PROCEEDINGS OF THE
FOURTH INTERNATIONAL
CONFERENCE ON

GENETIC ALGORITHMS

PROCEEDINGS OF THE
FOURTH INTERNATIONAL
CONFERENCE ON

GENETIC ALGORITHMS

University of California, San Diego
July 13–16, 1991

Editors/Program Co-Chairpersons:
Richard K. Belew
Lashon B. Booker

Supported By:
International Society for Genetic Algorithms
Office of Naval Research
Naval Research Laboratory

Morgan Kaufmann Publishers
San Mateo, California

Sponsoring Editor Bruce Spatz
Production Editor Yonie Overton
Cover Design Jo Jackson
Production Coordination Ocean View Technical Publications

Morgan Kaufmann Publishers, Inc.
Editorial Office:
2929 Campus Drive, Suite 260
San Mateo, CA 94403

94 93 92 91 5 4 3 2 1

Library of Congress Cataloging in Publication Data is available for this book.
ISBN 1-55860-208-9

CONTENTS

REPRESENTATION AND GENETIC OPERATORS

GENETIC ALGORITHM TECHNIQUES AND BEHAVIOR

FORMAL ANALYSIS OF GENETIC ALGORITHMS

PARALLEL GENETIC ALGORITHMS

CLASSIFIER SYSTEMS AND OTHER RULE-BASED APPROACHES

GENETIC ALGORITHMS IN HYBRID METHODS

GENETIC ALGORITHM APPLICATIONS

CONNECTIONISM AND ARTIFICIAL LIFE

PREFACE

This volume is the written record of the research to be presented at the Fourth International Conference on Genetic Algorithms (ICGA-91). The goal of the conference was to bring together an international community of scientists and engineers from academia and industry interested in algorithms suggested by the evolutionary process of natural selection. Topics of particular interest include genetic algorithms and classifier systems, evolution strategies, machine learning and optimization using these systems, and their relations to other paradigms such as connectionist networks and artificial life. The papers in this volume provide a capsule summary of current research on these topics.

As we write this preface, after having reviewed the many excellent papers contained in these proceedings but before the actual meeting, we cannot help but notice that the genetic algorithm (GA) community is experiencing a period of extremely rapid growth. Both of us remember that, only a decade or so ago, a meeting of the "GA community" meant a handful of people gathered in a classroom at the University of Michigan. In 1985 the gathering became a small international conference. Since then, the breadth of research activity and the number of papers submitted to the ICGA conferences has increased dramatically. More than 160 papers were submitted to ICGA-91, an increase of 60% over ICGA-89. The authors of accepted papers come from all over the United States and from nine foreign countries. Almost 30% of the authors come from outside the United States.

We had planned ahead for a modest increase in submissions by expanding the program committee to make sure each paper could receive at least two, and possibly three, reviews. The unexpectedly large number of papers submitted to ICGA-91 made this goal difficult to achieve. We succeeded, thanks only to a great deal of hard work by the entire program committee, who were operating under tight deadlines, and with the help of several additional reviewers whose efforts we gratefully acknowledge. Within these constraints, each paper was carefully reviewed. While the committee regrets having to reject more than half of the papers submitted, we are pleased with the outstanding quality of the papers that appear here.

We have done our best to organize these papers into thematic sections. The first section contains papers discussing genetic search operators and appropriate representations. One of the most exciting aspects of ICGA-91 will be the participation by German scientists who have studied the evolutionary techniques developed by Rechenberg and Schwefel known as *Evolutionsstrategie.* Their participation promises to focus attention on the fundamental trade-offs involved in using recombination versus mutation as the engine of genetic search. The next section considers variations on the "standard" GA originally formulated by John Holland, as well as characterizations of the behavior of these algorithms under specific conditions. The section on formal analysis describes a number of advances in our theoretical understanding of the GA, particularly regarding Walsh transforms and related characterizations of schemata processing. In large measure, it is the development of this theoretical core that has enabled the often diverging body of GA research to remain coherent.

The natural way in which population models map themselves onto massively parallel computer architectures continues to inspire many innovative uses of these new machines. The section on parallel GAs reports on the most recent experiments in this direction. Next, there is a set of papers describing classifier systems and related rule-based paradigms that use genetic techniques. Classifier systems continue to represent both an expressive rule-based knowledge representation amenable to GA adaptation and a class of computational models that is interesting in its own right. The section on hybrid methods highlights a new trend in the way GAs are being used as a problem-solving tool. As the relative strengths and weaknesses of GAs become more clearly understood, investigators have begun to use them as components in more elaborate, hybrid algorithmic designs. The GA applications section includes a diverse set of applied work, ranging from proof-of-principle studies to practical systems having immediate utility. Appropriately enough, the proceedings close with a series of papers that in one way or another relate back to the biological fountainhead from which evolutionary algorithms first sprang. Connectionist ("neural") networks continue to be another important class of biologically inspired adaptive algorithms, and "artificial life" is emerging as a label for the general class of computational investigations of living systems. The GA has already established an important role in the former and promises to make important contributions to the latter.

Because this volume contains so many papers and because we know our organization of them is but one of many, we have provided readers with two ancillary indices. The first is an initial attempt to organize the GA literature according to a taxonomic classification system, and the second is a simple "flat" index of important words and phrases. We generated these indices by combining our ideas with the classifications each author provided for his own work.

The ICGA-91 conference format will be a departure from the format of previous conferences. One of the frequent complaints many of us heard after ICGA-89 was that there was not enough opportunity for meaningful discussion and interaction. Consequently, we have attempted to balance the "lecture mode" constraints of plenary sessions with more lively interactions in small groups. We hope to achieve this objective by giving new emphasis to poster sessions and informal workshops.

The conference will begin with four invited talks that illustrate the breadth of issues related to GAs. The conference will then alternate between plenary presentations and poster and workshop sessions. The committee selected a small number of papers to be presented in plenary sessions to help "prime the pump" for the poster session and workshop discussions. Most papers will be presented in poster sessions. We hope this will promote considerable interaction between authors and those interested in their work. Moreover, it promises to be a more relaxing and enjoyable way to assimilate a large amount of information. Note also that all papers are uniform in appearance in this proceedings.

Informal workshops are another idea for increasing the amount of discussion at ICGA-91. Time has been set aside for special interest groups to discuss selected topics in depth, present preliminary results, exchange notes on work in progress, and the like. The workshops also provide a forum where those whose papers were not accepted for ICGA-91 can present their ideas to the rest of the community.

In summary, we thank the members of the Program Committee and everyone else whose dedication, cooperation, and labor made this endeavor possible. The results of their collective effort gives us every reason to believe that ICGA-91 will be a special event. It has been our special privilege to act as midwives for this collection. We hope this volume captures a fraction of the intellectual energy evident in the GA community at this important juncture in its evolution.

Richard K. Belew
Lashon B. Booker

Program Co-Chairpersons, ICGA-91

ICGA-91 CONFERENCE ORGANIZATION

ICGA-91 CONFERENCE COMMITTEE

Conference Co-Chairpersons:	Kenneth A. De Jong, *George Mason University* J. David Schaffer, *Philips Labs*
Vice Chairperson and Publicity:	David E. Goldberg, *University of Illinois*
Program Co-Chairpersons:	Richard K. Belew, *University of California, San Diego* Lashon B. Booker, *MITRE Corporation*
Financial Chairperson:	Gilbert Syswerda, *Bolt Beranek and Newman, Inc.*
Local Arrangements:	Richard K. Belew, *University of California, San Diego*

ICGA-91 PROGRAM COMMITTEE

Emile Aarts, *NV Philips*
Richard K. Belew, *University of California, San Diego*
Lashon B. Booker, *MITRE Corporation*
Yuval Davidor, *Weizmann Institute*
Lawrence Davis, *TICA Associates*
Kenneth A. De Jong, *George Mason University*
Larry Eshelman, *Philips Labs*
Stephanie Forrest, *University of New Mexico*
John J. Grefenstette, *Naval Research Laboratory*
David E. Goldberg, *University of Illinois*
Paulien Hogeweg, *University of Utrecht*
John H. Holland, *University of Michigan*
Gunar E. Liepins, *Oak Ridge National Laboratory*
Heinz Mühlenbein, *GMD, Germany*
John R. Koza, *Stanford University*
Gregory J.E. Rawlins, *Indiana University*
Rick L. Riolo, *University of Michigan*
George G. Robertson, *Xerox PARC*
J. David Schaffer, *Philips Labs*
Stephen F. Smith, *Carnegie Mellon University*
Gilbert Syswerda, *Bolt Beranek and Newman, Inc.*
Tom H. Westerdale, *University of London*
Stewart W. Wilson, *Rowland Institute for Science*
Darrell Whitley, *Colorado State University*

ACKNOWLEDGMENTS

ICGA-91 SUPPORTERS

The financial support provided by the following organizations is gratefully acknowledged.

International Society for Genetic Algorithms
Office of Naval Research
Naval Research Laboratory

ADDITIONAL REVIEWERS

We are grateful to the following individuals for their help in reviewing papers for ICGA-91.

David Ackley
Helen G. Cobb
David Demers
Terence C. Fogarty
Alexander Glockner
Michael Gordon
William Hart
Chris Langton
John McInerney
Jean-Arcady Meyer
Steven J. Nowlan
Connie Loggia Ramsey
Nicol Schraudolph
Alan C. Schultz
Stephen J. Smith
William M. Spears
Richard Sutton
Peter Todd

ICGA-91 GRAPHICS AND LOGO

The logograph on the conference poster and the cover of these proceedings was contributed by

Einat Delman

The ICGA-91 conference logo design was contributed by

Hugh Cartwright

The ICGA-91 conference t-shirt design was contributed by

Richard K. Belew

INVITED SPEAKERS

The ICGA-91 conference opened with a series of invited presentations by leading researchers from computer science, mathematical biology, economics and philosophy:

John Holland, University of Michigan
Complex Adaptive Systems

Marcus Feldman, Stanford University
Optimality and the Evolution of Recombination

John Miller, Santa Fe Institute
Artificial Adaptive Agents in Economics

William Wimsatt, University of Chicago
Developmental Constraints on Evolving Systems

Representation and Genetic Operators

A Survey of Evolution Strategies

Thomas Bäck* **Frank Hoffmeister†** **Hans–Paul Schwefel‡**

University of Dortmund
Department of Computer Science XI
P.O. Box 50 05 00 · D–4600 Dortmund 50 · Germany

Abstract

Similar to *Genetic Algorithms*, *Evolution Strategies* (ESs) are algorithms which imitate the principles of natural evolution as a method to solve parameter optimization problems. The development of Evolution Strategies from the first mutation–selection scheme to the refined (μ,λ)–ES including the general concept of self–adaptation of the strategy parameters for the mutation variances as well as their covariances are described.

1 Introduction

The idea to use principles of organic evolution processes as rules for optimum seeking procedures emerged independently on both sides of the Atlantic ocean more than two decades ago. Both approaches rely upon imitating the collective learning paradigm of natural populations, based upon Darwin's observations and the modern synthetic theory of evolution.

In the USA Holland introduced *Genetic Algorithms* in the 60ies, embedded into the general framework of adaptation [Hol75]. He also mentioned the applicability to parameter optimization which was first realized in the work of De Jong [Jon75].

This article focuses on the German development called *Evolution Strategies* (ESs), introduced by Rechenberg at Berlin in the 60ies as well [Rec73], and further developed by Schwefel [Sch75b]. ESs were applied first to experimental optimization problems with more or less continuously changeable parameters only. The first numerical applications were performed by Hartmann [Har74] and Höfler [Höf76], and a first attempt towards extending this strategy in order to solve discrete or even binary parameter optimization problems was made by Schwefel [Sch75a].

*baeck@lumpi.informatik.uni-dortmund.de
†iwan@lumpi.informatik.uni-dortmund.de
‡uin005@ddohrz11.bitnet

The aim of this paper is to give an overview of the development of ESs, beginning with the first simple mutation–selection mechanism with two individuals per generation only and stopping at the (μ,λ)–ES as used nowadays on single processor computers. The important idea of the on–line adaptation of the strategy parameters during the search process by incorporating them into the genetic representation of the individuals is also explained.

First a short introduction to the basic terminology concerning the parameter optimization problem is given. The overall goal of a parameter optimization problem $f : M \subseteq \mathsf{R}^n \rightarrow \mathsf{R}$, $M \neq \emptyset$, where f is called the *objective function*, is to find a vector $x^* \in M$ such that:

$$\forall x \in M \ : \ f(x) \geq f(x^*) = f^* \tag{1}$$

where f^* is called a *global minimum*; x^* is the *minimum location (point or set)*.

$$M = \{x \in \mathsf{R}^n \mid g_j(x) \geq 0 \ \forall j \in \{1, \ldots, q\}\} \tag{2}$$

is the set of feasible points for a problem with inequality constraints $g_j : \mathsf{R}^n \rightarrow \mathsf{R}$. For an unconstrained problem $M = \mathsf{R}^n$.

Since $\max\{f(x)\} = -\min\{-f(x)\}$, the restriction to minimization is without loss of generality. In the following a minimization problem is assumed. In general the optimization task is complicated by the existance of non–linear objective functions with multiple local optima. A *local minimum* $\hat{f} = f(\hat{x})$ is defined by the condition (3).

$$\exists \epsilon > 0 \ \forall x \in M \ : \ \|x - \hat{x}\| < \epsilon \Rightarrow \hat{f} \leq f(x) \tag{3}$$

Even if there is only one local optimum, it may be difficult to find a path towards it in case of discontinuities of the objective function or its derivatives. Simplification, e.g. linearization, may help to make things easier, but it can lead to results which are far away from the true optimum.

2 The Two Membered ES

According to Rechenberg [Rec73], the first efforts towards an evolution strategy took place in 1964 at the Technical University of Berlin (TUB). Then, the idea to imitate principles of organic evolution was applied in the field of experimental parameter optimization. The applications dealt with hydrodynamical problems like shape optimization of a bended pipe and of a flashing nozzle, or with control problems like the optimization of a PID regulator within a highly nonlinear system. Besides of simulating different versions of the strategy on the first available digital computer at the TUB, a Zuse Z23 [Sch65], computers soon were also used to solve numerical optimization problems by means of the first versions of simple ESs [Har74, Höf76].

The algorithm used in these applications was a simple mutation–selection scheme called *two membered* ES. It is based upon a "population" consisting of one parent individual (a real–valued vector), and one descendant, created by means of adding normally distributed random numbers. The better of both individuals then serves as the ancestor of the following iteration/generation. Such a (1+1)–ES can be described as the following 8–tuple:

$$(1+1)\text{–ES} = (P^0, m, s, c_d, c_i, f, g, t) \qquad (4)$$

where

P^0	$=$	$(x^0, \sigma^0) \in I$	population $I = \mathrm{R}^n \times \mathrm{R}^n$
m	$:$	$I \to I$	mutation operator
s	$:$	$I \times I \to I$	selection operator
c_d, c_i	$\in$	R	step-size control
f	$:$	$\mathrm{R}^n \to \mathrm{R}$	objective function
g_j	$:$	$\mathrm{R}^n \to \mathrm{R}$	constraint functions $j \in \{1, \ldots, q\}$
t	$:$	$I \times I \to \{0, 1\}$	termination criterion

P^0 denotes the initial "population" consisting of a single parent which produces by means of mutation a single offspring resulting in

$$\begin{aligned} P'^t &= (a_1'^t, a_2'^t) \in I \times I \\ a_1'^t &= P^t = (x^t, \sigma^t) \\ a_2'^t &= m(P^t) = (x'^t, \sigma^t) \end{aligned} \qquad (5)$$

The mutation operator is applied to all components of the object parameter x^t. According to the biological observation that offspring are similar to their parents and that smaller changes occur more often than larger ones, mutation is realized by normally distributed random numbers:

$$x'^t = x^t + \mathbf{N_0}(\sigma^t) \qquad (6)$$

where $\mathbf{N_0}$ denotes a vector of independent Gaussian random numbers with zero mean and standard deviations σ_i^t $(i = 1, \ldots, n)$. The selection operator then determines the fitter individual to become the parent of the next "generation":

$$P^{t+1} = s(P'^t) = \begin{cases} a_2'^t \text{ if } \begin{cases} f(x'^t) \leq f(x^t) \wedge \\ g_j(x'^t) \geq 0 \\ \forall j \in \{1, \ldots, q\} \end{cases} \\ a_1'^t \; = P^t \text{ else} \end{cases} \qquad (7)$$

in case of minimization. The iteration process $P^t \to P^{t+1}$ stops when the termination criterion $t(a_1'^t, a_2'^t) = 1$ holds. Function t depends on the implementation and may utilize elapsed CPU time, elapsed number of generations, absolute or relative progress per generation, etc.

In the description presented so far the standard deviations $\sigma^t \in \mathrm{R}^n$ remain constant over time. For theoretical considerations all components of σ^t are identical, i.e. $\forall i, j \in \{1, \ldots, n\} : \sigma_i^t = \sigma_j^t =: \sigma$. For a (1+1)–ES and a *regular* optimization problem a convergence property can be shown (see [Bor78]). The regularity of the optimization problem is specified by the criteria given in definition 1.

Definition 1

The optimization problem (1) is called *regular*, iff the following conditions are satisfied:

1. f is continuous.
2. M is a closed set.
3. $\forall \tilde{x} \in M \; : \; \tilde{M} := \{x \in M \mid f(x) \leq f(\tilde{x})\}$ is a closed set.
4. $\forall \epsilon > 0 \; : \; L^0_{f^*+\epsilon} := \mathrm{int}(L_{f^*+\epsilon}) \neq \emptyset$, where int denotes the set of all internal points and $L_{f^*+\epsilon} := \{x \in M \mid f(x) \leq f^* + \epsilon\}$ is a *niveau set* of f.

Then the global convergence of the (1+1)–ES can be shown:

Theorem 1

For $\sigma > 0$ and a regular optimization problem (1) with $f^* > -\infty$

$$p\left\{\lim_{t\to\infty} f(x^t) = f^*\right\} = 1$$

holds, i.e. the global optimum is found with probability 1 for sufficiently long search times.

The longish proof is omitted here; it can be found in [Bor78]. The basic idea is to use the monotone sequence $f(x^0) \geq f(x^1) \geq \ldots \geq f(x^i) \geq \ldots$ of objective function values generated by the process. Then, for the limit value $f^* \leq \tilde{f} = \lim_{t\to\infty} f(x^t)$ under the assumption $\tilde{f} > f^*$ a contradiction emerges, such that $\tilde{f} = f^*$ must be valid.

As is well known from similar theorems for *Simulated Annealing* [AK89] and *Genetic Algorithms* [EAH91], such results are not of much practical relevance due to the unlimited time condition. In fact, we are interested in the expectation of the *convergence rate* φ, which is given by the quotient of the distance covered towards the optimum and the number of trials needed for this distance.

Rechenberg calculated the convergence rates for the model functions

$$\begin{aligned} f_1(x) &= F(x_1) = c_0 + c_1 x_1 \\ & \quad \forall i \in \{2,\ldots,n\} : -b/2 \le x_i \le b/2 \\ f_2(x) &= \sum_{i=1}^{n} x_i^2 \end{aligned} \tag{8}$$

where $x = (x_1, \ldots, x_n)^T \in R^n$. f_1 is called the *corridor model* and represents a simple linear function with inequality constraints. Improvement of this objective function is only accomplished by moving along the first axis of the search space inside a corridor of width b. f_2 is called the *sphere model*. It comprises the simplest kind of non-linear, unimodal function. For these model functions the expectations of the rates of convergence are [Rec73]

$$\begin{aligned} \varphi_1 &= \frac{\sigma}{\sqrt{2\pi}} \left(1 - \sqrt{\frac{2}{\pi}\frac{\sigma}{b}}\right)^{n-1} \\ & \quad \text{for } n \gg 1 \\ \varphi_2 &= \frac{\sigma}{\sqrt{2\pi}} \left(\exp\left(-\left(\frac{n\sigma}{\sqrt{8}r}\right)^2\right)\right) \\ &- \frac{\sigma}{\sqrt{2\pi}} \left(\sqrt{\pi}\frac{n\sigma}{\sqrt{8}r}\left(1 - \operatorname{erf}\left(\frac{n\sigma}{\sqrt{8}r}\right)\right)\right) \\ & \quad \text{for } n \gg 1 \end{aligned} \tag{9}$$

where $\operatorname{erf}(x)$ refers to the well-known error function. The rate of convergence for the sphere model depends on the current location within the search space where r denotes the current euclidean distance from the optimum. From (9) it is possible to determine the optimum standard deviations σ_i^{opt} $(i = 1, 2)$ according to $\left.\frac{d\varphi_i}{d\sigma_i}\right|_{\sigma_i^{opt};\varphi_i^{max}} = 0$:

$$\begin{aligned} \sigma_1^{opt} &= \sqrt{\frac{\pi}{2}} \cdot \frac{b}{n} \quad ; \quad \varphi_1^{max} = \frac{1}{2e} \cdot \frac{b}{n} \\ \sigma_2^{opt} &\approx 1.224 \cdot \frac{r}{n} \quad ; \quad \varphi_2^{max} \approx 0.2025 \cdot \frac{r}{n} \end{aligned} \tag{10}$$

It should be noted, that in both cases the step size σ_i^{opt} is inversely proportional to the number of object variables n. Hence, the maximum rate of convergence is also inversely proportional to n.

The optimum standard deviations can be combined with the probabilities for a successful mutation:

$$\begin{aligned} p_1^t &= \frac{1}{2}\left(1 - \sqrt{\frac{2}{\pi}\frac{\sigma}{b}}\right) \quad \text{for } n \gg 1 \\ p_2^t &= \frac{1}{2}\left(1 - \operatorname{erf}\left(\frac{n\sigma}{\sqrt{8}r}\right)\right) \quad \text{for } n \gg 1 \end{aligned} \tag{11}$$

For optimum step sizes these probabilities turn to the values $p_1^{opt} = 1/(2e) \approx 0.184$ and $p_2^{opt} \approx 0.270$. From these findings Rechenberg postulated his *1/5 success rule*:

> *The ratio of successful mutations to all mutations should be 1/5. If it is greater than 1/5, increase the variance; if it is less, decrease the mutation variance.*

Though, in general, problems of interest may have characteristics different from those of the model functions used above, the following heuristic often helps to dynamically adjust the σ^t – not individually, but all at the same time, only. Hence, the mutation operator m is extended by the following equation:

$$\sigma^{t+1} = \begin{cases} c_d \cdot \sigma^t & , \text{ if } p_s^t < 1/5 \\ c_i \cdot \sigma^t & , \text{ if } p_s^t > 1/5 \\ \sigma^t & , \text{ if } p_s^t = 1/5 \end{cases} \tag{12}$$

where p_s^t is the frequency of successful mutations, measured e.g. over intervals of $10n$ trials. Schwefel [Sch81] gives reasons to use the factors $c_d = 0.82$ and $c_i = 1/0.82$ for the adjustment, which should take place every n mutations. It should be noted that with (6) and (12) the operator m consists of a random and a deterministic component, now.

As explained in [Sch81] the 1/5 success rule is a measure to increase the efficiency at the cost of effectiveness or robustness. It may lead the (1+1)–ES to premature termination even in the case of unimodal functions if there are discontinuities or active restrictions.

3 The First Multimembered ES

So far the population principle has not really been used. The (1+1)–ES can be designated as a kind of probabilistic gradient search technique – not, however, as a pure random or Monte Carlo method. To introduce the population concept into the algorithm, Rechenberg proposed the *multimembered* ES, where $\mu > 1$ parents can participate in the generation of one offspring individual. This has been denoted by Schwefel as $(\mu+1)$–ES and can be formalized this way:

$$(\mu+1)\text{–ES} = (P^0, \mu, r, m, s, c_d, c_i, f, g, t) \tag{13}$$

where

P^0	$=$	$(a_1^0, \ldots, a_\mu^0) \in I^\mu$	population
			$I = R^n \times R^n$
μ	$>$	1	number of parents
r	$:$	$I^\mu \to I$	recombination operator
m	$:$	$I \to I$	mutation operator
s	$:$	$I^{\mu+1} \to I^\mu$	selection operator

c_d, c_i	$\in$	R	step-size control
f	:	$\mathsf{R}^n \to \mathsf{R}$	objective function
g_j	:	$\mathsf{R}^n \to \mathsf{R}$	constraint functions $j \in \{1,\ldots,q\}$
t	:	$I^\mu \to \{0,1\}$	termination criterion

With the introduction of μ parents instead of only one, the imitation of sexual reproduction is possible, which is provided by the additional recombination operator r:

$$r(P^t) = a' = (x', \sigma') \in I \quad x' \in \mathsf{R}^n \ , \ \sigma' \in \mathsf{R}^n$$

$$\begin{aligned} x_i' &= \begin{cases} x_{a,i} \ , \ \mathcal{X} \leq 1/2 \\ x_{b,i} \ , \ \mathcal{X} > 1/2 \end{cases} \quad \forall i \in \{1,\ldots,n\} \\ \sigma_i' &= \begin{cases} \sigma_{a,i} \ , \ \mathcal{X} \leq 1/2 \\ \sigma_{b,i} \ , \ \mathcal{X} > 1/2 \end{cases} \quad \forall i \in \{1,\ldots,n\} \end{aligned} \tag{14}$$

where $a = (x_a, \sigma_a), b = (x_b, \sigma_b) \in I$ are two parents internally chosen by r. By convention all parents in a population have the same mating probabilities, i.e. the parents a and b are determined by uniform random numbers. $\mathcal{X}$ denotes a uniform random variable on the interval $[0,1]$, and it is sampled anew for each component of the vectors x' and σ'. r is called a *discrete* recombination operator due to the fact that component values are just copied from on of the parents.

According to the saying "survival of the fittest" (which was not coined by Darwin, but by one of his antagonists) the selection operator s removes the least fit individual – may it be the offspring or one of the parents – from the population before the next generation starts producing a new offspring.

$$\begin{aligned} P'^t &= (a_1'^t, \ldots, a_{\mu+1}'^t) = (a_1^t, \ldots, a_\mu^t, m(r(P^t))) \\ P^{t+1} &= s(P'^t) \quad \text{such that} \ \forall\, a_i^{t+1} = (x, \sigma) \\ & \quad \nexists a_j'^t = (x', \sigma') : f(x') < f(x) \end{aligned} \tag{15}$$

The mutation operator m and the adjustment of σ^t is realized in the same manner as for a (1+1)–ES (4,5,12). Self–adaptation of the step sizes has not been possible within the (μ+1)–ES scheme, since offspring with reduced mutation variances are always preferred.

4 (μ+λ)–ES and (μ,λ)–ES

The motivation to extend the (μ+1)–ES to a (μ+λ)–ES and (μ,λ)–ES has been twofold [Sch77, Sch81]: first, to make use of (at that time futuristic) parallel computers, and secondly, to enable self–adaptation of strategic parameters like the (even n different) standard deviations of the mutations. Instead of changing the σ^t by an exogenous heuristic in a deterministic manner, Schwefel completely viewed σ^t as a part of the genetic information of an individual. Consequently, it is subject to recombination and mutation as well. Those individuals with better adjusted strategy parameters are expected to perform better. Thus, selection will favour them accordingly and sooner or later those individuals will dominate the population, i.e. a better parameter setting will emerge by means of self-adaptation.

As the nomenclature (μ+λ)–ES suggests, μ parents produce λ offspring which are reduced again to the μ parents of the next generation. In all variants of (μ+λ)–ES selection operates on the joined set of parents and offspring. Thus, parents survive until they are superseded by better offspring. It might be even possible for very well adapted individuals to survive forever. This feature gives rise to some deficiencies of a (μ+λ)–ES:

1. On problems with an optimum moving over time a (μ+λ)–ES gets stuck at an out-dated good location if the internal parameter setting becomes unsuitable to jump to the new field of possible improvements.
2. The same happens if the measurement of the fitness (objective) or the adjustment of the object variables are subject to noise, e.g. in experimental settings.
3. For a (μ+λ)–ES with $\mu/\lambda \geq p_{f(x)}^{opt}$ (probability for a successful mutation) there is a deterministic selection advantage for those offspring which reduce some of their σ_i (9).

In order to avoid these effects Schwefel investigated the properties of a (μ,λ)–ES where only the offspring undergo selection, i.e. the life time of *every* individual is limited to one generation. The limited life span allows to forget inappropriate internal parameter settings. This may result in short phases of recession, but it avoids long stagnation phases due to misadapted strategy parameters [Sch87]. The (μ+λ)–ES and (μ,λ)–ES fit into the same formal framework with the only difference being the limited life time of individuals in (μ,λ)–ES. Thus, only a formal description of a (μ,λ)–ES is presented here:

$$(\mu,\lambda)\text{–ES} = (P^0, \mu, \lambda; r, m, s; \Delta\sigma; f, g, t) \tag{16}$$

where

P^0	=	$(a_1^0, \ldots, a_\mu^0) \in I^\mu$	population $I = \mathsf{R}^n \times \mathsf{R}^n$
μ	$\in$	N	number of parents
λ	$\in$	N	number of offspring $\lambda > \mu$
r	:	$I^\mu \to I$	recombination operator
m	:	$I \to I$	mutation operator
s	:	$I^\lambda \to I^\mu$	selection operator
$\Delta\sigma$	$\in$	R	step-size meta-control
f	:	$\mathsf{R}^n \to \mathsf{R}$	objective function
g_j	:	$\mathsf{R}^n \to \mathsf{R}$	constraint functions $j \in \{1,\ldots,q\}$
t	:	$I^\mu \to \{0,1\}$	termination criterion

(17)

In case of a $(\mu+\lambda)$–ES the selection operator s is modified to $s: I^{\mu+\lambda} \rightarrow I^{\mu}$. The major difference to the ES variants given before results from the handling of the internal strategy parameter σ^t, which now is incorporated into the genetic information of an individual $a_i^t = (x^t, \sigma^t) \in I$ and which is not controlled by some meta–level algorithm like the 1/5 success rule anymore. As a result, the mutation operator is different from before:

$$\begin{aligned} a_i'^t &= r(P^t) \\ m(a_i'^t) &= a_i''^t = (x''^t, \sigma''^t) \\ \sigma''^t &= \sigma'^t \exp \mathbf{N_0}(\Delta\sigma) \\ x''^t &= x'^t + \mathbf{N_0}(\sigma''^t) \end{aligned} \quad (18)$$

Mutation not only works on x^t but also on σ^t. The adjustment of σ^t by the 1/5 success rule has been replaced by a random modification, where unsuitable σ''^t are removed by means of selection. The operators r and s work as in (13).

Schwefel theoretically investigated the case of a $(1,\lambda)$–ES in a similar way as Rechenberg did before with respect to the (1+1)–ES. In particular, severe analytical problems arise as soon as one has to look for the stationary distribution of a population. More details are omitted here, but may be found in [Sch77, Sch81]. Like Rechenberg, Schwefel considered the corridor model and the sphere model (8). From an approximation of the maximum rates of convergence at optimum σ^t he deduced the corresponding best values for λ, the number of offspring per generation, for a $(1,\lambda)$–ES on a SISD computer with sequential evaluations of the offspring:

$$\begin{aligned} \lambda_1 &= 6 \\ \lambda_2 &= 5 \end{aligned} \quad (19)$$

On average at least one out of λ_1 (λ_2) offspring represents an improvement of the objective function, i.e. the probability for a successful offspring is approximately $1/\lambda_1$ ($1/\lambda_2$). These values are pretty close to Rechenberg's 1/5 success rule. On a MIMD computer, a $(1,\lambda)$–ES outperforms the simple (1+1)–ES by far and it has the additional advantage to be able to escape from local optima as soon as μ is increased beyond only 1 while $\mu/\lambda = \text{const}$. Like for the (1+1)–ES the maximum rate of convergence is inversely proportional to n, the number of object variables [Sch77, Sch81].

ESs which are operated with an optimum ratio of μ/λ for a maximum rate of convergence, are biased towards local search. As a result, such ESs tend to reduce their genetic diversity, i.e. the number of different alleles (specific parameter settings) in a population, as soon as they are attracted by some local optimum. In order to avoid the effect of missing alleles Born [Bor78] proposed the concept of a *genetic load* for some kind of $(\mu+1)$–ES. With genetic load a population is made up of an initially fixed, constant sub-population and a dynamic sub-population which evolves over time as described before: $P^t = P_l \cup P_a^t$ where P_l denotes the genetic load and P_a^t refers to the evolving (active) subpopulation. The genetic load may be used to introduce knowledge about suitable strategy parameters and object variables which are close to the optimum. Its major task is to maintain a minimum genetic diversity and a set of alleles that need not be learned by the algorithm itself. Recombination and mutation are extended and include the genetic load component as well as the dynamic part of the population. It is important to note that in this approach the genetic load is not subject to selection, i.e. no individual of the genetic load is replaced by offspring. For this case of a $(\mu+1)$–ES with genetic load Born could also prove the global convergence analogous to theorem 1 [Bor78].

4.1 Recombination Types

Intermediate recombination is motivated by the following Gedankenexperiment. When a population moves up-hill along a ridge or down-hill along a narrow valley, the individuals will have positions either on one or the other side of the ridge / ravine. In this case intermediate recombination of parents on different sides of the collective pathway (which is close to the gradient) may yield extraordinary successes.

$$\begin{aligned} r(P^t) &= a' = (x', \sigma') \in I \\ x_i' &= \frac{1}{2}(x_{a,i} + x_{b,i}) \qquad i = 1, \ldots, n \\ \sigma_i' &= \frac{1}{2}(\sigma_{a,i} + \sigma_{b,i}) \end{aligned} \quad (20)$$

Again, $a = (x_a, \sigma_a)$ and $b = (x_b, \sigma_b)$ are two parents chosen by r. Unfortunately, this type of recombination tends to reduce the genetic diversity of the population, but on the other hand it is a measure to avoid over–adaptation, especially with respect to the strategy parameters. In ESs the model functions suggest that the achievable rate of progress is inversely proportional to the number of object variables, hence individuals moving in a subspace of the object variables can exhibit a temporarily larger convergence rate than those moving in the full space. These individuals will reach a relative optimum only, and the whole evolution process may stagnate there for a while.

By intermediate recombination, a random but unsuitable extinction of a mutation step size σ_i is always reverted (increased again) as long as there is no mate with a similar step size adaptation. Actually, Schwefel's implementation of $(\mu+\lambda)$–ES and (μ,λ)–ES contains five types of recombination [Sch81]:

$$r(P^t) = a' = (x', \sigma') \in I$$

$$x_i' = \begin{cases} x_{a,i} & \text{(A) no recombination} \\ x_{a,i} \text{ or } x_{b,i} & \text{(B) discrete} \\ \frac{1}{2}(x_{a,i} + x_{b,i}) & \text{(C) intermediate} \\ x_{a,i} \text{ or } x_{b_i,i} & \text{(D) global, discrete} \\ \frac{1}{2}(x_{a,i} + x_{b_i,i}) & \text{(E) global, intermediate} \end{cases} \quad (21)$$

where $a, b, b_i \in P^t$ are parents chosen by r. Note that with global recombination the mating partners for the recombination of a *single* component x_i' are chosen anew from the population resulting in a higher mixing of the genetic information than in the standard case (B). An experimental comparison of most traditional, commonly used, direct optimization strategies to ESs on a set of 50 test functions, including multimodal as well as unimodal ones, restricted as well as unrestricted ones, has shown rather good results for ESs. Their convergence rate is comparable to other algorithms, but their reliability as well as their chance to find a low-dimensional global optimum was remarkably better than for the other strategies compared [Sch81]. Best results were obtained with different recombination types for the object variables (discrete) and the strategy parameters (intermediate).

4.2 Correlated Mutations

In ESs mutation realizes a kind of hill-climbing search procedure (12), when it is considered in combination with selection. With dedicated σ_i for each object variable x_i preferred directions of search can be established only along the axes of the coordinate system. In general, the best search direction (the gradient) is *not* aligned on those axes. Thus, an optimum rate of progress is achieved only by chance when suitable mutations coincide, i.e. when they are correlated. Otherwise, the trajectory of the population through the search space is zigzagging along the gradient. In order to avoid this reduction of the rate of progress, Schwefel [Sch81] extended the mutation operator to handle *correlated mutations* which require an additional strategy vector θ.

$$\begin{aligned} m(a_i''^t) &= a_i''^t = (x'', \sigma'', \theta'') \in I\ ,\ I = \mathrm{R}^n \times \mathrm{R}^n \times \mathrm{R}^w \\ \sigma'' &= \sigma'^t \exp \mathbf{N}_0(\Delta\sigma) \\ \theta'' &= \theta'^t + \mathbf{N}_0(\Delta\theta) \\ x'' &= x'^t + \mathbf{C}_0(\sigma'', \theta'') \end{aligned} \quad (22)$$

where $\mathbf{N}_0$ denotes a vector of independent Gaussian random numbers with expectation zero and standard deviations $\Delta\sigma_i$ and $\Delta\theta_i$, respectively. $\mathbf{C}_0(\cdot,\cdot)$ refers to a normally distributed random vector with expectation zero and probability density

$$p(\mathbf{C}_0(\cdot,\cdot)) = \sqrt{\frac{\det \mathbf{A}}{(2\pi)^n}} \exp\left(-\frac{1}{2}\mathbf{N}_0(\sigma'')^{\mathrm{T}} \mathbf{A}\, \mathbf{N}_0(\sigma'')\right) \quad (23)$$

The diagonal elements of the covariance matrix $\mathbf{A}^{-1}$ are the independent variances $\sigma_i''^2$ (squares of the mutation step sizes) of the object variables x_i^t, while the off-diagonal elements represent the covariances $c_{i,j}$ of the mutations. Schwefel restricts the space of equal probability density to the surface of n-dimensional rotating hyperellipsoids, which are realized by a set of inclination angles $\theta'' \in \mathrm{R}^w$ of the main axes of the hyperellipsoid, $w = n(n-1)/2$. This helps to keep the covariance matrix positive definite. The standard deviations σ_i'' serve as a kind of mean step size along those axes.

Like the strategy parameter σ^t, θ^t is also incorporated into the genetic representation of an individual and is modified in a similar way, i.e. the recombination operator is extended to work on the inclination angles θ^t, as it has been done before for the mutation step size σ^t (12). This way the ES may adapt to any preferred direction of search by means of self-learning. Due to the additional meta-parameter $\Delta\theta$ the signature of an ES with correlated mutations is defined as

$$(\mu,\lambda)\text{-ES} = (P^0, \mu, \lambda; r, m, s; \Delta\sigma, \Delta\theta; f, g, t) \quad (24)$$

The step sizes σ_i of an individual make up an ellipsoid of equal probability density to place an offspring if these step sizes are applied to x^t of the individual itself. The left part of the illustration in figure 1 shows some individuals with their corresponding ellipsoids if the step sizes are *not* correlated (simple mutations), while the illustration on the right hand shows the same individuals with correlated mutations. In long, narrow valleys the step sizes with simple mutations must be smaller than with correlated mutations, where a single step size may reach far into the valley if it is oriented appropriately. In such situations the resulting rate of convergence is much higher.

5 Conclusion

Evolution Strategies went through a long period of stepwise development since the formulation of the basic ideas in the middle of the 60ies. Some important milestones in their development were

- the analytical convergence rate calculations, which led to the development of the 1/5 success rule;
- the introduction of a population instead of one single individual, which also allowed for the sexual recombination process;
- the self-learning process of strategy parameters by means of incorporating them into the set of genetically inherited variables;
- the (μ,λ)-ES, which introduces a forgetting principle and is important in changing environments as well as a measure against over-adaptation, especially of the strategy parameters;
- the introduction of additional strategy parameters to allow for correlated mutations and thus self-learning of simple "natural laws" in the topological environment.

The third and fifth points establish a two-level learning process, since not only the object-variable population adapts according to the response surface, but

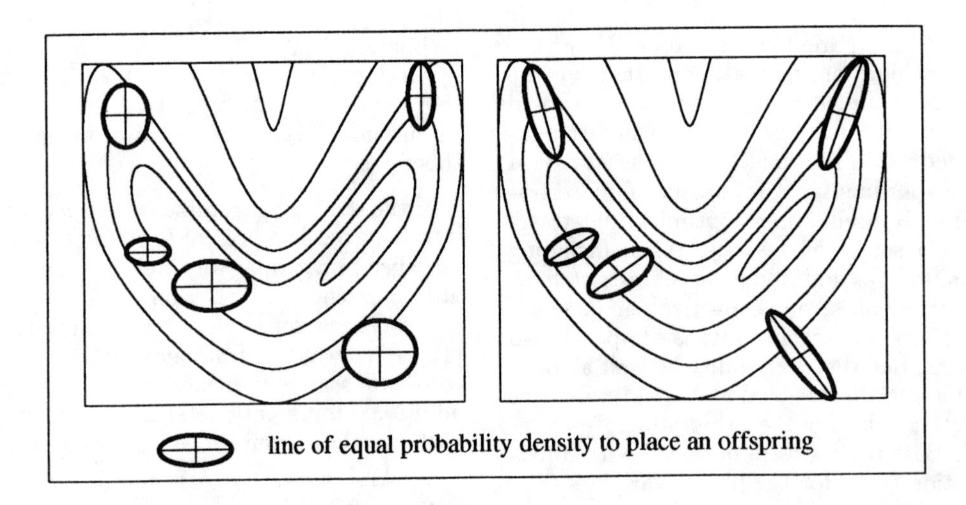

Figure 1: Searching with simple and correlated mutations

also the strategy parameters are changed with respect to the actual topolocigal requirements. Schwefel has pointed to the difficulties as well as opportunities of the two-level collective learning in [Sch87]. The strategy parameters make up an internal model of the objective function, which is learned on-line during the optimum seeking without any exogenous controlling instance or additional measure of fitness.

Besides of the different levels of genotypic/phenotypic information representation and of different selection mechanisms the two-level learning in ESs is the most striking difference between ESs and Genetic Algorithms [HB90, HB91].

Current research concerning ESs deals with applications like the travelling salesman problem [Her91, Rud91], girder-bridge optimization [Loh91], neural networks [Sal91], vector optimization [Kur91], and parameter optimization in general [BBK83, Bor78, BB79].

Furthermore the scalable parallelism of such an Evolutionary Algorithm [Hof91] is investigated with the help of implementations on MIMD-computers at the University of Dortmund, especially on a small Transputer network with 30 T800 processors [Bor89, Rud91].

References

[AK89] Emile H. L. Aarts and Jan Korst. *Simulated Annealing and Boltzmann Machines.* Wiley, Chichester, 1989.

[BB79] Joachim Born and Klaus Bellmann. *Numerische Parameteroptimierung in mathematischen Modellen mittels einer Evolutionsstrategie*, volume 18 of *Lecture Notes in Control and Information Sciences*, pages 157–167. Springer, Berlin, 1979.

[BBK83] U. Bernutat-Buchmann and J. Krieger. Evolution strategies in numerical optimization on vector computers. In Feilmeier, Joubert, and Schendel, editors, *Parallel Computing 83, Proceedings of the International Conference on Parallel Computing*, pages 99–105, FU Berlin, 26.–28.Sept. 1983.

[Bor78] Joachim Born. *Evolutionsstrategien zur numerischen Lösung von Adaptationsaufgaben.* Dissertation A, Humboldt-Universität, Berlin, GDR, 1978.

[Bor89] Andreas Bormann. *Parallelisierungsmöglichkeiten für direkte Optimierungsverfahren auf Transputersystemen.* Master thesis, University of Dortmund, Dortmund, Germany, April 1989.

[EAH91] A. E. Eiben, E. H. L. Aarts, and K. M. Van Hee. Global convergence of genetic algorithms: an infinite markov chain analysis. In Männer and Schwefel [MS91], pages 4–12.

[Har74] Dietrich Hartmann. *Optimierung balkenartiger Zylinderschalen aus Stahlbeton mit elastischem und plastischem Werkstoffverhalten.* PhD thesis, University of Dortmund, July 1974.

[HB90] Frank Hoffmeister and Thomas Bäck. Genetic algorithms and evolution strategies: Similarities and differences. Technical Report "Grüne Reihe" No. 365, Department of Computer Science, University of Dortmund, November 1990.

[HB91] Frank Hoffmeister and Thomas Bäck. Genetic algorithms and evolution strategies: Similarities and differences. In Männer and Schwefel [MS91], pages 455–470.

[Her91] Michael Herdy. Application of the 'Evolutionsstrategie' to discrete optimization problems. In Männer and Schwefel [MS91], pages 188–192. (In print).

[Höf76] A. Höfler. *Formoptimierung von Leichtbaufachwerken durch Einsatz einer Evolutionsstrategie.* PhD thesis, Technical University of Berlin, June 1976. Dept. Verkehrswesen.

[Hof91] Frank Hoffmeister. Scalable parallelism by evolutionary algorithms. In M. Grauer and D. B. Pressmar, editors, *Applied Parallel and Distributed Optimization*, Lecture Notes in Mathematical Systems and Economics. Springer, 1991.

[Hol75] John H. Holland. *Adaptation in natural and artificial systems.* The University of Michigan Press, Ann Arbor, 1975.

[Jon75] Kenneth De Jong. *An analysis of the behaviour of a class of genetic adaptive systems.* PhD thesis, University of Michigan, 1975. Diss. Abstr. Int. 36(10), 5140B, University Microfilms No. 76–9381.

[Kur91] Frank Kursawe. A variant of Evolution Strategies for vector optimization. In Männer and Schwefel [MS91], pages 193–197. (In print).

[Loh91] Reinhard Lohmann. Application of evolution strategy in parallel populations. In Männer and Schwefel [MS91], pages 198–208. (In print).

[MS91] Reinhard Männer and Hans-Paul Schwefel, editors. *Proceedings of the First International Conference on Parallel Problem Solving from Nature (PPSN), Dortmund, Germany, 1990*, Berlin, 1991. Springer. (In print).

[Rec73] Ingo Rechenberg. *Evolutionsstrategie: Optimierung technischer Systeme nach Prinzipien der biologischen Evolution.* Frommann–Holzboog Verlag, Stuttgart, 1973.

[Rud91] Günter Rudolph. Global optimization by means of evolution strategies. In Männer and Schwefel [MS91], pages 209–213. (In print).

[Sal91] R. Salomon. Improved convergence rate of back–propagation with dynamic adaptation of the learning rate. In Männer and Schwefel [MS91], pages 269–273. (In print).

[Sch65] Hans-Paul Schwefel. *Kybernetische Evolution als Strategie der experimentellen Forschung in der Strömungstechnik.* Diploma thesis, Technical University of Berlin, March 1965.

[Sch75a] Hans-Paul Schwefel. Binäre Optimierung durch somatische Mutation. Technical report, Technical University of Berlin and Medical University of Hannover, May 1975.

[Sch75b] Hans-Paul Schwefel. *Evolutionsstrategie und numerische Optimierung.* Dissertation, Technische Universität Berlin, Berlin, May 1975.

[Sch77] Hans-Paul Schwefel. *Numerische Optimierung von Computer-Modellen mittels der Evolutionsstrategie.* Interdisciplinary systems research; 26. Birkhäuser, Basel, 1977.

[Sch81] Hans-Paul Schwefel. *Numerical Optimization of Computer Models.* Wiley, Chichester, 1981.

[Sch87] Hans-Paul Schwefel. Collective phenomena in evolutionary systems. In *31st Annual Meeting of the International Society for General System Research, Budapest*, pages 1025–1033, June 1987.

Exploring Problem-Specific Recombination Operators for Job Shop Scheduling

Sugato Bagchi **Serdar Uckun** **Yutaka Miyabe*** **Kazuhiko Kawamura**

Center for Intelligent Systems,
Box 1804 Station B,
Vanderbilt University,
Nashville, TN 37235.

Abstract

In this paper, the development and implementation of a prototype job shop scheduling system based on a genetic algorithm is described. To enhance the performance of the algorithm and to expand the search space, a chromosome representation which stores problem-specific information is devised. Problem-specific recombination operators which take advantage of the additional information are also developed. The results of experimentation on three different representational schemes are discussed.

1 INTRODUCTION

Scheduling is an active area of research in applied artificial intelligence. Scheduling problems typically comprise several concurrent (and often conflicting) goals, and several resources which may be allocated in order to satisfy these goals. In many cases, the combination of goals and resources results in an exponentially growing problem space. As an immediate result, no deterministic methods exist for solving those problems in polynomial time. Such problems are called *NP-complete* problems, with respect to the exponential time and memory requirements necessary to reach optimal solutions. A common example of NP-complete problems is job shop scheduling (JSS). Since job shop scheduling problems challenge existing search methodologies, and since successful solution of real-world scheduling problems is an important focus in industry, JSS has been a favorite domain among researchers.

Among various search methodologies available, brute-force (blind) search algorithms are of no use for NP-complete problems due to exponential time and space requirements [6]. Heuristic search methods, when combined with appropriate abstractions to reduce the exponential complexity of the problem, offer near-optimal solutions. Methods such as ISIS [3] and OPIS [7] utilize domain-specific constraints to converge on near-optimal solutions using multiple layers of abstraction and other scheduling heuristics. Although such deterministic methods are quite effective, the use of abstractions constrains the boundaries of the search. For example, OPIS schedules bottleneck resources before other resources, and therefore limits the subsequent search to a subspace bounded by the constraints inflicted upon all other resources by the bottleneck resource allocation process.

Although genetic algorithms (GA's) may not resolve the optimization problem completely, they may serve as powerful search algorithms that can sample large search spaces randomly and efficiently. The assumption underlying the use of GA's for scheduling is that optimal solutions will be found in the neighborhood of good solutions. In other words, exploitation of favorable features in sub-optimal solutions will eventually lead to the discovery of near-optimal solutions. Once several near-optimal solutions are identified, a heuristic search method can rapidly traverse the remaining search space to optimize such solutions within their local neighborhood.

This paper describes the development and the implementation of a prototype job shop scheduling system based on a genetic algorithm. The next section covers the representational issues. Section 3 describes the implementation of domain-specific recombination operators that enhance the performance of the genetic algorithm. Section 4 details the results of experimentation, and the final section describes research issues that remain to be explored.

2 REPRESENTATIONAL ISSUES

Scheduling is a combinatorics problem where reusable resources are assigned a sequence of tasks necessary to

*Computer Systems Laboratory, Electronics R&D Laboratories, Nippon Steel Corporation, Sagamihara, Kanagawa 229, Japan

service orders. Optimization is performed to minimize the service time while maximizing machine utilization as well as satisfying other constraints such as due dates and customer priorities. Where an exhaustive search will fail due to time and memory constraints for all but the simplest of cases, genetic algorithms have the potential to discover near optimal schedules much faster. Theory and evidence suggest that genetic algorithms perform better when augmented with problem-specific knowledge and advanced recombination operators that exploit the added information [5]. This paper describes the development of a problem-specific representation for JSS, and the development of GA methods to explore the resulting search space.

2.1 BACKGROUND WORK

The use of genetic algorithms for job shop scheduling has been explored previously. Davis [2] uses an intermediate representation which is guaranteed to produce legal schedules when operated upon by generic recombination operators. However, the example used is not very complicated, and there are no significant results. Syswerda describes how a job shop scheduling problem may be represented similar to the traveling salesperson problem (TSP) [8]. He then demonstrates the use of different recombination operators on a hypothetical scheduling problem. Whitley and coworkers [9] define a new domain-independent recombination operator for TSP-like ordering problems. Cleveland and Smith [1] approach a flow shop release scheduling problem by representing it in a TSP-like fashion, using a variety of recombination operators designed to maintain the integrity of successful segments (subtours) of the chromosome. An interesting contribution of the paper is the use of heuristically guided recombination operators. The authors conclude that more research is necessary to devise more accurate representations of the problem domain.

2.2 THE JOB SHOP SCHEDULING PROBLEM

A job shop is a facility that produces goods according to prespecified process plans, under several domain-dependent and common-sense constraints. A brief description is as follows:

- A job shop may produce several parts.
- Each part has one or more alternative process plans.
- Each process plan consists of a sequence of operations.
- Each operation requires resources (machine, personnel, raw material, etc.)
- Each operation has a certain duration on a certain machine.
- Each order is for one or more units of the same part.
- Several orders are typically active at a given time.

The job–shop scheduling problem is the selection of sequences of operations in order to satisfy a number of orders, and the assignment of start times, end times, and resources for each operation [3]. There may be several constraints that further bound the scheduling problem, such as machine down-times, regular maintenance, machine setup times required between changes in operation, and preferences on the processing of orders. Figure 1 tabulates the information on the hypothetical job shop which is used for the proof-of-principle experiments described in the following sections. The figure also lists a set of hypothetical orders to be scheduled on this domain.

2.2.1 Search Spaces

In the job shop shown in Figure 1, three distinct search spaces can be identified:

Order permutations: The queue ordering of job orders is one of the most significant items on the search agenda. The example has 11 job orders, amounting to an order permutation search space of (11!), or roughly $4 * 10^7$ alternatives.

Process plans: Each part has two or three different process plans to choose from. However, each process plan is inherently related to the allocation of resources for its operations. Therefore, it is more convenient to estimate the size of the process plan and resource search spaces as a whole.

Resources: Each process plan contains two or three operations, which in turn may be performed on one or two different machines. Each process plan listed above has 2 to 8 alternative resource assignments. As a whole, 11 orders for the parts mentioned in the example represent a combined process plan/resource search space of approximately $1.3 * 10^{10}$ alternatives.

Since the permutation search space is orthogonal to the combined process plan/resource search space, the total search space size for this example is roughly $5 * 10^{17}$.

2.3 CHROMOSOME REPRESENTATION

Genetic algorithms typically require problem states to be represented in the form of strings. In addition, a diverse and random population of such strings must be created to represent an initial population. The following sections describe two different approaches that could be taken to tackle the problem of representation.

```
;; numbers indicate setup times
(machines
 (MachineM 5)
 (MachineN 15)
 (MachineP 10))

;; numbers indicate duration of operation
(operations
 (Op1
  (MachineM 40))
 (Op2
  (MachineM 7) (MachineN 10))
 (Op3
  (MachineN 20))
 (Op4
  (MachineM 12))
 (Op5
  (MachineN 30) (MachineP 5))
 (Op6
  (MachineN 16) (MachineP 8))
 (Op7
  (MachineN 15) (MachineP 12))
 (Op8
  (MachineM 5) (MachineP 8))
 (Op9
  (MachineM 4) (MachineN 6))
 (Op10
  (MachineM 8) (MachineP 3)))

(parts
 (partA (planA1 Op1 Op4 Op6)
        (planA2 Op1 Op5 Op6)
        (planA3 Op4 Op9 Op6))
 (partB (planB1 Op1 Op2 Op7)
        (planB2 Op7 Op9 Op10))
 (partC (planC1 Op1 Op3 Op8)
        (planC2 Op4 Op10)))

;; numbers indicate the number of units to
   be produced
(orders
 (order1 (batch1 partA #40))
 (order2 (batch1 partB #20))
 (order3 (batch2 partA #10))
 (order4 (batch2 partB #20))
 (order5 (batch2 partC #15))
 (order6 (batch3 partB #50))
 (order7 (batch3 partC #40))
 (order8 (batch4 partA #35))
 (order9 (batch4 partB #35))
 (order10 (batch5 partC #80))
 (order11 (batch5 partA #120)))
```

Figure 1: A hypothetical job shop

2.3.1 Direct Representation

One solution is to use the schedule itself as a chromosome, where a schedule for a particular machine is the time ordering of operations that will use that machine. This approach requires complicated recombination operators which should ensure that the sequence of operations needed to manufacture a certain part is not violated. These operators should also be able to inflict changes across schedules, since operations on the different machines are often interrelated. For example, when one operation on a certain machine is delayed, subsequent operations scheduled on other machines will be affected. Such constraints complicate the crossover and mutation operators to the extent that an operation on any string will affect the strings representing the schedules of other machines.

If one undertakes the burden of developing complicated genetic operators, this scheme has the advantage of being very direct. The chromosomes themselves represent the subject of optimization. New operators can be defined to perform domain specific operations such as inserting high priority orders into an existing near optimal schedule, or, replacing one manufacturing procedure with another (i.e., using a different sequence of tasks to create the same order). The advanced operators will perform the additional computation to guarantee that all produced schedules are legal.

2.3.2 Indirect Representation

An alternative scheme described in [8] is to represent the queue of job orders as a chromosome. In this scheme, the ordering of the chromosome represents the scheduling priority of the job order, as seen in Figure 2.a. A schedule may be built from this string by allocating resources to the tasks one by one, following the order in the string. The task of the genetic algorithm is to come up with different permutations on the list of orders. The advantage of this scheme is the simplicity of the chromosome structure and the operators. The added responsibility is to create a schedule builder that converts chromosomes into legal schedules. This representation will be referred to as **Representation 1**.

The drawback with this scheme is that the genetic algorithm is constrained to search only in the space of all permutations of job orders. However, as shown in Section 2.2.1, this is only a part of the complete search space. In addition to the ordering the job orders, other alternatives are available as well. These have been represented in the process and resource search spaces in Section 2.2.1. The representation described in [8] does not factor this information into the genetic algorithm. Hence, even after the genetic algorithm comes up with a sequence of job orders, a great deal of search remains to be done by the schedule builder. Furthermore, since the schedule builder has to be called for every member

order2	order1	order3
order1	order2	order3
order2	order3	order1

a. sample population of 3 strings representing order priorities only

order2	plan B1	order1	plan A1	order3	plan A2
order1	plan A2	order2	plan B1	order3	plan A1
order2	plan B2	order3	plan A1	order1	plan A1

b. sample population of 3 strings representing order priorities + randomly assigned process plans

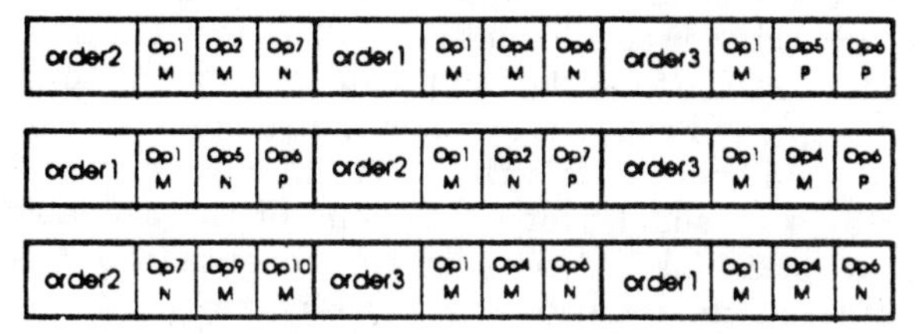

c. sample population of 3 strings representing order priorities + randomly assigned process plans + randomly assigned resources for each operation

Figure 2: Chromosome representations for three different representation schemes

of the population in every generation, the additional burden of search will slow down the overall optimization process.

2.3.3 Problem Specific Representation

In this research, an additional step is taken to represent more problem-specific information in the chromosome in order that a larger proportion of the search task may be addressed by the genetic algorithm.

The first step towards a representation that contains problem specific information in addition to job ordering is to include the selected process plan for each job order. Crossover and mutation operators that exploit this information have also been implemented. The search task is divided between the genetic algorithm and the schedule builder. The genetic algorithm searches the space comprising all the permutations of job orders along with the option to modify their processing plans. The schedule builder searches for the resources that can satisfy the process plans as early as possible. An example of the new representation, referred to as **Representation 2** is shown in Figure 2.b.

Figure 2.c shows **Representation 3**, which contains not only the process plan for each job order in the chromosome, but also the set of machines that will perform the various operations specified in the plan. This information comprises the entire search space, and the genetic algorithm is the only mechanism that performs search. The schedule builder is responsible only for creating legal schedules after factoring in parameters like machine setup and down times. Both the selection of a process and the allocation of resources are performed by the genetic algorithm.

The main difference between these methods is the extent of the search space explored by the genetic algorithm. **Representation 2** searches a smaller space and is therefore expected to converge to a solution faster. On the other hand, **Representation 3** is more desirable due to the homogeneity of its search. Therefore, genetic search with this representation is expected to yield more desirable results. All three representations have been implemented and analyzed. Section 4 illustrates the results obtained.

The introduction of an extended chromosome representation necessitates the definition of new crossover and mutation operators. These operators should operate along with the existing permutation-based operators. This issue is discussed in Section 3.1.

2.4 SCHEDULE BUILDER

The choice of an indirect chromosome representation necessitates a schedule builder which transforms a chromosome representation into a valid schedule. The activity of the schedule builder depends on the amount of information encoded in the chromosome representation. The schedule builder has three basic functions:

1. Local search for information not supplied by the chromosome. As described previously, there are alternate process plans and resources for each order. The schedule builder has to perform a local search if any of these options are omitted in the chromosome. The three representations described in the previous section have varying amounts of information encoded in the chromosomes. In **Representation 3**, all the search variables are encoded in the chromosome, and no additional search is required of the schedule builder. In **Representation 2**, the set of machines to perform the operations are not specified. The schedule builder is responsible for selecting the machine that can complete an operation at the earliest time, thereby enforcing a local decision. An exhaustive search of all machine combinations is not performed, since such a global deterministic search would require exponential time and memory to complete. **Representation 1** encodes only the sequence of job orders. In this case, the process plan and the set of resources for each order

are selected by the schedule builder. To avoid a complicated search, each order is assigned its default process plan and no further search is attempted. Resources are selected in the same manner as described for **Representation 2**.

2. Enforcing constraints, such as machine setup times and regularly scheduled down times for maintenance. Domain specific constraints may also be represented, such as a limit on the number of machines that can operate simultaneously.
3. Building the actual schedules for all the machines in the job shop.

The schedule builder introduces a deterministic aspect to the search which is otherwise stochastic. An attempt is made to minimize the deterministic component, so that the performance of the stochastic variants may be assessed and compared with minimal interference from a deterministic scheduling algorithm. As a result, the schedule builder described above is kept as simple as practicable.

2.5 FITNESS FUNCTIONS

The fitness of a chromosome can be determined after building the best possible schedule from it. Ideally, the schedule has to be evaluated to detect under-utilization of machines, response time of individual orders (with more weight given to high priority orders) and the overall response time. Other constraints like satisfaction of due dates must also be given appropriate weights. In the current implementation, the fitness function calculates overall machine utilization as the percent ratio of total operational time to the total time spent to service all job orders. This ensures that high fitness values will be given to schedules that distribute the load equitably among available machines.

3 PROBLEM SPECIFIC RECOMBINATION OPERATORS

As discussed above, the use of a problem specific chromosome representation requires the use of recombination operators designed to exploit the new representation. The new crossover and mutation operators designed for this purpose are described in the following sections.

3.1 CROSSOVER OPERATORS

The chromosome representations described in Section 2.3.3 contain information about both the ordering and value of the individual elements. In this implementation, the partially-mapped crossover (PMX) operation described in [4] is being used to exchange string ordering. However, as mentioned in Section 2.3.3, this operator by itself cannot search the entire problem space.

Another crossover operator is used to exchange problem specific information. This operator selects random substrings from two chromosomes, and for each element in a substring, exchanges the process plan with the corresponding element in the other chromosome. For **Representation 3**, selected resources (machines) are also exchanged. However, the ordering of the jobs in the chromosome is not modified.

An important issue arising from the presence of two crossover operators is the selection of the operator to be used when a crossover is to be performed. A user-defined selection probability is being used to make the decision. This will be discussed in Section 3.3. Results of experimentation with different selection probabilities are explained in Section 4.

3.2 MUTATION OPERATORS

As with the crossover operation, two separate operators are being used for mutation. The first operator is order based. It selects two points on a chromosome at random, and swaps the elements at these positions. The second operator selects an element at random and selects (with equal probability) one of the available process plans for the job. For **Representation 3**, a set of resources are also chosen at random.

Again, a decision has to be made to select the operator to be used when a mutation has to be performed. The same selection probability as described in the previous section is being used to make the decision. The selection procedure is detailed below.

3.3 OPERATOR SELECTION

With the presence of two operators each for crossover and mutation, the combination of these operators is an interesting issue. This implementation uses a selection probability to choose one recombination operator over the other, both for crossover and mutation. This is the probability by which the domain independent recombination operators (PMX for crossover and order swapping for mutation) will be used over the domain dependent ones. For example, if the selection probability is set to 50% when a crossover is to be performed, either PMX or the domain-specific operator will be chosen with equal probabilities. With the selection probability set to 80%, PMX will be chosen most of the time. When the selection probability is increased to 100%, the domain-specific operators will be defeated. The selection process applies similarly to the mutation operators.

The choice of a selection probability depends on the sizes of the search spaces due to the order permutation and the process plan/resources. As shown in Section 2.2.1, the order permutation space is smaller by three orders of magnitude. Thus, a smaller percentage should be assigned to the PMX operator. Experiments with different selection probability settings and the results are discussed below.

4 RESULTS

The recombination operators described in Section 3 have been implemented to operate on the chromosome representations described in Section 2.3.3. The scheduling example shown in Figure 1 was used for optimization. The first major objective of these experiments was a comparison of the representation schemes that were discussed in Sections 2.3.2 and 2.3.3. Another major objective was a comparison of various selection probabilities as described in Section 3.3. Due to the stochastic nature of the algorithm, the results were obtained by averaging the data obtained over 100 runs with the same parameters. Each run explored a population of size 50 for 1000 generations.

4.1 PERFORMANCE OF CHROMOSOME REPRESENTATIONS

The three chromosome representations discussed here are representation of the order priority alone (**Representation 1**), representation of the order priority and process plans (**Representation 2**), and representation of the order priority, process plan and resources (**Representation 3**). These schemes are illustrated in Figure 2.

Figure 3 depicts the performance of these representation modalities under the same selection probability of 20%. As seen in the figure, **Representation 1** prematurely converges to a sub-optimal value. This observation may be explained by the fact that the genetic algorithm is only searching the space defined by all permutations of job order priorities. For the example used, this is equal to (11!). The multiple options of process plans and resources are not being used. Instead, in this case, process plans and resources are being assigned deterministically by the schedule builder. This severely limits the space being searched by the algorithm. **Representation 2** fares better, because its search space includes not only the order permutations but also alternative process plans. It may be noted that in both these representations, the schedule builder performs a search to locate the resource (machine) that can perform an operation at the earliest. However, this does not seem as effective as **Representation 3**, where the genetic algorithm performs the search for the best resource as well. In this case the entire search task is handled by the genetic algorithm. As seen in the figure, the results are significantly better than the two other representations ($\alpha = 0.005$, for one sided *t-test* between **Representations 2** and **3**).

The results in Figure 3 suggest that the genetic algorithm should be allowed to operate on the entire search space. All alternatives and options that are specified in the scheduling problem should be represented in the chromosome, with domain dependent operators available to exploit them.

4.2 EFFECT OF SELECTION PROBABILITY

As discussed in Section 3.3, the introduction of the selection probability adds a new parameter to the genetic algorithm. The effect of varying this parameter has been investigated. Figure 4 shows the results obtained with the selection probability set at 5%, 20%, 50%, 80%, and 100%. As mentioned before, these percentage values indicate the probability with which the domain independent operators will be used over domain dependent operators. **Representation 3** has been used to experiment with different selection probabilities.

The results indicate that a selection probability of 20% performs better, on the average, than the other probabilities. The results also show that by changing the selection probability from 80% to 5%, the fitness percentage rises by a margin of only 2%. Therefore, the performance is not very sensitive to changes in the selection probability, as long as the probability remains within that range.

However, no independent conclusions can be made regarding the size and shape of search spaces vs. an optimal setting of the selection probability. As with other GA parameters (crossover and mutation probabilities, etc.), the selection probability must be tailored to the specific problem in hand through trial and error.

5 DISCUSSION

The success of genetic algorithms in any application area can only be determined by experimentation. Parameters like population size, crossover and mutation probabilities should be varied until satisfactory results are achieved. In addition, the operators themselves must be adjusted to suit the domain requirements. The chromosome representation should contain all the information that pertain to the optimization problem. As shown here, an expanded problem representation results in a better performance by the genetic algorithm. However, the increased search space may result in slower convergence.

One interesting topic for further research is the use of a genetic algorithm to rapidly reduce the search space to a size that can subsequently be handled by deterministic search algorithms [5]. The idea here is to use the genetic algorithm to identify the possible "hills" in the search space, which can later be climbed by conventional algorithms. Such a hybridization of search techniques, which could boost search performance in difficult tasks such as scheduling, is currently being

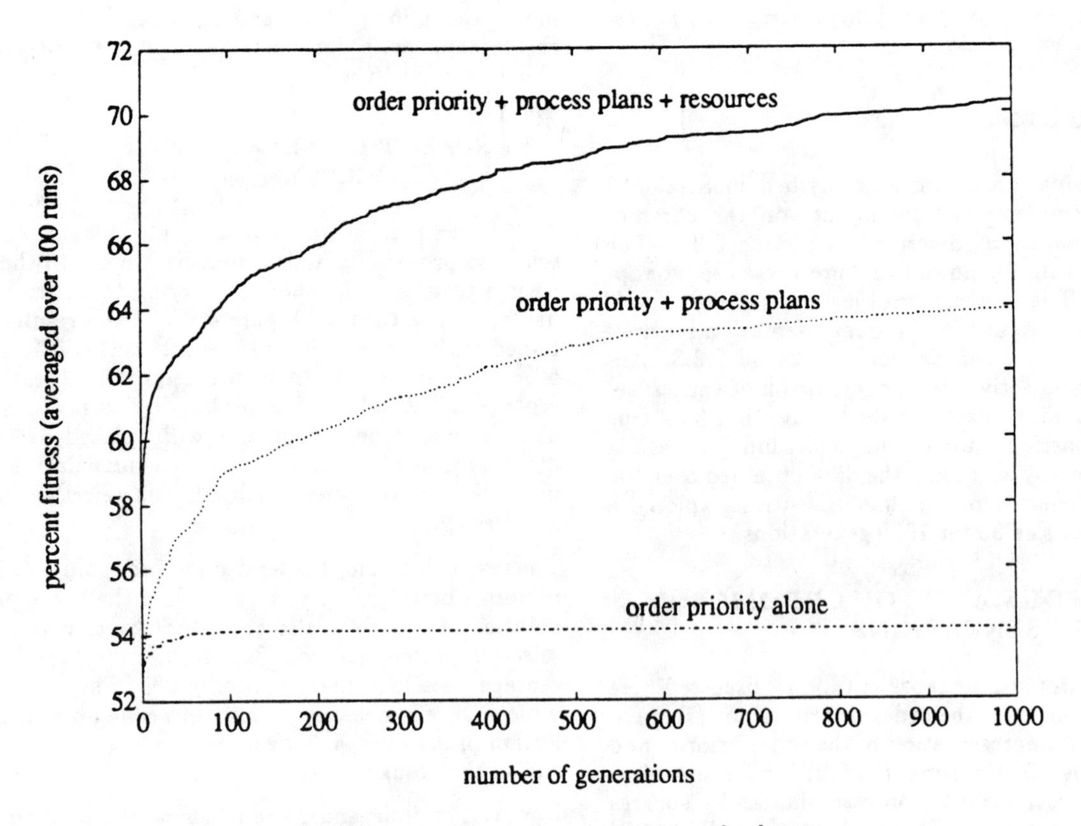

Figure 3: Effect of representational scheme

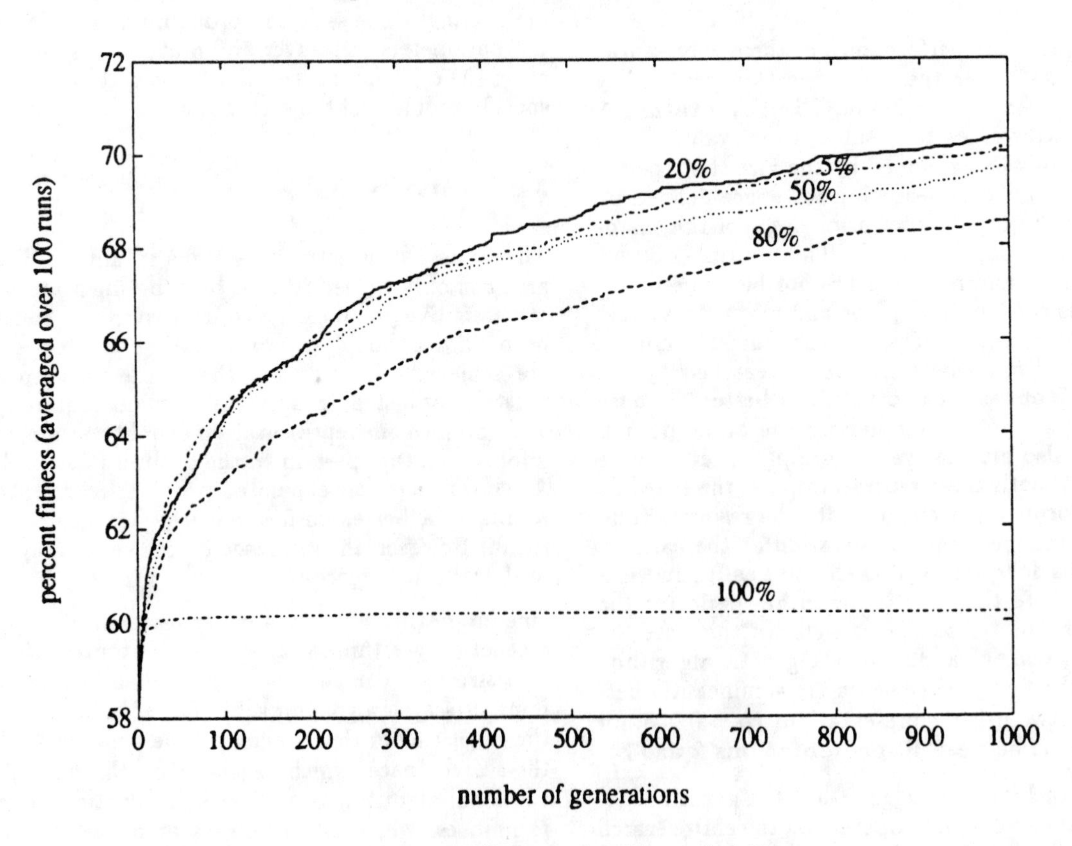

Figure 4: Effect of selection probability

researched.

Another issue is rescheduling and reactive scheduling. An operational scheduling system would never develop schedules from scratch. Instead, it will add incoming orders onto the existing schedule. Moreover, conflicts and problems arising in real time must be effectively dealt with by a reactive scheduler. The prototype scheduling system described in this paper is not capable of rescheduling or reactive scheduling. Therefore, improvements in search and scheduling methods are necessary in order to apply the existing prototype into dynamic, real world scheduling problems.

Acknowledgements

This research project is supported by a grant from the Computer Systems Laboratory of the Electronics R&D Laboratories of Nippon Steel Corporation, Sagamihara, Japan. The genetic algorithm used in this research is adapted from Splicer, developed by Steven E. Bayer for the MITRE Corporation and NASA. The authors would like to thank Mr. Bayer for his kind permission for the use of Splicer in this project.

References

[1] Cleveland, G. A. and S. F. Smith. *Using genetic algorithms to schedule flow shop releases,* Proceedings of the 3rd International Conference on Genetic Algorithms, 1989.

[2] Davis, L. Job shop scheduling with Genetic Algorithms, *Proceedings of an International Conference on Genetic Algorithms and their Applications,* Pittsburgh, PA, pp. 136–140, 1985.

[3] Fox, M. S. *Constraint-Directed Search: A Case Study of Job-Shop Scheduling,* Morgan Kaufmann Publishers, Inc., Los Altos, CA, 1987.

[4] Goldberg, D. E. and R. Lingle, Jr. Alleles, loci, and the traveling salesman problem, *Proceedings of an International Conference on Genetic Algorithms and their Applications,* Pittsburgh, PA, pp. 154–159, 1985.

[5] Grefenstette, J. J. Incorporating problem specific knowledge into Genetic Algorithms, *Genetic Algorithms and Simulated Annealing,* L. Davis, ed., Morgan Kaufmann Publishers, Inc., Los Altos, CA, 1987.

[6] Korf, R. E. Search: a survey of recent results, *Exploring Artificial Intelligence,* H. E. Shrobe, ed., Morgan Kaufmann Publishers, Inc., San Mateo, CA, 1988.

[7] Smith, S. F., M. S. Fox, and P. S. Ow. Constructing and maintaining detailed production plans: investigations into the development of knowledge-based factory scheduling systems, *AI Magazine,* Fall 1986, pp. 45–61, 1986.

[8] Syswerda, G. Schedule optimization using genetic algorithms. *The Genetic Algorithms Handbook,* L. Davis, ed., Van Nostrand Reinhold, 1991.

[9] Whitley, D., T. Starkweather, D. Fuquay. *Scheduling problems and the Traveling Salesmen: the genetic edge recombination operator,* Proceedings of the 3rd International Conference on Genetic Algorithms, 1989.

Bit-Climbing, Representational Bias, and Test Suite Design

Lawrence Davis
Tica Associates
36 Hampshire St.
Cambridge, MA 02139

Abstract

Genetic algorithms are often characterized as algorithms that are superior to hill-climbing algorithms in a number of domains because they manipulate higher-order combinations of bit strings called *schemata*. To develop, parameterize, and demonstrate the power of genetic algorithms, researchers have traditionally employed test suites of numerical function optimization problems. In this paper I show that these test suite problems do not require the ability to manipulate schemata for fast solution. I describe a hill-climbing algorithm that out-performs the genetic algorithm dramatically on the test suite problems. I discuss reasons for the success of the hill-climbing algorithm, and then apply the hill-climbing algorithm and the genetic algorithm to modified versions of the test suite problems that yield results more in line with theoretical expectations. I conclude by discussing the import of these results on our design of future test suite problems.

THE BIT-CLIMBING ALGORITHM

This paper describes a family of hill-climbing algorithms inspired by the traditional genetic algorithm. Like the traditional genetic algorithm, these hill-climbing algorithms use binary notation. Unlike the traditional genetic algorithm, they manipulate a single chromosome in a fashion I call *bit-climbing*.

Put intuitively, a bit-climbing algorithm works as follows. Begin with a chromosome composed of bits. Flip each bit of the chromosome in some order. Any time the evaluation of the new chromosome is the same as or better than the evaluation of the old, replace the old chromosome with the new one and continue testing bits. When the list of bits to check has run out, generate a new one. Whenever every bit has been tested with no improvement, stop and return the best chromosome.

This algorithm is more formally described in Figure 1. It should be noted that bit-climbing is different from a deterministic hill-climber, in which every bit is flipped and tested, and only then is the current chromosome altered by altering that bit giving the greatest improvement in the evaluation function. A bit-climber as defined here accepts any change yielding an identical or better evaluation when the change is found. This means that the behavior of RBC is critically determined by the order in which it tests the bits, whereas a deterministic hill-climber will always climb to the same point from a given starting point. It should also be noted that bit-climbing can be made more efficient if one notes which bit changes have been tested since the last change was accepted and avoids re-evaluating them. This refinement is not a feature of the algorithm in Figure 1, nor was it a feature of the bit-climbing algorithms whose performance is described in this paper.

Different versions of the bit-climbing algorithm may be obtained by altering the technique for producing the initial chromosome C and by altering the function F that decides in what order bit changes will be tested. One could set the initial chromosome to contain all 0's or all 1's, and test bits in a left-to-right or right-to-left direction. In this paper I will discuss only the performance of a bit-climber that begins with randomly-generated chromosomes, and that randomly generates a new flip-list whenever the current flip-list is empty. More formally, the algorithm considered here is such that, for each run, C is a randomly generated list of L bits and $F(1, 2, ..., L)$ is a random permutation that is recomputed with each call to F.

I call the bit-climbing algorithm used here "RBC", or "random bit-climber". It carries out multiple runs of random initialization and bit-flipping until a pre-allotted number of chromosome evaluations has been carried out. In some respects, RBC is reminiscent of simulated annealers, although RBC never accepts a change that lessens the evaluation of its current chromosome.

Figure 1: The Bit Climbing Algorithm

Given an initial chromosome *C* of length *L*, an evaluation function *E* that evaluates bit strings of length *L*, and a function *F* that returns a permutation of the list of numbers *(1, 2, ... L)*,

1. Set new-chromosome = *C*.

2. Set best-evaluation = *E(new-chromosome)*.
   Set flip-list = *F(1, 2, ... L)*.
   Set change = NO.

3. Set new-chromosome = new-chromosome with its nth bit flipped, where n is the first member of flip-list.
   Set new-evaluation = *E(new-chromosome)*.
   Set flip-list = flip-list with its first member removed.

4. If new-evaluation is better than best-evaluation,
   Set best-chromosome = new-chromosome,
   Set best-evaluation = new-evaluation,
   Set change = YES.

5. If flip-list is empty,
   If change = NO,
   return best-chromosome and halt.
   otherwise, go to 2.

6. Go to 3.

THE TEST PROBLEMS USED HERE

The first set of problems used in the experiments reported here consisted of the test suite assembled and/or designed by DeJong, described in [2]. The problems used also include two functions proposed as test suite functions by Schaffer et al in [3]. The DeJong test suite contained five functions -- F1 through F5. The first two functions in the Schaffer et al test suite are F6 and F7. Thus, the results reported here are based on runs of various optimization algorithms on F1 through F7 -- problems that have been fundamental in parameterizing and comparing the performance of traditional genetic algorithms.

The second set of problems used here is a modification of F1-F7. The modification involves shifting the values encoded on binary chromosomes, and so I call these problems "SF1-SF7" (for "shifted F1", etc.)

THE GENETIC ALGORITHMS

I compared RBC with two genetic algorithms. Both are taken from the OOGA system, a Common Lisp software system created to accompany the *Handbook of Genetic Algorithms* [1]. In all tests reported here, I used the double-precision arithmetic functions in GENERA(TM) 8.1 on a Symbolics 3630 Lisp Machine. The two genetic algorithms are as follows.

The Traditional Genetic Algorithm ("TGA") uses generational replacement; a population of 50 individuals; a linear ranked selection technique that assigns the best individual in the population a fitness of 100, the next-best a fitness of 99, and so forth; and a single crossover-and-mutate operator with mutation probability = .8% and crossover probability = 65%. The algorithm was run until 4000 individuals had been produced.

The Steady-State Genetic Algorithm ("SSGA") was like TGA except that no chromosomes that were duplicates of chromosomes already in the population were evaluated or inserted into the population; the fitnesses of population members were altered at 50-individual intervals, so that at the beginning of each run they were 100, 99.8, 99.6, ... and at the end of the run they were 100, 98.8, 97.6, 96.4, etc. SSGA was restarted whenever it had produced 4000 duplicates. (This generally happened on three of the problems in the test suite). SSGA was allowed restarts in order more fairly to compare it with RBC, an algorithm employing multiple runs. In the experiments reported here, allowing restarts improved the performance of SSGA to a minor degree on three of the problems; the amount of improvement does not alter the qualitative conclusions reported below.

The parameter settings for TGA and SSGA are fairly robust. I used both algorithms because the traditional genetic algorithm, while much slower, should do better with a noisy evaluation function and the test suite contains F4, a function with noise. As it turned out, given runs of 4000 individuals the steady-state algorithm did much better than the generational one, even on the noisy problem.

Although TGA produces some duplicate population members and for six of the seven functions such duplicate members require no re-evaluation, each new chromosome was evaluated in TGA even if it duplicated an earlier chromosome, and so each chromosome counted as an evaluation for TGA. This is the most robust version of the traditional genetic algorithm, and it handles noisy evaluation functions well when algorithms that do not re-evaluate their chromosomes may fail. Genetic algorithms that do not allow duplicates are represented by the SSGA.

THE EVALUATION METRIC

Most of the functions in the test set require extremely fine discrimination of results. For that reason, I used a place-counting metric for all functions except F3 and SF3. The place-counting metric is as follows: Given the optimal value for a problem, count the number of places in which an evaluation rounded off to that place matches the optimal value.

There is one parameter required for this evaluation metric, the number of places to compare to the left of the decimal point. I used the number of places in which the worst chromosome differs from the optimal result. Thus, the place-counting metric counts places starting this many places to the left of the decimal point: F1, 2; F2, 4; F4, 1; F5, 3; F6, 1; and F7, 6. The optimal value for all functions except F3 and F5 is 0. The place-counting metric measures with how much precision an optimization algorithm has found that value. The optimal value is -30 for F3 and ~1.0 for F5.

To illustrate the way this metric works, let us consider chromosome evaluations for F2 and their value using this metric. The optimal value is 0, and we begin counting four places to the left of the decimal. This metric scores 664.232 as 0, since the optimal value in the thousands place is 0 and 664 rounds off to 1 in the thousands place. The metric scores 4.256 as 3, since the thousands, hundreds, and tens places for this value match the optimal value. It should be noted that if the optimal value is found, this metric will produce an arbitrarily large result. I set a maximum value of 20 in producing the results described below.

The evaluation of chromosomes for F3 and SF3 is not done by place-counting. The optimal value of F3 is -30, and each evaluation is an integer. Unlike the other functions in the test suite, no logarithmically sensitive procedure is necessary for comparison of results. The results reported here for F3 and SF3 are simply the negation of each chromosome's evaluation.

The evaluation metric for F4 -- a function with noise -- was applied to the "true" evaluation of each chromosome -- the evaluation with noise omitted. When running any of the algorithms described in this paper on F4 I recorded both the noisy and "true" evaluation of each chromosome. The algorithms had access only to the noisy value. I applied the evaluation metric to the "true" value. Using this evaluation technique, optimization algorithms that handle noise badly may fail to find optimal points while those that can accommodate noise (like the traditional genetic algorithm) would be expected to do well.

The results reported on all examples here are ones in which a higher evaluation is better -- for every problem except F3 and SF3, because a higher evaluation means that more digits of the optimal answer have been found. These evaluation metrics were applied before averaging results to yield the performance measures described below. Each performance measure in the tables is the result of averaging evaluation metrics over 50 runs of an algorithm. The evaluation metric was always applied to the evaluation of the best chromosome in the population.

Figure 2: Average Final Evaluations For 50 Runs of 4000 Individuals on F1-F7.

	TGA	SSGA	RBC
F1	3.8	10.9	15.8
F2	4.3	6.9	6.7
F3	28.1	30/2250	30/1750
F4	5.2	7.6	14.0
F5	2.6	3.0/2000	3.0/900
F6	1.5	3.2	8.3
F7	2.7	6.0	13.3

INITIAL RESULTS

Figure 2 shows the average evaluation of the best individual in the population after 4000 individuals had been evaluated, using the evaluation metrics described above. The algorithms compared are TGA (the traditional genetic algorithm), SSGA (the steady-state genetic algorithm), and RBC (the random bit-climbing algorithm described above).

In Figure 2 we see that in every problem except F2 RBC exceeds the performance of TGA and SSGA. On F3 and F5 the performance of RBC at 4000 evaluations equals that of SSGA, but as the table shows, each run of RBC has found the optimal value after 1750 evaluations, as opposed to 2250 for SSGA. On F2 RBC is quite close to SSGA. The amount of improvement of RBC on SSGA is quite dramatic for several cases. In F1, F4, F6, and F7 RBC has found answers with five to seven more significant digits of precision than the genetic algorithms, in the same number of evaluations.

Figure 3 shows the number of evaluations required for each algorithm to equal the average evaluation of SGA after 4000 trials. Figure 2 gave us a feeling for the ultimate performance of each algorithm. Figure 3 gives us a feeling for the speed of each algorithm. The results

Figure 3: Number of Individuals Evaluated to Equal Evaluation of TGA at 4000 Individuals on F1-F7.

	TGA	SSGA	RBC
F1	3900	632	22
F2	3900	228	280
F3	4000	680	89
F4	4000	130	343
F5	3850	352	290
F6	3580	660	140
F7	4000	804	26
Average	3890	498	170

are not at all in keeping with our expectation, given our characterization of genetic algorithm performance. On average, SSGA is eight times faster than TGA. But RBC is about 23 times faster, and about three times faster than SSGA. A hill-climber has out-performed the genetic algorithm on all but one of the test suite problems, and its performance is quite close to that of the SSGA on that problem.

DISCUSSION OF INITIAL RESULTS

A central tenet of our field has been that genetic algorithms do well because they combine schemata, multiple-member subcomponents of good solutions. Yet RBC is an algorithm that manipulates one bit at a time and finds solutions from 3 to 23 times faster than the genetic algorithm, on the very problems that we have used to tune and compare genetic algorithms. Granted, it was never the explicit aim of the creators of the test suite problems that each problem require schemata recombination, but it was their belief that at least some of them require this capability. And in fact, I had expected that F6 was one of these. I created RBC for use on F6 when I was preparing the tutorial in [1]. I had intended to use RBC as a benchmark to display the power of the genetic algorithm as compared with a hill-climber. However, RBC is so obviously superior on F6 that my points would have been ill-served had I used it. I assumed that this phenomenon was a fluke and omitted discussions of RBC from the book.

As can be seen from the results reported above, the phenomenon was not a fluke at all. Each problem in the test suite displays the same sort of behavior: a simple hill-climbing algorithm nearly equals or greatly exceeds the performance of the genetic algorithms. What are we to make of this fact? Are we wrong about the power of genetic algorithms on numerical optimization problems? Is the strength of genetic algorithms their use of binary representation to encode numerical solutions rather than the use of population management and crossover? Can that strength be simply co-opted by a hill-climbing algorithm requiring none of the power of the GA?

The answer to these questions turns out to be "yes" for F1-F7, but "no" in general. Let us consider some points that bear on these questions.

One important feature of the runs of bit-climbing on F1-F7 is that they often require few bit changes to generate an optimal or near-optimal chromosome. Another way of putting this is that there is a high degree of linear separability in the solution fields in F1-F7. Another feature is that, because of the way the range of variable values has been set up, it is usually easy to determine by eye how close a chromosome is to the optimal chromosome. Either the optimal values are centered on 0 (as in F1, F6 and F7) and the fields of the optimal chromosome are of the form (1 0 0 0 0... 0 0) or (0 1 1 1 1 ... 1) or they are otherwise perched on very steep Hamming cliffs that may be strolled up if one approaches from the correct direction. RBC does just this: it approaches the cliffs from various sides in its multiple runs, and it rapidly finds their peaks when it is climbing from the correct side.

The patternedness of these optimal chromosomes leaps out at us humans (although it doesn't leap out at a genetic algorithm). This makes it easy to watch a GA run and determine at a glance how close the population is to a global optimum. But the fact that the problems have such patterned encodings appears to be correlated with ease in bit-climbing. This fact suggests that the problems in the test suite lack the property David E. Goldberg calls "GA-hardness". That is, there is very little deception in these problems. Put another way, although there appear to be cliffs that are difficult for hill-climbers to scale, it seems that these cliffs do not always lie between random bit-climbers and their goals.

To test this hypothesis, I created a second test suite containing SF1-SF7. Each problem is like its correlate in the original test suite. Each uses chromosomes of the same length and uses the same mathematical evaluation function. The only difference is that when each chromosome is decoded, the value of each decoded variable is shifted downward 10%. If this shift takes the value below the permitted range of the variable, the amount by which the value falls below the range is subtracted from the maximum value to yield the shifted value.

Figure 4: Average Final Evaluations For 50 Runs of 4000 Individuals on SF1-SF7.

	TGA	SSGA	RBC
SF1	2.2	5.8	3.9
SF2	4.4	6.5	4.5
SF3	25.3	26.9	25.3
SF4	2.6	4.0	3.2
SF5	3.7	5.0	3.7
SF6	1.3	2.1	2.0
SF7	1.9	2.8	2.5

Figure 5: Number of Individuals Evaluated to Equal Evaluation of SGA at 4000 Individuals on SF1-SF7.

	TGA	SSGA	RBC
SF1	3300	305	556
SF2	4000	212	3676
SF3	3600	513	3934
SF4	1850	234	2025
SF5	3900	380	680
SF6	3850	357	375
SF7	4000	716	850
Average	3500	388	1728

Shifting each value 10% in this way before applying the mathematical functions does not alter the size of the search space. Each set of chromosomes is mapped one-to-one onto the same set of values. Only the mapping function has changed. Thus, differences in the performance of our optimizers on the shifted test suite will not occur because of changes in the search space size.

Figure 4 shows the average final evaluations of the best population members on runs of SF1-SF7. Now the results are quite different. SSGA out-performs RBC in every case. And in nearly every case these evaluations for SSGA and TGA are much lower than those for F1-F7.

Figure 5 shows the number of evaluations required for each algorithm to equal or exceed the average best evaluation of TGA on SF1-SF7. Here again results are quite different. Now SSGA is four and a half times faster than RBC, SSGA is nine times faster than TGA, and RBC is only twice as fast as TGA. These results are more in keeping with our expectations. At a cost of lower final performance and slower optimization, we have produced a test suite in which RBC does not uniformly dominate a genetic algorithm.

CONCLUSIONS

There are a number of inferences to draw from these results. I should like to highlight several of them.

Bit climbing is a good test for GA-simplicity. A few runs of RBC on a function will provide a useful benchmark for comparison with a genetic algorithm. If the performance of bit climbers on a function exceeds that of the genetic algorithms, then the problem is probably not hard enough to require a genetic algorithm, and it may be that the problem is a simple one for any binary-based optimizer. RBC was able rapidly to determine the previously unknown fact that F1, F6 and F7 are uncharacteristically simple problems for a genetic algorithm using binary representation. I don't know of any other quick general test for binary bias. Of course, RBC is not a test for binary bias at the level of schemata of two or more bits, but it appears to be an effective diagnostic tool for the lowest level of binary bias.

Bit climbing may be a good optimizer in its own right. RBC and its relatives using left-to-right or right-to-left flip lists with all-zero or all-one initial chromosomes are very effective on some of the problems we have been considering. I believe it is worth communicating these algorithm to numerical function optimizers, for they merge the divide-and-conquer resolution strategy of the binary representation technique with the brute force hill-climbing methods that can be so effective in numerical function optimization. It seems likely that bit-climbing algorithms may be competitive hill climbers on problems that do not require the full range of genetic algorithm mechanisms.

The test suite is biased in favor of binary representation. The empirical results discussed above suggest that the problems in the test suite F1-F7 with optima centered on the range of variable values are inadvertently biased so that a binary genetic algorithm does better than it would over a range of similar problems without such centering. This conclusion is not *proven* by the empirical results. It could be that the performance of the genetic algorithm on F1-F7 is more typical than that for SG1-SF7. A set of experiments run by Gilbert Syswerda and me in the preparation of [4] confirms that F1-F7, rather than SF1-SF7, are atypical. On the basis of these and other experiments, I believe that a useful rule of thumb in test suite design should be that the optimal value of a variable should never lie on the midpoint of the range, as it does on F1, F6, and F7, since this is guaranteed to place the optimal values atop the second-highest Hamming cliffs in the problem world. Further, the range of values should not be manipulated so that the optima lie atop any high Hamming cliffs. If one fails to observe this design principle, one may not be equitably comparing the performance of an optimization algorithm using binary representation with that of algorithms using other representations. (And one may also be favoring algorithms that use mutation heavily as opposed to those that use crossover).

The test suite does not provide a good test for the handling of noise. F4 is the only problem in the suite that has a noisy evaluation function. However, the data show that SSGA is much more effective at finding good chromosomes on F4 than TGA, at least given 4000 evaluations. We need a noise-based example that discriminates algorithms that re-evaluate chromosomes from those that don't. A crude example of such a function would be F6 with an amount of noise uniformly distributed from 0 to 10.0 added to the evaluation. The amount of noise in this function is so great that genetic algorithms using re-evaluation techniques are bound to do better than those that do not.

The test suite does not provide a problem requiring long chromosomes. F4 is the only problem in the suite that has a long chromosome (if 240 fields counts as "long"). None of the other examples has a binary chromosome longer than 50. But F4 is also the only problem containing noise, and any test of our algorithms' ability to manipulate long chromosomes appears to have been swallowed up by the noise. A useful test suite problem would be one with long chromosomes and an evaluation problem that cannot be easily bit-climbed.

The standard parameters may not be representative. We have seen that the examples that have been used to find robust settings for a genetic algorithm are ones in which mutation will be much more effective than it might generally be. Will these values be robust when genetic algorithms are applied to more characteristic problems? Given the amount of time that can be (and has been) expended in finding robust parameter settings for F1-F5, it seems important to consider the generalizability of the results.

Gray coding may yield similar results. Hardness is not a property of problems. It is a relation between optimization algorithms and problems. What is a hard problem for a genetic algorithm using binary representation may be much simpler for one using Gray coding, and vice versa. An interesting question concerns the use of bit-climbing techniques on Gray-coded chromosomes to determine whether there is representational bias in favor of Gray-coded chromosomes in the test suite. The results we have produced in preparing [4] suggest to me that similar results will be obtained when Gray coding is used. However, this suggestion awaits further empirical test. Lawrence Eshelman, in personal communication, has reported similar results in empirical tests using Gray coding.

References

[1] Lawrence Davis. *Handbook of genetic Algorithms.* Van Nostrand Reinhold, New York, 1991.

[2] Kenneth A. DeJong. *An analysis of the behavior of a class of genetic adaptive systems.* Doctoral Dissertation, University of Michigan, 1975.

[3] J. David Schaffer, Richard A. Caruana, Larry J. Eshelman, and Rajarshi Das. A study of control parameters affecting online performance of genetic algorithms for function optimization. In J. David Schaffer, editor, *Proceedings of the Third International Conference on Genetic Algorithms.* Morgan Kaufmann, San Mateo, 1989.

[4] Gilbert Syswerda and Lawrence Davis. Binary bias in genetic algorithms. In preparation.

Don't Worry, Be Messy*

David E. Goldberg
Dept. of General Engineering
University of Illinois
Urbana, IL 61801

Kalyanmoy Deb
Dept. of Eng. Mechanics
University of Alabama
Tuscaloosa, AL 35487

Bradley Korb
McDonnell Douglas Space Systems Co.
Huntsville, AL 35806

Abstract

Foremost among the challenges facing genetic algorithms (GAs) is the so-called *linkage problem*. In difficult functions—so-called *deceptive* functions—if needed bits or features are not coded tightly on the artificial chromosome, simple genetic algorithms will bypass global solutions and will instead be misled toward local optima. Inversion and other reordering operators have not been successful in beating this problem, and straightforward theory has suggested that these operators are too slow to be of timely use. This problem has led to the invention of *messy genetic algorithms* (mGAs). Simply stated, messy genetic algorithms find and emphasize tightly coded substrings initially, juxtaposing them thereafter to find globally optimal structures. This paper reviews why mGAs should be investigated, how they differ from traditional GAs, why they work, what results have been obtained thus far, and which directions appear most fruitful for further work.

1 INTRODUCTION

Despite their empirical success, there has been a long-standing objection to the use of genetic algorithms (GAs). To assure convergence to global optima, strings in simple GAs must be coded so that *building blocks*—short, highly fit combinations of bits—can combine to form optima. If the linkage between necessary bit combinations is too weak, in certain types of problems called *deceptive* problems (Goldberg, 1987, 1989b, 1989c), genetic algorithms will converge to suboptimal points. A number of reordering operators have been suggested to recode strings on the fly, but these have not yet proved sufficiently powerful in empirical studies, and recent theoretical work (Goldberg & Bridges, 1990) has suggested that unary reordering operators are too slow to be of much use in searching for tight linkage.

To overcome this problem, a radically reorganized type of genetic algorithm called a *messy genetic algorithm* (mGA) has been devised and tested (Goldberg, Deb, & Korb, 1990; Goldberg, Korb, & Deb, 1989). Not only does the mGA respect a version of the schema theorem, it does so in a way that guarantees that appropriate building blocks will be formed and that they will be tight. In a number of problems of bounded deception, with mixed building-block size and fitness scaling, mGAs have repeatedly found global optima, and this convergence has taken place in a time that grows on a serial machine as a polynomial function of the number of variables.

This paper reviews the progress made to date in developing and applying messy genetic algorithms. Specifically, the differences between traditional and messy GAs are briefly outlined; elements of important mGA theory are then sketched, and computational results are quickly surveyed. The paper concludes with an examination of the next important steps in the development of these algorithms and their messy offspring.

2 TRADITIONAL GAs & mGAs

The details of simple genetic algorithms are covered in standard references (De Jong, 1975; Goldberg, 1989a; Holland, 1975), and messy GAs are described more fully elsewhere (Goldberg, Deb, & Korb, 1990; Goldberg, Korb, & Deb, 1989). Here, fundamental differences between traditional GAs and the messy approach are highlighted.

Messy GAs are different from simple GAs in four ways:

1. mGAs use variable-length codes that may be over- or underspecified with respect to the problem being solved;

*Portions of this paper are excerpted from a 1990 paper by the authors entitled Messy Genetic Algorithms Revisited: Studies in Mixed Size and Scale, *Complex Systems, 4*, 415–444.

2. mGAs use simple *cut* and *splice* operators in place of fixed-length crossover operators;
3. mGAs divide the evolutionary process into two phases: a primordial phase and a juxtapositional phase;
4. mGAs sometimes use competitive templates to accentuate salient building blocks.

Messy GAs are messy because they use variable-length strings that may be under- or overspecified with respect to the problem being solved. For example, the three-bit string 111 of a simple GA might be represented in a messy GA (using LISP-like notation) as `((1 1) (2 1) (3 1))`, where each bit is identified by its name and its value. In mGAs, since variable-length strings are allowed, interpretations must be found for strings with too few or too many bits. For example, the strings `((1 1) (2 1))` and `((1 1) (2 1) (3 1) (1 0))` are both valid mGA strings in a three-bit problem despite the lack of a third bit in the first and despite the extra first bit in the second. Note the conflict regarding bit one in the second string; in mGAs, simple first-come-first-served rules have been used with a left-to-right scan to arbitrate such conflicts. While overspecification may be handled with relatively simple rules, underspecification is the knottier problem, and its solution through the use of competitive templates is discussed in a moment.

To recombine the variable-length strings of mGAs, fixed-length, one- or two-cut crossover operators are abandoned in favor of separate cut and splice operators. Cut and splice are as simple as they sound. Cut severs a string with specified probability $p_c = (\lambda - 1)p_\kappa$ that grows as the string length λ, and splice joins two strings together with fixed probability p_s. Together, cut and splice have roughly the same potential disruption as simple crossover and the same juxtapositional power (Goldberg, Korb, & Deb, 1989). The detailed theoretical argument for cut and splice must also consider the continued expression of good building blocks. Such consideration has led to the division of the evolutionary process into two phases.

Simple GAs usually process strings in a homogeneous fashion. A population is generated initially (usually at random), and subsequent generations remain the same size, using reproduction, crossover, and other genetic operators to create the next population from the current population. By contrast, mGAs divide the genetic processing into two distinct phases: a primordial phase and a juxtapositional phase. In the primordial phase, the population is first initialized to contain all possible building blocks of a specified length, where the characteristic length is chosen to encompass possibly deceptive (misleading) building blocks. Thereafter, the proportion of good building blocks is enriched through a number of generations of reproduction without other genetic action. At the same time, the population size is usually reduced by halving the number of individuals in the population at specified intervals.

With the proportion of good building blocks so enriched, the juxtapositional phase proceeds with a fixed population size and the invocation of reproduction, cut, splice, and other genetic operators. Cut and splice act to recombine the enriched proportions of building blocks. In empirical tests to date, mGAs have always found globally optimal strings. Moreover, this convergence has been shown to occur in a time that grows no faster than a polynomial function of the number of decision variables.

This accomplishment was only made possible through the invention of a noise-free method of evaluating salient building blocks. This procedure, the method of *competitive templates*, solves the problem of underspecification alluded to earlier by filling in unnamed genes with a locally optimal structure. In this way, only salient building blocks obtain fitness values better than the template value and are thereby enriched during the primordial phase. In some problems, under some codings, it may be possible to evaluate partial strings without filling in missing positions from some competitive template. The original study demonstrated the effectiveness of mGAs in such problems in partial-string, partial-evaluation experiments.

Together, these four differences, messy strings, messy operators, a two-phase evolutionary process, and competitive templates, distinguish mGAs from their more traditional cousins. Yet, novelty for its own sake is no virtue. In the next section, we examine the key elements of mGA theory that have guided the design of mGAs in an effort to obtain more reliable building-block processing.

3 ELEMENTS OF mGA THEORY

Over the course of the investigation of mGAs, a number of critical theoretical conclusions have been drawn that have guided both the design of the mGA itself and the design of the computational experiments:

1. mGAs satisfy a form of the schema theorem and have similar juxtapositional power as compared to simple GAs.
2. Functions of bounded deception are the worst functions that GAs with reordering and recombination can hope to solve and are therefore the appropriate target for empirical and theoretical study.
3. Competitive templates achieve reliable building-block evaluation.
4. On a serial machine, mGAs require a number of function evaluations to convergence that grows only as a polynomial function of the number of

decision variables. On a parallel machine (with a polynomial number of processors), the convergence time grows only as the logarithm of the number of decision variables.

5. When processing kth-order building blocks, mGAs appear to converge with high probability to a solution no worse than the order-k truncated solution. For functions of problems with bounded deception, this leads to the conjecture that mGAs find global optima with high probability.
6. Thresholding and tie-breaking can handle the twin problems of mixed fitness scaling and length.

Each of these is considered in somewhat more detail.

3.1 SCHEMA THEOREM

The earlier paper (Goldberg, Korb, & Deb, 1989) developed the necessary extensions to the schema theorem to account for mGAs. The details will not be recounted here, but a primary difference between simple GAs and mGAs is that under messy GAs, both a schema's *physical presence* and its continued *expression* must be considered. When this is done, and when simple first-come-first-served expression rules are adopted with the cut and splice operators discussed earlier, and when these operators are applied *after* the population has been doped with the best building blocks of a particular order, straightforward analysis yields an interesting result: the probability of schema survival and continued expression is no worse than 0.5. In conjunction with binary tournament selection (which gives roughly two copies to the best schema), this arrangement of the algorithm means that the best schemata should be expected to do no worse than maintain their numbers in the population. With good, tight building blocks and growth of the best building blocks no worse than neutral, it is little wonder that results thus far have been encouraging.

3.2 DECEPTIVE FUNCTIONS

The use of deceptive functions (Goldberg, 1989b, 1989c; Liepins & Vose, 1990) as test functions is critical to understanding the convergence of mGAs, traditional GAs, or any other similarity-based search technique for that matter. The argument is straightforward. Deceptive functions are designed explicitly to mislead similarity-based methods away from global optima and toward the complement of the global optimum. Of course, it is unreasonable to expect a GA to find ℓ-deceptive optima (where ℓ is the length of the problem), because such optima are truly needles in a haystack. On the other hand, it is reasonable to expect GAs to solve problems of *bounded deception*, in which the bit combinations needed have order no higher than $k \ll \ell$, and in these first studies, test functions of bounded deception have been constructed from sums of disjoint deceptive subfunctions of known order. Subfunctions of different length have been adopted, and different scalings of fitness values have been tried in an attempt to deceive the mGA. In all cases, simple extensions of the algorithm have been sufficient to permit convergence to global optima. These extension will be discussed in a moment, but the ability to evaluate a part without benefit of the whole has turned out to be an important element of mGA theory.

3.3 COMPETITIVE TEMPLATES

Competitive templates are critical to mGAs, because they permit the accurate evaluation of partial strings. The argument is straightforward, yet subtle. Assume that order-k building blocks are being processed, and further assume that the competitive template that is locally optimal at level $k-1$ is used to fill in the missing $\ell-k$ positions of the substring (at level $k = 1$, any fixed template will do). If the function is deterministic and stationary, then the only structures that will achieve a function value better than that of the competitive template alone are those that are building blocks at the level k. Moreover, among directly competing gene combinations, the best building block at the level k will get the best increment over the competitive template value. In this way, mGAs are able to separate the value of a bit combination from the string without prior function knowledge and without large stochastic variations. Arguments have been made elsewhere (Goldberg, Korb, & Deb, 1989) against the use of other methods of handling the underspecification problem such as averaging techniques, because the amount of fitness variation or *collateral noise* is quite large. It is interesting that the same case argues against the use of traditional GAs unless proper accounting is made of the fitness variance in the sizing of GA populations (Goldberg & Rudnick, 1991). But assuming that collateral noise is properly accounted in simple GAs, it is still unclear why one should choose the noisy evaluation inherent in simple GAs over the quiet one designed into mGAs.

The idea of using locally optimal templates at the next lower level suggests the most practical way of using mGAs. Starting at the level $k = 1$, an order-1 optimal template can be found, which in turn is used to find a level-2 template, and so on. In this way, mGAs can climb the *ladder of deception* one rung at a time, obtaining useful intermediate results at the same time the solution is being refined. It is interesting to note that this ladder-climbing analogy carries over to the computational cost of solutions with increasing k, a matter addressed in the next subsection.

Table 1: Phase Duration and Time-complexity Estimates for an mGA

Phase	Duration	Serial	Parallel
Primordial	$O(\log \ell)$	$O(\ell^k)$	$O(1)$
Lengthening	$O(\log \ell)$	$O(\ell \log \ell)$	$O(\log \ell)$
Crossing	$O(\log \ell)$	$O(\ell \log \ell)$	$O(\log \ell)$
Overall mGA	$O(\log \ell)$	$O(\ell^k)$	$O(\log \ell)$

3.4 TIME COMPLEXITY OF mGAs

Assuming that function evaluations require much greater processing time than genetic operators, that is $t_f \gg t_{ga}$, and considering the use of either a serial machine that performs one function evaluation per unit time or a parallel processor that can simultaneously perform a polynomial number of function evaluations per unit time, the time complexity of the mGA may be considered by analyzing the mGA in three phases, the primordial phase, the lengthening portion of the juxtapositional phase, and the crossing portion of the juxtapositional phase. The details of the analysis are provided elsewhere (Goldberg, Deb, & Korb, 1990), but the duration, serial complexity, and parallel complexity of the mGA are shown in table 1 for each phase and for the overall mGA. On a serial machine, an mGA requires a number of function evaluations that grows as a polynomial function of the number of decision variables, $O(\ell^k)$. It is interesting that the calculation is dominated by the computation of the initialization phase. This suggests that if prior information is available regarding the function that would permit restriction of initialization to a limited number of building blocks (something less than $O(\ell \log \ell)$), then the overall serial complexity can be reduced to a svelte $O(\ell \log \ell)$. On a parallel machine with enough processors, initialization can be done in constant time, as can the generational function evaluations during lengthening and crossing. Thus, on a large enough parallel machine, the mGA requires computations that grow only as fast as a logarithmic function of the number of decision variables. These estimates are encouraging and bode well for the future of messy GAs in combinatorial function optimization, especially when considered in the light of the following conjecture.

3.5 CONJECTURE: mGAs FIND BEST

That mGAs converge in polynomial time or better is important, but polynomial convergence is no virtue if that convergence is incorrect. Empirically, mGAs have always found globally optimal results in problems of bounded deception, leading us to the following conjecture:

Conjecture 1 *With probability that can be made arbitrarily close to one, messy GAs converge to a solution at least as good as the truncated order-k solution. Moreover, this convergence occurs in a time that is* $O(\ell^k)$ *on a serial machine and* $O(\log \ell)$ *on a parallel machine.*

Previously, the truncated order-k solution was defined in terms of a truncated Walsh series, but this definition is incorrect because of recent results that construct fully deceptive functions of any order using Walsh coefficients no greater than order three (Goldberg, 1990). A better working definition considers the top ℓ/k nonoverlapping partitions and assumes that the mGA finds no worse a solution than the one that can be obtained through the independent mixing of the best and second-best bit combinations within each of the top nonoverlapping partitions. Regardless of this needed change in focus, the bottom line remains the same. It appears when a mGA attacks a problem of bounded (order-k) deception at level k that the mGA solves the problem to global optimality.

The plausibility of this conjecture can be seen quite readily. During the primordial phase, the best building blocks grow logistically as long as the fitness signal is reliable (and as long as apples are compared to apples, but we will have more to say about this in a moment when we discuss the need for genic thresholding). During the lengthening portion of the juxtapositional phase, the best building blocks will hold their own on average, because reproduction will continue to increase their number at a rate near doubling, and splicing must continue to express a building block no less than half of the time (because half the time a currently expressed building block will be placed at the left end of the string, guaranteeing continued expression under the first-come-first-served rule). Thereafter during the crossing portion of the juxtapositional phase, the mGA behaves very much like a simple GA with very tight building blocks, and continued convergence proceeds according to an inequality that looks very much like the standard schema theorem.

Although the conjecture is reasonable, taking it to theoremhood is nontrivial as it is insufficient to deal with the trajectory of the population in expectation. A method for turning the schema theorem into a rigorous decision-theoretic bound by including variance adjustments has been suggested elsewhere (Goldberg & Rudnick, 1991) and appears to be a promising road to rigorous convergence proofs for all GAs, messy or otherwise; however, a number of details of the analysis should not be underestimated. Nonetheless, the outline gives more than a hint of the convergence mechanism underlying mGAs and provides some explanation of the remarkable empirical results observed to date.

3.6 THRESHOLDING & TIE-BREAKING

In traditional fixed-length GAs, strings all have the same set of genes and, though the allele values may differ, all head-to-head comparisons have some meaning, because each string may be viewed as a full solution to the problem at hand. In mGAs, structures need not have any genes in common, and competition between arbitrary substrings is likely to be a case of comparing apples to oranges. Moreover, there are no guarantees that string lengths are matched to building-block size, and this opens the possibility that parasitic bits may tag along with good building blocks, later deactivating other needed substructures. These problems both result because mGAs are trying to solve many subproblems explicitly (and simultaneously), thereby giving an mGA a more ecological flavor than a traditional GA. Ecological GAs call for ecological solutions, and two mechanisms have been adopted with empirical success: *genic thresholding* and *length-based tie-breaking*.

Genic thresholding requires that two strings share some threshold θ number of *genes* before they are permitted to compete in tournament selection. The expected number of genes in common between randomly chosen strings of length λ_1 and λ_2 is hypergeometrically distributed. Thus, the expected number in common is $\theta = \frac{\lambda_1\lambda_2}{\ell}$, and this is the threshold value used in the mGA studies.

Tie-breaking is used to prevent parasitic bits—bits that agree with the competitive template, but currently serve no purpose other than acting as a placeholder—from tagging along with some low-order building block, thereafter preventing the expression of some other building block. An additional $k-1$ *null bits* (explicit placeholders) may be added to the functional genes, or all building blocks at the current level and less may be generated, but the key point is to prefer strings with least length (or least effective length in the case of null bits) when fitness ties occur between building blocks. In this way strings will be preferred that have the least amount of stray genetic material, thereby reducing the threat of blocked expression from parasitic bits.

4 RESULTS TO DATE

mGAs have been tested on a number of problems:

1. Base case: partial-string, partial-evaluation runs.
2. Base case: competitive template runs.
3. Scaling case: nine-up, one-down (27:1 scaling).
4. Scaling case: one-up, nine-down (27:1 scaling).
5. Scaling case: linear scaling (10:1 scaling).
6. Monkey-wrench case: $3+1$-bit monkey wrench.
7. Monkey-wrench case: $3+3$-bit monkey wrenches.
8. Mixed-size and scale case: 36-bit problem, with 3, 4, 5-bit building blocks (25:9 scaling).

In all cases, the mGA was run repeatedly, and in the base cases a simple GA was run as a control. In the base cases, a 30-bit function was created by summing 10 equally scaled copies of a three-bit deceptive function designed elsewhere (Goldberg, 1989b). Whether the simple GA could solve the problem depended upon the linkage condition assumed. With perfectly tight linkage the simple GA converged to global results, yet with maximally loose linkage the simple GA converged to the poorest local optimum (of 1024 total optima). The mGA partial-string, partial-evaluation runs sidestepped the issue of evaluating partial strings by only giving credit to subfunctions with full gene complements. So doing assumes knowledge of the function structure that is not always available, but the study demonstrated the utility of the mGA approach in those problems where such information may be available. It also set the stage for the competitive template runs. In the competitive template runs, the mGA was given the worst possible competitive template (all zeroes), and in all cases the algorithm converged to the global solution. This was the first time that any GA had found global results in a provably difficult problem without knowledge of appropriate linkage.

In the scaling runs, a 30-bit function was again constructed by adding ten copies of the 3-bit deceptive function, but in these runs subfunctions were scaled differently. Moreover, steps were taken to overwhelm the least-fit building blocks by biasing the population toward most highly fit building blocks. In all cases, the mGA was undeterred by these attempts to disrupt performance, and global optima were found quickly.

The problem of mixed subfunction size was first investigated in the monkey-wrench cases. Using the original 30-bit problem with level scaling, substrings (monkey wrenches) containing an optimal three-bit building block together with a single parasitic bit or a combination of three parasitic bits designed to block subsequent expression of other optimal building blocks were added to the additional population . This was done in an attempt to overwhelm the best building blocks during the juxtapositional phase. With a single parastic bit, it was interesting that convergence without tie-breaking was slowed, but it was not prevented. Convergence with tie-breaking was as rapid as when no monkey wrenches were used originally. With three parasitic bits added, the tie-breaking procedure was necessary to achieve global convergence, and that convergence was as rapid as that obtained originally.

A function with both mixed subfunction size and scale was created as the sum of three, 5-bit subfunctions, three, 4-bit subfunctions, and three, 3-bit subfunctions. Figure 1 shows the maximum number of opti-

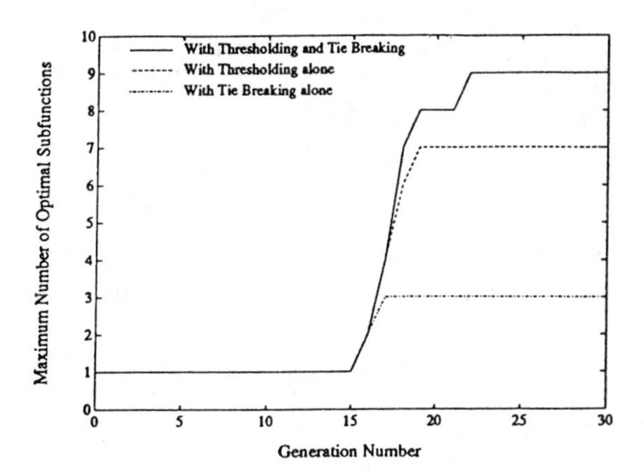

Figure 1: Maximum number of optimal subfunctions correct versus generation in a mixed size-scale problem.

mal building blocks versus generation for the function of mixed-size and scale. The mGA with tie breaking and thresholding gets all nine building blocks correct shortly after the beginning of the lengthening portion of the juxtapositional phase. Neither thresholding alone nor tie breaking alone is able to do as well.

Thus far, mGAs have always found global results in every run on all test functions investigated. Additional theoretical and empirical work is needed before this can be claimed as a general result, and the next section outlines some of the necessary steps.

5 NEXT STEPS

Further inquiry is needed and underway in a number of areas:

1. analyze overlapping subfunctions;
2. extend niching techniques to mGAs;
3. prove the fundamental conjecture of mGAs;
4. implement a parallel mGA;
5. develop other messy code types, including permutation codes, messy floating-point codes, and messy classifiers (rules);
6. extend mGAs to nondeterministic and nonstationary functions;

Some progress has been made in understanding the role of overlapping partitions. A recent paper (Goldberg, 1990) showed that high-order, fully deceptive functions can be created using overlapping low-order Walsh coefficients. This dashes early hopes of simple algebraic tests for deception, but mGAs remain a valid approach to such problems as long as the order of deception of the function is bounded and is a good bit less than the problem length.

The genic thresholding technique used in mGAs so far ensures that only meaningful gene combinations compete, but niching—the simulataneous discovery of multiple solutions—requires that the measure of similarity be extended to the allele level. Theoretical and empirical work is now pointed in this direction.

The variance results mentioned earlier (Goldberg & Rudnick, 1991) have taken us closer to a proof of the fundamental conjecture of mGAs. Thinking along variance lines may nail down bounding probabilities on the global convergence of both simple GAs and mGAs.

There are no immediately obvious serial bottlenecks preventing the use of mGAs on massively parallel machines. Tournament selection can be done in local processor neighborhoods, and cut, splice, and other genetic operators require no more than pairwise interactions between individuals. Actual parallel implementations and empirical tests should be undertaken to make sure these speculations are accurate.

Some work on messy floating-point codes has been attempted along the lines suggested in the earlier papers, and results from these studies will soon be reported. Messy permutations were also suggested previously, and these will be attempted shortly, but another possibility for encoding permutations in a somewhat indirect way involves the use of the Boolean satisfiability problem as suggested elsewhere (De Jong & Spears, 1989). Classifier systems lend themselves to messy codings, because only information-carrying positions need to be mentioned explicitly. Moreover, it is quite easy to imagine a syntactical means to Michigan-Pitt approach fusion by allowing messy rules to come together in bundled rule clusters, clusters that act as the locus of reward and reproduction. Careful experimental design is necessary to wade into these waters, and deception and its meaning in each problem domain needs special attention. Nonetheless, it is clear that mGAs are not restricted to functions over some number of Boolean decision variables.

Messy GAs have focused attention on the importance of *gene expression*, and mGAs may be extended to handle nonstationary functions if intrachromosomal dominance or more traditional dominance-diploidy mechanisms are adopted. Although mGAs filter out collateral noise through the use of competitive templates, functions that are inherently nondeterministic may be handled by mGAs if appropriate steps are taken to average multiple copies of each building block initially. The variance paper (Goldberg & Rudnick, 1991) may be especially useful in choosing appropriate duplication counts.

Clearly there are many avenues for further work, yet enough is known that mGAs may confidently be applied to difficult, blind combinatorial problems.

6 CONCLUSIONS

For some time, traditional genetic algorithms have promised broadly efficacious search behavior in fairly arbitrary environments, but GAs have sometimes failed to deliver because there have been no practical techniques for assuring the tight linkage necessary to solve tough problems to global optimality. By getting the linkage right, by using an expression operator that places a lower bound on the schema-loss term, by using competitive templates to get a low-noise evaluation of building blocks, and by invoking ecological operators to make most competitions meaningful, messy GAs appear to solve problems of bounded deception to global optimality in a time that grows no more quickly than a polynomial function of the number of decision variables on a serial machine and as a logarithmic function of the number of decision variables on a parallel machine with a polynomial number of processors.

Much work remains to be done. That mGA convergence is global remains to be proved, and important application areas remain to be tried. Yet, because the results to date have been so encouraging, we recommend the immediate application of mGAs to difficult combinatorial problems of practical import. Although several i's remain to be dotted and a number of t's are still there for the crossing, we believe that this technique is on the way to becoming an important weapon in the genetic algorithmist's arsenal to combat nontrivial blind combinatorial problems in a rational, efficient manner.

Acknowledgments

This material is based upon work supported by Subcontract No. 045 of Research Activity AI.12 under the auspices of the Research Institute for Computing and Information Systems (RICIS) at the University of Houston, Clearlake under NASA Cooperative Agreement NCC9-16 and by the National Science Foundation under Grants CTS-8451610 and ECS-9022007. Continued support of mGA research under U.S. Army Contract DASG60-90-C-0153 is also acknowledged. Hardware and software provided by the Digital Equipment Corporation and Texas Instruments Incorporated was used to perform some of the computations reported herein. Mr. Deb acknowledges research support under a University of Alabama Graduate Council Fellowship.

References

De Jong, K. A. (1975). An analysis of the behavior of a class of genetic adaptive systems (Doctoral dissertation, University of Michigan). *Dissertation Abstracts International, 36*(10), 5140B. (University Microfilms No. 76-9381)

De Jong, K. A., & Spears, W. M. (1989). Using genetic algorithms to solve NP-complete problems. *Proceedings of the Third International Conference on Genetic Algorithms*, 124–132.

Goldberg, D. E. (1987). Simple genetic algorithms and the minimal, deceptive problem. In L. Davis (Ed.), *Genetic algorithms and simulated annealing* (pp. 74–88). Los Altos, CA: Morgan Kaufmann.

Goldberg, D. E. (1989a). *Genetic algorithms in search, optimization, and machine learning.* Reading, MA: Addison-Wesley.

Goldberg, D. E. (1989b). Genetic algorithms and Walsh functions: Part I, a gentle introduction. *Complex Systems, 3*, 129–152.

Goldberg, D. E. (1989c). Genetic algorithms and Walsh functions: Part II, deception and its analysis. *Complex Systems, 3*, 153–171.

Goldberg, D. E. (1990). *Construction of high-order deceptive functions using low-order Walsh coefficients* (IlliGAL Report No. 90002). Urbana, IL: University of Illinois at Urbana-Champaign, The Illinois Genetic Algorithms Laboratory.

Goldberg, D. E., & Bridges, C. L. (1990). An analysis of a reordering operator on a GA-hard problem. *Biological Cybernetics, 62* (5), 397–405.

Goldberg, D. E., Deb, K., & Korb, B. (1990). Messy genetic algorithms revisited: Studies in mixed size and scale. *Complex Systems, 4*, 415–444.

Goldberg, D. E., Korb, B., & Deb, K. (1989). Messy genetic algorithms: Motivation, analysis, and first results. *Complex Systems, 3*, 493–530.

Goldberg, D. E., & Rudnick, M. (1991). *Genetic algorithms and the variance of fitness* (IlliGAL Report No. 91001). Urbana, IL: University of Illinois at Urbana-Champaign, The Illinois Genetic Algorithms Laboratory.

Holland, J. H. (1975). *Adaptation in natural and artificial systems.* Ann Arbor, MI: University of Michigan Press.

Liepins, G. E., & Vose, M. D. (1990). Representational issues in genetic optimization. *Journal of Experimental and Theoretical Artificial Intelligence, 2*(2), 4–30.

An Experimental Comparison of Binary and Floating Point Representations in Genetic Algorithms

Cezary Z. Janikow*
Department of Computer Science
University of North Carolina
Chapel Hill, NC 27599, USA

Zbigniew Michalewicz
Department of Computer Science
University of North Carolina
Charlotte, NC 28223, USA

Abstract

Genetic Algorithms (GAs) are innovative search algorithms based on natural phenomena, whose main advantages lie in great robustness and problem independence. So far, GAs were most successful in parameter optimization domains; however, even there certain problems, as lack of fine local tuning capabilities and severe time complexity, prohibit their wider use on most moderately and highly complex problems. Recently, there has been a growing interest in the floating point (FP) representation for genetic algorithms. In this paper we empirically study both FP and binary based GAs using a dynamic control problem — highly complex and quite difficult for any method. Results suggest that the well known advantages of low cardinality alphabets can be compensated for by designing new operators, and that such approach provides means for overcoming some of the mentioned disadvantages.

1 INTRODUCTION

The binary alphabet offers the maximum number of schemata per bit of information of any coding [6] and consequently the bit string representation of solutions has dominated genetic algorithm research. This coding also facilitates theoretical analysis and allows elegant genetic operators. But the 'implicit parallelism' result does not depend on using bit strings [1] and it may be worthwhile to experiment with large alphabets and (possibly) new genetic operators.

In [7] the author wrote:

> "The use of real-coded or floating-point genes has a long, if controversial, history in artificial genetic and evolutionary search schemes, and their use as of late seems to be on the rise. This rising usage has been somewhat surprising to researchers familiar with fundamental genetic algorithm (GA) theory ([6], [10]), because simple analyses seem to suggest that enhanced schema processing is obtained by using alphabets of low cardinality, a seemingly direct contradiction of empirical findings that real codings have worked well in a number of practical problems."

The same paper presents a theory of convergence for real-coded GAs that use floating point codings in their chromosomes, and discusses it further:

> "Although the theory helps suggest why many problems have been solved using real-coded GAs, it also suggests that real-coded GAs can be *blocked* from further progress in [some] situations."

However, we argue that some modifications of genetic operators on float point representation may result in a much better performance; these may be also very useful when the problem to be solved involves non-trivial constraints ([12], [13], [14]), and they can help in avoiding such "blocked" situations. Moreover, the search space of the floating point representation is (to a very close degree) equivalent to the problem space. This, in turn, allows a more conscious design of problem specific operators, and actually extends the idea of using a special coding (as Grey) to bring the two spaces together.

Subsequently, we empirically compare a binary implementation with a floating point implementation using various new operators. As a test case we selected a non–trivial, non–decomposable dynamic control problem [15]. The results, due to the limited context of such experiments, should be looked at as a case, rather than generalizable, study; more systematic experimentations must be performed to draw more convincing conclusions.

*Present address: Department of Mathematics and Computer Science, University of Missouri, St. Louis, Missouri 63121–4499

2 THE TEST CASE

For experiments we have selected the following dynamic control problem:

$$min \left(x_N^2 + \sum_{k=0}^{N-1} (x_k^2 + u_k^2) \right)$$

subject to

$$x_{k+1} = x_k + u_k, \; k = 0, 1, \ldots, N-1,$$

where x_0 is a given initial state, $x_k \in R$ is a state, and $\vec{u} \in R^N$ is the sought control vector. The optimal value can be analytically expressed as

$$J^* = K_0 x_0^2$$

where K_k is the solution of the Riccati equation

$$K_k = 1 + K_{k+1}/(1 + K_{k+1}) \text{ and } K_N = 1$$

During the experiments a chromosome represented a vector of the control states $\vec{u}$. We have also assumed a fixed domain $\langle -200, 200 \rangle$ for each u_i (actual solutions fall withing this range for the class of tests performed). For all subsequent experiments we used $x_0 = 100$ and (unless otherwise stated) $N = 45$. Therefore, a chromosome was represented by a vector $\vec{u} = \langle u_0, \ldots, u_{44} \rangle$, having the optimal value J^* 16180.4.

3 THE TWO IMPLEMENTATIONS

For the study we have selected two genetic algorithm implementations differing only by representation and applicable genetic operators, and equivalent otherwise. Such an approach gave us a better basis for a more direct comparison. Both implementations used the same selective mechanism: stochastic universal sampling [2].

3.1 THE BINARY IMPLEMENTATION

In the binary implementation each element of a chromosome vector was coded using the same number of bits. To facilitate a fast run time decoding, each element occupied its own word (in general it could occupied more than one if the number of bits per element exceeded the word size, but this case is an easy extension) of memory: this way, to read gene's values, elements could be accessed as unsigned integers, which removed the need for binary to decimal decoding (it still required representation range → domain scaling). Then, each chromosome was a vector of N words, with N equals the number of elements per chromosome (or again a multiple of such for cases where multiple words were required to represent desired number of bits).

The precision of such an approach depends (for a fixed domain size) on the number of bits actually used, and equals $(UB - LB)/(2^n - 1)$, where UB and LB are domain bounds and n is the number of bits per one element of a chromosome.

3.2 THE FLOATING POINT IMPLEMENTATION

In the floating point (FP) implementation each chromosome vector was coded as a vector of floating point numbers, of the same length as the solution vector. Each element was later initialized within the desired range, and the operators were carefully designed (closed) to preserve this requirement.

The precision of such an approach depends on the underlying machine, but is generally much better than that of the binary representation. Of course, we can always extend the precision of the binary representation by introducing more bits, but this considerably slows down the algorithm (see Section 5). In addition, the FP representation is capable of representing quite large domains (or cases of unknown domains). On the other hand, the binary representation must sacrifice the precision for an increase in domain size, given fixed binary length.

4 THE EXPERIMENTS

The experiments were conducted on a DEC3100 workstation. All presented results represent the average values obtained from 10 independent runs. During all experiments the population size was kept fixed at 60, and the number of iterations was set at 20,000. Unless otherwise stated, the binary representation was using $n = 30$ bits to code one variable (one element of the solution vector), needing $30 \cdot 45 = 1350$ bits for the whole vector.

Because of possible differences in interpretation of different operators, we accepted a probability of chromosomes' update as a fair measure of effort between the floating point and binary representations. Then, all experiments were conducted with individual operators probabilities set to achieve the same such update rates. Accordingly, we could compare runs of both implementations with approximately the same rate of function evaluations. (the number of function evaluations was approximately equal to population size × update rate × number of iterations).

4.1 RANDOM MUTATION AND CROSSOVER

In this part of the experiment we ran both implementations with operators which were equivalent (at least for the binary representation) to the traditional ones.

4.1.1 Binary

The binary implementation used traditional operators of mutation and crossover. However, for compatibility with the FP implementation, we allowed crossover points to fall between elements only. The probability of crossover was fixed at 0.25, while the probability of mutation varied to achieve desired rate of chromosome update (shown in Table 1).

Table 1: Relation between Probabilities of Chromosome's Update and Mutation Rate

	Probability of chromosome's update				
	0.6	0.7	0.8	0.9	0.95
Bin	0.00047	0.00068	0.00098	0.0015	0.0021
FP	0.014	0.02	0.03	0.045	0.061

4.1.2 FP

The crossover operators was analogous (and actually equivalent) to that of the binary implementation (split points between float numbers) and applied with the same probability (0.25). The mutation, which we call random, applies to a floating point number rather that to a bit, with an appropriate probability as to achieve the same rates of chromosome's updates (same number of function evaluations) as for the binary case (see Table 1); the result of such a mutation is a random value from the domain $\langle LB, UB \rangle$ with a uniform distribution (special non–uniform distributions will be used in so called "dynamic mutation" introduced in Section 4.2.1).

4.1.3 Results

Table 2: Average Results as a Function of Probability of Chromosome's Update

	Probability of chromosome's update					std.
	0.6	0.7	0.8	0.9	0.95	dev.
Bin	42179	46102	29290	52769	30573	31212
FP	46594	41806	47454	69624	82371	11275

The results (Table 2) are slightly better for the binary case; however, it is rather difficult to judge them better as all fell quite away from the optimal solution (16180.4). Moreover, an interesting pattern that emerged showed that the FP implementation was more stable, with much lower standard deviation.

In addition, it is interesting to note that the above experiment was not quite fair for the FP representation; its random mutation behaves "more" randomly than that of the binary implementation, where changing a random bit (with a uniform distribution) doesn't imply producing a totally random value from the domain. As an illustration let us consider the following question: what is the probability that after mutation an element will fall within δ% of the domain range (400, since the domain is $\langle -200, 200 \rangle$) from its old value? The answer is:

FP : Such probability clearly falls in the range $\langle \delta, 2 \cdot \delta \rangle$. For example, for $\delta = 0.05$ it is in $\langle 0.05, 0.1 \rangle$.

Binary : Here we need to consider the number of low order bits that can be safely changed. Assuming $n = 30$ as an element length and m as the length of permissible change, m must satisfy $m \leq n + \log_2 \delta$. Since m is an integer, then $m = \lfloor n + \log_2 \delta \rfloor$. Again, for $\delta = 0.05$, $m = 25$, and the sought probability is $m/n = 25/30 = 0.833$ — quite a different number.

Therefore, we will try to design a method of compensating for this drawback in the following subsection.

4.2 DYNAMIC MUTATION

In this part of the experiments we ran, in addition to the operators discussed in Section 4.1, a special dynamic mutation operator aimed at both improving a single element tuning and reducing the above disadvantage of random mutation in the FP implementation.

4.2.1 FP

The new operator is defined as follows: if $s_v^t = \langle v_1, \ldots, v_m \rangle$ is a chromosome (t is the generation number) and the element v_k was selected for this mutation, the result is a vector $s_v^{t+1} = \langle v_1, \ldots, v_k', \ldots, v_m \rangle$, where

$$v_k' = \begin{cases} v_k + \Delta(t, UB - v_k) & \text{if a random digit is 0} \\ v_k - \Delta(t, v_k - LB) & \text{if a random digit is 1} \end{cases}$$

The function $\Delta(t, y)$ returns a value in the range $[0, y]$ such that the probability of $\Delta(t, y)$ being close to 0 increases as t increases. This property causes this operator to search the space uniformly initially (when t is small), and very locally at later stages; thus increasing the probability of generating the new number closer to its successor than a random choice. We have used the following function:

$$\Delta(t, y) = y \cdot \left(1 - r^{(1 - \frac{t}{T})^b}\right),$$

where r is a uniform random number from $[0..1]$, T is the maximal generation number, and b is a system parameter determining the degree of dependency on iteration number (we used $b = 5$).

4.2.2 Binary

To be more than fair to the binary implementation, we modeled the dynamic operator into its space, even though it was primarily introduced to improve the FP mutation. Here, it is analogous to that of the FP, but with a differently defined v_k':

$$v_k' = mutate(v_k, \nabla(t, n)),$$

where $n = 30$ is the number of bits per one element of a chromosome; $mutate(v_k, pos)$ means: mutate value of the k-th element on *pos* bit (0 bit is the least significant), and

$$\nabla(t,n) = \begin{cases} \lfloor\Delta(t,n)\rfloor & \text{if a random digit is 0} \\ \lceil\Delta(t,n)\rceil & \text{if a random digit is 1} \end{cases}$$

with the b parameter of Δ adjusted appropriately if similar nonuniformity is desired (for examples see Figure 1).

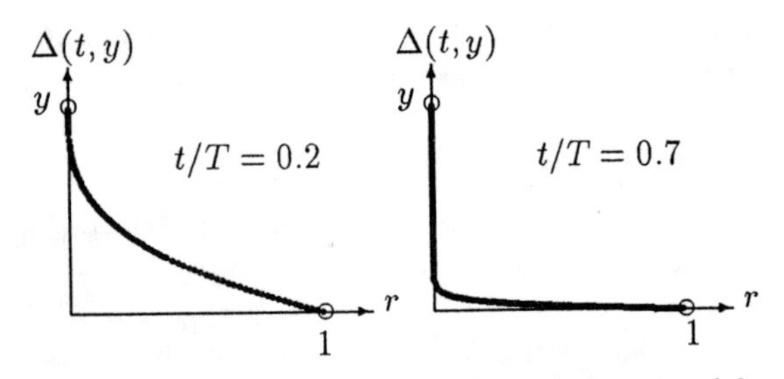

Figure 1: Δ function for two selected times and $b = 4$

4.2.3 Results

We repeated similar experiments to those of section 4.1.3 using also the dynamic mutations applied at the same rate as the previously defined mutations.

Table 3: Average Results as a Function of Probability of Chromosome's Update

	Probability of chromosome's update			std. dev.
	0.7	0.8	0.9	
Bin	32275	35265	30373	40256
FP	21098	20561	26164	2133

Now the FP implementation shows a better average performance (Table 3). In addition, again the binary's results were more unstable. However, it is interesting to note here that despite its high average, the binary implementation produced the two single best results for this round (16205 and 16189).

4.3 OTHER OPERATORS

In this part of the experiment we decided to implement and use some additional operators — those easy to implement in each space. Therefore, the purpose of this part of the experiment was not to compare both implementations in exactly the same context, but rather to see what level of quality could be obtained by using a set of easily implementable operators. Actually, this is where we start to distinguish between problem-independent and problem-dependent operators, to show that problem-specific operators are superior.

4.3.1 Binary

In the binary representation, the space is that of binary strings. This provides for the highly acclaimed operator problem-independence, since all operators can be defined in this space regardless of the underlying problem space. In addition to those previously described operators, we implemented a multi-point crossover, and also allowed for the classical crossovers (crossover points within bits of an element). The multi-point operator was introduced to set aside the single vs. muilt-point crossover debate; the probability of a crossover split was controlled by a system parameter (set at 0.3).

4.3.2 FP

In the floating point representation we deal directly (disregarding finite precision) with the problem space. Therefore, we can easily define new operators acting on real space vectors rather that some artificial agents. Accordingly, in addition to those previously described operators, we also implemented an analogous multi-point crossover, and single and multi-point arithmetical crossovers; They average values of two corresponding elements (rather that exchange them), at selected points. Such operators have the property that each element of the new chromosomes is still within the original domain. A version of such an arithmetical crossover averages two whole chromosomes along all dimensions, and simulates finding a midpoint between two points of a real space. For more details on these operators an interested reader is referred to [11], [13], [12], [18].

4.3.3 Results

Table 4: Average Results as a Function of Probability of Chromosome's Update

	Probability of chromosome's update			std. dev.	Best
	0.7	0.8	0.9		
Bin	23814	19234	27456	6078	16188.2
FP	16248	16798	16198	54	16182.1

Here the FP implementation shows an outstanding superiority (Table 4); Even though the best results are not so much different, only the FP was consistent in achieving that.

5 TIME PERFORMANCE

Many complain about the high time complexity of GAs on nontrivial problems. In this section we compare the time performance of both implementations using the mutation and crossover as defined in Section 4.1.

Table 5: CPU Time (sec) as a Function of Number of Elements

	Number of elements (N)				
	5	15	25	35	45
Bin	1080	3123	5137	7177	9221
FP	184	398	611	823	1072

Table 5 compares CPU time for both implementations on varying number of elements in the chromosome. The FP version is much faster, even for the moderate 30 bits per variable in the binary implementation; Both times are linear in the chromosome's length. Since we executed approximately the same number of function evaluations, and there was no need for binary decoding other than domain scaling (see Section 3.1), the major factor for these differences had to be the operator selection mechanisms. Furthermore, since the crossover operators were, in fact, identical, the mutation mechanisms had to contribute mostly. Actually, the reasons for such time disparities in these mechanisms are easily visible from Table 1: while seeking applicable mutation antities, the binary implementation iterates over all bits of a chromosome, but the floating point implementation iterates only over all elements of a chromosome. In other words, if other operations of the implementations were neglected, one would expect the binary one to run slower by the factor equal to the number of bits required to represent one gene (30 in the above run). The above fact was confirmed by measuring time spent in various parts of each algorithm. Note that the above holds only for our definitions of floating representation operators; if one wished to model these operators on the classical ones, the outcome might be quite different.

From the above discussion follows that the time disparity between binary and floating point implementations is directly proportional to the number of bits per one gene of the former. This implies that for problems which, due to some problem specific goals, require high precision, and, therefore, a longer bitwise gene representation, the difference should increase. This claim is exemplified in Table 6.

Table 6: CPU Time (sec) as a Function of Number of Bits Per Element; $N = 45$

	Number of bits per binary element					
	5	10	20	30	40	50
Bin	4426	5355	7438	9219	10981	12734
FP	1072 (constant)					

6 SUMMARY

The results of the conducted experiments indicate that the floating point representation is faster, more consistent from run to run, and provides higher precision (especially with large domains where binary coding would require prohibitively long representation). At the same time its performance can be enhanced by special operators to achieve high (even higher than that of the binary representation) performance accuracy. The design of such operators is, however, easy in the representation space approximately equivalent to the problem space. This approach abandons the idea of problem–independent operators; however, the floating point representation was introduced especially to deal with real parameter problems and we see no drawbacks of tailoring the operators to such domains.

These results support other studies praising the floating point representation, *e.g.* [7] gives the following reasons for such a preference: (1) comfort with one-gene-one-variable correspondence, (2) avoidance of Hamming clifs and other artifacts of mutation operating on bit strings treated as unsigned binary integers, (3) fewer generations to population conformity.

Acknowledgments: This research was supported by a grant from the North Carolina Supercomputing Center.

References

[1] Antonisse, J., *A New Interpretation of Schema Notation that Overturns the Binary Encoding Constraint*, in [17], pp. 86–91.

[2] Baker, J.E., *Reducing Bias and Inefficiency in the Selection Algorithm*, in [9].

[3] Bosworth, J., Foo, N., and Zeigler, B.P., *Comparison of Genetic Algorithms with Conjugate Gradient Methods*, Washington, DC, NASA (CR–2093), 1972.

[4] Davis, L., (Editor), *Genetic Algorithms and Simulated Annealing*, Pitman, London, 1987.

[5] De Jong, K.A., *Genetic Algorithms: A 10 Year Perspective*, in [8], pp.169–177.

[6] Goldberg, D.E., *Genetic Algorithms in Search, Optimization and Machine Learning*, Addison Wesley, 1989.

[7] Goldberg, D.E., *Real-coded Genetic Algorithms, Virtual Alphabets, and Blocking*, University of Illinois at Urbana-Champaign, Technical Report No. 90001, September 1990.

[8] Grefenstette, J.J., (Editor), Proceedings of the First International Conference on Genetic Algorithms, Pittsburg, July 24–26, Lawrence Erlbaum Associates, Publishers, 1985.

[9] Grefenstette, J.J., (Editor), Proceedings of the Second International Conference on Genetic Algorithms, MIT, Cambridge, July 28–31, Lawrence Erlbaum Associates, Publishers, 1987.

[10] Holland, J., *Adaptation in Natural and Artificial Systems*, Ann Arbor: University of Michigan Press, 1975.

[11] Janikow, C., and Michalewicz, Z., *Specialized Genetic Algorithms for Numerical Optimization Problems*, Proceedings of the International Conference on Tools for AI, Washington, November 6–9, pp.798–804, 1990.

[12] Michalewicz, Z. and Janikow, C., *GENOCOP: A Genetic Algorithms for Numerical Optimization Problems with Linear Constraints*, to appear in Communications of ACM, 1991.

[13] Michalewicz, Z. and Janikow, C., *Genetic Algorithms for Numerical Optimization*, Statistics and Computing, Vol.1, No.1, 1991.

[14] Michalewicz, Z. and Janikow, C., *Handling Constraints in Genetic Algorithms*, Proceedings of the 4th International Conference on Genetic Algorithms, San Diego, July 13–16, 1991.

[15] Michalewicz, Z., Krawczyk, J., Kazemi, M., Janikow, C., *Genetic Algorithms and Optimal Control Problems*, Proceedings of the 29th IEEE Conference on Decision and Control, Honolulu, pp.1664–1666, December 5–7, 1990.

[16] Schaffer, J., Caruana, R., Eshelman, L., and Das, R., *A Study of Control Parameters Affecting Online Performance of Genetic Algorithms for Function Optimization*, in [17], pp.51–60.

[17] Schaffer, J., (Editor), Proceedings of the Third International Conference on Genetic Algorithms, George Mason University, June 4–7, 1989, Morgan Kaufmann Publishers, 1989.

[18] Vignaux, G.A. and Michalewicz, Z., *A Genetic Algorithm for the Linear Transportation Problem*, IEEE Transactions on Systems, Man, and Cybernetics, Vol.21, No.2, 1991.

Evolving a Computer Program to Generate Random Numbers Using the Genetic Programming Paradigm

John R. Koza
Computer Science Department
Stanford University
Stanford, CA 94305 USA
Koza@Sunburn.Stanford.Edu
415-941-0336

ABSTRACT

This paper demonstrates that it is possible to genetically breed a computer program that is considered difficult to write, namely, a randomizer that converts a sequence of consecutive integers into pseudo-random bits with near maximal entropy.

1. INTRODUCTION AND OVERVIEW

"How can computers learn to solve problems without being explicitly programmed?" This question, which is a central question in the fields of artificial intelligence and machine learning, can be approached using an analogy to the evolutionary process in nature.

John Holland's pioneering 1975 *Adaptation in Natural and Artificial Systems* [3] described how the evolutionary process in nature can be applied to artificial systems using the "genetic algorithm" operating on fixed length character strings.

Representation is a key issue in genetic algorithm work because genetic algorithms directly manipulate the coded representation of the problem and because the representation scheme can severely limit the window by which the system observes its world [2]. For many problems in artificial intelligence, the most natural representation for solutions to problems are hierarchical computer programs of indeterminate size and shape, as opposed to structures whose size has been determined in advance. It is unnatural and difficult to represent hierarchies of dynamically varying size and shape with fixed length character strings.

In this paper, we show how to genetically breed a population of computer programs to convert a sequence of consecutive integers into a sequence of pseudo-random bits using the recently developed "genetic programming" paradigm.

2. DEFINITION OF RANDOMNESS

Numbers "chosen at random" are useful in a variety of scientific, mathematical, engineering, and industrial applications, including Monte Carlo simulations, decision theory, instant lottery ticket production, etc. Random numbers are difficult to create. Marsaglia [18] describes the difficulty of successfully randomizing numbers, particularly where a stream of several "independent" random integers are required to carry out related steps of one process.

When random numbers are required in computer programs, they are typically provided by a deterministic algorithm (as opposed to some non-algorithmic technique, such as neutron emissions). The algorithm is known to the programmer. Moreover, the program is typically written (using "seeds") so that its output is fully reproducible (to aid debugging, verification, etc.). Numbers which are produced by a deterministic, known, and reproducible algorithm are, of course, anything but random. Indeed, as John Von Neumann said, "Anyone who considers arithmetical methods of producing random digits is, of course, in a state of sin."

Not only is the term "random number" an oxymoron, but there is no generally accepted mathematical definition of a random number sequence. Knuth [17] describes a number of different specific tests, such as the equidistribution (frequency) test, gap test, run test, serial test, permutation test, coupon collector's test, poker test, etc. D.H.Lehmer (17) said "A random sequence is a vague notion embodying the idea of a sequence in which each term is unpredictable to the uninitiated and whose digits pass a certain number of tests, traditional with statisticians and *depending somewhat on the uses to which the sequence is to be put* ." (Italics added).

The notion of statistical independence can be used as the starting point for one possible rigorous mathematical definition of a sequence $\{u_i\}$ of random numbers. Anderson [1], for example, says that "each value of u_i is as likely as any other value and the value of u_i must be statistically independent of the value of u_j for $i \neq j$." The first part of this statement corresponds to the equidistribution (frequency) test described in Knuth [17] while the second part of this statement corresponds to a pairwise version of the serial test described in Knuth.

One can expand Anderson's conditions to a more general concept of statistical independence based on the intuitive notion of conditional probability that a specified sub-sequence predicts another sub-sequence. Let K be the number of values possible at a given position in the sequence. Let c be the number of consequent (i.e. predicted) positions within a sub-sequence of length N. Let g (where $g \leq N-c$) be the number of antecedent (i.e. given) values at the remaining N–c positions within the sub-sequence of length N. Then, for any integer N (where N runs from 1 to infinity), all K^c conditional probabilities of c particular specified values at the c consequent positions (given that g particular specified antecedent values have appeared at the g antecedent positions), are all equal to $\frac{1}{K^c}$ (within an acceptably small error $\varepsilon \geq 0$). Note that if g is strictly less than N–c, then there are N–g–c positions within the sequence which are part of neither the antecedent nor consequent part of the prediction. That is, they are "don't care" positions. For a given N, there are multinomial coefficient $\binom{N}{g \; c \; N-g-c}$ ways of picking the g given antecedent positions and c consequent positions out of the subsequence of length N. And then, after g such given antecedent positions are picked, there are then K^g particular assignments of values at the g antecedent positions. After those choices are made, there are then K^c particular assignments of values at the c consequent positions. The conditional probabilities of the K^c particular assignments of values that should be equal.

The above definition captures the intuitive notion based on conditional probabilities; however, it is particularly onerous combinatorially. The following simpler definition avoids conditional probabilities and is sufficient to guarantee satisfaction of the above conditional probability test. For any integer N (where N runs from 1 to infinity), the probabilities of each of the K^N possible sub-sequences of length N are all equal to $\frac{1}{K^N}$ (within an acceptably small error $\varepsilon \geq 0$).

No finite sequence can satisfy the above test. However, if N is then limited to some finite fixed integer N_{max}, then "only" $K^{N_{max}}$ probabilities must be estimated when $N = N_{max}$. Moreover, these $K^{N_{max}}$ separate probabilities can be conveniently summarized into a scalar quantity by using the concept of entropy for this set of events and probabilities. The entropy (which is measured in bits) is maximal when the probabilities of all the possible events are equal. Moreover, when K = 2, this maximal value of entropy happens to equal the length N (in bits) of the sequence. The entropy E_h for the set of K^h probabilities for the K^h possible sub-sequences of length h, equals

$$E_h = -\sum P_{hj} \log_2 P_{hj}.$$

Here j ranges over the K^h possible sub-sequences of length h. By convention, $\log_2 0$ is 0.

If K = 2, then this sum attains its maximum value of h precisely when the probabilities of all the K^h possible sub-sequences of length h are equal to $\frac{1}{2^h}$.

As h runs from 1 to N_{max}, it is convenient to further summarize the N_{max} separate scalar values of entropy into one single scalar value, namely, E_{total} as follows:

$$E_{total} = \sum_{h=1}^{N_{max}} \left[-\sum_j P_{hj} \log_2 P_{hj} \right]$$

When E_{total} attains the maximal value of

$$\sum_{h=1}^{N_{max}} h = N_{max}(N_{max} - 1),$$

then the sequence may be viewed as random.

To illustrate for K = 2: When h is 1, there are only two probabilities to consider, namely, the probability P_{10} of occurrence of zeroes and the probability P_{11} of occurrence of ones in the entire random sequence. These two probabilities are what are used in Knuth's equidistribution (frequency) test. For the worst randomizer (say one that always emits 1), P_{10} is 0.0 and P_{11} is 1.0 and the entropy is 0.0 bits. For a better randomizer, both P_{10} and P_{11} would equal approximately 0.5 and the entropy would approximately equal the maximal value of 1.0 bits. However, a defective randomizer such as one that emits

```
0101010101...
```

has entropy of 1.0 bits when only the two "singlet" probabilities associated with h = 1 are considered. However, when h = 2, the "pair" probabilities of the four possible pairs (i.e. 00, 01, 10, and 11) are not all equal to $\frac{1}{4}$. The two pairs 00 and 11 do not appear at all in the output of this defective randomizer. The two pairs 01 and 10 appear with probability $\frac{1}{2}$. Thus, the entropy for h = 2 for this defective randomizer is only 1.0, instead of the maximal value of 2.0 bits possible.

3. TYPES OF PSEUDO-RANDOM NUMBER GENERATORS

The common types of randomizers start with one or more seeds [1]. Multiplicative congruential randomizers start with a seed value x_0 and then produce subsequent elements of the sequence recursively as follows:

$$x_i = (a\, x_{i-1} + c) \mod M,$$

where a is the multiplier, c is the additive constant, and M is the modulus. Park and Miller [19] describe the especially simple and popular randomizer

$$x_i = 7^5\, x_{i-1} \mod [2^{31} - 1].$$

which provides especially good randomness by many tests for the low-order bits.

The popular URN08 randomizer came from IBM in 1970 and is widely known as "RANDU". IBM's RANDU is the multiplicative congruential randomizer

$$x_i = 65539\, x_{i-1} \mod 2^{31}.$$

Shift register randomizers start with a seed value x_0 and then produce subsequent elements of the sequence recursively in a shift register. In the popular SR[3,28,31] shift register randomizer (called "SHIFT REGISTER" herein), the numbers 3 and 28 specify the amount of shifting (end off, with zero fill) to the right or left (respectively) in a 31-bit shift register. With XOR being the exclusive-or operation,

```
temp =(XOR xi-1 (SHIFT-RIGHT xi 3))
xi =(XOR temp (SHIFT-LEFT temp 28).
```

Shuffling randomizers call on one or more other randomizers to shuffle numbers to produce random numbers. One two-sequence shuffling randomizer (called SHUFFLE herein) uses the Park-Miller randomizer described above to produce an initial set of uniformly distributed random numbers between 0.0 and 1.0 and then uses the shift register randomizer SR[3,28,31] to call out particular numbers from this set of numbers, while using additional calls on Park-Miller [19] to replace the numbers called out.

Texas Instruments supplies a randomizer called RANDOM with its Explorer™ computers.

Because of the size of tables needed for the sub-sequence probabilities needed to compute the total entropy, we will focus, for the remainder of this paper, on producing sequences of random binary digits (i.e. $K = 2$).

4. BACKGROUND ON GENETIC PROGRAMMING PARADIGM

We have recently shown that entire computer programs can be genetically bred to solve problems in a variety of different areas of artificial intelligence, machine learning, and symbolic processing. Specifically, this recently developed genetic programming paradigm has been successfully applied [4, 5, 12] to example problems in several different areas, including

- planning (e.g. navigating an artificial ant along a trail and developing a robotic plan for stacking blocks in to a desired order) [4, 6],
- emergent behavior (e.g. discovering a computer program for locating food, carrying food to the nest, and dropping pheromones, which, when executed by all the ants in an ant colony, produces interesting higher level "emergent" behavior) [10],
- machine learning of functions (e.g. learning the Boolean 11-multiplexer function) [11],
- automatic programming (e.g. solving pairs of linear equations, solving quadratic equations for complex roots, and discovering trigonometric identities) [4],
- discovering inverse kinematic equations (e.g. to move a robot arm to designated target points) [12, 16],
- optimal control (e.g. centering a cart and balancing a broom on a moving cart in minimal time by applying a "bang bang" force to the cart) [13, 14],
- pattern recognition (e.g. translation-invariant one-dimensional shape in a linear retina) [4],
- sequence induction (e.g. inducing a recursive procedure for generating sequences such as the Fibonacci and the Hofstadter sequences) [4],
- symbolic "data to function" regression, integration, differentiation, and symbolic solution to general functional equations (including differential equations with initial conditions, and integral equations) [4],
- empirical discovery (e.g. rediscovering Kepler's Third Law, rediscovering the well-known non-linear econometric "exchange equation" $MV = PQ$ from actual, noisy time series data for the money supply, the velocity of money, the price level, and the gross national product of an economy) [8],
- concept formation and decision tree induction [9],
- finding minimax strategies for games (e.g. differential pursuer-evader games, discrete games in extensive form) by both evolution and co-evolution [7], and
- simultaneous architectural design and training of neural networks [15].

A visualization of the application of the genetic programming paradigm to planning, emergent behavior, machine learning, empirical discovery, inverse kinematics, and game playing can be viewed in the *Artificial Life II Video Proceedings* videotape [16].

In the genetic programming paradigm, the individuals in the population are compositions of functions and terminals appropriate to the particular problem domain. The set of functions used typically includes arithmetic operations, mathematical functions, conditional logical operations, and domain-specific functions. The set of terminals used typically includes inputs (sensors) appropriate to the problem domain and various constants. Each function in the function set must be well defined for any combination of elements from the range of every function that it may encounter and every terminal that it may encounter.

The search space is the hyperspace of all possible compositions of functions that can be recursively composed of the available functions and terminals.

The symbolic expressions (S-expressions) of the LISP programming language are an especially convenient way to create and manipulate the compositions of functions and terminals described above. These S-expressions in LISP correspond directly to the "parse tree" that is internally created by most compilers.

The basic genetic operations for the genetic programming paradigm are fitness proportionate reproduction and crossover (recombination). The crossover (recombination) operation is a sexual operation that operates on two parental LISP S-expressions and produces two offspring S-expressions using parts of each parent. The crossover operation creates new offspring S-expressions by exchanging sub-trees (i.e. sub-lists) between the two parents. Because entire sub-trees are swapped, this crossover operation always produces syntactically and semantically valid LISP S-expressions as offspring regardless of the crossover points. For example, consider the two parental S-expressions:

```
(OR (NOT D1) (AND D0 D1))

(OR (OR D1 (NOT D0))
    (AND (NOT D0) (NOT D1)))
```

These two S-expressions are depicted as rooted, point-labeled trees with ordered branches in Fig. 1.

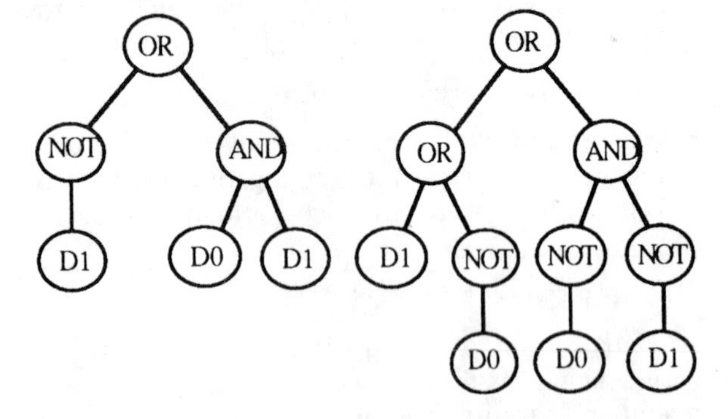

Figure 1: Two Parental LISP S-expressions shown as trees with ordered branches.

Assume that the points of both trees are numbered in a depth-first way starting at the left. Suppose that the second point (out of 6 points of the first parent) is randomly selected as the crossover point for the first parent and that the sixth point (out of 10 points of the second parent) is randomly selected as the crossover point of the second parent. The crossover points in the trees above are therefore the NOT in the first parent and the AND in the second parent. The two crossover fragments are two sub-trees shown in Figure 2 below:

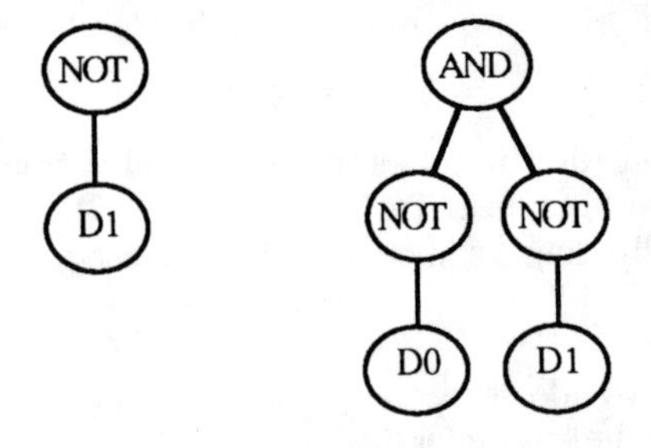

Figure 2: The Two Crossover Fragments

These two crossover fragments correspond to the bold, underlined sub-expressions (sub-lists) in the two parental LISP S-expressions shown above. The two offspring resulting from crossover are shown in Figure 3 below. Note that the first offspring above is an S-expression for the even-parity function, namely

```
(OR (AND (NOT D0) (NOT D1)) (AND D0 D1)).
```

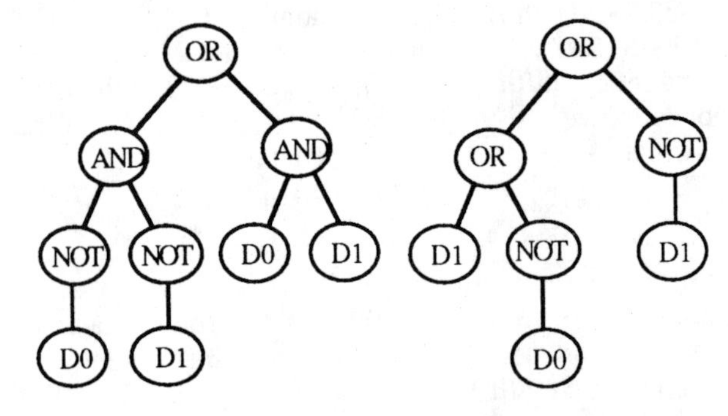

Figure 3: Offspring Resulting from Crossover

5. BREEDING A RANDOMIZER

Our goal is to genetically breed a computer program to convert a sequence of consecutive integers into a sequence of random binary digits. The input to our randomizer will merely be an argument J running consecutively from 1 to 16384 (2^{14}). Randomizers of this type can quickly reconstruct a particular single random number within a sequence without having to reconstruct the entire sequence. This type of randomizer typically has a complex structure because it does not rely on a recursive seed. A randomizer using a recursive seed as its input gains considerable computational leverage from the cascading of its own previous steps. For example, when the third call is made to a recursive randomizer with initial seed x_0 in order to compute x_3, the randomizer is, in effect, computing

$$x_3=(a[(a\{(a x_0+ c) \bmod M\}+c) \bmod M]+ c) \bmod M.$$

The first major step in using the genetic programming paradigm is to identify the set of terminals. The terminals in the genetic programming paradigm correspond to the inputs to the computer program being genetically bred. Thus, the terminal set for this problem is the set T = {J, $\Re$}, where $\Re$ represents small random integer constants between 0 and 3.

The second major step in using the genetic programming paradigm is to identify a sufficient set of functions for the problem. The protected modulus function MOD% and protected integer quotient function QUOT% uses the "protected" division function %, which returns one when division by zero is attempted, and, otherwise, returns the normal quotient. Since we anticipate creation of a randomizer involving congruential type steps, the function set for this problem

F = {+, –, *, QUOT%, MOD%}.

The third major step in using the genetic programming paradigm is identification of the fitness function for evaluating how good a given computer program is at solving the problem at hand. The raw fitness here is the sum of the entropy measures E_{total} (described earlier) for sub-sequences of length 1 through 7. The maximum raw fitness is therefore 28.

The fourth major step in using the genetic programming paradigm is selecting the values of certain parameters. The most important parameter is population. The population size is 500 here. Each new generation was created from the preceding generation by applying the fitness proportionate reproduction operation to 10% of the population and by applying the crossover operation to 90% of the population (with one parent selected proportionate to fitness). In selecting crossover points, 90% were internal (function) points of the tree and 10% were external (terminal) points of the tree. Mutation was not used. For the practical reason of conserving computer time, the depth of initial random S-expressions was limited to 4 and the depth of S-expressions created by crossover was limited to 15.

Finally, the fifth major step in using the genetic programming paradigm is the criterion for terminating a run and accepting a result. We will terminate a given run when either (1) the genetic programming paradigm produces computer program whose entropy of 27.990 or greater, or (2) 51 generations have been run. Double precision arithmetic was used in computing entropy.

Note that the first two of these five major steps in applying the genetic programming paradigm corresponds to the step (performed by the user) of determining the representation scheme in the conventional genetic algorithm operating on character strings (that is, determining the chromosome length, alphabet size, and the mapping between the problem and chromosomes).

In addition, note that the step (performed by the user) of determining the set of primitive functions in the genetic programming paradigm is equivalent to a similar step in other machine learning paradigms. For example, this same determination of primitive functions occurs in the induction of decision trees using ID3 [9] when the user selects the functions that can appear at the internal points of the decision tree. Similarly, this same determination occurs in neural net work when the user selects the external functions that are to be activated by the output of a neural network. The same user determination occurs in other machine learning paradigms (although the name given to this omnipresent determination varies and is often considered by the researcher to be implicit in the statement of his or her problem).

5.1. DESCRIPTION OF ONE RUN

The initial random generation of randomizers was, predictably, highly unfit. In one particular run, 56% of the 500 individuals had entropy 0.000 (usually because the S-expression merely emitted a constant). Another 14% of the population had entropy between 0.001 and 0.080 (out of a possible 28.000 bits). These individuals emitted a constant almost all of the time. Another 24% of the population had entropy between 7.000 and 7.056 bits. These S-expressions scored 7.000 because they mapped the consecutive sequence of integers J into another consecutive sequence.

The best 24 individuals from generation 0 scored between 10.428 and 20.920 bits of entropy. The best single S-expression scored 20.920 bits. It had 63 points, but can be simplified to 13 points, namely,

```
(+ J (* (* (MOD% J 3) 3) (QUOT% (+ J 1) 4))).
```

This individual attains its 20.920 bits (out of a possible 28.000) by getting a perfect 1.000 bits out of a possible 1.000 bits for sequences of length 1; getting 1.918 out of 2.000 bits for length 2; getting 2.792 out of 3.000 bits for length 3; and getting 3.268 out of 4.000 bits for length 4. Although this individual does a credible job of randomizing bits when the window is narrow, it does not do as well for the longer sequences. It gets only 3.689 bits out of 5.000 for length 5 (i.e. only 74% of the possible 5.000 bits); 4.002 bits out of 6.000 for length 6 (i.e. only 67% of the possible 6.000 bits); and 4.252 bits out of 7.000 for length 7 (i.e. only 61% of the possible 7.000 bits).

After 2 generations of one typical run, the entropy of the best-of-generation individual improved to 22.126 bits. After 4 generations, the entropy of the best-of-generation individual improved to 26.474 bits. Thereafter, entropy reached and slowly improved within the 27.800 to 27.900 area.

On generation 14, we obtained an individual S-expression that attained a nearly maximal entropy of 27.996. It had 153 points, but simplifies to

```
(- J (QUOT% (+ (+ (+ J J) J) (* (+ J 2) J))
(+ (MOD% (* (- 2 1) (QUOT% (QUOT% (+ (* J J)
(QUOT% (- (QUOT% (* J (MOD% (QUOT% J 3) (MOD%
J J))) (QUOT% (* 3 2) (QUOT% 2 1))) (- 3
(QUOT% (+ (* J J) (- 2 1)) 3))) (* 3 (+ (MOD%
1 0) J)))) 3) 3)) (+ (- 2 J) 1)) (+ (QUOT%
(MOD% J 3) (- (MOD% 2 0) (MOD% (MOD% 0 J)
J))) (- 3 3)))))
```

The simplified version of this individual (with only 41 points) is graphically depicted in Figure 4 below as a rooted, point-labeled tree with ordered branches:

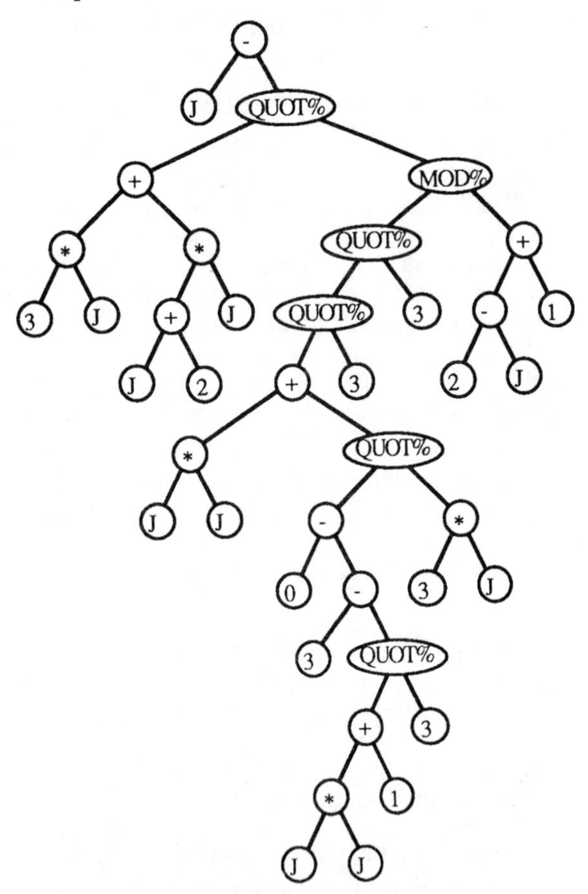

Figure 4: A Graphical Representation of the Genetically Bred Randomizer after Simplification

In scoring 27.996, this randomizer achieved a maximal value of entropy of 1.000, 2.000, 3.000, 4.000, 5.000, and 6.000 bits for sequences of lengths 1, 2, 3, 4, 5, and 6, respectively, and a near-maximal value of 6.996 for the 128 (2^7) possible sequences of length 7.

Note that the progressive change in size and shape of the individuals in the population is a characteristic of the genetic programming paradigm. The size (153 points) and shape of the best scoring individual from generation 14 differs from the size (63 points) and shape of the best scoring individual from generation 0. The size and particular hierarchical structure of the best scoring individual from generation 14 was not specified in advance. Instead, the entire structure evolved as a result of reproduction, crossover, and the relentless pressure of the fitness (i.e. entropy).

5.2. RESULTS OF SEVERAL RUNS

In the previous section we described one particular run of the genetic programming paradigm in which we obtained a randomizer with entropy of 27.996.

The genetic programming paradigm (as with genetic algorithms in general) contains probabilistic steps at several different points. As a result, we rarely obtain a solution to a problem in the precise way we anticipate and we rarely obtain the precise same solution twice.

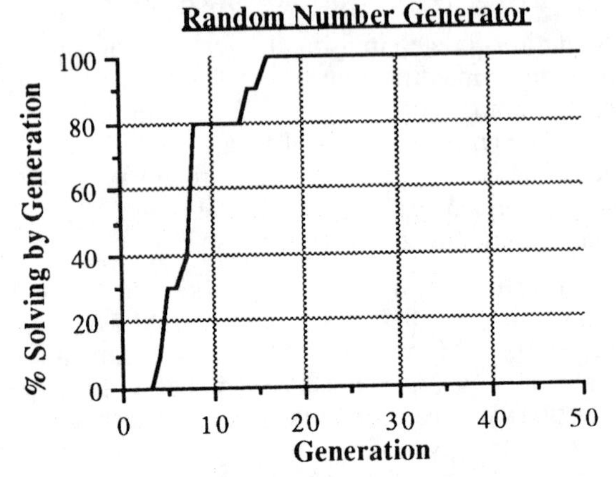

Figure 5: The Cumulative Percentage of Runs that Reached an entropy of 27.990 After the Execution of a Certain Number of Generations

We can measure the number of individuals that need to be processed by a genetic algorithm to produce a desired result (i.e. entropy of 27.990 or better) with a certain probability, say 99%. Suppose, for example, a particular run of a genetic algorithm produces the desired result with only a probability of success p_s after a specified choice (perhaps arbitrary and non-optimal) of number of generations N_{gen} and population of size N. Suppose also that we are seeking to achieve the desired result with a probability of, say, $z = 1 - \varepsilon = 99\%$. Then, the number K of independent runs required is

$$K = \frac{\log(1-z)}{\log(1-p_s)} = \frac{\log \varepsilon}{\log(1-p_s)}, \text{ where } \varepsilon = 1-z.$$

For example, we ran 10 runs of this problem with a population size of 500 and 51 generations. The graph in Figure 5, shows that the probability of success p_s of a run is 90% with 15 generations. With this probability of success p_s =0.90 on a particular single run, we need K = 2 independent runs with a population of 500 for 15 generations to assure a 99% probability of solving this problem on at least one run. That is, processing 15,000 individuals is sufficient.

5.3. DISCUSSION

Is the genetically bred computer program a good randomizer? The answer is a decisive "Yes and No."

Table 1, below, compares the shortfall in entropy from the maximal 28.000 bits for the genetically bred randomizer and the five commercial randomizers described earlier.

Table 1: Shortfall in Entropy for the Different Randomizers

Randomizer	Entropy Shortfall
Park-Miller	.009
IBM RANDU	.010
SHIFT REGISTER	.010
SHUFFLE	.015
TI RANDOM	.009
Genetic	.004

As can be seen, the genetically bred randomizer has precisely the characteristic for which it was bred (i.e. high entropy). With respect to that particular measure, it exceeded the performance of the other randomizers.

Table 2: Results of Equidistribution Test

Randomizer	# of 0's	# of 1's	χ^2
Park-Miller	8146	8238	0.52
IBM RANDU	8149	8235	0.45
Shift-Register	8234	8150	0.43
SHUFFLE	8234	8150	0.43
TI RANDOM	8235	8149	0.45
Genetic	8207	8177	0.05

If we apply the equidistribution (frequency) test from Knuth to the genetically bred randomizer and the five commercial randomizers, we get the results shown in Table 2. The genetic randomizer comes much closer to the uniform distribution of 8192 zeroes and 8192 ones in a random sequence of 16,384 binary digits than any of the five commercial randomizers. That is, the frequencies of zeroes and ones for the genetic randomizer are almost perfectly uniform. This is understandable, since the genetic randomizer was bred with the entropy fitness measure explicitly so as make single bit frequencies (and indeed sub-sequence frequencies up to length 7) uniform. The greater the deviation from uniformity, the worse the fitness (entropy).

But, in another sense, the frequency of zeroes and ones in the genetic randomizer leans in the direction of being too uniform. When we compute χ^2 (chi-square), we find that the five commercial randomizers have a χ^2 which, for one degree of freedom, is very close to the 50th percentile. In contrast, the χ^2 of the genetically bred randomizer is in about the 15th percentile. Whether the intentionally created uniform distribution or a less uniform distribution constitutes randomness depends on ones point of view. For example, some state lotteries explicitly enforce maximal uniformity (entropy) in the distribution of winners in preprinted instant lottery game tickets as opposed to "unstructured randomness."

If we apply the gap test from Knuth [17] to the genetic randomizer and the five other randomizers, we get the results shown in Table 3 below. The gap test counts the number of gaps of particular sizes in the sequence of 16,384 binary digits. For example, in the sequence 0010110 of length 7, we have one gap in the zeroes of length zero (i.e. one instance of consecutive zeroes), one gap of length one, and one gap of length two.

Table 3: Results of the Gap Test

Randomizer	χ^2 for 0's	χ^2 for 1's
Park-Miller	5.52	12.80
IBM RANDU	9.07	13.19
Shift-Register	11.21	4.05
SHUFFLE	9.22	10.72
TI RANDOM	4.50	9.44
Genetic	9.03	2.64

The 50th percentile of the χ^2 distribution for 10 degrees of freedom is at 9.342. The χ^2 for the zeroes for the genetic randomizer is 9.03. This is similar to most of the values for the five other randomizers. However, the χ^2 for the ones for the genetic randomizer is the rather low value of 2.64. That is, the gaps in the ones for the genetic randomizer almost perfectly match the expected number of gaps expected from a random source. That is, the genetic randomizer adheres too closely to the theoretical distribution of gaps. This match is especially strong for gap sizes of seven and below. This is understandable in view of how the genetic randomizer was bred. In contrast, the five other randomizers are "more random" because their performance deviates more sharply from the theoretical distribution of gaps expected from a random source. As Knuth said, "...the observed values are so close to the expected values, we cannot consider the result to be random!"

6. CONCLUSIONS

We demonstrated that it is possible to use the genetic programming paradigm to breed a computer program to perform the task of converting a sequence of consecutive integers into a sequence of pseudo-random bits with near maximal entropy. The size and shape of the initial, intermediate, and final programs were not specified in advance. Instead, the size, shape, and specific internal steps of the various computer programs emerged from a evolutionary process driven by the selective pressure applied by the fitness (entropy) measure. The better programs produced in the later generations of this evolutionary process were structurally more complex than the earlier programs.

7. REFERENCES

[1] S. L. Anderson. Random number generators on vector supercomputers and other advanced architectures. *SIAM Review*. 32(2). Pages 221-251. June 1990.

[2] K. A. De Jong. On using genetic algorithms to search program spaces. *Genetic Algorithms and Their Applications: Proceedings of the Second International Conference on Genetic Algorithms*. Lawrence Erlbaum Associates, Hillsdale, NJ 1987.

[3] J. H. Holland. *Adaptation in Natural and Artificial Systems*. University of Michigan Press, Ann Arbor, MI 1975.

[4] J. R. Koza. Hierarchical genetic algorithms operating on populations of computer programs. In *Proceedings of the 11th International Joint Conference on Artificial Intelligence*. San Mateo, CA: Morgan Kaufmann 1989.

[5] J. R. Koza. *Genetic Programming: A Paradigm for Genetically Breeding Populations of Computer Programs to Solve Problems*. Stanford University Computer Science Dept. Technical Report STAN-CS-90-1314. June 1990.

[6] J. R. Koza. Genetically breeding populations of computer programs to solve problems in artificial intelligence. In *Proceedings of the Second International Conference on Tools for AI*. Washington. November, 1990. IEEE Computer Society Press, Los Alamitos, CA 1990.

[7] J. R. Koza. Evolution and co-evolution of computer programs to control independent-acting agents. In Meyer, Jean-Arcady and Wilson, Stewart W. *From Animals to Animats: Proceedings of the First International Conference on Simulation of Adaptive Behavior*. Paris. September , 1990. MIT Press, Cambridge 1991.

[8] J. R. Koza. A genetic approach to econometric modeling. In Bourgine, Paul and Walliser, Bernard. *Proceedings of the 2nd International Conference on Economics and Artificial Intelligence*. Pergamon Press 1991.

[9] J. R. Koza. Concept formation and decision tree induction using the genetic programming paradigm. In Schwefel, Hans-Paul and Maenner, Reinhard (editors) *Parallel Problem Solving from Nature*. Springer-Verlag, Berlin, 1991.

[10] J. R. Koza. Genetic evolution and co-evolution of computer programs. In Farmer, Doyne, Langton, Christopher, Rasmussen, S., and Taylor, C. (editors) *Artificial Life II, SFI Studies in the Sciences of Complexity*. Volume XI. Addison-Wesley, Redwood City CA 1991.

[11] J. R. Koza. A hierarchical approach to learning the Boolean multiplexer function. In Rawlins, Gregory (editor). *Proceedings of Workshop on the Foundations of Genetic Algorithms and Classifier Systems. Bloomington, Indiana. July, 1990*. San Mateo, CA Morgan Kaufmann 1991.

[12] J. R. Koza. *Genetic Programming*. MIT Press, Cambridge, MA, 1991 (forthcoming).

[13] J. R. Koza and M. A. Keane. Genetic breeding of non-linear optimal control strategies for broom balancing. In *Proceedings of the Ninth International Conference on Analysis and Optimization of Systems. Antibes, June, 1990*. Pages 47-56. Springer-Verlag, Berlin, 1990.

[14] J. R. Koza and M. A. Keane. Cart centering and broom balancing by genetically breeding populations of control strategy programs. In *Proceedings of International Joint Conference on Neural Networks, Washington, January, 1990*. Volume I. Hillsdale, NJ: Lawrence Erlbaum 1990.

[15] J. R. Koza and J. P. Rice. Genetic generation of both the weights and architecture for a neural network. In *Proceedings of International Joint Conference on Neural Networks, Seattle, 1991*.

[16] J. R. Koza and J. P. Rice. A genetic approach to artificial intelligence. In C. G. Langton *Artificial Life II Video Proceedings*. Addison-Wesley 1991.

[17] D. E. Knuth. *The Art of Computer Programming*. Volume 2. Addison-Wesley, Reading, MA, 1981.

[18] G. Marsaglia. Random numbers fall mainly in the planes. *Proc. Nat. Acad. Sci. U.S.A.*, 61, pages 25-28. 1968.

[19] S. K. Park and K. W. Miller. Random number generators: Good ones are hard to find. *Comm. ACM*. 31, pages 1192-1201, 1988.

Intelligent Structural Operators for the k-way Graph Partitioning Problem

Gregor von Laszewski
The Ohio State University
Harold A. Bolz Hall

gregor@cis.ohio-state.edu
2036 Neil Avenue Mall
Columbus OH, 43210-1277

Abstract

A parallel genetic algorithm for the graph partitioning problem is presented, which combines general heuristic algorithms with techniques that are described in evolution theory. In the parallel genetic algorithm the selection of a mate is restricted to a *local neighborhood.* In addition, the parallel genetic algorithm executes an *adaptation* step after an individual is generated, with the genetic operators crossover and mutation. During the adaptation step the solution is improved by a common algorithm. Another selection step decides if the adapted descendant should *replace* the parent individual. Instead of using a uniform crossover operator a more *intelligent* crossover operator, which copies subsets of nodes, is used. Basic parameters of the parallel genetic algorithm are determined for different graphs. The algorithm found for a large sample instance a new unknown solution.

1 GENETIC ALGORITHMS

Genetic Algorithms are stochastic search algorithms introduced by J.Holland in the 70's [8]. These algorithms are based on ideas and techniques from genetic and evolutionary theory. Genetic algorithms simulate an evolutionary process with N individuals which represent points in a search space. Every individual is encoded as a string called a *genotype*. The value of the cost function which is defined for such a string is called a *phenotype*.

In each step of the genetic algorithm, called a generation, every individual is evaluated with regard to the entire population. This value is called the *relative fitness* of an individual. According to "natural evolution" offspring are produced using genetic operators. The *selection* operator chooses individuals with a probability that corresponds to the relative fitness. Two chosen individuals produce a descendant using the genetic operator *crossover*. The crossover operator exchanges substrings of the codes of the two chosen individuals. The descendant replaces an individual in the population after the generation step is complete. Another genetic operator, called *mutation*, changes the genotype of the descendant, with a small probability.

Mutation and crossover cause variation in the search process. The mutation operator allows a search close to a point in the search space, because only a small number of changes occur. Crossover causes longer jumps in the search space.

However, only selection leads the search in a specific direction. Substrings of individuals that are more fit than others are kept for the next generation. The search is successful if the search space has the property that a combination of two high valued points of the search space leads to a higher valued point with high probability [12].

Further information about genetic algorithms and their applications is provided in [3] and [5].

2 PARALLEL GENETIC ALGORITHMS

In Holland's genetic algorithm, selection occurs in the entire population, whereas in the parallel genetic algorithms the selection of a mate is restricted to a *local neighborhood.* In addition, the parallel genetic algorithm executes an *adaptation* step after an individual is generated, with the genetic operators crossover and mutation. During the adaptation step the solution is improved by a common algorithm. Another selection step decides if the adapted descendant should *replace* the parent individual.

The parallel genetic algorithm (PGA) can be described as follows: An environment consists of a set of locations $X = \{x_1, ..., x_N\}$ which are divided geographically. Connections between locations are described by a relation R on X. At each location x_k there exists an individual I_k^t at time t. At the beginning of the evolutionary process the initial individuals I_k^0 are randomly initialized. For each individual I_k^t, a set of neighbor individuals $\mathcal{N}^t(x_k)$ are determined by the relation R. Figure 1 shows the evolution process that runs on each location.

First, an individual chooses a partner for mating in its neighborhood and creates a descendant using the crossover operator. After the mutation operator is applied, the descendant is improved in the adaptation step. If the descendant is well adapted to the environment,[1] it replaces the parent individual. The algorithm is terminated when a termination constraint

[1] For example, if the descendant is better than the parents, or if it is better than the worst individual in the neighborhood, it is considered as well adapted.

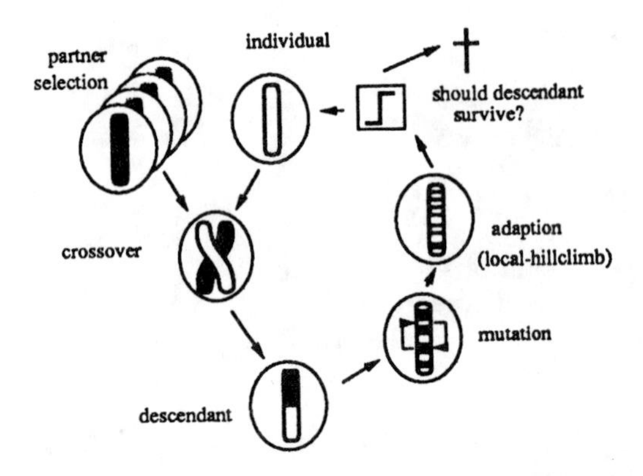

Figure 1: Evolution process

PGA $= (C, N, I^0, K, c, GO)$

C	is the set of genetic codings for the solutions.
N	is the number of locations. The locations are $X = \{x_1, ..., x_N\}$. At each time t there is a individual $I_k^t \in C$ on location x_k.
I^0	$= \{I_1^0, ..., I_N^0\}$ is the initial population at time $t = 0$.
K	$\subset X \times X$ is the communication relation.
c	is a cost function which determines the phenotype of the individual. A coding is evaluated.
GO	= {mutation, crossing-over, selection, parent replacement strategy} is the set of genetic operators.

Figure 2: Parameters of the parallel genetic algorithm

is fulfilled.[2]

Since the evolution process runs simultaneously on each location, this model can be mapped onto a multiprocessor system. Each processor must know the codings of the individuals living on its neighbor processors.

The parallel genetic algorithm has been successfully applied to the traveling salesman problem [14, 6]. In this paper it is demonstrated that the parallel genetic algorithm can also be applied to the complex k-way graph partitioning problem. A formal description of the k-way graph partitioning problem is given in the next section.

3 THE k-WAY GRAPH PARTITIONING PROBLEM

The k-way graph partitioning problem (k-GPP) is a fundamental combinatorial problem which has applications in many areas of computer science (e.g., design of electrical circuits, mapping) [10]. Mathematically we can formulate the k-way graph partitioning problem as follows:

Let $G = (V, E, w)$ be an undirected graph, where $V = \{v_1, v_2, ..., v_n\}$ is the set of nodes, $E \subseteq V \times V$ is the set of edges and $w : E \mapsto \mathbb{N}$ defines the weights of the edges. The k-way graph partitioning problem is to divide the graph into k disjoint subsets of nodes $P_1 ... P_k$, such that the sum of the weights of edges between nodes in different subsets is minimal, and the sizes of the subsets are nearly equal. The subsets are called *partitions*, and the set of edges between the partitions is called a *cut*.

Let $P = \{P_1, ..., P_k\}$ be the partitions. Then the string $(g_1 g_2 ... g_n)$ describes a solution:

$$g_i = a \iff v_i \in P_a \quad \forall i \in \{1, ..., n\}$$

With $a \in \{1, ..., k\}$. Node v_i is assigned to the partition specified by g_i. Instead of minimizing the cost of the cut we maximize the sum of the weights of all the edges between nodes in the same partitions. This is an equivalent problem because the total cost of edges is constant. This leads to a cost function of:

$$c(g_1 g_2 ... g_n) = \sum_{\substack{1 \leq i \leq j \leq n \\ g_i = g_j}} w(v_i, v_j) \ .$$

The advantage of this cost function is that a selection operator for a genetic algorithm can be easily formulated. Furthermore, the parallel genetic algorithm described in this paper does not change the sizes of the partitions during the computation. The equal size of the partitions is controlled by the variance

$$\sigma^2(P) \stackrel{\text{def}}{=} \frac{1}{m} \sum_{i=1}^{m} |P_i|^2 - \left(\frac{1}{m} \sum_{i=1}^{m} |P_i| \right)^2 .$$

4 PARALLEL GENETIC ALGORITHM APPLIED TO THE k-GPP

To apply the parallel genetic algorithm to the k-way graph partitioning problem, a representation of problem solutions has to be defined. Genetic operators which control the composition of two solutions or the modification of one solution have also to be defined. In addition, the values of the parameters used by the parallel genetic algorithm have to be determined (e.g. population size, relation between the locations, mutations, etc.).

4.1 Representation, Communication Relation, and Selection

Rather than a simple binary representation, the discrete string representation defined in section 3 is used

[2]For example, a time limit may be used as termination constraint.

to code solutions of the k-way graph partitioning problem. Therefore, a larger alphabet $\Sigma = \{1, .., k\}$ is used. To guarantee the constraint of the equal partition size, only a subset of all k^n possible strings is allowed. This straightforward representation implies that the phenotype of a string $g_1...g_n$ is given by the value $c(g_1...g_n)$.

For the experiments, the communication relation between the locations is determined by a ring:

$$x_k \text{ is neighbor of } x_l \Longleftrightarrow 0 < (l-k+N) \; mod \; N \leq A,$$

where A denotes the number of neighbors and N denotes the population size. Let $\mathcal{N}^t(x_k)$ be the set of individuals located in the neighborhood of the individual I_k^t. These individuals are called the *selection neighbors*. For example, let three be the size of the the selection neighborhood. Then those three individuals are in the selection neighborhood that lie in the ring directly before the individual itself (Figure 5).

The individual which is currently the best[3] can be added to the selection neighbors. This individual is called the *currently best* individual.

With the selection neighbors, the relative fitness of an individual in an environment is defined as follows:

$$f(I_l^t) = \frac{c(I_l^t)}{\sum\limits_{I_r^t \in \mathcal{N}^t(x_k)} c(I_r^t)} \qquad \forall I_l^t \in \mathcal{N}^t(x_k)$$

The relative fitness determines the probability of selecting an individual from the *selection neighbors* for mating. With the help of this fitness function, very good solutions can be found. Other selection strategies are described elsewhere [2, 5].

4.2 The Structural Crossover Operator

The crossover operator is very important for the success of the genetic algorithm. If a crossover operator destroys too much information already gained in the past, the genetic algorithm degenerates to a simple random search algorithm. To avoid losing too much information, an intelligent structural crossover operator is defined. It copies whole partitions from one solution into another.

Figure 3 depicts the recombination of two solutions. A grid with 4×4 nodes is to be divided into 4 partitions. To show the recombination step more clearly, colors are used in the figure instead of numbers to represent the different partitions.

First, a partition is randomly chosen in a parent solution (the light gray partition). Then this partition

[3]The currently best individual is the representation of the best solution found since the parallel genetic algorithm is started.

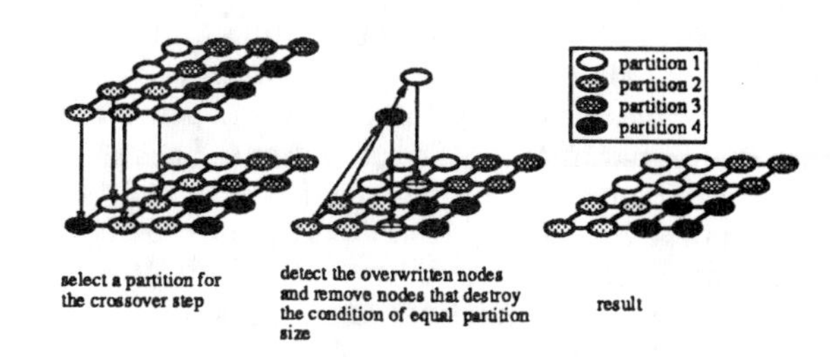

Figure 3: Recombination of two solutions

is copied into the other parent solution. Because this copying process may destroy the constraint of equal partition sizes, a repairing operator is applied. In the repairing step, all nodes in the temporary solution which are not elements of the copied partition, but have the same color as this partition, are detected. These nodes are marked in the second part of Figure 3 with horizontal lines.

To assign these nodes to a partition, they have to be marked (e.g. randomly) with the colors of those nodes which have been overwritten by the copied partition. In the example the white and the black partitions have one node too few. So the nodes marked with horizontal lines are relabeled with the colors white and black. A new code is generated which represents a valid solution for the problem instance.

Executing the crossover operator on arbitrary genotypes creates descendants which temporarily have a lot of open positions during the crossover process. In the extreme case, these positions could correspond to a whole partition. If the number of nodes in a partition is large in comparison to the number of nodes in the graph, a great disturbance of the old solutions will arise. In order to avoid losing too much information computed in the past, the codings are adapted before the crossover process starts. They are changed, in such a way that the difference between the two parent solutions is as small as possible. Let $(a_1...a_n), (b_1...b_n)$ denote the parent individuals. Then the difference of the two parent individuals is defined as follows:

$$\text{difference } (a_1...a_n, b_1...b_n) \stackrel{\text{def}}{=} \sum_{i=1}^{n} \begin{cases} 1 & \text{if } a_i \neq b_i \\ 0 & \text{otherwise} \end{cases}$$

4.3 The Structural Mutation Operator

A common mutation operator that replaces values in the string with an element randomly chosen out of Σ will destroy the condition of equal partition size. To avoid leaving the search space, a mutation is defined as the exchange of two numbers of the coding.

Because, at the beginning of the evolution process, the solutions generated with the crossover operator are very different from each other, there is no need to disturb them with a mutation operator. Mutations are only executed if the difference between a parent and the solution created by the crossover operator is smaller than a parameter called *mutations*. Let Δ denote the minimum of the difference between the two parents and its descendent generated by the crossover step. If this difference is smaller than the parameter *mutations*, then Δ - *mutations* swap operations are executed on the coding of the descendant.

4.4 The Adaptation Step

For large problem instances, it is important to restrict the solution space. This can be achieved by using a hill climbing algorithm to improve the solutions represented by the coding. Therefore, a variant of the 2-opt algorithm introduced by Kernighan and Lin is implemented [10]. For all pairs of nodes, the 2-opt algorithm exchanges these nodes if the solution can be improved by the exchange. This step is repeated until no further improvement can be made. Since one iteration step is done in $O(n^2)$ time, it is necessary to reduce the number of nodes on which this heuristic is used. Instead of trying the exchange over all pairs of nodes, the 2-opt algorithm is only executed on the nodes located at the border of the partitions.

5 RESULTS

The parallel genetic algorithm is implemented on a 64 node transputer system. Each evolution process is executed on one transputer. The maximal population size is 64.

This paper concentrates on two different problem instances. First, a graph whose edges are connected like a *grid* is used to demonstrate some basic effects of the parallel genetic algorithm. This graph is to be divided into four partitions. Therefore, the globally optimal solution for a grid with 100 nodes has a cost function value of 20. Without equivalent solutions[4], there exists only one globally optimal solution. The problem *grid* provides a test instance for determining the basic properties of the implemented algorithm.

Second, a graph called *beam* is used [4]. This graph has 918 nodes and 3233 edges, and is to be divided into 18 partitions.

There are only a few algorithms which can be compared with the PGA, because other algorithms are usually restricted to the 2-way graph partitioning problem. Two of these comparable algorithms are the round robin algorithm of Moore and the divide-and-conquer Kernighan-Lin algorithm of Zmijewski[13, 7]. These algorithms do not use the constraint of equal partition size, so that the partitioning problem is simpler.

[4] Solutions are equivalent to each other only if they are different in the names of the partitions

For the grid graph, parameters were found that allow the globally optimal solution to be generated in every case. Also, the PGA found the best known solution for the instance *beam*. Figure 4 shows the progress of this solution. The table 1 also shows the best known results found with the different algorithms.

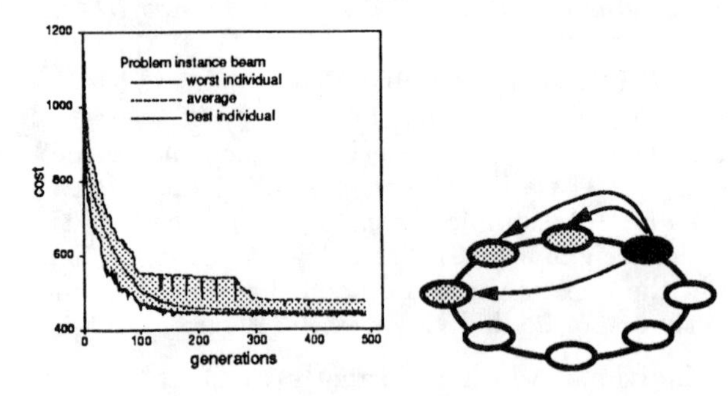

Figure 4: Problem beam, 64 individuals

Figure 5: Communication structure, Ring with 8 individuals

algorithm	minimal cost	$\sigma(P)$	running time
GZ87	587	0.99	78 rounds
Moore	453	0.99	78 rounds
PGA	430	0.00	500 generations, 28 min

Table 1: Comparison of the best solutions for the instance beam

6 PARENT REPLACEMENT STRATEGY

In the implementation of the parallel genetic algorithm for the k-way graph partitioning problem the convergence speed, is an important factor. To increase the convergence speed two special concepts are introduced:

1. The parent solution is only replaced if a specific condition is fulfilled.
2. The currently best individual is included in the selection neighborhood.

In this section, different strategies for deciding whether a descendant should survive and replace a parent individual are compared. The following replacement strategies are considered:

Strategy "each": Each parent individual is replaced by its descendant. The replacement is done regardless of the quality of the parent or the descendant solution.

Strategy "better": A parent individual is only replaced by its descendant if the descendant is better than the parent individual.

Strategy "locally better": This strategy is a combination of the previous strategies. The replacement of a parent individual is dependent on the cost of the neighbor individuals, the descendant, and the parent itself. A parent is replaced if the descendant is better than the parent solution, or if the descendant is better than the worst individual in the *local* selection neighborhood.

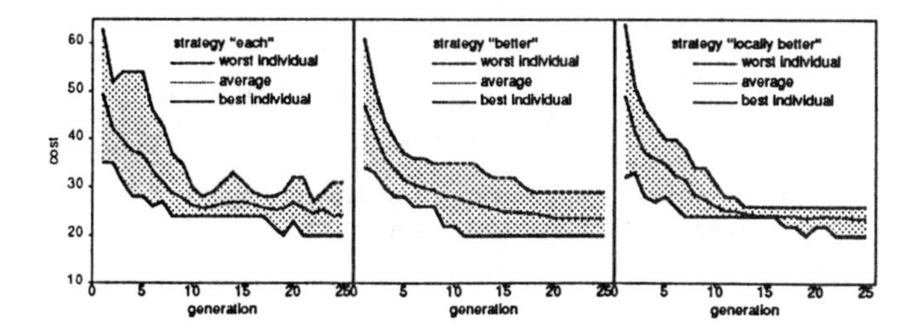

Figure 6: Replacement strategy for a parent without the currently best individual in the selection neighborhood

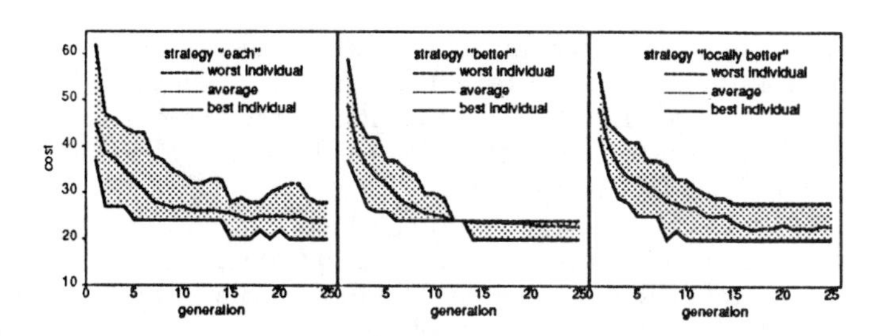

Figure 7: Replacement strategy for a parent with the currently best individual in the selection neighborhood

The same experiments are done with and without the currently best individual in the selection neighborhood. The problem instance *grid* is used. For the experiments with the problem instance grid, the population size is 16 and the size of the selection neighborhood is 4. Figure 6 and 7 display the range of the cost values of the population over generations. For each generation, the cost of the worst and best individuals are shown. Furthermore, the average cost of the individuals in the population is shown. The graphs shown in Figure 6 do not include the currently best individual in the selection neighborhood. Whereas, the graphs shown in figure 7 include the currently best individual in the selection neighborhood.

The experiments show that for the strategy "each," the cost range of the individuals in the population fluctuate heavily among the generations. A relative long time period is needed to find the minimal solution. If the currently best individual is included in the selection neighborhood, the convergence speed can be improved.

Using the strategy "better," the convergence speed is greater than with the strategy "each". If the currently best individual is also included in the selection neighborhood, then the algorithm often gets stuck in a locally optimal solution. The algorithm converges too quickly without discovering other possible solutions in the solution space. The second graph in Figure 7 shows that at a specific time period, all individuals have the same cost value. No new information is gained by the genetic algorithm.

Now the question arises, how convergence in low quality solutions can be avoided. One simple way to overcome this problem is to combine the two strategies with each other as described above. The replacement strategy "locally better", with the inclusion of the currently best individual in the selection neighborhood, enables the parallel genetic algorithm to find high quality solutions with a high convergence speed.

Another important factor for jumping out of stable but suboptimal solutions is a perturbation of the solution with the mutation operator. Using a sufficiently large mutation rate enables the parallel genetic algorithm to introduce new variation into the search process as shown in the second graph of Figure 7. Here, the mutation rate is $\frac{1}{5}$.

This result also holds for the larger problem instance *beam*. Figure 8 shows the range of the solutions generated with the strategy "locally better" and the inclusion of the currently best individual in the selection neighborhood. This approach produces the best results.

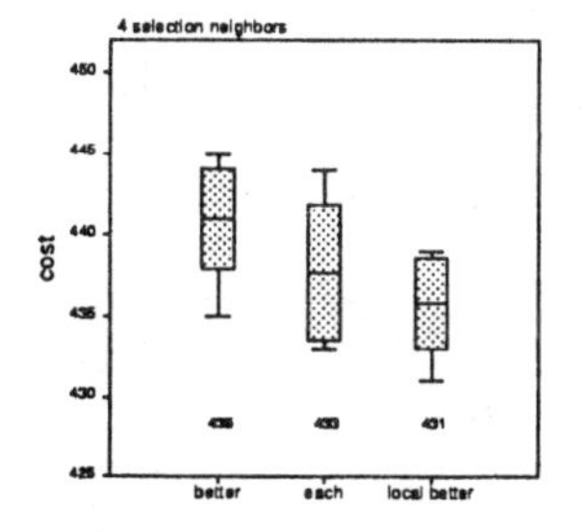

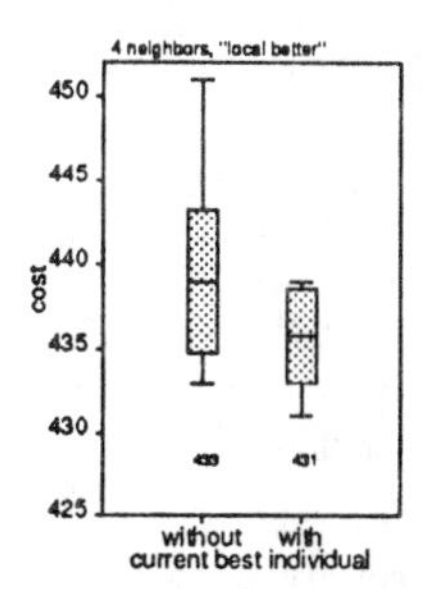

Figure 8: Replacement strategy. The problem instance *beam* is used.

7 MUTATION

In the last section, it was shown that the mutation operator is important for varying solutions when the genetic algorithm get stuck in locally optimal solutions. With a large experiment, the optimal number of mutations are determined for the problem instance *grid*. The optimal number of mutations achieves

1. that the average number of generations needed to find a very good solution[5] is minimal.
2. that the frequency of finding the globally optimal solution is maximal.

Figures 9 and 10 depict the result of the experiments used to find the optimal number of *mutations*. 100 experiments were done for each mutation in the interval from 0 to 50. Each experiment was terminated after the globally optimal solution is found, or the evolution cycle (Figure 1) is repeated more than 100 times. The strategy "locally better" is used for replacing the parent individual.

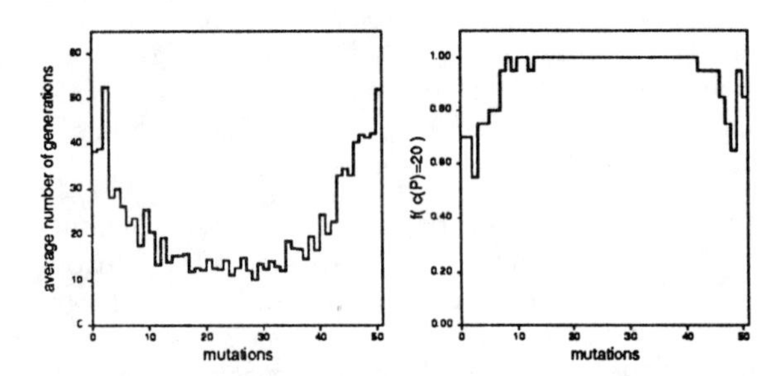

Figure 9: Analysis of the number of mutations. The currently best individual is not included in the selection neighborhood

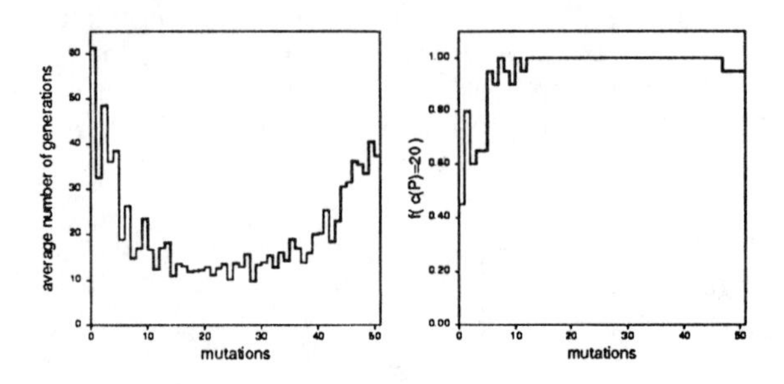

Figure 10: Analysis of the number of mutations. The currently best individual is included in the selection neighborhood

If the currently best individual is not included in the selection neighborhood, the optimal number of mutations is between 17 and 34. If fewer mutations are done, the algorithm often get stuck in suboptimal solutions. This can be avoided by increasing the number of mutations. Finding the globally optimal solution is prevented if too many mutations are executed. In this case the algorithm degenerates to a simple random search algorithm.

[5]For the problem instance *grid*, "very good" means "globally optimal"

If the currently best individual is included in the selection neighborhood, the same effects appear. However, the average number of generations which are needed to find the globally optimal solution is smaller.

7.1 The Correlation Between Mutation and the Adaptation Step

Common genetic algorithms use very small mutation rates. For the problem instance *grid* an optimal solution rate of about $\frac{1}{4}$ was observed. This section explains, why a high mutation rate is needed to find very good solutions quickly. To see the correlation between the mutation rate and the adaptation step, one has to remember that:

1. the adaptation step is executed after the recombination of a new descendant.
2. the mutation operator is applied when the crossover operator generates a descendant that is very similar to one of its parents.

The PGA is applied to the problem instance *grid*. At the end of the evolution process, a descendant is only slightly different from one of its parents. If the 2-opt algorithm were executed next, no new variation would introduced into the search process.

Furthermore, with advancing generations nearly optimal solutions are generated. Applying the 2-opt algorithm on a slightly disturbed solution near the optimum leads with high probability to the same old solution. This fact is displayed in Figure 11. Let a be a solution and b be the solution which is created by applying some mutations and the 2-opt algorithm on a. Figure 11 shows how often the solutions a and b are equivalent for different good solutions.

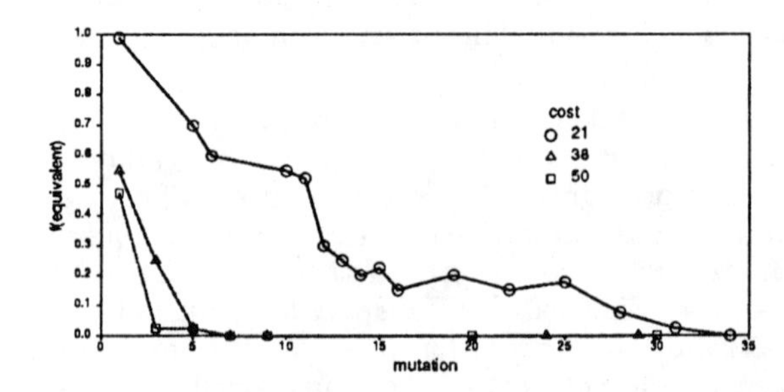

Figure 11: Relative frequency of generating an equivalent 2-optimal solution after applying a number of mutations and the 2-optimal algorithm for different good solutions

To prevent the parallel genetic algorithm from getting stuck, the mutation operator is used to disturb the descendants. In addition the number of mutations has to be sufficiently large.

The mutation operator is defined in such a way that it is only applied if the difference between the parent solutions and the descendant is smaller than the number of mutations. Because the solutions are so different at the beginning of the search process the mutation operator is only applied later. Therefore, the crossover and repairing operator are responsible for introducing variety early on in the search process.

Another factor for a large mutation rate is that also mutations are done regardless of the position of the nodes in a partition. If the nodes are in the same partition then the mutation operator generates obviously not a new solution. This can in future implementations avoided if only mutations between different partitions are allowed. Furthermore, the nodes which are allowed to mutate could be restricted to the borders of the partitions.

7.2 Population Size and Neighborhood Size

Using the problem instance *grid*, only slight differences occur if other parameters like the population size are modified. Therefore, the problem instance *beam* is focus of this section.

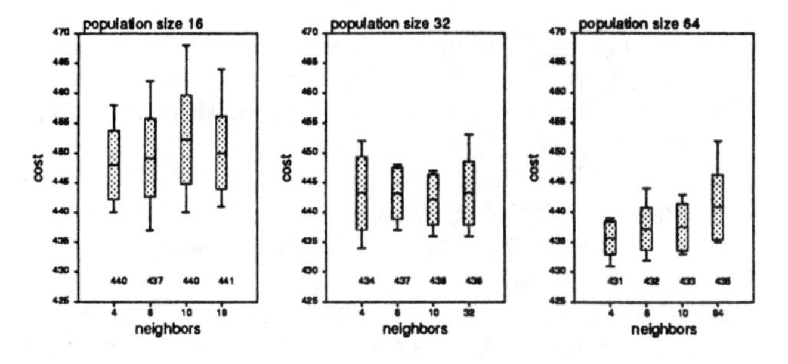

Figure 12: Analysis of the population size and the size of the selection neighborhood

Figure 12 compares the cost of solutions found with different population and neighborhood sizes. The selection neighborhood of each individual includes the currently best individual to increase the convergence speed of the algorithm. To get good results, the population size is more important than the size of the selection neighborhood. The best results are found using the largest population – i.e. 64 individuals. In addition, the size of the neighborhood should be small.

One important result is that, for large populations, the PGA algorithm produces better solutions with a restricted neighborhood than with a panmictic population.

Crossover and Mutation

In [1], a genetic algorithm for the graph partitioning problem can be found. This algorithm only generates solutions for the 2-way partitioning problem. Experimental results are only presented for graphs of up to 64 nodes. A uniform crossover operator is used to produce offspring. Each position of the offspring is randomly labeled by one of the two corresponding numbers in the parent genotypes.

In this paper the uniform crossover operator has been extended for the k-way graph partitioning problem. With this crossover operator, however, no solutions were found which were as good as those generated with subset crossover for different mutations (Figure 13).

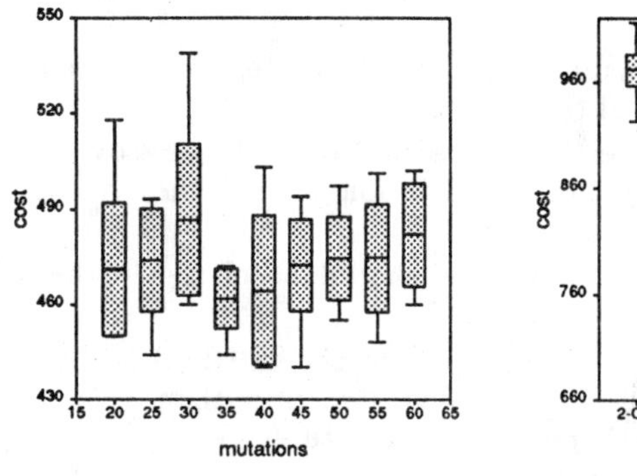

Figure 13: Uniform crossover operator with different mutations

Figure 14: Comparison of the cost of the cut for different heuristics on a random graph with 900 nodes

Random Graphs

Random graphs are defined so that the average degree of each node is $(n-1)p$, where n is the number of nodes, and p is the probability that a pair of nodes is connected by an edge. For a constant p and a large n, random graphs are dense. The graph partitioning problem is easier to solve for dense graphs, because the solutions have nearly the same cost of the cut. The GPP is more difficult for instances of restricted random graphs whose degree is bounded, e.g. by 4 [9].

To compare the PGA with other heuristics, the 2-opt and the KL-algorithm for the m-partitioning were also implemented. The KL-algorithm tries to exchange sequences of nodes instead of exchanging only two nodes in one step. A detailed description of these algorithms is provided in [10] and [11].

Figure 14 shows the comparison of the algorithms 2-opt, KL, and PGA terminated after 500 and 1000 generations. They are tested on a random graph with 900 nodes and maximum node degree of 4. Experiments with the problem instances with 900 and 918 nodes show that the PGA is much faster for regular graphs than for restricted random graphs.

8 CONCLUSIONS

The parallel genetic algorithm computes very good results for the graph partitioning problem. For a large problem instance the algorithm found a new unknown minimal solution. The search space has the property that a combination of two high valued points often leads to a higher valued point. Implanting a small, maximized subset of nodes from one solution into another and applying a local hill-climbing heuristic to this solution, often leads to a better partition.

Furthermore, this paper shows:

- that a parent replacement strategy improves the quality of the solutions.
- that mutation is needed only if the crossover operator produces a solution which is nearly equal to one of its parents.
- that the population should be chosen to be as large as possible.
- that better solutions are generated with the restricted neighborhood structure than with the panmictic population structure.
- that for the implementation presented in this paper, the selection neighborhood should have a size of 4, and should include the currently best individual to achieve the best results with a high convergence rate.
- that to restrict the solution space, a discrete problem representation and structural genetic operators are important.
- that the adaptation step is very important for restricting the solution space and improving the convergence rate of the algorithm.

This implementation of a parallel genetic algorithm shows that there exist two strategies for defining genetic algorithms. The first strategy uses a sophisticated representation and simple genetic operators onto the codings to generate good solutions. Sometimes it is difficult to find a sophisticated representation. Then it is easier to chose a simple straightforward representation and introduce intelligence into the algorithm by defining genetic operators which use the structure of the problem to generate offspring.

There are many opportunities for further research in this area. The most interesting would be to choose a larger population size to display more clearly the difference between the panmictic population and the neighborhood model. A more sophisticated mutation and selection operator may also be defined. The implementation environment makes it possible to run different adaptation strategies on different locations in order to inspect the solution space with different strategies. Also, different communication relations may be compared.

Acknowledgment

The author thanks Heinz Mühlenbein and Martina Gorges-Schleuter for providing the initial stimulation for the study and helpful discussions along the way, and Jim Buuck and Steve Edwards for suggestions that helped improving the final presentation.

References

[1] D. H. Ackley. A Connectionist Machine for Genetic Hill-climbing. The Kluwer international series in engineering and computer science SECS 28. Kluwer Academic Publisher, Norwell, Massachussets, 87.

[2] J.E. Baker. Adaptive Selection Methods for Genetic Algorithms. In 3rd Int. Conf. on Genetic Algorithms, Morgan Kaufmann, 85.

[3] L. Davis. Genetic Algorithms and Simulated Annealing. Morgan Kaufmann, Los Altos, 87.

[4] G.C. Everstine. A comparison of three resequencing algorithms for the reduction of matrix profile and wavefront. Int. J. Numer. Methods in Eng., Vol. 14, 837-853, 79.

[5] D. E. Goldberg. Genetic Algorithms in Search, Optimization, and Machine Learning. Addison-Wesley, 89.

[6] M. Gorges-Schleuter. ASPARAGOS: An Asynchronous Parallel Genetic Optimization Strategy. In 3rd Int. Conf. on Genetic Algorithms, San Mateo, Morgan Kaufmann, 89.

[7] J. R. Gilbert and E. Zmijewski. A Parallel Graph Partitioning Algorithm for a Message-Passing Multiprocessor. Techn. Report 87-803, Cornell University, 87.

[8] J. H. Holland. Adaptation in natural and artificial systems. Ann Arbor, University of Michigan Press, 75.

[9] D.S. Johnson, C.R. Aragon, L.A.McGeoch, C.Schevon. Optimization by Simulated Annealing: An Experimental Evaluation. Part I, Graph Partitioning. Operations Research, 37(6):865-892, Dec 89.

[10] B. W. Kernighan and S. Lin. An Efficient Heuristic Procedure for Partitioning Graphs. Technical report, Bell Syst. Techn. J., February 70.

[11] G. von Laszewski. Ein paralleler genetischer Algorithmus für das GPP. Master's thesis, Univerisität Bonn, 90.

[12] G. von Laszewski, H. Mühlenbein. Partitioning a Graph with a Parallel Genetic Algorithm. First Int. Workshop on Parallel Problem Solving from Nature, University Dortmund, West Germany, 90.

[13] D. Moore. A Round-Robin Parallel Partitioning Algorithm. Technical Report 88-916, Cornell University, Ithaca, NY, 88.

[14] H. Mühlenbein. Parallel Genetic Algorithm, Population Dynamics and Combinatorial Optimization. In 3rd Int. Conf. on Genetic Algorithms, San Mateo, Morgan Kaufmann, 89.

Designer Genetic Algorithms: Genetic Algorithms in Structure Design

Sushil J. Louis
Department of Computer Science
Indiana University, Bloomington, IN 47405
louis@iuvax.cs.indiana.edu

Gregory J. E. Rawlins
Department of Computer Science
Indiana University, Bloomington, IN 47405
rawlins@iuvax.cs.indiana.edu

Abstract

This paper considers the problem of using genetic algorithms to design structures. We relax one constraint on classical genetic algorithms and describe a genetic algorithm that uses differential information about *search direction* to design structures. This differential information is captured by a masked crossover operator which also removes the bias toward short schemas. We analyze performance and present some preliminary results. Further, consideration of this problem suggests a partial solution to the identification of the *deception problem.*

1 INTRODUCTION

The problem of designing structures is pervasive in science and engineering. The problem is:

> Given a function and some materials to work with, design a structure that performs this function subject to certain constraints.

As an example of the design problem, consider the *combinational circuit design problem*: Given a set of logic gates, design a circuit that performs a desired function. Two instantiations of this problem are the *parity problem* and the *adder problem.* A solution to these two problems is given in most introductory textbooks on digital design (see figures 1 and 2). Both problems are well-defined, unambiguous, easy to evaluate, and can be scaled in difficulty. In addition, we can change the number of solutions (the *footprint*) in the search space of a particular instantiation by varying the types of gates available. We therefore use them as a testbed and as a basis for performance comparison of various design strategies.

Design is traditionally considered a creative process and so difficult to automate. Expert systems that seek to codify knowledge are currently too brittle and not applicable across a broad range of domains. However, *natural selection* has been spectacularly successful in producing a broad range of robust structures that are efficient at performing a broad range of functions. Its success is evident from the abundance and diversity of life on this planet.

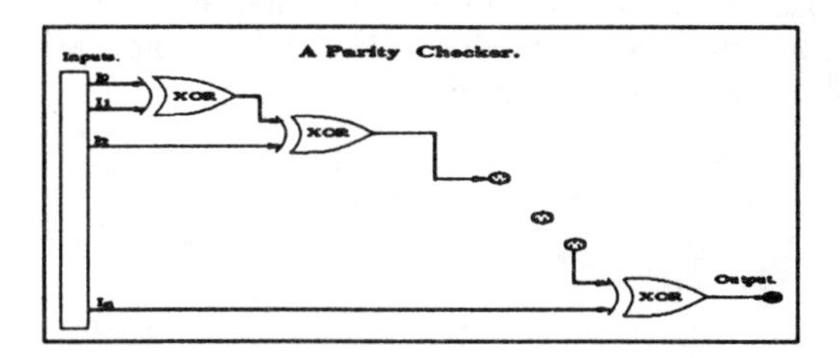

Figure 1: An n-bit parity checker.

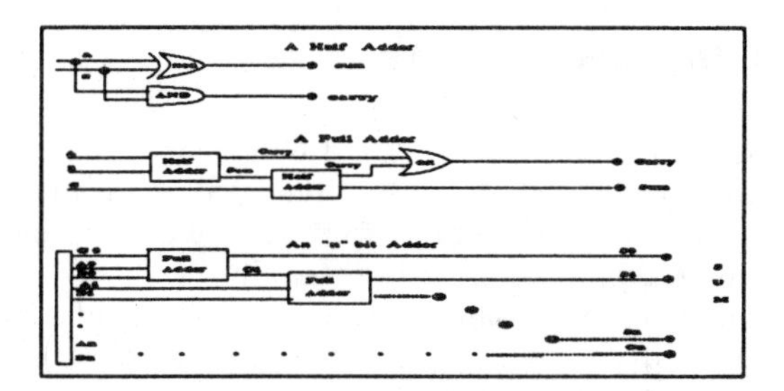

Figure 2: An n-bit adder.

Genetic Algorithms (GAs), based on natural selection should enjoy similar success in solving the problem of design. However when naively applied, their performance is less than encouraging. The difficulties lie in the enormous size of the problem, the interdependence among parts in the structure, and the biases inherent in current GAs. This interdependence is called *epistasis* and is an important aspect of well-designed structures. For example, if we use the elements of a two-dimensional array to represent gates in the textbook solution to the parity problem, the XOR gates must lie on the diagonal. Such interdependence is crucial in the design of a correct circuit [16].

The problem of structure design points out the major problem of choosing a good representation for GAs. A classical GA will do well, only if we artfully choose just the right encoding (non-epistatic), in essence, helping the search process. We solve this problem by using the fitness difference between parents and children to indicate good directions to bias search. Such directional information is easily available but cannot be explicitly stored and used in nature. Classical GAs mimicking nature also do not use this information. We however, can and do explicitly use this directional information to bias search toward high-performance schemas of ***arbitrary length*** thus reducing the dependence on encoding. In addition, we may identify deception by using this information alone or in conjunction with other methods and take remedial action. This approach is further developed in section three.

The next section defines a classical genetic algorithm and presents problems with using it for design. This leads to what we call a Designer Genetic Algorithm (DGA) described in section three. Preliminary results, presented in the fourth section, indicate the usefulness of DGAs. The last section covers conclusions and directions for future research.

2 CLASSICAL GAs

A genetic algorithm, first defined by Holland, is a randomized parallel search method modeled on evolution [15]. GAs are being applied to a variety of problems and are becoming an important tool in machine learning and function optimization [10]. Their beauty lies in their ability to model the robustness, flexibility and graceful degradation of biological systems. However, there has been little research on their applicability to design problems; much of the GA literature concerns function optimization. Any reference to design invariably means optimization of design parameters. In such problems the initial structure is fixed and the object is to optimize some associated cost [1, 17]. For example, in Goldberg and Samtani [9] a GA minimizes the weight of a 10-member plane truss, subject to maximum and minimum stress constraints on each member. Although such design parameter optimization is important, our problem is to design the initial structure itself.

A GA encodes each of a problem's parameters as a binary string. An encoded parameter can be thought of as a gene, the parameter's values, the gene's alleles. The string produced by the concatenation of all the encoded parameters forms a genotype. The basic algorithm, where $P(t)$ is the population of strings at generation t, is given below.

$t = 0$
initialize $P(t)$
evaluate $P(t)$
while termination condition false **do**
  select $P(t+1)$ **from** $P(t)$
  recombine $P(t+1)$
  evaluate $P(t+1)$
  $t = t + 1$

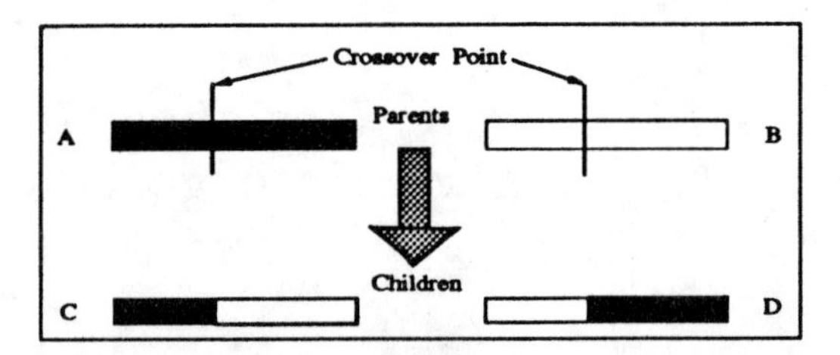

Figure 3: Crossover of the two parents A and B producing children C and D.

Selection is done on the basis of relative fitness and it probabilistically culls from the population those points which have relatively low fitness. Recombination, which consists of mutation and crossover, imitates sexual reproduction. Mutation probabilistically chooses a bit and flips it. Crossover (CX) is a structured yet randomized operator that allows information exchange between points. It is implemented by choosing a random point in the selected pair of strings and exchanging the substrings defined by that point. Figure 3 shows how crossover mixes information from two parent strings A and B, producing offspring C and D made up of parts from both parents. We note that this operator which does no table lookups or backtracking, is very efficient because of its simplicity.

Holland's ***schema theorem*** is fundamental to the theory of genetic algorithms [15]. A ***schema*** is a template that identifies a subset of strings with similarities at certain string positions. For example consider binary strings of length 6. The schema 1**0*1 describes the set of all strings of length 6 with 1*s* at positions 1 and 6 and a 0 at position 4. The "*" is a "don't care" symbol; positions 2, 3 and 5 can be either a 1 or a 0. The ***order*** of a schema is the number of fixed positions in the schema, while the ***length*** is the distance between the first and last specific positions. The order of 1**0*1 is 3 and its length is 5. The ***fitness*** of a schema is the average fitness of all strings matching the schema.

The schema theorem proves that relatively short, low-order, above-average schemas get an exponentially increasing number of trials in subsequent generations. Long, low-order, high-performance schema do not play a significant role in biasing genetic search.

2.1 GAs for STRUCTURE DESIGN

Using a genetic algorithm to design a structure is like playing with a child's construction kit. Given some low level building blocks, we have to put them together so that they perform a certain function. A GA used for design manipulates low-level "tools," or building blocks, playing with their arrangements, until it finds the required structure. But there are three problems:

First, a necessary condition for a GA to build a structure is that there should be at least one and preferably many evolutionary paths leading to the desired structure. A GA (or any search method) will perform poorly in optimizing a function that is zero at all points but one [3].

Second, the mapping from genotype to phenotype is now much more complex. We can compare the structure of an eye (a structure phenotype) with a point in the search space (a phenotype in function optimization) to get an idea of this complexity. Epistasis in phenotypic structures plays an important part in determining the suitability of classical genetic algorithms to structure design. Phenotypic epistasis may not be reflected in the genotype (unless it is very carefully encoded) and so will seriously degrade GA performance.

Finally, since we are working with structures, we often work in more than one dimension. Physical structures exist in three dimensions and may often be made up of many kinds of lower level building blocks. Higher dimensionality and a large alphabet increase the search space tremendously.

2.2 CROSSOVER BIAS

Long schemas tend to be disrupted by CX more often than shorter ones. Let H be a schema, $\delta(H)$ its length and $O(H)$ its order. Then the probability that the crossover point falls within the schema is $\delta(H)/(l-1)$ where l is the length of the string containing the schema. However, in epistatic domains, schemas of arbitrary length need to be preserved. If the encoding does not ensure that low-order schemas are short the GA will not make progress.

One way out of this is to use inversion. Inversion rearranges the bits in a string allowing linked bits to move close together. Inversion-like reordering operators have been implemented by Goldberg and others [8, 21] with some success. The problem with using inversion and inversion-like operators is the decrease in computational feasibility. If l is the length of a string, inversion increases the search space from 2^l to $2^l!$. Natural selection has geological time scales to work with and therefore inversion is sufficient to generate tight linkage. We do not have this amount of time nor the resources available to nature.

Another approach is to use a new crossover operator like punctuated crossover or uniform crossover. Punctuated crossover (PX) relies on a binary mask, carried along as part of the genotype, in which a 1 identifies a crossover point. Masks, being part of the genotypic string, change through crossover and mutation. Experimental results with punctuated crossover did not conclusively prove the usefulness of this operator or whether these masks adapt to an encoding [18, 19].

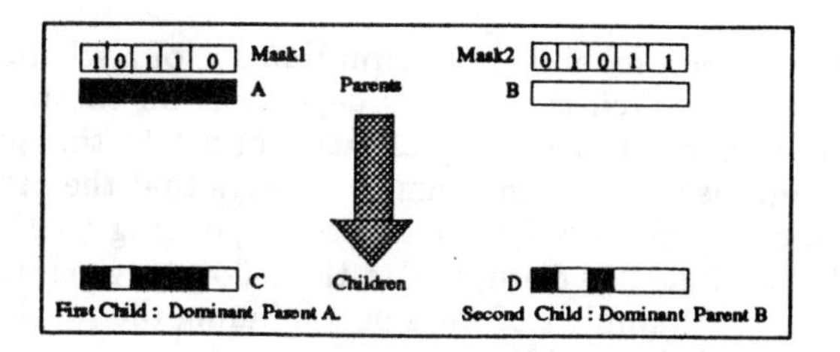

Figure 4: Masked crossover.

Uniform crossover (UX) exchanges corresponding bits with a probability of 0.5. The probability of disruption of a schema is now proportional to the order of the schema and independent of its length. Experimental results with uniform crossover suggest that this property is useful in some problems [20]. However, in design problems we would like *not* to disrupt highly fit schemas whatever their length.

None of these operators uses directional information. In the next section, we define a masked crossover operator that removes the bias toward short schemas by using directional information to efficiently bias search.

3 MASKED CROSSOVER

We define an operator that uses the relative fitness of the children with respect to their parents, to guide crossover. The *relative* fitness of the children indicates the desirability of proceeding in a particular search direction. The use of this information is not limited to our operator, and can be used in classical GAs with minor modifications [16].

Masked crossover (MX) uses binary masks to direct crossover. Let A and B be the two parent strings, and let C and D be the two children produced. $Mask1$ and $Mask2$ are a binary mask pair, where $Mask1$ is associated with A and $Mask2$ with B. A subscript indicates a bit position in a string. Masked crossover is shown in figure 4 and defined below:

copy A **to** C **and** B **to** D
for i **from** 1 **to** string-length
  if $Mask2_i = 1$ **and** $Mask1_i = 0$
    copy the ith bit **from** B **to** C
  if $Mask1_i = 1$ **and** $Mask2_i = 0$
    copy the ith bit **from** A **to** D

MX tries to preserve schemas identified by the masks. Call A the ***dominant parent*** with respect to C; C inherits A's bits unless B feels strongly ($Mask2_i = 1$) and A does not ($Mask1_i = 0$). The traditional way of analyzing a crossover operator is in terms of disruption. The probability of disruption P_d, of a schema H due to masked crossover is dependent on the masks. Assuming a random initialization of masks this probability is given by the number of ways that the bit positions in both parent masks corresponding to H can be combined to disrupt H in the following generation. The total number of ways of combining the mask bits corresponding to H is:

$$T_c = 2^{2 \times O(H)}$$

The number of ways of disrupting H is T_c minus the number of ways of preserving H, $\mathcal{P}_H$. For each bit position in H, there are three ways of preserving it, therefore:

$$\mathcal{P}_H = 3^{O(H)}$$

So the probability of disruption is:

$$P_d = \frac{T_c - \mathcal{P}_H}{T_c} = 1 - (\tfrac{3}{4})^{O(H)}$$

This probability of disruption does *not* depend on $\delta(H)$. Intuitively, 1's in the mask signify bits participating in schemas. MX preserves A's schemas in C while adding some schemas from B at those positions that A has not fixed. A similar process produces D. In addition, MX can combine overlapping schemas with less disruption than UX. This allows creation of schemas that would be impossible with one point crossover.[1] To ensure that the semantic interpretation of mask bits is correct, we modify masks in subsequent generations. Modifying masks will change the probability of disruption. Using fitness information to guide mask modification in subsequent generations, we would like to decrease the probability of disruption of highly-fit schemas independent of length. Instead of using genetic operators on masks, we use a set of rules that operate bitwise on parent masks to control future mask settings. Since crossover is controlled by masks, using meta-masks to control mask string crossover then leads to meta-meta masks and so on. To avoid this problem we use rules for mask propagation. Choosing the rule to be used depends on the fitness of the child relative to that of its parents. We define three types of children:

Good child: more fit than best parent.

Average child: fitness between that of the parents.

Bad child: less (or equally) fit than worst parent.

With two children produced by each crossover, and three types of children there are a six cases, with associated interpretations and possible actions on the masks (see figure 5).

[1] A simple example: the string 111 cannot be produced from 101 and 010 by one point crossover

Case	Rule
Both good	MF_{gg}
Both bad	MF_{bb}
Both average	MF_{aa}
One good, one bad	MF_{gb}
One good, one average	MF_{ga}
One average, one bad	MF_{ab}

Figure 5: Mask rules for the six ways of pairing children.

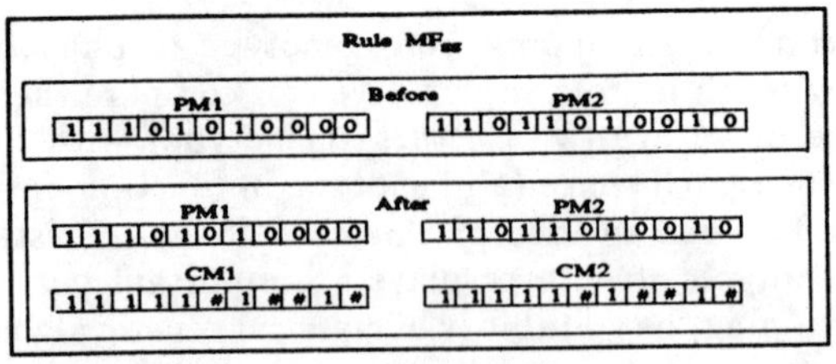

Figure 6: Mask rule MF_{gg}: Example of mask propagation when both C1 and C2 are good

3.1 MASK RULES

This section specifies rules for mask propagation. In each case a child's mask is a copy of the dominant parent's except for the changes the rules allow. The underlying premise guiding the rules is that when a child is less fit than its dominant parent, the recessive parent contributed bits deleterious to its fitness. We want to encourage search in the convex subspace defined by these loci. The idea is to search in areas close to one parent with information from the other parent providing some guidance. *Note that in MX, this is done without regard to length.* A mask mutation operator that flips a mask bit with low probability acts during mask propagation. We provide two representative mask functions rather than all, to give an intuitive understanding of their form. These are MF_{gg}, used when both children are good and MF_{bb} used when both children are bad (for more details see Louis and Rawlins [16]).

Let $P1$ and $P2$ be the two parents, $PM1$ and $PM2$ their respective masks. Similarly, $C1$ and $C2$ are the two children with masks $CM1$ and $CM2$. The modifications to masks depend on the relative ordering of $P1, P2, C1$ and $C2$. For the figures in this section, the "#" represents positions decided by tossing a coin.

1. MF_{gg}: Both children are good.

   Summary: Encouraging behavior. Parents' masks are OR'd to produce the children's masks, ensuring preservation of the contributions from both parents (see figure 6).

   Action:

   - $CM1$ and $CM2$: OR the masks of $PM1$ and $PM2$. If there are any 0's left in $CM1$, toss

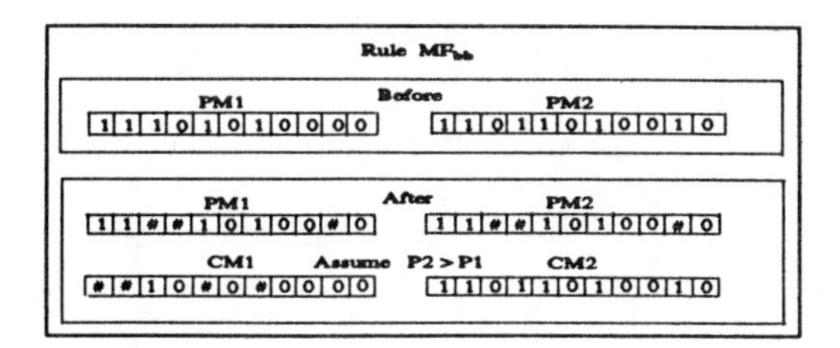

Figure 7: Mask rule MF_{bb}: Example of mask propagation when both C1 and C2 are bad.

a coin to decide their value.

- $PM1$ and $PM2$: No changes except for those produced by mutation.

2. MF_{bb}: Both children are bad.

   Summary: Discouraging behavior that must be guarded against in future. Each parent contributed bits that were detrimental to the children's fitness. MX has not set up the parents masks correctly. Changes are given below and shown in figure 7.

   Action:

   - $CM1$: This mask should reflect the undesirability of the current search direction. Contributions from $P2$ were detrimental, therefore $CM1$ should search in the area of $P2$'s contribution which is specified by the loci where $PM1_i$ is 0 and $PM2_i$ is 1. Set these loci in $CM1_i$ to 0. If $P2 > P1$ in fitness, toss a coin to set the bits of $CM1$ at those locations where both $PM1_i$ and $PM2_i$ are 1.
   - $CM2$: A similar rule applies to $CM2$.
   - $PM1$: $P2$'s contribution to $C1$ led to a bad child. The $C1$ positions copied from $P2$ need to be explored in $P1$. These loci are those for which $PM1_i$ was 0 and $PM2_i$ was 1. Therefore set these loci in $PM1_i$ by tossing a coin. In addition, $PM1$ specified loci that were detrimental to $C2$. Therefore when $PM1_i$ is 1 and $PM2_i$ is 0, set these locations by tossing a coin.
   - $PM2$: A similar rule applies for $PM2$.

With mask propagation through mask rules, directional information is explicitly stored in the masks and used by the crossover operator to bias search. Using the fitness of children relative to that of their parents we preserve and recombine high performance schema and disrupt low-performance schema, independent of schema length. The main features of MX are, storage and use of directional information, and independence from length of schemas. We think of masked crossover as a golden mean between the disruptiveness of UX and the bias toward short schemas of CX. Compared to the size of the search space when using inversion, $2^l!$, a genetic algorithm using MX searches only 2^{2l}.

Many sets of mask propagation rules can be defined. In fact a GA can search the space of mask rules to find a suitable set. This may be overkill, since the number of rules is usually quite small, simpler methods will suffice. Results, outlined in the next section, indicate that a significant performance increase is obtained from even the simple set of rules above.

MX presents a problem when using classical selection procedures. The classical strategy replaces the original population with the new children produced, but does not allow a genetic algorithm using masked crossover to converge. Masks will tend to disrupt the best individuals while searching for promising directions to explore because of the nature of the rules guiding mask propagation. Therefore our selection procedure is a modification of the CHC selection strategy [6]. If the population size is N, the children produced double the population to $2N$. From this, the N best individuals are chosen for further consideration. We use this *elitist* selection strategy to guarantee convergence. Another problem which may occur is that although MX preserves schemas of arbitrary length, the fitness information itself may be misleading. Such problems are called *deceptive* [10]. When fitness information is misleading we expect a GA using MX to perform worse than a GA using crossover operators that do not use such information. This is borne out by results from the adder problem. A Designer Genetic Algorithm (DGA) therefore differs from a Classical GA (CGA) in the crossover operator (MX) and in the selection strategy (elitist) used.

Identifying and overcoming deception, is an important area of research. Theoretically, deception is identifiable by mathematical analysis. However, from a practical standpoint, this analysis is prohibitively expensive. Messy genetic algorithms (MGAs), developed by Goldberg to handle deception, need to identify deceptive schemas to be applicable [13, 14]. We suggest an approach satisfying both criteria, using designer genetic algorithms.

Deception can be statically identified using the ANODE algorithm suggested by Goldberg [11, 12]. Recent results indicate that the Nonuniform Walsh-Schema Transform (NWST) [2] can dynamically analyze a GA. Using the NWST in concert with the normal operation of a GA, we can collect runtime statistics needed to identify deception. Furthermore, we can improve efficiency by removing some of the determinism in the ANODE algorithm. This will not significantly alter effectiveness as long as the probability of correctly identifying deception is greater than that of incorrectly identifying it. In other words, we propose to let a DGA collect runtime statistics on encoding (through the NWST) and use these statistics to set masks. Whenever the DGA detects deception either through a periodic check of these statistics and/or a decrease in rate of progress, the algorithm identifies *deceptive schema* with the help of the statistics collected and the masks. It then allows an MGA to work on just these schema and solve the deception at this level. The DGA then continues, appropriately seeded with the optimal schema produced by the MGA. Our current research follows this approach.

4 RESULTS

We compare a designer genetic algorithm's performance with that of a classical GA on the adder and parity problems. In all experiments, the population is made up of 30 genotypes. The probability of crossover is 0.7 and the probability of mutation for masks and genotypes is 0.04. These numbers were found to be optimal through a series of experiments using various population sizes and probabilities. The graphs in this section plot average fitness over ten runs.

Each genotype is a bit string that maps to a two-dimensional structure (phenotype) embodying a circuit. We need 3 bits to represent 8 possible gates. A gate has two inputs and one output. If we consider the phenotype as a two dimensional array of gates S, a gate S_{ij}, gets its first input from $S_{i,j-1}$ and its second from one of $S_{i+1,j-1}$ or $S_{i-1,j-1}$. An additional bit associated with each gate encodes this choice. If the gate is in the first or last rows, the row number for the second input is calculated modulo the number of rows. The gates in the first column, $S_{i,0}$ receive the input to the circuit. Connecting wires are simply gates that transfer their first input to their output. The other gates are AND, OR (inclusive OR), NOT and XOR (exclusive OR). We determine the fitness of a genotype by evaluating the associated phenotypic structure that specifies a circuit. If the number of bits is n, the circuit is tested on the 2^n possible combinations of n bits. The GA maximizes the sum of correct responses (For more detail see Louis and Rawlins [16]). It is also possible to use only a subset of the possible inputs, reverting to the complete set only when the population converges prematurely. This results in significant savings in time.

We compare the performance of a classical GA using elitist selection with a DGA on a 2-bit adder problem. The graph in figure 8 shows that the classical GA does better, although the difference is not great. This is not very encouraging. However, if we look at the solution space we see that solutions to the adder problem involve deception. As explained earlier, since MX uses fitness information to bias search, it is more easily mislead than traditional crossover. Even if a problem is deceptive, it does not mean that no solutions can be found. Figures 9 and 10 show solutions to the 2-bit adder problem found be a designer genetic algorithm and classical genetic algorithm. As wire gates ignore their second input, only one input is shown for such gates. The gate at position S_{33} is shown unconnected

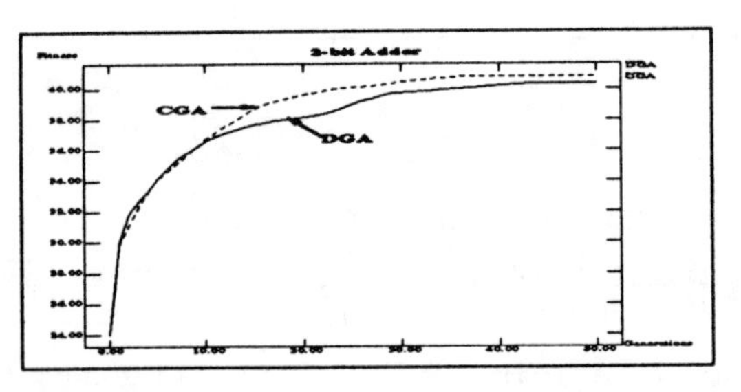

Figure 8: Performance comparison of average fitness per generation of a classical GA versus a DGA on a 2-bit adder.

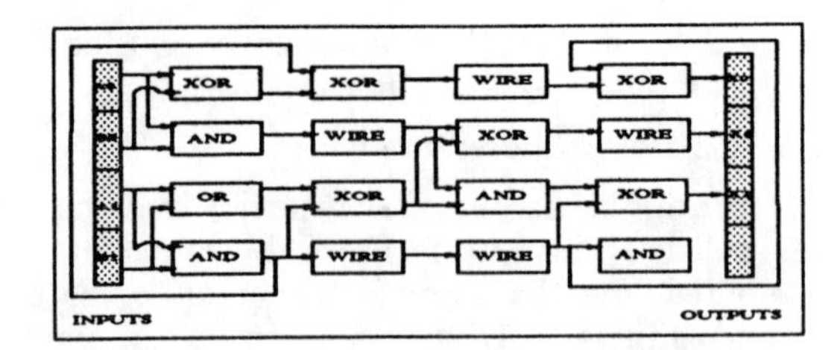

Figure 9: A 2-bit adder designed by a designer genetic algorithm.

because it does not affect the output. Although we have not done a rigorous study of the types of solutions found by both algorithms, we see that the circuit designed by the DGA depends on long schemas. For example S_{03} gets its input from S_{32} which is 11 units away, where 11 is large when compared with 16, the length of the genotype. This is in marked contrast to the CGA circuit.

We now consider the parity problem. The encoding described above will violate the "principle of meaningful building blocks" with regard to the solution to the parity problem as shown in figure 1 [10]. Since diagonal elements of S (the phenotype) are further apart in the string, any good subsolutions (highly fit, low-order schemas) found will tend to be disrupted by traditional crossover. MX however, will find and preserve these subsolutions as its performance is independent of length. To observe performance under these conditions, we restrict the number of gate types available to the GA to three and do not allow a choice of input (the second input is now always from the next row,

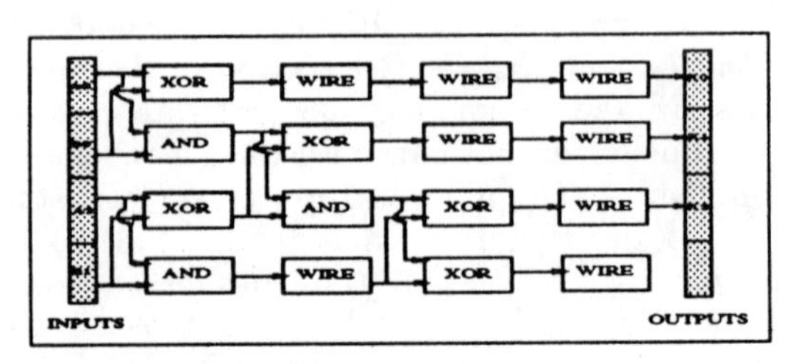

Figure 10: A 2-bit adder designed by a classical genetic algorithm.

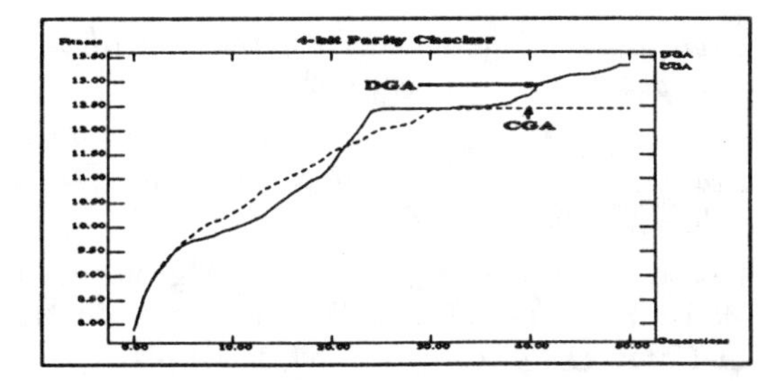

Figure 11: Performance comparison of average fitness per generation of a classical GA versus a DGA on a 4-bit parity checker.

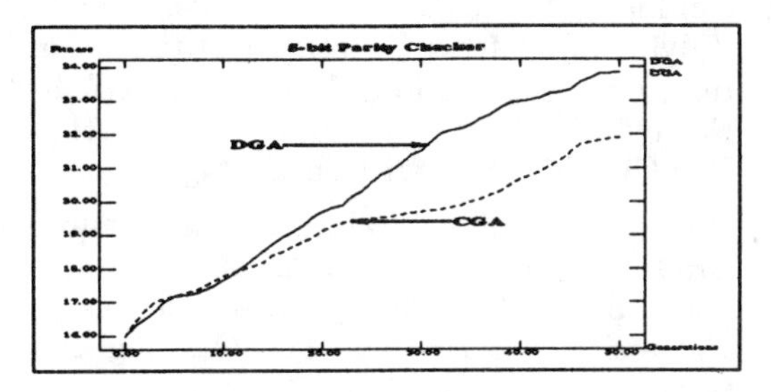

Figure 12: Performance comparison of average fitness per generation of a classical GA versus a DGA on a 5-bit parity checker.

modulo the number of rows). Although this reduces the size of the search space, CX disrupts low-order schemas and therefore performs worse than the DGA. Figure 11 shows this for a 4-bit parity checker. The difference in performance gets larger as the problem is scaled in size. Figure 12 compares the average fitness performance on a 5-bit problem. In the 5-bit experiments the choice of gates was still restricted to the same three as in the previous example. However, input choice was allowed, increasing the number of solutions in the search space. When allowed all possible gates, the performance difference is less, and is due to the large increase in the number of possible solutions and therefore a lesser degree of violation of the meaningful building block principle (see figure 13). However, as the problem becomes more epistatic, MX does better than CX. Hypothesis testing using the student's t test on our experimental data proves that MX is significantly better than CX at a confidence level greater than 95% [7].

In the comparisons above we ignored the effect of selection. Figure 13 compares the performance of: 1) a GA using traditional crossover and selection, 2) a GA using traditional crossover and elitist selection, and 3) a DGA on a 5-bit parity problem. The same parameter set as in the previous examples is used although we set the number of gate types to six, increasing the number of possible solutions. This was done in the hope of coaxing better performance from the GA using traditional selection and crossover. The figure clearly shows the importance of selection strategy. Finally, figure 14

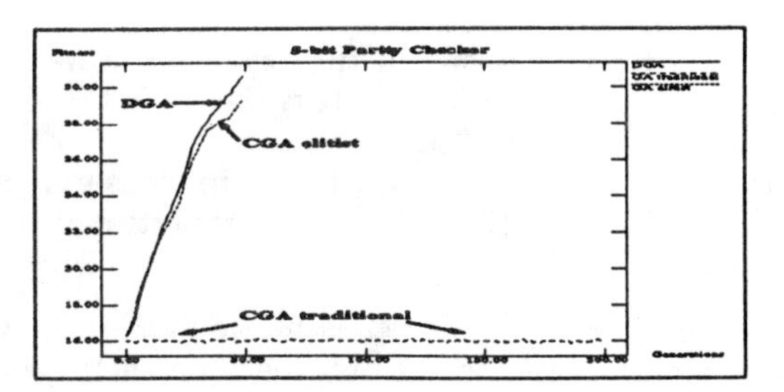

Figure 13: Performance comparison of average fitness per generation of a traditional CGA, an elitist CGA, and a DGA.

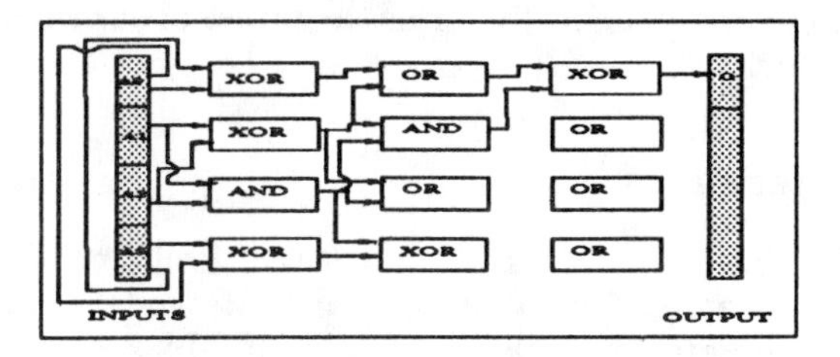

Figure 14: A Circuit designed by a DGA that solves the 4-bit parity problem.

shows a 4-bit parity checker produced by the DGA.

5 CONCLUSIONS

We have shown that a designer genetic algorithm relaxes the emphasis on schema of short length. This increases the domain of successful GA applications since a GA programmer no longer needs to follow the principle of meaningful building blocks. Using masked crossover mitigates the problem of epistasis while elitist selection is crucial to good performance on design problems. The increase in cost in using a DGA is by at most a constant factor per generation. This comes from the cost of sorting a population of size n ($n \log n$), and a constant cost for mask propagation. Comparing the performance of the two GAs on a problem also gives significant insights about properties of the search space.

Selection plays the largest part in biasing genetic search. We can think of a genetic algorithm as a search process at two levels. At the selection level, search is biased by fitness information. However, at the recombination level, search is essentially random (in classical GAs). Using fitness information to bias search at the recombination level allows a DGA to do better as is indicated by our results. However, if the fitness information is misleading, a GA will be led astray. The DGA will be misled at both levels, in contrast to the the CGA which will be mislead only at the selection level.

Experiments with the standard test suite of five functions first used by DeJong show no significant difference in performance [4]. In the experiments we used the same elitist selection strategy for both the DGA and the classical GA. These results were to be expected as DeJong's criteria for choosing his functions were not based on the epistatic or deceptive properties of these functions.

This paper uses a simple representation and only considers binary masks which piggyback on their associated strings. Mathematical analysis of the effects of mask rules on these simple masks is being done. We are also looking at more general representations and non-binary masks. Finally, identifying and handling deception dynamically, forms the thrust of our current research.

References

[1] Bramlette, Mark F., and Cusic, Rod., "A Comparative Evaluation of Search Methods Applied to Parametric Design of Aircraft." In *Proceedings of the Third International Conference on Genetic Algorithms.* Morgan Kauffman, 1989, 213-218.

[2] Bridges, Clayton L. and Goldberg, David E., "The Nonuniform Walsh-Schema Transform", in *Workshop on the Foundations of Genetic Algorithms and Classifier Systems* Morgan Kauffman, (to appear) 1991.

[3] Culberson, Joseph C., and Rawlins, Gregory J. E., "Genetic Algorithms as Function Optimizers." Unpublished Manuscript, Indiana University, Department of Computer Science. 1990.

[4] De Jong, K. A., "An Analysis of a the Behavior of a class of Genetic Adaptive Systems." Doctoral Dissertation, Dept. of Computer and Communication Sciences, University of Michigan, Ann Arbor.

[5] Eshelman, Larry J., Carauna, A. and Schaffer, J David. "Biases in the Crossover Landscape" In *Proceedings of the Third International Conference on Genetic Algorithms.* Morgan Kauffman, 1989, 10-19.

[6] Eshelman. L. J., "The CHC Adaptive Search Algorithm: How to have Safe Search When Engaging in Nontraditional Genetic Recombination." in *Workshop on the Foundations of Genetic Algorithms and Classifier Systems* Morgan Kauffman, (to appear) 1991.

[7] Freund. John. E., *Statistics A First Course,* Prentice-Hall, 1981.

[8] Goldberg, D. E., and Lingle, R., "Alleles, loci and the Traveling Salesman problem." in *Proceedings of the an International Conference on Genetic Algorithms and their Applications,* 1985. 154-159.

[9] Goldberg, David E., and Samtani, M. P., "Engineering Optimization via Genetic Algorithm." in *Proceedings of the Ninth Conference in Electronic Computation,* 1986, 471-482.

[10] Goldberg, David E., *Genetic Algorithms in Search, Optimization, and Machine Learning* Addison-Wesley, 1989.

[11] Goldberg, David E., "Genetic Algorithms and Walsh Functions: Part I, A Gentle Introduction", in *Complex Systems,* 3, 1989, 129-152.

[12] Goldberg, David E., "Genetic Algorithms and Walsh Functions: Part II, Deception and its Analysis, in *Complex Systems,* 3, 1989, 153-171.

[13] Goldberg, David E., Korb, Bradley., and Deb, Kalyanmoy. "Messy Genetic Algorithms: Motivation, Analysis, and First Results", TCGA Report No. 89002, Tuscaloosa: University of Alabama, The Clearinghouse for Genetic Algorithms, 1989.

[14] Goldberg, David E., Deb, Kalyanmoy, and Korb, Bradley. "An Investigation of Messy Genetic Algorithms," TCGA Report No. 90005, Tuscaloosa: University of Alabama, The Clearinghouse for Genetic Algorithms, 1990.

[15] Holland, John. H., *Adaptation In Natural and Artificial Systems.* Ann Arbor: The University of Michigan Press. 1975.

[16] Louis, Sushil. J., and Rawlins, Gregory J. E. "Using Genetic Algorithms to Design Structures," Technical Report No. 326, Department of Computer Science, Indiana University, 1990.

[17] Powell, D. J., Tong, S. S., and Skolnik, M. M., "EnGENEous Domain Independent, Machine Learning for Design Optimization." in *Proceedings of the Third International Conference on Genetic Algorithms and their Applications,* 1989, 151-159.

[18] Schaffer, David. J., and Morishima, Amy., "An Adaptive Crossover Distribution Mechanism for Genetic Algorithms" in *Proceedings of the Second International Conference on Genetic Algorithms.* Lawrence Erlbaum Associates, 1987, 36-40.

[19] Schaffer, J. David, and Morishima, Amy. "Adaptive Knowledge Representation: A Content Sensitive Recombination Mechanism for Genetic Algorithms" In *International Journal of Intelligent Systems* John Wiley & Sons Inc., 1988, Vol 3, 229-246

[20] Syswerda, Gilbert., "Uniform Crossover in Genetic Algorithms" In *Proceedings of the Third International Conference on Genetic Algorithms.* Morgan Kauffman, 1989, 2-8.

[21] Smith. D., "Bin Packing with Adaptive Search." in *Proceedings of an International Conference on Genetic Algorithms and Their Applications.* 1985. 202-206.

On Crossover as an Evolutionarily Viable Strategy

J. David Schaffer
Philips Labs
Briarcliff Manor, NY, USA
dsl@philabs.philips.com

Larry J. Eshelman
Philips Labs
Briarcliff Manor, NY, USA
lje@philabs.philips.com

Abstract

Genetic algorithm theory has long maintained that fitness proportional reproduction and crossover recombination yield a powerful search algorithm. The analyses behind this assertion has implied that highly disruptive crossover is likely to be inferior to a less disruptive one. Recent empirical results have provided two opposite challenges to these conclusions: one suggesting highly disruptive crossover can be more effective than less disruptive crossover, and one suggesting that crossover itself contributes nothing that mutation alone cannot provide.

To shed some light on these questions we present results from experiments in which a small subpopulation using crossover tries to invade a larger population that does not. We observe that crossover always takes over except when it is very disruptive. However, this takeover is not always to the benefit of the population; for some environments, mutation alone can eventually do better. We also find support for the assertion that crossover can exploit epistasis that mutation alone cannot and observe much faster search by crossover on tasks that mutation alone can solve.

1 Introduction

A popular model of evolutionary processes involves a repeated cycle of fitness biased selection and random mutation applied to a population of structures. Holland[10] has argued for the benefits of crossing over material from two parents. The claim is that there are non-linear interactions (epistasis) among the building blocks (genes, schemata) within these structures that mutation, proceeding as it does by single steps, could not properly exploit. Procedures employing these notions have become known as genetic algorithms (GAs). Nevertheless, to counter the premature convergence possibilities of crossover alone, GA prescription has been to use a low level of mutation in addition to crossover, but it has been believed that this was not the main driver of the search. Schaffer et al.[12] have suggested that contrary to this thinking, mutation was actually making a significant contribution to the search effectiveness, an observation supported by the higher mutation rates found by a meta-GA for a GA[8]. In addition, Davidor[2] has suggested that there are degrees of epistasis that are beyond even crossover's exploitive abilities. Fogel and Atmar[6] have gone so far as to argue that crossover cannot claim any generally useful advantage over mutation and selection alone, supporting their claim with experiments in which subpopulations using different combinations of genetic operators competed for dominance. Their results show mutation alone dominating the subpopulations using crossover.

Biological observations suggest that a very low level of mutation and crossover with low disruptiveness are what Nature uses and this is consistent with traditional GA theory. If one favors the Panglossian assumption — if Nature does it, it must be the best thing to do — then one must try to explain these new empirical observations.

To shed more light on these issues, we wish to address the question, "is crossover an evolutionarily viable strategy?" That is, suppose there is an evolutionary system in operation (i.e. a periodic reproduction and selection of the fittest cycle operating on a population of individuals), and suppose that a gene appears for recombination (crossover). Under what circumstances, if any, will this gene come to dominate in the population? And is the effect of crossover beneficial to the search process?

2 The Race

There will clearly be a race between the offspring of the crossover subpopulation and the rest of the population. Even in the case where the rest of the population has no method for introducing variation, selection will still be putting pressure on these offspring by increasing the instances of the best unchanging individuals, thus lowering everyone else's relative fitness. Once crossover produces an individual superior to the best unchanging individual, the takeover of the crossover gene is assured, so long as crossover is unlikely to quickly destroy him again. On the other hand, if crossover is highly disruptive of what has contributed to the survival of its possessors, they may not get enough tries to produce the wunderkind before their subpopulation dies out.

When mutation is also operating, the race is between the (possibly mutated) offspring of crossover, the offspring of mutation alone, and the unchanged individuals. All the while, selection will be tending to reduce the gene pool variation, mutation will be tending to maintain it, and crossover will be tending to redistribute it.

A critical factor in determining the outcome of this race seems to be the *riskiness* of the change operator (mutation or crossover). We define a *safety ratio* as the ratio of the probability that an offspring will be better than the parent(s) to the probability that it will be worse. These probabilities will be influenced by several factors, but unfortunately we know of no theory that identifies what they are. As a start toward building such a theory, we have chosen to examine empirically the epistasis in the problem[1], the disruptiveness of the operators to the schemata in the parents, and the rate of mutation.

3 Experimental Methods

Simulations of the invasion of a population by a small subpopulation of individuals possessing a crossover gene were conducted under varying conditions of problem epistasis, crossover disruptiveness and level of background mutation.

Following majority practice in GA work, the structures manipulated by our genetic search procedures were bit strings (chromosomes). All chromosomes contained 100 bits used to compute fitness and one more bit for a crossover gene. The simulations began with a random initial population of 500. This population size is larger than usually used for optimization, but was chosen to reduce the variance in the outcomes from stochastic effects (genetic drift). Furthermore, we are not interested here in discovering good optimization strategies per se, where minimizing the total computational effort to an answer of a given quality is the issue. We are assuming a Nature-like model in which all population members are evaluated in parallel and the race is conducted on a generation basis.

After evaluation of the initial population, the fittest members were selected using Baker's unbiased selection procedure[9]. A randomly selected 10% of this population had their crossover gene turned on and the others had theirs turned off. Thus, the invading subpopulation begins with 50 individuals, a size often used for genetic optimization tasks. A smaller subpopulation may not contain enough initial variation to be viable. No further direct manipulations of the crossover genes were done. Then followed generation cycles consisting of the following steps:

> Randomly mate the crossing individuals among themselves[2], and replace them by their offspring.
>
> If (mutation rate > 0) perform mutation on the entire gene pool (including the crossover gene).
>
> Evaluate the new offspring.
>
> Select the fittest.
>
> Terminate if (population converged, 2 generations with no new offspring (spinning) or 50,000 trials).

Twenty independent runs were executed with different random seeds, using each combination of the following experimental conditions.

Epistasis: We chose five problems with known amounts and types of epistasis, Onemax, Plateau, Plateau-d, Trap and Trap-d.

Onemax is a simple linear problem (i.e. zero epistasis). The evaluation returned for each 100-bit string is just the number of ones it contains. The optimum (minimum) value is zero and occurs for the all-zero string, the worst possible score is 100 and the expected score for a random 100-bit string is 50. The Plateau problem contains the maximum amount of epistasis within small segments (genes), and no epistasis between segments. For each five-bit segment, a value of five is given for the "00000" pattern (the optimum allele) and zero for all others.

The score for a string is 100 minus the sum of the segment scores. Thus the optimum score is zero and the worst possible score is 100. The expected score for a random 100-bit string is $100 - 20(5)(1/32) = 96.875$.

[1] The problem is taken to be the composite of the representation, usually a mapping of the instance space to be searched onto strings of some sort, and the evaluation function that is computed from these strings. It has been common to think of the problem as only the evaluations function, but recently more research attention has begun to be focused on the mapping as well[11,14,15].

[2] If the number of crossing individuals is odd, then one mating with a non-crossing individual will occur with probability 0.5.

The Trap problem is like the Plateau problem, but is deceptive[3]. A value of five is given for the "00000" allele, but for other patterns, a value of 0.5 times the number of ones is given. Unless the optimum allele is present, there is pressure away from it towards its complement. The optimum score is again zero, the worst possible score is 90 (for a string where all 20 segments have a single one and four zeros), and the expected score for a random string is

$$100 - 20\left[\frac{5}{32}+\binom{5}{1}\frac{0.5}{32}+\binom{5}{2}\frac{1.0}{32}+\binom{5}{3}\frac{1.5}{32}+\binom{5}{4}\frac{2.0}{32}+\frac{2.5}{32}\right]$$

$= 71.875$

Note that the epistasis in the Plateau and Trap problems is very local; it includes only the five adjacent bits of the segment. This situation favors the bias of the two-point crossover operator, but as Eshelman et al. [4] point out, this locality cannot generally be assumed. So, we included a variant of each of these problems, Plateau-d and Trap-d, in which the segments were uniformly intermixed along the chromosome. That is, the bits of segment 1 were at loci 1,21,41,61, and 81, segment 2 at 2,22,42,62 and 82, etc.

Disruptiveness of crossovers: We studied two crossover operators that differ on this dimension. Two-point crossover (2X) selects at random two loci and exchanges the parental segments between them. This operator shares the local bias of traditional one-point crossover, but is less disruptive and exhibits generally superior search behavior[3,12]. Uniform crossover (UX), simply exchanges the parental bit at each locus with an even probability. It is considerably more disruptive than 2X of short schemata, but does not share its local bias which makes it less exploitive of spurious correlations[13]. It is also a more vigorous recombiner of parental schemata[5].

Mutation rate: We chose two rates, zero (no mutation), and 0.0005. The latter rate was selected after some preliminary experiments, and represents an attempt to balance the expected number of improvers from the mutation and crossover subpopulations in the initial generation of the race. This balance could not be perfect for all of the problems; however, rates much higher than this lead to very poor search by mutation, and rates much lower lead to intolerably slow search.

[3] Deceptiveness is a concept introduced by Goldberg[7] and refers to a situation in which the average fitnesses of low order schemata lead away from the optimal solution.

4 Baseline Tests

Anticipating that insight into the riskiness of each operator would be useful for interpreting the race results, some baseline experiments were conducted. Each problem was run 20 times using selection with each change operator alone (i.e. mutation @ .0005, 100% 2X, or 100% UX). The population size was 500 in all cases. During these runs all offspring produced were compared to their parents and statistics were kept on the fraction of them that were better and worse than their parents (both parents in the case of crossover). The ratio of P(better)/P(worse) can be thought of as a *safety ratio* for the operator. The higher this ratio, the safer it is for a population to use the operator to produce offspring.

4.1 Mutation alone

The safety ratio curves for mutation are displayed in Figure 1. The Plateau-d and Trap-d were not tested for mutation or for UX since these operators have no local bias making these problems the same as their locally coded counterparts. These curves all reflect the increasing risk of using mutation as the population becomes more attuned to the problem (environment). On Onemax and Trap the ratio begins quite high, indicating that the likelihood of improving by random mutation is almost as high as the likelihood of worsening. Improvements are easy in the nearly random populations. The population on Trap converges more slowly than on Onemax because the selection pressure is less, particularly after the first 100 generations or so. Mutating an optimal allele yields a worse offspring, but mutating a non-optimal allele can still make small improvements. The Plateau problem represents a much riskier environment. After the very first selection step the population will consist only of individuals with one or more optimal alleles (those with none score zero and produce no offspring). Mutation is unlikely to add a new one, but is likely to destroy one. After 50 generations or so, mutation is very dangerous.

The view of the riskiness of using mutation from these curves is strictly local (i.e. offspring compared to parents). Showing how the population performs relative to the true global optimum, Table 1 presents the mean and variance (among the 20 runs) of the best individual found and the mean number of generations needed to find it. Note that in spite of the increasing risk associated with mutation, this is a fairly effective search algorithm. The global optimum is found every time for Onemax and most of the time (16/20) for Plateau and in a number of generations well beyond the point when the risk of destroying the parent's genetic material is very high. Even on Trap, the final solutions contain

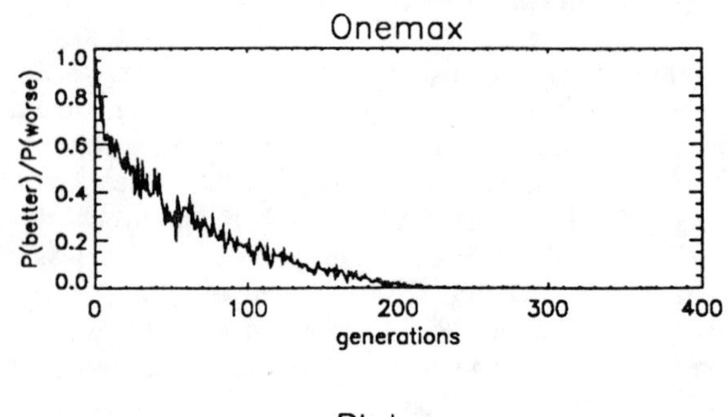

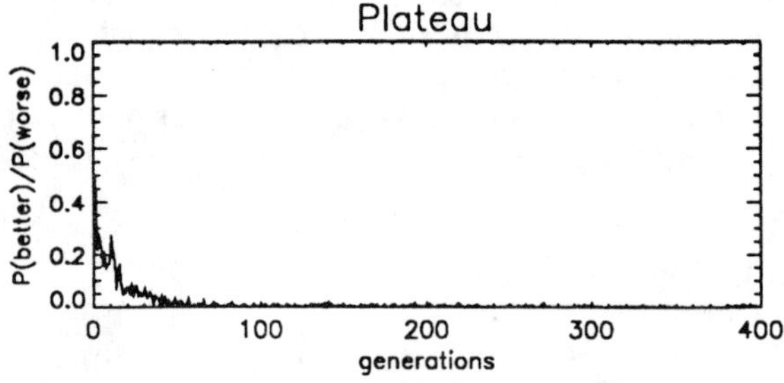

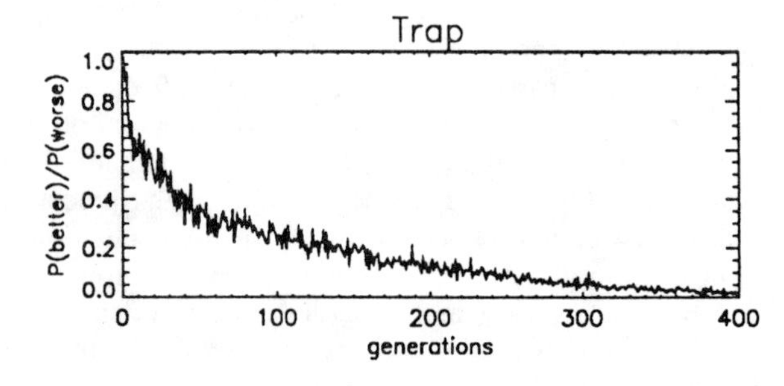

Figure 1. The *safety ratio* of using mutation alone

several optimal alleles (about 7/20) and the penoptimal alleles (the false peak) for the rest of the segments. This algorithm is somewhat analogous to simulated annealing or stochastic hillclimbing. In the early generations, improvements are easy and the survival threshold is low. Even offspring worse than their parent can survive so long as they're better than the worst in the population. Later, after much selection, only improvers are likely to survive (strictly descending).

Table 1: Performance of mutation alone

prob	mean best	sd best	mean generations @ best
Onemax	0	0	298
Plateau	0.015	0.033	1732
Trap	0.334	0.049	450

The good performance on the non-epistatic Onemax is no surprise, and climbing the false peaks of Trap is just a small version of Onemax, but Plateau seems to contradict the assertion of Holland[10, p. 110] that a reproductive plan using mutation alone is little better than an enumerative process. Plateau is highly epistatic, although the epistasis is low order (only 5 bits). There is a mechanism operating here that is not immediately obvious in Figure 1. Even after P(better)/P(worse) has dropped near zero, there is still a small chance of producing an improved offspring that is just perceptibly larger than for Onemax after it becomes essentially zero. To achieve this, mutation must be creating optimal alleles. For this to happen at the rates just visible in the figure, there must be a mechanism that favors 0s over 1s in the gene pool. This mechanism is the break-up of optimal alleles which selection has already concentrated, thus increasing the likelihood that they may be reconstructed by mutation. This experiment illustrates that, as Fogel and Atmar[6] observe, reproductive plans using only mutation can be very effective on surprisingly difficult problems.

4.2 2X alone

Risk curves for 2X are given in Figure 2. A curve for Onemax is not included, because its shape is obvious; by the nature of Onemax whenever an offspring is produced that is better than both parents, its sibling will be worse. Hence the P(better)/P(worse) ratio is forever 1.0. Note that these searches terminate much sooner than do those with mutation. As Holland has predicted, crossover is less effective than mutation at maintaining gene pool variation in the face of selection pressure to converge. On Plateau, the population after initial selection contains only individuals with at least one optimal allele and few with more (individuals with none must be the worst in the population, and do not reproduce). Hence disruption by 2X is not too likely relative to recombination, so the ratio is quite high. Very soon selection begins increasing the number of individuals with more than one optimal allele and the likelihood of disruption goes up so the ratio drops. Then convergence begins to make disruption less likely and creation by recombination more likely (0's are proliferating). The ratio climbs. A similar story is evident for Plateau-d, but now initial disruption is much more likely, so the ratio starts much lower. An important point here is that there are phases of the search when the risk of engaging in crossover may be high or low.

On Trap, the situation begins like Plateau, but here there is a force tending to proliferate ones instead of zeros. Even when convergence makes 2X less disruptive, it is also less creative, because the gene pool is not so rich in the building blocks of the optimal alleles. Unlike Plateau, there is no assurance that all the individuals in the initial population will have at least one optimal allele (not all those with none will be the population's worst individuals). If crossover does not disrupt any optimal alleles (either because there are none or because they're missed by crossover) the result is exactly like Onemax, one offspring is better than both parents and one is worse giving an instance of the ratio=1.0 that's added into the ratio sum. This occurs less often in Plateau where such events yield offspring that are equal to their parents unless there is recombination of optimal alleles. This accounts for the ratio initially being higher for Trap than Plateau and for Trap-d than Plateau-d. The curves for both Trap and Trap-d initially fall because of disruption of optimal alleles, but this disruption is more vigorous in Trap-d soon yielding a population in which there are few optimal alleles. Then the problem becomes a Onemax as the false peaks are climbed. The ratio looks like things are improving, but performance is poor. (See below.) On Trap, more optimal alleles are preserved than on Trap-d, so crossover remains risky.

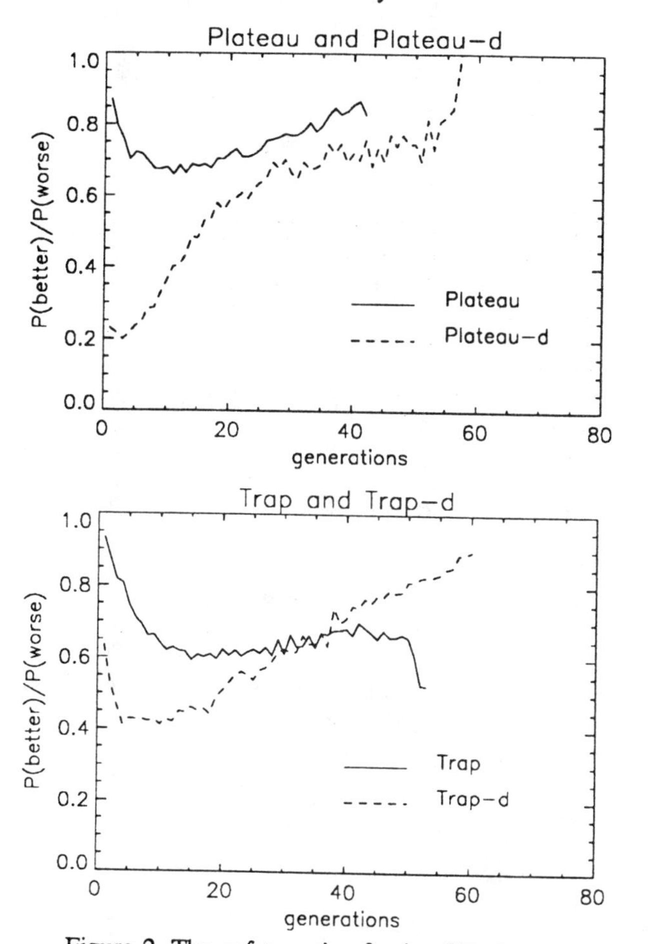

Figure 2. The *safety ratio* of using 2X alone

The performance is shown in Table 2. We see that 2X alone yields a very effective search algorithm when its local bias is beneficial. This algorithm finds the global optimum every time for Onemax and Plateau and most of the time (18/20) for Trap. Furthermore it does so more quickly than mutation alone. The poor performance on Trap-d is expected, but good performance on Plateau-d is a bit surprising. While 2X is highly disruptive of the distributed optimal alleles, the selection preference for them will yield a gene pool that is littered with their fragments even though there is no explicit selection preference for the fragments themselves. We observed above that mutation could exploit this situation, and here we observe that crossover, by recombining these fragments (building blocks), is even more effective (58 generations versus 1732).

Table 2: Performance of 2X alone

prob	mean best	sd best	mean generations @ best
Onemax	0	0	35
Plateau	0	0	30
Plateau-d	0.0025	0.011	58
Trap	0.003	0.008	40
Trap-d	0.434	0.049	59

4.3 UX alone

Risk curves for UX are shown in Figure 3. Again the Onemax curve is trivial and so not shown and the Plateau-d and Trap-d curves are the same as their non-distributed counterparts since UX has no local bias. These curves reflect the same mechanism discussed for 2X, but show the increased disruptiveness of UX.

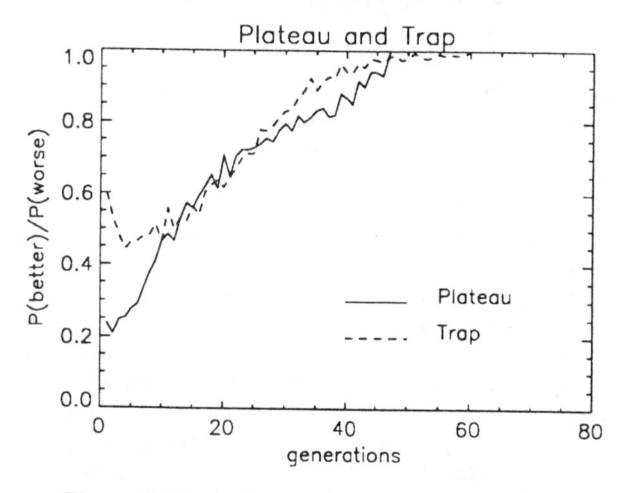

Figure 3. The *safety ratio* of using UX alone

The performance is shown in Table 3. This algorithm's performance is essentially the same as 2X on Onemax. On Plateau, its disruptiveness causes some trouble leading to slower search, but it's still able to find the global optimum. A longer epistatic segment would have caused more trouble. On Trap, UX is effectively trapped. It destroys all the optimal alleles and then climbs the false peaks as though this were a Onemax problem.

Table 3: Performance of UX alone

prob	mean best	sd best	mean generations @ best
Onemax	0	0	35
Plateau	0	0	37
Trap	0.491	0.013	41

5 Race Results

The fates of the crossover gene under the various race conditions are shown in Table 4. For each race, Table 4 presents the mean over 20 runs of the percent of the population with the crossover gene at its maximum point. This occurred at termination (convergence) for the runs without mutation, but with mutation the populations never converged and so ran to the specified computing limit (50,000 offspring). Once the best value ever found had essentially taken over the population, there was no longer any differential survival value to having the crossover gene, and its percent of the population tended to drift toward 50 under the action of mutation. Also shown are the mean best value and the average generation at which this value first appeared.

We can group the race outcomes into four classes.

Class 1: The crossover gene takes over and the population achieves a level of performance that mutation alone could not reach. Two cells in Table 4 fit this class: 2X (with and without mutation) on the Trap problem. This is the type of environment in which GA theory predicts success for crossover: there is significant but local epistasis. Not only is the performance better than mutation alone, this performance is achieved very rapidly. It is interesting to note that the performance of the population, once invaded by 2X, is worse than it was when 100% 2X was used from the beginning (see Table 2). When the subpopulation engaging in crossover is small, the gene pool variance can be lost too rapidly to sustain effective search. The effect of mutation may be to delay this convergence (56 vs. 62 generations to find its best) and to improve the performance slightly (.13 vs. .09). These differences are not statistically significant.

Class 2: The crossover gene takes over and the population performance is about the same as with mutation alone, but this performance level was reached much sooner. This class includes Onemax under all conditions, 2X (with or without mutation) on Plateau, 2X

Table 4: Race results (with mutation alone for comparison)

problem	MU		2X						UX					
			without MU			with MU			without MU			with MU		
	mean best	gen @ best	max % xo	mean best	gen @ best	max % xo	mean best	gen @ best	max % xo	mean best	gen @ best	max % xo	mean best	gen @ best
Onemax	0	298	100	.05	45	99	0	46	100	0	35	99	0	36
Plateau	.015	1732	100	.04	55	99	0	64	65	.49	26	99	.18	240
Plateau-d	*		40	.68	14	99	.16	482	*			*		
Trap	.33	450	100	.13	56	99	.09	62	5	.57	4	99	.33	173
Trap-d	*		6	.56	4	99	.34	191	*			*		

* Experiment not run. The expected results are known to be the same as the table entry immediately above.

with mutation on Trap-d, and UX with mutation on Trap and Trap-d. The Onemax results are no surprise, crossover provides more effective recombination than mutation alone, but the problem class is easy and could have been solved even more rapidly by hillclimbing. The surprise on Plateau is not the success of 2X (this was expected for the same reasons as Trap), but the success of mutation alone. With epistasis limited to five bits, mutation can accumulate beneficial changes even when they don't convey any immediate survival benefit. Another unexpected result is that the highly disruptive crossovers (2X on Trap-d and UX on Trap and Trap-d) could do so well when coupled with mutation. They did better with a small invading population than they did when they were used at 100% from the beginning (see Tables 2 and 3). When we examine these searches we note that often the invading subpopulation actually dies out, but is resurrected again by the action of mutation on the crossover gene. When this resurrection happens with the population in a more favorable state because of convergence (the *safety ratio* curves in Figures 2 and 3 rise), then the crossover gene can take over.

Class 3: The crossover gene takes over, but the population performance is worse than it was with mutation alone. This class includes 2X with mutation on Plateau-d and UX with mutation on Plateau and Plateau-d. Unlike the previous class, here there is a destructive interaction between crossover and mutation. Crossover is so disruptive that the beneficial mutations cannot survive long enough to accumulate.

Class 4: The crossover gene fails to take over consistently and performance is worse than with mutation alone. This class includes 2X on Plateau-d and Trap-d and UX on Plateau, Plateau-d, Trap and Trap-d, all without mutation. These are all circumstances where crossover is highly disruptive and there's no mutation to help it out. On the Trap problem the demise of the crossover gene is quite dramatic with the best individual ever found often appearing in the initial population. Convergence is rapid after the loss of any mechanism to combat it. On the Plateau problem, however, sometimes crossover survives (40% of the time for 2X and 65% for UX). The population performance is better when recombination survives (mean best scores of .55 in the 8 runs where 2X took over versus .77 for the 12 were it died and .35 in the 13 runs where UX took over versus .76 when it died), but still not as good as when the crossover was used 100% from the start (see Tables 2 and 3). As noted with class 1 above, the small population does not provide sufficient defense against selection's pressure for convergence.

6 Conclusions

A subpopulation producing offspring by recombination using crossover was unable to invade an otherwise static population undergoing fitness proportional reproduction when crossover was excessively disruptive. This was observed for the highly disruptive UX, but also for the less disruptive 2X when the critical schemata had long defining lengths. Under these circumstances, crossover is not an evolutionarily viable strategy. However, with mutation simultaneously operating at a small rate, the invasion was always successful. This takeover resulted in the production of high performance offspring much more quickly than was the case when mutation alone was operating. Furthermore, a limited amount of deception in the problem effectively trapped the mutation-only population, but the low disruptiveness of 2X allowed its population to produce superior offspring. We observed that this takeover was not always immediate; it was sensitive to the relative riskiness of the crossover and mutation operators. As selection pressure led to more converged and more fit populations, mutation tended to become more risky while crossover became less risky. In some cases, we observed the early demise of the invading group only to see them retry later by having mutation turn the crossover gene back on. If conditions were sufficiently favorable, the new invaders were quickly successful. Under these circumstances, crossover is an evolutionarily viable strategy. However, the takeover by crossover was not always associated with superior performance. Highly disruptive crossover on mildly epistatic non-deceptive problems yielded population performance that was worse than it would have been if mutation alone had been at work.

The introduction of crossover into a population slowly evolving under selection and mutation can be a mixed blessing. It can increase the speed of evolution many fold as well as overcome some epistasis that mutation alone could not. It can also disrupt schemata that mutation could have exploited leading to poorer performance for the population.

In these experiments, UX seemed generally to perform worse than did 2X and we would be remiss if we left this conclusion to be drawn. Because of its highly disruptive nature, UX does fair poorly under the selection scheme used in these experiments wherein the offspring replace the parents and the only way to preserve a parental structure into the next generation is to employ a reproduction without modification. UX can perform impressively with a more conservative selection scheme that allows parental structures to live for many generations, only being replaced when offspring (their own or others') appear who are superior[5,13]. Under these circumstances UX can cope successfully with problems that defy 2X because of non-local epistasis.

When Fogel and Atmar ran races similar to ours, they observed mutation winning and crossover losing. We speculate there are several reasons: 1) In their experiments, crossover was always linked with an inversion operator that was not true inversion as described by Holland[10]. Their inversion operator swapped alleles between genes instead of just reordering the genes. 2) Their representation coded the real parameter values and did not use a binary representation. Thus crossover was limited to recombining the alleles present and could do no within-parameter search.

We believe Fogel and Atmar are correct to point out the power of selection and mutation for search. This power is often underestimated in the GA community. We believe it is not correct to infer from their results that crossover provides no added value. We believe our experimental results begin to shed more light on the circumstances under which crossover will deliver the search power first described by Holland. However, these results also suggest circumstances in which crossover has a detrimental impact on search. It must be used with care.

References

[1] R. A. Caruana, J. D. Schaffer and L. J. Eshelman. Using Multiple Representations to Improve Inductive Bias: Gray and Binary Coding for Genetic Algorithms. *Proceedings of the 6th International Conference on Machine Learning*, 375-378. Morgan Kaufmann, San Mateo, CA, 1989.

[2] Y. Davidor. Epistasis Variance — Suitability of a Representation to Genetic Algorithms. In G. J. E. Rawlins (editor), *Foundations of Genetic Algorithms and Classifier Systems*, Morgan Kaufmann, San Mateo, CA, 1991.

[3] K. A. De Jong. Analysis of the Behavior of a Class of Genetic Adaptive Systems. Ph.D. Dissertation, Department of Computer and Communication Sciences, University of Michigan, Ann Arbor, MI, 1975.

[4] L. J. Eshelman , R. A. Caruana and J. D. Schaffer. Biases in the Crossover Landscape. *Proceedings of the Third International Conference on Genetic Algorithms*, 10-19. Morgan Kaufmann, San Mateo, CA, 1989.

[5] L. J. Eshelman. The CHC Adaptive Search Algorithm: How to Have Safe Search When Engaging in Nontraditional Genetic Recombination. In G. J. E. Rawlins (editor), *Foundations of Genetic Algorithms and Classifier Systems*, Morgan Kaufmann, San Mateo, CA, 1991.

[6] D. B. Fogel and J. W. Atmar. Comparing Genetic Operators with Gaussian Mutations in Simulated Evolutionary Processes Using Linear Systems. *Biological Cybernetics, 63* (1990), 111-114.

[7] D. E. Goldberg. Simple Genetic Algorithms and the Minimal, Deceptive Problem. In L. Davis (editor), *Genetic Algorithms and Simulated Annealing*, 74-88. Morgan Kaufmann, San Mateo, CA, 1987.

[8] J. J. Grefenstette. Optimization of Control Parameters for Genetic Algorithms. *IEEE Transactions on Systems, Man & Cybernetics, SMC-16*, 1, (January-February 1986), 122-128.

[9] J. J. Grefenstette and J. E. Baker. How Genetic Algorithms Work: A Critical Look at Implicit Parallelism. *Proceedings of the Third International Conference on Genetic Algorithms*, 20-27. Morgan Kaufmann, San Mateo, CA, 1989.

[10] J. H. Holland. *Adaptation in Natural and Artificial Systems*. University of Michigan Press, Ann Arbor, MI, 1975.

[11] G. E. Liepins and M. D. Vose. Representational Issues in Genetic Optimization. *Journal of Experimental and Theoretical AI*, May 1991.

[12] J. D. Schaffer, R. A. Caruana, L. J. Eshelman and R. Das. A Study of Control Parameters Affecting Online Performance of Genetic Algorithms for Function Optimization. *Proceedings of the Third International Conference on Genetic Algorithms*, 51-60. Morgan Kaufmann, San Mateo, CA, 1989.

[13] J. D. Schaffer, L. J. Eshelman and D. Offutt. Spurious Correlations and Premature Convergence in Genetic Algorithms. In G. J. E. Rawlins (editor), *Foundations of Genetic Algorithms and Classifier Systems*, Morgan Kaufmann, San Mateo, CA, 1991 forthcoming.

[14] N. N. Schraudolph and R. K. Belew. Dynamic Parameter Encoding for Genetic Algorithms. Technical Report CNLS, MS-B258, Los Alamos National Laboratory, Los Alamos, NM.

[15] C. G. Shaefer. The ARGOT Strategy: Adaptive Representation Genetic Optimizer Technique. *Genetic Algorithms and Their Applications: Proceedings of the Second International Conference on Genetic Algorithms*, 50-58. Lawrence Erlbaum Associates, Hillsdale, NJ, 1987.

A Comparison of Genetic Sequencing Operators

T. Starkweather, S. McDaniel,
K. Mathias, D. Whitley
Computer Science Dept.
Colorado State University
Fort Collins, CO 80523

C. Whitley
Mechanical Engineering Dept.
Colorado State University
Fort Collins, CO 80523

Abstract

This work compares six sequencing operators that have been developed for use with genetic algorithms. An improved version of the edge recombination operator is presented, the concepts of adjacency, order, and position are reviewed in the context of these operators, and results are compared for a 30 city "Blind" Traveling Salesman Problem and a real world warehouse/shipping scheduling application. Results indicate that the effectiveness of different operators is dependent on the problem domain; operators which work well in problems where adjacency is important (e.g., the Traveling Salesman) may not be effective for other types of sequencing problems. Operators which perform poorly on the Blind Traveling Salesman Problem work extremely well for the warehouse scheduling task.

1 INTRODUCTION

Gil Syswerda [5] conducted a study in which "edge recombination" (a genetic operator specifically designed for the Traveling Salesman Problem) performed poorly relative to other operators on a job sequence scheduling task. While the population size used by Syswerda was small (30 strings) and good results were obtained on this problem using mutations alone (no recombination), Syswerda's discussion of the relative importance of position, order and adjacency for different sequencing tasks raises an issue that has not been adequately addressed. Researchers, including ourselves [8] [10], seem to tacitly assume that all sequencing tasks are similar and that one genetic operator should suffice for all types of sequencing problems.

This paper compares six different operators on two different sequencing tasks. The comparisons include an improved version of "edge recombination." The problems are a 30 city "Blind" Traveling Salesman Problem and a 195 element sequencing task for a real world warehouse/shipping scheduling application. Our experiments show that different operators are better suited to different kinds of sequencing tasks. Edge recombination is only roughly competitive with operators such as PMX on the warehouse scheduling problem and the resulting search is an order of magnitude slower than operators which stress relative order as opposed to adjacency.

The genetic algorithm used in these experiments is *GENITOR,* which was developed at Colorado State University [7]; our results also suggest that *GENITOR* is a key part of our improved performance on the Traveling Salesman Problem. It uses a one-at-a-time replacement paradigm in which only one pair of strings reproduces during any given generation and only one offspring is generated. The new offspring replaces the worst string in the population rather than one of its parents. This ensures that the best string found so far will never be replaced in the population.

We do not offer comparative results in this paper to other approaches for the Traveling Salesman Problem. We do note, however, that *GENITOR* is really solving a more difficult version of the Traveling Salesman Problem than that solved by most other algorithms. Most algorithms use local edge information to do local improvements; *GENITOR* uses only the overall value of the total sequence. David Goldberg [2] points out that genetic algorithms are actually solving a "Blind" Traveling Salesman Problem. The Blind Traveling Salesman Problem is interesting because for certain types of sequence optimization tasks local information is not readily available. In a fair comparison on the "classic" Traveling Salesman Problem, the genetic algorithm would be allowed to also use local information about edge costs. In other words, parent and offspring tours could be improved by using local information. A group of European researchers [6] have used this type of strategy and have achieved impressive results on classic Traveling Salesman Problems with up to 666 cities. They found that by combining genetic operators and local search methods developed by Lin/Kernigan they obtained superior results to 5 other approaches on 8 different problems ranging in size from 48 to 666 cities (median size 318). Thus, it appears likely that genetic methods could become a basic part of the tool kit for solving Traveling Salesman Problems and that the genetic algorithm could be viewed as a structure in which to organize and apply many of the tools that already exist. The reader interested in comparison tests on the classic Traveling Salesman Problem should consult the results of Ulder et al. [6].

2 EMPHASIS OF DIFFERENT SEQUENCING OPERATORS

Six genetic sequencing operators are compared in this study: improved edge recombination, order crossover, variants proposed by Syswerda [5] which we shall refer to as order crossover #2 and position crossover, PMX crossover, and cycle crossover.

2.1 ENHANCED EDGE RECOMBINATION

The edge recombination operator is different from other genetic sequencing operators in that it emphasizes adjacency information instead of the order or position of the items in the sequence. The "edge table" used by the operator is really an adjacency table listing the connections into and out of a city found in the two parent sequences. The edges are then used to construct offspring in such a way that we avoid isolating cities or elements in the sequence.

For example, the tour [b a d f g e c j h i] contains the links [ba, ad, df, fg, ge, ec, cj, jh, hi, ib], when one considers the tour as a hamiltonian cycle. In order to preserve links present in the two parent sequences a table is built which contains all the links present in each parent tour. Building the offspring then proceeds as follows: (1) Select a starting element. This can be one of the starting elements from a parent, or can be chosen from the set of elements which have the fewest entries in the edge table. (2) Of the elements that have links to this previous element, choose the element which has the fewest number of links remaining in its edge table entry, breaking ties randomly. (3) Repeat step 2 until the new offspring sequence in complete.

An example is given of edge recombination is given in figure 1. Suppose element *a* is selected randomly to start the offspring tour. Since *a* has been used, all occurrences of *a* are removed from the right-hand side of the edge table. Element *a* has links to elements *b*, *f*, and *j*. Elements *b* and *f* both have 3 links remaining in their table entries, but element *j* has only 2 links remaining. Therefore, *j* is selected as the next element in the offspring, and all occurrences of *j* are removed from the right-hand side of the edge table. Element *j* has links to *i* and *h*, both of which have 3 links remaining. Therefore, one of these elements are selected at random (element *i* in figure 1) and the process continues until the child tour is complete.

```
Parent 1:     a b c d e f g h i j
Parent 2:     c f a j h d i g b e

Offspring:    a j i h d c f e b g

Edge table:   city   links
              a      b, f, j
              b      a, c, g, e
              c      b, d, e, f
              d      d, f, b, c
              e      d, f, b, c
              f      e, g, c, a
              g      f, h, i, b
              h      g, i, j, d
              i      h, j, d, g
              j      i, a, h
```

Figure 1: Edge Recombination

When the edge recombination operator was first implemented, we realized that it had no active mechanism to preserve "common subsequences" between the 2 parents. We have developed a simple solution to this problem. When the "edge table" is constructed, if an item is already in the edge table and we are trying to insert it again, that element of the sequence must be a common edge. The elements of a sequence are stored in the edge table as integers, so if an element is already present, the value is inverted: if A is already in the table, change the integer to -A. The sign acts as a flag. Consider the following sequences and edge table: [a b c d e f] and [c d e b f a].

```
a: b, -f, c          d: -c, -e
b: a, c, e, f        e: -d, f, b
c: b, -d, a          f: e, -a, b
```

The new edge table is the same as the old edge table, except for the flagged elements. One of three cases holds for an edge table entry. 1) If four elements are entered in the table as connections to a given table entry, that entry is not part of a common subsequence. 2) If three elements are entered as connections to a given table entry, then one of the first two elements will be negative and represents the *beginning* of a common subtour. 3) If only two elements are entered for a given table entry, both must be negative and that entry is an internal element in a common subsequence. Giving priority to negative entries when constructing offspring affects edge recombination for case 2 only. In case 1, no connecting elements have negative values, and in case 3 both connecting elements are negative, so edge recombination behaves just as before. In case 2, the negative element which represents the start of a common subtour is given first priority for being cho-

```
Parent 1:     a b c d e f g h i j
Parent 2:     c f a j h d i g b e
Cross Pts:        *       *
Offspring:    f g a j h d i b c e
```

Figure 2: Order Crossover #1

```
Parent 1:     a b c d e f g h i j
Parent 2:     c f a j h d i g b e
Cross Pts:        * *     *   *
Offspring:    a j c d e f g h i b
```

Figure 3: Order Crossover #2

sen. Once this common subsequence is started, each internal element (case 3) of the sequence has only one edge in and one edge out, so it is guaranteed that the common sections of the sequence will be preserved. The implementation of this idea (along with better mechanisms to ensure random choices when random choices are indicated) improved our performance on the Blind Traveling Salesman Problem. Using a single population of 1000 and a total of 30,000 recombinations *GENITOR* with the enhanced edge recombination operator finds the optimal solution on the 30 city problem described in Whitley et al. [8] on 30 out of 30 runs. On a 105 city problem the new operator finds the "best known" solution on 14/30 runs with no parameter tuning.

2.2 ORDER CROSSOVER

The original order crossover operator (which we refer to as order crossover) was developed by Davis [1] (also see [4]). The offspring inherits the elements between the two crossover points, inclusive, from the selected parent in the same order and position as they appeared in that parent. The remaining elements are inherited from the alternate parent in the order in which they appear in that parent, beginning with the first position following the second crossover point and skipping over all elements already present in the offspring. Thus, although the purported goal is to preserve the relative order of elements in the sequences to be combined, part of the offspring inherits the order, adjacency *and* absolute position of part of one parent string, and the other part of the offspring inherits the relative order of the remaining elements from the other parent, with disruption occurring whenever an element is present that has already been chosen.

An example is given in figure 2. The elements *a*, *j*, *h*, *d*, and *i* are inherited from P2 in the order and position in which they occur in P2. Then, starting from the first position after the second crossover point, the child tour inherits from P1. In this example, position 8 is this next position. P1[8] = h, which is already present in the offspring, so P1 is search until an element is found which is not already present in the child tour. Since *h*, *i*, and *j* are already present in the child, the search continues from the beginning of the string and Off[8] = P1[2] = b, Off[9] = P1[3] = c, Off[10] = P1[4] = e, and so on until the new tour is complete.

2.3 ORDER CROSSOVER #2

The operator which was developed by Syswerda [5] differs from the above order operator in that several key positions are chosen randomly and the order in which these elements appear in one parent is imposed on the other parent to produce two offspring; in our experiments we produce only one offspring.

In the example of figure 3, positions 3, 4, 7, and 9 have been selected as the key positions. The ordering of the elements in these positions from Parent 2 will be imposed on Parent 1. The elements (in order) from Parent 2 are a, j, i and b. In Parent 1 these same elements are found in positions 1, 2, 9 and 10. In Parent 1 P1[1] = a, P1[2] = b, P1[9] = i and P1[10] = j, where P1 is Parent 1 and the position is used as an index. In the offspring the elements in these positions (i.e., 1, 2, 9, 10) are reordered to match the order of the same elements found in Parent 2 (i.e., a, j, i, b). Therefore Off[1] = a, Off[2] = j, Off[9] = i and Off[10] = b, where Off is the offspring under construction. All other elements in the offspring are copied directly from Parent 1.

2.4 PARTIALLY MAPPED CROSSOVER (PMX)

This operator is described in detail by Goldberg and Lingle [3]. A parent and two crossover sites are selected randomly and the elements between the two starting positions in one of the parents are directly inherited by the offspring. Each element between the two crossover points in the alternate parent are mapped to the position held by this element in the first parent. Then the remaining elements are inherited from the alternate parent. Just as in the order crossover operator #1, the section of the first parent which is copied directly to the offspring preserves order, adjacency and position for that section. However, it seems that more disruption occurs when mapping the other elements from the unselected parent. In the first example (figure 4), the elements in positions 3, 4, 5 and 6 are inherited by the child from Parent 1. Then beginning with position 3, the element in P1 (c) is located in P2 (position 7) and this position in the offspring is filled with the element in Parent 2 at position 3: Off[7] = P2[3]. Moving to position 4 in Parent 1, we find a d and see that it occurs at position 1 in Parent 2, so Off[8] = P2[5] = a and f (P1[6]) is at P2[10] so Off[10] = P2[6] = g. The remaining elements

```
Parent 1:     a b c d e f g h i j
Cross Pts:      *       *
Parent 2:     d i j h a g c e b f
Offspring:    h i c d e f j a b g
```

Figure 4: PMX Crossover Example 1

```
Parent 1:     a b c d e f g h i j
Cross Pts:      *       *
Parent 2:     c f a j h d i g b e
Offspring:    a j c d e f i g b h
```

Figure 5: PMX Crossover Example 2

are inherited from P2: Off[2] = P2[2] = i, and Off[9] = P2[9] = b.

Since the segment of elements from the alternate parent does not contain any elements in the key segment of the first parent, both adjacency and relative order are preserved.

In the second example (figure 5), the mapping proceeds as above with Off[3 to 6] = P1[3 to 6]. Next Off[1] = P2[3] = a, since P1[3] = c and P2[1] = c. Next, we note that P1[4] = d and P2[4] = j. Since P2[6] = d, this is the preferred position for j in the offspring, but it has already been filled. City j is skipped over temporarily. Element h maps to element e which occupies position 10 in parent 2, so Off[10] = h. City d maps to element f which occupies position 2 in parent 2, so Off[2] = d; even though this is a duplicate it is left in the offspring temporarily. Elements i, g and b are then inherited from P2 leaving a sequence with no j element and two d elements. The element d which is outside the originally selected positions 3 through 6 is replaced with a j resulting in a complete and legal sequence. Note that when this substitution occurs, it results in a mutation where neither adjacency, position, or relative order is preserved by the substitution. Also note that PMX is influenced by position, especially in Example 2.

2.5 CYCLE CROSSOVER

Originally developed by Oliver et al. [4], this operator preserves the absolute position of elements in the parent sequence. A parent sequence and a cycle starting point are randomly selected. The element at the cycle starting point of the selected parent is inherited by the child. The element which is in the same position in the other parent cannot then be placed in this position so its position is found in the selected parent and is inherited from that position by the child. This continues until the cycle is completed by encountering the initial item in the unselected parent. Any elements which are not yet present in the offspring are inherited from the unselected parent. Note that cycle crossover always preserves the position of elements from one parent or the other without any disruption.

```
Parent 1:     a b c d e f g h i j
Cross Pts:          *
Parent 2:     c f a j h d i g b e
Offspring:    c b a d e f g h i j
```

Figure 6: Cycle Crossover

```
Parent 1:     a b c d e f g h i j
Cross Pts:      * *     *     *
Parent 2:     c f a j h d i g b e
Offspring:    a b c j h f d g i e
```

Figure 7: Position Based Crossover

In figure 6, position four in Parent 1 is the selected starting position for the cycle and Off[4] = P1[4] = d. Parent 2 is then searched until the position of element d is found (P2[6]) and the offspring tour at this position inherits the element in this position from Parent 1, Off[6] = P1[6] = f. f occurs in P2 at position 2, so Off[2] = P1[2] = b followed by Off[9] = p1[9] = i, Off[7] = P1[7] = h, Off[5] = P1[5] = e, and Off[10] = P1[10] = j. This completes a cycle, since P2[5] = j and P1[5] = d, which was the starting element in the cycle. Now any remaining elements are inherited from Parent 2: Off[1] = P2[1] = c and Off[3] = P2[3] = a.

2.6 POSITION BASED CROSSOVER

This operator, also proposed by Syswerda [5], is intended to preserve position information during the recombination process. Several random locations in the sequence are selected along with one parent; the elements in those positions are inherited from that parent. The remaining elements are inherited in the order in which they appear in the alternate parent, skipping over all elements which have already been included in the offspring. Thus, the operator appears to be similar to Davis' Order Crossover #1 operator except that the elements copied from the selected parent come from random locations in the sequence and not from adjacent locations; although designed as a "position" operator, it certainly is less effective at preserving position than cycle crossover and probably less effective at preserving position than PMX. We argue this is really another order operator.

In figure 7 the elements b, c, f and i are inherited from Parent 1 in positions 2, 3, 6 and 9 respectively. The remaining elements are inherited from Parent 2 as follows: Off[1] = P2[3] since P2[1] and P2[2] have already been included in the offspring. Then going in order, Off[4] = P2[4], Off[5] = P2[5], Off[7] = P2[6], Off[8] = P2[8] and Off[10] = P2[10].

Op	Bias	Trials	Pop	Best	Avg
Edge	1.5	50000	500	16/30	421.6
Order #1	1.5	50000	500	8/30	429.5
Order #2	1.5	50000	500	9/30	440.5
Position	1.5	50000	500	11/30	431.3
PMX	1.5	50000	500	437	514.6
Cycle	1.5	50000	500	459	519.9

Table 1: 30 City Results (untuned)

Op	Bias	Trials	Pop	Best	Avg
Edge	1.4	30000	1000	30/30	420.0
Order #1	1.1	100000	1000	25/30	420.7
Order #2	1.2	100000	1000	18/30	421.4
Position	1.2	120000	1000	18/30	423.2
PMX	1.2	120000	1400	1/30	452.8
Cycle	1.1	150000	1500	440	490.3

Table 2: 30 City Results (tuned)

3 THE 30 CITY BLIND TRAVELING SALESMAN

Each of the above operators was used to solve the 30 city Traveling Salesman Problem. In order to compare the performance on two levels they were each run using the same parameters for 30 experiments and then each was tuned for best results. The parameters for the first comparison were: selection bias of 1.5, population size of 500, no explicit mutation and 50,000 trials. The only exception to this is that the cycle crossover operator always mutates whenever the offspring and the selected parent are identical. Results appear in Table 1.

We attempted to optimize the performance of each operator by tuning the following parameters: bias, population size and number of trials. The results in able 2 are similar to those in Table 1, although PMX and cycle showed very little improvement despite the parameter tuning. The three order crossover operators (Order #1, Order #2 and Position) have similar performance. Using higher selection bias values in general gave poorer results for all operators except edge recombination. PMX and cycle crossover in particular converged too quickly in most cases to find the optimal solution. The improved edge recombination operator found the optimal solution 28 out of 30 times using a population of 650, bias of 1.7 and 30,000 recombinations. As shown, a larger population found the optimal solution on every attempt. Our results differ somewhat from the results cited by Oliver et. al.[4]. The ranking of the operators in terms of performance is the same (Order #1 is better than PMX which is better than Cycle). The main difference is that all of these operators produced much better results in the current study. Order crossover #1 and PMX both failed to find the optimal solution to this problem in the Oliver et al. study. The main difference in the two studies is the use of the *GENITOR* algorithm instead of the standard generational genetic algorithm. This strongly suggests that the use of *GENITOR* is partially responsible for the positive results we have obtained on this problem.

In a previous study we found that the old edge recombination operator coupled with a distributed genetic algorithm found the best known solution on 30 out of 30 attempts using 10 subpopulations of 200 individuals each, using up to a total of 70,000 evaluations/recombinations (7,000 per subpopulation). On the 105 city problem the old edge recombination operator operator coupled with a distributed genetic algorithm matched the best known solution on 15 out of 30 attempts using up to 2 million recombinations; all results were within 1% of the best known solution and 29/30 were within 0.5% of the best known. Using the enhanced edge recombination operator we found the best known solution on 14/30 runs using a single population algorithm (popsize: 5000) and only 1 million recombinations. These are first run results with no parameter tuning. While these results are not directly comparable, they do support the notion that the enhanced edge recombination operator is more effective than the original implementation.

3.1 DISCUSSION

The key difference between the operators is the information which each attempts to preserve during recombination. For the Traveling Salesman Problem the important information would seem to be the adjacency information. The edge recombination operator explicitly preserves adjacency information and clearly has the best performance on this problem. Information about absolute position appears to be relatively unimportant. None of the operators use mutation (except cycle crossover when the offspring is identical to one of the parents). We have done some experiments which suggest that the performance of some of the operators can be improved if mutation is used; resolving this issue requires further tests. Perhaps most surprising is the difference in performance between the order operators and among the position preserving operators. These differences can be explained by looking at how the operator preserves adjacency information, relative order and position. Adjacency information is clearly important, but the results obtained with the order operators (order crossover #1, order crossover #2, and the so-called position based crossover) suggests that order information is useful for solving this problem. PMX may produce a greater emphasis on absolute position than the other order operators; the cycle operator clearly stresses absolute position.

It is important to note that the performance of these operators on a given problem is directly related to the nature of that problem. In other problems such as scheduling, the important information may not be adjacency, but may have a higher correlation to the position in the string or the relative order among the encoded elements in the string.

4 A WAREHOUSE/SHIPPING SCHEDULER

A prototype scheduling system has been developed for the Coors brewery in Golden, Co., which uses a genetic algorithm to optimize the allocation of beer production and inventory to the orders at the plant. A simulator was constructed consisting of a representation for beer production, the contents of inventory, arrangement of truck and rail loading docks, and orders for a 24 hour period. Preliminary tests indicated that the system is viable and subsequent tests of the system used real data from the plant.

The objective of the Coors scheduling package is the efficient allocation of orders to loading docks in the plant based on a fixed production schedule. Beer production occurs on multiple production lines which operate 24 hours a day. Each line produces different product types. There are numerous product types which can be produced, based on type of beer as well as various packages and labels. The data which is available for each line includes flow-rate, start and stop times, and product-type. The scheduling simulator analyzes the production schedule for each line and creates a time-sorted list composed of the product-type, amount, and time available. This time-sorted production list is then examined during an event driven simulation. An input file to the scheduling simulator contains the contents of inventory at the start of the time-period which is to be scheduled. New orders enter a loading dock upon completion of a previous order; the inventory is initially checked for product needed by the new order. Minimizing the contents of inventory is an important aspect of this problem. Inventory impacts the physical work of moving product more than once on the plant floor, the physical space occupied by product in storage, as well as refrigeration costs, etc. The schedule simulation places orders in rail and truck loading docks and attempts to fill the orders with the product that comes out of production and inventory. An order consists of one or more product type and an associated amount. In the actual data for the test scheduling period 195 customer orders are present and waiting to be filled. The schedule simulator attempts an efficient mapping between the product that is available and these orders. A "good" schedule is one which minimizes the average inventory in the system, and fills as many orders as possible. Each individual in the population maintained by the genetic algorithm is a sequence of customer order numbers. This sequence is mapped to the loading docks by the schedule simulator and orders are filled and placed in the docks based strictly on this sequence. Initially these sequences are randomly created, and as genetic search progresses new sequences are created by the process of genetic recombination. For the genetic algorithm to work, an evaluation for the sequence is needed. The evaluation of the sequence of orders is obtained using a scheduling simulator, which models operation of the plant and creates a shipping schedule based on the sequence.

Our results indicate that for this sequencing problem, relative order of the items which make up the sequence is more important than adjacency. This is not surprising given the nature of the problem: the relative order in which product is used will clearly affect inventory more than adjacency. Adjacency would appear to be almost irrelevant in this domain. This means that genetic recombination operators which perform well on the Traveling Salesman because they stress adjacency will be poor for sequencing tasks where relative order is critical. Experiments with the same 6 recombination operators tested on the 30 city Blind Traveling Salesman Problem were conducted on this warehouse/shipping sequencing task.

Figure 8 gives two graphs comparing the six operators on both the Blind Traveling Salesman Problem and the warehouse/shipping scheduler. As the graphs show, the results of schedule optimization with the six operators are almost the opposite of the results for the Traveling Salesman Problem.

The graph of the warehouse/shipping scheduler shows the comparative results for the 6 operators with up to 30,000 recombinations. (The population size is 200, the selective bias is 1.7, the number of runs is 15; no parameter tuning was used.) Both of Syswerda's operators did extremely well on this problem; they also did relatively well on the Traveling Salesman Problem. When we use edge recombination with up to 200,000 recombinations it finds solutions comparable to those found by PMX in 20,000 recombinations; both results are inferior to order crossover #2 and the position based operator.

The difference in search speed displayed by the operators coupled with increased performance in workstations has allowed us to achieve 2 orders of magnitude improvement in execution speed on the scheduling application. We are currently using a 15 MIP workstation. This means that our scheduler now executes in minutes rather than hours. (e.g., 6 hours becomes 3.6 minutes given 100 times faster execution). Scheduling 195 jobs in under 5 minutes is close to real time in the context of this warehouse/shipping problem. This allows quick rescheduling in the event of line breakages, shortfalls in production, and other unforeseen circumstances. Our execution times are, of course, dependent on the complexity of the evaluation function.

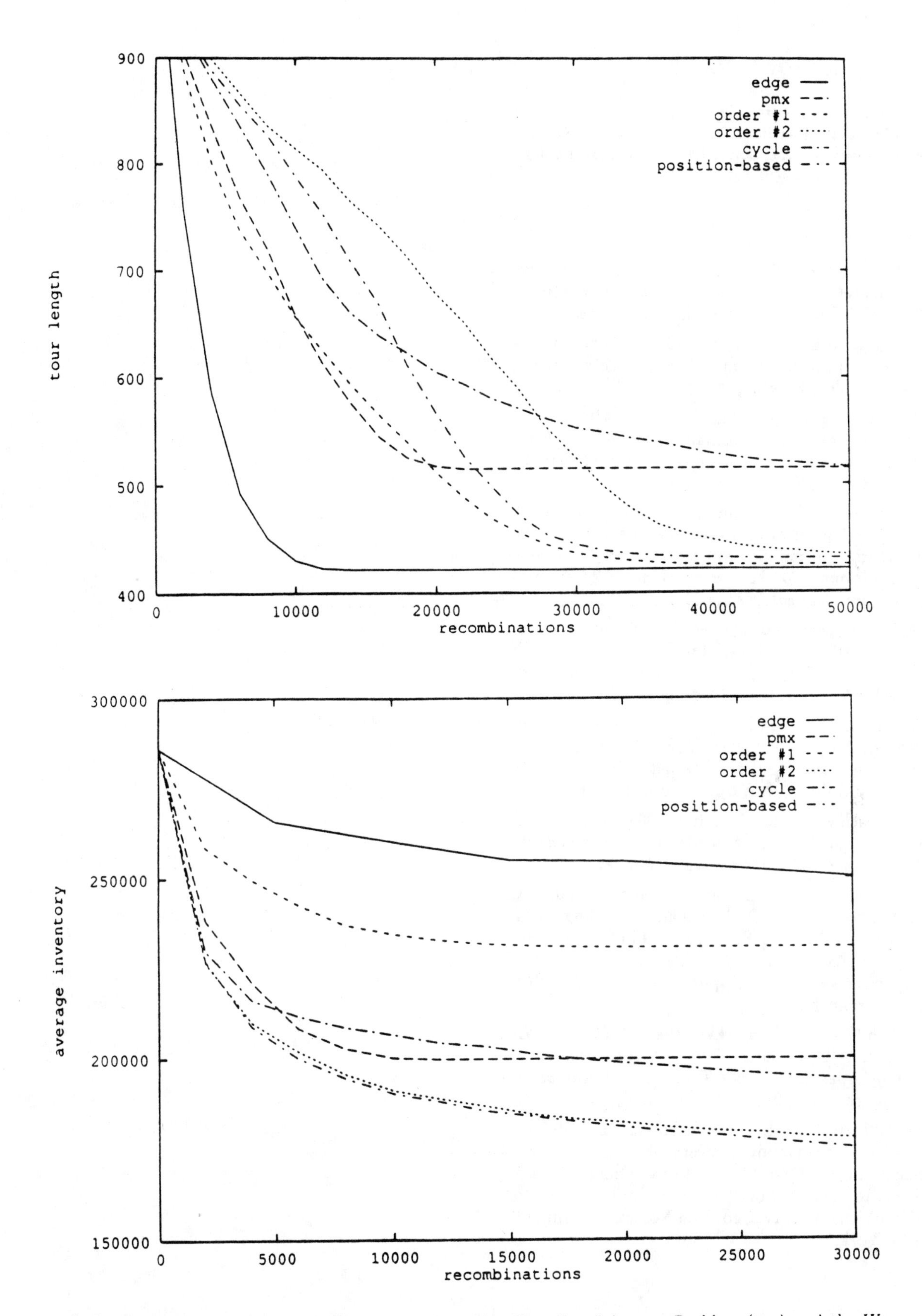

Figure 8: Graphs of 6 Operators for the Blind 30 City Traveling Salesman Problem (top) and the Warehouse/Shipping Scheduler (bottom).

ACKNOWLEDGEMENTS

This research was supported in part by a grant from the Colorado Institute of Artificial Intelligence (CIAI). CIAI is sponsored in part by the Colorado Advanced Technology Institute (CATI), an agency of the State of Colorado.

References

[1] L. Davis. (1985) "Applying Adaptive Algorithms to Epistatic Domains." In *Proc. International Joint Conference on Artificial Intelligence.*

[2] D. Goldberg. (1989) *Genetic Algorithms in Search, Optimization and Machine Learning.* Addison Wesley, Reading, MA.

[3] D. Goldberg and R. Lingle. (1985) "Alleles, loci, and the Traveling Salesman Problem." In *Proc. International Conference on Genetic Algorithms and their Applications.*

[4] I. Oliver, D. Smith, and J. Holland. (1987) "A Study of Permutation Crossover Operators on the Traveling Salesman Problem." In *Proc. Second International Conference on Genetic Algorithms and their Applications.*

[5] G. Syswerda. (1990) "Schedule Optimization Using Genetic Algorithms." In *Handbook of Genetic Algorithms.* l. Davis, ed. Van Nostrand Reinhold, New York.

[6] N. Ulder, E. Pesch, P. van Laarhoven, H. Bandelt, E. Aarts. (1990) "Improving TSP Exchange Heuristics by Population Genetics." In *Parallel Problem Solving In Nature.* Springer/Verlag.

[7] D. Whitley and J. Kauth (1988) "*GENITOR*: A Different Genetic Algorithm" In *Proc. Rocky Mountain Conf. on Artificial Intelligence.*

[8] D. Whitley, T. Starkweather, and D. Fuquay. (1989) "Scheduling Problems and Traveling Salesman: The Genetic Edge Recombination Operator." In *Proc. Third Int'l. Conference on Genetic Algorithms and their Applications.* J. D. Shaeffer, ed. Morgan Kaufmann.

[9] D. Whitley and T. Starkweather. (1990) "*GENITOR II*: A Distributed Genetic Algorithm." *Journal of Experimental and Theoretical Artificial Intelligence.* 2:189-214.

[10] D. Whitley, T. Starkweather, and D. Shaner. (1990) "Traveling Salesman and Sequence Scheduling: Quality Solutions Using Genetic Edge Recombination." In *Handbook of Genetic Algorithms.* L. Davis, ed. Van Nostrand Reinhold, New York.

Delta Coding: An Iterative Search Strategy for Genetic Algorithms

D. Whitley and K. Mathias
Computer Science Dept.
Colorado State University
Fort Collins, CO 80523

P. Fitzhorn
Mechanical Engineering Dept.
Colorado State University
Fort Collins, CO 80523

Abstract

A new search strategy for genetic algorithms is introduced which allows iterative searches with complete reinitialization of the population preserving the progress already made toward solving an optimization task. *Delta coding* is a simple search strategy based on the idea that the encoding used by a genetic algorithm can express a distance away from some previous partial solution. Delta values are added to a partial solution before evaluating the fitness; the delta encoding forms a new hypercube of equal or smaller size that is constructed around the most recent partial solution. Results are presented on two optimization problems involving geometric transformations; solving these problems with precision is difficult for conventional genetic algorithms as well as traditional mathematical optimization techniques. Tests using single population and distributed genetic algorithms are compared to *delta coding*. *Delta coding* is shown to produce more precise solutions while reducing the amount of work necessary to reach the solution.

1 INTRODUCTION

Genetic algorithms have been applied to many types of optimization problems. However, the limited success experienced on some problems, especially those requiring precise solutions, has driven researchers to look for new and more powerful strategies to augment the performance of genetic algorithms. *Delta coding* is introduced and shown to be capable of producing improved performance by yielding more precise solutions with less work. *Delta coding* treats population strings not as parameters, but as *delta values* ($\pm\delta$) which are added to a partial solution before being evaluated. The first run of a genetic algorithm (GA) using delta encoding is like any other run of the genetic algorithm. Subsequent runs are made by using the best solution obtained on the most recent run as a partial solution; the genetic algorithm is restarted with the substring coding for each parameter representing a distance or delta value ($\pm\delta$) away from the value of the corresponding parameter in the most recent partial solution forming a new hypercube around the previous solution. The "delta hypercube" can also be reduced or enlarged each time the genetic algorithm is restarted.

Delta coding effectively avoids what is usually a difficult trade off between achieving fast search and sustaining diversity (and thereby avoiding premature convergence). Diversity is a problem because the longer a genetic algorithm runs and the closer it gets to a solution, the less search power it has. *Delta coding* avoids the need to sustain diversity, since diversity is reintroduced by generating an entirely new and random population; previous work is preserved because the hypercube corresponding to the new encoding is defined with respect to the previous partial solution. By using smaller populations, *delta coding* reduces the amount of work it takes to reach a solution while at the same time improving the precision of the solution. Furthermore, if the precision of the resulting solution is more critical than search time, search can continue indefinitely without worrying about premature convergence, while normal genetic algorithms must be completely restarted when convergence occurs without reaching an adequate solution.

Delta coding preserves hyperplane sampling (and the theoretical foundations of genetic algorithms) since each individual iteration is a single run of a genetic algorithm; only the mapping strategy has changed. *Delta coding* makes mutation unnecessary. Furthermore, the "delta hypercubes" defined with respect to different partial solutions are never the same. Thus, Hamming cliffs are moved to different pairs of strings in the search space each time the hypercube is redefined. This has the potential to eliminate gray coding and may also be helpful in solving deceptive problems.

Delta coding is tested on two geometric transformation problems. The performance of GENITOR, GENITOR II, and *delta coding* are examined. While we use *delta coding* in conjunction with GENITOR, *delta coding* is an encoding strategy that can be used with any genetic algorithm. The two transformation problems involve a search for the set of optimal geometric transformation parameters (i.e. rotation, scale and/or translation) that will superimpose a set of test points in one geometric domain onto a given set of fixed points in another. The first problem can be stated as follows: Given two independent two dimensional representations of a single object, find a transformation which maps one representation (the primary image) onto the other (target image). The transformation is achieved by reducing the sum of the squared errors on two sets of n landmark points, one set of n from the primary image, and a similar set of n from the target, which must match when the two images are properly superimposed. In the second problem, a three dimensional primary representation must be mapped onto a target two dimensional image. As in the first problem, the error is minimized over a set of landmark points from the fixed two dimensional image.

2 BACKGROUND

Genetic algorithms work with a (typically binary) string encoding of a problem's parameters. It also employs an evaluation function to determine the relative performance of any particular set of parameters. The selection of strings for replication and reproduction is based on the performance measure. A string of parameters must be encoded in such a way that they can be broken apart and combined with other strings to form new parameter sets, which are referred to as offspring. The fitness (i.e. performance) of strings is used to allocate reproductive trials, such that a higher fitness value translates into an increased probability that a particular parent will be chosen to recombine with another parent string during reproduction. Recombination takes place between two parent strings by identifying two points randomly on the strings and choosing the information between the points from one parent to pass on to the offspring and the information outside of the points from the other parent to pass on to the offspring. The reduced surrogate crossover operator [2] used in these experiments, can ensure that offspring are not duplicates of the parents.

The goal of genetic algorithms is to exploit hyperplane information feedback by evaluating the relative fitness of strings, and then use this information to allocate reproductive opportunities so as to direct the search toward particular regions or partitions in hyperspace that contain above average competitive solutions. This method of search has been shown to be an effective optimization technique; genetic algorithms may be superior to other mathematical methods of search if the problem is complex with multiple local minima or if gradient information is difficult to obtain.

Sustaining diversity is a pervasive problem with genetic algorithms. Several strategies have been developed to support diversity maintenance such as mutation, adaptive mutation and distributed genetic algorithms[12] as well as more divergent variations on genetic search such as Ackley's connectionist genetic hill-climber [1] and Shaefer's ARGOT Strategy [10]. Diversity maintenance is critical, since the search reaches a plateau and stops if the entire population has converged to a single solution (or is populated with strings whose potential offspring are already members of the population). A typical search with a genetic algorithm shows significant progress toward better solutions when the population exhibits a large degree of diversity, but as search continues, progress becomes asymptotic to a limit imposed by the lack of diversity in the remaining population.

When first developing the *delta coding* idea, we did not relate it to any existing method. The motivation was based on the fact that solutions found by genetic algorithms were sometimes close to a precise solution such that we could complement a small number of bits and obtain a globally optimal solution. Also, recent theoretical work [3] shows that selective pressure on hyperplanes is not uniform during genetic search; bits that are selected against early on may actually be preferred as the search narrows to concentrate on specialized regions in hyperspace. But selective pressure cannot work during the later stages of search if there is not enough diversity in the population to allow alternate bit combinations to be tested.

After developing *delta coding*, we realized that it shared similarities with other strategies, especially ARGOT and Dynamic Parameter Encoding (DPE). Other strategies that have similarities that are not as obvious at first glance include micro-GAs. The next section briefly reviews GENITOR and the distributed GENITOR II, since these are compared to *delta coding*. ARGOT, DPE and micro-GAs are also discussed. Future work should include comparisons with these algorithms and should examine the merger of subsets of these strategies with *delta coding*.

2.1 THE GENITOR ALGORITHM

GENITOR is a genetic algorithm that uses "one at a time" recombination between two parents to form an offspring; each offspring is immediately evaluated and placed in a ranked population, displacing the string in the population with the worst fitness. This is different from the mass reproduction found in standard genetic algorithms where offsprings displace parents regardless of the fitness values of either string. One at a time reproduction guarantees that the best solutions found so far will be held undisturbed in the population until a better solution is located. GENITOR has displayed superior performance in many problem domains [15]. The perpetuation of superior strings guides the search by providing more bias or probability of commitment toward better solutions thereby introducing a kind of hill-climbing to genetic search, with only improved points in the search space being permitted to become part of the population. This can potentially be detrimental to the search in some instances since sampling biases can result from early commitment to strings that may not be representative of the general distribution of strings in hyperspace. Still, no one string dominates search and it has been shown that GENITOR does allocate increasing reproductive trials to hyperplanes in the population that contain above average strings [12].

2.2 GENITOR II: A DISTRIBUTED GA

GENITOR II is a version of GENITOR that attempts to improve search and maintain diversity by using distributed populations. Small populations, termed subpopulations, represent an independent search except that the subpopulations rendezvous at fixed intervals and exchange information by swapping copies of their "best" strings. The exchange interval, subpopulation size, number of subpopulations, and number of strings exchanged are user defined [12].

The algorithm used to exchange strings between subpopulations is a directional adjacent ring neighborhood distribution. This algorithm allows the subpopulation at position 1 to distribute copies of it's best n strings to the neighbor at position 2 at the first rendezvous. The new strings in the subpopulation at position 2 displace the n worst strings. This distribution occurs at all adjacent subpopulations in the ring at the same synchronized interval. During the next distribution the subpopulation at position 1 distributes copies of it's n best strings to the subpopulation at position 3, with corresponding distribution occurring throughout the ring. This method of distribution allows the best string from the subpopulation at position 1 to reach all positions in $log(X) + 1$ distributions, (where X is the number of subpopulations in the ring), assuming the string remains the best for the number of trials needed to perform $log(X) + 1$ swaps.

The GENITOR II algorithm has been shown to yield better results than GENITOR and standard GAs on many classes of problems. The distribution of the n best strings from a subpopulation provides a method to sustain diversity, share information and correct for sampling biases that may occur in individual populations. The GENITOR II algorithm not only shows improved performance in terms of more accurate solutions, it also requires less overall work [11].

2.3 THE ARGOT STRATEGY AND DYNAMIC PARAMETER ENCODING

The ARGOT strategy (Adaptive Representation Genetic Optimizer Technique) combines many ideas to improve optimization; the results appear to be successful, but the strategy is complex. The ARGOT strategy is designed to solve problems using a genetic algorithm, but several environmentally triggered operators are added to alter intermediate mappings that drive the search. Binary strings are used to represent search parameters. However, an intermediate mapping is performed between the bit strings and the search space. These intermediate mappings are based on internal measurements such as parameter convergence, parameter variance and parameter "positioning" within a possible range of parameter values. These measurements are used to drive strategies within ARGOT such as adjusting parameter resolution and roving parameter boundary locations [10].

Dynamic parameter encoding (DPE) is a strategy explored by Schraudoph and Belew [9] which has many similarities to ARGOT; the accuracy of the encoded parameters are dynamically adjusted to increase the resolution of the answer and to zoom in on the most promising area of the search space. DPE is designed to reduce the affects of the arbitrary experimenter assigned resolution inherent in solutions involving genetic optimization. At first the string encodes only the most significant bits of each parameter (e.g., 4 bits). This represents a coarse grain partitioning of the function space. As the genetic algorithm begins to converge, the most significant bit is recorded, dropped from the encoding, and a new bit is introduced which adds additional precision and creates a finer grain partitioning to the search space. Thus, while the number of bits used by the genetic encoding remains constant, the function is searched using an increasing level of detail.

Both of these approaches clearly have similarities to *delta coding* because bits are added or deleted (or remapped) as the range and location of parameter boundaries change. *Delta coding* does not allow each parameter to change independently (perhaps this could be done) while on the other hand ARGOT and DPE do not have discrete stages of search as does *delta coding*. The discrete stages of search may in fact be advantageous; the other strategy we have looked at that uses discrete stages of search is micro-GAs.

2.4 MICRO-GAs AND OTHER REINITIALIZATION METHODS

The micro-GA [7] uses a small population (e.g. five strings). The population is measured for convergence either by genotype convergence or phenotype convergence. If the population has converged the best string is kept and the rest of the population is randomly regenerated and the search is repeated. No mutation is used. This strategy has been shown to give good performance on certain test problems. Goldberg initially described this method of implementation for serial GAs [5]. The motivation is to maximize the number of schemata sampled during genetic search. Cataclysmic mutation, as described by Eshelman[4] also provides a method of restarting the GA search by reinitializing the population. During reinitialization, the best string in the population at the time of convergence is used as a template by randomly complementing some proportion of the bits; the resulting strings are used to reinitialize the population.

Micro-GAs and these other strategies are somewhat like *delta coding* in that the population is reinitialized and prior results are included in the population by saving the best string. One advantage of delta coding is that a completely new hypercube is defined, and the size and location of the new hypercube can be varied. One disadvantage of micro-GAs, for example, is that the small population prevents adequate hyperplane sampling and simple hill-climbing may dominate the search.

3 DELTA CODING

We now look at *delta coding* in more detail. Strategies such as micro-GAs are similar to *delta coding* in that diversity is *regenerated* in order to sustain search. Other strategies like ARGOT and DPE are similar in that they focus on promising areas of search. One of the advantages of *delta coding* is its simplicity: it is easy to understand, easy to implement and easy to analyze given the existing theory of genetic algorithms. Like ARGOT and DPE, *delta coding*, remaps hyperspace, can decrease the size of the search space by restricting the range of the *delta values* ($\pm\delta$) at each iteration and additionally eliminates diversity concerns.

Our *delta coding* implementation uses GENITOR as its basis. On the initial run, GENITOR evaluates the problem in a normal fashion except that the diversity of the population is checked on each trial. The population diversity is monitored by measuring the Hamming distance between each of the parameters in the best and worst strings in the population. If the sum of the parameter Hamming distances is more than one, the search continues. If the distance is less than or equal to one, the diversity is defined to be *sufficiently exploited* and the iteration is stopped. The best solution is saved as a starting point for the next *delta iteration*. The new values encoded on the string are recoded with a sign bit, thus defining a positive or negative delta value. Subsequent delta runs will often explore a smaller solution space. The number of bits used to encoded each delta value is reduced by a user determined number of bits. The population is reinitialized and the GENITOR algorithm is allowed to resume evaluation. Each delta parameter, when decoded for fitness evaluation, is applied as a delta value ($\pm\delta$) to the parameters saved from the previous iteration. This produces a refined search in a reduced hypercube space centered around the best initial solution.

Each *delta iteration* is allowed to proceed until diversity has been sufficiently exploited. The goal is to achieve solution improvement while the population is diverse. Typically, as a genetic algorithm begins to converge, more and more effort is expended to obtain smaller and smaller reductions in error. When using delta coding we want to stop search when the probable payoff for continued search begins to decline. In actual practice, this is not easy to determine, but the exact stopping criteria need not be too precise as long as the search is not stopped too quickly.

Once an iteration is terminated, the best result from that iteration is saved for the next *delta iteration*. Typically, at each iteration the *delta coding* is reduced by a single bit for each parameter; reductions are limited to some minimal number of bits for the *delta values* (e.g. 4 to 6 bits). If a delta run converges to exactly the same solution (i.e., all zeros), the number of bits used to encode the *delta values* ($\pm\delta$) is increased, thus expanding the hypercube. This search process continues until a solution meeting the user defined criteria is found or until the maximum number of trials (defined by the user) for the entire collection of iterations is exhausted. The algorithm is shown in Table 1.

One critical finding (that is somewhat counter-intuitive) is that *delta coding* works best when defined with respect to the immediate previous solution, ***not the best solution found so far.*** The reasons for this are not completely clear; however, a different (inferior) solution indicates something about the distribution of strings in hyperspace and provides a new direction in which to search. It is also consistent with the notion that search should not be exclusively driven by the single best string found so far. Furthermore, this strategy could help the search to escape local minima without having to expand the delta encoding.

4 PROBLEM DESCRIPTIONS

We attempted, unsuccessfully, to find techniques in traditional mathematical optimization methods that could be used to find optimal rigid body transformations for superimposing a set of test points onto a set of fixed target points. The only methods which appeared to be applicable for our 3-D problem were brute force methods that involved a great deal of ad hoc search. While we have not looked at methods such as simulated annealing, so far the genetic algorithm is the only approach that has produced adequate results using modest computational resources.

In the 2-D problem, four points are defined in two dimensional space and used as target points. Five transformation parameters are then applied to another set of four corresponding points in two dimensional space (the primary image) with the goal of overlaying them on the target points. The five transformation parameters used were translation along the X and Y axis,

PHASE 1: /* Apply GENITOR in Normal Mode */
Apply recombination/Evaluate/Insert Offspring
Check Hamming Distance Between Best and
Worst Members in Population
IF (Hamming Distance > 1) Continue PHASE 1
 	
TRANSITION PHASE:
Save Best Solution of Population as *PARTIAL*
Re-initialize Population
Encode Parameters Using $(X-1)$ Bits
(Plus Extra Bit Used as Sign)
 	
ITERATION PHASE:
DELTA: /* Apply GENITOR in Delta Mode */
Apply recombination
Add Decoded String Value to *PARTIAL*
Evaluate/Insert Offspring
Check Hamming Distance Between Best and
Worst Members in Population
IF (Hamming Distance > 1) Continue DELTA
 	
IF (Maximum Trials > Allocated Trials) OR
(*PARTIAL* < Threshhold) **HALT**
ELSE
Save New Best Solution as *PARTIAL*
Re-initialize Population
IF (*Delta Values* = 0) THEN
Encode Parameters with 1 Extra Bit
ELSE
Encode Parameters with 1 Less Bit
 	
Continue ITERATION PHASE

Table 1: *Delta Coding Algorithm*

scale in the X and Y directions, and rotation around the Z axis (restricting the points to the X-Y plane during rotation). A further constraint was imposed on the transformation in that each point was identified with a label such that *p1* of the primary image must overlay *p1'* of the target image for the transformation to be correct. The parameter set was encoded in binary using eight bits for each parameter, for a total of 40 bits. The fitness function was designed to measure the Euclidean distance between the corresponding target and transformed points, summing the squared distances for all point pairs. The goal was to minimize the error. (For implementation details see [8]).

The 3-D to 2-D transformational problem was defined by a specific application and motivated the 2-D test problems [6]. The goal in the 3-D transformation problem was to provide a set of transformation parameters that would map a set of 3-D points onto a target set of points in a 2-D domain. As in the 2-D transformation problem, the points were tagged so that the points undergoing transformation corresponded to specific 2-D target points. The evaluation function was used to measure Euclidean distances and report the sum of the squared distances as the error.

For the 3-D problem there were only three points on the object representation available to provide error information and eight transformation parameters were required [6]. The eight parameters were translation along the X, Y, Z axis, rotation around the X, Y, Z axis, a single uniform scale, and perspective. Parameters were encoded using 12 to 14 bits each, for a total of 104 bits. (For implementation details see [8]).

5 EXPERIMENT DESCRIPTIONS

The 2-D transformation problems were developed as a smaller test environment to see how genetic algorithms would perform on this class of problems. Several variants of the 2-D problem were developed and set up to run under GENITOR; each variant was known to have a solution of zero error. Several attempts at tuning parameters for optimal performance were performed. The problems were also run in the GENITOR II environment. The tunable parameters were kept the same in the GENITOR II runs as in the GENITOR runs except the population size and number of trials allocated to each experiment were altered to adjust for the use of multiple subpopulations. There were ten subpopulations used in the 2-D experiments. The swap interval was set to 5,000 and the number of strings distributed at each interval was five. With less than optimal results using GENITOR II the problem was encoded using the *delta coding* strategy. An effort was made to tune the population size and maximum overall trials to the lowest values possible while still achieving superior results. The bias value was kept at 2.0 in an effort to maintain an agressive level of search. Each experiment was repeated 30 times to obtain an adequate comparative sample.

5.1 REDUCTION AND EXPANSION STRATEGIES

Several types of reduction schemes were tested for use with *delta coding*. These reduction strategies involve user tunable values including (but not limited to) the smallest number of bits for a parameter during a run and the number of bits by which to reduce or expand each parameter during each new iteration. We tested numerous reduction and expansion strategies; only key examples used on the 2-D problem are reported here [8]. First, we reduced each parameter by 1 bit at each iteration, until the *delta coding* used only 2 bits. We found however, that we obtained better results if the *delta coding* was not allowed to use less that 4 bits; in other words, after reducing the *delta coding* to 4 bits per parameter, it stays at four bits (unless we expand the encoding). We thought that it was perhaps the 4 bit limit rather than the 1 bit reduction that was important, but this proved not to be the case in our test problems. We tried reducing by 2 bits with a 4 bit limit to the minimal encoding; this did not perform as

Genetic Method	Pop. Size	Sub-Pop.	Bias Value	Max. Trials	Avg. Error	Percent Solved
GENITOR	3,700	1	1.8	100,000	0.083	10%
GENITOR II	300	10	1.8	20,000	0.045	10%
GENITOR II	1,000	10	1.8	100,000	0.003	80%
Delta Coding	200	1	2.0	20,000	0.001	93%

Table 2: *Comparison of GENITOR and GENITOR II to* Delta Coding *for the 2-D Problem. Worst case results are reported for* delta coding, *while best case results are given for GENITOR and GENITOR II.*

Genetic Method	Pop. Size	Sub-Pop.	Bias Value	Max. Trials	Avg. Error	Best Found
GENITOR	5,000	1	1.75	500,000	1.45	0.4900
GENITOR II	500	10	1.75	500,000	1.22	0.3270
GENITOR II	50	100	1.75	500,000	0.88	0.1950
Delta Coding	500	1	2.00	500,000	0.10	0.0037

Table 3: *Comparison of GENITOR and GENITOR II to Delta Coding for the 3-D to 2-D Problem. For GENITOR II population size refers to subpopulation sizes.*

well. Reducing by 4 bits to more quickly reach the 4 bit limit further degraded performance.

Simple 1 bit reduction per parameter proved to be the best strategy on both the 2-D and 3-D problems; however, on the 3-D problem a 6 bit limit produced the best results, while on the 2-D problems the best results were obtained with a 4 bit limit. Solution curves for a few of the *delta coding* reduction strategies are presented in Figure 2. These results suggest that a conservative strategy which does not commit to a solution too quickly has an advantage over other strategies.

The other side of this issue is *delta coding* expansion. In our implementation of *delta coding* the criteria used to determine when the hypercube should be expanded was as follows: if a run converged to zero such that the solution was the same as that obtained on the previous iteration, then the hyperplane is enlarged by increasing the number of bits used in the *delta coding*. This "opens up" the search space when 1) further improvements seem unlikely or 2) the space defined by the current hypercube somehow favors the previous solution. The encoding is not expanded if the search converges to an inferior, but different solution. If a different solution is found, inferior or not, the encoding is reduced using the new inferior solution as the partial solution for delta encoding. This decision is based on empirical results and the intuition that a different but inferior solution shifts the direction of search.

6 RESULTS

Comparison of different approaches to the 2-D mapping problem are provided in Table 2 with performance curves in Figure 1. Several versions of the 2-D to 2-D problem were examined; on most of these "test cases" *delta coding* found zero error 100% of the time. Table 2 reports the worst test case results for *delta coding*, while it reports best test case results for GENITOR and GENITOR II. *Delta coding* out performs all other implementations compared in this context and provides a reliable mechanism to find a globally optimal set of transformations with zero error.

A summary of comparison experiments for the 3-D to 2-D transformation problem is provided in Table 3. The corresponding solution curves are presented in Figure 3. The swap interval for the GENITOR II experiments was set to 2500 and the number of strings distributed was set to five. The total number of trials allocated for each of the 3-D experiments was 500,000; for example, with 10 subpopulations the number of trials is 50,000 for each of the individual subpopulations in the GENITOR II environment. The population size for the 3-D problem was set to 500.

The delta strategy appears to out perform the other implementations on the 3-D mapping problem; it converged faster using an overall population size that is 10% the size of the others.

Figure 3 also shows that *delta coding* produces these more precise results using fewer trials; the *delta coding* experiments were largely converged by 100,000 trials. Furthermore the use of smaller population sizes in conjunction with *delta coding* in the GENITOR environment results in significant time savings [11]. The optimal solution was known to be an error of zero in all test cases and yet none of the experiments yielded a single run with an error of exactly zero. *Delta coding* made little progress after the initial 100,000 recombinations; ideally delta coding should have allowed the algorithm to continue to make progress, but in this case it did not. Nevertheless, the results are encouraging.

7 CONCLUSIONS

Delta coding has several attractive properties. It is conceptually simple to understand and simple to implement. It does not rely on forced mechanisms such as mutation to sustain diversity; it simply continues to search until the diversity in the current iteration is exhausted or reduced. It then saves the best solution parameters and reinitializes the population. This provides a new population with which to search again. This resurgence of diversity for each *delta iteration* functions to maintain a vigorous search at all times. With each *delta iteration* the number of bits used to

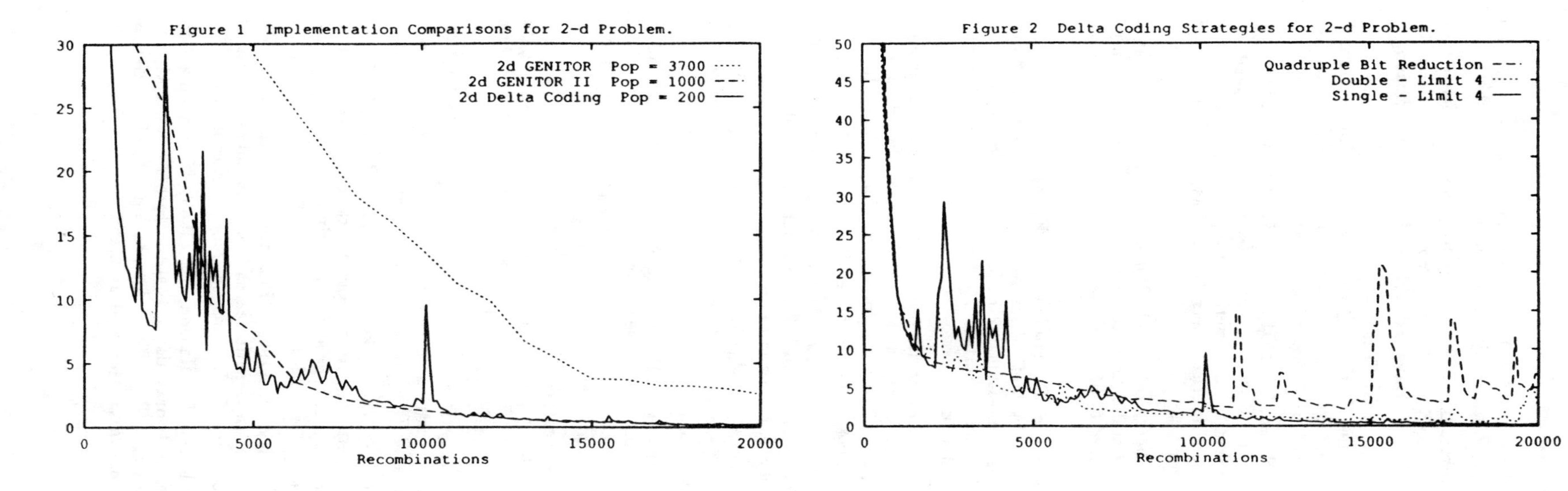

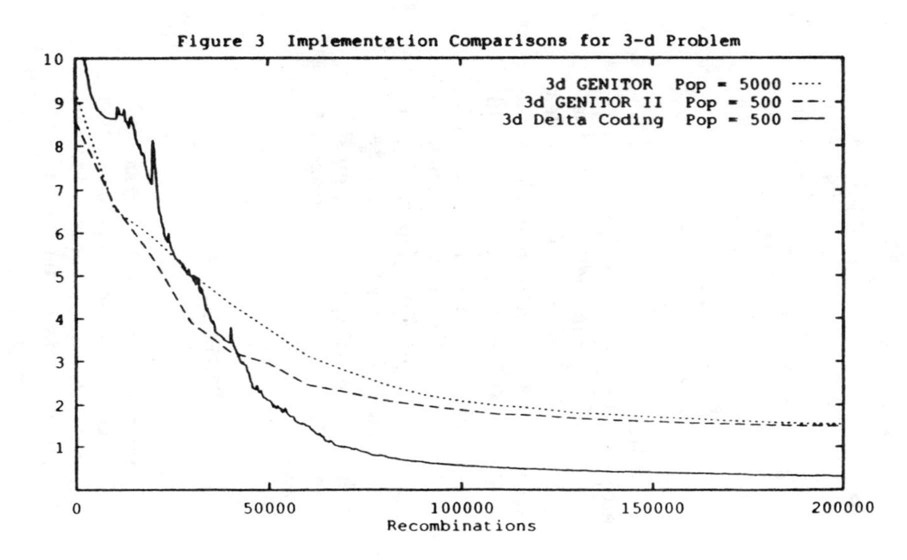

Figures 1-3. Figures 1 and 3 compare performance for GENITOR, GENITOR II, and GENITOR with delta coding on the 2-D to 2-D and the 3-D to 2-D mapping problems. Figure 2 compares performance for GENITOR with delta coding on the 2-D to 2-D problem using different reduction strategies. The "spiking" behavior in all three figures represents restarts of the algorithm using delta coding which are accompanied by changes in the encoding. The graphs do not represent the entire search; scales where choosen to highlight the "spiking" behavior and the early convergence of GENITOR using delta coding. Early "spiking" from vigorous restarts (such as that characterizing the single bit reduction in Figure 2) typically produces better optimization. The "spiking" behavior in Figure 3 is somewhat suppressed by the scale of the figure.

encode the parameters are typically reduced and thus, the solution space is made smaller. Since the parameters are *delta values* ($\pm\delta$) the new hypercube remaps the space, wrapping around on the opposite extreme of the parameter ranges and effectively remaps Hamming cliffs. The criteria for determining when to collapse the hypercube imposed on the solution space is a function of current population convergence.

Another advantage of *delta coding* is that smaller population sizes than those used in other implementations can be used; this can translate into significantly less work [11]. The reduction of the search space also potentially helps to reduce search time.

The failure to locate the optimal solution in the 3-D mapping problem indicates that there is more work to be done. Although these results are encouraging there are numerous ways in which *delta coding* might be enhanced. Some potential difficulties with *delta coding* are 1) additional user tunable parameters, 2) the definition of hyperplane expansion criteria, and 3) developing efficient but effective methods for determining when the population diversity has been *sufficiently exploited.*

While we have not yet carried out experiments to test these ideas, we also believe that *delta coding* may help in solving two types of problems that are known to cause difficulties for genetic algorithms. First, *delta coding* may help to solve deceptive problems. Deceptive problems exist because the relationship between specific hyperplanes is such that different hyperplane competitions have conflicting winners [14]; however, since *delta coding* changes the way hyperspace is defined with respect to the objective function at each iteration, the same partitions of hyperspace will not exist from one *delta iteration* to the next.

The second problem that *delta coding* could perhaps resolve is the recombination of *lethals. Lethals* refer to poor offspring that result from the recombination of two good, but dissimilar parents. For example, the bit strings 00000 and 11111 may represent local optima, while points in between (e.g. 00111) would result in very poor solutions. Therefore, recombination between the dissimilar parents would generally produce worse solutions providing inconsistent hyperplane feedback [13]. *Delta coding* may help resolve such difficulties by redefining new hyperplane mappings in N-space and redefining the hypercube with respect to only one of the competing solutions.

ACKNOWLEDGEMENTS

This research was supported in part by NSF grant IRI-9010546 and in part by a grant from the Colorado Institute of Artificial Intelligence (CIAI). CIAI is sponsored in part by the Colorado Advanced Technology Institute (CATI), an agency of the State of Colorado.

References

[1] Ackley, D. (1987) *A Connectionist Machine for Genetic Hillclimbing.* Kluwer Academic Publishers.

[2] Booker, L. (1987) "Improving Search in Genetic Algorithms," In, *Genetic Algorithms and Simulated Annealing.* L. Davis, ed. Pitman/Morgan Kaufmann.

[3] Baker, J., Greffenstette, J., (1989) "How Genetic Algorithms Work: A Critical Look at Implicit Parallelism." *Proc. Third International Conf. on Genetic Algorithms.* Morgan Kaufmann.

[4] Eshelman, L., (1991) "The CHC Adaptive Search Algorithm: How to have Safe Search When Engaging in Nontraditional Genetic Recombination." *Foundations of Genetic Algorithms.* G. Rawlins, ed. Morgan Kaufmann.

[5] Goldberg, D., (1989) "Sizing Populations for Serial and Parallel Genetic Algorithms." *Proc. Third International Conf. on Genetic Algorithms.* Morgan Kaufmann.

[6] Nickerson, B., Fitzhorn, P., Koch, S., Charney, M., (1991) "A Method for Near Optimal Computational Superimposition of 2D Digital Facial Photographs and 3D Cranial Surface Meshes." *J. Forensic Sciences.* 2:480-500.

[7] Krishnakumar, K., (1989) "Micro-Genetic Algorithms for Stationary and Non-stationary Function Optimization." *SPIE's Intelligent Control and Adaptive Systems Conf.*, Paper # 1196-32.

[8] Mathias, K., (1991) "Delta Coding Strategies for Genetic Algorithms." *Thesis.*, Department of Computer Science, Colorado State University, Fort Collins, CO.

[9] Schraudolph, N., Belew, R., (1990) "Dynamic Parameter Encoding for Genetic Algorithms." *CSE Technical Report* #CS 90-175.

[10] Shaefer, C., (1987) "The ARGOT Strategy: Adaptive Representative Genetic Optimizer Technique." *Genetic Algorithms and their Applications: Proc of the Second International Conf.*

[11] Starkweather, T., Whitley, D., Mathias, K., (1990) "Optimization Using Distributed Genetic Algorithms." *Parallel Problem Solving from Nature.* Springer/Verlag.

[12] Whitley, D. and Starkweather, T., (1990) "*GENITOR II*: A Distributed Genetic Algorithm." *J. Experimental and Theoretical Artificial Intelligence.* 2:189-214.

[13] Whitley, D., Starkweather, T., and Bogart, C., (1990) "Genetic Algorithms and Neural Networks: Optimizing Connections and Connectivity." *Parallel Computing.* 14:347-361.

[14] Whitley, D., (1991) "Fundamental Principles of Deception in Genetic Search." *Foundation of Genetic Algorithms.* G. Rawlins, ed. Morgan Kaufmann.

[15] Whitley, D., Hanson, T., (1989) "Optimizing Neural Networks Using Faster, More Accurate Genetic Search." *Proc. Third International Conf. on Genetic Algorithms.* Morgan Kaufmann.

GA-Easy Does Not Imply Steepest-Ascent Optimizable

Stewart W. Wilson
The Rowland Institute for Science
100 Cambridge Parkway
Cambridge, MA 02142
(wilson@think.com)

Abstract

It is shown that there are many functions which are GA-easy but not readily optimizable by a basic hill-climbing technique. The results, including a comparison of the genetic algorithm with a population of hill-climbers, provide insight into the operation of the GA and suggest further study of GA-easiness.

1 INTRODUCTION

Recent work beginning with Goldberg's [1] seminal paper on deceptiveness has aimed at characterizing functions (or more exactly, function-coding combinations) that are difficult to optimize using a genetic algorithm. The purpose of those investigations has been to explore the limits of GA power and, if possible, to exceed them via innovations in the algorithm. It would also be of interest to have good characterizations of functions that are *easy* for GAs, particularly if the functions cause difficulties for standard optimization methods. Identification of such functions and methods can give us a clearer picture of where and to what the GA is superior, and so aid in the practical selection of optimization techniques. But characterizing GA capability in terms of the methods it beats should also give insight into the still somewhat mysterious operation of the GA by suggesting—when we understand how the other methods work—which aspects of the GA could account for its superiority.

The present paper takes a few steps along this path by showing, through example problems, that there is a large class of functions which are *GA-easy* (according to a quite natural definition), but whose global optima cannot reliably be found by a basic hill-climbing technique. Further, an experiment is performed indicating that the GA is superior, on this class of functions, to a set of hill-climbers equal in number to the GA population size. These demonstrations contribute to the evidence that (1) the GA possesses local-optimum-avoiding capability exceeding that of basic hill-climbers, and (2) crossover gives the GA a distinct advantage over multiple independent local searches.

The rest of the paper is organized as follows. In the next section we warm up by defining what is perhaps the most elementary binary search technique, *bit-setting optimization*, and show that GA-easy does not imply bit-setting optimizable—in other words that there is a class of functions that are GA-easy but cannot be optimized by this technique. Then in Section 3 we strengthen this result by defining *steepest-ascent optimization* and showing that GA-easy does not imply steepest-ascent optimizable. In Section 4 we make sure this result applies to strings of arbitrary length. Then in Section 5 we demonstrate that on functions that are GA-easy but not steepest-ascent-optimizable, the GA should do better than an equal population of steepest-ascenders. Section 6 summarizes and comments on these results.

2 BIT-SETTING OPTIMIZATION AND GA-EASY

Let us consider an optimization problem in which the function f to be optimized is defined over a Hamming space so that points in the space are represented by binary strings. Suppose we begin with an arbitrary string S of length n and see which of the two possible settings of the kth bit gives the string a higher value for f (i.e., a higher fitness). We perform the same experiment, always starting with S, for each possible k, and then form a new string S' consisting of the better settings of each bit (if neither is better we pick one at random). If for any initial string S the value of $f(S')$ equals the highest possible value of f we shall say that f is *bit-setting optimizable* (BSO).

Now consider the function f_1 of three-bit strings shown in Figure 1. Each point of the lattice contains a bitstring together with the fitness (function value) for that string. The arrows indicate the direction of fitness change between points that differ by a single bit. By inspection, we see that f_1 is not bit-setting optimizable (NBSO) since if the optimizer happens to start with $S = 001$ it will end up with $S' = 110$, which is not the optimum. (Note in passing that if the fitnesses of 001 and 000 were changed to 1 and 0, respectively, the problem would be BSO.)

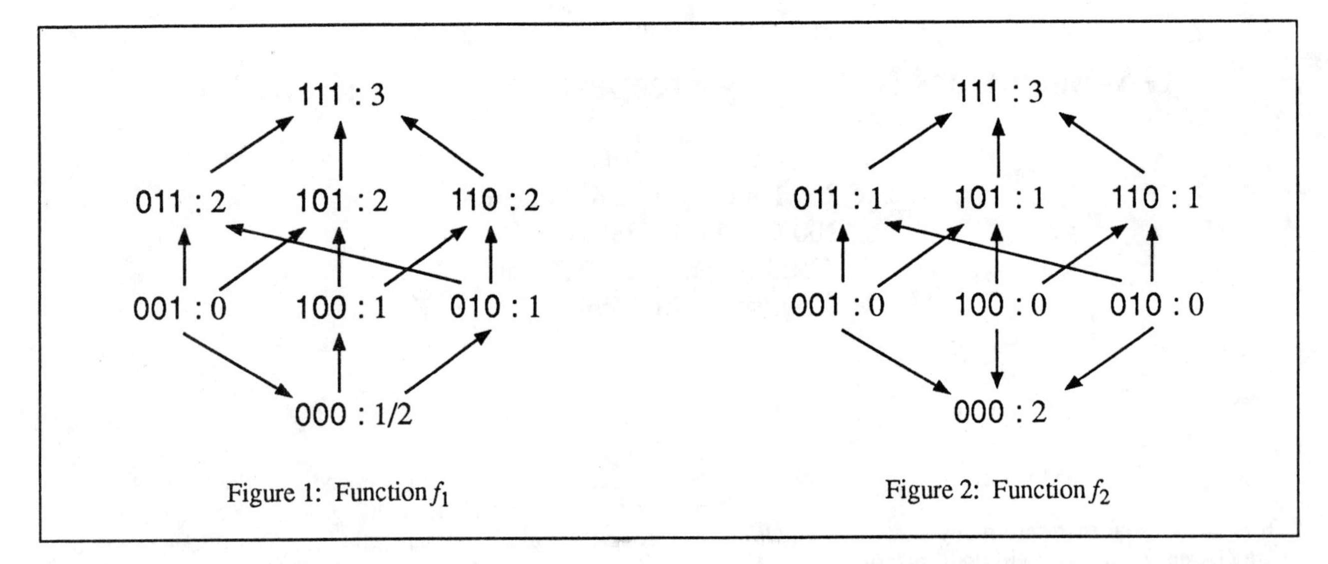

Figure 1: Function f_1

Figure 2: Function f_2

Let us now ask how hard f_1 is for a GA. Unfortunately, while we could frame a definition of "GA-optimizable" as being, say, optimizable by a GA starting from a random initial population of such-and-such a size, etc., it could not be usefully applied to a given problem because the precise action of a GA is not sufficiently well understood to allow prediction of the results. In effect, the only way currently to determine "GA-optimizable" is experimentally. However, optimization problems can be characterized in terms of their *deceptiveness* and there tends to be a correlation between deceptiveness and the difficulty in practice of optimization by a GA.

According to theory centered on the *schema theorem*, GAs "work well when *building blocks*—short, low-order schemata with above-average fitness values—combine to form optima or near-optima" [2]. Goldberg particularly has studied function-coding combinations that are deceptive to the GA in the sense that some or all short, low-order schemata whose fitness values are optimal within complete sets of *competing schemata* ([2], p. 39) do not in fact contain, or sample, the global optimum. For such functions, the schema theorem implies that the GA will be misled and may fail to find or retain the global optimum before converging to a suboptimal solution. Experimentally, this is the case (with some qualifications that are outside our present scope), and it can safely be said that deceptive problems will be difficult to optimize using a GA.

In this paper we focus on the converse proposition and suggest as a working hypothesis that the more *non*-deceptive a problem, the easier it will be to GA-optimize. In particular, let us define a problem to be *GA-easy* if the highest fitness schema in *every* complete set of competing schemata contains the optimum. Noting that our connection between GA-easy and GA-optimizable is reasonable but entirely heuristic, let us now ask whether of not f_1 is GA-easy. To do so we enumerate the problem's complete sets of competing schemata and see if the highest fitness schema in each set contains the optimum. The results are as follows (schema fitness values assume a uniform schema distribution, true on average at the start of search from a random initial population):

$f(0**)$	7/8	$f(*0*)$	7/8	$f(**0)$	9/8
$f(1**)$	2 <—	$f(*1*)$	2 <—	$f(**1)$	7/4 <—
$f(00*)$	1/4	$f(0*0)$	3/4	$f(*00)$	3/4
$f(01*)$	3/2	$f(0*1)$	1	$f(*01)$	1
$f(10*)$	3/2	$f(1*0)$	3/2	$f(*10)$	3/2
$f(11*)$	5/2 <—	$f(1*1)$	5/2 <—	$f(*11)$	5/2 <—

The maximum fitness schema (indicated by an arrow) in each competing set contains the optimum, 111, so that, according to our definition, f_1 is GA-easy. Earlier we found that f_1 is not bit-setting optimizable so that we have shown by example that the proposition "GA-easy does not imply bit-setting optimizable" is true.

This result is somewhat interesting, but not terribly surprising, since bit-setting optimization is a very weak method and should not often succeed. However, we now know that GA-easy, while making perhaps the strongest possible assumption about schema non-deceptiveness, nevertheless includes functions that cannot be bit-setting optimized. In the next section we develop a more interesting result.

3 STEEPEST-ASCENT OPTIMIZATION AND GA-EASY

Consider function f_2 (Figure 2) which is similar to f_1 except that it has a local maximum at 000. Like f_1, f_2 is not BSO. However, unlike f_1, f_2 cannot be optimized by a binary form of *steepest-ascent*. To define (binary) steepest-ascent, suppose we again start with an arbitary string S of length n and see which of the two possible settings of the kth bit gives the string a higher fitness. We perform this experiment, always starting with S, for each possible k, and determine the value of k and the bit value at k for which the highest fitness is found (breaking ties by some definite procedure). Call these k' and b', respectively. Next form a new string S' that is like S except that the bit at position k' is set to b'. Now replace S by S' and repeat this process until the fitness of S no longer increases. If for any initial string the fitness of the string finally arrived at equals the highest possible value of f we shall say that f is *steepest-ascent optimizable* (SAO).

Inspection of Figure 2 shows that f_2 is not steepest-ascent optimizable because 000, which is sub-optimal, will be reached starting from any of the four strings in the lower part of the figure. Let us next see if f_2 is GA-easy:

$f(0**)$ 3/4 $f(*0*)$ 3/4 $f(**0)$ 3/4
$f(1**)$ 5/4 <— $f(*1*)$ 5/4 <— $f(**1)$ 5/4 <—

$f(00*)$ 1 $f(0*0)$ 1 $f(*00)$ 1
$f(01*)$ 1/2 $f(0*1)$ 1/2 $f(*01)$ 1/2
$f(10*)$ 1/2 $f(1*0)$ 1/2 $f(*10)$ 1/2
$f(11*)$ 2 <— $f(1*1)$ 2 <— $f(*11)$ 2 <—

The function f_2 is indeed GA-easy. We have shown by this example that the proposition "GA-easy does not imply steepest-ascent optimizable" is true. This is stronger than our previous result, since apart from our intuition, it can be proved that functions that are bit-setting optimizable form a proper subset of functions that are steepest-ascent optimizable. Briefly, assume there exists a function f that is BSO but *not* SAO. If f is not SAO, it must have at least one local optimum (a maximum), where the steepest-ascender could get stuck. Now suppose the bit-setting optimizer happened to be started with initial string S precisely on one these local maxima. By definition of local maximum, any bit change in S would produce a string with non-increasing fitness, so that the bit-setting optimizer would also get stuck, contradicting the hypothesis that f is BSO. Thus BSO implies SAO. The converse is not true, since for example f_1 is SAO but not BSO.

4 LONG STRINGS

The previous main result, that GA-easy does not imply SAO, is certainly interesting but we need to check that it doesn't merely apply to short string problems like those used to prove it. Is the proposition also true for arbitrarily long strings? The answer is yes, as will now be shown.

Consider a function F of bit strings of length $n = 3m$ (m an integer >> 1). Further, let the value (fitness) of a string under F be the sum of the values obtained by applying f_2 to successive triples of the bits of the string. That is,

$$F(x_1, x_2, x_3, x_4, \ldots x_{3m}) = f_2(x_1, x_2, x_3) + f_2(x_4, x_5, x_6) + \ldots + f_2(x_{3m-2}, x_{3m-1}, x_{3m}).$$

Clearly, because f_2 is not steepest-ascent optimizable, F is also not SAO. To see this it is only necessary to assume a starting string for optimizing F in which at least one of the three-bit arguments to at least one of the f_2 terms above is a "bad" (unsuccessful) starting point for optimization of f_2. Then no matter how effective the optimizer on the rest of the string, those three bits will never arrive at 111, so the global optimum of F will not be reached.

Showing that F is GA-easy is more difficult. We shall again proceed by suggestive example, and not attempt a formal proof. First, shorten F, without loss of generality, to a function of six bits. We next need to be convinced that schema values are simply sums of the schema values for f_2, as in this example:

$$F(1***10) = f_2(1**) + f_2(*10).$$

But this equation is equivalent to the sum of the eight equations:

$$F(100010) = f_2(100) + f_2(010)$$
$$\ldots$$
$$F(111110) = f_2(111) + f_2(110),$$

all of which hold by definition of F. Therefore, we can say that F's schema fitnesses just sum f_2's schema fitnesses in the manner above.

Now, to demonstrate GA-easy, we must show that in every complete set of competing schemata, the schema with maximum fitness contains the optimum. In other words, e.g.,

$$F(1***11) > F(0***00), F(0***01), \ldots, F(0***11),$$

meaning that $F(1***11)$ is greater than any schema value on the right. Consider the first inequality. Since we have just found that F's schema values are sums of f_2's, we can write

$$F(1***11) = f_2(1**) + f_2(*11)$$
and
$$F(0***00) = f_2(0**) + f_2(*00).$$

We are therefore asking whether

$$f_2(1**) + f_2(*11) > f_2(0**) + f_2(*00).$$

But because f_2 is GA-easy, this must be true. Since there was nothing unique about our choice of schemata, we conclude that F is GA-easy. Further, If F is GA-easy but not SAO, then we have demonstrated that the proposition "GA-easy does not imply steepest-ascent optimizable" is true for strings of arbitrary length.

5 IS THIS FAIR TO HILL-CLIMBERS?

One might suggest that the genetic algorithm has a built-in unfair advantage over a steepest-ascender since the GA works from a population of N initial strings whereas the steepest-ascender uses just one. What if there were N steepest-ascenders each starting with a random string and searching independently? The probability of finding the optimum somewhere among N such steepest-ascenders certainly increases with N. How does this compare with a population of size N under the GA?

Here we are on somewhat more complicated ground. Let us start by assuming we are trying to optimize the function F using N steepest-ascenders. Since a steepest-ascender starting at four of the eight possible starting strings for f_2 will reach the optimum of f_2, the probability of reaching the optimum of F, starting with a random $3m$-bit string, is $(4/8)^m$. The probability of *not* finding the optimum in such a string is then $[1-(.5)^m]$. If we have N random strings, the probability that the steepest-ascender will not find the optimum in *any* of the strings is $[1-(.5)^m]^N$. Consequently, the probability of finding at least one optimum in a population of N strings, written in terms of string length $n = 3m$, is $1 - [1 - (.5)^{n/3}]^N$.

Let us go one step further and suppose this probability equals some criterion value, say one-half. We could then set the above expression equal to one-half and solve for N. The result $N(n)$ would be the minimum size population for which the steepest-ascender would have probability .5 of finding the optimum. The manner in which $N(n)$ increased with n would suggest how the steepest-ascender's ability scaled with increasing problem size.

Our main purpose is to compare the N-string steepest-ascender with a GA having population N. As noted in Section 2, there is at present no formal analysis that would tell us, say, how large N has to be for the GA, or even the GA on a GA-easy problem such as this, to have probability .5 of finding the optimum. Current formal work on optimal population size, while important, does not yet directly address the question of convergence to the optimum. It is therefore necessary to proceed experimentally.

Our approach is to take a range of values for n, calculate the corresponding values of $N(n)$, then see if a GA using these values for N finds F's optimum substantially more often than half the time. Specifically, we chose values for n of 15, 18, 21, 24, 27, and 30 bits. The resulting values of $N(n)$ are shown in the table below; in each case, the GA's population size was the next lower even integer. The GA employed tournament selection (2-string tournaments), single-point crossover with probability 0.6, and point mutation with probability equal to the reciprocal of population size. For each value of N, the GA was run 10 times for 30 generations. The measure of the GA's success was the proportion of the 10 runs for which the optimum was present in the final generation. Also tabulated are measures of computational effort for the steepest-ascender and the GA. In both cases we simply use the number of string evaluations. For the GA, this is just N times the number of generations (30). For the steepest-ascender, note that to determine the best bit change for a single string requires n evaluations and the total "climb" for that string should average on the order of $n/2$ steps. Since there are N strings, the total number of evaluations is therefore about $(Nn^2)/2$.

n	$N(n)$	GA-success	GA-effort	SA-effort
15	21.83	0.8	600	2250
18	44.01	0.9	1320	7128
21	88.38	1.0	2640	19404
24	177.10	1.0	5280	50688
27	354.54	1.0	10620	129033
30	709.44	1.0	21240	318600

These results suggest that the GA is more effective, with much less effort, than the N-string steepest-ascender on function F. With exponentially rising population sizes, the steepest-ascender has only an even chance of finding the optimum, whereas the GA found the optimum with little difficulty in the small populations, and with near certainty in the larger ones. The GA in fact found the optimum in earlier generations as n increased.

The tentative implication of the results of this and the previous section is that within the class of functions that are GA-easy, and therefore presumably *actually* rather easy for a GA, there exists a quite large subclass of problems that are not easily solvable by "steepest-ascent", a basic hill-climbing technique. Furthermore, on these problems the GA would appear to be more powerful than an equal population of steepest-ascenders. Further research along this particular direction might compare the GA, again on a GA-easy but not SAO problem, with a population of more sophisticated hill-climbers that employ stochastic methods to escape local optima.

6 CONCLUSION

In this paper we compared the GA with simple but well-defined optimization methods on relatively simple problems. We found, perhaps somewhat surprisingly, that "GA-easy" problems can be difficult for these other methods, but not for the GA. The results add to our confidence that "GA-easy" is a non-trivial characterization. Formal analysis of genetic algorithm action on GA-easy problems should be relatively straightforward. If such problems are actually moderately interesting, the analysis is doubly motivated.

Our approach proposed that careful comparisons with methods that the GA beats could yield insight into the GA.

The results of Section 5 are of this sort. One sometimes hears the question: Is a GA more powerful than an equal population of hill-climbers, and if so, why? The answer seems to be yes, as noted. But why?

Perhaps the basic reason is that the hill-climber can get tripped up by an individual case, whereas on a problem that is GA-easy but not SAO, the GA is carried over such cases by the statistical power of the schema theorem. Note also that each of a population of hill-climbers is likely to get stuck somewhere, but in different places than its neighbors. In a problem like the one constructed in this paper, each hill-climber will at the same time be *correct* in a number of places (in the string), again at different places than its neighbors. Crossover would permit the good parts to be communicated and accumulated. But only the GA has crossover.

References

[1] D. E. Goldberg. Simple genetic algorithms and the minimal, deceptive problem. In L. Davis, editor, *Genetic Algorithms and Simulated Annealing*, chapter 6, pages 74-88. Morgan Kaufmann, Los Altos, CA, 1987.

[2] D. E. Goldberg. *Genetic Algorithms in Search, Optimization, and Machine Learning*. Addison-Wesley, Reading, MA, 1989.

Genetic Algorithm Techniques and Behavior

Extended Selection Mechanisms in Genetic Algorithms

Thomas Bäck*

University of Dortmund
Department of Computer Science XI
P.O. Box 50 05 00 · D–4600 Dortmund 50

Frank Hoffmeister†

University of Dortmund
Department of Computer Science XI
P.O. Box 50 05 00 · D–4600 Dortmund 50

Abstract

Common selection mechanisms used in *Evolutionary Algorithms* are combined to form some generalized variants of selection. These are applied to a *Genetic Algorithm* and are subject to an experimental comparison. The feature of *extinctiveness* as introduced in *Evolution Strategies* is identified to be the main reason for a considerable speedup of the search in case of unimodal objective functions.

1 Introduction

Genetic Algorithms (GAs) [Hol75] and *Evolution Strategies* (ESs) [Rec73, Sch81] are two types of algorithms which try to imitate the mechanism of natural evolution. In this paper the generic term *Evolutionary Algorithms* is used to denote such algorithms with the common features of a population of individuals which undergo Darwinian selection of the fitter individuals and which are subject to mutation and sexual recombination processes [BH91, HB91]. The selection mechanism of such algorithms plays an important role for driving the search towards better individuals on the one hand and for maintaining a high genotypic diversity of the population on the other hand. This is directly related to the trade–off between high convergence velocity and high probability to find a global optimum in case of a multimodal problem, which is a well–known problem in current research concerning Evolutionary Algorithms [Bak85, Gol89, Sch81, Whi89].

Within this work we look at the selection techniques which are commonly used in Evolutionary Algorithms and describe a set of possible generalizations and recombinations of them in section 2. These new selection mechanisms are compared by experiments with respect to two simple but important topologies of objective functions in section 3.

*baeck@lumpi.informatik.uni-dortmund.de
†iwan@lumpi.informatik.uni-dortmund.de

2 Selection Schemes

Proportional selection [Hol75, Gol89] and ranking [Bak85, Whi89] are the main selection scheme used in GAs, while ESs are based on several variants of (μ,λ)–selection [Sch81].

To describe these techniques in a formal way the following notation is used: $P^t = (a_1^t, \ldots, a_\lambda^t) \in I^\lambda$ denotes the population at generation $t \in \mathbb{N}$, $\lambda > 1$ the population size, and I is the space of individuals a_i^t. The fitness function $f : I \to \mathbb{R}$ provides the environmental feedback for selection. Furthermore a mapping $rank : I \to \{1, \ldots, \lambda\}$ is given by the following definition:

$$\begin{aligned} & \forall i \in \{1,\ldots,\lambda\} : rank(a_i^t) = i \\ \iff & \forall j \in \{1,\ldots,\lambda-1\} : f(a_j^t) \square f(a_{j+1}^t) \end{aligned} \quad (1)$$

where $\square$ denotes the $\leq$ relation in case of a minimization task and $\geq$ in case of a maximization problem. Consequently, we can use the index i to denote the rank of an individual. In the following we assume that individuals are always sorted according to their fitness, with a_1^t being the best individual of P^t.

Selection in Evolutionary Algorithms is defined by selection (reproduction) probabilities $p_s(a_i^t)$ for each individual within a population. At present the following selection schemes exist:

- *Proportional Selection* [Hol75]:

$$p_s(a_i^t) = f(a_i^t) / \sum_{j=1}^{\lambda} f(a_j^t)$$

- *Linear Ranking* [Bak85]:

$$p_s(a_i^t) = \frac{1}{\lambda}\left(\eta_{max} - (\eta_{max} - \eta_{min})\frac{i-1}{\lambda-1}\right)$$

where $\eta_{min} = 2 - \eta_{max}$ and $1 \leq \eta_{max} \leq 2$.

- *(μ,λ)-Uniform Ranking* [Sch81]:

$$p_s(a_i^t) = \begin{cases} 1/\mu &, 1 \le i \le \mu \\ 0 &, \mu < i \le \lambda \end{cases}$$

While the latter two schemes are rank-based (i.e. instead of the actual fitness their rank-index i is used), proportional selection is directly based upon the fitness values of all individuals. When (μ,λ)–selection is also taken into account, a selection scheme can be classified with respect to the following criteria:

- *Dynamic* versus *static* selection:

  The selection probabilities can depend on the actual fitness-values (proportional selection) and hence they change between generations, or they can depend on the rank of the fitness-values only (linear ranking, (μ,λ)–selection) which results in fixed (static) values for all generations. This can easily be formalized as follows:

  DEFINITION 1 (Dynamic Selection)
  A selection scheme is called *dynamic*:

  $$\Longleftrightarrow \nexists i \in \{1, \ldots, \lambda\} \; \forall t \ge 0 : p_s(a_i^t) = c_i$$
  where the c_i are constants.

  DEFINITION 2 (Static Selection)
  A selection scheme is called *static*:

  $$\Longleftrightarrow \forall i \in \{1, \ldots, \lambda\} \; \forall t \ge 0 : p_s(a_i^t) = c_i$$
  where the c_i are constants.

- *Extinctive* versus *preservative* selection:

  The term preservative describes a selection scheme, which guarantees a non-zero selection probability for each individual, i.e. each individual has a chance to contribute offspring to the next generation. On the other hand, in an extinctive selection scheme some individuals are definitely not allowed to create any offspring, i.e. they have zero selection probabilities.

  DEFINITION 3 (Preservative Selection)
  A selection scheme is called *preservative*:

  $$\Longleftrightarrow \forall t \ge 0 \; \forall P^t = (a_1^t, \ldots, a_\lambda^t) \; \forall i \in \{1, \ldots, \lambda\} : \\ p_s(a_i^t) > 0$$

  DEFINITION 4 (Extinctive Selection)
  A selection scheme is called *extinctive*:

  $$\Longleftrightarrow \forall t \ge 0 \; \forall P^t = (a_1^t, \ldots, a_\lambda^t) \; \exists i \in \{1, \ldots, \lambda\} : \\ p_s(a_i^t) = 0$$

- *Left* versus *right extinctive* selection:

  In case of extinctive selection (def. 4) there is a major special case where the worst performing individuals have zero reproduction rates, i.e. do not reproduce. This situation is referred to as right extinctive selection. Although it might be of no practical relevance there may be also the opposite situation (left extinctive selection) where some of the best performing individuals are prevented from reproduction in order to avoid premature convergence due to super-individuals.

  DEFINITION 5 (Left Extinctive Selection)
  A selection scheme is called *left extinctive*:

  $$\Longleftrightarrow \forall t \ge 0 \; \forall P^t = (a_1^t, \ldots, a_\lambda^t) \\ \exists l \in \{1, \ldots, \lambda-1\} : i \le l \Longrightarrow p_s(a_i^t) = 0$$

  DEFINITION 6 (Right Extinctive Selection)
  A selection scheme is called *right extinctive*:

  $$\Longleftrightarrow \forall t \ge 0 \; \forall P^t = (a_1^t, \ldots, a_\lambda^t) \\ \exists l \in \{2, \ldots, \lambda\} : i \ge l \Longrightarrow p_s(a_i^t) = 0$$

Of course, in any case the condition $\sum_{i=1}^{\lambda} p_s(a_i^t) = 1$ must be satisfied. With regard to this classification proportional selection is a dynamic, preservative scheme, while linear ranking realizes a static, preservative scheme. (μ,λ)-uniform ranking is static and extinctive. Hence, the main difference is the preservativeness and extinctiveness of the selection schemes, respectively.

Apart from different assignments of reproduction rates there are other characteristics of selection:

- *Elitist* versus *pure* selection:

  Normally, parents are allowed to reproduce in one generation only. Then, they die out and are replaced by some offspring. A selection scheme which enforces a life time of just one generation for each individual regardless of its fitness is referred to as pure selection. In an elitist selection scheme some or all of the parents are allowed to undergo selection with their offspring [Jon75]. This might result in 'unlimited' life times of super-fit individuals.

  DEFINITION 7 (Elitist Selection)
  A selection scheme is called *elitist* or *k-elitist*:

  $$\Longleftrightarrow \exists k \in \{1, \ldots, \lambda\} \; \forall t > 0 \; \forall i \in \{1, \ldots, k\} : \\ f(a_i^t) \Box f(a_i^{t-1})$$

  DEFINITION 8 (Pure Selection)
  A selection scheme is called *pure* iff there is no $k \in \{1, \ldots, \lambda\}$ which satisfies the k-elitist property.

- *Generational* versus *steady-state* selection:

  With generational selection the set of parents is fixed until λ offspring, the members of the next generation, are completely produced. In case of selection on-the-fly or steady-state selection an offspring immediately replaces a parent if it performs better. Thus, the set of parents may change for every reproduction step [Whi89].

  It should be noted, that steady-state selection is a special variant of elitist selection (def. 7) where the set of parents incorporated into selection is larger than the set of offspring of size 1.

By "recombining" the major characteristics of the existing selection schemes, proportional selection and ranking can be generalized, which allows them to be also extinctive:

DEFINITION 9 ((μ,λ)-*Proportional Selection*)

$$p_s(a_i^t) = \begin{cases} f(a_i^t)/\sum_{j=1}^{\mu} f(a_j^t) & , 1 \leq i \leq \mu \\ 0 & , \mu < i \leq \lambda \end{cases} \quad (2)$$

DEFINITION 10 ((μ,λ)-*Linear Ranking*)

$$p_s(a_i^t) =$$

$$\begin{cases} \frac{1}{\mu}\left(\eta_{max} - 2(\eta_{max} - 1)\frac{i-1}{\mu-1}\right) & , 1 \leq i \leq \mu \\ 0 & , \mu < i \leq \lambda \end{cases} \quad (3)$$

Figure 1 tries to give an impression of the different selection schemes and their interdependencies. In each case selection probabilities versus the rank of the individual are sketched by step-functions. Remember the rank-ordering of individuals such that better individuals have lower ranks.

First, it should be noted, that each extinctive scheme turns into the corresponding preservative scheme for $\mu = \lambda$. For (μ,λ)-uniform ranking the case $\mu = \lambda$ should lead to random walk where the selective pressure towards better individuals is completely lost. The random walk variant is of no interest but mentioned here to complete the classification. (μ,λ)-uniform ranking in ESs is obviously a special case of the extinctive linear ranking selection ($\eta_{max} = 1$).

The selective pressure of the extinctive schemes can be guided by the exogeneous setting of μ. As μ approaches λ, the selective pressure towards the better individuals is decreasing continually and selection becomes "softer".

From theoretical investigations concerning (1,λ)-ES on a simple corridor and sphere model for the objective function, Schwefel derived values of $\lambda \approx 6.0$ for the corridor model and $\lambda \approx 4.7$ for the sphere model to achieve an optimum rate of convergence [Sch81]. The setting ($\mu/\lambda \approx 1/5$) emphasizes on the convergence speed for unimodal problems; for multimodal problems this ratio should be much higher in order to allow for the exploration of the search space to some extent.

With respect to super-individuals with a high fitness value or individuals with just a poor fitness proportional selection appears to be rather "hard", since the resulting rates of reproduction effectively discard the poor ones while high preference is given to the good ones, thus decreasing the genetic diversity quickly. With rank-based schemes the same situation is less drastic since the actual fitness does not influence the realized rate of reproduction, thus yielding a slower reduction of the genetic diversity. Hence, uniform and linear ranking appear to be "softer" than proportional selection.

3 Experimental Results

For the experimental comparison of the selection mechanisms it is concentrated on the two examples of objective functions given in table 1.

The functions f_1 and f_7 are representing the classes of unimodal as well as multimodal functions. For f_1 a high convergence velocity is expected to be sufficient to approach the optimum, while for f_7 a more explorative behaviour of the algorithm would give a chance to find the global optimum. To obtain the results, a modified version of Grefenstette's GENESIS–GA [Gre87] was used here. The GA is defined by the following parameter and configuration settings: Mutation rate $p_m = 0.001$; crossover rate $p_c = 0.6$; population size $\lambda = 50$; length of an individual $l = 32n$, where n denotes the dimension of the objective function[1]; two-point crossover; Gray code. For ranking the usual setting of $\eta_{max} = 1.1$ (maximum expected value) was chosen according to [Bak85].

Different values of $\mu \in \{5, 10, 15, 20, 30, 40, 50\}$ have been used for the test runs, and for a comparison the best values in each generation are plotted. The results are based on the averaged values of 10 runs in each case.

In the left parts of figures 2–4 the performances of (μ,λ)–proportional selection, (μ,λ)–linear ranking, and (μ,λ)–uniform ranking are shown for f_1. Obviously the performance is maximized for rather small values of $\mu \in \{5, 10, 15\}$. In each case performance decreases for growing values of μ, finally turning into the familiar normal ranking and proportional selection plots and a random walk wandering for (μ,λ)–uniform ranking, respectively.

A comparison of the different selection mechanisms for the same values of μ does not lead to a clear general statement, since no large differences exist. A tendency towards favouring (μ,λ)-linear ranking compared with proportional selection and the latter compared with (μ,λ)-uniform ranking can be deduced from a set of graphics not shown here. The major improvement is introduced by the idea of extinctive selection. This result can be interpreted as an indication of the validity of Schwefel's result for an optimum setting of the ratio μ/λ in case of f_1 [Sch81] not only for ESs, but also for GAs. Thus we can formulate the hypothesis, that the effect of an extinctive selection mechanism is

[1] A length of 32 bits per object variable is used for the representation of the real interval $[x_{min}, x_{max}]$ to which the bitstrings are mapped, in order to achieve a maximum resolution $\Delta x = (x_{max} - x_{min})/(2^{32} - 1)$.

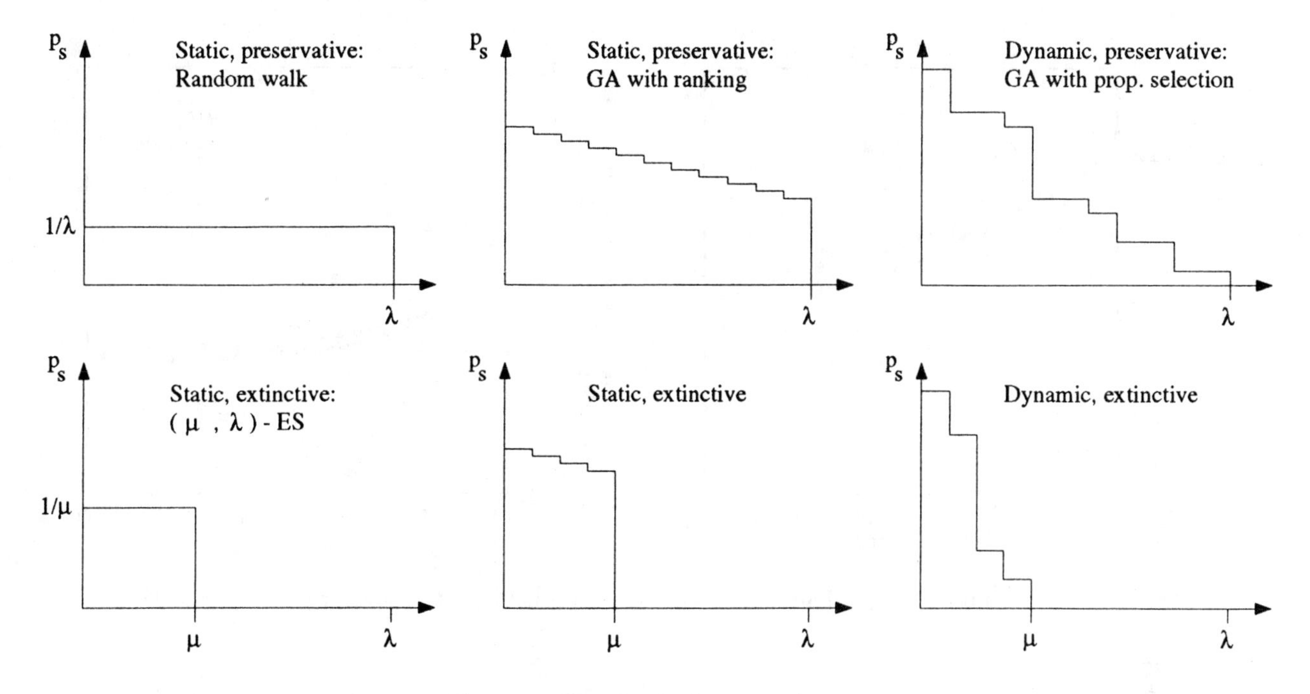

Figure 1: Sketch of selection schemes

Name	Description	Dim.	Characteristics	Ref.
f_1	sphere model $f_1(\vec{x}) = \sum_{i=1}^{n} x_i^2$ $-5.12 \leq x_i \leq 5.12$	$n = 30$	unimodal, high–dimensional	[HB91] [Jon75] [Sch81]
f_7	generalized Rastrigin's function $f_7(\vec{x}) = nA + \sum_{i=1}^{n} x_i^2 - A\cos(\omega x_i)$ $A = 10$; $\omega = 2\pi$; $-5.12 \leq x_i \leq 5.12$	$n = 20$	multimodal, high–dimensional, f_1 with sine wave superposition	[HB91] [TZ89]

Table 1: The set of test functions

very similar for both algorithms, independently of the representation of the individuals and the special kind of genetic operators.

For f_7 the results are completely different to the general results for f_1. The performance plots are given in the right parts of figures 2–4.

At first glance the similarity of all plots characterizing true extinctive selection mechanisms ($\mu < \lambda$) becomes apparent. Furthermore, all extinctive mechanisms as well as preservative proportional selection get stuck in local optima. Only preservative ranking, while progressing very slow in the early phase of a run, seems to promise better results in the long run. Besides random walk this is the only mechanism which behaves basically different compared to the rest. From these plots a cautious hypothesis about an optimum value of μ somewhere between 40 and 50 can be drawn up.

This completely inverse behaviour between f_1 and f_7 is expected to be caused by their topological differences solely.

The remarkably different shape of the preservative ranking mechanism for f_7 can be understood by looking at the genotypic diversity of the populations. This can be measured by the *bias* b ($0.5 \leq b \leq 1$) according to equation (4).

$$b = \frac{1}{l\lambda} \sum_{j=1}^{l} \max \left(\sum_{\substack{i=1 \\ \alpha_{i,j}^t = 0}}^{\lambda} (1 - \alpha_{i,j}^t) \; , \; \sum_{\substack{i=1 \\ \alpha_{i,j}^t = 1}}^{\lambda} \alpha_{i,j}^t \right) \quad (4)$$

$$\forall P^t = (a_1^t, \ldots, a_\lambda^t) \;\; \forall a_k^t = (\alpha_{k,1}^t, \ldots, \alpha_{k,l}^t) \;\; \forall t \geq 0$$

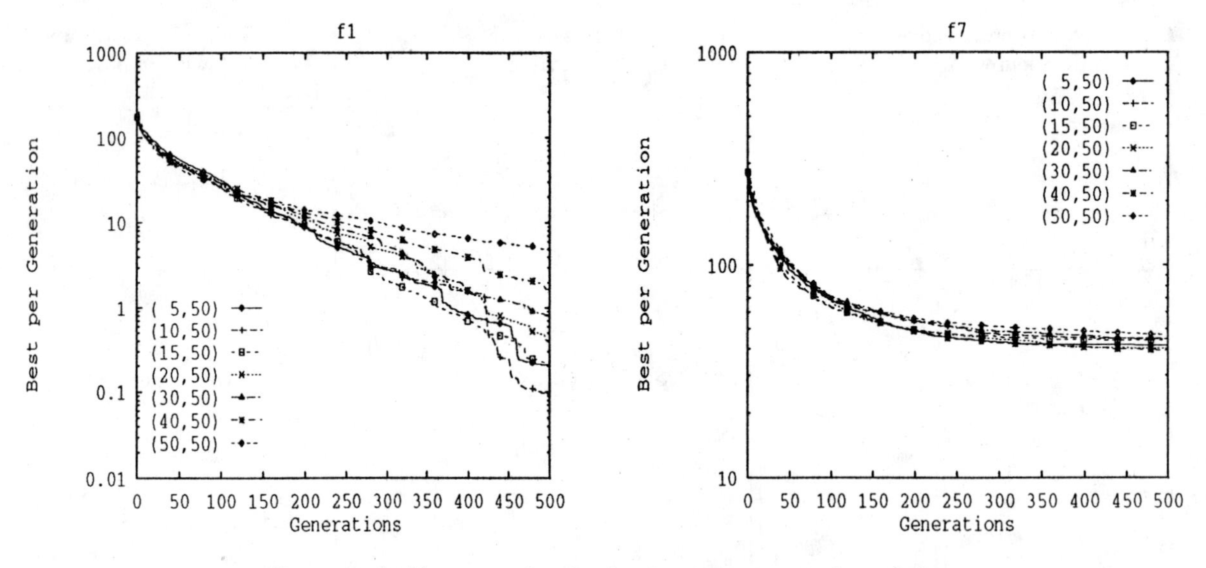

Figure 2: (μ,λ)–proportional selection schemes on f_1 and f_7

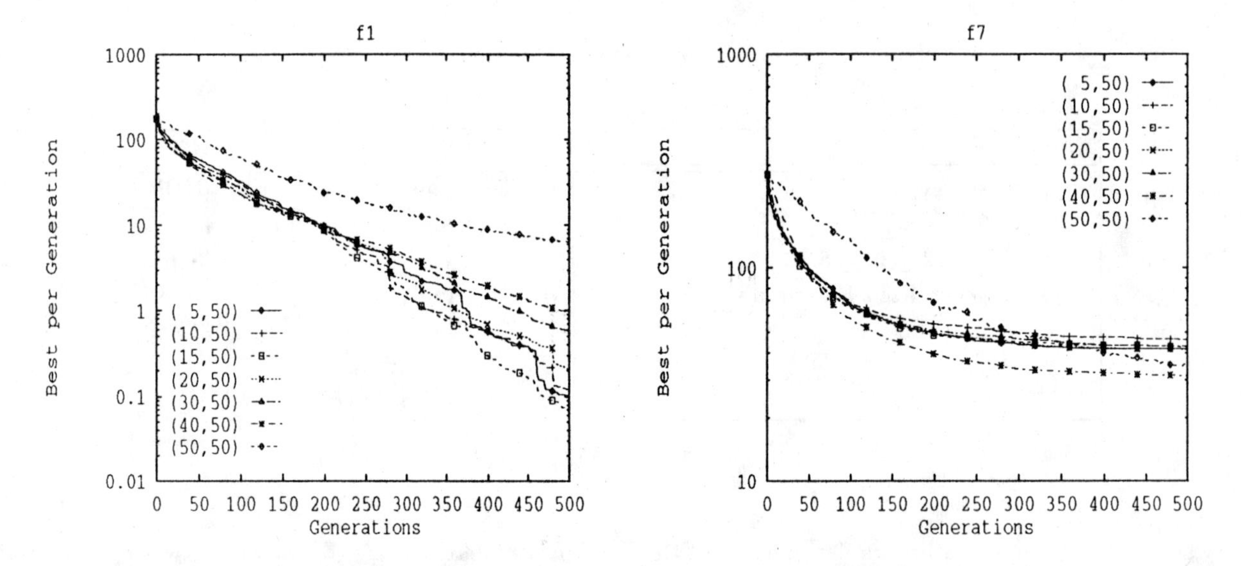

Figure 3: (μ,λ)–linear ranking schemes on f_1 and f_7

b indicates the average percentage of the most prominent value in each position of the individuals [Gre87]. Smaller (larger) values of b indicate higher (lower) genotypic diversity.

The preservative linear ranking mechanism shows a fundamentally larger genotypic diversity than the extinctive ones (left part of figure 5). This is the property which verifies the very slow convergence behaviour of such a selection mechanism. However, it is not experimentally checked whether such behaviour can lead to better solutions in the long run even for difficult surfaces like that of f_7.

A comparison of (50,50)-linear ranking and (40,50)-linear ranking for a longer run on f_7 seems to confirm this assumption, but the difference of the runs is rather small (right part of figure 5).

There are two major effects to observe for different degrees of extinctiveness ($0 < \mu < 50$) which depend on the number of optima. In general for a unimodal function like f_1 the best performance increases with "harder" selection, i.e. decreasing μ, while it stays on a similar level for most degrees of extinctiveness for multimodal functions like f_7. This is *not* a general fact. For different adaptation schemes the impact of selection varies as can be seen from figure 6 which summarizes the first results of a complete set of runs on f_1 and f_7 for all types of selection and various degrees

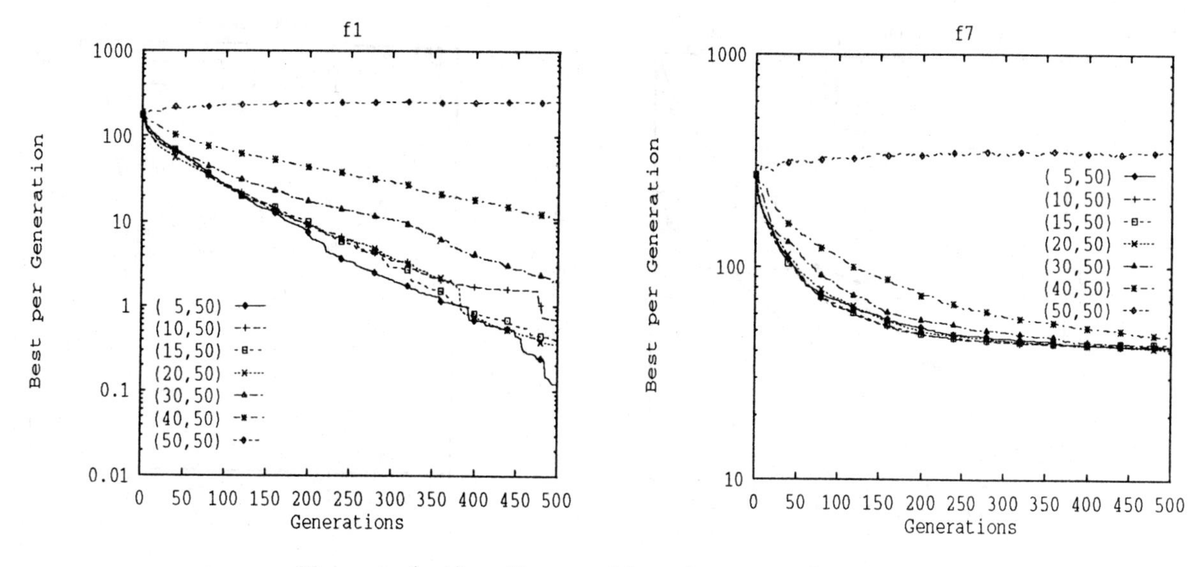

Figure 4: (μ,λ)–uniform ranking schemes on f_1 and f_7

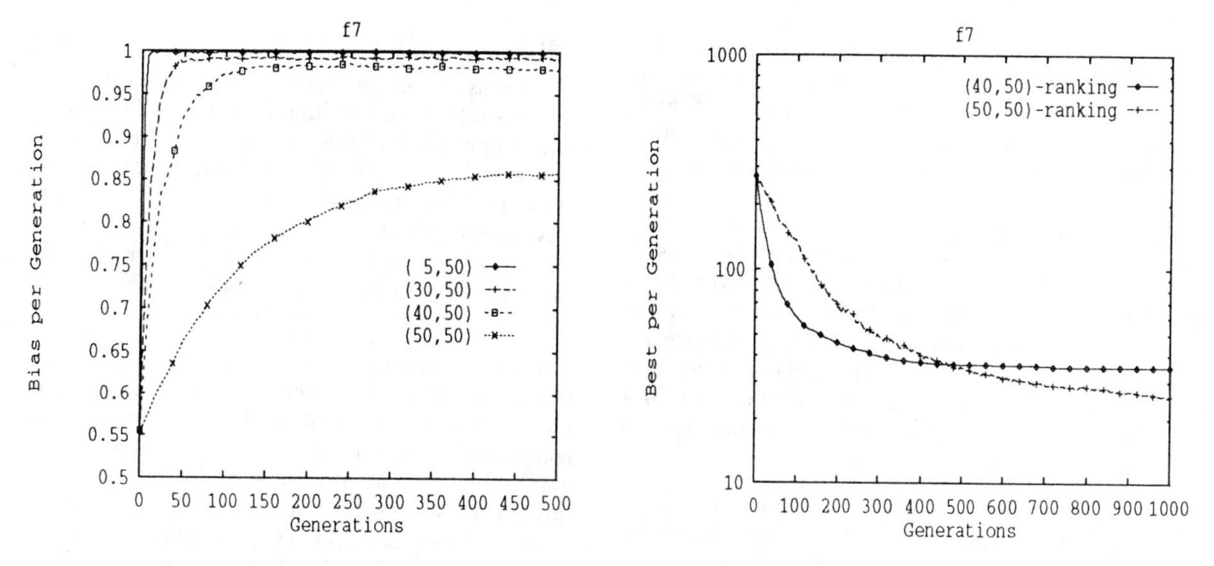

Figure 5: Bias and long runs for (μ,λ)–linear ranking selection schemes on f_7

of extinctiveness with an Evolution Strategy according to Schwefel [Sch81].

Evolution Strategies (ESs) work on a phenotypic level, i.e. they operate directly on the set of real-valued object variables x_i. Mutation is realized by adding to each x_i a normally distributed random number with expected value 0 and standard deviation σ_i. Recombination may be discrete or intermediate. Theoretical considerations for a maximum rate of convergence suggest that the optimal settings of the σ_i may depend on the distance from the optimum, i.e. they are a local feature of the response surface. Therefore, the genetic information of each individual not only consists of the x_i, but also of the strategy parameters σ_i which also undergo mutation and recombination *before* they are used to mutate the x_i. Better adapted settings of the σ_i are expected to result in a better performance of the x_i with respect to the objective function. Hence, selection automatically favours advantageous settings of the strategy parameters σ_i. A detailed description of ESs may be found in [BHS91].

Each curve in figure 6 shows the best solution obtained after 250 generations for a particular selection scheme with respect to various degrees of extinctiveness. Like for the other experiments a population size of $\lambda = 50$ was chosen; all values are averaged over 10 runs.

In case of the unimodal function f_1 a high rate of

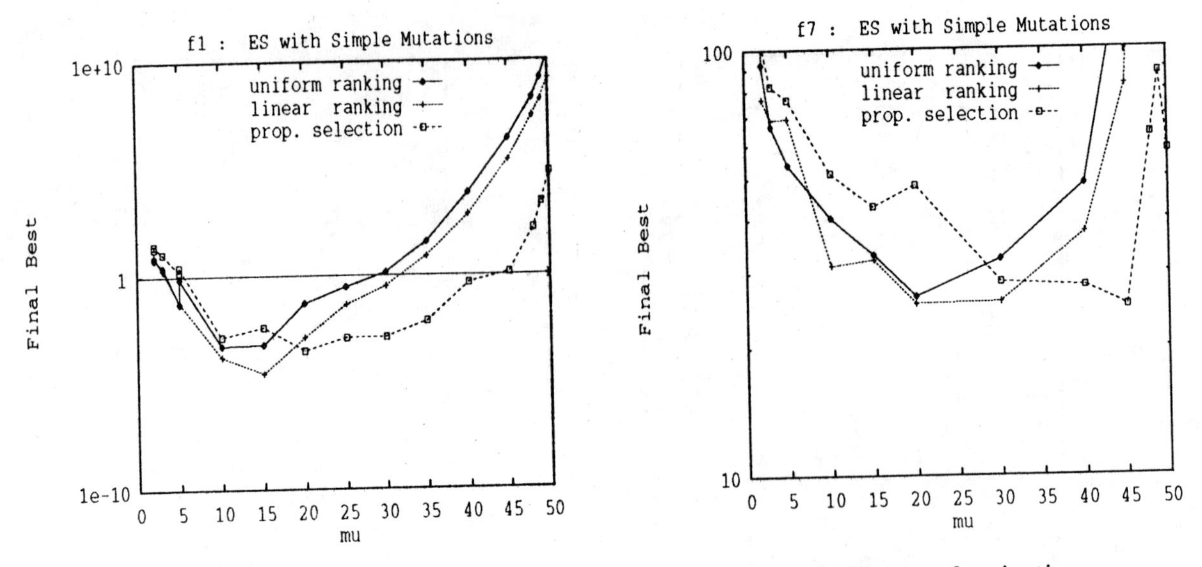

Figure 6: Best performance of ESs on f_1 and f_7 with respect to the degree of extinctiveness

convergence is required for optimal performance. For ranking this is achieved by a setting of $\mu/\lambda \approx 1/5$, which is pretty close to the theoretical results [Sch81]. The curves for f_1 also partially meet the expectation that a "harder" selection scheme (linear ranking) performs better than a "softer" one (uniform ranking). Proportional selection even results in a much "harder" selection, which for a high degree of extinctiveness, is not able to maintain a sufficient genetic diversity to allow for a rapid adaptation of the strategy parameters by means of recombination. This preference for the best is advantageous if extinctiveness is lowered. In that case the ranking schemes fail to maintain a proper setting of strategy parameters finally leading to a divergence from the optimum.

For a multimodal function like f_7 a high genetic diversity is required to explore the search space sufficiently. Hence, the "softer" selection schemes (uniform and linear ranking) perform best with respect to the quality of the final optimum. But if selection becomes too "soft" the population is unable to maintain partial solutions which may be used as a starting point for further improvements. This is why the right illustration in figure 6 shows an optimum at $\mu = 20$ for the ranking schemes while proportional selection performs better with growing μ. But even for proportional selection some degree of extinctiveness is required for an optimal performance.

4 Summary

Undoubtedly an extinctive selection mechanism produces a remarkable speedup for a unimodal function like f_1. This should be a sufficient reason to use such selection mechanisms as a further way of guiding the search of genetic algorithms.

In contrast by extinctiveness there is no improvement of the results for a multimodal surface. Due to this reason a superiority of selection mechanisms which maintain a high genotypic diversity can be concluded from the experimental runs. The question remains how to solve this contradiction concerning the algorithmic requirements caused by different topological surfaces of the actual optimization problem, which is noted as a list of characteristic properties in table 2.

The term convergence confidence is used to describe the probability to converge towards the global optimum. For a multimodal objective function a high convergence confidence is aspired, which requires an explorative character of the search. To achieve this behaviour a "soft" selection scheme can be used in order to maintain a large genotypic diversity of the population during the search. The resulting search process can be designated as volume oriented.

The corresponding appropriate properties for a unimodal problem aim at increasing the convergence velocity. A rather "hard" selection mechanism forces the search process into the gradient direction, resulting in a path oriented, exploiting search. Consequently, the genotypic diversity remains small.

Unfortunately, for a real-world application the user does not know anything about the objective function's properties. Besides the usual parameterization problem (which settings are appropriate for λ, l, p_m, p_c, η_{max}?) a additional parameter is introduced by extinctive selection.

To solve this problem at least two approaches can be thought of. As shown by Schwefel in the framework of

unimodal objective function	multimodal objective function
convergence velocity	convergence confidence
"hard" selection scheme	"soft" selection scheme
small genotypic diversity	large genotypic diversity
path oriented	volume oriented
exploitative character	explorative character

Table 2: Unimodal and multimodal search properties

Evolution Strategies [Sch81] the *self-learning* of strategy parameters provides a powerful mechanism of internal adaptation of the algorithm with respect to the objective function topology. This is often referred to as *second-level learning* and provides an alternative to the other approach of using a meta-level control algorithm as desribed in [Gre86, GBGK89].

References

[Bak85] James Edward Baker. Adaptive selection methods for genetic algorithms. In J. J. Grefenstette, editor, *Proceedings of the first international conference on genetic algorithms and their applications*, pages 101–111, Hillsdale, New Jersey, 1985. Lawrence Erlbaum Associates.

[BH91] Thomas Bäck and Frank Hoffmeister. Global optimization by means of evolutionary algorithms. In Alexander N. Antamoshkin, editor, *Random Search as a Method for Adaptation and Optimization of Complex Systems*, pages 17–21, Divnogorsk, UdSSR, March 1991. Krasnojarsk Space Technology University.

[BHS91] Thomas Bäck, Frank Hoffmeister, and Hans-Paul Schwefel. A survey of evolution strategies. In Richard K. Belew, editor, *Proceedings of the Fourth Internation Conference on Genetic Algorithms and their Application*, San Diego, California, USA, 1991. Morgan Kaufmann Publishers.

[GBGK89] Yu. V. Guliaev, I. L. Bukatova, L. N. Golubeva, and V. F. Krapivin. Evolutionary informatics and "intelligent" special processors. Academy of Sciences of the USSR, Institute of Radio Engineering and Electronics, 1989. Preprint.

[Gol89] David E. Goldberg. *Genetic algorithms in search, optimization and machine learning*. Addison Wesley, 1989.

[Gre86] John J. Grefenstette. Optimization of control parameters for genetic algorithms. *IEEE Transactions on Systems, Man and Cybernetics*, SMC–16(1):122–128, 1986.

[Gre87] John J. Grefenstette. *A User's Guide to GENESIS*. Navy Center for Applied Research in Artificial Intelligence, Washington, D. C., 1987.

[HB91] Frank Hoffmeister and Thomas Bäck. Genetic algorithms and evolution strategies: Similarities and differences. In Reinhard Männer and Hans-Paul Schwefel, editors, *Proceedings of the First International Conference on Parallel Problem Solving from Nature (PPSN), Dortmund, Germany, 1990*, pages 455–470, Berlin, 1991. Springer.

[Hol75] John H. Holland. *Adaptation in natural and artificial systems*. The University of Michigan Press, Ann Arbor, 1975.

[Jon75] Kenneth De Jong. *An analysis of the behaviour of a class of genetic adaptive systems*. PhD thesis, University of Michigan, 1975. Diss. Abstr. Int. 36(10), 5140B, University Microfilms No. 76–9381.

[Rec73] Ingo Rechenberg. *Evolutionsstrategie: Optimierung technischer Systeme nach Prinzipien der biologischen Evolution*. Frommann–Holzboog Verlag, Stuttgart, 1973.

[Sch81] Hans-Paul Schwefel. *Numerical Optimization of Computer Models*. Wiley, Chichester, 1981.

[Sch89] J. David Schaffer, editor. *Proceedings of the Third International Conference on Genetic Algorithms and Their Applications*, San Mateo, California, June 1989. Morgan Kaufmann Publishers.

[TZ89] A. Törn and A. Zilinskas. *Global Optimization*, volume 350 of *Lecture Notes in Computer Science*. Springer, Berlin, FRG, 1989.

[Whi89] Darrell Whitley. The GENITOR algorithm and selection pressure: Why rank-based allocation of reproductive trials is best. In Schaffer [Sch89], pages 116–121.

Initialization, Mutation and Selection Methods in Genetic Algorithms for Function Optimization

Mark F. Bramlette (818 782 6733)

Abstract

This paper presents and evaluates some modifications of the Genetic Algorithm for improving its performance in function optimization. The modified methods pertain to selection of initial populations, mutation, and competition in producing new generations. They were evaluated by using one Genetic Algorithm to select near-optimal values for parameters controlling another Genetic Algorithm. The values thus selected are evidence that the modifications improve performance in the context of function optimization with vector structures, relatively few fitness function evaluations, and the modified methods. Unusually high rates of mutation prevailed in the tuning competition. Optimal parameter selection was sensitive not only to the particular object function but also to a limit on the number of times it can be evaluated, derived from applications which limit search time severely.

1 INTRODUCTION

Modifications of the Genetic Algorithm for function optimization and a test that demonstrates their effectiveness are presented here. The modified methods are these: Initialization is accomplished by extended random search – each member of the first generation is taken as the best of some number of random trials. Structures in the population are represented as lists of integers, not binary strings. These lists are viewed as vectors in policy space; the integers are regarded as genes, with each gene having many possible values. Mutation uses exponential variates rounded to integers to determine the magnitude of changes to a gene value. The expected magnitude of these changes is gradually reduced as generations pass while the probability of mutation is constant. Using this mutation, search moves gradually from exploring policy space to refining known good solutions. Dynamic translation and scaling of fitness scores is used to establish a balance between intensity of competition and avoidance of premature convergence (the loss of genetic diversity without convergence on an optimum).

To evaluate these methods, an experiment was conducted using one Genetic Algorithm to tune another one, that is, to find near-optimal values for its control parameters. This tuning allows assessment of the value of the modified methods by showing the extent of their use. Evaluation of each set of control parameters was based on the best individual found; the higher-level Genetic Algorithm also addressed a function optimization. Each set of values for control parameters chosen by a higher-level Genetic Algorithm was used in a lower-level Genetic Algorithm limited to one thousand evaluations of its fitness function. This limit comes from applications in which the time available for search is only about a thousand times as large as the time required to evaluate a single alternative.

2 FUNCTION OPTIMIZATION

This section presents modifications to the Genetic Algorithm that are intended to improve its performance in function optimization tasks. In these tasks only the single best individual found at any time during search is relevant, not online or offline performance.

2.1 VECTOR REPRESENTATION

Although representation of individuals in a Genetic Algorithm as binary strings has been recommended [12], each individual is represented here as a list of (non-negative) integers [1] [4], seen as a vector in policy space. This seems more natural for function optimization. It facilitates comparing the Genetic Algorithm with combinatorial optimization techniques, such as simulated annealing [5] [18], which do not require binary representation. Integers were chosen, rather than real numbers, to support integer programming and other applications where only a finite number of values are possible in some variables.

2.2 INITIAL POPULATIONS

An initial population for a Genetic Algorithm is usually chosen at random; one random trial is made to produce each individual. In the present implementation each member of the initial population is chosen as the best of n randomly chosen individuals, where n has a value selected by the user of the Genetic Algorithm. This *extended random initialization* is a generalization of the Genetic Algorithm since setting n to 1 produces the usual initialization procedure. Some users of the Genetic Algorithm have seeded the initial population with some members that are known beforehand to have good scores or be in an interesting part of the search space. Extended random initialization differs from this in that all members of the initial population are chosen automatically (so the user does not intervene) and by the same procedure (so the expected score of each member of the initial population is the same). One purpose of the present experimentation was to determine whether extended random initialization leads to premature convergence.

2.3 EXPONENTIAL MUTATION

When individuals are represented as bit strings, mutation consists of reversing a randomly chosen bit. If a number is represented by a group of bits in the string, small changes in the number's value are unlikely to follow from such mutation. This results in difficulty for the Genetic Algorithm in refining solutions to find an optimum solution after discovering good solutions in its neighborhood. Gray scale coding may reduce this problem. Another approach uses different mutation probabilities for the bits in the string, increasing the probability of mutation exponentially with decreasing significance of the bits [10] so that small changes to a gene's value are more likely than large ones.

Instead of bit strings, individuals can be represented as lists of integers (genes) generally. With this vector representation, mutation consists of adding (or subtracting) some increment to a gene. Selecting this increment is a problem. If it is always large, the search cannot readily refine solutions and risks missing solutions that require values between the expressible values. If the increment is always small, the search cannot readily explore the search space, especially if there are many possible values for a gene. Intermediate values chosen to compromise on the goals of refinement and exploration may not do very well at reaching either one. So the present implementation of the Genetic Algorithm allows many possible values of each gene and avoids a fixed increment for mutational changes. It generates an exponentially-distributed pseudo-random variate and rounds it to an integer to use as an increment (or, with equal probability, decrement) to a gene being mutated. This distributes mutational changes widely while making small changes in a gene's value more likely than large ones.

The initial value of the mean of the exponential distribution used in mutation is a fraction of the range of the gene (variable) being mutated. Fraction of Range is a user-chosen control parameter in the present implementation. In this approach to "black box" optimization, the information available to the optimizer is both a fitness score for any selection of black-box inputs and a specified range and granularity of permissible values for each input variable. Since Genetic Algorithms would apparently benefit from improved ability to refine solutions, the mean of the exponential distributions used in each generation is reduced by a constant multiplier. This multiplier is computed to produce a value of 2 as the mean of the exponential distribution in the last generation of an iteration so that, near an iteration's end, small changes in a gene's value have high probability. (In this implementation, each iteration of the Genetic Algorithm consists of generating an initial population and then producing a user-selected number of new generations as described in Section 3.1; the user also selects the number of iterations.)

For example, if a variable's range is 10000, Fraction of Range is 0.75 and number of generations is 200, then the values used for the mean of the exponential distribution for the first three generations are 7500, 7197.65, and 6907.49. The mean decreases by 4.03% of its previous value between generations, reaching 2 in the two-hundredth generation. If Fraction of Range is changed to 0.25, the rate of decrease in the mean becomes 3.50% per generation and the means decline from 2500 in the first generation to 2 in the last. While the expected magnitude of mutational changes is gradually reduced as generations pass, the probability of mutation is constant. This automatically moves mutation from exploring the vector space to refining known good solutions.

2.4 SCALING AND CONVERGENCE

Genetic algorithms can suffer from premature convergence caused by early emergence of an individual that is much better than the others in the population, although far from optimal. Copies of this structure may quickly dominate the population. Search continues then but is concentrated in the vicinity of this structure and may miss much better solutions elsewhere in the search space. To assist in preventing premature convergence, another control parameter, Dynamic Range, has been introduced for scaling scores. The next generation is produced by copying a structure the number of times indicated by the integer part of its scaled score and providing one more copy with a probability indicated by the fractional part. The method used here translates and scales the scores linearly so that the resulting range of scaled scores divided by the best scaled score equals the Dynamic Range. Examples of

Table 1: Scaled Scores versus Dynamic-Range

Dynamic Range		0.25	0.50	0.75
Structure 1 Score	-10	12/14	4/6	4/10
Structure 2 Score	0	13/14	5/6	7/10
Structure 3 Score	10	14/14	6/6	10/10
Structure 4 Score	20	15/14	7/6	13/10
Structure 5 Score	30	16/14	8/6	16/10
Offset		130	50	23.3
Range	40	4/14	4/6	12/10
Range / Best		0.25	0.50	0.75

scaled scores are given as rational fractions to simplify interpretation (Table 1) of scaling in a maximization problem with various values for Dynamic Range. A below-average structure such as Structure 2 receives no guaranteed copies in the next generation but its probability of being copied once is 13/14, 5/6 or 7/10 for Dynamic Range of 0.25, 0.50, or 0.75. Structure 4, an example of an above-average structure, is guaranteed to be copied into the next generation once regardless of the value of Dynamic Range. The probability of copying Structure 4 once more is 1/14, 1/6, or 3/10 for Dynamic Range of 0.25, 0.50, or 0.75.

The scaled score of an individual is a function of (1) Dynamic Range, (2) score of the individual and (3) the largest score, smallest score and average score in the individual's generation. When maximizing, this function is

$$ScaledScore_i = (Score_i + Offset)/(AverageScore + Offset), \text{ where}$$

$$Offset = ((LargestScore - SmallestScore)/DynamicRange) - LargestScore.$$

When minimizing, this function is

$$ScaledScore_i = (Score_i - Offset)/(AverageScore - Offset), \text{ where}$$

$$Offset = ((LargestScore - SmallestScore)/DynamicRange) - SmallestScore.$$

Since scaling in each generation is a function of the current average score and range of scores, competitive pressure does not deteriorate as the population becomes more fit on average and less varied in fitness. Values of Dynamic Range near 1.0 produce a vigorous competition for reproduction with little concern for maintaining genetic diversity in the population by preserving individuals. Values of Dynamic Range near zero usually reduce competitive pressures and preserve genetic diversity.

3 TESTING EFFECTIVENESS

To evaluate the effectiveness of these modifications, one Genetic Algorithm, the "higher-level" one, was used to tune a "lower-level" Genetic Algorithm, that is, to find (nearly) optimal values for its control parameters. The lower-level GAs were run with values for eight control parameters provided by the higher-level GA. The lower-level GA performed a ten-dimensional function optimization. This experiment constitutes a test of extended random initialization since the higher-level GA can replace it with the usual initialization procedure by its setting of a control parameter, that is, by setting the number of random trials used in selecting each member of the initial population to 1. The usefulness of exponential mutation is tested by determining the extent to which tuning allocates evaluation of the fitness functions to mutants rather than to the other two sources of new individuals, initialization and crossover. The tradeoff between premature convergence and competitive pressure may be illuminated by the value chosen for Dynamic Range. Other studies of tuning GA, [6] [7] [8] [11] [14] [16], are not directly applicable to the present problem. This is because the modified methods presented here were not used, but also because online or offline performance is maximized rather than best single score, and binary representation is used rather than vectors, or many more than one thousand fitness function evaluations are made.

The parameters that control the modified methods – extended random initialization, exponential mutation, and dynamic range – were described in Section 2. The other five parameters selected by the higher-level Genetic Algorithm for controlling the lower-level Genetic Algorithm are discussed here along with the fitness functions used in experimentation.

Population Size. Population size is chosen from the integers between 4 and 100. These limits were chosen to be wide enough that there would be high probability that the derived optimal population size would be well away from both limits. This is intended to assure that the limits on population size are not binding in the best solutions found.

Producing a new generation stochastically can result in a generation that differs in size from the preceding generation due to random effects. Usually some special action is taken to assure constant population size in each generation. While the population size could be permitted a random walk or could be maintained by selecting individuals to copy or delete based on fitness rank, the present work preserves population size by random choice of individuals to duplicate or delete. Varying generation size is addressed in [2].

Number of Generations. The Genetic Algorithm tested here runs each iteration until a specified number of generations are done. The alternative of terminating when the population loses diversity is efficient when crossover is the primary source of variation in a population; crossover cannot produce any new individuals in homogeneous populations. However, a totally homogeneous population can be succeeded by populations with better fitness when using exponential mutation.

Probability of Mutation. Some authorities regard mutation as much less significant than crossover, serving merely as a background operator; other judgments of mutation's importance are higher. One purpose of the present work is to determine experimentally how best to allocate fitness function evaluations among the three sources of new individuals in a Genetic Algorithm — initialization, mutation and crossover — at least for a few problems.

Probability of Crossover. Crossover here consisted of using the first several entries from the list representing one individual in the population and the remaining entries from another individual. The single point of crossover occurs only at the boundary between two genes. Although [8] presents test results for binary representation indicating that other crossover operators do consistently better than one-point traditional crossover, this has not yet been tested for vector representation.

Iterations of Genetic Algorithm. Each iteration of the Genetic Algorithm here consists of establishing an initial population and then producing the specified number of generations using mutation, crossover, and reproduction with emphasis on individuals with good scores. Iterations of the Genetic Algorithm continues until the requested number of iterations have been done unless stopped sooner due to reaching the user-chosen limit on the total number of individuals evaluated. In many sets of values for control parameters, the value of iterations is irrelevant since the size of the population and the probabilities of mutation and crossover result in the limit of one thousand fitness function evaluations being reached before even the first iteration was finished.

Objective Functions. The objective functions chosen here are based on resemblances to applications whose occurrence motivated this experimentation, rather than the test suites in the literature of Genetic Algorithms. These resemblances pertain to number of dimensions of search and complexity of function form.

The first problem addressed in the lower-level Genetic Algorithms had an objective function of ten integers, each required to be between 0 and 10,000. The objective function, which is to be maximized, is the following:

$$
\begin{aligned}
Fitness(x_1 \ldots x_{10}) = 0.0001\times \\
(+1000x_1 + 0.50x_1^2 - 0.50x_1^{2.5} \\
+3000x_2 + 0.50x_2^2 - 0.50x_2^{2.5} \\
-2000x_3 + 0.30x_3^2 - 0.30x_3^{2.5} \\
+7000x_4 + 0.20x_4^2 - 0.20x_4^{2.5} \\
+2000x_5 + 0.20x_5^2 - 0.20x_5^{2.5} \\
+9000x_6 + 0.25x_6^2 - 0.25x_6^{2.5} \\
-4000x_7 + 0.60x_7^2 - 0.60x_7^{2.5} \\
-1000x_8 + 0.50x_8^2 - 0.50x_8^{2.5} \\
+9000x_9 + 0.20x_9^2 - 0.20x_9^{2.5} \\
+1000x_{10} + 0.40x_{10}^2 - 0.40x_{10}^{2.5})
\end{aligned}
$$

The second fitness function was produced by changing all exponents of 2.5 in this equation to 0.5. Each Genetic Algorithm was limited to one thousand evaluations of the objective function. This relatively low limit was derived from the application ultimately intended for function optimization. The present fitness function has the same dimensionality and number of possible values for each gene as that application, but is more quickly computed and hence more suitable for this tuning experiment.

4 EXPERIMENTAL RESULTS

The values for parameters controlling the higher-level Genetic Algorithm were chosen intuitively. Neither level of Genetic Algorithm used elitism. Single-point crossover was used. The higher-level Genetic Algorithm evaluated the fitness of a set of values for the control parameters in the lower-level Genetic Algorithm by using each set in one hundred iterations of the lower-level Genetic Algorithm and then averaging the resulting hundred best (maximum) values found. More efficient search can result from less accurate estimates of individuals fitness [9]; less effort is spent evaluating each individual so that more individuals can be evaluated. For this reason, smaller numbers of repetitions (one and ten) were used in earlier experiments for evaluating a set of control parameters. These results were so contaminated by random factors in the effectiveness of the lower-level Genetic Algorithm that the value of one hundred was used even though this greatly increased the required computer time.

Minimum values, maximum values, and resolution for each parameter, and the best combinations of control parameters found experimentally are shown (Table 2). When the best values for the control parameters found for Problem 2.5 are used, each iteration of

Table 2: Parameters Chosen by Higher-Level G. A.

	Limits	Units	Pr'm 2.5	Pr'm 0.5
Initial Trials.	1 to 20	1	18	20
Fraction of Range.	0.01 to 1.0	0.01	1.00	0.88
Dynamic Range.	0.01 to 1.0	0.01	0.96	0.97
Population Size.	4 to 100	1	8	7
Number of Generations.	1 to 200	1	191	169
Mutation Probability.	0.01 to 1.0	0.01	0.74	0.64
Crossover Probability.	0.01 to 1.0	0.01	0.12	0.27
Number of Iterations.	2 to 10	1	7	3

the lower-level Genetic Algorithm evaluates the fitness function 144 times (the product of Initialization Trials and Population Size, or 8×18) for initialization. The expected number of mutants evaluated is 1130.72 (the product of population size, probability of mutation, and number of generations, or $8 \times 0.74 \times 191$). The expected number of individuals produced by crossover is 366.72 (the product of population size, twice the probability of crossover, and number of generations, or $8 \times 2 \times 0.12 \times 191$). Each crossover results in two individuals being generated and evaluated. The limit of one thousand evaluations interrupts the search, so the 856 evaluations available after initialization were split approximately in the ratio 1130.72 to 366.72 between mutation and crossover. About 646 mutants and 210 products of crossover were evaluated. Since both tunings produced parameter values that left the first iteration unfinished when the limit on evaluations was reached, the value of Iterations does not matter.

In problem 0.5, 140 fitness evaluations were used in initialization, and there would be about 757 evaluations for mutants, and 639 for the products of crossover except that the limit on evaluations interrupts the search. The 860 evaluations left after initialization are split, using the ratio 757:639, between about 466 mutants and 394 products of crossover. Such different results in optimal tuning for problems that differ from each other so little suggest that no best tuning can be found for all problems, leaving open the question of whether there exists a tuning that is adequate (although suboptimal) over a broad range of problems.

Extended Random Initialization. The values of 18 and 20 for the number of random trials made in initializing populations are much higher than the value of 1 usually used in Genetic Algorithms. The value of 1 could have been chosen by the higher-level Genetic Algorithm (Table 2). Even more random search might have gone into initialization if the upper limit had been higher than 20. The best tunings found for the two problems spent about 14% of the evaluations (144 and 140) on initialization even though only a thousand fitness function evaluations are permitted. The results indicate that extended random initialization is preferable to standard initialization for these problems.

Exponential Mutation. The high values of Probability of Mutation (0.74 and 0.64) and of Fraction of Range (1.00 and 0.88) that prevailed in the tuning indicate that exponential mutation is beneficial also. In both problems, a major share (646 and 466) of the available thousand fitness evaluations are given to mutants. The value of "cooling" the mutation (reducing the mean of the exponential distribution used in mutation after each generation) is clear from the large amount of cooling that occurred. In both problems the mean of the exponential distribution declined substantially (from 10,000 to about 287 in about 109 generations for problem 2.5, from 8,800 to about 2.6 in about 104 generations for problem 0.5). It did not reach 2 because of being interrupted at one thousand fitness evaluations. Exponential mutation moved the search from exploration toward refinement, even though only a small number of alternatives were evaluated (10^3) out of a large search space (10^{40}). Reaching the minimum value for the mean of mutation's exponential distribution, 2, would accomplish more precise refinements of genes' values. This would occur if the values for the control parameters chosen by tuning allowed multiple iterations, as could happen if there were a higher limit on the number of evaluations of the fitness function so that at least one iteration could be completed.

Competition versus Convergence. The high values chosen for Dynamic Range (0.96 and 0.97) indicate that premature convergence is not a problem here and that unrestrained competition is permissible. This is plausible since so few fitness function evaluations are allowed that there appears to be inadequate time for convergence to occur. It may be that premature convergence is rarely a problem when exponential mutation is used or when the relatively few individuals are evaluated for fitness.

Comparison of Mutation and Crossover. In both problems, fewer fitness evaluations were allocated to the products of crossover (210 and 394) than to mutation (646 and 466). A relatively high rate of crossover seemed likely for these problems prior to the experiments since no terms of the fitness function involved more than one variable. Crossover and mutation rates are adapted by the higher-level Genetic Algorithm to work well with each other and the other six parameters controlling the lower-level Genetic Algorithms. Asserting that one is more important than another may therefore be like saying that genes for sharp teeth are more valuable to a predator than genes for a keen sense of smell, when actually they are co-adapted. The emphasis on mutation found here is sharply different from other results [7] [8] [11] [14] [16] in which mutation rates are set far below crossover rates. In those studies crossover was permitted at any point in a bit string. Thus, crossover produced not only two new individuals but also, when crossover occurred between the boundaries of a field defining a number, commonly also produced two new numbers in that field, a sort of mutation in itself. Crossover of bit strings has the effect of simultaneously accomplishing this mutation as well. In the present implementation of the Genetic Algorithm, crossover occurs only at a boundary between genes and therefore does not also accomplish a mutation. This result and the effects of exponential mutation may explain why mutation rates were set so much higher in the present work than in other experiments.

Comparison of Tuning Results. It might be that a wide variety of tunings give about the same performance on the problems described here. To test this, the 2.5 problem was tested using the tuning derived for it (the matched case) and then using the tuning derived for the 0.5 problem (the mismatched case). Both cases were run five times. The matched test produced scores of 17043.352, 16984.297, 16895.46, 16801.283, and 16638.467, with mean 16872.572 and standard deviation 129.89. The mismatched test produced scores of 15735.034, 15726.378, 15509.9, 15184.595 and 15091.787 with mean 15449.539 and standard deviation 124.30. All scores in the mismatched test are below all scores in the matched test. Consider the hypothesis that the two settings are equally effective on the 2.5 problem, that is, that the second tuning is among a number of settings that give about the same performance on that problem. For the present test results, this is quite unlikely: the probability that, in a random permutation of ten numbers, the larger five occur ahead of the other five is $(5 \times 4 \times 3 \times 2 \times 1)/(10 \times 9 \times 8 \times 7 \times 6)$ or 0.003968. The two means differ by about 8% of the larger mean or eleven times the larger standard deviation. Thus, for the 2.5 problem, the tuning found for the 2.5 problem works better than the tuning found for the 0.5 problem does, even though these two problems are so similar. In this case, the performance penalty resulting from not using the best known tuning is large enough to be significant in practical problems.

Table 3: Values Used and Produced by the G. A.

Tuning Problem 2.5	First Used	First Result	Second Used	Second Result
Initialization Trials.	10	18	18	9
Fraction of Range.	0.75	1.00	0.96	0.70
Dynamic Range.	0.75	0.96	0.96	0.97
Population Size.	30	8	9	25
Number of Generations.	200	191	184	196
Probability of Mutation.	0.10	0.74	0.79	0.28
Probability of Crossover.	0.05	0.12	0.12	0.06
Number of Iterations	1	7	7	7

The control parameters for the higher-level Genetic Algorithm were given intuitively-chosen values (Table 3, first column). Using these values for tuning the lower-level Genetic Algorithm produced the values shown in second column. These results are sharply different, especially in population size, number of initialization trials, and probability of mutation. To test whether the derived values at the lower level were more effective at the higher level than the originally chosen ones, the tuning experiment for Problem 2.5 was repeated using the control values shown in column 3, the best available values at the time this experiment was started and quite close to column 2. The resulting tuning, column 4, shows a large amount of time spent in initialization and a mutation rate considerably higher than the crossover rate, as before. To test whether the values in column 2 or column 4 were better, each was used six times. The intuitively-chosen values produced better results than the derived values every time, (17156.8, 17121.9, 16739.6, 17152.9, 16880.9, and 16722.5 to 16244.1, 16195.2, 15912.3, 16241.1, 16055.4, and 15712.2). The probability of this would be 0.00018 if the two tunings were equally effective for the higher-level Genetic Algorithm. So, the tuning that proved best for use on problem 2.5 was not also the best for *tuning* problem 2.5. Again, the loss in performance resulting from using a suboptimal tuning is large enough to be significant.

Significance of Limits on Number of Evaluations of Fitness Function. Limiting the number of evaluations of the fitness function in a Genetic Algorithm for function optimization may be expected to have a great ef-

Table 4: Tuning for Problem F6-10

Function Evaluations	10,000	100,000
Initialization Trials.	13	27
Fraction of Range.	0.74	0.79
Dynamic Range.	0.93	0.98
Population Size.	56	89
Number of Generations.	153	199
Probability of Mutation.	0.26	0.19
Probability of Crossover.	0.05	0.16
Number of Iterations.	6	7
Best Score	0.2356	0.0309

fect on tuning. When this limit is higher, larger populations and greater emphasis on exploring alternatives rather than refining them seem plausible. To test this prediction, experiments were conducted using a Genetic Algorithm to tune a lower-level Genetic Algorithm that has a fitness function called F6-10, derived from F-6 in [16] by extending it from two to ten dimensions. This fitness function to be minimized is

$$Fitness(x_1 \ldots x_{10}) =$$

$$0.5 + [(\sin^2 Sum^{0.5} - 0.5)/(1.0 + 0.001 Sum)^2]$$

where *Sum* is the sum of the squares of the ten variables. Each variable has 10^4 possible values, ranging between -100 and 100 with a resolution of 0.02. The search space has 10^{40} points. The original problem, F-6, has only 10^8 points when the variables have the same range and resolution. This provides some quantitative basis for asserting that F6-10 is much more difficult than F-6 is. The generalization of F-6 to ten dimensions is motivated by applications that have this dimensionality.

This problem was tuned with the limit on the number of evaluations of the fitness function set at 10,000, and at 100,000 (Table 4). These results indicate again that using extended random initialization and setting the probability of mutation greater than the probability of crossover are useful. The statistical significance of the differences in tuning shown in Table 4 were tested by using these two tunings for ten trials each on the problem in which 10,000 evaluations of the fitness function are permitted. In the matched case (the tuning derived for 10,000 function evaluations was used for 10,000 function evaluations) the scores were 0.2711, 0.2835, 0.2889, 0.2914, 0.3029, 0.3228, 0.3291, 0.3342, 0.3358 and 0.3408, averaging 0.3100. (The deterioration in average performance from 0.2356 in the tuning experiment to 0.3100 in the present experiment is typical of experience in using Genetic Algorithms with objective functions that have statistical variation in them. The structure that performs best has benefited not only from good genes but also from good luck in the particular sequence of pseudo-random variates used during the evaluation.) In the mismatched case (the tuning derived for 100,000 function evaluations was used for 10,000 function evaluations) the scores were 0.3355, 0.3449, 0.3596, 0.3683, 0.3708, 0.3853, 0.3921, 0.3974, 0.4033, and 0.3921, averaging 0.3764.

A time-saving method of testing the significance of these results is computer-intensive statistical hypothesis testing [15]. This method is usable when the standard t test of the statistical significance of differences in means is not valid due to small sample sizes. This method seeks to prove that the observed difference in two data sets is significant by reasoning from the "null" hypothesis that the difference is not significant to some low-probability conclusion. From this it is concluded with high confidence that the null hypothesis is incorrect and that the alternative hypothesis is therefore correct. The null hypothesis here is that the two tunings are equally good. This is tested by generating some large number of random permutations of the twenty scores produced by pooling the two samples of ten scores. The method computes the fraction of these permutations that show at least as large a difference in the mean of the first ten and the mean of the second ten as was observed originally.

The results of the test are these: The original difference in means is 0.0664. Out of 999 random permutations of the twenty scores, none had a difference of means as large as 0.0664. The observed fraction of permutations, counting the original one, with a difference in means as large as the original one is 0.001. Let ϕ be the true value of this fraction, that is, the value that is expected if testing continued indefinitely and the null hypothesis is true. The probability (estimated through a computation using [15]) that $\phi <= 0.01$ was found experimentally to be 0.999956. So, the null hypothesis is rejected with very high confidence. With the limit of 10,000 fitness function evaluations, the tuning derived for a limit of 100,000 fitness function evaluations is not as good as the tuning derived for 10,000 fitness function evaluations. The conclusion is that, for problem F6-10 with the number of evaluations of the fitness function limited to 10,000, performance when using the tuning derived for this limit significantly surpasses performance when using a tuning derived for another limit, even though the same objective function was used in finding both tunings. The more general conclusion is that optimal tuning can depend not only on the particular objective function being optimized but also on the number of times that it is possible to evaluate that function.

5 CONCLUSIONS

The modifications to Genetic Algorithm presented here are intended to improve its performance in limited search time and applied to difficult problems of fairly high dimensionality. For the problems under consideration, the total number of individuals that can be evaluated is severely limited because evaluating any one individual takes as much as 0.1% of the available search time. The experimental results provide evidence that the modified methods benefit the Genetic Algorithm's performance in the present context: function optimization with vector structures, relatively few fitness function evaluations, and the modified methods. The high values selected by the upper-level Genetic Algorithm for the number of random trials used in initialization, the probability of mutation and expected magnitude of mutation are evidence that the modified initialization and mutation methods are beneficial.

In the best tuning found by genetic search, about 14% of the function evaluations were allocated to initializing populations by random search even though only a thousand evaluations were permitted. While it may seem excessive to devote so much time to initializing the Genetic Algorithm, note that [3] found it cost-effective to use even more of the available computer time, about 80%, in initializing simulated annealing.

Optimal mutation probability was found to be much larger relative to crossover probability than was found in other contexts (optimizing online and offline performance, using binary strings and methods other than the modifications presented here). This may be partly because crossover here occurs only between genes. Further, the tuning results show that reducing competition in producing new generations (for the sake of preserving genetic diversity and avoiding premature convergence) is not necessary here. For the test problems the best results were invariably achieved with parameters setting that called for intense competition.

Finally, the question of how a limit on the number of individuals that can be evaluated influences tuning was investigated briefly. The experiments indicate that optimal tuning is different for limits of 10,000 and 100,000 on the number of individuals evaluated, at least for the problem demonstrated. It would be expected that when time permits a very large number of individuals to be evaluated, a very large number of tunings would perform about as well as the optimal tuning but that tuning is more significant when relatively few individuals are evaluated. In this context, even 100,000 individuals is relatively few.

References

[1] Antonisse, J., (1989) "A New Interpretation of Schema Notation that Overturns the Binary Encoding Constraint", in [17]

[2] Baker, J. E., (1985) "Adaptive Selection Methods for Genetic Algorithms", in [13]

[3] Basu, A. and L. N. Frazer, (1990) "Rapid Determination of the Critical Temperature in Simulated Annealing Inversion", *Science*, Volume 249, 21 September 1990, 1409-12

[4] Bramlette, M. F. and R. Cusic, (1989) "A Comparative Evaluation of Search Methods Applied to Parametric Design of Aircraft", in [17]

[5] Davis, L. D., (1987) *Genetic Algorithms and Simulated Annealing*, Los Altos, Morgan Kaufmann

[6] Davis, L. D., (1989) "Adapting Operator Probabilities in Genetic Algorithms", in [17].

[7] DeJong, K. A., "Analysis of the Behavior of a Class of Genetic Adaptive Systems", PhD dissertation, University of Michigan, Ann Arbor, 1975.

[8] Eschelman, L. J. et al., (1989) "Biases in the Crossover Landscape", in [17]

[9] Fitzpatrick, J. M. and Grefenstette, J. J., (1988) "Genetic Algorithms in Noisy Environments", Machine Learning, Volume 3, 101-120

[10] Fogarty, T. C., (1989) "Varying the Probability of Mutation in the Genetic Algorithm", in [17]

[11] Goldberg, D. E. (1985) "Optimal Initial Population Size for Binary-Coded Genetic Algorithms" TCGA Report Number 85001

[12] Goldberg, D. E. (1989) *Genetic Algorithms in Search, Optimization, and Machine Learning.* Menlo Park, Addison-Wesley

[13] Grefenstette, J. J., (1985) *Proceedings of an International Conference on Genetic Algorithms and Their Applications*, J. J. Grefenstette, Editor

[14] Grefenstette, J. J., (1986) "Optimization of Control Parameters for Genetic Algorithms", IEEE Transactions on Systems, Man and Cybernetics, volume SMC-16, Number 1 (January-February)

[15] Noreen, E. W., (1989) *Computer Intensive Methods for Testing Hypotheses.* New York, John Wiley and Sons.

[16] Schaffer, J. D., et alia, (1989) "A Study of Control Parameters Affecting Online Performance of Genetic Algorithms for Function Optimization", in [17]

[17] Schaffer, J. D., Editor, *Proceedings of the Third International Conference on Genetic Algorithms*, June 4-7, 1989 San Mateo, Morgan Kaufmann

[18] van Laarhoven, P. J. M., and E. H. L. Aarts (1987). *Simulated Annealing: Theory and Applications.* Dordrecht, Kluwer Academic

Looking Around: Using Clues from the Data Space to Guide Genetic Algorithm Searches

Hugh M. Cartwright and Gregory F. Mott

Physical Chemistry Laboratory, Oxford University,
South Parks Road, Oxford, England OX1 3QZ
e–mail: hcart@vax.oxford.ac.uk

Abstract

For some classes of problem, the most appropriate parameters in a GA calculation are a function not only of the problem type, but also of the data used. A method of assessing suitable population sizes and crossover rates for the GA solution of flowshop scheduling problems is described. Appropriate values of these parameters are shown to be related to the topology of the data hyperspace. Systems in which the surface of the hyperspace is relatively smooth appear to be searched more efficiently by GAs with smaller populations and higher crossover rates than data sets yielding corrugated surfaces. The method is illustrated by consideration of flowshop scheduling systems that generate surfaces of contrasting character.

1 Introduction

It is the ultimate goal of most Genetic Algorithm (GA) research to apply the technique to real–world problems. To be of value in these problems, the GA must clearly be able to both find an acceptable solution, and do this in reasonable time.

The ability to locate nearly–optimal solutions, and the rate of convergence to these solutions, are both strongly dependent upon the choice of adjustable parameters in the GA. Each of these parameters, such as population size, mutation rate, crossover rate, choice of crossover mechanism etc. may have a profound effect on the success of the procedure, and therefore whether the GA can be of value for the chosen problem.

In consequence, considerable effort has been devoted to investigating how the choice of each parameter affects performance of the GA [see, for example, 3, 4, 6]. There is broad agreement on suitable ranges for some parameters. The population, for example, is often chosen in the range 20 to 100. Very small populations cannot establish or maintain sufficient diversity, while in a large population good strings may be swamped by numerous inferior strings, or close copies of themselves. Similarly, while the mutation rate should be sufficient to prevent stagnation in the population, it must still be low enough that the propagation of schemata is not unduly hindered. This requirement is fundamental to the Schema Theorem, on which many of the theoretical constructs of the GA rest.

To an extent, such conclusions about parameter values apply to all GA calculations. As a result, values which workers have in the past found suitable, provide an appropriate starting point for any new problem. These values can then be refined by suitable means. Once an effective set of parameters has been found, that set may be used to direct the GA search.

This procedure of refining a starting set of generally accepted parameters, is widely employed. However, it is not without its limitations: even within a single problem type, parameters that are appropriate for one set of data may be inappropriate for a second. Further, the values chosen for parameters at the start of a calculation (for example, the relative rates of mutation and mating), may not remain ideal throughout its course. There exist some – perhaps many – applications of the GA for which the best values of parameters may only be determined once the calculation is underway [2, 6].

How, then, should calculations for these problems be tackled? It is the purpose of this paper to show that, by allowing the GA to look critically at the data space in which it will search, before the normal GA calculation begins, valuable information may be gained on suitable parameter values. This information can both improve the likelihood that the optimum will be located and accelerate convergence towards that optimum.

2 Flowshop Scheduling: A Problem in Chemical Engineering

High–value chemicals are routinely produced in chemical flowshops (figure 1).

During flowshop operation, batches of different starting materials are fed in succession into a chain of serially–linked reactors [5]. Each batch is in general chemically distinct, so will spend different lengths of time in a given reactor than other batches. In some batch sequences, the movements of the chemicals will mesh together efficiently, with batches moving in partial lock–step fashion from reactor to reactor. In other sequences, frequent mis–matches will occur: chemicals still occupying reactors towards the end of the reactor chain may prevent the progress of chemicals from reactors nearer the start.

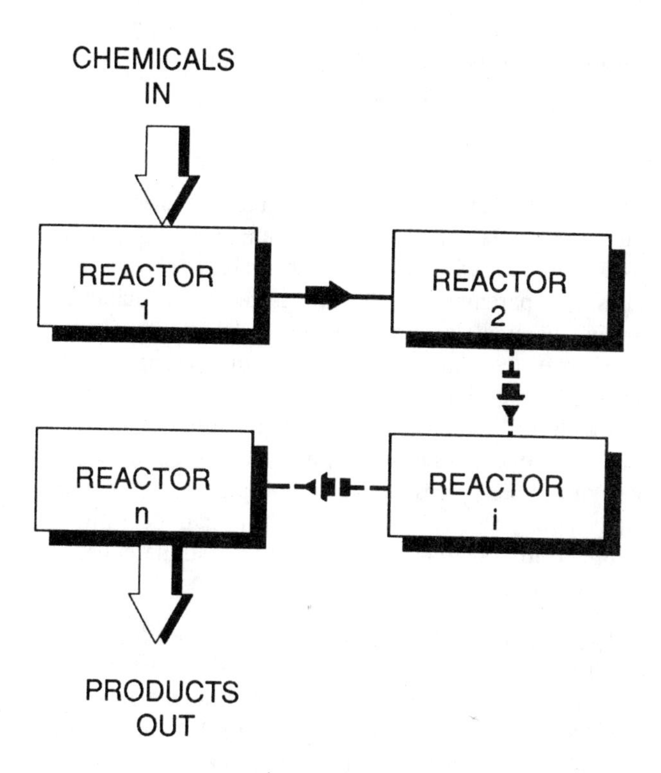

Figure 1: Schematic diagram of part of a chemical flowshop.

The total time taken for all batches to be processed through the flowshop (known as the "makespan"), is thus a critical function of both the sequence in which batches enter the reactors and the matrix defining the time spent by each batch in each reactor. The task of finding the optimum sequence – that of minimum makespan – has features in common with the Travelling Salesman Problem (TSP), but is both computationally more demanding than the TSP, and of considerable financial importance to the chemical process industry [8, 9].

The number of different sequences in which NB batches can enter the flowshop is NB!; the total number of sequences is thus considerable for non–trivial systems. A typical medium–sized plant may schedule 20 batches at one time, for which the number of possible sequences is of the order of 10^{18}. The problem then resolves itself into one of searching a very large data space for one or a few isolated points which represent the sequence(s) of minimum makespan. This is a problem ideally suited to the GA, involving as it does an NP–complete problem in which the schemata, consisting of sub–sequences of batches, have a clear physical interpretation [1].

We discuss below the performance of a benchmark GA in which parameters shown to be realistic by other work are used to direct the calculation. This is compared with the performance of a GA which starts by explicitly assessing the data space it is investigating, and, from its observations, advises on the most suitable GA parameters. As we shall show, the latter algorithm, which evaluates beforehand the nature of the data space in which it is to search, performs significantly better.

3 A Benchmark GA Calculation

It is useful first to perform a straightforward GA calculation to provide data which may be used as a benchmark by which to judge later results. For this calculation, the parameters shown in table 1 are used.

Table 1: Parameters For The Benchmark Calculation

Population: 30
Crossover rate: 0.6
Mutation rate (Pairswapping) 0.0015
Mutation rate (String regeneration) 0.0005

Throughout this work, decimal rather than binary coding is used; as other workers have noted, binary coding often hinders, rather than assists, GA solution of sequencing problems [10].

Figures 2 and 3 show how the best makespan for four representative systems improves as the benchmark calculations proceed. Results are shown for four systems: 8, 10 or 20 batches pass through a flowshop of 20 reactors. For NB=8 and NB=10, the size of the data space is such that the global optimum makespans can be found by exhaustive search of the data space; these are shown on the figure.

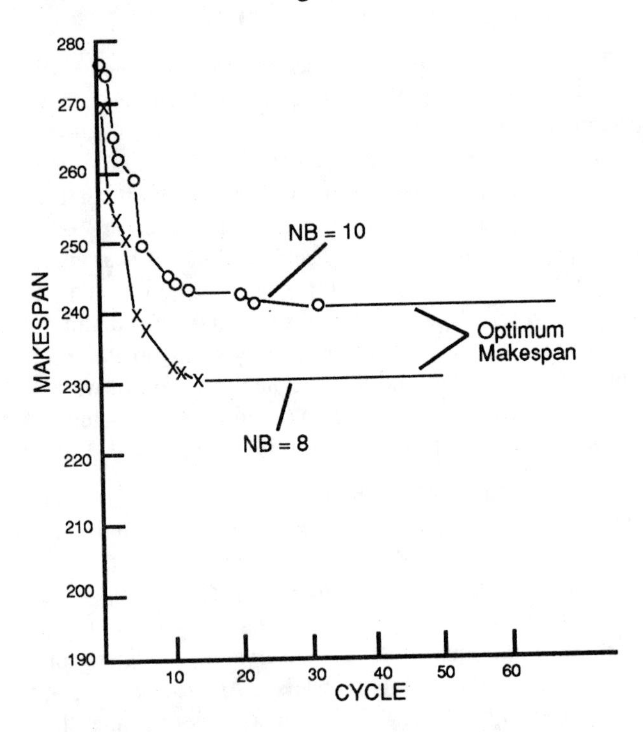

Figure 2: Best makespan found by the GA for systems of 8 and 10 batches. The optimum makespans were found by exhaustive search.

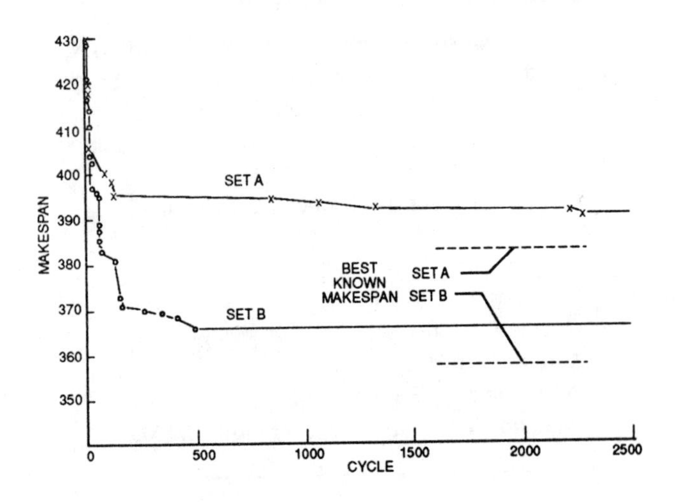

Figure 3: Best makespan found by the GA using benchmark parameters for systems of 20 batches. The "best known" makespans were found by later GA calculations.

The GA converges rapidly to the correct makespans for the smaller systems. Optimum sequences are found far more quickly than exhaustive search can locate them, as we would expect. (The ratios of cpu time required for exhaustive search to the time required for GA search are roughly 100 : 1 for NB=8, and 1800 : 1 for NB=10.) However, for NB=20, the performance of the algorithm is less satisfactory. Figure 3 is unlikely to persuade us that the sequence found after 5000 cycles is optimal; indeed, calculations for other trial systems in which NB=20 frequently reveal makespan improvement well beyond 5000 cycles. Searching – albeit effective searching – is evidently continuing, and many additional cycles may be necessary before convergence is apparently complete.

The results shown in figure 3 were obtained using realistic starting parameters. Various mating schemes, including some particularly suited to the TSP (for example, Whitley's edge recombination operator [10]) were tested. As a result of these tests, a variant of Goldberg's PMX mating scheme [7], which trials suggest works particularly well for this problem, was chosen. Despite this, performance of the GA, while encouraging, is not fully convincing. We need to consider what assistance may be available to the GA to enhance its performance.

4 The Nature of the Data Space

The central factor restricting progress of the GA in this problem is that location of, and convergence to good solutions are sensitive to values chosen for adjustable parameters. The GA is fed with "suitable" starting parameters, but these are suitable only in the sense that they have been found to be appropriate when tackling similar problems, or the same problem but with different data. However, using data that previous work has suggested might be appropriate does not guarantee success. To make progress, the GA needs some mechanism by which it can determine, before serious searching starts, what constitute suitable parameters.

The most productive way we have found to do this is to allow the algorithm to look around to assess the topology of the data space. To illustrate how this can be done, we consider results for two different sets of data for the flowshop scheduling problem: in each system, 20 different batches pass through 20 reactors; the number of possible sequences is thus 2.43×10^{18} for each set.

Set (A) Residence times of batches in each reactor are chosen at random in the range 1–15 hours.

Set (B) Residence times are chosen at random in the range 1–15 hours, with a strong bias towards times near the extrema of the range.

Results for the benchmark calculations on these sets (figure 3) show that the GA performs adequately in seeking to minimize the makespan, but fails to find the best known sequence for either set. We can assess whether the parameters chosen for the benchmark calculations are appropriate by sampling the data space mapped out by these two sets.

Figures 4 and 5 show how the makespan varies with batch sequence for a number of related sequences using data from these sets.

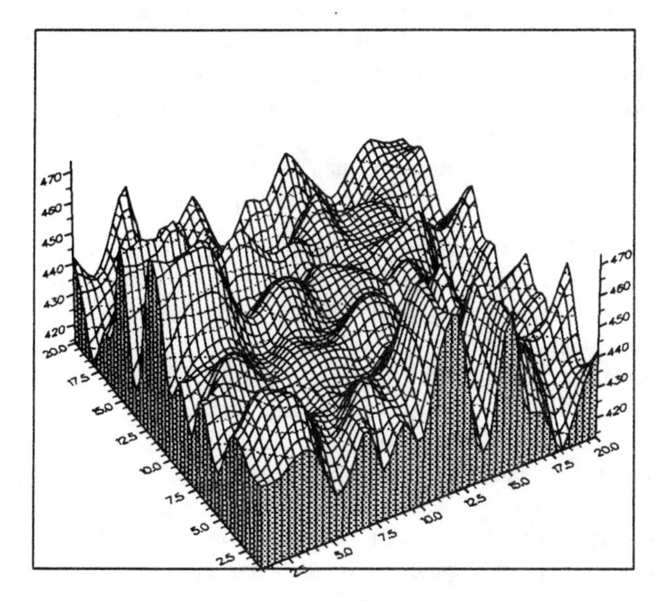

Figure 4: A portion of the data space showing makespans for some related sequences using set A.

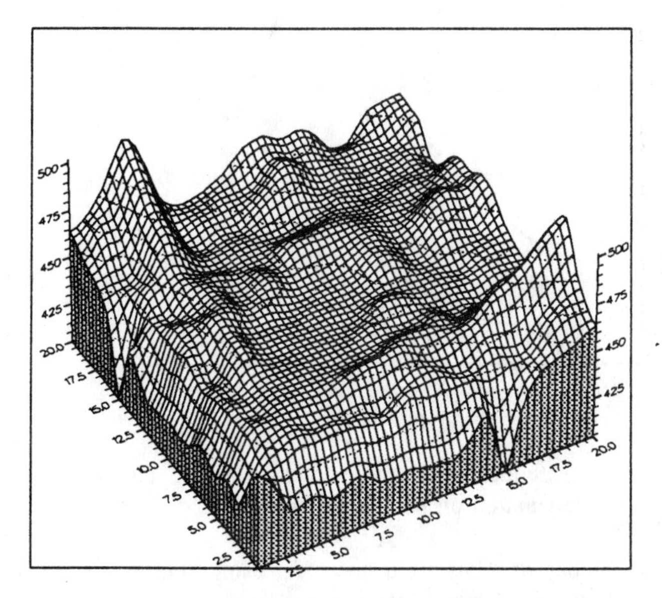

Figure 5: A portion of the data space showing makespans for some related sequences using set B.

As the data space is many–dimensional, some recipe is needed to visualize the data. We choose to display the makespan, plotted on the vertical axis, for a number of closely–related batch sequences. The sequences whose makespans are shown in the figures are generated by pairwise swaps of batch numbers in a single parent sequence, which lies at the lower left of the diagram.

Thus, if the parent sequence were 7, 1, 9, 16, 18, ..., the sequences lying along the line y=0 would be 7, 1, 9, 16, 18, ..., 1, 7, 9, 16, 18, ..., 9, 1, 7, 16, 18, ..., 16, 1, 9, 7, 18, ... and so on. Every sequence shown is related to the parent by a single pairwise swap, so the sequences share numerous schemata.

It is immediately clear that the surfaces in figures 4 and 5 are qualitatively different. That belonging to set A is uneven, lacking short–range order (apart from a mirror plane, which is a consequence of the way in which the surface was derived) and with abrupt changes in makespan for small changes in sequence. By contrast, the surface derived from set B is smooth and relatively featureless, with short–range regularity.

This marked difference between the two surfaces will help guide the GA towards a more suitable parameter choice. But before we argue how this might be done, we must be certain that, since each surface depicts only a tiny portion of very large surfaces, the difference between them is not just a statistical freak.

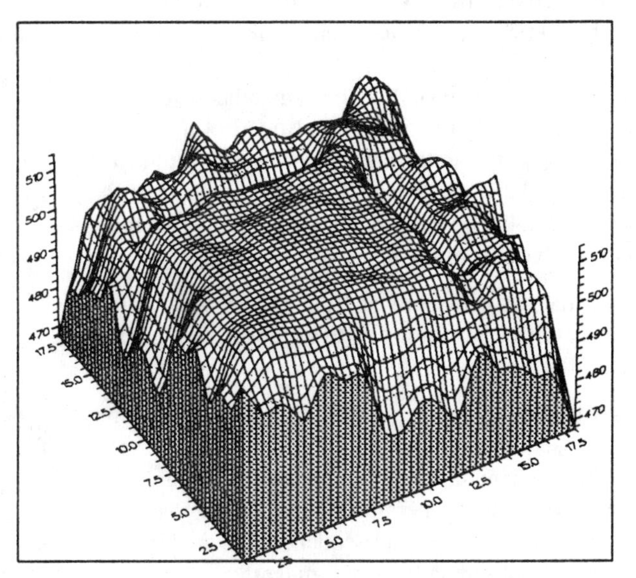

Figure 6: A portion of the data space showing makespans for sequences derived from a randomly–chosen parent sequence (18 batches, set B reactor times).

Surfaces in the region of other, randomly–selected sequences for the two sets can readily be calculated and displayed. Every surface we have observed for set A has the same irregular nature as that shown in figure 4. Surfaces from set B often (though not invariably) show large, quite uniform areas. Figures 6 and 7 illustrate this. These show portions of the surface generated when 18 batches pass through the same set of reactors (with the same residence times) used for set B. Though each surface displays its own individual character, the contrast with figure 4 remains.

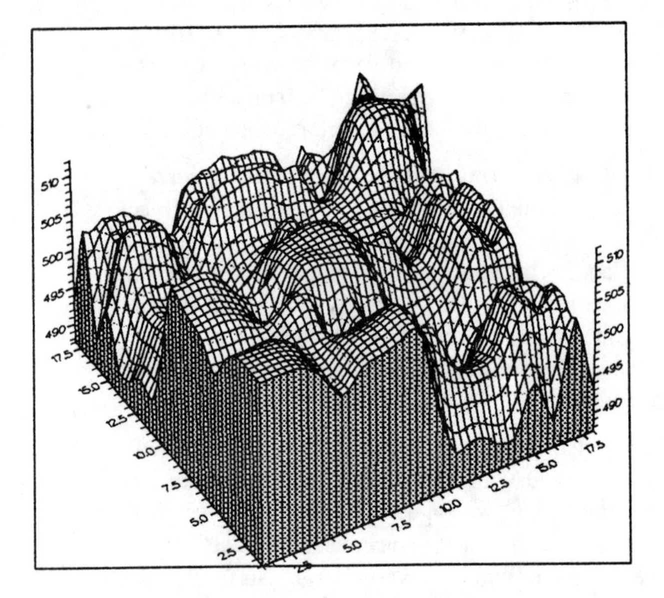

Figure 7: A portion of the data space showing makespans derived from a parent known to have a very high makespan (18 batches, set B reactor times).

(In passing, we may note that the smoothness of the surface shown in figure 5 might also arise if set B could somehow generate makespans only within a small range. This, however, is not the case; set B makespans actually cover a much wider range overall that those derived from set A.)

Given that the surfaces derived from sets A and B are qualitatively different, how can this information guide the GA search?

The basis of the operation of the GA is the evolution of good sequences by a cut-and-pasting of high-quality schemata. If the pairwise swapping procedure used to generate figures 4–7 disrupts important schemata in a good sequence, the makespan should change substantially (since makespans for both sets, and for set B in particular, cover a wide range). Yet, as figure 5 shows, most pairwise swaps leave the makespan for the set B sequence almost unchanged. If these swaps hardly affect the makespan, the quality of the sequence (as judged by its makespan) must be determined by a comparatively small group of batch numbers which are in the "right" positions in the sequence; the identity of batch numbers elsewhere in the sequence is evidently much less important.

In terms of schemata, this suggests:

> *The number and the length of high-quality schemata defining a family of good sequences in set B are probably both small. (If schemata were very long or numerous, pairwise swapping should frequently disrupt them, and lead to substantial irregularities in the surface).*

By contrast, figure 4 shows that almost every pairwise swap for a typical set A sequence produces a considerable change in makespan. Since the schemata in this good sequence are disrupted by pairswaps at so many different locations in the string, they must be more numerous, longer, or both, than the high-quality schemata required to define a good set B sequence.

We must now turn these observations to practical advantage in the GA. If the argument is correct that good set B sequences include fewer and shorter schemata, a smaller string population should be needed for calculations on set B than on set A. A small population will be sufficient to encompass all important schemata, and will give advantages in computational speed. Further, since schemata for set B evidently are less easily damaged, a higher crossover rate for this set can be used, which will allow more rapid investigation of the data space.

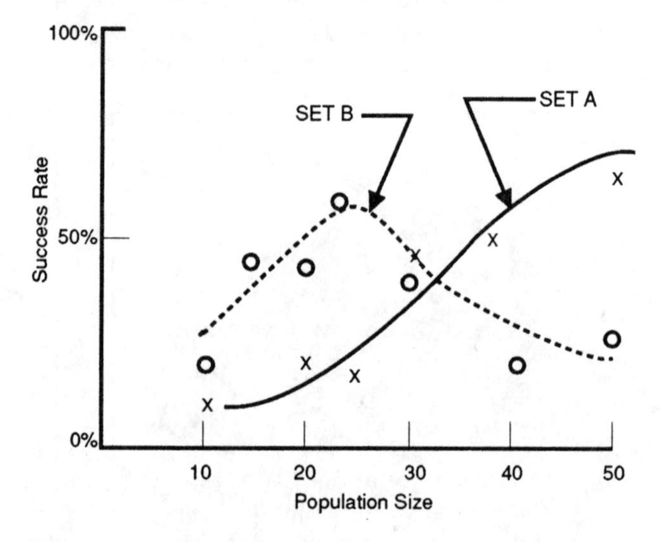

Figure 8: Dependence of the success rate of the calculation on population size.

A large number of trials have been run using sets A and B to test the validity of these propositions. Some representative results are given in figures 8–10. The figures use averaged data extracted from trials covering various population sizes, crossover rates and mutation rates.

Figure 8 shows how the success of a calculation depends upon population size. We define a "successful" calculation to be one which finds a sequence with makespan within 2% of the best known makespan within 5000 cycles; in a "near–optimum" calculation a makespan within 1% of the best known value is found within 5000 cycles. Although further improvement is not uncommon beyond 5000 cycles, the calculations generally are close to convergence at this point; thus the quality of the best makespan at 5000 cycles is a reasonable measure of the success of the calculation. It is evident that success depends in a rather different way on population size for the two sets. For set A calculations, success increases steadily with population size; "near–optimum" solutions, however, are found most readily using a population size around 30.

By contrast, success for set B is more dependent on choice of population. While small or large populations are moderately successful, population sizes in the range 15–25 are significantly better. This conclusion extends also to "near–optimum" solutions.

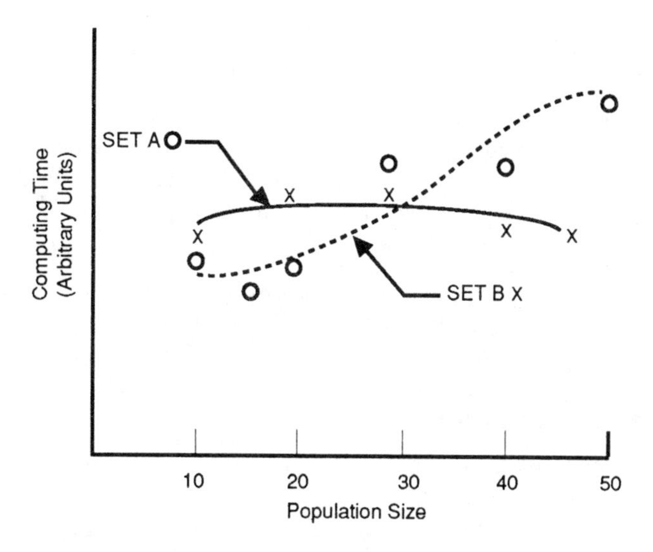

Figure 9: Computing time required to reach good solutions, as a function of population size.

Figure 9 shows the time taken by a successful calculation to locate a good sequence. Population size is of only minor importance for calculations on set A. A small population offers the advantage of a low cycle time, but this is counteracted by the need for a greater number of cycles to convergence than required for a large population. For set B, though, the time to convergence increases as the population increases. For this type of surface, small populations are about as successful as large, but reach their solutions in significantly less time. A small population in the range 15–25 is clearly appropriate for investigation of smooth surfaces of the type generated by this set, while a larger population is appropriate for the more corrugated surface.

We have also argued that different crossover rates might be appropriate for the two sets. Figure 10 shows how success of the calculations is affected by choice of crossover rate. Calculations for set B are largely unaffected by the crossover rate selected. In contrast, the success of set A calculations is strongly dependent upon the choice of crossover rate. This is particularly marked for "near–optimum" solutions. The frequency with which set A calculations find near–optimum solutions falls to almost zero for crossover rates > 0.6, while for set B, the optimum crossover rate seems to be in the region 0.55–0.7.

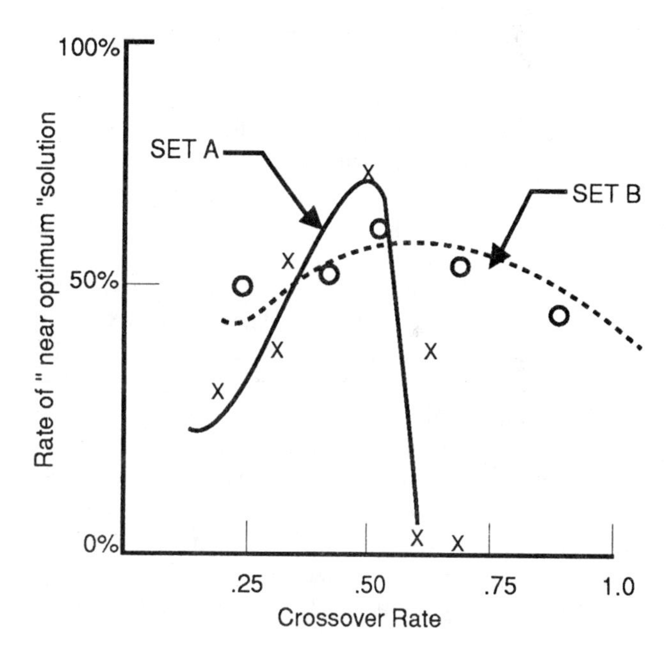

Figure 10: Dependence of the success of the calculation on crossover rate.

It is thus apparent that appropriate population sizes and crossover rates for the two sets are indeed different, and that we may be forewarned of this by a preliminary assessment of the surfaces. Figure 11 shows calculations on sets A and B using parameters suggested by consideration of figures 8–10; the improvement in performance, both in terms of convergence and the quality of the best sequence found over the benchmark calculations shown in figure 4, is apparent.

If preliminary assessment of the nature of the data

space can provide guidance on suitable population sizes and crossover rates, the question arises: how should this assessment be performed?

In view of the size and irregularity of the surfaces, it is clearly prudent to assess the surface around several sequences before drawing conclusions about the nature of the data space.

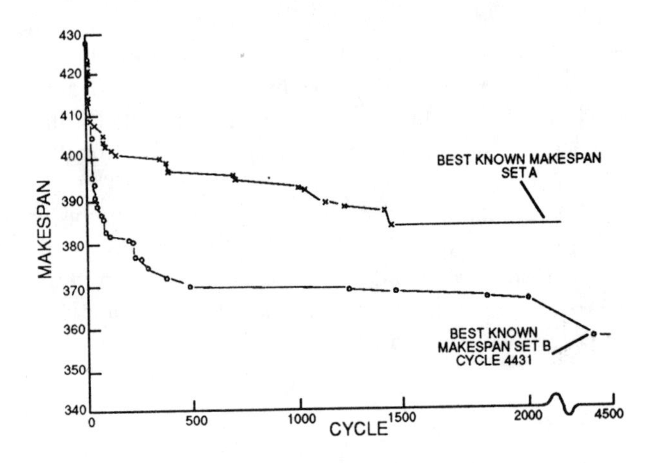

Figure 11: Best makespan found for sets A and B, using parameters suggested by figures 8–10. Set A: 28 strings, crossover rate = 0.5. Set B: 18 strings, crossover rate = 0.7.

Ideally, the GA should itself analyze a selection of surfaces chosen at random from those generated by a new data set, and assign a "crinkle factor" to each, determined by the extent of surface corrugation. The crinkle factor would then be used to set appropriate population size and crossover rate. We have investigated a number of ways in which the GA might do this, most involving some measure of the standard deviation of makespans on the surface from the mean. Such an approach can adequately interpret many surfaces, but no entirely satisfactory statistical measure of corrugation has yet been found for this purpose. In particular, it has proved difficult to categorize surfaces like that shown in figure 6, which occasionally arise from sets generating "smooth" data.

The most reliable means at present of judging the category into which new data fit is to display several surfaces and judge their corrugation by eye. This operator intervention conflicts somewhat with the goal of a completely automatic solution; nevertheless, the potential gains in quality of solution and rate of convergence suggest this is a valuable use of time early in a calculation.

References

[1] Bounds, D.G., 1987. New Optimization Methods from Physics and Biology, *Nature*, **329**, 215 (1987).

[2] Bramlette, M.F., and Cusic, R., 1989. A Comparative Evaluation of Search Methods Applied to the Parametric Design of Aircraft. *Proc. Intl. Conf. on genetic algorithms, 1989.*

[3] Eshelman, L.J., Caruana, R.A., & Schaffer, J.D., 1989. Biases in the Crossover Landscape. Philips Laboratories TN–89–021 (1989), and references therein.

[4] Fogarty, T.C., Varying the Probability of Mutation in the Genetic Algorithm. *Proc. Intl. Conf. on genetic algorithms, 1989.*

[5] Garetti, M., et al 1990. EXS: an Expert System for Detailed Production Scheduling *Autom. Strum.* 38, 115, (1990)

[6] Goldberg, D.E., 1989. *Genetic Algorithms in Search, Optimization and Machine Learning.* Addison–Wesley, Reading Mass, (1989).

[7] Goldberg, D.E., and Lingle, R., 1985, Alleles, loci and the Traveling Salesman Problem. *Proc. Intl. Conf. on genetic algorithms and their applications, 1985.*

[8] Ku, H–m., and Karimi, I.A., 1988, Scheduling in Serial Multiproduct Batch Processes with Finite Interstage Storage: A Mixed Integer Linear Program Formulation. *Ind. Eng. Chem. Res.* 27, 1840, (1988)

[9] Rajagopalan, D., and Karimi, I.A., 1989. Completion Times in Serial Mixed–Storage Multiproduct Processes with Transfer and Set–up Times. *Computers Chem. Engng.,* 13, 175, (1989)

[10] Whitley, D., Starkweather, T., & Fuquay, D'A., 1989. Scheduling Problems and Traveling Salesmen: The Genetic Edge Recombination Operator. *Proc. Intl. Conf. on genetic algorithms. 1989.*

Preventing Premature Convergence in Genetic Algorithms by Preventing Incest

Larry J. Eshelman
Philips Labs
Briarcliff Manor, NY, USA
lje@philabs.philips.com

J. David Schaffer
Philips Labs
Briarcliff Manor, NY, USA
ds1@philabs.philips.com

Abstract

A number of mechanisms have been suggested for combating premature convergence in genetic algorithms. Many of them, however, are subject to the objection that they tend to undermine the algorithm's implicit parallelism — the property that distinguishes genetic algorithms from other search methods. In this paper we examine several of these diversity-maintaining mechanisms, including a new one, *incest prevention*. We argue that by adopting a conservative selection strategy, *elitist selection*, the theoretical objection to these various mechanisms for combating premature convergence no longer applies. Finally, we present the results of an empirical study which suggests that these mechanisms, and especially incest prevention, are effective in preventing premature convergence, thereby significantly improving search.

1 Introduction

Premature convergence — loss of population diversity before optimal or at least satisfactory values have been found — has long been recognized as a serious failure mode for genetic algorithms (GAs). In this paper we examine three strategies for combating premature convergence: (1) a mating strategy, *incest prevention*, (2) a crossover strategy, *uniform crossover*, and (3) a population management strategy, *the weeding out of duplicates*. All these strategies, but especially the first two, are subject to the objection that they make crossover more disruptive of schemata, undermining the GAs implicit parallelism. Empirical evidence, however, seems to contradict GA theory on this point. Ackley [1], Syswerda [18], and Eshelman, et al. [7] have reported that a more disruptive crossover operator, uniform crossover, often outperforms traditional one- and two-point crossover. Syswerda [18] and Whitley [19] have reported improved results using the third strategy. Finally, Eshelman has compared CHC, a GA that uses the first two strategies, with a traditional GA and has reported that CHC outperforms the traditional GA on a suite of 10 test functions, even though the traditional GA used parameter sets tuned to each function whereas CHC used the same parameter set for all functions [8].

Our goal in this paper is to examine and compare these various strategies and to obtain a better understanding of the conditions under which they are effective. In the next section we briefly review several versions of these three strategies that have appeared in the GA literature. In the third section we address the objection that all these strategies, and especially the first two, have the effect, either directly or indirectly, of making crossover more disruptive of schemata. In sections four and five we present the results of an empirical study comparing and contrasting each of the three strategies. Finally, based on the empirical evidence, we make some recommendations for using these various mechanisms.

2 Strategies for Maintaining Population Diversity

Strategies for maintaining population diversity can naturally be grouped according to where they occur in the GA's reproduction-recombination-replacement cycle: (1) how mates are selected, (2) how children are created by recombination, (3) how parents are replaced.

2.1 Mating Strategies

Mates can be selected so as to maintain population diversity. Usually, mating strategies are considered in the context of speciation, where the goal is to prevent radically dissimilar individuals (e.g., classifiers with different functions) from mating [2,20]. Here we have a different goal. All other things being equal, children produced by crossover from diverse parents will tend to be more diverse. Goldberg's *sharing functions*, by reducing the fitness of individuals as a function of how similar they are to other individuals in the population, can be viewed as an indirect mating strategy [9]. Although Goldberg presents sharing functions in the context of speciation, they actually have the effect of inhibiting rather than promoting within-species matings, at least for the dominant "species." Instead of preventing diverse individuals from mating, similar individuals belonging to the largest species are less likely to be selected for reproduction with each other. Eshelman's *incest prevention* mechanism is a more direct approach for preventing similar individuals from mating [8]. Individuals are randomly paired for mating, but are only mated if their Hamming distance is above a certain threshold. The threshold is initially set to the expected average Hamming distance of the initial population, and then is allowed to drop as the population converges. Although this mechanism does not explicitly prevent siblings or near ancestors from mating, insofar as such individuals tend to be similar it has this effect. Mating strategies that pair diverse individuals, have the side effect that more schemata are disrupted by crossover, since fewer schemata are shared.

2.2 Recombination Strategies

The second strategy is to use a crossover operator that helps promote population diversity. To the extent that crossover tends to produce children that are different from both parents, the resulting population will tend to be more diverse. The most straightforward method for making crossover more vigorous is to increase the rate of crossover use. Another method for increasing the crossover rate is to cross over the reduced surrogates — i.e., make crossover operate on the differing bits [3]. So long as the parents differ by at least two bits, the children will be different from their parents. A more radical method for maintaining population diversity is to use a more disruptive crossover operator such as uniform crossover. Uniform crossover, by recombining at the bit level, is much less likely than traditional one- or two-point crossover to produce the same offspring twice from the same parents. Furthermore, uniform crossover tends to crossover close to half the differing bits, and so preserves fewer schemata than one-point or two-point crossover which on average produce children that differ from their nearest parent by only a quarter of the differing bits. CHC uses a crossover operator, HUX, that combines all three mechanisms — half of the differing bits are crossed over at random [8].

2.3 Replacement Strategies

A third strategy for maintaining population diversity is to monitor which individuals are allowed into the population. De Jong explored a crowding scheme in which new individuals are more likely to replace individuals in the parent population that are similar [6]. Syswerda and Whitley have taken a somewhat similar approach — their GAs only add a new individual to the population if it is not identical to any member already in the population [18, 19]. Such replacement strategies, simply by encouraging population diversity, have the side effect that more schemata will be disrupted by crossover, although fewer than in the case of the other strategies.

In the next section we address the issue of whether the disruptive side effects of these various strategies undermine the GA's implicit parallelism.

3 Implicit Parallelism and Schema Disruption

The distinguishing feature of genetic algorithms is that they are able to process many building blocks or schemata in parallel. This property is called implicit parallelism [14]. Typically, implicit parallelism is explicated in terms of the sampling behavior of the GA — i.e., a large number of the schemata in a population will be sampled in future trials in proportion to their fitness ratios (ratio of the average fitness of the chromosomes in the population that are members of the schema to the average fitness of the population), and so the instances in the population of above average schemata will grow exponentially [13]. The burden of this sampling behavior is carried by the selection algorithm. The selection algorithm, however, operates on individuals, whereas implicit parallelism is about the manipulation of schemata. Hence, it is critical that crossover be vigorous enough that schemata are tested in new contexts. Little is learned by increasing the market share of a schema appearing in above average individuals if the new instances appear in individuals that are nearly identical to their parents. On the other hand, implicit parallelism is undermined if the good schemata are destroyed in the process of vigorously testing schemata in new contexts. Implicit parallelism presupposes both that the good schemata will tend to survive crossover and that crossover will vigorously test schemata in new contexts.

Unfortunately, traditional GA theory has placed the burden of preservation as well as vigorous recombination completely upon crossover, forcing a tradeoff between preservation and exploration. This tradeoff, however, is not a serious constraint provided a very large population is used. Given a large population, it is unlikely that an individual will be crossed over with a sibling or a near ancestor, and so it is very likely that the schemata from any parent will be tested in a new context. When genetic algorithms are used for practical applications, however, such as optimization and machine learning and the cost of the algorithm is measured in terms of total evaluations rather than generations, large populations are inefficient. When small populations are used, on the other hand, operators that tend to preserve schemata also tend to lead to rapid convergence, and thus fail to satisfy the requirement that schemata be vigorously tested in new contexts. Premature convergence can be slowed by the various mechanisms described above, but to the extent that they make crossover more disruptive of good schemata, fewer schemata will be effectively processed in parallel.

Fortunately, there is a way out of this dilemma — some of the burden of preserving good schemata can be transferred from crossover to selection. The traditional GA uses two selection procedures: selection for reproduction and selection for replacement (or survival). Whereas selection for reproduction is biased according to fitness, selection for replacement is typically unbiased, i.e., uniform. Depending on the GA, *n* children are created each cycle (generation) where *n* ranges from 1 to the population size. (If the crossover and mutation rates are low enough, some of the *n* children may be identical to their parents.) At the replacement stage the *n* children replace *n* members of the parent population chosen at random. The limiting (and typical) case is when *n* is the population size, and the new population replaces the old population each cycle (i.e., a "generation gap" of 1.0 [12]). Under such a replacement procedure there is no way for a schema to be preserved except by being passed on to the children. This makes it necessary to use a crossover operator that is not very disruptive, often in conjunction with a fairly low rate of crossover. Replacement-selection, however, need not be unbiased. An alternative is a *population-elitist selection* strategy in which only those members of the parent population are replaced that are worse than the *n* offspring created. Although it remains the case that in order for a schema to be propagated (i.e., increase its market share) it must survive crossover, its preservation is independent of crossover. All schemata in surviving parents are preserved. In other words, elitist selection guarantees survival provided the number of new offspring better than this individual is less than this individual's rank in the parent population, counting from the worst. Crossover is no longer both the source of preservation and variation. Crossover can concentrate on variation, letting elitist selection ensure that good schemata are preserved.

The main theoretical objection to a biased replacement-selection policy is that together with a biased reproduction-selection policy there is too much pressure toward convergence [4]. Several researchers, however, have reported improved performance with replacement-selection [17-19]. Furthermore, Eshelman has shown that CHC, a GA that combines elitist replacement-selection with an *unbiased* reproduction-selection strategy and a very vigorous (disruptive) crossover operator, significantly outperforms a tradition GA over a suite of 10 test functions [8]. CHC is able to moderate selection pressure in two ways. First, it eliminates the traditional selection bias for reproduction, and relies only upon the fitness-bias of replacement-selection. Eshelman has argued that fitness proportional reproduction does occur with elitist selection, but over generations rather than at every generation [8].

Secondly, since the burden of preserving schemata no longer completely rests upon crossover, CHC is able to use mating and recombination strategies that help maintain diversity. In the remainder of this paper we shall attempt to tease out the contributions of several of the mechanisms that enable CHC to perform so well.

4 Experimental Design

Our goal is to test the effect of each of the three strategies both in the presence and the absence of the other strategies across a wide range of functions. We have chosen a diversity-maintaining mechanism from each of the three categories: (1) incest prevention versus no incest prevention (2) HUX versus two-point crossover, and (3) duplicate checking versus no duplicate checking. Various combinations of these mechanisms yield eight different algorithms (2×2×2). We tested each algorithm on thirteen functions using 50 replications. An analysis of variance (ANOVA) for each function was used to identify and quantify the influences of these factors.

4.1 The Algorithms Tested

All algorithms tested were variants of the CHC algorithm (see [8] for a detailed description). The following procedures were used in all variants: (1) Selection for reproduction was unbiased (i.e., one copy of each individual in the parent population was selected for mating). (2) Selection for replacement was elitist (i.e., the members of the parent and child populations were sorted and worst members of the parent population were replaced by better members (if any) of the child population). (3) There was no mutation during the recombination cycles; instead, when the search stopped making progress (three generations with no new members accepted into the population), the best member found so far was used as a template to reinitialize the population, by vigorous mutation, and the search continued. For the test problems described in § 4.5, vigorous mutation meant flipping 35% of the bits, with one copy of the best individual left unchanged. If, however, a total of three such reinitializations resulted in no improvements, the population was replaced with random strings and the search continued. The search was not halted until the global optimum was found or a maximum of 50000 evaluations. In all cases a population size of 50 was used.

Something needs to be said concerning the rationale for reinitialization. The various mechanisms that we are exploring affect how rapidly the GA converges. Once an algorithm has stopped making progress (usually because its population is highly converged), it seems a waste of time to let it continue churning away. By restarting an algorithm that converges faster than another algorithm, we give it a chance to effectively compete.

As Goldberg points out, "simply taking small population sizes and running them to convergence and letting them wiggle around due to mutation is not going to be very useful. On the other hand, if we restart populations when they converge ... we may be able to keep the rate of useful schema processing high" [11, p. 13]. The algorithms in this study allow both partial and full restarts. As long as improvements are being made, the algorithm only partially randomizes the population. After several failures to improve upon the best individual found so far, however, the algorithm does a full restart, completely randomizing the population. This process continues until the optimum is found or the maximum number of evaluations is reached.

4.2 Incest Prevention

Two parents are mated to produce two offspring via crossover only if their Hamming distance is above a threshold. Initially and at the beginning of a restart the threshold is set to the expected Hamming distance of any two individuals in the initial or reinitialized population. Whenever there is a generation with no individuals accepted into the parent population, the difference threshold is decremented. (The "no progress" criterion for reinitialization discussed above only applies once the difference threshold has dropped to zero.) For those algorithms without incest prevention, the potential parents produce two offspring provided they differ by at least two bits. Note that for all versions the number of children produced each generation can vary from zero up to the population size (50).

4.3 Crossover

Two-point crossover (2X) operates on the reduced surrogate — a randomly chosen segment of the differing bits is swapped to produce two children. HUX swaps half the differing bits chosen at random to produce two children.

4.4 Duplicate checking

Once a child is produced by crossover, it is matched against all the members of the parent population. If a duplicate is found, the child is discarded. Otherwise, the child is evaluated and included in the child population of potential candidates for replacing members in the parent population. (Note that duplicates are still possible, since no check is made to ensure that there are no duplicates in the child population.)

4.5 The Functions Tested

The thirteen test functions included F1-F10, previously studied by Schaffer, et al. [8,16]. In all cases a bit-string representation was used with Gray coded parameters. Functions F1-F5 are the five functions studied by De Jong [6]. Functions F6-F10 include two sine-wave-based functions, a FIR digital filter optimization task, a 30-city traveling salesman problem [15] with a sort order representation, and a 64-node graph partition task [1]. (In order to make functions F6 and F7 hillclimbing-hard the variables have been shifted a small amount so that the optimum is no longer at the origin [5].) Function F11 consists of 20, maximally epistatic, 5-bit segments — each segment scores a one for the "00000" pattern and zero for all others. The score for a string is 20 minus the sum of the segment scores (where the optimum is the minimum score). Function F12, like F11, consists of 20 five-bit segments, but each segment is deceptive (the average fitnesses of lower order schemata lead away from the optimum [10]). Each segment scores a one for the "00000" pattern and one-tenth for each "1". Again the score for a string is 20 minus the sum of the segment scores. The last function, F12-d, is a distributed version of F12. It is the same as F12 except that the bits defining a segment are maximally distributed instead of adjacent. That is, the bits of segment 1 are at loci 1, 21, 41, 61, and 81, segment 2 at 2, 22, 42, 62, and 82, etc.

Each of the eight algorithms was tested on each of twelve functions, F1-F12. Only the four algorithms using 2X, however, were tested on F12-d since the performance of HUX, unlike 2X, is not affected by the location of the bits. Each search was continued until either the optimum was found or until a maximum of 50000 evaluations was reached, and was replicated 50 times.

The functions were chosen so as to vary in difficulty from functions like F1, F3 and F5 that are relatively easy for an iterative hillclimber to optimize to functions like F8, F9, and F12-d that are not only nearly impossible for an iterative hillclimber to optimize, but are also challenging for a traditional GA (proportional selection with one- or two-point crossover). The difficulty of these latter functions is illustrated by the fact that we were not able to find any combination of parameter settings for a traditional GA so that for any of these functions it could consistently find (and in the case of F12-d, even once find) the global optimum in 50 runs of 50000 evaluations. (For each of functions F1-F10 we tested 840 different combinations of parameter settings [8,16] Our search for optimal parameter settings for F11, F12 and F12-d has not been so exhaustive, but we have tested at least 20 different combinations of parameter settings for each of these functions.)

5 Experimental Results

Table 1 summaries the results of the experiments in terms of the mean number of evaluations to find the global optimum. The only exceptions are F4 and F8. In the case of F4, the search was halted whenever the performance value before adding noise was less than two standard deviations of the noise function from the optimum. In the case of F8, a FIR digital filter optimization task, the global optimum is not known so the best value ever found was used as a halting criterion. For functions F6, F8-F12, and F12-d some of the means are underestimated because the search reached 50000 evaluations before the global optimum was found. (The search is halted at the end of the generation in which 50000 evaluations is reached. This explains why some of the means in Table 1 are somewhat greater than 50000.) Since HUX's performance is unaffected by bit position, the HUX values for F12 have also been used for F12-d. Table 2 shows the number of searches out of 50 that found the global optimum for those functions for which not every version of the algorithm consistently found the global optimum. (F1-F5 and F7 are not shown since all eight algorithms always found the global optimum.) Table 3 shows the mean performance values for the three functions, F9, F12, and F12-d, for which the searches rarely found the global optimum.

Table 1: Mean Number of Evaluations to Find the Global Optimum

Func	HUX		2X	
	No Inc	Inc	No Inc	Inc
	Duplicates			
F1	1089	2005	1189	1948
F2	9065	9937	9681	9103
F3	1169	1725	1135	1711
F4	1948†	2674†	5037	5179
F5	1396	2512	2207	3453
F6	6348	12261	15827	23331
F7	3462	11645	10737	14838
F8	7279	39295	15154	44046
F9	36567	50030	49576	50024
F10	16591	45164	29983	48502
F11	3827	13287	4776	19034
F12	46037	50023	35889	50022
F12-d	46037	50023	50009	50025
	No Duplicates			
F1	1104	1588	1214	1836
F2	8717	7928	8376	9066
F3	1169	1665	1277	1572
F4	1948†	2674†	4803	5256
F5	1268	2342	1884	3147
F6	6603	12018	15714	18245
F7	3436	10047	9420	13323
F8	7172	26349	15173	41369
F9	37816	49844	49670	50022
F10	14467	42467	28495	47425
F11	3633	8586	5434	11924
F12	45127	50025	33314	50023
F12-d	45127	50025	50008	50026

† No duplicates were encountered. Since all runs began with the same random seeds, these runs replicated exactly.

Table 2: Number of Runs out of 50 that Found the Global Optimum

Func	HUX		2X	
	No Inc	Inc	No Inc	Inc
	Duplicates			
F6	50	50	49	47
F8	50	25	50	13
F9	27	0	1	0
F10	50	13	37	4
F11	50	49	50	45
F12	12	0	29	0
F12-d	12	0	0	0
	No Duplicates			
F6	50	50	49	48
F8	50	44	50	17
F9	24	1	1	0
F10	50	15	40	8
F11	50	50	50	50
F12	14	0	30	0
F12-d	14	0	0	0

Table 3: Mean Performance

Func	HUX		2X	
	No Inc	Inc	No Inc	Inc
	Duplicates			
F9	425.4	454.6	448.4	486.4
F12	1.16	8.82	0.48	3.41
F12-d	1.16	8.82	6.69	11.52
	No Duplicates			
F9	427.5	451.8	451.7	491.1
F12	1.08	9.05	0.42	3.41
F12-d	1.08	9.05	6.95	11.13

It is obvious after a quick perusal of these three tables that HUX combined with incest prevention is, generally, a powerful algorithm, whether or not the mechanism for preventing duplicates is used. If incest is allowed, on the other hand, the prevention of duplicates becomes an important factor. A function by function analysis of variance (ANOVA) bears this out. Table 4 shows the main and interaction effects for the ANOVAs. (There were no significant interaction effects between duplicate checking and crossover nor among all three factors, so these are not included in the table.) The ANOVAs used as the measure of performance the number of trials to the optimum value plus an additional factor based on the best function value found to compensate for those cases where the optimum was not found before the maximum number of trials was reached:

$$num\_evals + (perf - opt) \times \frac{max\_evals - min\_evals}{max\_perf - min\_perf}$$

As is indicated by the column headed "incest," incest prevention has a statistically significant effect for twelve of the thirteen functions. In all cases incest prevention improved performance. The choice of the crossover operator made a significant difference for ten of the functions. Among these ten functions, HUX significantly outperformed 2X except on F12, the deceptive function with an adjacent representation. When the bits were distributed (F12-d), however, 2X performed significantly worse than HUX. Checking for duplicates has a significant effect for four functions — in all cases improving performance. In three of these cases there is a significant interaction with the incest prevention factor. In all three cases the improvement is only significant when there is no incest prevention. Function F5 is the exception. Duplicate checking helps with and without incest prevention. It helps somewhat more when there is incest prevention, but not enough to cause a significant interaction effect.

Table 4: ANOVA - Levels of Significance†

func	incest	x-over	dupes	incest x-over	incest dupes
F1	***		**		**
F2					
F3	***				
F4	*	***			
F5	***	***	*		
F6	***	***			
F7	***	***		**	
F8	***	***	***	***	***
F9	***	***		***	
F10	***	***		**	
F11	***	***	**	**	**
F12	***	***		**	
F12-d	***	***		**	

† '*', '**', and '***' indicate the 0.05, 0.01, and 0.001 significance levels, respectively.

The only other significant interactions indicated by the ANOVAs are between the choice of crossover and incest prevention. Of the seven functions for which there is this interaction, in two cases F8 and F11, incest prevention helps 2X more than it does HUX. The effect is the opposite for the other five functions. We have not been able to discover anything special about these functions that would account for these opposite interaction effects. It should also be noted that although these interaction effects are statistically significant, they are dominated by the main effects in terms of magnitude.

Our working hypothesis has been that the three mechanisms being tested slow convergence, and hence improve performance. This hypothesis is supported by the experimental results. The impact of the three factors is correlated with how much they slow convergence. Runs with incest prevention, the most effective factor, take twice as long before they converge — they have 2.19 times as many evaluations per restart as those that allow incest. Runs using HUX, the next most effective factor, have 1.33 times as many evaluations per restart as those using 2X. Runs using duplicate checking, the least effective factor, have only 1.06 times as many evaluations per restart as those without duplicate checking.

Function F2 (Rosenbrock's Saddle) requires some comment given that search was not significantly affected by the three mechanisms examined in this study. This function has some interesting properties that deserve further study. Although the surface defined by F2 is smooth and unimodal, in the vicinity of the global optimum it forms a shallow "ravine" that runs diagonal to the axes defining the parameters. This means that any string representing a point in the "ravine" can only be improved by changing both parameters simultaneously. Furthermore, because of the discrete sampling of the search space by the binary representation, local minima are introduced in this region, so that improvements sometimes lead search away from the global optimum. All the algorithms have no trouble quickly finding the region of the search space where the global optimum resides, but they all tend to get trapped by local minima in this region and typically require several restarts before they are able to find the global optimum. We conjecture that the schemata have no consistent differential value in this region and that all the algorithms find the global optimum by random search, through repeated restarts.

Finally, since none of the algorithms compared in this study is a traditional GA (using proportional selection and one- or two-point crossover), something should be said about how the traditional GA does in comparison to the best of the algorithms discussed in this study (the standard version of CHC using HUX and incest prevention). The answer is, not very well. For ten of the thirteen functions we were not able to find any combination of parameter settings so that the GA didn't perform significantly worse than CHC, even though we always used the same parameter settings for CHC for all thirteen functions (a population size of 50 and a divergence rate of 0.35). We were able to find parameter settings for only two functions, F1 and F12, such that the traditional GA outperformed CHC. For F2 we found parameter settings so that the traditional GA did about as well as CHC. Given that the parameters that enabled the GA to win for F1 in effect turned it into a hillclimber (population size of 10, no crossover, 0.023 mutation rate), the only significant win for the GA is F12, a deceptive function. By using a large population size (greater than 200), no mutation, and a crossover rate of at least 0.5, the traditional GA will do better than CHC (with its standard parameter settings). The GAs advantage, however, depends upon the deceptive schemata being of short defining length. When the deceptive schemata are distributed as in F12-d, the GA never finds the optimal solution whereas CHC finds it about 25% of the time. In brief, the only difficult functions (hillclimbing hard) that we have been able to find for which the traditional GA has an advantage over CHC are such benign deceptive problems as F12.

6 Conclusions

The results of this study indicate that incest prevention is always a good idea. The results also indicate that using a more disruptive crossover operator like HUX is usually a good idea. The exception to the rule is when the problem is deceptive and the deceptive schemata have a short defining length rather than being interspersed as in F12-d. Otherwise, HUX is a better crossover operator, even for deceptive problems. Duplicate checking is a good strategy if incest prevention is not used, but as an alternative to incest prevention, it is not as effective. Furthermore, incest prevention has a much lower overhead than duplicate checking. Whereas duplicate prevention is a global operation requiring that each potential child be compared to every member of the parent population, incest prevention is a local operation requiring that the two parents be compared in relation to a threshold. The payoff for duplicate checking in conjunction with incest prevention is marginal, but does not seem to hurt (discounting the extra overhead). This is not surprising. The number of duplicates discovered drops radically when incest prevention is used.

It should be kept in mind that all the strategies tested were used in conjunction with a fitness-biased replacement strategy — elitist selection. We can not infer that they would be as effective with a traditional GA. Furthermore, since incest prevention and duplicate checking have the effect of preventing some parents from having any offspring in a given generation, the traditional procedure of replacing the parent population with the children each generation needs to be modified in order to adopt either of these strategies.

A number of researchers have independently discovered the benefits of an biased replacement-selection strategy [1, 8, 17-19]. Typically, this has been in conjunction with a nongenerational GA that produces one or two offspring per cycle, instead of a completely new population. Syswerda and Whitley have both interpreted the resulting improvements as an outcome of the nongenerational aspect. We suggest, however, that the resulting improvements stem from the fitness-biased replacement-selection strategy, which, because of its conservative nature, allows the algorithm to use mechanisms for maintaining diversity, including higher crossover rates, crossing over of reduced surrogates, uniform crossover, duplicate checking, and incest prevention, without sacrificing implicit parallelism.

References

[1] D. H. Ackley. *A Connectionist Machine for Genetic Hillclimbing*. Kluwer Academic Publishers, Boston, MA, 1987.

[2] L. B. Booker. Intelligent Behavior as an Adaptation to the Task Environment. Ph.D. Dissertation, Department Computer and Communication Sciences, University of Michigan, Ann Arbor, MI, February 1982.

[3] L. Booker. Improving Search in Genetic Algorithms. In L. Davis (editor), *Genetic Algorithms and Simulated Annealing*, 61-73. Morgan Kaufmann, San Mateo, CA, 1987.

[4] A. Brindle. Genetic Algorithms for Function Optimization. Ph.D. Dissertation, Computer Science Department, University of Alberta, 1981.

[5] L. Davis. Bit-Climbing, Representational Bias, and Test Suite Design. *Proceedings of the Fourth International Conference on Genetic Algorithms.* Morgan Kaufmann, San Mateo, CA, 1991.

[6] K. A. De Jong. Analysis of the Behavior of a Class of Genetic Adaptive Systems. Ph.D. Dissertation, Department of Computer and Communication Sciences, University of Michigan, Ann Arbor, MI, 1975.

[7] L. J. Eshelman, R. A. Caruana and J. D. Schaffer. Biases in the Crossover Landscape. *Proceedings of the Third International Conference on Genetic Algorithms*, 10-19. Morgan Kaufmann, San Mateo, CA, 1989.

[8] L. J. Eshelman. The CHC Adaptive Search Algorithm: How to Have Safe Search When Engaging in Nontraditional Genetic Recombination. In G. J. E. Rawlins (editor), *Foundations of Genetic Algorithms and Classifier Systems.* Morgan Kaufmann, San Mateo, CA, 1991.

[9] D. E. Goldberg and J. Richardson. Genetic Algorithms with Sharing for Multimodal Function Optimization. *Genetic Algorithms and Their Applications: Proceedings of the Second International Conference on Genetic Algorithms*, 41-49. Lawrence Erlbaum Associates, Hillsdale, NJ, 1987.

[10] D. E. Goldberg. Simple Genetic Algorithms and

the Minimal, Deceptive Problem. In L. Davis (editor), *Genetic Algorithms and Simulated Annealing*, 74-88. Morgan Kaufmann, San Mateo, CA. 1987.

[11] D. E. Goldberg. Sizing Populations for Serial and Parallel Genetic Algorithms. TCGA Report Number 88004, Department Engineering Mechanics, University of Alabama, Tuscaloosa, AL, August 1988.

[12] J. J. Grefenstette. Optimization of Control Parameters for Genetic Algorithms. *IEEE Transactions on Systems, Man & Cybernetics SMC-16*, 1, (January-February 1986), 122-128.

[13] J. J. Grefenstette and J. E. Baker. How Genetic Algorithms Work: A Critical Look at Implicit Parallelism. *Proceedings of the Third International Conference on Genetic Algorithms*, 20-27. Morgan Kaufmann, San Mateo, CA, 1989.

[14] J. H. Holland. *Adaptation in Natural and Artificial Systems*. University of Michigan Press, Ann Arbor, MI, 1975.

[15] I. M. Oliver, D. J. Smith and J. R. C. Holland. A Study of Permutation Crossover Operators on the Traveling Salesman Problem. *Genetic Algorithms and Their Applications: Proceedings of the Second International Conference on Genetic Algorithms*, 224-230. Lawrence Erlbaum Associates, Hillsdale, NJ, 1987.

[16] J. D. Schaffer, R.A. Caruana, L. J. Eshelman and R. Das. A Study of Control Parameters Affecting Online Performance of Genetic Algorithms for Function Optimization. *Proceedings of the Third International Conference on Genetic Algorithms*, 51-60. Morgan Kaufmann, San Mateo, CA, 1989.

[17] J. D. Schaffer, L. J. Eshelman and D. Offutt. Spurious Correlations and Premature Convergence in Genetic Algorithms. In G. J. E. Rawlins (editor), *Foundations of Genetic Algorithms and Classifier Systems*. Morgan Kaufmann, San Mateo, CA, 1991.

[18] G. Syswerda. Uniform Crossover in Genetic Algorithms. *Proceedings of the Third International Conference on Genetic Algorithms*, 2-9. Morgan Kaufmann, San Mateo, CA, 1989.

[19] D. Whitley. The GENITOR Algorithm and Selection Pressure: Why Rank-Based Allocation of Reproductive Trials is Best. *Proceedings of the Third International Conference on Genetic Algorithms*, 116-121. Morgan Kaufmann, San Mateo, CA, 1989.

[20] S.W. Wilson. Classifier System Learning of a Boolean Function. RIS 27r, The Rowland Institute for Science, Cambridge, MA , February 9, 1986.

Inserting Introns Improves Genetic Algorithm Success Rate: Taking a Cue from Biology.

James R. Levenick
Computer Science Department
Willamette University
Salem, OR 97301
email - willamu!levenick@uunet.uu.net

Abstract

Molecular evolution may provide inspiration for genetic algorithm techniques. Molecular biologists have discovered non-functional sequences of DNA, called introns. This paper reports experiments demonstrating that the insertion of introns into bit strings can lead to dramatic increases in success rates of artificial evolution.

1 INTRODUCTION

Investigating the mechanisms of natural systems is a central activity of science. An understanding of a natural system's information handling can lead to more effective artificial methods. Molecular evolution is the biological counterpart of genetic algorithm (GA) research. Insights from genetics may suggest enhancements to GA techniques. In the best case GA research may offer insights into evolutionary biology in return. This paper presents evidence that inserting introns can improve the performance of a GA by a factor of ten. These results may shed light on why introns exist. Perhaps part of their function (in biological genetic material) is to make crossover more effective.

1.1 BACKGROUND

Two papers at the 1989 GA conference seemed to imply that GAs worked efficiently and well with small populations [4, 8]. There was also considerable discussion of the fact that GAs, although performing better than enumeration, did little better than iterated hill-climbing algorithms on both De Jong's test suite and several classes of Walsh polynomials [10, pages 68-69]. This prompted me to experiment with evaluation functions that were unlikely to be optimized by either small populations or iterated hill climbing.

One of these functions, containing five regions, had a single optimum and multiple plateaus. It had a total of thirty bits, so the probability of locating the optimum at random was slightly worse than one in a billion. Small populations seldom succeeded in optimizing this function (see Figure 3), and iterated hill-climbing succeeded less than one time in a thousand. The strange results mentioned above [8, 10] appear to have stemmed from using evaluation functions that were simply too easy; or in the case of the Walsh polynomials, pathological [3]. Disturbingly, the GA optimized this function less than one time in ten (on average) even with a population size of 1024. Analysis revealed that although all five regions typically existed in the initial population, they were often destroyed during crossover.

A method to ameliorate this problem was inspired by a Steven Gould column. Gould [5] discusses selection as a conservative force (as opposed to an agent for change). Among the evidence for this view is that the mutation rate of exons (regions of chromosomes that are expressed as proteins) exceeds the mutation rate of introns (regions that are not expressed, or intragenic regions). This is because mutations in exons tend to be selected against (since they are generally maladaptive) whereas mutations in introns are unaffected by selection.

Recently Monte Carlo simulation has been employed to estimate that all present-day genetic material stemmed from a pool of between 1000 and 7000 seminal exons [6]. That the tremendous diversity we see in the living world could have evolved from so few building blocks may encourage GA researchers to seek more potent operators. If all life on earth sprang from fewer than 10,000 exons, surely a few dozen artificial exons could generate interesting systems if created judiciously and then combined and modified appropriately.

1.2 ARTIFICIAL EVOLUTION

Assume that we observe a flying creature and seek to study the behavior of the underlying genetic operators that evolved the DNA that built this organism. Assume further that the phenotypic characteristics that impart the ability to fly in this creature are controlled by just five exons (this is a gross oversimplification, but may provide some insight into the interplay of crossover and introns). The strings in a GA may be thought of as encoding information about how to build individuals of that species; the various functional segments of a string (the exons) then code for different attributes of an individual.

As in the evaluation function discussed above, let the length of each exon region be six binary locations (again this is a tremendous oversimplification, real domains may include thousands of sites or base pairs). Call the exons A, B, C, D, and E. A string with appropriate values at every location in a particular exon will be referred to as having that gene; thus if a string has the correct values for attribute A at all six locations in that exon, it will be said to have gene A.

Having half a wing seems less adaptive than having none; and the chances of a whole wing evolving in one generation are prohibitively small. Thus, it may be difficult to understand how birds could have ever evolved "by random chance". This leads some people to mistakenly conclude that complex systems must have had a creator [1]. The following example addresses the puzzle of how flying creatures could have evolved, in a greatly simplified form. Imagine the following: A is a gene for elongated finger bones (advantageous for extracting food from crevices), B is a gene for feathers (for warmth), C is for skin flaps between fingers (again for warmth - widespread fingers with skin-flaps could be used to better cover young), D is for hind legs capable of a few good jumps (for catching prey), and E is for clutching feet (for holding prey).

The existence of any of these genes is somewhat advantageous, the coexistence of genes D and E is more advantageous (can jump from branch to branch and hang on), genes for A, B, and C together is even more advantageous (can wrap self or young in a feathered cover; can flap and so jump farther), and all five attributes together allows the creature to fly! This last case, where all five genes are present, is the solution for which the GA will be searching in the following experiments.

2 EXPERIMENTS

Introns commonly interrupt the encoding of a single protein [2] (thus enabling the production of variant proteins by a change of a few base pairs); the same effect occurs at the molar level. These experiments investigate whether inserting sequences of nonfunctional locations (introns) between functional regions (exons) affects the success rate of a GA. The GA will attempt to discover a string with all five genes (and thus evolve an organism that can fly). An evaluation function (called Mu, illustrated in Figure 1) was implemented to reflect the situation described above.

mu values	Genes occurring
mu0=1	None
mu1=5	A or B or C or D or E
mu2=50	D and E
mu3=100	A and B and C
mu5=1000	A and B and C and D and E

Figure 1: Evaluation Function Mu.

2.1 EXPERIMENT 1

A GA program was configured and run 50 times on each of 20 parameter settings both with and without the insertion of introns into the strings (see Figure 2).

S1→aaaaaaiiiiiibbbbbbiiiiiicccccciiiiiiddddddiiiiiieeeeee
S2→aaaaaabbbbbbccccccddddddeeeeeeiiiiiiiiiiiiiiiiiiiiiiii

Figure 2: Five exon regions, A, B, C, D, and E (each of 6 bits, denoted aaaaaa, etc) with (S1) and without (S2) intron regions between the exons.

Four different population sizes (16, 64, 256, 1024), and five rates of reproduction (10, 30, 50, 70, and 90 percent of the population per generation) were tested. Each new individual was created as a result of crossover and mutation. Selection for reproduction was without replacement (i.e., no individual may appear more than once in the reproductive pool; runs with replacement were not substantially different with these parameter settings) and probabalistic on the basis of mu. The probability of mutation was 0.003 per bit. Replacement of old individuals was by lowest mu. String length was 60 bits. Initial populations were random (every bit has probability 0.5 of being a 1).

Each run was continued until either 20000 mu evaluations were performed, or some individual had been evolved that could fly (i.e., had all five genes). Runs where flying creatures evolved were counted as successes, others as failures. The number of successes in each series (with and without introns) are shown in Figure 3.

Several things are clear from these results: 1) The

	Population size							
	16		64		256		1024	
Repro.	introns?		introns?		introns?		introns?	
Rate	no	yes	no	yes	no	yes	no	yes
10%	0	0	0	2	3	22	4	48
30%	0	0	0	2	2	16	4	28
50%	0	0	0	1	1	9	3	23
70%	0	0	0	0	0	8	1	11
90%	0	0	0	1	0	4	1	12

Figure 3: Successful evolutions of a flying creature (ABCDE) in 50 attempts, using the evaluation function in Figure 1. Selection is without replacement.

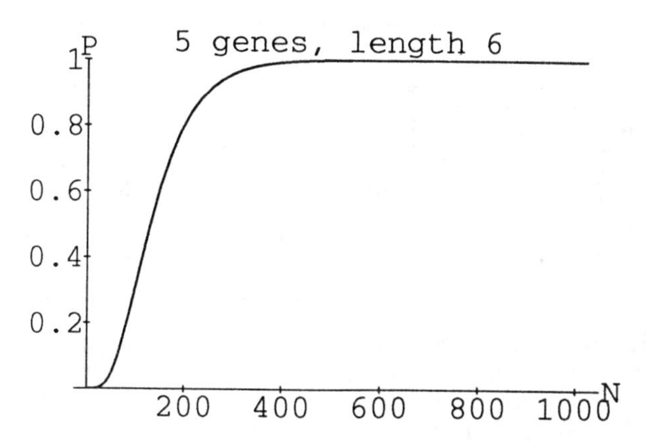

Figure 4: Population (N) versus the probability (P) that there is at least one instance of each of 5 different 6 bit genes in randomly generated populations.

	Population size 256				Population size 1024			
	replacement yes		replacement no		replacement yes		replacement no	
Repro. Rate	introns? no	introns? yes	introns? no	introns? yes	introns? no	introns? yes	introns? no	introns? yes
1%	0	0	0	6	1	8	1	12
3%	0	0	0	12	1	8	3	14
5%	1	1	0	7	2	16	0	21
10%	0	1	0	12	3	17	2	27
30%	0	0	0	30	2	23	5	48
50%	0	1	0	32	4	31	9	47
70%	0	0	2	38	5	37	6	50
90%	0	4	10	37	5	38	13	46

Figure 5: Successful evolutions of a flying creature (with the five genes arranged ADBEC) in 50 attempts. The evaluation function from Figure 1 is used.

population size is directly related to the number of successful runs. This would be true whether the implementation is serial or parallel (see [4]). 2) The rate of reproduction is inversely related to the number of successful runs. 3) The introduction of introns results in a nearly tenfold increase (187:19) in the number of successful evolutions of individuals with all five genes. These results are explained as follows.

1) Two things must happen to evolve a flying creature: i) all five genes must occur in the population, and, ii) they must be combined into a single individual. The latter is highly unlikely without the former. It is very improbable that a randomly generated initial population of size sixteen contains even one representative of each gene (the exact probability is 0.00055). This probability increases non-linearly with the size of the population; with population sizes of 64, 256, and 1024, the values are 0.103256, 0.91436, and 0.9999995, respectively (see Figure 4).

2) Even if all five genes occur in the initial population, there is no assurance that the GA will fortuitously combine them. Premature convergence is possible to either the gene triple, ABC, or the gene pair, DE. Lower reproduction rates help to avoid premature convergence. High rates of reproduction tend to eliminate strings containing only one of the five genes; these may be required to complete a five gene string. Lower reproduction rates tend to preserve copies of such strings longer.

3) The insertion of introns increases the number of crossover points that preserve genes. Without introns there are four crossover points that do not break genes, with seven bit introns between exons there are 32. This simple mechanism may provide part of the explanation of the existence of introns in real DNA. However, all locations are *not* equiprobable for crossing in DNA; various preferred locations in particular segments of DNA, called "hot spots", have been identified [9, 7].

The precise conditions that allow crossover are yet to be established.

2.2 EXPERIMENT 2: COMPLICATIONS

There is no reason why the five genes for flying must be arranged in the pattern, ABCDE; any permutation is equally likely. A more difficult problem is posed if the arrangement is ADBEC, since then it is less likely that an individual which can fly will be created as the result of crossing an individual with gene pair DE and another with gene triple ABC. Experiment 2 confirmed this expectation; the number of successes in each series (in each of four conditions) are shown in Figure 5.

Unlike Experiment 1, in Experiment 2 there was a large performance difference between the algorithms using selection for reproduction with and without replacement. In only one of 32 trials the GA with replacement during selection for reproduction performed better than the GA using selection without replacement. Overall there was a 9:1 advantage in the number of successful evolutions when selection was performed without replacement rather then with replacement. This is attributable to premature convergence to DE; the value for DE was high enough that it tended to proliferate and so crowd out the genes that might have combined to form ABC. Without replacement no string can produce more than one offspring per generation; this caused the population to remain heterogeneous longer and so lead to many more successes.

Reducing the ratio of mu2 to mu1 should slow the proliferation of high value strings and so lead to more successes. Experiment 2 was repeated using mu1=1.01, mu2=5 and mu3=10; thus the ratio of mu2 to mu1 is reduced from 10:1 to 5:1 and the ratio of mu2 to mu0 was reduced from 50:1 to 5:1. The predicted behavior was observed; runs without replacement succeeded less than twice as often as those with replacement. The

	Population size							
	256				1024			
	replacement				replacement			
	yes		no		yes		no	
Repro.	introns?		introns?		introns?		introns?	
Rate	no	yes	no	yes	no	yes	no	yes
10%	0	9	3	6	1	23	0	24
30%	0	12	4	18	1	27	4	40
50%	0	12	2	17	1	27	4	44
70%	1	10	0	31	2	26	12	36
90%	0	6	4	32	6	17	6	28

Figure 6: Successes with the five genes rearranged in the sequence ADBEC, using the evaluation function values mu0=1, mul=1.01, mu2=5, mu3=10.

success rates for GAs with introns show an advantage of approximately 9:1 over those without (see Figure 6).

2.2.1 Artificial and natural evolution

The increase in success rates with the introduction of introns is dependant on the evaluation function. With mu3=4, mu2=2, and mul=1.01, the strings with introns display an advantage of about 4:1; and with mul=25, mu2=50, mu3=100 the ratio decreases to about 3:1. With the combination mul=1.01, mu2=50 and mu3=100, successful evolution rarely occurred; only 11 of 1000 runs resulted in flying creatures with introns and only 4 of a 1000 without.

Do these results imply that a larger disparity between values of individual genes and their combinations causes it to be *more* difficult to evolve individuals with those combinations? I.e., does any great advantage conferred by having particular combinations of genes make larger, more advantageous combinations including those combinations more difficult to achieve? Or is this result an artifact of the operation of the artificial GA?

In a real population in the real world the fortuitous combination of complementary genes may enlarge an organism's niche, thus creating the opportunity to increase the population of that species. Admittedly, if some segment of the population has the gene triple ABC (and so is cold adapted), if the climate suddenly becomes much colder, only individuals with ABC may survive, thus precluding the possibility of evolving a flyer (at least in the short run); but, this is not a necessary result of ABC's high value. In other circumstances, ABC's high value could be instrumental in evolving ABCDE, or ADBEC. This suggests two hypotheses about GAs that will not be pursued here: i) constant population size is a constraint that might be beneficially relaxed, and ii) constant evaluation functions skew results relative to natural evolution.

	Population size							
	16		64		256		1024	
Repro.	introns?		introns?		introns?		introns?	
Rate	no	yes	no	yes	no	yes	no	yes
10%	0	0	0	0	0	10	2	21
30%	0	0	0	0	1	12	3	23
50%	0	0	0	3	1	13	1	29
70%	0	0	0	1	3	23	2	33
90%	0	0	0	0	2	12	0	12

Figure 7: Successes in evolving ADBEC, using the evaluation function values mu0=1, mul=1.01, mu2=1, mu3=2 (with replacement). Strings with DE had the same value as those with none of the desired genes.

2.3 EXPERIMENT 3: ANOMALOUS RESULTS?

In experimenting with different evaluation functions, I accidently set mul=1.01, mu2=1, and mu3=2, and went home for the night. Since there would be no advantage for DE, I expected no (ADBEC) flying creatures to evolve. To my surprise there were a fair number of successes (see Figure 7); in fact the results were very similar to those in Experiment 2 with replacement (see Figure 6). How can this be explained? As usual, apparently anomalous results indicate unrevealed pattern.

In Experiment 2 the five genes were reordered from ABCDE to ADBEC. Not only did this make a single cross between ABC and DE incapable of creating ADBEC, but no two crosses could accomplish it. Consider a case where X=A*B*C and Y=*D*E*, here * indicates something other than the correct bit values for the gene that might appear at that position. Crossing X and Y, we might obtain: AD*E* and **B*C, A**E* and *DB*C, A*BE* and *D**C, A*B** and *D*EC, or A*B*C and *D*E*, depending on where the cross occurred. None of these pairs can be crossed (either with each other, or with X or Y) to produce ADBEC. Because strings in *D*E* are not particularly helpful in evolving ADBEC, the algorithm is not hurt by considering a string in *D*E* as less fit than one with only D or E. On the other hand, the algorithm is not impacted by giving *D*E* a slightly higher value, since such strings contain two useful genes. Even though no single step (or even two step) recombination of X and Y can produce a flying creature, the repeated action of crossover generates a pool of individuals with fragments of the ADBEC string, which can eventually be assembled.

3 CONCLUSIONS

Several general conclusions my be drawn from the experiments reported herein. First, as Holland has long argued, large populations are required for nontrivial tasks. Second, the inclusion of introns in GA bit strings can afford important advantages. Third, limiting each string to one offspring per generation can increase success rates over unlimited reproduction. Fourth, certain kinds of premature convergence in GAs are artifacts of the impoverished environments in which they are tested.

More specifically; insertion of introns was demonstrated to produce as much as a ten-fold increase in successful evolution over strings without this superfluous material. It may seem bizarre that extraneous positions in a bit string should increase the efficiency of an algorithm by a factor of ten, but consideration of the action of the crossover operator makes the mechanism apparent.

Additionally, a fact was demonstrated that is initially counter-intuitive; genes whose existence singly provides even marginal advantage can be retained in the population even if their combination is disadvantageous. This makes possible their recombination into highly adaptive constellations with other marginally above average genes sometime in the future. On the other hand, combinations of genes that confer advantage (like *D*E*, using Mu) are helpful in evolving more complex combinations containing them even if that evolution cannot occur directly in one or even several steps; so long as the constituent genes proliferate, eventually they may be combined. The implications of these facts may help provide insight into the mechanisms of genetic algorithms and/or molecular biology.

Whether the improved operation of crossover with the insertion of introns explains (even in part) their existence in real genetic material remains to be seen. It is a plausible and intriguing probability.

Acknowledgements

I would like to thank John Holland, without whom I never would have thought to do any of this; and Frank Zizza, for aid and assistance with Mathematica.

References

[1] Dawkins - *The Blind Watchmaker.* Norton: New York, NY, 1987.

[2] R.L. Dorit, L. Schoenback and W. Gilbert. How Big Is the Universe of Exons? *Science*, 250, 1990.

[3] S. Forrest and M. Mitchell. The performance of genetic algorithms on Walsh polynomials: Some anomalous results and their explanation. In *Proceedings of the Fourth International Conference on Genetic Algorithms.* Kaufmann, San Mateo, CA, 1991.

[4] D.E. Goldberg. Sizing Populations for Serial and Parallel Genetic Algorithms. In *Proceedings of the Third International Conference on Genetic Algorithms*, pages 70-79. Kaufmann, San Mateo, CA, 1989.

[5] S.J. Gould. Through a Lens, Darkly: Do species change by random molecular shifts or natural selection? *Natural History*, 1989.

[6] W. Gregory. Why genes in pieces? *Nature*, 271, 1978.

[7] T.B. Grimer, M. Mueller, E. Dreier, T. Kind, T. Bettechen, G. Meng, and C. R. Mueller. Hot spot recombination within DXS164 in the Duchenne muscular dystrophy gene. *American Journal of Human Genetics* 45(3), 1989.

[8] J.D. Schaffer, R.A. Caruana, L.J. Eshelman, and R. Das. A Study of Control Parameters Affecting Online Performance of Genetic Algorithms for Function Optimization. In *Proceedings of the Third International Conference on Genetic Algorithms* pages 51-60. Kaufmann, San Mateo, CA, 1989.

[9] A.O. Sperry, V.C. Blasquez, and W.T. Garrard Dysfunction of chromosomal loop attachment sites: Illegitimate recombination linked to matrix association regions and topoisomerase II. *Proceedings of the National Academy of Science, U.S.A.* 86(14), 1989.

[10] R. Tanese. *Distributed Genetic Algorithms For Function Optimization.* Ph.D. Dissertation, Computer Science and Engineering; University of Michigan, 1989.

Adaptation on Rugged Landscapes Generated by Iterated Local Interactions of Neighboring Genes

Marc Lipsitch*
Santa Fe Institute
1850 Old Pecos Trail
Santa Fe, NM 87501
lipsitch@sfi.santafe.edu

Abstract

Fitness landscapes were constructed by creating phenotypes through 20 iterations of an elementary cellular automaton (CA), with the genotype determining initial state of the CA and bitwise conformity of the final state of the CA to a specified target string determining fitness. Landscapes were classified by the Li-Packard rule class of the CA rule that generated them and by several global measures of the landscape itself. A genetic algorithm (GA) simulated adaptation on these landscapes. Landscapes generated by locally chaotic (Class 3) CA rules had very infrequent local optima and were easy for GAs to adapt on. This suggests that bounded chaos is characteristic of GA representations that increase search effectiveness. Chaotic (Class 4) landscapes and landscapes with low correlation lengths (a global measure) were most difficult for all GAs. Compared to adaptation by mutation only, crossover reduced GA performance on low-correlation and chaotic fitness functions. Mean hill-climbing walk length to a peak is closely tied to GA performance: nearly all difficult landscapes were those with mean walk length near 1.

1 INTRODUCTION

Agents must adapt in a variety of environments that can be characterized by rugged fitness landscapes. One class of rugged landscapes are those generated by iterated local interactions among neighboring sites on the genotype. These landscapes are of particular interest for two reasons. First, such local interactions provide an idealized model of the chemical and physical interactions among adjacent genetic loci; three adjacent nucleotides interactively determine amino acid coding, and amino acids interact in determining the structure of the resulting protein. Second, these local interactions have been studied extensively in the context of cellular automata (CAs). This paper describes the relationships between the nature of these local interactions, characteristics of the fitness landscapes generated by the interactions, and the success of populations adapting on these landscapes.

Fitness landscapes based on local interactions can be implemented by the use of elementary cellular automata, which have been extensively studied by Wolfram, et al. (see, for example, [8]). The use of CAs provides a simple way to tune the degree of interaction among separate loci. By increasing the number of iterations of the CA, or the neighborhood size considered by CA rules, one can increase the number of neighbors with which each locus interacts. Furthermore, a significant amount of work has been done on classifying and describing the behavior of various classes of rules governing CAs. The relationship between known characteristics of certain classes of rules and the nature of adaptation on landscapes generated by those rules gives insight into the conditions under which various kinds of adaptation may work.

The process of adaptation on a rugged landscape can be modelled by the use of Genetic Algorithms (GAs). A genetic algorithm is a biologically inspired computer algorithm that models the processes of random mutation, recombination and natural selection. An important problem in the theory of genetic algorithms is to identify descriptive measures of a search space that can predict how successful the GA will be at such a search. Since GA's are simplified models of natural genetic processes, this question has important biological implications. Under what conditions will adapting populations "find" the optima in their fitness landscapes and be able to stay there? On the other hand, under what conditions is it unlikely that such a population would find and remain at these optima? Under what conditions is a sexually reproducing population, in which recombination is a source of genetic variability, likely to adapt more successfully than an asexually

*Letters sent to 431 West Wesley Road, Atlanta, GA 30305, may be answered more quickly.

reproducing counterpart, in which genetic variation is produced only by mutation?

This paper describes the results of numerical simulations on three sets of 256 different, deterministic fitness landscapes generated by local interactions among adjacent genes. It discusses the relationships among three factors: the kind of cellular automaton rule that generates a fitness landscape, a measure of correlation on the fitness landscape, and the performance of two kinds of genetic algorithms in adapting on the landscape. Finally, it suggests some implications of these findings for biology and for the theory of genetic algorithms.

2 DESCRIPTION OF EXPERIMENTS

2.1 Landscape formation and statistical measures to describe landscapes

The fitness landscapes used in these simulations are constructed by a two-step process. In the first step, the "genotype" encoded by the 30-bit GA string is converted to a phenotype by an idealized "environment" or "chemistry" represented by repeated iterations of an elementary cellular automaton (CA). The "phenotype" is the state of the CA after a set number of time steps. The fitness of the phenotype is then measured by its bitwise conformity to a specified target string.

A cellular automaton is an n-dimensional, ordered lattice of bits that are updated at successive time steps according to a deterministic rule. For *elementary* cellular automata, the lattice is 1-dimensional, and the update rule expresses the state of a bit at time step $t+1$ as a function of the states of that bit, and the states of its nearest neighbors to the left and to the right, at time step t. Thus each bit at time $t+1$ is determined by the state of 3 bits at time t. There are $2^3 = 8$ possible configurations of 3 bits, and each of these configurations may produce either 0 or 1 in the next time step. Thus there are $2^{2^3} = 256$ different elementary CA rules. Each simulation was run on CAs described by each of the 256 different CA rules.

In these simulations, the CA was 30 bits long, with periodic boundary conditions. The genotype encoded the starting condition (time step $t = 0$) for the 30 bits of the cellular automaton, and the CA rule was applied for 20 iterations. The phenotype corresponding to a given genotype was the state of the CA after the final time step. In each successive time step of the cellular automaton, a single bit can influence the state of those to the right and to the left of it. For each iteration, then, the influence of a single bit can spread by a maximum of two bits. After 20 time steps on a 30-bit lattice, each bit in the genotype could potentially influence every bit in the phenotype.

The second step in the construction of the fitness landscape was to assign a fitness to each phenotype. The fitness of a phenotype was determined by its conformity to a target string, measured by the number of bits they had in common. Three different target strings were tried, which will be referred to as ALLONES (1111 ...1111), ALTERN (1010 ...1010), and RANDOM, a "randomly" chosen string (010001100101100101011000000001). For a summary of the procedure for determining phenotype and fitness, see Figure 1.

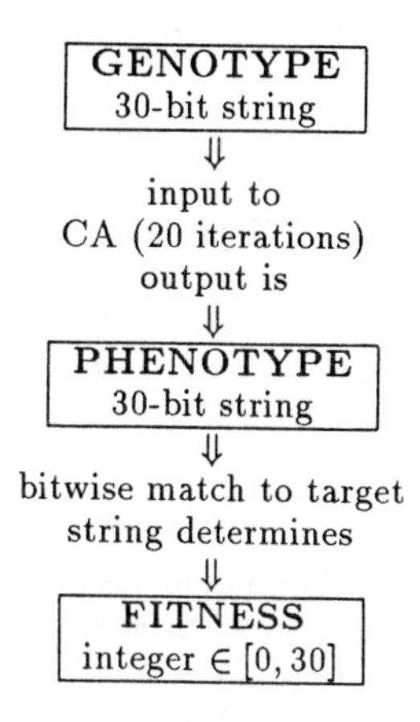

Figure 1: Construction of Fitness Landscapes

Fitness landscapes generated in this way were described by several methods. One set of measures describes the global features of the landscape. These included two statistics on the correlation of fitness values on the landscapes. Rugged landscapes show little correlation between the fitness values of neighboring points on a landscape, while smooth landscapes are more correlated. Landscape ruggedness was also measured by the mean length of an uphill walk from a random point to a local optimum. In addition to these measures of ruggedness, the landscapes were classified according to the type of local interactions that occur in the CA used to generate the landscape. Li and Packard [7] have classified CA rules in this way, and their scheme has been followed.

2.1.1 Global Measures of Landscapes

One statistical property of landscapes is the degree of correlation of "related" points. Correlation provides a measure of how "rugged" a given landscape is. Highly correlated landscapes are "smooth" because nearby points have similar fitnesses, while less correlated landscapes are more rugged, having larger fluctuations in fitness over short distances.

Measures of the correlation of nearby points on a fitness landscape are of interest because they should help to predict GA performance. Effective adaptation on a fitness landscape depends on the "building block hypothesis" — the condition that schemata that are highly fit on average combine to make highly fit genotypes [1, 2]. By contrast, in "deceptive" problems, highly fit schemata, when combined, may produce a less fit genotype. Highly correlated landscapes are those that are least deceptive, since correlation measures reflect the similarity in fitness of similar genotypes: those that share many schemata. On highly correlated landscapes, then, searching for highly fit schemata, which GAs do well, is an effective way to search for highly fit whole genotypes. We will expect, then, that adaptation will be easier on highly correlated landscapes.

Two statistical measures of correlation were devised for the landscapes in these experiments. The first, the correlation length, is analogous to a measure used in information theory and physics. This measure indicates the longest distance away from a given point in a space at which fitness is correlated with the fitness of the starting point. Highly correlated, "smooth" landscapes preserve this information over long distance, yielding long correlation lengths.

For these landscapes, the correlation length was measured as follows. On each landscape, 600 points were picked at random. For each such "initial point," one of its i-mutant neighbors[1] was picked at random, for each $i = 1$ to 30. For each i, the correlation coefficient

$$c_i = \rho[f(p), f(m_i(p))]$$

was calculated, where $f(p)$ is the fitness of point p, m_i is a randomly chosen i-mutant neighbor of point p, and ρ is the standard correlation coefficient.[2] This produces a range of values c_i for $i = 1$ to 30, where c_i is the correlation between the fitness of a point and its i-mutant neighbor. The correlation length, L, of a landscape is defined as one less than the value of i at which c_i first becomes nonpositive. It thus reflects the maximum hamming distance from a known point at which a "searcher" in the space retains some information about expected fitness value.

A second measure of correlation on a space was the correlation coefficient $\rho[h, \Delta f]$ where h is the hamming distance between two points $p1$ and $p2$ on the landscape, and $\Delta f = |f(p1) - f(p2)|$. For high values of this measure, long walks tend to yield large changes in fitness, while short walks give smaller changes in fitness. This measure and correlation length give similar information; however, this measure may be easier to compute.

A global measure of landscapes is the mean length of a hill-climbing walk from a random point on the landscape to a local fitness optimum. This figure is inversely related to the frequency of optima on the landscape. The upper bound on expected walk length (R) is $R = D/2$, where D is the dimension of the space (here $R_{max} = 30/2 = 15$). Kauffman [5] has shown that for a completely random landscape, $R = log_2(D-1)$; here, $R = 4.9$. "Perverse" landscapes may have even shorter mean walk lengths, approaching zero for flat landscapes or landscapes with isolated spikes or areas of ruggedness.

[1] An i-mutant neighbor is defined as a bit string formed by choosing i bits, without replacement, in the original string, and flipping them.

[2]
$$\rho[x, y] = \frac{\overline{(x - \overline{x}) \cdot (y - \overline{y})}}{\sigma_x \cdot \sigma_y}$$

ρ takes on values in $[-1, 1]$.

2.1.2 CA rule classification

In looking at landscapes generated by local interactions, it is of interest to consider the relationship between the kind of local interactions that generates the landscape and global properties of the landscape. Micro characteristics of the "underlying chemistry" affect the macro properties of the fitness landscape that depends on the chemistry.

Fitness landscapes were classified by the category of the CA rule that generated them, according to the classification developed by Li and Packard. They divide the elementary CA rules into five classes, according to the asymptotic attractors of the CA described by the rule. These classifications are based on the typical pattern of CA dynamics; classes 0–2 show periodic dynamics, while classes 3–4 show "chaotic" dynamics. Their categories are shown in Table 1:

Table 1: CA Rule Classes

0 *Fixed point.* All starting conditions go to the same fixed attractor. Small perturbation has no asymptotic effect.

1 *Nonhomogeneous fixed point.* Attractors are fixed points but depend on starting conditions. Small perturbation may or may not push the CA to a different asymptotic attractor.

2 *Periodic attractor.* Asymptotic behavior is a periodic cycle.

3 *Locally chaotic.* Impenetrable "barriers" of either fixed or periodic behavior separate regions of chaos. Small perturbations from a given starting condition spread in regions between the barriers.

4 *Chaotic.* Behavior of the entire CA is chaotic. Small perturbations spread throughout the CA. Cycle length (period) diverges exponentially with increased lattice length.

Li and Packard's classes are not exactly analogous to any property of landscapes in this experiment because they describe typical asymptotic behavior of cellular automata, while the landscapes considered here are based on the state of the automaton after a fixed number of steps, $t = 20$. However, because of the potential for complete "mixing" described above — each gene can affect every site in the phenotype — it is likely that the asymptotic chaotic behavior (bounded or global) described by Li and Packard will correspond to chaotic behavior in these experiments in the sense that phenotype and fitness will depend sensitively on the starting conditions, or genotype.

2.2 ADAPTATION ON THE LANDSCAPES: CHARACTERISTICS OF THE GA

Once these characteristics of the fitness landscapes were measured, further experiments determined the adaptability of various landscapes. These experiments were carried out using a simple version of a genetic algorithm [4]. 20 different genetic algorithm populations were allowed to evolve on landscapes produced by each rule for each target string. Unless otherwise specified, all figures for GA performance represent averages over these 20 replicates, on a given landscape. Population size in each GA generation was 40. Individuals were randomly selected for reproduction, with the chance of being selected proportional to an individual's renormalized fitness.[3] After selection, each individual had a 50% chance of being transferred directly into the following generation, and a 50% chance of undergoing modification by simple genetic transformations. The modification process consisted of either crossover[4] and mutation, or mutation only. The chance of mutation at each site was 3.33%, with each site independent, so that the expected number of mutations in a 30-bit string was 1.00. Genetic algorithms were allowed to run for 60 generations. This figure was chosen because in longer simulations, mean fitness within a generation had reached a plateau by that point.

3 RESULTS

Experiments were designed to answer two questions. First, what was the relationship between the characteristics of a fitness landscape and the nature of the underlying "chemistry," or CA rule that generated the landscape (Section 3.1)? Second, what was the relationship between these characteristics and the ease with which populations adapted on the landscapes. Answers to this second question fall into two categories. The first deals with the ease of adaptation on a given type of landscape in general (Section 3.2.1). The second looks at which landscape characteristics determine the relative success of adaptation by mutation only and adaptation by mutation and recombination (Section 3.2.2).

[3]Renormalization of fitness compensates for the effects of scaling by dividing fitness by the standard deviation of all fitness values in the generation.

[4]The GA's used circular crossover, rather than the standard linear crossover. Linear crossover cuts each parental genome at one point and combines the right half of one parental string with the left half of the other, and vice-versa. Circular crossover makes cuts at two points, and switches the portion between the two points. Linear crossover generates a boundary effect: genes near opposite edges of the genome are more often separated during crossover than genes near the middle. Under circular crossover, there are no "ends" of the genome. Therefore, while more distant genes are still more likely to be separated by crossover (with maximum probability of separation for genes separated by $L/2 = 15$ loci), the effect is symmetric.

3.1 CHARACTERISTICS OF FITNESS LANDSCAPES

Simulations showed a close relationship between the class of the CA rule that generated a landscape and the measures of landscape ruggedness. For all three target strings (fitness functions), *chaotic (Class 4) landscapes showed the shortest mean walk length to local optima*, $R = 2.9 - 3.0$,[5] compared to global averages of $R = 4.3 - 4.8$. *Chaotic landscapes also showed smaller values for each correlation function*, correlation lengths of $L = 4.7 - 6.0$, compared to global averages of $L = 10.4 - 11.3$, and hamming correlation coefficients of $\rho = 0.00 - 0.02$, compared to global averages of $\rho = 0.12 - 0.18$. *Locally chaotic rules (Class 3), had by far the longest mean walk length of any class*, ranging from $R = 7.0$ to $R = 7.8$. Statistical measures of landscapes, separated by CA rule class and target string, are reported in the Appendix, Table 2.[6]

As previously mentioned, the correspondence between Li and Packard's CA rule classification and the structure of these landscapes is only indirect. However, it seemed plausible that Class 4 chaotic rules would generate chaotic fitness landscapes, since small perturbations in the genotype should produce large changes in phenotype and thus in fitness. One would therefore expect that chaotic rules would generate fitness landscapes with many local optima. Since similar genotypes on these landscapes produce a wide spread of phenotypes, and therefore of fitness values, the chances that a random uphill walker will encounter a local fitness optimum are higher. Kauffman [5] shows that in completely uncorrelated landscapes, the expected length of an uphill walk is small and increases with $log_2(D - 1)$, where D is the dimension of the space. For the same reasons, one would expect chaotic rules to have short correlation lengths. The data on Class 4 rules confirm these predictions.

More surprising, perhaps, are the Class 3, locally chaotic CA rules. Although landscapes generated by these rules did not differ significantly from Class 1 or 2

[5]These ranges indicate the range of mean values across different target strings.

[6]Tables 2-6 and Figure 2 are in an Appendix, available on request.

rules in their correlation measures, they had by far the longest mean length of walks to local fitness optima of any rule class: $R = 7.0 - 7.8$, compared to global averages of $R = 4.3 - 4.8$. As we shall see, this rarity of local optima seems to have improved the performance of adapting populations substantially.

3.2 ADAPTATION ON CA-GENERATED LANDSCAPES

Two questions were asked about adaptation on these landscapes: first, on what kinds of landscapes was it difficult to adapt, and second, what parameters of the landscapes predicted whether recombination was helpful or harmful to genetic algorithm performance?

3.2.1 Ease of adaptation on landscapes

One measure of the ease of adaptation on a landscape was how close the population in the final generation got to the global fitness optimum in a space. Each space had $2^{30} \approx 10^9$ points, so sampling to find the global optimum was not practical. Instead, 20,000 points were sampled. Approximation to the global optimum was measured by percentile rank compared to this sample.[7] Numbers represent the average over 20 replications of the mean fitness of individuals in the final generation of the genetic algorithm. Table 3 shows the mean percentile rank of the final generation fitness of populations evolving on landscapes generated by cellular automata of Classes 0–4.

Populations adapting on landscapes generated by CA's in all classes except the globally chaotic (Classes 0–3) reached the high 99th percentile, as the data in Table 3 shows. The exception is Class 1 for the ALTERN target string, in runs both with and without crossover. This exception will be discussed below. In all runs, however, *Class 4, chaotic rules presented significantly greater challenges to adapting populations, with mean percentiles of 92.2–98.5.*

Locally chaotic (Class 3) landscapes produced the highest performance by adapting populations and the smallest standard deviation in performance of any of the four nontrivial classes (1–4), reaching 99.6–100.0 percentile with $\sigma = 0.0 - 0.6$ %ile. As reported above, Class 3 landscapes had much longer average walk lengths to local optima than those of any other class. This rarity of local optima facilitates adaptation on the landscapes in two ways. First, it results in a low probability of being "trapped" on unfavorable local optima. Second, it means that piecewise optimization of a genome is effective, so that good genetic material continues to be good as other parts of the genome change. In the language of schemata, such landscapes are susceptible to solution by building blocks; highly fit, short, low-order schemata combine to form highly fit genotypes [2].

Landscapes with low correlation lengths were the most difficult to adapt on, as Table 4 shows. This is not surprising, since the preceding section showed that Class 4 rules usually generated low-correlation landscapes. For each target string, above a certain threshold correlation length, between 4 and 7, adapting populations reliably reached nearly the 100th percentile, while performance fell off below that correlation length. However, that threshold correlation length varied between target strings and between adaptive strategies (mutation-only and mutation-and-crossover). For correlation lengths near the threshold, performance standard deviation was very high, indicating a transition at which some landscapes are easy to adapt on, and others are much more difficult. Low standard deviations ($\sigma \leq 0.5$%ile for correlation lengths higher than 11 for ALLONES and RANDOM target strings) in performance on high-correlation landscapes indicate that nearly every rule with high correlation length reached the high level of performance. Again, the ALTERN target string is an exception; standard deviations remain high at high correlation lengths, particularly in the runs with crossover.

Landscapes with short mean walk lengths to peak (R) were most difficult to adapt on. These landscapes, with their frequent local optima, presented greater risks of being trapped. Figure 2 (Appendix) shows the performance of GA's plotted against R. Except for the landscapes using the ALTERN target string, R was a very good predictor of GA success: GAs adapting on landscapes with R values greater than approximately 4 almost always reached near-100th %ile performance.

3.2.2 When does recombination help?

On each landscape, adapting populations with and without crossover were compared. Relative performance was measured, not by percentile, but by differ-

[7] Percentile rank of GA performance is the percentage of the random sample of the fitness values of 20,000 points whose fitness value was less than or equal to the average fitness attained by members of the final GA generation. Since all actual fitness values are integral, and average fitness attained by the GA is usually not, percentile rank of nonintegral averages was adjusted by linear interpolation. The GA ran for 60 generations, with a population size of 40, and with half of each generation, on average, passing unchanged into the next generation. Thus, the GA sampled approximately $40 + (59 \times 20) = 1220$ different genotypes. These were compared against 20,000 genotypes in the random sample. This discrepancy in sample size means that high percentile rank shows substantial improvement over random sampling. The percentile measure is sensitive for landscapes on which adapting populations do not reach the highest fitness value in the random sample. For landscapes on which it does, the measure is insensitive, since the maximum percentile is 100, and the measure cannot distinguish between populations that significantly exceed the best fitness found by random sample, and those that exceed it only slightly.

ence in performance, normalized by the standard deviation of fitness values in the 20,000-point random sample of the landscape: $P = \frac{\overline{G}}{\sigma_F}$, where P is the performance measure; $\overline{G}$ is the mean fitness of individuals in the final generation of the GA, and σ_F is the standard deviation of fitness values in the whole landscape. This measure provides sensitivity across the whole range of fitness values, not just those far removed from "optimal." The added sensitivity is important, since we are interested in relative performance, and percentile gives us no information about relative performance on "easy" landscapes, where both mutation and crossover may attain percentile ranks of 100 despite different absolute performance. This measure of performance was compared for runs using mutation only and those using both mutation and crossover.

We first consider the relative performance, shown in Table 5, of mutation-only adaptation and mutation-plus-crossover (hereafter referred to simply as crossover) adaptation on landscapes of varying correlation lengths. For the ALLONES and RANDOM target strings, *mutation-only adaptation performed better on less correlated landscapes, while populations evolving with crossover performed better on more correlated landscapes.* As with the absolute performance statistics, there is a transition in performance at a correlation length between $L = 4$ and $L = 8$, depending on the target string. As we have seen, correlation length is a good indicator of the effectiveness of optimization by building blocks, a process on which crossover crucially depends. Thus, it is to be expected that crossover becomes more helpful as landscapes become more correlated.

As in the previous section, the ALLONES and RANDOM data support what we would expect. Crossover generally represents a longer jump on the space than does mutation. Without accounting for the effects of population convergence in later generations of the GA, a mutated and crossed-over individual has an expected hamming distance of approximately 3.75 from its nearer parent.[8] It is not a completely random jump, since it is a jump "between" two highly fit points. However, on less correlated landscapes, where long jumps mean the loss of information about fitness, even such a "directed" long jump is more likely to land on an unfit point than the shorter jumps of mutation (expected length of jump under mutation was 1, since for those strings that were mutated, the probability of mutation was set to the reciprocal of the length of the gene string). The ALTERN target string once more presented an anomaly; crossover did better, on average, for nearly every correlation length.

Chaotic (Class 4) rules show the same characteristics as landscapes with low correlations: *mutation-only adaptation produced higher average fitnesses in the final generation on Class 4 rules*; in between 56 and 94% of all Class 4 landscapes, mutation-only performed better. This is unsurprising, since Class 4 CAs tend to generate uncorrelated landscapes (Table 2). Or in terms of schemata: since chaotic landscapes are highly "deceptive" — changes at one locus affect the fitness contributions of many other loci, so that optimization by schemata is difficult — crossing over, which works best when highly fit schemata combine effectively, is less likely to be helpful on these landscapes. *For all other classes, crossover produced better results on average,* outperforming mutation-only in 62 – 71% of all cases. These results, which omit flat landscapes (i.e., those for which $\sigma_F = 0$) appear in Table 6.

Locally chaotic (Class 3) rules show negligible differences in the performance of the two methods of adaptation. The extremely high performance of all populations on Class 3 landscapes, and the unusually low frequency of local optima in them, suggest that these landscapes were easy to adapt on regardless of whether mutation or crossover was used; hence the small difference between strategies.

R (mean walk length) is not a good predictor of the relative performance of mutation-only and mutation-and-crossover adaptation. There are significant numbers of landscapes at each R value for which each method performs better than the other, although the differences fall off substantially at higher R values.

3.2.3 The Mystery of the ALTERN target function.

Unlike the ALLONES and RANDOM target functions, the ALTERN target function consistently produced landscapes on which adapting populations did not act as expected. The reason for this difference is unclear. If ALTERN had produced landscapes that differed substantially in correlation or other measures of ruggedness from the other target functions, this should have shown up in the data. However, the differences between landscapes generated by this target function and those generated by other target functions do not show up in any of the statistics calculated here.

While this anomaly is disconcerting, it does point up an important feature of adaptation on rugged fitness

[8] The distance between the two points of crossover is chosen from an even distribution over $[0, 15]$, since a cut of size $s > 15$ is equivalent to a cut of size $30 - s$. Thus, the expected size of the crossed-over region is 7.5. On average, half of the transplanted genes will be the same allele in both parents. Therefore, the expected number of switched genes from crossover is 3.75. In half the cases, the one expected mutation will be on the portion of the genome coming from the nearer parent, adding 1 to the expected hamming distance. In the other half, it will be on the side coming from the further parent, having on average 0 effect on the distance from the nearer parent. This calculation is inexact because of the substitution of expected values for case-by-case analysis.

landscapes. Landscapes generated by all three fitness functions shared a number of characteristics; the distribution of correlation lengths was similar; the nature of local interactions (epistasis) was identical; and the "task" on each landscape was qualitatively the same: to match a given target string. The fact that one target string can be so anomalous suggests that these measures alone are of limited predictive value.

4 SUMMARY AND DISCUSSION

4.1 IMPLICATIONS FOR GA USERS

These experiments describe a set of easily measurable properties of fitness landscapes that help to predict GA performance. Landscapes generated by chaotic rules and those with low correlation lengths are more difficult to adapt on in general, and are more susceptible to adaptation by populations that do not use recombination than by those that do. Landscapes with many local optima are generally more difficult to adapt on than those with few local optima.

As we have seen, these results from a specific class of fitness landscapes provide empirical confirmation for predictions of the theory of rugged landscapes [5]. Both kinds of adaptation rely on correlation in fitness values of nearby points on the fitness landscape, and adaptive search will not reach the most fit points in the landscape if it does not have some minimal level of correlation with which to work. Chaotic rules tend to generate a high degree of sensitivity to perturbations, making adaptation difficult. Frequent local optima indicate a risk of trapping the genetic algorithm before it finds the global optimum.

Perhaps the most striking result in these experiments is that Class 3 landscapes generated the fewest local optima and made adaptation the easiest of any of the classes. The relevant feature of Class 3 cellular automata is that a perturbation in one bit at time t spreads chaotically in the succeeding time steps, but only within bounds. Thus, a mutation in the genotype can cause substantial but bounded exploration of phenotype space. The results here indicate that this kind of bounded exploration is very efficient.

This conclusion gives insight into the problem of representation: in general, how should the variables of a function to be optimized be represented on the GA string? The answer suggested here is that variables should be chosen so that a small change in the genotype makes a substantial but limited change in the phenotype.

An final result is insight into conditions under which crossover reduces GA performance. The results in Section 3.2.2 show that crossover often reduces GA performance on fitness functions that are highly sensitive to small changes in initial conditions and on very uncorrelated functions.

4.2 BIOLOGICAL IMPLICATIONS

The question that originally motivated these experiments was biological — under what conditions can we expect that an adapting population will reached optimal fitness levels in its space of possible genotypes? On the other hand, under what conditions will highly favorable genotypes be unable to spread and remain prevalent in a population? The results of these experiments give evidence in several biological contexts.

Brian Goodwin [3] has recently proposed a "structuralist" model of morphogenesis, in which genetic information encodes parameter values for a physical-chemical system that generates the organism's morphology, rather than directly encoding specific morphological traits. Goodwin and his colleagues have constructed such a physical-chemical model for *Acetabularia* algae, and the model predicts the existence of certain morphological features for which no satisfactory evolutionary explanation has been given. It accounts for these features as the outcomes of various chemical gradients and other complex interactions, parameterized by starting values and distributions of the chemical components of the system.

The interactions of genes in "morphogenesis" — here represented by iterations of the cellular automaton — are in the spirit of Goodwin's model. The genotype encodes starting parameters that interact in a mathematically complex but not random way to generate a phenotype. These simulations show that in cases where Goodwin's model is applicable, and phenotypic characteristics are dependent on these complex interactions, the ease of adaptation declines with increasing complexity of interactions. For highly chaotic, uncorrelated morphogenetic processes (CA rules), adapting populations are unlikely to settle at highly fit points in the space of possible genotypes.

The performance of adapting populations on Class 3 landscapes, by contrast, suggests a model for successful adaptation on fitness landscapes generated by complex interactions. One of the most notable results in these experiments was the contrast between the landscapes generated by Class 3 (locally chaotic) rules and those generated by Class 4 (chaotic) rules. Class 4, chaotic rules produced the least correlated landscapes, with the most local optima, and proved by far the hardest for adapting populations. Class 3 landscapes had the longest mean walk lengths to local optima, and their landscapes were among the most highly correlated. Populations adapting on Class 3 landscapes uniformly reached nearly the hundredth percentile. The locally chaotic nature of these rules allowed exploration of wide areas of phenotype space, since a change at a locus within one of these chaotic regions spread quickly within that region, but the bounds on the spreading of perturbation allowed containment of the changes. This is analogous to what Kauffman [6] calls "modularity." Small sets of parameters can be changed without changing the nature of the whole phenotype. By contrast, the chaotic fitness landscapes gave too little control; a small change in the genotype can destroy too much of the structure of the phenotype. This result suggests that bounded chaos may be a good compromise between the need for phenotypic stability and the need to generate substantial variation to "find" highly fit genetic configurations.

Bounded chaotic interaction is closely analogous to biological epistasis. Goodwin's model implies that epistasis is a widely prevalent phenomenon, since genes interact by setting the parameters of a single physical-chemical process. Biologically, we may expect that adaptation could be most successful if relatively small sets of genes interact epistatically in a given morphogenetic process, and if these sets of genes had small degrees of overlap. In that way, mutations or recombinations in those sets of genes could have substantial but limited effects; while generating significant variation, they would not be expected completely to change the character of an organism.[9]

The comparisons of sexually and asexually reproducing populations in this paper provide evidence on both sides of the debate over the claim that reproduction using recombination is evolutionarily advantageous because it generates genetic diversity. These results must be taken cautiously, since the model studied in this paper omits major features of sexual reproduction, such as diploidy. In the chaotic and less-correlated landscapes, crossover tends to disrupt highly fit but delicate genotypes, and that the combination of two highly fit genotypes may be much less fit. On such uncorrelated landscapes the genetic diversity generated by sexual reproduction tends to disrupt, rather than promote, adaptation. However, on the more correlated, and the non-chaotic landscapes, recombination appears to aid adaptation.

4.3 FURTHER WORK

One limitation of these results is that, although they show relationships between measures of a space and the success of adaptation on them, they do quantitatively predict the success of adaptation on an arbitrary space. An important next step would be to identify some generalized function of parameters of a space such as dimension, correlation measures, distribution of local optima, etc., that would be useful in predicting the adaptability of a landscape.

The case of the ALTERN target function, as already noted, calls into question the viability of such an enterprise using only the descriptive measures considered here. An understanding of exactly why this function caused such surprising results may help to uncover some other parameter of fitness landscapes that would help to make such predictions more accurate.

[9] The unusual results of the ALTERN target function again call into question the accuracy of measures of epistasis as predictors of adaptive success. Since the CA rule completely determines the rules of genetic interaction, epistatic interactions are independent of the target string. However, we have seen that the same epistatic interactions, when measured against different fitness functions, generated very different adaptive environments.

Acknowledgements

This paper is the result of an undergraduate internship at the Santa Fe Institute, supported by the Institute and by the Mory's Fund through the Yale College Dean's Office. The Yale Science & Engineering Computing Facility allowed work to continue in the "Zoo" at Yale during the academic year. John Miller of the Institute's Adaptive Computation Program patiently and helpfully advised this project from its conception onward. The genetic algorithm code used in the simulations came from his program *GAKERNAL*. I also wish to thank Stuart Kauffman, Wentian Li, Michael Palmer, Stephanie Forrest, and Stuart Newman for helpful discussions.

References

[1] A. D. Bethke. *Genetic Algorithms as Function Optimizers.* Doctoral Dissertation, University of Michigan. *Dissertation Abstracts International* 41(9), 3503B. University Microfilms No. 8106101, 1981.

[2] D. Goldberg. *Genetic Algorithms in Search, Optimization and Machine Learning,* pages 46–52. Addison Wesley, Reading, MA, 1989.

[3] B. C. Goodwin. Structuralism and biology. Paper circulated at the Santa Fe Institute, Conference on the Principles of Organization of Organisms, 1990.

[4] J. Holland. *Adaptation in Natural and Artificial Systems.* University of Michigan Press, Ann Arbor, MI, 1975.

[5] S. Kauffman. Adaptation on rugged fitness landscapes. In D. Stein, editor, *Lectures in the Sciences of Complexity,* Vol. I. Addison-Wesley, Redwood City, CA, 1989.

[6] S. Kauffman. *The Origins of Order: Self Organization and Selection in Evolution.* In press: Oxford University Press, 1991.

[7] W. Li and N. Packard. The structure of the elementary cellular automata rule space. Technical Report CCSR-89-8, Center for Complex Systems Research, University of Illinois at Urbana-Champaign, 1989.

[8] S. Wolfram, editor. *Theory and Applications of Cellular Automata.* World Scientific, 1986.

Adaptation in constant utility non-stationary environments

Michael L. Littman & David H. Ackley
Cognitive Science Research Group
Bell Communications Research
Morristown, NJ 07960

Abstract

Environments that vary over time present a fundamental problem to adaptive systems. Although in the worst case there is no hope of effective adaptation, some forms environmental variability do provide adaptive opportunities. We consider a broad class of non-stationary environments, those which combine a variable *result function* with an invariant *utility function*, and demonstrate via simulation that an adaptive strategy employing both evolution and learning can tolerate a much higher rate of environmental variation than an evolution-only strategy. We suggest that in many cases where stability has previously been assumed, the constant utility non-stationary environment may in fact be a more powerful viewpoint.

1 Non-stationary environments

An adaptive system within an environment performs two basic tasks. First, there is the search for, and the representation of, regularities in the history of interactions with the environment. Second, there is the attempt to gain some advantage from the constructed representation, by basing future actions on the assumption that those regularities will persist. If that fundamental adaptive assumption fails utterly, and the environment totally lacks such regularities, adaptation can provide no benefit.

Of course, not all non-stationary environments are so pathological. Even if the environment, when viewed as a monolithic function, may be observed to change over time in unpredictable ways, it may be possible to view that environment as being formed from interacting components some of which *do* possess persistent regularities. If that is possible, an adaptive system possessing such a view can gain advantage — even though the environment will always have some surprises in store — by finding, representing, and exploiting those component regularities.

1.1 Constant utility

In this paper we demonstrate one way this can occur within the context of evolutionary adaptation via the genetic algorithm [11, 8]. Consider this scenario in which an organism is faced with a series of interactions with an environment: In each interaction, the environment presents the organism with a *situation*, and then the organism responds with an *action*, and then the environment responds with a *result*. A *utility* function, hidden from the organism, reduces the result to a scalar value. The average utility of the results created by the organism over some given lifespan determines the organism's *fitness*. The nasty trick is this: At unpredictable times and in unpredictable ways, the environmental mapping from situation and action to result changes. As a consequence, if averaged over many environmental changes, the fitness produced by any single mapping will be at the chance level. The adaptive challenge is to produce organisms that yield high *online fitness* (i.e., fitness averaged over successive generations of organisms, see [6]), in the face of the changing environment.

Although this scenario is familiar enough to genetic algorithm researchers in some ways, it also has certain unusual aspects that are central to the present endeavor. To begin with, the environment is non-stationary: The fitness of a given situation-action map changes over time. Also, the fitness function can readily be decomposed into two sub-functions: One function mapping from situation and action to result, and another function mapping from result to utility. And finally, although the result function varies over time, the utility function does *not* change.

We call environments that possess these properties *constant utility non-stationary environments.* Although adaptation may not be beneficial in worst-case non-stationary environments, we can exploit the persistent regularity of the utility function to build an effective adaptation algorithm for this special class.

1.2 Evolutionary reinforcement learning

If an oracle were available to supply the organism with the utility of each result as it occurred, all that would be necessary is a learning algorithm to optimize the utility function during the lifetime of the organism. It could experiment with various action functions and see which lead to increased utility. Changes in the environmental result function would just mean some more learning to do. If such an oracle is unavailable, learning runs into trouble. In our scenario, the only information available is the lifetime average value of the utility function (i.e., the fitness), and that information is not available until end of the organism's lifespan, rendering moot the possibility of using it directly for learning.

The essential idea of *evolutionary reinforcement learning* (ERL) [3, 1] is that an evolutionary algorithm can supply learning with such an oracle. We can include in the genetic material a representation of a *goodness function*, and let learning proceed during an organism's lifetime under the *assumption* that the goodness function is an accurate representation of the utility function. When that goodness function is a poor predictor of utility, well then, learning won't help and the organism possessing it will have low fitness, but that's no worse than possessing a poor action function in an evolution-only algorithm. On the other hand, if the goodness function is a reasonable predictor of utility, then learning can be helpful and higher fitness can be the result, even when the environmental result function is changing.

By separating the stationary and varying portions of the environment, and genetically representing a predictor of the stationary part, and employing learning to compensate for the variable part, we can, at least in principle, and within limits imposed by the response time of the learning system, *stabilize* a non-stationary environment.

2 A simulation study

To demonstrate this effect, we require an abstract computational model of an organism facing a constant utility non-stationary environment, and an adaptive system employing both evolution and learning. Here, we employ very simple models in both of these roles. This choice helps isolate the phenomena of interest, and also reduces simulation costs, but it also rules out myriad effects — many probably harmful, some possibly useful — that could occur in more complex systems.

2.1 Notation

Let $T = 1 \ldots N$ index generations, and let $t = 1 \ldots n$ index interactions with the environment during one lifespan. Let $E = (r_t^T, U)$ denote an environment, consisting of a time-varying result function r and a constant utility function U. Let $S = (\Omega_T, G, L)$ denote an adaptive system, consisting of a sequence of populations $\Omega_T = \{x_p : 1 \leq p \leq P\}$ of binary strings $x = (0,1)^l$, a genetic algorithm G, and a learning algorithm L.

A string x yields an organism according to a constant genetic expression function $e(x) = o$. Each organism $o = (a_0, \{a_t\}, g)$ possesses an initial action function a_0, a goodness function g, and a set of action functions $a_1 \ldots a_n$ specified by the learning algorithm. During a run of the model, a set of PN organisms $\{o_p^T\}$ will be produced, performing a total of $I = PNn$ interactions with the environment.

The interaction of organism o_p^T with the environment at time t takes the following form: A uniform random binary situation vector v_s of length w — the "width" of the environment — is presented to the organism, which produces an action vector $v_a = a_{t-1}(v_s)$. The environment then produces a result vector $v_r = r_t^T(v_s, v_a)$. The organism then performs learning $L(a_{t-1}, g, v_s, v_a, v_r) = a_t$. The scalar-valued utility of the result is then determined $u_{pt}^T = U(v_r)$.

At each moment (T, t) in time, p interactions occur, one for each member of the population at time T. With some probability α — the environmental variation rate parameter — an alteration is made to r_t^T to produce r_{t+1}^T, otherwise $r_{t+1}^T = r_t^T$. To begin a new generation, $r_1^{T+1} = r_{n+1}^T$.

At the end of the lifespan of o_p^T, the fitness of the corresponding genetic code is evaluated

$$F(x_p^T) = \frac{1}{n} \sum_{t=1}^{n} u_{pt}^T.$$

After P lifespans, the genetic algorithm updates the population $\Omega_{T+1} = G(\Omega_T, \{F(x_p^T)\})$. The online system fitness

$$F_S = \frac{1}{NP} \sum_{T=1}^{N} \sum_{p=1}^{P} F_p^T.$$

is the quantity to be maximized by adaptation.

2.2 A bit-wise independent environment

Our test environment presents situations involving $w = 8$ bits, requiring 8 bit actions, and producing 8 bit results. There is a one-to-one correspondence between situation, action, and result bits, so that each bit of the result vector depends only on the corresponding situation and action bits. The adaptation problem is

At birth: Extract a_0 from x, giving weights w_{sa} and biases w_{0a}. Extract v_g from x. Let $u_* = 0$.

At time $1 \le t \le n$:

1. *(Situation)* Pick random v_s; compute output probabilities $v_o(i) = 1/\left(1 + e^{-(v_s(i)w_{ii} - w_{0i})}\right)$.
2. *(Action)* Given a uniform random variable $\xi \in [0,1]$, $v_a(i) = 1$ if $v_o(i) \ge \xi$ and 0 otherwise.
3. *(Result)* Compute result vector $v_r(i) = r_i^{(R(2i-1),R(2i))}(v_s(i), v_a(i))$. If $v_g = v_r$ go to 4, else go to 5.
4. *(Reward)* Update weights and biases: $\Delta w_{ji} = 10(v_o(i) - v_a(j))\max(0.1, v_s(j)(1 - v_s(j)))$. Go to 6.
5. *(Punish)* Update weights and biases: $\Delta w_{ji} = 0.5(1 - v_o(i) - v_a(j))\max(0.1, v_s(j)(1 - v_s(j)))$.
6. *(Utility)* Increment u_* by $U(v_r, v_d)$.

At death: Exit returning fitness score u_*/n.

Figure 1: The lifetime of an organism.

drastically simplified by building this "spatial" independence into the genetic representation of the action function, which specifies eight parallel functions each mapping one situation bit to the corresponding action bit.

The result function r_t^T consists of eight independent boolean functions each of which maps one bit pair $(v_s(i), v_a(i))$ to one result bit $v_r(i)$. The function is represented by a 16 bit vector R, interpreted as eight two bit values. Each value selects one of four boolean functions:

$$r_i^{(0,0)} = v_a(i), \quad r_i^{(0,1)} = \neg v_a(i),$$

$$r_i^{(1,0)} = v_s(i) \text{ eqv } v_a(i), \quad r_i^{(1,1)} = v_s(i) \text{ xor } v_a(i).$$

R_{t+1}^T is formed by first copying R_t^T, and then, with probability α, inverting a randomly chosen bit of R_{t+1}^T.

The utility function U is controlled by an eight bit desired result vector v_d:

$$U(v_r, v_d) = \frac{1}{8}\sum_{i=1}^{8} v_r(i) \text{ eqv } v_d(i).$$

v_d is unknown to the adaptive system at the outset but constant over all generations; it represents the stable component of the environment that is to be discovered by evolution and exploited by learning.

2.3 An ERL algorithm

In this study, we use a rather simple and generic form of evolutionary reinforcement learning. Here, the GENESIS genetic algorithm system [9] provides the evolutionary component, and a simplified form of the CRBP (complementary reinforcement back-propagation) algorithm [2] provides the learning component.

We adopted the default values for all the parameters in the GENESIS system (population size = 50, crossover rate = 0.6, mutation rate = 0.001, generation gap = 1.0, scaling window = 5, elitist strategy; see [9]), except for the "total trials" parameter (raised to 2,500), the "structure length" parameter (raised to 72 bits), and the use of the "a" option (since fitness is stochastic).

CRBP performs reinforcement learning using a back-propagation-style neural network [13], and a reinforcement function that returns a success or failure signal for each generated output. In this simulation, we used the simplest network — containing 8 input-output weights and 8 output biases — sufficient to represent all possibly optimal action functions.

The genetic representation of the initial action function a_0 occupies 64 bits, representing the 16 weights and biases in a range of ± 4, scaled into four bits each and gray-coded (see [9]). The reinforcement function compares each result vector to a genetically specified 8 bit goodness vector v_g, returns success if and only if they match exactly.[1] The overall length of the genetic material is $64 + 8 = 72$ bits. Figure 1 details the algorithm.

2.4 Comparative data

We compared the behavior of evolution-only and evolution-learning strategies in these environments. The evolution-only strategy is identical to the evolution-learning strategy except steps 4 and 5 in Figure 1 are skipped. Fitness is scored as the fraction of maximum utility achieved, averaged over 1000 situations per organism, 50 organisms per generation, 50 generations per run, and 10 runs. A fitness score near 1.0 indicates that almost every situation is handled correctly by almost every individual in almost every

[1] Thus, in this case, the reinforcement function, playing the role of the goodness function, attempts only to signal optimal values of the utility function, rather than to duplicate it.

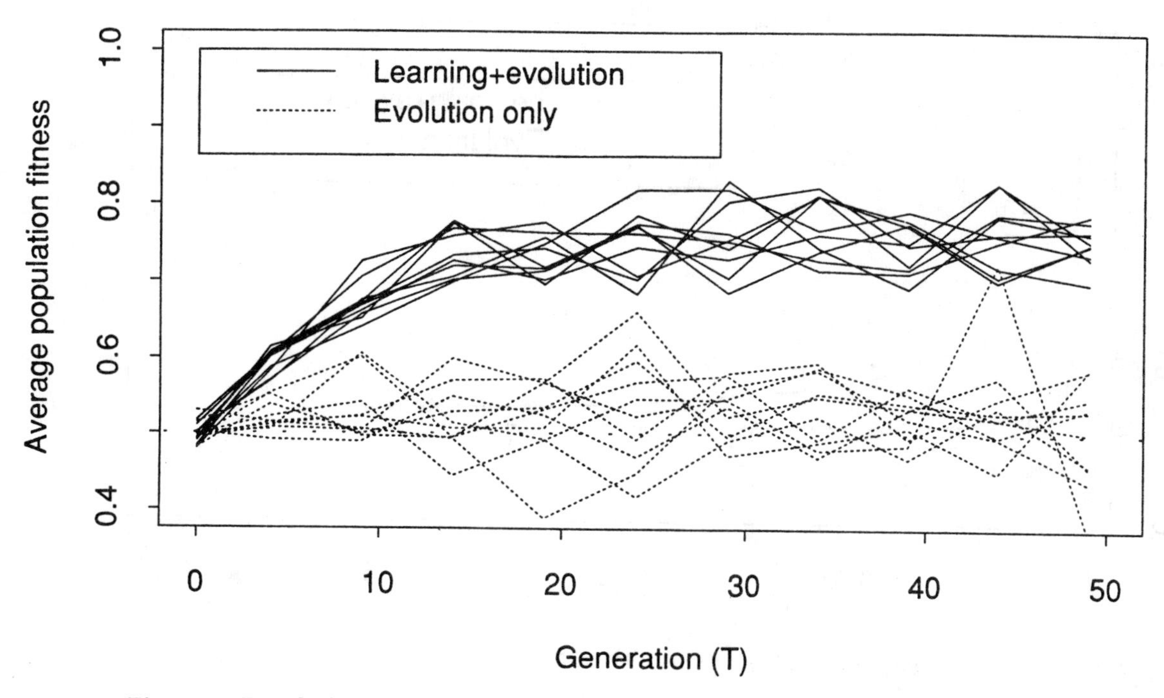

Figure 2: Population average fitness versus time of evolution-only and evolution-learning.

generation in all the runs.[2] A fitness of 0.5 indicates chance performance.

We employ the online fitness measure for F_S because other common performance measures — such as best structure found, offline fitness, and final population average — are either very noisy or dramatically misleading due to the varying environment. However, since the global averaging involved in online fitness can mask transient phenomena that occur during a run, in Figure 2 we display typical data produced by the evolution-only and evolution-learning strategies, as measured by average population fitness sampled at five generation intervals. Compared to typical optimizations in stationary environments, the population fitnesses over time are highly variable, but the overall effect is obvious: The evolution-learning strategy quickly attains and consistently maintains higher fitness compared to evolution alone.

The independent variable in our study is α, which determines the rate at which the result function r_t^T changes. When $\alpha = 0$, the environment is stationary. The data in Figure 2 were gathered at $\alpha = 0.005$. Figure 3 reports the system fitness F_S for the two algorithms, averaged over ten runs apiece for nine different values of α. Figure 4 displays the data from the lower α values in the form of fitness ratios of the two strategies, illustrating the relative advantage provided by learning.

As in other studies of evolution and learning (e.g., [12]), we observe an "inverted-U"-shape relating the independent parameter and the advantage of learning.

In a stationary environment, at the left of the graph, both algorithms perform almost equally well. This indicates that the populations using the evolution-only strategy were able to evolve an optimal action function reasonably quickly and therefore were able to produce the correct action in almost all of the cases. Similarly, the populations using the evolution-learning strategy evolved the correct goodness function (and perhaps the action function as well) and therefore were able to learn an effective action function and could spend most of their lives producing the correct action.

In a highly variable environment, at the right of Figure 3, neither strategy is able to find a useful action function. Performance for both strategies is at or very close to the chance level in these environments. As Figure 4 suggests, however, in between these extremes learning provides a substantial benefit. The evolutionary algorithm by itself tolerates only slow variation in the environment before its performance drops to chance. It seems that the genetic algorithm converges on an action function, and when environmental change necessitates an alternate action function, the population has difficulty responding quickly.

The evolution-learning algorithm, on the other hand, performs significantly better than chance even when the environment variation rate is so large that on average several changes occur in each lifespan. By "rolling with the punches," the evolution-learning strategy is able to stave off the decay towards chance even in quite variable environments.

Although not quite significant ($p < 0.015$ by the sign test), the evolution-learning strategy achieved slightly better mean performance even in a stationary environment. One might have expected that learning would

[2]Since GENESIS is configured to minimize rather than maximize, we provided function values based upon averaging $100(1 - U(v_r, v_d))$.

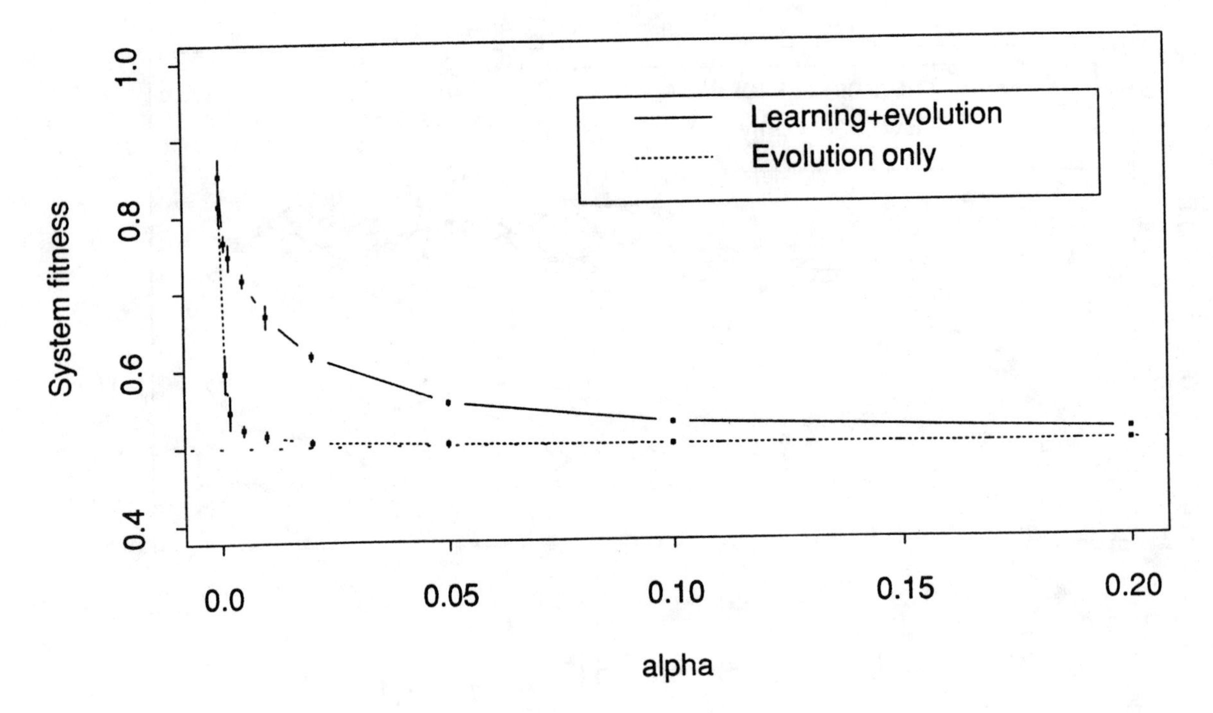

Figure 3: Online fitness scores of evolution-only and evolution-learning algorithms, as a function of the environmental variation rate α. Except when $\alpha = 0$, all differences between the two algorithms are significant ($p < .001$).

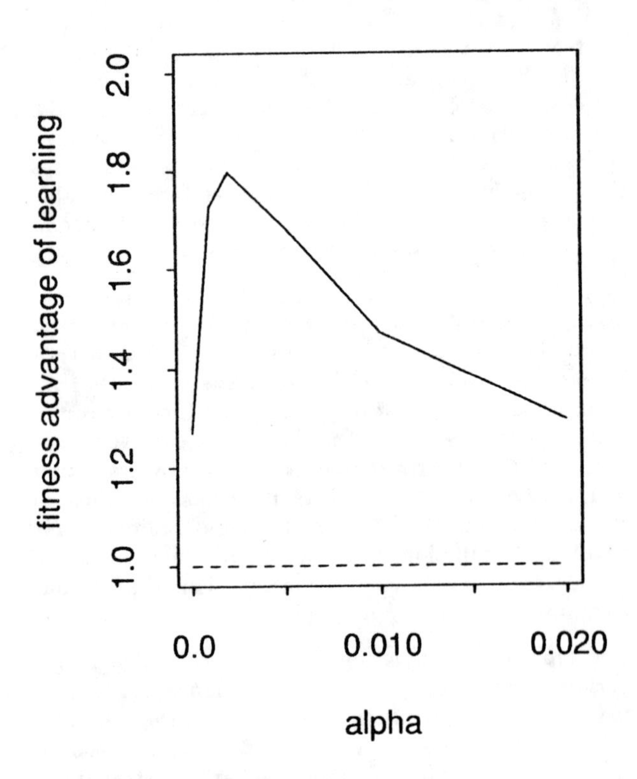

Figure 4: Ratios of evolution-learning to evolution-only system fitness at the lowest six tested values of α.

be a drawback in that case, since each generation would waste time relearning the same action function. Actually, learning can serve as a "scout" to help evolution find good directions in which to evolve [10, 15, 5, 7]. Such "Baldwin effects" may account for evolutionary reinforcement learning's edge in the stationary environment.

3 Analysis

At least in this abstract and simplified case, we have seen that evolution combined with learning can outperform evolution alone across a wide range of environmental variation rates. But how good is good? How well are the algorithms really doing?

To investigate, we wrote down a differential equation modelling instantaneous fitness growth per situation under the assumption that in a stationary environment the growth rate would be proportional to distance from maximum fitness: $\frac{df}{dt} \propto (1 - f)$. Although we lack a rigorous justification for this choice, it can be motivated at least in part by noting that under certain conditions a hillclimbing algorithm will display such a growth rate. Far from the maximum uphill directions are easy to find, but as the maximum is approached, more and more time is spent looking at directions that lead downhill, and progress slows.

In our specific scenario, the constant of proportionality involves at least two factors: An inverse factor of $2w$ representing the size of the search space and an inverse factor of n representing the number of situations per lifespan. In addition, the non-stationary case requires an additional term to reflect fitness changes due to the environment's random walk in the $2w$ dimensional result function space, leading to

$$\frac{df}{dt} = \frac{\rho}{2wn}(1-f) - \alpha\left(\frac{2f-1}{2w}\right),$$

with a scale factor ρ that depends on the effectiveness of the algorithm. We solved this differential equation and integrated to obtain a theoretical online fitness measure, and then fitted the experimental data to the measure to estimate ρ's for the two algorithms. The evolution-only data yielded a $\rho_e = 0.56$ with an excellent fit — a test to distinguish the empirical data from the theoretical curve via z-scores fails ($p > 0.5$). This result suggests that although the evolution-only algorithm is successfully hillclimbing over the course of generations, and in spite of non-zero α, its hillclimbing rate is too slow to keep up except when α is small.

Although fitting the evolution-learning data to this model yields an impressive $\rho_{el} = 11.52$, the model in fact misses the empirical data, on average, by almost four standard deviations (z-score $= 3.95$). It is clear that this model — which does not recognize the value of constant utility — is inappropriate to the combined algorithm. To account for the higher performance of the evolution-learning algorithm properly, more sophisticated models, incorporating both the timescale of learning as well as the timescale of evolution, are needed.

4 Discussion

There is a growing selection of model computational systems displaying interaction effects between learning and evolution, with "learning" taken to refer to just about any sort of intra-individual-lifetime adaptive process. In these models, according to various criteria, learning provides an improvement compared to evolution alone. In Hinton & Nowlan's model [10], learning guides an evolutionary search toward a well-hidden maximum. Smith [15] points out that learning can also help compensate for the disruptive effects of sexual recombination. Fontanari & Meir [7] provide simulations and analytic results of a model that shows how learning allows an evolutionary search to function successfully under much higher mutation rates. In Miller & Todd's model [12], learning speeds evolutionary search given noisy environmental cues. In the present model, the addition of learning allows evolution to track a basically deterministic but rapidly changing environment.

In one sense, all that these models do is show that learning definitely is helpful under conditions expressly chosen to highlight learning's ability to help. On that view, the successes of these evolutionary learning algorithms are all very obvious... *after* the natures of the chosen conditions are understood. The search for computational strategies is as much a search for useful problem settings as it is for effective problem solvers, and it is in that sense that these models are particularly relevant. It is all too easy to buy into some simple conceptualization and then overlook alternate or more complex approaches in the attempt to fit as much of reality as possible into the adopted view.

Viewing a model solely in terms of an adaptive system and an environment, we feel, can often be such an overly-simple perspective. In the present model we have, in effect, a three-way distinction between adaptive system, changeable environment, and constant (or very slowly changing) utility. If one forces this scenario into the simpler two-way view, the constancy of the utility function is obscured by the changeable environment, increasing the plausibility of the mistaken notion that non-stationary environments are fundamentally incompatible with adaptive systems simply by virtue of being non-stationary.

Although the "adaptive system versus environment" viewpoint is undeniably parsimonious, the proposed three-cornered approach may in reality be a much more useful perspective. After all, the nominal "environment" of many natural and artificial adaptive systems includes both an "external" environment and a privileged subsystem — a "body" — that controls and is controlled by the adaptive system. In such cases, although the utility function — some measure of reproductive success, perhaps — requires much of a lifetime to evaluate, the proximal needs of the body — hunger, thirst, and so forth — provide a goodness function that is a very useful first approximation to utility. With some ability to learn from the goodness function, new effective actions can be discovered fairly quickly in the face of all sorts of changes in the external environment, everything from the onset of an ice age or a river changing course down to the rise of home pizza delivery or the microwave oven.

A related circumstance occurs when a short-term-stable environment is significantly non-stationary over a longer term due to coevolution of multiple adaptive systems. The prey become harder to catch and the predators must change in response, or the predators become better at hunting and the prey must change in response. Of course, all such changes can simply be referred back for evolution to handle over the course of generations, and evidently that is what happens in many biological adaptive systems; what we have seen in this paper is that adding learning can provide a faster response than evolution can alone, allowing a greater rate of environmental change to be tolerated.

We believe that the perspective of the constant utility non-stationary environment perspective captures a wide range of natural and artificial adaptive systems. In such cases, evolutionary reinforcement learning is one obvious place to start. There is much to learn.

References

[1] Ackley, D.H. & Littman, M.L. (1990). Learning from natural selection in an artificial environment. In M. Caudhill, (ed.) *Proceedings of the International Joint Conference on Neural Networks, IJCNN-90-WASH-DC*, Volume I, 189–193, Lawrence Erlbaum Associates: Hillsdale, NJ.

[2] Ackley, D.H. & Littman, M.L. (1990). Generalization and scaling in reinforcement learning. In D. Touretzky, (ed.) *Advances in Neural Information Processing Systems – 2*, 550–557, Morgan Kaufmann: San Mateo, CA.

[3] Ackley, D.H. & Littman, M.L. (1991, in press). Interactions between evolution and learning. In C.G. Langton, (ed.) *Artificial Life II*, Morgan Kaufmann: San Mateo, CA.

[4] Baldwin, J. M. (1896). A new factor in evolution. *American Naturalist*, 30: 441-451, 536-553.

[5] Belew, R. (1989). Evolution, learning and culture: Computational metaphors for adaptive algorithms. University of California at San Diego, CSE Technical report CS89–156, La Jolla, CA.

[6] DeJong, K.A. (1980). Adaptive system design: a genetic approach. *IEEE Trans. Syst., Man, and Cyber., SMC-10*(9), 566–574.

[7] Fontanari, J.F., & Meir, R. (1990). The effect of learning on the evolution of asexual populations. *Complex Systems* **4**, 401–414.

[8] Goldberg, D. (1989). *Genetic algorithms in search, optimization, and machine learning.* Addison-Wesley, MA.

[9] Grefenstette, J.J. (1987). *A user's guide to GENESIS.* (Available with the GENESIS system via electronic mail to gref@NRL-AIC.ARPA).

[10] Hinton, G.E. & Nowlan, S.J. (1987). How learning can guide evolution. *Complex Systems*, **1**, 495–502.

[11] Holland, J.H. (1975). *Adaptation in Natural and Artificial Systems.* University of Michigan Press, Ann Arbor, MI.

[12] Miller, G.F. & Todd, P.M. (1990). Exploring adaptive agency I: Theory and methods for simulating the evolution of learning. In D.S. Touretzky, J.L. Elman, T.J. Sejnowski, & G.E. Hinton (Eds.), *Proceedings of the 1990 Connectionist Models Summer School.* San Mateo, CA: Morgan Kaufmann, 65–80.

[13] Rumelhart, D.E., Hinton, G.E., & Williams, R.J. (1986). Learning internal representations by error propagation. Chapter 8 of D.E. Rumelhart & J.L. McClelland (Eds.) *Parallel Distributed Processing: Explorations in the microstructures of cognition. Volume 1: Foundations.*

[14] Schull, J. (1990). Are species intelligent? *Behavioral and Brain Sciences*, 13:1, 61–73.

[15] Smith, J.M. (1987). When learning guides evolution. *Nature*, **32**, 761–762.

The Genetic Algorithm and the Structure of the Fitness Landscape

Bernard Manderick and **Mark de Weger** and **Piet Spiessens**
AI Laboratory
Free University of Brussels
Brussels B-1050

Abstract

In this paper we study the relation between the performance of a genetic algorithm (GA) and some statistical features of the fitness landscape that the GA has to optimize. On two problems, the NK-landscapes and the Traveling Salesman Problem, we show that there is a strong relationship between the fitness correlation coefficients of genetic operators and their roles in the overall GA-performance. In this way, the fitness correlation coefficients provide an easy way to assess the roles of and to tune the different components of the GA. Our results also suggest that the correlation length of the fitness landscape defines an exploratory horizon beyond which GA-exploration reduces to random search.

1 Introduction

Recently a number of attempts have been made to characterize how simple or hard a function is for a genetic algorithm. Firstly, the notion of deceptiveness has been described by [5,6]. Goldberg's analysis of deceptiveness is based on Walsh analysis and characterizes how low-order schemata lead the GA away from the global optimum. Secondly, epistasis variance measures the degree of chromosome interaction present in a sample of genotypes [3]. Davidor suggests that high epistatis variance characterizes problems that are hard for GAs.

In this paper we analyze the problem in a different way. The combination of a function and the discrete coding of its arguments defines a *fitness landscape* on some combinatorial space like the bit string space. Our goal is to show how statistical features of the fitness landscape reflect how easy or difficult it is for a GA to optimize the corresponding function.

The statistical features investigated are:

- The autocorrelation function of the landscape: This function expresses for each distance h how correlated the fitnesses of points are which separated by that distance h.
- The correlation length of the landscape: This corresponds roughly with the distance one can jump and still have some information about the fitness there given the fitness here. And,
- The fitness correlation coefficients of the different genetic operators. These correlation coefficients are related to the correlation length and express how correlated the landscape appears to the corresponding genetic operator. The distance between the parents and the children depends on the genetic operator. For mutation, the distance is always one. For crossover, the distance depends on the distance between the parents and the chosen crossover point. So, the landscape may appear more or less correlated depending on the operator.

On a number of problems we show that there is a strong relationship between the correlation length of the landscape, the fitness correlation coefficients of the genetic operators and the performance of the GA. This way, the fitness correlation coefficients provide an easy way to assess the roles of and to tune the different components of the GA. And, the correlation length provides a way to evaluate the coding used for the problem. Our results suggest that the correlation length of the fitness landscape defines an exploratory horizon beyond which GA-exploration reduces to random search.

In the next three sections we describe the basic idea and apply it to 1) the NK-landscapes, a family of landscapes whose ruggedness can be tuned [11], and 2) the Traveling Salesman Problem (TSP).

*Address: Pleinlaan 2 - B-1050 Brussels - Belgium. Email: bernard@arti.vub.ac.be

2 Statistical Features of the Fitness Landscape

Given a fitness landscape we calculate three simple statistical features:

1. for each genetic operator OP its coefficient of correlation ρ_{OP},
2. the autocorrelation function $\rho(h)$ of the landscape, and
3. the correlation length τ of the landscape.

2.1 The correlation coefficient of a genetic operator

Each genetic operator OP is applied to either one individual (e.g. mutation) or two individuals (e.g. crossover) and generates the same number of offspring. For each application of the operator OP, the random variables F_p and F_c either represent the fitness of the one parent and its child or the mean fitness (when OP is the mutation operator) of the two parents and the mean fitness of the two children (when OP is a crossover mutation operator). To calculate the *(fitness) correlation coefficient* ρ_{OP} of an operator OP we take a number of parents, apply that operator OP to get their offspring and calculate the correlation coefficient between the fitnesses F_p and F_c:

$$\rho_{OP}(F_p, F_c) = \frac{Cov(F_p, F_c)}{\sigma(F_p)\sigma(F_c)}, \quad (1)$$

where $Cov(F_p, F_c)$ is the covariance between the random variables F_p and F_c, and $\sigma(F_p)$ and $\sigma(F_c)$ are the standard deviations of the random variables F_p and F_c.

2.2 The correlation length of a landscape

The correlation length of a fitness landscape is defined in terms of the autocorrelation function of a process on that landscape.

The *autocorrelation function* $\rho(h)$ of a random process $\{X_t\}_{t\in Z}$, represents the correlation coefficient between pairs of values of $\{X_t\}_{t\in Z}$ separated by an interval of length h.

Random processes for which the autocorrelation function makes sense (see below) can only be defined for genetic operators that transform one point into another one, like mutation. Such an operator defines a neighborhood for the search space. The *neighborhood* of a point is defined as the set of points into which it can be transformed by applying the genetic operator once. We carry out a *random walk* through the search space by applying this operator repeatedly. This way, starting from a randomly chosen point with fitness F_0, we get a sequence of fitness values $F_1, F_2, \ldots, F_M$.

If we now assume that the fitness landscape is isotropic (i.e. the statistics of the random walk do not depend on the particular random walk used in the calculations) and that the process described by the random walk is stationary[1] then the obtained autocorrelation function and correlation length are independent from the particular random walk used to calculate the autocorrelation function and correlation length. So, the notions of autocorrelation function and correlation length *of the fitness landscape* make sense.

The autocorrelation function ρ is defined as

$$\rho(h) = \frac{1}{\sigma^2}R(h) = \frac{R(h)}{R(0)}, \quad (2)$$

where σ^2 is the variance and $R(h)$ is the *autocovariance function* of the stationary stochastic process. It is defined as

$$R(h) = E[(F_h - \mu_F)(F_0 - \mu_F)], \quad (3)$$

where E is the expectation or mean, F_0 is fitness of the starting point, F_h is the fitness at step h and μ_F is the mean fitness [10]. For practical purposes, $R(h)$ can be estimated by $R'(h)$ [15]:

$$R'(h) = \frac{1}{M}\sum_{i=0}^{M-|h|}(F_i - \bar{F})(F_{i+|h|} - \bar{F}), \quad (4)$$

$$\bar{F} = \frac{1}{M+1}\sum_{i=0}^{M} F_i \quad (5)$$

where $M > 0$ and $|h| < M$.

The *correlation length* τ is defined as the distance h where the autocorrelation function $\rho(h) = 1/2$.

3 The NK-landscapes

3.1 Introduction

Kauffman [11] describes a family of fitness landscapes determined by two parameters: N and K. The points of the NK-landscape are bit strings $\mathbf{b} = b_1 b_2 \ldots b_N$

[1] A stochastic process $\{X_t\}$ is stationary if for all $t_1, \ldots, t_k$ and h, the joint distributions of $\{X_{t_1}, \ldots, X_{t_k}\}$ and $\{X_{t_1+h}, \ldots, X_{t_k+h}\}$ are the same. This is a reasonable assumption for problems that are characterized by a large number of conflicting interacting constraints (like the TSP and the NK-landscapes). The assumptions have been verified for the TSP by Kirkpatrick and Toulouse [12] and for the NK-landscapes by Weinberger [17].

of length N. Each bit string $\mathbf{b}$ has the N strings at Hamming distance 1 as its neighbors. The parameter K specifies the *degree of epistatic interaction* between the individual bits i. The fitness landscape becomes more and more rugged as K increases. For a given N and K, the landscape is generated as follows:

1. For each bit i $(1 \leq i \leq N)$, pick K other bits to affect its fitness contribution to the total bit string fitness. Either K random bits are selected or the K closest neighbors of bit i are taken. This choice does not affect the resulting model [11].
2. Create for each bit i $(1 \leq i \leq N)$ a vector $\mathbf{w}_i$. The fitness contribution of the ith bit depends on its own value and the values of the K interacting bits. Hence, each vector $\mathbf{w}_i$ has 2^{K+1} entries, i.e. one entry for every possible combination of $K+1$ bit values. Each of these combinations is assigned a uniformly distributed random number between 0 and 1. This number represents the fitness contribution f_i of the ith bit b_i given the bit values of the K interacting bits.
3. The fitness $\mathbf{f}$ of the bit string $\mathbf{b} = b_1 b_2 \ldots b_N$ is obtained by taking the average of the fitness contributions f_i of the i bits, i.e. $f(\mathbf{b}) = \frac{1}{N}\sum_{i=1}^{N} f_i$.

An interesting property of the NK-landscapes is that the ruggedness of the fitness landscape can be tuned by changing the parameter K. This property allows us to experiment with the GA in a controlled way.

For $K = 0$, the fitness contribution f_i of bit b_i is independent of the values of all other bits. Starting with a given bit string, either it is the global optimum or we can find a fitter string by "flipping" one of its bits. The landscape has no local optima and a hillclimber will easily find the global optimum. Also, neighboring points have highly correlated fitnesses, as will be shown in the next section.

In contrast, for $K = N - 1$, the fitness contribution of any bit depends on the values of all other bits. If we flip one bit, not only its own fitness contribution changes, but also the contributions of all other bits. Since all of these contributions are drawn from a uniform random distribution, the fitnesses of the old and new strings are totally uncorrelated. As a consequence, the number of local optima is very high (the number is approximately $\frac{2^N}{N+1}$) and walks from any point to the nearest local optimum are very short. The ruggedness of this landscape makes it difficult to find the global optimum.

By gradually changing the degree of epistatic interaction K from 0 to $N - 1$, the fitness correlation between neighboring points of the landscape decreases, the number of local optima increases, the expected time to reach a local optimum decreases and the global optimum is harder to find [11].

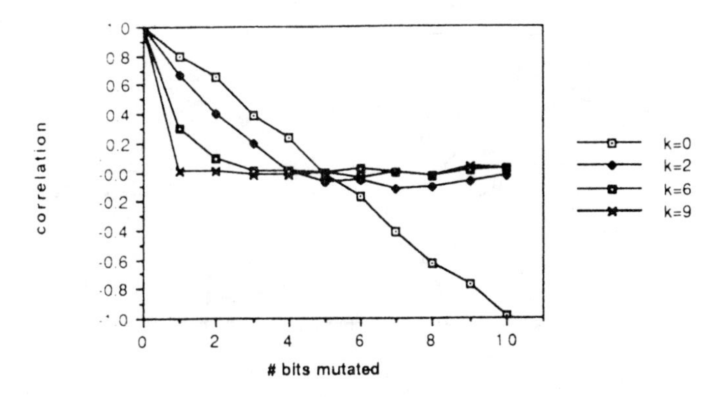

Figure 1: The correlation coefficient ρ_M as a function of the number of bits "flipped" by a mutation operator. 100 points and their mutants were used to calculate each value of ρ_M. $N = 10$, $K = 0,\ 2,\ 6,\ 9$.

3.2 Experimental results

First we describe two series of experiments that give insight in the correlation structure of the NK-landscape under mutation and crossover for different K. Then we investigate the relation between GA-hardness of an NK-landscape and its correlation length τ.

Mutation operators. In the first experiment we calculated the correlation coefficients ρ_M of different mutation operators M. Figure 1 plots this correlation ρ_M as a function of the number of bits that a mutation operator "flips". We took $N = 10$ and the correlation is plotted for $K = 0,\ 2,\ 6$ and 9.

For all K, the correlation ρ_M decreases approximately linearly from 1 to 0 with the number of bits flipped by a mutation operator. Only for $K = 0$, the correlation ρ_M continues to decrease to -1. This is a consequence of the linear character of the corresponding landscape. The steepness of each curve in Figure 1 is a good indication of the problems a landscape poses to a GA. For $K = 0$, the correlation coefficient $\rho_M = 0.81$ for a 1-bit mutation, whereas $\rho_M = 0$ for $K = 9$. The graph for $K = 0$ suggests that a high mutation rate or multiple-point mutations (i.e. in one mutation a fixed number of bits are flipped) may improve the GA-performance. There is still a high fitness correlation between points at Hamming distance 2 or 3. Multiple-point mutations might exploit these correlations. For $K = 9$, the mutation operator is unable to use the fitness information in the current generation of the GA to find better points. Neighboring points are totally uncorrelated.

Crossover operators. In the next experiments, we calculated the correlation coefficient ρ_X of the one-point crossover operator X. Figure 2 plots the correlation ρ_X for $K = 0, 2, 6$ and 9 $(N = 10)$ as a function of the Hamming distance between the parents.

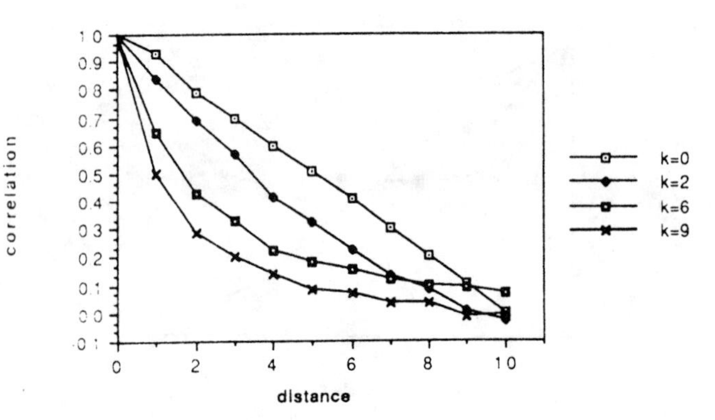

Figure 2: The correlation coefficient ρ_X of the one-point crossover operator as a function of the Hamming distance between the parents. 100 pairs of parents and their offspring were used to calculate each value of ρ_X. $N = 10,\ K = 0,\ 2,\ 6,\ 9$.

Again the fitness correlation ρ_X decreases from 1 to 0 and the higher the degree of epistatic interaction K is, the more rapidly it decreases. Even for small K, the correlation ρ_X becomes very small when the Hamming distance between the parents increases. As a consequence, the probability that the crossover operator produces highly fit offspring from highly fit but dissimilar parents becomes very low and crossover loses its exploratory effect. Moreover, the higher the degree of epistatic interaction K is, the closer the parents have to be.

This confirms the well-known heuristic in the GA-community that the crossover operator works best when restricted to similar parents. The statistical features of the landscape as we calculated them, indicate how similar the parents have to be. Our observations are also consistent with those of Kauffman who compared 1) crossover between points at a small Hamming distance, 2) crossover between randomly selected points, and 3) crossover between highly fit points. Kauffman concluded that crossover performs best in the first case [11].

Correlation length. In the last series of experiments, we looked at the relation between the correlation length τ of an NK-landscape and the GA-hardness of this landscape. For each K, we carried out random walks of length 2,048 and calculated the correlation length τ of the corresponding autocorrelation function. To measure the GA-hardness of a given landscape, we took the number of improvements in a fixed number of generations, i.e. the number of individuals with a higher fitness than in the starting population.

For a number of reasons, we used the GENITOR genetic algorithm [18]. Firstly, few parameters have to be tuned to optimize its performance. Only the selective pressure and the population diversity have to be controlled. And secondly, the fitness differences on the NK-landscape are so small that only rank-based allocation of trials works well.

The results of the experiments are shown in Table 1. The results show that the GA performs well on an NK-landscape when the correlation length τ of the landscape is high.

4 Traveling Salesman Problem

4.1 Introduction

In this section we analyze the TSP-landscape. For our GA-purposes, each possible tour is represented as a permutation of the numbers 1 to n. Each number represents a city and the cities are visited according to their order in the permutation. Mutation and crossover operators for the TSP are defined so that the resulting representation is again a permutation.

The particular TSP-landscape we analyzed is the standard 30-city problem (see [14]).

4.2 Experimental results

Next we describe experiments with three mutation and four crossover operators. Then we analyze the relation between the correlation length of the landscape and the overall GA-performance.

Mutation operators. We looked at three different mutation operators. Each mutation operator defines a different neighborhood relation for the TSP-landscape, resulting in different "topologies". We considered operators called Swap [14], Reverse [1] and Remove-and-Reinsert [4]. A mutation of a given tour is generated by randomly selecting two different cities and then respectively swapping the two cities, reversing the subtour between the two cities, or removing the first city from the tour and reinserting it after the second one. In each case, the neighborhood size is $n!/(n-2)!2!$ where n is the tour length. These neighborhoods are illustrated in Table 2 for a 10-city TSP.

Table 3 shows the correlation coefficients for swap mutation and reverse mutation. Table 4 shows the correlation lengths of landscapes under neighbourhoods defined by the different mutation operators. On the basis of these figures, we expected a GA using the Swap mutation to perform worse than the other ones.

Figure 3 compares the results of a GA with Swap mutation and Remove-and-Reinsert mutation on the 30-city problem.[2] Shown are the lengths of the short-

[2] GA-configuration: population size 50 for all experiments, mutation probability: 0.50. We used stochastic remainder sampling without replacement with a fitness scale factor of 2 as the selection procedure [7].

Table 1: The relation between the correlation length τ of an NK-landscape and the number of improvements Imp found by GENITOR during runs of 2048 generations. The dimension of the landscapes is $N = 96$ and the degree of epistatic interaction K takes the values $K = 0, 1, 2, 4, 8, 16, 32, 48, 95$. The results are averaged over 5 runs.

K	0	1	2	4	8	16	32	48	95
τ	29.96	24.37	19.51	14.15	7.06	3.90	1.72	1.00	0.52
$Imp.$	19.80	16.00	15.20	11.60	8.60	6.20	3.80	5.4	5.2

Table 2: The Swap, Reverse and Remove-and-Reinsert neighbors of the tour (1 2 3 4 5 6 7 8 9 10) when the 4th and 8th city are selected. The cities in the neighboring tour that differ from the original one are shown in bold.

Swap	Reverse	Remove-and-Reinsert
(1 2 3 **8** 5 6 7 **4** 9 10)	(1 2 3 **8 7 6 5 4** 9 10)	(1 2 3 **5 6 7 8 4** 9 10)

Table 3: The correlation coefficients of the 4 crossover and the 3 mutation operators.

ρ_{OX}	0.72
ρ_{PMX}	0.61
ρ_{CX}	0.57
ρ_{EX}	0.90

$\rho_{Reverse}$	0.86
$\rho_{Remove-and-Reinsert}$	0.80
ρ_{Swap}	0.77

est tour averaged over five runs. As expected, the Remove-and-Reinsert mutation performs slightly better than the Swap mutation.

Crossover operators. The four crossover operators considered here are: partially matched crossover PMX [8], cycle crossover CX [14], order crossover OX [14] and edge crossover EX [19]. Oliver *et al.* [14] compared the PMX-, CX- and OX-operators. They concluded that OX performs significantly better than PMX and that PMX performs slightly better than CX. EX was not included in their comparison. Whitley *et al.* [19] describe EX as performing impressively (for almost every tested TSP-problem, solutions were found that are at least as good as the best known solution).

We compared the four crossover operators on the 30-city problem. For each crossover operator we calculated its correlation coefficient. Table 3 shows the results.

Figure 4 shows the results of GA-runs with the different crossover operators and compares them with [14]. Figures 4 (a)-(c) (from [14]) give the best-so-far performance of one experiment with OX, PMX and CX, respectively. Figure 4 (d) gives the best-so-far performance (averaged over five experiments) for EX.[3]

It is clear that EX is far superior to any of the other crossover operators: after 1,000 generations it has already found a shorter tour (average tour length of 431) than OX after 50,000 generations (tour length of 449).

The GA-runs confirm our hypothesis that the better a GA-operator works, the higher its correlation coefficient is. Note also that even the quantitative differences—e.g. PMX works slightly better than CX, but OX a lot better than PMX—are reflected in the correlations.

Correlation length. The correlation length τ of the fitness landscape for the TSP depends on the used "topology". The correlation length τ of each landscape is shown in Table 4. As can be seen, the landscape is more correlated for the Reverse and the Remove-and-Reinsert neighborhoods than for the Swap neighborhood. This is also reflected in the correlation coefficients of the mutation operators corresponding with these neighborhoods (see Table 3). The test results of GA-runs were already shown in Figure 3.

On the TSP, the correlation length τ of the landscape and the correlation coefficients of the genetic operators are consistent with a number of observations reported in the literature [14,19]. Fogel [4] reported on experiments where a GA with Remove-and-Reinsert mutation but without crossover outperformed a GA without mutation but with PMX and questioned the role of crossover. Fogel's results are plausible if we take into account that 1) the TSP-landscape is more correlated using the Remove-and-Reinsert neighborhood than the Swap neighborhood, and 2) the correlation

[3]GA configuration: population size 50 for all experiments, probability of crossover: 0.50 (EX) or 0.80 (OX,CX,PMX), probability of mutation (swap): 0.30 (EX), 0.50 (OX,CX) or 0.60 (PMX). We used stochastic remainder sampling without replacement with a fitness scale factor of 2 as the selection procedure [7]. The settings are optimal for each experiment and were determined by experimentation.

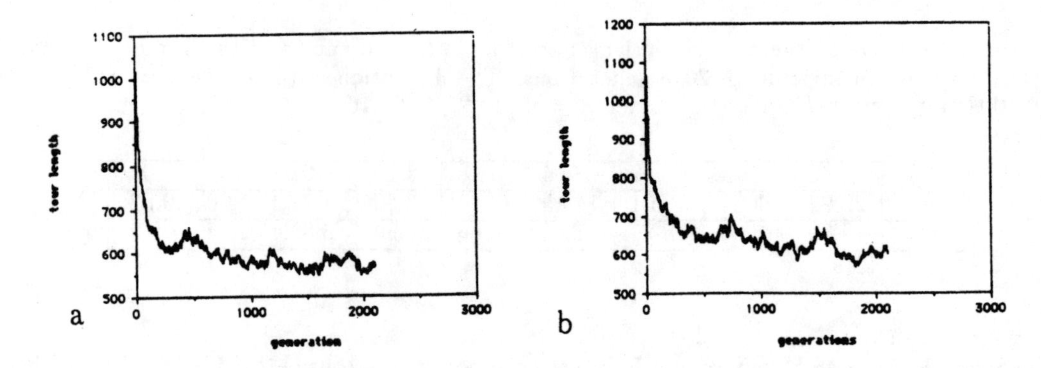

Figure 3: Best-so-far performance averaged over 5 runs on the 30-city TSP (no crossover). (a) Remove-and-Reinsert mutation, (b) Swap mutation.

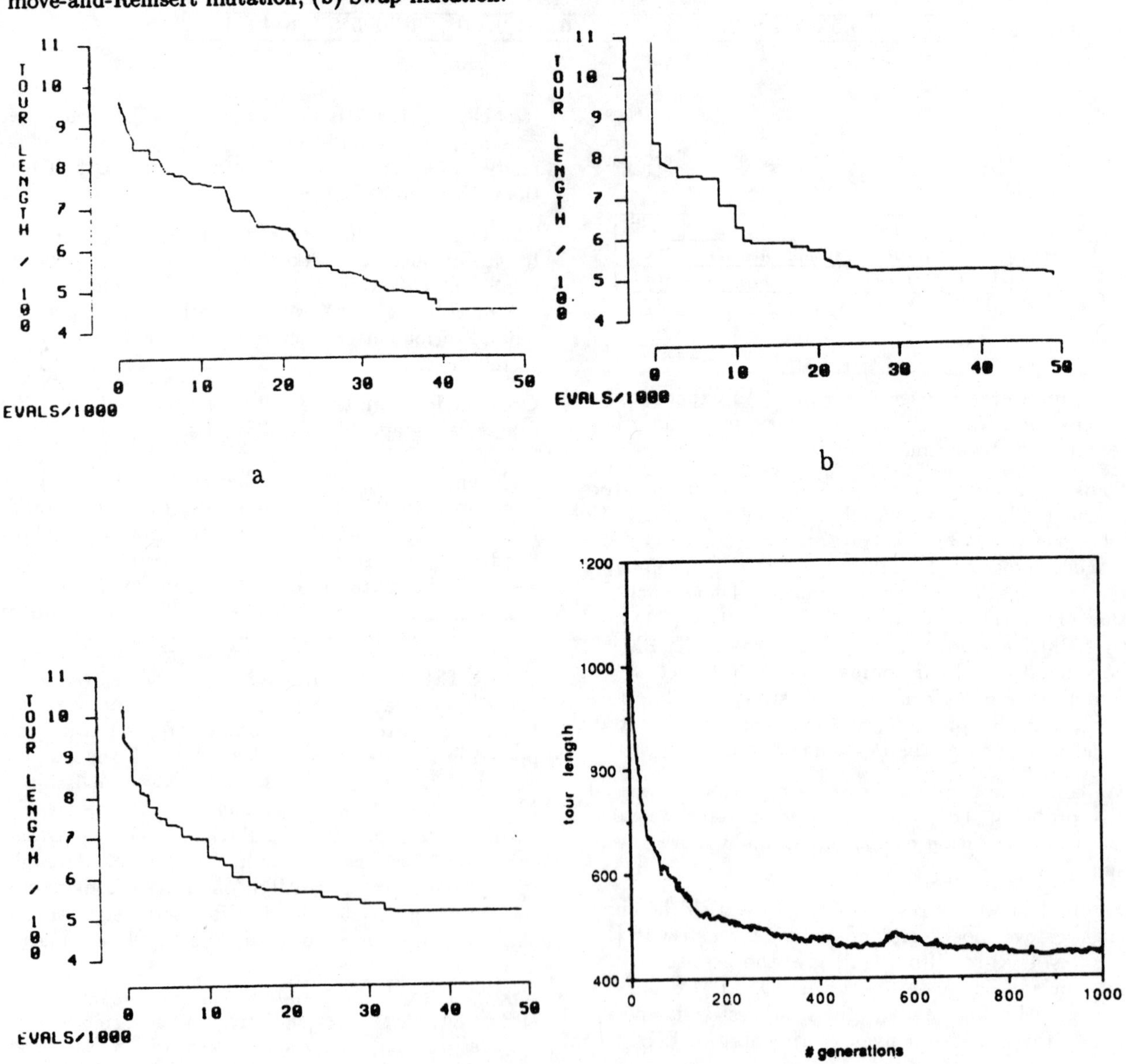

Figure 4: Results of a GA on the 30-city TSP. (a) OX, (b) PMX, (c) CX, (d) EX. (a)-(c) from [14].

Table 4: The correlation length τ of the TSP-landscape for the Swap, Reverse and Remove-and-Reinsert neighborhood of a 64-city problem. The cities are randomly distributed in a square with side 100. The results are averaged over 5 runs and the standard deviations are shown between brackets.

Swap	Reverse	Remove-and-Reinsert
10.44 (1.97)	20.43 (3.54)	14.53 (4.52)

coefficient ρ_{PMX} of the PMX-operator is rather small.

5 Conclusions

In this paper we have shown that for both the NK-landscapes and the TSP there exists a strong relation between the correlation coefficients of various operators and the performance of the GA.

A GA has to find a proper balance between exploitation and exploration. The correlation length τ of the fitness landscape and the correlation coefficients of the genetic operators may be used to control the exploration. Intuitively, the GA should explore potentially interesting new regions only when these regions are not too far (depending on the correlation length τ) from the currently exploited regions. A high degree of exploratory search (high mutation rate or multiple-point mutations and crossover between dissimilar parents) can be maintained if the correlation coefficients are high. In contrast, if the correlation length τ is small and the GA is too explorative then it will not perform much better than random search. So, τ defines an exploratory horizon beyond which genetic search degrades to random search.

Calculating the correlation coefficients of mutation and crossover operators also has some practical benefits. It is an easy way to evaluate the operators used in a GA and the coding used in the problem. The best alternative so far is running extensive GAs or meta-GAs to find optimal parameter values for the GA [9]. But these parameters are a compromise depending on the test suite and not tuned for the particular problem at hand. Finding optimal codings for a problem is a topic of active research [13]. Different codings result in different fitness landscapes for the same problem. Our results suggest that the correlation length τ of the landscape could be used to evaluate the coding since τ relates to the GA-hardness of the problem.

Both studied landscapes are isotropic. This implies that τ does not depend on the random walk carried out to calculate the autocorrelation function, and that the correlation coefficients do not depend on the sample used in the calculations. Fitness landscapes corresponding with simple mathematical functions or deceptive problems might not be isotropic. We do not know yet if our approach works in these cases. However, we already have a number of encouraging results. A comparison of binary and Gray coding on De Jong's testsuite using the correlation length τ matches the results of Caruana *et al.* [2] quite well. And experiments with two deceptive problems [13,16] indicate that the correlation length τ of the corresponding landscapes is small, as expected.

Acknowledgments

We would like to thank Luc Steels, the director of the AI Lab, for his continuous support. Thanks also to Didier Keymeulen and Charles Berger for many valuable comments. Bernard Manderick would also like to thank the ERBAS-group for providing a forum where the ideas presented in this paper could be discussed.

This research has been sponsored by the Belgian Government under contract "Incentive Program for Fundamental Research in Artificial Intelligence, Project: Self-Organization in Subsymbolic Computation".

References

[1] R. Brady. Optimization strategies gleaned from biological evolution. *Nature*, 317:804, 1985.

[2] R.A. Caruana and J.D. Schaffer. Representation and hidden bias vs. binary coding for genetic algorithms. *Proceedings of Fourth International Conference on Machine Learning*, pages 153–161, 1988.

[3] Y. Davidor. Epistasis variance: A viewpoint on representations, ga hardness, and deception. *Complex Systems*, 4:369–383, 1990.

[4] D.B. Fogel. An evolutionary approach to the traveling salesman problem. *Biological Cybernetics*, 60:139–144, 1988.

[5] D.E. Goldberg. Genetic algorithms and walsh functions: Part I, a gentle introduction. *Complex Systems*, 3:129–152, 1989.

[6] D.E. Goldberg. Genetic algorithms and walsh functions: Part II, deception and its analysis. *Complex Systems*, 3:153–171, 1989.

[7] D.E. Goldberg. *Genetic Algorithms in Search, Optimization, and Machine Learning.* Addison-Wesley, Reading, 1989.

[8] D.E. Goldberg and R. Lingle. Alleles, loci, and the traveling salesman problem. *Proceedings of an International Conference on Genetic Algorithms and their Applications*, pages 154–159, 1985.

[9] J.J. Grefenstette. Optimization of control parameters for genetic algorithms. *IEEE Transactions*

on Systems, Man & Cybernetics, SMC-16(1):122–128, 1986.

[10] S. Karlin and H.M. Taylor. *A First Course in Stochastic Processes.* Academic Press, New York, 1975.

[11] S.A. Kauffman. Adaptation on rugged fitness landscapes. In D.L. Stein, editor, *Lectures in the Sciences of Complexity*, pages 527–618, Reading, 1989. Addison-Wesley.

[12] S. Kirkpatrick and G. Toulouse. Configuration space analysis of the traveling salesman problems. *J. Phys.*, 46:1277, 1985.

[13] G.E. Liepins and M.D. Vose. Representational issues in genetic optimization. *Journal of Experimental and Theoretical AI*, 2:1–15, 1990.

[14] I.M. Oliver, D.J. Smith, and J.R.C. Holland. A study of permutation crossover operators on the traveling salesman problem. *Genetic algorithms and Their Applications: Proceedings of the Second International Conference on Genetic Algorithms*, pages 224–230, 1987.

[15] M. Priestly. *Spectral Analysis and Time Series.* Academic Press, London, 1981.

[16] T. Starkweather, D. Whitley, and K. Mathias. Optimization using distributed genetic algorithms. In H.-P. Schwefel and R. Maenner, editors, *Parallel Problem Solving from Nature*, Lecture Notes in Computer Science, pages 176–185, Berlin, 1991. Springer Verlag.

[17] E. Weinberger. Correlated and uncorrelated fitness landscapes and how to tell the difference. *Biological Cybernetics*, 63:325–336, 1990.

[18] D. Whitley. The GENITOR algorithm and selective pressure: Why rank-based allocation of reproductive trials is best. *Proceedings of the Third International Conference on Genetic Algorithms*, pages 124–132, 1989.

[19] D. Whitley, T. Starkweather, and D. Fuqua. Scheduling problems and the traveling salesmen: The genetic edge recombination operator. *Proceedings of the Third International Conference on Genetic Algorithms*, pages 133–140, 1989.

Handling Constraints in Genetic Algorithms

Zbigniew Michalewicz
Department of Computer Science
University of North Carolina
Charlotte, NC 28223, USA

Cezary Z. Janikow*
Department of Computer Science
University of North Carolina
Chapel Hill, NC 27599, USA

Abstract

The major difficulty in applicability of genetic algorithms to various optimization problems is the lack of general methodology for handling constraints. This paper discusses a new such methodology and presents results from the experimental system GENOCOP (for GEnetic algorithm for Numerical Optimization for COnstrainted Problems). The system not only handles any objective function with any set of linear constraints, but also effectively reduces the search space. The results indicate that this approach is superior to traditional methods when applied to the nonlinear transportation problem.

1 INTRODUCTION

In this paper we present a new approach to solving numerical optimization problems with linear constraints, based on genetic algorithms. Our new methodology seems to fit between the OR and AI approaches. First, it uses the OR technique for problem representation, *i.e.* the formulation of constraints is quantitative. On the other hand, our method uses a genetic algorithm, which is considered as an AI based search method.

Since the genetic approach is basically an accelerated search of the feasible solution space, introducing constraints can be potentially advantageous and can improve the behavior of the technique by limiting the space to be searched. However, all traditional GA approaches do not use this fact and rather emply techniques aimed at minimizing the negative effect of such constraints. This, in turn, often increases the search space by allowing some infeasible solutions outside the constrained solution space.

In the proposed approach, linear constraints are divided into equalities and inequalities. The equalities are eliminated at the start, together with an equal number of problem variables; this action removes also part of the space to be searched. The remaining constraints, in the form of linear inequalities, form a convex set which must be searched for a solution. Its convexity ensures that linear combinations of solutions always yield feasible solutions — a property used throughout this approach. The inequalities are used to generate bounds for any given variable: such bounds are dynamic as they depend on the values of the other variables and can be efficiently computed.

The full discussion on the proposed approach will appear elsewhere (see [15]); in this paper we explain the main idea of the proposed approach and present the first results.

2 THE CONSTRAINTS PROBLEM IN GENETIC ALGORITHMS

Three different approaches to the constraints problem in genetic algorithms have previously been proposed. The first two involve transforming potential solutions of the problem into a form suitable for a genetic algorithm and using penalty functions or applications of "decoders" or "repair" algorithms. The third approach involves modifying the genetic algorithm to suit the problem by using new data structures and new genetic operators.

In this section, after defining the class of linearly constrained optimization problems, we briefly discuss these three approaches in turn.

2.1 THE PROBLEM

We consider a class of optimization problems that can be formulated as follows:

Optimize a function $f(x_1, x_2, \ldots, x_q)$, subject to the following sets of linear constraints:

*Present address: Department of Mathematics and Computer Science, University of Missouri, St. Louis, Missouri 63121-4499

1. Domain constraints: $l_i \leq x_i \leq u_i$ for $i = 1, 2, \ldots, q$. We write $\vec{l} \leq \vec{x} \leq \vec{u}$, where $\vec{l} = \langle l_1, \ldots, l_q \rangle$, $\vec{u} = \langle u_1, \ldots, u_q \rangle$, $\vec{x} = \langle x_1, \ldots, x_q \rangle$.
2. Equalities: $A\vec{x} = \vec{b}$, where $\vec{x} = \langle x_1, \ldots, x_q \rangle$, $A = (a_{ij})$, $\vec{b} = \langle b_1, \ldots, b_p \rangle$, $1 \leq i \leq p$, and $1 \leq j \leq q$ (p is the number of equations).
3. Inequalities: $C\vec{x} \leq \vec{d}$, where $\vec{x} = \langle x_1, \ldots, x_q \rangle$, $C = (c_{ij})$, $\vec{d} = \langle d_1, \ldots, d_m \rangle$, $1 \leq i \leq m$, and $1 \leq j \leq q$ (m is the number of inequalities).

This formulation is general enough to handle a large class of standard Operations Research optimization problems with linear constraints and any objective function. The example considered later, the nonlinear transportation problem, is one of many problems in this class.

2.2 PENALTY FUNCTIONS

One way of dealing with candidate solutions that violate the constraints is to generate potential solutions without considering the constraints and then penalizing them by decreasing the "goodness" of the evaluation function. In other words, a constrained problem is transformed to an unconstrained problem by associating a penalty with all constraint violations and the penalties are included in the function evaluation. However, though the evaluation function is usually well defined, there is no accepted methodology for combining it with the penalty. Davis discusses this problem in [2] listing disadvantages of using high, moderate, or light penalties:

> "If one incorporates a high penalty into the evaluation routine and the domain is one in which production of an individual violating the constraint is likely, one runs the risk of creating a genetic algorithm that spends most of its time evaluating illegal individuals. Further, it can happen that when a legal individual is found, it drives the others out and the population converges on it without finding better individuals, since the likely paths to other legal individuals require the production of illegal individuals as intermediate structures, and the penalties for violating the constraint make it unlikely that such intermediate structure will reproduce. If one imposes moderate penalties, the system may evolve individuals that violate the constraint but are rated better than those that do not because the rest of the evaluation function can be satisfied better by accepting the moderate constraint penalty than by avoiding it".

In [18] and [16] the authors present the most recent approaches for using penalty functions in GAs for constrained optimization problems. However, the paper by Siedlecki and Sklansky [18] discusses a particular constrained optimization problem and the proposed method is problem specific. The paper by Richardson, Palmer, Liepins, and Hillard [16] examines the penalty approach discussing the strengths and weaknesses of various penalty function formulations and illustrate a technique for solving three dimensional constrained problem. However, in the last section of their article the authors say:

> "The technique used to solve the three dimensional problem described above can't be generalized since quantities like the trend and maximum derivative are seldom available".

We do not believe this to be a promising direction. For a heavily constrained problem, such as the transportation problem, the probability of generating an infeasible candidate is too great to be ignored. The technique based on penalty functions, at the best, seems to work reasonably well for narrow classes of problems and for few constraints, or for cases of non-essential constraints.

2.3 DECODERS AND REPAIR ALGORITHMS

Another approach concentrates on the use of special representation mappings (decoders) which guarantee (or at least increase the probability of) the generation of a feasible solution and on the application of special repair algorithms to "correct" any infeasible solutions so generated. However, decoders are frequently computationally intensive to run [2], not all constraints can be easily implemented this way, and the resulting algorithm must be tailored to the particular application. The same is true for repair algorithms. Again, we do not believe this is a promising direction for incorporating constraints into genetic algorithms. Repair algorithms and decoders may work reasonably well but are highly problem specific, and the chances of building a general genetic algorithm to handle different types of constraints based on this principle seem to be slim.

2.4 SPECIALIZED DATA STRUCTURES AND GENETIC OPERATORS

The last and relatively new approach for incorporating constraints in genetic algorithms is to introduce richer data structures together with an appropriate family of applicable "genetic" operators which can "hide" the constraints presented in the problem (see [10], [12]).

Several experiments ([7], [13], [19], [20]) indicate the usefulness of this approach, but it is not always possible, for an arbitrary set of constraints, to develop an efficient data structure hiding such constraints. In addition, such structures require specialized genetic operators to maintain feasibility. Such extensions lack also the theoretical basis enjoyed by the classical genetic operators. Despite these objections, experimental results suggest that this approach may be promising. However, the particular choice of representation and operators must still be tailored to the specific problem to be solved.

3 A NEW METHODOLOGY: THE GENOCOP SYSTEM

The proposed methodology provides a way of handling constraints that is both general and problem independent. It combines some of the ideas seen in the previous approaches, but in a totally new context. The main idea behind this approach lies in (1) an elimination of the equalities present in the set of constraints, and (2) careful design of special "genetic" operators, which guarantee to keep all "chromosomes" within the constrained solution space. This can be done very efficiently for linear constraints and while we do not claim these results extend easily to nonlinear constraints, the former class contains many interesting optimization problems.

A full description of the GENOCOP system is presented in [15]; below we give a small example which should provide some insight into the proposed methodology.

Let us assume we wish to minimize a function of six variables:

$$f(x_1, x_2, x_3, x_4, x_5, x_6)$$

subject to the following constraints:

$$\begin{aligned}
&x_1 + x_2 + x_3 = 5\\
&x_4 + x_5 + x_6 = 10\\
&x_1 + x_4 = 3\\
&x_2 + x_5 = 4\\
&x_1 \geq 0,\ x_2 \geq 0,\ x_3 \geq 0,\ x_4 \geq 0,\ x_5 \geq 0,\\
&x_6 \geq 0.
\end{aligned}$$

We can take an advantage from the presence of four independent equations and express four variables as functions of the remaining two:

$$\begin{aligned}
&x_3 = 5 - x_1 - x_2\\
&x_4 = 3 - x_1\\
&x_5 = 4 - x_2\\
&x_6 = 3 + x_1 + x_2
\end{aligned}$$

We have reduced the original problem to the optimization problem of a function of two variables x_1 and x_2:

$$\begin{aligned}
g(x_1, x_2) = f(x_1, x_2, &(5 - x_1 - x_2),\\
&(3 - x_1), (4 - x_2), (3 + x_1 + x_2)).
\end{aligned}$$

subject to the following constraints (inequalities only):

$$\begin{aligned}
&x_1 \geq 0,\ x_2 \geq 0\\
&5 - x_1 - x_2 \geq 0\\
&3 - x_1 \geq 0\\
&4 - x_2 \geq 0\\
&3 + x_1 + x_2 \geq 0
\end{aligned}$$

These inequalities can be further reduced to:

$$\begin{aligned}
&0 \leq x_1 \leq 3\\
&0 \leq x_2 \leq 4\\
&x_1 + x_2 \leq 5
\end{aligned}$$

This would complete the first step of our algorithm: elimination of equalities.

Now let us consider a single point from the search space, *e.g.* $\vec{x} = \langle x_1, x_2 \rangle = \langle 1.8, 2.3 \rangle$. If we try to change the value of variable x_1 without changing the value of x_2 (*uniform mutation*), the variable x_1 can take any value from the range: $[0,\ 5 - x_2] = [0,\ 2.7]$. Additionally, if we have two points within search space, $\vec{x} = \langle x_1, x_2 \rangle = \langle 1.8, 2.3 \rangle$ and $\vec{x'} = \langle x'_1, x'_2 \rangle = \langle 0.9, 3.5 \rangle$, then any linear combination $a\vec{x} + (1 - a)\vec{x'}$, $0 \leq a \leq 1$, would yield a point within search space, *i.e.* all constraints must be satisfied (*whole arithmetical crossover*). Therefore, both examples of operators would not move a vector outside the constrained solution space.

The above example explains the main idea behind the GENOCOP system. Linear constraints were of two types: equalities and inequalities. We first eliminated all the equalities, reducing the number of variables and appropriately modifying the inequalities. Reducing the set of variables both decreased the length of the representation vector and reduced the search space. Since we were left with only linear inequalities, the search space was convex — which, in the presence of closed operators, could be searched efficiently. The problem then became one of designing such closed operators. We achieved this by defining them as being context-dependent, that is dynamically adjusting to the current context. Such operators, used in GENOCOP system, are quite different from the classical ones. This is because:

1. We deal with a real valued space R^q, where a solution is coded as a vector with floating point type components,
2. The genetic operators are dynamic, *i.e.* a value of a vector component depends on the remaining values of the vector,
3. Some genetic operators are non-uniform, *i.e.* their action depends on the age of the population.

In the convex space, the value of the i-th component of a feasible solution $\vec{s} = \langle v_1, \ldots, v_m \rangle$ is always in some (dynamic) range $[l, u]$; the bounds l and u depend on the other vector's values $v_1, \ldots, v_{i-1}, v_{i+1}, \ldots, v_m$, and the set of inequalities.

Before we describe the operators, we present two important characteristics of convex spaces (due to the linearity of the constraints, the solution space is always a convex space $\mathcal{S}$), which play an essential role in the definition of these operators:

1. For any two points s_1 and s_2 in the solution space $\mathcal{S}$, the linear combination $a \cdot s_1 + (1-a) \cdot s_2$, where $a \in [0,1]$, is a point in $\mathcal{S}$.
2. For every point $s_0 \in \mathcal{S}$ and any line p such that $s_0 \in p$, p intersects the boundaries of $\mathcal{S}$ at precisely two points, say $l_p^{s_0}$ and $u_p^{s_0}$.

Since we are only interested in lines parallel to each axis, to simplify the notation we denote by $l_{(i)}^s$ and $u_{(i)}^s$ the i-th components of the vectors l_p^s and u_p^s, respectively, where the line p is parallel to the axis i. We assume further that $l_{(i)}^s \leq u_{(i)}^s$.

Because of intuitional similarities, we cluster the operators into the standard two classes: mutation and crossover. The proposed crossover and mutation operators use the two properties to ensure that the offspring of a point in the convex solution space $\mathcal{S}$ belongs to $\mathcal{S}$. However, some operators (*e.g.* non-uniform mutation) have little to do with GENOCOP methodology: they have other "responsibilities" like fine tuning and prevention of premature convergence. For a detailed discussion on these topics, the reader is referred to [9] and [11].

Mutation group: Mutations are quite different from the traditional one with respect to both the actual mutation (a gene, being a floating point number, is mutated in a dynamic range) and to the selection of an applicable gene. A traditional mutation is performed on static domains for all genes. In such a case the order of possible mutations on a chromosome does not influence the outcome. This is not true anymore with the dynamic domains. To solve the problem we proceed as follows: we randomly select $p_{um} \cdot pop\_size$ chromosomes for uniform mutation, $p_{bm} \cdot pop\_size$ chromosomes for boundary mutation , and $p_{nm} \cdot pop\_size$ chromosomes for non-uniform mutation (all with possible repetitions), where p_{um}, p_{bm}, and p_{nm} are probabilities of the three mutations defined below. Then, we perform these mutations in a random fashion on the selected chromosome.

- **uniform mutation** selects a random gene k of the chromosome $s_v^t = \langle v_1, \ldots, v_m \rangle$: the result of this mutation is a vector $s_v^{t+1} = \langle v_1, \ldots, v_k', \ldots, v_m \rangle$, where v_k' is a random value (uniform probability distribution) from the range $[l_{(k)}^{s_v^t}, u_{(k)}^{s_v^t}]$. The dynamic values $l_{(k)}^{s_v^t}$ and $u_{(k)}^{s_v^t}$ are easily calculated from the set of constraints (inequalities).
- **boundary mutation** is a variation of the uniform mutation with v_k' being either $l_{(k)}^{s_v^t}$ or $u_{(k)}^{s_v^t}$, each with equal probability.
- **non-uniform mutation** is one of the operators responsible for the fine tuning capabilities of the system. It is defined as follows: if $s_v^t = \langle v_1, \ldots, v_m \rangle$ is a chromosome and the element v_k was selected for this mutation from the set of movable genes, the result is a vector $s_v^{t+1} = \langle v_1, \ldots, v_k', \ldots, v_m \rangle$, with $k \in \{1, \ldots, n\}$, and

$$v_k' = \begin{cases} v_k + \Delta(t, u_{(k)}^{s_v^t} - v_k) & \text{if a random digit is 0} \\ v_k - \Delta(t, v_k - l_{(k)}^{s_v^t}) & \text{if a random digit is 1} \end{cases}$$

The function $\Delta(t, y)$ returns a value in the range $[0, y]$ such that the probability of $\Delta(t, y)$ being close to 0 increases as t increases. This property causes this operator to search the space uniformly initially (when t is small), and very locally at later stages. We have used the following function:

$$\Delta(t, y) = y \cdot \left(1 - r^{(1-\frac{t}{T})^b}\right),$$

where r is a random number from $[0..1]$, T is the maximal generation number, and b is a system parameter determining the degree of non-uniformity.

Crossover group: Chromosomes are randomly selected in pairs for application of the crossover operators according to appropriate probabilities.

- **simple crossover** is defined as follows: if $s_v^t = \langle v_1, \ldots, v_m \rangle$ and $s_w^t = \langle w_1, \ldots, w_m \rangle$ are crossed after the k-th position, the resulting offspring are: $s_v^{t+1} = \langle v_1, \ldots, v_k, w_{k+1}, \ldots, w_m \rangle$ and $s_w^{t+1} = \langle w_1, \ldots, w_k, v_{k+1}, \ldots, v_m \rangle$. Note that the only permissible split points are between individual floating points (using float representation it is impossible to split anywhere else).
However, such operator may produce offspring outside of the convex solution space $\mathcal{S}$. To avoid this problem, we use the property of the convex spaces saying, that there exist $a \in [0,1]$ such that

$$s_v^{t+1} = \langle v_1, \ldots, v_k, w_{k+1} \cdot a + v_{k+1} \cdot (1-a), \ldots, w_m \cdot a + v_m \cdot (1-a) \rangle \in \mathcal{S}$$

and

$$s_w^{t+1} = \langle w_1, \ldots, w_k, v_{k+1} \cdot a + w_{k+1} \cdot (1-a), \ldots, v_m \cdot a + w_m \cdot (1-a) \rangle \in \mathcal{S}$$

The only question to be answered yet is how to find the largest a to obtain the greatest possible information exchange: due to the real interval, we cannot perform an extensive search. In GENOCOP we implemented a binary search (to some depth only for efficiency). Then, a takes the largest appropriate value found, or 0 if no value satisfied the constraints. The necessity for such actions is small in general and decreases rapidly over the life of the population. Note, that the value of a is determined separately for each single arithmetical crossover and each gene.

- **single arithmetical crossover** is defined as follows: if $s_v^t = \langle v_1, \ldots, v_m \rangle$ and $s_w^t = \langle w_1, \ldots, w_m \rangle$ are to be crossed, the resulting offspring are $s_v^{t+1} = \langle v_1, \ldots, v_k', \ldots, v_m \rangle$ and $s_w^{t+1} = \langle w_1, \ldots, w_k', \ldots, w_m \rangle$, where $k \in [1, m]$, $v_k' = a \cdot w_k + (1-a) \cdot v_k$, and $w_k' = a \cdot v_k + (1-a) \cdot w_k$. Here, a is a dynamic parameter calculated in the given context (vectors s_v, s_w) so that the operator is closed (points x_v^{t+1} and x_w^{t+1} are in the convex constrained space $\mathcal{S}$). Actually, a is a random choice from the following range:

$$
\begin{array}{ll}
[max(\frac{l^{s_w}_{(k)} - w_k}{v_k - w_k}, \frac{u^{s_v}_{(k)} - v_k}{w_k - v_k}), min(\frac{l^{s_v}_{(k)} - v_k}{w_k - v_k}, \frac{u^{s_w}_{(k)} - w_k}{v_k - w_k})] & \\
 & \text{if } v_k > w_k \\
[0, 0] & \text{if } v_k = w_k \\
[max(\frac{l^{s_v}_{(k)} - v_k}{w_k - v_k}, \frac{u^{s_w}_{(k)} - w_k}{v_k - w_k}), min(\frac{l^{s_w}_{(k)} - w_k}{v_k - w_k}, \frac{u^{s_v}_{(k)} - v_k}{w_k - v_k})] & \\
 & \text{if } v_k < w_k
\end{array}
$$

To increase the applicability of this operator (to ensure that a will be non–zero, which actually always nullifies the results of the operator) it is wise to select the applicable gene as a random choice from the intersection of movable genes of both chromosomes. Note again, that the value of a is determined separately for each single arithmetical crossover and each gene.

- **whole arithmetical crossover** is defined as a linear combination of two vectors: if s_v^t and s_w^t are to be crossed, the resulting offspring are $s_v^{t+1} = a \cdot s_w^t + (1-a) \cdot s_v^t$ and $s_w^{t+1} = a \cdot s_v^t + (1-a) \cdot s_w^t$. This operator uses a simpler static system parameter $a \in [0..1]$, as it always guarantees closedness (according to characteristic (1) of convex spaces given earlier in this section).

4 EXPERIMENTS AND RESULTS

For experiments we selected the following problem (transportation problem) of forty nine variables:

$$\text{minimize } f(\vec{x}) = \sum_{i=1}^{49} g(x_i) \; ,$$

subject to the following equality constraints:

$$
\begin{array}{l}
x_1 + x_2 + x_3 + x_4 + x_5 + x_6 + x_7 = 27 \\
x_8 + x_9 + x_{10} + x_{11} + x_{12} + x_{13} + x_{14} = 28 \\
x_{15} + x_{16} + x_{17} + x_{18} + x_{19} + x_{20} + x_{21} = 25 \\
x_{22} + x_{23} + x_{24} + x_{25} + x_{26} + x_{27} + x_{28} = 20 \\
x_{29} + x_{30} + x_{31} + x_{32} + x_{33} + x_{34} + x_{35} = 20 \\
x_{36} + x_{37} + x_{38} + x_{39} + x_{40} + x_{41} + x_{42} = 20 \\
x_{43} + x_{44} + x_{45} + x_{46} + x_{47} + x_{48} + x_{49} = 20
\end{array}
$$

$$
\begin{array}{l}
x_1 + x_8 + x_{15} + x_{22} + x_{29} + x_{36} + x_{43} = 20 \\
x_2 + x_9 + x_{16} + x_{23} + x_{30} + x_{37} + x_{44} = 20 \\
x_3 + x_{10} + x_{17} + x_{24} + x_{31} + x_{38} + x_{45} = 20 \\
x_4 + x_{11} + x_{18} + x_{25} + x_{32} + x_{39} + x_{46} = 23 \\
x_5 + x_{12} + x_{19} + x_{26} + x_{33} + x_{40} + x_{47} = 26 \\
x_6 + x_{13} + x_{20} + x_{27} + x_{34} + x_{41} + x_{48} = 25 \\
x_7 + x_{14} + x_{21} + x_{28} + x_{35} + x_{42} + x_{49} = 26
\end{array}
$$

Experiments were made for six nonlinear cost functions g (for a full discussion on the selection and classification of these functions, see [13]); their graphs are presented in Figure 1.

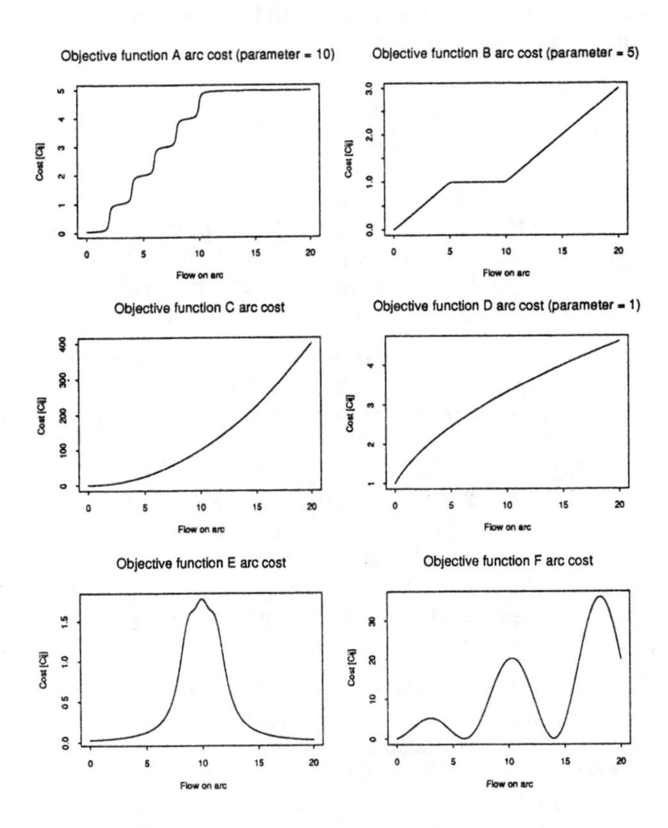

Figure 1: Six cost functions A – F.

There are thirteen independent and one dependent equations here; therefore, we eliminate thirteen variables: $x_1, \ldots, x_8, x_{15}, x_{22}, x_{29}, x_{36}, x_{44}$. All remaining variables are renamed $y_1, \ldots, y_{36}$. Each of these variables has to satisfy four two–sided inequalities, which result from the initial domains and our transformations. Now, each chromosome is a float vector $\langle y_1, \ldots, y_{36} \rangle$.

For comparative experiments we implemented all previously discussed GA approaches to the constraint problem (penalties, decoders, and specialized data structures), the GENOCOP system, and we used a version (the *student version*) of GAMS, a package for the construction and solution of large and complex mathematical programming models [1].

The experiments with penalty functions and decoders were not successful. For example, in experiments with penalty functions the evaluation function *Eval* was composed of the optimization function f and the penalty P:

$$Eval(\vec{x}) = f(\vec{x}) + P,$$

For our experiments we followed a suggestion (see [16]) to start with relaxed penalties and to tight them as the run progresses. We used

$$P = k \cdot (\frac{t}{T})^p \cdot \overline{f} \cdot \sum_{i=1}^{14} d_i$$

where $\overline{f}$ is the average fitness of the population at the given generation t, k and p are parameters, T is the maximum number of generations, and d_i returns the "degree of constraint violation".

We experimented with various values of p (close to 1), k (close to $\frac{1}{14}$, where 14 is total number of equality constraints; the static domain constraints were naturally satisfied by a proper representation), and $T = 8000$. However, this method did not lead to feasible solutions: in over 1200 runs (with different seeds for random number generator and various values for parameters k and p) the best chromosomes (after 8000 generations) violated at least 3 constraints in a significant way. The best solution was "far away" from the feasible solution for transportation problem.

Only GENOCOP system and a specialized data structures system (GENETIC-2, based on matrix data structure as a chromosome) gave feasible results: these are reported in [15]. In general, they are quite similar (GENETIC-2 was slightly better); however, note that the matrix approach was tailored to the specific (transportation) problem, whereas GENOCOP is problem independent and works without any hard-coded domain knowledge. In other words, while one might expect the GENOCOP to perform similarly well for other constrained problems, the GENETIC-2 cannot be used at all.

The results and comparison of GENOCOP and GAMS (summarized in Figure 2) indicate both usefulness of our method in the presence of many constraints and its superiority over some standard systems on nontrivial problems.

GENOCOP was run for 8000 iterations, with the population size equal 40. A single run of 8000 iterations took 2:28 CPU sec on Cray Y-MP. GAMS was run on an Olivetti 386 with a math co-processor and a single run took 1:20 sec.

Function	GAMS	GENOCOP	% difference
A	96.00	24.15	+297.52%
B	1141.60	205.60	+455.25%
C	2535.29	2571.04	−1.41%
D	565.15	480.16	+17.70%
E	208.25	204.82	+1.67%
F	43527.54	119.61	+36291.22%

Figure 2: GENOCOP versus GAMS: the results for the 7 × 7 problem.

5 CONCLUSIONS

After considering alternative ways for handling constraints in genetic algorithms for optimization problems, we proposed a new method for handling linear constraints. This new methodology should enable such constrained problems with difficult objective functions to be solved without incurring the heavy computational overhead associated with frequent constraint checking and without a need for designing a specific system's architecture.

The equality constraints are handled immediately by eliminating some variables, at one stroke removing constraints and reducing the search space. The inequalities are then processed to provide a set of bounds for each of the remaining variables considered in isolation. These bounds are dynamic in that they depend on the values of other variables of the current solution. Since the GA modifies each variable independently, this is not computationally complex.

Our results suggest that the method is useful as compared to the standard methods, and may lead towards the solution of some difficult Operations Research problems.

It is relatively easy to extend the GENOCOP system to handle discrete variables (nominal, linear, and boolean). Such variables would undergo different mutations and crossovers, to keep the solution vector within the constrained space. Also, it is possible to extend the GENOCOP system to handle nonlinear constraints provided that the search space is convex.

Acknowledgments: This research was supported by a grant from the North Carolina Supercomputing Center.

References

[1] Brooke, A., Kendrick, D., and Meeraus, A., *GAMS: A User's Guide*, The Scientific Press, 1988.

[2] Davis, L., (Editor), *Genetic Algorithms and Simulated Annealing*, Pitman, London, 1987.

[3] De Jong, K.A., *Genetic Algorithms: A 10 Year Perspective*, in [5], pp.169–177.

[4] Goldberg, D.E., *Genetic Algorithms in Search, Optimization and Machine Learning*, Addison Wesley, 1989.

[5] Grefenstette, J.J., (Editor), Proceedings of the First International Conference on Genetic Algorithms, Pittsburg, Lawrence Erlbaum Associates, Publishers, July 24–26, 1985.

[6] Grefenstette, J.J., (Editor), Proceedings of the Second International Conference on Genetic Algorithms, MIT, Cambridge, Lawrence Erlbaum Associates, Publishers, July 28–31, 1987.

[7] Groves, L., Michalewicz, Z., Elia, P., Janikow, C., *Genetic Algorithms for Drawing Directed Graphs*, Proceedings of the Fifth International Symposium on Methodologies of Intelligent Systems, Knoxville, pp.268–276, October 25–27, 1990.

[8] Holland, J., *Adaptation in Natural and Artificial Systems*, Ann Arbor: University of Michigan Press, 1975.

[9] Janikow, C., and Michalewicz, Z., *Specialized Genetic Algorithms for Numerical Optimization Problems*, Proceedings of the International Conference on Tools for AI, Washington, pp.798–804, November 6–9, 1990.

[10] Michalewicz, Z., Vignaux, G.A., Groves, L., *Genetic Algorithms for Optimization Problems*, Proceedings of the 11-th NZ Computer Conference, Wellington, New Zealand, pp.211–223, August 16–18, 1989.

[11] Michalewicz, Z. and Janikow, C., *Genetic Algorithms for Numerical Optimization*, Statistics and Computing, Vol.1, No.1, 1991.

[12] Michalewicz, Z., Schell, J., and Seniw, D., *Data Structures + Genetic Operators = Evolution Programs*, UNCC Technical Report, 1991.

[13] Michalewicz, Z., Vignaux, G.A., Hobbs, M., *A Non-standerd Genetic Algorithm for the Nonlinear Transportation Problem*, ORSA Journal on Computing, Vol.3, 1991.

[14] Michalewicz, Z., Krawczyk, J., Kazemi, M., Janikow, C., *Genetic Algorithms and Optimal Control Problems*, Proceedings of the 29th IEEE Conference on Decision and Control, Honolulu, pp.1664–1666, December 5–7, 1990.

[15] Michalewicz, Z. and Janikow, C., *GENOCOP: A Genetic Algorithm for Numerical Optimization Problems with Linear Constraints*, to appear in Communications of ACM, 1991.

[16] Richardson, J.T., Palmer, M.R., Liepins, G., and Hilliard, M., *Some Guidelines for Genetic Algorithms with Penalty Functions*, in [17], pp.191–197.

[17] Schaffer, J., (Editor), Proceedings of the Third International Conference on Genetic Algorithms, George Mason University, Morgan Kaufmann Publishers, June 4–7, 1989.

[18] Siedlecki, W. and Sklanski, J., *Constrained Genetic Optimization via Dynamic Reward–Penalty Balancing and Its Use in Pattern Recognition*, in [17], pp.141–150.

[19] Vignaux, G.A. and Michalewicz, Z., *Genetic Algorithms for the Transportation Problem*, Proceedings of the 4th International Symposium on Methodologies for Intelligent Systems, Charlotte, NC, pp.252–259, October 12–14, 1989.

[20] Vignaux, G.A. and Michalewicz, Z., *A Genetic Algorithm for the Linear Transportation Problem*, IEEE Transactions on Systems, Man, and Cybernetics, Vol.21, No.2, 1991.

Adapting the Evaluation Space to Improve Global Learning

Alan C. Schultz
Navy Center for Applied Research in Artificial Intelligence
Naval Research Laboratory
Washington, DC 20375-5000, U.S.A.
Email: schultz@aic.nrl.navy.mil (202) 767-2684

Abstract

In domains where a stochastic process is involved in the evaluation of a candidate solution, multiple evaluations are necessary to obtain a good estimate of the performance of an individual. This work shows that biasing the sampling of that problem configuration space can lead to better performance of the structure being learned given the same amount of effort.

1 Introduction

In many domains, particularly those where a stochastic or noisy evaluation process is involved, the evaluation of a candidate solution might require sampling, i.e. multiple evaluations, to get a good estimate of performance. This random sampling over the space of possible configurations of the problem environment is typically performed with a uniform distribution. In some cases, better performance can be achieved by using a non-uniform distribution of samples from this problem configuration space. Furthermore, instead of randomly choosing samples with a fixed distribution, the distribution can be altered *adaptively* over time to achieve a particular goal. We call this **adaptive sampling** of the problem configuration space.

It is important to note that the problem configuration space to which we refer is *not* the same space as the solution search space. The **solution search space** is the space that the genetic algorithm searches, i.e. the space where crossover and mutation are applied. This is the space of candidate solutions. The **problem configuration space** is the space of possible variations in the form of the problem being solved. For example, in the Evasive Maneuvers (EM) domain,[1] several parameters define the characteristics of the missile we are evading and the initial starting conditions of the missile in relation to the plane. Each evaluation, from the Genetic Algorithm (GA) point of view, requires multiple episodes of simulation, where the parameters of the missiles and starting conditions are randomly chosen each time. The performance is the average over these episodes. The space of possible parameter settings defines the problem configuration space, and adaptive sampling biases the distribution of samples from this space. The bias[2] is adjusted adaptively based on the performance of the samples seen. In essence, we are adaptively altering the evaluation function to learn a better global solution over the problem configuration space.

Many complex domains require multiple evaluations of the candidate solutions to get a good estimate of their performance. For example, some domains involve a simulation of an environment where the initial configuration of the environment is randomly chosen each episode (Schultz, 1991; Grefenstette, 1991; Selfridge, Sutton and Barto, 1985). In other domains, the evaluation process itself is noisy due to computational constraints. For example, in Fitzpatrick and Grefenstette (1988), an image registration process used statistical sampling of the image to reduce the computational complexity of the evaluation. The resulting noisy evaluation eliminated the need to examine all pixels in an image.

These complex domains typically use a uniform distribution in randomly selecting each sample from the space of problem configurations, in order to gain an accurate estimate of candidate solutions. What we are interested in is having the learning system choose the samples from the configuration space non-uniformly to maximize some aspect of the learning, e.g., the average performance over the entire configuration space. The learner, in this case, might be analogous to an active learner in that it chooses the specific training environment to maximize its learning.

There are several motivations for wanting to alter the selection of samples. In a general sense, we want our learning system to acquire knowledge structures that perform as well as possible in the domain of interest. In particular, we want the learned knowledge structures to be as generally useful as possible, while retaining high performance. It is well-known that randomly sampling the space of initial configurations of a problem yields more robust solutions (Sammut and Cribb, 1990). However, if this space of configurations is very large with much irregularity, then it is difficult to adequately sample enough of the space. Adaptive sampling tries to include the most productive samples so that the amount of sampling of the problem configuration space is reduced. Even if the space is not too large and has enough regularities to sam-

[1] The EM domain (Erickson and Zytkow, 1988) will be described in more detail in the next section.

[2] Here, we define bias to mean the non-uniform distribution.

ple adequately, adaptive sampling allows selecting a criteria over which the samples are chosen. The learned knowledge structures can be adapted to a particular use by biasing the sampling of the problem configuration space. Given an effort equal to uniform sampling, higher performance knowledge structures can be produced. This last use of adaptive sampling will be demonstrated in this study.

While the GA is sampling the solution space for good solutions to the problem, the adaptive sampling mechanism is sampling the space of possible configurations of the problem to be solved, in an effort to maximize what is learned by the GA. But what do we mean by "maximizing what is learned by the GA?" Depending on the goal, we might want to learn something that works as well as possible over the entire problem configuration space, or we might want a solution that produces a uniform performance over the area, i.e. we might be willing to accept a lower mean performance if the same performance is achieved at all points in the space. Another goal might be to maximize some subarea of the space.

To measure the above goals, we will use various statistics, such as the mean of performance over the area, the variance of the performance, the maximum or minimum of the performance, or the area of the space that achieves some level of performance.

This paper will show that without increasing the effort required, a higher performance solution can be found by biasing the sampling of the problem configuration space. This will be demonstrated empirically with the addition of an adaptive sampling mechanism to SAMUEL, a GA-based learning system that learns strategies for solving sequential decision problems.

Actively selecting training examples is not a new concept. In Scott & Markovitch (1989), a conceptual clustering system used a heuristic to guide the search of experience space, such that informative training examples could be generated. The heuristic was based on Shannon's uncertainty function. Although uncertainty is a useful heuristic when trying to maximize what is known about a region of the search space, in this study we base our biasing on the performance of regions of the space. The reason for this is two-fold. First, we assume that performance is the only feedback available to us, and we want to reduce the amount of explicit bookkeeping we must perform. Second, as will be shown later, we want to affect our sampling in ways that are related to the performance of the system.

Section 2 will identify the problem configuration space with respect to the Evasive Maneuvers domain. Section 3 will describe the adaptive sampling mechanism used in these experiments for sampling the configuration space. The experimental methodology will be explained in Section 4, and the results presented in Section 5. Finally, Section 6 will summarize the experiments and suggest better mechanisms for performing the sampling.

2 The EM Domain and Problem Configuration Space

The **Evasive Maneuvers** (EM) domain is a two-dimensional missile and plane problem where the object is for the plane to avoid being hit by the missile. The missile is initially much faster than the plane, but the plane is slightly more maneuverable. The missile will eventually exhaust its fuel and fall from the sky. The use of SAMUEL to learn plans for this domain was presented in Grefenstette, Ramsey, and Schultz (1990). The empirical results in this paper use the EM domain.

In this domain, missiles can have a wide variety of characteristics. In particular, the two main attributes for missiles are their initial speed and their maneuverability, and we have a rough idea of the minimum and maximum values for these attributes. Therefore, we will define our problem configuration space as a two dimensional space where one dimension corresponds to the missile speed, and the other dimension corresponds to the missile maneuverability.

Although we can learn high-performance strategies for evading specific instances of missiles from this space, we also want to learn a single strategy that has relatively high performance over as much of the space as possible. The system aboard the plane might then use the specialized strategies to defend against specific, *known* missiles, but will also have a general default strategy to use between the time a missile is first detected, and the time the missile is classified as a particular type.

Adaptive sampling will allow us to generate a high performance general strategy over a larger area of the problem configuration space. With uniform sampling, it would not be possible to learn to perform as well on the entire space.

3 An Adaptive Sampling Mechanism

This section describes the adaptive sampling mechanism in SAMUEL. Please note, however, that the specific mechanism used here is only one possible instantiation. Other mechanisms are possible, and will be discussed in the conclusion.

Outside the GA, there is a two-dimensional matrix. One dimension of the matrix corresponds to the missile speed and the other corresponds to the missile maneuverability. Each cell of this matrix represents a *gross* estimate of the performance for that combination of missile speed and maneuverability *over all episodes, over all members of the population, and across all generations.* The cells are initialized with the average possible payoff. This estimate is updated *every* episode with the following calculation:

$$CellValue = CellValue + B(reward - CellValue)$$

where B represents a rate of learning, *reward* is the performance from the episode, and *CellValue* is the current value of the cell in the matrix being updated. *CellValue* will converge to the mean payoff for the associated configuration.

The matrix is used *each* episode to bias the selection of the missile characteristics that will be used in that episode of the evaluation. Exactly how the matrix is used depends on the goal of using adaptive sampling. Many disciplines for biasing the distribution may be implemented. In this study, two biasing schemes are examined: **inverse bias**, and **contour bias**. In each case, the bias defines a weighting for the distribution of samples, and the samples are chosen stochastically based on the weighting.

The first criterion examined for biasing the distribution of samples, inverse bias,[3] can be stated as follows:

- Sample more heavily from areas of the space with lower performance, but never stop sampling the good areas.

This bias is shown in graphical form in Figure 1, where the X-axis is the value from the matrix (i.e., the input to the weighting function), and the Y-axis is the weighting for the selection of the sample.

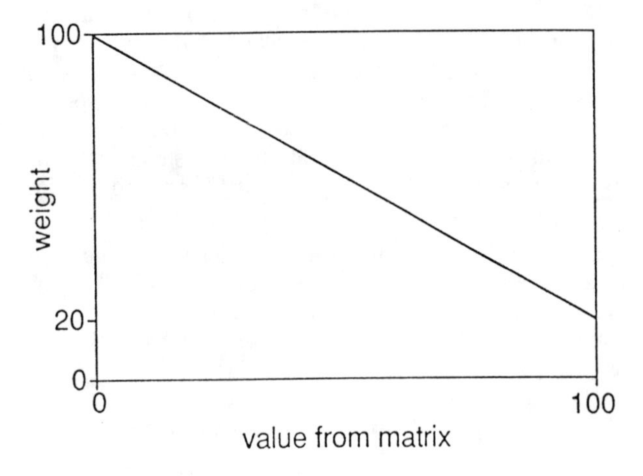

FIGURE 1: Inverse Bias function.

The intention here is that heavier sampling of the worst performing areas will force the system to "concentrate" on improving those areas of the problem configuration space. We want to continue to sample from the areas that already perform well so that we do not *forget* what we know for that area. One drawback of this approach is the underlying assumption that additional training on poor performing regions in the configuration space will improve the performance in these regions, or that the boundaries of the configuration space can be selected such that only "learnable" areas are included. If the total area of the problem configuration space includes a large area that can not be learned because of limited capabilities of the learning agent, then the overall performance will degrade. In the EM domain, if the missile is fast enough and manueverable enough, then no amount of learning will allow the aircraft to escape the missile. Therefore, it is important to limit the entire configuration space to areas known to be learnable. However, this might not be possible in practice. This observation was the motivation for the next criterion.

The next criterion used for biasing the distribution, contour bias,[4] can be stated as:

- Define a parabola shaped weighting around some performance value so that you sample more in areas that are close to that performance level, but never stop sampling at a fair rate in areas above that performance, and sample a little in areas below that performance.

This weighting function is illustrated in Figure 2.

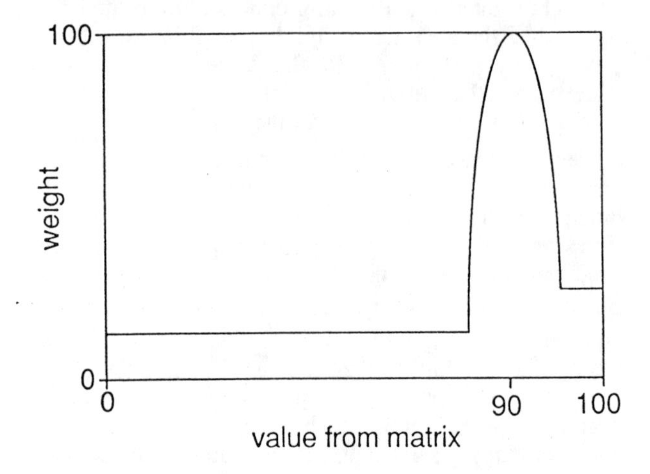

FIGURE 2: Contour Bias function.

the contour bias technique tends to *push* the area of at least the chosen performance out to cover a greater region, but does not suffer the problem of trying to sample areas where there is no hope of achieving any improvement. An improvement over the last criterion is that non-learnable regions are avoided. Whereas the last criterion tends to improve the mean performance over the entire area, this method is good for expanding the region of some given level of performance. This level of performance must be specified, and in Figure 2, as well as the reported experiments, is set at 90 percent.

Other disciplines are possible, depending on the objective of biasing the sampling. The results of applying these two adaptive sampling techniques is presented next, along with the results from the baseline (uniform distribution) experiment.

[3] So called because the distribution weighting is the inverse of the performance.

[4] This name refers to the distribution following a "contour" of a given performance level.

4 Experimental Method

In order to test the effectiveness of adaptive sampling, the following methodology was used. Each experimental run is composed of the learning stage, where an optimal plan is learned, followed by a testing stage, where that plan is evaluated.

TABLE 1: Statistics for uniform and adaptive sampling.

	uniform	inverse	contour
mean	86.86	91.17	92.13
variance	189.21	119.01	139.70
minimum	20.29	25.29	24.90
maximum	99.40	99.59	100.00
area above 95%	33%	51%	65%
area above 98%	13%	16%	35%
area above 99%	2%	2%	18%

During the learning phase, the adaptive sampling mechanism is enabled using one of the biases for choosing the samples from the problem configuration space. After 100 generations, the best plan is retrieved. In the testing phase, this best plan is subjected to extended evaluations on 256 combinations of values for the two parameters that define the problem configuration space (i.e. each parameter is divided into 16 values). The performance is presented as a contour plot where the x-axis is the missile speed and the y-axis is the maximum missile turning rate (maneuverability). The contours represent the level of performance with a given combination of speed and turning rate.

For comparative purposes, a baseline experiment was performed without adaptive sampling. In this experiment, random sampling with uniform distribution was performed over the problem configuration space. This would be equivalent to performing a Monte Carlo sampling of the space. The best plan from this experiment was again tested over the combination of values for the two parameters as described above.

In addition to the contour plots, which give a visual picture of the performance of the plan over the problem configuration space, various statistics can quantify the overall effect, as discussed earlier. For each experiment, we measure the mean performance over the space, the variance of the performance, the maximum and minimum performance, and the percentage of the area of the space where the performance was greater than or equal to 95, 98 and 99 percent. Table 1 summarizes the results for the baseline case of uniform distribution and for the two non-uniform distributions, inverse bias and contour bias.

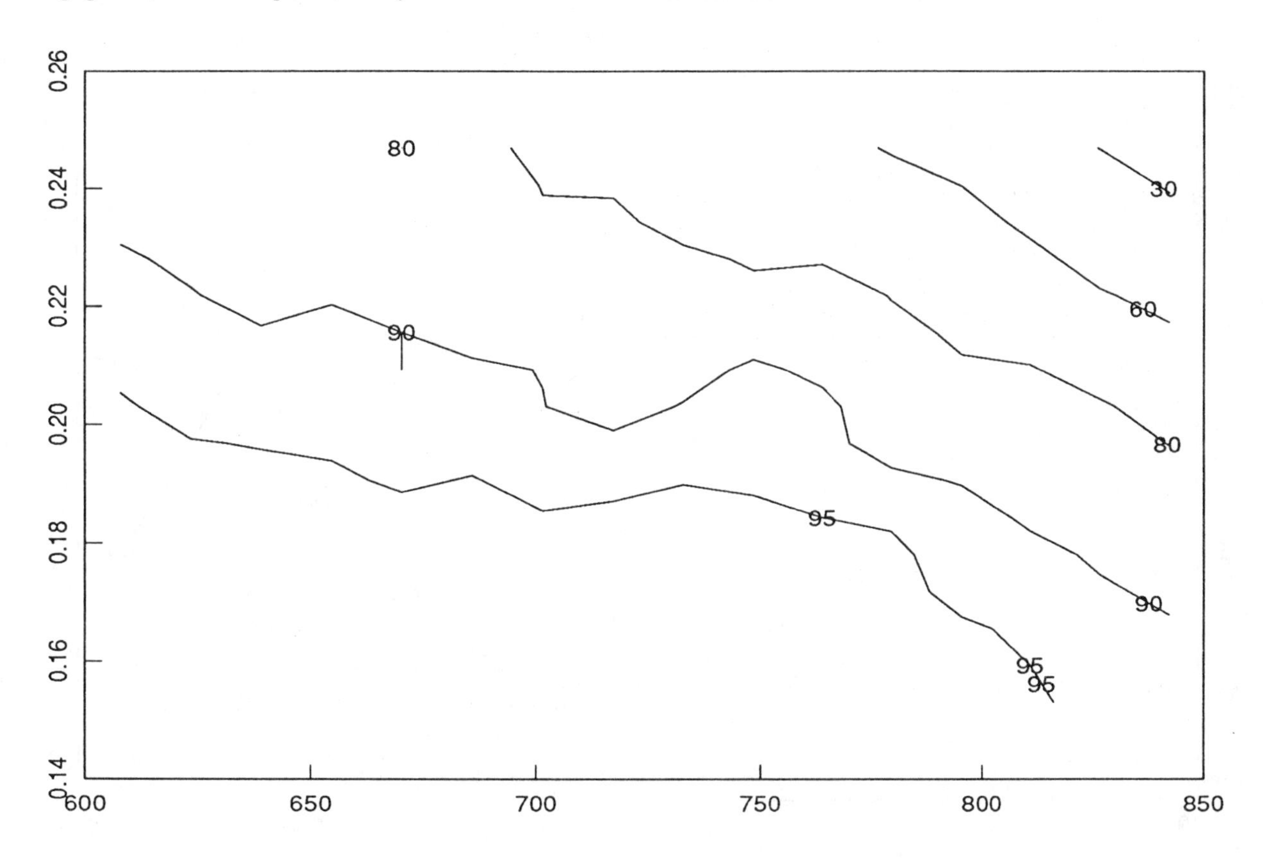

FIGURE 3: Baseline performance using uniform distribution.

5 Results

The results of the baseline experiment are shown in Table 1, under the heading *uniform*, and in Figure 3. Here we can see the effect of training over the entire problem configuration space using a uniform distribution of samples during the training. The results for the first adaptive sampling discipline, inverse bias, are shown in Table 1, under the heading *inverse*, and in Figure 4. These show the effect of sampling more where the performance of the samples is lower. The results from the second adaptive sampling discipline, contour bias, are shown in Table 1, under the heading *contour*, and in Figure 5. The results of biasing the samples towards a particular performance to expand its region is shown.

By comparing the columns in Table 1 and examining the three figures, the advantage of biasing the sampling becomes clear. The uniform sampling, as seen in Figure 3, only has acceptable performance on a small portion of the space. The inverse bias, by definition, strives to get a uniformly good performance over the entire area by concentrating on those areas where performance is lower. We can see from Figure 4 that the area of good performance is much larger. From the statistics, we see that the mean performance is better than in the uniform sampling case, and in particular, the variance in performance over the area is much lower. Also, the minimum performance in the area has risen significantly.

This indicates that the strategy learned is more robust and generally applicable to more of the situations it might encounter within the problem configuration space. Notice, however, that the performance at 98 percent and above did not increase significantly. Suppose that instead of wanting a strategy that performed relatively well, where the emphasis was on uniformity over the space, we wanted to emphasize a higher performance strategy, and find out how much of the space we could get it to cover.

With the contour bias, the emphasis is on making the area of very high performance as large as possible. This is achieved by concentrating on the area around a given, relatively high performance. This area should then "grow" to cover more area. In the table, we can see that the areas of very high performance are much greater when using the contour bias. In particular, the area with a performance greater than or equal to 98 percent has nearly tripled, while the area above 99% has increased by almost a factor of ten. A side effect, however, is that the variance is higher than in the inverse bias case. This is to be expected, since inverse bias strives for uniformity, while the contour bias attempts to enlarge the area of good performance, sometimes at the expense of other areas of the space. Note, however, that the variance is still better than when not using adaptive sampling at all. This bias also gave the highest maximum performance.

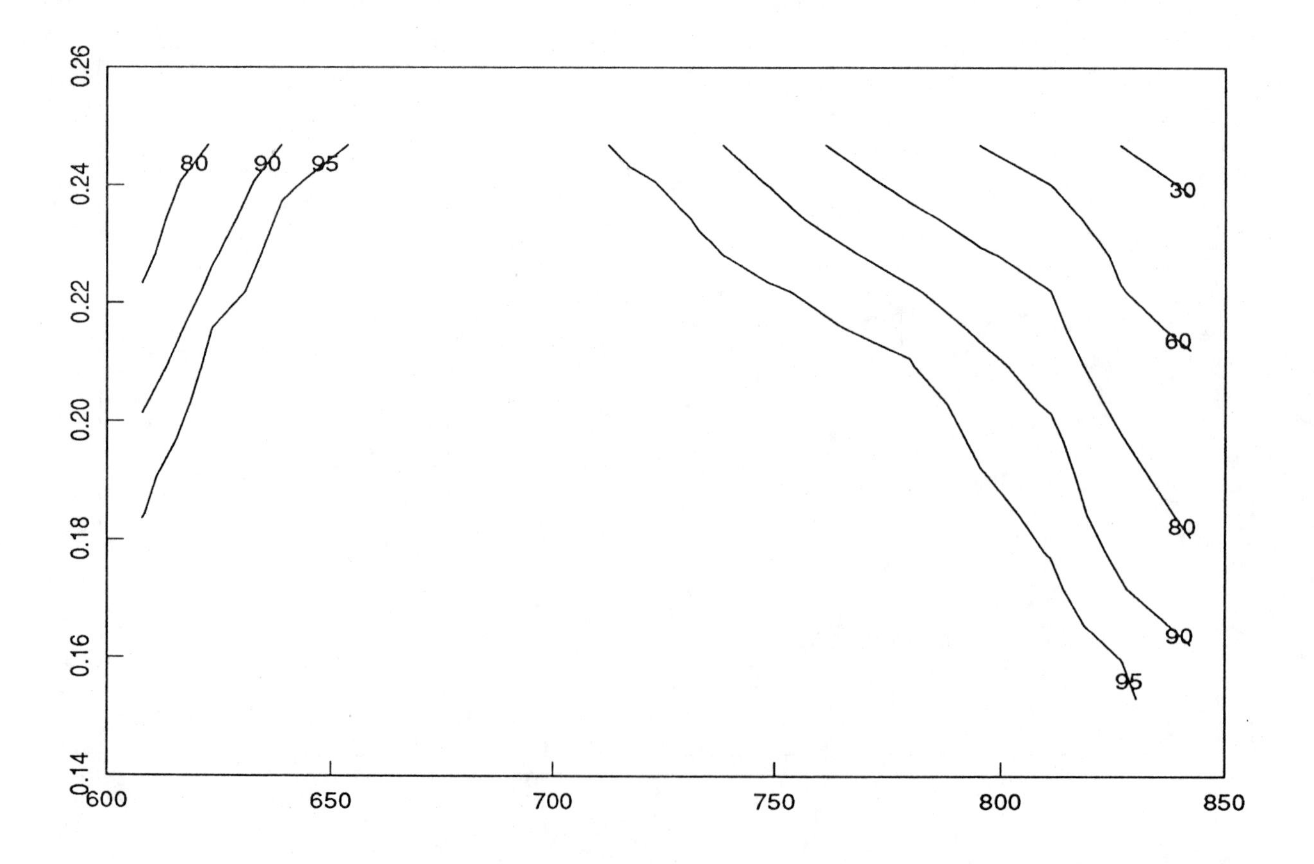

FIGURE 4: Performance using inverse bias.

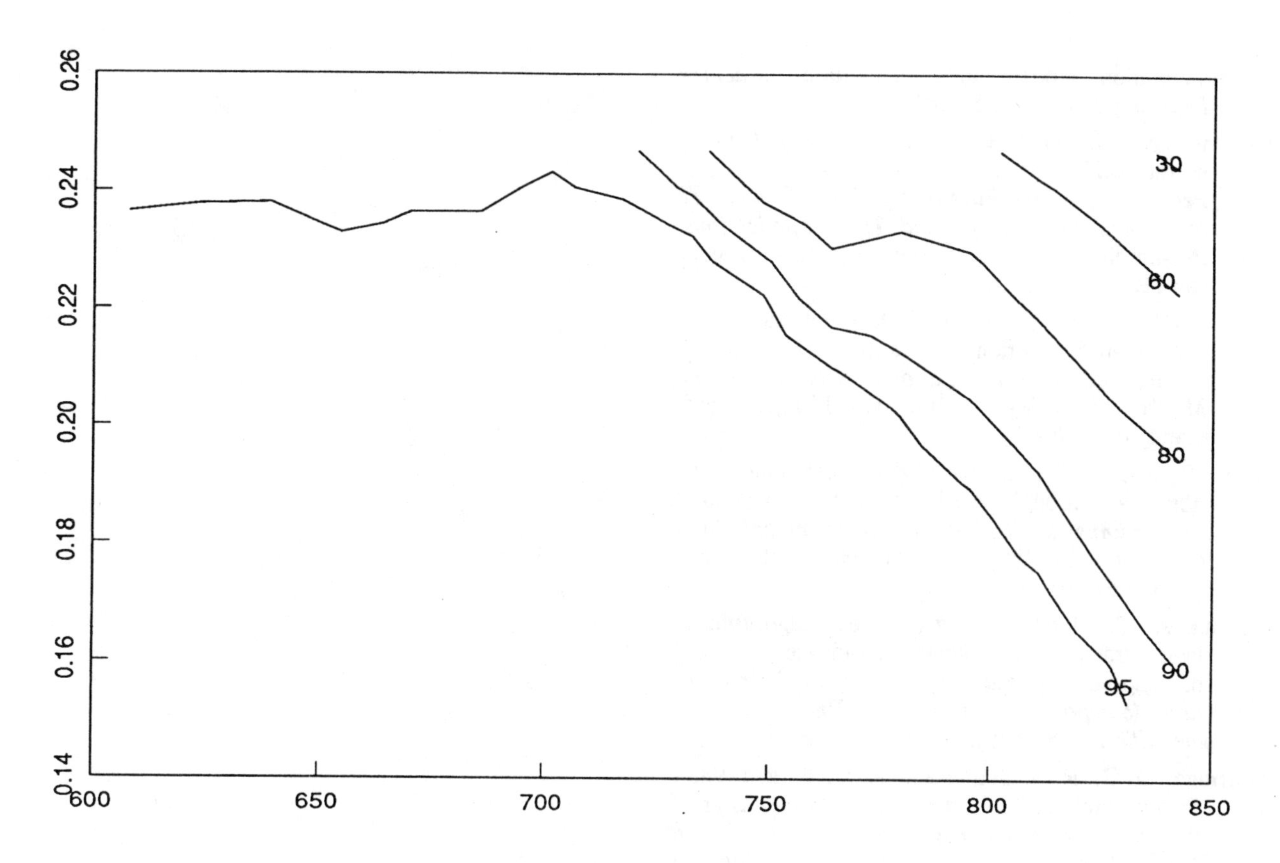

FIGURE 5: Performance using contour bias.

6 Conclusion

The empirical evidence presented in this paper shows that biasing the sampling of the problem configuration space can improve the overall performance (along some dimension) of the knowledge structures learned in problems involving stochastic evaluations. In the EM domain, the adaptive sampling is successfully used to generate a strategies with specific properties that are useful for a wide variety of opponents, by biasing the sampling of the space of opponents.

Future research will examine adaptive sampling in other domains, including domains where the motivation of reduced effort in sampling can be tested. Also, we would like to examine and characterize other biasing disciplines.

This paper's intention is to demonstrate the *idea* of adaptive sampling. The method used here (a two dimensional matrix) will not scale up if the problem configuration space is more than a few dimensions. In fact, the natural choice for the mechanism for adaptive sampling is a genetic algorithm, since it embodies the notion of implicit statistics without explicit bookkeeping. Therefore, another direction for further research is to develop adaptive sampling as a meta or cooperative genetic algorithm.

Acknowledgements

The author wishes to thank Helen Cobb and Connie Ramsey for their useful comments on the paper, the referees for their careful reviews, and the machine learning group at NCARAI for their comments on the idea of adaptive sampling, particularly John Grefenstette. This work is supported in part by ONR under Work Request N00014-91-WX24011.

References

Erickson, M. D. and J. M. Zytkow (1988). Utilizing experience for improving the tactical manager. *Proceedings of the Fifth International Conference on Machine Learning*. Ann Arbor, MI. (pp. 444-450).

Fitzpatrick, J. M. and J. J. Grefenstette (1988). Genetic algorithms in noisy environments. *Machine Learning, 3(2/3)*, (pp. 101-120).

Grefenstette, J. J. (1991). A lamarkian approach to learning in adversarial environments. *Proceedings of the Fourth International Conference on Genetic Algorithms*. San Diego, CA: Morgan Kaufmann.

Grefenstette, J. J., C. L. Ramsey, and A. C. Schultz (1990). Learning sequential decision rules using simulation models and competition. *Machine Learning, 5(4)*, (pp. 355-381).

Grefenstette, J. J. and J. M. Fitzpatrick (1985). Genetic search with approximate function evaluations. *Proceedings of the First International Conference on Genetic Algorithms and Their Applications*. Pittsburgh, PA: Lawrence Erlbaum Assoc. (pp.112-120).

Sammut, C. and J. Cribb (1990). Is learning rate a good performance criterion for learning? *Proceedings of the Seventh International Conference on Machine Learning*. Austin, TX: Morgan Kaufmann. (pp. 170-178).

Scott, P., and S. Markovitch (1989). Learning novel domains through curiosity and conjecture. *Proceedings of the Eleventh International Joint Conference on Artificial Intelligence*, Detroit, MI, pp. 669-674

Schultz, A. C. (1991). Using a genetic algorithm to learn strategies for collision avoidance and local navigation. *Proceedings of the Seventh International Symposium on Unmanned Untethered Submersible Technology*. Durham, NH: IEEE.

Selfridge, O. G., R. S. Sutton and A. G. Barto (1985). Training and tracking in robotics. *Proceedings of the Ninth International Joint Conference on Artificial Intelligence*. Los Angeles, CA: Morgan Kaufmann. August, 1985. (pp 670-672).

Formal Analysis of Genetic Algorithms

The Only Challenging Problems are Deceptive: Global Search by Solving Order-1 Hyperplanes

Rajarshi Das and Darrell Whitley
Department of Computer Science
Colorado State University
Fort Collins, CO 80523
whitley@cs.colostate.edu

Abstract

Empirical results are presented to support the observation that the only problems which pose challenging optimization tasks are problems that are at least partially deceptive. If no deception exists, the global solution to an optimization problem can theoretically be found by solving the order-1 hyperplanes, regardless of whether the problem is linear or nonlinear. In actual practice, we show that certain problems that have been used as test problems in the genetic algorithm literature are not deceptive (or are not *fundamentally* deceptive) and in fact the global solutions to these problems can be inferred by using statistical methods to determine (to some desired degree of reliability) the winners of the order-1 hyperplane competitions. In general, methods such as those described in this paper can be used to classify problems according to difficulty. Using problems that are not fundamentally deceptive as a test suite to compare different genetic algorithm implementations and operators can produce misleading results.

1 INTRODUCTION

Based on earlier work by Bethke (1980) Goldberg has investigated the notion of "deception" for several years now (1987, 1989a, 1989b, 1989c). The goal of these studies is to better understand what kind of situations are likely to create difficulty for a genetic algorithm when performing a function optimization task, and to create interesting and appropriate test problems for genetic algorithms. A few other researcher have also become interested in deception (e.g., Whitley 1991, Liepins and Vose 1990, Liepins and Vose 1991). Nevertheless, most researchers in the genetic algorithm community seem not to be particularly concerned with the notion of deception. This largely seems to be due to the perception that by studying deception we are concentrating on a set of aberrant or atypical problems that have little to do with the way genetic algorithms normally work. There are at least two major reasons why deception is critical to genetic algorithm research. First, deception interferes with "implicit parallelism." Because deception involves conflicting hyperplane competitions a genetic algorithm can only solve one hyperplane competition correctly when conflict is involved. Thus, deception is associated with a break down in implicit parallelism (Whitley 1991). Second, if no deception is present the problem can be solved by directly solving the order-1 hyperplane competitions. Thus, the only challenging problems for a genetic algorithm are problems that involve a significant degree of deception: one or more of the order-1 hyperplane competitions must be misleading.

A new search method and empirical results are presented which operationalize the idea that if there is no deception in a problem then the global solution for that problem can be found by correctly solving the order-1 hyperplanes. Simple statistical methods are used to infer the winners of order-1 hyperplane competitions. A problem is *fundamentally* deceptive if the globally optimal solution cannot be inferred by solving the order-1 hyperplanes either independently or in sequence of significance.

Finally, we show that the same problems which can be "solved" by looking only at order-1 hyperplane competitions are also problems where uniform crossover appears to outperform 1-point and 2-point recombination operators. In contrast, experiments indicate that on deceptive problems uniform crossover may be inferior to 1-point and 2-point crossover.

2 GLOBAL SOLUTIONS FROM ORDER-1 HYPERPLANES

The following theorem appears in Whitley (1991).

THEOREM 1: *Given a fitness function for a problem representing some optimization task with a binary encoding of length L,* **if** *1) no deception occurs in any of the hyperplanes associated with that particular binary* *encoding and 2) the winners of the L order-1 hyperplanes can be correctly determined,* **then** *the global optimum of the function is determined by the one string contained in the intersection of the L order-1 hyperplane competition winners.*

PROOF BY CONTRADICTION:

Assume that deception does not occur in any hyperplane of the fitness function and we determine the correct winners of each of the order-1 hyperplanes. The intersection of the order-1 schemata will identify exactly one string as the candidate for the global optimum. If the candidate is not the global optimum, then at least one of the order-1 hyperplanes is deceptive, thus generating a contradiction.

QED

While this theorem makes some idealistic assumptions, it can be operationalized. If no deception exists, then all hyperplane competitions are won by hyperplanes containing the global solution. We can therefore solve all hyperplane competitions by solving the order-1 hyperplane competitions. For each bit location we directly solve a competition involving the corresponding two schemata (e.g. *...*0*...* and *...*1*...*). If no deception is present, statistical methods can be used to determine (to some desired level of reliability) the probable value of the bit that belongs in the corresponding position in the binary string representing the global solution to the problem. The use of statistical methods also potentially allows us to make certain inferences about the role of sampling noise and the probability that we have correctly chosen the correct order-1 hyperplane winners.

Certain practical problems of course must be solved. For example, what if two competing order-1 hyperplanes have equal or near equal value? Note that if there is no deception, all subpartitions of the order-1 hyperplanes (i.e. the higher order hyperplanes it contains) must lead toward the global solution: to do otherwise would imply deception. Therefore, if we can solve *any* of the order-1 hyperplanes, we can use the solution to narrow the search space. If there is no deception, it does not matter in what sequence the order-1 hyperplanes are solved. Thus, if L winners cannot be decisively chosen when sampling from the entire space, the search space can be reduced by fixing the bits values for any winners that do emerge. If no deception exists, the sequence in which the order-1 hyperplanes are solved is irrelevant since no partition of hyperspace can support a winner other than the global optimum. This means that we can break ties and resolve ambiguous competitions if the winners of competing order-1 hyperplanes can be chosen in any relevant subpartition of the reduced search space.

This approach can also be used to solve problems that are partially deceptive if the order-1 hyperplanes are not misleading. However, if some deception exists and we try to narrow the search space by solving the order-1 hyperplanes sequentially, it is now possible that different sequences may produce different winners. In practice it will be hard to distinguish between problems in which no deception occurs and problems where we can only infer that the order-1 hyperplanes are not deceptive because the order-1 winners yield a solution that matches a known global solution. This is important because if higher order deception exists, then the sequence in which the order-1 hyperplanes are solved may be important. The following implementation is proposed. 1) Test for differences in all order-1 hyperplanes. 2) Select the order-1 competition characterized by the most significant difference to solve first. 3) Fix that bit and repeat the process until the number of strings remaining in the reduced search space can be exhaustively enumerated with easy. Any problem which *cannot* be solved in this way is said to be *fundamentally deceptive.* Note that this is an empirical class: in practice it may be impossible to distinguish problems that 1) are not deceptive in any hyperplane partition, 2) have no deception in the order-1 hyperplanes or 3) are not correctly solved because of sampling error.

To operationalize these ideas we need to compare two competing order-1 hyperplanes and determine average fitness. Unfortunately, a single direct comparison of their average fitness derived from a arbitrary population may or may not produce statistically significant results due to variations in hyperplane fitness and sampling error. This is especially true if the difference in the average fitness of two hyperplanes is relatively small. To overcome this problem, a running t-test was performed where the sample size is increased until the t-test indicates that the average fitness of the competing order-1 hyperplanes is significantly different. The t value is used to determine the confidence level of the result. We performed the running t-test on the order-1 hyperplanes for four different minimization problems. A 10 bit coding was used to represent each problem space with the global minimum at 1111111111.

The results in the following sections show that for problems with little or no deception, it is possible to perform simple statistical tests on the running average of the order-1 hyperplane fitness and infer the solution correctly. If a problem has a known optimum, then the method for determining if the problem is fundamentally deceptive are reasonably well defined. If a problem does not have a known optimum, the test to determine whether a problem is fundamentally deceptive can be used as a benchmark: if any search method finds a solution better than that which can be inferred from the order-1 hyperplanes, then the problem is by definition fundamentally deceptive. We believe that this test can be an important benchmark for test functions that are used in genetic algorithm research.

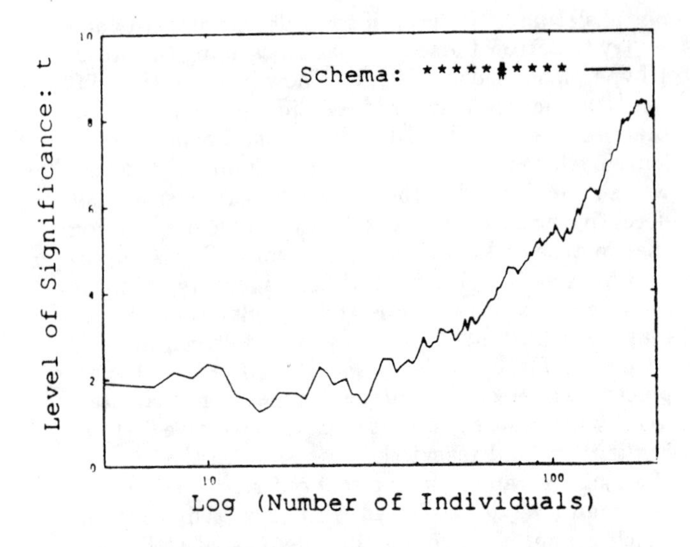

Figure 1: Significant difference between completing order-1 hyperplanes for the "one-max" problem as indicated by a t-test.

2.1 The "One Max" Problem

This is a simple bit counting problem described by Ackley [1987]. The total number of 1-bit in each string is counted and the value is subtracted from 10 to determine the individual's fitness. A full enumeration of all possible bit strings shows that in this bitwise linear function any order-1 hyperplane containing a 1-bit at a particular location is more fit than its competing hyperplane with a 0-bit at the same location. The One Max was a test problem used by Syswerda to compare the performance of uniform crossover with that of one-point and two-point crossover [1989].

Figure 1 shows the variation in t as a function of the population size for the hyperplane competition between *****1**** and *****0****. From this plot we can determine that the hyperplane *****1**** is more fit than the hyperplane *****0**** with a confidence level of 99.9% ($t > 3.37$) when the sample size is more than 75. The t-test for all other order-1 hyperplanes competitions lead to very similar results. The fact that all these tests were independent demonstrates that it is possible to solve each of the order-1 hyperplane competitions simultaneously.

2.2 A Unimodal Nonlinear Problem

This problem is the one dimensional version of De Jong's test function F1. As with the "one max" problem, a full enumeration of all possible bit strings shows that every order-1 hyperplane with a 1-bit is more fit than its competitor with a 0-bit. However a t-test shows some interesting differences. In Figure 2a the variation of t as a function of population size for the hyperplane competition 1********* and 0********* indicates $t > 5.0$ for a population size of 10 or more. In the hyperplane competition between *1******** and *0******** we find $t > 5.0$ for population size of more than 38 (Figure 2a). On the other hand, the value of t fluctuates between 0.0 and 2.0 when we consider the hyperplane competition between *********1 and *********0 (Figure 2a). Although we can statistically solve the order-1 hyperplanes competition quickly at the first two bit positions with a confidence level of more than 99.9%, the same is not possible for the last bit position. The first and the last bit are the most and least significant bit respectively under the binary encoding used here; furthermore, the contribution of fitness from each bit is correlated with its significance. Therefore, unlike the one max problem, it is difficult to solve all the order-1 hyperplane competitions independently in parallel in this case.

This problem can be solved using only order-1 hyperplane competitions if the competitions are solved sequentially. If at least one order-1 hyperplane competition is solved at a particular location, then that bit is set equal to the winning hyperplane before solving other competitions. This changes the sample average and variance, thus allowing a finer discrimination between order-1 hyperplanes in a smaller, restricted partition of hyperspace. As already noted, the sequence in which the order-1 hyperplanes are solved can influence the outcome if the problem is partially deceptive. The following discussion details steps in the general strategy described in Section 2 of this paper. Sample all of the order-1 hyperplanes. If they can be independently solved, stop. If they cannot be independently solved, increase the sample size until one of the order-1 hyperplane competitions has a winner. Use this winner to reduce the size of the search space by fixing the value of the relevant bit and repeat the process. The results of using this method to solve the unimodal nonlinear function is presented in Figure 2b.

The motivation for solving the order-1 hyperplane competitions in a sequential fashion is based on the fact that our empirical tests (Figure 2c) suggests that the genetic algorithm applies a similar strategy; our results show that on problems where fitness is correlated with the significance of the bits the genetic algorithm also solves the most significant bits first and then solves the less significant bits. This strategy is particularly relevant when the most significant bits in the parameter encoding(s) contribute most to the fitness value and the less significant bits contribute less to fitness. This same observation has been made by Schraudolph and Belew (1990) and employed in the implementation of *Dynamic Parameter Encoding* (DPE). While we will not describe the details of DPE in this paper, the similarities in the strategies does perhaps raise questions about the potential performance of DPE on deceptive problems.

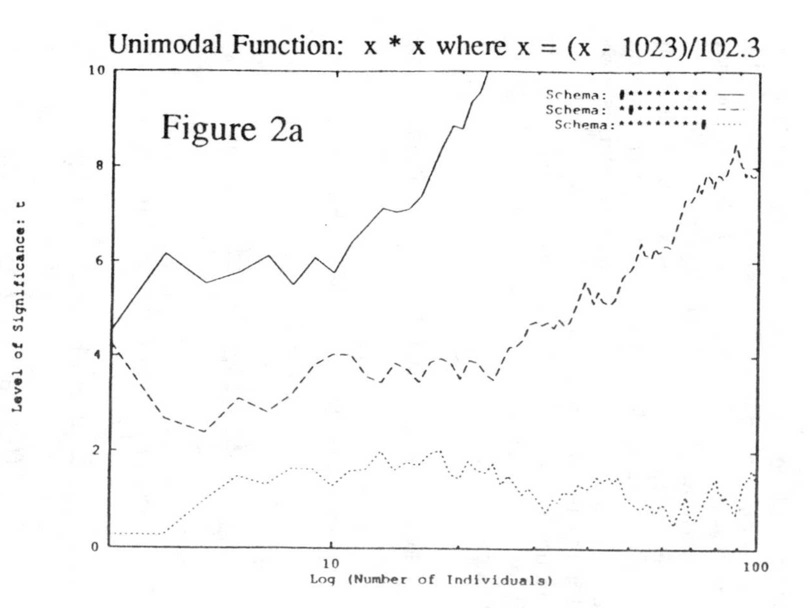

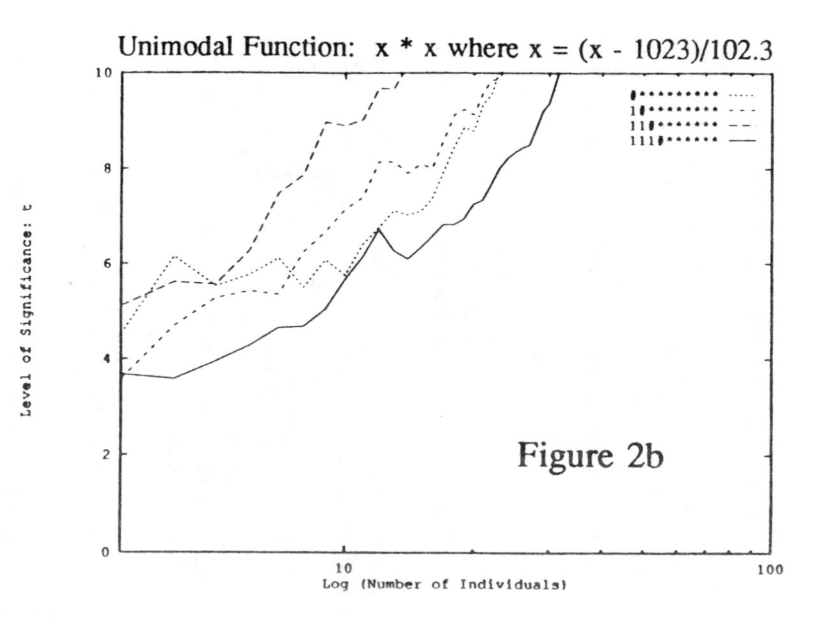

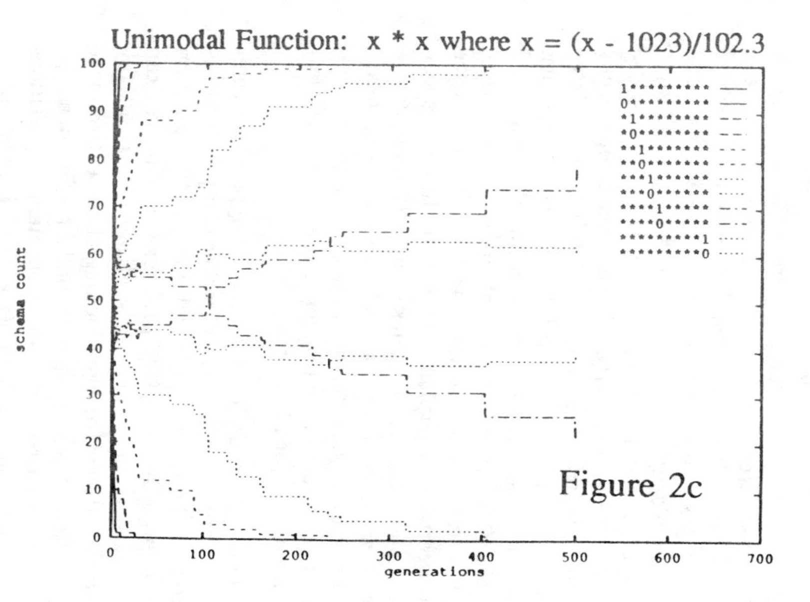

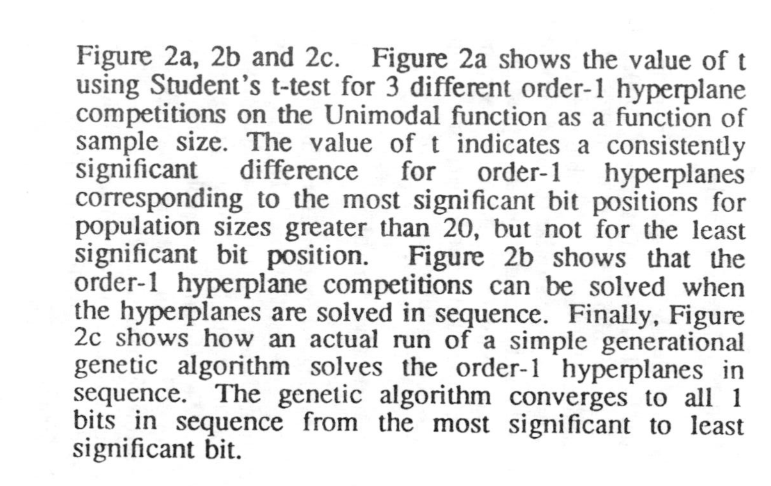

Figure 2a, 2b and 2c. Figure 2a shows the value of t using Student's t-test for 3 different order-1 hyperplane competitions on the Unimodal function as a function of sample size. The value of t indicates a consistently significant difference for order-1 hyperplanes corresponding to the most significant bit positions for population sizes greater than 20, but not for the least significant bit position. Figure 2b shows that the order-1 hyperplane competitions can be solved when the hyperplanes are solved in sequence. Finally, Figure 2c shows how an actual run of a simple generational genetic algorithm solves the order-1 hyperplanes in sequence. The genetic algorithm converges to all 1 bits in sequence from the most significant to least significant bit.

2.3 A Bimodal Nonlinear Problem

This quadratic function has a global minimum at 1111111111 and a local minimum at 0000000000. Although this problem is bimodal and nonlinear, it is in fact not fundamentally deceptive problem. The results from the t-test are very similar to those obtained from the unimodal nonlinear function. It is possible to independently solve the hyperplane competitions only for more significant bits. Thus the hyperplane competitions were solved sequentially.

2.4 A Fully Deceptive Problem

This function was designed by concatenating two fully deceptive five bit problems. Although 1111111111 was the global minimum, the problem has a "deceptive attractor" at 0000000000 which is a local minimum (see Whitley 1991). We attempted to solve the order-1 hyperplane competitions. As with the one max problem, it was possible to statistically solve all the order-1 competitions independently. In this case, however, the order-1 hyperplanes competitions lead to the string 0000000000.

3 GENETIC "HITCHHIKING"

The fact that order-1 hyperplanes either 1) cannot be distinguished when sampling from the space of all strings or 2) are deceptive when sampling from the space of all strings (but can be "correctly" solved when sampling from more specialized regions) results in a a phenomena referred to as genetic "hitchhiking." Schraudolph and Belew (1990) use this term to describe a process where certain hyperplane competitions are solved before other and hence influence the later competitions (i.e. the "hitch hikers.") As an example, consider the the unimodal nonlinear problem discussed in section 2.2. Although the problem is not deceptive, it is interesting because there is no clear preference toward certain hyperplanes when sampling the entire search space. But as the genetic algorithm biases the search toward particular values for the most significant bits the search space narrows. As the search narrows, preferences emerge with respect to the value of the less significant bits (See Figure 2c).

"Hitchhiking" occurs because certain hyperplane competitions are not solved until earlier competitions begin to bias the space from which the genetic algorithm is sampling. In deceptive problems hitchhiking can occur because an order-1 hyperplane that is initially preferred is deceptive when looking at the entire space and a different winner is eventually preferred as the search space narrows and more specialized partitions of hyperspace are explored (e.g. see Figure 3). In other problems, such as the unimodal nonlinear problem, hitchhiking occurs because no clear preference exists until the search space is narrowed. In both cases, the job of selecting a hyperplane winner is deferred until other competitions have started to be resolved. Clearly the genetic algorithm does not solve all of the order-1 hyperplane competitions simultaneously or at the same rate.

Consider the problem posed by Grefenstette and Baker (1989):

$$f(x) = \begin{cases} 2 & \text{if } x \in 111*...* \\ 1 & \text{if } x \in 0***...* \\ 0 & \text{otherwise} \end{cases}$$

We reduced this to a 3 bit problem and then used a set of equations to show the predicted changes for the various members the population. This problem is fundamentally deceptive since the order-1 hyperplane winners are 0**, *1* and **1, but the optimal solution is 111. The equations used are a variant of the equations used by Goldberg (1987) to study the minimally deceptive problem.

$$P(t+1,000) = P(t,000)\frac{f(000)}{\bar{f}}(1 - \text{losses}) + \text{gains}.$$

This is an idealized version of the genetic algorithm. P(t,000) is the proportion of the population that samples the string (schema) 000 at time t. The expression $f(000)/\bar{f}$ indicates the change in representation of 000 in the population due to fitness, where f(000) is the fitness associated with the string (schema) 000 and $\bar{f}$ is the average fitness of the population. Losses occur when a string (schema) crosses with another string and the resulting offspring fail to preserve the original string. Gains occur when two different strings cross and independently create a new copy of some string; for example 100 and 001 will always produce a new copy of 000. The probability of "losses" and "gains" are calculated as follows.

$$\text{losses} = \frac{f(111)}{\bar{f}}P(t,111) + \frac{f(101)}{\bar{f}}P(t,101)$$

$$+0.5\frac{f(110)}{\bar{f}}P(t,110) + 0.5\frac{f(011)}{\bar{f}}P(t,011).$$

$$\text{gains} = \frac{f(001)}{\bar{f}}P(t,001)\frac{f(100)}{\bar{f}}P(t,100)$$

$$+0.5\frac{f(010)}{\bar{f}}P(t,010)\frac{f(100)}{\bar{f}}P(t,100)$$

$$+0.5\frac{f(011)}{\bar{f}}P(t,011)\frac{f(100)}{\bar{f}}P(t,100)$$

$$+0.5\frac{f(001)}{\bar{f}}P(t,001)\frac{f(110)}{\bar{f}}P(t,110)$$

$$+0.5\frac{f(001)}{\bar{f}}P(t,001)\frac{f(010)}{\bar{f}}P(t,010).$$

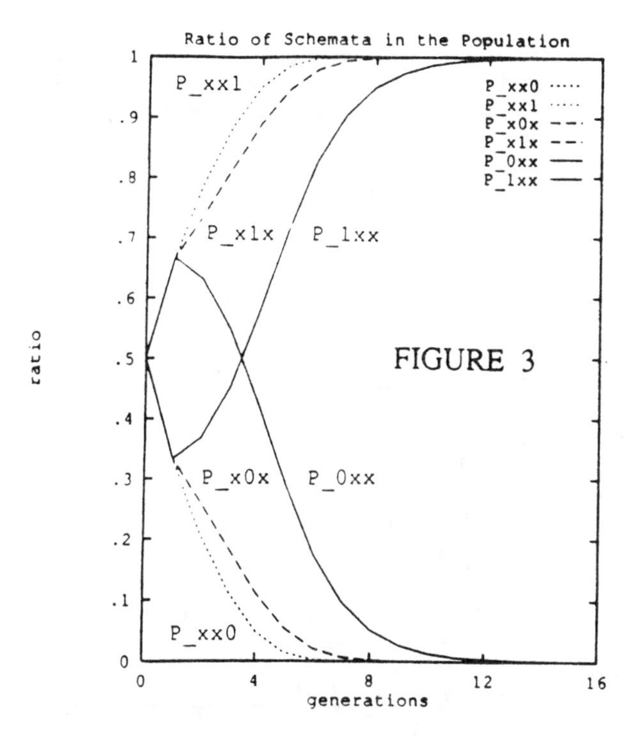

Figure 3: Plot of order-1 hyperplanes. Note that the hyperplane competition between 1** and 0** is deceptive in the early stages of the search.

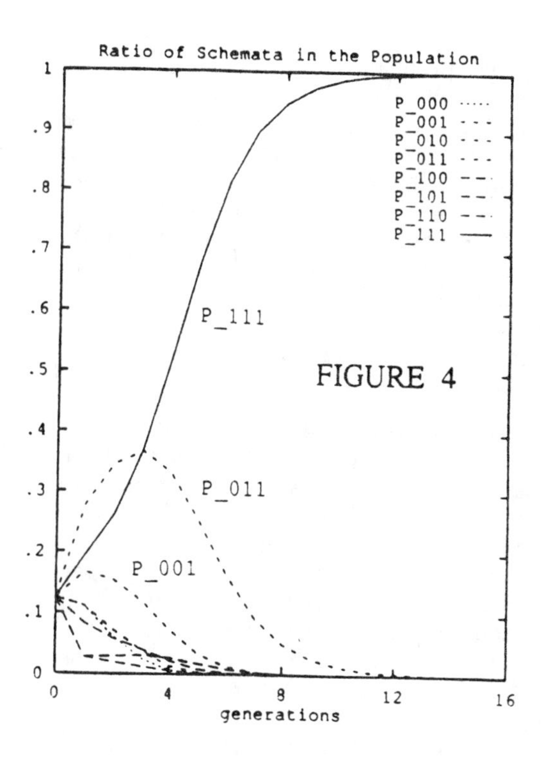

Figure 4: Plot of order-3 hyperplanes. By the fourth generation, *11 (represented by 111 and 011) dominates the search, and thus resolves the deception.

The equations can be generalized to cover the remaining 7 cases by using $(x \oplus c)$ to transform all the bit strings (represented by x) contained in the formula to the appropriate corresponding strings for computing the formula for P(t+1,c). The translation is accomplished using bitwise addition modulo 2 (i.e., a bitwise exclusive-or, here denoted by $\oplus$). Note that in the above equations we assume that crossover occurs at each time step (i.e., $p_c = 1.0$) and that disruption always occurs (i.e., $\delta(l)/(L-1) = 1.0$).

In this problem recombination is continually disrupting 111 since crossover always occurs within this 3 bit string. Nevertheless, if we look at the order-2 competition between *00, *01, *10 and *11 the clear winner is *11. As the second and third bit positions are set to the bit value 1, the search space is biased and the first position which initially favored 0 now favors 1. This can be clearly seen in Figure 3 which tracks the population representation of the order-1 hyperplanes using the above equations assuming that all strings start with an equal population representation of 12.5%. Figure 4 tracks the order-3 hyperplane competitions (or the 3 bit strings, depending on how one conceptualizes the problem).

4 CROSSOVER AND DECEPTION

Deception is a more serious problem (for most recombination operators) if it occurs across some random (and therefore, unknown) combination of bits in the encoding. To build a suitable test function that is likely to be difficult for a genetic algorithm to solve, Goldberg, Korb and Deb (1989) defined a 30 bit function made up of 10 copies of a fully deceptive 3-bit function. The problem is difficult (or "ugly") when the bits are arranged so that each of the 3 bits of the subfunctions are uniformly and maximally distributed across the encoding. Thus each 3 bit subfunction, i, has bits located at positions i, i+10 and i+20. In our tests we will use both 3-bit and 4-bit functions as subproblems of "ugly" functions. We report results in terms of the number of subproblems correctly solved by the genetic algorithm for both 1-point and uniform crossover. The genetic algorithm used in these tests was GENITOR, which is similar to the "steady state" algorithm used by Syswerda (1989) in his tests with uniform crossover.

Uniform crossover involves choosing each bit independently from the two parents, producing a random assortment of bits from the two parents. The potential advantage is that now the location of the bits on the encoding is irrelevant; since linkage is now irrelevant, there is no "ugly" version of the problem. Thus, there is no bias against schemata with long defining lengths. On the other hand, the advantages of tight linkage (i.e., functionally related bits that are physically close in the encoding) are lost, since bits can no longer can be inherited as a block. One might expect uniform crossover to be competitive on problems such as the "ugly" deceptive problem, since there is not a bias against schemata with long defining length. But when this is considered more carefully, it seems clear that uniform crossover breaks apart higher order schemata during recombination and thereby shifts the burden of search to the lower order (deceptive) schemata. If the problem is deceptive, this action will increase the probability that the deceptive hyperplanes will be able to mislead the genetic algorithm. This suggests that Syswerda's test problems (1989) for uniform crossover were not sufficiently deceptive to make them challenging tests problems.

Using a population of 2000 and 50,000 evaluations, uniform crossover solved 35% of the subproblems correctly on the ugly order-3 deceptive problem; note again that for uniform crossover that the "ugly" and tightly linked versions of the order-3 deceptive problem are exactly the same. The 1-point crossover operator solves 38% of the subproblems in its "ugly" form. When a random assortment of bit positions are used 1-point crossover solves over 50% of the subproblems. Other orderings of the bits produce no change in the performance of uniform crossover. On an "ugly" order-4 deceptive problem, uniform crossover solved only 3% of the subproblems correctly, while 1-point crossover solved 7% correctly.

In general uniform crossover may create difficulties when higher order hyperplane information is needed to guide the search toward global solutions.

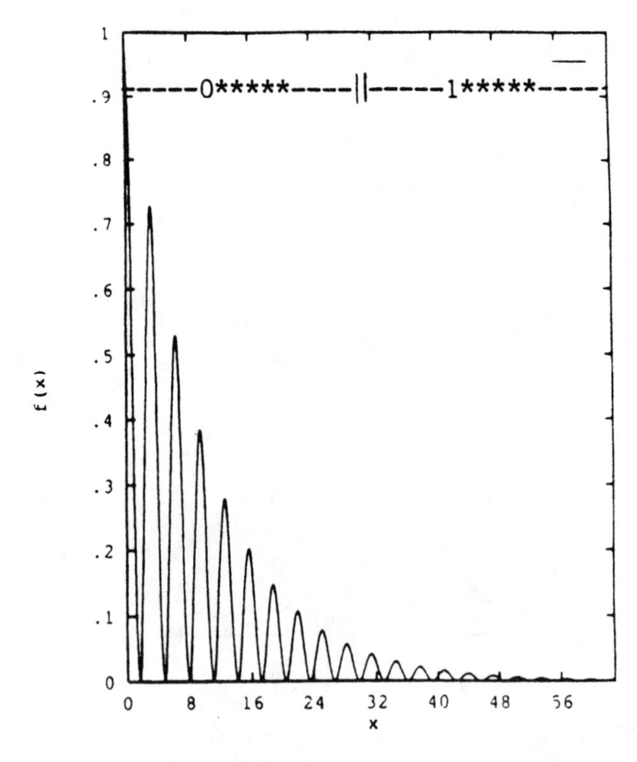

Figure 5:

4.1 Uniform Crossover Test Problems

The "One Max" problem appears in a paper by Gil Syswerda comparing "uniform" crossover to 1-point and 2-point crossover. Syswerda uses 6 test functions to motivate the argument that uniform crossover is superior to 1-point or 2-point crossover. Syswerda concludes, "The proof is in the empirical pudding: when in doubt, use uniform crossover" (1989:8).

While we recognize the danger of empirical arguments, our results suggest that the performance of uniform crossover may be correlated with the fact most of the test problems used by Syswerda display little or no deception: in other words these problems are not fundamentally deceptive. Because uniform crossover is highly disruptive to higher order hyperplanes, this means the search will be more strongly influenced by the order-1 hyperplanes, which obviously cannot be disrupted by any type of crossover.

Besides the "One Max" problem, Syswerda's second test problems was "Sparse One Max," which is a variant of One Max and which can also be solved by solving the order-1 hyperplanes. The third test problem is the "Contiguous Bits" problem, a lock and tumbler problem that can also be solved by solving the order-1 hyperplanes: 0 bits have no value and 1 bits have a value if they are contiguous to other 1-bits.

The fourth test problem is more interesting, but it can also be solved using information from the order-1 hyperplanes. The exponentially decreasing sine function (Figure 5) has clear order-1 hyperplane winners.

The fifth test problem is Shekel's Foxholes. Here 1 and 2 point crossover produced better results than uniform crossover. Oddly enough, Syswerda retests the problem using a different random shuffling of bits on each experiment. Clearly, this can disrupt any linkage that might exists, thereby degrading 1 and 2 point crossover; it has no effect on uniform crossover. Using a different shuffle pattern on each run, uniform crossover appears to be what Syswerda refers to as a "marginal" winner: in fact it is highly unlikely that the differences are statistically significant. This, coupled with the fact that shuffling unfairly affected the results and the best overall results were obtained using 2-point crossover discounts this case as being "supportive" of Syswerda's thesis.

Finally, the last problem was a traveling salesman problem using 16 cities uniformly distributed on a circle with radius of 45 units. We have not determined if this problem can be solved by solving the order-1 hyperplanes or not. We do note however, that this is particularly easy version of the Traveling Salesman problem since the cities are arranged as a circle and a greedy algorithm which always chooses the nearest neighbor will correctly solve this problem.

5 BUILDING A BETTER TEST PROBLEM

Our examination of Syswerda's results are largely intended to dramatize the following point: results based on problems that are not fundamentally deceptive have little empirical validity as measures of the overall search capability of genetic algorithms or the operators used during genetic. This is because it is not necessary to sample higher order hyperplanes if the problem can be solved by looking at only the order-1 hyperplanes. Mechanisms that do emphasize higher order hyperplanes are likely to result in a slower, more careful search. Furthermore, simple counting arguments indicate that larger populations are needed if we are to intelligently sample order-4, order-5 or order-6 hyperplanes since they are represented in only a small percentage of the total population.

We are not asserting that uniform crossover is inferior to 1 or 2 point crossover; but we do assert that, at this time, there is no solid empirical evidence to suggest that uniform crossover is superior to 1 or 2 point crossover on a challenging set of test problems. Furthermore, the theoretical evidence suggests (but does not conclusively show) that uniform crossover is typically inferior to 1 and 2 point crossover at preserving order-3 and higher order hyperplanes samples.

Tests on fully deceptive order-3 and order-4 problems using a "worst case" ordering of bits, shows that uniform crossover is not superior to a 1-point operator. When a more favorable ordering of the bits is used, then uniform crossover is inferior to 1-point crossover.

These ideas also have implications for other kinds of mechanisms which may be designed to "enhance" genetic search. For example if we wish to "seed" a population of binary strings by locally optimizing every string in the population before genetic search begins. Note that this is a weaker strategy than solving the order-1 hyperplane competitions because we are merely determining a bit preference for a single string, but the two are not completely unrelated. If a problem displays little or no deception, local optimization will most likely increase the rate of convergence to the global solution. If a problem is fundamentally deceptive the process of locally optimizing each string may bias the search toward the deceptive solution.

Not all challenging test problems must be fully deceptive problems; however, such test problems should be at least fundamentally deceptive. In this paper we outline some basic tests that can be used to determine whether a problem is fundamentally deceptive or not. Furthemore, the tests that we have used are simple and provide a form of measuring stick which can help us ensure that test problems are indeed interesting and challenging. At the very least, nondeceptive problems should be only a very small part of a larger test suite.

Acknowledgements

This research was supported in part by NSF grant IRI-9010546 and in part by a grant from the Colorado Institute of Artificial Intelligence (CIAI). CIAI is sponsored in part by the Colorado Advanced Technology Institute (CATI), an agency of the State of Colorado. CATI promotes advanced technology education and research at universities in Colorado for the purpose of economic development.

References

Bethke, A. (1980) Genetic Algorithms as Function Optimizers. Ph.D. Dissertation, Computer and Communication Sciences, University of Michigan, Ann Arbor.

Goldberg, D. (1987) Simple Genetic Algorithms and the Minimal, Deceptive Problem. In, *Genetic Algorithms and Simulated Annealing,* L. Davis, ed., pp: 74-88, Pitman/Morgan Kaufmann.

Goldberg, D. (1989a) *Genetic Algorithms in Search, Optimization and Machine Learning.* Reading, MA: Addison-Wesley.

Goldberg, D. (1989b) Genetic Algorithms and Walsh Functions: Part I, A Gentle Introduction. *Complex Systems* 3:129-152.

Goldberg, D. (1989c) Genetic Algorithms and Walsh Functions: Part II, Deception and its Analysis. *Complex Systems* 3:153-171.

Goldberg, D., Korb, B., and Deb, K. (1989) Messy Genetic Algorithms: Motivation, Analysis, and First Results. *Complex Systems* 4:415-444.

Grefenstette, J. and Baker, J. (1989) How genetic algorithms work: a critical look at implicit parallelism. *Proceeding of the Third International Conference on Genetic Algorithms, 1989.* Washington, D.C., Morgan Kaufmann.

Liepins, G. and Vose, M. (1990) Representation Issues in Genetic Optimization. *J. Experimental and Theoretical Art. Intell.* 2(1990)101-115.

Liepins, G. and Vose, M. (1991) Deceptiveness and Genetic Algorithm Dynamics. *Foundation of Genetic Algorithms,* G. Rawlins, ed. Morgan Kaufmann.

Schraudolph, N. and Belew, R. (1990) Dynamic Parameter Encoding for Genetic Algorithms. Computer Science and Engr. Dept. Univ. of Calif., San Diego. CSE Tech. Rprt. #CS 90-175.

Syswerda, G. (1989) Uniform Crossover in Genetic Algorithms. *Proc. Third International Conf. on Genetic Algorithms,* Morgan Kaufmann.

Whitley, D. (1991) Fundamental Principles of Deception in Genetic Search. *Foundations of Genetic Algorithms.* G. Rawlins, ed. Morgan Kaufmann.

A Simulated Annealing Like Convergence Theory for the Simple Genetic Algorithm

Thomas E. Davis
WL/MNGS
Eglin AFB
Fl 32542-5434

Jose C. Principe
Department of Electrical Engineering
University of Florida
Gainesville, Fl 32611-2024

Abstract

This paper synopsizes a substantial body of work whose goal is extrapolation of the existing theoretical foundation of the simulated annealing algorithm onto a Markov chain genetic algorithm model. Some key intermediate convergence results are reported, including a genetic algorithm mutation probability control parameter sequence bound analogous to the simulated annealing temperature schedule bounds.

1 INTRODUCTION

Both simulated annealing and the genetic algorithm are stochastic relaxation search techniques suitable for application to a wide variety of combinatorial complexity non-convex optimization problems [14, 15, 1, 13, 6, 9, 10, 11]. Each produces a sequence of candidate solutions (or in the case of the genetic algorithm a sequence of populations of candidate solutions) to the underlying optimization problem, and the purpose of both algorithms is to generate sequences with probability distribution biased toward candidates which optimize the objective function.

Simulated annealing exploits an analogy of combinatorial optimization to the annealing of crystalline solids [14], in which the solid is allowed to cool very gradually from some elevated temperature and thereby allowed to relax into one of its minimum energy states. The appeal of the algorithm class is that it provides asymptotic convergence to a globally optimal solution. A substantial body of knowledge exists for describing the algorithm convergence behavior based upon the asymptotic probability distribution associated with a non-stationary Markov chain algorithm model [15]. Section 2 below is a summary of the essential features of the simulated annealing convergence theory.

The genetic algorithm emulates the evolutionary behavior of biological systems [1, 6]. It generates a sequence of populations of candidate solutions to the underlying optimization problem by employing a set of genetically inspired stochastic state transition operators to transform each population of candidate solutions into a descendent population. A variety of distinct genetic operators are reported in the literature [1, 6, 9, 10, 11], the most important of which are (1) reproduction, (2) crossover and (3) mutation. A one, two or three operator genetic algorithm employing combinations of these operators and having specified algorithm parameters (i.e. population size, mutation and crossover probabilities), including the case in which the population size is fixed but the mutation and crossover probabilities are allowed to vary with the algorithm iteration index, is referred to herein as a simple genetic algorithm.

Other authors have employed Markov chains in examining genetic algorithm convergence behavior, notably [3] and [8] who use Markov chain methodology accompanied by approximate numerical analysis to examine genetic drift in finite population genetic algorithms, and [18] who develop their state transition matrix from the recombination and selection operators introduced in [21] and then employ it in examining steady state distributions as population size increases. Additionally, others have pursued unifying formalisms for simulated annealing and the genetic algorithm by introducing non-conventional GA operators [7, 4]. However, no convergence theory for the simple genetic algorithm comparable in scope to that of simulated annealing exists in the literature. The central theme of the work reported here is an attempt to develop a very general simple genetic algorithm model (and accompanying convergence theory) by directly extrapolating the simulated annealing convergence methodology.

An essential first step toward that goal is the development of a very general non-stationary Markov chain simple genetic algorithm model. That task is accomplished in the work synopsized here for simple genetic algorithm variants implementing three combinations of the primary operators. The resulting model provides a mechanism for employing the mutation probability parameter in a role analogous to the absolute temperature control parameter analog of simulated annealing. It is reviewed in Section 3 below.

Section 4 presents some empirical results generated from the algorithm model by computer simulation and Section 5 summarizes some key theoretical convergence results derived from the model. They include: (1) existence of a unique asymptotic probability distribution (stationary distribution) for the time-homogeneous algorithm variants which incorporate the mutation operator (the two and three-operator variants) with non-zero mutation probability, (2) formulation of the stationary distribution solution for the time-homogeneous two and three-operator algorithm variants in terms of the characteristic polynomials of matrices derived from the state transition matrix, (3) existence of a zero mutation probability limit for the time-homogeneous two and three-operator algorithm variants, (4) a mutation probability schedule bound (analogous to the annealing schedule bound of simulated annealing) sufficient for the non-stationary genetic algorithm variants to achieve (asymptotically) the limiting distribution and (5) a partially developed methodology for representing the stationary distribution components at all consistent values of mutation probability (including the zero mutation probability limit).

The simple genetic algorithm model and convergence results presented in this paper are synopsized from [2]. Space limitations prohibit replicating them in the generality developed there.

2 SIMULATED ANNEALING

Let a combinatorial optimization problem be represented by the pair (S, C) where S is the problem solution space and C its objective function, and assume without loss of generality that the problem requires minimization of C. Then, a simulated annealing algorithm executing on this problem generates a sequence of candidate solutions drawn from S by employing a stochastic state transition mechanism to transform each member of the sequence into a successor. The stochastic state transformation mechanism is selected such that the resulting asymptotic probability distribution over S is the Gibbs (or Boltzman) distribution

$$\forall i \in S : Pr\{i\} = \frac{\exp\{-C(i)/T\}}{\sum_{j \in S} \exp\{-C(j)/T\}} \quad (1)$$

with $C(i)$ playing the role of energy function for the analogous thermodynamic system and T a strictly positive algorithm parameter analogous to absolute temperature.

As the parameter T in Eq. 1 approaches zero, the probability distribution approaches a limit in which all states corresponding to sub-optimal solutions have zero probability, while the states corresponding to the optimal value of C are uniformly distributed, i.e.

$$\lim_{T \to 0+} Pr\{i\} = \begin{cases} \frac{1}{|S_{opt}|} & i \in S_{opt} \\ 0 & \text{otherwise} \end{cases} \quad (2)$$

where $S_{opt} = \{i \in S : C(i) = C_{opt} = C_{min}\}$. Thus, if sufficient conditions on the state transformation mechanism are invoked to ensure that Eq. 1 represents the algorithm asymptotic state probability distribution, and if the parameter T goes to zero asymptotically, then the candidate solution sequence probability distribution is (asymptotically) nonzero only for optimal solutions (Eq. 2).

Since the conditional dependence of each candidate solution in the sequence upon the sequence history is equal to its conditional dependence upon its immediate predecessor, the solution sequence evolves as a Markov chain. In fact, the algorithm can be represented by the quadruple $(S, i_0, \overline{P}_T, \tau)$ where S is as defined above, i_0 is an initial solution selected from S, $\overline{P}_T$ is a state transition matrix describing the stochastic state transition mechanism and $\tau = \{T_k\}$ is a sequence of parameter values. The state transition matrix at algorithm iteration k depends upon the parameter value T_k, so in general the Markov chain is non-stationary.

The essence of the simulated annealing convergence theory is a set of sufficient conditions on the functional form of $\overline{P}_T$ to ensure that the asymptotic probability distribution of the stationary Markov chain corresponding to every fixed, strictly positive value of T is given by Eq. 1, along with sufficient conditions on τ to ensure that the non-stationary algorithm achieves the limiting distribution (Eq. 2) asymptotically. Sufficient conditions on the functional form of P_T are established in [15, 16] and elsewhere, and are not reviewed here. The topic of interest here is sufficient conditions on τ, and consequently on the non-stationary algorithm behavior. The first condition is that the sequence have limit zero. Beyond that requirement, a sufficient condition on τ is obtained by enforcing strong ergodicity on the Markov chain and deducing a corresponding parameter sequence bound. Strong ergodicity of the non-stationary chain, along with the zero control parameter sequence limit, ensures asymptotic convergence to the zero temperature limit in Eq. 2 (i.e. convergence to global optimality). The methodology is thoroughly reviewed in [15]. Among the annealing schedule bounds so deduced are those of [5] and [17], both of which have the general form $T_k \geq K/\log(k)$.

3 THE MARKOV CHAIN GENETIC ALGORITHM MODEL

The sequence of candidate solution populations produced by the simple genetic algorithm is a realization of a stochastic process with finite state space, and the stochastic operators which emulate biological system behavior possess the property that the conditional dependence of each population upon its predecessors in the sequence is completely determined by its conditional dependence upon its immediate predecessor population. Thus, the sequence of populations evolves as a Markov chain. The work synopsized in this paper includes a Markov chain model of one, two and three-operator variants of the algorithm and exploration of the state transition behavior of each variant. The following paragraphs summarize the essential model results reported in [2].

Describing and analyzing the operation of the simple genetic algorithm is facilitated by assuming that the underlying optimization problem is defined over a bit-string solution space. This assumption is not essential and sacrifices very little generality.

Let a combinatorial optimization problem be characterized by the ordered pair (S, R) where $S = \{0,1\}^L$ is the problem's solution space and where R is a strictly positive reward function, and assume, with no loss of generality, that the problem requires maximization of R. Also, let a simple genetic algorithm designed to execute on this problem have fixed population size M, let $i \in S$ be interpreted as an unsigned integer $(0 \leq i \leq 2^L - 1)$, and let a generation be represented by $\overline{m} = (m(0), m(1), \ldots, m(2^L - 1))$ where $m(i)$ is the number of occurrences of solution $i \in S$ in the population. Thus, in the parlance of combinatorial mathematics, $\overline{m}$ is a distribution of M non-distinct objects over $N = 2^L$ bins [12, 19], and the set of all such distributions, $S' = \{\overline{m}\}$, is a suitable representation of the simple genetic algorithm search space. The cardinality of S' is given by

$$\begin{aligned} N' = \mathrm{card}(S') &= \binom{M + 2^L - 1}{M} \\ &= \binom{M + N - 1}{M}. \end{aligned}$$

Since both L and M are finite, so is N'.

Then, if $\overline{m}_0$ is selected as an initial population, the simple genetic algorithm can be represented by the quadruple $(S', \overline{m}_0, \overline{P_{\overline{Q}}}, \Gamma)$ where $\overline{P_{\overline{Q}}}$ is a state transition matrix (analogous to $\overline{P_T}$ of the simulated annealing model) and $\Gamma = \overline{Q}_k$ is a sequence of parameter vectors $\overline{Q}_k = (p_m(k), p_c(k))$. The algorithm parameters $p_m(k)$ and $p_c(k)$ are respectively the mutation and crossover probabilities. In the work reported here, the mutation probability sequence is employed in a role analogous to absolute temperature in simulated annealing, and consideration is limited to monotone non-increasing sequences. In general, the only limitation on the crossover probability sequence is that its values are probabilities. However, in [2], consideration is limited to constant crossover probability sequences.

The solution evolves as a sequence $\{\overline{m}_k\}$ of states $\overline{m}_k \in S'$ in which the conditional dependence of $\overline{m}_{k+1}$ on the sequence history is equivalent to its conditional dependence on $\overline{m}_k$, and thus the solution sequence is a Markov chain. In general, the chain is not time-homogeneous (i.e. it is non-stationary), however it is time-homogeneous if the parameter vectors are constant.

The functional forms of the state transition matrices corresponding to three combinations of the genetic algorithm operators are developed and their associated state behaviors explored in [2]. The first case consists of a one-operator algorithm which employs only reproduction. The second is a two-operator variant which employs reproduction with mutation. Finally a three-operator algorithm which includes crossover is developed. Following is a summary.

The proportional reproduction only model variant corresponds to the case $\forall k : \overline{Q}_k = (0,0)$. In this case, the conditional probability of selecting a solution $i \in S$ from a population described by the state vector $\overline{n} \in S'$ is

$$\forall i \in S, \forall \overline{n} \in S' : P_1(i|\overline{n}) = \frac{n(i) \times R(i)}{\sum_{j \in S} n(j) \times R(j)}. \quad (3)$$

Thus, the conditional probability of the successor generation $\overline{m}$, given that the present generation is $\overline{n}$, is a multinomial distribution,

$$\begin{aligned} &\forall \overline{m}, \overline{n} \in S' : \\ P_1(\overline{m}|\overline{n}) &= \frac{M!}{\prod_{i \in S} m(i)!} \times \prod_{i \in S} P_1(i|\overline{n})^{m(i)} \quad (4) \\ &= \binom{M}{\overline{m}} \times \prod_{i \in S} P_1(i|\overline{n})^{m(i)} \end{aligned}$$

where the symbol

$$\binom{M}{\overline{m}} = \frac{M!}{\prod_{i \in S} m(i)!}$$

denotes the indicated multinomial coefficient. The transition probability matrix of the Markov chain representing the one-operator algorithm is composed of the array of conditional probabilities defined by Eq. 4,

$$\overline{P} = [P_1(\overline{m}|\overline{n})]. \quad (5)$$

Since it is independent of the sequence index (i.e. the parameter vectors are constant), the one-operator Markov chain is time-homogeneous.

The set of states which represent uniform populations (i.e. the states $\overline{m}_A \in S'_A \subset S'$ in which one component is M and all others are zero) are absorbing states of the Markov chain, because for any such state, $P_1(\overline{m}_A|\overline{m}_A) = 1$ and the state cannot be escaped. $N = 2^L$ such uniform population states exist, one associated with each $i \in S$. It follows from Eq. 3 and 4 that $\forall \overline{n} \in S' - S'_A, \exists \overline{m}_A \in S'_A : P_1(\overline{m}_A|\overline{n}) > 0$, and consequently there are exactly $N = 2^L$ absorbing states. The corresponding rows of $\overline{P}$ contain 1 in the principal diagonal location and 0 elsewhere,

$$\forall \overline{n}_A \in S'_A : P_1(\overline{m}|\overline{n}_A) = \begin{cases} 1 & \overline{m} = \overline{n}_A \\ 0 & \overline{m} \in S' - \{\overline{n}_A\} \end{cases} . \quad (6)$$

It follows that the probability distribution given by the $N' \times 1$ vector $\overline{q}_{\overline{n}_A}$ whose $\overline{n}_A \in S'_A$ component is one is invariant with respect to the iteration index in the sense that

$$\overline{q}^T_{\overline{n}_A} \overline{P} = \overline{q}^T_{\overline{n}_A}.$$

Any such probability distribution is called a stationary distribution, or invariant distribution, of the Markov chain. In the case of the one-operator algorithm variant the stationary distribution is not unique because the corresponding vector for any of the $N = 2^L$ absorbing states satisfies the invariance requirement, as does any probability vector of the form $\overline{q} = \sum_{\overline{n}_A \in S'_A} \gamma_{\overline{n}_A} \overline{q}_{\overline{n}_A}$ where $\gamma_{\overline{n}_A} \geq 0$ and $\sum \gamma_{\overline{n}_A} = 1$.

The two-operator algorithm composed of reproduction and mutation corresponds to the case $\forall k : \overline{Q}_k = (p_m(k), 0)$ where $0 < p_m(k) < 1$. Results analogous to Eq. 3-5 are obtainable for this algorithm variant as follows. Let $P_2(i|\overline{n})$ and $P_2(\overline{m}|\overline{n})$ be the conditional distributions of the two-operator algorithm corresponding to the one-operator distributions defined by Eq. 3 and 4. Then, $P_2(i|\overline{n})$ can be expressed as a sum over all j of the corresponding $P_1(j|\overline{n})$ times a factor which accounts for the probability of the mutation event required to transform j into i. This probability can be expressed as $p_m^{H(i,j)}(1-p_m)^{L-H(i,j)}$ where $H(i,j) = H(j,i)$ is the Hamming distance of the pair i, j, and thus $P_2(i|\overline{n})$ can be written as

$$\forall i \in S, \forall \overline{n} \in S' :$$

$$\begin{aligned} P_2(i|\overline{n}) &= \sum_{j \in S} p_m^{H(i,j)}(1-p_m)^{L-H(i,j)} \\ &\quad \times P_1(j|\overline{n}) \qquad (7) \\ &= \frac{1}{(1+\alpha)^L} \sum_{j \in S} P_1(j|\overline{n}) \times \alpha^{H(i,j)} \end{aligned}$$

where

$$\alpha = \frac{p_m}{(1-p_m)}. \quad (8)$$

The two-operator analog of Eq. 4 is

$$\forall \overline{m}, \overline{n} \in S' :$$

$$P_2(\overline{m}|\overline{n}) = \binom{M}{\overline{m}} \times \prod_{i \in S} P_2(i|\overline{n})^{m(i)}. \quad (9)$$

The admissible range of p_m is $0 < p_m < 1$, and consequently that of α is $0 < \alpha < \infty$. The transition probability matrix of the Markov chain representing the two-operator algorithm is composed of the array of conditional probabilities defined by Eq. 9, i.e.

$$\overline{P} = [P_2(\overline{m}|\overline{n})]. \quad (10)$$

Since the elements of $\overline{P}$ depend on α (and hence by Eq. 8 on $p_m(k)$), the two-operator Markov chain is generally not time-homogeneous. It is time-homogeneous if the mutation probability is fixed. It also follows from Eq. 7 and 9 that

$$\begin{aligned} \lim_{\alpha \to 0^+} P_2(i|\overline{n}) &= P_1(i|\overline{n}) \\ \text{and} \qquad\qquad & \qquad\qquad (11) \\ \lim_{\alpha \to 0^+} P_2(\overline{m}|\overline{n}) &= P_1(\overline{m}|\overline{n}) \end{aligned}$$

Since the reward function, R, is strictly positive by hypothesis, and since $\forall i, j \in S : 0 \leq H(i,j) \leq L$, it follows that for α in the range $0 < \alpha \leq 1$, which corresponds to $0 < p_m \leq 1/2$, then

$$\begin{aligned} \alpha^L \sum_{j \in S} n(j) \times R(j) &\leq \sum_{j \in S} n(j) \times R(j) \times \alpha^{H(i,j)} \\ &\leq \sum_{j \in S} n(j) \times R(j) \end{aligned}$$

and consequently from Eq. 3 and 7 that

$$\forall i \in S, \forall \overline{n} \in S' :$$

$$\left(\frac{\alpha}{1+\alpha}\right)^L \leq P_2(i|\overline{n}) \leq \left(\frac{1}{1+\alpha}\right)^L. \quad (12)$$

Using Eq. 12 in Eq. 9 yields

$$\begin{aligned} \forall \overline{m}, \overline{n} \in S' : \quad & \binom{M}{\overline{m}} \left(\frac{\alpha}{1+\alpha}\right)^{ML} \\ \leq \;& P_2(\overline{m}|\overline{n}) \qquad\qquad (13) \\ \leq \;& \binom{M}{\overline{m}} \left(\frac{1}{1+\alpha}\right)^{ML} \end{aligned}$$

The lower bound in Eq. 13 can be used with the Perron-Frobenius theorem, which is fundamental to the study of non-negative matrices in general and stochastic state transition matrices in particular [20], to establish that the time-homogeneous two-operator algorithm variant possesses a unique stationary distribution [2]. That is, the vector $\overline{q}_\alpha$ given by

$$\begin{aligned} \overline{q}^T_\alpha \overline{P} &= \overline{q}^T_\alpha \qquad\qquad (14) \\ \overline{q}^T_\alpha \overline{1} &= 1 \end{aligned}$$

and having strictly positive components exists and is unique. Further, $\overline{q}_\alpha$ is the asymptotic state probability distribution of the corresponding Markov chain. It follows from Eq. 3 and Eq. 7-10 that it is completely determined by the objective function and the algorithm parameters, and is independent of the algorithm initial population, $\overline{m}_0$. The time-homogeneous two-operator stationary distribution is analogous to the asymptotic distribution in Eq. 1 for the simulated annealing algorithm.

The three-operator simple genetic algorithm corresponds to the case $\forall k : \overline{Q}_k = (p_m(k), p_c(k))$ with both

$p_m(k)$ and $p_c(k)$ non-zero. Results analogous to Eq. 7-11 are obtained in [2] by defining a crossover operator function which is similar in character to the Hamming distance function employed for the two-operator case. Further, the bounds in Eq. 12 and 13 apply without alteration, a consequence of which is that a unique stationary distribution exists for the three-operator algorithm also (Eq. 14). The details are omitted here in the interest of brevity.

The unique stationary distribution of the time-homogeneous two and three-operator algorithm variants (for all strictly positive values of the parameter α) and the fact that both algorithms degenerate into absorbing state behavior at zero mutation probability (Eq. 11 and its three-operator counterpart) suggests a mechanism for adapting the simulated annealing convergence theory onto the non-stationary simple genetic algorithm model. Some theoretical consequences of its behavior derived from the model and reported in [2] are summarized in Section 5 below, but first some empirical data generated from a computer simulation of the model is presented.

4 SOME EMPIRICAL RESULTS

This section reports the results of some computer simulations based upon the simple genetic algorithm Markov chain model summarized above. The results reported here concern the converged limiting (mutation probability approximately zero) stationary distribution of the three-operator algorithm variant executing on a four bit optimization problem at two selected values of the population size parameter, M. These results are a very small sample of the data reported in [2].

The underlying optimization problem for the data presented here is defined by the reward function presented in Figure 1. The solution state which maximizes the reward value is the $i \in S$ represented by the decimal integer value 12.

Figures 2 and 3 represent converged three-operator stationary distribution results at population sizes $M = 6$ and $M = 7$ respectively, and with crossover probability one. These results are for extremely small α (approaching zero). As shown in [2] and reviewed in Section 5 below, only the states corresponding to uniform populations (one-operator absorbing states) have non-zero probability in the $\alpha \to 0^+$ limit. Consequently, only the final probabilities for the uniform population states are displayed, with each such state indexed by the decimal integer value corresponding to the solution it represents. These results are generated by iteratively multiplying a starting probability vector by the state transition matrix until convergence (within a small tolerance) is obtained. In the two cases presented, 20 and 22 iterations respectively are required.

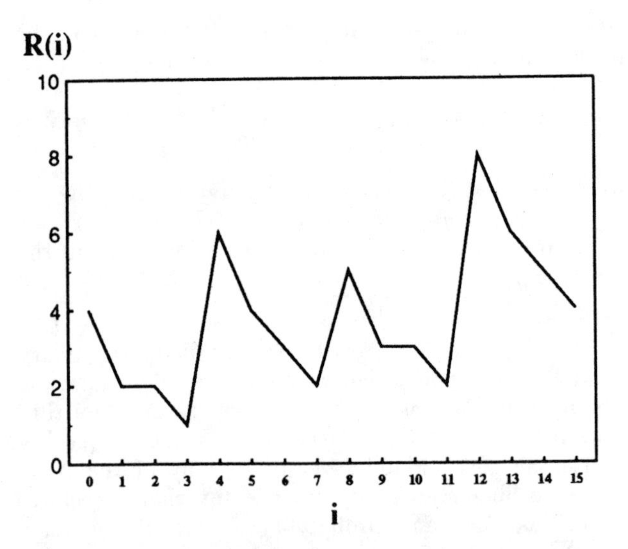

Figure 1: Four-bit Reward Function

A mildly surprising result suggested by these data is that the $\alpha \to 0^+$ limiting value of the stationary distribution is non-zero for all possible uniform states. This behavior, which is confirmed by theoretical results developed in [2] and reviewed in Section 5 below, has some consequences for the attempt to extrapolate the simulated annealing convergence theory onto the non-stationary simple genetic algorithm. It means that the algorithm cannot guarantee convergence to a globally optimal solution as does the simulated annealing algorithm, at least not for finite population size. However, as suggested by the data plotted in Figure 4, it may be possible to approach the desired limiting behavior as closely as required. That figure plots the computed limiting distribution entropy versus the algorithm population size parameter. The plots suggest that the entropy can be reduced arbitrarily close to zero. Results developed in [2] reinforce this premise, at least for the two-operator case.

5 SYNOPSIS OF KEY THEORETICAL RESULTS

Some key theoretical results required for extrapolating the simulated annealing convergence theory onto the simple genetic algorithm are obtained in [2] from the Markov chain model reviewed in Section 3 above. Fundamental to those developments is the formulation of the existence argument for the two and three-operator unique stationary distributions (Eq. 14) into a solution which allows a substantial degree of insight into the functional form of their components. The stationary distribution components are shown to be rational functions of both the objective function and the algorithm parameters. The rational functions are expressed in terms of the characteristic polynomials of a

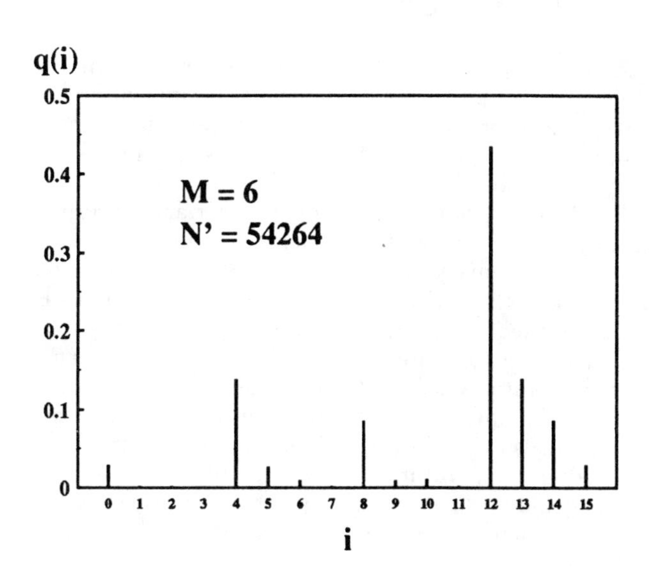

Figure 2: Three-Operator Stationary Distribution Limit ($M = 6$)

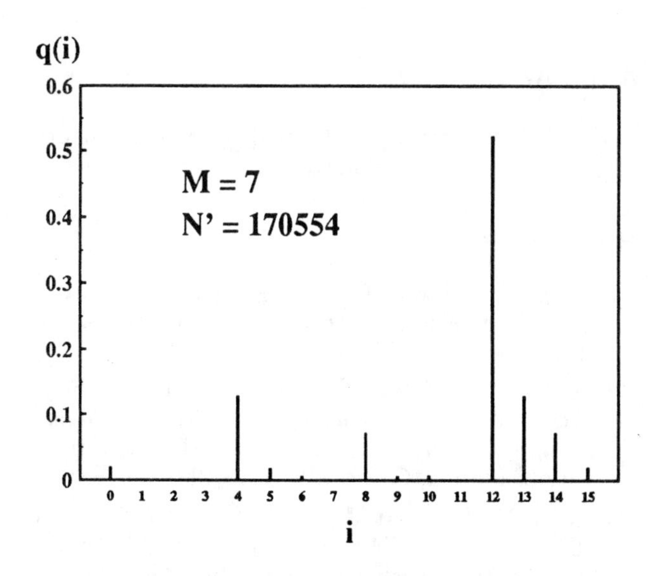

Figure 3: Three-Operator Stationary Distribution Limit ($M = 7$)

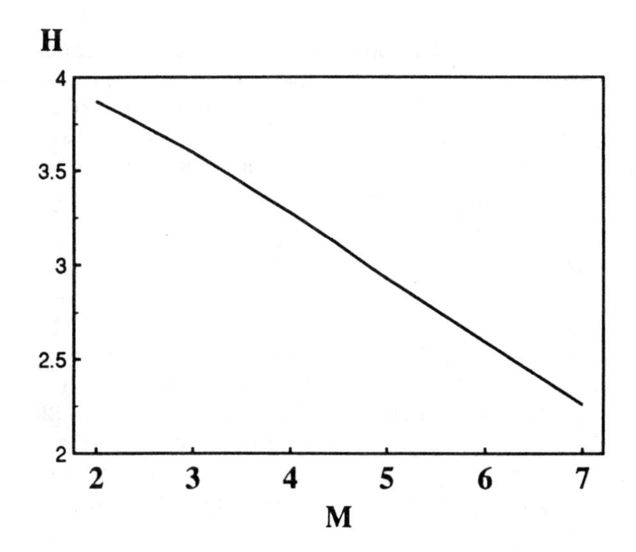

Figure 4: Three-Operator Limit Distribution Entropy vs M

set of matrices, each of which is derived from the state transition matrix by setting a selected row to zero. That is,

$$q(\overline{m}) = \left. \frac{\phi_{\overline{m}}(\lambda)}{\sum_{\overline{n} \in S'} \phi_{\overline{n}}(\lambda)} \right|_{\lambda=1} \tag{15}$$

where $q(\overline{m})$ is the $\overline{m}$ component of $\overline{q}$ and where $\phi_{\overline{m}}(\lambda)$ is the characteristic polynomial of the matrix $\overline{P}_{\overline{m}}$ derived from $\overline{P}$ by setting row $\overline{m}$ in $\overline{P}$ to the vector $\overline{0}^T$.

In terms of extrapolating the simulated annealing convergence theory onto the simple genetic algorithm model, the zero mutation probability limit of Eq. 15 is of extreme interest. Unfortunately, at $\alpha = 0$, Eq. 15 yields an indeterminate form because the characteristic polynomials of the matrices $\overline{P}_{\overline{m}}$ are all identically zero at zero mutation probability. That is, 1 is an eigenvalue of the zero mutation probability limit of each $\overline{P}_{\overline{m}}$. (The Eq. 15 result is developed on the premise that $\alpha > 0$). Nevertheless, by exploiting the functional form of the components of the elements in the $\overline{P}_{\overline{m}}$ matrices, [2] succeeds in translating Eq. 15 into a form which is determinate at $\alpha = 0$. The result is expressed in terms of matrix counterparts of the $\overline{P}_{\overline{m}}$ derived by coalescing the absorbing state rows of $\overline{P}_{\overline{m}}$ (i.e. the rows whose zero mutation probability limit is given by Eq. 6 and 11 for the two-operator case) into the set of adjacent states reachable by a single-bit mutation event. The result is of the form

$$\lim_{\alpha \to 0+} q_\alpha(\overline{m}) = \begin{cases} \left. \frac{\phi'_{\overline{m}_A}(\lambda)}{\sum_{\overline{n}_A \in S'_A} \phi'_{\overline{n}_A}(\lambda)} \right|_{\lambda=1} & \overline{m} = \overline{m}_A \in S'_A \\ 0 & \overline{m} \in S' - S'_A \end{cases} \tag{16}$$

where each of the $\phi'_{\overline{n}_A}$ is non-zero when evaluated at $\lambda = 1$ and each has the same algebraic sign.

Thus, the zero mutation probability limit of the time-homogeneous two and three-operator stationary distributions exists. Further, it has strictly positive components corresponding to each of the one-operator absorbing states and zero components elsewhere, confirming the empirical result noted in Section 4.

The non-zero value of the uniform population states in the zero mutation probability limiting stationary distribution means that the limiting condition necessary for extending the simulated annealing global optimality result does not obtain. However, the converged distribution entropy data presented in Section 4, and some further theoretical results developed in [2] suggest that the desired limiting behavior can be approached by adjusting the algorithm population size parameter, at least for the two-operator algorithm variant.

Sufficient conditions on the mutation probability control parameter to ensure that the non-stationary algorithm achieves the limiting distribution asymptotically are also developed in [2]. The methodology consists of exploiting the form of the matrices whose characteristic polynomials are represented in Eq. 15 and 16 to deduce some results concerning continuity of the stationary distribution components and their derivatives, and using those results along with the lower bound in Eq. 13 to replicate the methodology by which the simulated annealing temperature schedule bounds in [5] and [17] are developed. The product of that effort is a monotonic lower bound on the sequence $\{p_m(k)\}$ which, along with the condition that the sequence has limit zero, is sufficient to ensure that the non-stationary simple genetic algorithm attains (asymptotically) the limit distribution described in Eq. 16. The bound is given by

$$p_m(k) \geq \frac{1}{2} k^{-\frac{1}{ML}}$$

It is asymptotically superior to the $K/\log(k)$ bounds for the simulated annealing algorithm.

Also reported in [2] is a partially developed framework for representing the stationary distribution components at all consistent fixed mutation probability parameter values, including the zero mutation probability limit. The development is very extensive and is currently incomplete. However, it provides substantial insight into the functional form of the stationary distribution components. For example, the two-operator stationary distribution component corresponding to the uniform population state for solution $i \in S$ includes the numerator polynomial factor $R(i)^M$. Thus, the ratio of the two-operator stationary distribution components corresponding to the uniform population states for solutions $i, j \in S$ includes the factor $[R(i)/R(j)]^M$ which, if $R(i) < R(j)$, can be forced towards zero by selecting M sufficiently large.

6 CONCLUSIONS AND FUTURE DIRECTION

The work reported in [2] and synopsized here attempts to provide an algorithm model and accompanying convergence theory for the simple genetic algorithm comparable in scope to that which exists for simulated annealing. It includes a very general non-stationary Markov chain model for algorithm variants incorporating one, two and three-operator combinations of the three primary GA operators, and some key intermediate theoretical results derived from the model. Specific theoretical results include (1) stationary distribution existence for the time-homogeneous two and three-operator variants, (2) existence of a zero mutation probability limiting stationary distribution and (3) sufficient conditions, analogous to the simulated annealing temperature schedule bounds, on the mutation probability parameter sequence to ensure that the non-stationary algorithm achieves the limiting distribution.

The major incomplete task in extrapolating the simulated annealing convergence theory consists of reducing the stationary distribution existence results to an explicit solution in terms of the objective function and algorithm parameters (or at least to a result from which usable bounding values can be obtained). A substantial amount of effort is devoted to that task in [2] and considerable progress obtained, along with a proposed direction for continuing the work begun there.

Acknowledgements

The authors wish to acknowledge the generous support of the USAF Wright Laboratories to this work.

References

[1] L. Davis, editor. *Genetic Algorithms and Simulated Annealing.* Morgan Kaufman Publishers, Inc., Los Altos, CA., 1987.

[2] T. E. Davis. *Toward an Extrapolation of the Simulated Annealing Convergence Theory Onto the Simple Genetic Algorithm.* PhD thesis, University of Florida, Gainesville, FL., 1991.

[3] K. A. De Jong. *An Analysis of the Behavior of a Class of Genetic Adaptive Systems.* PhD thesis, University of Michigan, Ann Arbor, MI., 1975.

[4] A. E. Eiben, E. H. L. Aarts, and K. M. Van Hee. Global Convergence of Genetic Algorithms: On Infinite Markov Chain Analysis. First International Workshop on Problem Solving from Nature, Dortmund, Germany, 1990.

[5] S. Geman and D. Geman. Stochastic Relaxation, Gibbs Distributions and the Bayesian Restoration

of Images. *IEEE Trans. Patt. Anal. Mach. Intel.*, PAMI-6(6), Nov. 1984.

[6] D. Goldberg. *Genetic Algorithms in Search, Optimization and Machine Learning.* Addison-Wesley Publishing Company, Inc., Reading, Mass., 1989.

[7] D. Goldberg. *A Note on Boltzman Tournament Selection for Genetic Algorithms and Population Oriented Simulated Annealing.* TCGA Report 90003, University of Alabama, Tuscaloosa, AL., May 1990.

[8] D. E. Goldberg and P. Segrest. Finite Markov Chain Analysis of Genetic Algorithms. In J. J. Grefenstette, editor, *Proceedings of the Second International Conference on Genetic Algorithms*, pages 1–8, Lawrence Earlbaum Associates, Hillsdale, New Jersey, 1987.

[9] J. J. Grefenstette, editor. *Proceedings of an International Conference on Genetic Algorithms and Their Applications*, Lawrence Earlbaum Associates, Hillsdale, New Jersey, 1985.

[10] J. J. Grefenstette, editor. *Proceedings of the Second International Conference on Genetic Algorithms*, Lawrence Earlbaum Associates, Hillsdale, New Jersey, 1987.

[11] J. J. Grefenstette, editor. *Proceedings of the Third International Conference on Genetic Algorithms*, Lawrence Earlbaum Associates, Hillsdale, New Jersey, 1989.

[12] M. Hall. *Combinatorial Theory.* Blaisdell Publishing Company, Waltham, Mass., 1967.

[13] J. H. Holland. *Adaptation in Natural and Artificial Systems.* The University of Michigan Press, Ann Arbor, MI., 1975.

[14] S. Kirkpatrick, C. D. Gelatt, and M. P. Vecchi. Optimization by simulated annealing. *Science*, 220(4598), May 1983.

[15] P. J. M. Laarhoven and E. H. L. Aarts. *Simulated Annealing.* D. Reidel Publishing Company, Dordrecht, Holland, 1987.

[16] M. Lundy and A. Mees. Convergence of an annealing algorithm. *Mathematical Programming*, 34:111–124, 1986.

[17] D. Mitra, F. Romeo, and A. Sangiovanni-Vincentelli. Convergence and finite time behavior of simulated annealing. In *Proc 24 th Conference on Decision and Control*, Ft. Lauderdale, 1985.

[18] A. E. Nix and M. D. Vose. Modeling Genetic Algorithms with Markov Chains. Submitted to the Annals of Mathematics and Artificial Intelligence.

[19] J. Riordan. *An Introduction to Combinatorial Analysis.* John Wiley & Sons, New York, N. Y., 1958.

[20] E. Seneta. *Non-negative Matrices and Markov Chains.* Springer Verlag, New York, 1975.

[21] M. D. Vose and G. E. Liepins. Punctuated equilibria in genetic search. Submitted to Complex Systems.

The Performance of Genetic Algorithms on Walsh Polynomials: Some Anomalous Results and Their Explanation

Stephanie Forrest
Dept. of Computer Science
University of New Mexico
Albuquerque, N.M. 87131-1386

Melanie Mitchell
Artificial Intelligence Laboratory
University of Michigan
Ann Arbor, MI 48109-2110

Abstract

In this paper we discuss a number of seemingly anomalous results reported by Tanese concerning the performance of the genetic algorithm (GA) on a subclass of Walsh polynomials. Tanese found that the GA optimized these functions poorly, that a partitioning of a single large population into a number of smaller independent populations seemed to improve performance, and that hillclimbing outperformed both the original and partitioned forms of the GA on optimizing these functions. We reexamine these results experimentally and theoretically, and propose and evaluate some explanations. In addition, we examine the question of what are reasonable and appropriate ways to measure the performance of genetic algorithms.

1 Introduction

Tanese's 1989 doctoral dissertation reports experiments that apparently contradict some commonly held beliefs about genetic algorithms (GAs) as function optimizers [13, 12]. Specifically, Tanese's results show that for a certain class of Walsh polynomials, GAs are poor optimizers, perform worse than hillclimbing, and often perform best when the total population is split up into very small subpopulations. These results call into question several expectations about GAs — that they are good at optimizing Walsh polynomials [1], that they will routinely outperform hillclimbing and other gradient descent methods on hard problems such as those with nonlinear interactions [9], and that populations must be of a sufficient size to support effective schema processing [4].

Tanese's results are provocative and have caused some researchers to question the effectiveness of GAs in function optimization. Thus, they deserve a careful explanation. This paper reviews these apparently negative results, explains some of the anomalies, and discusses the general question of what criteria are appropriate to use in assessing the performance of GAs.

2 Experimental Setup

The experiments we report in this paper were performed with a similar GA and identical parameter values to those used by Tanese [2, 13]. All of Tanese's experiments used strings of length 32 and populations of 256 individuals. The population was sometimes subdivided into a number of smaller subpopulations. Tanese's algorithm used a *sigma scaling* method, in which the number of expected offspring allocated to each individual is a function of the individual's fitness, the mean fitness of the population, and the standard deviation from the mean. An individual with fitness one standard deviation above the population mean was allocated one and a half expected offspring. Multipoint crossover was used, with a crossover rate of 0.022 per bit (e.g., for 32-bit strings, there were on average 0.7 crossovers per pair of parents per generation). The crossover rate (per pair) was interpreted as the mean of a Poisson distribution from which the actual number of crosses was determined for each individual. The mutation probability was 0.005 per bit. With the exceptions of sigma scaling and multipoint crossover, Tanese's GA was conventional. Tanese ran each of her experiments for 500 generations, whereas we ran each of ours for 200 generations, although, as will be discussed later, this was not an important difference. For some experiments we altered certain parameter values; these exceptions are noted explicitly.

3 Walsh Polynomials

In her dissertation, Tanese describes experiments involving several fitness functions. In this paper we consider only the results with respect to Walsh polynomials. Walsh polynomials and their relevance to

GAs have been discussed by a number of researchers [1, 5, 9, 13]. A Walsh polynomial is defined as a sum of *Walsh functions*.

There is a correspondence between certain partionings of the GA's search space and Walsh functions. Such partitionings are induced by schemas. For example, the partitioning d**...* divides the search space into two halves, corresponding to the schemas 1**...* and 0**...*. Likewise, the partitioning dd**...* represents a division of the space into four quarters, each of which corresponds to a schema with the leftmost two bits defined. Each different partitioning of the search space can be indexed by a unique binary integer (bit string) in which 1's correspond to the partition's defined bits and 0's correspond to the non-defined bits. For example, under this enumeration, the partition d**...* has index j=100...0. The Walsh function corresponding to the jth partition is defined as follows (where both j and x are bit strings):

$$\psi_j(x) = \begin{cases} 1 & \text{if } x \wedge j \text{ has even parity} \\ -1 & \text{otherwise.} \end{cases}$$

A Walsh polynomial has the following form:

$$f(x) = \sum_{j=0}^{2^l - 1} \omega_j \psi_j(x)$$

where l is the length of the bit string x, and each ω_j is a real-valued coefficient. Walsh polynomials provide a basis for defining any real-valued function on bit strings.

From this general class of functions, specific subsets were selected as fitness functions, both for Tanese's work and for the experiments reported here. Tanese generated each fitness function, hereafter called *Tanese functions*, by randomly choosing 32 partition indices j (called the *defined partitions*), all of the same order — that is, all containing the same number of 1's. The coefficient ω_j for each of the 32 chosen partition indices was also chosen at random from the interval $(0.0, 5.0]$. The fitness function consisted of the sum of these 32 terms (all other coefficients were effectively set to 0).

Once the 32 partition indices were chosen, a point x' was chosen randomly to be the global optimum, and the sign of each of the 32 ω_j's was adjusted so that the fitness of x' would be $\sum |\omega_j|$.

This method of constructing functions had several advantages for comparing the performance of different GAs: (1) it was easy to construct random functions that on average were of similar difficulty (although Goldberg [6] has shown that it is possible to construct two functions of a given order that have different levels of difficulty for the GA); (2) functions of different degrees of difficulty could be constructed by varying the order, since low-order functions of this sort should on average be easier for the GA to optimize than high-order functions [13]; and (3) the global optimum was known, which made it possible to determine how close the GA came to the maximum value of the function.

Tanese conducted experiments on fitness functions with defined partitions at orders 4, 8, 16, and 20 (each function had all of its defined partitions at the same order). For each experiment she randomly generated 64 functions of a given order, and compared the performance of the *traditional* (single population) GA with a number of *partitioned* GA's, in which the population of 256 individuals was subdivided into various numbers of smaller populations. In this paper, we discuss results only for functions at order 8, since these were the functions Tanese analyzed in the greatest depth. All of our experiments involved manipulations of parameters on the traditional GA; we did not experiment with partitioned GA's. For each of our experiments, we ran the GA once on each of 20 different randomly generated functions, for 200 generations each. Tanese carried each run out to 500 generations, but in each of our runs that used strings of length 32, the population had reached a more or less steady state by about generation 100, and the results would not have changed significantly if the run had been extended. The shorter runs were sufficient for determining the comparative effects of the various manipulations we performed on the parameters.

4 Discussion of Anomalies

This section discusses and proposes explanations for the following anomalies in Tanese's results :

- The poor performance of all forms of the GA on optimizing both low-order and high-order Tanese functions.
- The improvement in optimization performance obtained by the partitioned GA over the traditional GA.
- The superior optimization performance of hill-climbing on the Tanese functions as compared with both the traditional and partitioned GA.

4.1 Why does the GA optimize Walsh polynomials poorly?

One of Tanese's most striking results was the poor performance of the GA (in both its traditional and partitioned-population form) when searching for maximum values of the functions described above. In Tanese's experiments, the GA was run 5 times on each of the 64 randomly generated functions; each set of 5 constitutes a *trial*. One of her primary criteria was the *success rate*: the number of trials on which the global optimum was found at least one out of five times. On the 64 trials (i.e., 320 runs total) on randomly generated order 4 Walsh polynomials, the success rate of the traditional GA was only 3 (the most successful partitioned algorithm's success rate was only 15). On 320 runs on randomly generated order 8 Walsh polynomials, neither the traditional nor the various partitioned GAs ever found the optimum.

The following possible explanations for the GA's poor performance will be discussed and evaluated in this section: (1) the average defining-lengths of schemas are very long and thus good schemas tend to be broken up by crossover; (2) randomly generating 32 partition-indices over strings of length 32 results in a large number of overlaps, which means that most of the significant loci are correlated, effectively making the functions very difficult; (3) crossover is ineffective on these functions because of the lack of lower-order building blocks; and (4) the fitness functions are deceptive.

4.1.1 Is the GA's poor performance due to long defining lengths of schemas?

Given a Tanese function F, the fitness of a string x under F depends on the parity of $x \wedge j$ for each j in the terms of F. Because of this parity calculation, a change of a single bit in x in any of the positions in which j has a 1 will produce an opposite value for $x \wedge j$ and thus reverse the contribution of that term to the total fitness. This implies that in general, for a Tanese function of order n, no schema s of order less than n will give the GA any useful information, since for any given j, half the instances of s will have even parity with respect to j, and half will have odd parity. This property is due to the parity properties of the Walsh functions and to the fact that in a Tanese function, all the terms are of the same order. This property and its implication for the effectiveness of crossover on these functions will be discussed in Section 4.1.3.

The expected defining length of a schema of order 8 in a 32-bit string is 26, a substantial proportion of the entire string [7]. Such long defining lengths could make the Tanese functions hard for the GA because of the high probability of crossover disruption, To what degree was this problem responsible for the poor performance of the GA on these functions?

To answer this question, we ran the traditional GA with Tanese's parameters on 20 randomly generated functions of order 8 whose partition indices were restricted to have a maximum defining length of 10. That is, for each of the 32 partition indices, a randomly positioned window of 10 contiguous loci was chosen, and all eight 1's were placed randomly inside this window.

Using the success-rate criterion described above, the performance was identical to Tanese's original results: the traditional GA never found the optimum. Other performance measures made it clear that limiting the defining length improved the GA's performance to some extent: the GA was able to find strings with slightly higher fitnesses (in terms of percent of optimum) and slightly higher mean-population fitnesses. These results, together with results from all experiments described in this paper, are summarized in Table 1 under the heading *Def-Len: 10*. They are to be compared with the values under *Original*, giving the results from our replication of Tanese's traditional GA runs on order-8 functions. We conclude that the contribution of long defining lengths to the GA's overall poor performance is not significant. As will be discussed, one reason for this is that crossover is not very effective on these functions in the first place.

4.1.2 Is the GA's poor performance due to overlap among significant loci in the partition indices?

Next, we considered the possibility that overlaps among significant loci in the partition-indices j (i.e., loci with 1's) were causing correlations among terms in the fitness functions, making the optimization problem effectively higher-order. As a simple example, suppose that 0011 and 0110 are two order-2 j's that have non-zero positive coefficients. There are eight strings that will cause $\psi_{0011}(x)$ to be positive, corresponding to the schemas ##00 and ##11. Likewise, there are eight strings that will cause $\psi_{0110}(x)$ to be positive, corresponding to the schemas #00# and #11#. So, in order to find a point that gets a positive value from both $\psi_{0110}(x)$ and $\psi_{0011}(x)$, the GA must discover either the schema #000 or the schema #111. The net effect of this overlap is that three bits are correlated instead of two, making the problem effectively order-3 instead of order-2. In the case of the Tanese functions, this is a likely source of difficulty; for example, with order-8 functions, where 32 order-8 terms were randomly generated, each 1 in any given j will on average be a member of 7 other different j's, and thus the effective linkage will be exceedingly high.

To assess the effect of overlaps, we ran the GA on strings of length 128 rather than 32, in order to reduce the number of overlaps. With a string length of 128, each defined locus with allele 1 would participate on average in only 2 of the 32 partition indices. As before, we generated 20 random functions and ran the GA for 200 generations on each. As shown in Table 1 (under *Str-Len: 128*), the GA's performance was remarkably improved. The GA found the optimum 19/20 times, compared with 0/20 for the 32-bit case, and came very close to finding the optimum on the other run. Of all the experiments we tried, this caused the most dramatic improvement, leading us to conclude that the principle reason the Tanese functions are difficult is because the short strings (32 bits) and relatively high number of terms (32) causes nearly all 32 bits to be correlated, thus effectively creating an order 32-problem. In the non-overlapping case, it is possible for the GA to optimize each term of the function almost independently.

The fact that overlap is much higher with 32-bit functions than with 128-bit functions explains the strikingly different dynamics between runs of the GA on the former and latter functions. Space constraints prevent us from reporting these data in detail (see [3]), but we describe them qualitatively.

A Walsh polynomial can be thought of as a "constraint satisfaction" problem in which each term produces an additional constraint. The goal is to find an individual that receives positive values from as many terms as possible, and preferably from terms with high coefficients. In a typical run on 32-bit strings, the GA very quickly finds its highest-fit individual, by around generation 30. This individual receives positive values from only a subset of the terms in the fitness function, leaving the rest of the terms "unsatisfied," that is, yielding negative values on that individual. However, because of the high degree of overlap, the constraint satisfaction problem here is very severe, and to discover an individual that receives additional positive values from other terms — without losing too many of the currently held positive values — is very difficult (especially since crossover is largely ineffective on these functions). This is particularly true since, once individuals of relatively high fitness are discovered, the diversity of the population falls very quickly. For example, on one typical run the number of different chromosomes in the population started out at 256, fell to 39 by generation 20, and stayed relatively constant thereafter[1]. Very quickly, the population subdivides itself into a small number of mutually exclusive sets: one large set of individuals receiving positive values from one large subset of the terms in the fitness function, and other, much smaller sets of individuals receiving positive values from the other, smaller subsets of terms in the fitness function.

To summarize, the GA stays stuck at a local optimum that is discovered early. It is possible that raising crossover and mutation rates might improve the GA's performance on such functions, since it would increase the amount of diversity in the population.

On a typical run involving strings of length 128, the situation is very different. In such a run, the GA tends not to discover its highest-fit individual until late in the run (on average, around generation 150), and the diversity of the population remains relatively high throughout the run. The continuing high diversity of the population seems to be a result of several factors, including the following: since the problem is less constrained, there are many ways in which an individual can achieve a relatively high fitness; and since the crossover rate was defined on a per-allele basis, there are on average a greater number of crossovers per chromosome here than in the 32-bit case. However, the relative lack of constraints was the major factor, since when crossover was turned off in the 128-bit case, the population diversity remained considerably higher than for the 32-bit run described above, although it was less than in the 128-bit case with crossover turned on. In the 128-bit case, the population does not segregate into mutually exclusive sets — throughout a typical run, each of the terms in the fitness function provide positive values to a significant fraction of the population. Because of the relative lack of constraints in the 128-bit case, the population does not quickly become locked into a local optimum, and is able both to continue exploring longer than in the 32-bit case and to optimize each term of the fitness function more independently.

[1] We counted two individuals as being different only if they were different at some significant locus — that is, at a locus in which one of the 32 js had a 1. As one would expect, with strings of length 32, every locus was significant, which is not generally the case for strings of length 128.

4.1.3 Is crossover effective on these functions?

As we mentioned earlier, the facts that (1) a Tanese function F evaluates points according to their parity with the j's in the terms of F, and (2) that a given F is created using partition indices of only a single order, imply that schemas of order lower than the order of F do not provide the GA with useful information. Thus crossover, one of the major strengths of the GA, is not a useful tool for recombining building blocks until schemas of order at least as high as that of the function have been discovered.

To verify this empirically, we ran the GA without crossover on 20 randomly generated 32-bit order 8 functions. The results are summarized in Table 1 under *No Xover 32*. The GA's performance was not impaired; the maximum fitness discovered and the mean population fitness are not significantly different from the runs in which crossover was used (*Original* in the table). To further verify the ineffectiveness of crossover, we ran the GA without crossover on 20 randomly generated 128-bit order 8 functions. The results are summarized under *No Xover 128* in Table 1: the performance of the GA was not significantly different from the 128-bit runs in which crossover was used (*Str Len 128*). With both the shorter and longer string-lengths (and thus with both large and small amounts of overlap), whether or not crossover is used does not seem to make any difference in the GA's performance on these functions.

	Times Optimum Found	Average Max. Fit. (% opt.)	Average Gen. of Max. Fit.	Average Max. Mean Fit. (% opt.)
Original	0	88.1 (2.9)	31 (33)	85.1 (3.8)
Def-Len: 10	0	92.3 (2.9)	41 (48)	89.2 (3.0)
Str-Len: 128	19	99.97 (0.13)	150 (30)	93.6 (1.3)
No Xover 32	0	88.4 (2.7)	22 (28)	86.2 (2.6)
No Xover 128	17	99.85 (0.45)	72 (41)	93.9 (0.6)
Hill 32	0	95.4 (1.5)	-	-
Hill 128	20	100.0 (0.0)	-	-

Table 1: Summary of results of all experiments. The experiments were all performed running the traditional GA (or, in one case, hillclimbing) on randomly generated order 8 Walsh polynomials. Each result summarizes 20 runs of 200 generations each. The experiments were: (1) *Original* (replicating Tanese's experiments); (2) *Def-Len: 10* (limiting the defining length of partition indices to 10); (3) *Str-Len: 128* (increasing the string length to 128 bits); (4) *No Xover 32* (same as *Original* but with no crossover); (5) *No Xover 128* (same as *Str-Len: 128* but with no crossover); (6) *Hill 32* (hillclimbing on 32 bits); and (7) *Hill 128* (hillclimbing on 128 bits) All runs except the 128-bit runs used strings of length 32. The values given are (1) the number of times the optimum was found; (2) the maximum fitness (% of optimum) found (averaged over 20 runs); (3) the average generation at which the maximum fitness was found; and (4) the maximum population mean (% of optimum) during a run (averaged over 20 runs). The numbers in parentheses are the standard deviations.

4.1.4 Is the GA's poor performance due to deceptive functions?

Goldberg has recently shown that it is possible to construct deceptive functions by exploiting interactions among low-order terms [6], and he proposes the possibility that the Tanese functions are deceptive. We believe that the Tanese functions are not fully deceptive and that even if they are partially deceptive, deception is not the major reason that the functions are difficult. The Tanese functions are not fully deceptive because there are some schemas that do not lead the GA away from the global optimum. For example, as we discussed in Section 4.1.3, schemas of lower order than the defined order of the function provide no information to direct the GA's search.

It should be pointed out that any even-order Tanese function will have the property that $f(x) = f(x_{complement})$. For example, consider an order-2 function with one defined partition-index, $j = 0011$. $\psi_j(x)$ will be positive for the schemas ##00 and ##11, and negative for ##01 and ##10. It is possible to show that this property will hold for any even-order positive term and for any linear combination of even-order terms.

This property implies that for all the Tanese functions there will be at least two global optima that are complements of one another. Additionally, there will be at least two second most fit points, again complements of one another, and so on. This fact may have implications for whether or not the functions are deceptive. One consequence of this property is that there will be mutually exclusive families of solutions (trees of schemas) and that the GA may have trouble selecting which peak to climb (see Section 4.1.2). As we have showed, the Tanese functions are effectively higher-order than the level at which they are defined, and this fact is sufficient to account for the functions' difficulty. The fact that the functions are effectively higher-order does not imply that they are deceptive.

4.2 Why does partitioning the population seem to improve performance?

Although both the traditional and partitioned GAs performed poorly on the Tanese functions with respect to the highest fitness found, the results reported by Tanese showed that the performance of most instances of the partitioned GA was better than that of the traditional GA. This was the case even when the population was subdivided into extremely small partitions, such as 64 partitions of 4 individuals each, and, in some cases, even for 128 partitions of 2 individuals each. For functions of order higher than 4, neither the traditional nor partitioned GA's ever found the function optimum, so the difference between the two was measured only in terms of proximity to the optimum of the best fitness discovered.

As Tanese points out, these results run against conventional wisdom about GAs: it has been thought that on difficult problems a large population is needed for processing a sufficient number of schemas [4].

Tanese proposes three main reasons for this surprising result.

1. Tanese functions have a large number of local optima and the GA tends to converge on one. Each of the smaller subpopulations of the partitioned GA will converge earlier than a larger single population, but the subpopulations tend to converge on different optima, and so are able to explore a greater number.

2. After the populations converge, the major adaptive force is mutation. Mutation will be more effective in a smaller population than in a large population, since in a large population, the greater number of mutations will drive up the standard deviation of fitnesses in the population, making it less likely that fit mutations will be able to produce enough offspring to spread in the population. In the smaller population, fewer mutations will occur and the standard deviation will be lower, allowing fit mutations to spread more effectively.

3. Since smaller populations tend to be more homogeneous (for the reasons given above), fit mutations are less likely to be destroyed through crossover.

Explanation 1 seems correct, especially in light of the dynamics that were discussed in Section 4.1.2. It seems that the traditional GA quickly finds a single local optimum that satisfies a subset of partitions, and cannot go beyond that point. Since the number of different chromosomes in the population falls so quickly, the traditional GA does not use the potential for diversity that its large population offers. This explains why a number of small populations running in parallel would likely result in at least one way of satisfying the constraints of the function that was superior to the single large-population GA.

Explanation 2 relies on the assumption that large populations tend to be more diverse and have higher standard deviations than smaller ones. Since in this case, the diversity of the large population with 32-bit strings falls very quickly, this may not be a correct assumption. It seems more likely that the main factor keeping successful mutations at bay is the degree to which the problem is constrained: once a local optimum is settled upon, it is very unlikely that a single mutation (or a small set of mutations) will yield a fitter individual.

A similar point could be made with respect to Explanation 3. Given the homogeneity that results even with a large population, it seems likely that this is not a significant factor in the relative poor performance of the traditional GA.

It is very important to point out that the effects discussed here come about because of the special characteristics of the fitness functions being used — in particular, the fact that the single-order functions prevent the GA from exploiting information from lower-order building blocks through crossover. For functions in which lower-order building blocks do yield useful information, these effects might not be seen. The results and analysis for these functions cannot be assumed to be applicable to all instances of Walsh polynomials.

5 Why does hillclimbing outperform the GA on these functions?

Another surprising result reported by Tanese is that iterated hillclimbing consistently outperformed the traditional GA on the Tanese functions. On 32-bit order-4 functions, the success rate of hillclimbing was 23 out of 64 — almost eight times the success rate of the traditional GA and more than one and a half times the success rate of the best partitioned GA. Hillclimbing, like both the traditional and partitioned GAs, had a success rate of 0 on all functions of higher order, but it did consistently achieve higher fitnesses than the traditional GA. We ran stochastic iterated hillclimbing [13] on 20 randomly generated order-8 functions using the same number of function evaluations as in the GA runs. The results are summarized under *Hill 32* in Table 1: on average, hillclimbing was able to find a string whose fitness was within about 5% of the maximum, whereas the original GA was able to find a string whose fitness was only within about 12% of the maximum.

Since the nature of the Tanese functions precluded the GA from using information from lower-order building blocks via crossover, the genetic operators of crossover and mutation served mainly as a means of generating random variation — in effect, an inefficient form of hillclimbing.

6 How should the GA's performance be measured?

A major criterion by which Tanese judged GA performance was success rate, i.e., the number of runs on which the global optimum was found. This is a very strict requirement, since it does not reflect the objective difficulty (independent of any specific optimization method) of the function or how close the algorithm came to the global optimum. One large question raised by her study is, what is a reasonable way to measure the performance of GAs?

Of course, any performance measure depends to some extent on the purpose for which the algorithm is being used. However, for many problems simply recording the number of times an algorithm finds the global optimum is not adequate. The global optimum may not be known or achieving it may be an infeasible goal; in such cases, success might be measured in terms of improvement over the previously known best solution or how rapidly the algorithm can discover a solution that is "good enough" (satisficing). Under these circumstances it would be unreasonable to expect the GA to find the exact optimum or get no credit for success.

Additionally, sampling procedures like the GA are inherently nondeterministic. This implies that even if the algorithm were performing well on a particular function (and in the theoretical limit could be guaranteed to find the optimum) it might be "unlucky" in some circumstances, not reaching the exact global optimum. To some extent nondeterminism can be accounted for by performing multiple runs with different initial populations. However, it seems that for nondeterministic procedures such as the GA, it would be appropriate to use a continuous measure of success.

We believe that more quantitative measures are appropriate for comparing the performance of different optimizers on different problems, and that they should have the following properties: (1) they should be normalized so that performance can be compared on different problems and with different optimizers, (2) they should reflect problem difficulty, and (3) they should indicate how far the optimizer got, not just whether or not it reached the global optimum.

We are currently studying a metric that computes the probability that a random point would have a higher fitness than that found by the optimizer [10]. If y is the highest value found by the optimizer, f is the fitness function, and X is a random variable, then we want to find

$$y' = P(f(X) < f(y)).$$

Using this metric, the optimizer is trying to maximize y'. P can be found by computing, either analytically or empirically, the distribution of $f(X)$.

In preliminary tests using this metric, the GA and hillclimbing both do extremely well on order 8 problems, and their performance is indistinguishable from one another. This suggests that the difference between the performance of hillclimbing and the GA is insignificant when compared with the difficulty of the order 8 Tanese functions.

7 Conclusions

After examining several possible causes for the GA's poor performance on the Tanese functions, we reach several conclusions:

- Overlaps in the defined loci between different terms of the function created functions of much higher effective order than their Walsh terms suggest. This was the principle reason for the difficulty of the functions.
- The lack of information from lower-order schemas hampered crossover from contributing to the search. This was a secondary cause of the GA's poor performance, and largely accounts for the superior performance of hillclimbing.
- Long defining lengths in the non-zero partitions contributed slightly to the GAs poor performance but were not a major factor.

In addition, we feel that the success-rate performance measure is inappropriate for comparative studies of the conventional GA with variants or with other optimization methods. Under another metric, the performance of hillclimbing and GA were indistinguishable on this problem.

Tanese's results could be erroneously interpreted to imply that, in general, Walsh polynomials are difficult for GAs to optimize, and that hillclimbing will generally outperform the GA on such functions. Such a result would be a very negative one for the GA, since the Walsh functions provide a basis for any real-valued function defined on bit strings, and thus any such function can be written as a Walsh polynomial. However, the experiments and analyses reported in this paper suggest that Tanese's results should not be interpreted as a general negative verdict on the efficacy of the GA in function optimization. The functions she studied are a highly restricted subset of the class of Walsh polynomials and have several peculiar properties that make them difficult to optimize. Her results could also mistakenly be taken to imply that partitioned GAs with smaller subpopulations will always outperform traditional GAs. This may not be true for functions in which recombination of lower-order building blocks plays a major role in the search.

These results raise the important question of what functions are well-suited for the GA, and more importantly, which of these functions distinguishes the GA from other optimization methods. One clearly important factor lacking in the Tanese functions is the availability of lower-order building blocks which can combine to produce fit higher-order schemas; such hierarchical schema-fitness structure is what makes crossover an effective operator. One hypothesis is that the degree to which a function contains such structure will in part determine the probability of better optimization success for the GA than for other methods. Another hypothesized contributing factor is the degree to which the fitness landscape contains different "mountainous" regions of high-fitness points that are separated by "deserts" of low-fitness points [8]. In order to travel from a low mountainous region to a higher one, a low-fitness desert must be crossed. Point-mutation methods such as hillclimbing can have very high expected waiting times to cross such deserts (if they can cross them at all); the hypothesis is that the GA will be much more likely to quickly cross such deserts via the crossover operation.

The degree to which these factors are present in a given function may depend to a large extent on the representation chosen for the function; the role of representation in GA function optimization has been recently discussed by Liepins and Vose [11]. We are currently studying the behavior of the GA and other optimization methods on a class of functions in which the degree to which these and other factors are present can be directly varied, and we are investigating ways in which the presence of such features can be detected in a given function with a given representation. We believe that this investigation will lead to a better understanding of the classes of functions for which the GA is likely to be a successful optimization method.

Acknowledgments

This work is a natural follow-up to Reiko Tanese's thesis work. We are grateful for the interesting results that she uncovered and the clarity with which she reported them in her thesis. We regret that she was unable to collaborate with us in this work and apologize for any misinterpretations or misrepresentations. John Holland and Quentin Stout made many valuable suggestions for this research. Partial support for this project was provided by the Santa Fe Institute, Santa Fe, NM and the Center for Nonlinear Studies, Los Alamos National Laboratory, Los Alamos, NM. Support for the second author was also provided by the Michigan Society of Fellows.

References

[1] A. D. Bethke. *Genetic Algorithms as Function Optimizers.* PhD thesis, The University of Michigan, Ann Arbor, MI, 1980. Dissertation Abstracts International, 41(9), 3503B (University Microfilms No. 8106101).

[2] Stephanie Forrest. Documentation for prisoner's dilemma and norms programs that use the genetic algorithm. Technical report, University of Michigan, 1985.

[3] Stephanie Forrest and Melanie Mitchell. The performance of genetic algorithms in function optimization: Some anomalous results and their explanation. Technical report, Santa Fe Institute, Santa Fe, New Mexico, 1991.

[4] David E. Goldberg. Optimal initial population size for binary-coded genetic algorithms. Technical Report TCGA Report No. 85001, The University of Alabama, Tuscaloosa, AL 35486-2908, 1985.

[5] David E. Goldberg. Genetic algorithms and Walsh functions: Part I, a gentle introduction. Technical Report TCGA-88006, The University of Alabama, Tuscaloosa, AL, 1988.

[6] David E. Goldberg. Construction of high-order deceptive functions using low-order Walsh coefficients. Technical Report 90002, University of Illinois at Urbana-Champaign, Dept. of General Engineering, U. of Illinois, Urbana, IL, 1990.

[7] David E. Goldberg, Bradley Korb, and Kalyanmoy Deb. Messy genetic algorithms: Motivation, analysis, and first results. *Complex Systems*, 3:493–530, 1990.

[8] John H. Holland. Personal communication.

[9] John H. Holland. The dynamics of searches directed by genetic algorithms. In Y. C. Lee, editor, *Evolution, Learning, and Cognition*, pages 111–128, Teaneck, NJ, 1988. World Scientific.

[10] David Lane. Personal communication.

[11] Gunar E. Liepins and Michael D. Vose. Representational issues in genetic optimization. *Journal of Experimental and Theoretical Artificial Intelligence*, 2:101–115, 1990.

[12] Reiko Tanese. Distributed genetic algorithms. In J. David Schaffer, editor, *Proceedings of the Third International Conference on Genetic Algorithms.* Morgan Kaufmann, 1989.

[13] Reiko Tanese. *Distributed Genetic Algorithms for Function Optimization.* PhD thesis, The University of Michigan, Ann Arbor, MI, 1989.

Optimizing an Arbitrary Function is Hard for the Genetic Algorithm

William E. Hart
whart@cs.ucsd.edu
Cognitive Computer Science Research Group
Computer Science and Engineering
University of California, San Diego

Richard K. Belew
rik@cs.ucsd.edu
Cognitive Computer Science Research Group
Computer Science and Engineering
University of California, San Diego

Abstract

The Genetic Algorithm (GA) is generally portrayed as a search procedure which can optimize pseudo-boolean functions based on a limited sample of the function's values. There have been many attempts to analyze the computational behavior of the GA. For the most part, these attempts have tacitly assumed that the algorithmic parameters of the GA (e.g. population size, choice of genetic operators, etc.) can be isolated from the characteristics of the class of functions being optimized. In the following, we demonstrate why this assumption is inappropriate. We consider the class, **F**, of all deterministic pseudo-boolean functions whose values range over the integers. We then consider the Genetic Algorithm as a combinatorial optimization problem over $\{0,1\}^l$ and demonstrate that the computational problem it attempts to solve is NP-hard relative to this class of functions. Using standard performance measures, we also give evidence that the Genetic Algorithm will not be able to efficiently approximate this optimization problem. These results imply that there does not exist a fixed set of algorithmic parameters which enable the GA to optimize an arbitrary function in **F**. We conclude that theoretical and experimental analyses of the GA which do not specify the class of functions being optimized can make few claims regarding the efficiency of the genetic algorithm for an arbitrary fitness function. When analyzing the computational complexity of the Genetic Algorithm, classes (or distributions) of functions should be analyzed relative to the algorithmic parameters chosen for the GA.

1 Introduction

The Genetic Algorithm [11,7] is a method of stochastic optimization which has attracted significant attention in recent years. There have been many attempts to analyze the computational behavior of the GA, with Holland's schema theorem [11] central to much of this analysis. Using it, we can justify how and why certain bit patterns (schemata) will be propagated from one generation to the next. This can be used to analyze the effectiveness of different genetic operators (see for example [15]). Related analysis with Walsh functions has also proven very rewarding. Walsh functions can be used to analyze the effectiveness of genetic operators, as well as analyze the difficulty of the function being optimized.[5,6]

While these approaches provide some understanding of the class of functions which the GA can efficiently optimize, they fall short of providing an analysis of the computational complexity of the GA. Holland's Building-Block Hypothesis suggests that the GA will do well on functions in which low order schemata correctly predict the values of high order schemata. Conversely, the work on deceptivity uses Walsh functions to measure how hard a function might be for the GA. Unfortunately, neither of these analyses have been able to specify exactly what class of functions the GA does efficiently optimize. Any discussion of the computational complexity of the GA must be relative to a specific class of functions. The assumptions that can be made about this class of functions are often critical to establishing interesting complexity bounds.

To illustrate the importance of selecting an appropriate class of functions, we analyze the computational complexity of the GA relative to a very broad class of functions. We consider **F**, the class of all deterministic pseudo-boolean functions, functions f such that $f : \{0,1\}^l \rightarrow \mathbf{R}$. To allow for our complexity analysis, we restrict **F** to functions assuming integer values. We present a formalism which theoretically describes the optimization problem which the GA is solving for functions in **F**. We then demonstrate that unless NP=RP, there is no version of the GA which allows it to efficiently optimize an arbitrary function from this class.

At this point, one could argue that while the GA is not able to solve the problem exactly, it might offer a reasonable approximation which performs acceptably well. However, we consider several standard performance guarantees and show that unless NP=RP, there can exist no version of the GA which can satisfy any of them. We conclude by noting that the GA must be analyzed relative to much smaller classes of functions.

Before proceeding, we introduce some notation and definitions. We let $B = \{0, 1\}$ and let B^l refer to binary strings of length l. We assume that the reader is familiar with formal language theory; we generally follow the notational conventions of [12]. Particularly notice that the expressions surrounded by $\langle\,\rangle$'s refer to encodings used when defining formal languages; for an integer k, $\langle k \rangle$ would refer to its encoding. We remind the reader that P refers to the class of formal languages which can be recognized by a deterministic Turing machine (TM) in polynomial time. Additionally, both NP and RP refer to the classes of formal languages which can be recognized by nondeterministic Turing machines in polynomial time. The distinction between the two is that for languages in NP there must exist at least one path of computation (sequence of machine states) which leads to an acceptance of the language, whereas for languages in RP at least half of all computation paths must lead to accepting states. It is known that $P \subseteq NP$, $RP \subseteq NP$ and $P \cap RP \neq \phi$, and it is widely believed that the two inclusions are proper.[1] We consider an algorithm to be **efficient** if it completes its computation in polynomial time. In other words, a TM M is efficient if the language it accepts is in P. A formal definition of the Genetic Algorithm is not necessary for the following presentation. We refer the interested reader to [7] for an excellent presentation of the GA and its applications.

2 Statement of the Problem

To analyze the computational complexity of the Genetic Algorithm, we need to formalize the task which it accomplishes. Intuitively, the Genetic Algorithm takes a fixed fitness function (usually chosen by the experimenter) and searches for a population which is most fit with respect to this function. There are three important elements to this rough description:

- the definition of the *class of functions* from which the fitness function might be chosen
- the definition of *search space* of populations which is being searched
- the definition of *fitness* for elements (populations) in that search space.

We consider each of these separately before formalizing the problem which the Genetic Algorithm attempts to solve.

As we noted before, we will consider the class of deterministic pseudo-boolean functions which assume integer values, **F**. We further restrict **F** to functions which are computable in polynomial time. It would be unreasonable to expect the GA to efficiently find optimal values for functions which couldn't compute those values efficiently themselves.

There are a number of ways in which we can characterize the GA's search space. Perhaps the most natural of these is the set of populations P, of binary strings of length l, such that $|P| = k$. This is a reasonable search space since both the population size and string length are typically fixed for a GA simulation. If, however, the fitness of the population is not dependent on the size of the population itself, the set of binary strings of length l may suffice.

The meaning of a 'most fit' population within this search space is somewhat ambiguous. Simulations of the GA are typically halted at the researcher's discretion and not at some well defined convergence criterion. Often several measures of the population's fitness are provided. Perhaps the two most common of these are (1) the value of the maximally fit individual in the population and (2) the average value of all individuals in the population.

In what follows, we define the fitness of a population to be the fitness of the maximally fit individual in that population (relative to a fixed fitness function). We use a search space consisting of binary strings of length l. This choice of a search space is appropriate since we have chosen a fitness criteria which is not dependent on the size of the population. Furthermore, it is worth noting that our choice of a search space does not preclude the generality of our results. An analysis using the first search space is virtually identical to ours. We choose the alternative search space for ease of presentation alone, since the population size would become an extraneous parameter in our presentation.

Having defined the search space for the Genetic Algorithm as well as the fitness criteria it uses for a population, we can now formalize the problem which the Genetic Algorithm attempts to solve as a combinatorial optimization problem, DGA-MAX (following the format of [13]).

[1] The reader is referred to [3] for an excellent discussion of the complexity differences between P and NP, and to [4] for an exposition of probabilistic computation.

Definition 1 DGA-MAX

The Genetic Algorithm combinatorial maximization problem which (1) uses a deterministic fitness function f and (2) assigns the fitness of the maximally fit individual in a population to the fitness of the population itself. An instance of DGA-MAX consists of the following two parts:
1) an integer l defining the combinatorial space B^l
2) an encoding of a TM M_f, which defines a function $f : B^l \to \mathbf{Z}$ ∎

3 Complexity Results

In order to determine the complexity of DGA-MAX , we need to define a version of this problem as a formal language (using the format of [3]).

Definition 2 DGA-MAX

INSTANCE: a string encoding integers l, and λ, and a Turing machine M_f which computes a function $f : B^l \to \mathbf{Z}$ in polynomial time.
QUESTION: Does there exist an $x \in B^l$ s.t. $f(x) > \lambda$? ∎

Note that the optimization version of DGA-MAX is more powerful than the formal language version of DGA-MAX . Given a TM which solves the optimization version, we can clearly solve the formal language version. However, it is unknown whether the opposite is true (see [13] for further details). Thus, the optimization version is at least as difficult as the formal language version of DGA-MAX . We now demonstrate that the formal language version of DGA-MAX is very difficult to solve. To do this, we use the following definition (recall that SAT is NP-complete).

Definition 3 SATISFIABILITY (SAT)

INSTANCE: Set U of variables, clauses C of clauses over U
QUESTION: Is there a satisfying truth assignment? ∎

Proposition 1 DGA-MAX is NP-complete.

<u>**Proof:**</u> First observe that DGA-MAX is in NP. We can construct a NTM M which does the following on input $w = \langle k, \lambda, M_f \rangle$: M randomly guesses a string x, evaluates it with the function f, and compares the result to λ. Since f is computed polynomially in l, M is in polynomial time in $|w|$.

Next we reduce SAT to DGA-MAX . We use a TM M' which outputs the encoding $\langle l, 0, M_f \rangle$ on input w. Here, l is the number of variables in U. On input $w' \in B^l$, the Turing machine M_f assigns the value of the ith bit of w' to the ith variable in U. It then evaluates the boolean expression Φ constructed from a conjunction of the clauses in C. This reduction only requires memory to store l. Clearly this requires no more than $O(log\,|w|)$, so M' is a log- space reduction.

Now if $M'(w) \in$ DGA-MAX , then there exists w' s.t. $f(w') = 1$. But this implies that w represents a set of satisfying assignments for Φ. Thus, $w \in SAT$. If $w \in SAT$ then there is a set of assignments to the l variables which satisfy Φ. Thus there exists a w' in B^l s.t. $f(w') = 1$. But if this is true, then $M'(w) \in$ DGA-MAX since B^l has an element (w') whose value is greater than zero (λ). We conclude that $w \in SAT \iff M'(w) \in$ DGA-MAX . ∎

If $P \neq NP$, as is widely suspected, this result implies that there does not exist an efficient Turing machine which recognizes DGA-MAX .

Corollary 1 The optimization version of DGA-MAX is NP-hard.

<u>**Proof:**</u> As we noted above, the optimization version of DGA-MAX is at least as difficult as the formal language version. Since the formal language version is NP-complete, the optimization version must be NP-hard. ∎

This last result indicates that there probably does not exist an efficient algorithm to solve the optimization version of DGA-MAX . However, this result only applies to the deterministic algorithms. Since the Genetic Algorithm is nondeterministic, it could be the case that its nondeterminism allows it to efficiently solve either of the versions of DGA-MAX . For example, it is known that there are languages which can be solved more efficiently by probabilistic Turing machines than by deterministic Turing machines.[4] The following corollary demonstrates that even though Genetic Algorithms are stochastic, they still require super-polynomial time to solve DGA-MAX unless $RP = NP$.

Corollary 2 If $RP \neq NP$, then DGA-MAX is not in RP.

<u>**Proof:**</u> This follows immediately since NP-hardness of DGA-MAX implies that DGA- MAX cannot be in in RP unless $RP = NP$. ∎

The remainder of the paper considers only the optimization version of DGA-MAX .

4 Performance Guarantees

The previous section demonstrated that it is highly unlikely that there exists an efficient algorithm which solves DGA-MAX , whether it be deterministic or nondeterministic. Given this, it is important to consider what other performance guarantees can or cannot be made for DGA-MAX . It is often the case that even if a problem is NP-complete one can still guarantee certain performance bounds which allow it to be effectively solved in practice. As an example, consider the Traveling Salesman Problem. Here the goal is to find a circuit/tour through n cities at a minimal cost. If the costs between the cities satisfy the triangle inequality (e.g. let the cost be the Euclidean distance between the cities) then the problem remains NP-complete. Even so, there exists an algorithm which can produce a tour through the cities which is at most twice the value of the optimal tour.[3]

Before continuing, we introduce some notation (from [3]). Let Opt(I) refer to the value of the optimal value for instance I, and let A(I) refer to the value that algorithm A returns for instance I (we assume that A is an efficient algorithm). We are considering a maximization problem, so $Opt(I) \geq A(I)$ for all algorithms A, and we assume that $A(I) \geq 0$ for all algorithms and for all instances.

There are a number of performance guarantees defined in the literature. Among them are the following (from [3,14]):

- Relative error : $|A(I) - Opt(I)| / Opt(I)$
- Absolute error : $|A(I) - Opt(I)|$
- Convergence rates
- Absolute and Asymptotic performance ratios

In the following, we consider the absolute and asymptotic performance ratios to analyze the difficulty of DGA-MAX . We take the following definitions from [3]. Let the ratio $R_A(I) = Opt(I)/A(I)$. We define

- Absolute Performance Ratio R_A:

$$R_A = \inf\{r \geq 1 \mid R_A(I) \leq r, \; \forall I \in \text{DGA-MAX}\}$$

- Asymptotic Performance Ratio R_A^∞:

$$R_A^\infty = \inf\{r \geq 1 \mid \exists N \in \mathbf{Z}^{>0} \text{ s.t. } \forall I \in \text{DGA-MAX}, \; Opt(I) \geq N, \; R_A(I) \leq r\}$$

- Best Achievable Asymptotic Performance Ratio R_{MIN}(DGA-MAX):

$$R_{MIN}(\text{DGA-MAX}) = \inf\{r \geq 1 \mid \textit{there exists a polynomial time algorithm } A \textit{ for } \text{DGA-MAX} \textit{ with } R_A^\infty = r\}$$

R_A^∞ indicates whether we can determine a bound on $R_A(I)$ above some value N, while R_A indicates whether we can determine a bound on $R_A(I)$ for values above $N = 0$. R_{MIN}(DGA-MAX) is the smallest value of R_A^∞ over all possible algorithms A. It is this last performance ratio that we analyze. In the following, we show that R_{MIN}(DGA-MAX) $= \infty$ which implies that no deterministic algorithm can provide a performance guarantee on $R_A(I)$ s.t. $R_A(I)$ is less than some fixed r. This is true even if we consider only instances which have optima above fixed thresholds.

Proposition 2 If $P \neq NP$, then R_{MIN}(DGA-MAX) $= \infty$.

Proof: We assume towards a contradiction that R_{MIN}(DGA-MAX) $< \infty$. Thus, there exists a polynomial time algorithm A, and a fixed K s.t. $R_A \leq K$.[3] We now show that we can solve SAT. Recall that an instance of SAT consists of U variables together with a set of J clauses, C, over these variables. Consider the TM M which does the following. On input w, M outputs a reduction of SAT to DGA-MAX on a spare input tape. Then M simulates A using this tape as its input tape.

The reduction used is very similar to that of Proposition 1. On input v, M' (1) Counts the number of variables in U, stores the value (as l), and outputs that value, (2) outputs an encoding for a Turing machine M_f; on input $w' \in B^l$, M_f assigns the ith variable the value of the ith bit of w', counts the number of satisfied boolean clauses in C, and then multiplies this value by K.

If the value generated by A is equal to J, M accepts, otherwise it rejects. The reduction M' only requires memory to store l. This is clearly no more than $O(log\,|w|)$ so M' is in polynomial time. Since we assume that A is polynomial in time, M is also polynomial in time.

Let $I = M'(w)$. Now, if w is satisfiable $Opt(I) = J \cdot K$, otherwise $Opt(I) < J \cdot K$. But A is guaranteed to get $R_A(I) = Opt(I)/A(I) \leq K$, so $A(I) \geq J$ implies that $Opt(I) \geq J \cdot K$ so w is satisfiable. Conversely, if w is satisfiable then $Opt(I) = J \cdot K$, so $A(I) \geq J$ and M accepts. Thus M accepts iff w is satisfiable. ∎

Now it is important to note that R_A^∞ is one of the weakest performance guarantees discussed in the literature. Hence, it is significant in that we have shown that even R_A^∞ cannot be satisfied by any algorithm. Given this result, we can easily demonstrate that other stronger performance results are not possible.

Corollary 3 If $P \neq NP$, then no polynomial time algorithm A can guarantee that

$$Opt(I) - A(I) \leq \delta, \; \forall I$$

for a constant $\delta \in \mathbf{R}^{\geq 0}$.

Proof: We assume towards a contradiction that such an algorithm A exists. Now

$$Opt(I)/A(I) \leq \delta/A(I) + 1.$$

Since $A(I)$ assumes only integer values,

$$R_A(I) = Opt(I)/A(I) \leq \delta + 1$$

for $A(I) > 0$. Thus $R_A^\infty = \delta + 1$ which contradicts the fact that R_{MIN}(DGA-MAX) $= \infty$. We conclude that no such TM A exists. ■

The previous two proofs considered deterministic performance guarantees. They were based on the assumption that $P \neq NP$ and are not directly applicable to an analysis of the Genetic Algorithm. Fortunately, the proofs for the probabilistic case are virtually identical to those above. The only difference we need to make in the definitions of our performance guarantees is that $A(I)$ is the value generated by A with probability greater than 1/2. The only difference in the proofs of the following lie in the justifications that the reductions created are in RP.

Proposition 3 If $RP \neq NP$, then R_{MIN}(DGA-MAX) $= \infty$.
■

Corollary 4 If $RP \neq NP$, then no probabilistic polynomial time algorithm A can guarantee that

$$Opt(I) - A(I) \leq \delta, \;\; \forall I$$

for a constant $\delta \in \mathbf{R}^{\geq 0}$.
■

5 Discussion

To analyze the computational complexity of the Genetic Algorithm, we have had to make a number of assumptions. While we have carefully cast the Genetic Algorithm as both an optimization problem and formal language, to do so required that we restrict our analysis to the class of pseudo-boolean functions with ranges over the integers. This is clearly undesirable, though for the moment it will have to suffice. Only recently have models been proposed in automata theory which can appropriately consider computation over real numbers (see for example [1]). Another important assumption is that we consider only deterministic functions. There is no particular reason our analysis should exclude the probabilistic case, though we expect it would require different proof techniques from those presented here.

The key to the hardness of DGA-MAX is that we have allowed the Turing machine M_f to be chosen from a very large class of Turing machines. Because we choose M_f from such a broad class, we can make few assumptions regarding the nature of f. Thus, it is very difficult to efficiently optimize an arbitrary function from this class. In some sense, our analysis of DGA-MAX provides a worst case scenario for a complexity analysis of the GA. In fact, it would be surprising to find that the GA, or any other algorithm, could efficiently optimize over such a broad class of functions. The conclusion we should draw here is that our complexity analyses of the GA must incorporate a class of functions for which the complexity is analyzed. As we noted earlier, this element is missing from current computational analyses of the Genetic Algorithm.

Finally, we comment on the performance guarantees we have considered. Each of the performance guarantees analyzed above is a 'worst case' guarantee. Each of these guarantees only bounds the performance of the worst instance of the class. Other performance measures which consider 'average case' performance may prove much more interesting. Empirically, the Genetic Algorithm seems to perform well on a broad range of functions. Thus we expect that average case analyses will prove a valuable technique for the GA.

6 Conclusion

We have considered the Genetic Algorithm as a combinatorial optimization problem and have demonstrated that the computational problem it attempts to solve is NP-hard relative to a very broad class of functions, **F**. Additionally, we have shown that a number of important performance guarantees cannot be be satisfied relative to this class. In light of these negative results, it should not be surprising that there are classes of deceptive functions[5,6] in **F** for which the Genetic Algorithm is a poor optimizer. In fact there will be subclasses of **F** which are deceptive relative to **any** set of algorithmic parameters, simply because the underlying combinatorial space is so difficult to optimize over. For example, while genetic operators like bitwise reordering may help the GA to optimize functions which it would otherwise find deceptive [8], there will necessarily exist other functions which this modified GA will not be able to efficiently optimize.

These results suggest that future analyses of the Genetic Algorithm must pay close attention to relationship between the the algorithmic parameters of the GA and the function space from which the fitness function is selected. Since any fixed set of algorithmic parameters cannot enable the GA to efficiently optimize an arbitrary function in a broad class like the one we have considered, we must consider smaller classes or distributions of functions for which those parameters are most appropriate. Alternatively, if we have a function to optimize, we must carefully select our algorithmic parameters if we wish to optimize the function efficiently with the Genetic Algorithm.

We conclude by noting that the literature on pseudo-boolean optimization has already examined the computational complexity of a large number of pseudo-boolean function classes.[10,2,9] This literature should prove a valuable resource to GA researchers since the computational complexity of many of these function classes has already been established.

Acknowledgements

This work was supported in part through INCOR Grant #486209-26961 through Los Alamos National Laboratory and the University of California. We would like to thank Christos Papadimitriou and Brian Bartell for their helpful discussions of these ideas. We would also like to thank our reviewers for their comments, particularly those pointing out the similarity between pseudo-boolean optimization and our presentation of DGA-MAX.

References

[1] Lenore Blum. Lectures on a theory of computation and complexity over the reals (or an arbitrary ring). In Erica Jen, editor, *1989 Lectures in Complex Systems*, pages 1–48. Addison-Wesley, 1990.

[2] Yves Crama. Recognition problems for special classes of polynomials in 0-1 variables. *Mathematical Programming*, 44:139–155, 1989.

[3] Michael R. Garey and David S. Johnson. *Computers and Intractability - A guide to the theory of NP-completeness.* W.H. Freeman and Co., 1979.

[4] John Gill. Computational complexity of probabilistic turing machines. *SIAM Journal of Computation*, 6(4):675–695, 1977.

[5] D.E. Goldberg. Genetic algorithms and walsh functions: Part I, a gentle introduction. *Complex Systems*, 3:129–152, 1989.

[6] D.E. Goldberg. Genetic algorithms and walsh functions: Part II, deception and its analysis. *Complex Systems*, 3:153–171, 1989.

[7] D.E Goldberg. *Genetic Algorithms in Search, Optimization, and Machine Learning.* Addison-Wesley Publishing Co., Inc., 1989.

[8] D.E. Goldberg and C.L. Bridges. An analysis of a reordering operator on a GA-hard problem. *Biological Cybernetics*, 62:397–405, 1990.

[9] P.L. Hammer, P. Hansen, and B. Simeone. Roof duality, complementation and persistency in quadratic 0-1 optimization. *Mathematical Programming*, 28:121–155, 1984.

[10] Pierre Hansen and Bruno Simeone. Unimodular functions. *Discrete Applied Mathematics*, 14:269–281, 1986.

[11] J.H. Holland. *Adaptation in Natural and Artificial Systems.* The University of Michigan Press, 1976.

[12] John E. Hopcroft and Jeffrey D. Ullman. *Introduction to Automata Theory, Languages, and Computation.* Addison-Wesley Pub. Co., 1979.

[13] Christos H. Papadimitriou and Kenneth Steiglitz. *Combinatorial Optimization - Algorithms and Complexity.* Prentice Hall, Inc., 1982.

[14] Alex Rinnooy Kan. Probabilistic analysis of algorithms. *Annals of Discrete Mathematics*, 31:365–384, 1987.

[15] Gilbert Syswerda. Uniform crossover in genetic algorithms. In *Proceedings of the Third International Conference on Genetic Algorithms*, pages 2–9, 1989.

ANALYSIS AND DESIGN OF A GENERAL GA DECEPTIVE PROBLEM

Abdollah Homaifar
Assistant Professor
Dept. Of Electrical Engineering
North Carolina A&T State University
Greensboro, NC 27411

Xiaoyun Qi
Research Assistant

John Fost
Professor

ABSTRACT

This paper presents the application of the Walsh/Hadamard transformation in generalizing the deceptive and optimal conditions of deceptive functions. This method is used in analyzing different coding combinations encountered in genetic algorithms(GAs). The effectiveness of this method is illustrated by designing a general deceptive function. Also, the regularities that exist among these functions are discussed.

INTRODUCTION

It has been shown that genetic algorithms work well when building blocks, short, low order(the number of fixed positions present in the similarity template) schemata with above average fitness values, combine to form optima or near-optima solutions [1]. Recently, a number of researchers have developed methods for analyzing the effect of combined coding operators which yield optima or near optima values. These methods are divided into two groups, dynamic and static methods [1]. The dynamic approach uses a full analysis of the propagation of competing species of schemata through the nonlinear difference equations resulting from the combined consideration of operators (such as reproduction, crossover, and mutation), coding and objective functions. Goldberg and Bridge [2] developed equations of motion under reproduction and crossover for general l-bit coding that permit dynamic analysis of higher-order problems. The static approach uses an efficient transform method (such as the Walsh transform devised by Bethke [3]) to calculate schema averages. These averages are then used to determine whether the building blocks combine to form longer, higher order schemata with below average fitness that will lead away from the optima; this condition forms the GA deceptive problem[4]. We focus on the second method, the static method.

The purpose of analyzing different coding function combinations is to study GA-hard problems (problems which diverge from the global optimum). It is shown that a GA hard problem is always a GA deceptive problem. To check whether a given function is GA deceptive, deceptive and optimal conditions must be determined. These conditions could be calculated according to the Walsh schema transform; however, it is a tedious computation. In this paper, the Hadamard transform is applied to overcome this difficulty, and its effectiveness is illustrated by designing general higher-degree deceptive functions. Once the deceptive functions are defined, simple calculations of schema fitness values by GA would show that on these functions GA is likely to converge to the second best point but not the global point for some cases.

ANALYSIS AND DESIGN OF GA DECEPTION

GA deception occurs when low order schemata lead away from the global optimum. The purpose of this study is to examine the conditions of GA deception and to establish a method of designing general GA deceptive functions.

For analyzing different coding function combinations, deceptive and optimal conditions are required. We obtain these conditions according to the Walsh/Hadamard schema transform. A brief review of Walsh/Hadamard transform and their functionalities is first introduced. For a complete review of Walsh/Hadamard transform see [5].

1. BASIC THEORY:

The Walsh Schema transform is used to calculate schema average fitness as the weighted sum of Walsh coefficients. Before doing this, we first establish the terminology used in this study. We shall assume a GA's process of λ-bit strings, where the bold face is used to denote an entire string and each bit is subscripted by its position.

$$\boldsymbol{X} = X_1X_2...X_\lambda \quad X_i \in (0,1) \tag{1}$$

We may write the fitness function based on Goldberg's work [6] as a linear combination of the Walsh monomials:

$$f(\boldsymbol{X}) = \sum_{j=0}^{2^\lambda-1} \omega_j \psi_j(\boldsymbol{X}) \tag{2}$$

$\psi_j(X)$ is a set of 2^λ monomials and takes a value $\varepsilon(-1,1)$; then, applying the orthogonality result of $\psi_j(X)$, the Walsh coefficients are obtained,

$$\omega_j = \frac{1}{2^\lambda} \sum_{X=0}^{2^\lambda-1} \psi_j(X) * f(X) \tag{3}$$

This enables us to write the schema average fitness as

$$f(S) = \sum_{j \in J(S)} \omega_j \psi_j(\beta(S)) \tag{4}$$

where the sum is taken over all partitions $j \in J(S)$ and $J(S)$ is a set of all subsets of the competing schemata defined by partition j. The β function is defined as

$$\beta(s_i) = \begin{cases} 0 & s_i=0,* \\ 1 & s_i=1 \end{cases} \tag{5}$$

From equation 4, we know that the schema average may be calculated as a partial, signed sum of the Walsh coefficients.

2. THE HADAMARD TRANSFORM:

The Hadamard transform matrix, HH_n is an N*N matrix[7], where N is equal to 2^n , and n=1,2,3,... This matrix can be generated by the core matrix,

$$HH_1 = \begin{pmatrix} 1 & 1 \\ 1 & -1 \end{pmatrix} \tag{6}$$

and the Kronecker product recursion

$$HH_n = HH_{n-1} \otimes HH_1 = \begin{pmatrix} HH_{n-1} & HH_{n-1} \\ HH_{n-1} & -HH_{n-1} \end{pmatrix} \tag{7}$$

The element of the basis vector of the Hadamard transform takes only the binary values (1, -1). These basis vectors can also be generated by sampling the Walsh functions since the elements of the basis vectors of the Walsh function take the binary values (1, -1) and form a complete orthogonal basis for square integral functions. The Hadamard transform just defined is also called the Walsh-Hadamard transform [7].

GENERATION OF DECEPTIVE AND OPTIMAL CONDITIONS BY HADAMARD TRANSFORM

To determine whether or not functions (linear or nonlinear) are GA deceptive, it is necessary to consider whether the low order schema leads away from the optimum. For analyzing this problem, it is necessary to obtain deceptive and optimal conditions. These conditions can be determined by the Hadamard transform.

In this section, we apply the Hadamard transform to obtain any arbitrary degree of deceptive conditions. We assume that f(1...1) is the best point for a schema of length λ. For full deception, it is required that all order one to λ-1 schemata lead away from the best point. Letting the schema order, O(S), be n and $1 \le n \le \lambda-1$, there are $\binom{\lambda}{n}$ different schemata of order n, and each order n schema has (2^n-1) number of deceptive conditions. Thus, there are $\binom{\lambda}{n}(2^n-1)$ deceptive conditions of order n.

Therefore, for a string of length λ, the total number of deceptive conditions is given as:

$$\sum_{i=1}^{\lambda-1} \binom{\lambda}{i}(2^i-1) \tag{8}$$

As an example, assume λ=3, then the total number of deceptive conditions is 12, which is in agreement with Goldberg's result [6]. Table 1 shows the total number of deceptive conditions for some selected string lengths. Therefore, a general method is required to form these conditions. The Hadamard transform matrix is the method used to obtain the overall deceptive conditions[5].

Tab.1. No.of deceptive condition versus string length

String Length	Number of Deceptive Conditions
3	12
4	50
:	:
8	6050
:	:
14	4,750,202

Suppose schema S=*ff**...f** is of length λ , then the deceptive conditions are determined from

$$HD_n * W > 0 \qquad (9)$$

$$HD_n = \frac{1}{2}\begin{pmatrix} \text{1st row-2nd row of } HH_n \\ \text{1st row-3rd row of } HH_n \\ \vdots \\ \text{1st row-}2^{o(H)}\text{ row of } HH_n \end{pmatrix} \qquad (10)$$

$$W = (\omega_{j_1}, \ldots, \omega_{j_{2^{o(H)}}})^T$$
$$j_i \in J(H) \text{ and } j_i < j_{i+1} \qquad (11)$$

where HD_n is the basis matrix of deceptive conditions, with the dimensions of $(2^n-1)*2^n$. Similarly, the optimal conditions are determined with the Walsh coefficients by using the Hadamard transform matrix. The optimal conditions are obtained by straightforward manipulation as[5]

$$HO_n * W > 0 \qquad (12)$$

$$HO_n = \frac{1}{2}\begin{pmatrix} \text{Last row-1st row of } HH_n \\ \text{,, - 2nd row of } HH_n \\ \vdots \\ \text{,,-2nd row of bottom of } HH_n \end{pmatrix} \qquad (13)$$

HO_n is the basis matrix of optimal conditions with dimension of $(2^\lambda-1)*2^\lambda$. Notice that the number of optimal conditions is $2^\lambda-1$. It is clear from Table 1 and eqns. 8 and 13 that both deceptive and optimal conditions are increasing dramatically as the string length increases. The following question may arise, among these number of conditions, is it possible to analyze the deceptive functions more rigorously? Is there any regularity among these conditions? These questions and the properties of a general deceptive function are discussed in the following sections.

ANALYSIS OF DECEPTIVE AND OPTIMAL CONDITIONS

OPTIMAL CONDITION

Theorem 1:
The first condition states that the sum of the Walsh coefficients of all odd order partition numbers must be less than zero. For the remaining $(2^\lambda-2)$ conditions, the partial sum of the Walsh coefficients of even order partition number is greater than the partial sum of the Walsh coefficients of odd order partition number.

Proof: Assume, f(X) represents the fitness function and f(11...1) and f(00...0) are the best and second best points respectively. In general, we can think of f(X) as a particular schema and its fitness can be written as a linear combination of the Walsh monomials (i.e., eqn. 2). Also, since each competing set of schema is numbered uniquely, depending on the number of even (or odd) 1's in the schema of the set *J(S)*, the sign of the corresponding Walsh coefficients are respectively positive (or negative). For f(11...1), the sign of Walsh coefficients of all odd order partition numbers are negative while the sign for all even order partition numbers are positive. For f(00...0), the sign of all Walsh coefficients is positive. Since the first condition requires f(11...1) - f(00...0) > 0, the Walsh coefficients of all even order partition numbers will cancel each other, and the equation reduces to the negative of the sum of all odd order Walsh coefficients. To satisfy the above requirement, this sum must be less than zero.
For the remaining $2^\lambda-2$ conditions, depending on the number of odd or even 1's in the schema of the set *J(S)* of the corresponding string, f(X) is the sum of the Walsh coefficients with negative and positive signs respectively. When they are subtracted from f(11...1), the corresponding conditions must be greater than zero. Those Walsh coefficients that have the same sign will cancel each other and we maintain the partial sum of the Walsh coefficients of the even and odd order partition number as positive and negative respectively.

Q.E.D.

In order to analyze what values of the Walsh coefficients will satisfy the first condition, we have considered the following cases:

i) λ = odd number
For the string **X** of length λ, the number of all odd order partition numbers except λ is $2^{\lambda-1}$-1. If we let the Walsh coefficient of partition number λ take some negative value and its corresponding absolute value be greater than the absolute sum of the Walsh coefficients of the remaining odd order partition numbers, then the first condition would not be violated. Therefore, if we let

$$\omega_{2^{\lambda}-1} = -2^{\lambda-1}$$
$$\omega_i = 1, i = \text{all odd order partition numbers except } \lambda \qquad (14)$$

then the condition is satisfied.

ii) λ = even number
The number of all odd order partition numbers is $2^{\lambda-1}$, of which λ of them are of order λ-1 (i.e. the highest odd order) Walsh coefficients. If the absolute sum of those λ Walsh coefficients are greater than the sum of the remaining $2^{\lambda-1}$-λ ones, it is then possible to meet the first condition. This subject will be discussed more in the section on design of the deceptive function. One must remember that the Walsh coefficient must be selected in such a way that f(11....11) and f(00....00) are the best and the second best points.

For the remaining 2^{λ}-2 conditions, if we choose each Walsh coefficient of odd order partition number to be smaller than that of the Walsh coefficients of even order partition number, then all the optimal conditions are satisfied. For now this requirement is sufficient to satisfy these optimal conditions. The details on the actual values of the Walsh coefficients assigned to odd and even order partition numbers are given in the section on design of general deceptive functions.

Deceptive Condition
As before, we assume f(11...1) is the best. For full GA deception, it is required that all 1 to λ-1 order schemata lead away from the optimal point. This implies that for deception to occur, all $\binom{\lambda}{n}$ combinations (O(S)=n) of f(**X**) of 1 to λ-1 order schemata with only zeros as bits must have a greater fitness than any corresponding schemata which uses ones and/or zeros as bits.

Theorem 2:

The partial sum of $\binom{\lambda}{n}(2^n-1)$ Walsh coefficients of the order n = 1 to λ-1 deceptive conditions must be greater than zero.

Proof: All combinations of the fitness function f(S) of string λ of order 1 to λ-1 of all 0's is the sum of positive Walsh coefficients in the set *J(S)*. Also, the corresponding f(H) with ones and/or zeros as bits in the set *J(S)* is the sum of positive and negative Walsh coefficients depending on the number of even 1's or odd 1's. Once the later fitness functions are subtracted from the former one, the Walsh coefficients with the same sign will cancel each other and only those terms with opposite sign will remain. Thus, the theorem is concluded. Q.E.D.
In order to study what values of the Walsh coefficients will satisfy both the deceptive and optimal conditions, two cases have been considered:
i) λ = odd number
If the Walsh coefficients of all order partition numbers except order λ take positive values, then all deceptive conditions are met. However, to satisfy optimal conditions at the same time, the partial sum of the Walsh coefficients of even order partition numbers must be greater than their counterparts(i.e., theorem 1).
ii) λ = even number
The deceptive conditions from order 1 to λ-2 lack the Walsh coefficients of the λ-1 order partition number. For order λ-1 deceptive conditions, each combination of Walsh coefficients has at most one Walsh coefficient of λ-1 order partition number. If we select the Walsh coefficients of λ-1 order partition numbers with appropriate negative values, and let the Walsh coefficients of other order partition numbers take positive values, then the deceptive and optimal conditions will not be violated. A more rigorous analysis is given in the next section.

METHODOLOGY FOR DESIGNING A GENERAL DECEPTIVE FUNCTION

As mentioned before, for string X of length λ, there are a total of $\sum_{i=1}^{\lambda-1}\binom{\lambda}{i}(2^i-1)$ deceptive and $2^{\lambda}-1$ optimal conditions. There are many sets of Walsh coefficients which could satisfy both the deceptive and optimal conditions. This implies that many deceptive functions can be designed for the particular string length when the Walsh coefficients are selected differently. In this section, we present the design of a particular deceptive function which satisfies all the constraints imposed by the deceptive and optimal conditions. We have considered the following two cases:

Case 1) λ = odd number
For this case we choose

$$\omega_i = \begin{cases} 1 & i = \text{odd order partition number except } 2^{\lambda}-1 \\ 2 & i = \text{even order partition number} \\ -2^{\lambda-1} & i = 2^{\lambda}-1 \end{cases} \quad (15)$$

where $\lambda \geq 3$. The deceptive functions which have been designed based upon the above criteria have the following properties. The fitness of all deceptive functions of odd numbers of ones in **X** except λ, after some straightforward manipulations, is given by

$$f(\mathbf{X}) = -1 + \| -2^{\lambda-1} \| + \omega_o \quad (16)$$

The fitness of all deceptive functions of even numbers of ones in **X** is given by

$$f(\mathbf{X}) = -3 + (-2^{\lambda-1}) + \omega_o \quad (17)$$

Also, f(11....1) and f(00....0) are given by

$$f(1..1) = \left(\sum_{i=1}^{\lambda-2} \binom{\lambda}{i}\right)(-\omega_m + \omega_n) + 2^{(\lambda-1)} + \omega_o \quad (18)$$
$$\text{for all } i = odd$$

$$f(00..0) = \left(\sum_{i=1}^{\lambda-2} \binom{\lambda}{i}\right)(\omega_m + \omega_n) + -2^{(\lambda-1)} + \omega_o \quad (19)$$
$$\text{for all } i = odd$$

where ω_m and ω_n are the Walsh coefficients associated with odd and even order partition numbers in eqn. 15. Note, the only difference between the best and the second best point of the deceptive function is in the sign of the Walsh coefficients of odd order partition numbers. We must also mention that based on eqns. 18 and 19, the difference between the best and the second best points is only two. Therefore, with a straightforward manipulation, we can show that

$$\sum_{i=1}^{\lambda-2} \binom{\lambda}{i} \omega_m + 1 = 2^{\lambda-1} \quad (20)$$
$$\text{for all } \quad i = odd$$

Thus, by the above method f(11....1) and f(00....0) are guaranteed to be the best and the second best points. To devise non-negative deceptive functions, the value of ω_0 must take some positive value which is chosen to make the fitness of all even numbers of ones in **X** greater than or equal to zero. In order to achieve non-negativity ,the value of ω_0 must satisfy the condition $\omega_0 \geq 3+2^{\lambda-1}$. For example, the minimum values of ω_0's = 7, 19, and 67 for a three-bit, five-bit, and a seven-bit, string length. For the aforementioned values of ω_0, f(**X**) of even numbers of ones are equal to zero.

Case 2) λ = even number
The Walsh coefficients are given in equ.21,

$$\omega_i = \begin{cases} 1 & i = \text{odd order partition number except } \lambda-1 \\ 2 & i = \text{even order partition number} \\ -\mu & i = \text{partition order } \lambda-1 \\ \lambda & i = 2^{\lambda} \end{cases} \quad (21)$$

where $\lambda \geq 4$. The Walsh coefficients of order λ-1 are selected to maintain f(11....1) and f(00....0) as the best and second best points. In order to achieve this requirement, the two constraint equations,the μ value is given in equ.22:

$$1. \quad \mu > \frac{\sum_{m=1}^{\lambda-3} \binom{\lambda}{m} \omega_m}{\lambda}$$
$$\text{for all } m \,/\, n = odd \,/\, even$$
$$2. \quad \mu < \frac{\sum_{m=1}^{\lambda-3} \binom{1}{m} \omega_m + \sum_{n=2}^{\lambda-2} \binom{\lambda}{n} \omega_n + (\lambda+2)}{2(\lambda-1)} \quad (22)$$

where the first equation is used to make f(11....1) the best point and the second equation is used to make f(00....0) the second best point. Based upon the above criteria the fitness functions have the following properties. The fitness of all even numbers of one in **X** except zero is given by

$$f(\boldsymbol{X}) = (2b-\lambda)(1+\mu)+\lambda-4+\omega_0 \quad (23)$$

where b is the number of one's in the string **X**. The fitness of all odd numbers of one in **X** except λ is given by

$$f(\boldsymbol{X}) = (\lambda-2b)(1+\mu)-\lambda+\omega_0 \quad (24)$$

Also, f(11....1) and f(00....0) are given by

$$f(11\ldots1) = \sum_{m=1}^{\lambda-3}\binom{\lambda}{m}(-\omega_m)+\sum_{n=2}^{\lambda-2}\binom{\lambda}{n}\omega_n+\lambda*\mu+\lambda+\omega_0$$
$$\text{for all } m\ /\ n = odd\ /\ even \quad (25)$$

$$f(0..0) = \sum_{m=1}^{\lambda-3}\binom{\lambda}{m}\omega_m+\sum_{n=2}^{\lambda-2}\binom{\lambda}{n}\omega_n+\lambda*(-\mu)+\lambda+\omega_0$$
$$\text{for all } m\ /\ n = odd\ /\ even \quad (26)$$

Again the only difference between the best and the second best points is in the sign of the odd order partition numbers.

As before, to devise non-negative deceptive functions, the value of ω_0 obtained from eq. 24 for b=λ-1.

$$\omega_0 \geq -((2-\lambda)(1+\mu)-\lambda) \quad (27)$$

From eqn. 22-1, the minimum integer values of μ are 2, 8, and 16 for the four-bit, six-bit, and eight-bit string lengths respectively. The corresponding values of the ω_0's = 10, 42, and 110. As an example, let λ=8 and μ=16, then f(0)=362, f(1)=204, f(3)=50, f(7)=136,..., f(255)=378. According to this regulation for setting ω_j, we have designed deceptive functions of order 3 to 14. Those Walsh coefficients not only satisfy both deceptive and optimal conditions but also make f(11...1) and f(00...0) the best and the second best points.

Based on the above design regulation, one might ask the following question: Given a deceptive function of length λ and deceptive order up to and including k=λ-1, can we have a reduced order deceptive function that satisfies both deceptive and optimal conditions? Or equivalently, can we set either all the odd order Walsh coefficients of order three up to and including λ(λ-1) to zero or only set order λ(λ-1) to zero when the string length is odd (even), and still have the deceptive function. The answer is no.

Lemma 1:
Given a deceptive function of length λ of order k=λ-1, it is not possible to have deceptive functions of reduced odd order partition numbers that satisfy both optimal and deceptive conditions.

Proof: Reducing the order implies that all the Walsh coefficients of odd order partition number must be set to zero for order three up to and including λ or only λ, when λ is odd, or for order three up to and including λ-1 or only λ-1, when λ is even. This condition has a direct bearing on the first optimal condition. The first optimal condition states that the sum of all odd order partition numbers must be less than zero. For λ =odd, setting the Walsh coefficients equal to zero for order three up to and including λ or only λ will set this condition equal to the sum of all one or λ-2, odd order partition numbers respectively. Likewise, for λ =even, setting the Walsh coefficients equal to zero for order three up to and including λ-1 or only λ-1 will set this condition equal to the sum of all one or λ-3, odd order partition numbers respectively. Since all of those Walsh coefficients are selected to be positive, hence their sum cannot be less than zero(i.e., theorem 1). Thus, it is not possible to have a deceptive function of order k with reduced odd order Walsh coefficients. Q.E.D.

EXAMPLES OF APPLYING SIMPLE GA ON THE DECEPTIVE FUNCTIONS

Deceptive functions of string length 3 up to 14 are designed based upon the methodology presented.In order to show the function designed are deceptive ones,the simulation results for the five-bit deceptive function are presented. Tab.2.shows the fitness.The GA parameters are chosen as

population size = 50
chromosome length = 5
maximum number of generation = 100
probability of crossover = 0.8
probability of mutation = 0.033

Depending on the initial population which is determined by a random seed number, the solution may converge to the optimal or to a suboptimal point. In Fig.1, where the seed = 0.467, the solution converges to the global optimal 11111 after 100 generations despite the function being deceptive. But, not all five-bit deceptive functions are like this. In Fig. 2, when the seed = 0.789, the final solution goes to the second best point 00000, thus illustrating that this function becomes difficult for GA to find the best point 11111.

Table 2 Five-bit deceptive function-

X (Decimal Number)	f(X)
0	48
1,2,4,,7,8,11,13,14,16,19,21,22,25,26,28	34
3,5,6,9,10,12,15,17,18,20,23,24,27,29,30	0
31	50

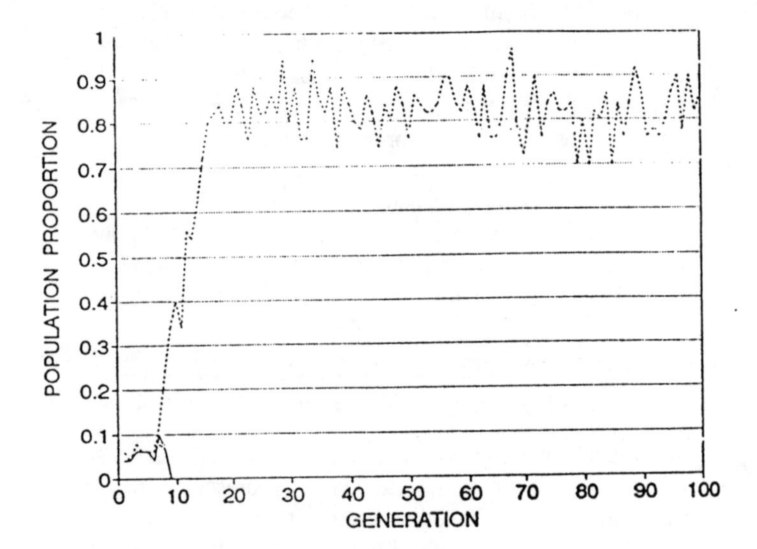

Fig.1 Numerical Solution of Five-bit Deceptive Function, Seed = 0.467

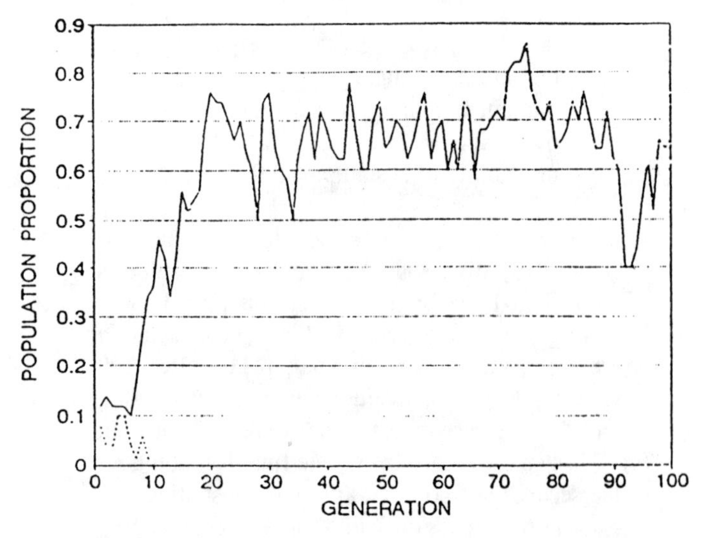

Fig.2 Numerical Solution of Five-bit Deceptive Function, Seed = 0.789

CONCLUSION

In this paper, the Hadamard transform has been used to generate higher-degree deceptive and optimal conditions in terms of Walsh coefficients. This transform is a powerful mathematical tool for analyzing deceptive problems. The method is used to analyze the deceptive and optimal conditions which are presented in the form of theorems 1 and 2. According to this analysis, a methodology for designing a general λ-bit deceptive function is given, and a number of examples and simulation results have demonstrated the usefulness of these methods. These examples are selected to make the difference between the best and the second best points minimal in order to make them harder GA deceptive problems.

The conclusion reached in Lemma 1 is based upon the particular design methodology introduced in this paper. As stated earlier there are many deceptive functions which could be designed that satisfy both the optimal and deceptive conditions. Hence, it is therefore possible to introduce a theorem which could contradict Lemma 1. Also, other issues such as the design of reduced deceptive functions of even order partition number is under investigation. This work enhances our understanding of deceptive problems and presents a more rigorous analysis of GA deception.

ACKNOWLEDGEMENTS

The material contributed is based upon work supported by the National Science Foundation under grant No.4-46009. We are indebted to Dr. David Goldberg of the University of Illinois at Urbana Champaign for the many stimulating discussions about this and other related materials.

REFERENCES

1. Goldberg DE, Genetic Algorithms in Search Optimization and Machine Learning, Addison Wesly, 1989.
2. Goldberg DE, and Bridges CL, An Analysis of Reproduction and Crossover in a Binary-Code Genetic Algorithm. Genetic Algorithms and Their Applications: Proceedings of the Second International Conference on Genetic Algorithms, 9-13,1987.
3. Bethke AD., "Genetic Algorithms as Function Optimizers" Ph.D. Dissertation, Ann Arbor: University of Michigan, 1980.
4. Goldberg DE, and Bridges CL, An Analysis of a Reordering Operator on a GA-Hard Problem.

Biological Cybernetics 62, 397-405, 1990.
5. Homaifar A., and Qi Xiaoyun, Analysis of GAs Deception by Hadamard Transform . IASTED International Symposium Machine Learning and Neural Networks, New York, October, 1990.
6. Goldberg DE, Genetic Algorithms and Walsh Function: Part I, A Gentle Introduction. Complex Syst. 3:129-152, 1989.
7. Anil K Jain, Fundamental of Digital Image Processing, Prentice, Englewood Cliffs, NJ07632,1989,155-159.

Analysis of Schema Distributions

Martin Hulin
Siemens AG, ZFE IS SYS 32
Otto-Hahn-Ring 6
8000 München 83, Germany,
Tel.: 089/636 49098, E-mail: mh%ara@ztivax.siemens.com

Abstract

This paper defines the notion of schema distribution (cost- or fitness distribution of a schema) which is a generalization of schema averages. The schema distributions of competing, short, low order schemata are used to judge problems whether they are deceptive for genetic algorithms (GAs). Different coding schemes for a special optimization task, the problem of circuit partitioning, are examined by this method to find the most suitable coding.

1 INTRODUCTION

Genetic algorithms (GAs) work well for optimization problems when building blocks - short, low order schemata with above average fitness values - combine to form optima or near optima [3]. In contrast problems are deceptive for GAs if many short, low order schemata lead away from the optima, i. e. are above average but do not contain optimal or near optimal solutions. Bethke [1] and Goldberg [3] use the average fitness of schemata to investigate them for deception and present an efficient method to calculate the fitness averages using Walsh functions. This paper refines the comparison of schema averages: schema distributions are compared. Section 2 defines the notion of schema distribution. Schema distributions are analysed and compared to find deceptive schemata. A small example illustrates the advantage of schema distributions to schema averages.

This new method is used in section 3 of the paper to judge two different coding schemes for a special application - the problem of circuit partitioning. Experimental results are given which support the theoretical investigations.

2 SCHEMA DISTRIBUTION

2.1 DEFINITION

For schemata the same terminology as in [3] is used. We assume that the GA processes bit strings of length l. A schema is defined as the set of all bit strings with a fixed value on certain given positions. It is normally described as a similarity template where the unfixed bit positions are marked by a wildcard symbol *. E. g. the schema {000, 001, 010, 011} ($l = 3$) is described by 0**.

For an optimization task to be solved by a GA a fitness or cost function $\Phi: \{0, 1\}^l \rightarrow \mathbb{R}$ is defined. Φ is a mapping of the l bit strings that represent solutions of the problem by a coding function into the reals. In this paper we concentrate on minimization problems without loss of generality and therefore only consider cost functions.

Now we can define the *cost distribution of a schema* H or simply *schema distribution*: $D_H: \mathbb{R} \rightarrow [0, 1]$. $D_H(x)$ is the fraction of elements of schema H with a cost value lower or equal to x or formally:

$$D_H(x) = |\{h \in H: \Phi(h) \le x\}| \, / \, |H|$$

Since every schema is a finite set it has at least one element with a minimal cost value min_H and at least one element with a maximal cost value max_H. It is evident that $D_H(x) = 0$ for $x < min_H$ and $D_H(x) = 1$ for $x \ge max_H$. D_H is a step function but for large l it looks like a continuous function. Like every distribution function D_H grows monotonically. Figure 1 shows a typical schema distribution.

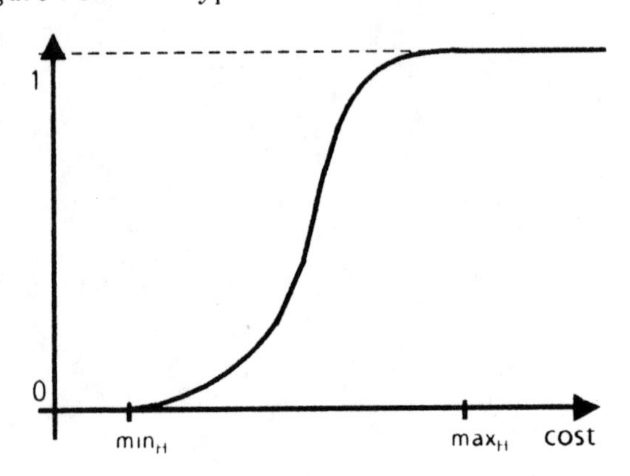

Figure 1: A Schema Distribution

2.2 DECEPTIVE PROBLEMS AND ANALYSIS OF SCHEMA DISTRIBUTIONS

Schema distributions can be used to judge whether a problem is deceptive for GAs and in which degree. Like the analysis of schema averages the analysis of schema distributions is relevant only at the beginning of the evolution process when the diversity of individuals in the population is high. Then it can be assumed that each element of a schema can be produced with equal chance. After relative short time (few thousand produced samples depending on problem size, cost function, population size, mutation rate and other parameters) the population concentrates on a relatively small area of the search space; the members of the population are very similar. In this situation only a very small and specific subset of a schema is likely to be produced by genetic operators and this subset will not represent the cost distribution of the whole schema.

To look for deceptive schemata we compare competing short, low order schemata that are defined on the same positions. Figures 2, 3, and 4 show three principal cases. Assume that schema H contains the minimum while schema K does not. The interesting part of schema distributions is the interval from about $min + (min + av)/3$ to av where min is the minimal cost value we are searching for and av is the average cost of all solutions. This is true because the average cost value of the population falls below av after the first three to five generations as De Jong has shown [2]. On the other hand experience shows that the population has already concentrated to a small area of the search space when the cost average of the population is within 30% to 40% of the minimum [4]. Then an analysis of schema distributions is no more relevant (see above).

Figure 2 shows the non deceptive case where a GA will work well. Samples of schema H will have above average cost values with high probability while most of the samples of schema K will have definitely worse cost values and be eliminated by selection. The number of samples of schema H in the population will increase the number of samples of schema K will decrease. Comparison of schema averages would give the same result in this case.

Figure 3 shows a deceptive case. Only few elements of H are very good or good most are rather bad. Schema K contains no optimal or near optimal elements but a large fraction of good elements. Thus during evolution the number of elements of K will increase in the population the number of elements of H will decrease. Schema H may become extinct. In this case again a comparison of schema averages would lead to the same result.

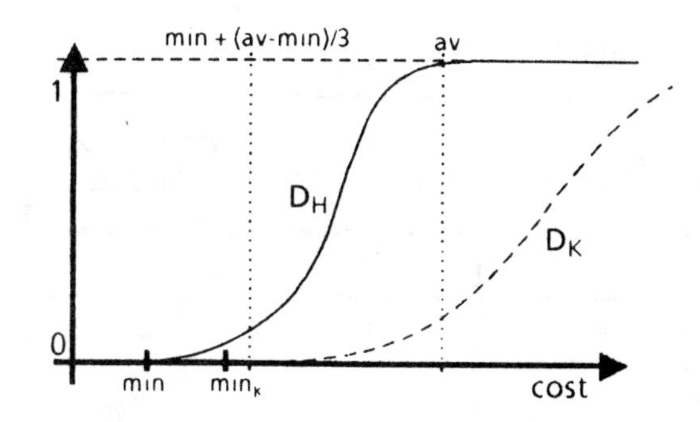

Figure 2: Non Deceptive Schema Distributions

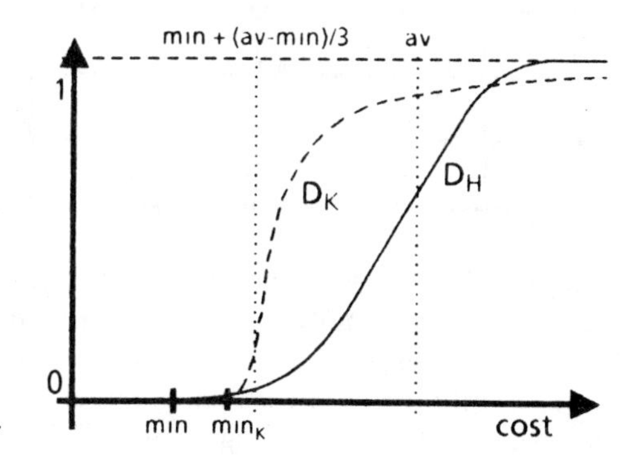

Figure 3: Deceptive Schemata Distributions

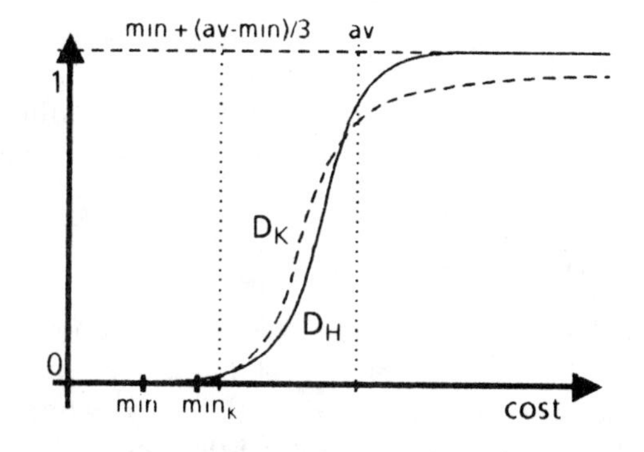

Figure 4: Deceptive Schemata Distributions although $\Phi(H) < \Phi(K)$

Case 3 (cf. Figure 4) is different. The average of schema H is better than that of schema K thus indicating no deception. But the analysis of schema distributions shows another result. In the interesting interval schema H and schema K have very similar distributions, schema K is even slighly better. If there are only few samples of H and K in the population, statistical sampling errors will determine whether H or K will win.

If a problem has many competing low order schemata where the schema distributions follow case 3, the problem is deceptive even if the schema

Table 1: Cost Functions of two Problems

Coding	Problem 1: Cost Φ1	Problem 2: Cost Φ2
000	1	1
001	6	6
010	6	6
011	5	5
100	11	7
101	4	4
110	3	3
111	2	2

averages do not indicate deception. It is unprobable that all schemata containing the optimum will win. The fact of sampling errors is the main disadvantage, if only averages are considered: For large problems, i. e. large l, memory limits allow only small populations. Thus at the beginning of the evolution process only few elements of a schema are created by chance even for low order schemata. E. g. for $l = 1000$, a population size of 1000 and $o(H) = \delta(H) = 8$ one can expect only about 4 from a total number of 2^{992} elements of H in the initial population. This low sampling rate does not allow to calculate with averages, the expected sampling error is too high.

2.3 EXAMPLE

We consider a problem with three variables ($l = 3$) similar to the fully deceptive problem in section 4.4 of [3] which is given by Table 1 or Figure 5. Schema $H = 0^{**}$ contains the optimum while schema $K = 1^{**}$ does not. $\Phi_1(0^{**}) = 4.5 < 5 = \Phi_1(1^{**})$ indicates that this problem is not fully deceptive, if we consider schema averages. On the other hand Figure 5 shows that there is a gradient to the suboptimal point 111 from all points except the optimum 000. This indicates a deceptive problem. Problem 1 seems to be as deceptive as problem 2 where $\Phi_2(0^{**}) = 4.5 > 4 = \Phi_2(1^{**})$. $\Phi_1(1^{**})$ is only worse because 1^{**} contains one extremely bad element 100 which will be eliminated by selection with high probability, if it is created.

Figure 6 shows the schema distributions. The situation is like that in case 3: Analysis of schema distribution shows deception while analysis of schema averages does not.

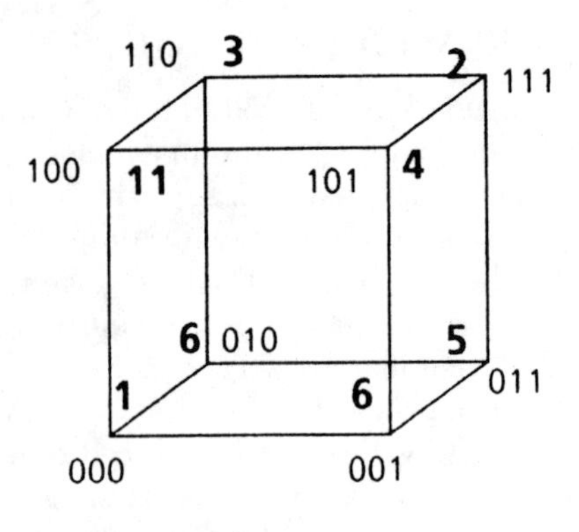

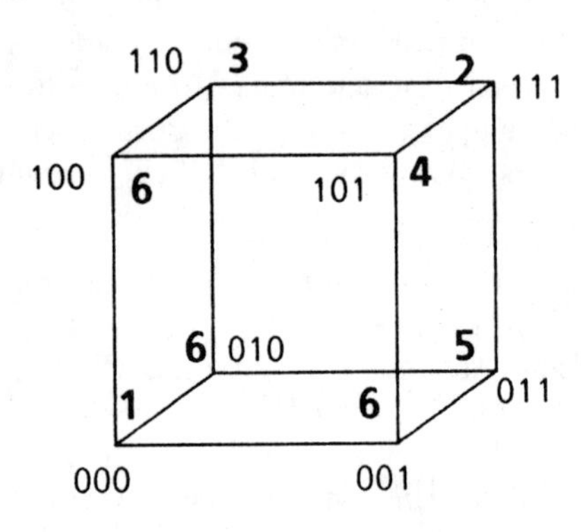

Figure 5: Two Example Problems

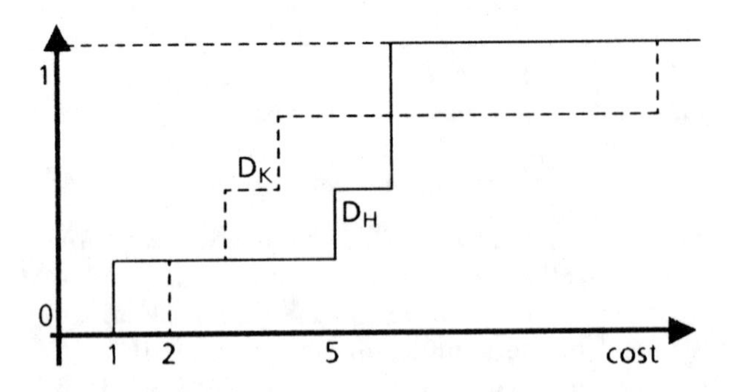

Figure 6: Schema Disributions for Problem 1 and Schemata $H = 0^{**}$ and $K = 1^{**}$

3 APPLICATION TO THE PROBLEM OF CIRCUIT PARTITIONING

The analysis of schema distributions can be used to select a proper coding scheme for the problem of circuit partitioning which arises during the design of electrical circuits and is a special case of hypergraph partitioning. The task is to partition the components of a circuit into subsets (henceforth called circuit-parts) with the goal to minimize the number of the subsets and the number of cut nets

(nets connecting components in different subsets) under the restriction of limited size of the subsets.

3.1 SIMPLE CODING

The coding of the parameters of a partition is done quite straight forward. A circuit is composed of atomic components, which are never split in the partitioning process, e.g. a 32 bit adder consists of 32 one bit full adders at register transfer level, while on gate level it consists of atoms like and-gates, or-gates, and inverters.

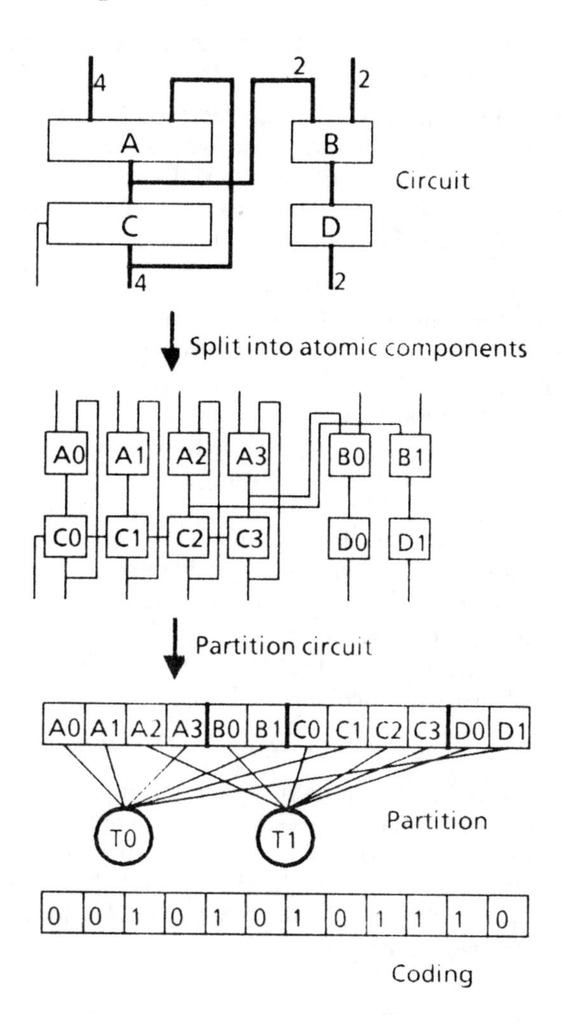

Figure 7: Small Example Circuit and Coding of a Partition

A partition is coded as an l-tuple $(P_1, \ldots P_l)$, where l is the number of atomic components. P_i = j means that component i is assigned to circuit-part j (Figure 7). Recombination works on these l-strings.

3.2 SPECIAL TWO STEP CODING

Complex components of a circuit are not split but remain integer. A two-step coding of a partition is used that supports a combination of horizontal and vertical slicing. The (complex) components are

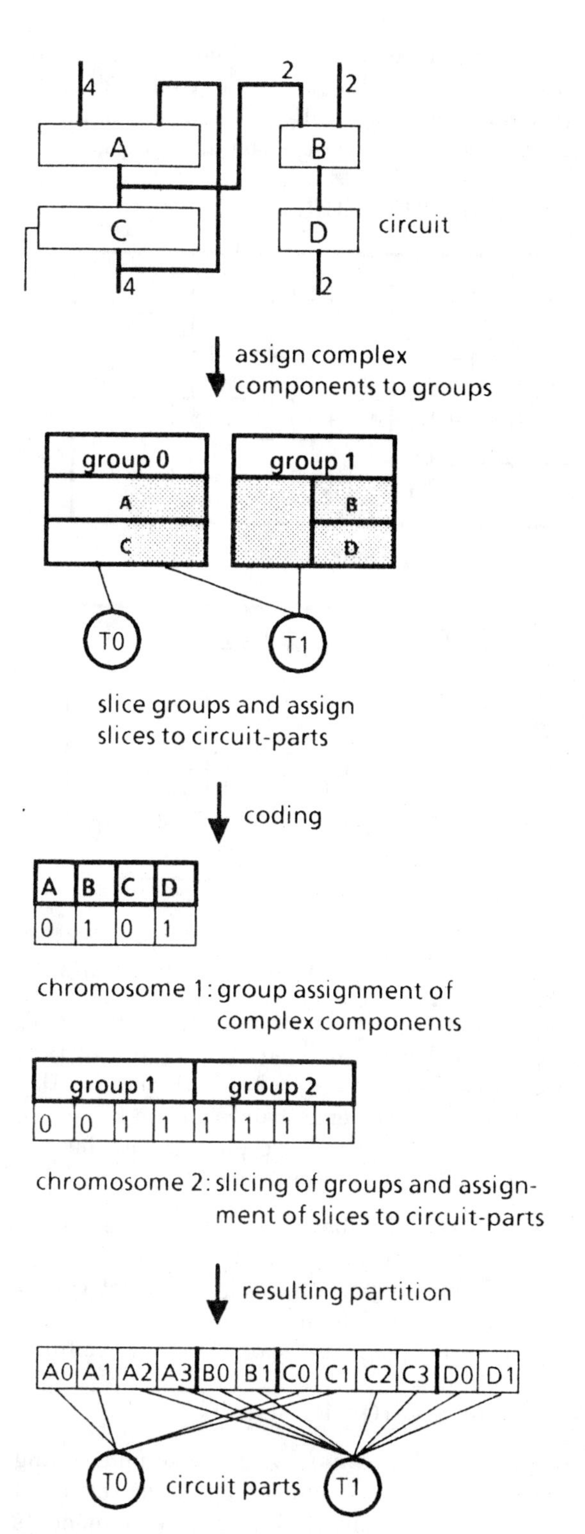

Figure 8: Two-step Coding of the Small Example Circuit

assigned to groups (horizontal slicing). A group is assumed to be built of bit-slices. Each group is sliced vertically, i. e. each bit-position of a group is assigned to a circuit-part. The result are two chromosomes: The first represents the assignment of complex components to groups, the second the slicing of the groups and the assigenement of the slices to the circuit-parts (Figure 8). Both parts of the information define a partition of the circuit clearly. The assignment of the components to groups is not done statically before partitioning, it evolves. Recombination works on the groups.

3.3 SCHEMA DISTRIBUTIONS

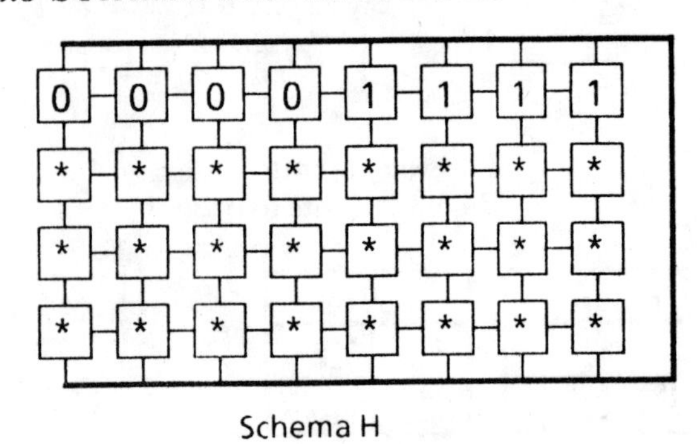

Schema H

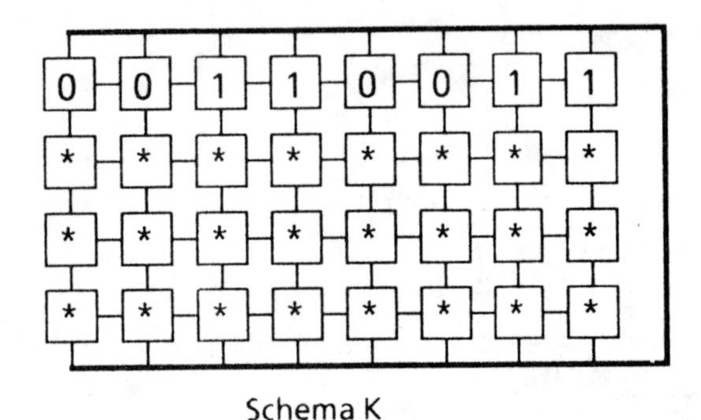

Schema K

Figure 9: Matrix-Circuit with two Schemata

Figure 9 shows a very regular circuit that is built like a matrix. The optimal partitioning is a vertical slicing in two 4-bit-slices. Schema H contains this optimum whereas schema K does not. In Figure 10 the schema distributions are plotted both for the simple coding and the special coding.

The schema distributions of H and K are very similar for the simple coding. This is like case 3 of section 2.2. For the special coding we detect case 1 without deception. The analysis of schema distributions clearly favours the special two step coding.

3.4 OPTIMISATION RESULTS

The results of partionings with the simple coding and the sophisticated coding support the results of 3.3. Two circuits - circuit 1 with 273 components and circuit 2 with 1170 components - have been partitioned using a GA. With the special coding better partitions (less pins) could be reached in less time (Table 2).

4 ACTUAL WORK, CONCLUSION

My aim is now to quantify the qualitative results described in this paper. There are four questions that have to be answered:

Table 2: Results of Partitonings for two Circuits with GAs (p.p. = produced partitions × 1000)

	circuit 1		circuit 2	
	p.p.	pins	p.p.	pins
simple cod.	4	70	55	134
soph. cod.	1.9	28	25	78

In which cost interval is the comparison of schema distributions appropriate?

As already explained in section 2.2 the upper bound ub of this cost area should be the average cost of the whole search space. Since the initial population is created at random the average cost of the first population is an estimate for ub.

To determine the lower bound lb of this cost interval is more difficult. If the population is too homogenuous, i. e. if the expected difference between parents and children due to the crossing over operator is less than the expected difference due to mutation, an analysis of schemata is not appropriate. A short calculation shows that this is true, if $d = 2\,p_m$, where d is the average Hamming-distance per gene of two individuals in the population. An empirical way to determine lb is to make some optimization runs for the problem and to record d and the best cost value in the actual population. The runs are continued until d falls below $2\,p_m$. The best cost value achieved so far is taken as lb.

How should two schema distributions be compared?

If the average cost value of a population is x, the quotient $D_H(x)/D_K(x)$ expresses the ratio of the probabilities to survive selection for members of the schemata H and K. A weighted average V(H, K) is taken over all these quotients between lb and ub, where the weight rises linearily:

$$V(H,K) = \frac{2}{ub-lb}\int_{lb}^{ub}\frac{D_H(x)}{D_K(x)}(x-lb)\,dx$$

Which schemata should be compared?

A metaschema M is a string M ∈ {▮, *}^{l} , e. g. M = **▮▮*▮*...* where a schema H ∈ M, if H is fixed at the positions ▮ in M and not fixed at the positions * in M. For every Metaschema the schema that contains the optimum (if there is only one optimum) is compared to all other schemata of the metaschema.

Irrelevant comparisons can be omitted, e. g. if the expected number of elements of a schema in the initial population is less than 1.

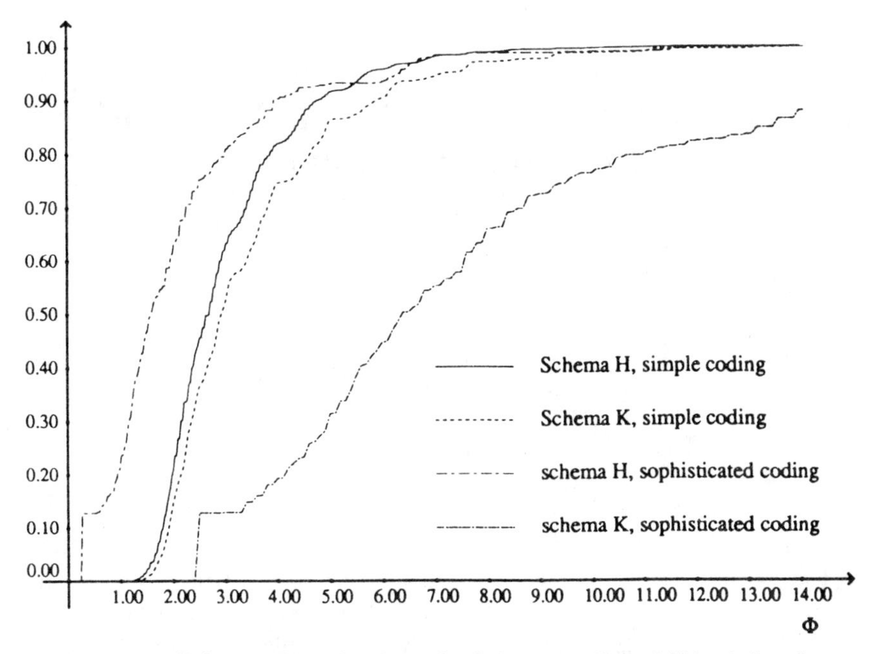

Figure 10: Schema Distributions for Schemata H and K both for the Simple Coding and the Sophisticated Coding

<u>What is the relevance of the comparison of two specific schemata?</u>

The more gene positions are fixed in a schema H, the more the building block H contributes to the optimum: the relevance increases by $o(H)/l$.

If population size is bounded, the relevance of a schema is proportional to its expected fraction in the initial population which is $1/2^{o(H)}$.

Crossing over and mutation do not destroy a schema with the probability

$$\frac{l-\delta(H)}{l-1}(1-p_m)^{o(H)}.$$

The relevance of a schema is proportional to this value.

These three factors are combined to a relevance $Rel(H, K)$ of the comparison of H and K.

$$Rel(H,K) = \frac{o(H)}{l}\,\frac{1}{2^{o(H)}}\,\frac{l-\delta(H)}{l-1}(1-p_m)^{o(H)}$$

The answers to the four questions above result in a function that calculates a suitability value for every given problem and its coding. This value is a measure for the suitability of GAs for this problem:

$$\frac{\sum\limits_{M\in\{\mathbf{I},*\}^n} Rel(M) \sum\limits_{I_{opt}\in H,\, K\in M,\, I_{opt}\notin K} V(H,K)}{\sum\limits_{M\in\{\mathbf{I},*\}^n} Rel(M)\,|M|}$$

The function is now tested for different problems and codings. It can be used to decide in advance whether GAs are a proper optimization method for a problem or a problem class with a specified coding scheme: The schema distributions for a sample problem like the matrix-circuit are calculated and the suitability value is established. If this value is greater than a threshold value, it is probable that GAs produce good solutions for the problem, and the effort to implement a GA for the problem seems to be worthwhile. The threshold value has to be gained by experience. If the suitability value is near 1 or lower than 1, a simple GA should not be used or a more appropriate coding has to be found.

References

[1] A. D. Bethke: "Genetic Algorithms as Function Optimizers", Doctoral dissertation, Umiversity of Michigan 1980. Dissertation Abstracts International, 41 (9), 3503B. (University Microfilms No. 8106101)

[2] K. De Jong, W. M. Spears: "An Analysis of the Interacting Roles of Population Size and Crossover in Genetic Algorithms", Proceedings of the first PPSN workshop, 1990, Dortmund (to be published)

[3] David E. Goldberg: "Genetic Algorithms and Walsh Functions: Part I, A Gentle Introduction", Complex Systems 3 (1989) 129 - 152

[4] David E. Goldberg, Philip Segrest: "Finite Markov Chain Analysis of Genetic Algorithms", Proceedings of the 2nd International Conference on Genetic Algorithms, Lawrence Erlbaum Associates, 1987, Hillsdale, New Jersey

Partition Coefficients, Static Deception and Deceptive Problems for Non-Binary Alphabets

Andrew J. Mason
Management Studies Group
University of Cambridge
Mill Lane, Cambridge
U.K. CB2 1RX
ajm19@uk.ac.cam.phx

Abstract

In this paper, we show how the calculation of partition coefficients for the analysis of Genetic Algorithms may be extended to non-binary alphabets. We demonstrate the use of such coefficients in the analysis of deception and the construction of fully deceptive problems of any given order. We also show how previous analysis of static deception may be extended to the non-binary case.

1 INTRODUCTION

The importance of building blocks – good, short, low-order schemata – to the workings of Genetic Algorithms has become widely known since Holland's initial work [8]. Bethke [3] has shown how *partition coefficients* may be used to analyse the processing of schemata. However, as has been common in GA literature, Bethke's and subsequent work [4,6,5] has concentrated on problems using binary coding.

The issue of binary versus non-binary alphabets currently remains open. Goldberg's 'principle of minimal alphabets' [6] has been questioned by Antonisse [1] who offered an interpretation of schemata that 'overturns the binary encoding constraint'. It is worth noting that Antonisse's argument may be extended until every possible subset of individuals is counted as a 'schema'; this brings into question the value of counting schemata when comparing coding schemes. Also, Grefenstette and Baker [7] have illustrated limitations in applying the k-armed bandits analogy (see [6]) to schema processing, and thereby questioned the independence of such processing assumed by schemata counting arguments.

In this paper, we extend partition coefficients to non-binary codings and discuss their use in locating deception. We also construct a non-binary problem that is fully deceptive to some given order. Recently Goldberg [5] has analysed and defined static deception for binary-coded problems; we extend his analysis to the non-binary case. These new analytical tools should prove useful in the debate over coding alphabets.

2 NOTATION

We firstly introduce our notation. An individual I, with fitness $f(I)$, is identified by a chromosome, a string of k genes. We write $I = i_1 i_2 \dots i_k$, where $i_j \in \aleph_j \; \forall \; j = 1, 2, \dots, k$, the $v_j = |\aleph_j|$ elements of $\aleph_j$ being the alleles (or alphabet) of gene i_j. If all genes are defined over the same alphabet we say $\aleph = \aleph_1 = \aleph_2 = \cdots = \aleph_k$, $v = v_1 = v_2 = \cdots = v_k$. Traditionally the binary alphabet $\aleph = \{0, 1\}$ is used, and this has been the focus of most research. However, in this paper we will not restrict ourselves to the binary case.

We represent a schema H as a string $H = h_1 h_2 \cdots h_k$, $h_j \in \aleph_i \cup \{*\}$, where $*$ means 'don't care'. We call $j : h_j \neq *$, a *defining position* in H. A schema H is defined as the set of individuals for which $i_j = h_j$ at each defining position in H. We write $H \ni I$ to mean H contains I. The fitness $f(H)$ of a schema is the average fitness of the individuals belonging to that schema.

We say that one schema H is contained within another H', $H \subseteq H'$ if $h_j = h'_j$ for each defining position in H'. For example, 010**1 $\subseteq$ 010***. When we write the statement $H' \supseteq H$ we say that H' is a *relaxation* of H.

We define the *order* $o(H)$ of a schema as the number of fixed positions in H, and the *defining length*, $\partial(H)$, as the distance between the first and last defining positions in H. We note that $H' \supset H \Rightarrow o(H') < o(H)$.

3 PARTITION COEFFICIENTS

Bethke [3] has shown how partition coefficients may be used in the analysis and construction of problems that are difficult to solve using a GA. Goldberg has repeated Bethke's analysis [4], and also used partition coefficients to characterise *static deception* in a problem [5]. These works have concentrated on binary-coded genes. In the following analysis this assumption is dropped.

For each schema H we introduce an associated partition coefficient $\epsilon(H)$ and write the following *general partition equation*.

$$f(H) = \sum_{H' \supseteq H} \epsilon(H') \tag{1}$$

Rearranging (1), we see that the partition coefficients are defined recursively by:

$$\epsilon(H) = f(H) - \sum_{H' \supset H} \epsilon(H'). \tag{2}$$

From (2), we see that $\epsilon(H)$ may be calculated immediately using only $f(H)$ and those $\epsilon(H')$ values for which $o(H') < o(H)$. Thus we may initially put $\epsilon(** \cdots **) = f(** \cdots **)$, and then calculate the partition coefficients associated with successively higher order schemata. Such a procedure will generate a unique set of partition coefficients that satisfy (1) for all H.

The partition coefficients quantify the non linearities associated with the schema fitness values. Each $\epsilon(H)$ gives the difference between the fitness of a schema H and a linear prediction of that fitness made by summing the ϵ associated with the lower order schemata containing H. For example, $\epsilon(**2)$ gives the difference between $f(**2)$ and $f(***)$, while $\epsilon(*62)$ is the difference between $f(*62)$ and the sum $\epsilon(***) + \epsilon(**2) + \epsilon(*6*)$.

As an example, if we have a tertiary alphabet $\aleph = \{0, 1, 2\}$ and fitness values

$$f(00) = 9, f(01) = 6, f(02) = 3,$$
$$f(10) = 3, f(11) = 3, f(12) = 5,$$
$$f(20) = 8, f(21) = 2, f(22) = 6$$

then we calculate the following coefficients as shown.

$$\begin{array}{lcll}
f(**) & = & \epsilon(**) = 5, & \Rightarrow \epsilon(**) = 5 \\
f(*0) & = & \epsilon(**) + \epsilon(*0) = 6\tfrac{2}{3} & \Rightarrow \epsilon(*0) = 1\tfrac{2}{3} \\
f(*1) & = & \epsilon(**) + \epsilon(*1) = 3\tfrac{2}{3} & \Rightarrow \epsilon(*1) = -1\tfrac{1}{3} \\
f(*2) & = & \epsilon(**) + \epsilon(*2) = 4\tfrac{2}{3} & \Rightarrow \epsilon(*2) = -\tfrac{1}{3} \\
f(0*) & = & \epsilon(**) + \epsilon(0*) = 6 & \Rightarrow \epsilon(0*) = 1 \\
f(00) & = & \epsilon(**) + \epsilon(0*) + \epsilon(*0) & \\
 & & +\epsilon(00) = 9 & \Rightarrow \epsilon(00) = 1\tfrac{1}{3}
\end{array}$$

Notice that $\epsilon(*0) + \epsilon(*1) + \epsilon(*2) = 0$. We now generalise this observation.

Given a schema $H = h_1 h_2 \cdots h_{j-1} * h_{j+1} \cdots h_k$, we define $H_{aj} = h_1 h_2 \cdots h_{j-1} a h_{j+1} \cdots h_k$ to be a schema derived from H by setting $h_j = a \in \aleph_j$. Note that $o(H_{aj}) = o(H) + 1$.

Lemma 1 *For* $H = h_1 h_2 \cdots h_{j-1} * h_{j+1} \cdots h_k$,

$$\sum_{H' \supseteq H} \sum_{a \in \aleph_j} \epsilon(H'_{aj}) = 0$$

Proof
From the definition of schema fitness, and using the partition equation, we may write

$$\begin{aligned}
v_j \cdot f(H) &= \sum_{a \in \aleph_j} f(H_{aj}) \\
&= \sum_{a \in \aleph_j} \sum_{H' \supseteq H_{aj}} \epsilon(H').
\end{aligned}$$

Consider the set $H' : H' \supseteq H_{aj}$. For each H' with a $*$ in position j (ie $h'_j = *$) there will be another schema with $h'_j = a$. Thus, we may write

$$\begin{aligned}
v_j \cdot f(H) &= \sum_{a \in \aleph_j} \sum_{H' \supseteq H} \epsilon(H') + \sum_{a \in \aleph_j} \sum_{H' \supseteq H} \epsilon(H'_{aj}) \\
&= v_j \cdot f(H) + \sum_{a \in \aleph_j} \sum_{H' \supseteq H} \epsilon(H'_{aj}) \\
\Rightarrow & \sum_{a \in \aleph_j} \sum_{H' \supseteq H} \epsilon(H'_{aj}) = 0 \quad \blacksquare
\end{aligned}$$

Theorem 1 *For* $H = h_1 h_2 \cdots h_{j-1} * h_{j+1} \cdots h_k$,

$$\sum_{a \in \aleph_j} \epsilon(H_{aj}) = 0 \tag{3}$$

Proof
We proceed by induction.
Basis: Consider $H = ** \cdots **$. Observing that $\{H' : H' \supseteq H\} = \{H\}$, Lemma 1 gives

$$\sum_{a \in \aleph_j} \epsilon(H_{aj}) = 0 \;\; \forall \;\; H : o(H) = 0$$

Inductive Step: Assume that the result holds for all $H : o(H) \leq m$, that is,

$$\sum_{a \in \aleph_j} \epsilon(H_{aj}) = 0 \;\; \forall \;\; H : o(H) \leq m.$$

Now, as we noted earlier, $o(H') \leq o(H) + 1 \; \forall \; H' \supset H$, and so from our assumption

$$\sum_{H' \supset H} \sum_{a \in \aleph_j} \epsilon(H'_{aj}) = 0 \;\; \forall \;\; H : o(H) \leq m + 1.$$

Using Lemma 1 we obtain

$$\sum_{a \in \aleph_j} \epsilon(H_{aj}) = 0 \;\; \forall \;\; H : o(H) \leq m + 1,$$

and so the result holds for all H by induction. $\blacksquare$

Theorem 2 *If there exists $\epsilon(\cdot)$ satisfying (3) then the Partition Equation is satisfied for the fitness function values*

$$f(I) = \sum_{H' \supseteq I} \epsilon(H'). \quad (4)$$

Proof
We proceed by induction.
Basis: For $o(H) = k$, (4) satisfies the Partition Equation by definition.

Inductive Step: Assume that the Partition Equation is satisfied for all $H : o(H) \geq m+1$. Now, for $H : o(H) < k$ and some j,

$$\begin{aligned} f(H) &\equiv \frac{1}{|H|}\sum_{I \in H} f(I) \\ &= \frac{1}{|H|}\sum_{a \in \aleph_j} |H_{aj}| f(H_{aj}) \\ &= \frac{1}{|H|}\sum_{a \in \aleph_j} \frac{|H|}{v_j} \sum_{H' \supseteq H_{aj}} \epsilon(H') \\ &= \frac{1}{v_j}\left[\sum_{H' \supseteq H}\sum_{a \in \aleph_j} \epsilon(H') + \sum_{H' \supseteq H}\sum_{a \in \aleph_j} \epsilon(H'_{aj})\right] \\ &= \sum_{H' \supseteq H} \epsilon(H'). \end{aligned}$$

The result holds for $H : o(H) \geq m$, and so for all H by induction. ■

Let us briefly consider the traditional case of binary genes. We first extend the standard terminology by defining a *mask* M as a set of all schemata with the same defining positions. A mask represents a set of parallel hyperplanes in $(\aleph_1, \aleph_2, \ldots, \aleph_k)$. It is identified using the notation $M = m_1 m_2 \cdots m_k$, $m_j \in \{\#, *\}$ where a # signifies a position as being defining. For example, $010{*}{*}1 \in \#\#\#{*}{*}\#$, $010{*}{*}1 \notin \#\#\#{*}{*}{*}$. The mask order $o(M)$ and defining length $\partial(M)$ are defined to be those of the schemata contained in the mask.

Now, in the binary case, it follows from Theorem 1 that the partition coefficients of schemata sharing the same fixed positions will differ only in sign. Thus, for $H \in M$, the Partition Equation may be defined in terms of partition coefficients $\epsilon'(M)$ and a Walsh function [2] $\mathrm{Wal}(M, H) = \pm 1$, where $\epsilon(H) = \mathrm{Wal}(M, H)\epsilon'(M)$. This is the approach taken by Bethke [3], although masks are not presented explicitly in his work.

We now see that the calculation of partition coefficients is not limited to problems with binary alleles. Bethke [3] has shown how partition coefficients may be used to construct problems that are difficult to solve using GAs. We next show how Bethke's methods may be extended for non-binary codings.

4 EASY AND DECEPTIVE PROBLEMS

For a problem to be efficiently solvable by a GA it should satisfy the building block hypothesis [6]; that is low-order schemata should 'point the way' towards an optimum. The most fit of those schemata contained in any mask should itself contain the optimum. A mask M is said to be deceptive if an optimal individual is not contained within (at least one of) the most fit schema(ta) in M. We say that deception exists at the p'th order if some p'th order mask is deceptive. A problem is said to be *fully deceptive* at order p if all masks of order p or less are deceptive.

Partition coefficients provide an easy means of calculating schema fitnesses, and thus checking masks for deception. In particular, we see from the Partition Equation that $f(H)$ depends only on those partition coefficients of order $o(H)$ or less. If all high order partition coefficients are small, then the optimum is likely to be well predicted by low-order schema, and we expect the problem to be easily solved by a GA. However, if the high order partition coefficients are large then the optimum may not be that predicted in the low-order masks; the low-order masks could contain deception. If a good solution could no longer be constructed from the short fit building blocks we would expect the problem to be difficult for a GA.

We next show how such a deceptive problem may be constructed. We assume strictly non-binary genes, $v_j \geq 3\ \forall j$, and construct the problem with a unique maximum at $I^* = i_1^* i_2^* \cdots i_k^*$. For $2 \leq p \leq k$, we put

$$\begin{array}{rll} o(H) = 1: & \epsilon(H) = -1, & H \ni I^* \\ & \epsilon(H) = -1/\Upsilon(H, I^*), & \text{otherwise} \\ 1 < o(H) < p: & \epsilon(H) = 0, & \\ o(H) = p: & \epsilon(H) = +3k, & H \ni I^* \\ & \epsilon(H) = +3k/\Upsilon(H, I^*), & \text{otherwise} \\ p < o(H) \leq k: & \epsilon(H) = 0, & \end{array} \quad (5)$$

where

$$\Upsilon(H, I^*) = \prod_{j \in \mathcal{D}(H, I^*)} (1 - v_j),$$

$$\mathcal{D}(H, I^*) = \{j : h_j \neq *, h_j \neq i_j^*\},$$

and the value of $\epsilon(** \cdots **)$ is chosen to ensure all $f(I) > 0$.

Firstly, it may be easily checked (see Mason [9]) that these partition coefficients satisfy (3), and so a fitness function may be constructed by Theorem 2.

Theorem 3 *The problem defined by (5) is fully deceptive up to order p-1, but no deception occurs in higher order masks.*

We refer to Mason [9] for full details of this proof. However, it is not difficult to see that for $o(H) < p$,

$f(H)$ will depend only on the partition coefficients $\epsilon(H') : H' \supseteq H, o(H) = 1$. Thus, the fitness will decrease by 1 for each position in H matching I^* and increase by $1/(v_j - 1)$ for each position j in which $i_j^* \neq h_j, h_j \neq *$. Therefore, for $H \in M$, $o(M) < p$, $f(H)$ will achieve a strict minimum when $H \ni I^*$, and thus M is deceptive.

A similar analysis is used to show that for $o(H) \geq p$, the order p partition values are large enough to dominate $f(H)$ and ensure no deception is present. That I^* is a unique optimum follows immediately from this result.

5 DECEPTION ANALYSIS

In the last section we discussed the presence of deception at some order within a problem. However, because of the overlap of positions between different schemata, it is not clear that deception in one or more masks will make the overall problem difficult to solve with a GA. (Note however that if the problem is fully deceptive to a high enough order, as results when using (5), then overall deception will result.)

In an attempt to be more rigorous, Goldberg [5] has defined the notion of *static deception*. He compares Ω', the set of individuals "likely to attract increasing GA samples", with the set of near optimal individuals $\Omega = \{I : f_{\max} - f(I) \leq \mu\}$ for some $\mu \geq 0$. The set Ω' is found by analysing the expected fitness of individuals after one application of the mutation and crossover operators.

Consider some individual I being changed to I' by mutation and crossover. We want to calculate $f'(I) = \mathrm{E}(f(I'))$, the *adjusted fitness* of I, knowing only I and the probabilities of performing mutation and crossover. $\mathrm{E}(f(I'))$ could be evaluated as $\sum_{I'} f(I')p(I, I')$ where $p(I, I')$ is the probability of I' being formed from I. Alternatively, the Partition Equation equation can be used: for any schema (or individual) the adjusted fitness $f'(H)$ can be calculated by considering separately the effects of crossover and mutation on each term $\epsilon(H')$, $H' \supseteq H$. In particular, Goldberg defines (presented here using our extended notation)

$$f'(H) = \sum_{H' \supseteq H} \epsilon'(H')$$

where $\epsilon'(H)$ is the apparent partition coefficient $\epsilon(H)$ after adjusting for the operators.

Goldberg defines $\Omega' = \{I : f'_{\max} - f'(I) \leq \mu'\}$, where

$$\mu' = \frac{f'_{\max} - \epsilon(** \cdots **)}{f_{\max} - \epsilon(** \cdots **)}\mu$$

is chosen to preserve the difference between maximum and average fitness in Ω and Ω'. A problem is said to be *statically easy* at level μ if $\Omega - \Omega' = \emptyset$, that is the optimal individuals are amongst those individuals the GA is increasingly sampling. Otherwise, the problem is termed *statically deceptive* at level μ.

Goldberg has calculated $\epsilon'(H)$ for binary coding with the standard mutation and 1-point crossover operators. We now show that his results extend immediately to the non-binary case.

Consider 1-point crossover (with a randomly chosen individual) being performed with probability p_c. A schema H will survive crossover with probability

$$1 - p_c\frac{\partial(H)}{k-1}.$$

However, if H is disrupted by crossover, it follows from Theorem 1 that $\mathrm{E}(\epsilon(H)) = 0$. Therefore, the partition coefficient corrected for crossover is

$$\epsilon'_c(H) = \epsilon(H)(1 - p_c\frac{\partial(H)}{k-1}).$$

Assume now that each gene in H is mutated with probability p_m to produce schema H'. We ignore the $o(p_m^2)$ possibility of 2 or more genes being mutated, and approximate the probability $1-(1-p_m)^{o(H)}$ of any mutation in H by $p_m o(H)$. Assuming that mutation results in a randomly chosen allele value differing from the initial value, then by Theorem 1, $\mathrm{E}(\epsilon(H')) = -\epsilon(H)$. This gives the mutation corrected partition coefficient

$$\begin{aligned} \epsilon'_m(H) &= \epsilon(H)(1 - p_m o(H)) - \epsilon(H)p_m o(H) \\ &= \epsilon(H)(1 - 2p_m o(H)). \end{aligned} \quad (6)$$

Combining the corrections for mutation and crossover, and ignoring any cross-product terms, we obtain Goldberg's result essentially unchanged:

$$\epsilon'(H) = \epsilon(H)\left[1 - p_c\frac{\partial(H)}{k-1} - 2p_m o(H)\right].$$

This result means that we have now defined static deception for general coding schemes.

Although we have shown that the calculations of apparent partition coefficients are identical under binary and non-binary codings, we have been assuming that the chromosomes are the same length under both codings. If both coding schemes were to distinguish the same number of individuals then the binary scheme would always give the longest chromosomes. Noting that the correction in ϵ increases with increasing schema order and defining length, we see that the coefficient corrections are potentially larger for the binary coding. Binary coding schemes may be expected to show greater deception than non-binary schemes. This is not surprising when we realise that in the extreme case $k = 1$ deception cannot occur. However, the greatest number of schemata (as they are traditionally defined in this paper) may be expressed when a binary coding is used, and hence low cardinality alphabets may be expected to encourage the formation of building blocks. It is the manipulation by crossover of such building blocks that distinguishes GA's from random search.

Acknowledgements

The author would like to thank Dr. Edward Anderson whose suggestions greatly improved this paper.

References

[1] J. Antonisse. A new interpretation of schema encoding that overturns the binary encoding constraint. In J. David Schaffer, editor, *Genetic Algorithms: Proceedings of the Third International Conference on Genetic Algorithms*, Morgan Kaufmann, San Mateo, CA, 1989.

[2] K.G. Beauchamp. *Applications of Walsh and Related Functions*, Academic Press, London, 1984.

[3] A.D. Bethke. Genetic Algorithms as Function Optimizers. Doctoral dissertation, University of Michigan. *Dissertation Abstracts International*, 41(9), 3503B, (University Microfilms No. 8106101) 1980.

[4] David E. Goldberg. Genetic Algorithms and Walsh Functions: Part I, A Gentle Introduction. TCGA Report No. 88006. University of Alabama, The Clearinghouse for Genetic Algorithms, Tuscaloosa, 1988.

[5] David E. Goldberg. Genetic Algorithms and Walsh Functions: Part II, Deception and its Analysis. TCGA Report No. 89001. University of Alabama, The Clearinghouse for Genetic Algorithms, Tuscaloosa, 1989.

[6] David E. Goldberg. *Genetic Algorithms in Search, Optimization and Machine Learning.* Addison-Wesley, Massachusetts, 1989.

[7] John J. Grefenstette, and James E. Baker. How Genetic Algorithms Work: A Critical Look at Implicit Parallelism. In J. David Schaffer, editor, *Genetic Algorithms: Proceedings of the Third International Conference on Genetic Algorithms*, Morgan Kaufmann, San Mateo, CA, 1989.

[8] John H. Holland. *Adaptation in natural and artificial systems*, The University of Michigan Press, Ann Arbor, MI, 1975.

[9] Andrew J. Mason. Non-Binary codings and Partition Coefficients for Genetic Algorithms. Management Studies Research Paper #3/91, Engineering Department, University of Cambridge, United Kingdom, 1991.

An Analysis of a Simple Genetic Algorithm

Yuri Rabinovich
Dept. of Computer Science,
Hebrew University,
Jerusalem, Israel 91904

Avi Wigderson
Dept. of Computer Science,
Hebrew University,
Jerusalem, Israel 91904

Abstract

The rate of convergence and the structure of stable populations are studied for a simple, and yet nontrivial, family of genetic algorithms.

1 INTRODUCTION

This paper originates in an attempt to use genetic algorithms as an alternative approach to theoretical problems of combinatorial optimization. In Holland's [1] pioneering work it is suggested that genetic algorithms are likely to work well in those cases where some short schemata have fitness exceeding the average and where these schemata combine well by the crossover operator. In this case the crossing-over of two well fitted structures usually results in a well fitted structure. In the context of combinatorial optimization this means that a genetic algorithm (with a genetic operator that is tailor-made for the problem) is likely to be effective when this operator usually merges two given structures of high fitness into a third good structure.

Consider for example the classical problem of finding large matchings in a given graph G. Genetic operators with the above-mentioned properties can be described, and it is therefore reasonable to believe that this nontrivial problem of finding a maximum size matching can be effectively solved (or at least approximated) by a genetic algorithm. This line of research quickly runs into serious mathematical difficulties we could not overcome. The present article grew out of our attempts to study similar, simpler systems.

The system under investigation has n independent binary attributes. The fitness of a structure is the number of positive attributes. Evolution in this system is governed by a simple shuffle operator, to be described below. This system stands in sharp contrast with the complicated situations genetic algorithms are usually applied in. Yet its analysis is by no means trivial. Our interest in this system is threefold: First, develop as much rigorous understanding of how it works, in the hope of creating a better basis for the study of similar, but more involved problems. Second, to show that at least the simplest genetic algorithms are amenable to exact mathematical analysis, and develop the suitable techniques. Third, the system turned out to have a measure of beauty.

2 THE SYSTEM AND ITS ANALYSIS

2.1 DESCRIPTION OF THE SYSTEM

Let n be a natural number. Every structure of our system has exactly n attributes. Each attribute has two possible values (0 and 1) so the set of n attribute may be represented by a binary string of length n. Different structures may have the same set of values for the n attributes.

The *population* will be regarded as a probability

distribution over the *kinds* of structures (i.e. over binary strings of length n). We shall not estimate in this paper the size needed in order that the population evolved with but a minor deviations from the corresponding distribution.

We define the fitness of a structure to be the number of 1-s it has. Later we shall discuss briefly other fitness functions.

The $\times$ operator (that plays the role of the ordinary crossover operator) acts as follows: given two structures s_1 and s_2, it produces an offspring structure, whose i-th position is determined by the i-th position of one of $\{s_1, s_2\}$, each with probability $\frac{1}{2}$.

Note that $\times$ is actually defined for the kinds of populations as well. It will be convenient to extend the domain of $\times$ to pairs of populations. It will be convenient to extend the the domain of $\times$ to the pairs of populations. Denoting symbolically a population $\mathcal{P}$ as $\sum p_i C_i$, and a population $\mathcal{Q}$ as $\sum q_j C_j$ where the sum runs over the kinds of structures, the natural definition for $\mathcal{P} \times \mathcal{Q}$ is

$$\mathcal{P} \times \mathcal{Q} = \sum p_i C_i \times \sum q_j C_j = \sum \sum p_i q_j (C_i \times C_j).$$

Define also $M(\mathcal{P})$ as $\mathcal{P} \times \mathcal{P}$.

Another operator we need is the operator of reevaluation W, which is an essential part of Holland's Reproductive Plans. Given a population $\mathcal{P} = \sum p_i C_i$, W is defined by

$$W\left(\sum p_i C_i\right) = \sum \left(\frac{f(i) p_i}{\sum f(j) p_j} \right) C_i.$$

where $f(i)$ stands for fitness of a structure of kind i. In other words, each kind in the population increases its part proportionally to its fitness. Note that the sum in the denominator is exactly the average fitness of the population; it will be denoted by $Av(\mathcal{P})$.

In this paper we confine ourselves to symmetrical initial populations (i.e. the populations which remain the same under any permutation of bits). Note that the operators $\times$ and W preserve symmetry. Of special interest to us is the system with the initial population being the uniform distribution over the singletones.

To finish the description of the system, we need to define how it evolves. Given the initial population $\mathcal{P}_0$, $\mathcal{P}_{n+1}$ is

$$\mathcal{P}_{n+1} = W(\mathcal{P}_n) \times W(\mathcal{P}_n) = MW(\mathcal{P}_n).$$

We follow the Reproductive Plan of Holland [1], with a generation replaced in a time-step by the generation of its offsprings. Since the system we have obtained has short above average schemata, there is every reason to believe that the population $\mathcal{P}_n$ converges to distribution having the whole weight on the string '11...111' (n 1-s). The main object of the following sections is to provide a mathematical foundation for this feeling.

Another object of interest are the populations arising from some iterative acting of M on some initial population, and the populations satisfying $\mathcal{P} \times \mathcal{P} = \mathcal{P}$. (Such populations will be called *stable*). Their importance will be revealed later.

2.2 HOMOGENEOUS REPRESENTATION

The representation of the the original system as given in section 2.1 is not convenient to work with. Therefore we shall use two other representations. The first is described in this section and the second in the next section. To every symmetric population $\mathcal{P}$ attach a vector $v = (a_0, a_1, ..., a_n)$ where a_i is a total weight of binary strings with exactly i 1-s. The a_i-s satisfy two conditions :
(a) $a_i \geq 0$; (b) $\sum_{i=0}^{n} a_i = 1$.

Call a $(n+1)$-vector *legal* if its entries satisfy the two conditions. There is a one-to-one correspondence between the symmetric populations and the legal $(n+1)$-vectors. Moreover, since the symmetric populations are closed under W and $\times$, our original system induces a new system over these vectors. The induced system is a genetic algorithm as well. When there is no place for confusion, we use for the induced operators the same symbols as for the original operators.

The original definition of $W(\mathcal{P})$ takes form
$W(v) = W(a_0, a_1, ..., a_n) =$

$(0, \frac{a_1}{Av(v)}, \frac{2a_2}{Av(v)}, ..., \frac{na_n}{Av(v)})$;
$Av(v) = Av(a_0, a_1, ..., a_n) = \sum_{i=0}^{n} ia_i$.

The introduced representation simplifies the investigation of our system, but it has its own drawbacks. The main problem lies in the description of the operator $\times$. Its direct definition would be hard to work with. This problem will be solved in the following section, using a different basis. Meanwhile we note that for two legal vectors v and u, $v \times u = (B_0(v,u), B_1(v,u), ..., B_n(v,u))$ where each $B_i(v,u)$ is a symmetric bilinear form with nonnegative coefficients. Their sum is defined by a $(n+1) \times (n+1)$ matrix with all entries equal 1.

2.3 CHANGE OF BASIS

The operator $\times$ is inconvenient to work with. In this section we present a basis, more suitable for both $\times$ and W.

Let us return to the original system. For a population $\mathcal{P}$ define the random variable X_i, $(i = 1, \ldots, n)$ to be the i-th coordinate of a random string in $\mathcal{P}$. Define e_k, $(k = 0, 1, ..., n)$ to be

$$e_0 = 1; \quad e_k = Pr(\wedge_{i=1}^{k}(X_i = 1)).$$

Note that because of the symmetry of $\mathcal{P}$ we could use in the definition of e_k any k-tuple of X_i-s with different indices.

We wish to express e_k-s in terms of a_i-s. Let A_k be a set of all strings in $\mathcal{P}$ having exactly k 1-s, and let s be a random string. For $k > 0$ we have:

$$e_k = Pr(\wedge_{i=1}^{k}(X_i = 1)) =$$

$$\sum_{j=0}^{n} Pr(\wedge_{i=1}^{k}(X_i = 1) |\ s \in A_j)\ Pr(s \in A_j) =$$

$$\sum_{j=k}^{n} \frac{\binom{n-k}{j-k}}{\binom{n}{j}} a_j.$$

Simplifying the above we obtain the equations $(k = 0, 1, ..., n)$

$$e_k = \sum_{i=0}^{n} \frac{i(i-1)...(i-k+1)}{n(n-1)...(n-k+1)} a_i = \frac{1}{\binom{n}{k}} \sum_{i=k}^{n} \binom{i}{k} a_i. \tag{1}$$

Thus the e_k-s are obtained from the a_k-s by a linear transformation. Since the transformation matrix is lower triangular with nonzero values on the diagonal, it is regular.

In what follows, we call the standard form of a vector the *a-form*, and its form in the new basis the *e-form*.

The following class of populations plays a key role in the investigation of our system:

Definition 2.1 *Let $\mathcal{P}(\alpha)$ be the symmetric population such that*
(1) $Pr(X_i = 1) = \alpha$ for $i = 1, 2, 3, ...n$.
(2) X_i are independent.

Such populations exist for every α between 0 and 1. They have a simple structure. For example, the probability of a binary string '1101...0' in $\mathcal{P}(\alpha)$ is $\alpha\alpha(1-\alpha)\alpha...(1-\alpha)$. If $v(\alpha)$ is a vector in the e-form corresponding to $\mathcal{P}(\alpha)$, then, of course, $v(\alpha) = (1, \alpha, \alpha^2, ..., \alpha^n)$.

Lemma 2.1 $\mathcal{P}(\alpha) \times \mathcal{P}(\beta) = \mathcal{P}(\frac{\alpha+\beta}{2})$ *where α, β are reals in $[0,1]$.*

Proof: First, note that since the $\times$ operator acts independently on different coordinates, it preserves their independence. Second, it is evident that $E(X_i)$ in $\mathcal{P} \times \mathcal{Q}$ always equals the average of $E(X_i)$-s in $\mathcal{P}$ and $\mathcal{Q}$. Since X_i is a binary random variable, $Pr(X_i) = E(X_i)$. ∎

Theorem 2.1 *Let $v = (e_0, e_1, ..., e_n)$, $u = (d_0, d_1, ..., d_n)$ be two vectors in the e-form, and $v \times u = (F_0(v,u), F_1(v,u), ..., F_n(v,u))$. Then $F_k(v,u)$ is a a bilinear form $\frac{1}{2^k}\sum_{i=0}^{k}\binom{k}{i} e_i d_{k-i}$.*

Proof: From the lemma 2.1 we know the statement is true if v and u are of the special form $v = (1, \alpha, \alpha^2, ..., \alpha^n), u = (1, \beta, ..., \beta^n)$. Indeed, in this case $F_k(v,u) = (\frac{\alpha+\beta}{2})^k = \frac{1}{2^k}\sum_{i=0}^{k}\binom{k}{i}\alpha^i\beta^{k-i}$. To conclude the proof it is sufficient to notice that a basis for R^{n+1} can be constructed of such vectors. Since any basis completely determines the coefficients of a bilinear form, the theorem is established. ∎

It remains to determine the form of the operator W in the new basis.

Proposition 2.1 *Let $v = (1, e_1, ..., e_n)$ be a vector in e-form. Then the k-th coordinate of $W(v)$ is given by $\left(\frac{n-k}{n}\frac{e_{k+1}}{e_1} + \frac{k}{n}\frac{e_k}{e_1}\right)$ for $k = 0, 1, ..., n$ (regard e_{n+1} as 0).*

Proof: The proposition can be verified directly by using the a-form of v, obtaining the a-form of $W(v)$, and going back to the e-form. ∎

Before we go on, we establish some additional properties of the e_k-s.

From equation 1 we see that $e_1 = \frac{Av(\mathcal{P})}{n}$. Since we are mainly interested in $Av(\mathcal{P})$, the e_1 is a very important parameter. Note that e_1 is preserved under M ($\times$ preserves average) and is changed to $(\frac{n-1}{n}\frac{e_2}{e_1} + \frac{1}{n})$ under W.

As follows from the above paragraph, e_2 has a big influence on the next population's average. Note that M changes e_2 to $\frac{1}{2}(e_2 + e_1^2)$.

¿From the definition of the e_k-s it follow that they constitute a nonnegative nonincreasing sequence. Therefore if $e_k = 0$ in some legal vector in the e-form, than $e_{k+1} = ... = e_n = 0$.

Proposition 2.2 *Let $v = (e_0, e_1, ..., e_n)$ be some legal vector in e-form, and denote $W(v)$ by $(e_0', e_1', ..., e_n')$. Then*

$$\frac{e_k'}{e_{k-1}'} \geq \frac{e_k}{e_{k-1}} \qquad (k = 1, 2, .., n).$$

(Regard $\frac{0}{0}$ as 0).

Proof: We can express (using proposition 2.1) the e_k'-s in terms of e_k-s. Our statement assumes the form :

$$\frac{(n-k)e_{k+1} + ke_k}{(n-k+1)e_k + (k-1)e_{k-1}} \geq \frac{e_k}{e_{k-1}}.$$

If $e_k = 0$ the inequality holds trivially ; otherwise we obtain an equivalent statement

$$(n-k)\frac{e_{k+1}}{e_k} + 1 \geq (n-k+1)\frac{e_k}{e_{k-1}}.$$

Expressing the e_k-s in terms of the a_i-s (via equation 1) and taking the common denominator in the left side, the inequality is equivalent to :

$$\frac{\sum_{i=0}^n i(i-1)...(i-k+1)(i-k+1)a_i}{\sum_{i=0}^n i(i-1)...(i-k+1)a_i} \geq$$

$$\frac{\sum_{i=0}^n i(i-1)...(i-k+2)(i-k+1)a_i}{\sum_{i=0}^n i(i-1)...(i-k+2)a_i}.$$

The numerator of the left side is the same as denominator of the right side. Observe that the ratio of the i-th terms in the left side ($i-k+1$ or 0) is equal to that of the right side. Therefore the last inequality is a direct consequence of the following lemma:

Lemma 2.2 *Let α, a_i, b_i, c_i be nonnegative reals $(i = 1, ..., n)$ and suppose that for every i, $\frac{a_i}{b_i} \geq \alpha\frac{b_i}{c_i}$ (regard $\frac{0}{0}$ as 0). Then*

$$\frac{\sum_{i=1}^n a_i}{\sum_{i=1}^n b_i} \geq \alpha\frac{\sum_{i=1}^n b_i}{\sum_{i=1}^n c_i}.$$

Proof: The statement follows from the Cauchy - Shwarz inequality with $x_i = \sqrt{\frac{a_i}{\alpha}}$, $y_i = \sqrt{\frac{c_i}{\alpha}}$. ∎

Corollary 2.1 *In the same notation of the proposition we have : $e'_1 \geq e_1$, $e'_2 \geq e_2, ..., e'_n \geq e_n$.*

2.4 A LOWER BOUND ON $Av(\mathcal{P}_r)$

The parameter $\frac{e_2}{e_1}$ is very important since the average of the next population depends only on it. The following lemma helps us to estimate its value. In this and the following sections only the e-form will be used.

Lemma 2.3 *Let $(e_0, e_1, ..., e_n)$ be the vector representing a nonzero population $\mathcal{P}$. Denote by $(e_0', e_1', ...e_n')$ the representation of $W(\mathcal{P})$, and by $(e_0^*, e_1^*, ...e_n^*)$ the representation of $MW(\mathcal{P})$. Then*

$$\frac{e_2^*}{e_1^*} \geq \left(\frac{2n-1}{2n}\right)\frac{e_2}{e_1} + \frac{1}{2n}.$$

Proof:

$$\frac{e_2^*}{e_1^*} \overset{(1)}{=} \frac{e_2' + e_1'^2}{2e_1'} = \frac{1}{2}e_1' + \frac{1}{2}\frac{e_2'}{e_1'} \overset{(2)}{\geq}$$

$$\frac{1}{2}e_1' + \frac{1}{2}\frac{e_2}{e_1} \overset{(3)}{=} \frac{1}{2}\left(\left(\frac{n-1}{n}\right)\frac{e_2}{e_1} + \frac{1}{n}\right) + \frac{1}{2}\frac{e_2}{e_1} =$$

$$\left(\frac{2n-1}{2n}\right)\frac{e_2}{e_1} + \frac{1}{2n}.$$

(1) is an expression of e_1 and e_2 in $MW(\mathcal{P})$ in terms of $W(\mathcal{P})$;
(2) is an application of the proposition 2.2;
(3) is an expression of e_1 in $W(\mathcal{P})$ in terms of $\mathcal{P}$. ∎

We can prove now the main theorem of this section. Call a population *zero free* if the weight of the string '00...000' is zero.

Theorem 2.2 *Let $\mathcal{Q}$ be some zero-free population and $\mathcal{Q}^+ = WM(\mathcal{Q})$. Then*

$$n - Av(\mathcal{Q}^+) \leq (n - Av(\mathcal{Q}))\left(1 - \frac{1}{2n}\right).$$

Proof: Since $n - Av(\mathcal{Q})$ equals $n(1-e_1)$, it is sufficient to prove the corresponding inequality for the e_1-s.

For every zero-free population $\mathcal{Q}$ there exists population $\mathcal{Q}^-$ such that $W(\mathcal{Q}^-) = \mathcal{Q}$.

Denote the e-form of $\mathcal{Q}^-$ by $(e_0, e_1, ..., e_n)$, the e-form of $\mathcal{Q}$ by $(e_0', e_1', ..., e_n')$, the e-form of $M(\mathcal{Q})$ by $(e_0^*, e_1^*, ..., e_n^*)$ and the e-form of $\mathcal{Q}^+$ by $(e_0^+, e_1^+, ..., e_n^+)$.

($\mathcal{Q}^+ = WM(\mathcal{Q}) = WMW(\mathcal{Q}^-)$).

Again we write a sequence of inequalities:

$$1 - e_1^+ \overset{(1)}{=} \left(\frac{n-1}{n}\right)\left(1 - \frac{e_2^*}{e_1^*}\right) \overset{(2)}{\leq}$$

$$\left(\frac{n-1}{n}\right)\left(1 - \frac{e_2}{e_1}\right)\left(1 - \frac{1}{2n}\right) \overset{(3)}{=}$$

$$(1 - e_1')\left(1 - \frac{1}{2n}\right).$$

Where
(1) is a representation of e_1^+ in $WM(\mathcal{Q})$ in terms of $M(\mathcal{Q})$;
(2) is an application of the previous lemma;
(3) is an expression of e_1 in $W(\mathcal{Q}^-)$ in terms of $\mathcal{Q}^-$. ∎

Corollary 2.2

$$Av(\mathcal{P}_r) \geq n - (n - Av(\mathcal{P}_1))\left(1 - \frac{1}{2n}\right)^{r-1} \geq$$

$$n - (n - Av(\mathcal{P}_1))\exp\left(-\frac{r-1}{2n}\right).$$

Proof: The proof follows directly from the theorem above and the observation that $Av(\mathcal{P}_{i+1}) = Av(\mathcal{Q}_i)$ for all natural i. (As usual, $\mathcal{Q}_i = W(\mathcal{P}_i), \mathcal{P}_i = M(\mathcal{Q}_{i-1})$.) ∎

The above corollary may be restated in the following form:
For every initial nonzero population $\mathcal{P}_0$, $2n \ln \frac{n}{\epsilon}$ generations shall always sufice to raise the average of the population to $1 - \epsilon$.

2.5 AN UPPER BOUND ON $Av(\mathcal{P}_r)$

Our aim now is to prove that there are populations which do not improve too fast. A class of *normal* populations will play a central role.

Definition 2.2 *A population is called normal if* $e_0 \geq \frac{e_1}{e_0} \geq \frac{e_2}{e_1} \geq \ldots \geq \frac{e_n}{e_{n-1}}$.
($\frac{0}{0}$ is regarded as 0 .)

Such populations exist. The simplest example is the uniform distribution on singletones, represented (in the e-form) by $(\frac{1}{n}, 0, 0, ..., 0)$.

Theorem 2.3 *The property of normality is preserved under the operators $\times$ and W.*

The proof for W is routine; one needs only to express the entries of $W(\mathcal{P})$ in terms of those of $\mathcal{P}$ and use the normality of $\mathcal{P}$.

Much more delicate is the proof that if $\mathcal{P}$ and $\mathcal{Q}$ are both normal, so is $\mathcal{P} \times \mathcal{Q}$. Unexpectedly enough, the result is actually proven by Walkup [2], in the paper on the binomial convolutions of Polya sequences. (The PF_2 sequences in this paper correspond to our normal sequences.)

Lemma 2.4 *Let* $(e_0, e_1, ...e_k)$ *be nonzero normal population* $\mathcal{P}$. *Then, in the notions of the lemma 2.3,*

$$\frac{e_2^*}{e_1^*} \leq \left(\frac{n-1}{n}\right)\frac{e_2}{e_1} + \frac{1}{n}.$$

Proof:

$$\frac{e_2^*}{e_1^*} \overset{(1)}{=} \frac{e_2' + e_1'^2}{2e_1'} = \frac{1}{2}e_1' + \frac{1}{2}\frac{e_2'}{e_1'} \overset{(2)}{\leq}$$

$$\frac{1}{2}e_1' + \frac{1}{2}e_1' = e_1' \overset{(3)}{=} \left(\frac{n-1}{n}\right)\frac{e_2}{e_1} + \frac{1}{n}.$$

(1) is an expression of e_1 and e_2 in $MW(\mathcal{P})$ in terms of $W(\mathcal{P})$;
(2) uses the normality preservation under W $(e_1'^2 \geq e_2')$;
(3) is an expression of e_1 in $W(\mathcal{P})$ in terms of $\mathcal{P}$. ∎

We are now in the position to prove the following theorem :

Theorem 2.4 *Let* $\mathcal{P}_0$ *be some nonzero normal population. Then all* $\mathcal{P}_i$ *and* $\mathcal{Q}_i$ *are normal, and for all natural* r

$$n - Av(\mathcal{Q}_{r+1}) \geq (n - Av(\mathcal{Q}_r))\left(1 - \frac{1}{n}\right).$$

Proof: Using notations analogous to those of the theorem 2.2, (e^+ for $\mathcal{Q}_{r+1}$, e^* for $\mathcal{P}_{r+1}$, e' for $\mathcal{Q}_r$, e for $\mathcal{P}_r$), we obtain

$$1 - e_1^+ \overset{(1)}{=} \left(\frac{n-1}{n}\right)\left(1 - \frac{e_2^*}{e_1^*}\right) \overset{(2)}{\geq}$$

$$\left(\frac{n-1}{n}\right)\left(1 - \frac{e_2}{e_1}\right)\left(1 - \frac{1}{n}\right) \overset{(3)}{=}$$

$$(1 - e_1')\left(1 - \frac{1}{n}\right).$$

Where
(1) is a representation of e_1^+ in $\mathcal{Q}_{r+1}$ in terms of $\mathcal{P}$;
(2) uses the normality preservation and the previous lemma;
(3) is an expression of e_1 in $\mathcal{Q}_r$ in terms of $\mathcal{P}_r$. ∎

Corollary 2.3 *Suppose that* $\mathcal{P}_0$ *is a normal population. Then*

$$Av(\mathcal{P}_r) \leq n - (n - Av(\mathcal{P}_1))\left(1 - \frac{1}{n}\right)^{r-1} \leq$$

$$n - (n - Av(\mathcal{P}_1))\exp\left(-\frac{r-1}{n+1}\right).$$

Proof: The statement follow from the theorem we just have proved, and the fact that $Av(\mathcal{Q}_{i-1}) = Av(\mathcal{P}_i)$ for all natural i. ∎

This corollary, together with the corollary 2.2, shows that our bounds are tight for systems with normal $\mathcal{P}_0$.

2.6 STABLE POPULATIONS AND ONE GENERAL REMARK ON THE LOWER BOUND

Theorem 2.5 *For any* α *in the interval [0,1]* $\mathcal{P}(\alpha)$ *is only stable population with* $e_1 = \alpha$. *Moreover, if we start with any* $\mathcal{P}$ *with* $e_1 = \alpha$ *and apply* M *repeatedly, the population converges to* $\mathcal{P}(\alpha)$. *(Recall that* M *preserves* e_1.*) In the metrics* $d(\mathcal{P}, \mathcal{Q}) = \sum_{k=0}^{n} |e_k - d_k|$ *this convergence is exponentially fast.*

Proof: We already know from theorem 2.1 that $\mathcal{P}(\alpha)$ is stable. If we prove the second part of the theorem, the uniqueness will be established.

We sketch the proof of the second part. Let $(e_0, e_1, ..., e_n)$ be our initial population. Denote by $e_k^{(r)}$ the e_k after r applications of M. Let α be e_1. Our goal is to prove that $|e_k^{(r)} - \alpha^k|$ tends to zero exponentially fast.

¿From the theorem 2.1 we know that $e_k^{(1)} = \sum_{i=0}^{k} \frac{\binom{k}{i}}{2^k} e_i e_{k-i}$. Observe that if $e_i = \alpha^i$ for $i = 0, 1, ..., (k-1)$ then $e_k^{(1)} = \frac{2^{k-1}-1}{2^{k-1}}\alpha^k + \frac{1}{2^{k-1}}e_k$. By above observation, after t applications of M, $|e_2 - \alpha^2|$ is $O(2^{-t})$. Since by now e_2 does not differ significantly from α^2, another $\frac{t}{2}$ applications of M cause $|e_3 - \alpha^3|$ to be $O(2^{-t})$. Meanwhile $|e_2 - \alpha^2|$ is already $O(2^{-1.5t})$. Advancing in the same fashion, after $(t + \frac{t}{2} + ...\frac{t}{2^n}) < 2t$

applications of M, for every $k \leq n$ $|e_k - \alpha^k|$ cannot be significantly bigger than $O(2^{-t})$. ∎

After we have found the stable populations and established a fast convergence to the stable population under action of M, we wish to outline a direction that might lead to a lower bound on $Av(\mathcal{P}_r)$ for other fitness functions.

We know that the operator W improves the population. It seems that in many cases a more general thing is true :

Conjecture 2.1 *Let $\mathcal{R}$ be a population, S is an operator obtained by a concatenation of the W-s and the M-s in a certain order, S^- is obtained exactly in the same way as S, with one W left out. Then* $Av(\ S(\mathcal{R})) \geq Av(\ S^-(\mathcal{R}))$.

Suppose that our fitness function f is such that the conjecture holds. Suppose, further, that W is not sensitive to small changes. Then we could obtain a lower bound on $Av(\mathcal{P}_r)$ in the following way:

Define a function F as $F(\alpha) = Av(\ W(\mathcal{P}(\alpha)))$.

Define k to be a natural number, such that k consecutive applications of M cause any population $\mathcal{P}$ to approach (in terms of the W's sensitivity) some $\mathcal{P}(\alpha)$. Then $Av(\mathcal{P}_r) \geq F^{*(\frac{r}{k})}(\beta) - \epsilon$, where $\beta = Av(\mathcal{P}_0)$ and F^{*m} is the m-th iteration of F.

Indeed, $\mathcal{P}_r = MWMW...MW(\mathcal{P}_0)$. By erasing all the W-s standing in the places not divisible by k, we cannot (by the conjecture) decrease the average. But now we have $\frac{r}{k}$ blocks of the type $WMMM...M$. The result follows from the definition of k.

3 CONCLUSION

The main focus of this paper is the applications of genetic algorithms to combinatorial optimization. The concrete system under study is very simple. Yet, investigating it hopefully teaches us something worthwhile about genetic algorithms in general.

A few interesting questions arise from our work. For example, what do stable populations look like? When does iterated applications of M converge ?

It is our hope to be able to investigate in the future some more complex genetic algorithms related to classical combinatorial problems. For example, genetic algorithms systems which produce maximum weight independent set of matroids by synthesizing complex structures from simple ones. We already know of a genetic algorithm that causes maximum size independent set of a matroid to evolve with a high probability in $O(n^2)$ steps.

Acknowledgments

The first author is partially supported by the Leibniz Center for Research in Computer Science.

We are very grateful to Profs. H. Furstenberg, I. Rinnot and B. Weiss for helpful discussions and concrete effective suggestions. We wish to thank also Dr. Yuval Davidor and Dr. Mario Szegedy for useful suggestions. Thanks to Prof. Nati Linial and Dr. Aviad Cohen who read the paper and made some necessary remarks.

References

[1] J. H. Holland, *Adaptation in natural and artificial systems*, University of Michigan Press, Ann Arbor, 1975.

[2] D. W. Walkup, Polya sequences, binomial convolution and Unions of Random Sets, *J. Appl. Prob.* 13(1976), pp. 76-85

Forma Analysis and Random Respectful Recombination

Nicholas J. Radcliffe

njr@castle.ed.ac.uk

Edinburgh Parallel Computing Centre, King's Buildings, University of Edinburgh, EH9 3JZ, Scotland

Abstract

Intrinsic parallelism is shown to have application beyond schemata and o-schemata. More general objects called *formae* are introduced and general operators which manipulate these are introduced and discussed. These include *random, respectful recombination*. The extended formalism is applied to various common representations and standard operators are analysed in the light of the formalism.

1 Introduction

The conventional understanding of genetic algorithms attributes much of their power to intrinsic parallelism, the phenomenon whereby each chromosome is an instance of many schemata (or o-schemata) and its measured performance contributes to an estimated fitness for each of these schemata. Efforts to maximise the level of intrinsic parallelism available are frequently in conflict with a desire to use natural representations and operators for the structures in space being searched. This paper demonstrates that intrinsic parallelism is a very general phenomenon, not restricted to schemata and o-schemata and explores the interaction between intrinsic parallelism, genetic representations and operators.

The paper begins (sections 2 & 3) with a review of earlier work which showed that more general partitions of the search space than schemata give rise to intrinsic parallelism. This motivates a shift of emphasis from schemata to more general kinds of regularities which may be present in the search space (sections 4 & 6), and allows the introduction of general-purpose operators—including so-called *random, respectful recombination*—which can be of both analytic and practical use (section 5). The generalisation of a schema is called a *forma*, and in sections 7 to 10 four different types of formae[1] are discussed, in conjunction with operators for their effective manipulation. Before concluding, there is a discussion of the future directions suggested by this work.

[1]Although Holland chose the neuter form for the Latin noun schema, there is no option but to choose the feminine form of its synonym, forma.

2 Analysis of Genetic Algorithms

In his seminal book on adaptation, John Holland [7] considered chromosomes which were k-ary[2] strings or similar and showed that genetic algorithms can usefully be analysed in terms of their effects on higher-order structures, variously known as *schemata, hyperplanes* or *similarity templates*. A schema specifies some set of alleles which the genes of a chromosome must express in order for that chromosome to be said to be an *instance* of the schema. Holland derived a simple but immensely significant expression which bounded the expected number of instances of any schema in the next generation of the population, commonly known as the "schema" or "fundamental" theorem. Later, David Goldberg [5, 4] introduced a number of classes of "o-schemata" which played much the same rôle for problems in which the chromosome was a member of the permutation group $\mathcal{P}_n$ (such as the travelling sales-rep problem, TSP) as did Holland's original schemata for k-ary string representations. The following analysis shows that schemata and o-schemata can both be regarded as examples of a more general kind of object which we shall term a *forma*.

For present purposes it will be convenient to identify a forma (currently a schema or an o-schema) with the set of all of its instances, so that if a chromosome η is an instance of a forma ξ we shall write $\eta \in \xi$. In this spirit, $3241 \in 3\Box 4\Box$ where $\Box$ is the "don't care" symbol.

Recall that the Fundamental Theorem bounds the expected number of instances $N_\xi(t+1)$ of each forma ξ in the population $\mathfrak{B}(t+1)$ by

$$\left\langle N_\xi(t+1) \right\rangle \geq N_\xi(t) \frac{\hat{\mu}_\xi(t)}{\bar{\mu}(t)} \left[1 - \sum_{\omega \in \Omega} p_\omega p_\omega^\xi \right],$$

where $\hat{\mu}_\xi(t)$ is the sample average for utility of ξ over all its instances in the population $\mathfrak{B}(t)$ and the terms $p_\omega p_\omega^\xi$ in the sum quantify the disruptive effect of each operator ω, drawn from a set Ω of genetic operators, on forma membership.

Recall also that the *defining positions* of a forma are those

[2]base k, for arbitrary k

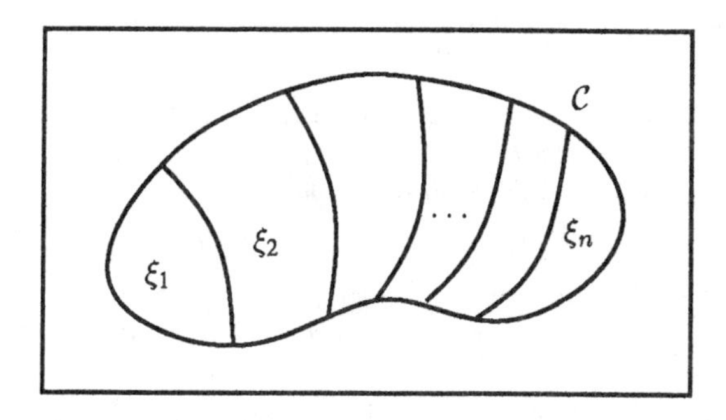

Figure 1: An equivalence relation partitions the space of chromosomes, $\mathcal{C}$, into a number of equivalence classes or *formae*, $\xi_1, \xi_2, \ldots, \xi_n$.

loci at which a value is specified, so that the forma $\xi = \Box a \Box b$, has defining positions at its second and fourth loci.

The generalisation sought in this paper requires us to introduce equivalence relations over the search space. Given any forma ξ, consider the equivalence relation which relates any pair of chromosomes having the same alleles at the forma's defining positions. We can choose to regard the forma as an equivalence class induced by this equivalence relation. Specifically, we can denote the equivalence relation which induces $\Box a \Box b$ by $\Box \blacksquare \Box \blacksquare$, which we understand to relate those chromosomes which have the same values at those loci marked with the $\blacksquare$ symbol, placing 1234 and 2224 in the forma $\Box 2 \Box 4$ but 1111 and 0101 in the forma $\Box 1 \Box 1$. We shall see examples of formae which are less like familiar schemata in section 10.

Having made this identification, the Fundamental Theorem can be seen to apply to *any* subset ξ of the space of chromosomes, $\mathcal{C}$, provided only that the disruption coefficient p_ω^ξ correctly bounds the disruptive effect of applying the operator ω to a chromosome which is a member of ξ. In practice we shall choose to regard these subsets as equivalence classes induced by some set of equivalence relations—a freedom we always have. For this reason we shall henceforth use the term "forma" to refer to an equivalence class of any equivalence relation over the space $\mathcal{C}$ of chromosomes. Holland's schemata and Goldberg's various o-schemata are then immediately seen to be special cases of formae. (See figure 1.)

3 Exploiting Intrinsic Parallelism

Holland observed that each evaluation of a chromosome can be regarded as a statistical sampling event which yields information about the sample averages for utility of *each* of the 2^n schemata of which it is an instance (the phenomenon referred to as intrinsic parallelism) but of course this applies equally to any formae we may choose to consider. The parallelism is exploited by using information gathered about these higher order structures, the formae, to guide the further exploration of the space. The critical tasks are thus finding formae which characterise solutions in meaningful ways and developing operators which usefully manipulate these formae.

It is important to notice that the primary factor governing the expected rate of increase of (instances of) any forma ξ is not the mean relative fitness of its members, $\mu_\xi / \bar{\mu}(t)$, but the *observed* relative fitness $\hat{\mu}_\xi(t)/\bar{\mu}(t)$ of its instances in the population $\mathfrak{B}(t)$. For this reason, our effective exploitation of information about the fitness of various formae is strictly limited by the reliability of the sample fitness $\hat{\mu}_\xi(t)$ as an estimator of the mean fitness μ_ξ of *all* instances of ξ. This suggests the unsurprising conclusion that we will only be able to exploit effectively information about formae whose instances display a low variance for fitness. More succinctly, only those formae which well-characterise solutions, identifying sets with broadly similar performance, will be of any value to the search. Thus the degree of intrinsic parallelism which can be effectively utilised by the search is limited to the number of formae which capture regularities in the performance of solutions over the space $\mathcal{C}$ of chromosomes.

These considerations and others (including Goldberg's principles of *minimal alphabets* and *meaningful building blocks*, [4]) led to the proposal of six *design principles* for constructing useful representations, formae and genetic operators (Radcliffe [10]). In the following, the number of formae induced by an equivalence relation will be referred to as the *precision* of both the relation and the formae it induces.[3] The set of equivalence relations which induce the formae (equivalence classes) under consideration will be written Ψ and the set of all formae induced by relations in Ψ will be denoted Ξ.

Two formae $\xi, \xi' \in \Xi$ will be said to be *compatible* if it is possible for a chromosome to be an instance of both ξ and ξ' (figure 2). In the familiar case of schemata, $1 \Box 1 \Box$ and $0 \Box \Box \Box$ are incompatible, because there is a conflict at the first locus, whereas $1 \Box 1 \Box$ and $11 \Box \Box$ are compatible.

In general, recombination operators take two chromosomes and produce different children depending on explicit or implicit control parameters such as the crosspoint used for one-point crossover and the binary mask used in uniform crossover. (See Syswerda [11] and Eshelman *et al* [3] for details of uniform crossover.) A generic recombination operator X will be taken to have an associated control set $\mathcal{A}_X$ and functional form

$$X : \mathcal{C} \times \mathcal{C} \times \mathcal{A}_X \longrightarrow \mathcal{C}.$$

The member of this control set chosen for some particular recombination completely determines which of the various possible children results from the cross.

[3] In the case of schemata and genes with k alleles, the precision is k^o, where o is the order of a schema.

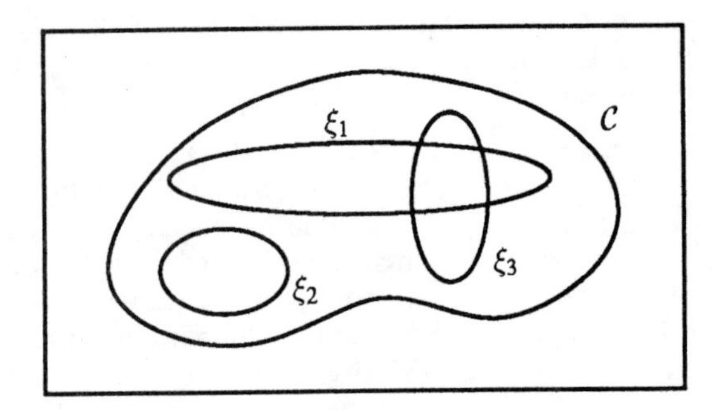

Figure 2: The formae ξ_1 and ξ_2 are *incompatible* because they have null intersection, whereas ξ_1 and ξ_3 are compatible because a single chromosome can be an instance of them both.

4 Design Principles

The first three design principles are general and suggest desirable characteristics of chromosomal representations and formae:

1. (Minimal redundancy) *The representation should have minimal redundancy; such redundancy as exists should be capable of being expressed in terms of the equivalence relations in* Ψ.
Ideally, each member of the space being searched should be represented by only one chromosome in $\mathcal{C}$. This is highly desirable in order to minimise the size of the search space. If redundant solutions are related by one of the equivalence relations used then the genetic algorithm should effectively be able to "fold out" the redundancy (see principle 4); otherwise it is doomed to treat redundant solutions as unrelated.

2. (Correlation within formae) *Some of the equivalence relations, including some of low precision, must relate chromosomes with correlated performance.*
This ensures that useful information can be gathered about the performance of a forma by sampling its instances. Such information is used to guide the search. The emphasis is placed on low-precision formae because these will generally be less likely to be disrupted by the application of genetic operators, and are also more likely to be compatible with one another.

3. (Closure) *The intersection of any pair of compatible formae should itself be a forma.*
This ensures that solutions can be specified with different degrees of accuracy and allows the search gradually to be refined. Clearly the precision of formae so constructed will be at least as high as that of the higher-precision of the intersecting formae.

The remaining three principles concern the way in which operators manipulate chromosomes and formae. It is helpful to use an informal analogy in which chromosomes specify people and some of the characteristics used to define a set of formae are hair colour and eye colour. The consequence of each design principle for these formae is given at the end of each principle.

4. (Respect) *Crossing two instances of any forma should produce another instance of that forma.*
Formally, it should be the case that

$$\forall \xi \in \Xi \ \forall \eta, \zeta \in \xi \ \forall a \in \mathcal{A}_X : \ X(\eta, \zeta, a) \in \xi,$$

where X is the crossover operator. In this case the crossover operator will be said to *respect* the equivalence relations (and their formae). This is necessary in order that the algorithm can converge on good formae, and implies, for example, that $X(\eta, \eta, a) \equiv \eta$, assuming that equivalence relations of maximum precision specify chromosomes completely.
[If both parents have blue eyes then all their children produced by recombination must have blue eyes.]

5. (Proper assortment) *Given instances of two compatible formae, it should be possible to cross them to produce a child which is an instance of both formae.*
Formally,

$$\forall \xi, \xi' \in \Xi \ (\xi \cap \xi' \neq \emptyset) \ \forall \eta \in \xi \ \forall \eta' \in \xi' \\ \exists a \in \mathcal{A}_X : \ X(\eta, \eta', a) \in \xi \cap \xi'. \qquad (1)$$

This relates to Goldberg's "meaningful building blocks", of which he writes ([4], p. 373)

> 'Effective processing by genetic algorithms occurs when *building blocks*—relatively short, low order schemata with above average fitness values—combine to form optima or near-optima.'

A crossover operator which obeys equation 1 seems very much more likely to be able to recombine "building blocks" usefully, and any crossover operator which obeys this principle will be said *properly to assort* formae.
[If one parent has blue eyes and the other has brown hair it must be possible to recombine them to produce a child with blue eyes and brown hair as the result of the cross.]

6. (Ergodicity) *It should be possible, through a finite sequence of applications of the genetic operators, to access any point in the search space* $\mathcal{C}$ *given any population* $\mathfrak{B}(t)$.
This provides the *raison d'être* for the mutation operator.
[Even if the whole population has blue eyes, it must be possible to produce a brown-eyed child. The mutation operator usually ensures this.]

5 Random, Respectful Recombination

Given a set Ξ of formae, an obvious question is whether it is possible to construct a recombination operator which simultaneously respects and properly assorts the formae, and if so, whether there is more than one such operator. It is simple to show that not all sets of formae can be so respected and properly assorted; a set which cannot be is described in section 9. Those which can be are said to be *separable*, and a recombination operator which respects and properly assorts a set of formae is said to *separate* them.

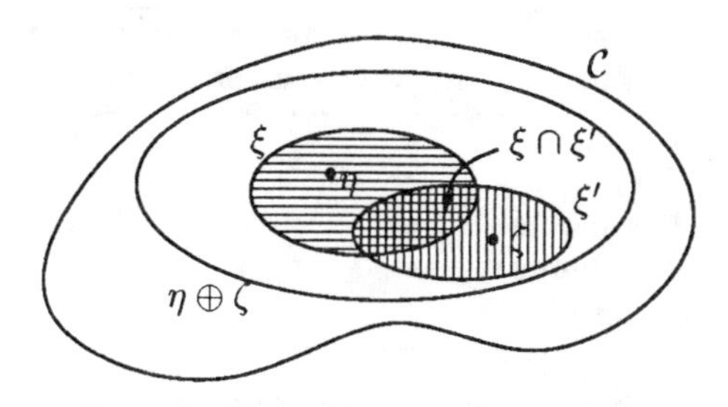

Figure 3: Any pair of chromosomes η and ζ have a *similarity set* denoted $\eta \oplus \zeta$, which is the smallest forma from Ξ containing them both. Respect requires that any children produced by recombining two solutions lie in their similarity set. The R^3 operator makes a uniform random choice of child from the parents' similarity set. If the set of formae, Ξ, is separable (i.e. capable of simultaneous respect and proper assortment) then R^3 will separate them.

The principle of respect amounts loosely to the requirement that characteristics shared by both parents are passed on to their children. It is useful to define the *similarity set* $\eta \oplus \zeta$ of chromosomes η and ζ as the highest precision forma which contains them both:

$$\eta \oplus \zeta \equiv \bigcap \{\xi \in \Xi \mid \eta, \zeta \in \xi\}.$$

In the familiar case of schemata, this is the schema having the alleles which η and ζ share at its (only) definition points. For example,

$$\begin{array}{r} 1\,0\,1\,1\,1\,0\,0\,1 \\ \oplus\,1\,0\,0\,1\,0\,0\,1\,0 \\ \hline 1\,0\,\square\,1\,\square\,0\,\square\,\square \end{array}$$

Respect then amounts to the requirement that all children produced by recombining η and ζ be members of their similarity set:

$$\forall \eta, \zeta \in \mathcal{C}\ \forall a \in \mathcal{A}_X : X(\eta, \zeta, a) \in \eta \oplus \zeta.$$

Clearly if Ξ is separable (that is, capable of simultaneous respect and proper assortment) the children required for proper assortment must also lie in the similarity set of the parents. It follows that a recombination operator which, given any pair of parents η and ζ, returns a randomly-selected member of $\eta \oplus \zeta$, is guaranteed both to respect and properly assort the formae. The recombination operator which makes a *uniform* random choice of children from the similarity set of the parents is called *random, respectful recombination*, which we abbreviate to R^3 (figure 3).

For example, when crossing the binary chromosomes 1010 and 0011, each of the four members of the similarity set □01□ (0010, 0011, 1010, and 1011) is chosen with probability one quarter by R^3.

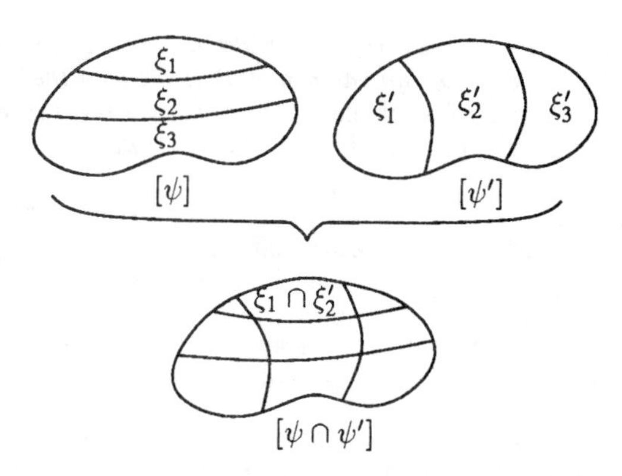

Figure 4: The set of formae induced by the equivalence relations ψ, ψ' and $\psi \cap \psi'$. The formae induced by $\psi \cap \psi'$ are intersections of those induced by ψ and ψ'.

6 Gene Transmission

It is instructive to notice that in the case of schemata for binary string representations the R^3 operator reduces to uniform crossover (Radcliffe [9]); see Syswerda [11] and Eshelman *al* [3] for details of uniform crossover) but that this is not so for k-ary representations with $k > 2$. For while uniform crossover requires each allele to be selected randomly from one of the parents, R^3, after copying all the shared alleles to the child, fills in the rest of the chromosome with genes randomly selected from the allele sets.

In constructing the *cycle* crossover operator for permutation representations and o-schemata, the principal aim of Oliver *et al* [8] was to ensure that every allele in the child was taken from one or other of the parents. More recently, Whitley *et al* [12] have used a similar criterion applied to *edges* rather than vertices of the graph to construct a highly-successful *edge recombination* operator for the TSP. It would be natural, therefore, to attempt to formulate a seventh design principle which specifies that all alleles present in the child are to be transmitted from one or other of its parents.

The difficulty with this for general representations and formae is that the notion of an allele is not necessarily well-defined (e.g. *locality formae*, introduced in section 10). We can, however, make some progress by introducing the notion of a *complete orthogonal basis* for a set of equivalence relations Ψ. In the case of schemata and o-schemata the basis E we shall seek to construct consists of all the relations with a single definition point. In the case of four-gene schemata, this gives

$$E = \{\, ■□□□,\ □■□□,\ □□■□,\ □□□■ \,\}.$$

We can then define the intersection of compatible relations in the obvious way (figure 4) so that

$$■□□\cdots□ \cap □■□\cdots□ = ■■□\cdots□.$$

This then allows us to propose the seventh design principle as follows:

7. (Strict Transmission) *Given a complete orthogonal basis $E \subset \Psi$ for a set of equivalence relations Ψ over the search space, under each equivalence relation in the basis E, every child produced by recombination must be equivalent to one of its parents.*
A recombination operator which obeys this principle will be said to be *strictly transmitting*. In the familiar case, this is precisely the requirement that every allele in the child come from one parent or the other.
[If one parent has blue eyes and the other has brown eyes, the child must have blue or brown eyes.]

Formalising the notion of a basis is quite hard, and the hurried reader can safely skip the rest of this section if uninterested in the formalism.

We begin by defining intersection for two equivalence relations. For these purposes an equivalence relation $\sim$ is best described by a binary function

$$\psi : C \times C \longrightarrow \{0, 1\}$$

which returns 1 if its arguments are equivalent and 0 if they are not:

$$\psi(\eta, \zeta) = \begin{cases} 1, & \text{if } \eta \sim \zeta, \\ 0, & \text{otherwise.} \end{cases}$$

We can then define the intersection of two equivalence relations $\psi, \psi' \in \Psi$ by

$$(\psi \cap \psi')(\eta, \zeta) = \begin{cases} 1, & \text{if } \psi(\eta, \zeta) = \psi'(\eta, \zeta) = 1, \\ 0, & \text{otherwise.} \end{cases}$$

Given this, a subset $E \subset \Psi$ will be said to be a complete orthogonal basis for Ψ provided that

- (Completeness) All relations $\psi \in \Psi$ can be constructed as as intersection of the basic relations:

$$\forall \psi \in \Psi \; \exists E_\psi \subset E : \bigcap E_\psi = \psi.$$

- (Orthogonality) Every forma F induced by every basic relation $\psi \in E$ is compatible with every forma F' induced by every other basic relation $\psi' \in E$:

$$\forall \psi, \psi' \in E \; (\psi \neq \psi') \quad \forall F \in [\psi] \; \forall F' \in [\psi'] : F \cap F' \neq \emptyset,$$

  where $[\psi]$ is the set of equivalence classes (formae) induced by ψ.

The transmission principle described above is now well-defined for a general set of formae provided that the equivalence relations inducing these formae have a complete orthogonal basis associated with them. We can define a *gene* as a basic equivalence relation and an allele as one of the equivalence classes (basic formae) induced by such a basic equivalence relation, motivating the term *gene transmission*.

7 Schemata

It is illuminating to classify the standard genetic operators used for recombining k-ary chromosomes and schemata.

(binary schemata)	respect	properly assort	strictly transmit
1-point	•		•
2-point	•		•
1-pt shuffle	•	•	•
2-pt shuffle	•	•	•
uniform/R^3	•	•	•

Table 1: Operators for binary schemata

(k-ary schemata)	respect	properly assort	strictly transmit
1-point	•		•
2-point	•		•
1-pt shuffle	•	•	•
2-pt shuffle	•	•	•
uniform	•	•	•
R^3	•	•	

Table 2: Operators for k-ary schemata

Tables 1 & 2 show which of the operators respect, properly assort, and strictly transmit schemata for binary and k-ary chromosomes respectively.

Notice that every standard crossover operator respects schemata and is strict in gene transmission, but that traditional 1- and 2–point crossover do not properly assort schemata. Plainly, just as R^3 for binary chromosomes is identical to uniform crossover, if one modifies it by enforcing strict gene transmission then uniform crossover is recovered for higher k-ary chromosomes also. These points are noteworthy principally because of the contrasting situation for (admittedly more complex) permutation representations and o-schemata described in the next section.

It should also be pointed out that while proper assortment and strict transmission have been defined as properties which recombination operators either possess (with respect to a given set of formae) or do not, in reality there are degrees of assortment and transmission. We might, for example, say that traditional crossover (with any number of cross-points) *weakly assorts* schemata on the basis that given a finite number of crosses between the parents and their various intermediate children it is possible to generate a child having any admixture of the parents' genes. Similarly, operators which do not enforce strict transmission of genes from parents to children will nevertheless pass on genes with some finite probability, giving rise to the notion of *partial transmission*.

8 O-schemata

In their paper on the travelling sales-rep problem, Goldberg & Lingle [5] introduced both the partially-mapped

o-schemata	respect	properly assort	strictly transmit
PMX	•		
cycle	•		•
order			
uniform			
R^3	•	•	

Table 3: Operators for o-schemata

crossover (PMX) operator for permutations and the original o-schemata. Subsequently Oliver *et al* [8] introduced a variation of o-schemata and Goldberg [4] introduced several others. For present purposes we shall consider only the original o-schemata, which are most similar to schemata; similar analysis is possible for the other types.

Goldberg's o-schemata are induced by equivalence relations which relate chromosomes having the same elements of the permutation in particular positions on the chromosome. This seems like a useful and appropriate relation when the absolute positions of the objects labelled by the permutation matter. In job-shop scheduling, for example, the numbers typically represent jobs and the positions on the chromosome specify on which machines and in which order the jobs should be placed. In these cases the absolute positions seem relevant and one might reasonably expect operators which reliably respect, assort and transmit o-schemata to be of great use. As Whitley *et al* [12] have argued, however, it is not apparent that absolute city position is of any great significance in the TSP; this is discussed further in the next section.

Table 3 shows the way in which o-schemata are manipulated by four standard crossover operators for permutations and the R^3 operator. The operators are Goldberg's PMX crossover [5], Oliver *et al*'s cycle crossover [8], Davis's order crossover operator, modified as described in [8], and what Davis calls *uniform (permutation) crossover,* which relates to order crossover in exactly the same way as uniform crossover relates to traditional 1- and 2-point crossover. (The elements from one parent are copied wherever the binary mask that acts as the control parameter has a 1, and the remaining elements are used to fill the gaps in the order that they occur in the other parent.)

The R^3 operator in this case acts simply by inserting the common genes straight into the child chromosome and then arranging the remaining elements of the permutation at random in the gaps.

9 Edge Formae

It seems clear, as Whitley *et al* [12] have argued, that the edges rather than the vertices of the graph are central to the TSP. While there might be some argument as to whether or not the edges should be taken to be directed, the symmetry of the euclidean metric used in the evaluation function suggests that undirected edges suffice.

If the towns (vertices) in an n-city TSP are numbered 1 to n, and the edges are described as non-ordered pairs of vertices (a, b), then apparently suitable *edge formae* are simply sets of edges, subject to the condition that no vertex appears in the description of more than two edges. Unfortunately, these formae are not separable. To see this, consider two tours η and ζ, with η containing the fragment 2–1–3 and ζ containing 4–1–3. Plainly these have the common edge $(1,3)$ $[\equiv (3,1)]$. We shall describe the formae by listing the edges they require to be present in angle-brackets, so that η is an instance of the forma $\langle(1,2)\rangle$ and ζ is an instance of the forma $\langle(1,4)\rangle$. These formae are clearly compatible, because any tour containing the fragment 2–1–4 is in their intersection[4]

$$\langle(1,2)\rangle \cap \langle(1,4)\rangle = \langle(1,2),(1,4)\rangle.$$

However, any recombination operator which respected the formae would be bound to include the common edge $(1,3)$ in all offspring from these parents, thus precluding generating a child in $\langle(1,2),(1,4)\rangle$. Since proper assortment requires that this child be capable of being generated this shows that these formae are not separable.

When Whitley *et al* introduced their powerful and attractive *edge recombination* operator they argued that 'there is no need for any new notion of "schema", with its own special schema theorem' because edge recombination manipulates an 'underlying binary representation' in the usual way. They went on to assert that 'where the parents have the same [edge], the offspring will have the same edge'. From the description in the paper, however, given tours containing the fragments shown above, it appears that while edge recombination *would* always generate a legal tour, there is no constraint which *requires* it always to transmit the common edge to the child. The operator does, however, provide a high rate of transmission of edges, this having been the major design criterion.

We can, of course, define the R^3 operator for the edge formae, even though they are not separable: it works simply by copying common edges into the child and then putting in random edges in such a way as to complete a legal tour. The lack of separability simply ensures that R^3 does not properly assort the formae.

10 Locality Formae

All of the formae discussed thus far have been fairly similar to traditional schemata. We now introduce *locality formae,* (Radcliffe [10]) which are rather different in character. Locality formae relate chromosomes on the basis of their closeness to each other. Suppose our function is

[4]Curiously, the intersection operation for these edge formae looks like the set *union* operation. This is because $\langle(1,3)\rangle$ is really an abbreviation for the set of chromosomes containing the 1–3 edge.

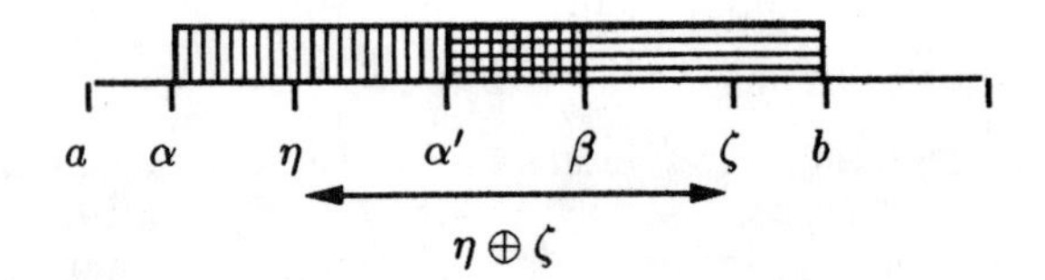

Figure 5: Given $\eta \in [\alpha, \beta)$ and $\zeta \in [\alpha', \beta')$, with $\zeta > \eta$, the formae are compatible only if $\beta > \alpha'$. The arrow shows the similarity set $\eta \oplus \zeta$.

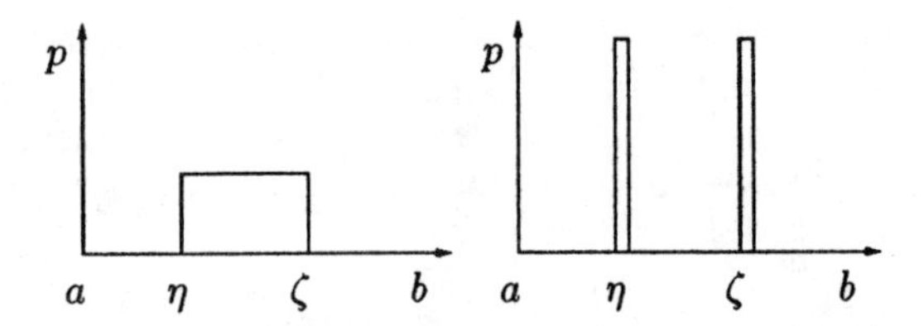

Figure 6: The left-hand graph shows (schematically) the probability of selecting each point along the axis under R^3 ("top hat"). The right-hand graph shows the corresponding diagram for standard crossover with real genes.

defined over a real interval $[a, b)$. We then define formae which divide the interval up into strips of arbitrary width. Thus, a forma is a half-open interval $[\alpha, \beta)$ with α and β both lying in the range $[0, b - a]$. These formae are separable. Respect requires that all children are instances of any formae which contain both parents η and ζ. Clearly the similarity set of η and ζ (the smallest interval which contains them both) is $[\eta, \zeta]$, where we have assumed, without loss of generality, that $\zeta \geq \eta$. Thus respect requires that all their children lie in $[\eta, \zeta]$. Similarly, if η is in some interval $\xi = [\alpha, \beta)$ and ζ lies in some other interval $\xi' = [\alpha', \beta')$, then for these formae to be compatible the intersection of the intervals that define them must be non-empty ($\beta > \alpha'$; figure 5) and so picking a random element from the similarity set $[\eta, \zeta]$ allows an element to be picked which lies in the intersection, showing that R^3 fulfils the requirements of proper assortment (figure 6).

The n-dimensional R^3 operator picks a random point in the n-dimensional hypercuboid with corners at the two chromosomes η and ζ (figure 7). This operator has been tested on De Jong's functions [2] which are not all obviously suitable for locality formae, and performed surprisingly well, apparently out-performing standard binary representations on four of the five functions. Full results are given in [9] and [10].

Both this operator and its natural analogue for k-ary string representations, which for each locus picks a random value in the range defined by the alleles from the two parents, suffer from a bias away from the ends of the interval. It is therefore necessary to introduce a mutation operator which offsets this bias in order to satisfy the ergodicity condition expressed in principle 6. An appropriate mutation operator

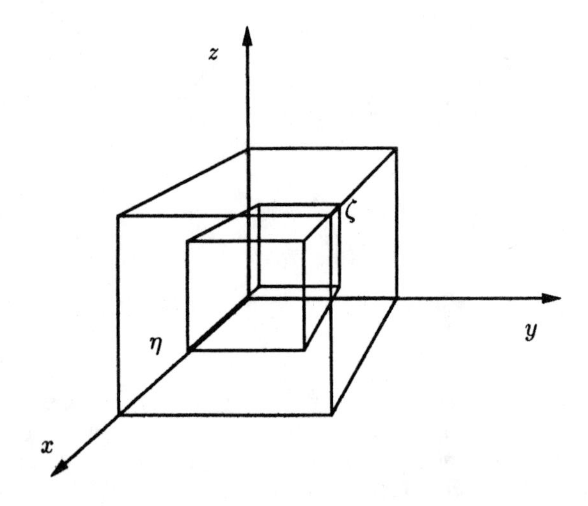

Figure 7: The n-dimensional R^3 operator for real genes picks any point in the hypercuboid with corners at the chromosomes being recombined, η and ζ.

acts with very low probability to introduce the extremal values at an arbitrary locus along the chromosome. In the one dimensional case this amounts to occasionally replacing the value of one of the chromosomes with an a or a b. The combination of R^3 and such *end-point* mutation appears to provide a surprisingly powerful set of genetic operators for some problems.

Locality formae are not, of course, the only alternatives to schemata which can be applied to real-valued problems, and there is no suggestion here that locality formae should be seen as a generic or definitive alternative to schemata. It would be interesting, for example, to attempt to construct formae and representations on the basis of fourier analysis, or some other complete orthonormal set of functions over the space being searched.

11 Future Directions

The random, respectful recombination operator discussed above has been introduced as one which automatically respects and properly assorts separable formae. In some circumstances R^3 is useful in its own right, but its principal utility seems likely to lie in providing a starting point from which to construct more sophisticated operators. The most obvious way to do this is to modify the flat probability distribution over the similarity set of the parents which R^3 uses. Such modification could either exclude some children entirely or simply bias the search towards some subset of the similarity set on the basis of other information or intuitions about the structure of the search space.

The problems which motivated the ideas in this paper are ones for which the author was unable to find traditional schemata which characterised the search spaces in useful ways—graph optimisation, neural networks and the travelling sales-rep problem among others. Little emphasis has been placed on gene transmission in this paper, even though this appears at first a very natural constraint to place on recombination operators. This lack of emphasis derives from a lack of clarity as to the meaning of the term "gene" in the context of problems such as those listed. The construction of a complete orthogonal basis for a set of formae provides a mechanism for defining genes rigorously, after which normal schema theory may be applied. There are, however, sets of formae for which no orthogonal basis exists. It is in areas such as these that the ideas in this paper are most likely be useful, through the development of equivalence relations (and their associated formae) which well characterise the regularities in the search space. This may be helpful even when these formae do not admit the construction of a complete orthogonal basis, and so do not allow a subsequent return to schema analysis.

12 Conclusion

Formae have been shown to be useful generalisations of schemata which help the exploitation of intrinsic parallelism in non string-based problems and extend the scope of the "fundamental" (schema) theorem. The random, respectful recombination operator (R^3) has been introduced as an operator which is sometimes useful in its own right and might often be a useful starting point for developing more sophisticated operators for a range of problems.

Acknowledgements

I would like to thank Mike Norman for many useful discussions about genetic algorithms, and also Andrew J. S. Wilson and Mike for careful criticism of a draft of this paper.

Some of this work was supported by the Edinburgh Parallel Computing Centre, a multidisciplinary project which receives major grants from the Department of Trade and Industry, the Computer Board and the Science and Engineering Research Council. The author acknowledges support from the University of Edinburgh and from Industrial Affiliates.

References

[1] J. David Schaffer (ed), *Proceedings of the Third International Conference on Genetic Algorithms*, Morgan Kaufmann (San Mateo) 1989.

[2] Kenneth A. De Jong, *A Genetic-Based Global Function Optimization Technique*, Technical Report 80-2, University of Pittsburgh, 1980.

[3] Larry J. Eshelman, Richard A. Caruna, & J. David Schaffer, *Biases in the Crossover Landscape*, in [1].

[4] D. E. Goldberg, *Genetic Algorithms in Search, Optimization & Machine Learning*, Addison-Wesley (Reading, Mass) 1989.

[5] D. E. Goldberg, & Robert Lingle Jr, *Alleles, Loci and the Travelling Salesman Problem*, in *Proceedings of an International Conference on Genetic Algorithms*, Lawrence Erlbaum Associates (Hillsdale) 1985.

[6] John Grefenstette, Rajeev Gopal, Brian Rosmaita, & Dirk Van Gucht, *Genetic Algorithms for the Travelling Salesman Problem*, in *Proceedings of an International Conference on Genetic Algorithms*, Lawrence Erlbaum Associates (Hillsdale) 1985.

[7] J. H. Holland, *Adaptation in Natural and Artificial Systems*, University of Michigan Press (Ann Arbor) 1975.

[8] I. M. Oliver, D. J. Smith & J. R. C. Holland, *A Study of Permutation Crossover Operators in the Traveling Salesman Problem*, in *Proceedings of the Second International Conference on Genetic Algorithms*, Lawrence Erlbaum Associates (Hillsdale, NJ) 1987.

[9] N. J. Radcliffe, *Genetic Neural Networks on MIMD Machines*, Ph.D. Thesis, Edinburgh University 1990.

[10] N. J. Radcliffe, *Equivalence Class Analysis of Genetic Algorithms*, to appear in *Complex Systems*.

[11] Gilbert Syswerda, *Uniform Crossover in Genetic Algorithms*, in [1].

[12] Darrell Whitley, Timothy Starkweather & D'Ann Fuquay, *Sheduling Problems and Traveling Salesmen: The Genetic Edge Recombination Operator* in [1].

On the Virtues of Parameterized Uniform Crossover

William M. Spears
Naval Research Laboratory
Washington, D.C. 20375 USA
spears@aic.nrl.navy.mil

Kenneth A. De Jong
George Mason University
Fairfax, VA 22030 USA
kdejong@aic.gmu.edu

Abstract

Traditionally, genetic algorithms have relied upon 1 and 2-point crossover operators. Many recent empirical studies, however, have shown the benefits of higher numbers of crossover points. Some of the most intriguing recent work has focused on uniform crossover, which involves on the average $L/2$ crossover points for strings of length L. Theoretical results suggest that, from the view of hyperplane sampling disruption, uniform crossover has few redeeming features. However, a growing body of experimental evidence suggests otherwise. In this paper, we attempt to reconcile these opposing views of uniform crossover and present a framework for understanding its virtues.

1 Introduction

One of the unique aspects of the work involving genetic algorithms (GAs) is the important role that recombination plays. In most GAs, recombination is implemented by means of a crossover operator which operates on pairs of individuals (parents) to produce new offspring by exchanging segments from the parents' genetic material. Traditionally, the number of crossover points (which determines how many segments are exchanged) has been fixed at a very low constant value of 1 or 2. Support for this decision came from early work of both a theoretical and empirical nature [Holland, 1975; DeJong, 1975]. However, there continue to be indications that there are situations in which having a higher number of crossover points is beneficial [Syswerda, 1989; Eschelman, 1989]. Perhaps the most surprising result (from a traditional perspective) is the effectiveness on some problems of uniform crossover, an operator which produces on the average $L/2$ crossings on strings of length L [Syswerda, 1989].

Recent work by [Spears and De Jong, 1990] has extended the theoretical analysis of n-point and uniform crossover with respect to disruption of sampling distributions. However, they pointed out that disruption analysis alone is not sufficient in general to predict and/or select optimal forms of crossover. In particular, they have shown that the population size must also be taken into account [DeJong and Spears, 1990]. This paper extends that work by looking at the properties of a parameterized uniform crossover operator and by considering two other aspects of crossover operators, namely, their recombination potential and their exploratory power. In this context, a surprisingly positive view of uniform crossover emerges.

2 Disruption Analysis

Holland provided the initial formal analysis of the behavior of GAs by showing how they allocate trials in a near optimal way to competing low order hyperplanes if the disruptive effects of the genetic operators used is not too severe [Holland, 1975]. Since mutation is typically run at a very low rate, it is generally ignored as a significant source of disruption. However, crossover is usually applied at a very high rate. So, considerable attention has been given to estimating P_d, the probability that a particular application of crossover will be disruptive.

Holland's initial analysis of the sampling disruption of 1-point crossover [Holland, 1975] has been extended to n-point and uniform crossover [DeJong, 1975; Spears and DeJong, 1990]. These results are in the form of estimates of the likelihood that the sampling of a kth order hyperplane (H_k) will be disrupted by a particular form of crossover. It turns out to be easier mathematically to estimate the complement of disruption: the likelihood of a sample surviving crossover (which we denote as P_s). As one might expect, the results are a function of both the order k of the hyperplane and its defining length (see the Appendix and [Spears and DeJong, 1990] for more precise details).

We provide in Figure 1 a graphical summary of a typical instance of these results for the case of 3rd order hyperplanes. The non-horizontal curves represent the survival of 3rd order hyperplanes under n-point crossover (n = 1...6). The horizontal line represents the probability of survival under uniform crossover. Figure 1 highlights two important points. First, if we interpret the area *above* a particular curve as a measure of the cumulative disruption potential of its associated crossover operator, then these curves suggest that 2-point crossover is the least disruptive of the n-point crossover family, and less disruptive than uniform crossover. Finally, unlike n-point crossover, uniform crossover disrupts all hyperplanes of order k with equal probability, regardless of how long or short their defining lengths are.

3 A Positive View of Crossover Disruption

A recurring theme in Holland's work is the importance of a proper balance between exploration and exploitation when adaptively searching an unknown space for high performance solutions [Holland, 1975]. The disruption analysis of the previous section implicitly assumes that disruption of the sampling distributions is a bad thing and to be avoided (e.g., a high disruption may stress exploration at the expense of exploitation). However, this is not always the case. There are important situations in which minimizing disruption hinders the adaptive search process by overemphasizing exploitation at the expense of needed exploration. One of the clearest examples of this is when the population size is too small to provide the necessary sampling accuracy for complex search spaces [DeJong and Spears, 1990].

To illustrate this we have selected a 30 bit problem with 6 peaks from [DeJong and Spears, 1990]. The measure of performance is simply the best individual found by the genetic algorithm. This is plotted every 100 evaluations. Since we are maximizing, higher curves represent better performance. Figures 2 and 3 illustrate the effect of population size on GA performance. Notice how uniform crossover dominates 2-point crossover on the 6-Peak problem with a small population, but just the opposite is true with a large population.

One conclusion of these results might be that we should maintain a portfolio of crossover operators and study the effects of various combinations. We have been examining another approach: achieving a better balance of exploration and exploitation using only uniform crossover. We are intrigued by this possibility for two reasons: its simplicity (only one crossover form) and its potential for increased robustness because the disruptive effect of uniform crossover is not influenced by hyperplane defining length.

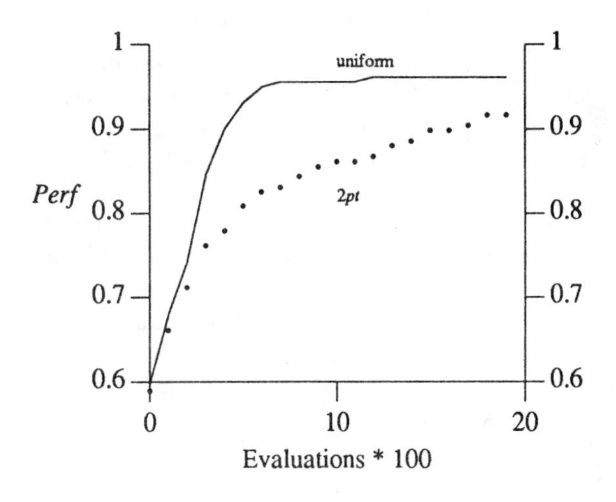

Figure 2: 6-Peak (30 bits) - Population 20

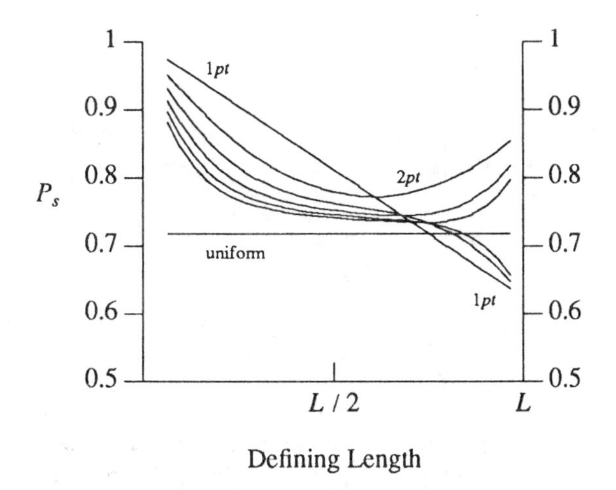

Figure 1. Survival of 3rd Order Hyperplanes

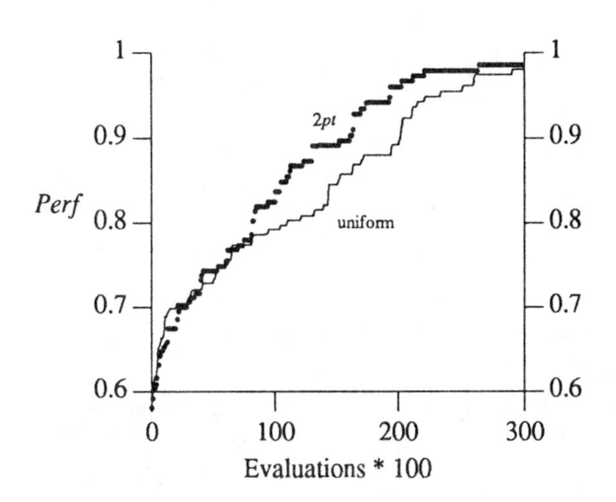

Figure 3: 6-Peak (30 bits) - Population 1000

4 A Closer Look at Uniform Crossover

It is clear that the level of disruption provided by uniform crossover is too high in many cases (e.g., when large populations are used). This standard form of uniform crossover swaps two parents' alleles with a probability of 0.5. Suppose, however, that we parameterize uniform crossover, where P_0 denotes the probability of swapping. We can now consider the effect of decreasing P_0.[1] Figure 4 illustrates this for 3rd order hyperplanes. Notice how the disruption of uniform crossover can be controlled by lowering P_0, *without* affecting the property that the disruption has no defining length bias. In particular, note that by simply lowering P_0 to .1, uniform crossover is less disruptive (overall) than 2-point crossover and has no defining length bias! This suggests a much more positive view of the potential of uniform crossover, namely, an unbiased recombination operator whose disruption potential can be easily controlled by a single parameter P_0.

To test this hypothesis, we have run a number of experiments in which P_0 varied. As expected, we can increase and decrease performance on a given problem with a fixed population size simply by varying P_0. Figure 5 illustrates this on the 6-Peak problem. Note that in this particular case, a value of $P_0 = 0.2$ produced the best results. Referring back to Figures 3 and 4, we can now see why. For the 6-Peak problem, a population size of 1000 has sufficient sampling capacity to require only the disruption level provided by 2-point crossover. Uniform crossover with $P_0 = 0.2$ provides approximately the same level of disruption but without the length bias.

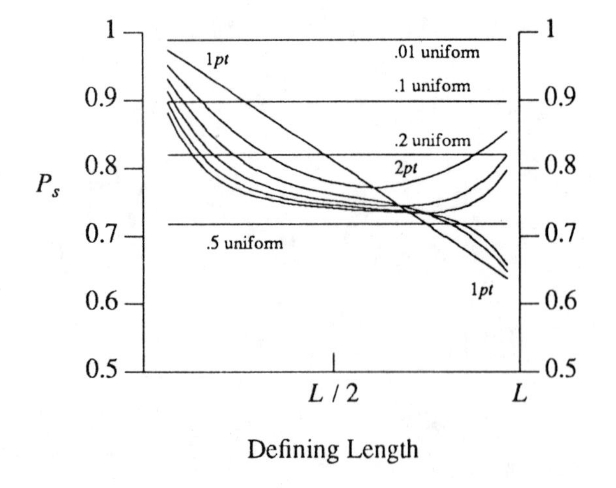

Figure 4. Survival of 3rd Order Hyperplanes

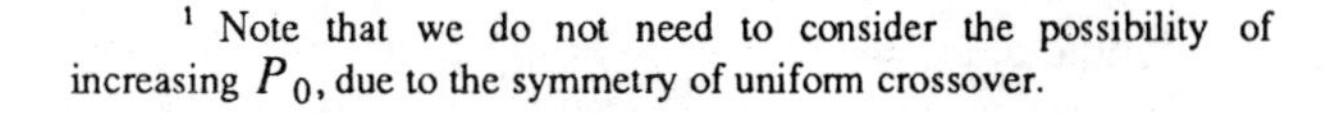

[1] Note that we do not need to consider the possibility of increasing P_0, due to the symmetry of uniform crossover.

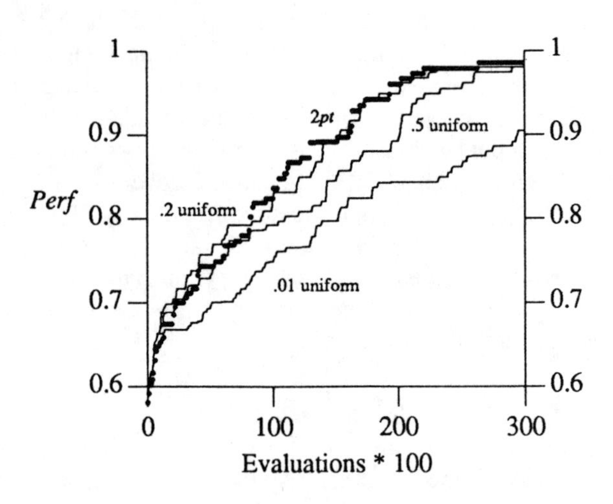

Figure 5: 6-Peak (30 bits) - Population 1000

Is this lack of length bias really important? Intuitively, it should help overcome representation problems in which important hyperplanes happen to have defining lengths which are adversely affected by the particular *n*-point crossover operator in use. Syswerda illustrated this clearly with his "sparse 1-max" problem in which 270 fake bits were appended to a 30-bit problem [Syswerda, 1989]. One can show similar results with almost any problem. Figure 6 illustrates this on our 6-Peak problem appended with 270 fake bits and the same evaluation function. Notice that, in comparison to the original 30-bit problem shown in Figure 5, the performance of 2-point crossover is worse, while the performance of uniform crossover ($P_0 = .2$) remains essentially unchanged.

How do we explain the drop in performance of 2-point crossover? In this case, the 30 important bits are all

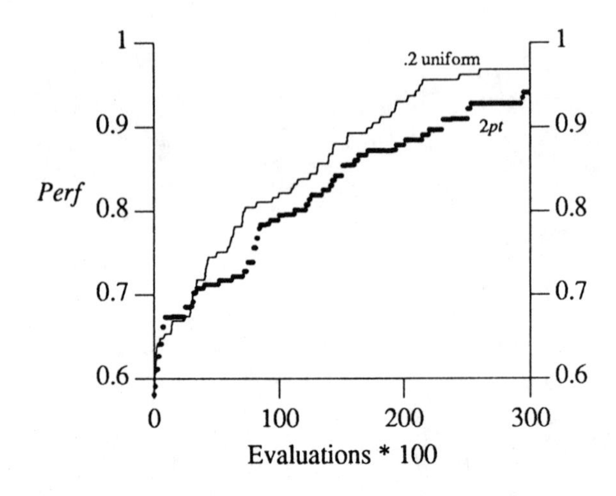

Figure 6: 6-Peak (300 bits) - Population 1000

within a distance of $L/10$ of each other (where L is the length of the string). If we examine Figure 4, we note that 2-point crossover is less disruptive within that range (0 to $L/10$) of defining lengths. In other words, the addition of 270 additional bits effectively decreases the disruption of the important hyperplanes under 2-point crossover. This effect is most obvious towards the end of the runs (see Figure 6), where disruption is increasingly useful (due to the increasing homogeneity of the population). Uniform crossover is not influenced by the added 270 bits, since it is insensitive to defining length.

In summary, we see two important virtues of uniform crossover. The first is the ease with which the disruptive effect of uniform crossover can be controlled by varying P_0. This is useful in achieving the proper balance between exploration and exploitation. The second virtue is that the disruptive potential of uniform crossover does not depend on the defining length of hyperplanes. This allows uniform crossover to perform equally well, regardless of the distribution of important alleles.

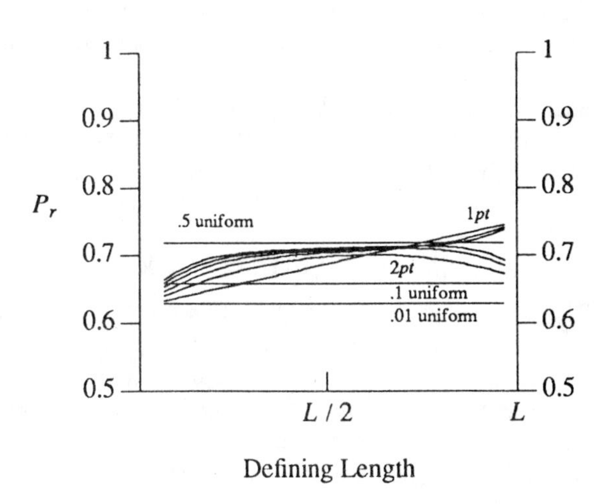

Figure 7: 3rd Order Hyperplane Recombination

5 Recombination Potential

Another possible virtue of uniform crossover that has been discussed in the literature is its recombination potential. In comparing uniform, 1 and 2-point crossover, Syswerda felt that uniform crossover gained significant advantage from its ability to combine small building blocks into larger ones [Syswerda, 1989]. He defined recombination potential as the ability of crossover to create higher order hyperplanes when the parents contain the necessary lower order hyperplanes. He provided an analysis showing uniform crossover ($P_0 = .5$) to have a higher recombination potential than 1 and 2-point crossover.

Syswerda pointed out that recombination can be considered to be a specialized form of survival, in which two lower order hyperplanes survive onto the same string, resulting in a higher order hyperplane. This observation allowed Syswerda to construct a recombination analysis from his survival analysis. However, since his survival analysis was limited to 1 and 2-point crossover, and to uniform crossover with a P_0 of .5, his recombination analysis was similarly limited. This motivated us to create a new recombination analysis in a similar vein, since our survival analysis includes all of n-point crossover and a parameterized uniform crossover.

In [Spears and DeJong, 1990], we developed a survival analysis for n-point crossover and a parameterized (P_0) uniform crossover. Details of this analysis, and our recombination analysis, are presented in the Appendix. Figure 7 illustrates the relationships of the crossover operators in terms of their recombination potential (we denote P_r as the probability of recombination). Note specifically that there is evidence to support the claim that uniform crossover ($P_0 = .5$) has a higher recombination potential than the other crossover operators. However, it is even more interesting to note that these relationships are qualitatively identical to those shown in Figure 4. In other words, if one operator is better than another for survival, it is worse for recombination (and vice versa). This observation appears to hold for all k, and suggests very strongly that the recombination analysis tells us nothing new about crossover.

6 Exploration Power

It has also been pointed out that disruption does not necessarily mean useful exploration. Crossover disruption simply implies that a hyperplane sample has been modified by crossover in some way so as to no longer be a member of that hyperplane, without any indication as to the possible forms that change might take. The potential number of ways in which a crossover operator can effect a change has been called its exploratory power. It has been pointed out that uniform crossover has the additional property that it has more exploratory power than n-point crossover [Eschelman, 1989].

To see that this is true, consider the extreme case in which one parent is a string of all 0s and the other all 1s. Clearly uniform crossover can produce offspring anywhere in the space while 1 and 2-point crossover are restricted to rather small subsets. In general, uniform crossover is much more likely to distribute its disruptive trials in an unbiased manner over larger portions of the space.

The difficulty comes in analyzing whether this exploratory power is a virtue. If we think of exploitation as the biased component of the adaptive search process, it makes sense to balance this with unbiased exploration. Clearly, this exploratory power can help in the early generations, particularly with smaller population sizes, to make sure the whole space is well sampled. At the same time, some of this exploratory power can be achieved over several generations via repeated applications of 1 and 2-point crossover. Unfortunately, our current analysis tools do not allow us to make comparisons of properties which span generations and are strongly affected by selection. Hopefully we will develop such tools and resolve questions of this type in the near future.

7 Conclusions and Further Work

The extensions to the analysis of n-point and uniform crossover presented in this paper open up an interesting and positive view of the usefulness of uniform crossover. There appear to be three potentially important virtues of uniform crossover. First, the disruption of hyperplane sampling under uniform crossover does not depend on the defining length of the hyperplanes. This reduces the possibility of representation effects, since there is no defining length bias. Second, the disruption potential is easily controlled via a single parameter P_0. This suggests the need for only one crossover form (uniform crossover), which is adapted to different situations by adjusting P_0. Finally, when a disruption does occur, uniform crossover results in a minimally biased exploration of the space being searched.

The first two virtues have been confirmed both theoretically and experimentally. At the same time, it should be emphasized that the empirical studies presented are limited to a carefully controlled experimental setting. The authors are currently working on expanding these experiments and on developing an exploration theory for recombination operators. Our goal is to understand these interactions well enough so that GAs can be designed to be self-selecting with respect to such decisions as optimal population size and level of disruption.

Acknowledgements

We would like to thank Diana Gordon for pointing out flaws in our preliminary recombination analysis.

References

De Jong, Kenneth A. (1975). *An Analysis of the Behavior of a Class of Genetic Adaptive Systems,* Doctoral Thesis, Department of Computer and Communication Sciences, University of Michigan, Ann Arbor.

De Jong, K. A. & Spears, W. (1990). An Analysis of of the Interacting Roles of Population Size and Crossover in Genetic Algorithms, *Proceedings of the First Int'l Conf. on Parallel Problem Solving from Nature*, Dortmund, Germany, October 1990.

Eschelman, L., Caruana, R. & Schaffer, D. (1989). Biases in the Crossover Landscape, *Proc. 3rd Int'l Conference on Genetic Algorithms,* Morgan Kaufman Publishing.

Holland, John H. (1975). *Adaptation in Natural and Artificial Systems,* The University of Michigan Press.

Spears, W. & De Jong, K. A. (1990). An Analysis of Multi-point Crossover, *Proceedings of the Foundations of Genetic Algorithms Workshop*, Indiana, July 1990.

Syswerda, Gilbert. (1989). Uniform Crossover in Genetic Algorithms, *Proc. 3rd Int'l Conference on Genetic Algorithms*, Morgan Kaufman Publishing.

Appendix

Summary of the Survival Analysis

For n-point crossover, P_s is expressed in the order dependent form ($P_{k,s}$):

$$P_{2,s}(n, L, L_1) =$$

$$\sum_{i=0}^{n} \binom{n}{i} \left[\frac{L_1}{L}\right]^i \left[\frac{L-L_1}{L}\right]^{n-i} C_s$$

and

$$P_{k,s}(n, L, L_1, \ldots, L_{k-1}) =$$

$$\sum_{i=0}^{n} \binom{n}{i} \left[\frac{L_1}{L}\right]^i \left[\frac{L-L_1}{L}\right]^{n-i} P_{k-1,s}(i, L_1, \ldots, L_{k-1})$$

Note that the survival of a kth order hyperplane under n-point crossover is recursively defined in terms of the survival of lower order hyperplanes. L refers to the length of the individuals. The $L_1 \cdots L_{k-1}$ refer to the defining lengths between the defining positions of the kth order hyperplane. The effect of the recursion and summation is to consider every possible placement of n crossover points within the kth order hyperplane. The correction factor C_s computes the probability that the hyperplane will survive, based on that placement of crossover points. Suppose that crossover results in x of the k defining positions being exchanged. Then the hyperplane will survive if: 1) the parents match on all x positions being exchanged, or 2) if they match on all $k-x$ positions not being exchanged, or 3) they match on all k defining positions. Hence, the general form of the correction is:

$$C_s = P_{eq}^{\,x} + P_{eq}^{\,k-x} - P_{eq}^{\,k}$$

where P_{eq} is the probability of two parents sharing an allele at each locus, and the $P_{eq}^{\,k}$ reflects an overlap within the 3 possibilities (and hence must be subtracted).

As an example, consider Figure 8. The two parents are denoted by P1 and P2. In this figure, we represent the survival of a 4th order hyperplane. The hyperplane defining positions are depicted with circles. Since 1 of the defining positions will be exchanged (under the 2-point crossover shown), the probability of survival is:

$$C_s = P_{eq}^{\ 1} + P_{eq}^{\ 3} - P_{eq}^{\ 4}$$

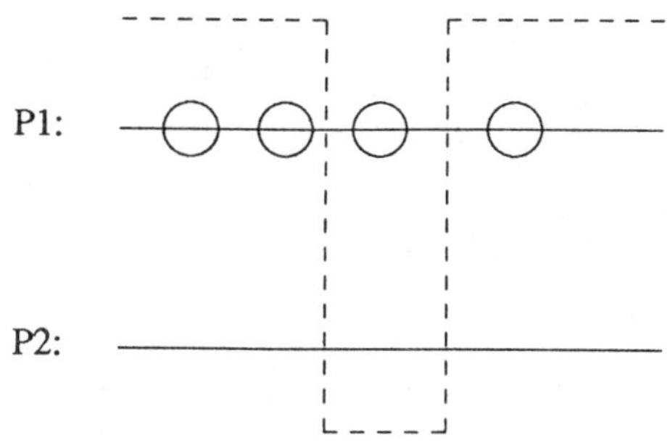

P1:

P2:

Figure 8: 4th Order Hyperplane Survival

For parameterized uniform crossover, P_s is also expressed in an order dependent form ($P_{k,s}$):

$$P_{k,s}(H_k) =$$

$$\sum_{i=0}^{k} \begin{bmatrix} k \\ i \end{bmatrix} (P_0)^i \ (1-P_0)^{k-i} \ (P_{eq}^{\ i} + P_{eq}^{\ k-i} - P_{eq}^{\ k})$$

where P_0 is the probability of swapping two parents' alleles at each locus. A graphical representation of these equations has been shown previously in Figure 4.

Recombination Analysis for *N*-Point Crossover

In our definition of survival, it is possible for a hyperplane to survive in either child. Recombination can be considered a restricted form of survival, in which two lower order hyperplanes survive to form a higher order hyperplane. The difference is that the two lower order hyperplanes (each of which exists in one parent) must survive in the same individual, in order for recombination to occur.

In the remaining discussion we will consider the creation of a *k*th order hyperplane from two hyperplanes of order *m* and *n*. We will restrict the situation such that the two lower order hyperplanes are non-overlapping, and $k = m + n$. Each lower order hyperplane is in a different parent. We denote the probability that the *k*th order hyperplane will be recombined from the two hyperplanes as $P_{k,r}$.

An analysis of recombination under *n*-point crossover is simple if one considers the correction factor C_s defined earlier for the survival analysis. Recall that recombination will occur if both lower order hyperplanes survive in the same individual. If an *n*-point crossover results in *x* of the *k* defining positions surviving in the same individual (i.e., *x* is a subset of the $m + n$ defining positions), then recombination will occur if: 1) the parents match on all of the *x* positions, or 2) if they match on all $k - x$ positions, or 3) they match on all *k* defining positions. Hence, the general form of the recombination correction C_r is:

$$C_r = P_{eq}^{\ x} + P_{eq}^{\ k-x} - P_{eq}^{\ k}$$

Note the similarity in description with the survival correction factor C_s (the only difference is in how *x* is defined). In other words, given a *k*th order hyperplane, and two hyperplanes of order *n* and *m*, $P_{k,r}$ is simply $P_{k,s}$ with the correction factor redefined as above.

As an example, consider Figure 9. In this figure, we represent the recombination of 2 2nd order hyperplanes. One hyperplane is depicted with circles, and the other with rectangles. Since 3 of the defining positions will survive onto the same individual (under the 2-point crossover shown), the probability of survival is:

$$C_r = P_{eq}^{\ 3} + P_{eq}^{\ 1} - P_{eq}^{\ 4}$$

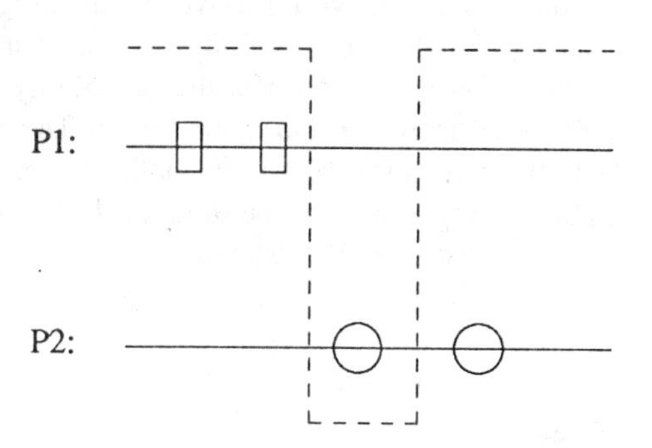

Figure 9: 2nd Order Hyperplane Recombination

Recombination Analysis for Uniform Crossover

The analysis of recombination under uniform crossover also involves the analysis of the original survival equation. Note that, due to the independence of the operator (each allele is swapped with probability P_0), the survival equation can be divided into three parts. The first part expresses the probability that a hyperplane will survive in the original string:

$$P_{k,s,orig}(H_k) = \sum_{i=0}^{k} \begin{bmatrix} k \\ i \end{bmatrix} (P_0)^i \ (1-P_0)^{k-i} \ (P_{eq}^{\ k-i})$$

The second part expresses the probability that a hyperplane will survive in the other string:

$$P_{k,s,other}(H_k) = \sum_{i=0}^{k} \begin{bmatrix} k \\ i \end{bmatrix} (P_0)^i \ (1-P_0)^{k-i} \ (P_{eq}^{\ i})$$

The final part expresses the probability that a hyperplane will exist in both strings:

$$P_{k,s,both}(H_k) = \sum_{i=0}^{k} \binom{k}{i} (P_0)^i \ (1-P_0)^{k-i} \ (P_{eq}{}^k) = P_{eq}{}^k$$

Then:

$$P_{k,s}(H_k) = P_{k,s,orig}(H_k) + P_{k,s,other}(H_k) - P_{k,s,both}(H_k)$$

Note, however, that this formulation allows us to express recombination under uniform crossover. Again, assuming the recombination of two non-overlapping hyperplanes of order n and m into a hyperplane of order k:

$$P_{k,r}(H_k) = P_{m,s,orig}(H_m)\, P_{n,s,other}(H_n) +$$

$$P_{m,s,other}(H_m)\, P_{n,s,orig}(H_n) -$$

$$P_{m,s,both}(H_m)\, P_{n,s,both}(H_n)$$

This equation reflects the decomposition of recombination into two independent survival events. The first term is the probability that H_m will survive on the original string, while H_n switches (i.e., both hyperplanes survive on one parent). The second term is the probability that both hyperplanes survive on the other parent. The third term reflects the joint probability that both hyperplanes survive on both strings, and must be subtracted. Finally, it is interesting to note that the last term is equivalent to $P_{eq}{}^m P_{eq}{}^n = P_{eq}{}^k$.

Schema Disruption

Michael D. Vose
The University of Tennessee
C. S. Dept., 107 Ayres Hall
Knoxville, TN 37996
(615) 974-5067 vose@cs.utk.edu

Gunar E. Liepins
Oak Ridge National Laboratory
MS 6360 Bldg. 6025, PO Box 2008
Oak Ridge, TN 37831-6360
(615) 574-6640 gxl@msr.epm.ornl.gov

Abstract

We use results of Battle, Liepins, and Vose to motivate a reexamination of schemata disruption. We formalize a "building block hypothesis" in which the relation of building blocks to utilities is not prominent. What does matter is how the crossover operator interacts with schemata. We argue that this interaction determines what the building blocks are, and once determined, these building blocks are the appropriate objects on which to conduct schemata analysis.

1 INTRODUCTION

Genetic algorithms (GAs) sometimes perform poorly as function optimizers. We will call any such function which is not well optimized a GA difficult problem.

Various reasons have been identified and blamed for the failure of genetic search on GA difficult problems (see for example [4]), among which are:

1. Deceptiveness
2. Sampling Error
3. Schema disruption

The essence of deceptiveness stems from the observation that in some sense schemata represent the direction of genetic search. It follows from the Schema Theorem that the number of instances of a schema is expected to increase in the next generation if it is of above average utility and is not disrupted by crossover. Therefore, such schemata indicate the area within the search space which the GA explores. Hence it is important that, at some stage, these schemata contain the object of search; problems for which this is not true are called deceptive.

Sampling error occurs when a schema of above average utility has less than average fitness when computed with respect to the current population. In this case, reproductive pressure may force an important building block to die out.

Schema disruption is usually regarded as an orthogonal mode of failure. Even though a problem is not deceptive, if crossover is incompatible with the progression from low order to high order schemata then important schemata could be prevented from guiding genetic search.

The common thread above is *schemata*. But what are schemata? Increasingly, applications of GAs to combinatorial problems involve modifications to simple crossover. When alternate mixing operations are used, what are the building blocks? Moreover, if Holland's schemata are not always appropriate then how should "schemata analysis" proceed?

We touch on all these questions, although indirectly, by considering the relationship between a GA difficult problem and our archetypical easy problem which we refer to as "counting 1's":

$$x \longmapsto \text{the number of 1's in the binary string } x$$

2 REPRESENTATIONS

It is a common practice to choose among alternate representations. For example, consider the objective function

$$f(x) \;=\; 4 + \frac{11}{6}x - 4x^2 + \frac{7}{6}x^3$$

It may readily be checked that f is a type − II minimal deceptive problem [2] over the integer interval $[0, 3]$ when integers are given their natural binary encoding:

string	fitness
00	4
01	3
10	1
11	5

However, if a Gray code is used then the problem gets easier; it becomes type – I:

value	string	fitness
0	00	4
1	01	3
2	11	1
3	10	5

The point is that values in the domain of f may be represented by binary strings in several ways, and some are better than others (we do not suggest that Gray coding is on average better than natural coding; it is an instructive exercise to prove that all ***fixed*** coding schemes do equally well when all functions are averaged over).

It was noticed by Liepins and Vose [4] that an entire class of deceptive problems would be totally easy for genetic optimization if an appropriate representation were used.

Of course, this by itself is trivial; the nontrivial result of Liepins and Vose is that a suitable representation is induced by a matrix. Assuming that f is injective over the finite optimization domain[1], simply represent values in the domain with nondecreasing fitness by binary strings having a nondecreasing number of 1's:

value	string	fitness
0	10	4
1	01	3
2	00	1
3	11	5

In this representation, optimization of f corresponds to that of counting 1's.

Suppose that genetic search is implemented with selection based on ranking so that all monotonic functions are treated the same. Then, from the perspective of genetic optimization,

Every function has a representation in which it is essentially the counting 1's problem.

3 SCHEMATA

We will refer to what is usually called a schema as a Holland schema. Since every Holland schema describes a subset of strings, one might therefore consider an *arbitrary* set of strings to be a schema.

This perspective was considered by Vose [5] who showed that these generalized schemata, which we call *predicates*, satisfy a Schema Theorem. Moreover, he classified predicates according to their susceptibility to sampling error by introducing the concept of globality. We denote the set of all predicates by $\mathcal{S}$.

A predicate is global if given that it is of above average fitness in some population, it will then never have below average fitness in any population.

A partial explanation of why counting 1's is easy for a GA is that even though the global predicates are not Holland schemata, the union of Holland schemata with fixed order containing $\vec{1}$ is global (see appendix). Therefore, these unions are immune to sampling error.

Moreover, in every pair of competing Holland schemata, if one of them contains the optimum then it has the greater utility. Therefore, counting 1's is antipodal to deceptive.

To summarize:

- *Predicates are arbitrary sets of strings, and they satisfy a Schema Theorem.*
- *From the perspectives of sampling error and deceptiveness, Holland schemata explain why counting 1's is easy.*

4 DUALITY

A duality between representations and schemata was discovered by Battle and Vose [1]. Their premise was that the role of Holland schemata in directing genetic search should be granted both as a matter of empirical fact and as a natural consequence of the (classical) Schema Theorem. Their conclusion was that schemata other than Holland's (i.e., predicates) can also be made to direct genetic search, and that a duality exists between problem representations and which predicates are relevant for their optimization.

Let f be an arbitrary objective function, and let M be an invertible mapping of binary strings. Battle and Vose showed that the genetic optimization of $f \circ M$ when guided by Holland schemata is equivalent to the genetic optimization of f if directed by the images under M of Holland schemata.

To illustrate this principle, let f be the deceptive function from the representation section, and let R map values in the domain of f with nondecreasing fitness to binary strings having a nondecreasing number of 1's:

x	R(x)	f(x)
0	10	4
1	01	3
2	00	1
3	11	5

In other words, R is such that genetic optimization of $f \circ R^{-1}$ is essentially that of counting 1's. By the duality principle, genetic optimization of counting 1's

[1] We will continue to make this special assumption throughout this paper.

when guided by Holland schemata (success) is equivalent to genetic optimization of f if directed by the images under R^{-1} of Holland schemata (hence also successful).

This example raises an interesting question. On the one hand:

- *If f is any GA difficult problem, then genetic optimization is likely to fail.*
- *A representation R exists such that $f \circ R^{-1}$ is essentially counting 1's.*
- *By duality, genetic search succeeds if guided by the images $\mathcal{I}$ under R^{-1} of Holland schemata.*
- *Therefore, $\mathcal{I}$ does not direct the optimization of f.*

On the other hand, the predicates $\mathcal{I}$ satisfy a Schema Theorem. So why shouldn't an arbitrary problem f be as easy to optimize as counting 1's? After all, what was said in the previous section concerning the suitability of Holland schemata for optimizing $f \circ R^{-1}$ (i.e., counting 1's) transfers directly (see [1] for details) to the suitability of $\mathcal{I}$ for optimizing f !

What can be keeping the predicates $\mathcal{I}$ from directing genetic search?

5 CROSSOVER

Let χ denote simple crossover, and let M be an invertible mapping of binary strings. Define the mixing operator $\mathcal{M}$ by

$$\mathcal{M} = M \circ \chi \circ M^{-1}$$

The duality principle of Battle and Vose [1] was proved by establishing an isomorphism between the genetic optimization of f using χ and that of $f \circ M^{-1}$ where $\mathcal{M}$ is used for crossover instead of χ (actually, the mutation operator was also changed, but the effects of mutation need have no significant influence).

Consequently, if we choose M to be the representation R of the previous section, then we see that genetic optimization of the counting 1's problem when $\mathcal{M}$ is used for crossover is equivalent to ordinary genetic optimization of f. Conversely, genetic optimization of f with an appropriate mixing operator is equivalent to ordinary genetic optimization of counting 1's. Therefore, the mixing operation chosen for crossover has a staggering effect on genetic search:

The difference between counting 1's and a GA difficult problem lies in the choice of crossover.

Therefore, one cannot explain the failure of $\mathcal{I}$ in directing genetic search by claiming Holland schemata are categorically the only relevant predicates. With appropriate crossover operator, f is equivalent to perhaps the easiest – from the perspective of Holland schemata – problem there is, counting 1's !

The answer clearly lies with the choice of crossover operator, and the natural question is:

What is the relationship between crossover and the appropriate building blocks?

6 THE LATTICE OF PREDICATES

A binary crossover operator $\mathcal{M}$ may produce children which depend on crossover position(s). So as to simplify both thinking and exposition, it is convenient to regard $\mathcal{M}$ as nondeterministicly producing one child. Both "demonic" and "angelic" nondeterminism will be used; the former means that statements are implicitly quantified over all possibilities (the devil may do his worst), the latter means that statements should be existentially quantified. We will make the choice of nondeterminism clear by using $\overline{\mathcal{M}}$ in demonic situations.

A lattice is a partially ordered set in which each pair of elements x and y has a greatest lower bound $x \wedge y$ and a least upper bound $x \vee y$. Since the collection $\mathcal{S}$ of predicates is simply a collection of sets of binary strings, it forms a lattice when partially ordered by set inclusion. In fact, greatest lower bound is intersection and least upper bound is union within the lattice $\mathcal{S}$.

We define a predicate $S \in \mathcal{S}$ as invariant under a binary operator $\mathcal{M}$ if $\overline{\mathcal{M}} : S \times S \to S$. A sublattice $\mathcal{H}$ of $\mathcal{S}$ is defined as *invariant* under $\mathcal{M}$ if every predicate of $\mathcal{H}$ is invariant under $\mathcal{M}$.

An invariant sublattice $\mathcal{H}$ is closed under $\mathcal{M}$ in the sense that if $s_1 \in S_1 \in \mathcal{H}$ and $s_2 \in S_2 \in \mathcal{H}$, then $s_1 \overline{\mathcal{M}} s_2$ belongs to some predicate of $\mathcal{H}$. This follows from the diagram

$$\begin{array}{ccc} (s_1, s_2) \in S_1 \times S_2 & \subset & (S_1 \vee S_2) \times (S_1 \vee S_2) \\ & & \Big\downarrow \overline{\mathcal{M}} \\ & & S_1 \vee S_2 \in \mathcal{H} \end{array}$$

Hence $\mathcal{M}$ may be regarded as a search operator on $\mathcal{H}$. Thinking of the object of search as some small predicate, we see that if S_1 and S_2 are close to the goal and $\mathcal{H}$ is an invariant lattice, then $S_1 \overline{\mathcal{M}} S_2$ cannot move too far away; it must be within their least upper bound.[2]

[2] The notation $S_1 \overline{\mathcal{M}} S_2$ represents the image of $S_1 \times S_2$ under the binary operator $\overline{\mathcal{M}}$. From this section on, $\mathcal{M}$ is intended to be general; it is not necessarily as restricted in the previous section on crossover.

For this view of $\mathcal{M}$ to be reasonable, it must be compatible with the progression from large to small predicates. This condition might be phrased using a condition like: for all $S_1 \in \mathcal{H}$ there exists S_2 such that

$$S_1 \mathcal{M} S_2 \subset S_1 \wedge S_2 \qquad (*)$$

A solution S_2 to $(*)$ is called trivial when $S_1 \wedge S_2 = S_1$ or $S_1 \wedge S_2 = S_2$. In an invariant lattice, the trivial solution $S_2 = S_1$ always exists. Because we are interested in genuine progress toward smaller predicates, trivial solutions are not interesting. By a solution of $(*)$ we therefore mean a nontrivial solution.

We define an invariant sublattice $\mathcal{H}$ as *complete* if for all $S_1 \in \mathcal{H}$

$$\text{A solution } S_2 \in \mathcal{S} \text{ of } (*) \text{ exists} \implies \text{A solution } S_2 \in \mathcal{H} \text{ of } (*) \text{ exists}$$

A complete sublattice $\mathcal{H}$ is *closed* if it is maximal, i.e., if

$$\mathcal{H} \subset \mathcal{H}' \text{ and } \mathcal{H}' \text{ is complete} \implies \mathcal{H} = \mathcal{H}'$$

To summarize,

- *If $\mathcal{H}$ is invariant, then $\mathcal{M}$ may be regarded as a search operator which traverses it.*
- *This search is constrained in the sense that $S_1 \overline{\mathcal{M}} S_2$ cannot move too far away from S_1 or S_2; it must be within their least upper bound.*
- *Moreover, $\mathcal{M}$ is compatible with the progression from large to small predicates of $\mathcal{H}$ when $\mathcal{H}$ is complete.*

7 SCHEMA DISRUPTION

It is revealing to consider the problem of counting 1's with normal crossover from the perspective of schema disruption.

Since those predicates immune to sampling error (the global predicates) are exact unions of Holland schema, and since the Holland schemata form an invariant sublattice (see appendix), it follows that the global predicates are also invariant. Therefore, the predicates relevant for counting 1's are disrupted less than others; global predicates are always important [5], and Holland schemata are relevant because they form a complete sublattice with respect to 1-point crossover. In fact, if crossover is 2-point or uniform, then the lattice of Holland schemata is closed (see appendix).

When some alternate operator $\mathcal{M}$ is used, Holland schemata may not be relevant. We argue that if corresponding to $\mathcal{M}$ is a complete or closed sublattice $\mathcal{H}$, then $\mathcal{H}$ represents predicates which direct genetic search. In this case we call the predicates of $\mathcal{H}$ *schemata* (hence schemata are Holland schemata when $\mathcal{M}$ is ordinary crossover).

The Schema Theorem for arbitrary predicates [5] is essentially identical to the classical one. Let P be a finite population drawn from some universe Ω,[3] and let $f : \Omega \to \Re$ be a fitness function. For any predicate H, define its representation in P as

$$|H|_P = \sum_{p \in P \cap H} 1$$

and define the utility of H with respect to P as

$$f_P(H) = \frac{1}{|H|_P} \sum_{p \in P \cap H} f(p)$$

Regarding P as changing under the influence of a genetic algorithm, let P_t denote the generation under consideration. The Schema Theorem is the statement that

$$\mathcal{E}\,|H|_{P_{t+1}} \geq |H|_{P_t} \frac{f_{P_t}(H)}{f_{P_t}(\Omega)} [1 - \alpha(H, P_t) - \beta(H, P_t)]$$

where $\mathcal{E}$ is an expectation operator, and the functions α and β approximate the probabilities that H will be disrupted by crossover and mutation (respectively). This theorem is quite general and does not require any specific assumptions about which crossover or mutation operators are used.

According to the Schema Theorem, the extent to which predicates represent the direction of genetic search is related to size of the functions α and β. Since mutation rates are usually small, so is β.

As far as α is concerned, a schema $S \in \mathcal{H}$ has an advantage over an arbitrary predicate of Ω. When $\mathcal{H}$ is invariant, crossover maps $S \times S$ into S. For this reason,

- *schemata are less likely to be disrupted by crossover, and their corresponding α are typically smaller.*

Moreover, when $\mathcal{H}$ is complete,

- *the disruption of two schemata S_1 and S_2 can be thought of as moving towards a larger schema since the resulting children are both instances of the least upper bound $S_1 \vee S_2$. However, this disruption is less likely to be irreparable. Crossover is compatible with the progression from large to small schemata; genetic search may return from $S_1 \vee S_2$ to a smaller schema (perhaps even to $S_1 \wedge S_2$).*

[3] The population P is intended to be a multiset.

We propose that schemata (i.e., predicates from a complete or closed sublattice) corresponding to an operator $\mathcal{M}$ form the objects appropriate to schemata analysis. Furthermore, we call schemata S_2 with above average fitness which are nontrivial solutions to

$$S_1 \mathcal{M} S_2 \subset S_1 \wedge S_2$$

building blocks, and argue for their relevance to genetic optimization. Consider the question raised at the close of the section on duality:

What can be keeping the predicates $\mathcal{I}$ from directing genetic search?

First note that for any Holland schema S,

$$\sum_{x \in R^{-1}S} f(x) = \sum_{x \in S} f \circ R^{-1}(x)$$

It follows that the utility of $R^{-1}S$ with respect to f is essentially the utility of S with respect to counting 1's. Hence *utilities* cannot be blamed for the failure of $\mathcal{I}$ to direct genetic search of f.

But what about disruption? If $\mathcal{I}$ is not invariant under crossover, then it is disadvantaged in the competition to direct genetic search. The sublattice of invariant schemata (i.e., Holland schemata) are less ravaged by crossover disruption, and are therefore favored. Moreover, if $\mathcal{I}$ is not a complete sublattice, then the progression from large to small predicates via crossover is hindered by the absence of appropriate building blocks.

Emphasizing this point in a different way explains the conclusion reached at the close of the section on crossover. The mixing operation $\mathcal{M}$ chosen for crossover has a profound effect on a GA. Since it is through $\mathcal{M}$ that search proceeds, it determines which predicates are invariant and how they combine, thereby choosing which schemata (i.e., which sublattice) will guide genetic search.

Given any function f, a sublattice with correctly aligned schemata utilities exists for its optimization (i.e., the images under R^{-1} of Holland schemata). Given the Schema Theorem, it is therefore quite natural that chosing this sublattice to guide genetic search (by chosing a crossover operator which makes it complete) makes f just as easy as counting 1's. On the other hand, if this sublattice is not complete with respect to the chosen crossover operator, then how its utilities align is irrelevant; the building blocks of the problem will lie elsewhere.

8 CONCLUSION

We have used results of Battle, Liepins, and Vose to motivate a reexamination of schemata disruption. Increasingly, applications of GAs to combinatorial problems involve modifications to simple crossover, and, when alternate mixing operations are used, what the building blocks of the problem are may not be clear.

We address this situation by formalizing a "building block hypothesis" in which the relation of building blocks to utilities is not prominent. What is critically important is how the crossover operator interacts with schemata. We argue that this interaction determines what the building blocks are, and once determined, these building blocks are the appropriate objects on which to conduct schemata analysis.

We have introduced the lattice of schemata to better understand how the mixing operator (crossover) conducts exploration. Through the concepts of invariance, completeness, and closure, a coherent perspective has emerged from which the relationship between crossover operators and building blocks may be better understood.

The role of schema disruption is central in our view. There is always a collection of schemata with correctly aligned utilities, the important question is: are they schemata which are invariant, and do they contain building blocks? About the schemata of a complete or closed sublattice, one can answer "yes"; such schemata are those which direct genetic search.

9 APPENDIX

Global schemata were introduced in the context of injective functions [5], and therefore are not directly appropriate to counting 1's. However, ties in fitness may be broken to make counting 1's injective (with tie breaking increments some multiple of t), and we can take the limit as the tie breaking scheme disappears (i.e., as $t \to 0$).

Using this method together with Theorem 1 of [5] leads to characterizing global schemata as sets of strings containing at least j 1's. Note that the union of Holland schemata containing $\vec{1}$ of order j is exactly the set of strings containing at least j 1's. Hence the global predicates are exact unions of Holland schemata.

For technical reasons, we regard $\emptyset$ (the empty set) as a Holland schemata. This way, Holland schemata form a lattice partially ordered by set inclusion (otherwise, the greatest lower bound of differing strings would not exist).

Lemma : The Holland schemata are an invariant sublattice.

Proof: Strings which agree at specified string positions have children which at those same positions agree with their parents. □

Theorem 1: The Holland schemata are a complete sublattice.

Proof: Holland schemata are invariant under crossover by the previous lemma. Let S_1 be a Holland schema. Note that if every position is fixed or no position is fixed, then a nontrivial solution to $S_1 \chi S_2 \subset S_1 \wedge S_2$ can not exist. Therefore, let i and j be such that $\{i, j\}$ contains both a fixed and nonfixed position. Let S_2 be obtained from S_1 by changing the fixed position of $\{i, j\}$ to $*$, and the nonfixed position of $\{i, j\}$ to a specific value. Since the crossover position may fall between positions i and j, S_2 is a nontrivial solution. □

Theorem 2: The Holland schemata are a closed sublattice under 2-point crossover.

Proof: Let $\mathcal{H}$ denote the Holland schemata, and suppose $\mathcal{H} \subset \mathcal{H}'$ for some invariant sublattice $\mathcal{H}'$. Let $S \in \mathcal{H}'$ and suppose i is a nonfixed position of S. If $s_1, s_2 \in S$ are strings which differ in the i th position and χ is 2-point crossover, then a possibility is that

$$s_1 \chi s_2 = \begin{cases} s_1 \text{ at all positions except the } i \text{ th} \\ s_2 \text{ at the } i \text{ th position} \end{cases}$$

Since S is invariant under χ, it follows that if a string position in S is not fixed, it is arbitrary. Therefore S is in fact a Holland schemata, and hence $\mathcal{H} = \mathcal{H}'$. □

Note that the requisite point in the proof of Theorem 2 is that 2-point crossover has the ability to modify a single position in a string. Hence this theorem is also valid for other crossover operators (like uniform crossover).

Acknowledgements

This research was supported by the National Science Foundation (IRI-8917545), the Air Force Office of Scientific Research, and the Office of Naval Research (F 49620-90-C-0033).

References

[1] D. Battle & M. D. Vose, Isomorphisms Of Genetic Algorithms. Workshop on the Foundations of Genetic Algorithms & Classifier Systems. Bloomington Indiana, July 1990.

[2] D. E. Goldberg, *Genetic Algorithms in Search, Optimization, and Machine Learning*, Addison-Wesley, 1989.

[3] J. H. Holland, *Adaptation in natural and artificial systems*, Ann Arbor, The University of Michigan Press, 1975.

[4] G. E. Liepins & M. D. Vose, Representational Issues in Genetic Optimization. *Journal of Experimental and Theoretical Artificial Intelligence*, 1990, 2(2), pp. 4-30.

[5] M. D. Vose, Generalizing The Notion Of Schema In Genetic Algorithms. *Artificial Intelligence*, in press.

Parallel Genetic Algorithms

A Multi-population Genetic Algorithm for Solving the K-Partition Problem on Hyper-cubes

J.P. Cohoon W.N. Martin D.S. Richards

Department of Computer Science
Thornton Hall
University of Virginia
Charlottesville, VA 22903-2442

Abstract

We have derived a multi-population formulation of genetic algorithms [1] that provides a means of investigating the effects of environment diversity on the evolution process. The empirical study reported here involves a VLSI application, namely, the K-partition problem. The results are derived from a system implemented on an INTEL i860 at Oak Ridge National Laboratory.

1 Introduction

Early in this century, as the fields of biological evolution and genetics were just beginning to be merged, Sewall Wright [12] developed the important conceptualization of the *adaptive landscape*. The original conceptualization proposes an underlying space (two-dimensional for discussion purposes) of possible genetic combinations and at each point in that space an "adaptive value" is determined and specified as a scalar quantity. The surface thus specified is referred to as the "adaptive landscape." A population of organisms can be mapped to the landscape by taking each member of the population, determining the point in the underlying space that its genetic codes specifies and marking the associated surface point. The figure used repeatedly by Wright shows the adaptive landscape as a standard topographic map with contour lines of equal adaptive value instead of altitude. A population is then depicted by a shaded region overlaid on the map.

There are several reasons that we used the word "conceptualization" in the previous paragraph. First and foremost is that it is not clear what the topology of the underlying space should be. Wright [12] considers initially the discrete gene sequences and connects genetic codes that are "one remove" from each other, implying that the space is actually an undirected graph. He then turns immediately to a continuous space with each gene locus specifying a dimension and with units along each dimension being the possible allelomorphs at the given locus. Specifying the underlying space to be an n-dimensional Euclidean space determines the topology. However, if one is to attempt to make inferences from the character of the adaptive landscape the ordering of the units along the various dimensions is crucial. The metric notions of "near by" and "distance" have no clearcut meaning; similar ambiguities occur in many discrete optimization problems. For instance, given two tours in a travelling salesperson problem, what is the proper measure of their closeness? This is also a divergent point from functional optimization problems in which the underlying space is in fact Euclidean n-space[8].

In any case, the concept of the adaptive landscape has had a powerful effect on both micro- and macroevolutionary theory. As Wright states, "The problem of evolution as I see it is that of a mechanism by which the species may continually find its way from lower to higher peaks in such a field. In order that this may occur, there must be some trial and error mechanism on a grand scale ... [12]."

Wright also used the adaptive landscape concept to explain his mechanism, "the shifting balance theory." In the shifting balance theory the ability for a species to "search" and not be forced to remain at lower adaptive peaks by strong selection pressure is provided through a population structure that allows the species to take advantage of ecological opportunities. The population structure is based upon *demes*, as Wright describes, "Most species contain numerous small, random breeding local populations (demes) that are sufficiently isolated (if only by distance) to permit differentiation ... [13]."

Wright conceives the shifting balance to be a microevolutionary mechanism, that is, a mechanism for evolution within a species. For him the emergence of a new species is a corollary to the general operation and progress of the shifting balance. Eldredge and Gould [3] (and more recently Eldredge [4]) have proposed that macroevolutionary mechanisms are important and see the emergence of a new species to be associated very often with extremely rapid evolutionary development of diverse organisms. As Eldredge states [4, page 119]:

> Other authors have gone further, suggesting that SMRS[1] disruption actually may *induce* [his emphasis] economic adaptive change, i.e., rather than merely occur in concert with it, ... [Eldredge and Gould] have argued that small populations near the periphery of the range of an ancestral population may be ideally suited to rapid adaptive change following the onset of reproductive isolation. ... Thus SMRS disruption under such conditions may readily be imagined to act as a 'release,' or 'trigger' to further adaptive change the better to fit the particular ecological conditions at the periphery of the parental species's range.

Our GA formulation has been strongly influenced by this theory of punctuated equilibria [3], so we have dubbed the developed system **GAPE** [1]. The important aspect of the Eldredge and Gould theory is that one should look to small disjoint populations, i.e., peripheral isolates, for extremely rapid evolutionary change. For our analogy to discrete optimization problems, such as the K-partition problem, we regard the rapid evolutionary change to be indicative of extensive search of the adaptive landscape. Thus, as we will describe in the following section, our formulation of the genetic algorithm is based on a population structure that involves subpopulations which have their isolated evolution occasionally punctuated by inter-population communication.

The application problem will be presented in Section 3 and the empirical study discussed in Section 4. The intent of this study, as well as our continuing research, is to investigate the effects on the evolution process of having a diversity of environmental characteristics across the populations.

2 The GAPE Formulation

Our basic model of a *genetic algorithm with punctuated equilibria* [1] assigns a set of n individuals to each of N processors, for a total population of size $n \times N$. Each individual is a possible solution to the discrete optimization problem being analyzed. The set of individuals assigned to each processor is its **subpopulation**. The processors are connected by an interconnection network. In practice, we might expect a conventional topology to be used, e.g., a mesh. For this study we have used the "native" hyper-cube topology of the INTEL i860, though the implemented system allows the user to specify an arbitrary adjacency graph for the topology. The network, however, should have sufficient connectivity and small diameter to ensure adequate mixing as time progresses.

The overall structure of the process comprises E major iterations called **epochs**. During an epoch each processor disjointly executes the genetic algorithm on its subpopulation. Theoretically each processor continues until it reaches equilibrium. Since, as yet, we know of no adequate stopping criteria we have used a fixed number, G, of generations per epoch. After each processor has stopped there is a phase during which each processor copies randomly selected subsets (of size S) of its population to neighboring processors. Each processor now has acquired a surplus of individuals and must probabilistically (with respect to fitness value) select a set of n individuals to survive to be its initial subpopulation at the beginning of the next epoch. In this way each subpopulation undergoes G generations of isolated evolution then intermixes its individuals with those at neighboring subpopulations, and repeats the cycle for E epochs. The high-level code for the overall **GAPE** system is given in Figure 1.

Each processor executes a GA for G generations in isolation. In each generation offspring are created at a rate, $0 \leq C \leq 1$, relative to the subpopulation size. The *parents* are chosen probabilistically (by fitness) with replacement. The crossover itself, and other problem specific details, are discussed in Section 3. The fitnesses are recalculated, relative to the new larger population. Then, probabilistically (by fitness) without replacement, the next population is selected. Finally, in a uniformly random manner elements are mutated at a rate, $0 \leq M \leq 1$, relative to the subpopulation size. The high-level code for the GA used by each processor is given in Figure 2.

The configuration of **GAPE** used in this study is as follows:

- number of subpopulations $\rightarrow$ 16;
- topology $\rightarrow$ hyper-cube of degree 4;
- number of individuals per subpopulation $\rightarrow$ 80;
- number of epochs $\rightarrow$ 30;
- generations per epoch $\rightarrow$ 50;
- $C \rightarrow 0.5$;
- $M \rightarrow 0.1$; and
- number of individuals communicated $\rightarrow$ 9.

[1] SMRS denotes "specific mate recognition system," the disruption of which is presumed to cause reproductive isolation.

3 The Partition Problem

We typically take our applications from the realm of VLSI problems, and here we will discuss the *K-partition problem*. In Figure 3 we display the components of an example problem. The seven small rectangles are the "units," shown with their relative areas. The three large rectangles are the "modules" into which the units are to be placed, shown with their relative capacities. The five "nets" that must connect the units are indicated by the line (dotted and dashed lines are used for visual clarity.)

In Figure 4 we show two possible partitions. On the left, the three modules are shown with the units assigned placed inside. Since the capacity of each module is larger that the summed area of the assigned units, this partition is called "feasible." If a net must connect units that are in t distinct modules it forms $(t-1)$ "cuts."[2] The left partition has four nets that exit modules and each of those nets connect units assigned to 2 distinct modules, so we say that the partition has a *totalcut* count of 4. For each module we also count the number of nets that must exit the module and call the maximum of those counts, *maxcut*. For the given partition, the module on the right has the maximum count of exiting nets, so *maxcut* is 4.

The importance of the two measures, *totalcut* and *maxcut*, derives from the VLSI context for the problem. *Totalcut* relates to the total wire length needed to realize the circuits specified by the nets under the unit-to-module assignment of the partition. *Maxcut* relates to the number of input/output "pads" required per module (in a worst case sense). *Maxcut* also relates to the degree of wire congestion that can be expected in local areas, i.e., immediately around each module. Note that in VLSI design the overall process is separated into several phases, and here the partition problem does not include geometric placement. Without that placement it is not possible to have direct measures of the necessary wire length and congestion implied by a given partition. Symmetrically, our work on placement [1] assumed the partition had been specified previously and the derived placement was with respect to the given partition.

Returning to Figure 4, on the right we display a partition that has both *totalcut* and *maxcut* of 0, since all nets connect only units that are assigned to the same module. However, the partition assigns (to the middle module) a set of units that has a total area requirement exceeding the capacity of that module. Such partitions are referred to as "infeasible," and as will be discussed later, we believe that it is important that **GAPE** be allowed to operate on infeasible solutions in intermediate stages. As an optimization problem we allow this possibility by including a penalty proportional to the over-capacity area.

[2] Here we are assuming that the connection between modules can be laid out as a *tree*.

To state the K-partition problem rigorously we make the following definitions. $\mathbf{U}$ is the set of units. Each u_i in $\mathbf{U}$ has an area a_i. $\mathbf{M}$ is the set of K modules. Each m_i in $\mathbf{M}$ has a capacity c_i. $\mathbf{N}$ is the set of nets. Each member $\mathbf{n}_i$ of $\mathbf{N}$ is a set of units, i.e., each $\mathbf{n}_i$ is a subset $\mathbf{U}$, indicating the units that are connected by the given net. Note that any unit may appear in several or none of the nets. A solution to the K-partition problem (given $\mathbf{U}$, $\mathbf{M}$ and $\mathbf{N}$) is $\mathbf{P}$, where $\mathbf{P}$ contains K elements. Each $\mathbf{p}_i$ in $\mathbf{P}$ is a set of units, indicating the units assigned to module m_i. Here $\mathbf{P}$ is a strict set partition of $\mathbf{U}$, in that,

$$\bigcup_{i=1}^{K} \mathbf{p}_i = \mathbf{U} \quad \text{and} \quad (\forall i \neq j)\ \mathbf{p}_i \cap \mathbf{p}_j = \emptyset.$$

For partition $\mathbf{P}$ we define the following measures. For each net $\mathbf{n}_i$ of $\mathbf{N}$ we define the function $NETCUTS$ to be the number of distinct modules that the net enters, and for each module m_i of $\mathbf{M}$ we define the function $MODULECUTS$ to be the number of nets that exit the module. Thus, for partition $\mathbf{P}$ we have

$$totalcut(\mathbf{P}) = \sum_{j=1}^{|\mathbf{N}|} (NETCUTS(\mathbf{n}_j) - 1)$$

and

$$maxcut(\mathbf{P}) = \max_{j \in 1 \ldots K} \{ MODULECUTS(m_j) \} .$$

The total area of the units assigned to a module is given by the function A

$$A(\mathbf{p}_i) = \sum_{u_j \in \mathbf{p}_i} a_j .$$

Thus, partition $\mathbf{P}$ is *feasible* if

$$(\forall \mathbf{p}_i \in \mathbf{P})\ A(\mathbf{p}_i) \leq c_i ,$$

otherwise it is *infeasible*. With regard to area we define the function $EXP(b, y)$, which is 0 if $y \leq 0$ and b^y, otherwise. Then our area penalty for partition $\mathbf{P}$ is given by:

$$areapenalty(\mathbf{P}) = \sum_{i=1}^{K} EXP(\, expbase,\ (A(\mathbf{p}_i) - c_i)\,) ,$$

where *expbase* is a system specified real value. Finally, our objective function to be minimized is defined as follows:

$$\begin{aligned} SCORE(\mathbf{P}) &= \mu \times areapenalty(\mathbf{P}) \\ &+ \lambda \times totalcut(\mathbf{P}) + (1-\lambda) \times maxcut(\mathbf{P}) , \end{aligned}$$

where λ is a tradeoff parameter for the two cut objectives and μ is a scaling factor for the area penalty (which is strongly affected by the value of *expbase* in the EXP function). For the study presented in the next section the following settings were used:

- $\lambda \rightarrow 0.5$; $\mu \rightarrow 100.0$; $expbase \rightarrow 1.02$.

A straightforward representation for an individual solution is as an array, *soln*, of $|\mathbf{U}|$ integers, where $soln[i]$ indicates the module to which unit u_i is assigned. In our implementation we have added an "ordering" array that specifies a permutation of the unit indices, i.e., $ordg[i]$ contains the current index of unit u_i. The module to which u_i is assigned is thus given by $soln[\ ordg[\ i\ ]]$.

Each subpopulation has its own ordering array. The motivation for including the ordering array is that building blocks are assumed to be formed by consecutive elements in the solution array from the point of view of the single-point crossover that we are using. Yet for the K-partition problem, we believe that the building blocks are the sets of units in each net. Note that the objective function of the K-partition problem is based on "cut" measures (considering only feasible solutions in which the area penalty is zero) and "cuts" are created by units from the same net being assigned to different modules. Thus, if a net has units on both sides of the crossover point (more likely for non-consecutive elements) then a "good" assignment of that net in either of the parents is more likely to be disrupted. The use of the ordering allows the **GAPE** system to explore different building block patterns, i.e., via the permutation in *ordg*.

Having the subpopulation-wide ordering array has a different dynamic behavior than using a "uniform" [10] crossover operator. Clearly, the latter has the effect of attempting new building block patterns, but the choice of the permutation is done independently for each new offspring pair. We contend that the stability, through numerous generations, of the ordering array is required in order for building blocks to be formed and exploited. The existence of different ordering arrays across the subpopulations provides a variation conducive to exploration. The computational cost of using the array is minor since it consists of simple level of indirection in the scoring function, and for **GAPE**, a simple translation during inter-epoch communication of individuals.

As we mentioned above, the crossover operator is a single-point crossover. No translation using the ordering array is necessary for creating the offspring. The mutation operator is also a single-point operator. In particular, the mutation operator randomly selects a position in the solution array and changes the entry to a randomly selected (though distinct) module. That is, the mutation operator picks a unit and assigns it to a new module.

The fitness function for individual **P** in subpopulation **S** is defined by

$$fitness(\mathbf{P}) = \frac{\hat{\mu} + \alpha\hat{\sigma} - SCORE(\mathbf{P})}{2\alpha\hat{\sigma}}$$

where $\hat{\mu}$ and $\hat{\sigma}$ are the mean and standard deviation over the scores in $\hat{\mathbf{S}}$, while $\hat{\mathbf{S}} \subseteq \mathbf{S}$ is defined by

$$\hat{\mathbf{S}} = \{\mathbf{P} : SCORE(\mathbf{P}) \leq med + (med - minimum)\}$$

with *med* and *minimum* taken over **S**. In previous studies [1] we used the mean and standard deviation taken over the whole set **S** instead of this selected subset. We made this change because the scoring function we defined for the K-partition problem involves an exponentially increasing factor, i.e., $areapenalty(\mathbf{P})$. As mentioned in Section 3, the scoring function has been defined with this factor in order to allow, but heavily penalize, infeasible solutions. Our variation operators, i.e., crossover and mutation, however, do not attempt to assure feasibility, so a newly derived solution may be infeasible and may substantially overassign a module. The penalty is exponential in the degree of overassignment resulting in an extremely large score. A subpopulation with even one substantially infeasible individual will yield a σ that is large relative to the vast majority of the individuals in **S**. Such a large σ causes almost all individuals to have a fitness of 0.5, making our random selections uniform instead of proportional to score.

4 Empirical Study

We have chosen a "random" problem instance of the K-partition problem for investigation. The specifics of the problem instance for our study are as follows:

- number of units $\rightarrow$ 200; number of nets $\rightarrow$ 300; K, the number of modules $\rightarrow$ 15
- a_i, the area of each unit $\rightarrow$ randomly selected from [25,75]
- c_i, the capacity of each module $\rightarrow$ $1.1 \times \sum_{u_j \in \mathbf{P}_i} a_j \ / \ K$
- length of each net $\rightarrow$ randomly selected from [2,6]; net elements randomly selected from **U**.

Our empirical study compares two GA configurations: the multi-population **GAPE**, and a single GA. For each configuration we made 5 individual runs, i.e., distinct seeds for the random number generator. Both configurations are derived from our implementation of **GAPE** on the INTEL i860 hyper-cube. The single GA is just a 1-subpopulation **GAPE** having no inter-epoch communication. In order to compare final results, we allowed each configuration to evolve through the same number of total generations [1] before recording the best. This means that the single GA runs sequentially for 16 times the number of generations of any one of the individual subpopulations. In this particular case, each **GAPE** subpopulation evolves for a total of 1500 generations, while the single GA evolves for 24000 generations.

In each run **GAPE** records the best score observed in any generation of any epoch. Table I summarizes the results with entries for the best of the 5 "best score observed" values, the average of the 5, and the standard deviation. From Table I one can see that **GAPE** both found the best score overall and did better in its average. We attribute this performance enhancement to the multi-population structure of **GAPE** and the environmental diversity attained through that structure. In addition, our implementation is able to make full use of the multi-processor configuration of the i860. That is, the 16-subpopulation **GAPE** completes in 1/16th the "wall-clock" time as the single GA.

	avg.	best	worst	std-dev
GAPE	303.6	293.0	312.0	6.5
single GA	308.7	304.5	316.5	4.5

Table I. Results for 5 runs of each system.

Acknowledgement

Our access to the INTEL i860 operated by Oak Ridge National Laboratory was made possible through the Institute for Parallel Computation at the University of Virginia.

References

[1] J. P. Cohoon, S. U. Hegde, W.N. Martin and D.Richards, "Distributed Genetic Algorithms for the Floorplan Design Problem," *IEEE Trans. on Computer-Aided Design*, vol. 10, no. 4, April 1991.

[2] J. P. Cohoon, W.N. Martin and D.Richards, "Genetic Algorithms and Punctuated Equilibria in VLSI," *Workshop on Parallel Problem Solving in Nature*, Dortmund, Germany, October, 1990.

[3] N. Eldredge and S.J. Gould, "Punctuated Equilibria: An Alternative to Phyletic Gradualism," in *Models in Paleobiology*, T.J.M. Schopf (ed.), Freeman, Cooper, and Co., San Fransico, 1972, 82-115.

[4] N. Eldredge, *Macro-evolutionary Dynamics: Species, Niches, & Adaptive Peaks*, McGraw-Hill, New York, 1989.

[5] D.E. Goldberg, "Genetic Algorithms and Walsh Functions: Part II, Deception and Its Analysis," *Complex Systems, 3*, 1989, pp. 153-171.

[6] J. H. Holland, *Adaptation in Natural and Artificial Systems*, University of Michigan Press, Ann Arbor, MI, 1975.

[7] I. Rechenberg, *Evolutionsstrategie*, Frommann-Hozboog, Stuttgart, 1973.

[8] H.-P. Schwefel, *Numerical Optimization of Computer Models*, Wiley, New York, 1981.

[9] C. L. Seitz, "The Cosmic Cube," *Communications of the ACM*, 28(1), 1985, 22-33.

[10] G. Syswerda, "Uniform Crossover in Genetic Algorithms," *International Conference on Genetic Algorithms*, 1989, 2-9.

[11] S. Wright, "Character Change, Speciation, and the Higher Taxa," *Evolution 36*, no. 3, 1982, pp. 427-443.

[12] S. Wright, "The Roles of Mutation, Inbreeding, Crossbreeding and Selection in Evolution," *Proceedings of the Sixth International Congress of Genetics 1* 1932, pp. 356-366.

[13] S. Wright, "Stochastic Processes in Evolution," in *Stochastic Models in Medicine and Biology*, J. Gurland, ed., University of Wisconsin Press, 1964, pp. 199-241.

```
initialize N subpopulations on N processors
for E iterations do
        parallel-for each processor i do
                run GA for G generations
                endfor
        parallel-for each processor i do
                for each neighbor j of i do
                        send a set of solutions from i to j
                        endfor
                endfor
        parallel-for each processor i do
                select n individuals to start next epoch
                endfor
        endfor
```

Figure 1. High-level code for overall **GAPE** process.

```
for G iterations do
        while number of offspring created < n × C do
                select two solutions
                create two offspring from the selected pair
                endwhile
        add offspring to subpopulation
        calculate fitnesses
        select n individuals to start next generation
        endfor
```

Figure 2. High-level code for GA executed at each processor.

Selection in Massively Parallel Genetic Algorithms

Robert J. Collins
David R. Jefferson
Artificial Life Laboratory
Department of Computer Science
University of California, Los Angeles
Los Angeles, CA 90024

Abstract

The availability of massively parallel computers makes it possible to apply genetic algorithms to large populations and very complex applications. Among these applications are studies of natural evolution in the emerging field of artificial life, which place special demands on the genetic algorithm. In this paper, we characterize the difference between panmictic and local selection/mating schemes in terms of diversity of alleles, diversity of genotypes, the inbreeding coefficient, and the speed and robustness of the genetic algorithm. Based on these metrics, local mating appears to not only be superior to panmictic for artificial evolutionary simulations, but also for more traditional applications of genetic algorithms.

1 INTRODUCTION

The availability of powerful supercomputers such as the Connection Machine [14] means that genetic algorithms are now applied to larger and more difficult optimization problems (e.g. [4], where the search space consists of 2^{25590} points). Some of our recent artificial life work [15, 5, 3, 4] has involved massively parallel genetic algorithms characterized by large populations, enormous search spaces, and fitness functions that change through time.

These simulated evolution applications place special demands on the genetic algorithm. The simulations generally attempt to model the evolution of populations of tens of thousands of artificial organisms in a simulated environment over a period of thousands of generations. The ecosystem in which the fitness of each individual is evaluated can potentially include both direct and indirect interactions with other members of the population, members of coevolving populations, the background environment, etc. In addition, the environment and selection criteria may change both during a generation and over a period of many generations (and may be different in different parts of the simulated world). Such applications require a genetic algorithm that is able to simultaneously explore a wide range of genotypes and can maintain enough genetic diversity to respond to changing conditions.

Genetic algorithms that use panmictic selection and mating (where any individual can potentially mate with any other) typically converge on a single peak of multimodal functions, even when several solutions of equal quality exist [8]. Genetic convergence is a serious problem when the adaptive landscape is constantly changing as it does in both natural and artificial ecosystems. Crowding, sharing, and restrictive mating are modifications to panmictic selection schemes that have been proposed to deal with the problem of convergence, and thus allow the population to simultaneously contain individuals on more than one peak in the adaptive landscape [7, 10, 8]. These modifications are motivated by the natural phenomena of niches, species, and assortative mating, but they make use of global knowledge of the population, phenotypic distance measures, and global selection and mating, and thus are not well suited for parallel implementation. Rather than attempting to directly implement these natural phenomena, we exploit the fact that they are emergent properties of local mating.

In this paper, we introduce several metrics for evolving populations, and use them to characterize the differences between local and panmictic selection/mating schemes. Although our target applications involve the evolution of artificial organisms, for computational convenience we have performed this study using the optimization of graph partitions as the evolutionary task.

The results of this empirical study indicate that local mating is more appropriate for artificial evolution than the panmictic mating schemes that are usually used in genetic algorithms. In addition, local mating appears to be superior to panmictic mating, even when considering traditional applications. Local mating (a) finds optimal solutions in faster; (b) typically finds multiple optimal solutions in the same run; and (c) is much more robust. These results are encouraging, but are based on a single optimization problem, a single recombination rate, a single mutation rate, one local mating algorithm, and large populations. Further investigation seems both appropriate and necessary.

2 GRAPH PARTITIONING

The graph partitioning problem [1] is to find two subsets V_0 and V_1 of the set V of vertices of a fixed graph G such that $V = V_0 \cup V_1$, $V_0 \cap V_1 = \emptyset$, $|V_0| = |V_1|$ (which assumes that $|V|$ is even), and the cut size (the number of edges with one endpoint in V_0 and the other in V_1) is minimized. We represent a partition of G by a string S of $|V|$ bits, such that vertex $i \in V_{S[i]}$ (the value of each bit indicates to which subset the corresponding vertex belongs).

In order to incorporate this problem statement into a genetic algorithm, we use the fitness function defined by Ackley [1], which has a penalty that is quadratic in the imbalance of a partition:

$$f(x) = -cutsize(x) - 0.1(Z(x) - O(x))^2$$

where x is an arbitrary bit string, $Z(x)$ is the number of 0's in x, and $O(x)$ is the number of 1's in x. The cutsize and penalty term are given negative signs to make this a function maximization problem (with maximum value of 0).

There is little structure in small random graphs, so good partitions are relatively easy to find by simple hill climbing methods [1]. To make the problem harder, a "clumpy" or *multilevel* graph is used (a clump is a strongly connected group of vertices). We adopt Ackley's multilevel graphs, where a clump consists of four fully connected vertices, and each graph consists of two identical, disconnected pieces (so the optimal partition has fitness 0). Each of the connected components of the graph connects its clumps in a hypercube. For example, a 64 vertex multilevel graph consists of 16 clumps. Each of the two connected components is cube with a clump in each of the 8 corners.

We place the vertices within each clump and each connected component consecutively on the string S. This means that there are two optimal solutions to the multilevel graph partitioning problem: $\frac{|V|}{2}$ 0's followed by $\frac{|V|}{2}$ 1's, and the bitwise complement.

To the genetic algorithm, the clumps act as short, low-order and highly fit schemata, thus the search is quickly focused on partitions that do not cut clumps. In order to continue the search for an optimal partition, it is necessary to move clumps across the partition via recombination. Recombination is necessary because moving a clump across the cut one vertex at a time (i.e. by point mutations) results in a dramatic decrease in fitness. The result is that the multilevel graph partitioning problem is difficult for genetic algorithms, unless convergence can be avoided (recombination requires diversity in order to have an impact, and convergence implies little diversity).

3 EVOLUTION METRICS

In this section, we introduce four metrics that we use to characterize evolving populations. These particular metrics were chosen in order to measure and highlight the differences in the evolutionary dynamics of structured (locally mating) and panmictic (globally mating) populations.

3.1 DIVERSITY OF ALLELES

We measure the allele diversity of a population by comparing the observed allele frequencies to the expected allele frequencies of a maximally diverse population. In this paper, we consider loci with exactly 2 alleles, so the maximum diversity occurs when each allele frequency is 0.5. We define the diversity of alleles of a population at locus (bit position) i as

$$D_i = 1 - 4(0.5 - observed\_freq_i)^2$$

where $observed\_freq_i$ is the frequency of the 0 allele at locus i. D_i can range from 0, which indicates complete fixation (convergence), to 1 which indicates the maximum possible genetic diversity. D, the genetic diversity of the population for the entire genome S, is

$$D = \frac{\sum D_i}{|S|}$$

which also ranges from 0 (fixation) to 1 (maximal diversity).

3.2 DIVERSITY OF GENOTYPES

We measure genotypic diversity by choosing a random sample (without replacement) of 10 loci and count the number of unique genotypes (with respect to these loci) represented in the population. A new set of loci is sampled each generation. This provides a measure of the breadth of the genetic search.

3.3 INBREEDING COEFFICIENT

We define inbreeding to be the mating of two individuals who are more similar to each other than would be expected if mates were chosen randomly. The primary effect of inbreeding is to decrease the frequency of heterozygous genotypes [13]. A (diploid) genotype is heterozygous at a locus if the two haplotypes contain different alleles at that locus. The inbreeding coefficient F is calculated by comparing the actual proportion of heterozygous genotypes in the population with the proportion that would occur under random mating. Heterozygosity (and thus the inbreeding coefficient) is defined in terms of diploid organisms, but with few exceptions genetic algorithms use haploid strings. Only during sexual reproduction are our individuals diploid, so this is when we measure F.

Let H be the observed proportion of heterozygous genotypes, and H_0 be the expected heterozygosity in a randomly mating population with the same allele frequencies. The standard form for F is

$$F = \frac{H_0 - H}{H_0}$$

In this paper, there are two alleles (0 and 1) at each locus, so (from the Hardy–Weinberg principle) $H_0 = 2p(1-p)$, where p is the frequency of the 0 allele.

3.4 SPEED AND ROBUSTNESS

We measure the speed of evolution achieved by a genetic algorithm in two ways.

- Number of generations required to discover an acceptable solution. This measure allows implementation independent comparisons.
- Computational time required to discover an acceptable solution. This measure takes into account the varying computational costs of the reproduction portion of the genetic algorithm.

We define the robustness to be the fraction of runs that find at least one acceptable solution. In this paper, we define an acceptable solution as either of the two optimal graph partitions. Since we do not always discover acceptable solutions, we stop such runs at 1000 generations. We report speed in terms of the median run (of all runs, including those that do not find optimal partitions).

4 SELECTION SCHEMES

4.1 LOCAL SELECTION

One of the basic assumptions of Wright's shifting balance theory of evolution is that spatial structure exists in large populations [19, 6]. The structure is in the form of *demes*, or semi–isolated subpopulations, with relatively thorough gene mixing within a deme, but restricted gene flow between demes. One way that demes can form in a continuous population and environment is *isolation by distance*: the probability that a given parent will produce an offspring at a given location is a fast–declining function of the geographical distance between the parent and offspring locations. Wright's theory of evolution has played a role in the design of several parallel genetic algorithms [17, 11, 16, 4].

To simulate isolation by distance in the selection and mating process, we place the individuals on a toroidal, 1 or 2 dimensional grid, with one organism per grid location. Selection and mating take place locally on this grid, with each individual competing and mating with its nearby neighbors. In our local mating scheme, the two parents of each offspring are the highest scoring individual encountered during two random walks that begin at the offspring location, one parent per walk. The parents are chosen with replacement, so it possible for the same high–scoring individual to be encountered during both random walks, in which case it would act as both parents for the offspring. Deme size (and thus the rate of gene flow) is a function of R, the length of the random walks.

4.2 STOCHASTIC SELECTION

The most typical panmictic selection strategy that is used in genetic algorithms is known as stochastic selection with replacement (or roulette wheel selection) [9]. We define the probability that individual x is chosen as a parent as

$$P(x) = \frac{f_x}{\sum f_i}$$

The stochastic selection algorithm assumes that all fitness values are non–negative. Because $f_x \leq 0$ for the graph partitioning problem, we adjust the fitness scores for the stochastic selection algorithm

$$f'_x = f_x + |\min f_i|$$

where f_x is the fitness of partition x.

4.3 LINEAR RANK SELECTION

Another panmictic selection algorithm is the linear rank method [12]. The linear rank selection algorithm defines the *target sampling rate* (TSR) of an individual x as

$$TSR(x) = Min + (Max - Min)\frac{rank(x)}{N-1}$$

where $rank(x)$ is the index of x when the population is sorted in increasing order based on fitness, and N is the population size. Also, the constraints that $0 \leq TSR(x)$, $\sum TSR(x) = N$, $1 \leq Max \leq 2$, and $Min + Max = 2$ are imposed. The TSR is the number of times an individual should be chosen as a parent for every N sampling operations.

5 EMPIRICAL RESULTS

We have compared various genetic algorithms on the multilevel graph partitioning problem. The selection algorithms that we used were local mating in both 1 and 2 dimensional geometries (for $R = \{1, 2, 3, 5, 10, 20, 30\}$), linear rank selection (with $Min = 0.0$ and $Max = 2.0$), and stochastic selection with replacement. The optimization problem is the partitioning of the 64–vertex multilevel graph. The population size N varies over a range from $2^{13} = 8,192$ to $2^{19} = 524,288$ individuals in each generation. In this section, we present only a representative sample of our results, because space constraints prevent us from presenting all of our data here.

The genetic operators include both recombination and mutation. During recombination, crossovers occur with a constant probability of 0.02 between each pair of consecutive bits, so most matings (64%) will result in zero or one crossovers, but some matings may result in many crossovers. In a similar way, point mutations (bit flips) occur with a constant probability of 0.001 per bit, with only 6% of the individuals experiencing one or more mutations.

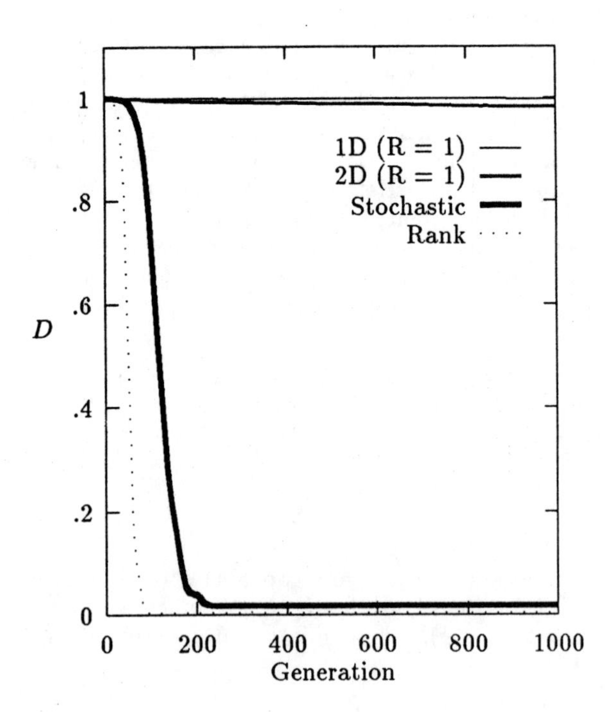

Figure 1: The diversity of alleles D maintained by the four selection algorithms. $N = 2^{16} = 65,536$ and the task is the 64 vertex multilevel graph. Each curve is the average of 7 runs.

5.1 DIVERSITY OF ALLELES

The diversity of alleles over time that is maintained by the four selection algorithms is plotted in Figure 1. Both of the panmictic selection algorithms quickly lose diversity, while both of the local selection schemes maintain a high degree of variation. Although the 1 dimensional local scheme maintains nearly perfect variation, the 2 dimensional algorithm loses a small amount over time. Of the two panmictic selection algorithms, linear ranking loses diversity sooner, and stabilizes at a lower level.

5.2 DIVERSITY OF GENOTYPES

The genotypic diversity for the four selection algorithms is plotted in Figure 2. This data is based on a random sample (with a new sample each generation) of 10 of the 64 loci in the genome. In all four cases, the genotypic diversity begins to fall after only a few generations. Both of the local mating schemes maintain a count of about 150 genotypes (out of a possible 1024), while stochastic selection remains around 40, and linear rank selection around 15. With the exception of 1 dimensional local mating, all of the algorithms quickly reach their stable values.

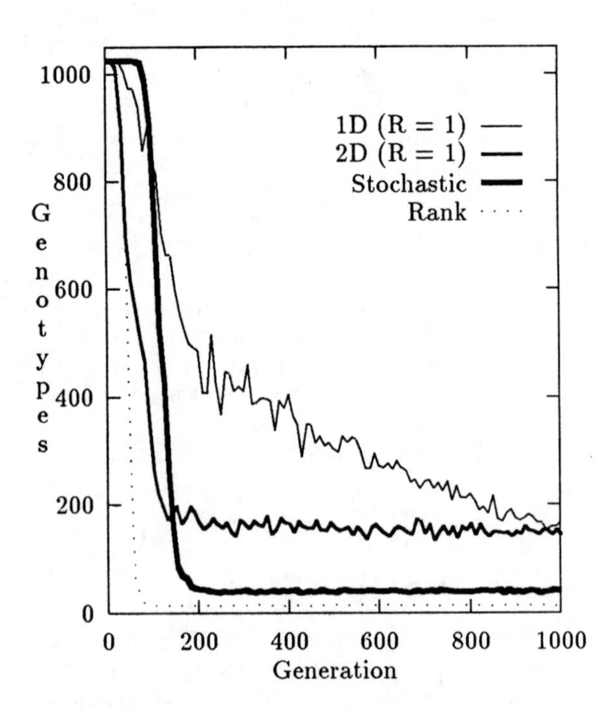

Figure 2: The genotypic diversity for the four selection algorithms, based on sampling 10 loci per generation. $N = 2^{16} = 65,536$ and the task is the 64 vertex multilevel graph. Each curve is the average of 7 runs.

5.3 INBREEDING COEFFICIENT

The inbreeding coefficient for the four selection algorithms is plotted in Figure 3. In the early gen-

Algorithm	log$_2$ Population Size						
	13	14	15	16	17	18	19
Stochastic	†	†	151	137	119	†	136
Linear Rank	†	57	66	52	35	32	31
1D ($R=1$)	156	148	142	124	126	108	114
1D ($R=5$)	85	79	59	63	57	49	50
1D ($R=10$)	56	50	47	50	43	40	41
1D ($R=20$)	42	40	37	37	34	30	27
1D ($R=30$)	41	39	32	32	29	26	24
2D ($R=1$)	48	43	41	42	40	40	38
2D ($R=5$)	23	20	16	17	16	16	15
2D ($R=10$)	14	13	13	12	12	11	11
2D ($R=20$)	13	11	10	11	9	9	9
2D ($R=30$)	11	8	9	9	8	8	8

Table 1: Generation of first appearance of an optimal partition (median of 11 runs) on the 64 vertex multilevel graph problem. † indicates that the median run did not find an optimal solution within 1000 generations.

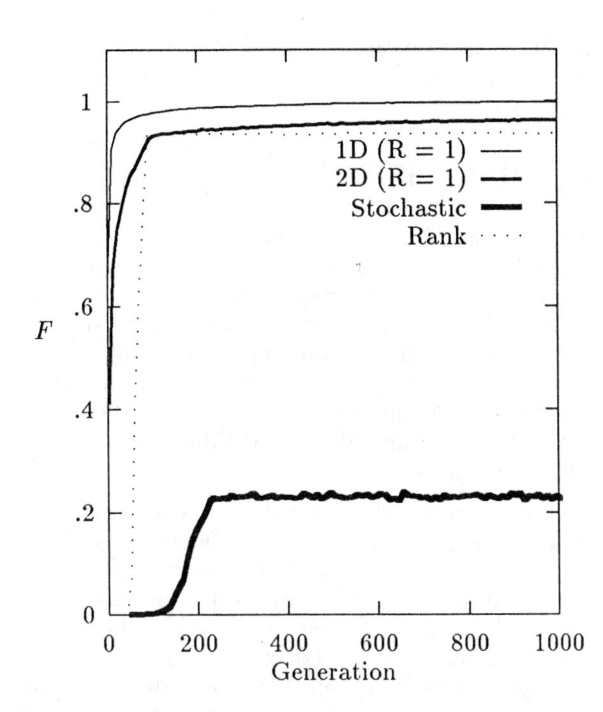

Figure 3: The inbreeding coefficient F for the four selection algorithms. $N = 2^{16} = 65,536$ and the task is the 64 vertex multilevel graph. Each curve is the average of 7 runs.

erations, both of the panmictic selection algorithms show no excess homozygosity (F near 0), while both local algorithms immediately show significant and increasing excess homozygosity. A significant degree of inbreeding is observed in later generations with both panmictic algorithms, and throughout the experiments for both local selection schemes. In later generations, stochastic selection is characterized by a much lower inbreeding coefficient than the others.

5.4 SPEED AND ROBUSTNESS

The first speed comparison is based on the number of generations to the first appearance of an optimal partitioning of the 64 vertex graph (Table 1). Of the two panmictic selection algorithms, linear rank selection finds solutions nearly twice as fast as stochastic selection. We also note that both panmictic algorithms do not reliably find optimal solutions when applied to the smaller populations. For both local mating algorithms, longer random walks result in faster evolution, and for the same R value, 2 dimensional local mating is faster than 1 dimensional local mating. With the exception of the most constrained local mating (1 dimensional with $R = 1$), local mating always beats panmictic, and in some cases by about a factor of 7.

Across all of the genetic algorithms, increasing the population size causes only slight speed improvements (in terms of number of generations to an optimal solution). In addition, we found that an optimal solution is either found within a couple hundred generations, or the run times out (across all selection algorithms). We never observed a run that first discovered an optimal solution between generation 200 and 1000.

The second speed comparison is based on the actual time required to find an optimal solution (Table 2). The run-time measurements reported here are based on implementations in C++/CM++ [2] that differ only in the selection/mate choice code. The data was gathered on an 16K processor Connection Machine–2 equipped with 64K bits of memory per processor and 32-bit floating point accelerators, with a Sun 4/330 front end running SunOS 4.1.1 and Connection Machine software version 6.0.

Although the run time per generation for stochastic selection is less than linear rank selection, it still re-

Algorithm	$\log_2$ Population Size					
	14	15	16	17	18	19
Stochastic	† (0.57)	165 (1.09)	285 (2.08)	552 (4.64)	† (9.47)	2652 (19.50)
Linear Rank	40 (0.70)	83 (1.26)	132 (2.54)	200 (5.72)	379 (11.84)	755 (24.34)
1D ($R=1$)	24 (0.16)	41 (0.29)	70 (0.56)	139 (1.10)	264 (2.44)	540 (4.74)
1D ($R=5$)	15 (0.19)	21 (0.35)	42 (0.67)	81 (1.42)	133 (2.72)	265 (5.29)
1D ($R=10$)	13 (0.25)	20 (0.42)	37 (0.74)	65 (1.50)	122 (3.06)	252 (6.15)
1D ($R=20$)	16 (0.39)	20 (0.54)	35 (0.95)	72 (2.13)	116 (3.85)	199 (7.38)
1D ($R=30$)	16 (0.42)	22 (0.69)	38 (1.18)	71(2.44)	120 (4.61)	216 (9.00)
2D ($R=1$)	7 (0.16)	12 (0.29)	25 (0.59)	51 (1.28)	98 (2.46)	185 (4.87)
2D ($R=5$)	4 (0.22)	6 (0.38)	12 (0.73)	26 (1.62)	48 (3.01)	86 (5.72)
2D ($R=10$)	3 (0.26)	6 (0.47)	10 (0.87)	22 (1.86)	39 (3.54)	75 (6.86)
2D ($R=20$)	4 (0.39)	7 (0.67)	13 (1.22)	23 (2.58)	43 (4.82)	83 (9.27)
2D ($R=30$)	4 (0.53)	8 (0.91)	14 (1.60)	26 (3.29)	49 (6.14)	93 (11.67)

Table 2: Computation time seconds to the first appearance of an optimal partition (median of 11 runs) on the 64 vertex multilevel graph problem. The time (seconds) per generation is shown in parentheses. † indicates that the median run did not find an optimal solution within 1000 generations.

$\log_2 N$	Linear Rank	Stochastic
13	0.45	0.27
14	0.55	0.36
15	0.64	0.64
16	0.82	0.55
17	1.0	0.55
18	1.0	0.27
19	1.0	0.73
overall	0.78	0.48

Table 3: Fraction of runs finding an optimal solution within 1000 generations for the panmictic selection algorithms.

quires more than twice as much time to find optimal solutions. For local mating, although long random walks require significant computation, the fastest evolution occurs when R is in the range $5 < R < 20$. Even with long random walks, the local mating algorithms run significantly faster (per generation) than the panmictic algorithms, which accentuates the fact that local mating optimizes in fewer generations.

When we compare the fastest panmictic algorithm that we implemented (linear rank) to our slowest local mating algorithm (1 dimensional with $R = 1$), we find that local mating is faster by about a factor of 2. When compared to the fastest local mating algorithm (2 dimensional with $R = 10$), linear rank is slower by more than an order of magnitude and stochastic selection is slower by about a factor of 25.

We measure the robustness of the various genetic algorithms in terms of the fraction of runs that find one of the two optimal solutions within 1000 generations. None of the local mating runs, across all population sizes and R values, failed to find an optimal solution. Unlike the local algorithms, the panmictic algorithms are not 100% robust (Table 3). Of the two panmictic algorithms, linear rank selection is more robust.

6 DISCUSSION

Simulations of the evolution of artificial organisms operate on large populations in complex and changing ecosystems. The adaptive landscapes are generally enormous (hundreds or thousands of orders of magnitude larger than the population size) and constantly changing. Artificial life simulations require a genetic algorithm that is resistant to convergence and can simultaneously explore different parts of the adaptive landscape.

Local mating should be resistant to convergence, because each deme can explore different peaks in the adaptive landscape. Small demes allow effective exploration of the adaptive landscape, because temporary reductions in fitness are possible, due to random genetic drift (selection would prevent this in a large population) [18]. When a deme discovers a higher adaptive peak, the new genotype will spread and deliver more genes to the nearby demes. This results in the gradual spread of the favorable alleles and allele combinations throughout the population, by means of interdeme selection.

In our empirical experiments, we have observed that the panmictic selection algorithms become focused towards one of the two optimal solutions in the very early generations, and eventually converge on or near that solution. In fact, we have not observed a single run in which a panmictic selection algorithm discovered both optimal solutions. In sharp contrast, the local mating algorithms consistently discover both optimal solutions. For all but the smallest population sizes, both optimal solutions are usu-

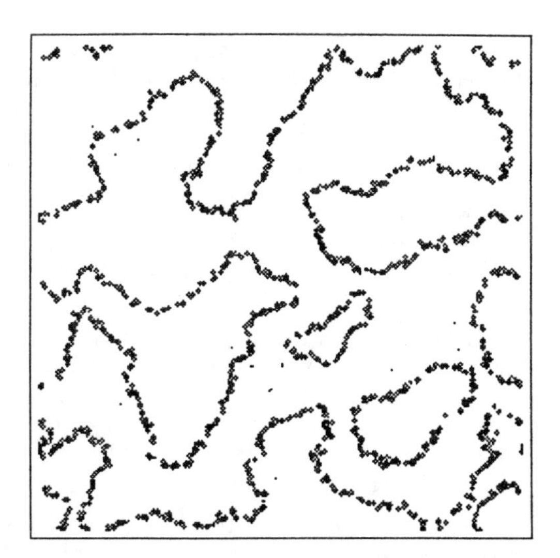

Figure 4: The geographical distribution of fitness scores in generation 150 a run using the 2 dimensional local mating algorithm ($R = 1$) with a population size of $2^{16} = 65,536$ on the 64–vertex multilevel graph partitioning problem. Individuals scoring less than -3 are represented by black pixels.

ally discovered many times in any one run by geographically separated demes. Every run with a large, locally mating population resulted in a nearly even mixture of both optimal solutions, and formed genetically homogeneous regions that are separated by boundaries (composed of low–fitness hybrids) that are stable for thousands of generations (Figure 4).

The genetic diversity data (Figure 1) demonstrates quite dramatically how panmictic selection converges towards only one of the solutions, while local selection contains a nearly equal mix of both. (Remember that the two solutions are bitwise complements of each other.) Local selection also maintains much greater diversity of genotypes than panmictic selection (Figure 2). What we expect to see is a cloud of mutants surrounding (in the adaptive landscape) an optimal solution. Because local mating finds and maintains both optimal solutions but panmictic mating finds only one, we would expect local mating to have something like twice the genotypic diversity of panmictic selection. This is clearly not the case—we observe a factor of 4 difference from stochastic selection and a factor of 10 from linear rank selection. We find this difference to be particularly important, because it demonstrates that at equilibrium, local selection explores many more genotypes in each generation.

The inbreeding coefficient results (Figure 3) are quite interesting, because they show two different sources of inbreeding. As we expected, local mating is characterized by a high degree of inbreeding throughout the experiment. The inbreeding is a result of fixation on a particular genotype within each deme due to selection pressure and random genetic drift. The fixation is local—diversity is maintained because each deme fixes on a different genotype.

On the other hand, panmictic selection results in almost no inbreeding until the population begins to converge on a particular solution. When this occurs, we start to observe excess homozygosity, which is due to selection for the optimal genotype. Linear rank selection shows a very high degree of inbreeding, indicating that very few sub–optimal genotypes are chosen as parents for the next generation. In contrast, stochastic selection shows a much lower degree of inbreeding. This shows that the stochastic selection algorithm allows sub–optimal genotypes to be sampled with a higher frequency (which is consistent with the greater allele and genotype diversity).

The diversity and inbreeding data demonstrate the dramatic dynamical differences between local and panmictic mating. It is clear that local mating is more appropriate for artificial evolution, because it maintains a broad genetic search for thousands of generations. If the adaptive landscape or fitness function were changing over time, this diversity would allow the population to exploit the genotypes which suddenly have higher relative fitness values. Local mating would be more appropriate than panmictic mating for artificial life applications, even if local mating results in slower evolution. Fortunately, this is not the case.

The data on the speed of evolution shows that the dynamics of local mating results in a faster and more robust genetic algorithm. Local mating is characterized by faster evolution both in terms of the number of fitness function evaluations and time required to find an optimal solution. The robustness of local mating is rather surprising: across all population sizes, both 1 and 2 dimensional geometries, and all R values, local mating *never* failed to find an optimal solution, while both of the panmictic algorithms had a significant fraction of their runs "time out" (no optimal solution found within 1000 generations).

The fact genetic algorithms that use local mating can be significantly faster and more robust than traditional genetic algorithms is an important result, and suggests that further investigation is in order. While many important theoretical results have been developed for panmictic mating algorithms, it is not clear that any of these results can be applied directly to local mating. Also, it is not at all clear how local mating will perform with small populations, although the difference in robustness between local and panmictic mating is greatest for the smaller population sizes that we examined. In addition, we have only examined one very simple local mating algorithm, and it is almost certainly not the best that can be (or has been) found.

Acknowledgments

We would like to thank Ernst Mayr, Joe Pemberton, Chuck Taylor, Greg Werner, and Alexis Wieland for their valuable input. We also acknowledge the three anonymous reviewers for their helpful comments. This work is supported in part by W. M. Keck Foundation grant number W880615, University of California Los Alamos National Laboratory award number CNLS/89-427, and University of California Los Alamos National Laboratory award number UC-90-4-A-88. The empirical data was gathered in part on a CM-2 computer at UCLA under the auspices of National Science Foundation Biological Facilities, grant number BBS 87 14206, and a CM-2 computer at the Advanced Computing Laboratory of Los Alamos National Laboratory under the auspices of the U.S. Department of Energy, contract W-7405-ENG-36.

References

[1] David H. Ackley. *Stochastic iterated genetic hillclimbing.* PhD thesis, Department of Computer Science, Carnegie Mellon University, 1987.

[2] Robert J. Collins. CM++: A C++ interface to the Connection Machine. In *Proceedings of the Symposium on Object Oriented Programming Emphasizing Practical Applications.* Marist College, September 1990.

[3] Robert J. Collins and David R. Jefferson. Representations for artificial organisms. In Jean-Arcady Meyer and Stewart Wilson, editors, *Proceedings of Simulation of Adaptive Behavior.* The MIT Press/Bradford Books, 1990.

[4] Robert J. Collins and David R. Jefferson. AntFarm: Towards simulated evolution. In Christopher Langton, J. Doyne Farmer, Steen Rasmussen, and Charles Taylor, editors, *Artificial Life II.* Addison-Wesley Publishing Company, (in press).

[5] Robert J. Collins and David R. Jefferson. An artificial neural representation for artificial organisms. In Reinhard Männer and David E. Goldberg, editors, *Proceedings of Parallel Problem Solving from Nature.* Springer-Verlag, (in press).

[6] James F. Crow. *Basic Concepts in Population, Quantitative, and Evolutionary Genetics.* W. H. Freeman and Company, New York, 1986.

[7] Kenneth A. De Jong. *An analysis of the behavior of a class of genetic adaptive systems.* PhD thesis, Department of Computer and Communication Sciences, University of Michigan, 1975.

[8] Kalyanmoy Deb and David E. Goldberg. An investigation of niche and species formation in genetic function optimization. In *Proceedings of the Third International Conference on Genetic Algorithms.* Morgan Kaufmann, June 1989.

[9] David E. Goldberg. *Genetic Algorithms in Search, Optimization and Machine Learning.* Addison-Wesley Publishing Company, Inc., Reading, Massachusetts, 1989.

[10] David E. Goldberg and Jon T. Richardson. Genetic algorithms with sharing for multimodal function optimization. In *Genetic algorithms and their applications: Proceedings of the Second International Conference on Genetic Algorithms.* Lawrence Erlbaum Associates, 1987.

[11] Martina Gorges-Schleuter. ASPARAGOS an asynchronous parallel genetic optimization strategy. In *Proceedings of the Third International Conference on Genetic Algorithms.* Morgan Kaufmann, June 1989.

[12] John J. Grefenstette and James E. Baker. How genetic algorithms work: A critical look at implicit parallelism. In *Proceedings of the Third International Conference on Genetic Algorithms.* Morgan Kaufmann, June 1989.

[13] Daniel L. Hartl and Andrew G. Clark. *Principles of Population Genetics.* Sinauer Associates, Inc., Sunderland, Massachusetts, 1989.

[14] W. Daniel Hillis. *The Connection Machine.* The MIT Press, Cambridge, Massachusetts, 1985.

[15] David Jefferson, Robert Collins, Claus Cooper, Michael Dyer, Margot Flowers, Richard Korf, Charles Taylor, and Alan Wang. The Genesys System: Evolution as a theme in artificial life. In Christopher Langton, J. Doyne Farmer, Steen Rasmussen, and Charles Taylor, editors, *Artificial Life II.* Addison-Wesley Publishing Company, (in press).

[16] Bernard Manderick and Piet Spiessens. Fine-grained parallel genetic algorithms. In *Proceedings of the Third International Conference on Genetic Algorithms.* Morgan Kaufmann, June 1989.

[17] Heinz Mühlenbein. Parallel genetic algorithms, population genetics and combinatorial optimization. In *Proceedings of the Third International Conference on Genetic Algorithms.* Morgan Kaufmann, June 1989.

[18] William B. Provine. *Sewall Wright and Evolutionary Biology.* University of Chicago Press, 1986.

[19] Sewall Wright. *Evolution and the Genetics of Populations. Volume 2: The Theory of Gene Frequencies.* University of Chicago Press, 1969.

A Naturally Occuring Niche & Species Phenomenon: The Model and First Results

Yuval Davidor
Department of Applied Mathematics and Computer Science, The Weizmann Institute, Rehovot 76100, Israel.
yuval@wisdom.weizmann.bitnet

Abstract

A new concept for synthesizing the genetic operators is introduced in this paper. The novelty of the proposed model lies primarily in the way that the traditional genetic operators interact among themselves to optimize their effect. The model deals with three fundamental problems of GA applications and attempts: to offer an optimal algorithmic solution for controlled convergence, to achieve a natural emergence of a niche & species phenomenon on both the genotypical and fitness levels, and to separate the choice of representation from the applicability of controlled convergence operators. In other words, to provide a GA model which is based entirely on local, low computational cost operators which provide a more optimal information flow in a GA.

1 A Brief History Of GA Operators For The Control Of Convergence

Since the early days of the '70's, two important aspects about GAs were realized. One, that controlled convergence is an illusive goal, difficult to achieve. Two, that fine tuning of the GA operators is not a robust solution to the controlled convergence problem. In other words, it was realized that the operators themselves, through their mutual interaction, should provide an adaptive, and hopefully optimal, balance between the exploration and exploitation of the search space. Novel mechanisms to provide such a desirable controlled convergence property were suggested in recent years with varying degrees of success. The main idea is that there should be a mechanism that will control and prevent an unbalanced proliferation of genotypes. It is clear that such an operator can be imposed through an explicit testing of genotype diversity, but that such a mechanism is computationally prohibitive. The 'battle' for controlled convergence focuses on the attempt to obtain this property indirectly in order to minimize computational cost. The following is a brief history of the mechanisms that were suggested as an answer to the control of convergence issue.

Cavicchio, in his Doctoral dissertation , suggested a *preselection* mechanism as a mean of enhancing the maintenance of high genotype diversity. The preselection mechanism replaces parent members in the population with their offspring [Cavicchio, 1970].

De Jong's *crowding* scheme is an elaboration on the preselection mechanism. In the crowding scheme, an offspring replaces the most similar string in bit terms (hamming distance) from a randomly drawn subpopulation of size CF (crowding factor) from the main population. The more similar a member of the population becomes to other members in the population, it experiences a heavier selection pressure [DeJong, 1975]. The experimentally optimal size of the subpopulation was typically 3 when applied to De Jong's five-function test bed.

Booker implemented a *sharing* idea in a classifier system environment [Booker, 1982]. Booker's idea was that if related rules share payments (his classifier system used the bucket brigade mechanism), subpopulations of rules will form naturally in the system. However, it is difficult to apply this mechanism to standard GAs. Schaffer has extended the idea of subpopulations in his VEGA model in which each fitness element had its own subpopulation [Schaffer, 1984].

A different approach to help maintain high genotype diversity was introduced by Mauldin in his *uniqueness* operator [Mauldin, 1984]. The uniqueness operator helped to maintain diversity by incorporating a 'censorship' operator with which the insertion of an offspring into the population is possible only if the offspring is genotypically different from all members of the population at a specified number of loci (hamming distance).

Recently Goldberg incorporated some of the ideas mentioned above in a mechanism he called a *sharing function* [Deb and Goldberg, 1989; Goldberg and Richardson, 1987]. This mechanism determines the reproduction probability according to the average fitness of similar strings in the population. The similarity criterion can be either specified in terms of hamming distance in the genotype space, a metric distance in the phenotypic or fitness spaces. Experimental results from GAs running Goldberg's sharing function mechanism on function optimization problems exhibit good performance. However, sharing functions suffer from two problems. The major one being that genotypic sharing, which most effectively helps to maintain diversity, is computationally very expensive. The second drawback is the fact that the similarity criterion is not natural to the search space and therefore cannot emerge naturally from the interaction of the reproduction and selection mechanisms. The latter point becomes important when the search space is extensively multi-modal or incorporates some deception.

Although the above mechanisms partly improve the convergence, they suffer from the fact that they employ additional operators, sometimes at heavy additional computing costs, and they are not applicable to all problem domains. In other words, these models are effective for specific domains, but usually add computation complexity and, above all, do not provide a robust solution.

A novel attempt to put to use effectively some of the ideas mentioned above, in a parallel computer architecture environment led, Muhlenbein and Gorges-Schleuter to explore a particular hardware architecture called ASPARAGOS [Muhlenbein, 1989; Schleuter, 1989]. The underlying idea of ASPAROGOS is that subpopulations are held on a network of transputers, and act as interacting subpopulations. This topological arrangement of communicating subpopulations deals with an aspect of nature which is called speciation. Speciation refers to the occurrence of local optimization concurrent with global optimization. Members of a subpopulation adapt to a local optimum and then interact with other locally optimal members of other subpopulations. The main drawback of ASPAROGOS lies in its very merit – the fact that it relies on a particular (and expensive) hardware. There are other relevant works involving parallel implementations, but they are not referred to here because they recapitulate the essence of what was discussed earlier [Cohoon, et al., 1987; Jog and VanGucht, 1987; Pettey and Leuze, 1989; Tanese, 1989].

To summarize the state of the art in the field of co-evolution and controlled convergence, one has to acknowledge that there is no model which is hardware independent, capable of handling arbitrary representation formats (variable length, real value genes, etc.), and allows controlled convergence without substantially increasing computation overhead. This paper introduces a new GA model which addresses the controlled convergence problem through the introduction of a topological GA model in which speciation occurs naturally with minimal computation overhead.

2 From Population Dynamics To Controlled Convergence

There are certain assumptions regarding population genetics and population dynamics which affect the interactions between the genetic operators and the environment. Since these interactions have a major role in shaping evolution and its robust adaptive behaviour, they shall guide this work. A simplistic list of these assumptions is given below:

1. An individual's life can be characterized by three fundamental activities: foraging, mating, and winning conflicts.
2. The cost of foraging is proportional to search efforts (distance of travel, etc.).
3. Matting partners are selected from the local environment, and proportionally to fitness.
4. Offspring remain in the geographical vicinity of their parents.
5. Good habitats are more likely to be inhabited by stronger individuals.
6. The frequency of aggressive conflicts is inversely proportional to resource availability.
7. Agresive conflicts are resolved probabilistically depending on the relative strengths of the opponents.

It is clear from the above list that the behaviour of the individual is a result of mechanisms which rely only on local information. Mechanisms which at best have complete information about the individual's local environment, but in many cases only have partial view even of their immediate surroundings.

3 The Basic Idea Of The ECO Genetic Algorithm

Rather than being held as an indistinguishable (apart from fitness) collection of strings in the traditional GA models, the population of strings is held on a 2-D grid having its opposite edges connected together so that each grid element has 8 adjacent elements (Figure 1). At initialization, strings are placed on the grid at random, but no more than one string per grid element.

This GA model is a steady-state GA [Syswerda, 1989; Syswerda, 1990; Whitley, 1989], and its reproduction part is slightly different than what commonly used. Nevertheless, and as will be shown latter, the ECO GA reproduction follows the schema theorem ideology. A grid element is selected at random, and defines a 9-element sub-population around it (Figure 2). For this sub-population, the reproduction cycle proceeds as in a simple GA having the population size set to 9. Two strings are selected probabilistically from this sub-population according to their relative fitness. Mutations, if turned on, occur at the duplication phase with probability P_m per locus. The offspring are put back into the 9-element sub-population, so that they are more likely to stay in the vicinity of their parents. Placing an offspring on a grid element which is already occupied by an 'adult' string initiates a conflict between that string and the offspring. Once initiated, conflicts are resolved probabilistically according to relative fitness of the opponents so that the probability of string i to survive a conflict with string j is proportional to its relative strength.

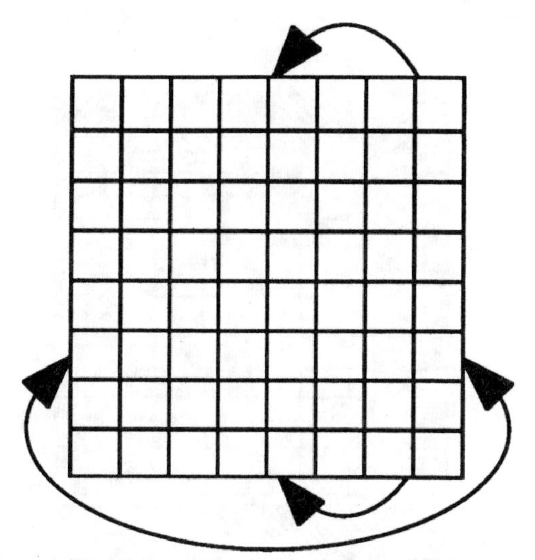

Figure 1 – The 2-D grid on which the population is stored. The opposite edges are connected and form a torus.

The schema theorem for this model becomes,

$$m(H,t+1) \geq m(H,t) + \qquad \text{(1a)}$$

$$+\, m(H,t)\frac{f(H)}{\bar{f}} - \qquad \text{(due to replication)} \quad \text{(1b)}$$

$$-\, m(H,t)\frac{f(H)}{\bar{f}} P_m\, o(H) - \qquad \text{(due to mutations)} \quad \text{(1c)}$$

$$-\, m(H,t)\frac{f(H)}{\bar{f}} P_c P_d P_h - \qquad \text{(due to crossover)} \quad \text{(1d)}$$

$$-\, m(H,t)\frac{f(H)}{\bar{f}} [P_m o(H) + P_c P_d P_h][1 - P_h]\frac{\bar{f}}{f(H)+\bar{f}}$$

$$\text{(due to conflicts)} \quad \text{(1e)}$$

where all arguments follow the conventional notation apart from the generation time, the probability P_h , and the average fitness $\bar{f}$.
The first term (Eq. 1a) involves the future frequency of schema H and its present frequency. The notation here for generation time is based on a steady-state GA model which designates an update of the adult population reproduction probability profile [Davidor, in preparation]. The second term (Eq. 1b) designates the growth due to replication. Since the reproduction potency $R_{i,j}$ of a string $S_{i,j}$ is based only on the string's fitness $f_{i,j}$ in its local sub-population grid environment so that,

$$R_{i,j} = \frac{f_{i,j}}{\frac{1}{9}\sum_{k=i-1}^{i+1}\sum_{m=j-1}^{j+1} f_{k,m}} ,$$

the average fitness of strings in the vicinity of strings containing schema H is,

$$\bar{f} = \frac{1}{m(H,t)} \sum_{\forall s_{i,j} \in H} R_{i,j} ,$$

which is similar to the conventional average fitness $\bar{f}$ used in the schema theorem, but converges more quickly (to either 1 or 0) as the search progresses. This aspect – quick, but local convergence – is the central theme of this work. It is based on the rational (and on widely documented experiments [Goldberg, 1989; Schaffer, et al., 1989]) that a simple GA operating on a finite population converges in finite time due to irreversible loss of diversity. As convergence time in a simple GA is proportional primarily to the population size, one can view the ECO GA as a model which uses a very small population size, and hence, converges quickly (see the discussion on optimal population sizes in [Goldberg, 1989]). Nevertheless, the ECO GA is not just a multi subpopulation GA. It has an implicit parallel overlapping subpopulations which evolve locally, but allow information to flow in the form of migration to adjacent grid elements.

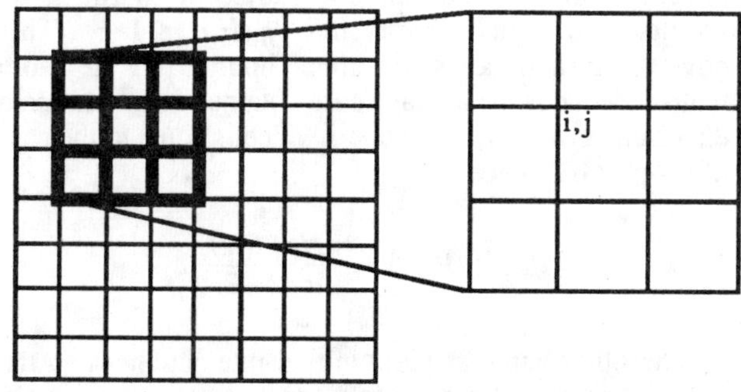

Figure 2 – The local environment of the grid element (i,j) which determines its replication potency, natural selection as a result of conflicts, and the gene pool for mating.

The third term on the right (Eq. 1c) designates the loss of schema instances due to mutations, and is the same as given in the basic schema theorem.

The fourth term on the right (1d) is the loss of schema instances due to recombination. This amounts to the number of offspring containing the given schema, times the probability a crossover was used (P_c), times the probability to disrupt the schema due to the type of crossover mechanism used (P_d), times the probability that the schema is mated with a string which does not contain the schema (P_h) [Spears and DeJong, in press]. Both P_c and P_d are as used in the schema theorem (for example, for a single-point uniform distributed crossover $P_d = \frac{\delta(H)}{l-1}$. See also [Bridges and Goldberg, 1987]).

The new probability element, P_h, designates the probability that the string which is selected as a mating partner contains a different schema than that under consideration. Though missing in Holland's original formulation, this probability is important for generating a more accurate prediction on schemata sampling, and in particular for this model. The probability of mating with a specific schema was already partly addressed in the context of a simple GA [Bridges and Goldberg, 1987]. Estimating P_h for this model has an added importance since it explains the emergence of niches, and other properties mentioned earlier.

P_h is dependent on the distribution of the $m(H,t)$ in the grid. The P_h probability of any schema at any time interval, $p_h(H,t)$, is bounded by the two extreme arrangements possible for its $m(H,t)$ members. One which packs the schema members together resulting in a minimal interaction with other schemata, and one which spreads the schema members as far apart as possible, an arrangement which maximizes the interaction with other schemata. These two extreme arrangements correspond to minimal and maximal $p_h(H,t)$ values. A conservative value of $p_h(H,t)$ is obtained when all $m(H,t)$ members have no other $m(H,t)$ member adjacent to them, an arrangement which results in $p_h(H,t) \leq 1$. This however is a rather strict upper bound and the more realistic value is the average case obtained from a random distribution of $m(H,t)$ members which is equal to the case of simple GA, hence

$$p_h(H,t) \leq 1 - \frac{m(H,t)}{n}\frac{f(H)}{\bar{f}} .$$

On the other hand, at the minimal interaction when the $m(H,t)$ members are tightly packed together, only members on the perimeter of the homogeneous pack (the interface ring of grid elements) can interact with a different schemata. Hence, the lower bound of $p_h(H,t)$ is,

$$\left(4\left(\sqrt{m(H,t)} - 2\right)\frac{3}{8} + 4\,\frac{5}{8}\right) \leq p_h(H,t) \quad , \quad m(H,t) < \left(\sqrt{n} - 2\right)^2$$

which is a good approximation for the lower bound (derived from the number of $m(H,t)$ members on the perimeter of a $\sqrt{m(H,t)} \times \sqrt{m(H,t)}$ cube and the number of different schemata they can interact with).

The last term (Eq. 1d) is the schema loss due to conflicts. This loss is the compound effect of the total number of offspring produced at the current time step, times the probability that an offspring contains a different schema, times the probability it is conflicting with a string containing schema H, times the probability it wins the conflict.

Figure 3 – The emergence of genotypically similar islands in the grid system. Above, at an early stage of the simulation, while on the bottom can be seen an advanced stage of island growth.

4 Niche & species emerge naturally in the ECO GA

At initial stages of the search, schemata sampling rate in the ECO GA is very similar to that in a SGA. However, islands of growing genotypic and fitness homogeneity emerge relatively quickly in comparison to the convergence in a SGA of the same population size. These islands represent near local optimum. As the search advances, These islands of near optimum strings mature and approach local optimality. Certain islands of highly fit strings (schemata) take-over lower fitness islands which are in their vicinity (Figure 3). Thus, an hierarchical growth and convergence is maintained until the entire grid is occupied by one schema.

It is important to notice the effect of the different operators while local genotypically/fitness islands emerge. One, proliferation of the dominating string/schemata within an island is reduced since its potency factor $R_{i,j} \rightarrow 1$. This leads to a global control over highly fit schemata. Two, the disruption due to recombination is reduced substantially within islands since $p_h \rightarrow 0$, and so does the probability of schema elimination due to conflicts. At this stage, a balance between schema growth and elimination within the islands is reached, and mutations introduce low probability perturbations for greedy optimization.

This process enables the concurrent assessment of highly fit schemata, without having to pay the price of global operators or unnecessarily risking premature convergence. Furthermore, it provides a natural balance between the effect of mutations recombination, and selection.

5 Initial trials

Initial experiments with the ECO GA model are presented in this section primarily to demonstrate the quality of naturally emerging niches. The model is applied to a demonstrative multi-modal function optimization problem with an standard binary representation. The function, $f(x) = e^{-\left(\frac{x-0.1}{0.8}\right)^2} \sin^6(5\pi x)$, is similar to the type of functions used to analyze and demonstrate the effect of controlled convergence operators [Goldberg and Richardson, 1987].

The ECO GA model used here was not tailored to achieve high performance. The main purpose is to demonstrate the workings of the new model, and primarily the emergence of niche & species phenomenon. To this end, the results presented here intend only to show the basic behaviour of the ECO GA.

When the distribution of strings in the grid is chaotic and rugged, islands of similar genotypes/phenotypes/fitness are expected to be present. Such a distribution is expected to be present during initial stages of ECO GA run. Figure 4 presents a contour map of the phenotype distribution in the grid after 500 evaluations. This map shows the beginning of island formation, though in general the distribution of phenotypic values is still quite chaotic.

On the other hand, after 5000 evaluations were performed, the grid contains phenotypically flat regions (phenotypic value plateaus). This is demonstrated in Figure 5.

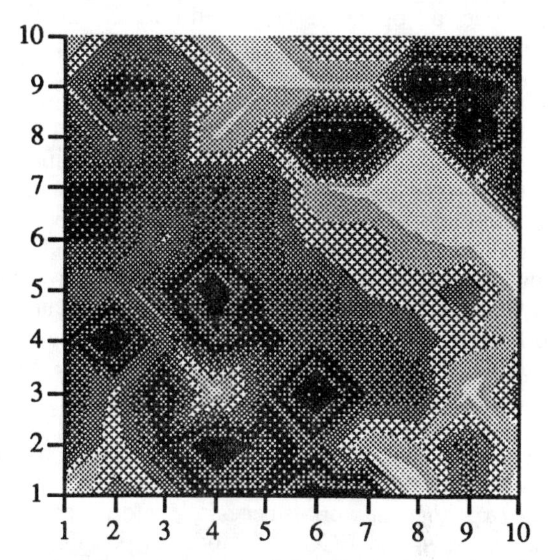

Figure 4 – A contour map of the phenotypic values in the 2D grid system after 500 evaluations (the light colours indicate low values and solid black indicates values between 0.9 and 1.0. The surface is quite rugged indicating that island of similar strings have not been formed yet, or that they are relatively small.

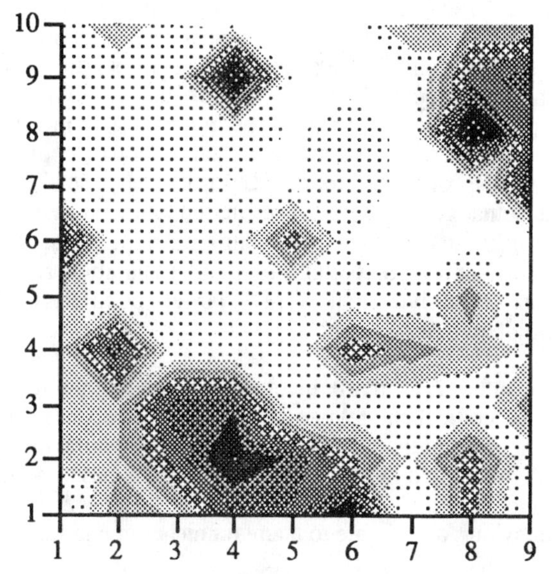

Figure 5 – After 5,000 evaluations one can easily observe large regions of phenotypically similar strings indicating that islands of similar strings have been formed. Also, the dominating phenotypic values concentrate at low values which corresponds to the global optimum of the fitness function.

6 Summary

The ECO GA model discussed in this paper presents a new synthesis of the conventional genetic operators. In this model all operators are based on local interaction among strings in a 2D grid system. The decision on a 2D topology rather than an n-dimensional grid system is not arbitrary. The number of dimensions in the topology determines the size of the sub-population (3 for 1D, 9 for 2D, 27 for 3D, and so forth). The 2D configuration was chosen to achieve a balance between the overall number of sub-populations and their size. This new population interaction arrangement results in a GA model which allows a rapid local convergence. The local convergence produces a niche & species like behaviour. Concurrent to the local convergence within niches, global optimum is obtained through the interaction among the locally optimized strings.

This work did not include a performance comparison study between conventional GA and the ECO GA. Though more experimental work is necessary before a conclusive assessment of the ECO GA model is possible, initial results indicate that this model improves on the problems associated with controlling the convergence. Furthermore, this model can be immediately applied to most parallel architectures.

The ECO GA model as presented, does not claim to maintain niche & species *ad infinitum*. Rather, it allows niches to emerge in order to enhance the efficiency and robustness of a GA search. To achieve a complete niche & species equilibrium, additional operators are needed. Another aspect of the model is the implicit small population size used for local convergence. Since all of the genetic operators are local, the effective population size to be considered for convergence is much smaller than the actual grid capacity. The small population size aspect conforms to recent experimental and theoretical results which suggest that multiple small population runs are more effective than a few larger population runs [Goldberg, 1989].

The author is currently involved in extending the research on ecological models of GAs in two directions. One, gathering additional experimental data comparing the ECO GA model to a SGA. Two, extending the basic model to a model which exhibits a complete niche & species behaviour, and is able to maintain niches at equilibrium.

Acknowledgements

Preparation of this article was supported in part by the Centre for Absorption in Science, The Ministry of Immigrant Absorption, The State of Israel. Y. Davidor is a recipient of a Sir Charles Clore Fellowship.

References

Booker, L. B. (1982) Intelligent behavior as an adaption to the task environment, (Doctoral dissertation, University of Michigan) *Dissertation Abstracts International, 43(2), 469B*

Bridges, C. L. and Goldberg, D. E. (1987) An analysis of reproduction and crossover in a binary-coded genetic algorithms, *2nd International Conference on Genetic Algorithms*, 9-13

Cavicchio, D. J. (1970) Adaptive search using simulated evolution, (Doctoral dissertation, University of Michigan) *Unpublished doctoral dissertation*

Cohoon, J. P., *et al.* (1987) Punctuated equilibria: A parallel genetic algorithm, *2nd International Conference on Genetic Algorithms*, 148-154

Davidor, Y., and Syswerda, G., (in preparation). *Analytical and experimental comparison between generational and steady-state genetic algorithm.*

Deb, K. and Goldberg, D. E. (1989) An investigation of niche and species formation in genetic function optimization, *3rd International Conference on Genetic Algorithms*, 42-50

DeJong, K. (1975) An analysis of the behavior of a class of genetic adaptive systems, (Doctoral dissertation, University of Michigan) *Dissertation Abstracts International 36(10), 5140B*

Goldberg, D. E. (1989) Sizing population for serial and parallel genetic algorithms, *3rd International Conference on Genetic Algorithms*, 70-79

Goldberg, D. E. and Richardson, J. (1987) Genetic algorithms with sharing for multimodal function optimization, *2nd International Conference on Genetic Algorithms*, 41-49

Jog, P. and VanGucht, D. (1987) Parallelisation of probabilistic sequential search algorithms, *2nd International Conference on Genetic Algorithms*, 170-176

Mauldin, M. L. (1984) Maintaining diversity in genetic search, *National Conference on Artificial Intelligence*, 247-250

Muhlenbein, H. (1989) Parallel genetic algorithms, population genetics and combinatorial optimization, *3rd International Conference on Genetic Algorithms*, 416-421

Pettey, C. C. and Leuze, M. R. (1989) A theoretical investigation of a parallel genetic algorithm, *3rd International Conference on Genetic Algorithms*, 398-405

Schaffer, J. D. (1984) Some experiments in machine learning using vector evaluated genetic algorithms, (Doctoral dissertation, Vanderbilt University) *Unpublished*

Schaffer, J. D., *et al.* (1989) A study of control parameters affecting online performance of genetic algorithms for function optimization, *3rd International Conference on Genetic Algorithms*, 51-60

Schleuter, M. G. (1989) ASPARAGOS, An asynchronous parallel genetic optimization strategy, *3rd International Conference on Genetic Algorithms*, 422-427

Spears, W. M. and DeJong, K. A. (in press) An analysis of multi-point crossover, *International Workshop on the Foundations of Genetic Algorithms*,

Syswerda, G. (1989) Uniform crossover in genetic algorithms, *3rd International Conference on Genetic Algorithms*, 2-9

Syswerda, G. (1990) A study of reproduction in generational and steady state genetic algorithms, *International Workshop on the Foundations of Genetic Algorithms*,

Tanese, R. (1989) Distributed genetic algorithms, *3rd International Conference on Genetic Algorithms*, 434-439

Whitley, D. (1989) The GEnitor algorithm and selection pressure: Why rank-based allocation of reproductive trials is best, *3rd International Conference on Genetic Algorithms*, 116-123

Simulated Co-Evolution as The Mechanism for Emergent Planning and Scheduling

Philip Husbands
School of Cognitive & Computing Sciences
University of Sussex
Falmer, Brighton, England, UK, BN1 9QH

Frank Mill
Dept. Mechanical Engineering
University of Edinburgh
The King's Buildings, Edinburgh EH9 3JL, Scotland

Abstract

The underlying structure of many combinatorial optimisation problems of practical interest is highly parallel. However, traditional approaches to these problems tend to use mathematical characterisations that obscure this inherent parallelism. By contrast, the use of biologically inspired models casts fresh light on a problem and may lead to a more general characterisation which clearly indicates how to exploit parallelism and gain better solutions. This paper describes a model based on simulated co-evolution that has been applied to a highly generalised version of the manufacturing scheduling problem, a problem previously regarded as too complex to tackle. Results from an implementation on a parallel computer are given.

1 Introduction

There is a very large body of work on solving planning and scheduling problems, mainly emanating from the fields of Artificial Intelligence and Operations Research. Traditional AI approaches have had limited success in real-world applications, indeed their shortcomings have been thoroughly explored and documented [2]. The general resource planning, or scheduling, problem is well known to be NP-Complete [5]. Consequently OR techniques have been developed to give exact solutions to restricted versions of the problem, but in general there is a reliance on heuristic-based methods. Because of the complexity and size of the search spaces involved, a number of simplifying assumptions are always used in practical applications. These assumptions are now implicit in what have become the standard problem formulations. The authors hold the view that in many instances this has led to the most general underlying optimisation problem being ignored or, more often, not even being acknowledged as existing at all.

This paper will concentrate on the domain of manufacturing planning and scheduling. In this domain the true relationship between planning and scheduling appears to have been lost sight of long ago. Scheduling is essentially seen as the task of finding an optimal way of interleaving a number of plans which are to be executed concurrently and which must share resources. The implicit assumption is that once planning has finished scheduling takes over. In fact there are many possible choices for the sub-operations in most planning problems. Very often the real optimisation problem is to *simultaneously* optimise the individual plans and the overall schedule. This paper describes how manufacturing planning has been radically recast to allow solutions to the simultaneous plan and schedule optimisation problem, a problem previously considered too hard to tackle at all. A model based on simulated co-evolution is described and it is shown how complex interactions are handled in an emergent way. Results from an implementation on a parallel machine are reported.

Although this paper is largely focused on one particular optimisation problem, it should be noted that the model presented can be generalised. This work is concerned with using parallel GA search in a form of distributed problem solving. Most previous parallel GA work has been concerned with speed up and devising parallel implementations which provide a more robust search [14,15]. In contrast, the work reported here is concerned with using parallel GA search to simultaneously solve interacting subproblems. From this emerges the solution to some wider more complex problem. The idea is to recast a highly complex problem in terms of the cooperative and simultaneous solution of a number of simpler interacting subproblems. Using this variation of divide and conquer, the inherent parallelism in a problem is brought out and thoroughly exploited. The model involves a number of separate populations each evolving using a GA. The genotype for each population is *different* and represents a solution to one of the subproblems. Because the fitness of any individual in any population takes into account the interactions with members of other populations, the separate species co-evolve in a shared world. In this model, possible conflicts between species (e.g. disputes over shared resources) are decided by a further *co-evolving* species, the Arbitrators. The Arbitrators evolve under a pressure to make decisions that benefit the whole ecosystem (cooperative distributed solution to the overall problem). Without explicitly encoding the overall problem, the Arbitrators are used to try and adhere to the global constraints demanded by the problem.

Further details should become clear on reading about the specific example described here. A drive behind this work was to find robust techniques for tackling extremely large combinatorial optimisation problems. Heuristic-free search (as provided by GAs) is of great interest because non-brittle heuristics are very hard to come by in many practical problems. It should be noted that we are describing problems with unimaginably huge search spaces (far larger than the number of particles in the universe) so 'solution' does not necessarily been global optimum. Of course, there are no know guaranteed methods for finding optimums in these sorts of problems.

2 Domain of Application: Manufacturing Planning and Scheduling

Consider a manufacturing environment in which *n* jobs or items are to be processed by *m* machines. Each job will have a set of constraints on the order in which machines can be used and a given processing time on each machine. The jobs may well be of different lengths and involve different subsets of the *m* machines. The job-shop scheduling problem is to find the sequence of jobs on each machine in order to minimise a given objective function. The latter will be a function of such things as total elapsed time, weighted mean completion time and weighted mean lateness under the given due dates for each job [3]. In the standard model process planning directly proceeds the scheduling. A process plan is a detailed set of instructions on how to manufacture each part (process each job). This is when decisions are made about the appropriate machines for each operation and any constraints on the order in which operations can be performed [1]. Very often completed process plans are presented as the raw data for the scheduler. However, in many manufacturing environments there are a vast number of legal plans for each component. These vary in the orderings between operations, the machines used, the tools used on any given machine and the orientation of the work-piece on any given machine. They will also vary enormously in their costs. Instead of just generating a reasonable plan to send off to the scheduler, it is desirable to generate a near optimal one. Clearly this cannot be done in isolation from the scheduling: a number of separately optimal plans for different components might well interact to cause serious bottle-necks. Because of the complexity of the overall optimisation problem, that is simultaneously optimising the individual plans and the schedule, and for the reasons outlined in the introduction, up until now very little work has been done on it. However, recasting the problem to fit an 'ecosystem' model of co-evolving organisms has provided a solution. Partly because of the power of the central optimisation technique (genetic algorithms) and partly because the recasting has allowed many of the complex interactions inherent in the problem to be represented in a simple and natural way.

3 Overview of Approach

This paper concentrates on one core aspect of a complete framework for dealing with a certain class of planning problems. To give the reader a clearer understanding of the context of the work described here, the overall approach is briefly presented. This is captured, at a very high level, in figure 1.

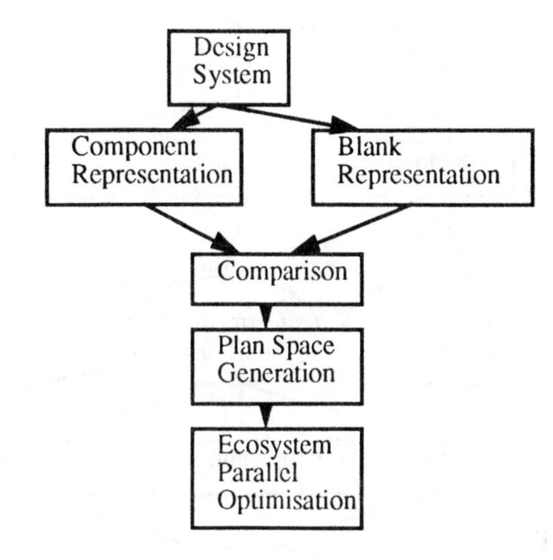

Figure 1: Information Processing Flow in System

A design system, whose description is outside the scope of this paper, produces component and blank representations. These representations are compared in order to find out which component features are to be machined and which, if any, already exist in the blank. The complete space of plans for each component is implicitly generated. These spaces are searched in parallel, taking into account interactions between and within plans, using an ecosystem model. From this emerges a solution to the simultaneously optimal plans and schedule problem. The earlier, knowledge based, parts of the system determine the boundaries and structure of the search space that the emergent optimisation techniques work in. For further details of other aspects of the system see [11,12].

In order to understand the interpretation of the genomes described later, a few more words need to be said about the plan space generation. This is done by a knowledge-based system, which breaks down the manufacture of a component into a number of nearly independent operations. The entire space of possible plans can then be generated by finding all the possible ways to carry out each operation, along with ordering constraints. The execution of an operation is defined in terms of a <machine/process/tool/setup> tuple. The first three fields indicate how to use the machine and the fourth refers to the orientation of the work-piece (partially completed component). The output from this process is a large number of interconnected networks like the one in figure 2. A manufacturing process for the sub-goal described by the fragment of network shown is a route from the starting conditions node to the goal conditions node. Implicit in the representation are functional dependencies and ordering constraints between sub-operations.

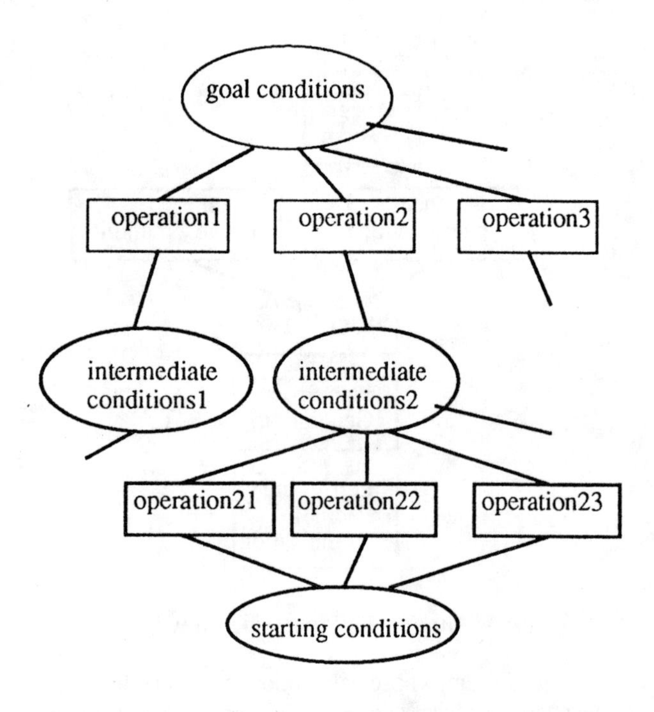

Figure 2: Planning Network

4 The Co-Evolving Species Model

4.1 Overview

The idea behind the co-evolving species model is shown in figure 3. The genotype of each specie represents a feasible process plan for a particular component to be manufactured in the machine shop. Separate populations evolve under the pressure of selection to find near-optimal process plans for each of the components. However, their fitness functions take into account the use of shared resources in their common world (a model of the machine shop). This means that without the need for an explicit scheduling stage, a low cost schedule will emerge at the same time as the plans are being optimised.

The data provided by the plan space generator, and depicted in figure 2, is used to randomly generate populations of structures representing possible plans, one population for each component to be manufactured. An important part of this model is the population of Arbitrators, again initially randomly generated. The Arbitrators' job is to resolve conflicts between members of the other populations; their fitness depends on how well they achieve this. Each population, including the Arbitrators, evolve under the influence of standard Holland-type genetic algorithms [10,6], using crossover and mutation and employed as the breeding stage of the core algorithm of figure 3. It is important to note that the

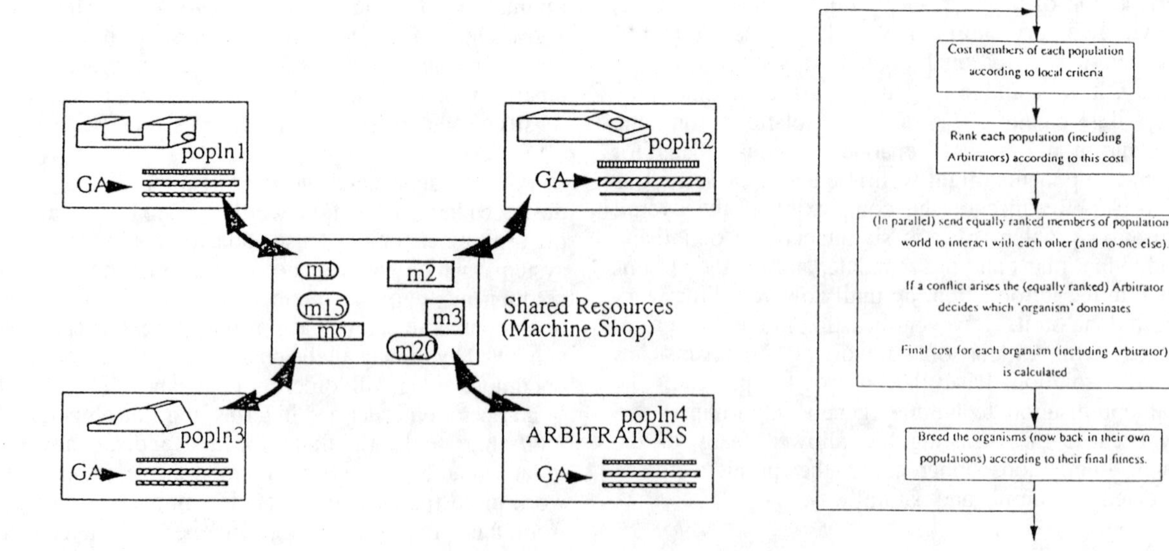

Figure 3. Co-Evolving Species Model for Simultaneous Plan/Schedule Optimisation

environment of each population includes the influence of all the other populations.

4.2 The Genotypes

The genotype of a process plan organism can be represented as follows:

$op_1 m_1 s_1 op_2 m_2 s_2 G op_3 m_3 s_3 op_4 m_4 s_4 op_5 m_5 s_5 G$ ....

Where op_i refers to the ith operation in a plan, m_i to the machine to use for that operation and s_i to the setup. Operations with interdependencies are grouped together, each group being terminated by a special symbol (G in above example). As long as the group terminators are the only legal crossover points, the crossover operation will always produce legal plans. If crossover were to occur within a group, data for dependent operations would be split up and illegal plans would probably occur on recombination. The mutation operator is also fairly involved because the gene values are context sensitive due to the dependencies. This encoding encapsulates the network structures of the data produced by the plan space generator. Each op_i, m_i and s_i have associated with them finite sets of possible integer coded values. Because these sets are all quite different, bit string representations would be awkward and unnatural, hence so called real valued codes are used.

The genotype is transformed into another form for interpretation by the fitness function. This is to take into account the ordering aspect of the problem. There is a network of partial ordering constraints associated with each genotype (specie), the operations must be ordered in accordance with these. Several methods have been used to represent the ordering part of the problem: extra genes to denote the relative orderings, a separate chromosome that holds the equivalent information and the use of a separate species of parasites who performed the translation. The first two required PMX type crossover operators [7]. Further work is required on the latter, but it appears to be the most promising.

The Arbitrators' genotype is a bit string which encodes a table indicating which population should have precedence at any particular stage of the execution of a plan, should a conflict over a shared resource occur. There is one bit for each possible population pairing at each possible stage. Hence the Arbitrator genome is a bit string of length S.N.(N-1)/2, where S = maximum number of stages in a possible plan (defined later) and N = number of process plan organism populations. Each bit is uniquely identified with a particular population pairing and is interpreted according to the following function:

$$f(n_1, n_2, k) = g\left[\frac{kN(N-1)}{2} + n_1(N-1) - \left(\frac{n_1(n_1+1)}{2}\right) + n_2 - 1\right]$$

Where n_1 and n_2 are unique labels for particular populations, $n_1 < n_2$, k refers to the stage of the plan and g[i] refers to the value of the ith gene on the Arbitrator genome. If $f(n_1, n_2, k) = 1$ then n_1 dominates, else n_2 dominates. By using pair wise filtering the Arbitrator can be used to resolve conflicts between any number of different species.

4.3 Cost Functions

As indicated in the algorithm shown in figure 3, the cost functions for all species involve two stages. The first stage involves local criteria and the second stage takes into account interactions between populations. The first stage cost function for the process plan organisms, COST1 shown below, is applied to the genotype shown above *after it has first been translated into a linearised format that can be interpreted sequentially.*

$$\mathrm{COST1}(\mathrm{plan}) = \sum_{i=1}^{N} \left(M(m_i, i) + S(s_i, i, m_i)\right)$$

Where s_i = setup used while processing ith operation, m_i = machine used for processing ith operation, $S(s_i, i, m_i)$ = setup cost for ith operation, $M(m_i, i)$ = machining cost for ith operation, N = number of operations to be processed and $M(m_i, i)$ has been previously calculated and is looked up in a table. Note that a setup cost is incurred every time a component is moved to a new machine or its orientation on the same machine changes. This function performs a basic simulation of the execution of the plan. Its input data is an ordered set of (machine,setup) pairs, one for each operation. The operations must be ordered in such a way that none of the constraints laid down by the planner are violated. Ordered sets of operations to be processed using a particular machine/setup combination are (effectively) built up on a 2D grid. $S(s_i, i, m_i)$ governs the way in which the sets are built up on the grid. The operations in any set can be performed in isolation from those in any other set. Such a set is referred to as a *stage* of a job throughout this paper. These sets themselves are ordered and the outcome is a process plan like the one shown below, where the integers in the sets refer to particular operations.

1) machine: 6 setup: 5 [0,3,5,7]
2) machine: 2 setup: 21 [1,8,12,19]
3) machine: 11 setup: 4 [2,4,6,9,13,15] ...etc

In fact COST1 provides a mapping from the process plan genotype to its phenotype: one of the plans illustrated above. Note that the setup cost is often considerably more (orders of magnitude) than the basic machining costs. The essential workings of COST1 is to sequentially process the transformed genome in order to group operations together in clusters which can then be scheduled as single units (stages). At the same time the final executable ordering of the operations is found, as well as the basic machining costs.

The definition of $S(s_i, i, m_i)$ is given below:

$$S(s_i, i, m_i) = \begin{cases} f(s_i, m_i) \text{ ,if s,m combo not previously encountered} \\ f(s_i, m_i) \text{ ,if i causes break-constraint in all grid sets} \\ f(s_i, m_i) \text{ ,if i causes set-incompatibility on all grid sets} \\ 0, \quad \text{otherwise} \end{cases}$$

$f(s_i,m_i)$ is a simple table look-up function, the pre-calculated cost of performing operation i with the particular machine and setup.

Full details of the simulation functions would take up too much space, but the following gives a brief explanation of break-constraint and set-incompatibility, as mentioned in the function definition above. Suppose operation x is being processed by COST1(plan) . It has associated with it <machine/setup> combination (m_x,s_x). Also suppose this combination has already been encountered by the objective function and so a corresponding set, S_0, exists on the simulation grid. A break-constraint occurs, in an attempt to add x to S_0, under the following condition:

$\exists z \exists y\, (y \in S_0 \wedge z \in S_n \wedge S_n \neq S_0 \wedge y \rightarrow z \wedge z \rightarrow x)$
Where, -> represents a partial ordering constraint, y and z are operations and S_n is any set on the grid.

That is, a break-constraint occurs for operation x with the <m_x , s_x > combination when there exists some operation y which has already used this same combination, i.e $s_y = s_x$ and $m_y = m_x$, and has the following additional property: due to the ordering constraints on the problem there exists a third operation, z , which must be processed after y but before x and does not use the same < setup , machine > combination, i.e. (m_z , s_z)$\neq$ (m_x, s_x). When a break-constraint occurs it is not possible to process all those operations linked to a particular <machine, setup > combination without changing machine and/or setup part way through to process some other operation. Obviously the setup cost is incurred again when processing moves back to the original machine. As far as the mechanism of the simulation is concerned, if a break-constraint occurs in an attempt to add operation x to grid set S_0, a new set, with x as the first member, is started at the same grid location as S_0 .

The set-incompatibility condition is slightly more subtle and is defined as follows: feature x causes a set-incompatibility if, when we are trying to add it to some set, S_0, of operations on the grid,

$\exists y\, (y \in S_n \wedge S_n \neq S_0 \wedge y \rightarrow x)$
$\& \; \exists z \exists w\, (z \in S_n \wedge w \in S_0 \wedge w \rightarrow z)$
Where, -> represents a partial ordering constraint, y, w and z are operations and S_n is any set on the grid.

The sets S_0 and S_n are incompatible as it is not possible to order them in relation to each other. We must start a new set at the same position as S_0 on the grid and with x as the first member. Every time a new set is created on the grid its number (order in which sets are created) and grid position are added to the end of a 'sets_so_far' list. The set ordering algorithm is based on the well known bubble-sort method; it uses this list as the array to sort. The crucial test in any sorting algorithm is one for deciding whether two adjacent members of the array are in the correct order. The action of the setup function S(s_i , i ,m_i) , particularly the set-incompatibility condition, ensures that it is possible to order any two grid sets. The ordering condition is simple. If any member of set S_i has any member of set S_j in its extended after-constraints list, then S_i is ordered before S_j. Formally, S_i -> S_j if:

$(\exists x \exists y\, (x \in S_i \wedge y \in S_j \wedge y \in A_x^{ext}))$ (i.e. x->y)

Where, S_i and S_j are grid sets, x and y are operations, -> represents a partial ordering constraint and A^{ext}_x is the complete set of all operations lying after x in the overall partial ordering.

It should be clear from the above that COST1 involves a fairly complex *interpretation* of the genotype, there is quite a high level of epistasis and the mechanisms for genes to influence each other's contributions to the overall genome fitness are complex. Despite this, genetic search performs very well.

The local cost criteria for the Arbitrators is derived from the final fitness of their parents. The function is given below, the section of the offspring genome up to *cp* (crossover point) was copied from parent1 and the section after *cp* was copied from parent2.

If cp > ALEN, cost = cost of parent1 (active part of genome was inherited solely from parent1)
Else, cost = (cp/ALEN)(cost of parent1) + (1 - cp/ALEN)(cost of parent2)

W here ALEN is the average useful Arbitrator length. This is a dynamic quantity as fitter plans tend to become shorter meaning that arbitrating decisions are not needed for later stages as the system evolves. This is only used because a fully dynamic implementation of the Arbitrators has not yet been completed.

The second phase of the cost function involves simulating the simultaneous execution of plans derived from stage one. Additional cost are incurred for waiting and going over due dates. There are a number of interesting problems here. We are working towards a set of optimal plans, one for each component, which when executed simultaneously will provide an optimal schedule. This means that most of the possible interactions between members of one population and all the members of another population are largely irrelevant. The solution used here was to rank each population according to the local cost functions described above and to run the simulation of phase two for equally ranked organisms. What happens when two plans want the same resource at the same time? Fixed precedences would be far too inflexible and random choices would be of no help. As already indicated, the most general and powerful solution developed was to introduce a new species, the Arbitrators, whose genetic code holds a table indicating which population had precedence at any stage. The Arbitrators are costed according to the amount of waiting and the total elapsed time for a given simulation. The smaller these two values, the fitter the Arbitrator. Hence the Arbitrators, initially randomly generated, are allowed to co-evolve with the plan organisms. Again, the Arbitrators are ranked and a simulation involves equal ranking members from each population, including the Arbitrators. If there is a conflict the Arbitrator resolves it. This scheme allows the evolution of sensible priorities at the various stages of the simulation. After the second phase each individual's fitness is calculated according to its *total* cost. This means that selection pressure takes account of both optimisation problems: interactions during phase two that increase an individuals cost will reduce its chances of reproduction, just as will a poor result from phase one of the costing. In general, a population of co-evolving Arbitrators could be used to resolve conflicts due to a number of different types of operational constraint. The cost of an Arbitrator after the second phase simulation is a function of the total schedule length and the weighted penalties incurred by the various plan populations.

5 Results

Figure 4 shows results from an implementation on a transputer based parallel machine. Typical results for a two job problem are shown. The graph shows how the machining costs (COST1) of the best individual in each population reduce with time, and also how the Arbitrator costs reduce. It also shows how the total elapsed time reduces. The gantt charts show how the emergent schedule evolves. The vastly reduced number of stages in the lower chart reflects the fact that machining costs can be decreased by putting more operations into a single stage. Clearly both optimisation problems have been tackled simultaneously. Note that there is some tension between the various objectives, one cost may momentarily rise while others drop, but the overall trend is down. A model of a real job-shop is used and the components planned for are of medium to high complexity needing 25-60 operations to manufacture. Each job has a number of internal partial ordering constraints but is by no means strongly constrained. Typically each operation has 8 candidate machines and each of these machines has 6 possible setups To simplify matters, tool changes and machine transfer costs have not been modelled in great detail. However, it is a simple matter to include them and future versions of the model will be complete in that respect. Experiments with up to 4 jobs have been conducted. Very promising results have been obtained for this extremely complex optimisation problem, never before attempted.

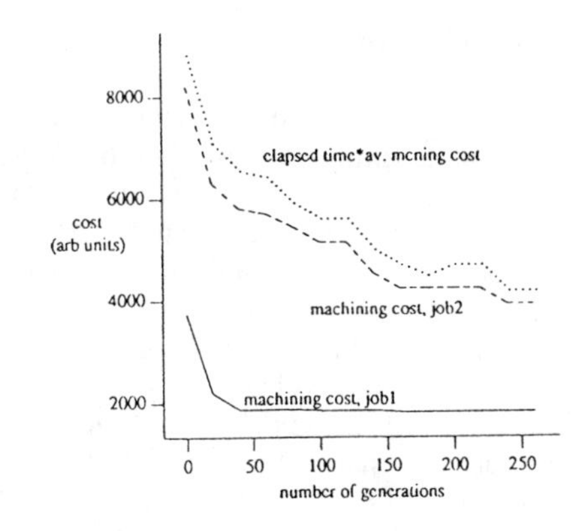

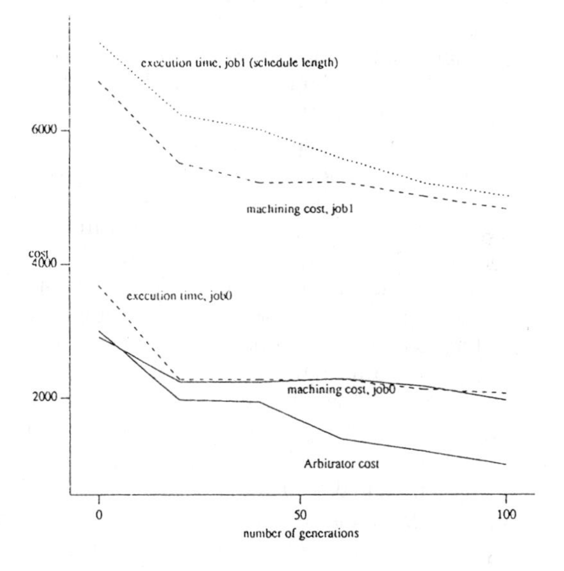

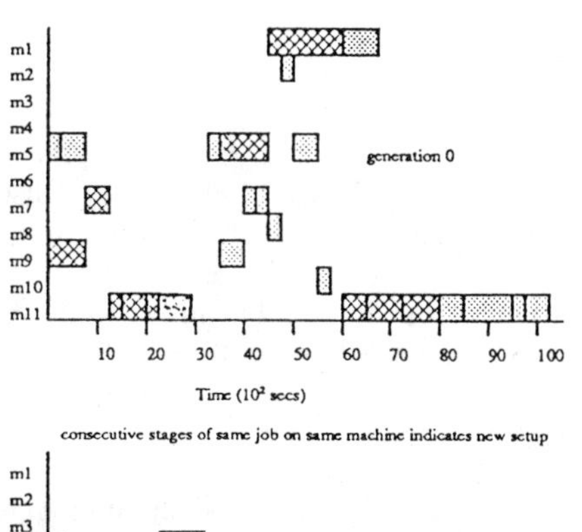

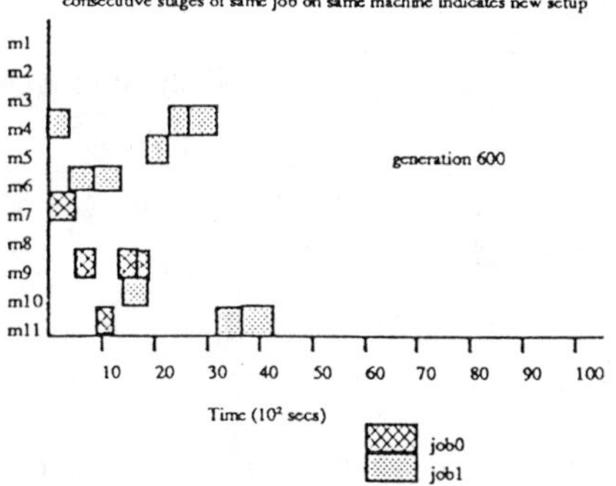

Figure 4. Results

6 Discussion

Davis has done some work on using GAs to solve job-shop scheduling problems [4], but his solution was for the simplified problem that does not take into account the proper relationship between planning and scheduling. Each genome represented an entire schedule, that approach cannot exploit the inherent parallelism of the problem in the same way that the work described here has. Hilliard et al [8] have used a classifier system to discover scheduling heuristics. That work may possibly tie in with ongoing research on enabling the Arbitrators to learn how to resolve a number of different type of conflicts, there is no reason why the Arbitrators should not become fully blown classifier systems. Because this system runs on a powerful parallel machine (500 transputers) very good solutions are found within a few minutes, because of this not much effort has yet been put into making the system react to sudden changes in the manufacturing environment. However, this is an area for future research. One possible scenario that is envisaged is that the system will run in the background and be continuously updated with feedback from the job-shop, in other word the simulated environment will dynamically mirror the actual manufacturing environment. Hillis and Koza [9,13] have previously used co-evolution but in quite different contexts. The work has used a very straightforward implementation of the actual GAs on the transputers, it may benefit from some more work in that direction. Certainly work is ongoing in extending the model and running it with 50-60 jobs rather than the handful used to date. There is a great deal of work in setting up the planning data for further jobs. However, an interactive system that should make that much easier is near completion.

Acknowledgements

This work was in part supported by SERC grants GR/D 63103 and GR/E 04837. It has benefited from discussion with many people including Stephen Warrington, Inman Harvey and Phil Agre. Acknowledgements are also due to The Edinburgh Manufacturing Planning Group and The Edinburgh Parallel Computing Centre.

References

[1] Chang, T. & Wysk,R. "An Introduction to Automated Process Planning Systems", Prentice-Hall, 1985.

[2] Chapman, D. "Planning for Conjunctive Goals", Tech. report AI-TR-802, MIT AI Lab, 1985.

[3] Christophedes, N. "Combinatorial Optimisation", Wiley, 1979.

[4] Davis, L. 'Job Shop Scheduling with Genetic Algorithms', in J. Grefenstette (ed), *Proc. Int. Conf. on Genetic Algorithms and their Applications*, Lawrence Erlbaum,1985.

[5] Garey, M. & Johnson, D. "Computers and Intractibility: A Guide to the Theory of NP-Completeness", W.H. Freeman, 1979.

[6] Goldberg, D. "Genetic Algorithms", Addison Wesley, 1989.

[7] Goldberg, D. & Lingle, R. "Alleles, Loci and The TRavelling Salesman Problem", in J. Grefenstette (ed), *Proc. Int. Conf. on GAs and their Applications, Lawrence Erlbaum*, 1985.

[8] Hilliard, M et al. 'A Classifier based system for discovering scheduling heuristics', in J. Grefenstette (ed), *Proc. 2nd Int. Conf. on GAs*, Lawrence Erlbaum,1987.

[9] Hillis, W.D. 'Co-Evolving Parasites Improve Simulated Evolution As an Optimisation Procedure', Physica D 42,228-234, 1990.

[10] Holland, J. "Adaptation in Natural and Artificial Systems", University of Michigan Press, 1976.

[11] Husbands,P., Mill,F.G. & Warrington,S.W., 'Representation, Reasoning and Decision Making in Process Planning with Complex Components',in "*Geometric Reasoning*", Woodwark, J (ed),203-215, Oxford University Press, 1989.

[12] Husbands,P., Mill,F.G. & Warrington,S.W., 'Generating Optimal Process Plans from First Principles', in *"Expert Systems for Management and Engineering"*, Balagurasamy, E. & Howe, J. (eds),130-153, Ellis Horwood, 1990.

[13] Koza, J. 'Genetic programming: A paradigm for genetically breeding populations of computer programs to solve problems', Tech. Report STAN-CS-90-1314, Dept. Compt. Sci., Stanford University, 1990.

[14] Pettey, C., Leuze, M., Grefenstette, J. "A Parallel Genetic Algorithm", in J. Grefenstette (ed), *Proc. 2nd INt. Conf. on GAs*, Lawrence Erlbaum, 1987.

[15] Muhlenbein, H. "Parallel Genetic Algorithms, Population Genetics and Combinatorial Optimisation", in J. Schaffer (ed), *Proc. 3rd INt. Conf. on GAs*, Morgan Kaufmann, 1989.

The Parallel Genetic Algorithm as Function Optimizer

H. Mühlenbein, M. Schomisch, J.Born
Gesellschaft für Mathematik und Datenverarbeitung mbH
Postfach 1240
D-5205 Sankt Augustin 1

Abstract

In this paper, the parallel genetic algorithm PGA is applied to the optimization of continuous functions. The PGA uses a mixed strategy. Subpopulations try to locate good local minima. If a subpopulation does not progress after a number of generations, hill-climbing is done. Good local minima found by a subpopulation are diffused to neighboring subpopulations. Many simulation results are given with popular test functions. The PGA performs better than other genetic algorithms on simple problems. A comparison with mathematical optimization methods is done for very large problems. Here a breakthrough can be reported. The PGA is able to find the global minimum of Rastrigin's function of dimension 400 on a 64 processor system! Furthermore, we give an example of a superlinear speedup.

Keywords: Structured populations, function optimization, parallel computation

1 Introduction

The parallel genetic algorithm PGA was introduced in 1987 ([9],[10]). It is an adaptation of the genetic algorithm GA to parallel computers. The PGA is totally asynchronous with distributed control. This has been achieved by the introduction of a spatial population structure and active individuals.

We have successfully applied the PGA to a number of problems, including function optimization and combinatorial optimization. In this paper we summarize the results for function optimization. An extended version of this paper has been published in [12]. The results in combinatorial optimization can be found in ([7],[2], [8]).

In all these applications we have used the same algorithm with only slight modifications. We have taken the largest published problem instances known to us. The PGA found solutions, which are comparable or even better than any other solution found by other heuristics. This is proof of its power by experiment.

We hope that future researchers will start were we have left off. The evaluation of heuristics for difficult optimization problems has to be done with large problems. Results on small toy problems cannot be extrapolated to large problems.

The most difficult part of a random search method is to explain why and when it will work. We will make a comparison with other random search methods and explain the more tricky parts of the algorithm. These are the spatial structure of the population, the selection schedule and the genetic operators. A more detailed discussion can be found in [8].

Throughout the paper we will use biological terms. We believe that this is a source of inspiration and helps one to understand the PGA intuitively.

2 Description of the PGA

The PGA can be informally described as follows. Subgroups of individuals live on a ladder-like 2-D world. New offspring are created by genetic operators within a subpopulation. Every k generations, the best individual of a subpopulation is sent to its neighbors. If the subpopulation does not progress for a number of generations, one individual will try local hill-climbing.

We will now describe the PGA more formally. The PGA is a black-box solver, which can be applied to the class of problems

$$min\{f(x) \mid x \in X\} \ where \ X \subseteq R^n, f : R^n \mapsto R.$$

$$X = \{x \in R^n \mid a_i \leq x_i \leq b_i, i = 1, ..., n\},$$
$$where \ a_i < b_i, i = 1, ..., n.$$

Note that f is not required to be convex, differentiable, continuous or unimodal.

The PGA operates with a set (a population) of vectors (individuals). The *phenotype* of an individual is given by a real vector x with $x \in X$. The *genotype* of an individual is given by the bit representation y of x

according to the floating point format of the computer. A *gene locus* is a single bit position. The bits of the floating point representation define a *chromosome*. The genetic representation of an individual consists of a *set of n chromosomes*. This is an important difference to other GA's, which combine the floating point representations into a single chromosome.

The PGA can be described as an eight-tuple

$PGA = (\mathbf{P}^0, \lambda, \mu, \delta, \tau, GA, \Lambda, t)$

where

$\mathbf{P}^0$ - initial population

λ - number of subpopulations

μ - number of individuals of a subpopulation

δ - number of neighbors in a neighborhood relation among subpopulations

τ: - isolation time in generations

GA - Genetic Algorithm applied to a subpopulation

Λ - local optimizer

$t : \{0,1\}^\lambda \mapsto \{0,1\}$ - termination criterion

In the following we call the application of the Genetic Algorithm GA to a subpopulation P_i^0, $i \in \{1, ..., \lambda\}$ an evolution process $GA(P_i^k)$, $k = 1, 2, ...$.

Each subpopulation is mapped onto a different processor. The above definition includes the special case $\mu = 1$. Then the subpopulation consists of one member only. This case has been used extensively for combinatorical problems (see [8]).

There are many neighborhood relations possible. We have used for all our applications a neighborhood relation which can be called a ladder. An example is shown in Figure 2. It is outside the scope of this paper to describe why such a population structure is promising (see [8] for details).

The PGA allows the independent evolution of the subpopulations for a certain time. This "isolation" time is given in generations. Migration to the neighbors takes place at generation $\tau, 2\tau, 3\tau, ...$. In the PGA only the best individuals migrate.

The termination criterion t operates on a set of termination flags generated by the evolution processes $GA(P_i^k)$. The PGA will be terminated if all λ flags are on.

The Genetic Algorithm GA can be described as a nine-tuple:

$GA = (P^0, \mu, \Omega, \Gamma, \Delta, \Theta, l, \Lambda, t)$

where P^0 is the initial population, μ is the population size, Ω is the selection operator, Γ the crossing-over operator, Δ the mutation operator, and Θ the recombination operator. Λ is the local hillclimbing method,

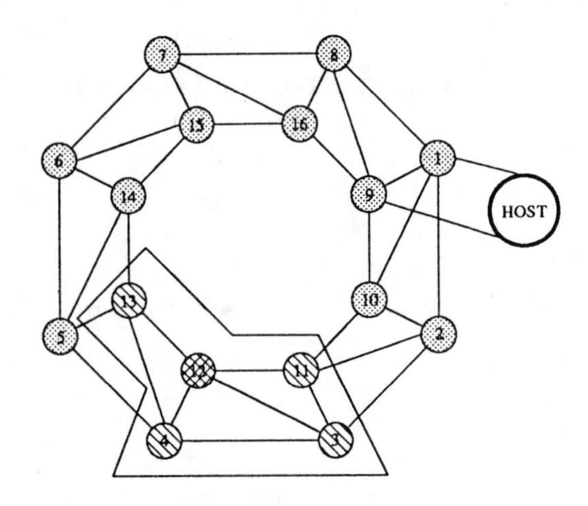

Every node represents a subpopulation $P_i, i = 1, ..., \lambda$. Population P_{12} has four populations in its neighborhood.

Figure 1: 16-node version of the ladder-population

l the criterion when to start hillclimbing and t is the termination flag. Ω, Γ, Δ, Λ use probability distributions.

In the following we will describe the GA in more detail. It has to be mentioned that our intention is a robust algorithm, which can be applied to a large variety of problems without adjustment of the parameters.

2.1 Selection

The selection operator works as follows. At each stage of the algorithm the subpopulation $P_j^k = (y_1^k, ..., y_\mu^k)$ is sorted in increasing order by $f(y_i^k)$.

For individuals with an odd index $i = 1, 3, 5, ...$ a mate is selected by a ranking scheme. The individual is selected according to the formula:

$$i_s = \frac{\mu}{2(a-1)}(a - \sqrt{a^2 - 4(a-1) * rnd(0,1)})$$

We use a high selection pressure by setting $a = 2.5$ as default. In this case onethird of the bad individuals are never selected.

The main purpose of this scheme is to select as severely as possible without destroying the diversity of the population too much. Each couple produces two offspring, therefore only $\mu/2$ couples are selected.

In addition, the GA uses the elitist strategy. The elitist strategy guaranties that the best individual of P_j^k survives to P_j^{k+1}.

2.2 Mutation, crossing-over and recombination

After the selection step has been done, the genetic operators Δ, Γ and Θ are applied to the two selected individuals y'^k and y''^k. All genetic operators, which the PGA uses, work in the space of the genotypes. In order to understand the effect of these operators, it is necessary to describe the genetic representation, in our case the ANSI-IEEE floating point representation.

In this representation the genotype g has the following form

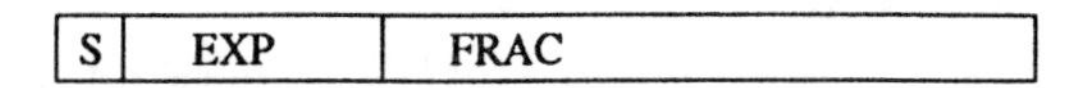

The value of g ,the phenotype p, is given by

$$p = (-1)^s \times 1.frac \times 2^{exp-bias}$$

In single precision, g is 32 bits long. s is coded by 1 bit, exp by 8 bits and $frac$ by 23 bits. $Bias$ is equal to 127 in decimal.

The PGA mutation scheme works as follows. First, with probability p_m a chromosome is mutated at all. Second, one bit is chosen randomly in the $frac$ part and flipped. If bit k is flipped then the mutated phenotype y' is given by

$$y' - y = (-1)^a 2^{-k} y$$

where $a = 1$ if bit k was on and $a = 0$ otherwise.

The mutation operator samples the neighborhood of y exponentially more frequently than regions a larger distance from y. If the feasible set is symmetric around the origin, we also flip the sign bit s.

The crossing-over operator Γ is the traditional one point crossover applied to every homologous chromosome in the mutated individuals y'^k and y''^k with probability p_c.

The recombination operator Θ exchanges homologous chromosomes with probability p_r.

The genetic operators are explained in figure 2.

In the PGA the most important operator is the recombination opearator. Recombination between $y'^k = (y_1'^k, ..., y_n'^k)$ and $y''^k = (y_1''^k, ..., y_n''^k)$ gives new phenotypes at the corners of the parallelepiped given by y'^k and y''^k.

Figure 3 shows the effect of recombination in the case of two dimensions. The square denotes the area, where new points are sampled by the genetic operators.

2.3 Local hill-climbing

The feature "local hill- climbing" has been very important in the application domain of combinatorial optimization (see [2], [8]). There are two reasons for this. First, hill-climbing is often much faster than applying genetic operators and second, the hills often have a "building block feature" i.e. each local hill has some information in common with the global minimum.

The situation is different in the application domain of function optimization. First, the cost of the evaluation of a new sample point is the same for a local hill-climbing method or a genetic operator. Furthermore, hill-climbing in the early stages of the search process does not pay off because no information about the global minimum is obtained. Therefore we decided to start the hill-climbing in the last stages of the algorithm (when we believe that the subpopulation has reached a region of attraction of a good minimum).

The decision to do local hill-climbing is done dynamically. An individual of a subpopulation is selected for local hill-climbing if one of the following criteria are fulfilled

$$f_1^k - f_1^{k-\Delta k_1} = 0$$

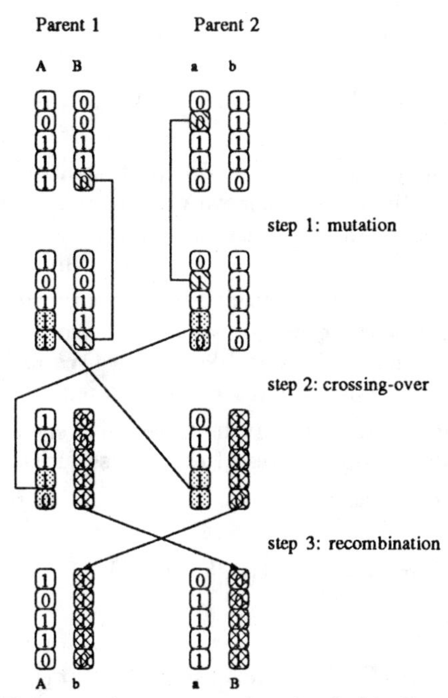

The mutation operator is a simple bit-flip. Crossing over can be described as an exchange of bit-sequences between homologous chromosomes, while recombination exchanges pairs of homologous chromosomes whithout destroying their binary structure.

Figure 2: How the PGA produces offspring

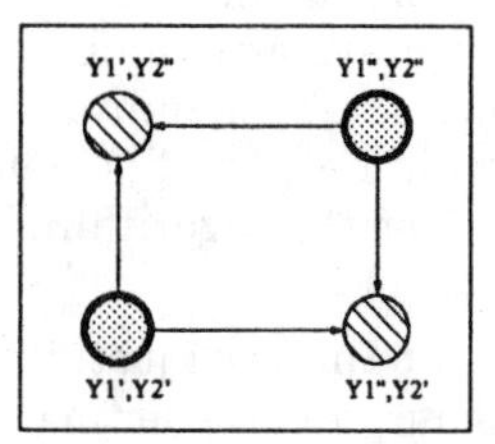

Figure 3: Recombination in phenotype space

$$||x_1^k - x_\mu^k|| \leq \alpha||b-a||$$

f_i^k is the fitness value of an individual with rank i and phenotype x_i^k, $i \in \{1, ..., \mu\}$.

The subpopulations may use different strategies for hill-climbing. In this manner we want to make the algorithm robust with regard to assumptions about (unknown) regions of attraction of the global minimum.

One of the implemented local hill-climbing methods is a random search technique with adaptive step size control proposed by Solis and Wets [14] (We use "Algorithm 1" with normally distributed steps).

We have chosen this strategy because it is very robust and does not use derivatives of the fitness function.

The vector of start step sizes σ^0 is determined by:

$$\sigma^0 = \frac{1}{\sqrt{n}} min\{\alpha||b-a||, \delta\}$$

$$\delta = max\{||x_1^k - x_m^k||, max\{10^{-3}, 10^{-3}||x_1^k||\}\}$$

x_1^k is the individual with the best fitness value, x_m^k is the individual with median fitness value in P_j^k.)

The control parameters of this hill-climbing method are explained in [14].

2.4 Termination

The PGA is a totally distributed algorithm. In distributed algorithms the problem of termination is difficult. In the PGA, the host program decides about the global termination (see figure 1). Every process $GA(P_i^k)$ sets a termination flag which is used by the termination criterion of the host program. The termination flag of a subpopulation is set, if one of the following criteria is fulfilled:

$$|f_1^k - f_1^{k-\Delta k_2}| \leq \varepsilon |f_1^k|$$

These criteria are tested every $0 mod \Delta k_2$ generations.

3 Performance evaluation

Performance evaluation of probabilistic search algorithms is a difficult task by itself. To do the evaluation, at least a performance measure and a representative suite of test functions are needed. We will use as performance measure the average number of function evaluations to compute the optimum as well as the time needed for computation. The first number measures the "intelligence" of the algorithm, the second the speed.

The difficulties in constructing well justified test functions are similar to those in constructing mathematical models of "typical" real life functions. Since no mathematical models for constructing test functions are known, they are constructed heuristically. Normally they consist of a number of combinations of elementary functions whose local minima are known.

In broad terms the complexity of a function optimization problem depends on

- the number of local minima
- distribution of the local minima
- the distribution of the function values of the local minima
- the domain of attraction for local minima

We will present performance data from eigth functions, which vary the above variables (see table 1).

Functions F1-F5 have been proposed by De Jong [5]. They are extensively used in the Genetic Algorithms community. The test environment includes functions which are discontinuous (F3), nonconvex (F2), multimodal (F5) and stochastic (F4).

Functions F6 - F8 are used in "mainstream" optimization. Each function tests the performance of a global search method on a specific aspect of the optimization problem.

F6 was proposed by Rastrigin [15]. It is highly multimodal. The local minima are located at a rectangular grid with size 1. The global minimum is at $x_i = 0, i = 1, ..., n$, giving $f(x) = 0$. Grid points with $x_i = 0$ except one coordinate, where $x_j = 1$, give $f = 1$, the second best minimum. With increasing distance to the global minimum the fitness values of local minima become larger.

Function F7 was proposed by Schwefel [13]. Here we use the function in arbitrary dimensions. The global minimum is at $x_i = 420.9687, i = 1, ..., n$. The local minima are located at the points $x_k \approx (\pi(0.5 + k))^2, k = 0, 2, 4, 6$ and $x_k \approx -(\pi(0.5+k))^2, k = 1, 3, 5$. The second best minimum is at $x_i = 420.9687, i = 1, ..n, i \neq j, x_j = -302.5232$ - far away from the global minimum. Therefore the search algorithm may be trapped in the wrong region.

Function F8 has been suggested by Griewangk. The function has its global minimum $f = 0$ at $x_i = 0, i = 1, ..., n$. The local minima are located approximately at $x_i(k) \approx^{+}_{-} k * \pi\sqrt{i}, i = 1, ..., n, k = 0, 1, 2, 3, ...$ In ten dimensions the function has four suboptimal minima $f(x) \approx 0.0074$ at $x \approx (^{+}_{-}\pi, ^{+}_{-}\pi\sqrt{2}, 0, ..., 0)$. This function is the most difficult to optimize because it is non separable. Furthermore the search algorithm has to climb a hill to get to the next valley.

4 Numerical results

The PGA has many internal control parameters, which have been derived from the evolution metaphor. It is not necessary to tune all these parameters for a specific function. The following numerical results have been obtained with a single set of parameters. This

Function Number	Function	Limits
1	$f_1(x) = \sum_1^3 x_i^2$	$-5.12 \le x_i \le 5.12$
2	$f_2(x) = 100(x_1^2 - x_2^2)^2 + (1 - x_1)^2$	$-2.048 \le x_i \le 2.048$
3	$f_3(x) = \sum_1^5 integer(x_i)$	$-5.12 \le x_i \le 5.12$
4	$f_4(x) = \sum_1^{30} ix_i^4 + Gauss(0,1)$	$-1.28 \le x_i \le 1.28$
5	$f_5(x) = 0.002 + \sum_{j=1}^{25} \frac{1}{j+\sum_{i=1}^{2}(x_i - a_{ij})^6}$	$-65.536 \le x_i \le 65.536$
6	$f_6(x) = nA + \sum_1^{20} x_i^2 - Acos(2\pi x_i)$	$-5.12 \le x_i \le 5.12$
7	$f_7(x) = \sum_1^{10} -x_i sin(\sqrt{\lvert x_i \rvert})$	$-500 \le x_i \le 500$
8	$f_8(x) = \sum_1^{10} x_i^2/4000 - \prod_1^{10} cos(x_i/\sqrt{i}) + 1$	$-600 \le x_i \le 600$

Table 1: Eight function test bed

F	n	Best		Average	
		time	feval	time	feval
F1	3	0.48	340	1.44	1170
F2	2	0.69	580	1.48	1235
F3	5	1.75	1360	3.44	3481
F4	30	7.52	2000	12.08	3194
F5	2	1.25	400	3.33	1256

F5: The global optimum was not found in 3 of the 50 runs.

Table 2: Results for $\lambda = 4$

F	n	Best		Average	
		time	feval	time	feval
F1	3	0.69	780	1.57	1526
F2	2	0.70	920	1.68	1671
F3	5	1.43	1480	3.43	3634
F4	30	6.84	3600	9.92	5243
F5	2	1.09	740	2.79	2076
F6	20	7.55	6980	9.41	9900
F7	10	4.38	3780	6.61	8699
F8	10	11.24	39240	16.84	59520

F7: The global optimum was not found in 4 of the 50 runs
F8: $\lambda = 16, \mu = 40; p_m = 0.3$

Table 3: Results for $\lambda = 8$

demonstrates the robustness of this search method. The parameters, which have been used for these runs, are given below.

number of subpopulations $\lambda = 4, 8$
number of individuals of a subpopulation $\mu = 20$
isolation time $\tau = \mu/2$
mutation probability $p_m = 1/n$ $(= 3/n$ *for F8*$)$
crossing over probability $p_c = 0.65$
recombination probability $p_r = 0.5$

Numerical results are shown in table 2 for the case $\lambda = 4, \mu = 20$ and in table 3 for the case $\lambda = 8, \mu = 20$.

The avarage is based on 50 runs. In all runs the global minimum has been found to at least three digits of accuracy. The time needed for a run is the time at the host between starting the computation and arriving of the minimum function value at the host. The number of function evaluations is not as accurate because of the termination problem mentioned earlier. When the minimal function value arrives at the host computer, it sends a termination command to all processors. Before terminating, these processors send the number of function evaluations done so far to the host which sums this number up.

It is instructive to compare our results with the results of Eshelman et. al [1], which are shown In table 4. G is the average number of function evaluations with optimal parameter setting by Grefenstette [3] and E is the average number of function evaluations with optimal parameter setting by Eshelman (We present the results for the best version: "eight-point traditional crossover").

The non-optimized PGA for F1 and F2 needs less function evaluations than these optimized Genetic Algorithms which use binary coding and a simple panmictic population. The most dramatic improvement is with F2. The discontinuous function F3 is solved better by the standard GA. This may indicate that the PGA is tuned more to continuous functions than to discontinuous ones. The results for F4 are in the same order for $\lambda = 4$. Shekel's foxholes problem F5 is also solved much better by the PGA. We found these results encouraging, because the problems are too small for the PGA to work ideally.

F	n	G feval	E feval
F1	3	2210	1538
F2	2	14229	9477
F3	5	2259	1740
F4	30	3070	4137
F5	2	4334	3004

Table 4: Results of Grefenstette(G) and of Eshelman(E)

We turn now to the more complex functions. F6 has been used by Hoffmeister et. al [4] to compare Genetic Algorithms with Evolution Strategies. Their methods did not find the global minimum within 250 generations. In a population of size 50 this gives 12500 function evaluations. After 250 generations they arrived at a minimum value of $f = 10$. The PGA found the optimum in all runs with an average of 9900 evaluations.

The PGA always found the optimum. In the beginning there is very good progress until a fitness value of about 20 is reached. In the last part of the algorithm the subpopulations get similar and concentrate around suboptimal minima. The global minimum is found by recombination only. A typical run is shown in figure 4.

There are no results available for function F7 in higher dimensions. Griewangk's function F8 is considered to be one of the most difficult test functions. The following results for F8 have been compiled in Törn and Zielinkas ([15]).

Griewangk's algorithm needed 6600 function evaluations (average number), but only one of the four suboptimal minima was found. Snyman and Fatti found the global minimum in 23399 function evaluations. Rinnooy Kan and Timmer [6] also tried function F8. They wrote: "For the two-dimensional problem, our method never really got started. Only one minimum was found, after which the method terminated. For the ten-dimensional prblem, the global minimum was found in six of the ten runs. The method terminated at an early stage using 700 function evaluations on the average. In the other four runs the method was stopped after 4000 points were sampled. At that time 17000 function evaluations had been used."

A performance evaluation based on time is not possible because of lack of data. In mainstream optimization a normalization of the computing time is done in order to compare the results for different machines. The normalized time unit is 1000 computations of Shekel's function S5 [15]. For our machine, the MEGAFRAME HyperCluster [11], the time unit is 1.7640 on one processor.

We will now turn to very large problems, which have not been dealt before. The PGA is an asynchronous parallel algorithm and therefore ideally suited to parallel computers. The behavior of the PGA on parallel processors will be investigated in the next section.

5 Superlinear speedup

Measurements of the speedup of the PGA on parallel processors is very difficult. First, each individual run is different because of the probabilistic nature of the algorithm. Second, a different number of processors needs a different parameter λ, which changes the search strategy.

The data in tables 2 and 3 show a small speedup for the functions F4 and F5 only. But the global minimum of F1-F3 is computed faster on four processors than on eight processors. This situation is typical for small problems. Parallel search pays off only, if the search space is large and complex.

The largest speedup is to be expected with function F7. Here the suboptima are far away from the optimum, so that a parallel search with subpopulations seems to be promising. Table 5 shows the result.

λ	μ	Best		Average	
		time	feval	time	feval
4	80	68.7	37280	84.6	50720
8	40	27.4	28640	32.3	37573
16	20	15.1	31680	17.2	39773

Table 5: Superlinear speedup F7 (n=30)

In the table a superlinear speedup can be observed when going from four to eight processors. The speedup is 2.6. This can be explained easily. Eight processors need less function evaluations than four processors to locate the minimum. The speedup between one processor and four processors would have been still more. Unfortunately we cannot make a comparison, because a single panmictic population (we tried up to $\mu = 640$) was not able to find the minimum at all. This shows the advantage of a spatial population structure. Similar results have been obtained with combinatorial optimization problems([8]).

All optimization problems so far have been solved by a small number of processors. We claim that the PGA is a breakthrough in parallel search methods. We hope to demonstrate this with the next examples. In these examples we will try to solve very large problem instances with the maximum number of processors we have available . The number is 64 [11].

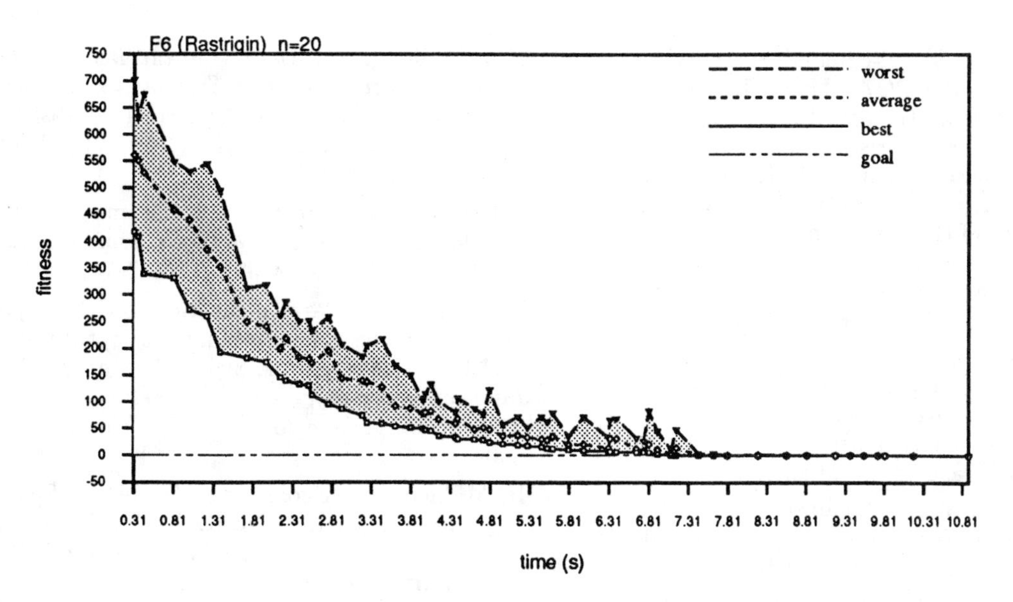

Figure 4: Average run for test-function F6

F	n	λ	μ	Average time	Average feval
F6	50	8	20	57.40	42753
	100	16	20	129.69	109072
	200	32	40	404.26	390768
	400	64	40	4849.458	7964400
F7	50	32	20	34.96	119316
	100	64	20	213.93	1262228
	150	64	40	1635.56	7041440

Table 6: Benchmarks for F6 and F7 (5 runs)

The number of function evaluations for F6 increases approximately like $n * \sqrt{n}$ in the range from $n = 20$ until $n = 200$. The same is true for F7 in the range from $n = 10$ until $n = 50$. Then the number of function evaluations increases rapidly. It is an open question, if a much larger number of processors could reduce the above numbers substantially. $F6(n = 400)$ and $F7(n = 150)$ are single runs only.

6 The PGA and mathematical optimization

There exists no unique optimal search method. Most of the proposed search methods for global optimization use heuristics to generate the sample points as well as possible, given the information at hand. The search strategy of the PGA can be explained as follows.

λ random samples of μ points are placed uniformly within the feasible set. Within each subgroup additional points are tested by recombination. These points are drawn probabilistically out of the n-dimensional grid defined by the μ sample points. Selection concentrates the sample points at promising regions. Therefore the size of the grid, i.e. the region tested, gets smaller. If the points are too close to each other, local hill-climbing will be done. So each subgroup ends up at one or more local minima.

By migration to the neighboring subgroups, the information of good points is spread. In the last stage of the algorithm the grid defined by the local minima found is also tested by recombination. This is the reason why subpopulations are so important in the PGA. The subpopulations do not need to converge to the global minimum, this can be found later by recombination.

It is instructive to compare the PGA heuristic with more standard optimization techniques. The most similar heuristic is used by clustering methods. For grouping points around minima, two strategies have been used. The first strategy tries to form groups around *some* of the most promising local minima, the second tries to produce groups at *all* local minima. In both cases a cluster analysis is applied to prevent redetermination of already known local minima.

The second strategy cannot be applied to large optimization problems. Rastrigin's function F6 has n^{11} local minima, where n is the dimension! We have to use strategies which concentrate the point in promising areas. A discussion of the different strategies can be found in [15].

The similarity to strategies used by the PGA is obvious. The major difference is the cluster analysis to prevent redetermination of already known local minima.

7 Conclusion

The PGA is no miracle, it is a very simple parallel search which concentrates the search in promising regions. Recombination gives an automatic step size control. The nearer the parent points are, the smaller will be the sampling area. Hill-climbing is a very important component of the PGA. The PGA outperforms more traditional genetic algorithms , also if the "intelligence" of the search heuristic is considered - the number of function evaluations needed. The PGA like any other *clean* genetic algorithm suffers from the fact that in the last stage of the algorithm the same function evaluations are done over and over again. Nevertheless, the *clean* PGA is able to solve complex search problems in very high dimensions which have not been solved before.

It has to be proven, that search methods based on *mathematical* heuristics are able to solve such large problems. We hope that researchers in mathematical optimization methods take up our challenge!

An objection often mentioned against genetic algorithm is, that there is no convergence theorem. But any algorithm which tries to estimate the probability to find the global minimum in a high dimensional problem will suffer from the *curse of dimensionality*. This can be easily demonstrated. In n dimensions, a hypercube of size a has 2^n hypercubes of size $1/2a$. With statistical information only, we need an exponential number of sample points to be sure that the global minimum is in one of the subcubes!

The most important feature of the PGA is that it is a totally asynchronous parallel algorithm. It runs with 100% efficiency on any number of processors. Many extensions of the *clean* PGA are straightforward . We only want to mention the following: Subpopulations could do different search strategies, the population structure could change during the run, the population could consist of different species etc...These extensions will be implemented in the course of applying the PGA to more and more challenging applications.

References

[1] L.J. Eshelman, R.A. Caruana, and J.D. Schaffer. Biases in the Crossover Landscape. In J.D. Schaffer, editor, *Proceedings of the Third International Conference on Genetic Algorithms*, pages 10–19, 1989.

[2] M. Gorges-Schleuter. *Genetic algorithms and population structure - A massively parallel algorithm* . PhD thesis, University of Dortmund, 1990.

[3] J.J. Grefenstette. Optimization of Control Parameters for Genetic Algorithms. *IEEE Transactions on Systems, Man and Cybernetics*, 16:122–128, 1986.

[4] F. Hoffmeister and Th. Bäck. Genetic Algorithms and Evolution Strategies: Similarities and Differences. In H.-P Schwefel, editor, *PPSN - First International Workshop on Parallel Problem Solving from Nature. October 1-3, 1990. Dortmund, FRG. Preprints*, 1990.

[5] K.A. De Jong. *An Analysis of the Behavior of a class of Genetic Adaptive Systems.* PhD thesis, University of Michigan, Dissertation Abstracts International 36(10), 5140B. (University Microfilms No. 76-9381), 1975.

[6] A.H.G. Rinnooy Kan and G.T. Timmer. Stochastic Global Optimization Methods, I and II. *Mathematical Programming*, 39:27–26,57–78, 1987.

[7] H. Mühlenbein. Parallel genetic algorithms, population genetics and combinatorial optimization. In J.D. Schaffer, editor, *proceedings of the Third International Conference on Genetic Algorithms*, pages 416–421, San Mateo, 1989. Morgan Kaufman.

[8] H. Mühlenbein. Evolution in Time and Space - the Parallel Genetic Algorithm. In G. Rawlins, editor, *Foundations of Genetic Algorithms.* Morgan Kaufman, 1991.

[9] H. Mühlenbein, M. Gorges-Schleuter, and O. Krämer. New solutions to the mapping problem of parallel systems - the evolution approach. *Parallel Computing*, 4:269–279, 1987.

[10] H. Mühlenbein, M. Gorges-Schleuter, and O. Krämer. Evolution algorithm in combinatorial optimization. *Parallel Computing*, 7:65–85, 1988.

[11] H. Mühlenbein, O. Krämer, G. Peise, and R. Rinn. The MEGAFRAME HyperCluster - a reconfigurable architecture for massively parallel computers. In *PARCELLA 90*, pages 143–156, Berlin, 1990. Akademie Verlag.

[12] H. Mühlenbein, M. Schomisch, and J. Born. The parallel genetic algorithm as function optimizer. *Parallel Computing*, 17, 1991.

[13] H.P. Schwefel. *Numerische Optimierung von Computer-Modellen mittels der Evolutionsstrategie*, volume 26 of *Interdisciplinary System Research.* Birkhäuser, 1977.

[14] F.J. Solis and R.J-B. Wets. Minimization by Random Search Techniques. *Mathematics of Operation Research*, 6:19–30, 1981.

[15] A. Törn and A. Zilinskas. *Global Optimization*, volume 350 of *Lecture Notes in Computer Sciences.* Springer, Berlin, 1989.

A Massively Parallel Genetic Algorithm
Implementation and First Analysis

Piet Spiessens **Bernard Manderick**

Artificial Intelligence Laboratory
Free University of Brussels
Pleinlaan 2, 1050 Brussels
Belgium
e-mail: piet@arti.vub.ac.be

Abstract

In this paper we describe an implementation of a fine-grained parallel genetic algorithm, the FG-algorithm, on the DAP, a massively parallel computer. We develop a model of the behavior of clusters of individuals by analyzing the parallel selection scheme used in FG. A difficult optimization problem, an "ugly" deceptive problem, is used to experiment with selection schemes, neighborhood sizes and crossover operators.

1 INTRODUCTION

Ever since Holland's seminal work [8], it is known that genetic algorithms (GAs) work by processing similarity subsets (schemata) in a powerful, implicitly parallel manner. According to Holland, the number of schemata processed effectively in one generation is proportional to the cube of the population size - the often cited $O(n^3)$ estimate. It follows directly from this work that the larger the population size of a GA, the better it will perform. However, if the population size increases, the number of generations a GA needs to converge also increases. This makes it infeasible to work with large populations on a serial computer. It takes too long to process one generation, and not enough search can be done per unit of computation time [5]. This explains why it is worthwhile to parallelize GAs.

In a previous paper we introduced a fine-grained parallel GA, called the FG-algorithm (for Fine-Grained) [9]. An outline of this algorithm is given in figure 1. Its main characteristics are that the individuals of the population are placed on a planar grid and that selection and crossover are restricted to an individual's immediate neighborhood on that grid (an extra parameter denotes the size of the neighborhood). This way, the global control structures present in a standard GA are eliminated and an effective parallelization is possible. A similar algorithm was introduced by Mühlenbein [10] who achieved very impressive results on hard optimization problems like the traveling salesman problem.

```
PARFOR each grid element el in the grid DO

  Initialization: Randomly generate an individual to occupy el.
  Evaluation: Compute the fitness of individual(el)

  REPEAT

    Selection: Calculate the fitness-distribution of neighborhood size
               individuals in the neighborhood of el. Select according
               to this distribution an individual from the neighbor-
               hood to become the new individual(el).
    Crossover: Randomly select an individual from the neighbor-
               hood. Crossover this individual and individual(el)
               with probability crossover rate. Choose an individual
               from the offspring as the new individual(el).
    Mutation: Mutate individual(el) with probability mutation rate.
    Evaluation: Compute the fitness of individual(el).

  UNTIL <some end-criterion>
```

Figure 1: Outline of the FG-algorithm

In this paper, we describe and analyze the implementation of the FG-algorithm on the DAP, a massively parallel computer. The parallel implementation makes it possible to observe the evolution of a very large population in "real-time". Pictures 1 and 2 show snapshots of the genotypes of a large population (16,384 individuals) after 10 and after 60 generations. As can be seen, within a few generations small clusters of identical genotypes are formed. As the search continues, the clusters grow or shrink depending on their surroundings. The diffusion of a cluster makes it come in contact with another one and new regions of the search space are explored through the recombination (crossover) of individuals at the edges. The "alien" individuals within a cluster are due to the mutation operator.

Because the evolution of a population under the FG-algorithm can be described at the level of clusters of individuals, we developed a model of the behavior of clusters by analyzing the parallel selection scheme used in FG. This model proved helpful in understanding how the FG-algorithm works and therefore enabled us to purposefully experiment with various components of the algorithm. All experiments are carried out on a difficult optimization problem, known as an "ugly" deceptive problem [11].

The remainder of this paper is organized as follows. The next section describes the implementation of FG on the DAP. Part 1 of this section describes the DAP itself and part 2 the actual implementation. In part 3 we calculate the time complexity of the FG-algorithm and present some execution times. In section 3, we present a model of the behavior of clusters, and in section 4 some experiments and results are described.

2 IMPLEMENTATION OF FG ON THE DAP

2.1 DESCRIPTION OF THE DAP

The DAP (standing for Distributed Array of Processors) from Active Memory Technology Ltd is a fine-grained, massively parallel SIMD computer. The processors are arranged in a 2-dimensional cyclic mesh. The size of the dimensions is 32x32 for the DAP 510 (which we are using) and 64x64 for the DAP 610. Each processor has a direct connection to its own local memory. The DAP attaches to a host computer which is used for program development, debugging, loading and controlling DAP programs. A fast data channel can be used to drive a high resolution color display, enabling run-time visualization of the data of an application.

The DAP is mainly programmed in a version of the FORTRAN language which includes extensions for dealing with vectors and arrays as single objects. This language, called FORTRAN-PLUS, is a forerunner of the ISO/ANSI standard FORTRAN 8X.

2.2 IMPLEMENTATION OF FG

In general, an efficient parallel program should have the following characteristics: no global control to avoid serial bottlenecks, and only local interactions between processors to reduce the communication cost. This last requirement is especially true for a mesh-connected system like the DAP, where the communication time between arbitrarily chosen processors is $O(\sqrt{n})$, where n is the total number of processors.

In the FG-algorithm, both characteristics needed for an efficient parallel implementation are present. There is no global control, which would be needed if, e.g., a global selection selection scheme was used, and each processor only needs to communicate with a small number of processors in its immediate neighborhood.

Consequently, the implementation of the FG-algorithm on the DAP is straightforward. It is no problem at all to parallelize the evaluation and the mutation steps of the algorithm, since they require no interprocess communication. For the selection and the crossover steps it is necessary to get the individuals of the neighboring processors to the processor itself. This is done with very efficient machine-level *shift* operations.

The fast data channel of the DAP is used to display the genotypes and their fitness values on a high-resolution color monitor. A graphical user interface on the host computer completes the implementation. The interface provides an easy way to inspect individuals, to alter mutation and crossover rates, to set end-conditions etc.

2.3 TIME COMPLEXITY AND CPU TIME

In this section we look at the time complexity of the FG algorithm and the CPU time needed to process one generation on the DAP. The following notations will be used: n for the population size, s for the neighborhood size, and l for the genotype length.

First we calculate the time complexity of the standard GA. This requires the time complexities of the various steps of the algorithm (selection, crossover and mutation). The evaluation step is left out since its time complexity is of course application-dependent.

Various selection schemes are possible in a GA, e.g., proportionate selection, ranking selection and tournament selection. The time complexity of the selection step depends on the selection scheme used and on the way it is implemented. Except for one selection scheme implementation (roulette wheel proportionate selection, which is $O(n^2)$), the time complexity of the aforementioned schemes is $O(n.logn)$ or $O(n)$ [7]. These figures only take into account the time needed to select the individuals. Additional time is needed to copy these individuals from one generation to the next. To copy one individual requires $O(l)$ steps. So, to fill the whole population $O(n.l)$ steps are needed. Overall, this gives a time complexity for the selection step of $O(n.logn+n.l)$ or $O(n+n.l)$ - which is the same as $O(n.l)$.

Commonly used crossover operators (1-point, 2-point, uniform) require $O(l)$ steps, which gives $O(n.l)$ steps for the whole population. Similarly, a pass through the population to mutate the individuals requires $O(n.l)$ steps.

For the complete serial GA this gives a time complexity of $O(n.logn+n.l)$ or $O(n.l)$, depending on the selection scheme used.

To calculate the time complexity of the FG-algorithm, we first have to describe some local selection schemes. Recall that in a local selection scheme only one individual is chosen from the immediate neighborhood of the individual.

Local proportionate selection assigns to each individual in the neighborhood a probability of selection according to its fitness value. Then, a roulette wheel spin picks one individual. This makes local proportionate selection $O(s)$ (s is the size of the neighborhood). Local ranking selection first sorts the individuals in the neighborhood from best to worst, then assigns probabilities of selection according to rank, and finally applies local proportionate selection using these probabilities. The sorting can be performed in $O(s.logs)$ and the rest is $O(s)$, which makes local ranking selection $O(s.logs)$. Local tournament selection chooses some number of genotypes randomly from the neighborhood and selects the best individual from this group. Instead of first choosing some number of individuals randomly, the whole neighborhood can be used as the tournament group. In either case, local tournament selection is $O(s)$. Again, the time to copy the selected individuals has to be added to the time complexity of the local selection schemes. This makes the selection step $O(s+l)$ or $O(s.logs+l)$.

The crossover step in the FG-algorithm proceeds in two steps. First a mate is chosen randomly from the neighborhood ($O(s)$) and then the actual crossover takes place ($O(l)$). Overall, this gives a time complexity of $O(s+l)$ for the crossover step. The mutation step is simply $O(l)$.

The complete FG-algorithm has a time complexity of $O(s+l)$ or $O(s.logs+l)$, depending on the local selection scheme used.

Table 1 summarizes the time complexities for the serial GA and the FG-algorithm. The overall time complexities of the algorithms resemble each other. Both are linear in l, and the roles of n and s are more or less interchangeable in the respective estimates. The speedup of the parallel algorithm is a consequence of the fact that when the genotype length (l) increases, the population size (n) has to be increased. For the serial GA this means that the execution time increases polynomially, whereas for the FG-algorithm, the execution time increases only linearly (the neighborhood size s is independent of the population size).

Table 1: Summary of time complexities

	Serial GA	FG-algorithm
Selection	$O(n.l)$	$O(s+l)$
	$O(n.logn+n.l)$	$O(s.logs+l)$
Crossover	$O(n.l)$	$O(s+l)$
Mutation	$O(n.l)$	$O(l)$
Total	$O(n.l)$	$O(s+l)$
	$O(n.logn+n.l)$	$O(s.logs+l)$

Figure 2 shows the CPU time needed to process one generation of 1,024 (32x32) individuals. The selection scheme used was local proportionate selection (overall time complexity $O(s+l)$). The graph gives the CPU time for genotype lengths from 3 to 100 bits and for neighborhood sizes of 4, 8 and 12 individuals. It is clear that the execution time is a linear function of both l and s. The CPU times include the execution of the evaluation step which was also $O(l)$. This way, a good estimate of the execution time of the FG-algorithm applied to a typical optimization problem can be calculated.

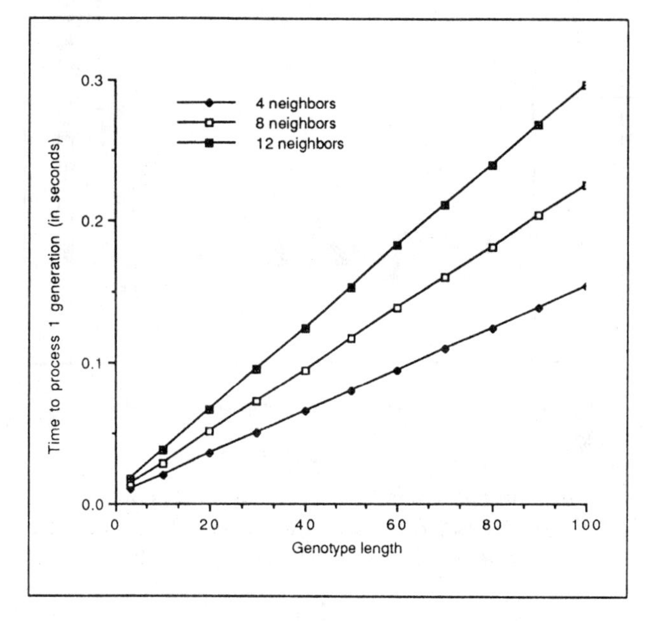

Figure 2: Processing time for one generation of 1,024 individuals

The fact that the DAP 510 has 1,024 processors does not mean that we are limited to populations of this size. Using a technique called slicing, a larger than 32x32 grid is cut into subgrids, which are then mapped on the physical processor grid and processed sequentially. There is only a slight overhead in dealing with the edges of the subgrids, which means that the execution time is approximately the execution time for a population of 1,024 individuals times the number of subgrids.

The execution times for various population sizes are shown in figure 3 (the evaluation of a 30-bit individual is included). A block diagram is used to stress that the CPU time remains the same for a population that is up to 1,024 individuals smaller than indicated. As can be seen, populations as large as 65,536 individuals can still be processed in a reasonable amount of time.

3 A CLOSER LOOK AT THE BEHAVIOR OF CLUSTERS

The behavior of a population under the FG-algorithm can be described at a higher level than that of its individuals. As mentioned before and as shown in pictures 1 and 2, clusters of individuals grow and shrink as the search continues. In this section we will take a closer look at the selection step of the FG-algorithm, which

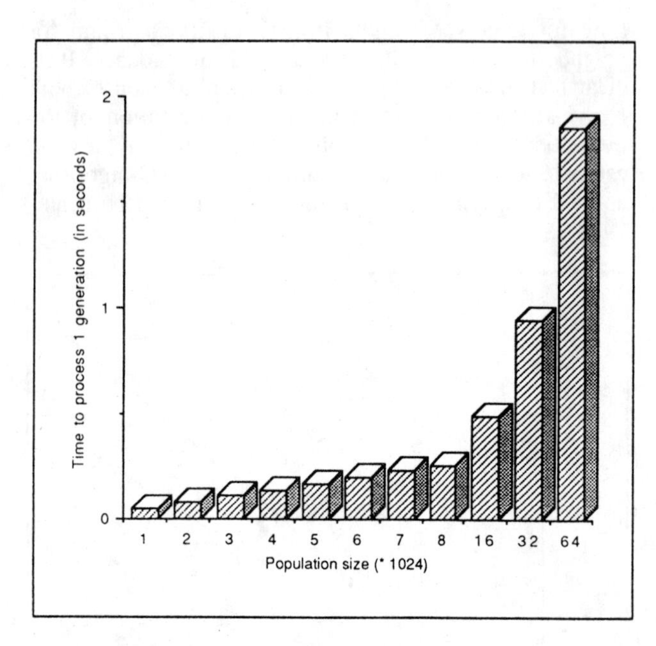

Figure 3: Execution times for various population sizes

mainly determines the behavior of clusters.

In local proportionate selection, the probability that a single individual i with fitness f_i is chosen from a neighborhood with s individuals is

$$\frac{f_i}{\sum_{j=1}^{s} f_j}$$

If the neighborhood contains $m_{i,t}$ copies of individual i at generation t, then the probability of selection of individual i at generation $t+1$ is

$$p_{i,t+1} = m_{i,t} \frac{f_i}{\sum_{j=1}^{s} f_j}$$

If only copies of individuals i and j are present in the neighborhood, the equation becomes

$$p_{i,t+1} = \frac{m_{i,t} f_i}{m_{i,t} f_i + m_{j,t} f_j}$$

If we denote f'_{ij} as the fitness ratio f_i/f_j and substitute $m_{j,t}$ by $s-m_{i,t}$, we get

$$p_{i,t+1} = \frac{m_{i,t} f'_{ij}}{m_{i,t}(f'_{ij} - 1) + s}$$

We define the growth $G_{i,t+1}$, or the "amount" by which an individual i will increase its presence in a particular grid location, as the probability that it is selected minus the probability that the other individual j is selected.

$$G_{i,t+1} = p_{i,t+1} - p_{j,t+1} = 2p_{i,t+1} - 1$$

In general, this equation is useless since we do not know how the individuals are distributed across the grid. Therefore we will make the following (reasonable) assumptions: (1) all copies of an individual i are located in a cluster, (2) a circle is a good approximation of the shape of this cluster, (3) growth or decay only takes place at the edge of the circle.

Now, suppose that a cluster of individuals i is surrounded by copies of individual j with the same fitness (f'_{ij}=1), then the growth of this cluster is

$$G_{i,t+1} = \frac{2m_{i,t}}{s}$$

Because of the convexity of the cluster, we have that $m_{i,t}<s/2$ for all grid locations on the edge, and therefore $G_{i,t+1} < 0$. This means that a cluster with the same fitness as its surroundings will eventually disappear. The form of niching [4] where equally good (suboptimal) individuals remain in the population thus seems impossible. It has to be noted however that it can take a fairly long time before such clusters disappear completely.

Next, we will look at how the total amount $M_{i,t}$ of individuals in a cluster changes. The number of individuals $m_{i,t}$ in a neighborhood on the edge of a circle (cluster) depends on the radius of this circle. If there are $M_{i,t}$ individuals in the circle, this radius is $r_{i,t}=\sqrt{M_{i,t}/\pi}$.

In a neighborhood of size s on the edge of a cluster, $m_{i,t}$ can be approximated by

$$m_{i,t} = \frac{s}{2}(s-r_{i,t}) + \int_{-s/2}^{s/2} \sqrt{r_{i,t}-x}.dx$$

The expected number of individuals in the next generation is approximately

$$M_{i,t+1} = M_{i,t} + 2\sqrt{\pi M_{i,t}}\, G_{i,t+1}$$

An exact solution of this equation is nontrivial, but it is clear that the solution has to follow quadratic growth. Figure 4 shows a number of iterations of the equation for various fitness ratios. In the same way as above, growth curves can be derived for local proportionate selection with scaling, local ranking selection and local tournament selection. All these schemes have the effect of a higher selection pressure and thus of steeper growth curves than those of local proportionate selection.

4 EXPERIMENTS AND RESULTS

In the previous section we have seen that a cluster with the same fitness value as its surroundings will eventually disappear. Unfortunately, this will also happen if this fitness value is only marginally better. In the first experiments presented in this section, we tried to find out if techniques to prevent such clusters from disap-

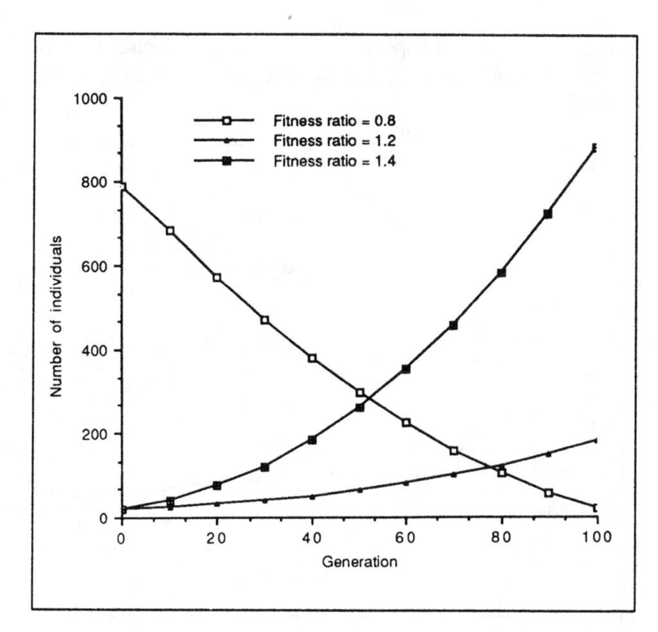

Figure 4: Growth curves for clusters with various fitness ratios

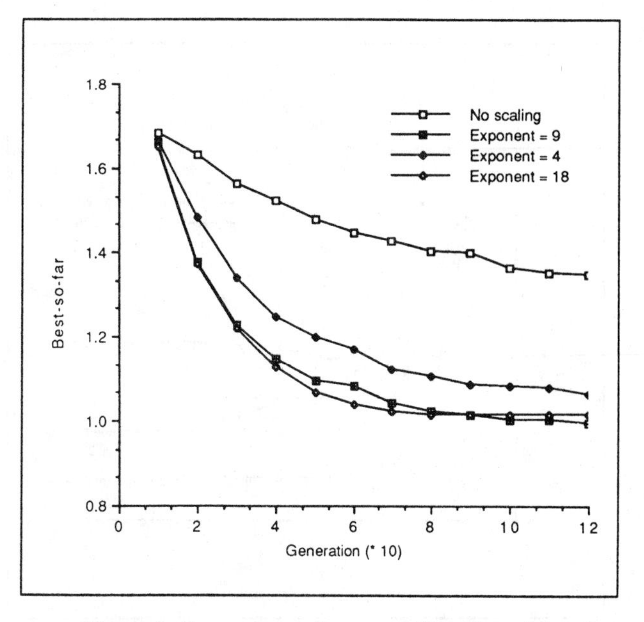

Figure 5: Best-so-far performance with polynomial scaling (population size = 4,096, neighborhood size = 4, mutation rate = 0.001, crossover rate = 0.6)

pearing would also improve the performance of the FG-algorithm.

A first technique is to add polynomial scaling to the proportionate selection scheme. The equations of the previous section were used to calculate minimal exponents for which small, marginally better clusters would not disappear.

The test function for this experiment (and all others in this section) was an "ugly" deceptive problem, like the one found in [11]. This problem is called "ugly" because it consists of a sum of 10 *disjoint* order-4 deceptive problems. We selected this problem because it is sufficiently difficult and because it is a useful test case since the solution of deceptive problems is believed to be crucial for the solution of all combinatorial optimization problems [6].

For the "ugly" deceptive problem, the smallest fitness ratio (f_i/f_j) of individuals i and j with $f_i > f_j$ is around 1.01 (the fitness value is an integer value between 0 and 300). According to our calculations, a scaling exponent equal to 9 was needed in this case. Figure 5 shows the best-so-far curves for various scaling exponents. The performance with a scaling exponent equal to 9 is better than the performance of proportionate selection without scaling and also better than proportionate selection with a scaling exponent equal to 4. A scaling exponent greater than 9 does not seem to further improve performance. As with all other experiments in this section, the results are averaged over 30 runs.

Since performance seemed to improve with the use of polynomial scaling, we set up a second experiment where the FG-algorithm used local tournament selection (the tournament group was formed by all individuals in the neighborhood). Local tournament selection also has the effect of a harder selection, because the local tournaments make sure that the best individual in the neighborhood always survives. This makes it possible to use higher mutation rates because there are enough copies of the (locally) best individual and the probability that they simultaneously disappear is very low. Higher mutation rates result in a more thorough exploration of the search space and also reduce the chance of premature convergence, a possible danger when using a hard selection scheme.

Figure 6 shows some results of local tournament selection with various mutation rates. With a mutation rate of 0.001, performance is comparable to the best performance obtained in the previous experiment. With a mutation rate of 0.025 performance increases. If the mutation rate is set too high, the search degenerates to random search after a while.

In a third experiment, we tried to assess the influence of the neighborhood size on the local selection schemes used in the previous experiments. It is easy to show and it is intuitively clear that an increase of the neighborhood size results in a higher selection pressure. For the local proportionate selection schemes, this means that marginally better individuals will have a higher chance of surviving in their neighborhood. For local tournament selection, this means that a better individual will spread much faster across the grid.

In figure 7 the performance curves for neighborhood sizes of 4, 8 and 12 individuals are compared for local proportionate selection with and without polynomial scaling. In both cases the performance slightly increases

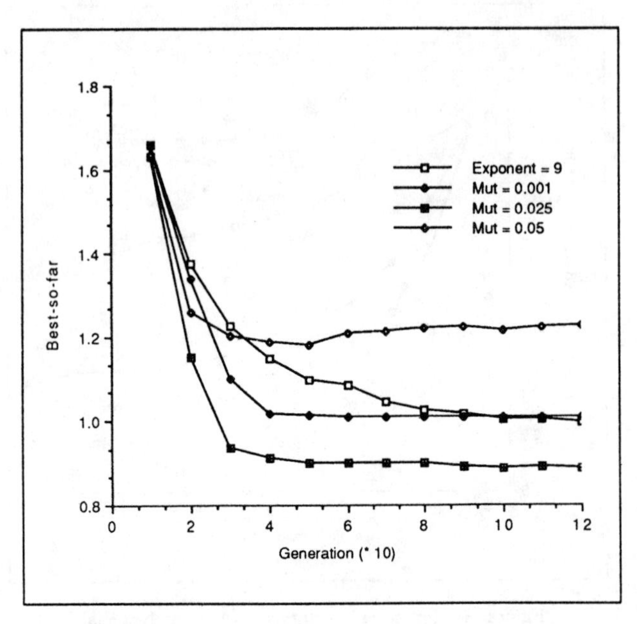

Figure 6: Best-so-far performance with local tournament selection (population size = 4,096, neighborhood size = 4, crossover rate = 0.6)

with the neighborhood size and the improvement is less pronounced if scaling is used.

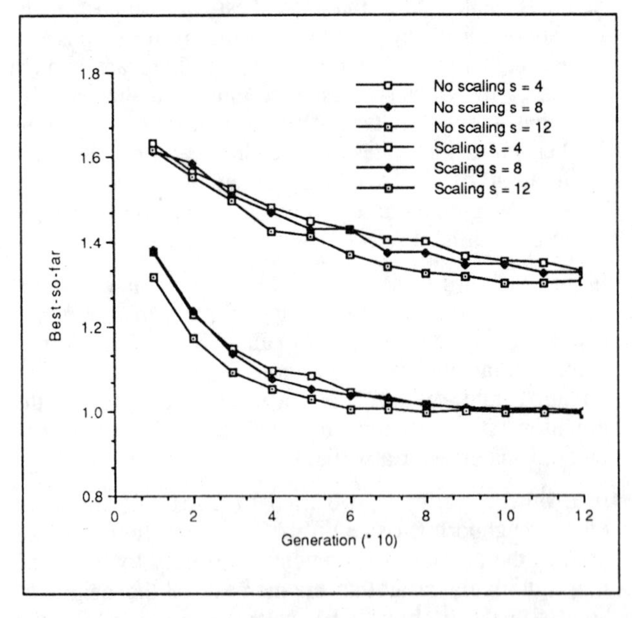

Figure 7: Influence of the neighborhood size on local proportionate selection (population size = 4,096, mutation rate = 0.001, crossover rate = 0.6)

Figure 8 shows the results for local tournament selection. Here, performance *decreases* with the size of the neighborhoods. A possible explanation is that the extra selection pressure amounts in a too severe selection which reduces the diversity in the population too quickly. The fact that local tournament selection performs best with small neighborhoods is another advantage over local proportionate selection (recall that the time complexity is linear in the size of the neighborhoods).

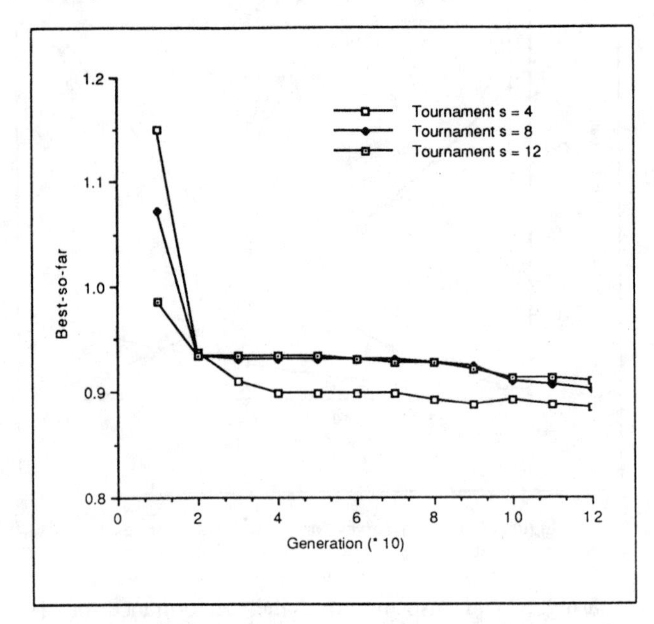

Figure 8: Influence of the neighborhood size on local tournament selection (population size = 4,096, mutation rate = 0.025, crossover rate = 0.6)

In all experiments above, we used the 2-point crossover operator. It has been known for some time that this operator performs better than the traditional 1-point crossover operator [2]. More recently, a number of researchers have reported even better results using the uniform crossover operator first introduced by Ackley [1]. Uniform crossover is more disruptive than the traditional crossover operators, and this is believed to be beneficial in the later stages of the search when the population becomes more homogeneous [12][3].

The results of our last experiment where we compared the three aforementioned operators seems to confirm this hypothesis. The graph in figure 9 shows that initially, the performance of the three operators is almost identical. Halfway through the search however, uniform crossover takes the lead and ends much better than the others. Note that the difference in performance only shows up after 120 generations, the maximum number of generations for the other experiments. Incidentally, this graph also shows that the use of local tournament selection does not lead to premature convergence. The FG-algorithm continues to find better solutions and eventually finds the optimum (even with 1-point and 2-point crossover operators).

5 CONCLUSION

This paper has described the implementation of the FG-algorithm on the DAP, a massively parallel computer. Execution times have shown that it is possible to

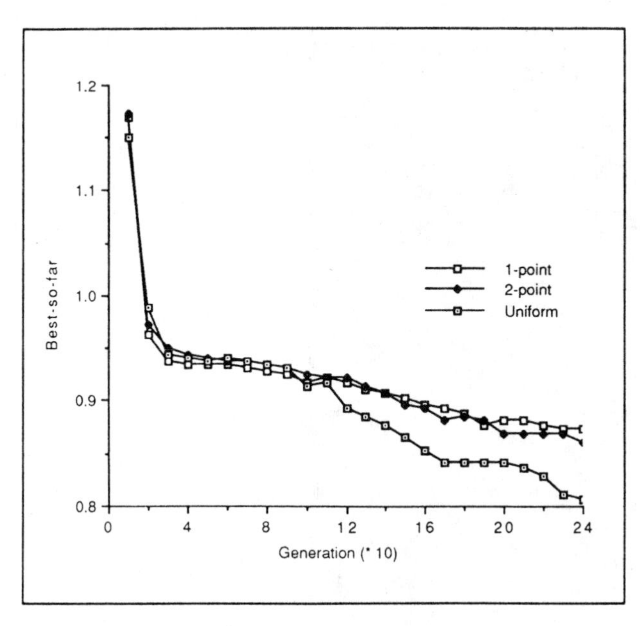

Figure 9: Best-so-far performance with different crossover operators (selection scheme : local tournament selection, population size = 4,096, neighborhood size = 4, mutation rate = 0.025, crossover rate = 0.6

work with very large populations. The time complexities of the parallel algorithm and the standard GA have been calculated and compared to each other. A model has been developed of the behavior of clusters by analyzing the parallel selection scheme used in FG. This model enabled us to purposefully experiment with the FG-algorithm. Experiments carried out on an "ugly" deceptive problem have shown a preference for local tournament selection over local proportionate selection with or without scaling. In contrast to these schemes, the performance of local tournament selection was found to be best with small neighborhood sizes. Additional support was also found for the thesis that in the later stages of the search, the uniform crossover operator performs better than 2-point or 1-point crossover. It is acknowledged however, that tests have been limited and further research is required to validate these results.

Acknowledgements

We would like to thank Didier Keymeulen and Jo Decuyper for their assistance in programming the DAP. Jan Torreele and Tony Bell provided valuable comments on earlier versions of this paper. This research is sponsored by the Belgian Government under IMPULS contract no. RFO/AI/10.

References

1. D.H. Ackley, Stochastic Iterated Genetic Hillclimbing, Doctoral Dissertation CMU, 1987.

2. L. Booker, "Improving Search in Genetic Algorithms," in *Genetic Algorithms and Simulated Annealing*, ed. L. Davis, Morgan Kaufmann, Los Altos, CA, 1987.

3. K. De Jong and W. Spears, "An Analysis of Multi-point Crossover," in *Theoretical Foundations of Genetic Algorithms*, ed. G. Rawlins, Morgan Kaufmann, Los Altos, CA, 1991.

4. D.E. Goldberg and J. Richardson, "Genetic Algorithms with Sharing for Multimodal Function Optimization," in *Proceedings of the Second International Conference on Genetic Algorithms*, ed. J.J. Grefenstette, Lawrence Erlbaum, Hillsdale, NJ, 1987.

5. D.E. Goldberg, "Sizing Populations for Serial and Parallel Genetic Algorithms," in *Proceedings of the Third International Conference on Genetic Algorithms*, ed. J.D. Schaffer, Morgan Kaufmann, San Mateo, CA, 1989.

6. D.E. Goldberg, "Genetic Algorithms and Walsh Functions," *Complex Systems*, vol. 3, no. 2, 1989.

7. D.E. Goldberg and K. Deb, "A Comparative Analysis of Selection Schemes Used in Genetic Algorithms," in *Theoretical Foundations of Genetic Algorithms*, ed. G. Rawlins, Morgan Kaufmann, Los Altos, CA, 1991.

8. J.H. Holland, *Adaptation in Natural and Artificial Systems*, University of Michigan Press, Ann Arbor, 1975.

9. B. Manderick and P. Spiessens, "Fine-grained Parallel Genetic Algorithms," in *Proceedings of the Third International Conference on Genetic Algorithms*, ed. J.D. Schaffer, Morgan Kaufmann, San Mateo, CA, 1989.

10. H. Mühlenbein, "Parallel Genetic Algorithms, Population Genetics and Combinatorial Optimization," in *Proceedings of the Third International Conference on Genetic Algorithms*, ed. J.D. Schaffer, Morgan Kaufmann, San Mateo, CA, 1989.

11. T. Starkweather, D. Whitley, and K. Mathias, "Optimization Using Distributed Genetic Algorithms," in *Parallel Problem Solving from Nature*, ed. H. Schwefel and R. Maenner, Springer Verlag, Berlin, Germany, 1991.

12. G. Syswerda, "Uniform Crossover in Genetic Algorithms," in *Proceedings of the Third International Conference on Genetic Algorithms*, ed. J.D. Schaffer, Morgan Kaufmann, San Mateo, CA, 1989.

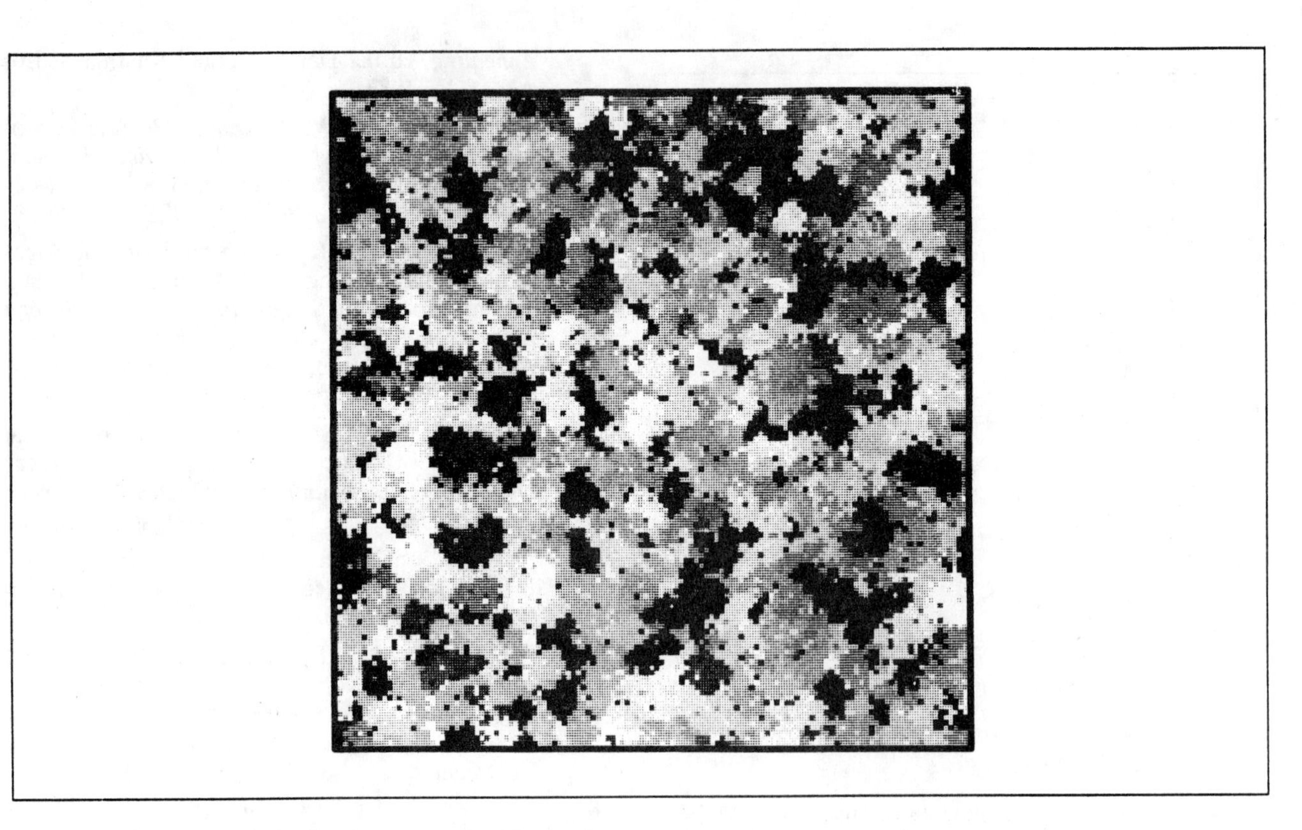

Picture 1: Typical distribution of genotypes after 10 generations

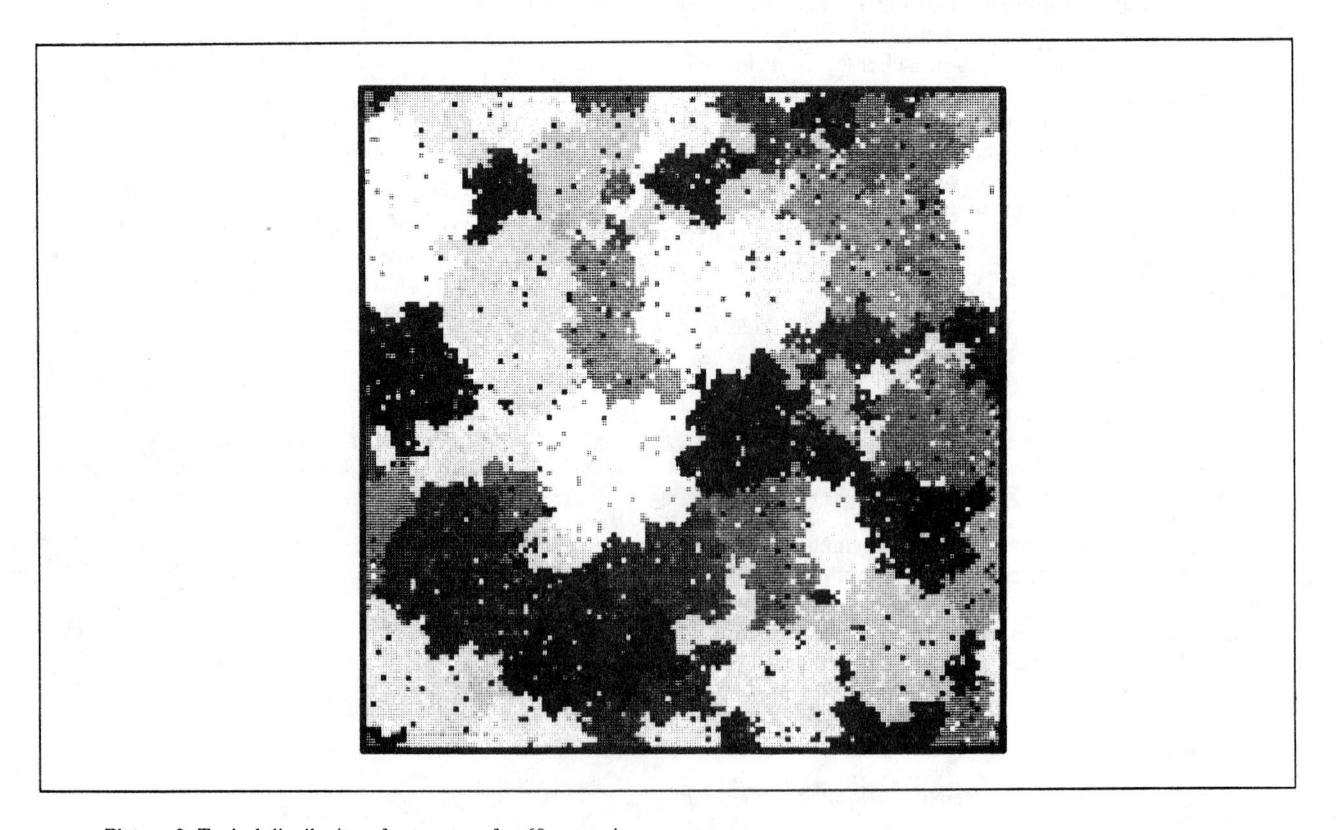

Picture 2: Typical distribution of genotypes after 60 generations

Classifier Systems and Other Rule-based Approaches

An Efficient Classifier System and its Experimental Comparison with two Representative learning methods on three medical domains.

Pierre Bonelli
Equipe Inférence et Apprentissage
Laboratoire de Recherche en Informatique (bat. 490)
Domaine universitaire de Paris-Sud
91405 Orsay Cedex, France
E-mail: bonelli@lri.lri.fr

Alexandre Parodi
INFODYNE SARL
10, rue de la Paix
75002 Paris, France.
E-mail: parodi@lri.lri.fr

Abstract

In this paper, we describe a Classifier System, **Newboole**, and we present its experimental comparison with two widely used learning algorithms, CN2 (logic reduction system) and Back Propagation (neural net), on three medical domains. The experimental results, obtained in the context of learning from preclassified examples, demonstrate two main points: firstly, that all three systems perform very closely on the induction tasks with a slight advantage for the Back Propagation algorithm. Secondly, that a Classifier System can provide comprehensive solutions in the form of a reasonable number of "symbolic" decision rules, which is not the case for a neural net using Back Propagation.

1. Introduction

Performing concept learning from preclassified examples is a learning task that many Machine Learning (ML) systems have undertaken. The majority of these systems are descendants from either one (or even both) of two benchmark algorithms, namely ID3 (decision tree classifier) [11] and AQ (logical reduction classifier) [9]. However, more recently, a new family of learning systems has come to life, known as Genetics Based Machine Learning (GBML), which derives its principles from Darwin's concept of evolution. Much of the progress in GBML is due to the emergence of a class of inductive learning systems, called Classifier Systems (CS) [7], which acquire rules incrementally.

In this paper, we describe a CS, **Newboole**, and we present its empirical comparison with two widely used learning algorithms, CN2 [2] and Back Propagation (BP) [12], on three medical domains which have been extensively used throughout the ML litterature. This study provides an evaluation of the symbolic concept acquisition capability of a CS versus already existing techniques.

1.1 A few good reasons for learning with genetic CS.

Most ML systems learn by being told, i.e. they require examples including the input and the expected output; learning is done through a teacher who presents problems and their solutions. Therefore such systems have limited intelligence: their task merely consists in the compression of given information, and thus in the discovery of underlying theories that pre-exist in the teacher's mind and are sampled by the examples.

Such paradigms can greatly improve knowledge acquisition; however they are not satisfactory when faced with learning tasks for which no teacher is available. For example, autonomous agents such as robot rovers cannot be given explicit knowledge regarding the adaptive behavior corresponding to each situation they encounter: they must infer it from past experience, using as only source of knowledge reward (payoff or penalty) received from the environment. Furthermore, since this knowledge is not available all at once, they must learn it *incrementally*.

It seems at the present time, that CS, which somehow mimic the way living species improve the structure of their organisms by adapting it to the environment, are best suited to fulfill these needs. However, it is also interesting to evaluate the performance of reinforcement learners such as CS in the restricted context of concept learning from *preclassified* examples. Indeed, knowing the correct answer in each situation can be easily taken into account by a CS: the system will heavily reward rules which classify the examples correctly and symmetrically penalize rules which give wrong answers. Since many real world learning tasks involve concept learning from preclassified examples, and since most other ML systems perform supervised learning, the comparison presented in this paper is restricted to learning from preclassified examples. It is important to keep in mind that we explore here only one facet of CS learning, which are usually presented as reinforcement learners that receive payoff from the environment.

A further quality of CS is their ability to extract knowledge in the form of comprehensible rules which can be understood and exploited by the human user. This is not yet the case with neural nets using BP.

A last argument that holds not only for classifier systems but also for GBML in general, is their capacity of supporting and processing complex knowledge representations. In the context of CS, Forrest [3] implemented in 1985 a system which translates KL-ONE semantic networks representing instances into CS bit string notation, thus enabling the application of classical genetic operators. However, such a preliminary low level coding does not seem necessary in many cases and might even turn out as being unsuitable. For example, Grefenstette et al. [5] use a more natural representation for their planning problem: genetic operators are directly applied on chains of condition-action rules representing plans.

In certain cases, it might even be better to adapt genetic operators to the natural knowledge representation rather than the opposite as Forrest did. For example, Koza [8] directly applies a specially designed crossover operator to programs in the form of LISP S-expressions.

1.2 A Quick Overview of Newboole, CN2 and Back Propagation.

Newboole is a modified version of Wilson's CS BOOLE, which performs (un)supervised symbolic concept learning from examples. Like any CS, **Newboole** maintains a population of classifiers according to Darwinian evolution principles. The algorithm and knowledge representation will be detailed in section 2.

When developing a new inductive learning system, it is important to evaluate its performance on real world problems in comparison with widely used algorithms. We decided to compare **Newboole** with two recent algorithms which represent the two main computer based learning methodologies: symbolic concept acquisition and connectionist learning.

CN2 is a system that was created by Clark & Niblett in 1987, and which is inspired from both the AQ & ID3 algorithms. It is extensively described in the ML literature (for example in [2]), and thus will not be detailed in the present paper. It is important to mention that CN2 induces rules in the form of a conjunction of selectors in the form of (attribute *selector* value) followed by a class description, using logic reduction techniques. We chose CN2 as the representative of the "mainstream" inductive learning algorithms for two reasons:

- its design was influenced both by ID3 and AQ.

- CN2 obtained the best classification accuracy rates in the many experimental papers in which it was compared to other rule induction methods such as decision tree classifiers.

Back Propagation [12] is the most widely used algorithm employed by a multi-layer perceptron to adjust the weights on its connections. It is certainly the most widely used *incremental* learning algorithm, and is very efficient when learning from preclassified patterns (examples). Since **Newboole** is an incremental algorithm, it seemed important to compare its performance with an "inherently" incremental algorithm.

The paper is organized as follows: in section 2.0, we describe **Newboole's** knowledge representation and algorithm. In section 3.0, we present and discuss the experimental results obtained using three medical domain testsets. Finally, section 4.0 includes a short summary of results, and outlines future research.

2 Overview of the Newboole system

In this section, we describe the **Newboole** system, in order to support the experiments which are presented in the next section.

A more detailed description can be found in previous articles [1], [10].

2.1 Knowledge representation

Historically, CS employ a simple, sub-symbolic representation for both instances and rules based on Holland's Schema Theory. The usual knowledge representation consists of fixed length strings of alleles over the simple {0, 1, #} alphabet. This representational scheme reflects the so called "basic principles for choosing a GA coding" cited by Goldberg in [4]: the principle of meaningful building blocks and the principle of minimal alphabets.

For **Newboole**, we chose to use a *natural* representation for instances and rules throughout the system rather than code them in the form of binary strings. We assume that the examples are described with a collection of attributes. Each attribute can take a set of discrete mutually exclusive values.

The attribute-value representation that we propose is not optimal according to Holland's schema theory, since it uses a larger alphabet: each attribute can take many values, not only zero or one.

However, it has some advantages. Firstly, the number of necessary attributes does not have to be known in advance; in this case, the machine only processes the rule attributes which are really used. Also, the example attributes that have unknown values do not cause any problem: these attributes are simply not present in the example description. Furthermore, this representation is more powerful for future enhancements, especially for connections with background knowledge.

2.2 The Newboole algorithm

Newboole has been largely inspired from BOOLE, a CS that was described by Wilson in [14], and which learns complex disjunctive Boolean functions.

Newboole improves a population of P attribute-value rules by using selection and genetic breeding. Each rule has a strength which somehow measures its utility (ability to be matched, provide good answers and be general). This system is composed of three components:

1/ At each cycle n, the **performance component** takes one example description (set of (attribute, value) couples) and attempts to match it to rule conditions (complexes of (attribute, value)'s). The (match)set M of all rules whose conditions match the input example description is formed, and the system output is set to the conclusion of the strongest rule in M.

2/ The **reinforcement component** updates the strength $s_i(n)$ of every rule
$R_i \in M$ at cycle n according to:

conclusion(R_i) = class(example(n))
=> $s_i(n+1) = (1-e)\, s_i(n) + r_i$
conclusion(R_i) $\neq$ class(example(n))
=> $s_i(n+1) = (1-p)\, s_i(n)$

p is the penalty coefficient, e is a "tax" coefficient and r_i is the "good answer reward" for rule R_i which is higher if the rule is more general, depending on the number N_i of attributes in the rule premise, according to:

$$r_i = \frac{1-gN_i}{\sum_i (1-gN_i)} R$$

g can be considered as a "generality enforcement" coefficient, and R is the total reward to be provided to M at each cycle.

Please note that in the case of reinforcement learning without knowledge of the correct answer, we would have:

conclusion(R_i) = system output
=> $s_i(n+1) = (1-e)\, s_i(n) + r_i$
conclusion(R_i) $\neq$ system output
=> $s_i(n+1) = (1-p)\, s_i(n)$

R then depends on how "valuable" the system output is for this particular cycle.

3/ Genetic Symbolic discovery

As stated above, the algorithm would only select pre-existing rules, which conceptually looks like Samuel's Checkers [13]. However, the discovery component discovers new rules which will then be utilized, measured and selected.

The discovery component is based on Holland's [6] genetic algorithm and employs reproduction, genetic operators, and deletion. At each learning cycle, r rules ("parents") are selected probabilistically according to strength and copied. Genetic operators are then applied to these copies with a different probability for each operator. The resulting offspring are then added to the population and a like number of weak classifiers are deleted.

The genetic operators we use are crossover and mutation. In crossover, randomly selected information is exchanged between a pair of rules. In mutation, some of the information contained in a rule is altered randomly.

4/ Population initialization algorithm

Any Genetic Algorithm based system needs a routine to create an initial population of solutions. Since **Newboole** learns from a set of available pre-classified examples, the "Learning Base", it seemed logical to use this Learning Base as the basis for the population initialization algorithm. Indeed, the search space to be explored is often far too vast to permit a random initialization process since the population of classifiers is supposed to recognize examples from the Learning Base which usually represents a microscopic fraction of the search space. Therefore, each initial rule $R_i(0)$ for $i \in \{1,2,...,P\}$ is obtained as follows:

* an example from the learning base is chosen at random;

* The condition part of the new rule is a copy of the attribute-value part of the example, except that fictitious "don't care" symbols are inserted with a probability of 0.5. This is done so that the initial classifiers are not too specific.

*The conclusion part of the new classifier is chosen randomly among the possible conclusion

5/ Knowledge Base Compaction algorithm

The knowledge base is the rule population. However, since the population is numerous, many extraneous rules exist. At the beginning, all the rules have the same strength; as the examples are presented, the overall population strength tends to be packed into a few very useful rules, and there still exists a number of rarely used weak ones. Only useful rules should be kept in the final version of the knowledge base.

One way to prune them out is to order the rules according to their strength from i=1 for the strongest to i=P for the weakest, and compute the cumulated strength:

$$S_k = \sum_{i=1}^{k} s_i$$

When S_k/S_P reaches a predefined number s, the first k rules represent the smallest set that contains a proportion s of the total "usefulness" of the population: these k rules can be considered as the final knowledge base.

However, in order to avoid to set s by hand, another method is used: it consists in forming the set of rules which are actually utilized as classifiers when the examples from the training set are presented; it is assumed that only those will be needed to classify future data (unseen examples).

3 Experiments and discussion of results

We tested **Newboole**, CN2 and BP's respective abilities to perform medical diagnosis using data from three medical domains, provided by the Institute of Oncology in Ljubljana, Yugoslavia. We chose these data sets because numerous other inductive learning systems have acquired their reputation from testing on these particular bases.

3.1 Description of the medical domains

Lymphography. This example base consists of 148 examples separated into 4 distinct diagnoses. Each example is completely described by 18 discretely valued attributes representing the symptoms and one of 4 classes representing the corresponding diagnosis. The estimated performance of a human expert (specialist) is 85 % of correct diagnoses (60 % for an intern).

Prognosis of Breast cancer recurrence. This example base consists of 288 examples separated into 2 distinct diagnoses. Each example is completely or *incompletely* described by 10 discretely valued attributes representing the symptoms and one of 2 classes representing the corresponding diagnosis. The quality of human expert diagnosis is unknown for this domain.

Location of Primary tumor . This example base consists of 339 examples separated into 22 possible locations of primary tumor. Each example is completely or *incompletely* described by 17 discretely valued attributes representing the symptoms and one of 22 classes representing the corresponding diagnosis. Four oncologists were tested on this domain, and obtained an average rate of 42 % of correct diagnoses (four interns obtained 32 %).

3.2 Conditions of the experiments

Our main goals were on the one hand to test the induction capability of the three algorithms, and on the second to study the comprehensibility of the solutions they provided.

For each experiment, 70% of the examples were randomly chosen for learning, and the remaining 30 % for testing . The results shown in the following section represent the average over 10 experiments. One important point is that we compared all three systems on exactly the ten same [Learning Set, Testing Set] couples for each domain, thus insuring a fair comparison.

3.3 Experiments and results

The three tables below describe the experimental parameters used for each learning system:

Parameters / **Medical Domains**	e	c cross over rate	m mutation rate	g gene-rality enf.	R reward	p penalty coef.	r renewing	P popu. size
Lymphography	0.1	0.5	0.001	0.044	1000	0.95	4	1000
Breast Cancer	0.1	0.5	0.001	0.08	1000	0.95	4	1000
Primary Tumor	0.1	0.5	0.001	0.047	1000	0.95	4	5000

Table 1: Experimental parameters in NEWBOOLE

Parameters / Domains	Star Size	Significance Threshold
Lymphography	5	1.00
Breast Cancer	5	1.00
Primary Tumor	5	1.00

Table 2: Experimental parameters in CN2

	architecture	learning rate	initialization
Lymphography	18:30:20:4	$0.1/sqrt(N_{in})$	$[-1.5/N_{in}; 1.5/N_{in}]$
Breast Cancer	10:20:10:2	$0.1/sqrt(N_{in})$	$[-1.5/N_{in}; 1.5/N_{in}]$
Primary Tumor	17:30:20:22	$0.1/sqrt(N_{in})$	$[-1.5/N_{in}; 1.5/N_{in}]$

Table 3: Experimental parameters for BP

The following graphs plot Newboole's respective performances on both the Learning and Testing bases for each medical domain.

Since **Newboole** and BP are both incremental algorithms, the learning process needs some explanation. During each learning cycle, an example from the Learning base is chosen at random and presented as input to either system. Each system then makes its decision according to its past experience and applies its learning algorithm. For both systems, learning was performed over 10000 cycles to ensure proper induction had been achieved. Figures 1,3 and 4 plot the moving average of the percentage of correct decisions over the past 50 cycles versus the number of cycles since the experiment began. As CN2 is not incremental, its performance is plotted as a horizontal line.

The testing procedure is the following. Every 100 cycles, Newboole's reinforcement and discovery components are *turned off* (as well as the neural net's), and the rule population is tested on the entire Testing base. Figures 2, 4, 6 plot the percentage of correct decicsions over the entire Tesing base. Please note that both Newboole's and the neural net's learning algorithms are turned off during testing. As CN2's learning procedure can not be "interrupted" in the same way, its test performance is plotted as a horizontal line.

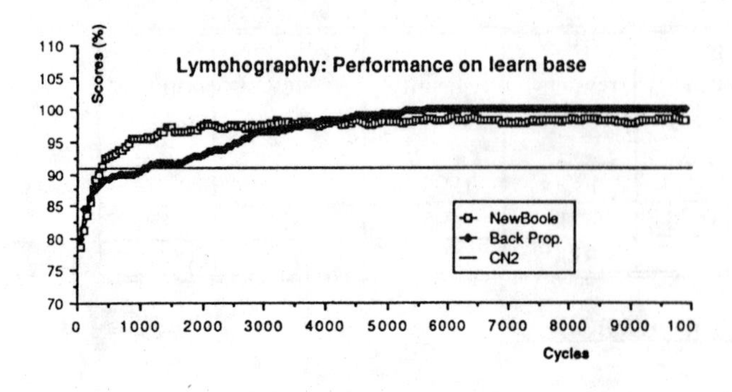

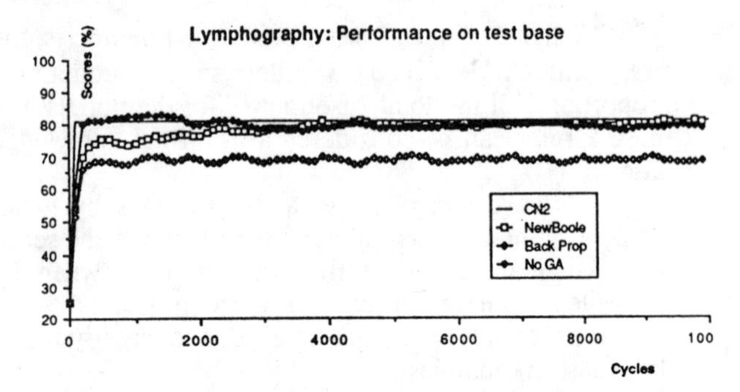

Figures 1, 2: Performance on Lymphography

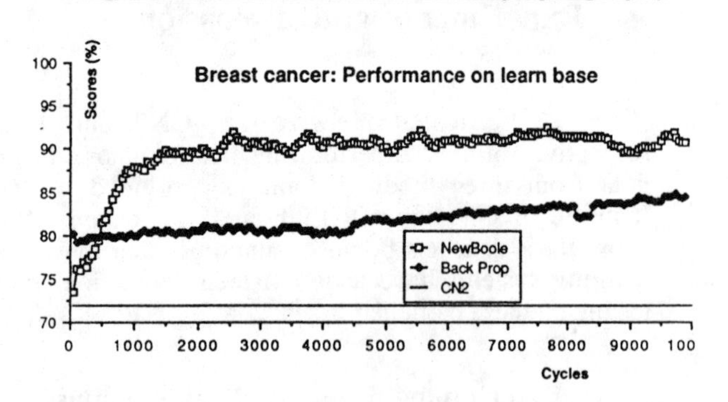

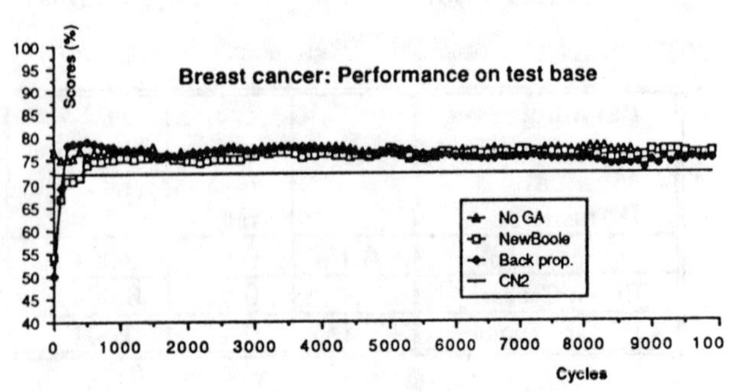

Figures 3, 4: Performance on Breast Cancer

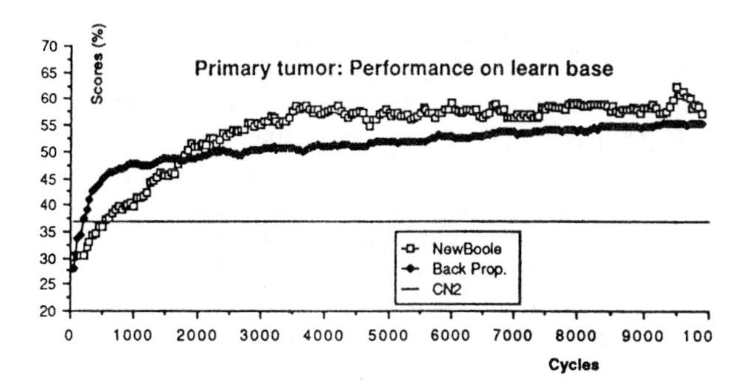

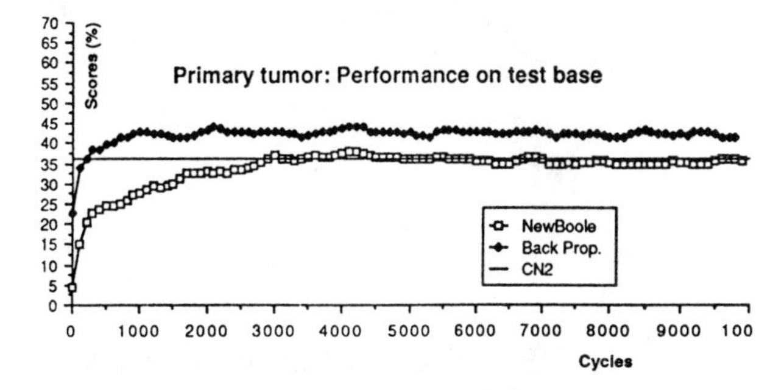

Figures 5, 6: Performance on Primary Tumor

3.4 Discussion of results

In this section we carefully analyze the experimental results by addressing different comparative aspects.

a/ Accuracy of performed induction.

On all three tasks, **Newboole**, CN2 and BP performed closely as far as the induction rate is concerned (see graphs above). This suggests that there is a limit to the quality of induction that a system can achieve from learning on a given database, which would correspond to the implicit knowledge contained in the examples. This is reinforced by the fact that human experts themselves don't obtain significantly different results: 85 % on lymphography, 42 % on primary tumor for specialists.

However, the results obtained in the three medical domains do not have similar significance. Indeed, a more careful examination of the test sets and results clearly show that the Breast Cancer data set is not appropriate for evaluating a ML system's ability to perform induction.

Lymphography is the data set which contains the most complete implicit information. For instance, examples from two different classes differ by the values of at least two attributes. In other words, the example attribute set is discriminant for the classes, and the examples are pertinent. Indeed, our experiments with the Genetic Algorithm (GA) turned off clearly showed that the GA is needed in order to fully explore the search space. For this experiment, we have an estimated mean square root error of 2 %.

As far as Breast Cancer is concerned, we were surprised to find that **Newboole** performed a little better without the GA than with the GA! This proves that the data does not contain much implicit information concerning the laws of the domain, and thus does not constitute proper testing grounds. A closer look at the examples reveals that one of the two possible diagnoses is greatly prevalant in the base as it corresponds to 80 % of the examples. Since it is very difficult to obtain more than 80% of correct answers in this kind of domain anyway, the systems merely find the appropriate default rule, which classifies each example as the most probable diagnosis a priori. The mean square root error in this experiment was of 1.7 %.

Finally, the Primary Tumor domain is difficult. Indeed, many patients are described by the same exact symptoms but are diagnosed differently. The performance of a learning system is thus limited since the similarities to be exploited result in contradictory rules. However, although attributes are not very discriminant for the classes, the default rule method would not obtain more than 25% accuracy. Since all three learning systems performed over 35 % accuracy, this data set tests their ability to extract knowledge from unreliable data. It is no surprise that the BP algorithm performs better than the other two. In this experiment, the mean square root error was very low: 0.65 %.

Overall, it seems that BP performs slightly better than the other two algorithms; however, a neural net using BP has the great disadvantage of merely providing as "solutions" sets of numerical coefficients without any semantic meaning.

b/ Time needed for learning

CN2 provides a clear advantage from this point of view. Indeed, the CN2 algorithm takes less than a couple of minutes on a Sparc station, whereas **Newboole** and BP take approximately 20 to 30 minutes. However, both **Newboole** and BP are essentially parallel algorithms which are not designed to run most efficiently on conventional sequential computers. It is also interesting to note that **Newboole** can be more easily implemented on parallel architectures than BP, since there is much less computational coupling between classifiers than there is between neurons: each rule is evaluated and selected very much independantly from the others, which is not the case for heavily connected neural nets.

c/ Comprehensibilty of provided solutions

Unlike a neural net using BP, both CN2 and **Newboole** provide comprehensible solutions in the form of "symbolic" diagnosis rules.

However, it is important to compress the acquired knowledge within the least number of rules possible. The experiments described below show that the large rule population **Newboole** comprises can be greatly reduced by an appropriate pruning method. These experiments were conducted over the lymphography domain.

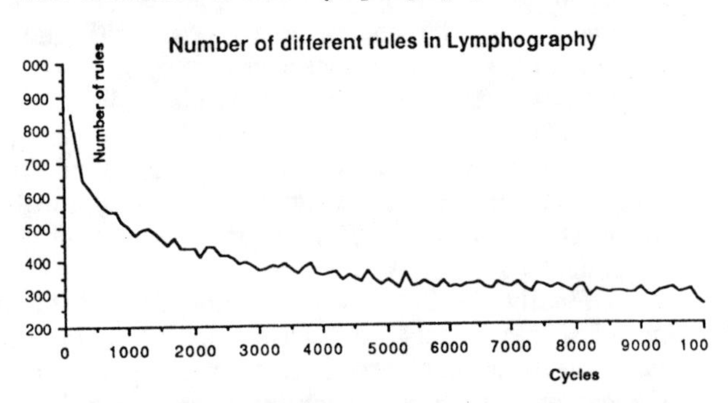

Figure 7: number of different rules in the population

If we look at the evolution of the number of *different* rules in NEWBOOLE's Knowledge Base (KB) over learning (Figure 7), we find that although this number decreases, it is still very large (over 200) after 10000 learning cycles. This number seems tremendous when compared to the number of rules CN2 provides after learning (15 on the average).

We defined the set of *useful* rules as the following:

Useful rules = { rules in the Population which are fired when the KB is tested over the learning set of examples}

It seemed as a good guess that these rules would be sufficient to classify unseen examples.

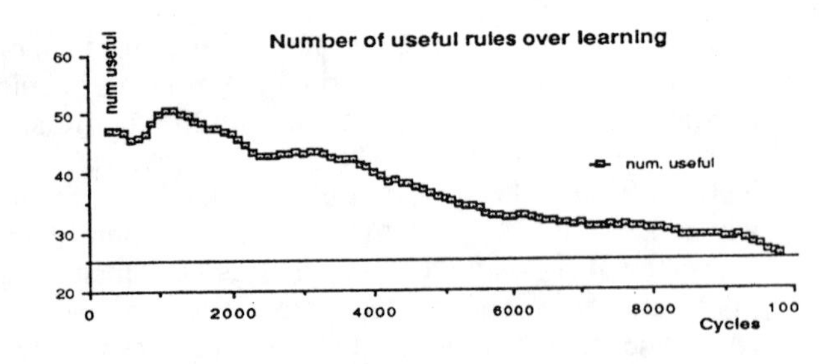

Figure 8: number of *useful* rules in the population

Figure 8 plots the evolution of the number of useful rules as defined above; one can notice that this number decreases strongly over learning until 25 after 10000 learning cycles (to be compared with 15 rules for CN2).

Are these rules truly useful? Figure 9 shows a comparison between the induction score over the unseen examples obtained respectively by the total rule population and the set of *Useful rules*.

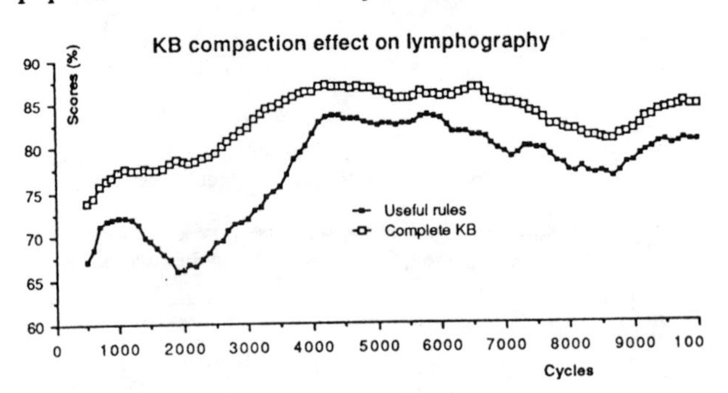

Figure 9: comparison of induction rate with useful rules vs total population

As predicted, the set of *Useful rules* compares favorably with the complete population: the loss of performance decreases down to about 2,5 % when learning is achieved, whereas the number of decision rules was reduced by a factor 10.

Furthermore, an increasing amount of the Population total strength (up to 60% after 10000 learning cycles) is packed into the *Useful set* during learning (Figure 10).

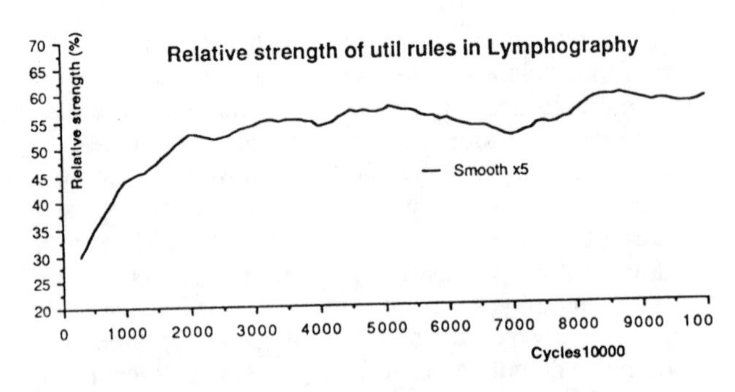

Figure 10: relative strength of useful rules vs entire population

Finally, *useful* rules are quite general: they each comprise an average of 4.9 attributes (18 attributes were available at the start). Here is an example of a high strength *useful* rule :

block of affere = *yes*
& **by pass** = *yes*
& **regeneration of extrasavates** = *yes*
& **changes in nodes** = *lacunar margin*

==> **diagnosis** = *metastases*

d/ General Remark

These experiments illustrated the need for construction of valid learning datasets that could serve as validation tools for new learning systems. It would be very interesting to gather up and characterize a set of data bases that could be used to evaluate different qualities of learning systems. Such data sets would permit researchers to hold experiments in more rigorous conditions, which is very important since ML is by essence a highly experimental field.

4 Conclusion

In this paper, we presented an empirical comparison of a genetic CS, **Newboole**, with two widely used learning algorithms, CN2 and BP, on three medical domains. The experimental results, obtained in the context of learning from preclassified examples, demonstrated two main points:

- firstly, that all three systems perform very closely on the induction tasks with a slight advantage for the BP algorithm.
- secondly, that a CS can provide comprehensive solutions in the form of a limited number of "symbolic" decision rules, which is not the case for a neural net using BP.

Some extensions to this work might be to experiment with **Newboole** on more complex problems such as problems where the correct answer is not always available, and to explore the possibility of incorporating domain specific knowledge into CS in order to guide genetic search more intelligently.

Acknowledgements

The data for all three medical domains was obtained from the University Medical Center in Ljubljana. We thank Hakim Lounis from "l' équipe Inférence et Apprentissage" for helping us with the CN2 experiments.

References

[1] P. Bonelli, A. Parodi, S. Sen and S. Wilson. Newboole: A Fast GBML System. In Proceedings of the international Conference on Machine Learning, June 1990, Austin, Texas.

[2] P. Clark and P. Niblett. Induction in noisy domains. In I. Bratko & N Lavrac (Editors), Progress in ML, 11-30, Sigma Press, 1987.

[3] S. Forrest. Implementing semantic network structure using the classifier system. In J.J. Grefenstette, editor, Proc. ICGA, 1985.

[4] D. Goldberg. Genetic Algorithms in search, Optimization, and Machine Learning. Addison Wesley, 1989.

[5] J. Grefenstette, C. Ramsey, and A. Schultz. Learning Sequential Decision Rules Using Simulation Models and Competition. In ML Journal 5, p. 335-381, Kluwer Academic Publishers, Boston, 1990.

[6] J. Holland. Adaptation in natural and artificial systems. University of Michigan Press, 1975.

[7] J. Holland. Escaping brittleness: the possibilities of general-purpose learning algorithms applied to parallel rule-based systems. In Machine Learning: an Artificial Intelligence Approach, Vol 2, Michalski R.S., Carbonell J.G., Mitchell T.M. eds, Morgan Kaufmann, Los Altos (CA), 1986.

[8] J. Koza. Evolution and co-evolution of computer programs to control independently-acting agents. In Proc. of the first conference on the Simulation of Adaptive Behavior, the MIT Press, 1991.

[9] R. Michalski, I. Mozetic, Jiarong Hong, and Nada Lavrac. The multi-purpose incremental learning system AQ15 and its testing application to three medical domains. In Proc. of AAAI 86, p. 1041-1045.

[10] A. Parodi and P. Bonelli. The Animat and the Physician. In Proc. of the first conference on the Simulation of Adaptive Behavior, the MIT Press, 1990.

[11] J. Quinlan. Induction of decision trees. In ML journal 1:81-106, Kluwer Academic Publishers, Boston, 1986.

[12] D. Rumelhart, G. Hinton, and R. Williams. Learning Internal Representations by error Propagation. In Parallel distributed processing, vol. 1, the MIT Press, 1986.

[13] A. Samuel. Some studies in machine learning using the game of checkers. In E.A. Feigenbaum and J. Feldman (Eds.), Computer and thought. McGraw-Hill, 1963.

[14] S. Wilson. Classifier Systems and the Animat Problem. In Machine Learning Journal 2:199-228, Kluwer Academic Publishers, Boston, 1987.

Alecsys: A Parallel Laboratory for Learning Classifier Systems

Marco Dorigo and Enrico Sirtori

MP-AI Project
Dipartimento di Elettronica
Politecnico di Milano
Piazza Leonardo da Vinci 32
20133 Milano - Italy
E-mail: dorigo@ipmel1.elet.polimi.it

Abstract

A major problem with learning systems is how to tackle real world problems. A distinctive characteristic of many real world problems is that they present a complexity that cannot be "user-defined", and which is generally orders of magnitude higher than in toy systems. The use of more powerful, parallel machines, is a way to attack this problem from two sides: through an increase in the performance of standard algorithms, and by design of a new structural organization of the learning system - organization that should allow a better control on the environmental complexity. In order to explore these potentialities we have built a tool, ALECSYS, that can be used to implement parallel learning classifier systems in a modular fashion. In ALECSYS parallelism is used both to increase the system performance, by what we call *low-level* parallelization, and to allow the use of many different learning classifier systems simultaneously, by what we call *high-level* parallelization. In the paper we first present the system organization and the algorithms used, then we report some simulation results and finally we give some hints for further work.

1 INTRODUCTION

Learning Classifier Systems (LCS) are a class of adaptive systems considered to be a very interesting and promising approach to reinforcement learning problems, i.e. learning situations where the only information available for the learning system is a payoff saying how useful a given action was with respect to the system goal. In the past years they have often been proposed as a powerful model to solve learning problems, but due to their computing power requirements, they met only limited success in facing real world problems. As parallel machines like the transputer and the Connection Machine are getting more and more widely used, new algorithms for exploiting parallel computing power in LCS applications have been developed by many researchers (e.g. [12], [13], [8], [15]).

ALECSYS (A Laboratory for Enhanced Classifier SYStems) is an environment for studying different forms of concurrency in LCSs. The system implements both a *low-level* parallelism within the structure of a single LCS, and a *high-level* type of concurrency which allows various LCSs to work together. We believe that this approach can lead to a powerful enhancement of the possibilities of a LCS. The paper is organized as follows: in section 2 some motivations to use parallelism in LCSs are given, sections 3 and 4 deal with low-level and high-level parallelism, while section 5 reports on some experiments. In the last section some conclusions and ideas about further work are sketched.

We assume the reader is familiar with classic LCS structures and algorithms. Shouldn't it be the case, various publications can be consulted to get a more detailed description (see for example [11], [9], [2])

2 REASONS FOR PARALLELIZING A LCS

One of the main problems faced by LCSs trying to solve real problems is the presence of heavy limitations both to the number of rules which can be employed and to their length. These limits arise from the present availability of computing power, which allows acceptable[1] elaboration times only when using populations of reduced size, e.g. sets of 100-1000 classifiers, with a 16-32 bits length.

One possible solution to increase the amount of processed information without slowing down the basic elaboration cycle, comes from the use of new parallel

---

[1] By "acceptable" we mean short enough for a machine learning researcher not to grow old while waiting for his experimentations to end...

architectures, such as the Connection Machine and the Transputer. Though each of the units composing such machines may have no greater computing power than other already existing hardware, their capability of independent, concurrent operations makes them more powerful than sequential Von Neumann architectures.

A parallel implementation of the LCS on the Connection Machine has been proposed by Robertson [13]. That work has demonstrated the power of such a solution, but still retained, in our opinion, a basic limit: as the Connection Machine is a SIMD architecture (i.e. it allows the use of a unique flow of control, applied to many data [7]), the resulting implementation was only a more powerful but still classic learning classifier system. To implement ALECSYS we have used the transputer system that, because of its MIMD architecture, permits the presence of many simultaneously active control flows operating on different data sets. This architectural organization allowed us to distinguish between low- and high-level software forms of parallelism in a way that directly maps on the hardware architecture.

3 LOW-LEVEL PARALLELIZATION OF A LCS

We will consider in the following the *micro-structure* of ALECSYS, i.e. the way we used parallelism to enhance the performance of the classic LCS model. The classic LCS is usually depicted as a set of three interacting systems (see Fig.1). The performance system is a kind of parallel, rule-based, production system; the rule-discovery (genetic algorithm) and the credit apportionment (bucket brigade algorithm) systems implement the two levels of learning that can be found in a LCS. We have maintained this basic organization, but, in order to simplify the presentation of the low-level parallelization algorithms, we discuss the various systems separately. In section A we show how to parallelize the LCS performance and credit apportionment subsystems; in B we will make similar considerations about the genetic algorithm.

3.1 THE PERFORMANCE AND CREDIT APPORTIONMENT SUBSYSTEMS

A basic execution cycle of the sequential LCS can be looked at as the result of the interaction between two data structures: the list of messages ML and the set of classifiers CF. Therefore, we decompose a basic execution cycle into two concurrent processes, MLprocess (Message List) and CFprocess (ClassiFiers set), interfaced to input and output processes (DTprocess and EFprocess, DeTector and EFfector processes), see Fig.2.

The processes communicate by explicit synchronization, with the following sequence of steps:

1 - MLprocess receives messages from DTprocess and places them in the message list ML;
2 - MLprocess sends ML to CFprocess;
3 - CFprocess "matches" ML and CF, calculating bids for each triggered rule;
4 - CFprocess sends MLprocess the list of triggered rules;
5 - MLprocess erases the old message list and makes an auction among triggered rules; the winners, selected with respect to their bid, are allowed to post their own messages, thus composing a new message list ML;
6 - MLprocess sends ML to EFprocess;
7 - EFprocess chooses the action to apply and, if necessary, discards from ML conflicting messages; EFprocess is then able to calculate the rewards owed to each message in ML; this list of rewards is sent back to MLprocess, together with the remaining ML;
8 - MLprocess sends to CFprocess the set of messages and rewards;
9 - CFprocess modifies the strengths of CF elements, assigning rewards and/or collecting taxes;
10 - while not stopped, go back to step 1;

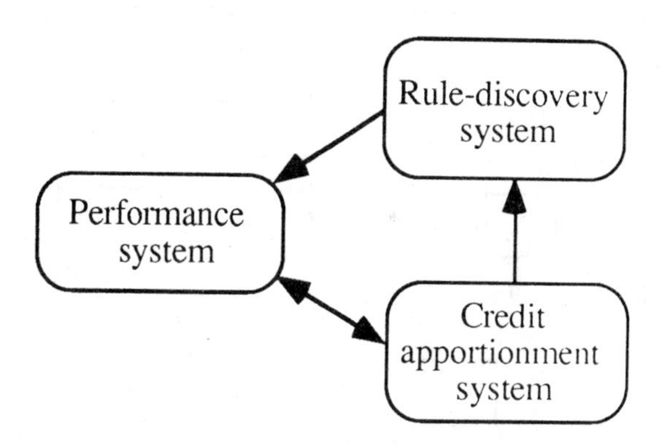

Fig. 1 - Structure of a classic learning classifier system

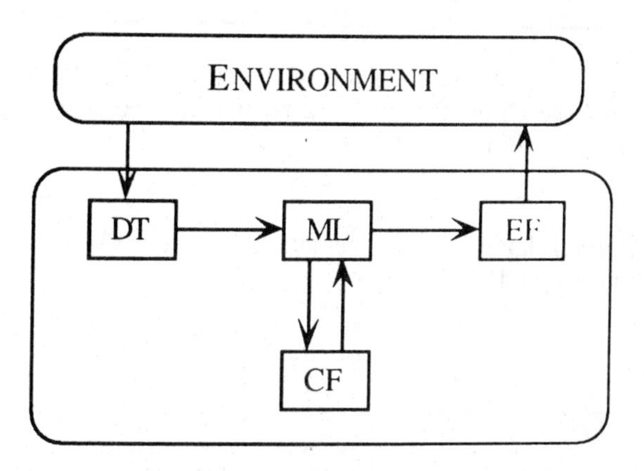

Fig. 2 - Concurrent processes in a learning classifier system

If we consider steps 3+4+5 (matching and message-production), we may see that they can be executed on each classifier independently. So, it is natural to split CFprocess into an array of concurrent subprocesses CFprocess.1, ... , CFprocess.i, ... , CFprocess.n, each one taking care of 1/n-th fraction of CF. The higher goes n, the more intensive is the concurrency. When n is equal to the size of CF, each CFprocess.i manages a single classifier: this is the typical Connection Machine version of a concurrent LCS [13]. In transputer-based machines we instead allocated about 100-500 rules to each processor (see section V for an empirical motivation to this number). CFprocesses can be organized in hierarchical structures such as for example, a tree (see Fig.3) or a toroidal grid[2].

As a matter of fact, many other steps can be parallelized: we propagate the message list ML from MLprocess to i-th CFprocess and back, obtaining concurrent processing of credit assignment on each CFprocess.i (the auction among triggered rules is subject to a hierarchical distribution mechanism, and the same hierarchical approach is applied to reward distribution).

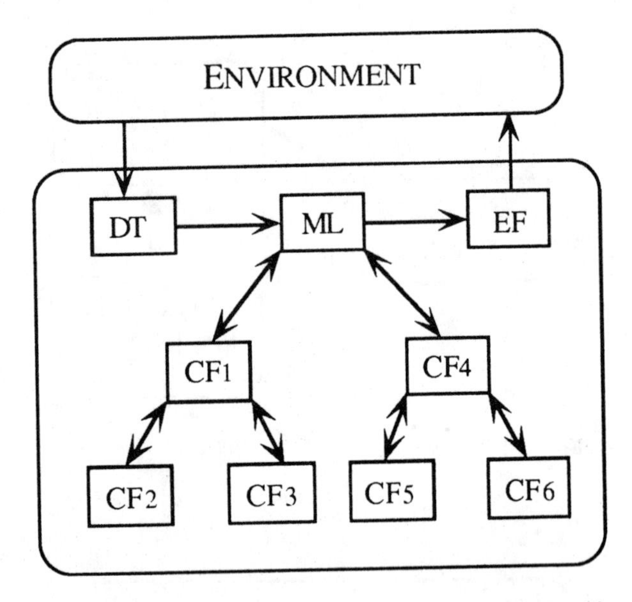

Fig. 3 - A parallel version of the classic LCS. In this example CFprocess of Fig.2 is split in six concurrent processes CF_1, ... , CF_6.

Further details on implementation issues can be found in [14] and [6].

[2]It should be noted that the structure actually chosen deeply influences the distribution of computational loads and therefore, the computational efficiency of the overall system.

3.2 THE RULE-DISCOVERY SUBSYSTEM

Let us now see how to distribute Genetic Algorithms on a transputer network. A first process, named GAprocess, can be assigned the duty to select, among CF elements, those individuals which are to be either replicated or discarded. It will be up to the (split) CFprocess, after receiving GAprocess decisions, to apply genetic operators, each single CFprocess.i focusing upon its own fraction of CF population (see Fig.4).

MLprocess stays idle during GA operations, as it could affect CF strengths, upon which genetic selection is based; and vice versa, GAprocess is "dormant" when MLprocess works. It is then quite natural to merge MLprocess and GAprocess into one single *manager* process.

A typical background genetic algorithm works as follows:

1 - within CF two sets of rules are selected, one built with rules to be replicated (*parents* classifiers), the other with those to be replaced (*offsprings* positions);
2 - *parents* are mated two by two;
3 - to each couple created at step 2 is applied a genetic crossover operator, thus generating a new couple of rules (*offsprings*);
4 - *offsprings* undergo a stochastic mutation operator;

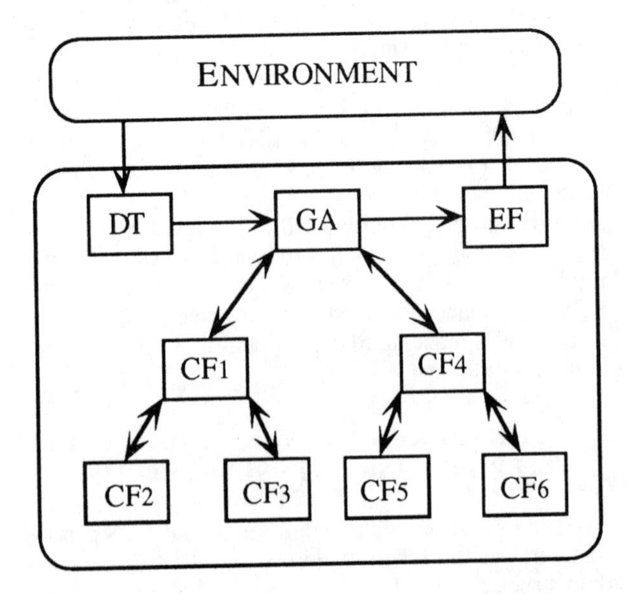

Fig.4 - The same type of architecture used for the apportionment of credit and performance systems parallelization is employed to parallelize the genetic algorithm.

Considering our MIMD parallel implementation, step 1 may be seen as an auction, which can be distributed over the processor network by a "hierarchical gathering and broadcasting" mechanism, similar to the one we used to propagate the message list. Step 2 (mating of rules) is not easy to parallelize, as it requires a central management unit. Luckly, the number of couples is usually low and concurrency seems to be unnecessary (at least for not very big populations). Step 3 is the hardest to parallelize, because of the many communications it requires, both between MLprocess and the array of split CFprocesses, and among CFprocesses themselves. Step 4 is a typical example of local data processing, extremely suited to concurrent distribution.

The parallel GA is then:

1 - each CFprocess.*i* selects, within its own subset of CF, m rules to replicate and m to replace (note: m is a system parameter);
2 - each CFprocess.*i* sends MLprocess some data about each selected classifier, enabling MLprocess to set up a hierarchical auction based on strength values; this process results in selecting 2m individuals within the overall CF population;
3 - MLprocess sends the following data to every i-th CFprocess containing a *parent* classifier:
identifier of the *parent* itself;
identifiers of the two *offsprings*;
CrossOver Point (COP);
at the same time MLprocess sends to the i-th CFprocess containing a position for any *offspring* the following data:
identifier of the *offspring*;
identifiers of the two *parents*;
COP;
4 - all the CFprocesses which have *parents* in their own fraction of CF send a copy of the *parent* rule to the CFprocess.*i* which has the corresponding *offspring* position; this process will apply crossover and mutation operators.

By similar procedures we have implemented parallelized standard operators such as Cover Detector, Cover Effector and Triggered Chaining.

4 HIGH-LEVEL PARALLELIZATION OF A LCS

In the preceding section we have analyzed a way to parallelize a single LCS, with the goal of obtaining computing speed improvements. What we did (just like other researchers, see for example [13], [3], [15]) was to parallelize the classic LCS model, introducing no significant improvement except for average execution times. This approach shows its weakness when a LCS is applied to multigoal problems - as unluckily seems to be the case in most of real problems. In these cases some mechanisms, like restricted mating [1], can be used. A different solution can be to dedicate different LCS to the solution of the different problems. This approach requires the introduction of a stronger form of concurrence, that we call high-level parallelism.
Moreover, scalability problems arise in a low-level parallelized LCS: adding a node to the transputer network implementing the LCS makes the communication load grow faster than the computational power, obtaining a less than linear speed-up. Therefore, adding processors to an existing network is less and less effective.

As already said, a better way to deal with complex problems could be to code them as a set of easier subproblems, each one committed to a LCS.

In ALECSYS the processor network is partitioned into subsets, each having its own size and topology. Every subset is allocated a single LCS, distributed by means of a low-level parallelization whenever the number of nodes used by each single LCS is greater than one (see Fig.5).

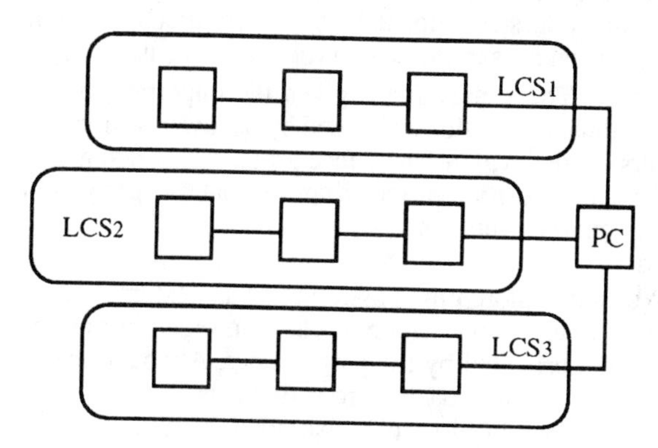

Fig.5 - Using high-level and low-level parallelism simultaneously: a transputer network in which each one of the three classifier systems (high-level parallelization) is mapped on three nodes (low-level parallelization).

Each of these LCSs learns to solve a specific subgoal, depending on the inputs it receives: each LCS perceives the external environment by its own detectors while the output interface is obviously common to all LCSs, thus requiring some type of interaction among LCSs proposing actions.

As an example we assigned the task of learning rules for tracing a light source to LCS_1 while a second learning classifier LCS_2 learned how to avoid heat sources; these tasks are simple components of a more complex goal, i.e. the system must simultaneously learn to follow the light, to avoid the heat sources, and how to behave in conflicting situations (e.g. when the light source goes too near to a heat source for the system to continue to follow it without being injured).

When proposing different actions, concurrent LCSs may interact either by *competition* or by *cooperation*.

In competitive interaction each LCS associates to every effector-directed message an activation value, which depends on the reliability of the rule inside the LCS, on the learning level achieved (i.e. how far the system is from steady-state performance), and on the relative importance of the LCS with respect to the others. Each action has an activation value that is the sum of all the activation values of effector messages that activate it; the action with the highest value is applied to the environment.

In cooperative interaction all suggested actions are taken into account to determine which action to apply. In this case, the LCSs may communicate by direct links (*direct communication*), thus influencing each other's decisions, or they may instead communicate their choices to an upper mediator (*indirect communication*), which will compose them into a proper action.

We investigated indirect, cooperative interaction. Indirect communication mode allows to create an easily expandable structure, though this architecture has its main weakness in the central role of the mediator-supervisor[3]. The supervisor is given the important duty to coordinate the actions proposed by its lower-degree partners. In this way it has to face a new task, arising from the decomposition of the primitive goal: the problem of subgoals coordination.

We first explored the possibility of an "external supervisor", i.e. a coordination mechanism set according to a scheme provided by the programmer. Although the results of this approach were satisfying, we resolved to assign the coordination task to a learning system, namely another LCS. The superiority of this approach clearly is in the fact that, in this way, the solution of the original task is completely delegated to a set of concurrent LCSs. We have called the coordinating LCS "internal supervisor".

We are now ready to present the general algorithm for ALECSYS:

1 - each LCS receives from the environment its own input messages; by processing them each LCS deduces a suggested action to actuate (usually not the same for all the LCSs);
2 - proposed actions are communicated to the supervisor, together with their associated bids;
3 - the supervisor deduces a resulting action, which is then actuated by means of the output interface;
4 - the environment assigns each LCS a reward;
5 - go back to step 1;

5 HALE: A CASE STUDY FOR ALECSYS.

We now depict the complete task used to test ALECSYS capabilities in dealing with multigoal problems, and report some preliminary experiments we made about both unimodal and multimodal tasks.

The acronym HALE stands for Heat And Light Environment and summarizes the characteristics of the problem: a vehicle (the HALE), placed within a rectangular boundary, should learn to:

a) follow a mobile light source
b) avoid one or more fixed heat sources

The tasks may be presented either separately (unimodal task) or simultaneously (multimodal task). In the case of multimodal problems the objectives may be set to be more or less conflicting, by choosing the smallest distance of heat sources from the path traced by the light source. Both light and heat sources are perceived by detectors operating in four cardinal directions while the system can move in eight directions.

A first set of experiments examined the effects of some parameters upon the system performance in the case of unimodal task (tracing the light source). We investigated the optimal value of the following parameters:

1) number of classifiers
2) rate of activation of genetic operators

In experiment 1) we found a peak in system performance when using 200 classifiers on each one of 9 processors (see Fig.6). This seems to support Goldberg hypothesis about the existence of an optimal size for the set of classifiers [10] (but it is counter the Riolo-Robertson [12] work).

Steady-state performance
(norm. avg. distance from light source)

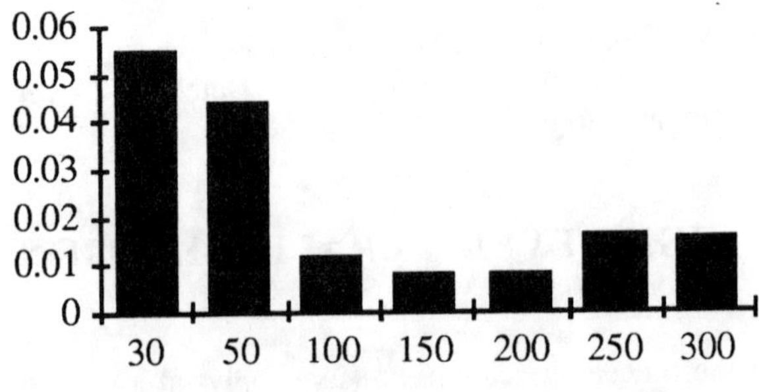

Fig.6 - Steady-state performance measured as normalized average distance of the HALE from the light source (averaged on the last 5000 cycles over 20000) as a function of the number of classifier on each node

In experiment 2) we tested the GA performance as a function of the rate of activation of background genetics (crossover and mutation) and of triggered chaining (see Fig.7). Our system obtained its best results when using a Triggered Chaining activation rate equal to 0.1, with a background-genetic rate of 0.01 (equivalent to explore 10 new rules every 1,000 cycles).

---

[3] By means of a proper structural organization this problem may be overcome (see [5]).

A second set of experiments was devoted to find out if the system was able to develop an internal model of the world. Experiments have been done putting the learning system in different external environments (unimodal and multimodal tasks) and testing different sets of system parameters; unluckily the results do not permit any definitive assertion to be drawn. Some tests showed a system behaviour based on *stimulus-response* rules, some others seemed instead to demonstrate memory-based actions.

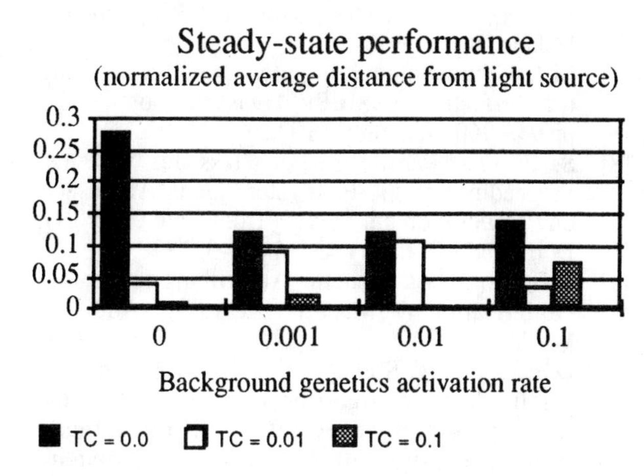

Fig.7 - Steady-state performance measured as normalized average distance of the HALE from the light source (averaged on the last 5000 cycles over 20000).

Most of these behaviours are probably better perceived when watching at the simulations. Yet, we present here a sample of classifiers taken from a steady-state population, originated as final output of an ALECSYS simulation when applied to an unimodal task (see table 1). Each classifier is made of two conditions and one action. Each message is 6 bits long. The first two bits indicate whether the messages are coming from Detectors (00), directed to Effectors (01) or aimed at internal use only (10 or 11). The others four bits have the following meaning: for detector-originated messages they encode the presence of a light source in the four cardinal directions, i.e. North, East, South and West; each sensor detects light when placed in a semicircle facing the sensor itself. For instance, 001001 means that light is perceived from North and West (see Fig.8).

In effector-directed messages the four bits have quite a different meaning. The first of them encodes whether the system should move or not, the remaining three set one of eight possible motion directions. For example, 011010 means that the system should move eastward.

Examining the population we found high-strength rules of the kind *stimulus-response*, and rules that send internal messages. An example of stimulus-response rule is

00#100.00#100.011001

that can be translated in words as:

"when light is perceived in direction East and/or North, then go North-East".

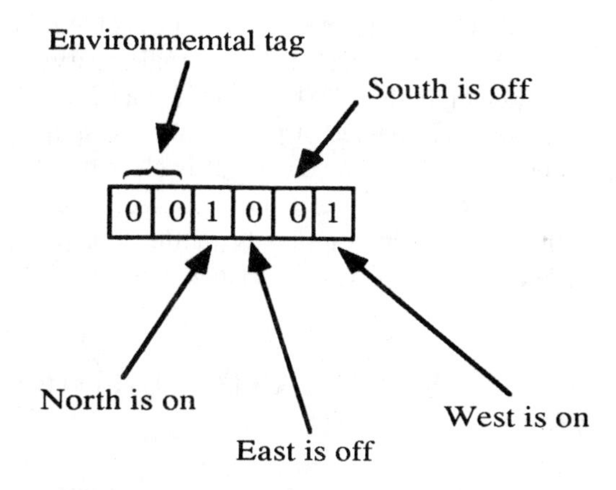

Fig.8 - Example of environmental message

We also observed many rules that send internal messages: their relation to the system behaviour is not completely clear yet. Their presence can probably justify the results of some experiments in which memory-based actions seemed to be used: the learning system developed a set of rules capable to trace a light moving on a circular path and another set capable to trace a light moving on straight segments with randomly changing orientation. The steady-state performance in the experiment with the circular path resulted to be the highest. We interpreted this behaviour as a possible clue of internal model development (the system was someway able to predict the light next movement), but further experimentation is necessary and is going on (see [4]).

A third set of experiments analyzed the performance of the system when applied to the HALE task, comparing the results obtained with the two different supervisor models (*internal-external*).

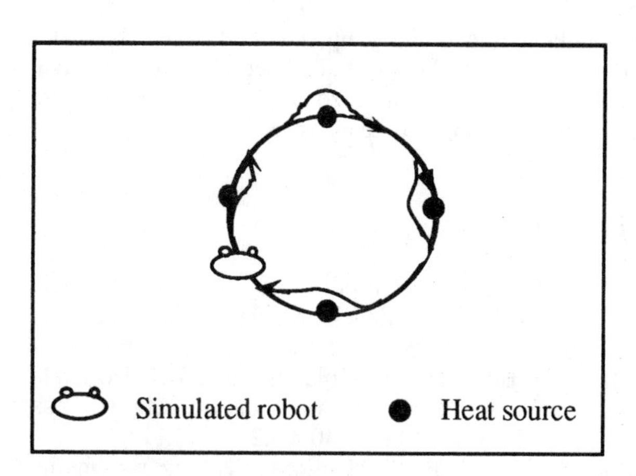

Fig.9 - Example of multimodal task: the simulated robot learns to follow the light moving on the circular path and at the same time to avoid the heat sources

The external-supervisor model could reach an acceptable performance within 20-40,000 cycles (30-60 minutes on our 9-transputer architecture), even when facing complex environments (up to 4 heat sources along the light source path - see Fig.9); the internal-supervisor model required more than 80,000 cycles (about 2 hours) to reach a sufficient degree of performance in simpler environments (only 1 heat source). It is clear that the longer time required by the internal supervisor model is due to the increase in complexity of the task to be learned (both simple behaviours and their coordination).

6 CONCLUSIONS AND FURTHER WORK

In this paper we have presented ALECSYS, a parallel version of the classic LCS model. We have distinguished between low- and high-level parallelism and have implemented a modular environment that can be used to test LCSs applied to complex problems. The proposed architecture will allow a researcher to investigate the validity of organizational models in which coordination of competing and cooperating agents is important. A first set of experiments, aimed at system tuning, has been presented: they seem to confirm previous experimental evidence proposed by other researchers. We have also begun to investigate coordination strategies and first results are given in the case of a multimodal task. Further work will consider the possibility of new coordination policies, such as for example experimentation with ethological models of behaviour coordination [5].

Acknowledgements

This research was supported in part by a grant from CNR - Progetto finalizzato sistemi informatici e calcolo parallelo - Sottoprogetto 2 - Tema: Processori dedicati and from CNR - Progetto finalizzato robotica - Sottoobiettivo 2 - Tema: ALPI. The system ALECSYS is available free of charge upon request to any of the authors. I would like to thank Uwe Schnepf, Andrea Bonarini, Vittorio Maniezzo, Domenico Sorrenti and Vincenzo Caglioti for the many useful discussions.

References

[1] L.B. Booker. Classifier Systems that Learn Internal World Models. Machine Learning, 3(3), p.161-192,1988.

[2] L.B. Booker, D.E. Goldberg, and J.H. Holland. Classifier Systems and Genetic Algorithms. Artificial Intelligence,40, pp.235-282,1989.

[3] A. Camilli, R. Di Meglio. Classifiers systems in massively parallel architectures (in Italian). Master thesis, University of Pisa, 1990.

[4] M. Dorigo, U. Schnepf. A Bootstrapping Approach to Robot Intelligence: First Results. Technical Report No. 90-068, MP-AI Project, Department of Electronics, Politecnico di Milano, Italy, December 1990.

[5] M. Dorigo, U. Schnepf. Organisation of Robot Behaviour Through Genetic Learning Processes. Proceedings of the Fifth IEEE International Conference on Advanced Robotics (in press) - Pisa - Italy - June 20-22, 1991.

[6] M. Dorigo, E. Sirtori. A Parallel Distributed Environment for Genetics-based Machine Learning. Technical Report No. 91-015, MP-AI Project, Department of Electronics, Politecnico di Milano, Italy,1991.

[7] M.J. Flynn. Some Computer Organisations and their Effectiveness. IEEE Transaction on Comp., pp.948-960, September 1972.

[8] S. Forrest. Parallelism in Classifier Systems. Proceedings of the First International Workshop on Parallel Problem Solving from Nature (PPSN), Dortmund - Germany, October 1-3, 1990.

[9] D.E. Goldberg. Genetic Algorithms in Search, Optimization & Machine Learning. Addison-Wesley, 1989.

[10] D.E. Goldberg. Sizing populations for serial and parallel Genetic Algorirhms. Proceedings of the Third International Conference on Genetic Algorithms, pp. 70-79, Morgan Kaufmann, George Mason Univ. CA, June 4-7,1989.

[11] J.H. Holland. Escaping brittleness: The possibilities of general purpose learning algorithms applied to parallel rule-based systems. In R.S. Michalski, J.G. Carbonell, and T.M. Mitchell (Eds.), Machine Learning II, Los Altos, CA: Morgan Kaufmann, 1986.

[12] R.L. Riolo, G.G. Robertson. A Tale of Two Classifier Systems. Machine Learning 3, pp.139-159, Kluwer Academic Publishers, Netherlands, 1988.

[13] G.G. Robertson. Parallel Implementation of Genetic Algorithms in a Classifier System. Proceedings of the Second International Conference on Genetic Algorithms, Lawrence Erlbaum, July 28-31, 1987.

[14] E. Sirtori. ALECSYS - A parallel architecture for Machine Learning (in Italian). Master thesis, MP-AI Project, Department of Electronics, Politecnico di Milano, 1991.

[15] A. Camilli, R. Di Meglio, F. Baiardi, M. Vanneschi, D.Montanari, R. Serra. Classifier System Parallelization on MIMD architectures. CNR Technical Report N.3/17 (in Italian), Progetto Finalizzato Sistemi Informatici e Calcolo Parallelo, March 1990.

Lamarckian Learning in Multi-agent Environments

John J. Grefenstette
Navy Center for Applied Research in Artificial Intelligence
Code 5514
Naval Research Laboratory
Washington, DC 20375-5000
E-mail: GREF@AIC.NRL.NAVY.MIL

Abstract

Genetic algorithms gain much of their power from mechanisms derived from the field of population genetics. However, it is possible, and in some cases desirable, to augment the standard mechanisms with additional features not available in biological systems. In this paper, we examine the use of Lamarckian learning operators in the SAMUEL architecture. The use of the operators is illustrated on three tasks in multi-agent environments.

1 INTRODUCTION

The goal of this work is to explore the application of machine learning techniques to reactive control problems arising in competitive, multi-agent domains. In such domains, traditional AI planning approaches are usually infeasible, because of the complexity of the multi-agent interactions and the inherent uncertainty about the future actions of other agents. On the other hand, genetic algorithms [11] appear to be a promising approach to developing high performance control strategies. SAMUEL is our platform for exploring the use of genetic algorithms to learn control strategies, expressed as sets of condition/action rules, for sequential decision problems.

Several new features have recently been added to SAMUEL that improve significantly both the quality of the rules that are learned and the computational cost of learning those rules. These improvements have been achieved by complementing the Darwinian principles embodied in SAMUEL with a number of mechanisms that are more Lamarckian in flavor. While such mechanisms may not be appropriate for systems intended to accurately model biological adaptive systems, they are appropriate for systems like SAMUEL whose primary motivation is the practical demonstration of genetic learning in interesting domains. This report will emphasize the new mechanisms in SAMUEL, and illustrate their utility in learning some interesting behaviors in multi-agent environments.

The reactive systems we consider here may be characterized by the following general scenario: The decision making agent interacts with a discrete-time dynamical system in an iterative fashion. At the beginning of each time step, the agent observes a representation of the current state and selects one of a finite set of actions, based on the agent's decision rules. As a result, the dynamical system enters a new state (perhaps based on the actions of other agents in the environment) and returns a (perhaps null) payoff. This cycle repeats indefinitely. The objective is to find a set of decision rules that maximizes the expected total payoff.[1] Several tasks for which reactive systems are appropriate have been investigated in the machine learning literature, including pole balancing [15], gas pipeline control [3], and the animat problem [18]. For many interesting problems, including those considered here, payoff is delayed in the sense that non-null payoff occurs only at the end of an episode that may span several decision steps.

SAMUEL is a genetic learning system designed for sequential decision problems. The design of SAMUEL builds on De Jong and Smith's LS-1 approach [17] as well as our own previous system called RUDI [7]. Some of the key features of SAMUEL are:

- A flexible and natural language for expressing rules.
- Incremental rule-level credit assignment.
- Competition both at the rule level and at the strategy level.
- A genetic algorithm for search the space of strategies.
- A set of heuristic rule learning operators that are integrated with the genetic operators.

Initial studies on an evasive maneuvers task have demonstrated that

- SAMUEL can learn general strategies for evasion that are effective against adversaries with a broad range of maneuverability characteristics, and under a variety

[1] Barto, Sutton and Watkins [1] give a good discussion of broad applicability of this general model.

of initial conditions (e.g., initial speed and range) [9].

- SAMUEL can learn high-performance strategies even with with noisy sensors [13].
- SAMUEL can effectively use existing knowledge to speed up learning [16].

Rather than detail a particular application, this report will try to illustrate how SAMUEL can be applied to learn several different behaviors, with a focus on tasks for an autonomous agent in a hostile environment. New results with three such environments are presented. We believe that these illustrations convey something of the generality of the approach.

The remainder of the paper is organized as follows: Section 2 offers a brief overview of the SAMUEL architecture. The next section presents the newest enhancements to SAMUEL, including a more strongly biased conflict resolution algorithm, an extension of the rule language to include symbolic attributes organized into a generalization hierarchy, and heuristic rule learning operators, including SPECIALIZE, GENERALIZE, MERGE, and DELETE. Section 4 presents a test suite of environments that each requires the system to learn a different kind of behavior. This is followed by some empirical studies of SAMUEL's performance on these environments. The last section contains a few final comments.

2 OVERVIEW OF SAMUEL

SAMUEL adopts a number of assumptions that make the system potentially applicable to real-world problems. First, the learning agent's perception facilities are limited to a fixed set of discrete, possibly noisy, sensors. There is also a fixed set of control variables that may be set by the decision making agent. Reflecting our primary interest in rapidly changing and uncertain environments, the agent's decision rules are limited to simple condition/action rules of the form

```
if   (and c1  . . .  cn)
then (and a1  . . .  am)
```

where each condition c_i specifies a set of values for one of the sensors and each action a_j specifies a setting for one of the control variables. We call a set of such decision rules is called a *reactive control strategy.*

One of the key features of SAMUEL is that, unlike many previous genetic learning systems, the knowledge representation consists of symbolic condition-action rules, rather than low-level binary pattern matching primitives.[2] The use of a symbolic language offers several advantages. First, it is easier to transfer the knowledge learned to human operators. Second, it makes it easier to combine genetic algorithms with analytic learning methods that explain the success of the empirically derived rules [5]. Finally, it makes it easier to incorporate existing knowledge. A recent study [16] addressed this final point by comparing two mechanisms for initializing the knowledge structures in SAMUEL. The results show that genetic algorithms can be used to improve partially correct strategies, as well as to learn strategies given no initial knowledge.

[2] Booker [2] shows how more expressive encodings could be implemented in a classifier system, but experimental results are not yet available.

The genetic algorithm in SAMUEL is generational and includes the standard genetic operators, SELECTION, CROSSOVER, and MUTATION. SAMUEL uses a proportional selection algorithm and Baker's SUS sampling algorithm. CROSSOVER consists of exchanging rules between two selected parents. (CROSSOVER occurs on rule boundaries only.) MUTATION consists of making a random change to a value in a rule. For example, MUTATION might change the condition

```
(time is [5 .. 10])
```

to

```
(time is [1 .. 10]).
```

A restricted form of mutation, called CREEP, makes a minimal change to a value in a rule. For example, CREEP might change the condition

```
(time is [5 .. 10])
```

to

```
(time is [6 .. 10])
```

but not to

```
(time is [7 .. 10]).
```

For further details on the operation of SAMUEL, see [9] and [10].

3 LAMARCKIAN ASPECTS OF SAMUEL

This section describes some of the more recently developed features of SAMUEL, especially those which modify the internal knowledge structures of a strategy as a direct result of the strategy's experience with the task domain. These changes are subsequently passed along to the strategy's offspring, reflecting an evolutionary theory most often connected with the work of Jean Baptiste Lamarck [6]. Lamarck developed a theory that stressed the inheritance of acquired characteristics, in particular acquired characteristics that are well adapted to the surrounding environment. Of course, Lamarck's theory was superseded by Darwin's emphasis on two-stage adaptation: undirected variation followed by selection. Research has generally failed to substantiate any Lamarckian mechanisms in biological systems. Fortunately, in artificial systems, we can easily provide that which nature cannot.[3] The primary Lamarckian feature SAMUEL is the association of strengths with individual rules, and the use of this information for conflict resolution. In addition, many of the rule modification operators are Lamarckian in that they are triggered either directly by a strategy's interaction with the environment or indirectly by the strength of the rules within a strategy.

[3] It might be mentioned that human cultural evolution is highly Lamarckian, and subsequently much more rapid than biological evolution [6].

The next two sections describe these mechanisms in detail.

3.1 CREDIT ASSIGNMENT AND CONFLICT RESOLUTION

Each rule in SAMUEL has an associated *strength* that estimates the rule's utility for the learning task [7].[4] When a rule is inherited by a newly formed strategy, the rule's strength is passed along as well. The primary way that rule strength is used in SAMUEL is in conflict resolution, which runs as follows:

1. Find the *match set*, consisting of all rules that most nearly match the current sensor readings.
2. For each possible action, define the action's *bid* as the maximum strength of any rule in the match set that specifies that action.
3. Raise each (non-null) bid to a power specified by a parameter called the *bid bias.*
4. Select an action by sampling from the probability distribution defined by the modified bids.

The bid bias serves as kind of a Lamarckian control knob.[5] For example, if the bid bias $= 0$, all non-null bids are considered equal, and the impact of the inherited strength information on conflict resolution is nullified. Any non-zero value for the bid bias results in a Lamarckian system, with a varying degree of greediness. For example, if the bid bias $= 1$, we get a roulette wheel selection based on the maximum strength associated with each action.[6] If the bid bias > 1, we get a bias toward the higher bids. (Any value of bid bias ≥ 10 is treated as infinite -- all bids less than the maximum bid are deleted.) Initial experience indicates that the best performance is obtained with the maximum value for the bid bias. This is the default value used in the experiments described below.

3.2 LEARNING OPERATORS

SAMUEL has one binary recombination operator, CROSSOVER. CROSSOVER exchanges rules between two strategies.[7] In its default mode, CROSSOVER first *clusters* rules so that rules that fire in sequence within a high-payoff environment tend to be assigned to the same offspring. The idea is to promote the inheritance of behavior associated with high payoff. This form of CROSSOVER is Lamarckian since the clustering depends directly on the strategy's past experience with the environment.

SAMUEL currently includes six unary operators that modify the rules within a single strategy: MUTATION, CREEP, SPECIALIZE, GENERALIZE, MERGE, and DELETE. Unlike previous versions of the system, SAMUEL has now adopted the policy, common in classifier systems [12], that all of these operators (except DELETE, of course) are *creative*, i.e., any modifications are made on a new copy of the original rule. Once created, a rule survives intact unless it is explicitly deleted or lost when its strategy is not selected for reproduction. We have found that this policy allows a much more aggressive application of rule modification operators with little damage if the changes are maladaptive.

A little more detail on the rule language is necessary before we discuss the new rule creation operators. Each sensor and control variable has an *attribute type* declared by the user. There are four type of attributes: *linear, cyclic, structured, pattern.* Linear and cyclic attributes take on values from a linearly or cyclicly ordered numeric range, respectively. For example, the sensor *time* might be a linear attribute with values between 0 and 20. A condition for this attribute might be

```
(time is [5 .. 10])
```

which would be matched if $5 \leq time \leq 10$. The cyclic attribute *direction* might take on value between 0 and 360, and a condition for this attribute might be

```
(direction is [270 .. 90])
```

which would be satisfied if ($270 \leq direction \leq 360$ or $0 \leq direction \leq 90$). A pattern attribute is associated with a fixed-length string over the alphabet { 0, 1, # }, like classifiers in classifier systems [12]. For example, the pattern attribute *visual-field* might be defined as a six-bit string, and a condition for this attribute might be

```
(visual-field is 0####1)
```

which would be matched if the first bit of *visual-field* is 0 and the last bit is 1. A structured attribute can assume values from a hierarchy of symbolic values specified by the user. For example, an attribute called *distance* might be defined as shown in Figure 1. Conditions for structured sensors specify a list of values, and the condition matches if the sensor's current value occurs in a subtree labeled by one of the values in the list. A condition for the *distance* sensor might be

```
(distance is [close, 400])
```

This would match if the sensor *distance* had the value *close, medium-close, very-close,* 100, 200, 300, or 400.

The user also specifies an *ordering* for the structured hierarchy that indicates the order relationship among the nodes at each level of the hierarchy. The order may be *linear, cyclic,* or *none*. For example, the *distance* attribute above has a linear order among the leaves, as well

---

[4] A rule's strength increases as a function of the mean of the expected payoff and decreases with the variance of the expected payoff, so that high strength indicates both high utility and high confidence in the rule [8].

[5] The bid bias was inspired by a similar mechanism in Riolo's classifier system CFS-C [14], and is similar in effect to the notion of using bidding noise based on a classifier's variance [4].

[6] In all previously reported results with SAMUEL, the bid bias was set to 1.

[7] We prefer to rely on explicit mutation operators, rather than overload CROSSOVER with the additional task of introducing new rules by crossing with rule boundaries.

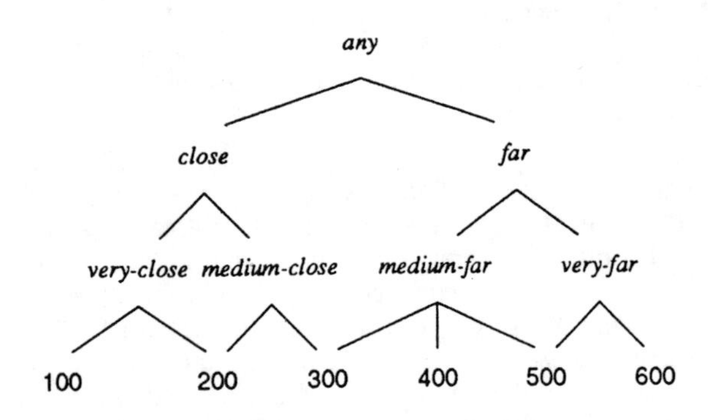

Figure 1: A Structured Attribute

as among the nodes at each higher level. On the other hand, a hierarchy having to do with an object's color may have no inherent order among the nodes at a given level of generality. An example of a hierarchy with cyclic order would be one based on compass direction, with leaf values such as *due-north, north-east, due-east, north-west.* The structured type allows the user to bias the learning operators to reflect the semantics of the sensor. We will now discuss the new rule creation operators, emphasizing their action on the structured type.

SPECIALIZE

The SPECIALIZE operator can be applied when a general rule fires in a high payoff episode. (The generality threshold and the payoff threshold for the operator are run-time parameters). The operator creates a new rule whose left-hand-side more closely matches the current sensor values and whose right-hand-side more closely matches the current action value. For numeric conditions (i.e., linear or cyclic), the operator creates a new condition with roughly half the generality of the previous condition by moving each endpoint half way toward the sensor reading. For example, if the original condition is

```
(speed is [100 .. 1500])
```

and the sensor reading is

```
speed = 500
```

then the new condition would be

```
(speed is [300 .. 1000])
```

For structured conditions, SPECIALIZE replaces each value in the disjunct by each of its children that covers the current sensor reading. For example, if the original condition is

```
(distance is [close, far])
```

and the sensor reading is

```
distance = 300
```

then the new condition would be

```
(distance is [medium-close, medium-far])
```

If the sensor reading is

```
distance = 400
```

then the new condition would be

```
(distance is [medium-far])
```

since there is no specialization of *close* that covers the sensor reading.

GENERALIZE

The GENERALIZE operator can be applied when a rule fires due to a partial match, during a high payoff episode. (The payoff threshold for the operator is a run-time parameter). A partial match occurs when there is no rule that completely matches all the current sensor readings. The operator creates a new rule whose left-hand-side is generalized enough to match the current sensor values. For numeric conditions, the operator creates a new condition with one of the end points set to the sensor reading. For example, if the original condition is

```
(speed is [700 .. 1500])
```

and the sensor reading is

```
speed = 500
```

then the new condition would be

```
(speed is [500 .. 1500])
```

For structured conditions, GENERALIZE adds the current sensor value to the disjunct and generalizes up the hierarchy if all the children of a given node are present in the disjunct. For example, if the original condition is

```
(distance is [very-close, very-far])
```

and the sensor reading is

```
distance = 300
```

then the new condition would be

```
(distance is [close, very-far])
```

MERGE

The MERGE operator creates a new rule from two existing high-strength rules that have identical right-hand-sides. The new rule will match any sensor value matched by either of the original rules. The right-hand-side of the new rule is the same as both of the original rules. For example, the result of MERGE of two rules:

```
if    ((time is [1 .. 5])
       (distance is [very-close]))
then ((turn is [right]))
```

and

```
if    ((time is [3 .. 8])
       (distance is [medium-close, medium-far]))
then  ((turn is [right]))
```

would be

```
if    ((time is [1 .. 8])
       (distance is [close, medium-far]))
then  ((turn is [right]))
```

The MERGE operator, in combination with the DELETE operator below, helps to eliminate overspecialized rules from the strategy.

DELETE

The DELETE operator is the only mechanism for removing rules from a strategy. A rule may be deleted if it meets one or more of the following criteria: (1) the rule has low activity level (hasn't fired recently); (2) the rule has low strength; or (3) the rule is subsumed by another rule with higher strength. All of these criteria are controlled by run-time parameters.

The operators SPECIALIZE and GENERALIZE are clearly Lamarckian in the sense that they are triggered only by successful experiences and they change a strategy to more closely reflect this experience. MERGE and DELETE are indirectly Lamarckian in the sense that they are sensitive to the strength or activity level of the rules, and these statistics directly reflect the rule's past experience with the environment. It should be noted, however, that with the proper selection of run-time parameters, the degree of Lamarckism in all these operators can be reduced or eliminated. Future studies will explored the effects of Lamarckism in SAMUEL in more depth.

4 A TEST SUITE OF COMPETITIVE ENVIRONMENTS

This section describes three rather challenging testing environments we have designed for SAMUEL. In each environment there is one learning agent and another adversary agent. This adversary may behave unpredictably (within certain bounds). The learning agent is the same for all three environments, but the adversaries and the performance tasks differ. This arrangement provides an interesting test of SAMUEL's ability to learn various tasks with an agent with only general purpose sensors and effectors.[8]

We first describe the learning agent. The agent has a fixed set of sensors, namely: *time* (since the beginning of the episode), *last-turn* (by the agent), *bearing* (direction to adversary's position), *heading* (relative direction of adversary's motion), *speed* (of adversary), and *range* (to adversary). Each sensor has a fixed granularity that is fairly large.[9] That is, the mapping from the true world state to observed world state is many-to-one. The sensors are also noisy, and may report incorrect values. The agent has two actions: it can change its own direction and speed. (In two cases, the agent only learns to directly control its turning rate, and its speed is determined by its turning rate.) Finally, the agent's own actions are noisy. That is, the agent may select a 90 degree turn, but in fact, it may turn a little less or a little more than it had indicated. Unlike an agent in a typical AI planning program, our agent generally cannot accurately predict the next state on the basis of the current observed state and the action it selects. These assumptions, which are intended to capture some of the flavor of robotic interactions with the real world, preclude the use of traditional AI planning techniques, and argue in favor of SAMUEL's more reactive approach. We now describe the three test environments.

4.1 EVASION

The first environment is a model of predator-prey situation in which the learning agent plays the role of prey.[10] The adversary, or predator, can track the motion of the prey and steer toward the prey's anticipated position. In this environment, the agent learns only its turning rate; its speed is determined by the turning rate. The process is divided into episodes that begin with the predator approaching the prey from a randomly chosen direction. The predator initially travels at a far greater speed but is less maneuverable than the prey (i.e., the predator has a greater turning radius than the prey) and gradually loses energy (i.e., speed) as it maneuvers. The episode ends when either the predator captures the prey or the predator's energy drops below a threshold and it gives up. This requires between 2 and 20 decision steps, depending on how many turns the predator performs while tracking the prey. At the end of each episode, the critic provides full payoff if the agent evades the adversary, and partial payoff otherwise, proportional to the amount of time before the agent's capture.

4.2 TRACKING

The second environment is a slightly different predator-prey model in which the learning agent plays the predator. In this model, the goal is to stalk the prey at a distance. The adversary (the prey) follows a random course and speed. The tracker must learn to control both its speed and its direction. It is assumed that the tracker has sensors that operate at a greater distance than the prey's sensors. The object is to keep the prey within range of

[8] These environments can be made available to other researchers who wish to experiment with classifier systems or other learning architectures. We would be happy to participate in comparative studies.

[9] In the experiments described here, all sensors are structured attributes.

[10] This environment differs from the EM problem in previous papers [8, 9, 13, 16]. In this paper, we introduce noise into both the agent's sensors and actions, and we vary both the initial state and the maneuverability characteristics of the adversary. As a result, the task is more realistic and more challenging.

the tracker's sensors, without being detected by the prey. If the tracker enters the range of the prey's sensors, it will be detected and captured with a probability that depends on the tracker's distance and speed. At the end of each episode, the critic provides full payoff if the tracker keeps within a certain average range of the prey, proportionately less payoff if the average range exceeds the threshold, and 0 payoff if the tracker is captured by the prey.

4.3 DOGFIGHT

The final environment pits the learning agent against a rule-based adversary with identical sensor and action capabilities. Like the learner in the *Evasion* environment, each agent controls its own turning rate, but its speed is a deterministic function of its turning rate. Each agent has a weapon that allows it to destroy the opponent if the agent is heading toward the opponent and is within the weapon's range. The object, therefore, is both to evade the opponent's fire while getting in position to make an attack. The learner receives full payoff for an episode in which the adversary is destroyed, partial payoff for a draw, and 0 payoff if the learner is destroyed. The adversary operates according to a fixed set of rules, and does not learn during these experiments.[11]

5 PERFORMANCE OF SAMUEL ON TEST ENVIRONMENTS

This section presents some initial empirical studies of the performance of SAMUEL on the test environments. At intervals of five generations, a single strategy is extracted by running extended tests on the top 20% of the current population. The performance of the extracted strategy is shown in the graph. All graphs represent the mean performance over 10 independent runs of the system, each run using a different seed for the random number generator. The error bars indicate one standard deviation across the runs. All experiments used a common set of parameters.[12]

In Figure 2 the solid line shows the performance of the current version of SAMUEL on the *Evasion* environment. The initial strategy (a random walk) evades the adversary about 31% of the time. After 50 generations, the final strategy evades the adversary about 82% of the time. Due to differences in the rule representation language, a direct comparison with the previous version of SAMUEL could not be performed. However, a good approximation of the previous behavior of SAMUEL can be obtained by lowering the bid bias to 1, disabling the GENERALIZE, MERGE, and CREEP operators, and restricting SPECIALIZE

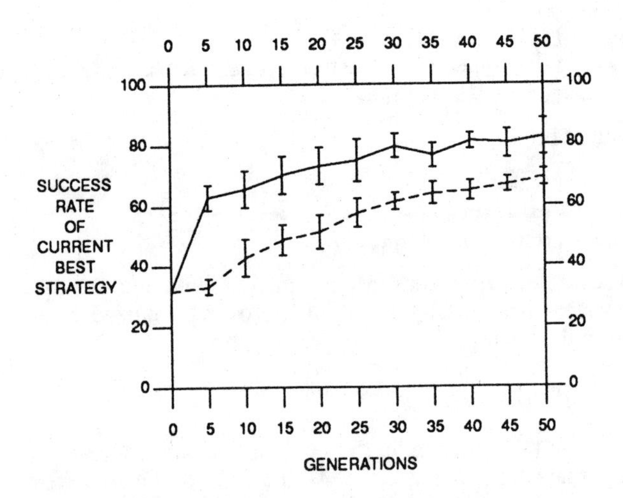

Figure 2: SAMUEL on *Evasion* Environment

to the maximally general rules. The resulting learning rate is shown by the dashed line in Figure 2. The mechanisms in the current version appear to yield significantly better performance, particularly in the early stages of learning. Note again that this environment is much more challenging than our earlier studies of the EM problem [8, 9].

Figure 3 shows a typical learning curve for the *Tracking* environment.

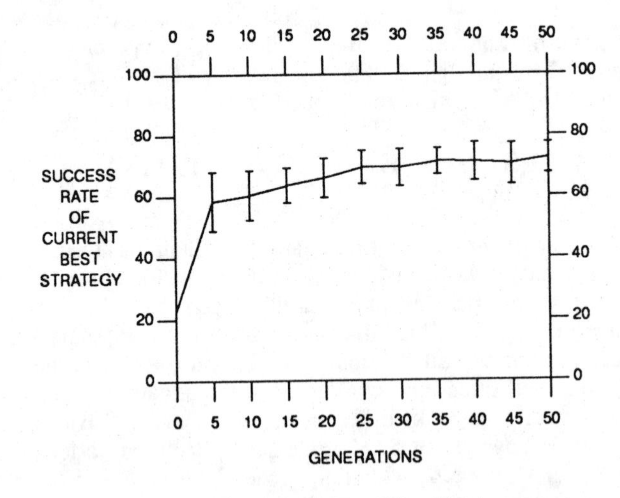

Figure 3: SAMUEL on *Tracking* Environment

This environment is more difficult than the *Evasion* environment in the sense that a random walk has very little chance of producing acceptable behavior. An initial plausible strategy, shown in Figure 4, provides an overgeneral but plausible initial starting point. The initial strategy successfully tracks the adversary approximately 22% of the time. After 50 generations, the final strategy evades the adversary over 72% of the time. It is not currently known whether there exists a completely successful strategy for this environment. Since the adver-

[11] We plan to address adaptive adversaries in future experiments.

[12] Population size = 100; crossover rate = 0.6; maximum number of rules per strategy = 64; noisy sensors and actions for the learning agent, 50 generations per run. After each evaluation, the remaining space in each strategy was allocated equally to the rule creation operators: MUTATION, CREEP, SPECIALIZE, GENERALIZE, and MERGE. Investigation of optimal parameter settings awaits future studies.

```
if    (and (bearing is [directly-ahead])
           (range is [high]))
then  (and (turn is [straight])
           (speed is [medium high]))

if    (and (bearing is [hard-right, behind-right])
           (range is [high]))
then  (and (turn is [soft-right])
           (speed is [medium, high]))

if    (and (bearing is [directly-behind])
           (range is [high]))
then  (and (turn is [hard-right])
           (speed is [medium, high]))

if    (and (bearing is [hard-left behind-left)
           (range is [high]))
then  (and (turn is [soft-left])
           (speed is [medium, high]))

if    (and (range is [close low medium]))
then  (and (turn is [straight])
           (speed is [low, medium]))
```

Figure 4: Initial Strategy for *Tracking* Environment

sary follows a random route, it can, and often does, turn directly toward the tracker and approach at high speed. Since the probability of detection depends in part on the tracker's own speed, it can easily be surprised and trapped by the adversary. Future studies will shed more light on the ultimate level of performance that can be obtained in this setting.

Figure 5 shows a typical learning curve for the *Dogfight* environment.

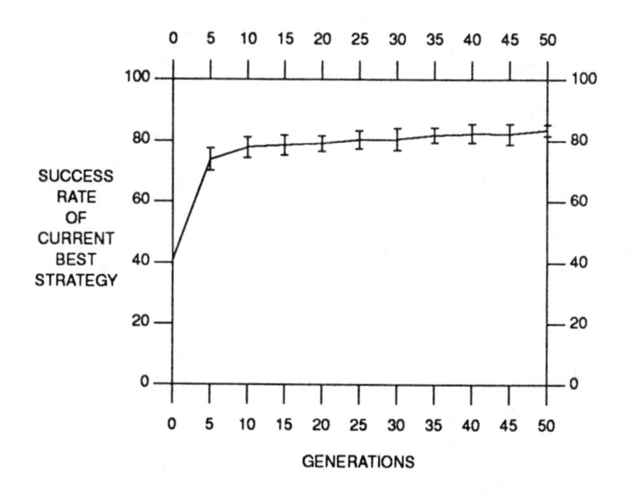

Figure 5: SAMUEL on *Dogfight* Environment

The initial strategy (a random walk) defeats the adversary approximately 40% of the time. After 50 generations, the final strategy evades the adversary about 83% of the time. Again, it is not currently known whether there exists a completely successful strategy for this environment.

SAMUEL appears to perform well in these initial studies on the new environments. Although the current version represents a significant improvement in learning speed over previous versions, some limitations of the system remain. There seems to be a window of environmental complexity in which SAMUEL performs best. If the environment is too simple, other methods such as traditional control theory or explanation-based learning may be much more efficient ways to develop high performance control rules. If the environment is too complex, SAMUEL flounders badly. As an example, the *Tracking* environment requires some initial knowledge in order to provide a minimum level of successful experience upon which SAMUEL can build better strategies. The user should not expect SAMUEL to develop strategies for a difficult environment on its own. Nonetheless, we believe that SAMUEL can be part of a methodology that combines knowledge engineering and machine learning in a way that significantly reduces the overall development effort for systems that exhibit expert performance in complex environments.

6 SUMMARY

This paper has presented a number of recent enhancements to SAMUEL, emphasizing the enhanced rule representation language and learning operators that take advantage of this new representation. It is expected that the inclusion of these operators will present new opportunities to merge to power of genetic algorithms with traditional machine learning approaches.

The performance of the system has been illustrated on three competitive environments. We encourage others in the GA community to explore learning in environments of at least this complexity. Complex, uncertain environments offer a promising niche for genetic learning approaches, a niche that has not been addressed adequately by traditional learning methods.

Finally, SAMUEL represents an integration of the two major genetic approaches to machine learning, the Michigan approach (i.e., Holland's classifier systems [12]) and the Pittsburgh approach (i.e., De Jong and Smith's LS-1 approach [17]). It is interesting to note that the more Lamarckian features of SAMUEL — using rule strengths for conflict resolution, and the triggered rule creation operators — were inspired by mechanisms in classifier systems. This suggests a fascinating question: Is John Holland a Lamarckian?

Acknowledgments

I want to acknowledge the contributions toward the development of SAMUEL by the members of the Machine Learning Group at NRL, especially Alan Schultz, Connie Ramsey, Diana Gordon, Helen Cobb, and Ken De Jong. This work is supported in part by ONR under Work Request N00014-91-WX24011.

References

[1] Barto, A. G., R. S. Sutton and C. J. C. H. Watkins (1989). Learning and sequential decision making. COINS Technical Report, University of Massachusetts, Amherst.

[2] Booker, L. B. (1991). The classifier system concept description language. *Proceedings of the 1990 Foundations of Genetic Algorithms Workshop.* Bloomington, IN: Morgan Kaufmann.

[3] Goldberg, D. E. (1983). *Computer-aided gas pipeline operation using genetic algorithms and machine learning,* Doctoral dissertation, Department Civil Engineering, University of Michigan, Ann Arbor.

[4] Goldberg, D. E. (1988). Probability matching, the magnitude of reinforcement, and classifier system bidding. (TCGA Report No. 88002). Tuscaloosa: University of Alabama, Department of Engineering Mechanics.

[5] Gordon, D. G and J. J. Grefenstette (1990). Explanations of empirically derived reactive plans. *Proceedings of the Seventh International Conference on Machine Learning.* Austin, TX: Morgan Kaufmann (pp. 198-203).

[6] Gould, S. J. (1980). *The Panda's Thumb.* New York: Norton & Co.

[7] Grefenstette, J. J. (1988). Credit assignment in rule discovery system based on genetic algorithms. *Machine Learning, 3(2/3),* (pp. 225-245).

[8] Grefenstette, J. J. (1989). A system for learning control strategies with genetic algorithms. *Proceedings of the Third International Conference on Genetic Algorithms.* Fairfax, VA: Morgan Kaufmann (pp. 183-190)

[9] Grefenstette, J. J., C. L. Ramsey, and A. C. Schultz (1990). Learning sequential decision rules using simulation models and competition. *Machine Learning, 5(4),* (pp. 355-381).

[10] Grefenstette, J. J., and H. C. Cobb (1991). User's guide for SAMUEL. NRL Report, Naval Research Lab, Washington, DC.

[11] Holland, J. H. (1975). *Adaptation in natural and artificial systems.* Ann Arbor: University Michigan Press.

[12] Holland J. H. (1986). Escaping brittleness: The possibilities of general-purpose learning algorithms applied to parallel rule-based systems. In R. S. Michalski, J. G. Carbonell, & T. M. Mitchell (Eds.), *Machine learning: An artificial intelligence approach* (Vol. 2). Los Altos, CA: Morgan Kaufmann.

[13] Ramsey, C. L., A. C. Schultz and J. J. Grefenstette (1990). Simulation-assisted learning by competition: Effects of noise differences between training model and target environment. *Proceedings of the Seventh International Conference on Machine Learning.* Austin, TX: Morgan Kaufmann (pp. 211-215).

[14] Riolo, R. L. (1987). Bucket brigade performance II: Default hierarchies. *Proceedings of the Second International Conference on Genetic Algorithms.* Cambridge, MA: Lawrence Erlbaum Assoc. (pp. 196-201)

[15] Selfridge, O., R. S. Sutton and A. G. Barto (1985). Training and tracking in robotics. *Proceedings of the Ninth International Conference on Artificial Intelligence.* Los Angeles, CA. August, 1985.

[16] Schultz, A. C. and J. J. Grefenstette (1990). Improving tactical plans with genetic algorithms. *Proceeding of IEEE Conference on Tools for AI 90,* Washington, DC: IEEE (pp. 328-334).

[17] Smith, S. F. (1980). *A learning system based on genetic adaptive algorithms,* Doctoral dissertation, Department of Computer Science, University of Pittsburgh.

[18] Wilson, S. W. (1987). Classifier systems and the animat problem. *Machine Learning, 2(3),* (pp. 199-228).

GA-1: A Parallel Associative Memory Processor for Rule Learning with Genetic Algorithms

Hiroaki Kitano[1], Stephen F. Smith[1], Tetsuya Higuchi[2]
[1]School of Computer Science
Carnegie Mellon University
Pittsburgh, PA 15213
[2]Electrotechnical Laboratory
1-1-4 Umezono, Tsukuba
Ibaraki, 305 Japan

Abstract

In this paper, we discuss the underlying hardware and supportable learning paradigms provided by the GA-1 system. GA-1 is a system currently under development which offers unique opportunities for research into large-scale rule learning with genetic algorithms (GAs). The base hardware is the IXM2 parallel associative memory machine which enables high performance processing by using 64 T800 transputers and associative memories providing 256K parallelism. Various population/subpopulation models, mating strategies, and generation models can be implemented to investigate architectures for high performance GA-based systems. Regardless of these options, however, GA-based rule learning takes maximum advantage of the hardware through extensive use of associative memory for bit-vector matching. Preliminary experiments indicate that GA-1 exhibits high execution speeds for such an approach.

1. Introduction

Genetic algorithms (GAs), and the rule-based systems typically assumed as targets for GA-based learning systems, inherently possess a high degree of parallelism. As such learning systems are applied to real-world tasks, the need to effectively exploit this parallelism becomes fundamental.

The principal motivation underlying our current development the GA-1 system is to offer a suitable parallel processing environment for research and development of GA-based learning systems for real-world and real-time applications. GA-1 is intended not only as a basis for approaching complex, real-time applications, but also as a research tool for investigating issues of scale previously considered to be computationally prohibitive. Our goal is to provide a flexible software environment in which specific GA-based learning systems composed of various generation models, migration models, and mating and reproduction strategies are easily configurable. In this paper we discuss the learning paradigms supportable within GA-1, and give some preliminary performance results with the currently implemented system.

Previous research has produced parallel implementations of GAs [Tanase, 1989], and classifier systems [Robertson, 1987], as well as supporting hardware [Gorges-Schleuter, 1989] [Twardowski, 1990]. [Tanase, 1989], [Gorges-Schleuter, 1989], and [Grefenstette et. al., 1990] have implemented GAs on coarse-grain processors to speed up evaluations. [Robertson, 1987] implemented a classifier system on the connection machine [Hillis, 1985]. [Twardowski, 1990] used an associative memory for matching of classifiers. In relation to these previously developed systems, GA-1 offers unique opportunities for parallel and distributed implementation of GA-based rule learning through use of 64 T800 transputers and associative memories. Unlike other parallel machines, IXM2, the base machine of GA-1, is equipped with a large associative memory chips. Due to these associative memory chips, GA-1 is capable of carrying out 256K parallel bit-vector matching when all 64 PEs are used. GA-1 exploits 4K parallel bit-vector matching at each of 64 PEs. This provides the capability to implement both the "Pitt Approach" to rule learning [Smith, 1980] and the grammatical encoding method [Kitano, 1990b], as well as the "Michigan" (or classifier system) approach [Holland and Reitman, 1978], with direct support from the hardware level. Since the GA-1 offers extremely high performance, as we will discuss in section 5, large-scale GA-based learning systems with the real-time performance would appear to be practical.

This paper is organized into four parts. First, we describe a basic hardware system of GA-1. The base machine is called IXM2 developed at the Electrotechnical laboratories in Japan. Second, we discuss some critical issues in designing GA-based systems on the base machine. These issues involves (1) population mappings, (2) processor idling, (3) communication bottleneck, and (4) parallelization of evaluation. Third, we consider the support for alternative parallel models for GA-based learning provided by GA-1, including some hybrid Pitt/Michigan schemes. Fourth, we report some basic performance evaluation results obtained with GA-1.

2. The Base Machine: The IXM2 Associative Memory Processor

IXM2 is a massively parallel associative processor designed and developed at the Electrotechnical Laboratory [Higuchi et. al., 1991] (figure 1). It is dedicated to semantic network

Figure 1: The IXM2 Associative Memory Processor

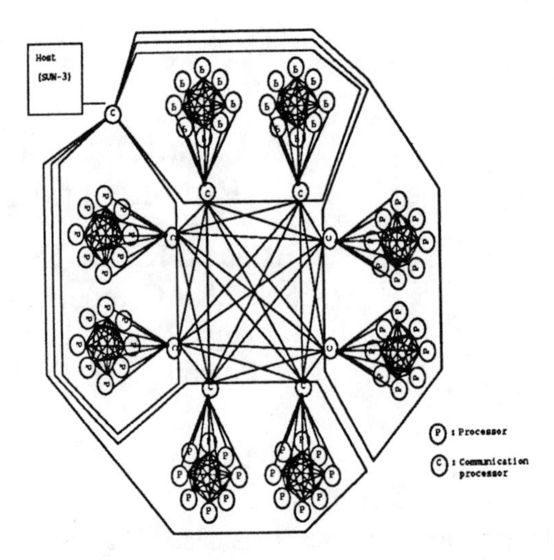

Figure 2: Architecture of the IXM2 Associative Memory Processor

No. of PE	IXM2	hypercube	torus
4	1	1.33	1.33
8	1	1.71	—
16	2.06	2.13	2.13
32	2.54	2.61	—
36	—	—	3.08
64	2.77	3.04	4.03

Table 1: Average Distance between PEs

processing using marker-passing.

IXM2 consists of 64 processors, called *associative processors*, which operate with associative memory each of which has a memory capacity of 4K words by 40 bits. Each associative processor is connected to other associative processors through network processors.

An associative processor consists of an IMS T800 transputer, 8 associative memory chips, RAM, link adapters, and associated logic. When operated at 20 MHz clock, T800 attains 10 MIPS [Inmos, 1987]. Each associative memory chip is a 20 Kbit CAM (512 words $\times$ 40 bits) manufactured by NTT [Ogura et. al., 1989]. The IXM2 has 64 such processors, thus attaining 256K parallelism which is far larger than 64K parallel of the Connection Machine [Hillis, 1985]. This high level of parallelism allows us to implement practical memory-based systems. The design decision to use associate memory chips driven by 32 bit CPUs, instead of having thousands of 1-bit CPUs, is the major contributing factor for performance, processor efficiency, and cost performance.

Network processors are used to handle communication between associative processors. There is one top-level network processor which deals with communication among the lower-level network processors, and 8 lower-level network processors each of which is connected to 8 associative processors. Unlike most other massively parallel architectures which use N-cube connections or mesh connections, IXM2 employs a full connection so that communication between any two processors can be attained by going through only 2 network processors. Table 1 shows average distance between two processors in various connections. IXM2's full connection is the best among others up to, at least, 64 PEs. This full connection architecture ensures high communication bandwidth and expandability which are critical factors in implementing real-time applications. Each interconnection attains high speed serial links (20 Mbits/sec), which enables a maximum transfer rate per link of 2.4 Mbytes/sec.

3. Mapping GAs onto Parallel Hardware

In this section, we focus on some design issues in the GA-1 parallel genetic algorithm system. We discuss parallel and distributed models on GA-1 from the view point of the parallel processing. Several issues must be addressed in order to design GA systems on parallel machines. These are: (1) Speeding Up Rule Interpretation, (2) Population Mapping to PEs, (3) Mating and Generation Model.

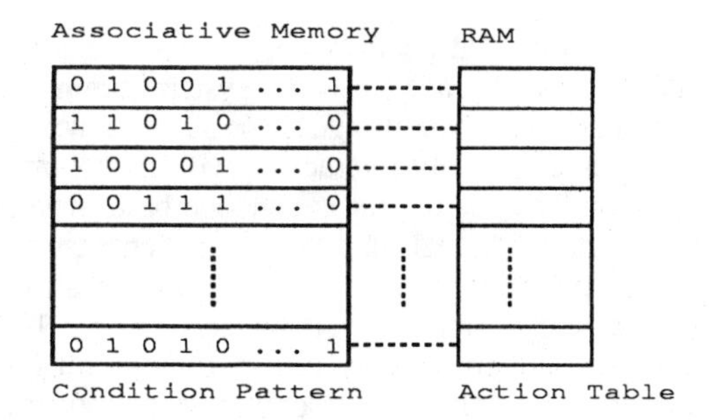

Figure 3: Loading Rules in Associative Memory

3.1. Speeding Up Rule Interpretation

For the most part, the classifier system variants that have appeared in the literature [Holland and Reitman, 1978] [Booker, 1982] [Riolo, 1986] [Wilson, 1985] and the rule-base systems that have been investigated within the Pitt approach [Smith, 1983] [Schaffer, 1985] [Grefenstette, 1988] [Greene, 1987], share a common low-level rule representation with capabilities to match fixed-length binary strings (encoding external and/or internal problem solving states) and post fixed-length binary messages (encoding external responses to be issued to the environment and/or aspects of the systems internal state). The grammatical encoding method of [Kitano, 1990b] assumes a similar condition action structure.

Within GA-1, such rules are contained in associative memory on each PE to determine which rules are capable of being executed at any point. At this level of implementation there is no distinction relative to the model under which this rules are being manipulated by the GA (e.g. "Pitt" or "Michigan" approach). Condition parts are stored in associative memory and action parts are stored in RAM (figure 3). Each PE has an associative memory which can load 4,000 rules with a 40 bits condition part. When 64 PEs are used collectively as one huge rule interpreter, GA-1 provides 256K parallelism.

To search for rules to be executed on a given cycle, an appropriate search mask and search data is formulated. A search mask specifies a set of memory bits to be considered. A portion of a program for search and retrieval of rules is shown below:

```
AM.Search.Mask:=literal.search.mask – specify bits to be searched
AM.Search:=literal.search.data    – initiate search operation

classifier[0]:=AM.Read.IN        – retrieve one classifier just hit
address:=AM.Addr.Reg ∧ #FFFF      – get the address of the classifier
WHILE (address <> End.OF.Hit)    – repeat until all the matched classifiers
 SEQ                            – are retrieved
  classifier[i]:=AM.Read.IN
  address:=AM.Addr.Reg ∧ #FFFF
  i:=i+1
```

3.2. Population Mapping

There are several ways to assign a population to processors (table 2). MMD stands for *Maximum Mating Distance*. First, we can assign one individual per one processor. This would be necessary when evaluation of each individual is computationally expensive. In this case, the population may consist of 64 processors, or be configured into subpopulations to exploit the 8 processor clusters. An alternative approach is to assign one population per processor. At maximum, 64 populations can be evaluated in parallel.

Population	Mapping to PE	MMD
Single population	Individual	Inter-Cluster
		Intra-Cluster
	Subpopulation	Inter-Cluster
		Intra-Cluster
Multiple population	Individual	Intra-Cluster
	Subpopulation	Inter-Cluster
		Intra-Cluster
	Population	Intra-PE

Table 2: Population mapping and Mating in GA-1

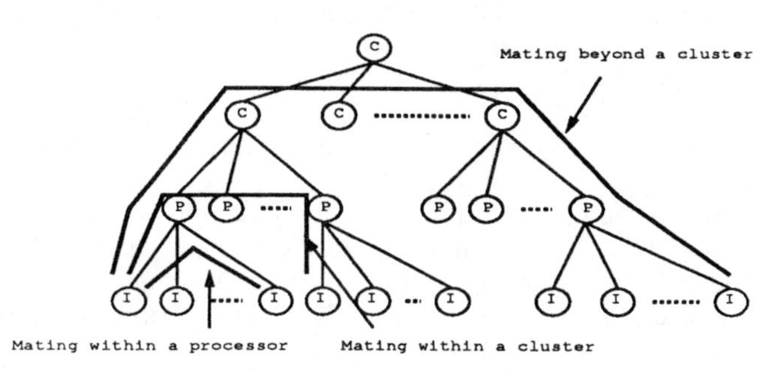

Figure 4: Mating and Migration in the GA-1

Depending upon how the population is mapped to processors, different mating and migration strategies can be used. GA-1 provides support for three layers of mating: (1) intra-processor, (2) intra-cluster, and (3) inter-cluster mating. This is shown in figure 4. The relationship between population mapping and maximum mating distance is shown in table 2.

When implementing the "Pitt" approach, one basic mapping strategy is one individual per PE with a single population. Under this mapping, 64 individuals each having maximum of 4K rules can be processed in parallel. For the "Michigan" approach, mappings are possible where each PE manages a separate subpopulation or where each PE contains components of a single large population. These mappings are by no means exhaustive; other possibilities exist under either paradigm. It is also possible to consider a multiple population mapping as a model of multiple species, or even a co-evolution model.

3.3. Mating and Generation Model

Decisions regarding which mating and generation model to use for a specific application can have a significant effect on performance in terms of both processor idle time and communication bottleneck. The traditional central control of mating under the discrete generation model (synchronous mating) is the worst option for two reasons. First, when computing cost for evaluation is different in each individual, a significant portion of total processor time is wasted while waiting other individuals to complete evaluation. [Kitano, 1991] estimated that over 80% of total CPU time can be wasted in some cases. Second, synchronous mating requires simultaneous exchange of all parent chromosomes. The next evaluation stage cannot begin until the exchange of chromosomes and application of genetic operations are completed in all processors. If several (or all) processors start sending chromosomes to other processors at once, large amounts of message collision and access queueing will occur. Loss of the computation time due to the communication bottleneck cannot be overlooked, particularly when each evaluation is relatively computationally cheap, i.e. evaluation time is comparable to communication time. Localized mating schemes reduce this problem of communication bottleneck; processor idle time still exists however and is generally far more significant problem.

Although, GA-1 supports use of a global synchronous mating strategy if necessary, the principal strategy advocated in GA-1 is the *continuous generation model* proposed in [Kitano, 1991]. In the continuous generation model, each individual performs mating with no synchronization using either a global or a local mating scheme. By allowing mating whenever evaluation of an individual is complete, processor idle can be minimized. For example, if we adopt an individual per PE mapping with a single population, the continuous generation model carries out 5.5 times more reproductions than the discrete generation model counterpart. Processor idle time is virtually eliminated.

Since the continuous generation model does not require synchronization, there is also far less chance of message collision and access queuing. Furthermore, the IXM2's connection network structure (which is full connection) ensures that only minimum number of network processors are involved in passing message to other PEs. As shown in table 1, IXM2 has less average distance between PEs than other architectures.

4. Parallel Models for Rule-Based Learning in GA-1

Rule-based learning with GAs has been considered within two basic paradigms: the so called "Michigan" (Classifier Systems) and "Pitt" approaches to which we have been refering above. At an operational level, these organizations can be distinguished by the role given to the GA in the learning process. The Classifier Systems approach assumes a "population" of individual rules (classifiers) which collectively constitute the knowledge of a problem solving agent. Here the GA serves as the basic mechanism for generating new rules, to fill in gaps in the knowledge and improve deficiencies in the rule base. The development and reinforcement of rule linkages is considered separable and handled via separate mechanisms (the bucket brigade algorithm, triggered operators, etc.) The Pitt Approach, alternatively, is more directly based on the standard GA-based optimization framework. Each individual in the population represents a complete candidate solution to the problem (i.e. a complete set of rules). The scope of genetic search is extended to directly manipulate building blocks representing subsets of high performance rules as well as building blocks representing high performance rule components.

From a task-oriented perspective, we can further differentiate the two paradigms. Classifier systems are motivated largely from an "autonomous agent" point of view: a problem solver that is actively interacting with its environment and incrementally updating its knowledge base in response to these interactions in real time. Here, the benefits of parallelization relate directly to real-time responsiveness and the size/complexity of problems that can be tackled. The Pitt Approach, alternatively, can be seen more as an off-line approach, where a training phase is conducted to produce a rule base for subsequent use by a real time problem solver. In problems requiring extraction of the regularities in existing data sets for future predictive use (e.g. [Greene, 1987]), as well as problems for which minimum problem solver competency requirements must be ensured prior to operational use [Grefenstette et. al., 1990], this off-line view is quite natural. Here parallelism dircetly impacts training time, allowing application to larger/more complex problems, exploration with larger populations, and more extensive evaluation of alternatives, etc.

In the subsections below, we consider the suitability of GA-1 first as a platform for parallel implementation of classifier-like systems and then as a basis for parallelizing the Pitt approach to rule learning.

4.1. Parallelizing Classifier Systems

Within serial implementations of either GA-based learning paradigm, the computational performance bottleneck is invariably the rule interpretation process. Section 3 above has indicated the advantages of IXM2's associative memory processors in eliminating this bottleneck, yielding a match time that is independent of the size of the rule base. Associative memory hardware has been previously exploited by Twardowski to produce a parallel implementation of a simple classifier system [Twardowski, 1990]. His Coherent Processor provides a single associative memory of 4K. The cummulative associative memory of the IXM2's 64 PEs, in contrast, provides support for much larger rule bases; of the size likely to be required for complex, real-world applications. In fact, IXM2 is capable of supporting rules bases comparable in size to those possible in a separate parallel implementation of a classifier system on the Connection Machine [Robertson, 1987], however with a much faster match cycle time (see Section 5).

With respect to the larger issue of implementing learning classifier systems (i.e. the full Michigan Approach), the selection and reproduction algorithms developed in [Twardowski, 1990] are equally applicable within GA-1. At the same time, GA-1 provides additional capabilities for further parallelization of more complex classifier systems that exploit problem decomposibility and make use of partitioned rule sets. For example, the rule base might be partitioned into subsets responsible for the settings of distinct external effectors (as suggested in [Holland and Reitman, 1978]). Alternatively, rule set partitions might be assigned responsibility for distinct problem solving contexts. By associating selection and reproduction logic with individual PEs, and assuming replacement of a fixed percentage of rules in each rule base partition at the end of each problem solving episode, up to 64 rule base partitions of size comparable to those of the Coherent Processor can be updated in parallel at the end of each problem solving epoch. Smaller partitions are of course possible through assignment of multiple partitions to each processor.

4.2. Parallelizing the Pitt Approach

The Pitt Approach to rule learning obviously introduces much more severe computational demands into the learning cycle, due to need to execute and derive fitness measures for populations of rule sets. With their SAMUEL system, [Grefenstette et. al., 1990] provide one example of a parallel implementation designed to minimize this evaluation time through distribution of rule set evaluation over a 128 node BBN butterfly machine. In this case, a standard globally synchronized generational model of the GA is employed, but individuals in the population are evaluated in parallel. Overall evaluation time is reduced (in this case by over two orders of magnitude) but the process of evaluating individuals remains costly.

A similar model is straightforwardly realized in GA-1 by distributing individuals in the population to each of its 64 PEs. While the number of individuals that can be evaluated in parallel is roughly half of that possible on the butterfly, this is overwhelmingly compensated for by the reduction in time required for each single evaluation.

If the time necessary to evaluate individuals is not constant (implying processor idle time) then opportunities for further speedup are possible though use of the continuous generation model described in Section 3. The most appropriate strategy here ultimately depends on the nature of the task domain and the defined evaluation procedure. For example, in the "map building" domains explored by ANIMAT [Wilson, 1985] and CS-1 [Holland and Reitman, 1978], evaluation time is inversely proportional to the individual's fitness. In this case, use of local selection and mating not only minimizes processor idle time, but actually increases the amount of effective reproduction that is performed [Kitano, 1991]. In other domains, where this relationship between evaluation time and fitness does not hold (e.g. SAMUEL's evasive maneuvers domain), global selection and mating would be the appropriate strategy. In comparison to other coarse-grain parallel architectures, the processor interconnection structure within GA-1 is best suited for minimizing communication bottlenecks across this range of strategies.

As in any implementation of the Pitt Approach, there are scale limitations in relation to what is possible under the Michigan Approach. There is the pragmatic need to allocate a portion of each processor's memory to the evaluation procedure and/or training examples. In the case of GA-1, these limits enable investigation of applications that far exceed the requirements of any domain considered to date. For example, if we assume that at most 2 individuals are associated to each processor (i.e. population size of 128) and that half of each processor's associative memory is devoted to storage of an training example data base (e.g. 2000 1 word examples), then the maximum allowable number of rules per individual would still be roughly an order of magnitude larger than the number required by any study utilizing the Pitt appraoch performed to date. The control store available to each processor (128K) is a bit more severe of a constraint, although again it would appear sufficient to accommodate all existing simulation-based learners that have been studied thus far. Note, however, that this constraint derives not from the GA-1 architecture but from the currently available IXM2 machine. Processor RAM could be expanded to 4 Gigabyte (logical address space) at additional cost.

Scale-up of the Pitt Approach to large-scale real-world problem domains is in fact, really more of a methodolgical issue than an architectural issue. It is reasonable to assume that applications requiring large rule bases are decomposable in the same sense as previously discussed above with respect to the Michigan Approach. If this is the case, then knowledge base components can be developed independently for later integration into the operational problem solver. The performance capabilities of GA-1 make investigation of such methodological issues possible.

4.3. Hybrid Models and Extensions

Several models for integrating elements of the Michigan and Pitt approaches to rule learning are imaginable. Here we briefly discuss two which appear to exploit the relative advantages of each and can be practically investigated within GA-1.

Considering the two paradigms generally, the Michigan approach places a premium on maintaining continuity in the rule base over time (i.e. gradual rule base refinement without sharp discontinuities in performance). The Pitt Approach, alternatively, emphasizes more global search for the best possible solution with limited consideration of the "refinement potential" of each candidate solution. Given these properties, an obvious model for integration (reflected to some extent in the work of [Grefenstette, 1988]) is one where a population of Michigan-style classifier systems are manipulated by a Pitt-style organization. The result would be a sort of Lamarkian framework where candidate solutions undergo a refinement process prior to evaluation of their overall worth.

A second model for coupling the two approaches relates to the above discussion of methodologies for scaling the Pitt Approach to large-scale domains. Complex problems are, in fact, rarely totally decomposable, which makes integration of independently developed rule base components becomes an important concern. One natural approach to this integration problem is to assume a Michigan organization as the operational problem solver. Rule base components derived under the Pitt Approach provide substantial building blocks, whose behavioral interactions can be subsequently sorted out in standard learning classifier system fashion (by full scale training in advance of execution if necessary). Overall system behavior can then be adapted as necessary in response to operational experiences.

One other new area for which GA-1 provides opportunities is in integration of GAs and self-reproducing automata. As discussed in [Kitano, 1990b], use of a graph generation system (which is an extension of the L-system [Lindenmayer, 1968]) enables systems which integrate evolutionary, developmental, and learning stages in a consistent manner. In the developmental stage, a single cell divides itself and creates daughter cells. Each daughter cells further subdivide themself. Finally, it stabilizes at a certain point. This process is analogoues to morphogenesis of living organs. One of the difficulties with this model is the computational cost of interpreting the grammar at each cell. Even with a population size of 100, if each individual eventually consists of 100 cells, a grammar interpretation has to be performed on 10,000 cells. Since the graph generation grammar is a set of rules, GA-1's associative memory can carry out efficient interpretation of the grammar. GA-1 allows us to investigate such systems of much larger-size. In addition, when a grammar is defined in a way that takes into account messages from other individuals, the entire system (with 64 PE of GA-1) would be a highly context-sensitive set of automata. This is yet another dimension in which GA-1 can be utilized for research of complex system, or artificial life.

5. Performance

In this section, we report some of the preliminary results of the basic performance data we have obtained with GA-1.

Figure 5 shows the time required for matching and retrieval of rules in associative memories. Depending upon the length of the rule, time required to search and retrieve a matching rule varies. The time required for matching is constant regardless of number of condition parts in the associative memory. However, retrieval is a serial operation within each PE, and 64 parallel as a whole system. For each hit, it takes 10 micro seconds when the condition part is 32 bits. Suppose we have a rule with a 32 bits condition parts, and an average hit per cycle is around 10, In this case, GA-1 attains 8,929 cycles per second (without bit-competition operations) – 0.112 millisecond per cycle.

This cycle rate is magnitude faster than that reported for the connection machine (5 milliseconds for one match and retreive cycle up to 65,536 classifiers [Robertson, 1987]). This performance advantage was independently confirmed even with data using a fully optimized program (using C* with PARIS Parallel Instruction Set) by ourself as shown in figure 5. The data for CM-2 shows time to take (1) broadcast of a bit-vector, (2) matching at each PE, and (3) retrieval of matched PE number to the host. This is equivallent to GA-1's memory search and retrieval phase. Data on CM-2 and GA-1 shows match-and-retrieve without bit-competition operations. GA-1 is consistently faster than CM-2. In order to attain GA-1's matching performance on a serial machine, it must carry out 2286 million bit-vector matching per seconds. Matching cycle per rule need to be less than a nano second. Obviously, this is physically impossible speed at current device technologies.

Figure 6 shows the time required to search for a matching rule for various rule bit widths. Data for CM-2 was obtained using programs written in C* with PARIS. While GA-1 fully utilizes associative memory of IXM2 hardware, CM-2 carries out bit-serial operation in each 1-bit PE. GA-1 is faster than CM-2 by the factor of 2, even at the bit-vector matching.

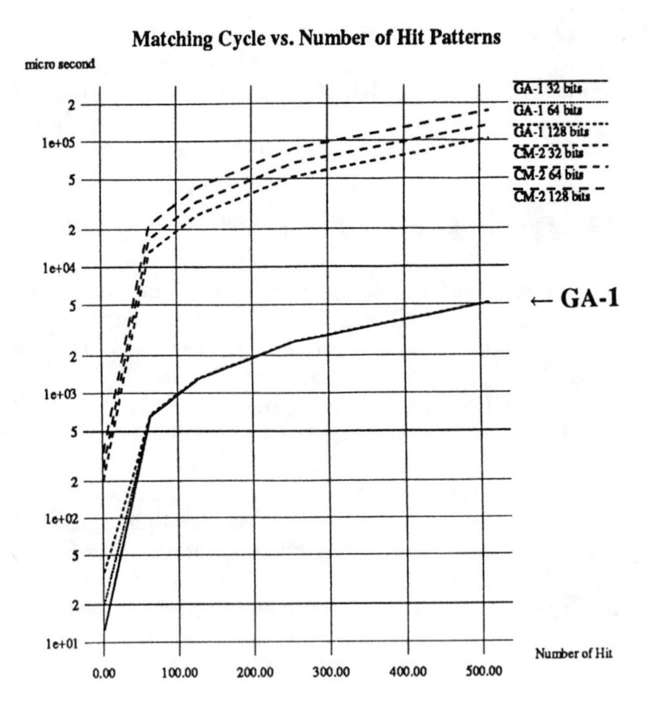

Figure 5: Match-and-Retrieve Time of GA-1 and CM-2

Due to high communication band-width (2.4 Mbyte/second), exchange of chromosome is carried out without causing substantial loss in the execution speed. Sending an entire chromosome in a PE (stored in an associative memory) to other PE in the same cluster requires 8.3 milliseconds. This is sufficiently fast for most real-time tasks. However, even this level of communication contention prevents GA-1 from exhibiting its full potential. Since GA-1's matching cycle is operating on micro-seconds order, 8.3 milliseconds is equivalent to a few hundreds matching cycles or a few matching cycle depends upon its number of hits. It is plausible that there are tasks for which evaluation only requires a few matching cycles so that the major computational bottleneck would be the communication for mating. In these cases, synchronous mating significantly degrades GA-1's performance from its full potential. As discussed previously, the continuous generation model further mitigates communication bottleneck. Benchmark experiments to measure effects of the continuous generation model on communication bottleneck is underway.

6. Conclusion

This paper discussed the GA-1 architecture, some of the design concerns, and new opportunities provided by GA-1 for GA-based rule learning. GA-1 is a parallel and distributed GA system based on IXM2 associative memory processor. The central feature of the IXM2 is extensive use of associative memory which enables GA-1 to perform highly parallel bit-vector matching. This capability allows development of

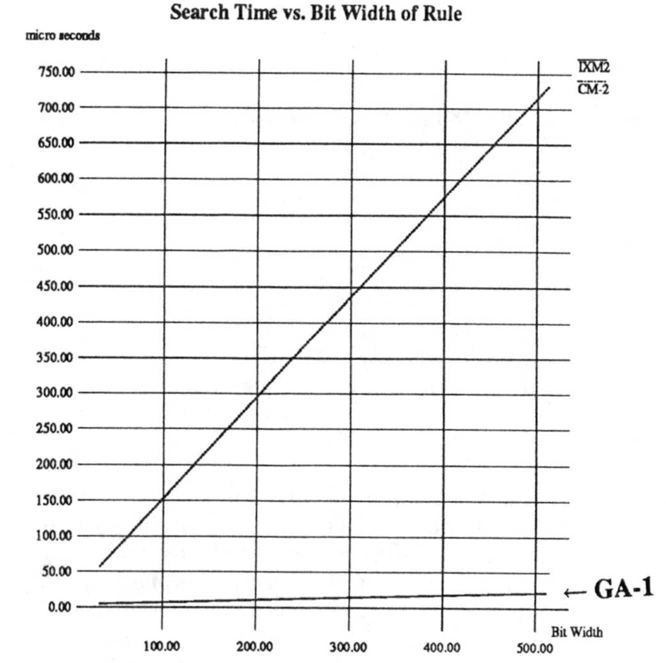

Figure 6: Bit Vector Matching Time of GA-1 and CM-2

large scale, real-time GA-based rule learning systems.

Preliminary performance data demonstrates that use of associative memory provides orders of magnitude of speed up in training and executing GA-based rule learning system comparing with other machines such as CM-2 connection machine. We discovered that regardless of implementation and mapping, some sort of central or local control at coarse-grain level processor is necessary to carry out GA operations. Thus, a simple fine-grain machine such as CM-2 is not really suitable for GA implementation. In GA-1, we avoid this problem by tightly coupled 32-bit transputers with 4K associative memory chips. This architecture attains both coarse-grain control and massively parallel bit-vector matching.

Given its high performance, GA-1 opens new research opportunities in the area of GA-based rule learning. We are currently engaged in full implementation of a Pitt-based approach, and intend to use this implementation to first explore the benefits of continuous generation strategies.

References

[Booker, 1982] Booker, L. B., *Intelligent Behavior as an Adaptation to the Task Environment*, Ph.D. Thesis, Depart of Computer Science, University of Michigan, 1982.

[De Jong, 1975] De Jong, K., *An analysis of the behavior of a class of genetic adaptive systems*, Doctoral dissertation, University of Michigan, 1975.

[Fitzpatrick and Grefenstette, 1988] Fitzpatrick, M., and Grefenstette, J., "Genetic Algorithms in Noisy Environments," *Machine Learning* 3, 2/3, 1988.

[Gorges-Schleuter, 1989] Gorges-Schleuter, M., "ASPARAGOS An Asynchronous Parallel Genetic Optimization Strategy," *Proceedings of ICGA-89*, 1989.

[Greene, 1987] Greene, D., Smith, S. F., "A Genetic System for Learning Models of Consumer Choice," *Proceedings of ICGA-87*, 1987.

[Grefenstette et. al., 1990] Grefenstette, J., Ramsey, C., and Schultz, A., "Learning Sequential Decision Rules Using Simulation Models and Competition," *Machine Learning*, 1990.

[Grefenstette, 1988] Grefenstette, J., "Credit Assignment in Rule Discovery Systems Based on Genetic Algorithms," *Machine Learning*, 3, 1988.

[Higuchi et. al., 1991] Higuchi, T., Kitano, H., Handa, K., Furuya, T., Takahashi, N., and Kokubu, A., "IXM2: A Parallel Associate Processor for Knowledge Processing," *Proceeding of the National Conference on Artificial Intelligence (AAAI-91)*, 1991.

[Hillis, 1985] D. Hillis, *Connection Machine*, MIT Press, 1985.

[Holland and Reitman, 1978] Holland, J., and Reitman, J., "Cognitive systems based on adaptive algorithms," Waterman, D., and Hayes-Roth, F., (Eds.) *Pattern directed inference systems*, New York, Academic Press, 1978.

[Inmos, 1987] Inmos, *IMS T800 Transputer*, April 1987.

[Kitano, 1990a] Kitano, H., "Empirical Studies on the Speed of Convergence of Neural Network Training by Genetic Algorithms," *Proc. of AAAI-90*, 1990.

[Kitano, 1990b] Kitano, H., "Designing Neural Networks Using Genetic Algorithms with Graph Generation Systems," *Complex Systems*, Vol. 4, No. 4., 1990.

[Kitano, 1991] Kitano, H., "Continuous Generation Genetic Algorithms," Carnegie Mellon University, 1991.

[Lindenmayer, 1968] Lindenmayer, A., "Mathematical Models for Cellular Interactions in Development," *J. theor. Biol.*, 18, 280-299, 1968.

[Ogura et. al., 1989] T. Ogura, J. Yamada, S. Yamada and M. Tanno, "A 20-Kbit Associative Memory LSI for Artificial Intelligence Machines ", *IEEE Journal of Solid-State Circuits*, Vol.24, No.4, August 1989.

[Riolo, 1986] Riolo, R. L., *CFS-C: A package of domain independent subroutines for implementing classifier systems in arbitrary user-defined environments*, Technical report, University of Michigan, Logic of Computers Group, 1986.

[Robertson, 1987] Robertson, G., "Parallel Implementation of Genetic Algorithms in a Classifier System," Davis, L., (Ed.) *Genetic Algorithms and Simulated Annealing*, Morgan Kaufmann Publishers, 1987.

[Schaffer, 1985] Schaffer, J. D., "Learning Multiclass Pattern Discriminator," *Proceedings of ICGA-85*, 1985.

[Smith, 1983] Smith, S. F., "Flexible Learning of Problem-Solving Heuristics through Adaptive Search," *Proceedings of IJCAI-83*, 1983.

[Smith, 1980] Smith, S. F., *A Learning System Based on Genetic Adaptive Algorithms*, Ph. D. Thesis, University of Pittsburgh, 1980.

[Tanase, 1989] Tanase, R., "Distributed Genetic Algorithms," *Proceedings of ICGA-89*, 1989.

[Twardowski, 1990] Twardowski, K. E., "Implementation of a Genetic Algorithm based Associative Classifier System (ACS)," *Proceedings of International Conference on Tools for Artificial Intelligence*, 1990.

[Wilson, 1985] Wilson, S., "Knowledge Growth in an Artificial Animal," *Proceedings of International Conference on Genetic Algorithms*, 1985.

Classifier System Learning of Boolean Concepts

Gunar E. Liepins
MS6360 Bldg. 6025
Oak Ridge National Laboratory
PO Box 2008
Oak Ridge, TN 37831-6360
(615) 574-6640 gxl@msr.epm.ornl.gov

Lori A. Wang
MS6109 Bldg. 4500S
PO Box 2008
Oak Ridge, TN 37831-6109
(615) 574-6576 wang@cs.utk.edu

Abstract

We investigate classifier system learning of Boolean concepts. We introduce a symmetric reward-penalty mechanism, speciation, generality thresholds and rule evaluation by queries. These enable the classifier system to learn the twenty multiplexor significantly faster than previously reported for classifier systems. Conversely, we provide theoretical analyses that suggest that classifier systems are not competitive with the best known learning algorithms for stationary deterministic Boolean problems. We suggest instead that they are particularly well suited to non-stationary problems for which the target concept evolves over time.

1 INTRODUCTION

In this paper we focus on classifier system learning of Boolean concepts. We introduce speciation, generality thresholds and rule evaluation by queries to empirically generate solutions to the twenty multiplexor significantly faster than previously reported. Conversely, we provide theoretical motivation that suggests that classifier systems are inefficient for deterministic stationary Boolean concept learning. Instead, we suggest that they are the discovery mechanism of choice for non-stationary problems.

Our work builds on that of Wilson [15], Sen [12], Bonelli et. al. [3], and borrows heavily from Wang [14] all of whom studied classifier system applications to the multiplexor problem. Wilson's empirical results demonstrated that a classifier system would solve the six multiplexor to approximately 94% accuracy in slightly more than 5,000 examples. For the twenty multiplexor, he reported better than 90% accuracy in roughly 70,000 examples. Sen [12] achieved approximately a factor of five speed-up for the six multiplexor by using a heavy handed penalty for incorrect classifiers. Bonelli et. al. modified Sen's approach by treating reward and punishment symmetrically and using deterministic selection of the acting classifier. With appropriate tuning, they achieved another factor of two improvement over Sen's results for the six multiplexor, and demonstrated approximately 94% accuracy for the eleven multiplexor in somewhat over 5,000 examples. Wang [14] performed an extensive systematic study of a wide variety of classifier formulations for the multiplexor problem. She introduced speciation, limited the specificity of classifiers, and investigated specialized search operators such as probabilistically flipping the output bit and specializing incorrect general classifiers by use of a "relative-complement" operator. Her results for the six and eleven multiplexor are comparable to those of Bonelli et. al. [3]. Moreover, for the twenty multiplexor, she was able to demonstrate approximately 90% accuracy within 25,000 examples and high 90% performance as early as 50,000 examples, where performance was measured as the average number of correct classifications of the most recent 50 examples.

Other machine learning approaches have also been applied to the multiplexor problem, including neural networks (Anderson, [1]; Jacobs, [7]; Bonelli et. al., [3]), decision trees (Quinlan, [10]; Pagallo and Haussler, [9]) and GPAC, a variant of Valiant's [13] learning algorithm, Oblow, [8]. Anderson [1] used a neural network with one hidden layer of four nodes for the six multiplexor and achieved his best results using Rumelhart-Hinton-Williams [11] backpropagation: accuracy in the high 90% after approximately 15,000 cycles through all the (2^6) examples. For the six multiplexor, Jacobs [7] used six hidden units together with steepest descent, momentum, delta-bar-delta, and a hybrid learning algorithm. His fastest learning time (to achieve mean square error of 0.64) was an average of about 106 cycles through the data — about 7,000 examples. Bonelli et. al. reported neural network learning times approximately a factor of two slower than those for their classifier system for each of the six and eleven multiplexors.

Quinlan [10] applied $C4$, a variant of $ID3$, to each of the six, eleven, and twenty multiplexors. Representative number of examples required to achieve low to mid 90% accuracy were 100, 200 and 400, respectively. Oblow [8] reports approximately a factor of two speedup (over $C4$) with his GPAC algorithm. Pagallo and Haussler [9] investigated augmented six and eleven multiplexor problems which they solved to high 90% accuracy in approximately 500 and 1,000 examples, respectively. Oblow [8] reports that his GPAC can solve these same augmented problems in about a factor of three fewer examples.

The results of Quinlan, Pagallo and Haussler, and Oblow beg the question of whether classifier systems can be made competitive with the best of the inductive learning approaches for stationary Boolean problems. Although our empirical results show substantial improvement over previously reported classifier system approaches to the twenty multiplexor, our analyses suggest that they are not the algorithms of choice for stationary Boolean concept learning.

2 BOOLEAN CONCEPT LEARNING

Boolean concept learning is the problem of generating a pre-specified Boolean function from a sequence of examples, each labeled according to whether the output is t or f (1 or 0). Computational learning theory, a recent branch of artificial intelligence, has devoted considerable attention to learnability of Boolean concepts. Valiant published his seminal paper in 1985 and Angluin [2] introduced the term Probably Approximately Correct (PAC) learning in 1988. The realistic goal of Boolean concept learning is not to learn arbitrary functions (concepts) perfectly, but rather with high probability to learn functions to a high degree of accuracy. Valiant [13] formalized this objective and began the characterization of sample complexity (minimum sample size) as a function solely of the concept class, prespecified representation, and required confidence and accuracy. Loosely stated, computational learning theory strives to provide lower and upper bounds, L and U respectively, to the number of examples required by an algorithm to approximate a concept. The bounds are to be independent of the underlying distribution from which the examples are chosen, that is, regardless of the underlying distribution of examples, there exist learning algorithms that with high degree of confidence, closely approximate any fixed target concept after examination of at most U examples, and no algorithm can guarantee an adequate approximation with the required confidence with less than L examples.

Formally, let (X, μ) be a measure space from which examples are randomly chosen. Let $\delta, \epsilon > 0$. Let $\mathcal{C}$ be a class of concepts in 2^X, the power set of $X, \bar{x}_m$ a sequence of m examples from $X, \mid \bar{x} \mid$ the cardinality of the sequence $\bar{x}$, and $\mathcal{A}$ a family of algorithms $A(\bar{x}) -> 2^X$ that associate a concept (hypothesis) to every sequence $\bar{x}$. For concepts C_1 and C_2 let $\Delta(C_1, C_2)$ be the measure of the symmetric difference between the sets. (For purposes of this exposition, we assume that all sets considered are measurable.) A major emphasis of computational learning theory is to determine sample complexity, bounds L and U such that for every fixed concept $C \in \mathcal{C}$ and every $A \in \mathcal{A}$, $\Delta(A(\bar{x}), C) < \epsilon$ with probability $1 - \delta$ whenever $\mid \bar{x} \mid > U$ and with probability $< \delta$ whenever $\mid \bar{x} \mid < L$. That is, more than U examples are likely to yield a good approximation to any chosen concept of $\mathcal{C}$. For fewer than L examples, it is unlikely (for some distribution) that all concepts can be uniformly well approximated.

3 THE MULTIPLEXOR AND ITS SAMPLE COMPLEXITY

Multiplexor problems are a class (indexed by k) of $k + 2^k$ Boolean variable problems whose solutions can be succinctly represented as a disjunction of 2^k terms, each of which is a conjunction of $k + 1$ variables. (An s-fold disjunction of terms, each of which is a conjunction of no more than k variables is defined to be an s-term $k - DNF$ - - disjunctive normal form- -. If the number of terms is not limited, then the expression is simply a $k - DNF$.) In particular, for any fixed k, the first k "address" bits when interpreted as the binary representation of an integer $0, ..., 2^k - 1$, point to the "data" bit which specifies the associated output. For example, the six multiplexor ($k = 2 => k + 2^k = 6$) can be represented by the following 4-term $3 - DNF$:

$$\bar{a}_0\bar{a}_1d_0 + a_0\bar{a}_1d_1 + \bar{a}_0a_1d_2 + a_0a_1d_3$$

Thus, if $a_0 = a_1 = d_0 = 0$ the function value is 0, whereas for $a_0 = 0, a_1 = 1$ and $d_2 = 1$, the output is 1.

Valiant [13], Haussler [5] and others have estimated the sample complexity for $k - DNF$s and s-term $k - DNF$s. For $h = 1/\epsilon = 1/\delta$ Valiant's estimate is that $U < h(ln \mid S \mid + ln\ h)$, where $\mid S \mid$ is the cardinality of the hypothesis space. For the case of $k - DNF$s, $\mid S \mid$ is less than or equal to the cardinality of the set of all conjunctions of k or fewer variables. For a universe of discourse with n variables with $k << n$, the set of all such conjunctions has $\sum_{i=0}^{k} \binom{n}{k} = O(n^k)$ distinct elements, and hence for the class of multiplexor problems $U < O(2^k)^k = O(2^{k^2})$.

Conversely, Haussler [5] demonstrated that $c_0(log\ h) + ksh\ log(n/[ks]^{1/k}) \leq L$. Again for the class of multiplexor problems, this implies that L is bounded below by a term that is $O(2^k)$, that is, the number of examples required to learn the k-multiplexor grows exponentially in k, but linearly in the coding of the k-multiplexor.

4 INEFFICIENCY OF CLASSIFIER SYSTEMS

Classifier systems are adaptive production rule systems that use the genetic algorithm to discover new rules. Because Boolean concept learning problems are stimulus-response learning problems (an oracle provides immediate feedback about whether any given example corresponds to t or f), no bucket brigade is required. Matching classifiers can be immediately rewarded (punished) depending on the correctness of their classification. The reader is referred to Bonelli et. al. [3] and Wilson [15] for details of NewBoole and Boole, classifier systems for stimulus-response problems. General background about classifier systems is available in Holland [6] and Goldberg [4]. For purposes of this section it suffices to note that classifier systems have interleaved discovery and evaluation phases; the genetic algorithm generates new rules, and the reward-punishment of matching rules modifies rule strengths. Thus, classifier systems cannot learn any more quickly than they can differentiate between good and bad rules in any population.

We do not provide a rigorous proof of classifier system inefficiency (for stationary concepts), but rather sketch an argument that we feel is compelling and could be made rigorous. First, consider the classifier system during the evaluation phase. Assume furthermore that for some $n >> k' > k$, the sets S^t and S^f of all conjuncts of k' or fewer variables that predict an outcome t and f, respectively, can be kept in memory.[1] Consider a classifier system population $S = S^t \cup S^f$ of all classifiers with no more than k' specific bits in the matching condition. Note that a classifier system cannot reliably evaluate individual rules unless those rules match some example(s). Thus, rule evaluation of the population S is (within an average factor of 2) no more efficient than Valiant's striking algorithm on the two sets S^t and S^f, separately:

"Striking Algorithm"

> Randomly select an example x. If it is not a counterexample to some conjunct, select again. Otherwise, without loss of generality, it is a counterexample to S^t. Strike from S^t those conjuncts which match x. Continue.

Valiant [13] provides bounds on how many examples need to be drawn before the "striking algorithm" yields a close approximation to the true concept with high probability.

It follows that any claim to classifier system efficiency must rest on discovery by the genetic algorithm.

Moreover, since $n >> k' > k$ implies

$$\binom{n}{k'} > \binom{n}{k},$$

classifier efficiency would seem to require that the genetic algorithm be able to quickly specialize a small population of overly general classifiers.

While the classifier system seems to be ill suited to learning stationary Boolean concepts, the "striking algorithm" is totally inappropriate for non-stationary concepts.

5 CLASSIFIER SYSTEM MODIFICATIONS

Although we have argued that classifier systems are inherently inefficient for stationary deterministic Boolean concept learning, we nonetheless feel that there is much useful that can be learned in this environment and can be carried over to non-stationary problems and problems that require chaining. Because the learning speeds of our modifications to classifier systems are comparable to Bonnelli et. al.'s [3] for the six and the eleven multiplexor, we concentrate primarily on the twenty and the lessons that are likely to carry over to other problems.

The basic classifier system that we used in our investigations is highly similar to NewBoole (Bonelli, et. al., [3]): it incorporates a steady state GA, symmetric reward-punishment, Wilson's [15] biased reward distribution function that favors general classifiers, and Wilson's [15] CREATE routine that generates classifiers when no classifier in the population matches the current example. We made three additional modifications that considerably improved performance.

First, we used exponentially increasing population sizes (400 for the six, 800 for the eleven, and 1600 for the twenty multiplexor) with the classifiers satisfying certain minimal generality conditions. The exponentially increasing sample sizes are consistent with the exponentially increasing number of disjuncts in the solution. The minimal generality conditions were motivated by our efficiency analysis.

Knowing the form of the maximally general, fully correct classifiers led us to conjecture that the six multiplexor should have 50% "#"s ("#" signs are "don't cares" that allow the rule to match with a {0,1} at that bit position), the eleven 64%, and the twenty 75% in their initial populations. A coarse grained experiment

[1] Incremental techniques that require less memory are also known.

with different percentages of "#" in the initial population suggested that 33%, 50%, and 64%, respectively were good initial population generalities. These generalities were imposed in two different ways. First, we constrained the average density of "#"s in the initial population to meet the target level (*tg*). Second, the generality level was imposed on each and every classifier in the initial population (*igl*). What was not tried but could be easily implemented was to further restrict the generation of new classifiers to also satisfy some minimal generality threshold.

These types of generality thresholds are motivated by the recognition that the unrestricted search space is exponential; for n variables, there are 2^n $n - DNF$s. If the target concept requires a great many highly specific terms, learning it (without additional special structure) is virtually hopeless. Instead, a reasonable expectation would be to try to learn an approximation to the target concept. Clearly, at this time we offer little guidance about how to choose appropriately general initial populations; our determination was made on the basis of coarse grained experiments.

Our second modification was a weak form of speciation (sp); we segregated classifiers into separate subpopulations depending on whether they predicted 0 or 1. Mating was allowed only within subpopulations. Moreover, we sometimes imposed additional mating restrictions; only unique offspring are allowed into the population (sing). If the offspring are not unique to the population, the parents are recombined. If the current parents fail to generate unique offspring within a preset time threshold, another set of parents is selected.

The final modification was made to the example selection mechanism. We implemented what in the computational learning literature would be called a partial query (pq); examples were alternately selected at random or based on the classifiers in the current population.[2] (The reader should note the previously cited results from Valiant [13] and Haussler [5] do not apply to queries. In fact, query systems generally learn far faster than systems depending on random selection of examples.)

Our use of pq was motivated by the observation that with random selection of examples, some classifiers might rarely match and the estimates of their strengths would rarely be updated. Conversely, if all examples were generated based on templates of the classifiers in the population, there would be no pressure for coverage; that is, for some examples, the system might not develop classifiers that match them. In some cases, we took one further step; we put classifiers in a FIFO stack and used this stack to generate the alternate templates for examples (pq-FIFO).

[2]Performance was measured exclusively on randomly selected examples.

6 EXPERIMENTAL RESULTS

In the interest of space and because small problems submit to search strategies that are useless for larger problems (the scaling issue), we only present performance graphs for the twenty multiplexor; the reader is referred to Wang [14] for details of the other experiments. Figure 1 presents the performance (on a single run of up to 15,000 examples) for five variations of Wilson's [15] basic algorithm. The bottom curve (in these graphs it's equal to zero) represents the number of "solution set" classifiers, or perfect rules, in the population. Perfect rules are maximally general rules that always answer correctly when matched. The middle curve (the smooth one) shows the population's hash percentage. This is simply the relative frequency of "#"s found in each bit location of each rule in the population. The top curve (irregular and jagged) shows the average score of the system. This measures the percentage of correct system answers over the last 50 running trials. In Figure 1 systems c and d use combinations of options tried earlier. These systems showed improved performance at the end of 15,000 trials and were run longer to see if average score would continue to improve.

Figure 2 presents the performance of systems c and d when trained on 70,000 examples. System c shows learning at 40,000 trials. The number of solution set classifiers also starts to grow at 40,000th trial. The hash percent initially drops off and then rises. System d shows learning occurring earlier at 20,000 trials. The major difference between systems c and d is that system d used FIFO stack with partial query to generate exemplars.

Although the results are not presented here, the six and eleven multiplexor problems can be easily and quickly solved using a variety of classifier system implementations; the twenty multiplexor is a much more difficult problem. As concept complexity increases the classifier system needs to be more carefully crafted to avoid exponential growth in the number of examples required to learn. Our modifications have yielded learning complexity in the order of 500-1000, 5000 and 20,000-40,000 examples for the six, eleven and twenty multiplexors, respectively. We make no claims for the thirty-seven multiplexor. Certainly, scaling issues need to be more thoroughly addressed before classifier systems can be routinely applied to "real world" problems.

7 SUMMARY AND FUTURE RESEARCH

We have provided a summary of previous machine learning applications to the multiplexor problem, and have suggested that the classifier system cannot be competitive with the most efficient known learning al-

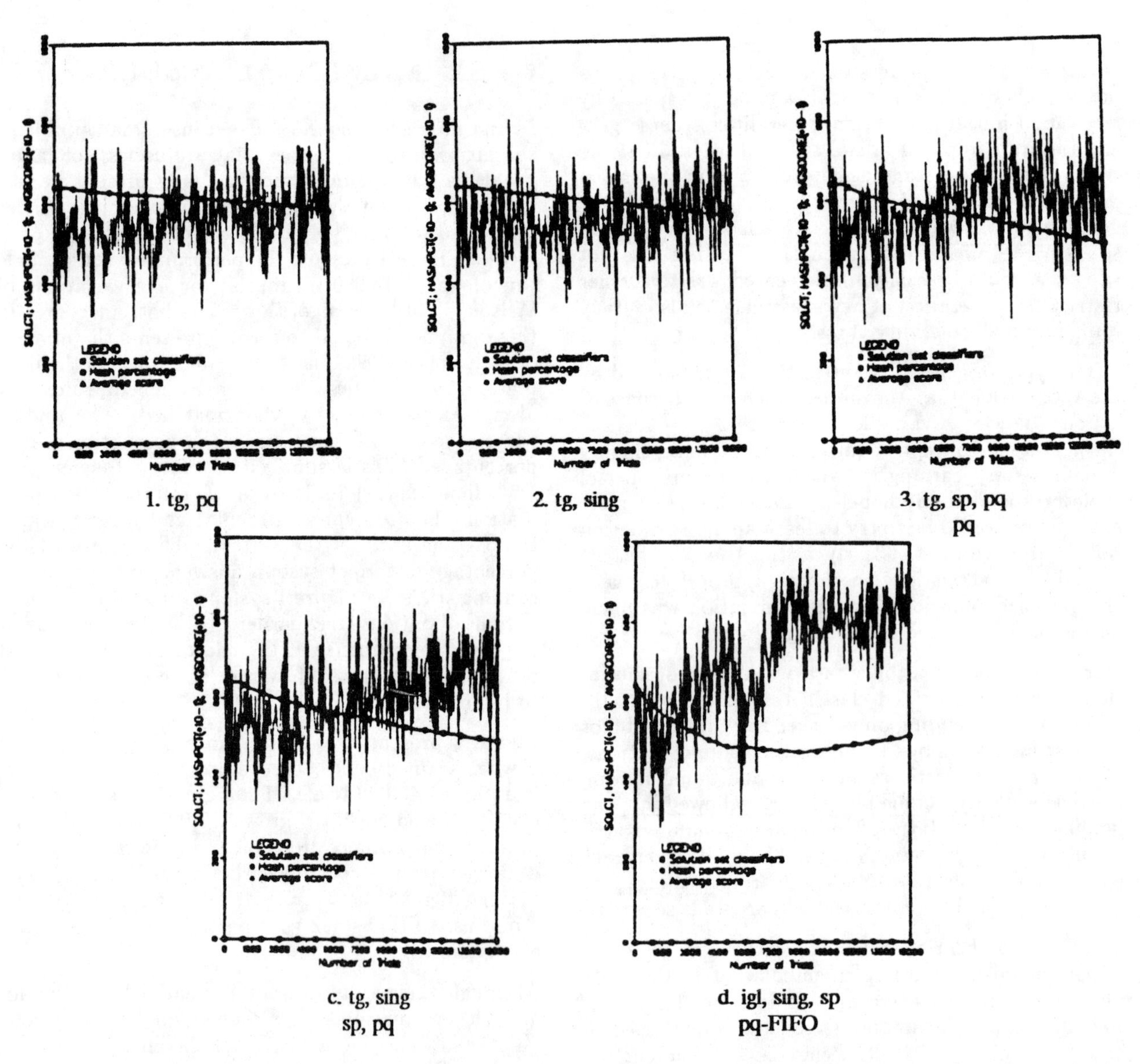

1. tg, pq 2. tg, sing 3. tg, sp, pq pq

c. tg, sing sp, pq d. igl, sing, sp pq-FIFO

Figure 1. Typical Single Run Performance of Five Variants of the Classifier on the Twenty Multiplexor (15,000 Generations).

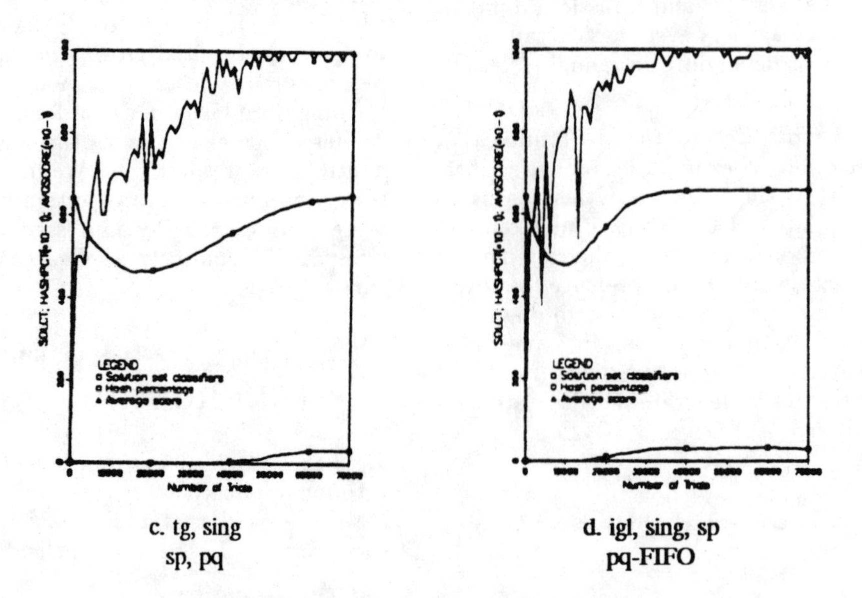

c. tg, sing sp, pq d. igl, sing, sp pq-FIFO

Figure 2. Performance of Variants c and d When Trained on 70,000 Examples.

gorithms for stationary Boolean concept learning. At the same time, we have investigated a variety of modifications to the classifier system that enabled it to learn the twenty multiplexor faster than previously reported (for classifier systems) in the literature. Our results suggest that for difficult (large) learning problems, it is useful to limit the specificity of the classifiers generated by the system, speciation is important to the success of the system, and that the system should be designed so that it can accurately estimate the relative merit of the classifiers in any population.

We did not address learning with noise nor learning of nonstationary concepts. In the first case, we suspect that the classifier system will continue to be dominated by other learning approaches. Conversely, for learning of nonstationary concepts, we feel that the classifier system is uniquely well suited. Care will need to be taken to insure that the system is capable of perpetually novel populations, while still converging to the current concept, but this is an issue for future research.

References

[1] C. W. Anderson. *Learning and problem solving with multilayer connectionist systems.* PhD thesis, University of Massachusetts, 1986.

[2] D. Angluin. Queries and concept learning. *Machine Learning*, 2:319–342, 1988.

[3] P. Bonelli, A. Parodi, S. Sen, and S. W. Wilson. Newboole: a fast gbml system. In *Proceedings of the Seventh International Conference on Machine Learning*, pages 153–159, Morgan Kaufmann, Palo Alto, CA, 1990.

[4] D. E. Goldberg. *Genetic Algorithms in Search, Optimization, and Machine Learning.* Addison-Wesley Pub. Co., Reading, Massachusetts, 1989.

[5] D. Haussler. Quantifying inductive bias. *Artificial Intelligence*, 36:177–221, 1988.

[6] J. H. Holland. Escaping brittleness: the possibilities of general-purpose learning algorithms applied to parallel rule-based systems. In R. S. Michalski, J. G. Carbonell, and T. M. Mitchell, editors, *Machine Learning*, Morgan Kaufmann, Los Angeles, CA, 1986. Volume 2.

[7] R. A. Jacobs. Increased rates of convergence through learning rate adaptation. *Neural Networks*, 1:295–307, 1988.

[8] E. M. Oblow. *Implementing Valiant's Learnability Theory using Random Sets.* Technical Report ORNL/TM-11512R, Oak Ridge National Laboratory, Oak Ridge, TN, 1990.

[9] G. Pagallo and D. Haussler. Boolean feature discovery in empirical learning. *Machine Learning*, 5(1):71–100, 1990.

[10] J. R. Quinlan. An empirical comparison of genetic and decision-tree classifiers. In *Proceedings of the Fifth International Machine Learning Conference*, pages 135–141, 1988.

[11] D. E. Rumelhart, G. E. Hinton, and R. Williams. Learning internal representations by error propagation. In J. L. McClelland D. E. Rumelhart and the PDP Research Group, editors, *Parallel Distributed Processing: Explorations in the Microstructure of Cognition*, Bradford Books, Cambridge, MA, 1986.

[12] S. Sen. *Classifier system learning of multiplexer function.* The University of Alabama, Tuscaloosa, Alabama, 1988. Class Project.

[13] L. G. Valiant. Learning disjunctions and conjunctions. In *Proceedings of the Ninth International Joint Conference on Artificial Intelligence*, pages 560–566, Morgan Kaufmann, Los Angeles, CA, 1985.

[14] L. A. Wang. *Classifier System Learning of the Boolean Multiplexer Function.* Master's thesis, Computer Science Department, University of Tennessee, Knoxville, TN, 1990.

[15] S. W. Wilson. Classifier systems and the animat problem. *Machine Learning*, 2:199–228, 1987.

Modeling Simple Human Category Learning with a Classifier System

Rick L. Riolo
The University of Michigan
Ann Arbor, MI 48109 USA

Abstract

A classifier system is used to model human performance on a simple deterministic discrimination task in which subjects must aquire categories based on their experience. The performance of the classifier system is compared to data from experiments with humans and to the performance of an adaptive neural net model described by Gluck and Bower. The classifier system is able to replicate data on human performance, including one observation not replicated by the neural net model. The classifier system also misses one prediction the neural net makes correctly. Three keys to the classifier system's performance are: (1) default hierarchies in which exceptions usually overrule more general rules; (2) a bucket brigade algorithm in which each classifier pays the average bid made by co-winning rules that produce the same message it does (rather than just paying its own bid) and receives an equal share of any payoff; and (3) the use of a bid tax.

1 INTRODUCTION

Classifier systems (CSs) have been used to understand, through metaphor and simulation, the behavior of adaptive organisms and robots from animats ([2], [26]) to rabbits [11] to humans ([10], [3]). In this paper, I will describe a CS model of human performance on a simple deterministic discrimination task in which subjects must acquire categories based on their experience. The performance of the CS is compared to data from experiments with humans described by [13] and to the performance of an adaptive neural net model described by [6] and [5].

Section 2 describes the basic human performance to be modeled. Section 3 briefly describes the CS used in these studies, CFS-C/FSW1. Section 4 describes the results obtained when a simple "flat" model of six component rules is used to model this task, first using a roulette wheel bucket brigade algorithm (BBA) and then using the noisy-auction, variance-sensitive scheme suggested by [9]. Section 5 describes the results obtained when a default hierarchy is used to model the task. Section 6 describes results obtained when the CS employs a BBA in which each classifier pays the average bid made by co-winning rules that produce the same message it does (rather than just paying its own bid) and receives an equal share of any payoff, first using just a flat model, then using a default hierarchy and finally using a default hierarchy and applying a bid tax. Section 7 is a general discussion of these results and their implications.

2 THE TASK: PERFORMANCE BY HUMANS

The basic classification task to be modeled, as described by Medin and Edelson [13], involves training non-medical students to perform a simple medical diagnosis task. The following is a simplified version of the task presented to the human subjects which captures the main relationships between the features and categories; the task used in this study is identical to that used by Gluck and Bower with their simple neural network model ([7],[4],[5]).[1] There were three symptoms, *a*, *k*, and *r*, and two diseases, K (common) and R (rare). As the names suggest, the *k* and *r* symptoms are defining features for diseases K and R, respectively: the disease is indicated if and only if those features are present. On the other hand, the symptom *a* is ambiguous—it is associated with both diseases. (Of course the subjects don't know any of this initially.) On each trial during the training phase a subject is shown a list of two symptoms, either *ak* or *ar*, and then asked to indicate which disease is expected; the subject is then told the correct answer. During train-

[1] In the experiments with humans, there were 9 symptoms and 6 diseases, but they all fell into the 3 symptom types and two disease types described here.

ing, the subjects were shown symptoms *ak* three times as often as symptoms *ar*; thus K is the common disease and R is rare. The primary interest of Medin and Edelson was the effect on classification performance of the different "base rate frequencies" of the diseases (and symptoms). In particular, they wanted to know if humans were rational (in a Baysian sense) when they used the information they observed to make classification decisions.

After a subject reached a performance criterion (16 trials without an error) or after a total of about 200 trials, the training was stopped. Essentially all subjects reached the criterion; thus the ability to discrimination $ak \Rightarrow$ K from $ar \Rightarrow$ R almost 100% of the time is the first behavior (B1) a model of these results must exhibit.

After the training phase was over, the subjects were given a number of test trials, during which no feedback was given. These tests were of three types:

- B2. Generalization to the ambiguous feature *a*. When presented with symptom *a* alone, subjects showed a strong bias toward classifying this as disease K. In fact, the ratio of K-responses to R-responses when presented with *a* roughly reflected the base rate of the two diseases, i.e., 3 to 1.
- B3. Generalization to the conflicting case, i.e., symptoms *kr* presented together. In this case, subjects tended to respond R instead of K. Medin and Edelson called this *base rate neglect*, since a Baysian use of the basic frequencies would lead one to expect disease K. [1] called this the *relative novelty effect*, suggesting that people put more weight on the novel symptom *r* exactly because it is rare.
- B4. Generalization to the combined case, i.e., symptoms *akr*. In this case, subjects tended to respond K more often than R, thus using the base rates to make their decision.

The results on these three tests, B2 to B4, along with the ability to classify the training samples very well (B1), make up the four basic behaviors a good model of this type of human discrimination and category learning must exhibit. Table 1 summarizes these results.

In addition, Medin and Edelson carried out one experiment in which they varied the instructions given to two groups. This experiment is described in section 6.

Table 1. The Behavior of Human Subjects.

	Symptoms	Ratio K / R
B1	*ak* (training)	100% K
	ar (training)	100% R
B2	*a* (ambiguous)	>1 (favor K)
B3	*kr* (conflicting)	<1 (favor R)
B4	*akr* (combined)	>1 (favor K)

Figure 1 shows the basic execution cycle of CFS-C/FSW1. For this simple classification domain, one detector message is added at the start of each trial (step 1). The detector message indicates which features (symptoms) are present for this trial (by having 1's in the appropriate loci)[2]. Classifiers match detector messages (sets of symptoms) and post messages that support one disease or the other as the system's guess.

In step 2, the bid, $B_i(t)$, of classifier i at step t is

$$B_i(t) = k * \sigma_i * S_i(t) \tag{1}$$

k is a small constant (0.04), which acts as a "risk factor" (learning rate), and $S_i(t)$ is the classifier's strength at step t. σ_i is a number between 0 and 1 that is a measure of the classifier's *specificity*; more specific classifiers having higher σ_i and more general classifiers have lower σ_i.

When a competition is run to determine which classifiers are to post messages, a "roulette wheel" selection procedure is used to determine which rules win, i.e., the probability that a given (satisfied) classifier i will win is

$$Prob(\ i\ wins\ ) = \beta_i(t) / \sum_j \beta_j(t) \tag{2}$$

where $\beta_i(t)$, the *effective bid* of classifier i at step t, is

$$\beta_i(t) = ({\sigma_i}^b * B_i(t))^2 \tag{3}$$

where σ_i and B_i are the specificity and bid as defined above, and b is the *effective bidratio power*.[3] For $b = 0$, the effective bid is just the bid squared, which gives a bias toward higher bidding classifiers winning the competition. For $b > 0$, e.g., $b = 8$, the effective bids of

3 THE CFS-C/FSW1 SYSTEM

All experiments were carried out with the CFS-C/FSW1 classifier system, which has been described in detail in [16] [17] [18]. This section briefly summarizes the important features of the CFS-C/FSW1 system.

[2]The messages are actually 16 bits long: 2 bits indicate whether the message is from the detectors or to the effectors, 3 bits are for the three symptoms, and the rest are just 0.

[3]Effective bids were first used by Goldberg, though they were used for a different reason than they are used here. I think he also was first to distinguish "roulette wheel" from "noisy auction" bid competition procedures; cf. [8].

1. Add the detector message indicating the current state (symptoms) to the message list.
2. Compare all messages to all classifiers and record matches.
3. Generate new messges by activating satisfied classifiers. If activating all the satisfied classifiers would produce more messages than will fit on the message list, a competition is run to determine which classifiers are to be activated. Classifiers are chosen probabilistically, without replacement, until the message list is full. The probability that a given classifier is activated is proportional to its *effective bid.*
4. Process the new messages through the output interface, resolving conflicts and selecting one response, i.e., disease K or R, to be used for the current time step.
5. Apply the *bucket brigade algorithm,* to redistribute strength from the environment to the system and from classifiers to other classifiers.
6. Apply *discovery algorithms,* to create new classifiers and remove classifiers that have not been useful. (This is not done in the experiments described in this paper: all rules are predefined.)
7. Replace the contents of the message list with the new messages, and return to step 1.

Figure 1: The basic CFS-C/FSW1 execution cycle.

more general rules (with $\sigma < 1$) are reduced relative to those of more specific rules (with σ closer to 1). Thus increasing b biases the competition to more specific rules, i.e., exception rules overrule defaults when they are competing [15].

In step 4, each effector message posted by a winning classifier supports a classification of the instance into disease K or disease R. If the effector interface must resolve conflicts, e.g., when one classifier produces a message "Disease K" and another produces a message "Disease R," the system's response r is chosen with probability:

$$\sum_{m_r} \beta_{m_r}(t) / \sum_{m} \beta_m(t) \qquad (4)$$

where m_r ranges over the effector messages that support response r, m ranges over all effector messages, and the β's are the effective bids of the classifier that posted the messages. Once a response is chosen, all messages that are inconsistent with it are deleted from the new message list. If the response is correct, the system reward is 10, otherwise it is 0.

There are two ways CSs can learn, one a short term and one a longer term learning process. In the short term, the strengths of existing rules are adjusted so that rules that lead to good performance are more likely to win the competition. Adjusting strengths is done by the *bucket brigade algorithm* (BBA). In the longer term, new rules can be added to the system, and rules that are not useful can be removed. Creating and removing rules is done by the *rule discovery algorithms.* In the experiments described in this paper, all rules are loaded into the system at the start of each run; no rules are created or destroyed. Thus learning only occurs as a result of the re-allocation of strength by the BBA.

The BBA is basically a type of temporal difference method [23] for updating rule strengths.[4] When a classifier posts a message, its strength is reduced by its bid. When a reward is received, all classifiers that posted effector messages (after any effector conflicts are resolved) have an equal share of the payoff added to their strengths.[5] The BBA also redistributes strength from classifiers to other classifiers but in the experiments described in this paper no classifiers are coupled, so strength is not transfered between classifiers. Thus for the classifiers in these experiments, S'_i, the strength at $t+1$ of a classifier i that posted an effector message at t is

$$S'_i = S_i + (R/n) - B_i \qquad (5)$$

where S_i is the strength of classifier i at t, R is the reward from the environment at step t, B_i is the bid of i at t, and n is the number of coactive rules that share the reward with classifier i.[6] Rules that do not win the competition to post messages or control the system response do not update their strengths. The change in strength is thus

$$\Delta S_i = (R/n) - B_i \qquad (6)$$

A rule's strength will stabilize when its income equals its payout (i.e., $\Delta S = 0$); if the average reward for rule i is R, substituting for B_i in the above equation gives

$$S_i^* = R/(n * k * \sigma_i) \qquad (7)$$

as the "equilibrium" strength of rule i. This is the maximum strength a rule can attain, given its specifity and a constant reward rate (and constant n).

4 MODEL I: SIX SIMPLE RULES

To see how well the CFS-C/FSW1 system can model the human performance described in section 2, the system was run with the following six rules:

[4]The BBA is also related to the Q-Learning techniques described by [25].

[5]In some systems, the entire payoff is added to each rule's strength.

[6]Note that within any experiments described in this paper, all rules share reward with the same number of rules, so n is not a determiner of relative strengths.

Table 2. A Simple Six-Rule Model.

(1) $a/$ K	(3) $k/$ K	(5) $r/$ K
(2) $a/$ R	(4) $k/$ R	(6) $r/$ R

For example, rule 1 means "if symptom a is present, guess disease K". The specificity of all six rules is 0.81. These six rules cover all possible relationships between individual symptoms and diseases in this task. As such they implement a very simple "component only" model of the task [7]. This model is analogous to the simple two layer network they describe which has three input nodes (one for each symptom) connected to two output nodes (one for each disease). The strengths of the six classifiers correspond to the weights between input and output layer nodes.

The CS with these six rules was trained for 2000 trials of ak(K) and ar(R) instances in a ratio of three K to one R (randomly distributed), at which point the strengths had been stable for many trials. At the end of the training period, the marginal performance on the K versus R discrimination was 78%,[7] i.e., the system was not able to pass test B1, perhaps the most fundemental of all the tests.

As described in section 3, the system's response to a given set of stimuli that support conflicting diseases is chosen probabilistically in proportion to the effective bids made to post messages supporting the different diseases. Therefore, the ratio of sums of effective bids for K to sums for R is used to estimate the expected K to R response ratio to a test stimuli (hereafter the "KR-ratio"). For example for test B2, the KR-ratio for symptom a can be estimated by β_1/β_2, the ratio of the effective bid of rule 1 to that of 2. After 2000 trials, this ratio was 1.52; thus the rules pass test B2, since the ratio should be greater than 1. Similarly for test B3, the KR-ratio for the conflicting symptoms kr can be estimated by $\beta_3+\beta_5/\beta_4+\beta_6$; for this experiment, the ratio was found to be 1.00, i.e., K to R responses will be 1/1, so test B3 (ratio less than 1) is not satisfied. And for test B4, the KR-ratio for the combined symptoms akr was estimated as $\beta_1+\beta_3+\beta_5/\beta_2+\beta_4+\beta_6$; this was found to be 1.19, which is greater than 1, so B4 was passed.[8]

The basic reason this set of rules fails on tests B1 and B3 is that the BBA with a roulette wheel auction leads to a kind of *probability matching* behavior by rules 1 and 2 [9]. For example, on ak trials, rules 1 to 4 are matched. If the system responds correctly (K), rules 1 and 3 are reinforced. Since rules 3 and 6 never make mistakes, each of their strengths soon approaches the same asymptote ($R/k*\sigma$) and remains there; thus the KR-ratio for kr is 1.0, failing test B3. However, when the strength of rule 1 increases, it causes the system to sometimes respond K on ar trials; similarly, rule 2 causes errors on ak trials when its strength increases. Thus discrimination is below 100%, failing B1. However, because ak trials occur three times more often than ar trials, the average strength of rule 1 is greater than that of rule 2, so B2 and B4 are satisfied. The behavior of the CS was found to be analogous to the behavior of *stimulus sampling* models ([1], [5]), i.e., they both probability match instead of discriminate, and they are able to pass tests B2 and B4 but not B1 and B3.

To solve probability matching problems in another context, Goldberg suggested a *variance sensitive bidding* (VSB) algorithm [9]. With this algorithm, the effective bid of equation 3 is:

$$\beta_i(t) = B_i(t) + G[1.35 * \sqrt{V_i(t)}] \quad (8)$$

where $G[\ldots]$ returns Gaussian noise with mean 0 and standard deviation $1.35 * \sqrt{V_i(t)}$. Then the winner of a competition is just the rule that bids the most (or for effector control, the response with the largest support). V_i is a recency weighted measure of the variance of rewards received by the rule, updated by:

$$V_i(t+1) = (1-c) * V_i(t) + c * (B_i(t) - R(t))^2 \quad (9)$$

where c is a constant (0.04), B_i is i's bid at t and R is the reward received at t. With the VSB scheme, if the variances of rewards paid to some rules are high, rules with lower bids (strengths) will have a better chance of winning the competitions and the system's behavior will approach probability matching, whereas with low reward variance, the largest bidders always win.

To see if a CS with variance sensitive bidding could exhibit behaviors B1 to B4, the CFS-C/FSW1 system was trained on 2000 (ak, ar) trials using the VSB algorithm just described. The system rapidly ($<$ 200 trials) reached 100% discrimination, satisfying B1. However, the KR-ratio was 1.0 for each of the B2 through B4 tests, thus failing them all. The basic reason for these results is that rules 3 and 6, which respond to the defining symptoms k and r, rapidly come to dominate system behavior as they achieve maximum strength and near zero variance, and hence maximum effective bids. Thus B1 is satisfied and B3 is not. At the same time rules 4 and 5 rapidly approach 0 strength, so for the ak case, for example, rules 1 and 3 always have a higher total effective bid than do rules 2 and 4, so the system never makes a mistake. Since the system never makes mistakes, the ambiguous rules 1 and 2 then are also able to approach the same maximum strength as

[7] All results are averages from five runs started with different pseudo-random number generator seeds.

[8] In fact, in all experiments the strengths and bids of rules 4 and 5 rapidly went to zero, since those rules were never correct.

rules 3 and 6, as defined by equation 7, leading to KR-ratios of 1.0 for B2 and B4.

In summary, a CS using the simple set of six rules in Table 2 does not seem to be able to replicate the performance of humans on the simple medical diagnosis task described in section 2. If the roulette wheel auction of CFS-C/FSW1 is used, the basic problem is that the system behaves more like a probability matcher than a discriminator. On the other hand, if a VSB scheme with noisy auction is used, the system is able to maximize discrimination but then the BBA allocates equal strength to all rules (except those always wrong), leading to behavior that is not like that exhibited by humans on the ambiguous (B2), conflicting (B3) and combined (B4) transfer tests. Its possible other parameter settings for the VSB scheme might have led to better performance, but no attempt was made to find them.

5 MODEL II. A SIMPLE DEFAULT HIERARCHY

Another way to construct models in CSs is to use *default hierarchies* ([10], [18], [11]). A simple default hierarchy for the medical diagnosis task consists of the six rules in Table 2 plus the four rules in Table 3.

Table 3. Exceptions for a Default Hierarchy with the Rules in Table 2.

(7) *ak*/ K	(9) *ar*/ K
(8) *ak*/ R	(10) *ar*/ R

For example, for the *ak* instance, classifier 1 is a very general *default* rule which is activated by that instance, the other training instance, *ar*, and by the test cases B2 (*a* alone) and B4 (*akr*). On the other hand rule 7 is an *exception* to the general default classifiers, activated only by the *ak* instance and the combined test, B4. For the experiments described in this paper, the specificity of rules 1 to 6 was 0.81, whereas the specificity of rules 7 to 10 was 0.97.

To see if this simple default hierarchy could replicate the performance of humans, CFS-C/FSW1 was run with the 10 rules for 2000 (*ak*, *ar*) training trials (as always, with a base rate of 3 to 1 K to R). All parameters were the same as in the experiment with only the 6 generalist rules, except that $b = 8$ (instead of 0) in equation 3. That is, there is a strong bias toward exceptions overiding conflicing default rules during the competitions to post messages and control the system's responses. This bias toward exceptions overriding defaults has been found to lead to better performance in a number of classifier system studies ([18], [8], [11]). There is also psychologically based arguments for having more specific rules override more general ones ([10], [22], [3], [11]).

With this simple default hierarchy, the system was able to discriminate *ak* from *ar* 96% of the time, performance almost as high as humans on test B1. The KR-ratio for B2 (*a* alone) was 1.11, (weakly) satisfying B2. The KR-ratio for B3 (*kr*) was 1.0, not less than 1 as B3 requires, and for the combined *akr* test the ratio was 1.04, just barely satisfying B4.

In summary, the default hierarchy replicated human behavior better than the simple six-rule, one level model, but it still did not satisfy all four tests. The improved discrimination was a result of the exception rules 7 and 10 overriding the ambiguous default rules 1 and 2 and supporting the defining defaults 3 and 6. However, the system still made 1 error every 20 trials, a much higher rate than exhibited by humans (even on somewhat harder problems). Because rules 3 and 7 (for *ak*⇒K) and rules 6 and 10 (for *ar*⇒R) never make mistakes, they all approach their maximum strengths as specified by equation 7. Since the specificity of rule 3 is the same as 6, and that of rule 7 is the same as 10, the ratio $(\beta_3+\beta_7)/(\beta_6+\beta_{10})$ will be 1.0 at equilibrium, thus failing test B3.

6 A MORE COOPERATIVE BBA

[11] showed how CSs with default hierarchies could replicate a wide variety of phenomena observed in classical conditioning. However, they used a slightly different version of the BBA. In particular, instead of each (winning) rule paying out its bid, co-winning rules pay out the *average* of their bids. Thus equation 6, the change in strength of rule i, becomes

$$\begin{aligned} \Delta S_i &= (R/n) - (\sum_j B_j/n) \\ &= k/n(R/k - \sum_j S_j) \end{aligned} \tag{10}$$

where R is the reward, n is the number of co-winners, k is the risk factor (learning rate), B_j is the bid, and j ranges over all co-winners. A rule's strength will stabilize when its income equals its payout, so setting $\Delta S = 0$, substituting for B_j, rearranging, and solving for S_i gives

$$S_i^* = R/(k * \sigma_i) - \sum_{j'} \sigma_{j'} S_{j'}/\sigma_i \tag{11}$$

as the "equilibrium" (asymptotic) strength of rule i, where j' ranges over co-winners other than i.

[11] points out that these equations for updating and asyptotic strengths are essentially the same as

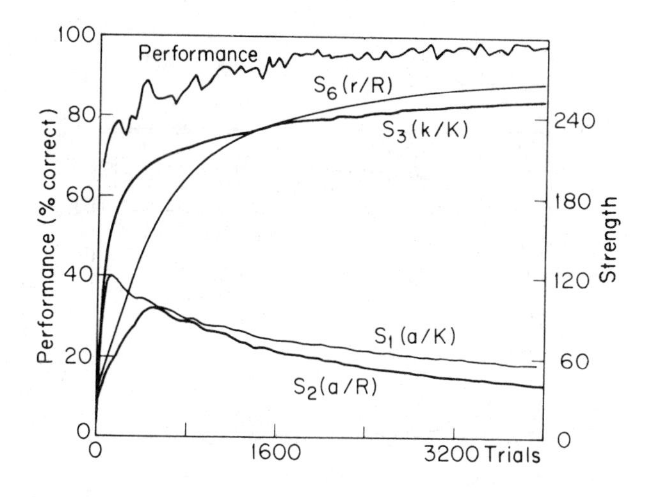

Figure 2: Strength and performance of a flat model using the CBBA.

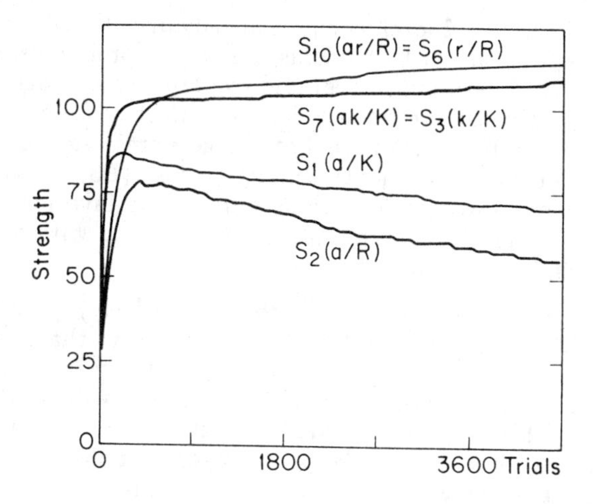

Figure 3: Strength and performance of a default hierarchy using the CBBA.

those derived from the equations proposed by Resorla/Wagner (the "delta" rule) for updating weights in neural net models of classical conditioning ([14], [24]). One difference is that in neural nets, *all* weights from active nodes are updated, whereas in a classifier system a classifier may be "active" (bidding) but not have its weight updated if it fails to win the competitions to control the system's response.

The key feature of the *cooperative*[9] BBA (CBBA) defined by Equations 10 and 11 is that co-winning rules collectively predict the expected reward, so that at equilibrium it is the *sum* of their strengths that comes to equal the reward. This contrasts with the traditional BBA described by equations 5–7, under which each *individual* rule's strength comes to equal the reward (ignoring the k and σ factors). The summative predictions made by the CBBA are a key reason [11] were able to replicate classicial condition phenomena like blocking, overshadowing, and so on.

To see if the CBBA would enable a CS to replicate the category learning of humans described in section 2, CFS-C/FSW1 was modified to use a CBBA. Figure 2 shows the performance of the system when trained on 4000 (*ak*, *ar*) trials using the six rules in Table 2 which implement a "flat" model of the world. The strengths of rules 1 ($a \Rightarrow$K), 2 ($a \Rightarrow$R), 3 ($k \Rightarrow$K), and 6 ($r \Rightarrow$R) are also shown. (The strength of rules 4 and 5 rapidly dropped to 0.)

Discrimination performance rapidly rises, reaching an average of 98.9% at the end of the run, which is almost as high as humans (B1). The strengths of rules 1 and 3 rapidly rise, since there are three times as many *ak* trials as *ar* trials. But then S_1 begins to fall, a result of its mistakes combined with the sharing of total predictive value with rule 3. Because rule 2 can make three times as many mistakes as rule 1, and get only (a share of) one third as many rewards, S_2 does not rise or remain as high as S_1. This results in the KR-ratio for a of 1.57, satisfying B2. Further, because rule 1 shares total predictive value with rule 3, and rule 2 shares with rule 6, $S_1 > S_2$ means that $S_3 < S_6$, giving a KR-ratio for *kr* of 0.94, satisfying B3.

However, because the total predictive value (i.e., reward) shared by 1 and 3 is equal to that shared by 2 and 6, the KR-ratio for *akr* is 1.0, failing B4.

Figure 3 shows the results obtained when the system with the CBBA was trained on 4500 (*ak*, *ar*) trials using the six rules in Table 2 plus the 4 rules in Table 3 which together implement a simple default hierarchy. Figure 3 show the strengths of rules 1 ($a \Rightarrow$K), 2 ($a \Rightarrow$R), 7 ($ak \Rightarrow$K), and 10 ($ar \Rightarrow$R); $S_3 = S_7$ and $S_6 = S_{10}$ throughout, and S_4 and S_5 rapidly dropped to 0. Discrimination performance rapidly rose to above 95% by 200 trials, reaching 99.0% at then end of the run. S_7 and S_1 rapidly rise, followed by S_{10} and S_2. The KR-ratios for a, *kr*, and *akr* are 1.36, 0.94, and 1.04, respectively, satisfying B2 through B4 as well. Thus using a CBBA along with a simple default hierarchy can replicate, qualitatively, human behavior on tests B1 to B4 for this classification task.

These results are comparable to those obtained by [4] and [5] using the Rescorla/Wagner delta rule to update

[9] In the 1980's this might have been called a "competitive" BBA.

strengths in a 2 level, component-only distributed network, i.e., one with "stimulus sampling" (probabilistic activation of input nodes) and overlapping activation of input nodes for different features. Both models satisfy B1 to B4; in addition, both make a prediction that the KR-ratio for the *kr* conflicting test (B3) should start out favoring K over R, but end up favoring R over K. This prediction has not been tested with humans, to my knowledge.

While the CS model is able to (qualitatively) predict B1 through B4, the 1% error rate means that the strengths of rules 1 and 2 will (very) slowly fall, and so the strengths of rules 3, 6, 7, and 10 will slowly rise. This slow change, coupled with the minimal differences of KR-ratios from 1.0 means that this model only tenuously satisfies the four criteria.[10] The underlying problem for the CS is that the KR-ratio for symptom *a*, 1.36, is low compared to the ratio humans exhibit, which is close to the base rate of K to R training samples, i.e. 3/1[11].

One way to make a CS more sensitive to the base rates in this task is to apply a *bid tax*, i.e., to reduce the strength of each bidding rule by some constant fraction (e.g., 0.01) whether it wins the competition or not. Such a tax introduces a bias against general rules that bid often but actually win the competition (and receive rewards) less often than other general or more specific rules ([18], [19]). Put another way, a bid tax penalizes a rule like $a \Rightarrow R$ relative to a rule like $a \Rightarrow K$ because they both bid the same number of times but the former receives fewer rewards than the latter, given the 3/1 K to R base rate.

Figure 4 shows the strengths of rules 1 ($a \Rightarrow K$), 2 ($a \Rightarrow R$), 7 ($ak \Rightarrow K$), and 10 ($ar \Rightarrow R$) when trained on 2000 (*ak*, *ar*) trials using the CBBA and a bid-tax of 0.01. These rules are again part of a default hierarchy implemented by the rules in Tables 1 and 2; $S_3 = S_7$ and $S_6 = S_{10}$ throughout, and the S_4 and S_5 rapidly dropped to 0.

Discrimination performance rapidly rose to above 95% by 150 trials, reaching 99.4% for the rest of the run. As in Figure 3, S_7 and S_1 rapidly rise, followed by S_{10} and S_2. The KR-ratios for *a*, *kr*, and *akr* are 2.68, 0.87, and 1.18, respectively. Thus B1 to B4 are all satisfied, and by a greater margin than when no bid tax is used.

To better understand the effects of a bid tax, consider experiment 4 carried out by [13], in which two groups

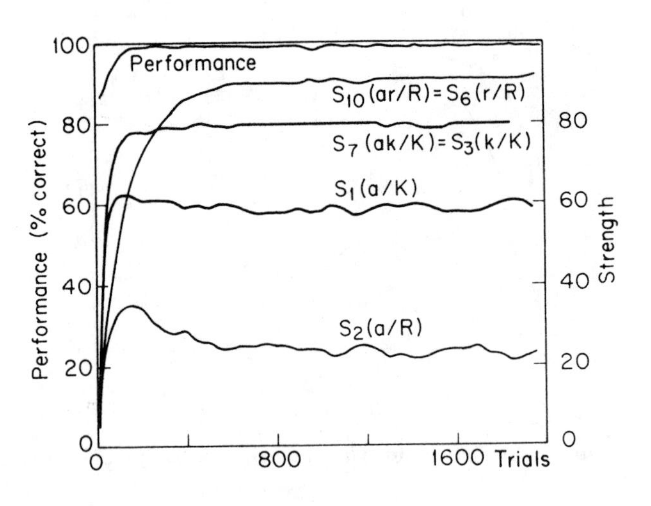

Figure 4: Strength and performance of a default hierarchy using the CBBA with a bid tax.

of subjects were given different instructions prior to the training trials of the medical diagnosis task. Both groups were told that some symptoms were better predictors than others. Then the *focus* group was told to learn about *only* the best predictors. The *complete* group was told they should try to learn about *all* the symptoms, because they may be tested on cases with only partial information available. Medin and Edelson noted three differences between the two groups on the three tests B2–B4, all basically reflecting an increase in K responses by the complete group as compared to the focus group. Thus for B2 (*a*) and B4 (*akr*), the KR-ratio for the complete group were closer to the base rate of K to R than were those in the focus group. For B3 (*rk*), the higher number of K responses in the complete group means the KR-ratio for the conflicting *kr* case was greater (closer to 1); in other words, the instructions given to the complete group *decreased* the "relative novelty effect" relative to the focus group.

If we consider the no-bid-tax simulation to be the focus case and the bid-tax simulation to be the complete case, the results with a CS default hierarchy are in agreement with two out of three of Medin and Edelson's results in their experiment 4: for the complete (bid tax) group relative to the focus group, the KR-ratio for *a* and for *akr* went up, replicating those results; on the other hand, for the *kr* case, the bid tax also *decreased* the KR-ratio from 0.94 to 0.87, i.e., *increasing* the relative novelty effect, contrary to the results reported by Medin and Edelson.

My interpretation of these results is that for the focus group, the instructions to learn about *only* the best predictors has the additional effect of instructions to

[10] The adaptive network model of Gluck and Bower also exhibits KR-ratios for B2 to B4 that are only slightly different from 1.0.

[11] As mentioned in section 2, its difficult to compare results quantitatively with those of Medin and Edelson, since (a) they used a more complex design, and (b) we have no idea what parameters human subjects are using. However, the ratios for B2 to B4 in their data are on the order of 3, 0.6, and 2.

not learn about the ambiguous cue *a*. Thus the subjects do not learn much about the fact that the *a*⇒K rule is useful (i.e., agrees with the correct answer) three times more often than the *a*⇒R rule, so these two rules are considered about equally valid (with just a slight bias toward the K rule). On the other hand the focus group is still learning that *r* signals a rare event, so this concentration on reliable predictors may enhance the increase in the relative novelty effect. Similarly, without the bid tax the CS does not penalize the R rule relative to the K rule, i.e., it does not learn as much about this relationship. On the other hand, the bid tax alters the relative strengths of rules 1 and 2 (and so indirectly, the rest of the rules) so that they do include information about the base rates of the features and diseases. This is just the kind of information the subject's in the complete case seem to have learned. This interpretation does not explain, however, why the relative novelty effect was found to decrease in the complete group.

Note that the distributed adaptive network model of [4] also replicates not only B1 to B4, but also two out of three of the key differences observed between the complete and focus groups. They argue that the focus group is analogous to their adaptive network with no overlapping activation of symptom nodes, because there is no "stimulus confusion" for these subjects. On the other hand, the complete group corresponds to their distributed network in which there is an overlap in activation of symptom nodes (i.e., if the stimulus is *a*, 0.2 of the *a* nodes will be activated, and 0.07 of the nodes for *k* and *r* will also be activated). They found that introducing overlapping activation did decrease the KR-ratio for the *kr* test and increase the ratio for *akr*, as Medin and Edelson found for the complete group relative to the focus (no overlap) group. However, the overlap model also *decreased* the KR-ratio for the ambiguous *a* test, which does *not* agree with Medin and Edelson's results.

In summary, a CS with a default hierarchy model, using the CBBA and a bid tax, can replicate almost all of the behaviors of human subjects on a medical diagnosis task as reported by Medin and Edelson both qualitatively and to a certain extent quantitatively. The CS also produced one result observed in humans that was not predicted by the adaptive network model of Gluck and Bower (and their network predicted one result correctly that the CS did not get). Both the CS model and Gluck and Bower model also make a prediction that has not yet been tested.

7 DISCUSSION

The results presented in this paper show how a CS can serve as a good model of the acquisition of simple categories from experience by humans. The key problem presented by the data of [13] was how to attain essentially 100% discrimination of the (*ak*, *ar*) cases and then transfer (generalize) this experience to the ambiguous *a*, conflicting *kr*, and combined *akr* cases in the same ways that human subjects do. A CS with a roulette wheel auction and a stardard BBA was not able to discriminate well; instead, it acted in a probability matching mode as a result of the ambiguous cue *a*, biased by the base rate differences in the training trials. This happens because (a) under the BBA each individual rule in effect adjusts its strength to predict the expected reward, independent of other co-active rules, and (b) the roulette wheel auction just transforms the strength ratios into response ratios.

It could be suggested that if there were rule discovery operators being used by the CS, e.g., the genetic algorithm and other triggered operators suggested by [10] and others, rules for an ambiguous cue might be removed because their strengths would be less that rules that just use defining cues. This would allow the system to achieve near perfect discrimination. This solution is similar to the "selective attention" mechanisms proposed as part of stimulus sampling models ([1], [5]). The problem for a CS is the same as that for stimulus sampling models: if the rules that respond to ambiguous cues like *a* are removed, then when a test on that cue is done later, a subject should show no bias toward one response or another. But this is not what is found by [13] and others: instead, a subject's response to an ambiguous cue is biased by the base rates that have been experienced for that cue in the training trials.

Goldberg has suggested the VSB scheme to solve other kinds of probability matching problems [9]. That variance-sensitive noisy-auction scheme does indeed enable a CS to discriminate 100% in the tasks described in this paper, but it does not lead to any bias in response to the transfer cases B2 to B4. The basic problem is that the defining features, which have zero variance, rapidly dominate performance, allowing other co-active rules that respond to ambiguous cues to increase their strengths to equal asymptotic values as allowed by the standard BBA defined in Equations 5 to 7. This was found to be the case in simulations using either the simple flat model of six rules described in section 4, or using the default hierarchies of ten rules described in section 5. It may be that other parameter settings would enable a VSB scheme to replicate human performance—no attempt was made to explore the effects of altering the VSB parameters. (The VSB scheme was also tried with the CBBA; it did not pass the B2 through B4 transfer tests, again because the rules responding to defining cues dominated system behavior.)

Two factors were required for modeling human behavior on this task:

1. A default hierarchy was used to represent system

knowledge. Because exception rules overrule generalists when they disagree (other things being equal), a default hierarchy is able to achieve good discrimination while still retaining and protecting the strength of the generalists. The generalists will then have strengths that reflect the system's training experience, and so they can influence the system's behavior when various transfer tests are carrried out.

2. A cooperative BBA (CBBA) like that proposed by [11] was used instead of a standard BBA. Under the CBBA, co-winning rules that agree on a response share the rewards they receive and pay out the average of their bids, rather than each paying its own bid. [11] showed that under a CBBA co-winning rules cooperate to predict rewards, and so it is the *sum* of their strengths that is proportional to the rewards received, rather than each rule's strength alone. They also showed that the updating of strength under a CBBA is analogous to the updating of weights between nodes in an adaptive network using a Rescorla/Wagner delta rule.

The results presented in section 6 showed that a CS using a CBBA to update the strengths of rules in a simple default hierarchy was able to replicate, at least qualitatively, all four of the basic discrimination and transfer behaviors observed when human subject carryied out a similar (but more complicated) task.

The results in section 6 also showed that applying a bid tax led to rule strengths that better reflect the base rates found for ambiguous cues in the training data. The application of a bid tax produced behavior that was qualitatively the same as and quantitatively similar to that exhibited by subjects who were told to learn about all cues, not just about the best predictors, for two out of the three changes observed in humans who were given different task instructions. The results obtained with the CS showed that when a bid tax is applied, the KR-ratio of responses to the ambiguous *a* symptom and the combined *akr* test increased, just as is found by Medin and Edelson for the "complete" group. This is the one prediction which the adaptive network model of Gluck and Bower does *not* make correctly. On the other hand, their network model with stimulus sampling from overlapping input nodes was able to predict that the relative novelty effect would be decreased for subjects in the complete group. For the four basic criteria B1–B4, the CS and adaptive networks make similar, correct predictions.

In [11] and in the studies reported here, the CBBA is applied to CSs in which all rules produce effector messages. One key issue is how to apply a CBBA when rules are coupled, e.g., in the CFSC2 system [20]. The main problem is how to decide which rules should have their bids averaged to determine how much their strengths should be decremented. In general terms, the basic idea of the CBBA is to average bids over those rules that are predicting the same event or outcome. In the case of rules that produce only effector messages, the solution is to average bids over rules that support the same action. One simple solution for the more general case is to average the bids of all rules that produce *identical messages*. A more general solution might be to average the bids of rules that are coupled to the *same rules*, i.e., that produce messages that are used by the same rule on the next step. This would mean there would be a lag in updating strengths until new messages were matched to classifiers on the next step. It is not obvious that this would create any significant problems, but this should be considered in more detail in future studies.

A second question that should be addressed in future studies is whether a CBBA should completely replace the traditional BBA, or in what circumstances one should be employed rather than the other. Some answers to this question relevant to modeling animal and human learning may be suggested by the studies of [21], [22] and others, which suggest that the summation of bids (i.e., the calculation of an aggregate prediction) is carried out in the hippocampus. They cite numerous studies which show that just those learning phenomena that require the ability to sum predictions and feed them back to modify strengths are disrupted in animals with hippocampal lesions, including the blocking, latent inhibition, negative patterning (the Exclusive-OR problem), and many higher cognitive functions. [22] also proposes a two level system, one a simple associative system, which uses a strength updating rule similar to the traditional non-summative bucket brigade algorithm, and a second configural associative system, which uses an updating rule that is like the Rescorla/Wagner delta rule or the CBBA. It may be possible to mix the traditional BBA and the CBBA in a single CS in ways that makes it possible to formally model some of these proposals made in the psychological literature.

Acknowledgements

This work was supported in part by National Science Foundation grant IRI-8904203 and by a University of Michigan Kellog Presidential Initiative Fund grant to the BACH research group.

References

[1] Binder, Arnold and Estes, William K. "Transfer of Response in Visual Recognition Situations as a Function of Frequency Variables." *Psych. Monographs, 80 (23)*, p 1-26, 1966.

[2] Booker, L.B. "Intelligent Behavior as an Adap-

tion to the Task Environment." Ph.D. Dissertation. University of Michigan, 1982.

[3] Druhan, Barry B. and Mathews, Robert C. "THIYOS: A Classifier System Model of Implicit Knowledge in Artificial Grammars." In *Proc. Ann. Cog. Sci. Soc.*, 1989.

[4] Gluck, Mark "Stimulus Sampling in a Distributed Network Model: Effects of Category Frequency on Generalization." In *Proc.11th Ann.Conf.Cog.Sci.Soc.*, Ann Arbor MI, 1989.

[5] Gluck, Mark. "Stimulus Sampling and Distributed Representations in Adaptive Network Theories of Learning." In *Festschrift for W.K.Estes,* A.Healy, S.Kosslyn, R.Shiffrin (eds). Lawrence Erlbaum Assoc, NJ, 1991 (in press).

[6] Gluck, Mark and Bower, Gordon H. "Evaluating an Adaptive Network Model of Human Learning." *J. Memory and Language, 27,* 166-195, 1988.

[7] Gluck, Mark and Bower, Gordon H. "From Conditioning to Category Learning: An Adaptive Network Model." *J. Exp. Psych: General, 117,* 227-247, 1988.

[8] Goldberg, David E. *Genetic Algorithms in Search, Optimization, and Machine Learning.* Addison-Wesley, Reading MA, 1989.

[9] Goldberg, David E. "Probability Matching, the Magnitude of Reinforcement, and Classifier System Bidding." *Machine Learning, 5,* 407-426, 1990.

[10] Holland, John H., Holyoak, Keith J., Nisbett, Richard E. and Paul A. Thagard. *Induction. Processes of Inference, Learning, and Discovery.* The MIT Press, Cambridge, MA, 1986.

[11] Holyoak, K.J., Koh, K. and Nisbett, R.E. "A Theory of Conditioning: Inductive Learning within Rule-Based Default Hierarchies." *Psych. Review 96,* 315-340, 1990.

[12] Kehoe, E. James. "A Layered Network Model of Associative Learning: Learning to Learn and Configuration." *Pysch. Rev., 95,* 411-433, 1988.

[13] Medin, Douglas L. and Edelson, Stephen M. "Problem Structure and the Use of Base-Rate Information From Experience." *J. Exp. Psych.: General, 117,* 68-85, 1988.

[14] Rescorla, R. A., and Wagner, A.R. "A Theory of Pavlovian Conditioning: Variations in the Effectiveness of reinforcement and non-reinforcement." In *Classical Conditioning II: Current Research and Theory,* A.H. Black and W.F. Prokasy (eds). Appleton-Century-Crofts, NY, 1972.

[15] Riolo, Rick L. "Bucket Brigade Performance II: Simple Default Hierarchies." In *Proc. Second Intern. Conf. on Genetic Alg. and their Applications,* 196-201. John J. Grefenstette (ed). Cambridge MA, 1987.

[16] Riolo, Rick L. "CFS-C: A Package of Domain Independent Subroutines for Implementing Classifier Systems in Arbitrary, User-Defined Environments." Logic of Computers Group, Division of Computer Science and Engineering, University of Michigan, Ann Arbor, 1988.

[17] Riolo, Rick L. "CFS-C/FSW1: An Implementation of the CFS-C Classifier System in a Domain that Involves Learning to Control a Markov Process." Logic of Computers Group, Division of Computer Science and Engineering, University of Michigan, Ann Arbor, 1988.

[18] Riolo, Rick L. "Empirical Studies of Default Hierarchies and Sequences of Rules in Learning Classifier Systems." Ph.D. Dissertation, University of Michigan, 1988.

[19] Riolo, R.L. "The Emergence of Default Hierarchies in Learning Classifier Systems." In *Proc. Third Internat. Conf. On Genetic Alogirithms,* 322-326. J.D.Schaeffer (ed). 1989.

[20] Riolo, Rick L. "Lookahead Planning and Latent Learning in a Classifier System." In *Proc. Conf. Simulation of Animal Behavior: From Animals to Animats.* Jean-Arkady Meyer and Stewart Wilson, eds. MIT Press, Paris, 1990.

[21] Schmajuk, Nestor A. "Role of the hippocampus in temporal and spatial navigation: an adaptive neural network." *Behav. Brain Res., 39,* 205-229, 1990.

[22] Sutherland, R. J. and Rudy, J. W. "Configural association theory: The role of the hippocampal formation in learning, memory, and amnesia." *Psychobiology 17,* 129-144, 1989.

[23] Sutton, R.S. "Learning to Predict by the Methods of Temporal Differences." *Machine Learning, 3,* 9-44, 1988.

[24] Sutton, Richard S. and Barto, Andrew G. "Toward a Modern Theory of Adaptive Networks: Expectations and Prediction." *Pysch. Rev., 88,* 135-170, 1981.

[25] Watkins, Christopher J.C.H. "Learning from Delayed Rewards." Ph.D. Thesis, King's College, 1989.

[26] Wilson, Stewart W. "Knowledge growth in an artificial animal." In *Proceedings of the First International Conference on Genetic Algorithms and their Applications,* 196-201. John J. Grefenstette (Ed.). Pittsburgh, PA, July 24-26, 1985.

Application of a Genetic Classifier for Patient Triage

Tod A. Sedbrook
CIS Department
University of
Northern Colorado
Greeley, Colorado

Haviland Wright
Avalanche Development
Corporation
Boulder, Colorado

Richard Wright
Director of Public Health
Denver Dept. of Health
and Hospitals
Denver, Colorado

Abstract

This research develops and applies a genetic classifier system (CS) to triage patients presenting with symptoms of upper respiratory infections (URIs). The CS searches among 66 dichotomous patient signs and symptoms to evolve classifiers that best explain care provider triage decisions. The systems search is directed by specifying relative costs of false positives and false negatives. The model achieved a sensitivity and specificity of 100% and 42%, respectively, when applied to a triage case base of URI patients. A split-sample validation of the system shows its accuracy is comparable to that achieved with a triage protocol developed by Infectious Disease Specialists.

1.0 Introduction

Discriminant and cluster analysis are familiar multivariate statistical approaches for analyzing features to classify subgroups. These techniques require feature spaces that conform to assumptions of multidimensional normal density; however, such approaches may not be appropriate when these assumptions are violated.

Of particular interest are techniques for analysis of dichotomous feature spaces. Dichotomies are common in medical diagnostics where patient signs and symptoms are recorded as either present or absent and laboratory test results are noted as either normal or abnormal. Classification techniques, applicable to dichotomous features, such as logistic regression, kernel methods and classification trees have been criticized for excessive computation times, poor intuitive interpretations, sensitivity to noisy data and difficulties with multimodal and multidimensional spaces [3,6].

A genetic classifier system (CS) is an alternative method to discriminate between patient groups on the basis of dichotomous patient features. This paper reports results of applying a CS model to detect dichotomous features which best discriminate between care provider triage classifications for patients presenting with acute, upper respiratory infections (URIs).

2.0 Background and Motivation

Triage is a medical term for assigning degrees of medical severity to a patient's condition to economically allocate appropriate levels of medical resources. This study applies the CS model to triage acute upper respiratory infections (URIs). The majority of URI patients present with conditions caused by viruses such as common cold syndrome or influenza. Other causes of URI episodes are bacteria related (e.g. streptococcal pharyngitis) and result in complications (e.g. otitis media and sinusitis). The CS is challenged to classify patients on the basis of symptom complexes.

Patients with URI's rarely manifest prototypical symptom patterns expected from a complicated or uncomplicated URI. Some patients may exhibit symptom patterns less extreme than complicated but more extreme than uncomplicated. Other patients may demonstrate symptoms uncharacteristic of either complicated or uncomplicated because the patient does not have a URI or has another condition in addition to a URI.

For this investigation, care-providers classified patients into one of two operationally defined triage classes - complicated or uncomplicated. The operational definition of uncomplicated was a patient who was judged not to need a same day appointment; whereas, a complicated patient was defined as one who required a same day appointment.

The CS uses genetic search procedures to develop classifier strings to 'best predict' care providers' triage decisions, where the meaning of 'best predict' depends on clinical concerns for patient safety. There are two aspects to consider in rewarding effective classifiers: (1) the sample counts of agreements and disagreements between the CS and care providers, and (2) the model's success in avoiding risks to patient safety. Patient safety is threatened by false negatives (FNs) when the model inappropriately assigns a patient to a less urgent triage class and needed care is delayed. Patient safety is also threatened by false positives (FPs) when resources are inappropriately allocated to patients and unavailable to those truly in need. The CS explicitly considers trade-offs between FPs and FNs through proportionate reduction in error (PRE) measures.

Cohen suggested a weighted kappa measure to reflect that some types of misclassifications are more serious than others [1]. Hildebrand, Laing and Roesenthal proposed weighted proportionate reduction in error (PRE) measures to assess prediction success according to an a priori proposition incorporating implicit costs of misclassification [4].

3.0 Methods

3.1 Genetic Classifier Model

The CS's applies iterative genetic search to find discriminators for triage classifications. Classifier strings represent patterns that trigger URI triage classifications according to degree of match to a patient's symptom profile. Classifiers with below average match post complicated triage messages. Those at or above average post uncomplicated messages. Figures 1 and 2 illustrate methods for determining a classifier's response to patients' symptom patters.

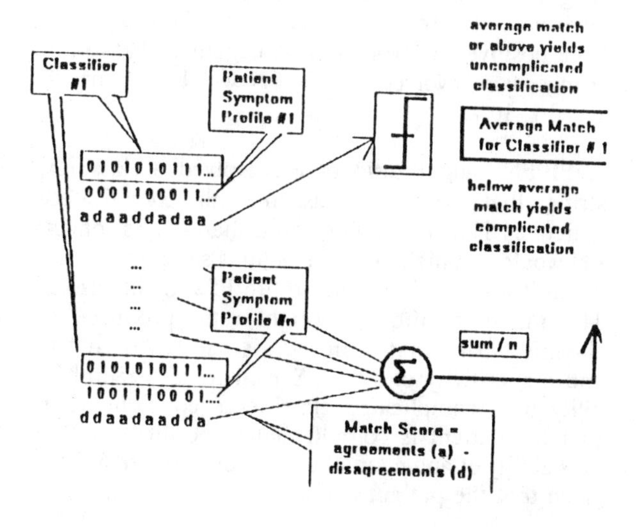

Figure 1: Classifier Response to Patient Symptom Profiles

$$\text{triage class}_{ik} = \begin{cases} 0 & \text{if } w_{ik} - \overline{w}_k >= 0 \\ 1 & \text{otherwise} \end{cases}$$

$$\text{where: } \overline{w}_k = \frac{\sum_i w_{ik}}{n}$$

$$w_{ik} = \sum_j y_{ikj}$$

$$y_{ikj} = \begin{cases} 1 & \text{if } s_{ij} = c_{kj} \\ -1 & \text{otherwise} \end{cases}$$

s_{ij} is symptom j for patient i
c_{kj} is bit j for classifier k

Figure 2. Classifier (k) prediction message for patient(i) (triage class $_{ik}$, 0 = uncomplicated, 1 = complicated)

The CS model allows the decision makers to engineer the decision environment by setting relative costs of false positives and false negatives. Classifier strings are rewarded and evolve according to their ability to avoid costly errors as reflected in the magnitude of their PRE measures (see Figure 3).

The CS's search is directed by steady state genetic algorithms as described by Syswerda and by Whitley [5, 7]. Classifiers are ranked according to their PRE measures and at each iteration two parents are selected with a probability proportional to their PRE value for a crossover operation. During crossover, each bit within the new child string has a .5 percent chance for random mutation (bit flipping). The child solution then joins the population which forces removal of a lower ranking solution. The removed string is selected from the lower ranking one-third of the population on the basis of degree of hamming similarity to the new child. By removing similar strings and introducing mutations, the population avoids genetic drift, maintains diversity and allows solution niches to form [2].

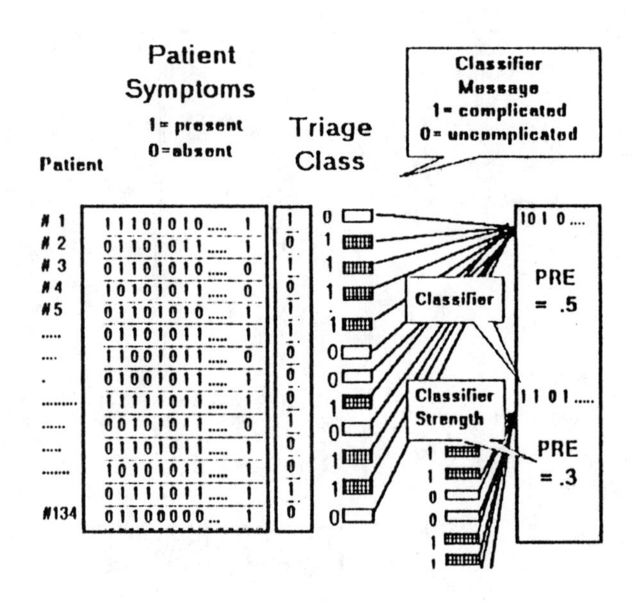

Figure 3. Classifiers are ranked according to PRE measures.

The genetic search procedure initiated from a population of 100 classifier strings each represented by 66 bits randomly set to either 0 or 1. Ranking and selecting a pair of parent classifier strings, and producing a new child proceeded for 4000 offspring. At the end of the iterations, the highest ranking classifier string represents the best of the evolved schemata. Exploration of lower iterations of 1000, 2000, and 3000 revealed prediction levels stabilized between 3000 and 4000 iterations.

4.0 Field Study Methods for Developing the Triage Case Base

The data for evaluating the CS model were collected from a clinical trial conducted at the Eastside Health Center of the Denver Department of Health and Hospitals. The study's sample patients were selected from a low socioeconomic population more likely to have chronic diseases and multiple health problems which complicates triage. The patient population adds to the diversity of the triage case base to provide a rigorous test of the CS model and strengthens the model's generalizability to other clinical settings.

Consenting patients were included in the study if they presented with any one or more of the following chief complaints: "cold", stuffy nose, cough, earache, stuffy ears, "flu", water/burning eyes, nasal discharge, runny nose, sinus pain, sneezing, sore throat, swollen glands, or hoarseness. Care providers recorded the presence or absence of patient symptoms and recorded a triage judgement of complicated or uncomplicated. As a follow-up, an independent physician reviewed medical charts for study patients. The chart review was conducted approximately two months after the patient's visit to determine if the patient returned for complications. The independent physician's retrospective opinion, in all but one case, agreed with the patient's triage class at the time of the original visit. The disagreement concerned a patient originally classified as uncomplicated, who presented with symptoms of bronchitis. This patient was re-classified as complicated for model validation.

The case base developed from the field study consists of 66 binary symptom attributes for each of 134 patients along with a triage classification from either a physician or a nurse practioner. An Infectious Disease Specialist developed the list of patient features. Each feature was labeled as either typical of a complicated or uncomplicated URI and the list of symptoms and associated labels were subjected to critical review by several Infectious Disease and Primary Care physicians.

5.0 Results

5.1 Patient Classification

The following presents results achieved by classifier strings in discriminating between complicated and uncomplicated patients at various levels of weight ratios representing implicit costs of false positives (FPs) and false negatives (FNs). For example, a weight ratio of 8:1 allows the model to adjust its environmental reward structure to allow up to 8 false positives to avoid 1 false negative. Classifier strings were evolved for FP:FN weight rations of 1:1, 8:1, 10:1, 50:1.

A FP:FN weight ratio of 8:1 evolved a classifier string that correctly classified all complicated patients. That is, trading more than 8 FPs for an FN would not make the predictions safer and could make it less efficient since more FPs might result. The model at the 8:1 weight ratio produced a classifier string that achieved a sensitivity of 100% and a specificity of 42%. Sensitivity is the probability of a complicated classifier prediction given that the patient is complicated. Specificity is the probability of an uncomplicated classifier prediction given that the patient is uncomplicated.

Table 1 represents the cross classifications of predictions found at the several FP:FN weight ratios. The table shows that classifier strings evolve according to levels of bias against FNs. The table also shows that increasing the FP:FN ratio from 5:1 to 7:1 results in trading one FP to avoid one FN.

Table 1. Cross-classification of agreements between care-providers and the CS's model derived classifier string for various weight ratios of false positives to false negatives.

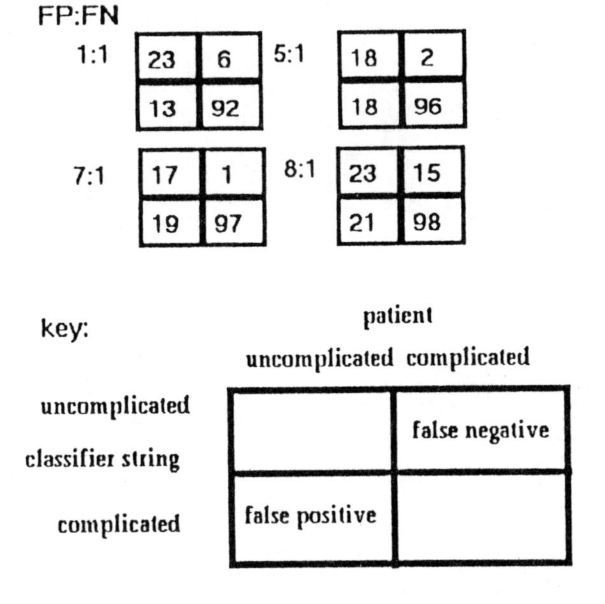

5.2 Model Validity

A split-sample validation test estimated the misclassification rate for classifier strings at four FP:FN ratios. Thirty different training sets of 100 patients and test sets of 34 patients were randomly constructed from the triage case base. Classifier strings with weight ratios of FP:FN of 1:1, 8:1, 10:1 and 50:1 were evolved from each of the 30 training samples of 100 patients. The top classifier string was then used to classify triage categories for the test set of 34 patients.

Table 2 presents the results of 30 split-sample experiments conducted for each FP:FN weight ratio. The training set classifier strings achieved high sensitivity in test sets and resulted in usually less than two FNs. The split-sample results compared favorably with a classification protocol developed from consultations with Infectious Disease specialists. The specialists achieved a sensitivity of 99% and a specificity of 11% with one false negative.

The above split-sample experiment does not eliminate the effects of any biases in patient selection and classification or data collection. Follow-up studies in a new clinical setting would provide the best way to prospectively evaluate the misclassification rates associated with the derived classifier strings of the CS model.

Table 2: Results of Split-Sample Validation of Classifiers

Weight Ratio FP:FN	Test Set Results (30 trials) Sensitivity	Specificity	Number of FN's
	Average (range)		
1:1	.88 (.78 - 1)	.31 (0.0 - 0.8)	2.73 (0 - 5)
8:1	.91 (.73 - 1)	.22 (.09 - .36)	1.96 (0 - 6)
10:1	.93 (.83 - 1)	.13 (0.0 - .45)	1.56 (0 - 6)
50:1	.93 (.74 - 1)	.18 (0.0 - .65)	1.83 (0 - 6)
Specialist Protocol	.99	.11	1

6.0 Conclusion

The results indicate the CS model responds appropriately to a range of FP:FN weight ratio, across a diverse set of URI patients and care-provider classifications. When so directed, the CS found a classifier string that correctly classified all complicated patients to avoid FNs even if more FPs result. When provided with a training set that was representative of the test set, the CS learned to classify cases at a level comparable to a specialist designed protocol.

References

[1] J. Cohen. Weighted Kappa: Nominal Scale Agreement with Provision for Scaled Disagreement or Partial Credit. *Psychological Bulletin.* October 1968: 213-220.

[2] D. E. Goldberg. *Genetic Algorithms in Search Optimization and Machine Learning.* New York: Addison Wesley, 1989.

[3] D. Greene and S. Smith. A Genetic Study for Learning Models of Consumer Choice. In J. Grefenstette, editor, *Second International Conference on Genetic Algorithms and Their Applications.* Hillsdale, NJ: Lawrence Erlbaum Associates, 1987.

[4] D.K. Hildebrand, J.D. Laing and H. Roesenthal. *Prediction Analysis of Cross Classifications*. New York: John Wiley & Sons, 1977.

[5] G. Syswerda. Uniform Crossover in Genetic Algorithms. In, J. D. Schaffer, editor, *Proceedings of the Third International Conference on Genetic Algorithms*. San Mateo, CA: Morgan Kaufmann, 1989.

[6] P.E. Utgoff and C.E. Brodley. An Incremental Method for Finding Multivariate Splits for Decision Trees. In J. Porter and R. Mooney, editors, *Proceedings of the Seventh International Conference on Machine Learning*. Palo Alto, CA.: Morgan Kaufmann, 1990.

[7] Whitley, D. The GENITOR Algorithm and Selection Pressure: Why Rank-Based Allocation of Reproductive Trials is Best. In J. D. Schaffer, editor, *Proceedings of the Third International Conference on Genetic Algorithms*. San Mateo, CA: Morgan Kaufmann, 1989.

HCS: Adding Hierarchies to Classifier Systems

Lingyan Shu † shu@cs.UAlberta.CA
Jonathan Schaeffer jonathan@cs.UAlberta.CA

Computing Science Dept.
University of Alberta
Edmonton, Alberta
Canada T6G 2H1

ABSTRACT

Genetic-algorithms-based learning classifier systems suffer from a number of problems that cause system instability, resulting in poor performance. These problems include genetic operation disruptions and difficulties in maintaining good classifiers and classifier structures in the population. A method is proposed in which structural ties are used to achieve coherence, impose cooperation and encourage co-adaptation among classifiers. A hierarchically structured classifier system (*HCS*) has been implemented to show the effect of this structuring. At the lowest level, classifiers (*individuals*) are grouped into *families*. Higher-order structures, such as a *community of families*, can be introduced if necessary. The experimental results show a significant improvement in system performance and stability. The relationships between the *HCS* framework and the Michigan and Pittsburgh approaches are discussed.

1. INTRODUCTION

Classifier systems have recently emerged as an exciting area of research into genetic-algorithm-based learning [4-7,10,12]. Although they show great promise, they are plagued with a number of serious difficulties. The presence of good classifiers and classifier structures in a population are vulnerable to genetic operation disruptions. This difficulty is reflected by the problems of rule clustering and rule association.

The rule clustering problem refers to the phenomenon that all the classifiers in a population may converge to similar patterns, but these patterns do not include all the sub-solutions. A similar problem has been pointed out with genetic algorithms being used for function optimization [3]. DeJong has proposed a method called *crowding* to limit the number of strings for each sub-solution in the population [3]. However, one obvious weakness of this method is that the stronger classifiers face the danger of being deleted, possibly affecting the stability of the system. Also, one classifier belonging to one sub-solution may be deleted for the sake of other sub-solutions. Booker proposed a method called *restricted mating strategies* to solve the rule clustering problem and used this strategy in his GOPHER-1 system [1,2]. In GOPHER-1, classifiers that are excited by the same message are called a classifier cluster. The genetic operations are restricted to classifiers in the same cluster. This reduced the clustering problem in the sense that, globally, classifiers for different messages would not disrupt each other. However, locally, in each classifier cluster, premature convergence could be a serious problem when the search space is much larger than the sample space.

The rule association problem refers to the difficulty in achieving cooperation among a set of classifiers. Cooperation includes the building of classifier chains and default hierarchies. The behavior of the rule association affects the robustness and the stability of genetic-algorithm-based systems. However, conventional classifier systems lack effective mechanisms for forming and maintaining rule associations. For example, several efforts have been made to solve this problem in conjunction with default hierarchies. One way to form and maintain default hierarchies is to favor more specific classifiers when bidding to fire classifiers [8,11]. However this scheme encourages the survival of more specific classifiers over more general ones. This phenomenon is described as *starving the generals* [8,11]. The recent scheme of *necessity auction and separate priority factor* would not work well in complicated situations [9]. The difficulty here lies in the lack of a strong relationship among the exception classifiers and default classifiers. Therefore, there is no distinction between a real exception classifier and a more specific classifier that is covered by the default classifier.

A method is proposed in this paper in which structural ties are used to achieve coherence, impose cooperation and encourage co-adaptation among classifiers. At the lowest level, classifiers (*individuals*) are grouped into *families*. Members of a family cooperate to maximize both the family's strength and the individual member's strength.

† Current address is Bell-Northern Research, P.O. Box 3511, Station C, Ottawa, Ontario, Canada K1Y 4H7.

As in conventional classifier systems, matching is done at the classifier level. However, the bid of a matched classifier is determined not only by the strength of the classifier and its specificity, but also the strength of the family that it belongs to. Therefore, the fates of the family members are bound together. Most genetic operations are performed at the family level, meaning that the basic unit for a genetic operation is the family instead of the classifier.

A hierarchically structured classifier system, *HCS*, has been implemented to show the effect of imposing structural ties. The experimental results show significant improvements in system performance and stability for *HCS* over conventional classifier systems.

The concept of a family superficially appears similar to the idea of *rule sets* in the Pittsburgh method [4]. The fundamental differences between the HCS and the the Pittsburgh and Michigan methods are discussed.

2. HIERARCHICAL CLASSIFIER SYSTEMS

Establishing relationships among a collection of entities usually results in common interests and mutual benefits. Therefore the behaviors of the entities tend to be for each other instead of against each other. Gaining coherence through establishing structural relationships is proposed as a method for achieving stability in a classifier system.

Structuring is superimposed on classifier systems by the introduction of hierarchies. Hierarchical Classifier Systems (*HCS*) retain the basic structure of classifier systems, but join several classifiers together as a family. The probability of a genetic operation between families is much higher than that between classifiers in a family. The utility of a rule is in direct proportion to the sum of the utilities of the classifiers in the family. Matching and firing are still performed at the classifier level. However, the bid of a matched classifier is determined not only by the utility of the classifier but also by the utility of the family. Hence, families that contain good classifiers and classifier structures would survive over those containing improper classifiers.

2.1. HCS FRAMEWORK

The syntax of HCS is defined by four-tuple $<E, F, G, R>$ where:

- $E = \{e_i\}$, where e_i $(i = 1, 2, ..., n)$ is an environment state.
- $F = \{f_i\}$, where f_i $(i = 1, 2, ..., m)$ is a family. f_i consists of a set of *fs* (family size) classifiers.
- $G = \{g_i\}$, where g_i $(i = 1, 2, \cdots j)$ is a category.
- $R = \{r_i\}$, where r_i is a relation over f_i. In other words, r_i is a relation over all the classifiers contained in a family.

A classifier contains a condition part and an action part. Both parts are strings over the alphabet {'0', '1', '#'}. Family members are bound together by relation r_i. r_i could be classifier chains or default hierarchies.

The computational behavior of *HCS* is defined by four-tuple $<GA, CA, SA, P>$ where:

- *GA* is a set of the genetic operations.
- *CA* is a set of credit assignment algorithms.
- *SA* is a set of selection algorithms used for choosing classifiers for firing, families for deletion, and families or classifiers for the genetic operations.
- *P* is a set of application-dependent parameters that includes the crossover rate, population size, the number of classifiers contained in a family (*fs*), etc.

In HCS, there are two kinds of crossover. The first one is crossing over two classifiers from different families. The second one is crossing over two families. The latter has two situations, swapping family members of the two selected families and swapping part of each family member with that of the corresponding member in the other family.

Swapping family members is done through the following steps:

1) Randomly choose a number n between 0 and the size of the families (exclusive) as the number of family members to be swapped.

2) Swap the first n members of the two families.

For example, assuming the following two three-member families are selected for a crossover operation:

$family_1$: { 00011#/0,
00111#/1,
11####/1 }

$family_2$: { 101111/0,
010000/0,
010100/1 },

and 2 is chosen as the number of members to be swapped, meaning the first two members in the two families are going to be swapped. A crossover between these two families by swapping family members would produce the following two families:

$family_3$: { 101111/0,

010000/0,
11####/1 },

family$_4$: { 00011#/0,
00111#/1,
010100/1}.

Swapping part of each family member between two families are done through the following steps:

1) Randomly select a position between two characters of a classifier. Therefore the classifier is split into two segments.

2) Swap the two segments with the corresponding classifier in the other family.

3) Repeat Steps 1 and 2 on every member of the families.

The following is an example of this kind of crossover. Consider *family*$_1$ and *family*$_2$. If the position between the third and fourth left characters is chosen, a crossover of this kind would produce

family$_5$: { 10111#/0,
01011#/1,
010###/1 },

family$_6$: { 000111/0,
001000/0,
11#100/1 }.

Considering a family as a matrix of classifiers, the above two kinds of crossover can be viewed as swapping rows and swapping columns of the matrices, respectively.

In *HCS*, classifiers are bound into families. Families, instead of classifiers, are the main objects of genetic operations. Within a family, classifiers are related to each other by a relation r_i and also by the common fate imposed on them. Naturally, the competitions among these classifiers are restricted, and the competitions inside a family are different from those among the families. These characteristics of *HCS* make it possible to reduce the effect of the rule clustering and rule association problems.

2.2. REDUCING THE RULE CLUSTERING PROBLEM

The competitions among useful patterns or schemata, and the disruption of useful patterns by genetic operators are the main causes of the clustering problem. For example, assume classifiers "###1/0" and "0001/1" are members of the solution set. In conventional classifier systems, crossover between these two classifiers may produce children "##01/0" and "00#1/1" which are not compatible with the solution set. However, in *HCS*, by imposing a relation r_i over the two classifiers, they may stay in the same family with high probability. If so, genetic operations and competitions between the two classifiers would be restricted, reducing the likelihood of producing incompatible classifiers.

Premature convergence is another reason for the clustering problem. If a solution classifier c is included in the initial population and other useful classifiers appear much later, c would become a super-classifier with a much higher strength in the population. Consequently, the population tends to converge to this classifier prematurely. The probability that at least a correct family in *HCS* is contained in an initial population is much smaller than the probability that at least one solution classifier is included in the initial population in a conventional classifier system. Hence, the chance that one family dominates the population is smaller.

2.3. REDUCING THE RULE ASSOCIATION PROBLEM

In *HCS*, by selecting proper relations r_i, classifiers can be grouped into families that contain reasoning chains or default hierarchies. Thus these associated classifiers are bound together explicitly. Therefore, their relationships can be considered locally inside the family. For example, consider a classifier system with a bidding scheme that favors the exception classifiers. With a default hierarchy of "###1/0" and "0001/1", the system would also favor classifiers such as "0011/0" and "0101/0", which are covered by the default classifier "###1/0". In *HCS*, if the bidding scheme favors the exception classifier only locally in the default hierarchy family {"###1/0", "0001/1"}, the *starving general classifiers* phenomenon would not occur as a result of giving higher priority to the exception classifiers.

3. EXPERIMENTS WITH HCS

3.1. IMPLEMENTATION OF HCS

HCS has been implemented as follows:

- $E = \{e_i\}$, where e_i $(i = 1, 2, ..., 2^l)$ is the binary string representation of a number between 0 and $2^l - 1$, l being the length of the strings.

- $F = \{f_i\}$, where f_i $(i = 1, 2, ..., m)$ is a family that contains *fs* (family size) classifiers.

- Category $G = \{0, 1\}$.

The relation over the classifiers in a family is not specified. Classifiers are grouped into families randomly. The only genetic operation used was crossover. The basic credit assignment algorithm is used, i.e. when a classifier responded correctly, some credit is given to the classifier. When a classifier responded incorrectly, some credit is taken from the classifier. Two levels of credit measurements are used; the utility of a classifier and the utility of a family. The utility of a family is the sum of the utilities of all the classifiers contained in the family. When any of the classifiers in a family gains, the family gains. Whether a matched classifier would be fired is determined by its bid, which is calculated by the following formula:

$$bid = \frac{k \times u_c \times u_f \times sp}{2^{l-sp}},$$

where k is a constant, u_c is the utility of the classifier, u_f is the utility of the family containing the classifier and sp (specificity) is the number of 'non-#'s in the condition part of the classifier.

The crossover operations are usually conducted at the family level, with a small possibility of operating at the classifier level (0.001 was used in our experiments). The so-called *roulette wheel selection* [5] was used to select candidates for crossover operations. Families (or classifiers in the case of crossing over at the classifier level) with higher utilities have a greater chance of being selected. In each generation, families with the lowest utilities are selected as candidates for deletion.

3.2. RESULTS AND ANALYSIS

HCS was tested by having the system learning Boolean functions. Many functions were used in our experiments to check the effectiveness of *HCS* in achieving better and more stable performances. The following two functions are used in this paper for illustrative purposes:

$f_1(x_0, x_1, x_2, x_3, x_4, x_5) =$
$(x_0$ *AND* x_1 *AND* $x_2)$ *OR* $(x_3$ *AND* x_4 *AND* $x_5)$ *OR*
$(x_1$ *AND* x_2 *AND* $x_4)$ *OR* $(x_0$ *AND* x_3 *AND* $x_5)$ *OR*
$(x_0$ *AND* x_2 *AND* $x_4)$ *OR* $(x_0$ *AND* x_1 *AND* $x_4)$ *OR*
$(x_1$ *AND* x_2 *AND* $x_3)$ *OR* $(x_2$ *AND* x_4 *AND* $x_5)$ *OR*
$(x_2$ *AND* x_3 *AND* $x_5)$ *OR* $(x_2$ *AND* x_3 *AND* $x_4)$ *OR*
$(x_0$ *AND* x_4 *AND* $x_5)$ *OR* $(x_0$ *AND* x_3 *AND* $x_4)$ *OR*
$(x_0$ *AND* x_1 *AND* $x_5)$.

$f_2(x_0, x_1, x_2, x_3, x_4) =$
$(x_0$ *AND* $x_1)$ *OR* $(x_2$ *AND* x_3 *AND* $x_4)$.

f_3 is a minimal Boolean function of 8 variables and is the sum of 58 products (too detailed to show here).

The performance of *HCS* has been examined with different family and population sizes. The family size refers to the number of classifiers contained in a family. The population size here refers to the number of families multiplied by the family size. Figure 1 and 2 show the results of learning function f_1 and f_2 with family sizes varying from 1 to 4. Each performance line shows the average of five runs generated with different random number seeds. The horizontal axis measures the number of generations elapsed. The vertical scale in the figures represents the performance, which is defined as

$$P = \sum_{t=1}^{T} (C_t - I_t) / T$$

where C_t is the number of the correct responses at generation t, I_t is the number of the incorrect responses at generation t and T is the current generation.

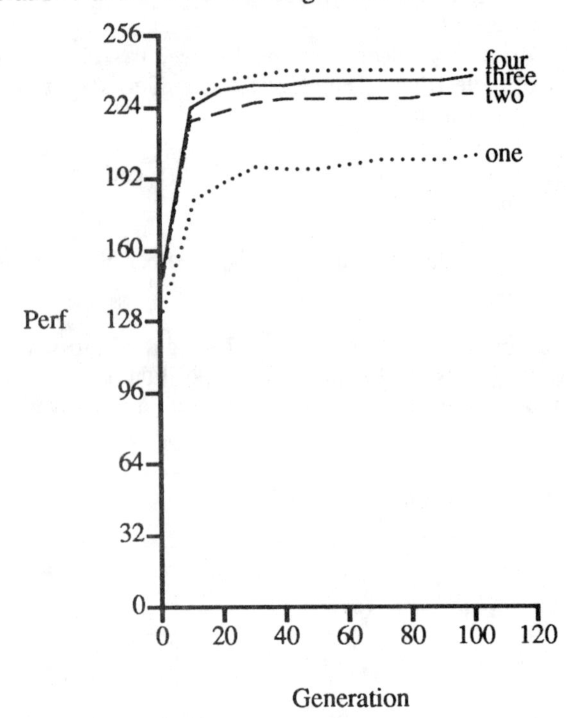

Figure 1: f_2 Performance, ps = 240.

These two figures show consistent improvements with *HCS* ($fs = 2, 3, 4$) compared with traditional classifier systems ($fs = 1$). These cases are representative of most of our experimental runs. In Figure 1, the performances of family sizes 2, 3 and 4 are not significantly different. The major benefits are obtained using families of size 2, in part because the population sizes are too big relative to the complexity of the problem being solved. This point is addressed later in this section.

In Figure 2, the performance for all family sizes is significantly less than that shown in Figures 1. For this problem, a population of 720 is too small given the complexity of the solution set. Absolute performance is not important, as the size of the solution set (complexity of the problem being solved) and the population size are important factors in determining this. Relative performance, however, clearly demonstrates the significant advantages to be gained by using families.

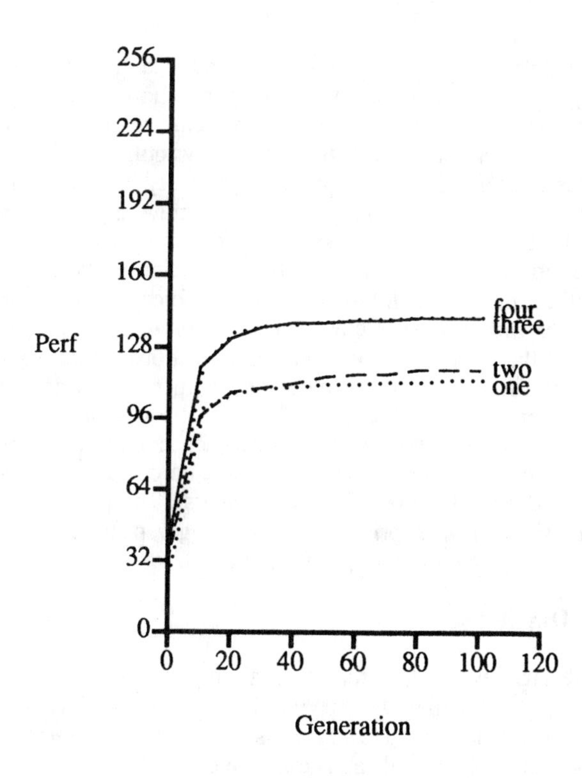

Figure 2: f_3 Performance, ps = 720.

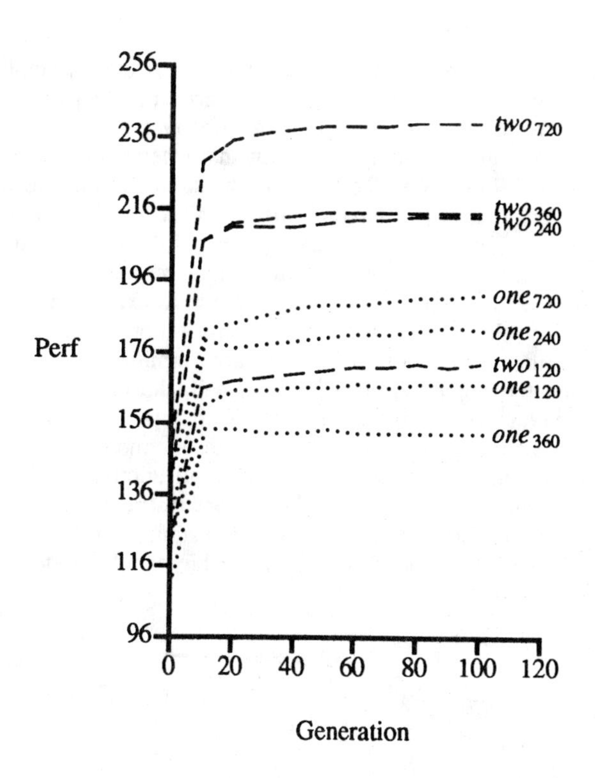

Figure 3: f_1 Performance.

The experiments show that within a range (determined experimentally per problem), the larger the population size, the better the performance of *HCS*. Classifier systems ($fs = 1$) generally show the same trend as well, but appear to be not as stable as *HCS*. Figures 3 and 4 illustrate the performance of families as a function of population size ($ps = 120, 240, 360, 720, 1440$). An example of classifier system instability is shown in Figure 3 where classifier system performance does not necessarily increase as the population size grows. For example, performance for $ps = 360$ is *less* than that for populations of size 120 and 240.

An interesting point to note is that in Figures 3 and 4, often a smaller population of families can out-perform a classifier system with more classifiers. For example, in Figure 3, a population of 120 families ($ps = 240$) *out-performs* classifier systems with up to 720 classifiers. This example is not an isolated experience.

Larger populations help reduce the premature convergence problem in the early generations. Classifier rules representing different sub-solutions can be established in parallel, accounting for the better performance of larger populations. However, for classifier systems, in the later generations these co-existing sub-solutions may disrupt each other. In *HCS*, if a (sub-)solution is contained in a family, then family ties makes the probability of crossover within that (sub-)solution small.

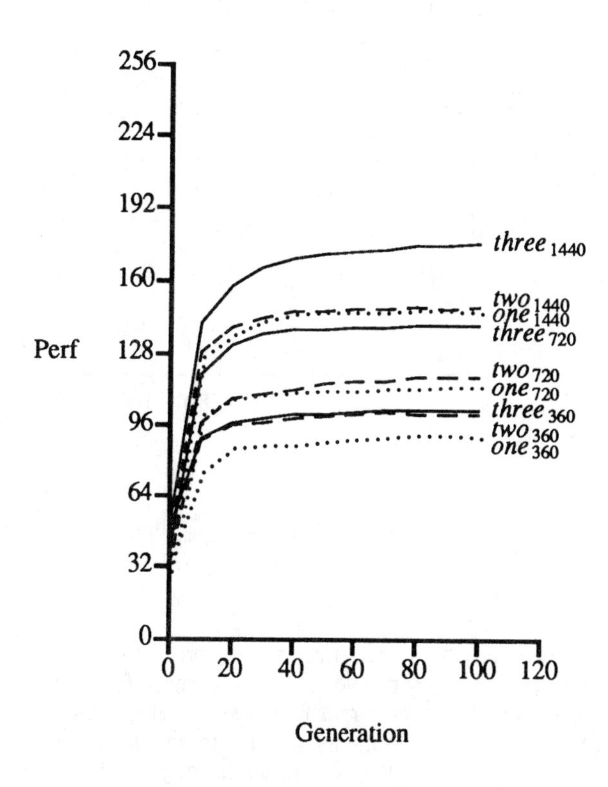

Figure 4: f_3 Performance.

Consider two populations, one of which is relatively small and the other relatively large in comparison to the population required to solve the problem effectively. Comparisons between Figures 1 and 2 show that when a population is relatively small to the problem, the differences between the performances of larger family sizes (*fs* = 4, for example) and smaller family sizes (*fs* = 2, for example) are more significant. There are two major reasons for this. First, when the population size is larger, as explained previously, there is more room for sub-solutions to evolve, negating the benefits of having large families. Second, when the population size is too small, the problems of premature convergence and genetic disruption are more severe. In this case, larger family sizes are more effective in reducing these problems. The implication of this is that, when the sizes of populations are restricted in practice, larger family sizes can be used to achieve higher performance for problems that require a big population to be solved effectively.

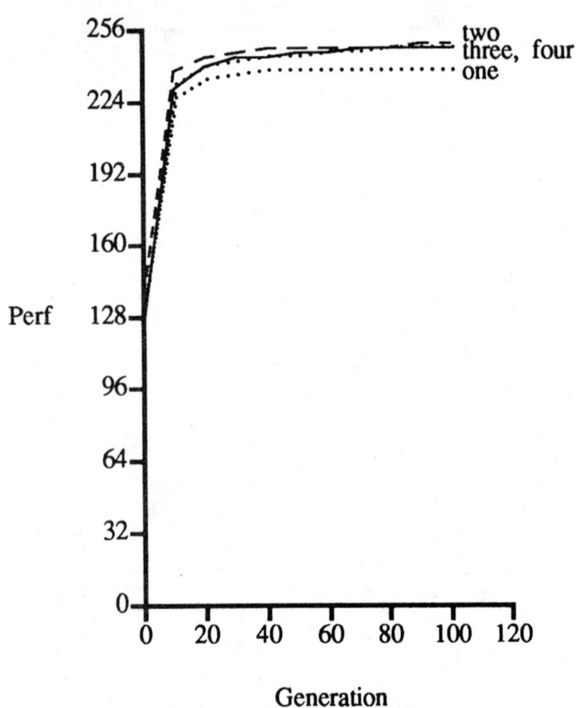

Figure 5: f_2 Performance, ps = 720.

The necessity for large family sizes is problem dependent. Generally, bigger problems (in this case, functions of more variables) need larger family sizes. However, larger families may not bring any benefits to smaller problems. Figure 5, in which the performances for *fs* = 2, 3 and 4 are roughly identical, shows this. In this case, larger families would increase the search space dramatically and therefore probably slow down the learning process. Our experiments also shown that if the family size is too large relative to the population size, the number of resources available in the population is limited and performance degrades. Another reason for the degradation is related to the parasite problem. This problem refers to the phenomenon that a classifier with a low utility is able to offer a high bid because of the high utility of the family that the classifier belongs to. This may happen in a multiple-member family system for two reasons. First, the bid of a classifier is in direct proportion to the utility of the family it resides in. Second, the objects for deletion are families instead of classifiers. Therefore, if the majority of the family members have high utilities, some weak members would survive over the competitions with the members of other families. This problem is more severe when the family size is large since each family may have more than one parasite and each parasite can take advantage of more than one family members. The problem of choosing an appropriate family size, given an arbitrary problem, is an interesting future research topic.

4. DISCUSSION

The HCS was implemented in a simple way. The relations r_i are actually not specified. Classifiers are grouped together randomly. Nevertheless some default hierarchies are still established and maintained well. The main benefit of the structuring lies in the ability to maintain good structures and classifier sets. Once a correct default hierarchy is built in a family, the chances that the hierarchy is well maintained in *HCS* is much higher than in conventional classifier systems. However, without specifying the relations of the family members r_i, the system may take a long time to overcome some wrong groupings of family members. For example, classifiers "1##0/1" and "1110/0" form a default hierarchy. But they may be split into two families. If the other members of the families have high utilities, these two classifiers may become parasites and disrupt performance. Specifying a relation r_i as a default hierarchy relation can help solve this problem because then default hierarchies are likely grouped into the same family. Further studies are needed on this issue.

Allowing varied family sizes in a population is another interesting issue. The coexistence of different sized families would be amenable to building families according to relations r_i and help in reducing the parasite problem.

There are only two levels (classifiers and families) in *HCS*. For problems with large search and solution spaces, *HCS* can be generalized into a hierarchy with more than two levels. The main motivation for building hierarchies is to gain better organizations. Too many hierarchies could put an extra burden on a system.

5. TOWARDS A UNIFICATION OF THE MICHIGAN AND PITTSBURGH METHODS

The previous discussions are based on the comparisons between *HCS* and the Michigan type classifier systems

[4], where single classifiers are the basic units for genetic operations. Another type of genetic-algorithm-based machine learning system is the Pittsburgh type systems [4]. In these systems, the basic units for genetic operations are production system programs. A fitness measure called strength is associated with each program and used to guide the selections for genetic operations. The strength of a program is changed by evaluating the performance of the program on a learning task. The main disadvantage of the Pittsburgh method is its computational complexity in both time and space. The distinct advantage of the Pittsburgh method over the Michigan method is its ability to maintain rule structures and emphasis on the cooperations among classifiers. *HCS* inherits this property of the Pittsburgh method while reducing the computational complexity dramatically by organizing classifiers into small structures instead of complete programs.

There are several important differences between *HCS* and the Pittsburgh method. In *HCS*, a family consists of a set of classifiers constrained by a pre-defined relation. Usually, those classifiers do not constitute a complete program. The idea here is to organize classifiers. This changes the traditional one-level classifier system architecture (classifiers) to multiple-level (individuals and families) architecture. Also, in *HCS*, the credit assignment process is based on the evaluations on individual classifiers. Families gain credit through the gain of family members. This allows for two levels of competition among classifiers. One is the competitions among classifiers in different families. The other is the competitions among classifiers in same families. This illustrates one of the important differences between *HCS* and the Pittsburgh method: *HCS* inherits the implicit parallelism of the Michigan method in the sense that each evaluation of a classifier affects many classifiers and their families.

6. CONCLUSIONS

A method for achieving coherence and cooperation among a classifier population is proposed. A hierarchical approach is used to impose structural ties in classifier systems, which has the benefit of reducing the difficulties of rule clustering and rule association. Using the hierarchical approach, classifier individuals are grouped into "families". The structural family ties encourage cooperation among family members, effectively enhancing the stability of the system. Experiments with having a system learn Boolean functions have shown significant improvement in performance and stability of hierarchical classifier systems over conventional classifier systems. *HCS* is a model towards a unification of the Michigan and Pittsburgh approaches.

Acknowledgments

The financial support of the Canadian Natural Sciences and Engineering Research Council is appreciated.

References

1. L. B. Booker, *Intelligent Behavior as an Adaptation in the Environment*, Ph.D. Thesis, Technical Report, University of Michigan, 1982.

2. L. B. Booker, Triggered Rule Discovery in Classifier System, *Third International Conference on Genetic Algorithms*, 1989, 265-274.

3. K. A. DeJong, *An Analysis of the Behavior of a Class of Genetic Adaptive Systems*, Ph.D. Thesis, University of Michigan, 1975.

4. K. A. DeJong, Learning With Genetic Algorithms: An Overview, *Machine Learning 3*, (1988), 121-138, Kluwer Academic Publishers.

5. D. E. Goldberg, *Genetic Algorithms in Search, Optimization & Machine Learning*, Addison-Wesley Publishing Company, 1989.

6. J. H. Holland, K. J. Holyoak, R. E. Nisbett and P. R. Thagard, *Induction: Processes of Inference, Learning, and Discovery*, MIT Press, Cambridge, 1986.

7. J. H. Holland, Escaping Brittleness: The Possibilities of General Purpose Learning Algorithms Applied to Parallel Rule-Based Systems, in *Machine Learning II*, R. S. Michalski, J. G. Carbonell and T. M. Mitchell (ed.), 1986.

8. R. L. Riolo, The Emergence of Default Hierarchies in Learning Classifier Systems, *Proceedings of the Third International Conference on Genetic Algorithms*, 1989, 322 - 327.

9. R. E. Smith and D. E. Goldberg, Reinforcement Learning With Classifier Systems, *AI, Simulation and Planning in High Autonomy Systems*, Los Alamitos, 1990, 184-192.

10. S. W. Wilson, Knowledge Growth in an Artificial Animal, in *Adaptive and Learning Systems: Theory and Applications*, K. S. Narendra (ed.), Plenum, New York, 1986, 255-264.

11. S. W. Wilson, Classifier Systems and the Animat Problem, *Machine Learning 2*, (1987), 199-228.

12. H. H. Zhou and J. J. Grefenstette, Learning by Analogy in Genetic Classifier Systems, *Proceedings of the Third International Conference on Genetic Algorithms*, 1989, 291-298.

The Fuzzy Classifier System: A Classifier System for Continuously Varying Variables

Manuel Valenzuela-Rendón
mvalenzu@mtecv2.mty.itesm.mx
Centro de Inteligencia Artificial
Instituto Tecnológico y de Estudios Superiores de Monterrey
Sucursal de Correos "J," C.P. 64849 Monterrey, N.L., Mexico

Abstract

This paper presents the fuzzy classifier system which merges the ideas behind classifier systems and fuzzy controllers. The fuzzy classifier system learns by creating fuzzy rules which relate the values of the input variables to internal or output variables. It has credit assignment mechanisms which reassemble those of common classifier systems, but with a fuzzy nature. The fuzzy classifier system employs a genetic algorithm to evolve adequate fuzzy rules. Preliminary results show that the fuzzy classifier system can effectively create fuzzy rules that imitate the behavior of simple static systems.

1 INTRODUCTION

In spite of the potential of classifier systems as a learning paradigm, they have found little application in the adaptive control of processes. It can be said that this is due in part to the limitations of the classifier syntax when representing continuously varying variables. Fuzzy controllers are a controller design approach which is not based on a mathematical description of the process being controlled. A fuzzy controller models the knowledge used by human operators as a set of rules in which variables take linguistic values [13].

In a fuzzy controller, relations between inputs and outputs are expressed as fuzzy rules. The fuzzy controller implements a fuzzy relation between all the possible values of the inputs and the indicated values of the outputs. This concept has been successfully applied in many occasions: control of a cement kiln [5], traffic control [9], robot arm control [11], and temperature control of a heated air-stream [6], are only a few examples. In a typical fuzzy controller [7], fuzzy rules are derived either from manuals of operation or from human operators that have successfully controlled the system and which have acquired their rules through experience. Algorithms that modify the fuzzy relation implemented by the fuzzy controller have been proposed [8, 10]. The fuzzy classifier system (FCS) is motivated by the ideas of fuzzy controllers, but departs from previous implementations in the manner in which it creates new rules and adjusts the contribution of the existing rules to the system outputs.

The FCS merges the credit assignment mechanisms of common classifier systems [4, 2] and the use of fuzzy logic of fuzzy controllers. It represents its fuzzy rules as binary strings on which a genetic algorithm operates, therefore allowing for the evolution of adapted rule sets. The FCS allows inputs, outputs, and internal variables to take continuous values over given ranges, thus it could be applied for the identification and control of dynamic systems. In the rest of this paper, the FCS will be described. Preliminary results will be presented that show that the FCS can successfully identify simple static systems.

2 FUZZY LOGIC

Similarly to fuzzy controllers, the FCS represents all its knowledge by means of fuzzy rules. To introduce the concept of fuzzy rules, let us first review fuzzy sets and their operations.

2.1 FUZZY SETS AND OPERATIONS

Fuzzy set theory [1, 13] can be defined as a generalization of common set theory in which the membership of an element to a set is defined by a membership function. In this way, an element can partially belong to a set. In fuzzy set parlance, common sets are also called *crisp*.

Set operations can be defined over fuzzy sets analogously to crisp set operations. The generalization of these crisp set operations to fuzzy sets can be performed in different ways. The following definitions are one of such possible manners, they constitute a consistent framework, and they are the most commonly used.

Let A and B be fuzzy sets over the variable $x \in [x_0, x_f]$ where $x_0, x_f \in \Re$. Let A and B be defined by the membership functions $\mu_A(x) \in [0,1]$ and $\mu_B(x) \in [0,1]$, respectively.

Definition 1 *The union of A and B, denoted by $A \cup B$, is defined by the following membership function:*

$$\mu_{A \cup B}(x) = \max(\mu_A(x), \mu_B(x)).$$

Definition 2 *The intersection of A and B, denoted by $A \cap B$, is defined by the following membership function:*

$$\mu_{A \cap B}(x) = \min(\mu_A(x), \mu_B(x)).$$

Definition 3 *The complement of A, denoted by A', is defined by the following membership function:*

$$\mu_{A'}(x) = 1 - \mu_A(x).$$

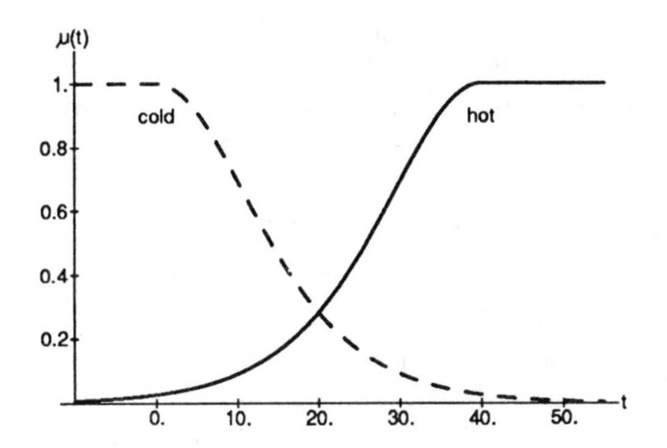

Figure 1: Membership functions of sets *cold* and *hot*.

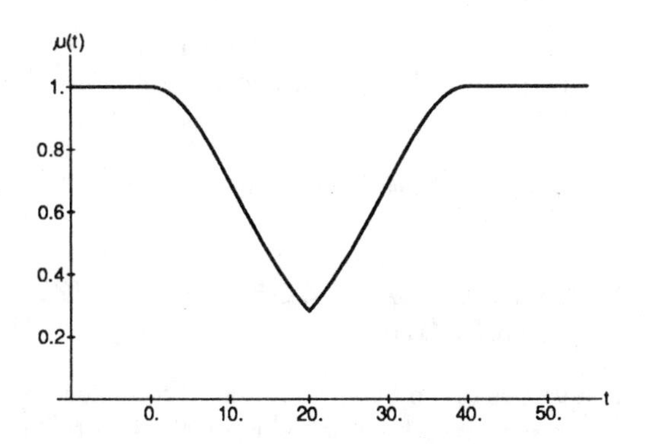

Figure 2: Fuzzy union of sets *cold* and *hot*.

For example, consider two fuzzy sets, *cold* and *hot*, over the variable t (temperature in Celsius degrees), and defined by the membership functions shown in Figure 1. The membership functions of the union and the intersection of these sets are shown in Figures 2 and 3.

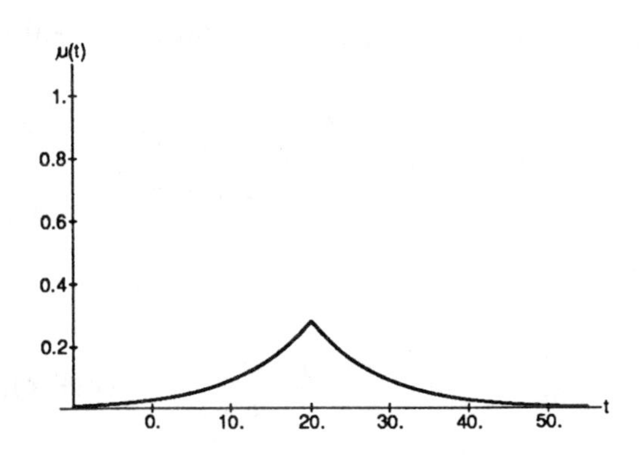

Figure 3: Fuzzy intersection of sets *cold* and *hot*.

2.2 FUZZY RULES AND RELATIONS

A fuzzy rule is an *if-then* expression in which conditions and action are fuzzy sets over given variables. Fuzzy rules are also called linguistic rules, because they represent the way in which people usually formulate their knowledge about a given process. The following are examples of fuzzy rules:

if the temperature is high,
then slightly reduce the gas intake;

if the pavement is very wet,
then moderately reduce your driving speed.

A fuzzy relation maps one or more independent variables into a dependent variable. A set of fuzzy rules, in which the antecedents refer to the same independent variables and the consequences refer to a same dependent variable, form a fuzzy relation. In other words, a way of expressing a fuzzy relation is by a set of fuzzy rules. Expressing a fuzzy relation as a set of fuzzy rules is computationally convenient because in general it requires less memory storage than expressing the relation in tabular form.

2.3 FUZZY RULES IN THE FCS

The FCS operates over variables that can be inputs, outputs, or internal. The FCS relates the values of the inputs and the internal variables and generates outputs according to fuzzy rules, similarly to a fuzzy controller. These rules or classifiers are represented as binary strings that encode the membership functions of the fuzzy sets involved in the fuzzy rule. To allow for a uniform procedure, the FCS linearly maps all variables to the range [0,1]. For each variable, n component fuzzy sets are defined so that their membership functions span the interval [0,1]. The number of these component sets is defined by the user according to the precision required. The peaks of the membership functions of the component sets that span a variable are equally spaced. The following expression defines the membership function of the i-th fuzzy set for the variable $x \in [x_0, x_f]$:

$$\lambda(x, h_i) = \frac{4e^{-(x-h_i)/\sigma}}{\left(1 + e^{-(x-h_i)/\sigma}\right)^2}, \qquad (1)$$

where

$$h_i = (i-1)\frac{(x_f + \delta) - (x_0 - \delta)}{n-1} + (x_0 - \delta),$$

for $i = 1, 2, \ldots, n$. The parameter σ controls the width of the component sets. The user must choose a value of σ according to the value of n so that the interval is adequately covered. The parameter δ expands the effect of the component sets outside the range of x thus increasing the precision for values near x_0 and x_f. The user must choose a value of δ according to the sensitivity of his application to values of x close to x_0 or x_f.

All the membership function used by the FCS are generated from Equation 1 with different values of h and σ. Conditions and recommended actions are binary coded so that the number of bits in a condition or an action is the number of fuzzy sets defined over the given variable. A "1" indicates that the corresponding fuzzy set is part of the condition or action. Additionally, a non-fuzzy binary tag is attached. This tag indicates to which variable the condition or action is referring to. Tags are followed by a colon. Consider for example the classifier 0:110/1:001 in which three component sets are defined over variables x_0 and x_1. The membership functions of condition 0:110 and action 1:001 are shown in Figures 4 and 5. The rule represented by these sets can be expressed in words as "*if* x_0 is low or medium *then* x_1 should be high." Notice that the FCS syntax does not include the wildcard character #.

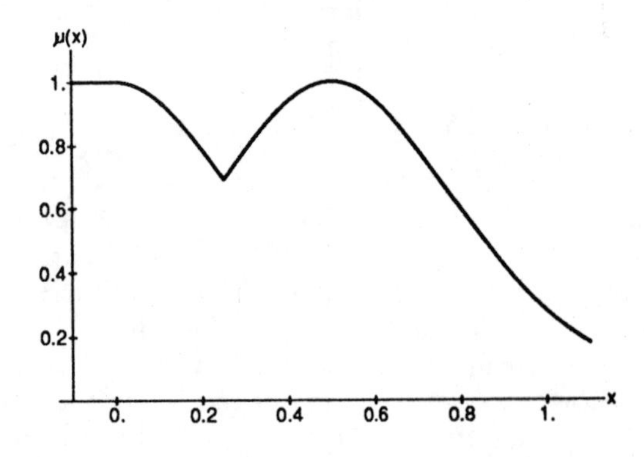

Figure 4: Membership function of condition 0:110.

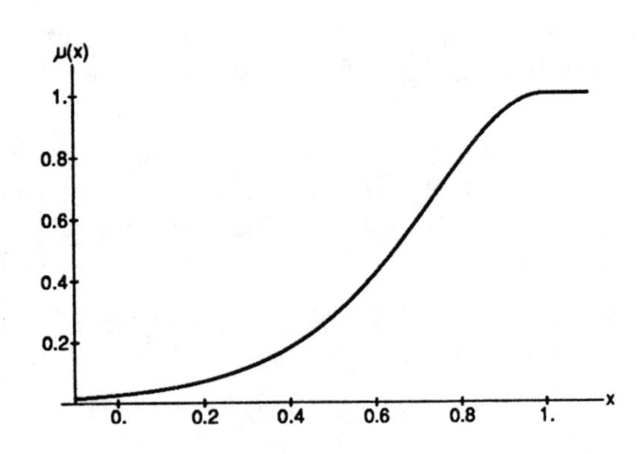

Figure 5: Membership function of action 1:001.

3 OPERATION OF THE FCS

The operation of the FCS is similar to that of a common classifier system. Figure 6 presents a block diagram of an FCS. The basic cycle of a FCS is as follows:

1. The input unit receives input values, encodes them into fuzzy messages, and adds these messages to the message list.
2. The classifier list is scanned to find all classifiers whose conditions are satisfied by the messages in the message list.
3. The message list is erased.
4. All matched classifiers are fired, and the messages produced are stored in the message list.
5. The output unit detects the output messages, and erases these messages from the message list.
6. In the output unit, output messages are decomposed into minimal messages.
7. Minimal messages are *defuzzified* and transformed into output values.
8. Payoff from the environment and classifiers is transmitted through the messages to the classifiers.

The following subsections explain the operation details of the FCS.

3.1 FUZZY MESSAGES AND FUZZY MATCHING

The values taken by variables are broadcast to the classifiers as *messages*. Each classifier will verify if its conditions are matched, i.e. satisfied, by the messages, and if so, will post a new message according to its indicated action.

There are two alternative and equivalent ways in which the process of a fuzzy rule being matched can be viewed. First, a fuzzy rule can be thought of as receiving real valued variables, and performing a fuzzy

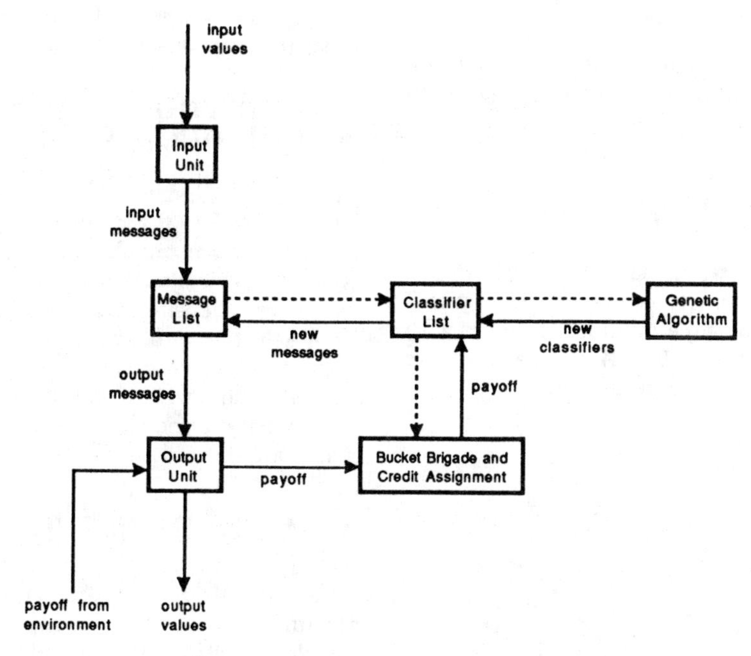

Figure 6: Block diagram of the FCS.

match between the values of these variables and the fuzzy sets defined in its condition. A second view is possible, one more convenient when handling sets of rules. A fuzzy rule can be thought of as receiving fuzzy values, and performing a perfect match between these fuzzy values and its condition.

Suppose for example two classifiers, 0:010/1:100 and 0:001/1:001, each with a condition over the input variable x_0 and an action over the output variable x_1; also suppose that x_0 takes the value of 0.3. It is equivalent to say that classifier 0:010/1:100 is matched with a degree of $\lambda(0.3, h_2)$ and classifier 0:001/1:001 is matched with a degree of $\lambda(0.3, h_3)$, or to say that the messages 0:010 and 0:001 are present with *activity levels* of $\lambda(0.3, h_2)$ and $\lambda(0.3, h_3)$ respectively, and that classifiers 0:010/1:100 and 0:001/1:001 are perfectly matched by these messages.

The FCS implements the second approach by having an input unit that fuzzifies inputs into fuzzy messages. Each message has an associated activity level which measures the degree of belonging of the input variable to the fuzzy set defined by the membership function represented by the message. Fuzzy classifiers match these messages and generate new messages with activity levels that correspond to the degree in which the conditions of the classifier are satisfied. This *fuzzification* of inputs is accomplished by creating *minimal messages* one for each fuzzy set defined over the variable. A minimal message has a single "1."

The matching of a condition in a classifier by a message is done in two steps. First, the tags of the message and condition are compared; if they are the same they refer to the same variable, if so, the rest of the message and condition are compared. Second, if at least there is one position in which the condition has a "1" and the message also has a "1," then the condition is satisfied.[1] The satisfaction level of a condition is equal to the maximum activity level of the messages that match this condition; this implements a fuzzy union. The activity level of a classifier is equal to the minimum of the satisfaction levels of all its conditions; therefore, implementing a fuzzy intersection. When a classifier fires, it generates a new message with an activity level proportional to the classifier's own activity level.

3.2 DEFUZZIFICATION IN THE OUTPUT UNIT

The FCS must translate messages referring to output variables into real values. This process of *defuzzification* is accomplished in the *output unit* by decomposing each output message into its corresponding minimal messages. The activity levels of minimal messages corresponding to the same variable and component set are added up and the messages are substituted by a single message. For each output variable, a fuzzy union is performed over the component sets represented by the minimal messages multiplied by their activity levels. Then, the gravity center of the union is taken. This

[1] This can be explained by recalling that a condition is a fuzzy union of the fuzzy sets that correspond to the bits that are "1." A condition as 0:101 in words would be "variable 0 takes the value low or high." This is satisfied by any of the messages 0:100, 0:001, 0:101, 0:110, 0:011, or 0:111.

gravity center is the output value.

The total gravity center can be obtained by first calculating the gravity center of the contribution of each minimal message. Then the total gravity center G is calculated as:

$$G = \frac{\sum A_i g_i}{\sum A_i},$$

where A_i is the area contribution of the i-th minimal message, and g_i is the gravity center of A_i. Figure 7 shows the intersection of three minimal messages with activity levels of 1.0, 0.6, and 0.8. A_1, A_2, and A_3 are the area contributions of these messages.

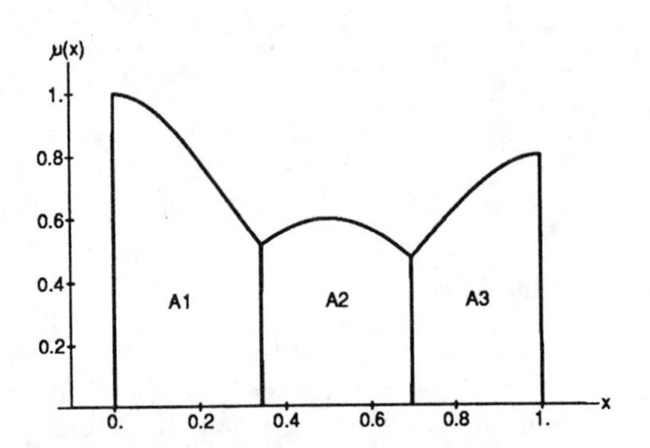

Figure 7: Intersection of three minimal messages and their area contributions.

3.3 CREDIT ASSIGNMENT AND FUZZY BUCKET BRIGADE

After the output unit has produced an output value, the environment judges this output, and accordingly, gives payoff to the output unit. The output unit distributes this payoff through the minimal messages to the classifiers that produced the output messages. The payoff is distributed in such a manner that classifiers that contributed more to the output taking a specific value receive a larger share of the payoff. In this way, classifiers that are directly involved in producing correct outputs receive increases in their strength.

Payoff to other classifiers that do not produce outputs, but post messages that allow for others to fire and receive payoff, is distributed following the *bucket brigade algorithm* of common classifier systems. According to the bucket brigade algorithm, matched classifiers *bid* a small portion of their strength for the right to fire. Firing classifiers pay their bids to those that posted the messages that allowed them to fire. In this way, a competitive economic system is established in which classifiers that produce or help produce good outputs have their strengths increased, and other have their strengths decreased. A basis against classifiers that do not participate in the competition is achieved through a living tax by which all classifiers have a small portion of their strength deducted every time step.

3.4 CREATION OF NEW RULES THROUGH A GENETIC ALGORITHM

In a FCS, classifiers are selected by a genetic algorithm (GA) [2, 3] for reproduction according to their strength: stronger classifiers are selected more than weaker ones. Pairs of classifiers for mating are obtained by choosing classifiers randomly according to their selection probabilities. Each couple reproduces, creating new rules by *crossover*. After crossover, mutation occurs with a specified small probability. To keep constant the size of the classifier list, weak classifiers are deleted.

4 A LEARNING TASK

As a first test for the FCS, a simplified task was chosen: the imitation of static one-input one-output systems in which the output depends only on the present input and not on past states of the system.

4.1 IDENTIFICATION WITH A STIMULUS-RESPONSE FCS

For this task we only require a stimulus-response FCS. A stimulus-response FCS is one in which classifiers post only output messages, and classifiers only respond to input messages. In this way, the bucket brigade does not need to operate. The FCS and the system are setup as shown in Figure 8. The FCS and the system to be identified receive the same input generated randomly among a range of possible inputs by an *input generator*. The output of the FCS is compared to that of the system, and a payoff is assigned to the FCS by a *performance evaluator*. The FCS receives higher payoff when it best imitates the behavior of the system. In this way, the FCS constructs a fuzzy model of the system.

The performance evaluator assigns payoff to the FCS output according to the following equation:

$$P = P_0 |u - y|,$$

where P_0 is a constant, u is the FCS output, and y is the system output. The output unit distributes this payoff to the minimal output messages. Each minimal message receives a portion according to its contribution to the output.

The i-th minimal message receives a payoff given by the following expression:

$$p_i = \phi(-\text{sgn}(\varepsilon_o)\varepsilon_i) \cdot P,$$

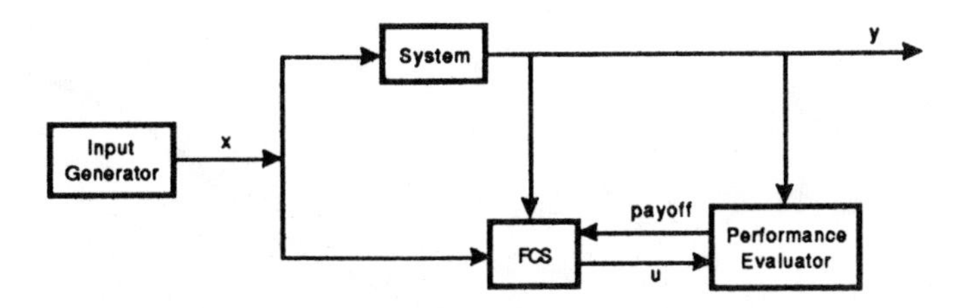

Figure 8: FCS in the identification setup.

where

$$\phi(x) = \begin{cases} -\eta_1 & \text{if } x \leq -\eta_2; \\ 1 & \text{if } x \geq \eta_2; \\ (1+\eta_1)(x+\eta_2)^2/4\eta_2^2 - \eta_1 & \text{otherwise.} \end{cases}$$

The output error ε_o and message error ε_i are calculated as follows:

$$\begin{aligned} \varepsilon_o &= u - y; \\ \varepsilon_i &= h_i - y. \end{aligned}$$

To understand these expressions, note that the product $-\text{sgn}(\varepsilon_o)\varepsilon_i$ is positive if, and only if, ε_0 and ε_i have opposite signs. Furthermore, note that this implies that one of the following is true: $u < y < h_i$ or $h_i < y < u$. Therefore, an adequate increase in the i-th minimal message activity level could reduce the output error to zero. Function $\phi(x)$ gives a large positive payoff to messages satisfying the above conditions, and gives a lower payoff to all others. The constants η_1 and η_2 must be chosen according to the problem.

This payoff distribution scheme assures that minimal messages that would contribute to minimize the error will receive a higher payoff.[2]

4.2 PAYOFF DISTRIBUTION IN THE STIMULUS-RESPONSE FCS

In the stimulus-response FCS, the specificity of the classifiers is involved in the payoff distribution scheme. A specificity ρ is defined only for the action of a classifier in the following manner:

$$\rho = \left(\frac{1 + \text{number of 0's}}{\text{action length}}\right)^2.$$

A message has a specificity equal to that of the classifier that posted it.

Each minimal message distributes its payoff to its corresponding messages according to their activity level and to their specificity. The payoff to a message m is the sum of all the payoffs it receives from its minimal messages:

$$\text{payoff to message } m = \sum_{i \in \mathcal{M}} \frac{p_i\, \rho_m\, (\text{activity level}_m)}{\text{activity level}_i},$$

where $\mathcal{M}$ is the set of minimal messages posted by message m. The activity level of a message is equal to the sum of the activity level multiplied by the bid of the classifiers who posted it.

Firing classifiers pay a bid B equal to a small fraction k of their strength S. Each classifier receives a payment proportional to its specificity and activity level, the payoff received by the message it posted, and divided by the number of classifiers that posted the same message.

4.3 RESULTS

The stimulus-response FCS just described was trained to imitate a system in which the output is equal to the input, i.e., $y = x$. The value of δ was set to 0 for the input and 0.2 for the output. The parameters $\sigma = 0.11$, $\eta_1 = 0.15$, and $\eta_2 = 1.0$ were obtained by trial and error. The genetic algorithm ran over a population of 40 classifiers.

After learning, the payoff distribution scheme and the genetic algorithm were turned off, and the input range was scanned to obtain the input-output behavior of the FCS. Figure 9 shows the performance obtained from the fuzzy rules found by the FCS using four fuzzy intervals after 64,000 cycles. The FCS approximates the straight line with an absolute error[3] of 1.72%.

[2] This payoff distribution scheme involves information about the correct output, and thus, it is not a blind reinforcement scheme.

[3] The absolute error was calculated as

$$\text{absolute error} = \frac{1}{x_f - x_0}\int_{x_0}^{x_f} |u - y|\, dx.$$

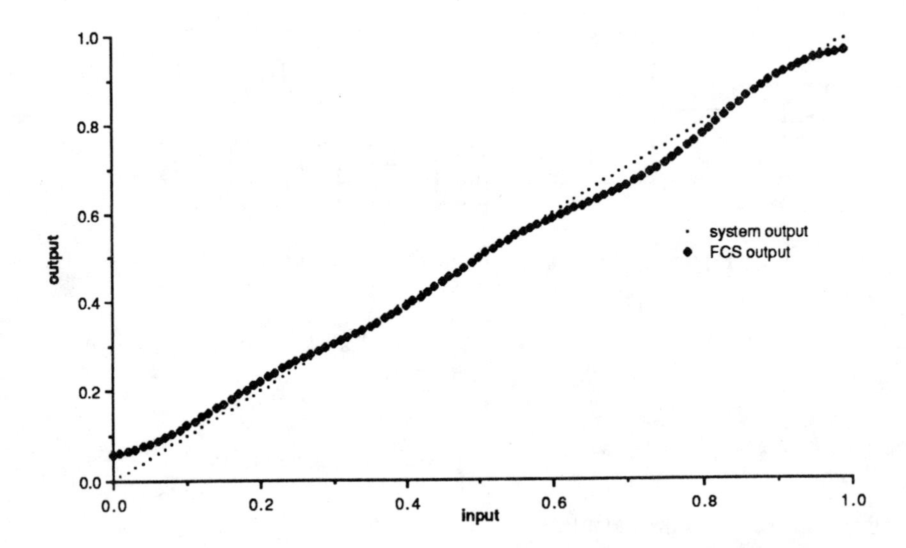

Figure 9: FCS imitation of the straight line $y = x$ using four fuzzy component sets.

When analyzing this result, we must consider the following step-ladder function:

$$\gamma(x) = \begin{cases} 1/2n & \text{if } 0 \leq x < 1/n; \\ 3/2n & \text{if } 1/n \leq x < 2/n; \\ \vdots & \\ (2n-1)/2n & \text{if } (n-1)/n \leq x < 1. \end{cases}$$

A learning system based on crisp intervals would tend to approximate the function $y = x$ with the step-ladder function. For a given number of intervals, the FCS finds a set of fuzzy rules that produces a smaller error than the step-ladder function.

The FCS was also setup to imitate the function $y = 4(x - 0.5)^2$. Figure 10 shows the performance of the FCS after 53,000 cycles. The absolute error was of 3.76%.

The previous results indicate that the FCS can learn to imitate simple, static systems. More experiments, involving other more complex systems, are required.

5 FUTURE WORK

Many issues remain open for research. Among the most important are increasing speed convergence while retaining stability; implementing pure reinforcement learning, so control as opposed to identification can be performed; allowing rule chaining, so that dynamic systems can be identified and controlled; allowing the FCS to adaptively change the membership functions of the component sets, so that greater sensitivity can be achieved where required; and implementing a scheme for continuous time output.

Efforts are currently being made to improve the convergence speed of the FCS while retaining stability. The following enhancements are now under evaluation.

- Mating restrictions
  For each classifier to be crossed, the genetic algorithm selects n classifiers and then chooses the one which is most similar to the classifier to be crossed.

- Replacement restriction
  When a classifier is created, the genetic algorithm selects the m weakest classifiers, then it deletes the one most similar to the new born classifier.

- Aging classifiers
  Classifiers have an age equal to the number of time cycles since their creation. Age limits for reproduction and death are defined. Classifiers under age are not selected for reproduction or deletion by the genetic algorithm.

- Limiting the weights of young classifiers
  Classifiers under a given age have a reduced weight in the decision of an output. The reduction is proportional to the difference between the age of the classifiers and a given age limit. This is possible because all classifiers help produce a single output. Reducing the weight of young classifiers on the output should improve the stability of the system.

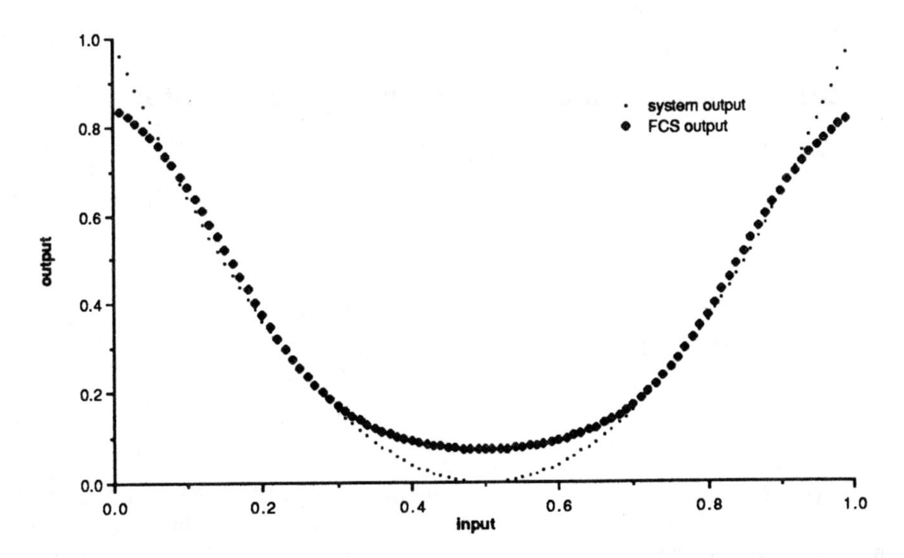

Figure 10: FCS imitation of the parabola $y = 4(x - 0.5)^2$ using four fuzzy component sets.

Acknowledgements

The author gratefully acknowledges the useful comments received from the reviewers.

References

[1] D. Dubois and H. Prade. *Fuzzy Sets and Systems: Theory and Applications.* Academic Press, New York, 1980.

[2] D. E. Goldberg. *Genetic Algorithms in Search, Optimization, and Machine Learning.* Addison-Wesley, Reading, MA, 1989.

[3] J. H. Holland. *Adaptation in Natural and Artificial Systems.* University of Michigan Press, Ann Arbor, MI, 1975.

[4] J. H. Holland, K. J. Holyoak, R. E. Nisbett, and P. R. Thagard. *Induction: Processes of Inference, Learning, and Discovery.* MIT Press, Cambridge, 1986.

[5] J.-J. Holmblad, L. P. Østergaad. Control of a cement kiln by fuzzy logic. In M. M. Gupta and E. Sánchez, editors, *Fuzzy Information and Decision Processes*, pages 389–399. North-Holland, Amsterdam, 1982.

[6] A. Ollero and A. J. García-Cerezo. Direct digital control, auto-tuning and supervision using fuzzy logic. *Fuzzy Sets and Systems*, 30:135–153, 1988.

[7] W. Pedrycz. *Fuzzy Control and Fuzzy Systems.* John Wiley, New York, 1989.

[8] T. J. Procyk and E. H. Mamdani. A linguistic self-organizing process controller. *Automatica*, 15:15–30, 1979.

[9] T. Saski and T. Akiyama. Traffic control process of expressway by fuzzy logic. *Fuzzy Sets and Systems*, 26:165–178, 1988.

[10] S. Shao. Fuzzy self-organizing controller and its application for dynamic processes. *Fuzzy Sets and Systems*, 26:151–164, 1988.

[11] R. Tanscheit and E. M. Scharf. Experiments with the use of a ruled-based self-organising controller for robotics applications. *Fuzzy Sets and Systems*, 26:195–215, 1988.

[12] M. Valenzuela-Rendón. The fuzzy classifier system: motivations and first results. In H.-P. Schwefel and R. Männer, editors, *Parallel Problem Solving from Nature*, 330–334, Springer (Verlag), Berlin, 1991.

[13] H. J. Zimmermann. *Fuzzy Set Theory—and Its Applications.* Kluwer, Boston, 1988.

Redundant Classifiers and Prokaryote Genomes

T. H. Westerdale
Department of Computer Science, Birkbeck College,
University of London, Malet Street, London WC1E 7HX,
England

Abstract

Redundant classifiers waste space. This paper suggests an approach to the problem of classifier redundancy based on the cost of gene replication. The approach stems from the view that reduction and simplification are the essence of the evolutionary creative process, and that the most advanced organisms are Prokaryotes, not Eukaryotes.

1 The Redundancy Problem

In this paper, we shall be concerned with reducing the space taken up by classifiers. In the classifier systems we shall discuss, there is attached to each classifier an explicit adjustable parameter called the classifier's availability.[1] Eligible classifiers fire with probabilities proportional to their availabilities. Rewarding a classifier consists in raising its availability. The scheme by which this is done is called the classifier system's *reward scheme.* Classifiers with very low availability are physically removed from the system.

A redundant[2] classifier is one whose effect, when it fires, is identical to or very similar to that of some other classifier.[3] We wish to reduce the availabilities of redundant classifiers so they can be physically eliminated. If two classifiers have similar effects, we want the reward scheme to reward the one with the lower availability less. However, in designing such a reward scheme we would prefer to avoid introducing detrimental biases.

2 Safe Profit Sharing

Safe [12] reward schemes are those that have no detrimental biases. Where improvement is possible, they tend towards improvement.[4] Fisher's fundamental theorem of natural selection[4] says roughly that if, in a population, reward is equal to value, then the rate of change of average value equals the variance of the value, the type of variance depending on the recombination used. Thus if reward is proportional to value, the rate of change is non-negative and the scheme is safe. By regarding a classifier firing sequence as a genotype and each classifier as an allele, Fisher's theorem can be extended to profit sharing [11] reward schemes. Thus properly formulated profit sharing schemes are safe. (See [12] for details of the argument.) We now mention two profit sharing schemes and discuss their safety. We will then ask how they can be modified to penalize redundancy without destroying safety.

In the *fixed length scheme* we let K be a positive real constant and n a positive integer constant. Every time payoff μ is received we look back at the sequence of the last n classifier firings and give reward $K\mu$ to each classifier in the sequence. n is called the profit sharing span. To implement the reward scheme, the system keeps track, for each classifier x, of a parameter N_x called the fire count. This is the number of times the classifier has fired in the last n firings. When payoff μ is received, each classifier x receives reward $K\mu N_x$. Thus the sequence of the last n classifiers to fire plays the role of the genotype and is rewarded.

It is useful to define $\Delta(x, i)$ to be 1 if classifier x fired i time units ago, and 0 otherwise. Then
$N_x = \sum_{i=1}^{n} \Delta(x, i)$. Updating the N_x parameters can be difficult.

It is easier to update the N_x's in the *exponential scheme.* In this scheme, c is a positive real slightly less than 1, and $N_x = \sum_{i=1}^{\infty} c^{i-1}\Delta(x, i)$. One way to describe this is to say that instead of a single rewarded genotype, we have a whole population of rewarded genotypes, of various lengths. Those of length i are composed of the last i classifiers to fire. The proportion of the population of length i is $c^{i-1} - c^{i}$. The reward is spread over the whole population.

If the scheme is constructed correctly, Fisher's theorem ap-

[1] Holding availabilities implicitly (see[11]) is perhaps not suitable when we are concerned with space.

[2] In this paper, "redundant" means mostly redundant, not necessarily completely redundant.

[3] Triggered chaining often produces such redundant classifiers.

[4] Bucket brigade schemes are notoriously unsafe, allowing freeloaders undeserved reward. In [11] I called freeloaders "cannibals". Goldberg's term *freeloader* is better.

plies and the whole scheme is safe if c is close enough to 1. Correct construction is a bit fiddly. The whole scheme comes out as follows. (The details are discussed and the safety proved in [12].)

For each classifier x, keep a fire count N_x, an eligibility count C_x, and an availability v_x. [5] K is a positive constant and c is a positive constant slightly less than one. For each time unit, do the following:

(a) Determine the set of eligible classifiers D.

(b) Calculate $\bar{v} = \sum_{x \in D} v_x$.

(c) Select a member $\hat{x}$ of D probabilistically, selecting x with probability $(v_x/\bar{v})$.

(d) Fire $\hat{x}$ and record payoff μ.

(e) Multiply all N_x and all C_x by c.

(f) Add 1 to $N_{\hat{x}}$.

(g) Add $(v_x/\bar{v})$ to C_x for each x in D.

(h) To each v_x add $K\mu(N_x - C_x)$.

Note that as c approaches 1, the length of the genotypes (current sequences, what is rewarded) increases.

Now suppose we start with such a safe scheme and then institute an *ad hoc* transfer of availability between classifiers in an attempt to penalize redundant classifiers. Obviously there is a grave danger we will lose safety. Is there any scheme modification that penalizes redundancy without damaging safety? Let us ask how organisms have tackled the redundancy problem in the biological world.

In doing so we shall feel free to make wild evolutionary speculations such as a biologist would be ashamed to put in print. For a biologist is searching for truth, whereas we are only searching for ideas. If we can implement the ideas and so make a better classifier system, then that is justification enough for us. Whether the ideas contribute toward a better biological understanding is not our concern here. Ask not what classifier systems can do for biology; ask what biology can do for classifier systems. Of course we hope to repay the debt some day, but perhaps not now, and certainly not in this paper.

3 Prokaryote Genomes

The *Magnolia* is a primitive flower. It has 6-9 large showy petals, 3 sepals, many stamens, and many pistils. The weeping willow (*Salix babylonica*) is an advanced flower. It has no petals, no sepals, and either one pistil and no stamens or no pistils and a couple of stamens. It is traditional to derive all flowering plants from *Magnolia* like organisms by a process of reduction, that is of fusion and elimination of parts.[6] Some flowering plants have lost distinct roots, leaves, or stems as well. (E.g. *Wolffia* has lost all three.) Presumably the *Magnolia* flower itself arose by the loss of photosynthetic ability in certain leaves.

Reduction and simplification are the essence of the evolutionary creative process. Here are a few examples:[7] Horses' feet originated by loss of toes. The aortic arch structure of mammals was built from that of fish by eliminating arches 1, 2, and 5 and eliminating the dorsal aorta between arches 3 and 4 on both sides and caudal to the subclavian artery on the right. Insect body plan was built from that of Annelids by eliminating appendages on most segments. Vertebrates originated from Tunicates by elimination of the entire adult phase.[8] Homo sapiens originated from an earlier primate by elimination of the adult phase.[9] Many lines of animals originated from photosynthetic flagellates by loss of chloroplasts. Ulotrichales originated from a flagellate by loss of flagella. This last example is the pattern followed in the origin of most multicellular organisms (perhaps not all; e.g., perhaps not red Algae). *Brauniellula* originated from a *Gomphidius*-like mushroom by a suppression of stipe expansion.[10] This example may show the way most or all Gastromycete lines originated. Seed plants originated by the reduction of the free living gametophyte. The green Alga *Protococcus* usually grows as solitary cells or in colonies of a few cells. It is found on the north side of some tree trunks and in high school level biology texts, where it appears as a supposedly simple plant. By botanists it "is generally interpreted as a reduced form from a branching Ulotrichaceous ancestor." ([8] vol. 1. p. 53).

Evidently von Wettstein first proposed the reduction scenario for Liverwort evolution. (See comments in [8] vol. 2 p. 16.) Schuster felt that it was "rather difficult to attempt to maintain the largely discredited notion that the Bryophytes 'led' to the higher plants. ...It would be more sensible, in many ways, to say that the Bryophytes developed from forms near the Psilophytales ...the entire viewpoint of reduction as the major evolutionary tendency has been accepted for the Bryophytes." ([7] p. 266). Though his position is controversial, note that if the fossil Naiadita is one of the Sphaerocarpales, then it would seem to fit into a reduction scenario for this group at least.

The Bryophyte example is fairly typical. It is always easier to see how something might be lost in evolution than how it might be gained. Hence when taxonomists draw evolutionary trees (or hedges) they tend to draw lines in which reduction and elimination predominate.

The above comprises only a few selected examples illustrative of the general pattern. The above examples are all

[5] The reason for the eligibility count C_x and its biological analogue are given in [12]. The fixed length scheme needs a similar eligibility count for safety. This paper does not discuss the eligibility count.

[6] but see [9] for a questioning of tradition.

[7] Of course most evolutionary conjectures are controversial. Not all these examples are generally accepted. They are, however, all plausible.

[8] For discussion and counter argument see [6] p. 280.

[9] An oversimplification obviously. See [6] chapter 10.

[10] See A.H. Smith in [1] vol. IV B, pp. 421,422, 442.

of free living organisms. Most[11] organisms are parasites or symbionts. The view that the evolution of parasites and symbionts is dominated by reduction and elimination is traditional. The process by which the endosymbiotic mitochondrion lost its unnecessary genes is the same as the process by which the tapeworm lost its unnecessary digestive system. Thus Eukaryotes originated by a process of reduction and elimination of unnecessary or redundant parts or alleles.

The elimination of unnecessary or redundant parts and the consequent streamlining of the organism is the central process of evolution. The more advanced organisms have carried the process further than the more primitive ones.

In which groups do we find the most advanced organisms? That Eukaryotes appeared more recently than Prokaryotes does not argue that they are more advanced. Tilden, referring to Algae, argued that "— the simplest plants in any group have been most recently evolved and most nearly resemble the ancestral form." ([10] p. 107). Though she unwarrantedly assumes the ancestral primitive form is "simple", she argues that more recent plants are more primitive. Following a similar argument would make Eukaryotes more primitive, not more advanced. In fact, the evolutionary history of all present day organisms is presumably of equal length, so any such argument is questionable. On the other hand, it is true that the evolutionary history of a present day Eukaryote contains fewer *generations* than that of a present day Prokaryote, so perhaps one would expect Eukaryotes to be primitive.

In fact, Eukaryotes *are* primitive. If we look at the truly important characters, we see that the most advanced successful streamlined organisms on earth, the Eubacteria, are Prokaryotes. So successful are they that when we think of Prokaryotes it is usually Eubacteria that we think of. So streamlined are they that they have eliminated almost all the useless DNA that clogs the genomes of primitive organisms like vertebrates. Virtually the entire Eubacterial genome is used either to code for proteins or essential RNA or to implement control of gene expression. Even the introns have been virtually eliminated. Redundant alleles are nearly gone.

It is to these advanced organisms that we need to look to see how pressure has been applied toward elimination of redundancy. How do they differ from their more primitive Eukaryote aunts and uncles? Perhaps the difference will help account for their success.

The most obvious difference is size. Eukaryote cells are usually between 10 and 100 microns in diameter, and Prokaryote cells between 1 and 10 microns in diameter. ([2] p. 19). Though there are so many exceptions to these figures they must be questionable, it seems reasonable to think of Eukaryote cells as 10 times the diameter, 1000 times the volume of Prokaryote cells.

[11]depending of course on what and how you count. What proportion of free living individuals contain at least one parasite?

Let's compare haploid genome size. Bacteria have between 10^6 and 10^7 base pairs. Mammals, Reptiles, and Cartilaginous Fish have between 10^9 and 10^{10} base pairs. Many other Eukaryotic groups have a roughly similar range, though the range for insects and vascular plants extends down to 10^8, and the range for vascular plants and amphibians extends up to 10^{11}. The range for Fungi falls outside these figures, being from 10^7 to 10^8. The figures ([2] p. 486) again are highly suspect, but generally it is true that Eukaryote genomes are larger than Prokaryote, perhaps 1000 times as large in Metazoa and vascular plants. To be more specific, a mammal cell has about 1000 times the volume of *Escherichia coli* and about 1000 times the genome size.

Although the Mammalian genome is 1000 times the size of the *E. coli* genome, this is not because there is 1000 times as much information encoded there, for much, presumably most, of the mammalian genome is non coding junk DNA. How can Eukaryotes withstand the obvious cost of carrying around so much junk?

Actually the cost is not all that great. Since the junk DNA is not translated into protein, the energy cost is chiefly that of replication. This is small compared to the energy of expression. Expression of a gene usually involves transcribing it many times into RNA and translating each of the RNA's many times into proteins. Replication of the gene involves making only one DNA copy. This replication cost is miniscule compared to expression cost. So selection pressure in Eukaryotes is exerted chiefly against the expression, and selection pressure against unexpressed or untranslated DNA is small compared to other selection pressures operating. No wonder Eukaryotes have been so lazy in eliminating their junk DNA.

It would be quite another story if Eukaryotes had to keep 1000 copies of each gene in each cell. Then the replication cost would be 1000 times as great and selection pressure would tend to eliminate unexpressed DNA. This is the situation in Prokaryotes. In place of a Eukaryote cell, the Prokaryotes provide (for the same volume of protoplasm) 1000 Prokaryote cells, each with a copy of every gene. The thousand fold increase in replication cost evidently means that junk DNA is selected against significantly. Doolittle and Sapienza commented, "The intensity of phenotypic selection pressure to eliminate excess DNA [is] greatest in organisms for which DNA replication comprises the greatest fraction of total energy expenditure. Prokaryotes in general are smaller and replicate themselves and their DNA more often than eukaryotes (especially complex multicellular eukaryotes). Phenotypic selection pressure for small 'streamlined' prokaryotic genomes with little excess DNA may be very strong." [5]

For the unicellular organism, expression cost presumably depends roughly on the organism volume, whereas replication cost depends on genome size. If replication cost is a significant portion of total cost, selection pressure will reduce genome size until replication cost becomes less damaging. This will happen when the ratio of genome size to cell volume reaches a small enough number. It seems that selection pressure has achieved roughly the same ratio in both Eukaryotes and Prokaryotes. (Multicellularity does not affect this argument.)[12]

In the Eukaryotes this ratio leaves room for junk DNA. In the Prokaryotes it doesn't.[13] (And notice that the Eubacteria, being more advanced than Eukaryotes, have dispensed with any wasteful diploidy nonsense, as a result of replication cost selection pressure.)

So the lesson is that if we wish to safely eliminate redundant alleles (junk DNA) then we can use the method of the most advanced organisms, the Eubacteria, and simply increase replication cost. They did it by reducing volume. We need not necessarily use that method.

4 Replication Cost and Profit Sharing

In the profit sharing schemes of section 2, the allele is the classifier and the genome is the classifier firing sequence. Firing sequences are rewarded on the basis of payoff received during the sequence. The rate of increase of genotype probability is (or would be if there were no recombination) the probability times the excess genotype value. This excess value can be thought of as the payoff minus the average value of all the genotypes. Actually, I should say payoff minus $\bar{m}i$, where $\bar{m}$ is the average payoff per unit time (per classifier firing) and i is the genotype length. In the fixed length scheme, $\bar{m}i$ is the average value of all the genotypes, but in the exponential scheme, where genotypes differ in length, it is the average weighted by length. This $\bar{m}i$ can be thought of as the cost of the sequence. Thus in the exponential scheme, the cost is proportional to the length. This makes sense. The cost is the *time* wasted in using that sequence rather than some other. The longer sequences waste more time.

We want to introduce a space cost via a replication cost. Space cost is the number of *distinct* classifiers in the sequence. We can think of the sequence of classifiers as a sequence not of genes, but of gene expressions, of manufactured proteins if you like.[14] So the expression cost is proportional to the length of the sequence. But the replication cost is proportional to the number of genes, the number of distinct classifiers in the sequence. If i is the sequence length and j is the number of *distinct* classifiers in the sequence, then the cost is $\bar{m}(i+rj)$, where r is a non-negative constant that is the same for all sequences. $\bar{m}i$ is the expression cost and $\bar{m}rj$ is the replication cost. $\bar{m}r$ is the cost of replicating a single gene. If r is small, the system behaves like a Eukaryote, like a primitive organism. If r is large, the system behaves like a Prokaryote, like an advanced organism, and tries to streamline itself by eliminating redundant classifiers.

So how do we change our schemes to tidily embody these ideas?

In the fixed length scheme above, all sequences (genotypes) were the same length and hence the same cost. In modifying the fixed length scheme, it is tidy to retain the notion that these genotypes are all the same total cost, though now this means that some sequences must be shorter than others. The profit sharing span will have to vary over time.

In the exponential scheme, we have a whole population of genotypes or strings. The safety argument in [12] for the exponential scheme relies on the fact that in the population, the proportion of strings with lengths in the half open interval $[\alpha, \beta)$ is constant through time. If α and β are integers, the proportion is $c^{-1}(c^{\alpha}-c^{\beta})$. If they are not then it is still close to this proportion and constant. In our modified exponential scheme, we want the proportion of strings with *costs* in the interval $[\alpha, \beta)$ to be constant through time. I don't at present see a reasonable way of doing this exactly, but we can certainly arrange that the proportion is approximately $c^{-1-r}(c^{\alpha}-c^{\beta})$.[15] Let us see how we can arrange this.[16]

Our population changes over time, but at any given time, all the strings of a given length are identical. We define $f(i)$ to be $i+rj$, where j is the number of different classifiers in the strings of length i. Thus $\bar{m}\, f(i)$ is the cost of any string of length i. We let $F(i)$ be the current proportion of the population that is of length $\geq i$. We modify the exponential scheme by setting $F(i) = c^{f(i)-1-r}$. If $r = 0$, this is the unmodified exponential scheme. Note that, for example, $f(5)$ and $F(5)$ will vary over time.

We let $L(x)$ be the number of time units ago that the last x fired. We define $n(x,i) = \sum_{j=1}^{i} \Delta(x,j)$. This is the number of times x fired in the last i time units. ($n(x,0) = 0$.) We define

[12] I admit it is surprising that organisms as different as the mammals (primitive organisms) and *E. coli* (a very advanced organism) should have ratios as similar as they do. The *extreme* similarity must be partly coincidence.

[13] Cavalier-Smith [3] has proposed a nucleoskeletal role for Eukaryote junk DNA. Thus junk DNA would have some value for Eukaryotes, just as showy petals have some value for *Magnolia*. The balance between the replication cost selection pressure (weak in Eukaryotes) and the weak selection for the thus required junk would then help explain the observed rough proportionality of cell volume and genome size in Eukaryotes (a proportionality we have not mentioned), particularly as the amount of junk DNA thus required would depend on cell volume. Evidently advanced organisms (Eubacteria) have evolved a structure that makes junk DNA as redundant to their needs as petals evidently are to a willow's. Thus it is these advanced Prokaryotes that have used replication cost selection pressure to eliminate the redundant DNA that is evidently continually created in any evolving population.

[14] In the most obvious analogy, the classifier is an enzyme, its left side the regulatory site and its right side the catalytic site. The various 1's, 0's, and #'s are individual codons. The messages are metabolytes. It doesn't quite work, but it's suggestive.

[15] It will be exactly this proportion if there are, in the population, strings with costs exactly α and β. But whether there are or not changes over time. When there are not, then we will have what we can think of as rounding errors.

[16] The safety proof in [12] has not yet been extended to cover this case, so at this stage it is only a reasonable guess that the errors in the approximation will not ruin safety.

$P(i) = F(i) - F(i+1)$. This is the proportion of the population that is length i. We write $Q(x)$ to mean the proportion of the population that contains at least one x, so $Q(x) = F(L(x))$. Now we define

$$s(x,y) = \sum_{i=L(y)}^{\infty} P(i)n(x,i).$$

Then

$$s(x,y) = Q(y)n(x,L(y)) + \sum_{i=L(y)+1}^{\infty} F(i)\Delta(x,i).$$

It follows that

$$s(x,x) = \sum_{i=1}^{\infty} P(i)n(x,i) = \sum_{i=1}^{\infty} F(i)\Delta(x,i),$$

and we see that $s(x,x)$ is the fire count of x. $s(x,y)$ is the portion of the fire count of x due to strings that contain at least one y.

Suppose the system holds in its memory the current values of the parameters $L(x)$, $Q(x)$, $n(x,L(y))$, and $s(x,y)$ for all classifiers x and y. Suppose then that the classifier z fires. Let us see how the system can correctly update the values it holds. We will indicate with a bar, parameter values after the firing of z, to distinguish them from the values before the firing. Thus for example $\bar{\Delta}(x,i+1) = \Delta(x,i)$ for all $i \geq 1$.
$\bar{L}(z) = 1$, and if $x \neq z$, then $\bar{L}(x) = L(x) + 1$.
$\bar{n}(x,\bar{L}(z)) = \bar{\Delta}(x,1)$, and if $y \neq z$ then
$\bar{n}(x,\bar{L}(y)) = \bar{\Delta}(x,1) + n(x,L(y))$.

If $L(z) \leq i$ then
$\bar{f}(i+1) = f(i) + 1$ and $\bar{F}(i+1) = cF(i)$.
If $1 \leq i < L(z)$ then
$\bar{f}(i+1) = f(i) + 1 + r$ and $\bar{F}(i+1) = c^{1+r}F(i)$.
$\bar{Q}(z) = 1$.
If $L(z) < L(x)$ then $\bar{Q}(x) = cQ(x)$.
If $L(x) < L(z)$ then $\bar{Q}(x) = c^{1+r}Q(x)$.

If we define $R_1(x,y,z) = s(x,y) + Q(y)\bar{\Delta}(x,1)$ and
$R_2(x,z) = (c - c^{1+r})(s(x,z) + Q(z)(\Delta(x,L(z)) - n(x,L(z))))$,
then it is easy to show that $\bar{s}(x,y)$ can be written as follows:

$$\begin{cases} c^{1+r}R_1(x,y,z) + R_2(x,z) & \text{if } L(y) < L(z); \\ c^{1+r}s(x,x) + \bar{\Delta}(x,1) + R_2(x,z) & \text{if } L(y) = L(z); \\ cR_1(x,y,z) & \text{if } L(y) > L(z). \end{cases}$$

Since $\bar{\Delta}(x,1)$ and $\Delta(x,L(z))$ are both 1 if $x = z$ and zero otherwise, the above formulae show how the system can update the values of $L(x)$, $Q(x)$, $n(x,L(y))$, and $s(x,y)$. It thus can obtain the current values of the firing counts $s(x,x)$ for use in our modified exponential scheme.

The system also needs to be able to obtain the current values of the eligibility count. This can be done in a way similar to the way just sketched for the firing count. We need only redefine $\Delta(x,i)$ as follows. If x was eligible i time units ago, then $\Delta(x,i)$ is the availability of x divided by the sum of the availabilities of all the classifiers that were eligible then. If x was not eligible i time units ago, then $\Delta(x,i) = 0$. The definitions of the other quantities are as before except that they use the re-defined Δ. Then $s(x,x)$ becomes the eligibility count of x, and the formulae above give a method for updating the values of these new quantities so that the current eligibility counts are always available. This method differs from that for firing counts in that $\bar{\Delta}(x,1)$ and $\Delta(x,L(z))$ are determined differently. The system will have to hold the parameters $\Delta(x,L(z))$ in its memory too, and update them in the obvious way.

So though we have only given illustrative unproved suggestions, it looks likely that replication cost can be safely incorporated in profit sharing in an implementable algorithm. This would provide safe selection pressure against redundant classifiers. However, if there are N classifiers in the system, the algorithm we have discussed requires the system to hold and update order N^2 different parameters, whereas the system using profit sharing with no replication cost needs only order N. It may be that there is a safe profit sharing algorithm with replication cost that uses only order N parameters, but I am not hopeful.

5 The Bucket Brigade with Replication Cost

In this respect, a bucket brigade with replication cost looks more hopeful. In the simplified bucket bucket brigade described elsewhere [11], with each classifier is held both an availability parameter and a cash balance parameter. The reward scheme works roughly as follows. When a classifier fires, its availability is increased by an amount proportional to its cash balance. Also a fixed proportion of its cash balance (say a hundredth) is passed back to the classifier that fired immediately previously. Any immediate payoff is added to the cash balance of the classifier that just fired.

If we normalize the cash balances by subtracting from each cash balance the sum of the cash balances, then each normalized balance is the system's estimate of the excess true value of the classifier. (Actually, it is the estimate of excess true value divided by the proportion of cash the bucket brigade passes back.) Once normalized, the estimates will tend to drift up if $\bar{m} > 0$, but this drift would be counteracted if every time a classifier fires, $\bar{m}$ is subtracted from its cash balance. We can think of this $\bar{m}$ as the cost of using that classifier rather than another one.

Looking at the same genotypes we looked at in profit sharing, we see that for fixed length genotypes of length l, what we can call the rough cost of firing x is $1+r$ if $L(x) > l$ and 1 if $L(x) \leq l$. If we let $\bar{\kappa}$ be the average rough cost per time unit, then $\bar{\kappa}$ is positive and $\bar{m}/\bar{\kappa}$ is the average payoff per rough cost unit. Thus the cost of firing x (the payoff you don't get because you're not doing something else) is $(\bar{m}/\bar{\kappa})(1+r)$ if $L(x) > l$ and $\bar{m}/\bar{\kappa}$ if $L(x) \leq l$.

For a population of genotypes like we had in the exponential scheme, the rough cost of firing x is
$1 + (1 - Q(x))r$, and the firing cost is
$(\bar{m}/\bar{\kappa})(1 + (1 - Q(x))r)$. We merely need to subtract this cost from the cash balance of x every time x fires.

Since the system can easily update the parameters $L(x)$ and $Q(x)$, the bucket brigade with replication cost can be implemented holding only order N parameters.

So what is the catch? Well, the cost of firing x is supposed to contribute only to firing sequences containing x, but the bucket brigade only does this approximately, so safety is not guaranteed. The problem is the two platform stations. [11]

Here is a particularly pathological case in which a two platform station destroys safety even though payoff per time unit is a positive constant $\bar{m}$. A two platform station simply misdirects the cost charges. Suppose there are five classifiers, a, b, d, t, and e. Suppose n is a non-negative integer constant and suppose eligibility of classifiers is such that they fire either in sequence *eatd* (the *first sequence*) or in sequence *ebte*^{n}, (the *second sequence*) where e^n is a string of n e's. When a sequence is completed, either of the two sequences may follow, the first sequence with probability p and the second with probability $q = 1 - p$. The size of p is clearly determined by the availabilities of a and b.
t is a two platform station.

We examine the value of $Q(a)$ just before a fires. The firing sequence back to the previous a is $atde(bte^ne)^j$ with probability pq^j. Then

$$f(L(a)) = \begin{cases} 4+4r & \text{if } j=0 \\ 4+(n+3)j+5r & \text{if } j>0. \end{cases}$$

Now $Q(a) = F(L(a)) = c^{f(L(a))-1-r}$. We will call this A_j. It is c^{3+3r} if $j = 0$ and $c^{3+(n+3)j+4r}$ if $j > 0$. Thus it is never greater than c^{3+3r} and on average it is
$\bar{A} = \sum_{j=0}^{\infty} pq^jA_j = pc^{3+3r} + pqc^{n+6+4r}(1 - qc^{n+3})^{-1}$.

We similarly examine the value of $Q(b)$ before b fires. The firing sequence back to the previous b is $bte^ne(atde)^j$ with probability p^jq.

$$f(L(b)) = \begin{cases} n+3+3r & \text{if } j=0 \\ n+3+4j+5r & \text{if } j>0. \end{cases}$$

$B_j = Q(b) = F(L(b)) = c^{f(L(b))-1-r}$ and this is
c^{n+2+2r} if $j = 0$ and
$c^{n+2+4j+4r}$ if $j>0$.
The average is
$\bar{B} = \sum_{j=0}^{\infty} p^jqB_j = qc^{n+2+2r} + pqc^{n+6+4r}(1 - pc^4)^{-1}$.

Now

$$(1 - qc^{n+3})^{-1} - (1 - pc^4)^{-1} = (-pc^4 + qc^{n+3})(1 - qc^{n+3})^{-1}(1 - pc^4)^{-1} > (-pc^4)(1 - c)^{-2}.$$

Thus $\bar{A} - \bar{B} > pc^{3+3r} - qc^{n+2+2r} - p^2qc^{10+4r}c^n(1-c)^{-2}$. Suppose we choose n so large that $c^n(1 - c)^{-2} < \frac{1}{2}$ and $c^{n-1-r} < \frac{1}{4}$. Suppose also that $p \geq q$. Then $\bar{A} - \bar{B} > \frac{1}{8}c^{3+3r}$.

Using the formula $(\bar{m}/\bar{\kappa})(1 + r - Q(x)\,r)$ for the cost of firing x, we see that the cost of firing a is not less than $(\bar{m}/\bar{\kappa})(1+r-c^{3+3r}r)$, since $A_j \leq c^{3+3r}$. Since the cost of firing x is not less than $\bar{m}/\bar{\kappa}$, the cost of firing the first sequence *eatd* is not less than $(\bar{m}/\bar{\kappa})(4 + r - c^{3+3r}r)$, giving a cost per time unit not less than $(\bar{m}/\bar{\kappa}) + (1 - c^{3+3r})(r\bar{m})(4\bar{\kappa})^{-1}$. In the second sequence *ebte*^{n}, the cost of firing each of the last $n - 1$ e's is $(\bar{m}/\bar{\kappa})$ since at that point $Q(e) = 1$. So the cost of firing the second sequence *ebte*^{n} is not more than $(\bar{m}/\bar{\kappa})(n + 3 + 4r)$, giving a cost per time unit of not more than $(\bar{m}/\bar{\kappa}) + (4r\bar{m})(\bar{\kappa}(n + 3))^{-1}$. Thus if n is large enough, the first sequence is more costly per time unit than the second, whatever the values of p and q, and so the system is better off with a high q.

But if we start with $q \leq \frac{1}{2}$, the bucket brigade tends to decrease q, for consider the cost of firing a. On average it is $(\bar{m}/\bar{\kappa})(1 + r - \bar{A}r)$. Thus when a fires, the average inflow of cash (positive or negative) to a is $\bar{m} - (\bar{m}/\bar{\kappa})(1 + r - \bar{A}r)$ plus the cash passed back from t. When b fires, the average inflow of cash to b is $\bar{m} - (\bar{m}/\bar{\kappa})(1 + r - \bar{B}r)$ plus the cash passed back from t. Since the cash passed back from t depends only on the cash balance of t, it is the same for a and b, so if n is large enough that $c^n(1 - c)^{-2} < \frac{1}{2}$ and $c^{n-1-r} < \frac{1}{4}$, then for any $q \leq \frac{1}{2}$ there is more average inflow to a than to b since $\bar{A} - \bar{B} > \frac{1}{8}c^{3+3r}$. Since $\bar{\kappa} \leq 1 + r$, the amount by which the average inflow to a exceeds that to b depends on q, but is never less than $(\bar{m}/(1 + r))\frac{1}{8}c^{3+3r}$ for any $q \leq \frac{1}{2}$. Thus the cash balance of a tends to become larger than that of b, and q tends to decrease.

6 General Considerations

Notice that selection against redundancy is stronger if r is higher. But if the classifier population size is too small then a high r will produce premature convergence, so we must be careful. The fact is that classifier populations that are too small are too small. Clever costings don't change this fact.

This paper does not discuss classifier crossover. It would be nice if our scheme were a scheme friendly to classifier crossover (crossover within a gene, in our biological metaphor). For it to be friendly, the notion of the cost of a schema must make sense (a schema being part of a classifier here, not part of a genome). It does, at least somewhat. A high cost schema is one found in classifiers that make eligible those firing sequences of following classifiers that have high cost. One can imagine such high cost schemata, and our costing schemes should exert at least some selection pressure to eliminate them.

7 Summary

Since we still lack a satisfactory theoretical foundation for Genetic Algorithms, we are usually guided by biological metaphor. It is important, then, that we not base our work on a prejudiced or parochial view of evolution. This paper is based on the view that reduction and simplification are the essence of the evolutionary creative process, and that the Eubacteria are the most advanced organisms, far more advanced than insects or vertebrates. This is only one of many reasonable views. We have seen how this view suggests a method for tackling the problem of classifier redundancy. Other views may suggest methods for tackling other problems, or indeed different methods for tackling this one.

References

[1] Ainsworth, Sparrow, and Sussman. *The Fungi.* 4 vols.,Acad. Press, London, 1973.

[2] Alberts, Bray, Lewis, Raff, Roberts, and Watson. *Molecular Biology of the Cell.* 2nd ed.,Garland Publishing, New York and London, 1308 pp., 1989.

[3] T. Cavalier-Smith. *The Evolution of Genome Size.* Wiley, New York, 1985.

[4] J.F. Crow and M. Kimura. *An Introduction to Population Genetics Theory.* Harper and Row, New York, 1970.

[5] C. Doolittle, F.W. and Sapienza. Selfish genes, the phenotype paradigm and genome evolution. *Nature*, 284:601–603, 1980.

[6] Stephen Jay Gould. *Ontogeny and Phylogeny.* Harvard Univ. Press, Cambridge, Mass. (501 pp), 1977.

[7] Rudolph M. Schuster. Boreal hepaticae, a manual of the liverworts of minnesota and adjacent regions. *The American Midland Naturalist*, 49(2):257–684, 1953.

[8] Gilbert M. Smith. *Cryptogamic Botany*, volume 1 (545 pp.) and 2 (397 pp.). McGraw Hill, New York, 1955.

[9] L.J. Taylor, D.W. and Hickey. An aptian plant with attached leaves and flowers. *Science*, 247:702–704, 1990.

[10] Josephine E. Tilden. *The Algae and Their Life Relations.* 2nd ed.,Oxford Univ. Press, London (550 pp.), 1937.

[11] T.H. Westerdale. A defense of the bucket brigade. In J. David Schaffer, editor, *Proceedings of the Third International Converence on Genetic Algorithms*, pages 282–290, San Mateo, California, 1989. Morgan Kaufmann.

[12] T.H. Westerdale. A reward scheme for production systems with overlapping conflict sets. *IEEE Trans. Syst., Man, Cybern.*, SMC-16(3):369–383, May/June 1986.

Genetic Algorithms in Hybrid Methods

Self-Optimizing Image Segmentation System Using a Genetic Algorithm

Bir Bhanu
College of Engineering
University of California
Riverside, CA 92521

Sungkee Lee
Dept. of Computer Sc.
University of Utah
Salt Lake City, UT 84112

John Ming
Human Interface Tech. Ctr.
NCR Corp., 500 Tech Parkway
Atlanta, GA 30313

Abstract

One of the fundamental weaknesses of current computer vision systems to be used in practical outdoor applications is their inability to adapt the segmentation process as real-world changes occur in the image. We present the first closed loop image segmentation system which incorporates a genetic algorithm to adapt the segmentation process to changes in image characteristics caused by variable environmental conditions such as time of day, time of year, clouds, etc. The segmentation problem is formulated as an optimization problem and the genetic algorithm efficiently searches the hyperspace of segmentation parameter combinations to determine the parameter set which maximizes the segmentation quality criteria. The goals of our adaptive image segmentation system are to provide continuous adaptation to normal environmental variations, to exhibit machine learning capabilities, and to provide robust performance when interacting with a dynamic environment. We present experimental results which demonstrate that genetic algorithm can be successfully used to adapt the segmentation performance automatically in outdoor color imagery.

1. INTRODUCTION

Image segmentation is typically the first, and most difficult task of any automated image understanding process. All subsequent interpretation tasks including object detection, feature extraction, object recognition, and classification rely heavily on the quality of the segmentation process. Despite the large number of segmentation techniques presently available [3,6], no general methods have been found that perform adequately across a diverse set of imagery. Only after many modifications to its control parameter set can any current segmentation technique be used to process the diversity of images found in real world applications. When presented with a new image, selecting the appropriate set of algorithm parameters is the key to effectively segmenting the image. The image segmentation problem can be characterized by several factors which make the parameter selection process very difficult. *First*, most segmentation techniques contain numerous control parameters which must be adjusted to obtain optimal performance. The size of the parameter search space in these systems can be prohibitively large, unless it is traversed in a highly efficient manner. *Second*, the parameters within most segmentation algorithms typically interact in a complex, non-linear fashion, which makes it difficult or impossible to model the parameters' behavior in an algorithmic or rule-based fashion. *Third*, since variations between images cause changes in the segmentation results, the objective function that represents segmentation quality varies from image to image. The search technique used to optimize the objective function must be able to adapt to these variations. *Finally*, the definition of the objective function itself can be subject to debate because there are no universally accepted measures of image segmentation quality.

Hence, we must apply a technique that can efficiently search the complex space of parameter combinations and locate the values which yield optimal results. The approach should not be dependent on the particular application domain nor should it have to rely on detailed knowledge pertinent to the selected segmentation algorithm. The key elements of our adaptive image segmentation system are: (1) A closed-loop feedback control technique that consists of a genetic learning component, an image segmentation component, and a segmented image evaluation component; (2) A genetic learning system that optimizes segmentation performance of each image and accumulates segmentation experience over time to reduce the effort needed to optimize the segmentation quality of succeeding images; (3) Image characteristics and external image variables are represented and manipulated using both numeric and symbolic forms within the genetic knowledge structure, only the segmentation parameters are represented and manipulated in binary strings; (4) Image segmentation performance is evaluated using five measures of segmentation quality that measure *global* characteristics of the entire image as well as *local* features of individual object regions; (5) The adaptive segmentation system is not dependent on any specific segmentation algorithm or type of sensor. The performance of the adaptive algorithm will be limited by the capabilities of the segmentation algorithm, but the results will be optimal for a given image based on our evaluation criteria.

To date, no segmentation algorithm has been developed which can automatically generate an "ideal" segmentation result in one pass (or in an open loop manner) over a range of scenarios encountered in practical outdoor applications. While there are adaptive threshold selection techniques

[12,14] for segmentation, these techniques do not accomplish any learning from experience to improve the performance of the system. Any technique, no matter how "sophisticated" it may be, will eventually yield poor performance if it can not adapt to the variations in outdoor scenes. Therefore, in this paper we attempt to address this fundamental bottleneck in developing "useful" computer vision systems for practical scenarios by developing a closed-loop system that incorporates a genetic algorithm and automatically adapts the segmentation algorithm's performance by changing its control parameters and will be valid across a wide diversity of image characteristics and application scenarios.

2. ADAPTIVE IMAGE SEGMENTATION SYSTEM

2.1 SEGMENTATION AS A SEARCH PROBLEM

Fig. 1 shows an outdoor image and the typical segmentation quality surface (discussed in Section 2.2.4) associated with the image in which only two segmentation parameters [8,13] are being varied. Because of the large number of potential parameter combinations and the subtle interaction of the algorithm parameters, the objective function is complex, multimodal, and presents problems for many commonly used search and optimization techniques. The drawbacks to some of these methodologies for the segmentation optimization problem are summarized by Lee [9].

Genetic algorithms [1,2,4,5,7] which are designed to efficiently locate an approximate global maximum in a search space show great promise in solving the parameter selection problem encountered in the image segmentation task. Since they use simple recombinations of existing high quality individuals and a method of measuring current performance, they do not require complex surface descriptions, domain specific knowledge, or measures of goal distance. Moreover, due to the generality of the genetic process, they are independent of the segmentation technique used, requiring only a measure of performance (which we refer to as segmentation quality) for any given parameter combination.

2.2 ADAPTIVE IMAGE SEGMENTATION

Adaptive image segmentation requires the ability to modify control parameters in order to respond to changes that occur in the image as a result of varying environmental conditions. The block diagram of our approach is shown in Fig. 2. After acquiring an input image, the system analyzes the image characteristics and passes this information, in conjunction with the observed external variables, to the genetic learning component. Using this data, the genetic learning system selects an appropriate parameter combination, which is passed to the image segmentation process. After the image has been segmented, the results are evaluated and an appropriate reward is generated and passed back to the genetic algorithm. This process continues until a segmentation result of acceptable quality is produced. The details of each component in this procedure will be described in the following subsections.

2.2.1 Image Characteristics

We compute twelve first order image properties for each color component (red, green, and blue) of the image. These features include mean, variance, skewness, kurtosis, energy, entropy, x intensity centroid, y intensity centroid, maximum peak height, maximum peak location, interval set score, and interval set size. Since we use a black/white version of the image to compute edge information and object contrast during the evaluation process, we also compute the twelve features for the Y (luminance component) image as well. Combining the image characteristic data from these four components yields a list of 48 elements. In addition, we utilize two external variables, time of day and weather conditions. The external variables are represented symbolically in the list structure (e.g., time = 9am, 10am, etc. and weather conditions = sunny, cloudy, hazy, etc). The distances between these values are computed symbolically when measuring image similarity. The two external variables are added to the list to create an image characteristic list of 50 elements. A system consisting of knowledge structures is used to store the image characteristics and the associated segmentation parameters that are generated by the genetic learning system.

2.2.2 Genetic Learning System

Fig. 3 shows a simple example of our genetic learning system. The image characteristics for a new image are compared with the individuals in the global population to obtain the initial seed for the local population. The global population represents the accumulated segmentation experience for all images that the system has processed whereas the local population contains the set of segmentation parameters processed by the genetic algorithm during the optimization of the current image. To obtain the initial local population (seed population) for a new image from the global population, a normalized Euclidean feature distance is computed from the new image to every member of the global population and this distance is used along with the fitness of each individual in the global population for selecting the closest individuals. Although we have limited the seed population to 3 in this example, our experiments utilize a seed population of 10 individuals. The global population holds 100 knowledge structures in order to maintain a diverse collection of segmentation experience. The parameter sets in the seed population are used to segment the image and the results are evaluated to generate a fitness for each individual. The fitness value (leftmost value in the list) varies from 0.0 to 1.0 and measures the quality of the segmentation parameter set. Note that only the fitness value and the action portion (segmentation parameters) of the knowledge structure are subject to genetic adaptation; the conditions (image characteristics) remain fixed for the life of the structure. If the fitness values are not acceptable, the individuals are recombined and the process repeats. Each pass through the loop (segmentation-evaluation-recombination), is known as a generation. The cycle continues until the maximum fitness achieved at the end of a generation exceeds some threshold. The global population is updated using the high quality members of the local population from the current image and the system is then ready to process another image.

2.2.3 Segmentation Algorithm

Since we are working with color imagery in our experiments, we have selected the well known Phoenix segmentation algorithm developed at Carnegie Mellon University [8,10,13]. Phoenix, which was the subject of several PhD dissertations, has been widely used, refined, and documented. The algorithm, which is based on a recursive region-splitting approach, has been extensively tested on color imagery. Phoenix [8] contains fourteen different control parameters which are used to control the thresholds and termination conditions used within the algorithm. There are 10^{33} conceivable parameter combinations using these fourteen values. Of the fourteen values, we have selected two of the most critical parameters that affect the overall results of the segmentation process, *maxmin* and *hsmooth*. Maxmin specifies the lowest acceptable peak-to-valley-height ratio used when deciding whether or not to split a large region into two or more smaller parts. Hsmooth controls the width of the window used to smooth the histogram of each image region during segmentation. The use of only two parameters for the initial tests aids in the *visualization* of the optimization process since we can plot the associated segmentation quality corresponding to each parameter combination using a 3D plotting technique. Future research will incorporate a larger number of modifiable parameters.

2.2.4 Segmentation Evaluation

There are a large number of segmentation quality measures that have been suggested, although none have achieved widespread acceptance as a universal measure of segmentation quality. In order to overcome the drawbacks of using only a single quality measure, we have incorporated an evaluation technique that uses five different quality measures to determine the overall fitness for a particular parameter set. The five segmentation quality measures are:

(1) *Edge-Border Coincidence:* Measures the overlap of the region borders in the image acquired from the segmentation algorithm relative to an edge image.
(2) *Boundary Consistency:* Similar to edge-border coincidence, except that region borders which do not exactly overlap edges can be matched with each other. Also, region borders which do not match with any edges are used to penalize the segmentation quality [9].
(3) *Pixel Classification*: This measure is based on the number of object pixels classified as background pixels and the number of background pixels classified as object pixels.
(4) *Object Overlap:* Measures the area of intersection between the object region in the ground truth image and the segmented image.
(5) *Object Contrast*: Measures the contrast between the object and the background in the segmented image, relative to the object contrast in the ground truth image.

The maximum and minimum values for each of the five segmentation quality measures are 1.0 and 0.0, respectively. The first two quality measures are *global* measures since they evaluate the segmentation quality of the whole image with respect to edge information. Conversely, the last three quality measures are *local* measures since they only evaluate the segmentation quality for the object regions of interest in the image. When an object is broken up into smaller parts during the segmentation process, only the largest region which overlaps the actual object in the image is used in computing the local quality measures. The three local measures require the availability of object ground truth information in order to correctly evaluate segmentation quality. Since we desire good object regions as well as high quality overall segmentation results, we have combined global and local quality measures (with equal weighting) to obtain a *combined* segmentation quality measure that maximizes overall performance of the system. Fig. 4 shows the surfaces defined for the five individual quality measures that are used to create the combined quality measure surface shown in Fig 1.

3. EXPERIMENTAL RESULTS

An initial database of outdoor imagery was collected to demonstrate the system's ability to adapt to real world conditions and produce the best segmentation result based on our evaluation criteria. The database consists of twenty frames that were collected approximately every 15 minutes over a 5 hour period (1:30 pm to 6:30 pm) using a JVC GXF700U color video camera. A representative subset of these images is shown in Fig. 5. This database will be used to describe the experimental results. Weather conditions in our image database varied from bright sun to overcast skies. Varying light level is the most prominent change throughout the image sequence, although the environmental conditions also created varying object highlights, moving shadows, and many subtle contrast changes between the objects in the image. The car in the image is the object of interest. The auto-iris mechanism in the camera was functioning, which causes a similar appearance in the background foliage throughout the image sequence. Even with the auto-iris capability built into the camera, there is still a wide variation in image characteristics across the image sequence. This variation requires the use of an adaptive segmentation approach to compensate for these changes.

To precisely evaluate the effectiveness of the adaptive image segmentation system, we exhaustively defined the segmentation quality surfaces for each frame. The segmentation quality surfaces were defined for preselected ranges of maxmin and hsmooth parameters Maxmin values, which affect segmentation performance in a non-linear fashion, were sampled exponentially over a range of values from 100 to 471. Values near 100 were spaced closer together than values at the upper end of the range. Hsmooth values were sampled linearly using numbers between 1 and 63. By selecting 32 discrete values (5 bits of resolution) for each of these parameter ranges, the search space contained 1024 different parameter combinations.

3.1 BASIC EXPERIMENTS

The first set of experiments with the adaptive segmentation system was divided into two separate phases: 1) a training phase where the optimization capabilities of the genetic algorithm were measured; and 2) a testing phase where we evaluated the reduction in effort achieved by utilizing previous segmentation experience. The image data was separated into two halves, 10 images (1,3,...,19) for training and 10 images (2,4,...,20) for testing. During the training phase, seed populations were selected using random locations on the combined segmentation quality surface for each image. The genetic system was then invoked using the seed population for each image and the convergence rate of the process was measured. Each training image was processed 100 times, each with a different collection of random starting points. These results were combined to compute the average number of generations needed to optimize each surface. The genetic component used a local population size of 10, a crossover rate of 0.8, and mutation rate of 0.01. A crossover rate of 0.8 indicates that, on average, 8 out of 10 members of the population will be selected for recombination during each generation. The mutation rate of 0.01 implies that on average, 1 out of 100 bits is mutated during the crossover operation to insure diversity in the local population.

The stopping criteria for the genetic process contains three tests. First, since the global maximum for each segmentation quality surface was known a priori (recall that the entire surface was precomputed), the first stopping criteria was the location of a parameter combination with 95% segmentation quality or higher. In experiments where the entire surface is not precomputed, this stopping criteria would be discarded. Second, the process terminates if 3 consecutive generations produce a decrease in the average population fitness for the local population. Third, if 5 consecutive generations fail to produce a new maximum value for the average population fitness, the genetic process terminates. If any one of these three conditions is met, the processing of the current image is stopped and the maximum segmentation quality currently in the local population is reported.

Fig. 6 shows the combined segmentation quality surfaces for the images shown in Fig. 5. Note that due to the complexity of these surfaces, most commonly-used search techniques [9] would not be effective at optimizing the segmentation quality.

At the end of training phase, the final local population from each of the training images (1,3,...,19) was combined to create a global population of 100 individuals. From this global population, the 10 initial seed members of each local population for the testing images (2,4,...,20) were selected. The testing was performed in a parallel fashion; the final local population for each of the testing images was not placed back into the global population for these tests. The alternative approach to testing, which processes each frame in the outdoor imagery database in a sequential manner and integrates the results into the global population, has been discussed in detail in [9].

For testing phase since the fitness of each seed population is based on previous segmentation experience, the genetic process is able to converge to the global maximum much faster during the testing phase. During the training experiments, the maximum number of generations was 13, the minimum number was 5, and the average number of generations was 9. By combining the information accumulated during training in the global population, the average number of generations was reduced from 9 during training to 3 during testing. This represents a considerable improvement in the adaptive system's efficiency. On average, the adaptive segmentation system visits approximately 2.5% of the search space (i.e., ~ 2.5 generations) for the experiments described here.

Since there are no other known adaptive segmentation techniques in the computer vision field to compare our system with, we measured the performance of the adaptive image segmentation system relative to the set of default Phoenix segmentation parameters [8,13] and a traditional optimization approach. The *default parameters* have been suggested after extensive amounts of testing by various researchers who developed the Phoenix algorithm [8]. The parameters for *traditional approach* are obtained by manually optimizing the segmentation algorithm on the first image in the database and then utilizing that parameter set for the remainder of the experiments. This approach to segmentation quality optimization is currently standard practice in state-of-the-art computer vision systems. Fig. 7 presents the comparison of these three approaches. The average segmentation quality for the adaptive segmentation technique was 95.8% (average of 100 experiments). In contrast, the performance of the default parameters was only 55.6% while the traditional approach provided 63.2% accuracy. As the figure shows, the performance of both of these alternative approaches was highly erratic throughout the sequence. Fig. 8 illustrates the quality of the segmentation results associated with the adaptive system, the default parameters, and the traditional approach. Each result corresponds to the average segmentation performance produced by each technique for the first frame in the database. By comparing the extracted car region in each of these images, as well as the overall segmentation of the entire image, it is clear that the adaptive segmentation results are superior to the other methods.

3.2 COMPARISON OF THE ADPATIVE SYSTEM WITH RANDOM SEARCH

Several tests were performed to compare the optimization capabilities of the adaptive segmentation system with a simple random walk through the search space. This experiment used only the training images (1,3,...,19) from the outdoor image database so that the adaptive system would not benefit from the reuse of segmentation experience from one image to the next. The intent of this restriction was to measure the efficiency of the genetic algorithm in optimizing a complex surface. In addition, the stopping criteria for the adaptive system was simplified so that when a surface point with 95% segmentation quality or better was located, the optimization process would terminate. The random walk algorithm searched the segmentation quality surface by visiting points randomly and used the same 95% stopping criteria. Finally, in order to insure correctness of the results, each segmentation quality surface was optimized by each technique 100 times and the results are averaged to create the performance figures.

Fig. 9 presents a comparison of the efficiency for the two techniques described above. The bars represent the total number of points visited on the surface using each technique for each of the images and the average number of points visited for each approach. As the average values show, the adaptive technique is far superior to the random walk approach. In addition, the average number of points visited by the adaptive approach is ~ 6.9% of the total number of points on the surface, compared to the earlier experiments where we processed ~ 2.5% of the surface, since we have not reused any segmentation experience gained from processing earlier images. Fig. 10 contrasts the segmentation quality achieved by the two techniques. Since the adaptive segmentation technique insures the achievement of a near global maximum for each image, we modified the random walk approach so that it would terminate after the same number of visited locations required by the adaptive technique. The maximum segmentation quality achieved by the random approach was then compared with the adaptive system. On the average, the adaptive system achieved 99.3% segmentation quality after the number of segmentations shown in Fig. 9. In comparison, the random walk achieved only 81.4% of the maximum quality for the same number of segmentations for each image.

3.3 THE EFFECTIVENESS OF THE REPRODUCTION AND CROSSOVER OPERATORS

A number of tests were performed to demonstrate the effectiveness of the reproduction and crossover operators in the adaptive image segmentation system. The optimization capability of the pure genetic algorithm was compared with two variations of the genetic algorithm. The first variation of the pure genetic algorithm was implemented without a reproduction operator. Instead of reproducing individuals according to their fitness values, the algorithm selected the individuals at random for further genetic operator action with the restriction that any individual be selected only once. The second variation of the genetic algorithm simply skipped a crossover operator. To ensure that this approach generates about the same number of offsprings as the pure genetic algorithm, the mutation rate of this approach was increased to the crossover rate (0.8) of the genetic process. The stopping criteria for each technique is to locate a surface point with 95% or higher segmentation quality. In order to ensure correctness of the results each image was tested by each technique 100 times and the results were averaged to create the performance figures. Fig. 11 presents the comparison of the optimization capability for three techniques. As the histograms show, the pure genetic algorithm results are much better than the results of the other two approaches for both the training and testing experiments. This demonstrates that the reproduction and crossover operators are critical for the success of genetic algorithms.

4. CONCLUSIONS

We have shown the ability of the adaptive image segmentation system using genetic algorithm to provide high quality (> 95%) segmentation results in a minimal number of segmentation cycles. The performance improvement provided by the adaptive system was consistently greater than ~33% over the traditional approach or the default segmentation parameters [8,13]. Using outdoor data, for the first time, we have shown that learning from experience can be used to improve the performance of the segmentation process. There are many more experiments that we have performed when the segmentation quality is a vector valued function [11] and optimization technique is either a pure genetic algorithm or a combination of genetic algorithm and hill climbing [9]. Although the segmentation and interpretation processes are interlinked, we have investigated how much the segmentation performance can be improved without complicating the adaptive segmentation process with the effects of recognition system performance. There are many ways in which adaptive image segmentation system described here can be used in practical computer vision systems. These research topics are currently under investigation.

Acknowledgements

The authors are grateful to Honeywell Inc. where part of this work was performed under an initiative grant.

References

1. K. DeJong, "Learning with Genetic Algorithms: An Overview," *Machine Learning* **3** pp. 121-138 (1988).

2. J.M. Fitzpatrick, J.J. Grefenstette, and D. Van Gucht, "Image Registration by Genetic Search," Proceeding of IEEE Southeastern Conference, pp. 460-464 (1984).

3. K.S. Fu and J.K. Mui, "A Survey on Image Segmentation," *Pattern Recognition* **13** pp. 3-16 (1981).

4. A.M. Gillies "Machine Learning Procedures for Generating Image Domain Feature Detectors," Ph.D. Thesis, dept. of Computer and Communication Sciences, University of Michigan, (April 1985).

5 D.E. Goldberg, "Computer-Aided gas Pipeline Operation Using Genetic Algorithms and Rule Learning," Ph.D. Thesis, Dept. of Civil Engineering, University of Michigan (1983).

6. R.M. Haralick and L.G. Shapiro, "Image Segmentation Techniques," *Computer Vision, Graphics and Image Processing* **29** pp. 100-132 (1985).

7. J.H. Holland, "Escaping Brittleness: The Possibilities of General-Purpose Learning Algorithms Applied to Parallel Rule-Based Systems," pp. 593-623 in *Machine Learning: An Artificial Intelligence Approach, Vol. II*, ed. R.S. Michalski, J.G. Carbonell, and T.M. Mitchell, Morgan Kaufmann Publishers, Inc.)1986).

8. K.I. Laws, "The Phoenix Image Segmentation System: Description and Evaluation," SRI International Technical Note No. 289 (December 1982).

9. S.K. Lee, "Adaptive Image Segmentation," Ph.D. Thesis, Dept. of Computer Science, University of Utah (July 1990).

10. R. Ohlander, K. Price, and D.R. Reddy, "Picture Segmentation Using a Recursive Region Splitting Method," *Computer Graphics and Image Processing* **8** pp. 313-333 (1978).

11. J.D. Schaffer, "Multiple Objective Optimization and Vector Evaluated Genetic Algorithms," Proceedings of International Conference on Genetic Algorithms and Their Applications, pp. 93-100 (1985).

12. P.G. Selfridge, "Reasoning About Success and Failure in Aerial Image Understanding," Ph.D. Thesis, University of Rochester (1982).

13. S. Shafer and T. Kanade, "Recursive Region Segmentation by Analysis of Histograms," Proceedings of IEEE International Conference on Acoustics, Speech and Signal Processing, pp. 1166-1171 (1982).

14. J.S. Weszka, "A survey of Threshold Selection Techniques," *Computer Graphics and Image Processing* **7** pp. 259-265 (1978).

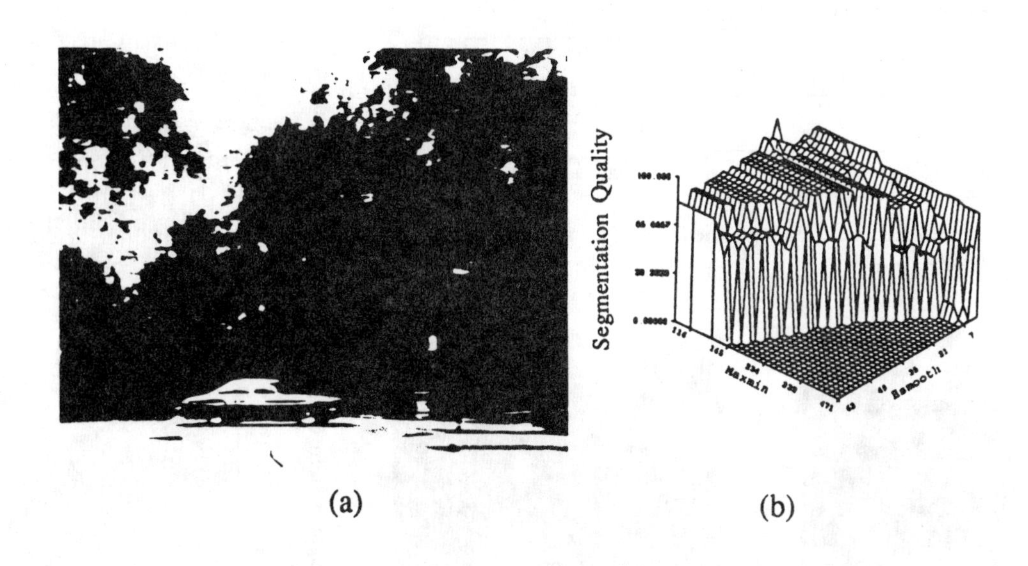

Figure 1: An outdoor image and its associated segmentation quality surface. *(a)* Frame 1 of outdoor image database. *(b)* Combined segmentation quality surface.

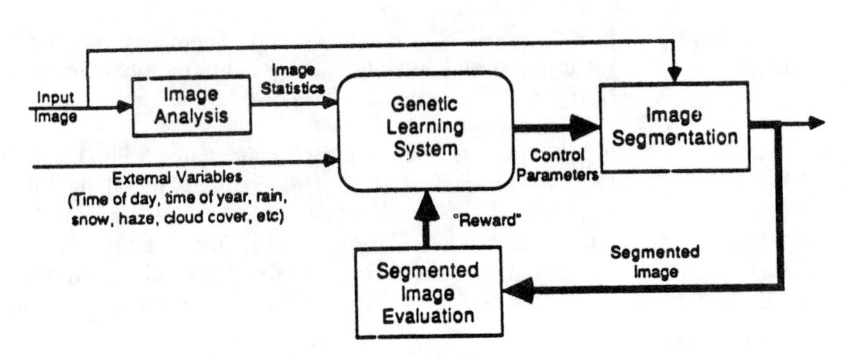

Figure 2: Block diagram of the adaptive image segmentation system.

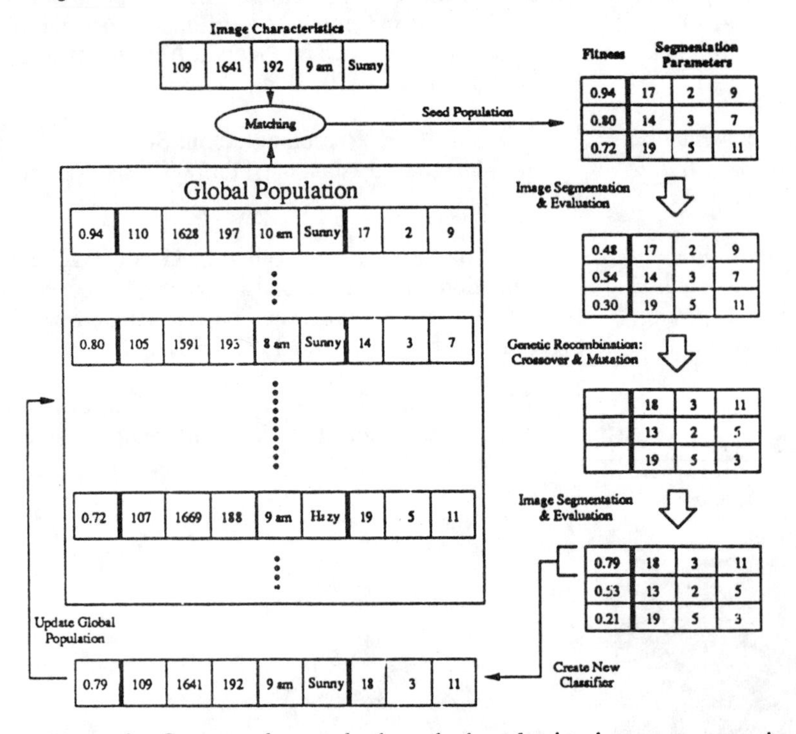

Figure 3: One complete cycle through the adaptive image segmentation

Figure 5: Selected color images from the outdoor experiments.

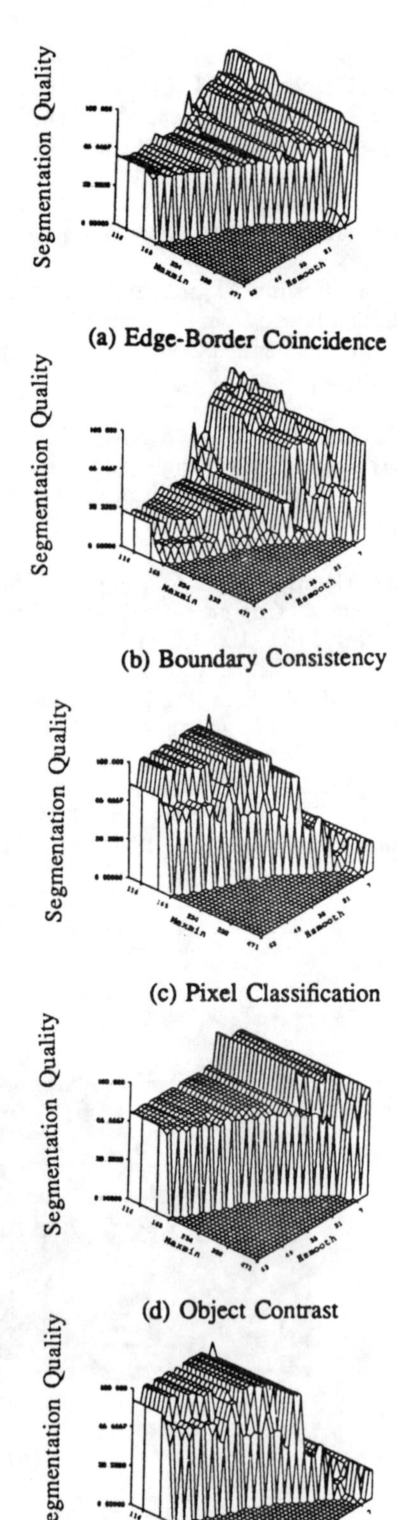

Figure 4: Individual quality surfaces

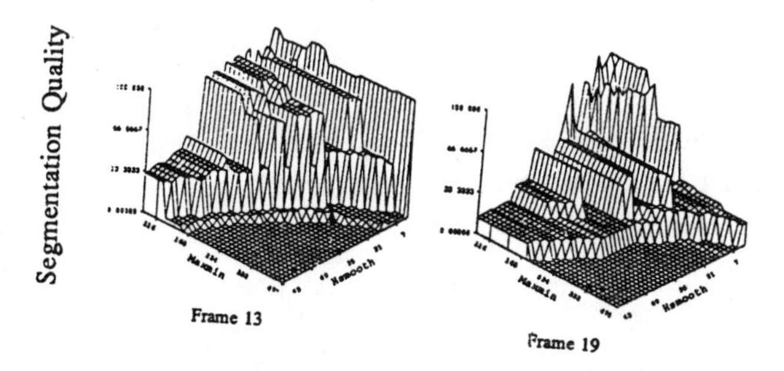

Figure 6: Quality Surfaces for the images shown in Fig. 5

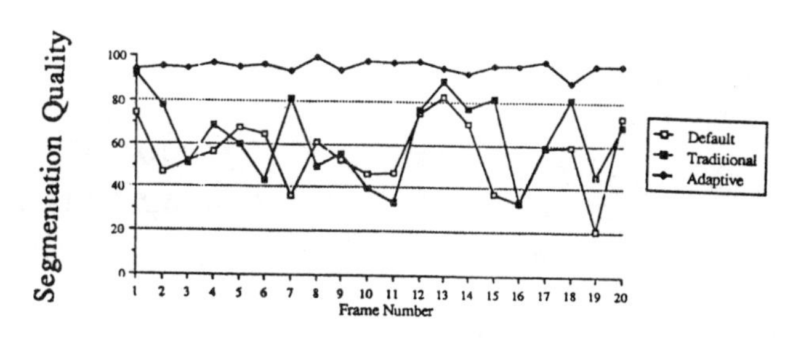

Figure 7: Comparison of the adaptive image segmentation system with default Phoenix parameters and the traditional segmentation approach commonly used in the computer vision field.

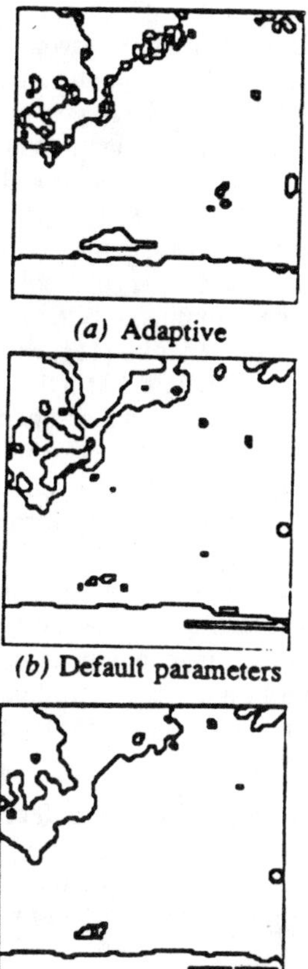

Figure 8: Segmentation performance comparison.

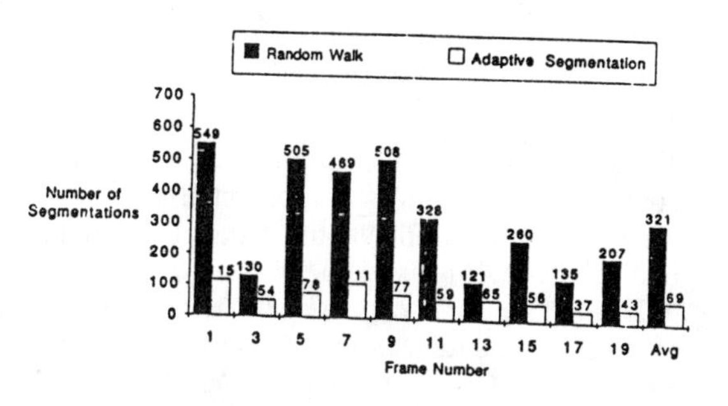

Figure 9: Performance comparison of the adaptive segmentation technique and a random walk approach.

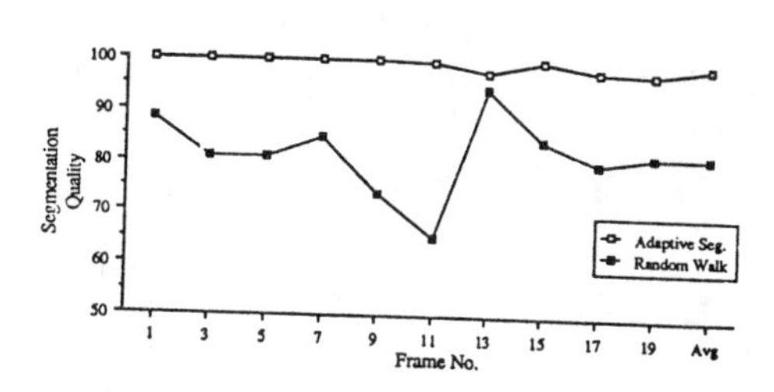

Figure 10: Segmentation quality performance for the adaptive segmentation technique and the random walk approach.

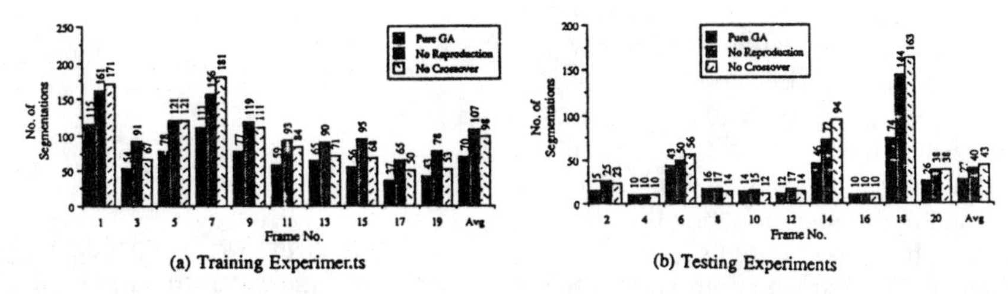

Figure 11 Performance comparison of the pure genetic algorithm and its variations. The superior performance of the pure genetic algorithm demonstrates the effectiveness of the reproduction and crossover operators.

System Identification with Evolving Polynomial Networks

Hillol Kargupta and **R. E. Smith**
Department of Engineering Mechanics
University of Alabama
Tuscaloosa, Alabama 35487

Abstract

The construction of models for prediction and control of initially unknown, potentially nonlinear systems is a difficult, fundamental problem in machine learning and engineering control. In this paper, a *genetic algorithm* (GA) based technique is used to iteratively form polynomial networks that model the behavior of nonlinear systems. This approach is motivated by the *group method of data handling* (GMDH) [11], but attempts to overcome the computational overhead and locality associated with the original GMDH. The approach presented here uses a multi-modal GA [4] to select nodes for a network based on an information-theoretic fitness measure. Preliminary results show that the GA is successful in modeling continuous-time and discrete-time chaotic systems. Implications and extensions of this work are discussed.

1 Introduction

Many problems in signal processing, learning control, and machine learning can be cast as *system identification* problems, where one must gather data from an system whose structure is initially unknown, and build a model of the system's structure. The resulting model can be used for prediction and control of the initially unknown system. System identification problems have been examined for decades. A substantial amount of work has been done in the identification of systems whose dependent variables (i.e. outputs) are known to vary as a linear function of independent variables (i.e. inputs). However, the identification of nonlinear systems remains a difficult problem. This problem compounds when no initial information is available about the system's structure. Once we assume the structure of the system, the problem reduces to a comparatively trivial job of identifying parameters. Identifying and modeling the structure of potentially nonlinear, black-box systems is a problem of particular significance to engineering control and machine learning research.

Connectionist systems [18] have shown great potential in their ability to model nonlinear input-output relationships. Genetic algorithms (GAs) [7] have proven themselves effective for search in complex spaces with little prior information. Several studies have attempted to marry these two naturally-motivated techniques [17]. This study takes a novel approach to nonlinear system identification by using a genetic algorithm to form layers of a polynomial network to model an unknown, nonlinear system. This approach is motivated work of Ivakhnenko [11], who introduces the *group method of data handling* (GMDH). The GMDH is a procedure that attempts to model an unknown system by iteratively connecting layers of nodes that compute polynomial functions. Unfortunately, this method suffers from huge computational overhead and the locality of its search procedure. This paper improves on the GMDH by using a GA that forms nodes in the network layer-by-layer, and employs an information-theoretic measure of effectiveness as each nodes fitness. This technique is tested in the identification of continuous and discrete time chaotic systems. Results show that the GA forms effective models of these systems.

The remainder of this paper introduces the GMDH, GA-based modifications to this scheme, an information-theoretic fitness measure, and experimental results. Conclusions discuss implications and possible extensions of this research.

2 Polynomial Networks and the GMDH

The Kolmogorov-Gabor theorem [11] shows that any function $y = f(\vec{x})$ can be represented as

$$y = a_0 + \sum_i a_i x_i + \sum_i \sum_j a_{ij} x_i x_j$$

$$+ \sum_i \sum_j \sum_k a_{ijk} x_i x_j x_k + \ldots$$

where x_i is the ith independent variable in the vector $\vec{x}$. However, the determination of the correct terms and coefficients for this expansion is a difficult task. Ivakhnenko

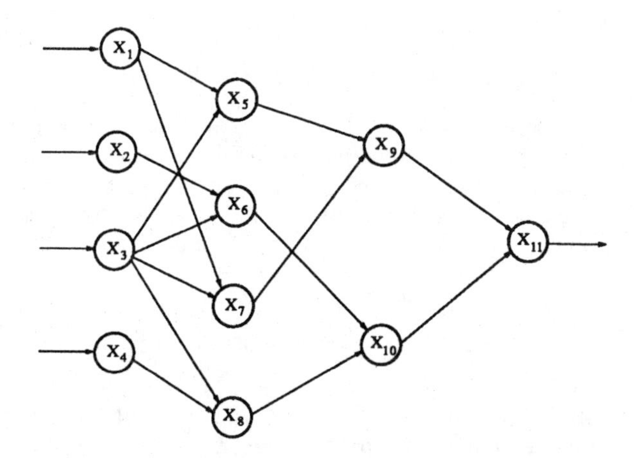

Figure 1: An illustration of a typical GMDH network. Nodes are labeled with their output variable names. Each node computes its output as an elementary function of its input variables, e.g. $x_5 = f(x_1, x_3), x_6 = f(x_2, x_3)$, etc.

suggested the GMDH procedure for determining the appropriate polynomial for a given set of (x, y) data by constructing simple computational networks. GMDH constructs the network's first layer assigning an input node to each independent variable x_i. Subsequent layers are constructed as follows:

1. Construct $\binom{n}{2} = n^2 - n$ nodes for the new layer, where n is the number of nodes in the previous layer.
2. Connect the outputs of pairs of nodes from the previous layer to nodes in the new layer. Each new node receives output from one of the $n^2 - n$ pairings of node outputs from the previous layer.
3. Determine the output of each node in the new layer by fitting an *elementary function* of the node's inputs to the y data. One possible elementary function is a quadratic of the nodes input signals:

$$y' = a + bx_i + cx_j + dx_ix_j + ex_i^2 + fx_j^2,$$

   where the parameters a, b, c, d, e, and f are determined by curve fitting to the y data.
4. Prune high-error nodes from the new layer. If the error associated with a node's output (as compared to the y data) is above a threshold value, that node is eliminated. The original GMDH used the root-mean-square (RMS) error as a basis for pruning new network layers.
5. Repeat this process for each new layer.

This process continues until performance ceases to improve with the addition of new layers. Figure 1 shows a typical GMDH network.

Versions of the GMDH have been applied with some success to economic and environmental modeling problems [5, 10, 11]. However, the technique suffers from two major drawbacks;

- the huge computational overhead associated with enumerating all the possible nodes in each layer, and
- the typical ills of local search techniques.

The limitations implied by the selection of elementary functions are also a target of criticism in the GMDH. Tenorio and Lee [20] have used multiple elementary functions and other modifications to improve the performance of the GMDH. However, the use of multiple elementary functions only increases the computational difficulties associated with GMDH.

One can also criticize the comparison of nodes in intermediate layers with the optimal final output. Discarding nodes at in an intermediate layer based on a final-output criteria could mislead the system to a sub-optimal network. A more suitable approach could be to maintain several nodes in each layer based on an intermediate measure of their relative merit in that layer.

The following section introduces a genetics-based polynomial network construction algorithm attempts to correct the deficiencies of the GMDH procedure.

3 Building Polynomial Networks with a Genetic Algorithm

This section introduces a *genetics-based self-organizing network* (GBSON) as a procedure for constructing polynomial networks. In implementing a genetic algorithm for this purpose, one must consider the multi-modal character of the search problem, and an appropriate measure of a network node's fitness. These issues are considered in the following sections.

3.1 Multi-modal Search with GAs

In constructing a new layer, the GMDH procedure generates hypersurfaces based on elementary functions of all possible pairings of the previous layer's outputs. The procedure then selects amongst these hypersurfaces based on an error criterion. Clearly, this is an optimization procedure over the space of possible hypersurfaces at a given layer. Once hypersurfaces are selected, they are combined to form a new hypersurface space, and, therefore, a new optimization problem. As the layering process continues, previously selected hypersurfaces act as *building blocks* for subsequent layers and the final network output.

Genetic algorithms (GAs) are search procedures that are based on mechanics of natural genetics. Although most GAs are structurally simple (consisting only of primitive selection, recombination, and mutation operators), mathematical analyses and many empirical studies have shown that GAs are robust and effective in linear and nonlinear search domains [7]. Therefore, the GA seems a likely candidate for the complex, nonlinear task of determining hypersurfaces in the GMDH layering process. However, the traditional GA drives towards a single structure as the solution to a given search problem. In forming a layer in a GMDH network, it is necessary to search for several hypersurfaces that may serve as building blocks for a satisfactory final output. Therefore, it is necessary to modify the GA such that it searches for several peaks in the space of possible hypersurfaces.

Niching and *speciation* techniques have been suggested to allow the GA to locate several peaks in a multi-modal optimization problem [8]. Deb [4] presents a detailed examination of the use of GA niching and speciation techniques. The work presented here follows Deb's guidelines on the use of *fitness sharing* and *mating restrictions* for multimodal genetic search.

3.2 An Information-Theoretic Fitness Function

To effectively form polynomial networks with a GA, it is necessary to assign each node a fitness value that adequately reflects that node's utility in the system identification task. One approach would be to simply follow the lead of the original GMDH, and use an objective function based on the node's RMS error. However, this scheme is somewhat myopic, in that it only consider's each nodes utility relative to the network's final goal, without consideration of the node's role as an intermediate structure. Clearly, the problem of assigning utility measures to intermediate structures is directly related to the well-known problems of credit assignment encountered in machine learning systems [13]. This study does not address these problems directly, but it does suggest a fitness measure that reflects both the utility of a node in a polynomial network relative to the network's final goal, and the node's role as an intermediate structure. This fitness measure is based on a information-theoretic measure that describes both the nodes utility and the complexity of the function represented by the node and its connections.

Prediction and system identification problems are intimately interrelated. In prediction problems, the objective is to determine future output of a system based on data from the system's past. The accuracy of such predications depends on how well one can estimate the system's underlying mechanism. This can be viewed as the problem of identifying the system and subsequently determining a model that describes the system's future behavior.

Consider a probability distribution over the a time series of data, $\vec{x}$ generated by an unknown system, $P(\vec{x})$. Also consider an approximation of this function, $P(\vec{x}, m, \vec{\theta})$, given by a model with m parameters $\vec{\theta}$. In general, a prediction problem requires the determination of both m and $\vec{\theta}$. The performance of a prediction model depends on both the model's ability to predict accurately, and its degree of complexity. This leads to the concept of a model's *description length, I*:

$$I_{m,\vec{\theta}}(\vec{x}) = -\log P(\vec{x}, m, \vec{\theta}) + 0.5m \log n.$$

This expression shows that I is influenced by two factors - the code length required to represent the information obtained from the observed data and the code length needed to describe the model itself. For models defined by linear polynomial regression this expression reduces to

$$I = 0.5n \log D_n^2 + 0.5m \log n$$

where D_n^2 is the model's mean-square error [20].

The GA-modified version of the GMDH procedure will use I as its objective function. The GA will attempt to minimize I in each subsequent layer. In effect, the GA is attempting to find a *minimum description length* (MDL). MDL determination is well known problem [1, 15, 16]. The following section describes the network-building algorithm in detail.

4 The GBSON Procedure

The GBSON procedure treats the formation of each layer of a polynomial network as a separate multi-modal function optimization problem. For each layer, GBSON proceeds as follows:

Generate new nodes The first step is to generate GA structures that represent new network nodes. Each node is represented with eight fields of a bit string. The first two fields identify the two nodes from the previous layer to which the represented node is connected. The last six fields represent the coefficients of a quadratic function that determines the node's output:

$$a + bz_1 + cz_2 + dz_1z_2 + ez_1^2 + fz_2^2,$$

where z_1 and z_2 are the outputs of the connected nodes in the previous layer. Note that this differs from the original GMDH procedure in that it is not necessary to generate all possible nodes.

Calculate description lengths For each new node, the description length of the function represented is determined. The number of parameters m is determined by recursively descending the network. The description length is used as a basis for the node's fitness in the GA.

Apply a multi-modal GA A pre-determined number of iterations of a GA are used to search the space of possible nodes for the current layer. A simple GA with tournament selection, single-point crossover, and point mutation is used. The tournament selection scheme favors low description length nodes. Genotypic fitness sharing [4] insures that the GA locates several viable nodes. Under the fitness sharing scheme, shared fitness is given by

$$f_{si} = \frac{f_i}{m_i}$$

where

$$m_i = \sum_{j=1}^{N} Sh(d_{ij})$$

where N is the population size and

$$Sh(d_{ij}) = \begin{cases} 1 - \left(\frac{d_{ij}}{\sigma_s}\right)^{\alpha} & \text{if } d_{ij} < \sigma_s \\ 0 & \text{otherwise.} \end{cases}$$

In the experiments presented here, the distance measure d_{ij} between structures i and j is the number of bits that differ between the two strings. The critical parameter in this scheme is σ_s. Deb [4] presents analyses that show how this parameter can be set based on the number of peaks, q, that are expected in the search space. Specifically,

$$\frac{1}{2^{\ell}} \sum_{i=0}^{\sigma_s} \binom{\ell}{i} = \frac{\ell}{q},$$

where ℓ is the string length. In the experiments presented here, σ_s is determined by assuming that q is the number of nodes in the previous network layer. This is clearly an arbitrary heuristic for determining the number of nodes in a given layer. This procedure could be inappropriate in problems that require extensive fan-out in the polynomial network. However, the heuristic has proven effective in the problems examined here.

Deb [4] suggests that *mating restrictions* can help to stabilize convergence of a multi-modal GA. To implement this a mating restriction in GBSON, the usual crossover procedure is modified as follows: For all structures i that have been selected for reproduction;

1. If i is to be crossed, select its mate j (also from the reproduced group) such that $d_{ij} < \sigma_m$.
2. If no such structure is found, select structure j (from the reproduced group) at random.
3. Cross structures i and j.

In the experiments presented here $\sigma_m = \sigma_s$.

Select nodes After the GA is applied, most of the structures in the population cluster around peaks in the space of possible nodes. Peak nodes are selected to form the new network layer.

The process repeats for subsequent layers until the GA converges to a layer with a single node. The resulting network is taken as a model of the input data.

5 Test Problems: Predicting Chaotic Time Series

Simple chaotic systems are useful for testing GBSON, since they exhibit complex behavior that is difficult to predict, but have structures that are easy to examine and describe. The following sections examine two chaotic time series prediction problems as test cases for GBSON.

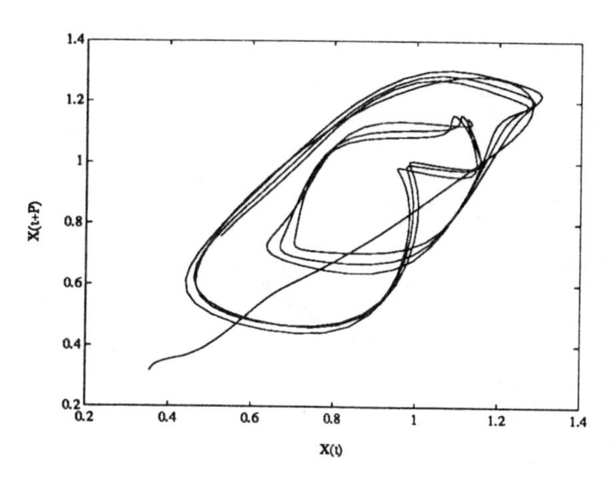

Figure 2: Geometry of $x(t+P)$ versus $x(t)$ for prediction time P = 6.

5.1 A Continuous-Time System: The Glass-Mackey Equation

Chaotic data is generated by solving Glass-Mackey delay equation [12]:

$$\frac{dx}{dt} = \frac{ax(t-\tau)}{1+x^{10}(t-\tau)} - bx(t)$$

where a, b, τ are constants.

Takens [19] shows that past states of a chaotic system can be used to form the state-space representation of the system's dynamics. Taken's theorem states that for a system with an attractor dimension D, $x(t+P)$ can be predicted based on $x(t), x(t-\delta), x(t-2\delta), \ldots, x(t-m\delta)$, where

$$D < m + 1 < 2D + 1.$$

GBSON's goal in this test problem is to determine $x(t+P)$ based on m previous x values.

For τ = 30, a =0.2 and b = 0.1, the equation generates chaotic data with an attractor dimension of 3.5. Figures 2 and 3 show the time series geometries for $P = 6$ and $P = 1$, respectively. The figures show that for short prediction times, local, linear approximation based on small m can work well. However, as the prediction time increases a nonlinear prediction model becomes necessary.

In the test case presented here, τ = 30, a =0.2, b = 0.1, P = 6, m = 4 and δ = 6. GBSON must determine a network that defines a function f such that

$$x(t+6) = f(x(t), x(t-6), x(t-12), x(t-18)).$$

In the tests presented here, 300 points along the Glass-Mackey time series are used as training data. GA parameters are shown in Table 1.

GBSON converges after creating the eight-node network shown in Figure 4. Node outputs are determined as follows:

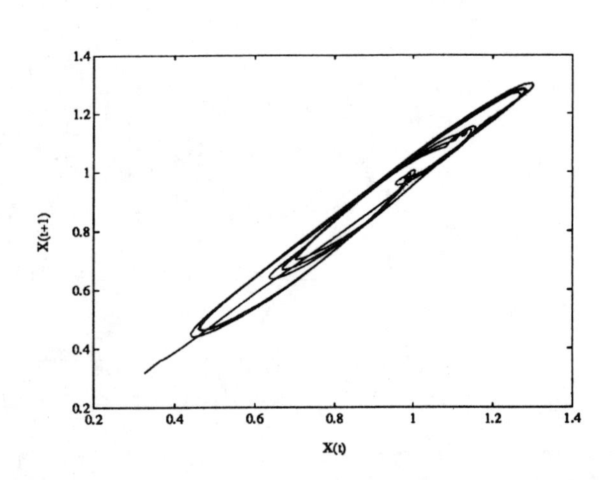

Figure 3: Geometry of $x(t+P)$ versus $x(t)$ for prediction time $P = 1$.

Population size	100
Probability of crossover	0.88
Probability of mutation	0.001
Tournament size	2
Generations per layer	30

Table 1: Parameters used in GBSON test problems.

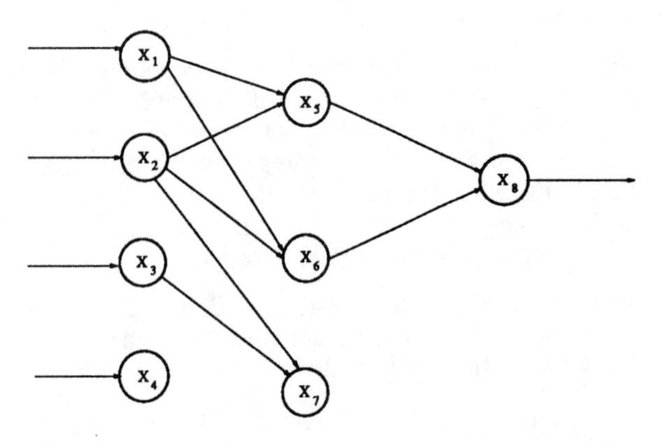

Figure 4: A network generated by GBSON for the Glass-Mackey equation.

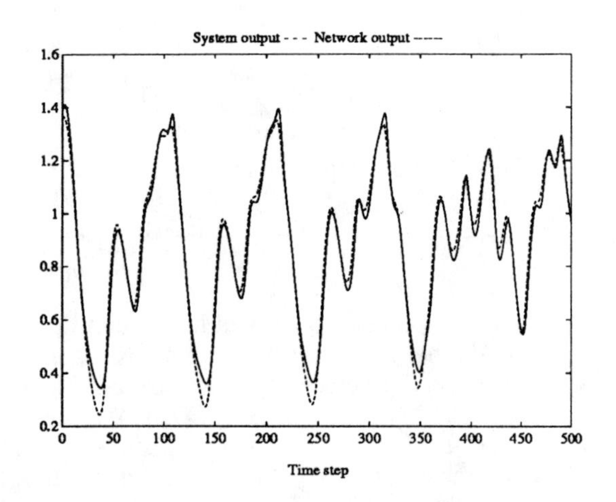

Figure 5: Chaotic data generated by the Glass-Mackey equation along with output generated by a network constructed by GBSON.

$$
\begin{aligned}
x_1 &= x(t) \\
x_2 &= x(t-6) \\
x_3 &= x(t-12) \\
x_4 &= x(t-18) \\
x_5 &= -2.54412x_1^2 - 2.12354x_2^2 + 1.23214x_1 \\
x_6 &= -1.99823x_2^2 + 2.88567x_1 - 0.00125 \\
x_8 &= 0.45547x_6^2 + 0.06563x_5^2 - 0.72348x_6x_5 \\
&\quad +0.78965x_5 - 0.11234
\end{aligned}
$$

Output of the network is shown alongside the actual time series in Figure 5.

5.2 A Discrete-Time System: The Logistic Equation

In this test GBSON models the chaotic time series given by the well-known logistic equation:

$$x(t+1) = \lambda x(t)(1 - x(t))$$

with $\lambda = 3.9$. Four previous x values are used to predict $x(t+1)$. GA parameters kept are the same as in the previous test case. Figure 6 shows the network determined by GBSON. The node outputs are given by:

$$
\begin{aligned}
x_1 &= x(t) \\
x_2 &= x(t-1) \\
x_3 &= x(t-2) \\
x_4 &= x(t-3) \\
x_5 &= -2.14902x_1^2 - 1.77255x_1^2 + 0.894118x_1x_1 + \\
&\quad 1.89804x_1 + 0.67451x_1 + 0.4549
\end{aligned}
$$

Figure 7 shows the output of the logistic equation alongside the output of the network generated by GBSON. It is interesting to note that although the form of equation determined by the GBSON network differs from that of the logistic function, the time-series behavior is similar.

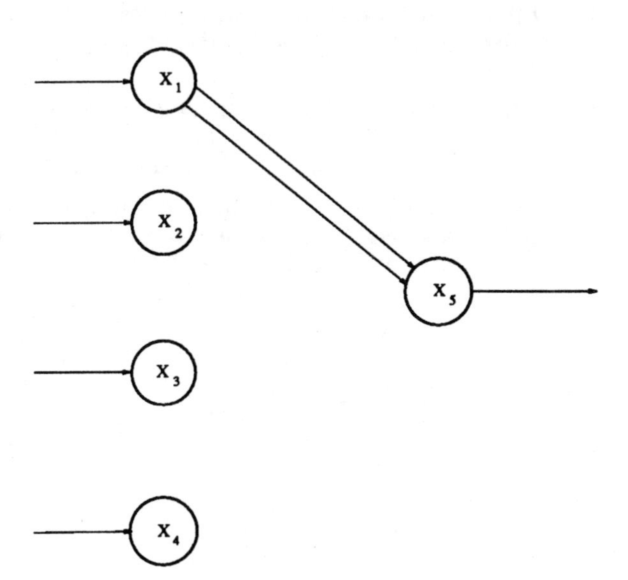

Figure 6: A network generated by GBSON for the logistic equation.

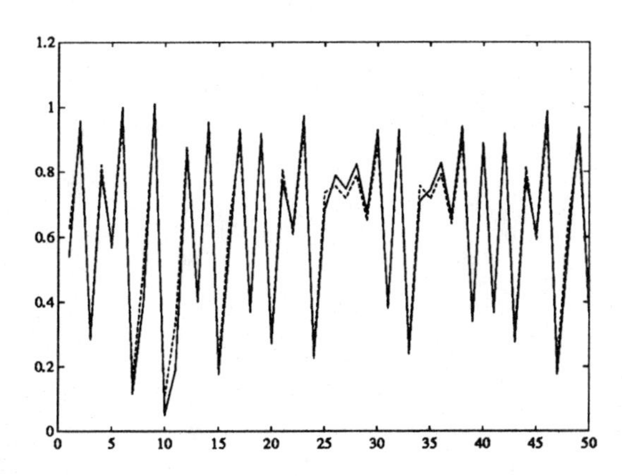

Figure 7: Chaotic data generated by the logistic equation along with the output of a network generated by GBSON.

The prediction of $x(t+1)$ in the logistic problem simply requires the identification of the equation itself. Long term prediction in the logistic equation is a more difficult problem, because of the equation's chaotic behavior. The results presented here are preliminary, and experiments with longer-term prediction in the logistic equation are currently underway.

6 Conclusions and Final Comments

The results presented here are preliminary, but encouraging. The computational leverage of the GA provides a ready means for reducing the computational overhead associated with the GMDH procedure. Results show that the GBSON procedure is successful in short-term prediction of chaotic system dynamics. However, GBSON also points out the two primary difficulties associated with the iterative formation of complex systems with interacting elements. The first of these is the well-known credit assignment problem [13]. GBSON assigns a node's fitness based on its description length, which combines information on the node's utility with information on the complexity of the node's functional role in the network. This technique has proven effective, and may be useful in other complex, evolving systems that are composed of interacting elements. The second difficulty in such systems is the dilemma of cooperation versus competition. In GBSON, this dilemma manifests itself in the need to prune high error nodes from the network, while insuring that sub-optimal nodes can combine their effects in subsequent layers to form an effective nonlinear model. GBSON uses fitness sharing and mating restriction techniques to insure the formation of several nodes at each network layer, despite selection competition between nodes. The key parameter in this approach is σ_s, which determines the number of potentially cooperative nodes that the GA can discover in each layer. GBSON uses the heuristic approach of determining σ_s based on the number of GA-determined nodes in the previous layer. This approach has proven effective in the test problems considered here. However, this heuristic is likely to fail in problems that require considerable network fan-out. The determination of appropriate criteria for eliminating intermediate structures as an evolving network expands is an important issue that deserves further consideration.

Clearly, the problems of credit assignment and cooperation versus competition remain difficult, key issues in complex, evolving, systems with many interacting elements. However, GBSON has shown that complex systems can be modeled with evolving networks that are constructed layer-by-layer. Similar approaches may prove useful in other complex, evolving systems, including other evolutionary-connectionist systems and learning classifier systems [9].

References

[1] H. Akaike. A new look at the statistical model identification. *IEEE Transactions on Automatic Control*,

19:716–723, 1974.

[2] V. V. Chalam. *Adaptive Control Systems: Techniques and Applications*. Dekker, New York, 1987.

[3] R. Das and D. E. Goldberg. Discrete–time parameter estimation with genetic algorithms. *The Modelling and Simulation Conference Proceedings*, 19:2391–2395, 1989.

[4] K. Deb. Genetic algorithms in multimodal function optimization. TCGA Report No. 89002, The University of Alabama, The Clearinghouse for Genetic Algorithms, Tuscaloosa, 1989.

[5] J. J. Duffy and M. A. Franklin. A learning identification algorithm and its application to an environmental system. *IEEE Transactions on Systems, Man, Cybernetics*, 5(2):226–240, 1975.

[6] D. E. Goldberg. System identification via genetic algorithm. 1981.

[7] D. E. Goldberg. *Genetic algorithms in search, optimization, and machine learning*. Addison–Wesley, Reading, MA, 1989.

[8] D. E. Goldberg and J. Richardson. Genetic algorithms with sharing for multimodal function optimization. *Proceedings of the Second International Conference on Genetic Algorithms*, pages 41–49, 1987.

[9] J. H. Holland, K. J. Holyoak, R. E. Nisbett, and P. R. Thagard. *Induction: Processes of inference, learning, and discovery*. MIT Press, Cambridge, 1986.

[10] M. O. Ikeda and Y. Sawarogi. Sequential gmdh algorithm and its application to river flow prediction. *IEEE Transactions on Systems, Man, Cybernetics*, 6(7):473–479, 1976.

[11] A. G. Ivakhnenko. Polynomial theory of complex systems. *IEEE Transactions on Systems, Man, Cybernetics*, 1(4):364–378, 1971.

[12] M. Mackey and L. Glass. Oscillation and chaos in physiological control system. *Science*, pages 197–287, 1977.

[13] M. L. Minsky. Steps toward artificial intelligence. In E. A. Feigenbaum and J. Feldman, editors, *Computers and Thought*, pages 406–450. McGraw-Hill, New York, 1963.

[14] N. H. Packard. A genetic learning algorithm for the analysis of complex data. Tech. Rep. CCSR–89–10, University of Illinois, Beckman Institute - Center for Complex Systems Research and Physics Department, Urbana, 1989.

[15] J. Rissanen. Modeling by shortest data description. *Automatica*, 14:465–471, 1978.

[16] J. Rissanen. Universal coding, information, prediction and estimation. *IEEE Transactions on Information Theory*, 30(4):629–636, 1984.

[17] M. Rudnick. A bibliography of the intersection of genetic search and artificial neural networks. Technical Report No. CS/E 90–001, Oregon Graduate Center, Beaverton, 1990.

[18] D. Rumelhart and J. L. McClelland. *Parallel Distributed Processing: Explorations in the Microstructure of Cognition*, volume 1. MIT Press, Cambridge, 1986.

[19] F. Takens. Detecting strange attractor in turbulence. In D. Rand and L. Young, editors, *Lecture notes in mathematics*, page 366. Springer Verlag, Berlin, 1981.

[20] M. Tenorio and W. T. Lee. Self-organizing network for optimum supervised learning. *IEEE Transactions on Neural Networks*, 1(1):100–109, 1990.

Hybridizing the Genetic Algorithm and the K Nearest Neighbors Classification Algorithm

James D. Kelly, Jr.
Bolt Beranek and Newman, Inc.
10 Moulton Street
Cambridge, MA 02138
jkelly@bbn.com

Lawrence Davis
Tica Associates
36 Hampshire Street
Cambridge, MA 02139

Abstract

In this paper we describe GA-RWKNN, a hybrid of a genetic algorithm and a classification algorithm called k nearest neighbors. GA-RWKNN uses the genetic algorithm and a training set of data records to learn rotations of the data set members and scaling factors (real-valued weights) for each attribute. The purpose of the rotation is to align the data in a manner such that the differential scaling along the new axes diminishes intraclass differences and magnifies interclass differences among data set members. The GA-RWKNN algorithm uses the k nearest neighbors algorithm to classify new data records based on their weighted distance from the rotated members of the training set. We describe experiments in which the classification results obtained with the hybrid algorithm exceed the performance of the simple k nearest neighbors algorithm and ID3. We conclude by describing fertile areas for future work.

1 INTRODUCTION

The field of automated classification has recently seen a number of advances, many of them stemming from the combination of statistical approaches with other approaches. In earlier work, we showed how the genetic algorithm may be used to improve the performance of k nearest neighbors, a statistical classification technique. In this paper, we describe a more powerful way to combine the genetic algorithm with k nearest neighbors techniques, with strong increases in classification performance.

Below, we describe the weighted k nearest neighbors algorithm and we describe the algorithm developed in our earlier work. We then describe types of problems for which weighted k nearest neighbors algorithms are not effective and we describe new techniques for using the search power of the genetic algorithm to produce an algorithm with classification potential greater than that of the k nearest neighbors, weighted k nearest neighbors, and ID3 algorithms. We conclude with descriptions of experiments we have run comparing these types of algorithms.

2 THE WEIGHTED K NEAREST NEIGHBORS CLASSIFICATION ALGORITHM (WKNN)

This section summarizes prior work in which we developed a hybrid of the k nearest neighbors algorithm and a genetic algorithm. The hybrid out-performs the k nearest neighbors algorithm, but for reasons we discuss in the next section, it has weaknesses stemming from arbitrary features of the training set.

The type of classification algorithm we will be discussing deals with data points. Each data point has a value for each of an antecedently known set of attributes, and each data point has a value for another field, called its *classification*. The performance of a classification algorithm is typically assessed by first partitioning the data into two exclusive subsets, a *training set* and a *test set*. After the classification algorithm has been trained, or tuned, using the training set, the performance is measured by then having it classify the members of the test set using the tuned algorithm. Given the knowledge of the correct classification of each test datum, one can compute a classification algorithm's performance by computing the ratio of cases in which the classification algorithm assigns the correct classifications to the test set members.

A *nearest neighbor* classification algorithm (NN) is one that uses the training set as a database of examples. Given a member of a test set, the nearest neighbor algorithm assigns it to the class of its nearest neighbor in the training set. The measurement of nearness, or similarity, between two data points has been researched extensively (e.g., [12]). We believe the results of our research are independent of the particular nearness metric used by such algorithms. The similarity metric we used was a Euclidean distance metric, which computes the distance between two data points i and j as follows:

$$d_{ij} = \{\sum_{a=1}^{n}(X_{ia} - X_{ja})^2\}^{\frac{1}{2}} \qquad (1)$$

where X_{ia} is the value of the ath attribute for the ith datum.[1]

[1]The reader should note that we preprocessed all data to stan-

A commonly-employed extension to the nearest neighbor algorithm is that of classifying a new data point by considering the classification of its *k nearest neighbors*, where k > 1. Some resolution of conflicts among the neighbors is made, and the result is a single classification of the new point. We call such algorithms KNN classification algorithms. As is the case for the computation of similarity, we believe the techniques developed by us will be effective whatever resolution mechanism is employed by a KNN algorithm.

KNN algorithms are easy to use. Some classification algorithms (neural networks, for example) require an intensive "learning" period. Some (decision tree algorithms like ID3, for example) require a good deal of processing. KNN algorithms require no preprocessing in order to determine the classification of a new data point. KNN algorithms are often highly effective although, as is the case for all classification algorithms, their effectiveness is critically determined by the nature of the training data and the nature of the domain.

KNN algorithms are particularly weak when the attributes that are important in the classification process vary in significance, since KNN algorithms effectively treat each attribute as equally important in classification. In [9], we discussed a class of algorithms we called *weighted k nearest neighbor algorithms* (WKKN algorithms). A WKNN algorithm uses a vector of weights, one associated with each data point attribute, to determine the nearest neighbors of a new point. These weights are used in determining neighbor distances. Ideally, a WKNN algorithm would weight the most important classification attributes most strongly.

It is important in what follows to understand the effect of weighting the attributes in a nearness computation. Figure 1 displays an example of the effect of attribute weighting. In the left-hand diagram in Figure 1 if the class of the middle datum was not known to be "unfilled box" it would be incorrectly classified as "filled circle". This misclassification is corrected by when the weighting of an attribute is decreased. For this example the weight, or scale, associated with attribute Y is decreased from 1 to 0.5. Thus in the right-hand diagram the class of the nearest neighbor to the middle datum becomes "unfilled box" and it would be correctly classified.

The distance metric we used for our WKNN algorithms was a slight variant of the Euclidean metric (Equation 1, above), where w_a are attribute weights:

$$d_{ij} = \{\sum_{a=1}^{n} w_a(X_{ia} - X_{ja})^2\}^{\frac{1}{2}} \quad (2)$$

The principal problem in using WKNN algorithms lies in ascertaining an optimal vector of weights, since the range of possibilities is extremely large, and the domain of possible vectors is not a unimodal one. The purpose of our earlier work was to show that when a genetic algorithm is used to search for high-performance vectors, it finds them

dard units in our experiments, so that the preprocessed attribute values had 0 mean and unit standard deviation.

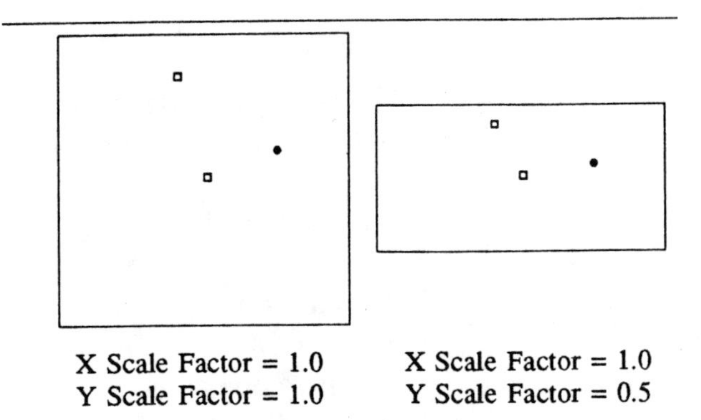

X Scale Factor = 1.0 Y Scale Factor = 1.0 — X Scale Factor = 1.0 Y Scale Factor = 0.5

Figure 1: Example of the Effect of Attribute Weighting.

quickly enough to improve the performance of the KNN algorithm on our test problems, and to make the WKNN algorithm a candidate for real-world use. We called hybrid algorithms using a genetic algorithm to find weight vectors and the WKNN classification technique GA-WKNN algorithms.

In our earlier work, we also investigated variably weighting the effect of the k neighbors in the decision mechanism. For some problems and training sets, the k neighbors are of equal importance in deciding on the classification of a new data point. For some problems, the nearest neighbor is much more important. And so on. We computed the voting strength V for a class i in this way:

$$V_i = \sum_{j=1}^{k} \begin{cases} W_j & \text{if } C_j = i \\ 0 & \text{if } C_j \neq i \end{cases} \quad (3)$$

where W_j is the weight associated with the jth nearest neighbor and C_j is the jth neighbor's class. The class of the focus datum is determined to be that class with the largest V. Again, the genetic algorithm was able to find high-performance weights for the decision process when finding an optimal set of neighbor weights in conjunction with an optimal set of attribute weights would be a prohibitively time-consuming process.

3 ROTATED, WEIGHTED K NEAREST NEIGHBORS ALGORITHMS

The use of a GA-WKNN algorithm extracts information from the training set about the relative importance of the various data point attributes in the classification process. But as we experimented with GA-WKNN algorithms we came to realize that such algorithms are critically affected by an irrelevant feature of the data in the training set – its orientation. This irrelevant feature affects many other classification algorithms as well. In what follows, we describe the problem, our solution in the form of a new algorithm, and empirical comparisons of our algorithm with others.

The "irrelevant" feature of the data we are discussing is its orientation. In fact, the performance of WKNN algorithms depends critically on the orientation of the data in the training set. The right-hand data set pictured in Figure 2 (shown in Section 7, where we describe our experimentation) demonstrates this fact. (The figure illustrates a data set with two classes, "unfilled box" and "filled circle". Each datum is described by two variables, "X" and "Y".) If these data were rotated 45 degrees to the left or right, a WKNN algorithm could compress the y or x dimension and greatly improve classification performance. In its current orientation, however, the training data set will lead to nearly identical performance for a KNN and WKNN algorithm even if the weights for the WKNN algorithm are optimal. This feature of the data set affects the performance of binary tree algorithms like ID3 as well. Since ID3 makes cuts parallel to the principal axes, poorer performance results when the significant attributes do not lie parallel to these axes.

We would like the orientation of the training data to be irrelevant to the performance of the classification algorithm, as it is for the KNN algorithm. Accordingly, we have experimented with rotating data sets before applying weight vectors to them. Assuming one can find the optimal rotation for a data set, then a WKNN algorithm with an optimal weight vector will have identical performance independently of the initial orientation of the data. We call an algorithm that rotates the data before applying attribute weights a *rotated, weighted k nearest neighbors algorithm*, or RWKNN.

In the next two sections we describe an approach to learning rotation and attribute weight vectors. Our technique uses a genetic algorithm to learn vectors of rotation parameters and attribute weights. As is the case for WKNN algorithms, the genetic algorithm does not necessarily find the optimal rotation and weight vectors for a problem. The significant result of our research is that a genetic algorithm does generate vectors which produce levels of classification performance superior to that obtained with the KNN and WKNN algorithms, and it does so quickly enough to make the algorithm a candidate for real-world classification problems.

4 OUR GENETIC ALGORITHM

The discovery of effective rotation and weight vectors for a RWKNN algorithm is a hard optimization problem with a very large search space. This is just the sort of problem that genetic algorithms have been shown to be good at, and so we hypothesized that the hybridization of a GA with a RWKNN algorithm was a promising way to improve the performance of the RWKNN algorithm.

The genetic algorithm we used is very similar to the real-number genetic algorithm described in [3]. The parameter settings and operator set of that algorithm were not tailored to the RWKNN domain. It is possible that domain-based heuristics could be added to the suite of GA-RWKNN operators, particularly given knowledge of the classification domains. We did not do that in this stage of our research, preferring instead to use the more generic version of a genetic algorithm.

5 THE GA-RWKNN ALGORITHM

The GA-RWKNN algorithm combines the search power of a genetic algorithm with the classification capability of a RWKNN. The goal of the algorithm is to learn rotation and attribute weight vectors that improve KNN classification. Specifics of the GA-RWKNN algorithm are:

- Chromosomes are lists of real numbers. Where there are n attributes associated with each datum in the database and one is considering k neighbors, a chromosome will have 2n+k-1 fields:
  - The first n-1 fields of each chromosome describe a rotation of the data in the attribute space. The rotation fields of the chromosome contain real values between 0 and 90.
  - The next n fields of the chromosome are weights, one associated with each attribute. These fields contain real values between 0 and 1 inclusive. These weights are used in the evaluation and classification phases of the algorithm as described above.
  - The final k fields of a chromosome contain weights, one for each of the k neighbors. These fields contain real values between 0 and 1 inclusive. It is a simple matter to produce problems with k = 3, for instance, in which one ought to weight the nearest neighbor most heavily and the next two neighbors very low. One can also produce problems in which one ought to weight all three neighbors equally. More interestingly, one can produce problems in which the second and third neighbor should be weighted more heavily than the first. These k fields of the chromosome evolve to determine what type of combination of the neighbor measures best fits the problem being solved.
- Chromosomes are evaluated by iterating through each data set element and classifying each data point by using its associated weights in equation 2 to determine the k closest neighbors, and then using the weights in equation 3 to make the class determination. When inserting new chromosomes into the population, we did not determine an absolute fitness. Instead we used a ranking technique. One could use any of a number of ranking criteria. The one we used is as follows:

  The first criterion in ranking chromosomes is the number of number of data which were assigned to an incorrect class by the GA-RWKNN algorithm. Chromosomes with a lower number of misclassifications rank higher than ones with a greater number. Among chromosomes which generate equal numbers of misclassifications there are finer degrees of difference. This function orders such chromosomes by additionally considering 1) the number of k neighbors which are of the same class (*k same neighbors*), 2) the total distance to the *k same neighbors*, 3) the number of k neighbors which are not of the same class (*k different neighbors*), and 4) the total distance to the *k different neighbors*. While there are a number of plausible ways to rank chromosomes using these criteria, we settled on the following cascaded method, which ranks chromosome X higher than chromosome Y if:

$$nm_X < nm_Y \quad \textbf{or;}$$

$$\begin{aligned} nm_X &= nm_Y \quad and \\ nks_X &\geq nks_Y \quad and \\ dks_X &< dks_Y \quad and \\ dkd_X &> dkd_Y \quad \textbf{or;} \end{aligned}$$

$$\begin{aligned} nm_X &= nm_Y \quad and \\ (dks_X/nks_X + dkd_X/nkd_X) &< (dks_Y/nks_Y + dkd_Y/nkd_Y) \end{aligned}$$

Where :

nm = *total number of misclassifications,*

nks = *total number of k same neighbors (maximum equals k × number of examples),*

nkd = *total number of k different neighbors,*

dks = *total distance to the k same neighbors, and,*

dkd = *total distance to the k different neighbors.*

In the experiments reported here, the GA-RWKNN algorithm evolves a single chromosome which is used to discriminate two classes of data. A natural extension of our technique is to train a chromosome for each class, normalize the neighbor and attribute fields of each chromosome so that the sum of the neighbor fields and the sum of the attribute fields is a constant, and then assign new instances to the class whose GA-derived chromosome produces the closest neighbors.

6 EXPERIMENTATION

Experimentation focussed on demonstrating the operation and utility of the GA-RWKNN algorithm and comparing the classification performance of the GA-RWKNN algorithm to that of the KNN and ID3 algorithms. In this section we discuss our data sets, the specific GA-RWKNN operating parameter settings used in our experiments, testing methods, and our results.

6.1 DATA SETS

In the development and testing of the GA-RWKNN algorithm we used artificial data sets, generated to test features of the algorithm in controlled settings. In this paper we present the results of two artificial data sets. In [9], we presented results with other data sets.

- **Data Set 1: Disjunctive with Additional Random Values**

  This data set is comprised of 200 data points, 5 attributes, and 2 decision classes. It is partially illustrated in Figure 2A. In the figure each symbol represents a datum from one of the two classes. Each datum is displayed based on its values for two of the five attributes, X and Y. The values for these two attributes were generated randomly (using the Common Lisp random function) within disjoint rectangular areas, the only constraint being that each of the 4 areas contain a quarter of the total data points. The other three attribute values which describe each datum were generated randomly, irrespective of the datum's class.

- **Data Set 2: Diagonal Disjunctive**

  This data set is comprised of 210 data points, 2 attributes, and 2 decision classes. It is illustrated in Figure 2B. Data points were generated randomly within disjoint rectangular areas, the only constraint being that each of the 6 areas contain 35 points. As illustrated in the figure the areas are rectangular, with boundaries rotated 45 degrees from the x-y axes. (Note that Data Set 2 was generated independently of Data Set 1. It is not a modification of Data Set 1.)

6.2 GA-RWKNN PARAMETERS

The population size of the GA-RWKNN algorithm was 50 in each test. Runs terminated after 600 individuals had been produced. We hypothesize that as the number of attributes increases, the population size and run length should probably increase as well to achieve good performance. We used the steady-state without duplicates reproduction technique [14; 11].

The operators used were *uniform crossover*, *average values*, *real number mutation*, and two real number creep operators, one with a large creep range and one with a smaller creep range. The parameter settings of these operators were as follows. Real number mutation replaced a field on a chromosome with a 10% probability. The new number was randomly generated from the interval between 0 and 1. The first creep operator (*large creep*) altered a chromosome field with 10% probability. The amount of the alteration was a randomly generated number between 0 and 0.25 in magnitude. The second creep operator (*small creep*) altered a chromosome field with 5% probability. The amount of the alteration was a randomly generated number between 0 and 0.1 in magnitude. The creep operator altered values up or down with equal probability.

Only one of these operators was used in any reproduction event. The number of parents used and children created in a reproduction event depended on the operator employed. The uniform crossover and average values operators used two parents. Uniform crossover produced two children and average values produced one. The other three operators, real number mutation and large and small creep, used one parent and produced one child. The relative probabilities that these operators would be selected for use in a reproduction event was held constant at 30%, 6%, 20%, 29%, and 15%, respectively, over the course of the run. These values had been found to perform well over a range of real number optimization problems during the research

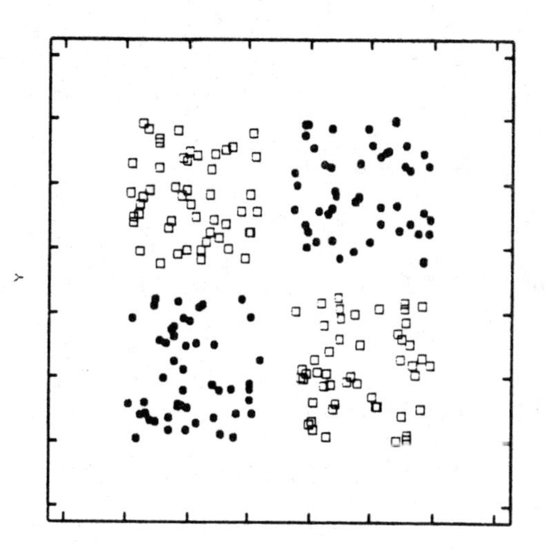

2A: Data Set 1

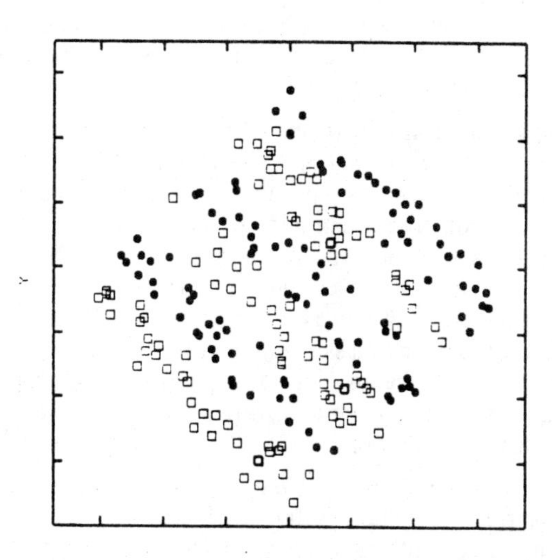

2A: Data Set 2

Figure 2: Data Sets

reported in [3].

Fitness was assigned in the following way. The population was rank ordered using one of the two ranking functions described above. Then each member was assigned a fitness from the series $[1000c^0, 1000c^1, 1000c^2, \ldots, 1000c^{50}]$, except that where any of these values fell below 1 it was replaced by 1. The value of c was not held constant during the runs. At the beginning of each run c was set to 0.95. At the end, c was set to 0.7. Intermediate values were the result of interpolating between 0.95 and 0.7. The effect of this technique is to produce mild pressure in favor of the best population members when the run begins. The curve steepens over the course of the run as the population of solutions converges on similar individuals. The steeper curve increases the selection pressure, causing the algorithm to focus more and more on the best individuals in the population.

While we experimented with various values of k throughout this research, we did not do an exhaustive search for optimal settings. For the sake of consistency we set k = 3 during our experiments.

Our genetic algorithm was created using an early version of OOGA (the Object-Oriented Genetic Algorithm). OOGA was written in Common Lisp and the Common Lisp Object System to accompany [3] and is described in [6].

6.3 TRAINING

We used the *cross validation* error estimation technique, described in [1]. Each data set was divided into five partitions. We generated five training/test sets for each data set, merging four out of the five partitions as the training data and using the remaining partition for testing. Thus, all testing was done with data not seen in the preceding training run. The results presented here are based on ten training runs (2 data sets, five distinct training/test sets for each) for each of the three classification techniques.

Our experiments were run on a Symbolics Ivory-based Lisp machine. Training times were on the order of fifteen minutes for Data Set 1 (without rotation) and one hour for Data Set 2. The minimization of operating time was not a goal of this experimentation and certain algorithmic and software optimizations will decrease operating times (we did not, for example, cache a list of potentially closest neighbors to each test point but computed instead the distance of a test point to every member of the data set). There are many ways to cut our run times significantly. But even the aforementioned times are reasonable for many applications and would not prohibit the use of the GA-RWKNN algorithm for real problems.

6.4 RESULTS AND DISCUSSION

The first part of this section discusses error rates for the three classification techniques we compared and the second part discusses several details of the GA-RWKNN performance.

Figure 3 summarizes the test set error rates for GA-RWKNN, KNN (unweighted k nearest neighbors), and

	GA-RWKNN	KNN	ID3
DATA SET 1	4%	20%	14%
DATA SET 2	13%	42%	29%

Figure 3: Classification Error Rates.

ID3 for the two data sets described above.[2] Each cell contains the test set classification error rates, averaged across the five partitions.[3]

To summarize Figure 3, the GA-RWKNN algorithm significantly out-performs the two other classification techniques in all 4 comparisons. The genetic algorithm has indeed found weighting vectors that, while not optimal, nonetheless produce better classification performance given the parameter settings above.

Additional insight can be gained by reviewing the chromosomes evolved during each training run. Figure 4 contains the mean and standard deviation of the top ranked chromosome in each training run for each data set.[4]

The results presented in Figure 4 demonstrate several features of our algorithm. The first is that weights evolved for noisy attributes, which hinder classification accuracy, are lower than weights learned for "important" attributes. This is illustrated by the chromosomes evolved for Data Set 1. As can be seen in Figure 4, the weights associated with the three attributes whose values were randomly generated without regard for class membership (R1, R2, R3) are considerably less than the weights associated with the attributes whose values do correlate with class membership (X, Y). The algorithm worked correctly in diminishing the effect of the three randomly generated variables.

The chromosomes evolved from Data Set 2 demonstrate both the utility of a rotation parameter as well as the differential weighting of attributes. The GA-RWKNN algorithm achieved superior classification performance in this data set by evolving a rotation magnitude which served to align class boundaries with the attribute axes. Corresponding with the rotation, weighting factors evolved which could take advantage of the new alignment to minimize the boundaries between classes. If one visualizes the points in Figure 2B as rotated approximately 45 degrees clockwise, one can see that confusions at class boundaries of this rotated data set will be reduced by associating a greater scale value with the Y axis than with the X. Such rotation and scaling is evident in the evolved chromosomes from Data Set 2.

Data Set 1:

	X	Y	R1	R2	R3	K1	K2	K3
mean:	**0.64**	**0.76**	**0.06**	**0.01**	**0.15**	**0.67**	**0.32**	**0.31**
dev:	0.06	0.14	0.06	0.01	0.24	0.12	0.25	0.23

Data Set 2:

	Θ	X	Y	K1	K2	K3
mean:	**39.60**	**0.12**	**0.91**	**0.78**	**0.45**	**0.38**
dev:	7.55	0.13	0.13	0.07	0.29	0.19

Figure 4: Evolved Chromosome Fields (Average Across 5 Training Runs)

Mention should also be made of the weights evolved for the three nearest neighbors. The average k weights for Data Set 1 suggest that the closest neighbor is the sole determinant of class membership, as the sum of the average values evolved for neighbors two and three is less than the average value evolved for neighbor one. Data Set 2 results suggest that any two out of the three closest neighbors determine class membership. The setting of the k parameter and the analysis of the performance of GA-RWKNN in evolving optimal weights for the neighbors is one area for future work.

Another note of interest is the number of nodes in the decision trees generated by ID3. Prior to any pruning, ID3 generated trees with an average of 29 nodes for Data Set 1 and 80 nodes for Data Set 2. In general, MDL pruning [10] decreased the number of nodes in the range of 30% to 50%, but with corresponding increases in classification errors. Aside from classification performance differences, it seems to us that the weight vectors evolved by the GA-RWKNN algorithm provide better characterizations of the data sets than decision trees with these number of nodes.

7 CONCLUSIONS

We have described a new type of classification algorithm that improves the performance of the k nearest neighbors algorithms given additional computer time. We have

[2]ID3 error rates are for unpruned decision trees. Pruned trees yielded higher error rates.

[3]For example, the value in the upper left cell, 4%, is determined by totaling the number of misclassifications in each of the five test runs (8) and dividing by the total number of all test data points (200).

[4]Note that the chromosome for Data Set 1 does not include rotation factors. This data set was set up early in the development of our algorithm to examine just the differential scaling mechanism; experiments with Data Set 1 were run prior to the implementation of the rotation mechanism. Given the structure of the data a strong hypothesis can be made that learned rotation weights will be random and will not effect classification performance. This hypothesis will be tested in further experimentation.

shown that the new algorithm, GA-RWKNN, is superior to KNN and ID3 on two problems created by us.

It is a simple matter to create problems for which ID3 will be superior to GA-RWKNN. Each type of classification algorithm has problems for which it is well-suited and problems for which it is ill-suited. One important feature of GA-RWKNN, in our view, is that it does not have the sensitivity to data orientation that affects ID3 and WKNN. In fact, one could use a genetic algorithm to find rotation parameters that improve the performance of ID3, just as we did for WKNN.

A point we would like to stress again is that there is a place for the GA-RWKNN algorithm in the real world. There are many well-known classification problems for which the performance of KNN algorithms exceeds the performance of other standard classification algorithms. What we have done here is to show how to improve the performance of WKNN algorithms on such problems. Weighting attributes and rotating data have not been used in a realistic classification technique before because finding an optimal weight vector and an optimal rotation is a prohibitively time-consuming process. The genetic algorithm does not find optimal weight vectors or rotations. It does, however, find vectors that significantly improve the performance of the KNN algorithm, and in periods of time that are appropriate for a wide range of classification tasks.

We would like to note a related area of research, that of using a genetic algorithm to find improvements in the data set stored in an associative memory of the type described by Kanerva. David Rogers, a genetic algorithm researcher working with Kanerva, has employed genetic algorithms to generate new data points [Rogers, personal communication]. Rogers believes that using similar weighting techniques may improve the effectiveness of Kanerva networks for classification.

We find this research suggestive, and expect that we will discover further connections as our work develops. There are a number of issues that one might address in future work, including these:

- Alternative distance/similarity metrics;
- Alternative k values. One interesting idea is to have algorithm evolve this setting rather than it being statically set a priori;
- Formal characterization of the example sets on which this algorithm out-performs other classification techniques such as ID3 and CART;
- Comparison of the genetic algorithm as a technique for finding attribute weight, neighbor weight, and rotation vectors with other search techniques;
- Class-based derivation of attribute weight vectors and rotation vectors;
- Incorporation of domain knowledge into the genetic algorithm in the form of heuristic operators; and
- Improvements in computation time.

Acknowledgements

We would like to thank Ken Anderson, Albert Boulanger, Herb Gish, William Salter, and Gil Syswerda for comments which have helped us in working out our research program and presentation of results. Bolt Beranek and Newman Inc., Systems and Technologies Division provided the facilities in which we conducted our research.

References

[1] L. Breiman, J. Freidman, R. Olshen, and C. Stone. *Classification and Regression Trees*. Wadsworth, Monterrey, CA, 1984.

[2] W. L. Buntine. *A Theory of Classification Rules*. Ph. D. Thesis, School of Computing Science, University of Technology, Sydney, 1990.

[3] L. Davis (editor). *The Handbook of Genetic Algorithms*. Van Nostrand Reinhold, New York, 1991.

[4] R. Duda and P. Hart. *Pattern Classification and Scene Analysis*. Wiley, New York, 1973.

[5] B. Everitt. *Cluster Analysis*. Heinemann, London, 1974.

[6] J. J. Grefenstette, L. Davis, and D. J. Cerys GENESIS and OOGA: Two Genetic Algorithm Systems TSP Publications, P. O. Box 991, Melrose, MA, 1991.

[7] J. Holland. *Adaptation in Natural and Artificial Systems*. University of Michigan Press, Ann Arbor, 1975.

[8] I.T. Jolliffe. *Principal Component Analysis*. Springer-Verlag, New York, 1986.

[9] J. Kelly and L. Davis. A Hybrid Genetic Algorithm for Classification. In *Proceedings of the Twelfth International Joint Conference on Artificial Intelligence*, Sydney, August, 1991.

[10] J.R. Quinlan and R.L. Rivest. Inferring Decision Trees Using the Minimum Description Length Principle. *Information Processing and Computation*, 80:227-248, 1989.

[11] G. Syswerda. Uniform Crossover in Genetic Algorithms. In *Proceedings of the Third International Conference on Genetic Algorithms*, Morgan Kaufman, San Mateo, CA, 1989.

[12] S. Vosniadou and A. Ortony (editors). *Similarity and Analogical Reasoning*. Cambridge University Press, Cambridge, 1989.

[13] S. Weiss and I. Kapouleas. An Empirical Comparison of Pattern Recognition, Neural Nets, and Machine Learning Classification Methods. In *Proceedings of the Eleventh International Joint Conference on Artificial Intelligence*, Detroit, MI, August, 1989.

[14] D. Whitley. GENITOR: A Different Genetic Algorithm. In *Proceedings of the Rocky Mountain Conference on Artificial Intelligence*, Denver, CO, 1988.

G/SPLINES: A Hybrid of Friedman's Multivariate Adaptive Regression Splines (MARS) Algorithm with Holland's Genetic Algorithm

David Rogers
Research Institute for Advanced Computer Science
MS Ellis, NASA Ames Research Center
Moffett Field, CA 94035
(415) 604-6363

Abstract

G/SPLINES are a hybrid of Friedman's Multivariable Adaptive Regression Splines (MARS) algorithm with Holland's Genetic Algorithm. In this hybrid, the incremental search is replaced by a genetic search. The G/SPLINE algorithm exhibits performance comparable to that of the MARS algorithm, requires fewer least-squares computations, and allows significantly larger problems to be considered.

1 INTRODUCTION

Many problems in diverse fields of study can be formulated into the problem of approximating a function from a set of sample points. For functions of few variables a large body of statistical methodology exists; these methods offer robust and effective approximations. For functions of many variables, relatively fewer techniques are available, and these techniques may not perform adequately in the desired high-dimensional setting. The interest in so-called neural-network models is due in part to their performance in these high-dimensional multivariate environments.

One class of algorithms proposed for high-dimensional environments rely on local variable selection to reduce the number of input dimensions during model construction. These methods approximate the desired function locally using only a subset of the large number of possible input dimensions. Some of the members of this class of algorithms are k-d Trees [1], CART [2], and Basis Function Trees [10]. These algorithms build an approximation model starting with the constant model, and refine the model incrementally by adding new basis functions.

Recently Friedman proposed another algorithm in this class, the Multivariate Adaptive Regression Splines (MARS) algorithm [3]. This statistical approach performs quite favorably with respect to many neural-network models. Unfortunately, the algorithm is too computationally intensive for use in problems that involve large (>1000) sample sizes or extremely high (>20) dimensions. This behavior is caused by the structure of the MARS algorithm, which builds models incrementally by testing a large class of possible extensions to a partially-constructed spline regression model, then adding the best extension.

G/SPLINES are a hybrid of Friedman's Multivariable Adaptive Regression Splines (MARS) algorithm with Holland's Genetic Algorithm. In this hybrid, the incremental search is replaced by a genetic search. The G/SPLINE algorithm exhibits performance comparable to that of the MARS algorithm, requires less computation, and allows significantly larger problems to be considered.

In this paper I begin with a discussion of the problem of functional approximation models, and the use of splines in these models. I then describe the MARS algorithm and estimate the number of least-squares regressions it requires. I follow with a description of the G/SPLINE algorithm. I conclude with experiments to illustrate its performance relative to the MARS algorithm and to study properties unique to G/SPLINES.

2 THE PROBLEM

We are given a set of N data samples $\{X_i\}$, with each data sample X_i being a n-dimensional vector of predictor variables $<x_{i1}, x_{i2}, \ldots, x_{in}>$. We are also given a set of N responses $\{y_i\}$. We assume that these samples are derived from an underlying system of the form:

$$y_i = f(X_i) + \text{error} = f(x_{i1}, \ldots, x_{in}) + \text{error}$$

The goal is to develop a model G(X) which minimizes some error criterion, such as the least-squares error:

$$LSE(G) = \frac{1}{N}\sum_{i=1}^{N} (y_i - G(X_i))^2$$

The model G is commonly constructed as a linear combination using some set of basis functions:

$$G(X) = a_0 + \sum_{k=1}^{M} a_k \phi_k(X)$$

Given an appropriate set of basis functions, standard least-squares regression techniques can be used to find a set of coefficients $\{a_k\}$ which minimizes the least-squared error [7]. This process suffers from two major weaknesses. First, if the basis functions for G do not reflect the underlying global structure of the function F, the accuracy of G is likely to be poor. Second, if too many basis functions are used in the approximation, the model may suffer from *overfitting*; while it generates reasonable approximations for F when given a data sample in $\{X_i\}$, previously unseen data samples may generate large errors. See Figure 1.

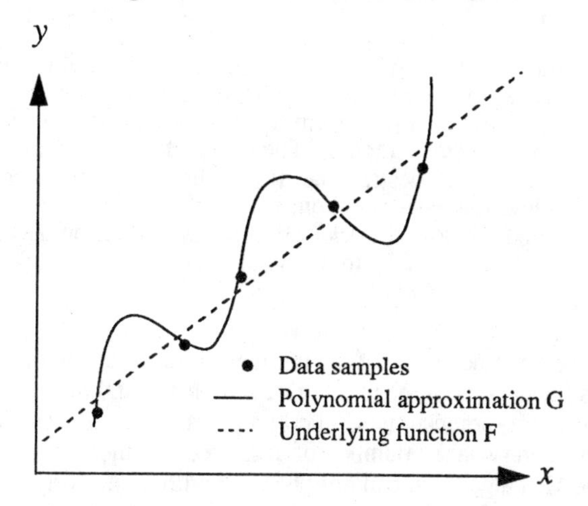

Figure 1: Overfitting. Using polynomials as the basis functions in constructing G, we create an approximation which exactly fits the data sample points but does not approximate the underlying function F well in other regions of the domain.

3 SPLINE APPROXIMATIONS

Spline functions have been used to address some of the difficulties mentioned in the previous section. The basic idea is that if global models are difficult to construct and often poorly behaved, it may be preferable to build a model piecewise using linear or low-order polynomials, each defined locally over some subregion of the domain. Because they are nonzero only in a part of the domain, they can represent local structure of functions that may not have easily-modeled global structure [2].

Such a set of spline basis functions in one dimension is given by:

$$1, x, (x - t_1)_+, (x - t_2)_+, \ldots, (x - t_K)_+$$

which leads to models of the form:

$$G(x) = a_0 + a_1 x + \sum_{k=1}^{K} a_{k+1} (x - t_k)_+$$

(In this notation, the subscript "+" means that the expression is assigned a value of zero if the argument is negative.) This type of spline is called a *truncated power spline*. The variables t_k are called "knots"; they are the locations where the spline functions subdivide the domain. The full basis set has a size (K + 2). A graph of one of these basis functions is shown in Figure 2.

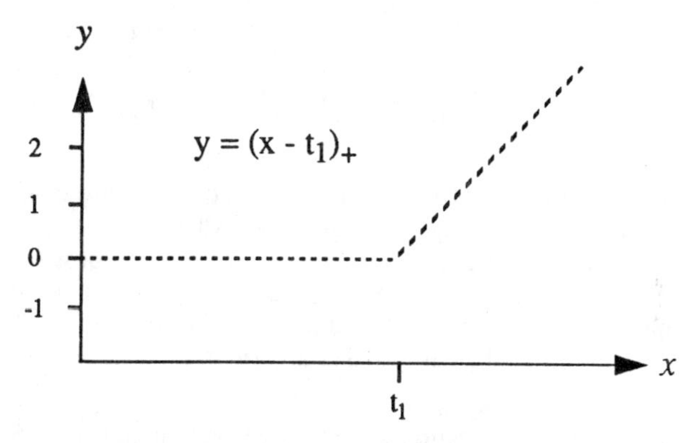

Figure 2: Spline Function. A spline function is zero over part of a domain, and a low-order polynomial over the remainder of the domain. This 1-power spline is continuous but has a discontinuous derivative. A q-power spline is continuous and has (q - 1) continuous derivatives.

Splines perform quite successfully in building low-dimensional models, but the extension to higher dimensions has proven, in the understated words of Friedman, "straightforward in principle but difficult in practice." Specifically, the standard extension of splines to higher dimensions requires $(K + q + 1)^n$ basis functions and the calculation of a corresponding number of coefficients; here, *n* is the number of input dimensions, *K* is the number of knots per dimension, and *q* is the order of the splines. Even for a relatively small number of dimensions, the computational costs of calculating the coefficients and the large number of data samples needed makes the procedure prohibitive.

4 THE MARS ALGORITHM

The MARS algorithm was developed to allow spline approximations in high-dimensional settings. The basic idea is to build the model using only a small subset of the $(K + q + 1)^n$ proposed basis functions. This is done by extending a partial model using an incremental search for the best new partition of the domain. This partitioning is repeated until a model with the desired number of terms is developed.

The algorithm begins with the linear model:

$$G_0(X) = a_0$$

At each partitioning step, the current model is extended by selecting: a basis function currently in the model; a dimension not currently partitioned in that basis function; and a knot location, assigned by selecting in turn the value for that dimension in each data sample. This triple (b, v, t) defines a possible extension to the current model:

$$G_{m+2}(X) = G_m(X) + a_{m+1} BF_b(X)(x_v - t)_+ + a_{m+2} BF_b(X)(t - x_v)_+$$

The coefficients of the newly generated model are computed using least-squares regression. All possible triples of (b, v, t) are tried; the model $G_{m+2}(X)$ which best fits the data samples is selected, and becomes the current model for further partitioning. A more detailed "C" description of the core MARS algorithm is given in Figure 3.[1]

The most computationally-intensive part of the MARS algorithm is the calculation of the least-squares coefficients for the newly proposed model. Thus, one estimate of the cost of building the final model is the number of least-squares regressions that must be performed. The upper limit on the number of models the MARS algorithm must generate and test at a given step is (N*m*n), where N is the number of data samples, m is the current number of basis functions in the model, and n is the number of input dimensions. If the number of basis functions in the final model is M_{max}, the maximum number of models generated is:

$$\text{max models} = (N \times n) \times \sum_{m=1}^{\frac{M_{max}}{2}} (2m+1) = (N \times n)\left(\frac{M_{max}^2}{4} + M_{max}\right)$$

1. Figure 3 contains what Friedman calls the *forward stepwise* portion of the algorithm; I do not describe the *backwards stepwise* portion in this paper. That procedure is a pruning process conducted on the model discovered by the forward stepwise algorithm; it gives some minor additional optimization. (In the experimental, however, both forward and backward stepwise sections were used before comparison with G/SPLINES.)

```
1   Model = constant_model ();
2   for (size = 1; size <= number_of_BFs; size += 2) {
3      lowest_score = INF;
4      for (m = 1; m <= size; m++) {
5         var = NONE;
6         while (var = next_unused_var(BF (m), var)) {
7            for (s = 1; s <= number_of_samples; s++) {
8               if (basis_function_nonzero(BF(m), data[s])) {
9                  knot = data[s][var];
10                 new = partition (Model, BF(m), var, knot);
11                 least_squares_fit (new);
12                 new_score = LOF(new, data);
13                 if (new_score < lowest_score) {
14                    best_model = new;
15                    lowest_score = new_score;
16                 }
17              }
18           }
19        }
20     }
21     Model = best_model;
22  }
```

Figure 3: The MARS Algorithm. The MARS algorithm is an incremental search to find the best new partition to the current spline model, starting with the constant model. The outer loop is over the number of desired partitions; the inner three loops choose a basis function, variable, and knot. The new model is tested after least-squares regression using a lack of fit (LOF) function, and the best new model is used for the next partitioning.

The good news is that the number of intermediate models is linear in both the number of samples N and the number of input dimensions n. The bad news is that this can still be a very large number of intermediate models, so the MARS algorithm can only be used with a relatively small number of data samples and input dimensions. Friedman claims the algorithm is effective for up to 1,000 data samples and 20 input dimensions; this places many interesting problems out of reach of the procedure.

Further, while the MARS algorithm is effective at creating well-performing models for many problems, models which cannot be reached through an incremental search are not discovered. Thus, functions which have a large number of non-linear interactions between the input variables may not be well-suited to modeling using the MARS algorithm.

5 G/SPLINES

The idea behind G/SPLINES is to use a genetic algorithm to do a search using full-size models rather than using the incremental search. The G/SPLINE algorithm starts with a collection of functional models, generated randomly. The coefficients for each model are determined using least-squares regression, and the lack of fit (LOF) over the data set is measured. The inverse of that lack of fit score is used as the fitness criterion. The main cycle begins by probabilistically choosing two parent models based on their inverse LOF score. Crossover is used to generate a new model. Mutation operators may be used to add additional terms to the model. The worst scoring function in the collection is then replaced by this new model. A more detailed "C" description of the G/SPLINE algorithm is given in Figure 4.

```
1   for (i = 0; i < number_of_functions; i++) {
2       fcn[i] = random_function();
3       least_squares_fit(fcn[i]);
4       function_score[i] = 1.0 / LOF(fcn[i], data);
5   }
6   for (i = 1; i < number_of_cycles; i++) {
7       select_parents(function_score, &par1, &par2);
8       child = worst_function(function_score);
9       crossover(fcn[par1], fcn[par2], fcn[child]);
10      if (random_new_mutation())
11          add_new_BF(fcn[child]);
12      if (random_merged_mutation())
13          add_merged_BF(fcn[par1], fcn[par2], fcn[child]);
14      least_squares_fit(fcn[i]);
15      function_score[i] = 1.0 / LOF(fcn[child], data);
16  }
17  model = best_function(function_score);
```

Figure 4: The G/SPLINE Algorithm. The G/SPLINE algorithm is a genetic search over a set of models, replacing the worst model with a crossover of two highly-rated models, using an initial setup of random functions. The models are tested after least-squares regression using the inverse of a lack of fit (LOF) function as the fitness score.

5.1 CROSSOVER

The core process in the G/SPLINE algorithm is a crossover algorithm. In G/SPLINE, two well-performing models are chosen, and the worst-performing model in the system is chosen for replacement. (While there are a number of possible procedures that could be used, basis functions seemed natural as the atomic unit in the crossover algorithm.)

The fitness function used was the inverse of the lack-of-fit function developed by Friedman in his MARS research. The lack of fit function is based on the least-squared error, with an additional penalty term related to the size of the model and the number of data samples. Without this penalty, the size of the models grows without bound, resulting in increased computational costs and increased risk of overfitting. Since both MARS and G/SPLINES use the same lack-of-fit function as their error measure, comparisons between MARS and G/SPLINES are more informative.

The process begins by probabilistically selecting two parents based upon the inverse lack-of-fit score. The two model parents are in this form:

$$\text{Model-1}(X) = a_0 + \sum_{k=1}^{M_1} a_k \phi_k(X)$$

$$\text{Model-2}(X) = b_0 + \sum_{k=1}^{M_2} b_k \beta_k(X)$$

In each parent model, we randomly choose a cut point, and select one of the generated two segments for inclusion into the child. We denote the selected segments of each models with the inclusive sets [Start_1, End_1] and [Start_2, End_2]. We then construct the new child model as:

$$\text{Child-Model}(X) = c + \sum_{k=\text{Start}_1}^{\text{End}_1} d_k \phi_k(X) + \sum_{k=\text{Start}_2}^{\text{End}_2} e_k \beta_k(X)$$

The child model is a linear combination of basis functions derived from each parent. (Some genetic algorithms use the unselected segments to create an additional child; in this initial work, I found that creating both children did not appreciably improve the performance of the G/SPLINES algorithm.) Once the basis functions for the child are determined, the coefficients are derived using a standard least-squares regression.

5.2 MUTATION OPERATORS

As the crossover process proceeds, two effects are seen. First, the number of different basis functions is reduced as combinations of better-approximating basis functions propagate through the models. Second, the models often contain basis functions which contribute no benefit to the quality of the model and increases the cost of the least-squares computation. To counteract these effects I used three mutation operators: NEW, MERGE, and DELETION. After the standard crossover is performed, there is a probability that one or more of these mutation operators may be applied, resulting in the addition or removal of basis functions.

The NEW operator creates a new basis function by randomly choosing an input variable v, a sign s (+1 or -1), and a data sample $\langle t_1, \ldots, t_n \rangle$. These parameters are used to create a basis function of the form:

$$BF_{new}(X) = (s \cdot (x_v - t_v))_+$$

The MERGE operator takes a random basis function from each parent, and creates a new basis function by multiplying them together, that is:

$$BF_{new}(X) = randBF(X, par1) \cdot randBF(X, par2)$$

It is through the MERGE operator that basis functions containing multiplicative terms are introduced.

For both ADD and MERGE, the newly generated basis function is added to the new model generated by the crossover process. This has a cost for functions, which find the crossover search slowed by the additional variance caused by this additional factor. However, in the longer term, this keeps the pool of basis functions from becoming dangerously small, and aids the process in finding high-quality approximations.

The DELETION operator ranks the basis functions in order of minimum maximum contribution to the generated model. Unlike the other two mutation operators, DELETION requires an additional least-squares operation to calculate the coefficients of the generated model. However, while it doubles the number of least-squared operations, it also speeds convergence and encourages compact models.

5.3 ALPHABET CARDINALITY

Considerable study has gone into developing effective codings for genetic algorithms, but coding design remains an art. In this work, an alphabet of high cardinality seemed the most appropriate: the basis functions are the atomic unit in the crossover process, and there is an extremely large number of possible basis functions. However, this choice places the work in the middle of an ongoing debate regarding the size of the alphabet best suited for genetic algorithms.

One school, of which Goldberg [4] is representative, is "almost obsessed with the idea of binary codings." This is not simply a preference for binary representations, but rather a rejection of other representations: "... in general the use of high-cardinality alphabets so severely reduces implicit parallelism that it is inappropriate to call these schemes genetic algorithms in the sense of Holland."

In reality, this argument is a poorly-disguised version of the holism/reductionism debate; what is the "appropriate" level of description? [5] Goldberg may be semantically correct in stating that high-cardinality alphabets do not result in "genetic algorithms in the sense of Holland," but he is mistaken to assume that the use of a high-cardinality alphabet "severely reduces implicit parallelism"; by using a higher-level description, the search (and resulting implicit parallelism) is simply being conducted on a different level of representation, not eliminated. Arguments can certainly be given for and against the use of different alphabets in different situations, but across-the-board claims of superiority for one side or another should be viewed with great suspicion.

6 EXPERIMENTAL

A training and a test data set were created for the experiments. Each data set contained 200 data samples. The function modeled was from Friedman [3]:

$$\begin{aligned} f(X) = {} & 10 \cdot \sin(\pi x_1 x_2) \\ & + 20 \cdot (x_3 - 1/2)^2 \\ & + 10 \cdot x_4 + 5 \cdot x_5 \\ & + \sum_{n=6}^{10} 0 \cdot x_n \end{aligned}$$

The data contained 10 predictor variables; the response is dependent on the first five variables, and independent of the next five variables. Noise was added to the response so that the signal/noise ratio was 4.8/1.0.

The domain for G/SPLINE was a population of 200 functions. Each function was initialized with 10 basis functions generated using the NEW mutation operator. After each crossover, there was some probability that one or more of the mutation operators would be applied to the child. The model with the lowest error on the training data set was chosen for testing against the testing data set.

The model generated by the MARS algorithm was reduced using his backward-stepwise algorithm, then applied to the testing data set.

6.1 Experiment 1

The first experiment was designed to see if the G/SPLINE algorithm could compete with the MARS algorithm in being able to generate models with comparable quality in an environment that favored the MARS algorithm.

The least-squared error versus the number of least-squared regression operations was graphed. The results of this experiment are shown in Figure 5.

This experiment illustrates the rapid learning capability of the G/SPLINE algorithm relative to the MARS algorithm. After 5000 least-squares operations, the MARS model has placed its first knot; the G/SPLINE algorithm is already nearing its asymptote. The final MARS model slightly outperformed the G/SPLINES model, but only after doing an order-of-magnitude more least-squares operations.

Did the model found by the genetic search use relationships in the data set reflective of the underlying function, or only discover easily-modeled patterns in random data? Figure 6 addresses this issue; it demonstrates how the variable use in the set of discovered models indeed reflects the underlying structure of the generating function.

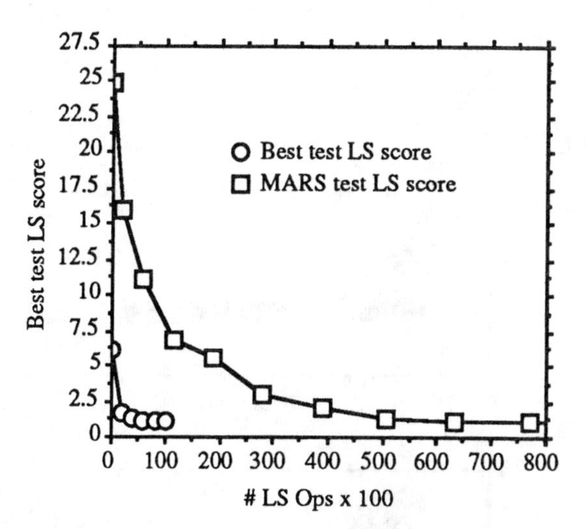

Figure 5: MARS vs. G/SPLINE. Measured in the number of least squares regression operations that must be performed, the G/SPLINE algorithm performs significantly better than the MARS algorithm. The G/SPLINES algorithm was close to convergence after 4,000 least-squared operations (LS ops), and showed no further improvement after 10,000 LS ops. The MARS algorithm was close to convergence after 50,000 LS ops, and showed no further improvement after 80,000 least-squares operations. The final least-squared error of the best G/SPLINES model was 1.17; the final least-squared error of the MARS model was 1.12. The optimal model would have a least-squared error of 1.08 on the test set.

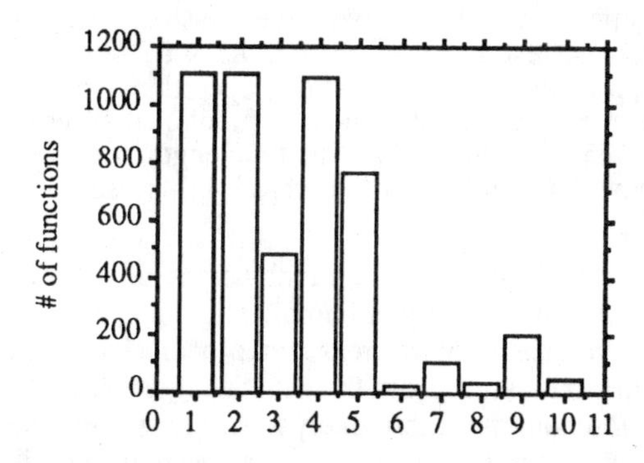

Figure 6: Variable use. The graph is the index versus the number of times the variable is used in some basis function in a discovered model. The variables were counted after 5,000 genetic crossover operations. The underlying function depends on the first five variables, and does not depend on the next five variables. In this case, the use of the variables reflects the underlying function.

The performance of G/SPLINE is reminiscent of the performance curves comparing Genetic Algorithm-based systems with Backpropagation-based neural networks [1]. In these systems, the genetic search is rapid at the beginning of the process, far outperforming the neural network. It is only after the problem is well-developed that the backpropagation algorithm begins to compete with the genetic search; eventually, the Backpropagation-based neural network slightly outperforms an algorithm based solely on genetic search. It is possible that the MARS algorithm and the G/SPLINE algorithm have a similar relationship with respect to their performance and speed.

6.2 Experiment 2

This experiment is identical to the first experiment, with the exception that the function is changed to have 5 dependent and 95 dependent variables, for a total of 100 predictor variables. The size of the data sets was unchanged, with each containing 200 samples. (Note that this change increased the problem size beyond Friedman's stated capabilities of the MARS algorithm.)

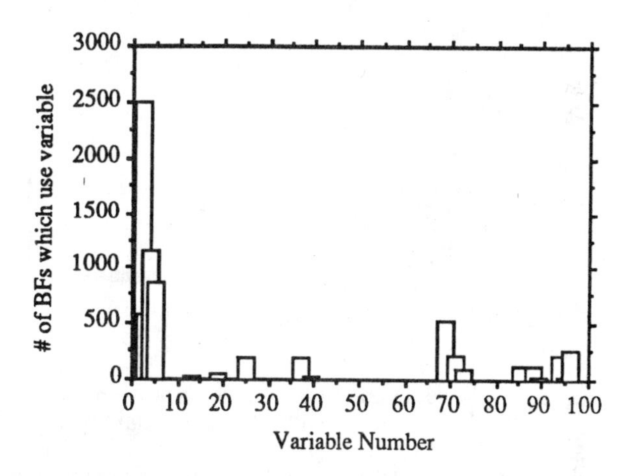

Figure 7: Variable elimination. The graph is the index versus the number of times the variable is used in some basis function. The variables were counted after 10,000 genetic crossover operations. The underlying function depends on the first five variables, and does not depend on the other 95 variables. The five dependent variables were the top five variables in terms of actual use in basis functions.

As Figure 7 shows, even with only 200 samples of data, the G/SPLINES algorithm still discovers the relative importance of the 5 dependent variables amidst the 95 independent variables.

6.3 Experiment 3

In this experiment I wanted to study the effect of sample set size on the rate of elimination of the independent variables.

This experiment is identical to the first experiment, except that two different data set sample sizes were used. The first set was the standard 200-sample set, and the second a 50-sample set.

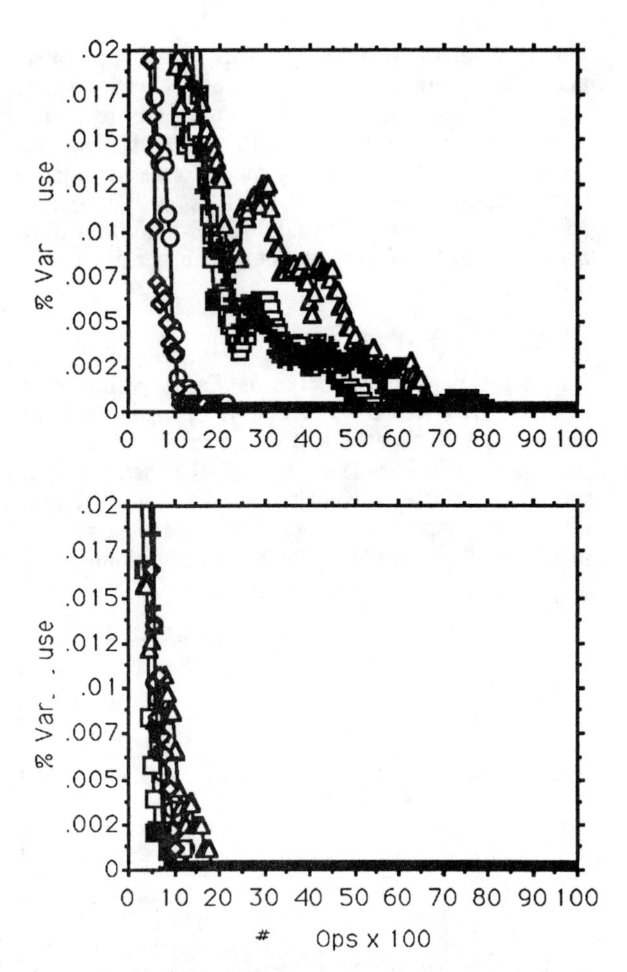

Figure 8: Variable elimination and sample size. The graphs are the number of genetic operations versus the percentage use of the five independent variables. The top graph is for a sample size of 200; the bottom graph for a sample size of 50. In the top graph, the variables are gradually eliminated. For the smaller sample size, the variables are eliminated rapidly.

Figure 8 shows the surprising result that the independent variables are eliminated from consideration *faster* when we have only 50 samples of data rather than 200. This is because G/SPLINES does not eliminate variables through positive identification of them as independent of the input, but rather, these variables get left behind as variables with more predictive power are preferentially selected for crossover. With smaller data sets, there is more pressure for smaller models, which causes a fiercer competition and earlier removal of basis functions which do not substantially contribute to a model's performance.

6.4 Experiment 4

In this experiment I wanted to study the effect of the genetic algorithm on the size of the generated models. The experimental conditions were identical to those of the first experiment.

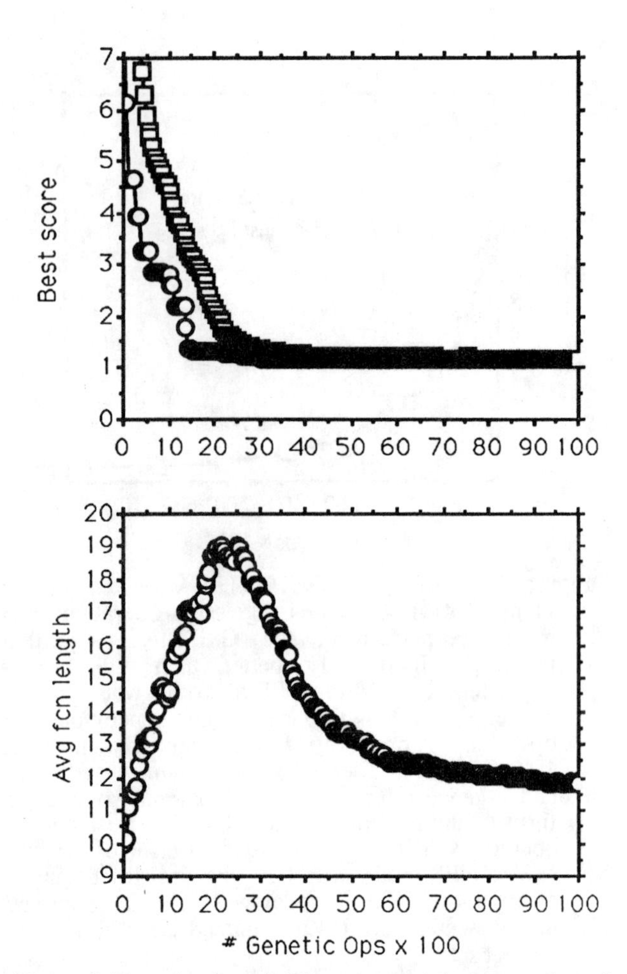

Figure 9: Function length. The top graph is the number of genetic operations versus the least-squared error (squares: test set, circles: training set). The bottom graph is the number of genetic operations versus the average model length (in basis functions). There is a rapid increase in average function length, followed by a slower decrease after the score is minimized.

Figure 9 shows that there is a rapid increase in model size until the score is nearly minimal; after that, there is a slower but consistent decrease in model size. This is likely due to pressure from the genetic algorithm; a compact representation is more likely to survive the crossover operation without loss. Thus, the genetic algorithm encourages compactness of representation in addition to the advantages such compactness affords during scoring. Since we are often interested in compact models (everything else being equal), this is a beneficial effect.

7 CONCLUSIONS

It is difficult to properly compare the utility of two algorithms, even when they share many similar features or are applied to the same data sets. Such comparisons are too often uninformative and unneccessarily harsh to the losing algorithm (which is usually *not* the algorithm developed by the author). My goal in this work was not to supplant the MARS algorithm, which I find elegant in concept and has a proven record of success in practice. Rather, my goal was to extend the reach of the algorithm by proposing a variant that may not suffer some of the limitations of the procedure as proposed by Friedman, yet may retain most of its advantages. I believe I was successful, but will leave the final judgement to the reader.

In this paper, the problems we selected were relatively small, and generated relatively small models, containing perhaps a dozen basis functions. As the complexity of the problems grow, the necessary size of the model will also grow; a genetic approach such as G/SPLINE may be the most effective technique for deriving such models. Similarly, the MARS model is most effective when most of the predictor variables have additive effects on the response; the G/SPLINE model, since it is not based on the incremental search, may be better suited to discover appropriate models in this case.

The number of least-squares regressions performed was proposed as a measure of the inherent efficiency of the algorithms. The structure of the MARS algorithm is such that the cost of the least-squares regression can be greatly reduced, so direct comparisons of the number of least-squares regressions is not truly a measure of the amount of computation involved in executing the algorithms. Still, even with these improvements, the MARS algorithm cannot effectively handle large (>1000) data samples or large (>20) numbers of input variables. It is appropriate to look for algorithms which keep the many advantages of the MARS approach while overcoming its limitations.

Finally, while the G/SPLINE algorithm uses linear splines as its basis functions, there is no reason the algorithm could not use non-linear splines and non-spline basis functions. While splines have attractive properties that make them generate useful models in many circumstances, there is no reason that other functions, which may perform well in circumstances where splines fail, should not be included in the set of possible basis functions. Work on this extension to G/SPLINE is ongoing.

Program Availability

The WOLF program implements the G/SPLINE algorithm. This program currently runs under UNIX on Sun and Silicon Graphics minicomputers, and under THINK/C 4.0 on the Apple Macintosh II microcomputer. The UNIX version of the software and data is available by FTP on the INTERNET by sending mail to drogers@riacs.edu. The Macintosh version is available on floppy disk for a $20 copying fee.

It is my goal that the timely sharing of both the software and the data sets will encourage comparison of this work to other work, speed the dissemination of the algorithm, and encourage others to similarly share their algorithms and data. My rapid progress in developing this research program was due in part to Dr. Friedman's policy of openly releasing his software (admittedly comment-free...) for distribution; I encourage others to join me in following that excellent precedent.

Acknowledgments

This work was supported in part by Cooperative Agreements NCC 2-387 and NCC 2-408 between the National Aeronautics and Space Administration (NASA) and the Universities Space Research Association (USRA). Special thanks to Doug Brockman (who shared my enthusiasm even though he didn't know what the hell I was up to), and to my father Philip Rogers, who made me want to become a scientist.

References

[1] Bentley, J., "Multidimensional Binary Search Trees used for Associative Searching," *Communications ACM*, 18, pp. 509-517, 1975.

[2] Breiman, L., Friedman., J., Olshen., R., and Stone, C., *Classification and Regression Trees*, Wadsworth, Belmont, CA, 1984.

[3] Davis, L., *Genetic Algorithms and Simulated Annealing*, Morgan Kaufmann, Los Altos, CA, 1987.

[4] deBoor, C., *A Practical Guide to Splines*, Springer-Verlag, New York, NY, 1979.

[5] Friedman, J., "Multivariate Adaptive Regression Splines," Technical Report No. 102, Laboratory for Computational Statistics, Department of Statistics, Stanford University, November 1988 (revised August 1990).

[6] Goldberg, D., *Genetic Algorithms in Search, Optimization, and Machine Learning*, Addison-Wesley, Reading, MA, 1989.

[7] Hofstadter, D., *Gödel, Escher, Bach: an Eternal Golden Braid*, Basic Books, New York, NY, 1979. [Chapter X has a formal discussion of levels of description, but I recommend the section titled "...Ant Fugue" for a more informal and inspired discussion on reductionism.]

[8] Holland, J., *Adaptation in Artificial and Natural Systems*, University of Michigan Press, Ann Arbor, MI, 1975.

[9] Ralston, A., and Rabinowitz, P., *A First Course in Numerical Analysis*, McGraw-Hill, New York, NY, 1978.

[10] Sanger, T., "Basis-Function Trees as a Generalization of Location Variable Selection Methods for Function Approximation," Proceedings of the 1990 Neural Information Processing System Conference, Denver, CO, 1991.

Genetic Algorithms in Asynchronous Teams

Pedro S. de Souza and **Sarosh N. Talukdar**
Engineering Design Research Center
Carnegie Mellon University
Pittsburgh, PA - 15213

Abstract

An A-team (asynchronous team) is a simple and flexible way of combining algorithms so they can interact synergistically. This paper illustrates how genetic algorithms and more traditional processes, such as Newton's Method, can be formed into A-teams for solving sets of nonlinear algebraic equations. We feel that similar combinations could be developed and used with advantage in other problem domains.

1. INTRODUCTION

Consider a class of problems for which many algorithms are available, such as nonlinear programming, integer programming or solving sets of nonlinear algebraic equations. When faced with a member of such a class, one has three options:

1. Select and apply one of the available algorithms.
2. Select several algorithms. Apply each separately and independently to the problem. Select the best solution obtained.
3. Select several algorithms. Make arrangements for them to cooperate so that better solutions are obtained more quickly than in option 2.

How can algorithms be made to cooperate? This paper illustrates one way, called an asynchronous team or A-team, with the aid of examples from the domain of nonlinear equation solving.

2. SOME TERMINOLOGY

Consider a set of nonlinear algebraic equations:

$$F(X) = 0 \quad (1)$$

where X is an N-dimensional vector of unknowns and F is an N-dimensional vector of functions. Let X_* be a solution to (1), and let $\{X_n\} = X_1, X_2, X_3, \ldots$ be a sequence of points produced by an iterative algorithm of the form:

$$X_n = PX_{n-1} = P^n X_0 \quad (2)$$

where n=1, 2, 3, ..., P is an operator and X_0 is called the starting value of the iteration. Suppose that the sequence $\{X_n\}$ converges to X_* for all X_0 in the set $B(X_*,P)$. Then B is called the basin of attraction for X_* and P. In other words, if an algorithm is to find a solution, then it must be started from a point within its basin of attraction for that solution. The form and size of that basin depend on both the problem and the algorithm.

Of the many algorithm available for solving sets of nonlinear algebraic equations, none is completely free of weaknesses. Some are fast (have high rates of convergence) but have small basins of attraction. Other are more robust (have larger basins of attractions) but tend to be slow. Newton's Method [1] is a good example of the former type; Steepest Descent [2] is a good example of the latter. Let P_{NM} denote the operator used by Newton's Method, and P_{SD} the operator used by Steepest Descent.

3. A-TEAMS

To obtain a process that is both fast and robust, Newton's Method can be combined with Steepest Descent as follows [4], [5], [6]:

$$X_n = w_n P_{NM} X_{n-1} + (1 - w_n) P_{SD} X_{n-1} \quad (3)$$

where n=1, 2, 3, ..., and the weight, w_n, lies in the interval [0, 1]. As X_{n-1} moves into the basin of attraction of Newton's Method, w_n is changed in value from 0 to 1, causing the overall process to change from Steepest Descent to Newton's Method.

A generic scheme for running Newton's Method and the Steepest Descent algorithm in parallel, called a supervised group or blackboard [5], is shown in Fig. 1. The incumbent estimate to the solution, X_{n-1}, is stored in the shared memory. The terms: $P_{SD}X_{n-1}$ and $P_{NM}X_{n-1}$ are computed concurrently. Then, the supervisor chooses a value for w_n and calculates the new estimate: X_n.

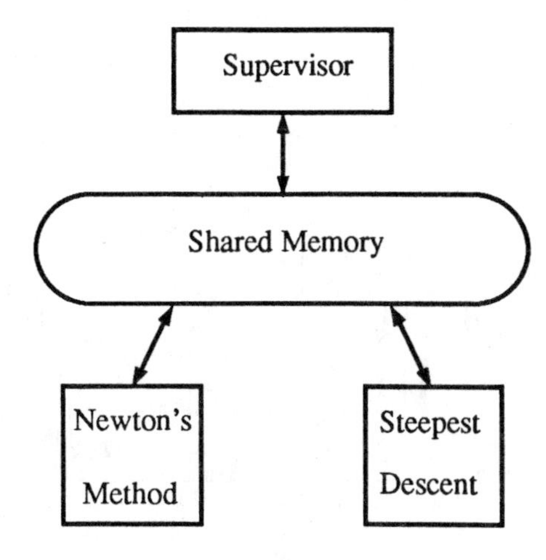

Figure 1: A Supervised Group or Blackboard for Implementing (3).

The main failing of this scheme is its inflexibility. Before new algorithms can be added, one must redesign the supervisor so it can choose good weights for these algorithms.

A more flexible arrangement, called an A-team (for asynchronous team)[6], is obtained by adopting the organizational structures of social insects [5]. These insects work without supervision and select their own tasks. As a result, their organizations are very flexible - members can be added or removed effortlessly. One might imagine that the price of this flexibility would be confusion and poor performance. In fact, at the beginning of a major undertaking, the actions of the insects often appear to be confused and in conflict. Actually, this "confusion" allows for a variety of alternatives to be identified and considered - an insect version of brain-storming. With the passage of time, alternatives are eliminated and eventually a consensus develops. The overall effects is one of synergistic cooperation; groups of insects seem able to complete tasks that exceed the sum of their individual capabilities [7].

To emulate the organizations of social insects in software, we make two modifications to the structure of a supervised group or blackboard (Fig. 1). First, we eliminate the supervisor. Second, we eliminate the need for supervision by encapsulating each member in a small, rule-based program that gives it consensus-building capabilities. The result is an unsupervised group of the form shown in Fig. 2. We call this structure an A-Team to emphasize that members interacts asynchronously, that is, they exchange information when they find it convenient to do so, rather than at predetermined points in time. These exchanges occurs through additions to, or modifications of, the contents of the shared memory.

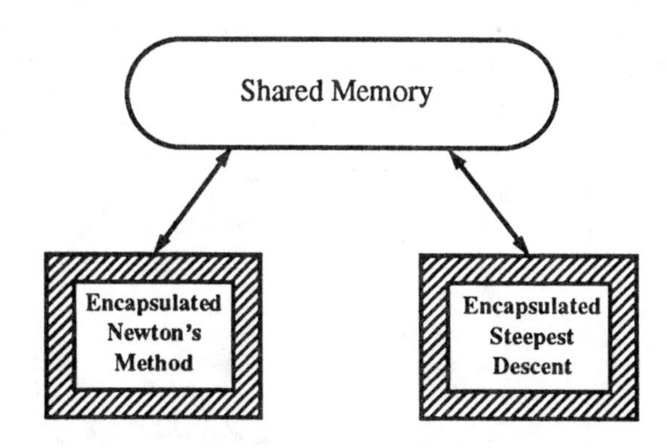

Figure 2: An A-Team with two members. The encapsulations provide the members with consensus building capability.

To illustrate the assembly of an A-Team, consider the pair of algorithms mentioned before, namely, Newton's Method and Steepest Descent. Suppose that each algorithm is assigned a computer and provided with a rule-based encapsulation that performs the following tasks:

- fetches a given starting point, X_0, from the shared memory.
- starts the algorithm from this point and monitors its progress, i.e. monitors the values of $\|F(X_n)\|$, where $\|\bullet\|$ is any vector norm.
- posts the results, X_n and $\|F(X_n)\|$, in the shared memory periodically.
- if progress is good, (i.e. if $\|F(X_n)\|$ is decreasing faster than some given threshold), then the encapsulation allows its algorithm to continue. Otherwise, it restarts its algorithm from the best point recorded in shared memory. (As a result, an algorithm that is performing poorly is restarted from the best point found by a more successful algorithm.)

Note that this A-Team will converge to a solution if it is started from any point that is within the basin of attraction of any one of its algorithms, as illustrated in Fig. 3. As such, it performs at least as well as the team of (3) but is much more flexible - new algorithms can be added to the A-Team merely by providing them with encapsulations of the type described above.

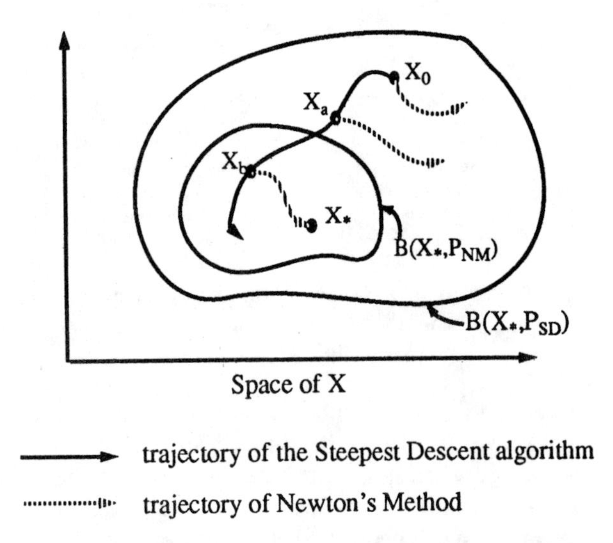

Figure 3: Both algorithm are started from X_0, a point that is inside the basin of attraction for Steepest Descent (SD) but outside the basin for Newton's Method (NM). As a result Newton's Method diverges. It is restarted from point X_a and again diverges. Finally, it is restarted from X_b which is within its basin of attraction, and it proceeds to find the solution X_* well before SD (SD is notoriously slow when close to a solution).

Genetic Algorithms can be loosely thought of as having very wide basins of attraction for global solutions and therefore would make good additions to A-Teams. In the next section we point out how genetic algorithms may be used to solve algebraic equations. Subsequent sections illustrate how A-teams containing genetic algorithms may be assembled for a few different types of nonlinear algebraic equations. Finally, we mention some of the implications of this work for domains other than algebraic equations.

4. USING GENETIC ALGORITHMS TO SOLVE NONLINEAR ALGEBRAIC EQUATIONS

Consider the single equation:

$$F(x) = x^2(\sin(8x)\sin(x) + 1.1) \quad (4)$$

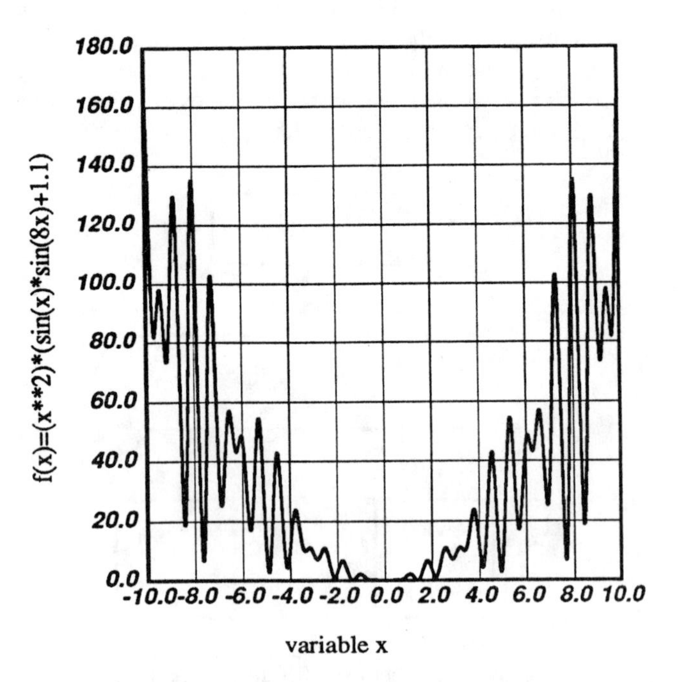

Figure 4: A plot of F(x) from equation (4)

A plot of F(x) is shown in Fig. 4. The multiple minima make (4) a difficult equation to solve. Conventional algorithms tend to get stuck in local minima instead of finding the global minimum which is the solution to (4). A GA, however, does not suffer from this limitation, as indicated by the results shown in Table 1. (In this GA we used a population of 25 strings, each string being 16 bits long [8]. Gray coding was employed for mapping the continuous variable, x, into each string [9][10]. The fitness function was chosen to be 1/|F(x)|. The best 8 members of each population were used to form a mating pool. Reproduction was through simple crossover and modifications through mutation were allowed.)

Table 1: Global minimum detection by GA in a multimodal function

Search Interval	Solution Found
$-10 \le x \le 10$	0.0
$1 \le x \le 10$	1.4
$3 \le x \le 10$	4.9
$6 \le x \le 10$	7.7

The results of Table 1 suggest that it would be beneficial to add GAs to A-Teams for solving sets of nonlinear equations. In the next two sections we illustrate how this might be done.

5. WEAKLY COUPLED EQUATIONS

A set of equations is said to be weakly coupled if it can be partitioned into subsets such that each subset contains a set of variables (called local variables) that appear in no other subset. The remaining variables are called global variables. For example:

$$x_1^2 \sin x_3 + x_2^2 \cos x_3 + (x_1 + x_2) e^{-(x_2 + x_3)} = 0 \quad \text{(a)}$$

$$x_1^2 + x_2^2 + x_3^2 - 16 = 0 \quad \text{(b)}$$

$$x_4 \tan(x_3) + x_5^2 \tan(x_6) = 0 \quad \text{(c)}$$

$$x_4^2 + x_3 x_5 - \pi = 0 \quad \text{(d)}$$

$$x_8 \cos(x_6) + x_7^3 = 0 \quad \text{(e)}$$

$$x_8^2 + x_7^2 \log(x_6 + 2) - 10 = 0 \quad \text{(f)}$$

Ex. 1: Weakly Coupled Nonlinear Equations (Set 1)

The set of equations above can be divided into subset I with equations (a) and (b), subset II with equations (c) and (d), and subset III with equations (e) and (f). Variables x_1 and x_2 are local to subset I, x_4 and x_5 are local to subset II, and x_7 and x_8 are local to subset III. Variables x_3 and x_6 are global.

A GA could be applied to the entire set of equations, but it would be very slow. Instead, we apply a GA only to the global variables and use conventional algorithms (Newton's Method) for the local variables.

The chromosomes of the GA are composed of the global variables. Their values are sent to NM algorithm to find the corresponding values of the local variables.

The fitness of the i-th chromosome is:

$$\text{Fitness}(i) = -\sqrt{\sum_{j=1}^{S} \{\text{Min}[g_j(x_{G_j}^i)]\}^2}$$

where:

- S : is the number of subsets;
- g_j : is the norm of subset j, and
- $x_{G_j}^i$: :are the global variables of subset j specified in the chromosome i.

A solution is expressed by:

$$\bigcup_{j=1}^{S} \{\hat{x}_{G_j}, \hat{x}_{L_j}\}$$

and is found when

$$|\text{Fitness}(i)| \le \varepsilon$$

where:

- ε : is a given threshold;
- $\hat{x}_{G_j}$: are the global variables of subset j, and
- $\hat{x}_{L_j}$: are the local variables of subset j.

The parameter used to measure the amount of work was the number of functions evaluated until a solution was found. In example 1 we need six function evaluations for each fitness value. Table 2 shows the average of twenty experiments for finding solutions of the above-mentioned equation set.

Table 2: Performance of GA combined with Newton's Method (NM).

Method	Function Evaluation (10^3)
GA	170
GA + Newton's Method	24
Work Reduction	**86%**

The work reduction comes from the fact that the number of variables manipulated by GA is reduced when compared to a pure GA procedure, which means that the space of probabilistic search is reduced. Once the global variables are fixed in a subset the local variables can be more readily evaluated by conventional techniques than by a pure GA.

Note that instead of Newton's Method other techniques could be used to combine with GA, as we see in section 6.

6. ANOTHER EXAMPLE

High precision is not a GAs strong point. It requires long strings and makes the algorithm expensive. This suggests that a GA be combined with algorithms that provides high precision more cheaply. These combinations can be made in a number of ways. One is indicated in Fig. 5. In this combination, the GA finds promising solutions which it posts in the shared memory. High accuracy Levenberg-Marquardt algorithms [11] refine those solutions.

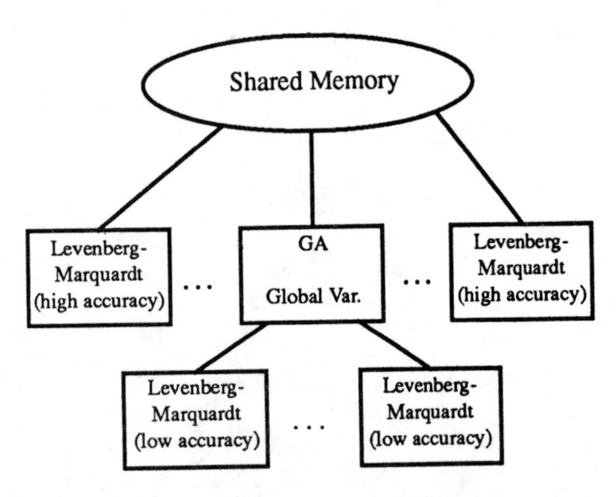

Figure 5: An A-Team Containing a GA and several copies of a Levenberg-Marquardt Algorithm.

Using the above combination we solved the equation set presented in the Ex. 2. The results are shown in Table 3.

$$x_1 \sin(x_2)\cos(x_{10}) - x_2 x_3 \sin(x_{10}-1) = 0 \quad \text{(a)}$$

$$x_1(x_1-2) - x_3\cos(x_2) + x_6 + 1 = 0 \quad \text{(b)}$$

$$x_4\cos(x_2) - x_5\sin(x_6) = 0 \quad \text{(c)}$$

$$x_4 x_2 - x_5^2(x_6-2) - 1 = 0 \quad \text{(d)}$$

$$(x_5-1)x_2 + x_6\cos(x_4) + x_{10} = 0 \quad \text{(e)}$$

$$x_7 x_8 \cos(x_2)\cos(x_6) - \frac{x_2 x_6 x_7^2}{10} + 2 = 0 \quad \text{(f)}$$

$$\frac{x_8^2}{1+(\sin(x_2))^2} - \frac{x_7^2}{1+(\cos(x_6))^2} = 0 \quad \text{(g)}$$

$$\frac{(x_9-1)^3\log(x_{10}^2+3)}{5} + x_{10}\cos(x_{10}) - x_6 - 2\pi = 0 \quad \text{(h)}$$

$$x_9 x_{10} + x_6^2 = 0 \quad \text{(i)}$$

$$x_9^2(\sin(x_6)-1)^2 + x_{10}^2 - 1 = 0 \quad \text{(j)}$$

Ex. 2: Weakly Coupled Nonlinear Equations (Set 2)

Line 1 of Table 3 is the average number of function evaluations to find a solution for example 2 using only Levenberg-Marquardt method. Every time the method converged, it was restarted until we got twenty solutions

Table 3: A-Team Performance

Method	Function Evaluation (10^3)
Levenberg-Marquardt	599
A-Team	349
Work Reduction	**42%**

for the equations set. Line 2 presents the average number of function evaluations for solving the same example using an A-team. This team had one GA-based procedure and one Levenberg-Marquardt procedure as its members. The experiment was repeated until twenty solutions were reached. Line 3 shows a 42% reduction of work is achieved by using an A-team to solve this problem.

7. CONCLUSIONS

We have demonstrated that genetic algorithms can be combined with other algorithms to form A-teams for solving sets of nonlinear algebraic equations. These teams result in considerable savings in the amount of computational effort (measured in terms of function evaluations) required to find solutions. In the combinations we have discussed, the genetic algorithms provide starting values for less robust but faster algorithms. This model of computation does not have to be confined to solving algebraic equations. Other domains, such as nonlinear and mixed integer programming, could benefit from its use. In our continuing work, we are exploring some of these possibilities.

Acknowledgments

The work reported here has been supported by Fundação de Amparo à Pesquisa do Estado de São Paulo (Brazil) and by the National Science Foundation through its program for Engineering Research Centers.

References

[1] J. M. Ortega, and W. C. Rheinboldt. "Iterative Solution of Nonlinear Equations in Several Variables", New York, NY: Academic Press, 1970.

[2] A. W. Westerberg, and S. W. Director. "A Modified Least Squares Algorithm for Solving Sparse NxN

Sets of Nonlinear Equations", Computers and Chemical Eng., Vol. 2, 1978.

[3] E. Levenberg. "A Method for the Solution of Certain Nonlinear Problems in Least Squares", Q. Appl. Math., Vol. 2, 1944, pp. 164-168.

[4] D. W. Marquardt. "An Algorithm for Least Square Estimation of Nonlinear Parameters", Jour. Soc. Ind. Appl. Math., Vol. 11, 1963, pp. 431-441.

[5] S. N. Talukdar, R. Quadrel, and E. Subrahmanian. "Organizations for Large-Scale Problem Solving," to appear in IEEE Proceedings.

[6] S. N. Talukdar, S. S. Pyo, and R. Mehrotra. "Designing Algorithms and Assignments for Distributed Processing", Electric Power Research Institute, Palo Alto, CA, 1983.

[7] E. O. Wilson. "The Insect Societies", Belnap Press of Harvard University, Cambridge, MA,1971.

[8] D. E. Goldberg. "Genetic Algorithms in Search, Optimization & Machine Learning", Reading, MA: Addison-Wesley, 1989.

[9] D. E. Goldberg and J. Richardson. "Genetic Algorithms with Sharing for Multimodal Function Optimization", Proceedings of the Second International Conference on Genetic Algorithms, Massachusetts Institute of Technology, Cambridge, MA, 1987.

[10] R. A. Caruana and J. D. Schaffer. "Representation and Hidden Bias: Gray vs. Binary Coding for Genetic Algorithms", Proceedings of the 5th International Conference on Machine Learning, Morgan Kaufmann, Los Altos, CA, June 12-14, 1988.

[11] R. Fletcher. "Practical Methods of Optimization", Chirchester, UK: John Wiley & Sons, 1987.

Genetic Algorithm Applications

A Genetic Algorithm for Database Query Optimization

Kristin Bennett **Michael C. Ferris** **Yannis E. Ioannidis**
Computer Sciences Department
University of Wisconsin
1210 West Dayton Street
Madison, Wisconsin 53706

Abstract

Current query optimization techniques are inadequate to support some of the emerging database applications. In this paper, we outline a database query optimization problem and describe the adaptation of a genetic algorithm to the problem. We present a method for encoding arbitrary binary trees as chromosomes and describe several crossover operators for such chromosomes. Preliminary computational comparisons with the current best-known method for query optimization indicate this to be a promising approach. In particular, the output quality and the time needed to produce such solutions is comparable to and in general better than the current method.

1 INTRODUCTION

Genetic algorithms [4, 6] are becoming a widely used and accepted method for very difficult optimization problems. In this paper, we describe the implementation of a genetic algorithm (GA) for a problem in database query optimization. In order to give a careful formulation of our GA, we first give a broad outline of this particular application.

The key to the success of a Database Management System (DBMS), especially of one based on the relational model [3], is the effectiveness of the query optimization module of the system. The input to this module is some internal representation of a query q given to the DBMS by the user. Its purpose is to select the most efficient *strategy* (algorithm) to access the relevant data and answer the query. Let $\mathcal{S}$ be the set of all strategies appropriate to answer a query q. Each member s of $\mathcal{S}$ has an associated cost $c(s)$ (measured in terms of CPU and/or I/O time). The goal of any optimization algorithm is to find a member s_0 of $\mathcal{S}$ that satisfies

$$c(s_0) = \min_{s \in \mathcal{S}} c(s).$$

Query optimization has been an active area of research ever since the beginning of the development of relational DBMSs. Good surveys on query optimization and other related issues can be found elsewhere [9, 10].

In the relational model, data is organized in *relations*, i.e., collections of similar pieces of information called *tuples*. Relations are the data units that are referenced by queries and processed internally. A *strategy* to answer a query q is a sequence of relational algebra operators applied to the relations in the database that eventually produces the answer to q. The cost of a strategy is the sum of the costs of processing each individual operator. Among these operators, the most difficult one to process and optimize is the *join*, denoted by $\bowtie$. It essentially takes as input two relations, combines their tuples one-by-one based on certain criteria, and produces a new relation as output. Join is associative and commutative, so the number of alternative strategies to answer a query grows exponentially with the number of joins in it. Moreover, a DBMS usually supports a variety of *join methods* (algorithms) for processing individual joins and a variety of *indices* (data structures) for accessing individual relations, which increase the options even further. Thus, all query optimization algorithms primarily deal with join queries. These are the focus of this paper as well.

In current applications, each query usually involves a small number of relations, e.g., less than 10. Hence, although exponential in the number of joins, the size of the strategy space is manageable. Most commercial database systems use variations of the same query optimization algorithm, which performs an exhaustive search over the space of alternative strategies, and whenever possible, uses heuristics to reduce the size of that space. This algorithm was first proposed for the System–R prototype DBMS [14], so we refer to it as the System–R algorithm.

Current query optimization techniques are inadequate

to support the needs of some of the newest database application domains, such as artificial intelligence (e.g., expert and deductive DBMSs), CAD/CAM (e.g., engineering DBMSs), and other disciplines (e.g., scientific DBMSs). Simply put, queries are much more complex both in the number of operands and in the diversity and complexity of operators in the query. This greatly exacerbates the difficulty of exploring the space of strategies and demands that new techniques be developed.

One of the proposed solutions is to use randomized algorithms. Simulated Annealing, Iterative Improvement, and Two-Phase Optimization (a combination of the first two) have already been successfully tried on query optimization [8, 15, 7], giving ample reason to believe that a GA will perform well also. Many of the operators used in these studies can be adapted for use in a GA and incorporated into a standard GA code. An advantage of our version of the GA [1], is that it is designed for a parallel architecture and significant computational savings over the other randomized methods can be obtained by a parallel implementation.

This paper is organized as follows. Section 2 defines two strategy spaces that are of interest to query optimization, which were used in our experiments. It also contains a description of the System–R algorithm, which is used as a basis for comparison of our results. Section 3 describes the specific genetic algorithm that we developed, including the representation that we used for the chromosomes for the two strategy spaces and the adopted crossover operators. Section 4 contains the results of our experiments. Finally, Section 5 gives a summary and provides some direction for future work.

2 QUERY OPTIMIZATION SPECIFICS

2.1 STRATEGY SPACES

Most query optimizers do not search the complete strategy space $\mathcal{S}$, but a subset of it, which is expected to contain the optimum strategy or at least one with similar cost. To understand the various options, we need some definitions related to databases. In a slight abuse of notation, consider the following query:

$$\begin{array}{c}(A \bowtie C) \text{ and } (B \bowtie C) \text{ and } (C \bowtie D) \\ \text{and } (D \bowtie E) \text{ and } (D \bowtie F)\end{array} \quad (1)$$

Each join is associated with a constraint (omitted for clarity of presentation) that specifies precisely which tuples of the joined relations are to appear in the result. Query (1) can be represented by a *query graph* [16], which has the query relations as nodes and the joins between relations as undirected edges, as shown in Figure 1. Throughout, we use capital letters to denote relations and numbers to represent joins. In this

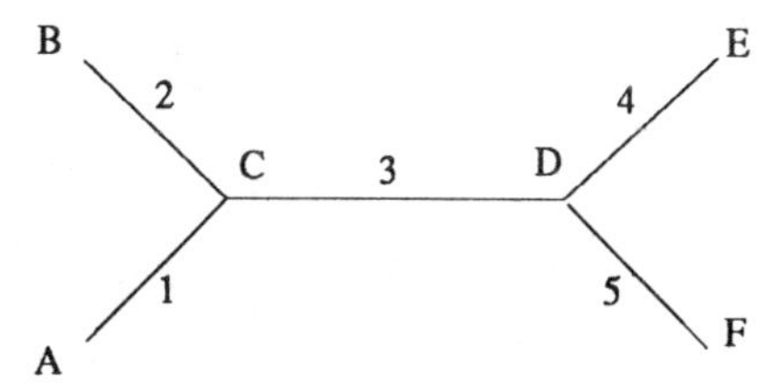

Figure 1: Query Graph

paper, we study tree queries, i.e., queries whose query graph is a tree. The answer to a given query is constructed by combining the tuples of all the relations in a query based on the constraints imposed by the specified joins. This is done in a step-wise fashion, each step involving a join between a pair of relations whose tuples are combined. These can be relations originally stored in the database or results of operations from previous steps (called *intermediate relations*). As a very strong and effective heuristic, database systems never combine relations that are not connected with a join in the original query. This is because such an operation produces the cartesian product of the tuples in the two relations. Not only is this an expensive operation, but its result is also very large, thus increasing the cost of subsequent operations. Most query optimizers confine themselves into searching the subspace of $\mathcal{S}$ of strategies with no cartesian products. This heuristic is adopted in the work presented in this paper as well.

Given the above, each strategy to answer a query can be represented as a *join processing tree*. This is a tree whose leaves are database relations, internal nodes are join operators, and edges indicate the flow of data from bottom-up. In addition, the chosen index for each database relation and the chosen join method for each join is specified. If all internal nodes of such a tree have at least one leaf as a child, then the tree is called *linear*. Otherwise, it is called *bushy*. Most join methods distinguish the two join operands, one being the *outer* (left) relation and the other being the *inner* (right) relation. An *outer linear join processing tree (left–deep tree)* is a linear join processing tree whose inner relations of all joins are base relations. In this study, we deal with two strategy spaces: one that includes only left-deep trees, which is denoted by $\mathcal{L}$, and one that includes both linear and bushy ones, which is denoted by $\mathcal{A}$. Examples of a left-deep tree and a bushy tree for query (1) are shown in Figure 2 (avoiding the details of the join constraints and the join methods). The interest in $\mathcal{L}$ stems from the fact that many DBMSs are using it as their strategy space, and is the one on which the System–R algorithm can be applied. We experiment with $\mathcal{A}$ as well, because quite often the optimum strategy is not in $\mathcal{L}$. We present results for applying a genetic algorithm on both spaces and compare them with the results of applying the System–R algorithm on $\mathcal{L}$.

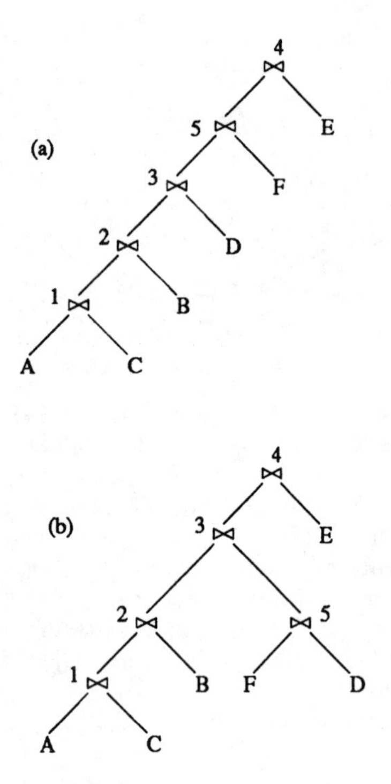

Figure 2: (a) Left-deep Tree; (b) Bushy Tree

We should emphasize at this point that, even after ignoring all strategies that contain cartesian products, the number of the remaining strategies still grows exponentially with the number of joins in the query. The above is true even for the $\mathcal{L}$ space, let alone for the $\mathcal{A}$ space, which is a superset of the former. The size of these spaces depends not only on the specific query which is being optimized (being dependent on the shape of its query graph), but also on the number of join methods supported by the system and the index structures that exist on the query relations. Hence, although cartesian products are excluded from strategies, even for the small queries with which we deal in this paper (up to 16 joins), the corresponding $\mathcal{L}$ and $\mathcal{A}$ spaces are very large and the associated optimization problem is computationally very difficult.

2.2 THE SYSTEM–R ALGORITHM

The System–R algorithm is based on dynamic programming. Specifically, the complete space $\mathcal{L}$ is constructed, occasionally pruning parts of it that are identified as suboptimal. The space is constructed by iteration on the number of relations joined so far. That is, at the k-th iteration, the best strategy to join k relations from the query is found, for all such sets of k relations. In the next iteration, strategies of $k+1$ relations are constructed, by combining each strategy from the previous collection with the appropriate remaining relations. For each set of $k+1$ relations, multiple strategies are usually constructed, of which only the one with the least cost is kept, since it can be shown that all the rest cannot be part of the optimum final strategy. This process needs as many iterations as there are relations in the query to complete. The main disadvantage of the algorithm is that it needs to maintain and process a very large number of strategies during its execution, a number that grows exponentially with the number of joins in the query. Especially towards the later iterations, this implies both a significant cpu cost for processing and a significant I/O cost due to increased page faults. This makes the algorithm inapplicable to queries with more than about 16 relations.

The above informal description of the algorithm is slightly simplified. In the interest of space, we have not discussed various complications that arise from side-effects that the use of specific types of indices can have on the desirability of a strategy. However, the version of the algorithm that was used in our experiments did address all these complications as well. The interested reader can find further details in the original paper on the System–R algorithm [14].

3 GENETIC ALGORITHM

In this section, we describe the implementation of a genetic algorithm to solve the problem outlined above. For completeness, we briefly review our terminology, details of which can be found in [1]. Our GA works with a population of chromosomes, each of which can be decoded into a solution of the problem. For each chromosome i in the population, a measure of its quality is calculated, called its fitness, $f(i)$. Chromosomes are selected from the population to become parents, based on fitness. Then, reproduction occurs between pairs of chromosomes to produce offspring. The newly created population becomes the next generation and the process is repeated. The model that we use has a fixed population size N.

Our GA [1] uses a neighborhood scheme in which the fitness information is only transmitted within a local neighborhood, see for example [12]. A model algorithm for such a scheme is as follows.

Local Neighborhood Algorithm:

repeat
    for each chromosome i *do*
        evaluate f(i)
        broadcast f(i) in the neighborhood of i
        receive f(j) for all chromosomes j
            in the neighborhood
        select chromosome k to mate from the
            neighborhood of i based on fitness
        reproduce using chromosomes i and k
        replace chromosome i with one
            of the offspring
until population variance is small

In the experiments that we report below, we have used a neighborhood structure we call *ring6*, which considers chromosomes i and j to be neighbors if

$$\min(|i-j|, |i+N-j|, |i-N-j|) \leq 3$$

This can be viewed as each chromosome residing on a ring with neighbors that are chromosomes no further than three links away. This neighborhood structure was chosen since it has proven very effective in other problem instances [1]. For query optimization, the aim is to minimize the cost of the strategy, so to generate a fitness distribution in a neighborhood, we take the negative of the cost (to convert to maximization) and use a linear scaling of these values in each neighborhood so that the maximum fitness is some proportion (user supplied) of the average fitness. Details of this scaling routine are found in [1]. Since each chromosome i only selects one mating partner, selection is carried out by choosing chromosome k from the neighborhood with probability

$$f_k / \sum_{j \in \mathrm{nhd}(i)} f_j.$$

Reproduction produces two offspring. The current chromosome is replaced with its best offspring provided this offspring is better than the worst chromosome in the neighborhood (see [12]).

For a GA to be an efficient optimization technique, we believe that reasonable parameter and algorithm choices (such as those detailed above) can be made without reference to a particular problem. However, the effectiveness of the overall algorithm depends crucially on the representation of the problem and the operators we use to exchange genetic information. Thus, we now specialize to the particular problem of query optimization. We describe two ways of encoding this problem which correspond to the two strategy spaces $\mathcal{L}$ and $\mathcal{A}$. We attempt to incorporate as much problem specific information as possible (see [5]), and show how mutation, initial population choice and crossover are carried out. Our discussion is broken into three parts, the first dealing with left–deep strategies, the second with bushy strategies, and the third dealing with crossover operators.

3.1 LEFT–DEEP STRATEGIES

Left–deep strategies ($\mathcal{L}$) are a relatively small subset of the strategy space $\mathcal{A}$. However, it has been observed that frequently a strategy exists in $\mathcal{L}$ whose cost is very close to the optimal cost over $\mathcal{A}$. System–R uses $\mathcal{L}$ as its strategy space, so it is appropriate to apply a GA in $\mathcal{L}$ and compare the results of these two methods. The advantage of choosing $\mathcal{L}$ over $\mathcal{A}$ is that the search space is much smaller (although still large in absolute terms); the disadvantage is that we cannot beat System–R in terms of output quality using a GA in $\mathcal{L}$.

Each chromosome represents a left–deep strategy. The chromosome is an ordered list of genes, where each gene consists of a relation and a join method. For example,

$$^JA \; ^JC \; ^JB \; ^JD \; ^JF \; ^JE$$

represents the left–deep strategy in Figure 2(a), with J representing some join method.

In our code, we associate the join method with the inner (right) relation of each join. Thus, to recreate the join processing tree, join the first and second relations using the method associated with the second relation. Then join the resulting intermediate relation with the next relation according to the specified method. Repeat until no relations remain. At each step verify that an edge exists in the query graph (Figure 1) between the current relation and one of the relations that occurred previously in the chromosome. If no such edge exists, then the query strategy contains a cartesian product, and the chromosome is penalized with an infinite cost. Note that the join method associated with the first relation is ignored and that if the crossover method produces many cartesian products, then the GA will not perform well.

Using this encoding, the problem is similar to a constrained traveling salesman problem (TSP) with a choice of methods of transport between the cities. In fact we use this analogy to motivate our choices of crossover operator. However, the query optimization function is much more expensive to evaluate than typical TSP functions, since each join, or equivalently the cost of traveling *directly* between cities, can be dependent on the route previously taken and/or the future cities to be visited. As an indication of the above complexity, we want to emphasize that the cost of a join between two relations is a function of their sizes. That size depends directly on the precise set of joins that have occurred previously. It also depends on the joins that remain to be processed later, because much of the data that is necessary for their execution is contained in the two relations.

Mutation is a secondary operator used to guarantee connectedness of the search pace. In our implementation it is of two types. The first type changes the join method randomly, and the second swaps the order of two adjacent genes. A left–deep strategy generator ensures that the initial population contains no cartesian products. This is achieved by cycling through a randomly generated permutation of the relations, only adding a relation to our chromosome if this can be done without introducing a cartesian product. The join methods are generated randomly.

3.2 BUSHY STRATEGIES

Quite often the best strategy for a query is in $\mathcal{L}$. However, in order to produce better solutions that System–R, we must consider a larger strategy space, ($\mathcal{A}$), which contains both linear and bushy strategies. We encode such strategies into chromosomes by considering each join as a gene, so that k_o^J represents join k with some join method J and its constituent relations (found on the query graph) in o orientation (for instance (a)lphabetically or (r)everse-alphabetically). A chromosome is then an ordered list of these genes. An example is

$$1_a^J\ 5_r^J\ 2_r^J\ 3_a^J\ 4_a^J$$

which represents the strategy given in Figure 2(b).

The decoding of the list into a solution is more costly than the left-deep decoding but has the ability to represent many more strategies. The decoding process grows the bushy tree from the bottom up. It maintains a list of intermediate relations waiting to be joined. Scanning the chromosome from left to right, it finds the constituent relations in each join (gene) by examining the query graph. The orientation of the gene indicates which relation is the outer (left) and inner (right) relation. If the right relation has been used in the formation of some intermediate relation in the list, the latter is substituted as the right relation of the join and the intermediate relation is removed from the list. The same process is done for the left relation. The left and right relations are joined according to the method in the gene, and the resulting relation is added to the list. After all the genes are processed, one intermediate relation remains in the list. This corresponds to the root of the bushy tree.

Note that although the representation is similar to the one described for left-deep trees (it is an ordered list of genes), there are three important differences. First, the decoding scheme guarantees that the corresponding query strategy has no cartesian products. Thus, we only consider "feasible" strategies of our problem. Second, our representation is based on labeling joins, not relations. This is somewhat natural since there may be several intermediate relations in use at any given step of the decoding scheme and these relations can be easily associated with a join. This association is critical for simple computation. In the left-deep case, only one intermediate relation is constructed at each stage, so the problem of handling these relations does not occur. Third, the representation is not uniquely defined, i.e., several chromosomes can decode into the same tree.

The coding described above has the advantage that it may now be possible to beat the System-R solution. However, the search spacc has been greatly increased giving the GA a more difficult task. We believe that the decoding scheme is very important for this GA to perform well: the extra work that we carry out in decoding guarantees that the algorithm only considers feasible solutions of the problem, and furthermore, that we can use standard crossover schemes motivated by GA's for TSP and other database work to generate good chromosomes.

Mutation is carried out in two ways. The first is to randomly change the join method or the orientation. The second is to perform reordering of genes on the chromosome by transposing a gene with its neighbor. Together, these guarantee that the search space is connected. The initial population is generated randomly.

3.3 CROSSOVER

In order to complete the discussion of our method, we describe the two crossover operators that we investigated. In each of the encodings above, the chromosome is an ordered list of genes. As outlined above, in the left-deep case the genes can be identified by their relation letter and in the bushy case by their join number. We describe the crossover operators solely in terms of the genes.

The first method, modified two swap (M2S), modifies the local improvement algorithm given in [11] to incorporate information from both parents and was designed primarily for the left-deep case. It can be described as follows. Given two parent chromosomes, $\mathcal{X}$ and $\mathcal{Y}$, randomly choose two genes in $\mathcal{X}$ and replace them by the corresponding genes from $\mathcal{Y}$, retaining their order from $\mathcal{Y}$, to create one offspring. For example, in the left-deep case, suppose the parent chromosomes $\mathcal{X}$ and $\mathcal{Y}$ are given by

$$\mathcal{X} = {}^mA\ {}^nC\ {}^mB\ {}^mD\ {}^nF\ {}^nE\ ,$$
$$\mathcal{Y} = {}^mB\ {}^nC\ {}^mD\ {}^mF\ {}^mE\ {}^nA$$

where m and n represent particular join methods and we randomly choose genes labeled A and D. The resulting chromosome is

$${}^mD\ {}^nC\ {}^mB\ {}^nA\ {}^nF\ {}^nE$$

We interchange the roles of $\mathcal{X}$ and $\mathcal{Y}$ to create another offspring. The use of M2S was partly motivated by the *Swap* transformation that has been successfully used for database query optimization in the context of other randomized algorithms [7, 15]. The two transformations are quite similar, except that, as a crossover, the transformation takes into account two strategies, whereas in its previous use it simply operates on one. Note that most of the ordering information from one of the chromosomes is retained, and it is this information that is of primary importance in both decoding schemes outlined above. Therefore, the crossover operator retains most of the ordering information from one chromosome, and uses order information from the other chromosome to exchange two genes. In the bushy case, the modified chromosome can be very different from both of its parents (depending on the locus of the genes being exchanged) due to the decoding scheme; in the left-deep case the solution will look very similar to one parent, although there is a small possibility of introducing a cartesian product.

The second method, which we refer to as CHUNK, is adapted from [2, 12] and was designed primarily for the bushy case. Here, we generate a random chunk of the chromosome as follows. Suppose the number of genes in the chromosome is l. The start of the chunk (of genes) is a uniformly generated random integer in $[0, l/2]$ and the length of the chunk is uniformly generated from $[l/4, l/2]$. Suppose we randomly generate the chunk $[3, 4]$, then one resulting chromosome copies the third and fourth genes of $\mathcal{X}$ into the same position in the offspring, then deletes the corresponding genes of $\mathcal{Y}$, using the remainder of $\mathcal{Y}$'s genes to fill up the remaining positions of the offspring. For example, if $\mathcal{X}$ and $\mathcal{Y}$ are given by

$$\mathcal{X} = 1_a^n\ 5_r^n\ 2_r^m\ 3_a^m\ 4_a^n\ ,$$
$$\mathcal{Y} = 3_r^n\ 5_a^n\ 1_r^m\ 4_a^m\ 2_a^m$$

where m and n represent particular join methods, the resulting chromosome is

$$5_a^n\ 1_r^m\ 2_r^m\ 3_a^m\ 4_a^m$$

Again, another chromosome is created by interchanging the roles of $\mathcal{X}$ and $\mathcal{Y}$. We arrived at this method after some experimentation. The essential motivation behind the above scheme is to force a reasonably sized subtree (between a quarter and a half of the size of the original tree) from one parent to be incorporated into the other parent with minimal disruption to the latter. Of course, the chunk may or may not correspond to a subtree, but to avoid excessive computation in determining a chunk, the above scheme was used. The ordering information on the chromosome crucially determines the strategy in our decoding scheme, so the crossover operator attempts to minimize ordering changes. Since our representation is many to one, forcing the chunk to have a large size generally results in some genetic information being exchanged, enabling the algorithm to look at a larger variety of solutions and hence generate better solutions in reasonable times.

Although the crossover operators were designed with particular strategy spaces in mind, both operators can be applied in the two strategy spaces since they essentially consider the chromosomes as an ordered list of genes. Thus, we experimented with all combinations of crossover operators and spaces.

4 PERFORMANCE RESULTS

In this section, we report on an experimental evaluation of the performance and behavior of the above genetic algorithm on query optimization compared to the System–R algorithm. First, we describe the testbed that we used for our experiments, and then we discuss the obtained results.

4.1 TESTBED

For our experiments we assumed a DBMS that supports the *nested–loops* and *merge–scan* join methods [14]. Tree queries were generated randomly whose size ranged from 5 to 16 joins. The limit on the query size was due to the inability of the System–R algorithm to run with larger queries, primarily because of its huge memory requirements. Moreover, not all generated 16-join queries were runnable by the System–R algorithm. Thus, for large queries, the genetic algorithm is clearly superior to the traditional algorithm.

In the interest of space, we do not present the precise cost formulas that were used in this study. They capture the I/O cost of the various join methods and indices used and can be found in any textbook on databases. We also avoid presenting any details on the assumed physical design of the database. The specifics are exactly as in previous studies [7].

We implemented all algorithms in C, and tested them on a dedicated DecStation 3100 workstation. All experiments were conducted with a population size of $N = 64$. Ten different queries were tested for each size up to 16 joins. However, System–R managed to terminate on only seven of the 16-join queries, so the results for that case represent a smaller number of queries. For each query, each algorithm was run five times.

4.2 OUTPUT QUALITY

We compare output quality based on the lowest cost chromosome in the final generation. The cost of the average output strategy produced by the algorithms as a function of the query size is shown in Figure 3(a). The x–axis is the number of joins in the query. The y–axis represents scaled cost, i.e., the ratio of the output strategy cost over the cost found by the System-R algorithm. For each size, the average over all queries of that size, of the average scaled output cost over all five runs of each query is shown.

The results are rather interesting. We observe that on the average, when GA is applied to $\mathcal{L}$, it fails to find the optimum strategy, but except for the largest queries (16 joins), it is clearly within a small range (10%) of optimality. For small queries (4-6 joins), both crossovers always find the optimum. As the query size grows, however, the algorithm becomes less stable and the quality of its output deteriorates, primarily due to the dramatic increase in the size of the strategy space.

When GA is applied to $\mathcal{A}$ the results improve. On the average, for all three sizes, the algorithm found a better strategy than the best left-deep tree, with the gains ranging up to 13%. Note that as query size grows, GA becomes relatively better than System–R. This is because, with increased query size, the relative difference between the best bushy strategies and the best left-deep strategies increases as well, so by searching a richer space, GA is able to improve on the output quality. A small set of experiments with an increased population size has given very promising results for further improving the output quality in large queries.

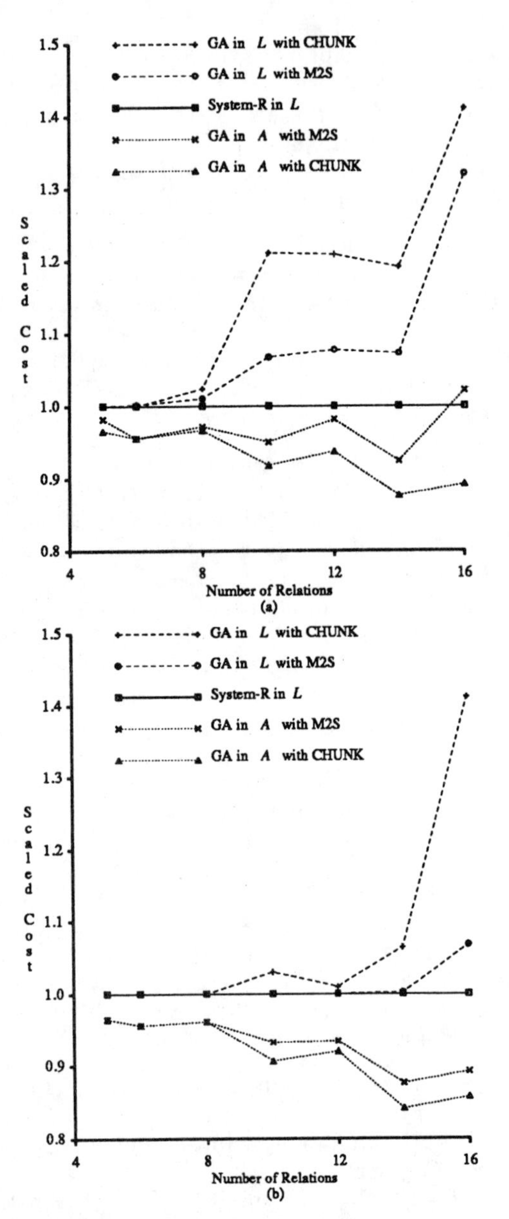

Figure 3: Scaled Cost of Strategy at Convergence: (a) average of 5 runs and (b) best of 5 runs

Another interesting comparison is that between the two crossovers. When GA is applied to $\mathcal{L}$, M2S is the preferred crossover, with CHUNK having much worse performance. This is due to the fact that, when the relations are the genes of the chromosome, applying CHUNK produces many offspring with cartesian products. Therefore, in that case, the algorithm spends much time in useless matings, thus failing to converge to a good strategy. On the other hand, M2S produces much fewer strategies with cartesian products and is the overall winner. Exactly the opposite happens when GA is applied in $\mathcal{A}$. CHUNK is the best performer, since it generates a large variety of strategies, thus looking at a larger area of the space and using its time much more effectively.

To overcome some of the inherent problems of randomized algorithms, it is occasionally proposed that such algorithms are run multiple times on a given instance problem, and the best solution among those found be chosen. With that in mind, we also compare the best output found among the five runs of each version of the GA algorithm for each query. We show the average of that over all queries of a given size in Figure 3(b). We now see that in $\mathcal{L}$, M2S is perfect, almost always finding the optimum strategy. CHUNK is considerably improved as well, but is still has inferior performance for the reasons explained above. Similar improvements are seen in the $\mathcal{A}$ space as well. Especially in the large joins, both crossovers find very good strategies. All these results indicate that multiple runs of the GA algorithm may be a plausible way to avoid some of its potential instabilities and produce high quality results.

4.3 TIME

The average time results are presented in Figure 4, where the x–axis represents the number of joins in the query, and the y–axis represents the processing time in seconds. For the 16-join queries on which System–R failed to finish, we use the time-to-failure in this figure. The results are as follows. System–R performs faster for queries of size up to 14, but the GA in $\mathcal{L}$ is much faster for queries of size 16. The increase in times for GA in $\mathcal{L}$ is almost linear. There is a larger increase in the time for GA in $\mathcal{A}$ than in $\mathcal{L}$. This is to be expected since $\mathcal{L}$ is a much smaller space than $\mathcal{A}$. As it is obvious from Figure 4, this increase is much less steep than the corresponding increase in time for System–R. We believe that even if System–R was runnable for queries beyond 16 joins, it would require much more time than GA.

A final comment that we want to make is that this version of GA is designed to be ported to a parallel machine. In a parallel implementation, each chromosome in the population resides on a processor and communication is carried out by message passing. The total communication overhead is thus minimal. Based on results on other optimization problems [1] where the evaluation of the fitness function dominates the processing time, as is the case with query optimization, we expect linear speedups in execution time. Since only limited parallelism can be incorporated into System–R, the time to execute the parallel GA should become much smaller than that of System–R.

5 CONCLUSIONS

We have presented a genetic algorithm for database query optimization. In doing so we have intro-

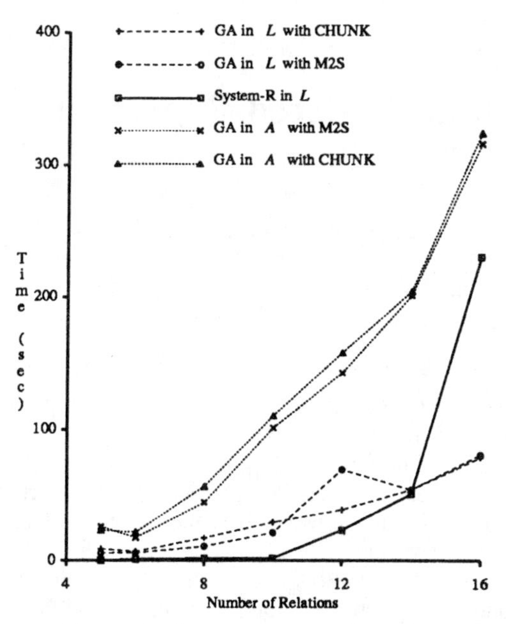

Figure 4: Average Processing Time

duced a novel encoding/decoding of chromosomes that represent binary trees together with associated new crossover operators. Although we did not exploit it in this paper, an important characteristic of the algorithm is its efficient parallelization. Our computational experiments with sequential implementations of the algorithm have shown the method to be a viable alternative to the commercially established algorithm. In fact, for large queries, one implementation of the GA found comparable solutions in much better time, whereas a different implementation found better quality solutions at the expense of additional time. Moreover, the GA was capable of optimizing large size problems on which the established algorithm fails.

In the future, we plan to adapt our parallel implementation of the GA to query optimization and verify our claims on its superiority over the System-R algorithm. In addition, we plan to investigate its applicability to query optimization in more complex database environments, e.g., parallel database machines.

Acknowledgements

The work described in this paper has been partially supported by The Air Force Laboratory Graduate Fellowship Program, the Air Force Office of Scientific Research under Grant AFOSR-89-0410 and the National Science Foundation under Grant IRI-8703592.

References

[1] E.J. Anderson and M.C. Ferris. A genetic algorithm for the assembly line balancing problem. In *Proceedings of the Integer Programming / Combinatorial Optimization Conference, Waterloo, September 1990*, Ontario, Canada, 1990. University of Waterloo Press.

[2] G.A. Cleveland and S.F. Smith. Using genetic algorithms to schedule flow shop releases. In Schaeffer [13], pages 160–169.

[3] E. F. Codd. A relational model of data for large shared data banks. *CACM*, 13(6):377–387, 1970.

[4] D.E. Goldberg. *Genetic Algorithms in Search, Optimization and Machine Learning.* Addison–Wesley, Reading MA, 1989.

[5] J.J. Grefenstette. Incorporating problem specific knowledge into genetic algorithms. In L.D. Davis, editor, *Genetic Algorithms and Simulated Annealing.* Pitman, London, 1987.

[6] J. Holland. *Adaptation in Natural and Artificial Systems.* The University of Michigan Press, Ann Arbor, Michigan, 1975.

[7] Y. E. Ioannidis and Y. Kang. Randomized algorithms for optimizing large join queries. In *Proc. of the 1990 ACM-SIGMOD Conference on the Management of Data*, pages 312–321, Atlantic City, NJ, May 1990.

[8] Y. E. Ioannidis and E. Wong. Query optimization by simulated annealing. In *Proc. of the 1987 ACM-SIGMOD Conference on the Management of Data*, pages 9–22, San Francisco, CA, May 1987.

[9] M. Jarke and J. Koch. Query optimization in database systems. *ACM Computing Surveys*, 16(2):111–152, June 1984.

[10] W. Kim, D. Reiner, and D. Batory. *Query Processing in Database Systems.* Springer Verlag, New York, N.Y., 1986.

[11] S. Lin and B.W. Kernighan. An efficient heuristic algorithm for the traveling salesman problem. *Operations Research*, 21:498–516, 1973.

[12] H. Mühlenbein. Parallel genetic algorithms, population genetics and combinatorial optimization. In Schaeffer [13], pages 416–421.

[13] J.D. Schaeffer, editor. *Proceedings of the Third International Conference on Genetic Algorithms*, San Mateo, California, 1989. Morgan Kaufmann Publishers, Inc.

[14] P. Selinger et al. Access path selection in a relational data base system. In *Proc. of the 1979 ACM-SIGMOD Conference on the Management of Data*, pages 23–34, Boston, MA, June 1979.

[15] A. Swami and A. Gupta. Optimization of large join queries. In *Proc. of the 1988 ACM-SIGMOD Conference on the Management of Data*, pages 8–17, Chicago, IL, June 1988.

[16] J. D. Ullman. *Principles of Database Systems.* Computer Science Press, Rockville, MD, 1982.

Genetic Algorithm for Clustering with an Ordered Representation

Jay N. Bhuyan
Computer Science Department
Tuskegee University
Tuskegee, AL 36088

Vijay V. Raghavan, Venkatesh K. Elayavalli
The Center for Advanced Computer Studies
University of Southwestern Louisiana
Louisiana, LA 70504-4330

Abstract

This paper considers the problem of partitioning N objects into M disjoint, nonempty clusters. Existing algorithms that optimally solve this problem employ branch-and-bound or dynamic programming techniques. Some heuristic algorithms, such as the K-MEANS, C-MEANS and basic ISODATA, are available to find an approximate solution. In this paper, we apply genetic algorithm to find a good ordered representation (permutation) of the objects. The fitness of a particular ordered representation is determined through a dynamic programming algorithm having a polynomial time-complexity. Two efficient problem specific crossover operators that generate a new representation are developed and compared. The performance of the algorithm is enhanced by the use of *local-tuning* (which improves the fitness of the members of the current generation before using them as parents to define the next generation). Experimental evaluation shows that the proposed strategies represent a compromise, in terms of the complexity and accuracy, between the best heuristic algorithm known to us and the traditional branch-and-bound or dynamic programming techniques for clustering.

1 INTRODUCTION

Clustering is an important technique used in the simplification of data or in discovering some inherent structure present in the set of objects. More specifically, the purpose of cluster analysis is to partition a given object set into a number of groups (clusters) such that objects in a particular cluster are more similar to each other than objects in different clusters. In clustering environment, one of the problem formulations involves the distribution of N objects among M clusters such that some extrinsic optimization criterion, which is additive over all the clusters, is minimized. When such an optimization criterion is given, the clustering problem is reduced to that of providing an efficient algorithm for searching the space of all possible classifications to find one that minimizes the optimization function. Clustering algorithms are categorized as either constructive or iterative. In the constructive method the allocation of objects to various clusters is determined by extending a partial solution to a complete feasible solution. Branch-and-bound [1] and solution through dynamic programming [2] belong to this category. Although the Branch-and-bound approach tries to reduce the complexity through pruning and the dynamic programming approach tries to avoid some redundant calculations present in the total enumeration, the complexity of both the algorithms is still too high. In particular, if the data collection is large, it is almost impossible to get results within a reasonable time.

In an iterative method, an initial solution is always feasible and complete and an attempt is made to improve this solution through various means from one iteration to the next. Popular clustering algorithms such as the K-MEANS (C-MEANS or basic ISODATA) method [3,4,5], belong to this category. Another improvement over this algorithm has recently been developed by Ismail et. al. [11], (called ABF and AFB) which alternates between a depth-first search and breadth-first search to effectively minimize the objective function. According to the various experimental evaluations done by the authors, ABF and AFB algorithms are the best available clustering algorithms if complexity is given more importance than accuracy. In these algorithms, in every iteration, objects are systematically moved to different clusters if there is a decrease in the value of the objective function. This greedy nature of these algorithms may cause them to get stuck at a local minima. They try to avoid this problem by taking several different random initial configurations and by applying the transformation procedure to each configuration. This type of evaluation is too ad hoc and the quality of the results strongly depends on the type of data and objective function.

Recently Klein et al. [6] have applied simulated annealing, a global combinatorial optimization technique, which uses probabilistic hill climbing (initially proposed by Kirkpatrick et al. [17]), for the clustering problem. But the disadvantages of simulated annealing are that a tremendous amount of execution time is needed and the determination of an efficient annealing schedule is difficult. In this paper, we consider a solution to the clustering problem based on genetic algorithm [13], where the optimal solution is searched through simultaneous consideration and manipulation of a *set* of possible solutions (population). In our approach, a solution is an ordered representation of the objects in a single dimension. Since for a permutation of the objects a considerable constraint is implied on possible clusterings, a polynomial solution for finding the optimal clusters from a permutation is used to find the fitness of each string in the population. Two types of problem-specific operators, which are applied on two parent solutions to give a new solution, are constructed. According to our experiments, our solution to the clustering problem through the use of genetic algorithm is promising.

1.1 REVIEW OF COST FUNCTION

The cost function we use is the sum of the squared Euclidean distance of each object from the centroid of its cluster. Suppose there are N objects with L dimensions and the objects are to be clustered into M classes. Gordon and Henderson [7] formalize a squared error clustering problem as follows.

$X = [x_{ij}]$, is a $N \times L$ object matrix.

$Y = [y_{ik}]$, is a $N \times M$ cluster membership matrix.

where $y_{ik} \in \{0,1\}$; $y_{ik} = 1$, if object i is in cluster k and $y_{ik} = 0$, otherwise.

$Z = [z_{kj}]$ is $M \times L$ matrix of cluster centers, where $z_{kj} = \sum_{i=1}^{N} y_{ik} \times x_{ij} / \sum_{i=1}^{N} y_{ik}$.

$$F(y_{11}, y_{12}, \ldots, y_{NM}) = \sum_{i=1}^{N}\sum_{k=1}^{M} y_{ik} \sum_{j=1}^{L} (x_{ij} - z_{k})^2.$$

The problem is to minimize F subject to the constraint $\sum_{k=1}^{M} y_{ik} = 1$, where $y_{ik} = 0$ or 1.

The number S(N,M) of ways of partitioning N objects into M clusters is given by

$$S(N,M) = \frac{1}{M!} \sum_{j=1}^{M} (-1)^{M-j} \binom{M}{j} j^N$$

where S(N,M) is defined as

$S(N,M) = M\, S(N-1,M) + S(N-1, M-1)$,

with $S(1,1) = 1$ and $S(1,M) = 0$, for $M \neq 1$

For example, if we have 100 objects and we want to partition them into 5 clusters then $S(100,5) = 10^{68}$. Hence, in an optimal solution, an exhaustive enumeration is impossible but for small values of N and M. Though branch-and-bound and dynamic programming reduce the number of cases to be explored, the complexity is still too high. Hence a genetic algorithm, which represents a compromise between the branch-and-bound and greedy algorithms, should be a viable alternative.

1.2 OUTLINE OF THE PAPER

In Section 2, the greedy algorithms for clustering called ABF and AFB developed by Ismail et al.[11], which are used for comparing the performance of our algorithm to earlier work, are discussed in detail. The specific implementation details of our schemes to solve the clustering problem are given in Section 3. Some experimental results comparing our work with the existing methods are given in Section 4. The last section deals with conclusions and directions of future work.

2 GREEDY ALGORITHM (AFB AND ABF)

The algorithm developed by Ismail et al. [11] is outlined in Figure 1. The algorithm starts with an initial configuration by randomly distributing the objects among different clusters. In the procedure DHF, each object from 1 to N is moved to the first cluster in the sequence of 1 to M, where there is a decrease in the objective function F. In the procedure DHB, each object is moved to a cluster which results in a maximum decrease in the value of the objective function F. An object can be removed from a cluster only if the latter has more than one object. For any object, if there can not be any decrease in the value of the objective function by moving it into any of the clusters, then the position of the object is not disturbed. The methods ABF and AFB, shown below, alternate between the algorithms DHB and DHF in each iteration until there is no decrease in the value of the objective function for the N objects.

```
Algorithm ABF:
  begin
     Select an initial configuration arbitrarily
     repeat
        if (iteration number is odd) then DHB();
        else
              DHF();
     until (there is no change in the objective
              function)
  end.
Algorithm AFB:
  begin
     Select an initial configuration arbitrarily
     repeat
        if (iteration number is odd) then DHF();
        else
              DHB();
     until (there is no change in the objective
              function)
  end.
```

Figure 1: Greedy Algorithm for Clustering

3 GENETIC CLUSTERING

This section describes in detail the genetic algorithms that we propose for clustering. We first give the details about the representation for each string in the population. Then we describe the strategies for initial population construction, for the selection of parents and for the various genetic operations.

3.1 REPRESENTATION OF CLUSTERING SOLUTIONS

We have explored several string representations for the solution space. In our view in choosing a representation one should look for two objectives:

1. To find an encoding that will permit the application of traditional crossover operators developed by Holland [13] and

2. The crossover operator should identify subunits of the parents associated with good performance and incorporate these subunits (from the parents) into the offsprings generated.

We next describe the different options we considered, and the representation we settled on for our algorithm.

3.1.1 Binary String Representation

In the binary string representation a particular partition of N objects into M groups is assumed to be represented by an undirected graph of N nodes. Each object corresponds to a node in the graph. Two nodes are connected in the graph if the objects they represent belong to the same cluster. Assuming that we have N objects, we will have N(N-1)/2 possible distinct edges where the existence of an edge between two objects represents that they belong to the same cluster. Let us assume that we have 6 objects 1, 2, 3, . . . , 6 and we want to put them into 3 clusters. Then the number of possible edges are (12, 13, 14, 15, 16, 23, 24, 25, 26, 34, 35, 36, 45, 46, and 56). Each possible solution to the clustering problem (chromosome) is an ordered representation of these edges by means of a binary string of length 15, where 1 indicates the presence of and 0 the absence of the corresponding edge. For example, suppose one of the parents called parent1 has {1, 3, 5}, {2, 4} and {6} as clusters, while the other parent called parent2 has {3, 5, 6}, {2, 4} and {1} as clusters. Then parent1 is represented as (0 1 0 1 0 0 1 0 0 0 1 0 0 0 0) and parent2 as (0 0 0 0 0 0 1 0 0 0 1 1 0 0 1). Now the objective of a crossover operator is to create an offspring which will have 1 in those bits positions where both the parents have 1 and pick up some additional 1s in bits positions selected at random, such that either of the parents has a 1 in those positions. The obtained offspring should consist of 3 complete graphs, to make it a legal (that which has same number of clusters as the parents) solution. We are still exploring the possibility of developing a crossover algorithm, similar to that of Whitley et al. [15], but we do not adopt this representation in this paper because of the high time complexity associated with operations on this representation. We recently came across an edge based representation developed by Jones and Beltramo[20] which is similar to this representation and in which they claim to have obtained a crossover operator with time complexity $\Theta(M^4)$.

3.1.2 Ordinal Representation

Another possibility for representation is a variant of the ordinal function introduced by Grefenstette et al. [16]. Given N objects and M clusters, each chromosome is represented by means of a string of length (N+M−1) containing each integer between 1 and (N+M−1) exactly once. While the numbers 1 to N represent the objects present in the problem, the numbers from (N+1) to (N+M−1) represent a boundary (an indication that two adjacent objects belong to different clusters) between objects. For example, let us take the previous example of having 6 objects, labeled 1 through 6, and 3 clusters. Let the clusters {2 3 5}, {4 1} and {6} be one of the solutions. Then one of the string representations of this solution is (2 3 5 7 4 1 8 6). Here, the digits 7 and 8 mark the boundaries between clusters. In order to create an ordinal representation of the above string we maintain two lists called ordinal list and free list. While, initially, the ordinal list is empty, the free list will contain all the numbers from 1 to (N+M−1) in an ascending order. In order to construct the ordinal string for each object or boundary in the solution string its position in the free list is appended to the ordinal list and the object or the boundary is deleted from the free list. The construction of the ordinal string from a solution string, for this example, is shown below.

Ordinal list	Free list
( )	(1 2 3 4 5 6 7 8)
(2)	(1 3 4 5 6 7 8)
(2 2)	(1 4 5 6 7 8)
(2 2 3)	(1 4 6 7 8)
(2 2 3 4)	(1 4 6 8)
(2 2 3 4 2)	(1 6 8)
(2 2 3 4 2 1)	(6 8)
(2 2 3 4 2 1 2)	(6)
(2 2 3 4 2 1 2 1)	( )

Through a procedure similar to the one shown above, we can recover the solution string back from an ordinal string. This encoding produces a legal solution after the application of the traditional crossover operator, except in some rare cases where an offspring being created has numbers corresponding to the boundaries in one of the following three configurations: two numbers are adjacent, one of them is in the beginning of the string, or one of them is at the end of the string. Although we are able to achieve our first objective, from the point of view of the second objective this type of encoding is not efficient. This is because while the string corresponding to the left of the crossover point is a sub-unit of one of the parents, the one to the right side of the crossover point is essentially random. Thus no useful information is passed on to the offspring from the second parent. We therefore discard this representation from further consideration.

3.1.3 Ordered Representation

In this representation each partition is represented by a permutation of objects. This type of ordered representation gives information about the similarity between objects rather than specify the optimal clusters directly. For example, let us assume that the chromosomal representation of objects 1 through 6 is the permutation (2 3 5 1 6 4). We assume that, in a permutation which leads to an optimal partition, objects which are similar to each other are placed near each other. With this assumption, the order (2 3 5 1 6 4) of objects indicates 2 is more similar to 3 than 5, and 5 cannot be included in the same cluster as 2, without also including 3. Thus there are several possibilities for identifying 3 clusters from the above permutation. Some of the possible combinations of clusters are {2} {3 5 1} {6 4}, {2 3} {5 1 6} {4} and {2 3} {5 1} {6 4}. Of all these combinations, the order (2 3 5 1 6 4) represents that combination which has an optimal value for the objective function.

In case of partioning N objects into M clusters, let us assume that $(O_1, O_2, ..., O_N)$ is a particular permutation of N objects. Then each cluster consists of an interval of objects $(O_i, O_{i+1}, ..., O_{j-1}, O_j)$ where $1 \leq i, j \leq N$. With this constraint the number of possible distinct clusters is $\frac{1}{2} \times N(N+1)$. Each interval can be evaluated separately to determine its contribution to the optimization function F. The value of the optimization function F of the ordered representation is determined by summing up the contributions of certain optimally selected non-overlapping intervals. The selected intervals span the whole interval $(O_1, O_2, ..., O_N)$ and the sum of their contribution is the lowest. Because of this additive property, a dynamic programming algorithm developed by Fisher[8,9], having a polynomial time complexity of $\Theta(N^2M)$, should be invoked each time we need to evaluate the fitness of a string in a population. We decide to use this representation for our clustering problem.

3.2 INITIAL POPULATION CONSTRUCTORS

In this section, we explore three different initial population constructors, namely A, B and C, for the clustering algorithms based on GAs. The constructor A randomly places the objects into an ordered list. The use of this strategy to obtain an initial population enables the complete testing of the GA in most challenging circumstances. In other words, the GA is made completely ignorant of knowledge about good regions of the search space.

Constructor B uses a heuristic algorithm based on a minimum distance between two objects. It generates a object label at random, and puts it in the first position of the ordered list. Then it looks for all objects which do not exist in the ordered list in order to find a object nearest to the most recently included one (w.r.t. squared euclidean distance) in the ordered list. This procedure is repeated until we have included all the objects into the ordered list. This process is repeated until all the individuals of the initial population are generated. The complexity of this constructor is $\Theta(N^2)$.

Constructor C uses a probabilistic greedy heuristic for constructing the initial population. The primary intention of this constructor is to maintain a balance between randomness and problem specific knowledge. The difference between this and the constructor B is in finding the object nearest to the most recently included one in the ordered list. Instead of searching for all the objects which are not included in the ordered list, we search only for k (a constant defined by the user) number of objects. If this k is equal to the total number of objects, then this constructor is same as constructor B. By this method, the time complexity of the constructor is considerably reduced.

3.3 PARENT SELECTION OPERATOR

As our problem is one of minimization, the objective function F(x) has to be mapped to a fitness function f(x) (where the probability of a string x_i being selected for crossover is given by $\frac{f(x_i)}{\sum_{i=1}^{p} f(x_i)}$, where p is the population size). If f(x) is taken as the inverse of F(x) there would not be much difference between a good string and a bad string. This problem is handled by the use of another transformation f(x) = Cmax − F (x), where Cmax corresponds to the value of the poorest string in the population. In addition to the transformation, a scaling function (e.g. f(x) = a f(x) + b), similar to that of Goldberg [18], is used. The select operator then selects parent strings according to the scaled fitness of the strings. It may be noted that the values of a and b are determined in such a way that each member of the population, having an average cost, contributes one expected offspring to the next generation and Cmult number of copies for the best population member, where Cmult is a parameter decided by the user.

3.4 CROSSOVER OPERATORS

In this section we explore two types of crossover operators. The operator 1 does a cut and paste type operation similar to the operator developed by Coohen et al. [10]. In this operation one of the parents is randomly assigned as the dominating parent, while the other is considered to be a supporting parent. A portion of the string is cut from the supporting parent and pasted to the dominating parent. The dominating parent is rearranged with as minimal a disturbance of its alleles as possible to make it a legal solution. The other operator (Operator 2) gives equal importance to both the parents. The location of each object with respect to other objects is studied in both the parents. This adjacency information is exploited in the creation of an offspring. Both the operators are explained below in detail.

3.4.1 Operator 1

Let X_T be a set of all objects to be clustered. A window size x_w and a object $x \in X_T$ is selected at random. Let x_{p1} be the position of the object in the dominating parent (parent1) and x_{p2} the position of the corresponding object in the supporting parent (parent2). Let X be the set of all objects between the positions $x_{p2}-x_w$ and $x_{p2}+x_w$ in the supporting parent. The object x is put into the position x_{p1} in the child. Now the objective is to put the objects of $X-\{x\}$ around x_{p1} and the objects of (X_T-X) in the remaining position of the child so that the order of the objects of X_T-X and X in parent1 is maintained in the child. This is done by a *sliding* process in 3 stages. In the first stage it stores the objects of $X_T-X \cap$ {parent1[1], parent1[2], ..., parent1[x_{p1}]} of parent1 in the child by storing the first object in the first position of the child and maintaining the sequence of these objects existing in parent1. Then, in the second stage, the objects of X are stored in the child from its left most empty position with the same sequence as in parent1. In the third stage $X_T-X \cap$ {parent1[x_{p1}], parent1[x_{p1}+1], ..., parent1[N]}, where N is the length of the string, are filled in the vacant position of the child while maintaining its sequence in parent1. It can be easily seen that the complexity of this algorithm is $\Theta(N)$. Let us take an example of 8 objects (Figure 2) and let $X_T = \{1, 2, 3, 4, 5, 6, 7, 8\}$, with parent1 as (5, 3, 8, 1, 2, 7, 6, 4) and parent2 as (6, 5, 1, 2, 4, 7, 8, 3). Suppose x = 2 and x_w = 2 are selected at random. Then the corresponding portion to be cut from parent2 is X = {5, 1, 2, 4, 7}. The sequence corresponding to X present in parent1 is (5, 1, 2, 7, 4). In the first stage the objects 3 and 8 are placed in 1st and 2nd positions of the child. In the second stage objects 5, 1, 2, 7, and 4 are copied to the child starting from its 3rd position. In the final stage object 6 is copied to the last position of the child resulting in the offspring (3, 8, 5, 1, 2, 7, 4, 6).

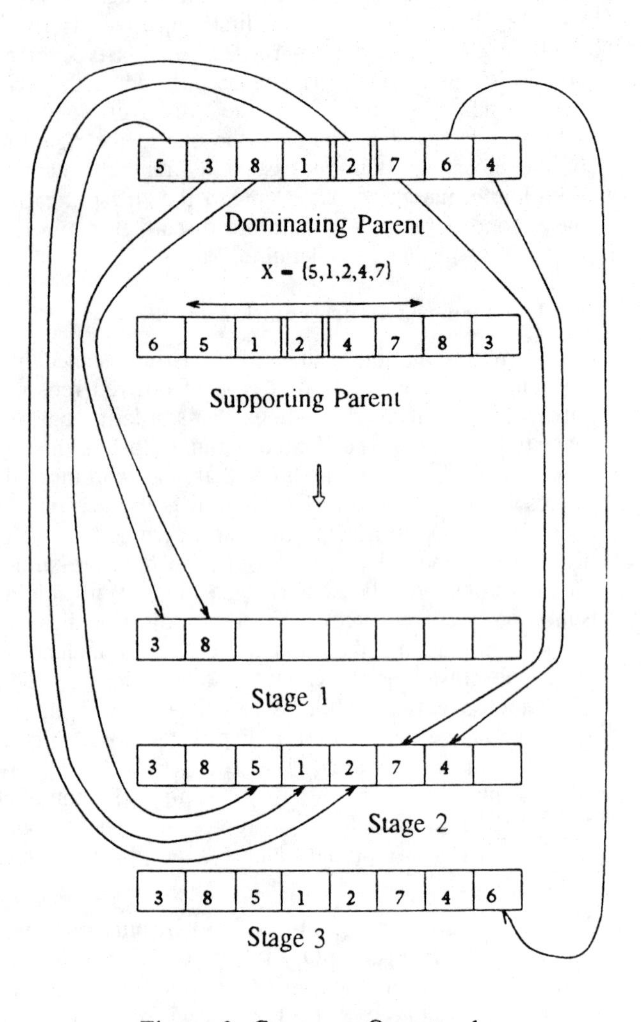

Figure 2: Crossover Operator 1

3.4.2 Operator 2

In this crossover operation the new child is initiated with the value from the first position of either of the two parents. The distance of the nearest object (not yet included in the child) to the recently included object is determined in both the parents. The object having the shortest distance is inserted in the left most empty position of the child. In case of a tie the object from either of the parents is picked up at random. This process continues until all the positions of the child are filled. Let us trace this algorithm through an example (Figure 3) of 8 objects with parent P_1 as (5, 3, 8, 1, 2, 7, 6, 4) and parent P_2 as (6, 5, 1, 2, 4, 7, 8, 3). The first position of the child is randomly picked from the first position of either parent, and let us assume in this case parent P_2 is selected at random resulting in object 6 being put in the first position of the child. The nearest object from 6 within a distance of 1 in parent P_1 is {7, 4} while that in parent P_2 is 5. Hence one of the objects out of 4, 5, and 7 is randomly selected. Let this be the 7. This object occupies the second position in the child. The objects within distance 1 from the position of object 7 in parent P_1 and parent P_2 are {2, 6} and {4, 8} respectively. Since object 6 is already included, an object is picked up out of 2, 4, 8 for the third position of the child (say, 2). In a similar fashion all the remaining positions of the child are fixed, and we may obtain the offspring as (6, 7, 2, 4, 8, 1, 5, 3).

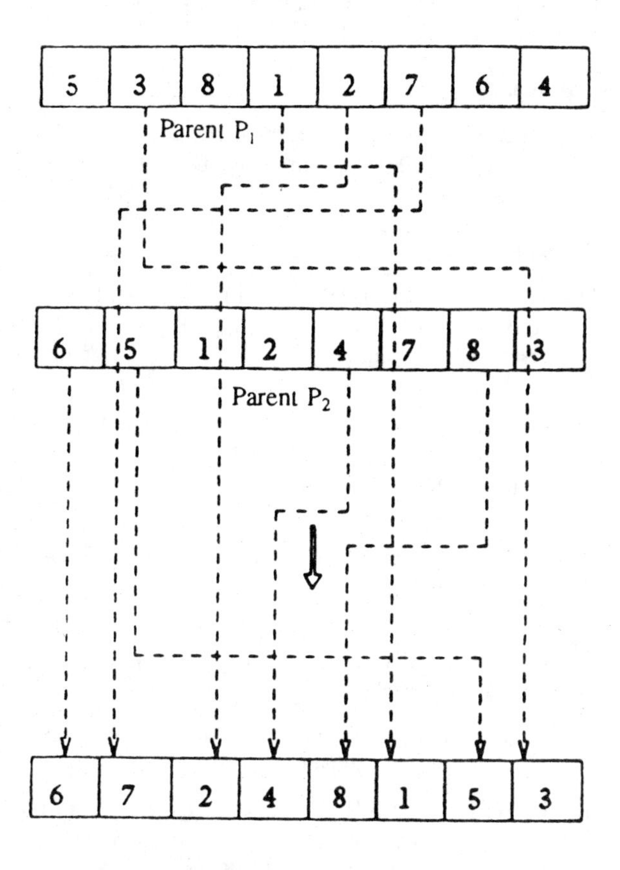

Figure 3: Crossover Operator 2

3.5 REPLACEMENT OPERATOR

The *creat* operator is used to select a population of fixed size from the old population P(old) and offspring P(offspring) (which is created by the crossover operator). A parameter X, provided by the user, indicates that a new population P(new) is formed by picking the X best strings from the combination of P(old) and P(offspring). The rest of the strings of the new population P(new) are selected at random from P(offspring). It may be noted that when X is 0, the new population P(new) consists of strings selected only from P(offspring). On the other hand, if the value of X is equal to the size of the population, then P(new) is constituted solely of the X best strings amongst the strings present in P(old) and P(offspring).

3.6 LOCAL TUNING AND MUTATION

Grefenstette[12] observed some problems in genetic algorithm with respect to local search. GAs focus on overall population instead of identifying one best individual. As in natural genetic systems, GAs progress by identifying high performance regions of the search space. GAs would be more useful for combinatorial optimization problems if they were adapted to invoke a local search strategy to optimize members of the final population. Brown et al. [14] recently applied simulated annealing to every individual in each generation of the genetic algorithm in the context of the Quadratic Assignment problem. But in our opinion, even if some inherent parallelism is present in genetic algorithm, it is too expensive to apply an algorithm like simulated annealing to each structure during every generation. The local tuning operation is proposed in this work as an alternative to Brown's strategy. This operator is greedy in nature and has a very low complexity. The Local-tuning operation is used only after a predetermined (user specified) number of generations have elapsed. When we try to evaluate each individual of a population through dynamic programming, the objects belonging to each cluster are determined. Objects are considered from 1 to N and each is moved to the first encountered cluster that is different from the current one and leads to a decrease in the value of the objective function. The complexity of this operator is found to be $\Theta(NM)$, which is very low compared to that of simulated annealing. In case of mutation, we randomly select two objects and exchange their positions.

4 EXPERIMENTAL RESULTS AND ANALYSIS

Experimental evaluation of the proposed algorithms were conducted on two sets of data. One was the classical German Towns data [19] which contains 59 two-dimensional objects. The other set of data consists of several sets of pseudo-random vectors. These vectors were chosen from normally distributed populations with the specified mean and variance-covariance matrices. The data sets 1, 2 and 3 (Table 1) contain 40 2-dimensional objects each and the other three, data sets 4, 5 and 6 (Table 1) contain 40 3-dimensional objects each.

The algorithms ABF and AFB were implemented and the experiments were conducted for 40 initial cluster configurations generated at random. The lowest values obtained are displayed in the second and third columns of Table 1 (for German Towns collection) and Table 2 (for the normalized data sets). These results were used to compare the performance of our genetic algorithm.

For the experiments of our genetic algorithm, crossover rate and mutation rate were fixed at 1.0 and 0.2, respectively. The population size in all the experiments was taken to be 40. The experiments were conducted using constructors A, B, and C. The constructor A was not found attractive because of its poor convergence time. The constructor B led to a fast convergence in some cases, but unfortunately to a local minimum many times. Hence we only report results based on constructor C.

Experiments on the *creat* operator were conducted for different values of X. When the value of X was chosen to be 0, we observed that the method often lost the best solution and took a long time to converge. When the value of X was chosen as the size of the population (i.e. 40) we observed that our method lost its diversity too soon and converged to a local minimum. The reported results are for experiments conducted with the value of X as 5.

The algorithm was run for 40 generations, with the local-tune operator being applied after 30 generations. However, even without local-tune (not shown in the tables) the algorithm produced better results than Greedy.

The Tables 1 and 2 show the performance of the two crossover operators. For the German Towns data, the results produced by our algorithm were as good as those of ABF and AFB for 4 to 7 clusters and better in case of 8 to 10 clusters (Table 1). In case of the normalized data sets, our algorithm performs better than AFB and ABF algorithms in three cases (Data sets 1, 2 and 3) and performs as good as the latter for the other 3 data sets (Table 2). The performance of both the operators were found to be equally good except in the case when 9 clusters were formed in the German Towns data collection (Table 1). In this case the performance of crossover operator 2 was same as that of ABF algorithm and operator 1 performed better. We also observed that performance of crossover operator 1 was better than that of operator 3 in many cases before the application of local tune up operator (not shown in the table). We could not determine any specific reason for this and we are still investigating this behavior.

5 CONCLUSION

Earlier work on the clustering problem has concentrated mainly on simplified greedy heuristics which sacrifice accuracy for time complexity, or on branch and bound along with dynamic programming which does the reverse. In this paper, we have explored the application of genetic algorithm to the clustering problem to demonstrate that a balance between time complexity and accuracy can be maintained. Preliminary experimental results reflect the superiority of genetic algorithm over the best available heuristics. In addition, by proper tuning of parameters, we can achieve further improvement on the accuracy, thereby completely justifying our claim. We also need to explore a different evaluation function for our ordered representation that has reduced time complexity (although the present one has polynomial time complexity), in order to provide the ability to handle large collections of objects. Our future work will concentrate on these issues and also on developing a crossover operator on a binary string representation.

Acknowledgements

This paper is based on research supported by the NSF under grant IRI-8805875 and by Digital Equipment Corporation.

References

[1] W.L.G. Koonz, P.M. Narendra, and K. Fukunga. A Branch and bound clustering Algorithm. *IEEE Tran. on Computers*, Vol. C-24, No.9 , September 1975.

[2] R.E. Jensen. A dynamic programming Algorithm for cluster Analysis. *Journal of Op. Res. Society*, Vol 7 pp 1034-1057,1969.

[3] R. Dubes and A. K. Jain. Clustering methodologies in exploratory data analysis. *Adv. Comput.*, 19 pp 213-228, 1980.

[4] R. Dubes and A. K. Jain. Clustering Techniques: the user Dilemma. *Pattern Recognition*, 8 pp 247-268, 1976.

[5] J. Tou and R. Gonzalez. *Pattern Recognition*. Reading, M.A: Addison-Wesley, 1974.

[6] R. W. Klein and R. C. Dubes. Experiments in projection and clustering by simulated Annealing. *Pattern Recognition*, vol 22 pp 213-220, 1989.

[7] A. D. Gordon and J. T. Henderson. An Algorithm for Euclidean sum of square classification. *Biometric*, 33, pp 355-362 , 1977.

[8] W.D. Fisher. On Grouping for maximum homogeneity. *Journal of American Stat. Asoc.*, 53, pp 789,798, 1958.

[9] J.A. Hartigan. *Clustering Algorithms*, John Wiley & Sons, Inc., 1975.

[10] J.P. Coohen and W.D. Paris. Genetic Placement. *IEEE Trans. on Computer-Aided Design*, Vol. 6, No.6, pp 956-964, Nov 1987.

[11] M.A. Ismail and M.S. Kamel. Multidimensional data clustering utilizing hybrid search strategies. *Pattern Recognition*, Vol 22 No. 1, pp 75-89, 1989.

[12] J.J. Grefenstette. Incorporating problem specific knowledge into genetic Algorithms. *Genetic Algorithms and Simulated Annealing*, Morgan Kaufman Publishers, pp 42-60, 1987.

[13] J.H. Holland. Adaptation in Natural and Artificial System. *The University Of Michigan Press*, Ann Arbour, 1975.

[14] D.E. Brown, C.L. Huntley, and R. Spillane. A Parallel Genetic Heuristic for the Quadratic Assignment Problem. *Proceedings of third International Conference on Genetic Algorithm*, pp 406-415, 1989.

[15] D. Whitley, T. Starkweather, and D. Fuquary. Scheduling Problem and Traveling Salesman : The genetic edge recombination operator. *Proceedings of third International Conference on Genetic Algorithm*, pp 406-415, 1989.

[16] J. Grefenstette, R. Gopal, B. Rosamita, and D.V. Gucht. Genetic Algorithm for the Traveling Salesman Problem. *Proceedings of first International Conference on Genetic Algorithm*, pp 160-165, 1987.

[17] S. Kirkpatrick, C.D. Gelatt, and M.P. Vecchi. Optimization by Simulated Annealing. *Science*, 220 pp 671-680.

[18] D.E. Goldberg. *Genetic Algorithms in Search Optimization & Machine Learning*. Addison-Wesley Publishing Company, Inc. pp 75-79, 1989.

[19] H. Spath. *Cluster Analysis Algorithm* John Wiley & Sons, 1980.

[20] D.R. Jones and Mark A. Beltramo. Solving Partitioning Problems with Genetic Algorithms. *Fourth International Conference on Genetic Algorithm*,1991.

Table 1: Results (Values of optimization function F) for German Towns Data Collection

#of clusters	ABF	AFB	Cross-over Operator 1	Cross-over Operator 2
4	49,600	49,600	49,600	49,600
5	38,716	38,716	38,716	38,716
6	30,535	30,535	30,535	30,535
7	24,432	24,432	24,432	24,432
8	21,499	21,499	21,483	21,483
9	18,970	19,042	18,550	18,970
10	16,711	16,864	16,353	16,516

Table 2: Results (Values of optimization function F) for Normally Distributed Data Sets

Data Sets	ABF	AFB	Cross-over Operator 1	Cross-over Operator 2
1	161	161	148	148
2	642	642	507	507
3	2714	2714	2710	2710
4	3302	3302	3302	3302
5	1549	1549	1549	1549
6	3966	3966	3966	3966

Tracking a Criminal Suspect through "Face-Space" with a Genetic Algorithm

Craig Caldwell
Victor S. Johnston
Psychology Department
New Mexico State University
Las Cruces, NM 88003

Abstract

The genetic algorithm can rapidly search a "face-space" containing over 34 billion possible facial composites. Using subjective estimates of resemblance to a culprit as a measure of fitness, a witness can exert selection pressure on the search to evolve the likeness of a criminal suspect in as few as ten generations. This multivariate approach may adapt to any individual recognition strategy. Sequential selections also present complete phenotypic composites, mapped from the feature-coded genotypic information, promoting natural recognition ability, rather than subjective recall, to evolve a target face. Methods for increasing the efficiency of the GA are examined. Some early results and future directions of this research are discussed.

A novel attribute of the genetic algorithm (GA) is its ability to dynamically interact with a human user while conducting a multivariate search. Because it is patterned after natural selection, evolution toward a desired configuration proceeds through the empirical selection of the most fit solution. This selection can be exploited to reflect human preferences, by relying on direct interaction with a human user to subjectively establish the measures of fitness that guide the evolving search. One area where this human interaction with the GA may be utilized fully is in assisting a witness to build a facial composite of a criminal suspect.

Humans have excellent facial recognition ability. They can distinguish among an "infinity" of faces seen over a lifetime, while recognizing large numbers of unfamiliar faces early in their development [4]. In contrast, witnesses frequently have great difficulty in recalling facial characteristics with sufficient detail to provide an accurate composite of a criminal suspect. As a consequence, current identification procedures, which depend heavily on feature recall, often yield unsatisfactory results.

The field of criminology provides most of the research on the systematic study of face representation in various media. The extensive use of the Photofit system has produced the largest body of research on recognition of composite facial images. Penry developed Photofit in Britain, between 1968 and 1974 [11]. This technique uses over 600 interchangeable photographs of facial parts, picturing five basic features: forehead and hair, eyes and eyebrows, mouth and lips, nose, and chin. With additional accessories, such as beards and eyeglasses, combinations can produce approximately fifteen billion different full-face views of Caucasian males [9]. Alternatives to Photofit have since been developed. They include the Multiple Image-Maker and Identification Compositor (MIMIC), which uses film strip projections; Identikit, which uses plastic overlays of drawn features to produce a composite resembling a sketch, and Compusketch, a computerized version of the Identikit process.

The success of current systems, such as Photofit, depends upon the ability of a witness to accurately recall each feature of a suspect, and to simultaneously be aware of which features and feature positions in the generated composite require modification. The success of such procedures has also been shown to be a function of age, sex, cognitive strategy and hemispheric advantage of the witness [4,5,7,8,10,12,13,14]. The inefficiency of a linear search and the need for recognition of isolated features is compounded with the need for further recall of both features and feature positions to produce a final composite.

An approach utilizing the Genetic Algorithm relies on recognition rather than recall. Unlike current procedures, the GA is capable of efficiently searching an extremely large sample space of alternative faces, and of finding an accurate likeness in a relatively short period of time. Since the implementation makes no assumptions concerning the age, gender, hemispheric advantage or cognitive strategy of a witness, it can find an adequate solution irrespective of these influences. Additionally, an examination of the processes used to locate a facial likeness in the multi-dimensional space of feature variables can provide insights into the cognitive strategies used in face recognition.

The development of a composite face is a saltant process, where only a limited number of configurations are considered. The GA, as first described by Holland, is the serialized recipe for the evolutionary process, implemented by the random influences of chromosome crossover and gene mutations, and steered by fitness for survival. When translated into computer software, the GA provides a unique search strategy that obtains the most "fit" outcome from a choice of evolutionary paths [6]. When used to implement a selection routine for facial identification, the GA provides a selection strategy that performs the double function of preserving an evolving facial composite while introducing non-destructive random changes. By representing facial variables as a form of genetic code, both cephalometric measurements and specific feature elements of a composite face may be varied simultaneously, using crossover and mutation operators.

In the current design, a series of twenty faces (phenotypes) are generated from a random series of binary number strings (genotypes) according to a standard developmental program. During decoding, the first seven bits of the genotype specify the type and position of one of 32 foreheads, the next seven bits specify the eyes and their separation, and the remaining three sets of seven designate the the shape and position of nose, mouth and chin, respectively. Combinations of these parts and positions allow over 34 billion composite faces to be evolved An initial twenty random faces can be viewed as single points spread throughout a multi-dimensional face space. The function of the GA is to search this space and find the best possible composite in the shortest possible time.

The first step in the GA is the "selection of the fittest" from the first generation of faces. This is achieved by having a witness view all twenty faces and rate each one on a nine point scale, according to its resemblance to a culprit. This measure does not depend upon the identification of any specific features shared by the culprit and the composite face, and the witness need not be aware of why any perceived resemblance exists. After such fitness ratings are made, a selection operator assigns genotype strings for breeding the next generation, in proportion to the relative fitness values. Selection according to phenotypic fitness is achieved using stochastic universal sampling [1]. The next step in the process is sexual reproduction between random pairs of the selected genotypes. The breeding of genotypes employs crossover and mutation operators. When two genotypes mate, they exchange portions of their bit strings according to a specified crossover rate, and mutate according to a selected mutation rate. Following selection, crossover, and mutation, the next generation of faces is developed from the new genotypes and subsequently rated by the witness, as before. This procedure continues until a satisfactory picture of the culprit is evolved.

The GA asssembles a target sequence from multiple subtargets by optimizing segments of bit strings and preserving them through repeated selections over generations, based on their viability. Variables that control feature configurations and inter-feature distances are represented as string segments, and can be equated to a set of coordinate values located in a manifold of five feature dimensions. Evolving facial constructs demonstrate emerging phenotypic properties that are functionally dependent upon the gene values. Each set of five coordinates, decoded from five segments, comprises a point that specifies a composite face. Evolutionary progress over generations can be represented conceptually as a track through the "face-space", from a random initial location to a "region-of-recognition" encompassing the target point. This progress appears as a stochastic walk, connecting the most fit point of each generation, and is diagramed in Figure 1.

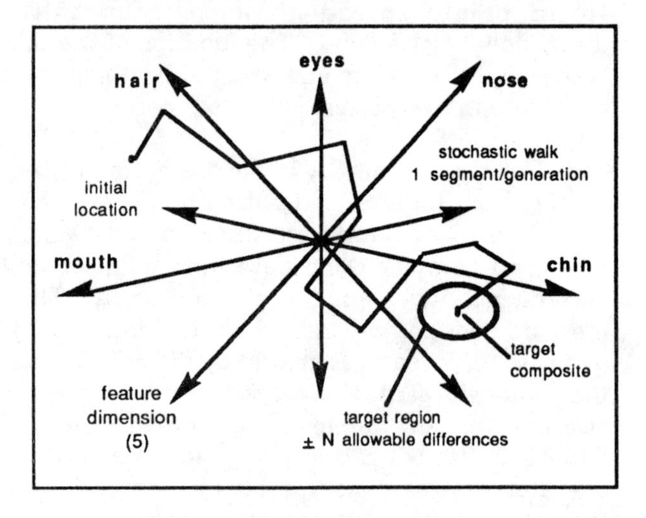

Figure 1. A stochastic walk over ten generations in "Face-Space".

Improving the GA based search can be equated with streamlining movement through "face-space". Such movement may be enhanced by several procedures which speed or direct the trajectory towards the target region. These procedures are enumerated below.

1. The starting location may be optimized. Initially, a set of coordinates is randomly generated. The set of points specified by these coordinates constitutes the first generation of composites. A high fitness rating for any composite represents the witness' desire to remain in that region of face space. The starting locations may be improved by generating a new randomized set of starting points when first generation fitness values are very low.

2. Different genotypic coding systems influence the rate of convergence on a known target. In binary, a change from number 011 (base ten: 3, gray code: 110) to number 100 (base ten: 4, gray code: 010), requires three simultaneous bit mutations, a "Hamming distance" of three. Some research indicates that gray code may facillitate search since there is a constant Hamming distance of one between adjacent values [2,3]. However, convergence tests using binary, gray and bingray (a composite with the first four digits decoded as binary code and the last three decoded as gray code), indicate a superiority for binary code under the current search conditions. (The rationale proposed for examining bingray code is that altering the binary part can jump the decoded value to the region of interest more rapidly, and the gray segment can mutate smoothly to then refine the local search.) A curve showing the performance of all three codes is presented in Figure 2. Over multiple simulations, the performance at generation 20 using binary code (88.6%) has proved superior to gray code (81.1%) and also superior to bingray code (87.1%). The problem with using gray code appears to arise from an inconsistent base ten value associated with any bit position. A statistical analysis of the data, using the method of contrasts, also shows that a binary coded GA converges significantly faster.

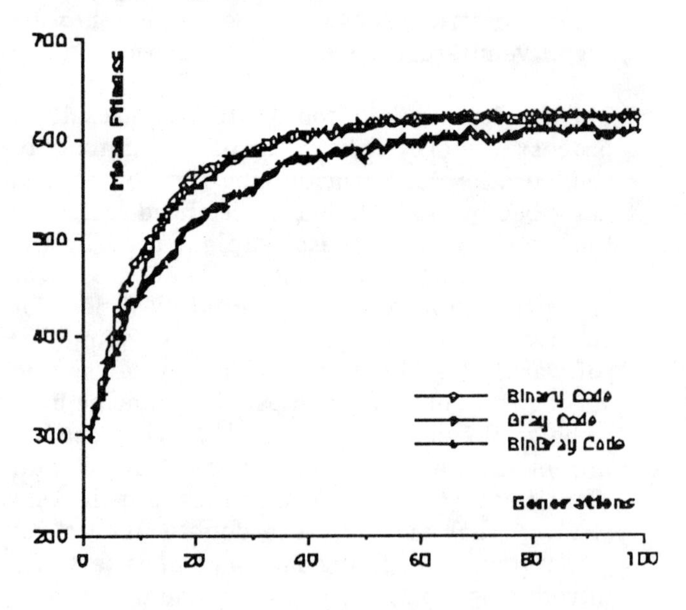

Figure 2. Binary, gray and bingray code performance.

3. The crossover rate determines how quickly the probability density distribution of points narrows around sequential segments of the walk through "face-space". The efficiency in convergence is controlled by the likelihood that an exchange of genes occurs, and by the consequence of this exchange altering the value (coordinate) of a feature specifying segment. For a string of length L, when only one crossover

occurs at random, only L-1 strings may be formed from the original, or 2(L-1) for both breeding strings. When multiple crossovers are allowed, occuring sequentially at each bit where the probablilty dictates, 2^L-1 reconfigurations are possible. This method of uniform crossover permits a wider and more thorough search, augmenting mutation effects by also allowing a single bit crossing to occur.

4. The mutation rate controls the breadth of (regional) variance that occurs at each step in the converging walk to the target. The largest dimension in this region of variance is constrained by the probability of mutation of each successive bit, but is limited in its extremum to one less than the upper maximum of the bit value (e.g. 0-63). [This region is analogous to the shell of a topological "ball" formed in a continuum. However, since the dimensional points consist of discrete feature coordinates, rather than a true continuum, the defined space is actually a decagon.]

5. To optimize the crossover and mutation rates, a meta-program has been constructed with a meta-level GA, where binary, meta-level strings can be decoded to provide rates for the base level GA. The meta-level GA evaluates each meta-string by determining how well a simulation evolves a composite face using the crossover and mutation rates specified by this string. The meta-level population of strings is evolved over a series of generations to breed the optimum percentages for the particular constraints of this task. This has provided the parameters for the current implementation of the facial composite search program.

6. The search may be further facilitated by permitting a witness to lock a highly desirable feature of a displayed composite. Such locking of a feature is implemented by the sequential flooding of the corresponding gene segments of the entire population with the appropriate code, in all future generations. This immediately confines all future search to the remaining dimensions of the "face-space". An alternative to this permanent locking of a feature (freeze) is to flood the population with the feature code but permit mutations to alter the sequence (flood). Results from running the meta-level program indicate that both the flood and freeze options produce a marked improvement in the performance of the algorithm. At the 20 generation mark, the mean performance of the standard GA has been equal to 88.6% of the maximum fitness possible, with mean performance using flooding equal to 93.1%, and freezing equal to 96.4%. The superior performance of freezing over flooding appears to be a consequence of the harmful effects of mutations as composites begin to converge to the likeness of the culprit. Mutations in early generations may enhance performance by exploring more regions of face space, but in later generations these mutations have a higher probability of being disruptive. The freeze option has been included in the current implementation of the GA program to increase the efficiency of the search.

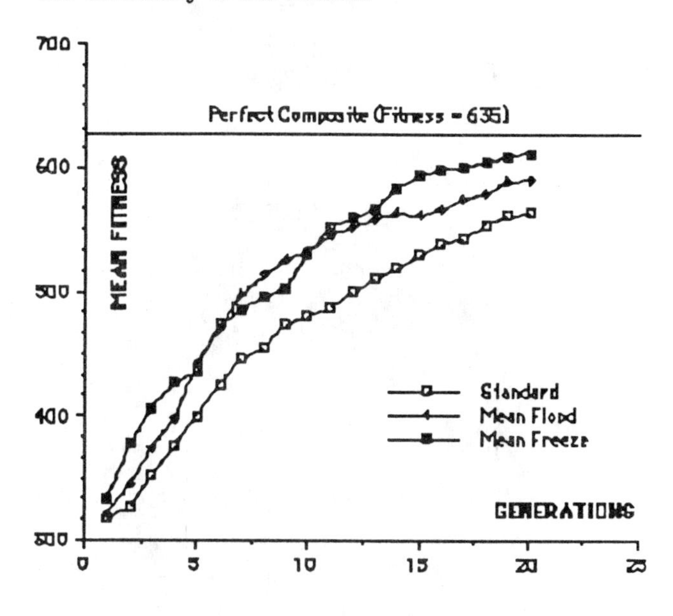

Figure 3. Flooding and freezing effects on code performance.

The current facial composite process has incorporated all of the above modifications. This implementation is being tested for its ability to generate an accurate facial composite. For this evaluation, witnesses have been exposed to a simulated crime and have then been required to use the GA based search to evolve the likeness of the culprit at varying times after exposure to the crime. Accuracy has been measured by subjective evaluations performed by independent judges. An example of a digitized target face, and a witness' composite, produced three days later is shown in Figure 4.

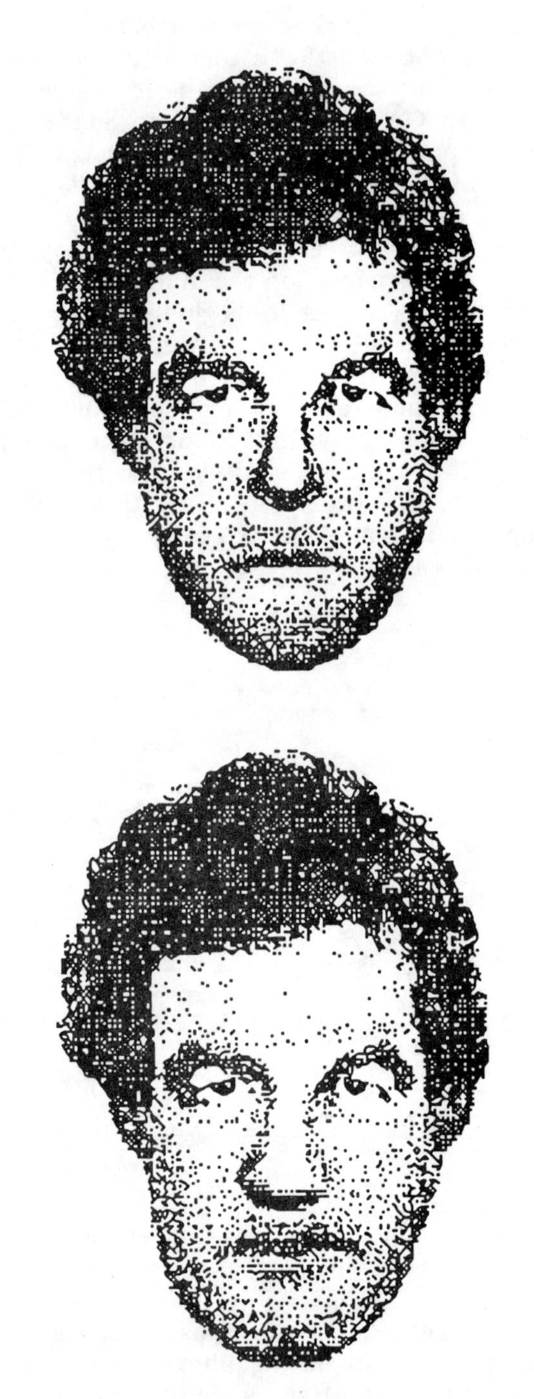

Figure 4. Target criminal (top) and composite (bottom), produced after 10 generations.

The GA may also provide a sensitive technique for examining the criteria used to generate and refine composite faces. Tracking the history of choices made by witnesses, as they search the multi-dimensional space of facial features, may also provide specific information on the salient cues employed in recognition tasks. Because of the flexibility in accommodating individual cognitive styles, face recognition with stimuli generated by the GA is sufficiently robust to ensure that the effects of elapsed time do not appear to substantially decrease the accuracy, or increase the number of decisions needed to produce it. Generating a composite face also provides a single genotype for that composite. This genotype can serve as a code for that face, not unlike a fingerprint. Thus, these genotypes are potentially useful codes for comparing composite faces with stored records. Genotypes generated by several witnesses may be combined and used to generate a new composite face. Such a composite may be more reliable than single source composites.

Acknowledgements

This research was supported under award #90-IJ-CX-0025 from the National Institute of Justice, Office of Justice Programs, U.S. Department of Justice. Points of view in this document are those of the authors and do not necessarily represent the official position of the U.S. Department of Justice.

References

[1] Baker, J.E. Reducing bias and inefficiency in the selection algorithm. GAs and their applications: *Proceedings of the Second International Conference on Genetic Algorithms*, pages 14-21. MIT, MA, 1987.

[2] Bethke, A.D. Genetic algorithms as function optimizers. Dissertation, Department of Computer and Communication Sciences, Univ. Michigan. Ann Arbor, MI, 1981.

[3] Caruana, R.A. and Schaffer, J.D. Representation and hidden bias: Gray vs. binary coding for genetic algorithms. *Proceedings of the 5th International Conference on Machine Learning*, pages 153-161. Morgan Kaufmann, Los Altos, CA, 1988.

[4] Ellis, H.D., Davies, G.M. and Shepherd, J.W. Introduction: Processes underlying face recognition. In R. Bruyer, editor, *The Neuropsychology of face perception and facial expression*, pages 1-38. Lawrence Erlbaum Associates, NJ, 1978.

[5] Going, M. and Read, J.D. Effects of uniqueness, sex of subject, and sex of photograph on facial recognition. *Perceptual & Motor Skills*, 39(10), pages 109-110, 1974.

[6] Goldberg, D.E. *Genetic algorithms in search, optimization & machine learning*, Addison Wesley, MA, 1989.

[7] Hall, D.F. Obtaining eyewitness identification in criminal investigations - applications of social and experimental psychology. Dissertation: Ohio State University, OH, 1976.

[8] Hines, D., Jordan-Brown, L. and Juzwin, K.R. Hemispheric visual processing in face recognition. *Brain & Cognition*, 6(1), pages 91-100, 1987.

[9] Kitson, T., Darnbrough, M. and Shields, E. Let's face it. *Police Research Bulletin*, Spring (30), pages 7-13, 1978.

[10] Miller, L.K. and Barg, M.D. Dissociation of feature vs. configural properties in the discrimination of faces. *Bulletin of the Psychonomic Society*, 21(6), pages 453-455, 1983.

[11] Penry, J. Photo-Fit. *Forensic Photography*, 3(7), pages 4-10, 1974.

[12] Ross-Kossak, P. and Turkewitz, G. A micro and macro developmental view of the nature of changes in complex information processing: A consideration of changes in hemispheric advantage during familiarization. In R. Bruyer, editor, *The neuropsychology of face perception and facial expression*, pages 125-145. Lawrence Erlbaum Associates, NJ, 1986.

[13] Solso, R.L. and McCarthy, J.E.. Prototype formation of faces. *British Journal of Psychology*, 72, pages 499-503, 1981.

[14] Yarmey, A.D. and Kent, J. Eyewitness identification by elderly and young adults. *Law and Human Behavior*, 4(3), special issue, pages 359-371, 1980.

Using a Genetic Algorithm to Design Binary Phase-Only Filters for Pattern Recognition

David L. Calloway
Armament Directorate
Wright Laboratory
Eglin AFB, FL 32542

Abstract

A genetic algorithm is applied to the task of designing binary phase-only filters in a pattern recognition application. Binary phase-only filters have traditionally been designed using the classical matched filter as a baseline and then setting the magnitude portion of the filter to unity and binarizing the phase information. The resulting filter has much of its original information content, but is represented with a greatly reduced set of elements. Such filters have been shown to exceed the pattern recognition ability of the classical matched filter on which they are based. However, binary phase-only filters designed using this method are not optimal for discrimination or invariance to pattern changes and several different researchers have investigated various optimization techniques. This paper introduces a new technique for designing binary phase-only filters using a genetic algorithm. A four operator genetic algorithm is used consisting of a stochastic remainder selection operator, a two-dimensional crossover operator, a diversity-based mutation operator, and a survival operator. The fitness function used in the selection and survival operators is based on the ability of the binary phase-only filter represented by an individual's chromosome to recognize one class of character while rejecting members of a different, but similar class. The fitness function consists of comparing the signal-to-noise ratio in the output (correlation) planes for the image(s) to be recognized to the image(s) to be rejected.

1 BINARY PHASE-ONLY FILTER BACKGROUND

A common technique in pattern recognition is to perform a cross-correlation between an unknown image and a set of pattern images. The pattern that is closest to the reference image produces a higher Signal-to-Noise Ratio (SNR) in the output (correlation) plane than the other patterns. The correlation operation is efficiently performed in the frequency domain by Fourier transforming the image and the pattern and then inverse Fourier transforming their mutual power spectrum. This approach is known as the Classical Matched Filter. It is possible to perform such correlations in real time using an optical processing system consisting of a set of Fourier optics to implement the transform operations and spatial light modulators to apply the filters.

In an IEEE Proceedings survey paper [3], Oppenheim and Lim demonstrated the importance of phase information in visual images. They showed that while many images could be reconstructed from phase-only information, magnitude-only information was typically not sufficient to reconstruct most visual images. Gianino and Horner [4] used this information to produce a modified filter in which all magnitude information was set to unity. The resulting filter contained only phase information and was thus called a phase-only filter. Surprisingly, the phase-only filter produced by normalizing all magnitude information is more effective (in terms of producing high SNR correlations) than the matched filter on which it is based. In addition, phase-only filters are especially useful in optical image recognition systems since they pass nearly 100% of the incident energy while matched filters attenuate much of the input signal.

Most real-world images have comparatively little high frequency energy. However, the features that we use to differentiate one image from another are contained primarily in this relatively small amount of high frequency information. The phase-only filter effectively acts as a high-pass filter since setting all frequency magnitudes to unity emphasizes the high frequencies more than the corresponding matched filter. This explains, in part, why the phase-only filter outperforms the matched filter in typical pattern recognition tasks.

A Binary Phase-Only Filter (BPOF) is a phase-only filter in which the phase information is compressed to only two phase values. This is typically accomplished by starting with the phase-only filter described above and setting all positive phases to a value of π radians while setting all negative phases to zero:

$$\theta_B(i) = \begin{cases} \pi & ,0 < \theta(i) \leq \pi \\ 0 & ,0 \geq \theta(i) > -\pi \end{cases}$$

The resulting filter contains a very compact representation of the original filter's information content—the magnitude information is set to unity at all frequencies and the phase information is represented by only a single bit at each frequency. These binary phase-only filters are nearly as proficient as the corresponding classical matched filter in correlating unknown images with stored patterns. They are especially useful in optical systems using binary spatial light modulators.

The BPOF design approach described above is straightforward, but not optimal for discriminating between patterns. Several different researchers have demonstrated improved methods of designing binary phase-only filters for pattern recognition. For example, Bartelt, Horner, and Kallman applied various iterative techniques to improve the performance of binary phase-only filters [5][6], Kumar and Bahri optimized a phase-only filter based on the magnitude information in the Fourier transform [7], Kim and Guest used simulated annealing techniques to search for an optimal BPOF [1][2], and various other researchers have identified the advantages of cosine and Hartley transforms as opposed to Fourier transforms in implementing binary phase-only filters [8][9]. The effort described in this paper is based on applying a genetic algorithm as a stochastic search technique to generate an optimal binary phase-only filter in a typical pattern recognition application.

2 IMPLEMENTATION OF THE GENETIC ALGORITHM

The goal of this study was to use a genetic algorithm to generate a pair of binary phase-only filters capable of discriminating between the letter "P" and the letter "R". This pair of letters is very difficult for conventional matched filters to discriminate because the bit pattern for the letter "P" is wholly contained in the image of the letter "R". This causes the "P" matched filter to produce a high correlation when applied to an image containing an "R". The images used in this study consisted of the 16×16 bit patterns shown in Figures 1 and 2.

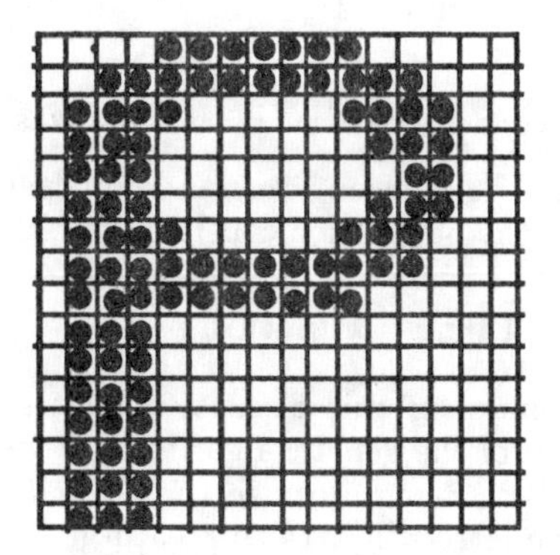

Figure 1: Image of the Letter "P"

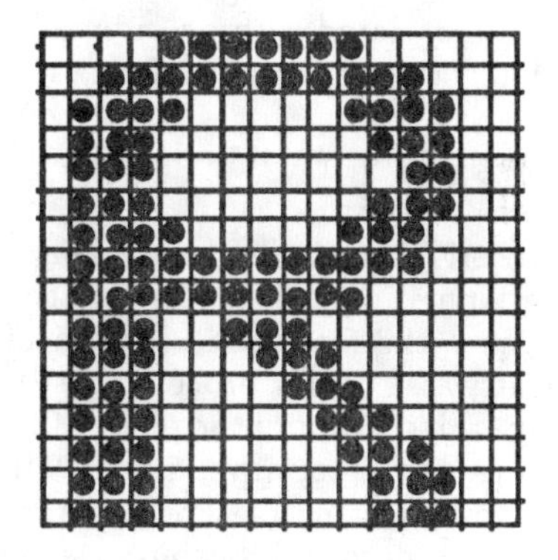

Figure 2: Image of the Letter "R"

The genetic algorithm was applied twice, once to generate a filter to recognize a "P" while rejecting an "R" and a second time to generate a filter to recognize an "R" while rejecting a "P".

A genetic algorithm is characterized by the data structure which defines the individual members of a population, the fitness function which serves as a figure of merit for comparing one individual to another, and a set of operators to convert one population into a new population.

2.1 The Data Structure

The data which was manipulated by the genetic algorithm consisted of a population of individuals. Each individual was represented by a chromosome consisting of a 256-element string taken from a binary alphabet:

$$x = (\alpha_1, \ldots, \alpha_{256}), \alpha \in \{0, 1\}$$

The chromosome was decoded to produce a 16×16-element matrix of 1's and 0's. A two-dimensional BPOF was produced from this chromosome by setting the corresponding BPOF phase to π radians if the allele was a one and setting the phase to zero if the allele was zero:

$$\theta(i) = \begin{cases} 0 & , \quad \alpha_i = 0 \\ \pi & , \quad \alpha_i = 1 \end{cases}$$

All magnitude information in the filter was set to unity.

The initial population was randomly generated. In this study the population consisted of 100 individuals.

2.2 The Fitness Function

The remainder of this discussion relates to the first application of the genetic algorithm, in which we generated a BPOF to recognize the letter "P" while rejecting the letter "R". During each generation, every individual in the current population was evaluated to determine how well the filter it represented differentiated between the two images described above. The fitness value associated with an individual member of a population was determined as follows:

1. The chromosome was decoded to produce a binary phase-only filter.
2. The BPOF was applied to the Fourier transform of the binary, 16×16-element image of the letter "P". Since the magnitude portion of every filter element was 1.0, this process consisted of simply adding each element of the phase of the "P"-image transform to the corresponding BPOF phase. The result was then inverse Fourier transformed and the magnitude of the output was squared. This was equivalent to forming the auto-correlation of the image of the letter "P" with the inverse Fourier transform of the BPOF.
3. The BPOF was then similarly applied to the letter "R", producing the cross-correlation between the inverse Fourier transform of the BPOF and the letter "R".
4. The Signal-to-Noise Ratios (SNRs) in each of the two correlation planes were then calculated.
5. Finally, the fitness value was calculated by dividing the SNR resulting from applying the BPOF to the letter "P" by the SNR resulting from applying the BPOF to the letter "R":

$$F = \mathrm{SNR}_P / \mathrm{SNR}_R$$

The SNR was defined as the ratio of the correlation peak maximum amplitude to the root-mean-square (rms) noise response elsewhere in the correlation plane:

$$\mathrm{SNR} = [R(x_i)]_{max} / N_{rms}$$

$$N_{rms} = \frac{\sqrt{\sum_{i \in (A-A')} |R(x_i)|^2}}{(N_A - N_{A'})}$$

where A is the entire correlation plane, $A' \subset A$ is a smaller region surrounding the correlation peak, N_A is the total number of pixels in the correlation plane, and $N_{A'}$ is the number of pixels considered to be in the correlation peak region [10]. In this study, A consisted of a 16×16 array of real numbers and A' was a 2×2 subset of this region. Thus, $N_A = 256$ and $N_{A'} = 4$.

To calculate the SNR for a BPOF applied to the target letter ("P"), the correlation peak region was defined as the four pixels in center of the correlation plane. This restriction encouraged the correlation peak to lie in the center of the correlation plane, as it would for a conventional BPOF created using the matched filter design approach. To calculate the SNR for the BPOF applied to a non-target letter ("R" in this case), the correlation peak region was centered on the location of the highest peak in the correlation plane (not necessarily in the center of the correlation plane). This procedure enhanced the robustness of the BPOFs produced by this GA.

2.3 The Genetic Operators

The genetic operators are responsible for combining or modifying members of an initial population to produce a new population. The four genetic operators used in this study consisted of selection, crossover, mutation, and survival. Each of these operators is described below.

2.3.1 Selection Operator

The selection operator used in this study was based on the stochastic remainder without replacement method described by Goldberg [11]. In this method, a trial population is constructed from the base population and individuals are then selected from the trial population for mating. The trial population is constructed in four steps:

1. The average fitness value in the base population is calculated:

$$F_{avg} = [\sum_{i=1}^{N} F(i)]/N$$

where F_{avg} is the average fitness value, $F(i)$ is the fitness value for the ith member of the population, and N is the total number of individuals in the population.

2. The fitness value for each member in the base population is compared to the average value:

$$F_{trial}(i) = F(i)/F_{avg}$$

3. An individual receives one guaranteed slot in the trial population for each whole multiple that the individual's fitness is above the average fitness:

$$N_{guaranteed} = \mathrm{Int}(F_{trial})$$

4. The remaining slots in the trial population are filled stochastically based on the remainders of each individual's fitness value after subtracting one for each guaranteed slot that was awarded:

$$P(N_{additional}) = F_{trial} - N_{guaranteed}$$

For example, if the average fitness value was 1.0 and an individual had a fitness value of 3.6, then this individual would recieve three guaranteed slots and have a 60% chance of receiving an additional slot in the trial population.

After filling the trial population, members are randomly selected from it for mating. Each copy selected for mating is removed from the trial population. Thus, each individual in the trial population is used once and only once in each generation.

2.3.2 Two-Dimensional Crossover Operator

The crossover operator is responsible for the mating process in which two members from an initial population are combined to produce two offspring in the new population. Eshelman et al discuss the importance of positional bias in the crossover landscape, and they introduce the concept of multi-point (ring) and uniform (bit-by-bit) crossover operations [12]. In this application the underlying structure of a BPOF filter is two-dimensional. Thus, positional biases exist between both east-west and north-south neighbors of any particular allele in the chromosome. A two-dimensional crossover operation was introduced to overcome these positional biases without resorting to the highly disruptive uniform crossover.

The two-dimensional crossover operation begins by randomly selecting two horizontal and two vertical crossover sites: H_1, H_2, V_1, and V_2. If the chromosome contains m rows and n columns, then the crossover points are restricted to the following values:

$$\begin{array}{ccccc} 1 & \leq & H_1 & < & m \\ H_1 & < & H_2 & \leq & m \\ 1 & \leq & V_1 & < & n \\ V_1 & < & V_2 & \leq & n \end{array}$$

This divides the two-dimensional chromosome into nine rectangular sections. The crossover points are identical for the two parents to be mated. For example, Figures 3 and 4 show two hypothetical parent chromosomes. Each small rectangle represents a single allele. A blank rectangle represents an allele with a value of zero and a filled-in rectangle represents an allele with a value of one. The two dark horizontal and vertical lines define the crossover points for this pair.

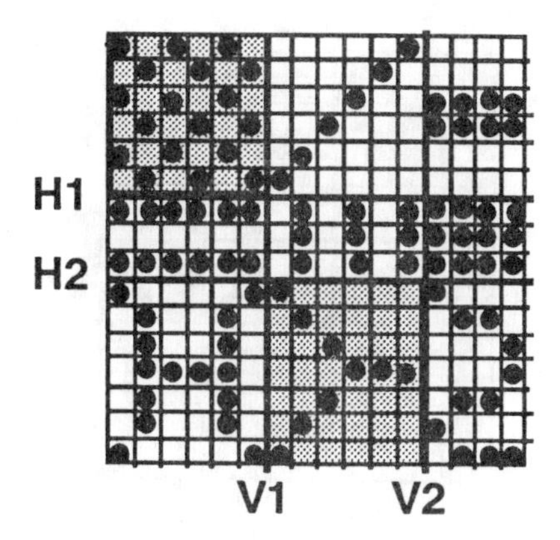

Figure 3: Parent 1

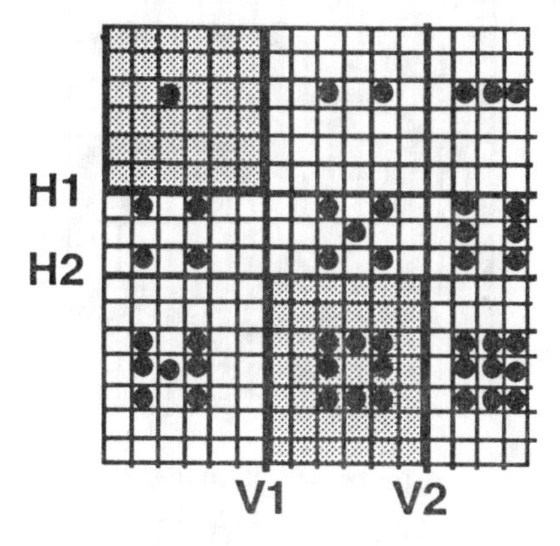

Figure 4: Parent 2

Two offspring were generated from each pair of parents. Each section in the parent chromosome was crossed independent of the other eight sections based on a predefined probability of crossover parameter. In this study, the probability of crossover was defined to be 0.2 (or, equivalently, 0.8). Thus, the first offspring was expected to receive 80% of its alleles from the first parent and 20% from the other parent. The second offspring received all alleles not included in the first offspring. In this example, the shaded portions from each parent were selected for crossover. This produced the two offspring shown in Figures 5 and 6.

2.3.3 Mutation Operator

After applying the crossover operator, each allele in an individual is given the opportunity to mutate. Fogarty introduced the concept of varying the mutation rate over time and across integer representation rather than leaving it at a constant value [13]. In this study, we varied the mutation rate based on the diversity in the population.

At the beginning of a generation the diversity of each allele in the population is determined. The diversity array contains one real number for each allele in a chromosome. The diversity array thus consisted of a 16×16 array of real numbers. For a problem such as this one, with a two-element alphabet consisting of zero or one,

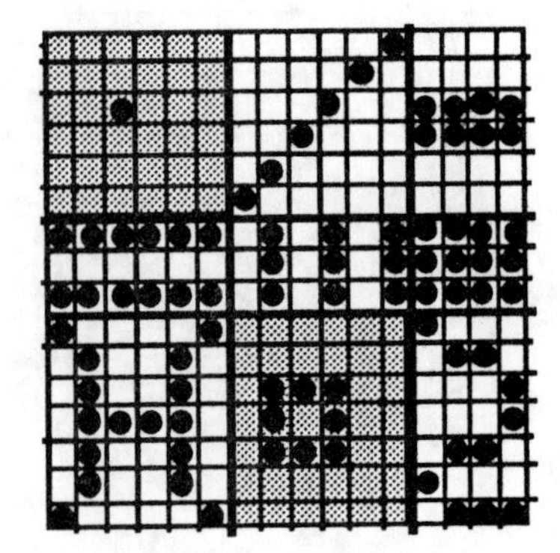

Figure 5: Offspring 1

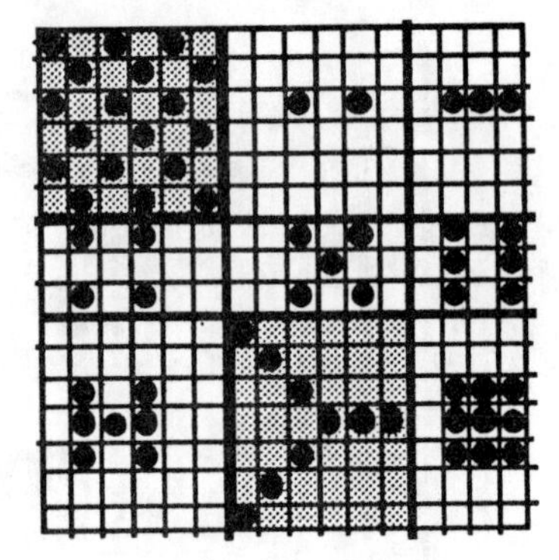

Figure 6: Offspring 2

the diversity is simply the average value over the entire population at each allele position in the chromosomes:

$$D(i) = [\sum_{j=1}^{N} \alpha_i(j)]/N$$

where $D(i)$ is the diversity at location i (allele i) in the chromosomes, $\alpha_i(j)$ is the value of the ith allele in the jth member of the population, and N is the total number of individuals in the population. A diversity of 0.0 means that all alleles in the population at the corresponding position are zero, a diversity of 1.0 means that all alleles are one, and a value of 0.5 implies that half the alleles are zero and half are one (the most diverse case).

The mutation rate is controlled by two parameters:

1. P_{min} – The minimum mutation rate, and
2. P_{max} – The maximum mutation rate.

The probability of mutation at each allele in a chromosome is then specified by a parabola which passes through P_{max} when the diversity is 0.0 or 1.0 and has a minimum value at P_{min} when the diversity is 0.5:

$$P_{mutation}(i) = 4.0 * (P_{max} - P_{min}) * D(i)^2 - 4.0 * (P_{max} - P_{min}) * D(i) + P_{max}$$

In other words, the higher the diversity at a particular location in the chromosome, the lower the probability of mutation for alleles at that location.

2.3.4 Survival

The final GA operator applied in this study was a survival operator. Note that each mating process starts with two parents and produces two offspring. The fitness function for each of these individuals is calculated as described in Section 2.2. The survival operator chooses for survival the two individuals from each parent/offspring set with the highest fitness function and these individuals then become candidates for selection. This process is similar to a tournament selection applied before the stochastic remainder selection operator, except that a separate tournament is held for each set of two parents and their two offspring, rather than to the entire population as a whole.

3 RESULTS

We applied the techniques described in the previous section to the task of designing a specific binary phase-only filter capable of discriminating between the letter "P" and the letter "R". As mentioned earlier, these two letters are notoriously difficult for traditional correlation filters to differentiate because the letter "P" is wholly contained in a subset of the pixels which make up the letter "R". The genetic algorithm was executed for 100 generations and the individual in the final population with the highest fitness value was selected to produce the resulting optimal filter.

We produced four filters in this study: two were designed using the conventional matched filter design approach, and the other two were designed using the genetic algorithm. These four filters were applied to the images of the letter "P" shown in Figure 1 and the letter "R" shown in Figure 2. Table 1 summarizes the results of this experiment.

This table shows that the SNRs produced by the GA-based filters are significantly less than for conventionally-designed filters. However, the SNR resulting from applying the filter to a non-target image was reduced significantly more than the SNR produced by applying the filter to a target image. This discrepancy is the key to the ability of these filters to discriminate between the two images.

A discrimination ratio can be defined as a figure of merit with which to compare these two sets of filters. Given an image to be recognized by one filter and rejected by another, the discrimination ratio is simply the SNR obtained by applying the first filter to the image divided by the SNR obtained by applying the second filter:

$$\mathcal{D} = \mathrm{SNR}_{RecognizingFilter}/\mathrm{SNR}_{RejectingFilter}$$

Table 1: Summary of Results

Filter	Image "P" SNR	Image "R" SNR	Discrimination Ratio (dB)
Conventional "P" BPOF	35.1	23.4	**1.17**
Conventional "R" BPOF	26.8	31.9	**1.34**
GA "P" BPOF	12.5	2.7	**5.76**
GA "R" BPOF	3.3	12.6	**6.67**

Figure 7 graphically illustrates the comparitive advantage of the GA-produced BPOFs with their conventional counterparts.

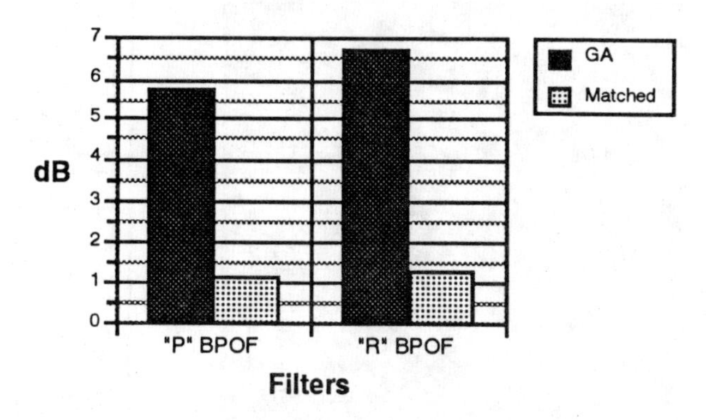

Figure 7: Discrimination Ratios

Figure 8 shows the output (correlation plane) of the GA BPOF trained to recognize the letter "R" when it is applied to this letter while Figure 10 shows the output when this filter is applied to the letter "P". The top half of these figures provide a three-dimensional view of the output while the lower half provides a shaded contour image. Figure 8 shows a definite peak in the center of the correlation plane with very little noise elsewhere in the plane while Figure 10 shows the absence of a strong signal anywhere in the plane.

Figures 9 and 11 provide a two-dimensional illustration of the discrimination capability of the GA-designed filters. Figure 9 compares the recognition capabilities of the "R" and "P" filters when applied to the letter "R". The dark line in this graph is a slice taken through the peak of the output correlation plane after applying the "P" BPOF to the letter "R". The lighter line is a similar slice taken through the peak in the correlation plane produced by applying the "R" BPOF to this same image. This figure illustrates the ability of the "R" BPOF to recognize that a letter "R" is present in the input plane and the ability of the "P" BPOF to recognize the fact that this letter is *not* a "P". Similarly, Figure 11 illustrates the ability of the "P" BPOF to recognize the letter "P" and the corresponding ability of the "R" BPOF to indicate that this image is *not* an "R".

4 SUMMARY

We have demonstrated the ability of a genetic algorithm to design binary phase-only filters capable of discriminating between two similar images. We intend to extend this application to larger images (128×128) in the near future and verify the simulated responses using spatial light modulators to implement the filters in an optical image processing system.

Acknowledgements

The author wishes to thank Dr. Dennis Goldstein for the original idea of applying the genetic algorithm to the design of binary phase-only filters, Tom Davis for many interesting discussions and expert guidance, and the U.S. Air Force Reserve program and the Air-to-Air Guidance Branch of the Air Force Armament Directorate for providing an excellent environment in which to pursue this research.

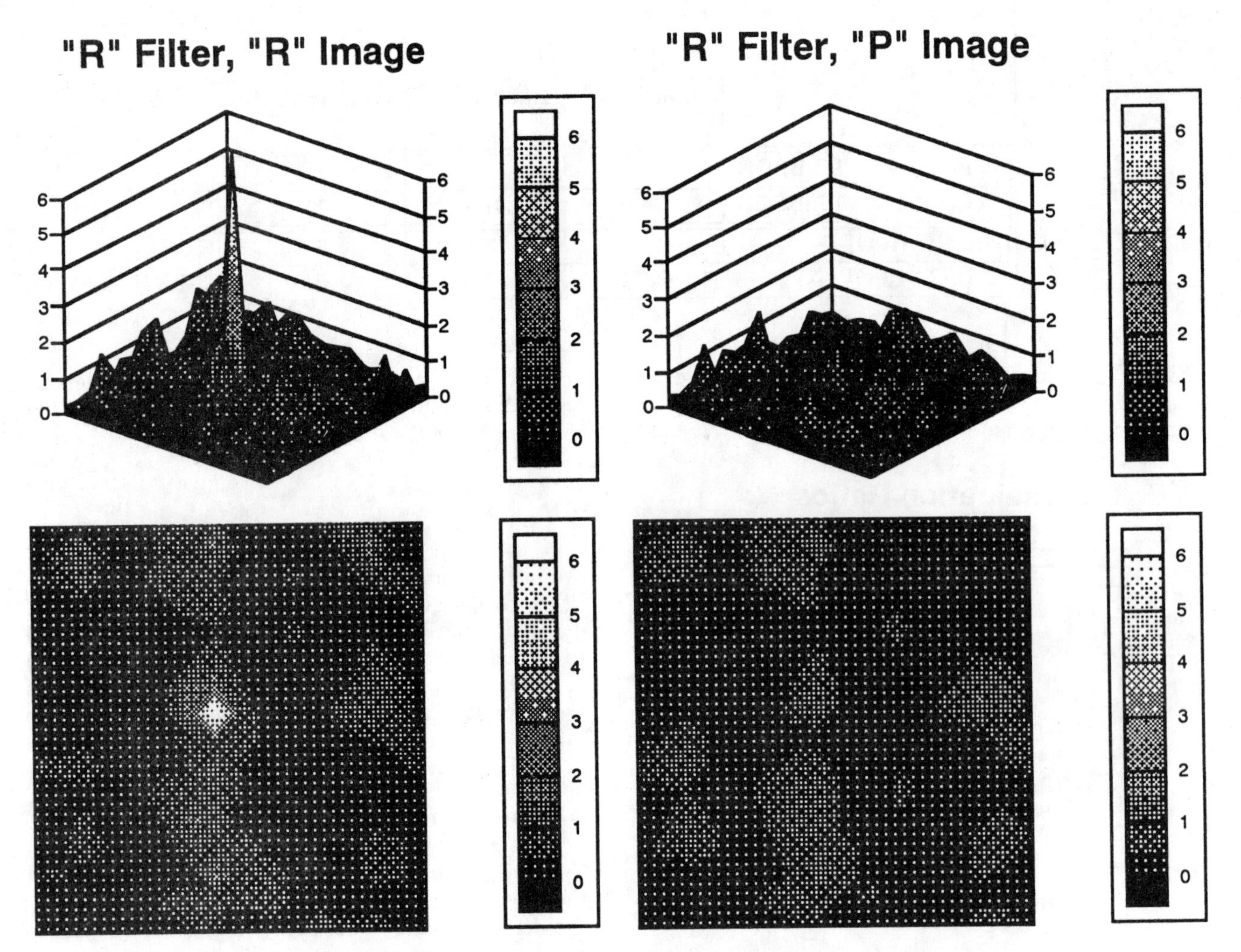

Figure 8: "R" Filter Applied to Letter "R"

Figure 10: "R" Filter Applied to Letter "P"

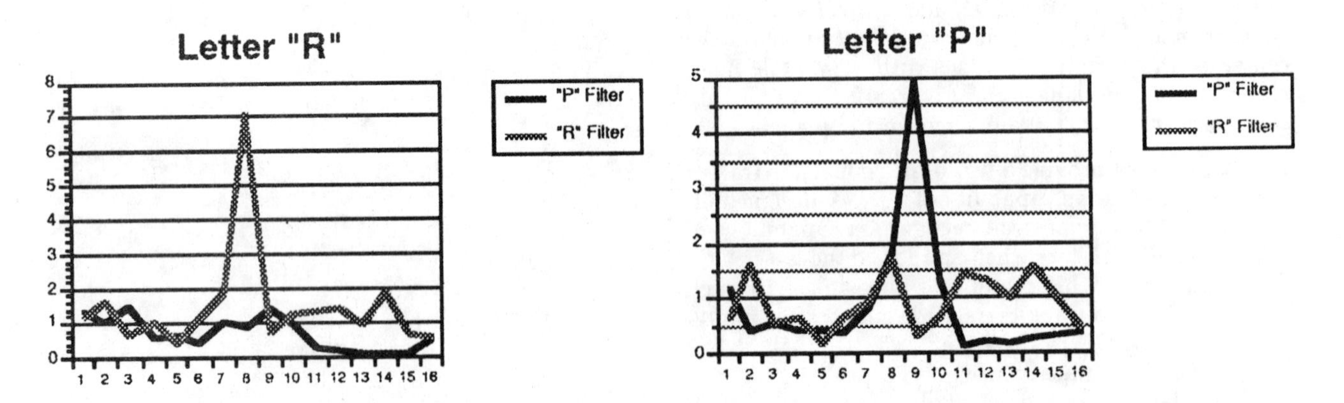

Figure 9: Filters Applied to Letter "R"

Figure 11: Filters Applied to Letter "P"

References

[1] Kim, Myung Soo and Guest, Clark C., "Simulated Annealing Algorithm for Binary Phase Only Filters in Pattern Classification", *Applied optics*, Vol. 29, No. 8, March 10, 1990, pp. 1203–1208.

[2] Kim, Myung Soo and Guest, Clark C., "Experiments on Annelaed Binary Phase Only Filters Fabricated with Electron Beam Lighography",*Applied Optics*, Vol. 29, No. 23, August 10, 1990, pp. 3380–3386.

[3] Oppenheim, Alan V. and Lim, Jae S., "The Importance of Phase in Signals",*Proceedings of the IEEE*, Vol. 69, No. 5, May 1981, pp. 529–541.

[4] Gianino, Peter D. and Horner, Joseph L., "Additional Properties of the Phase-Only Correlation Filter", *Optical Engineering*, Vol. 23, No. 6, November/December 1984, pp 695–697.

[5] Bartelt, Hartmut and Horner, Joseph, "Improving Binary Phase Correlation Filters Using Iterative Techniques", *Applied Optics*, Vol. 24, No. 16, September 15, 1985, pp. 2984–2897.

[6] Kallman, Robert R., "Optimal Low Noise Phase-Only and Binary Phase-Only Optical Correlation Filters for Threshold Detectors", *Applied Optics*, Vol. 25, No. 23, December 1, 1986, pp. 4216–4217.

[7] Kumar, B. V. k. Vijaya and Bahri, Zouhir, "Phase-Only Filters with Improved Signal to Noise Ratio", *Applied Optics*, Vol. 28, No. 2, January 15, 1989, pp. 250–257.

[8] Cottrell, Don M., Lilly, Roger A., Davis, Jeffrey A., and Day, Timothy, "Optical Correlator Performance of Binary Phase-Only Filters Using Fourier and Hartley Transforms", *Applied Optics*, Vol. 26, No. 18, September 15, 1987, pp. 3755-3761.

[9] Davis, Jeffrey A., Cottrell, Don M., Bach, Glenn W., and Lilly, Roger A., "Phase-Encoded Binary Filters for Optical Pattern Recognition", *Applied Optics*, Vol. 28, No. 2, January 15, 1989, pp. 258–261.

[10] Horner, Joseph L. and Bartelt, Hamut O., "Two-bit Correlation", *Applied Optics*, Vol. 24, No. 18, September 15, 1985, pp. 2889–2893.

[11] Goldberg, David E., *Genetic Algorithms in Search, Optimization, and Machine Learning*, Addison-Wesley Publishing Company, Inc., Reading, MA,1989.

[12] Eshelman, Larry J., Caruana, Richard A. and and Schaffer, David J., "Biases in the Crossover Landscape", *Proceedings of the Third International Conference on Genetic Algorithms*, 1989, pp. 10–19.

[13] Fogarty, Terence C.," Varying the Probability of Mutation in the Genetic Algorithm", *Proceedings of the Third International Conference on Genetic Algorithms*, 1989, pp. 104–109.

A System for Learning Routes and Schedules with Genetic Algorithms

Paula S. Gabbert
Systems Engineering Department
University of Virginia
Olsson Hall
Charlottesville, VA 22901

Donald E. Brown
Systems Engineering Department
Institute for Parallel Computation

Christopher L. Huntley
Systems Engineering Department
Institute for Parallel Computation

Bernard P. Markowicz
CSX Transportation
100 North Charles St.
Baltimore, MD 21201

David E. Sappington
Institute for Parallel Computation

Abstract

A Genetic Algorithm approach to routing and scheduling trains along a rail network is shown to be a feasible and useful technique. Routing and scheduling in the context of a rail freight transportation network for CSX is discussed. A representation of this problem for a Genetic Algorithm approach is proposed and results of the technique are presented.

1 INTRODUCTION

Virtually all transportation services seek low cost routing and scheduling strategies. However, routing and scheduling are among the most difficult combinatorial optimization problems. Difficulties arise because of both nonlinearities in cost functions and interactions between routes and schedules. Traditionally, these difficulties are assumed away (i.e. costs are assumed linear and routing and scheduling are considered independently). In the best cases these assumptions lead to good solutions to the actual problem. However, there are many cases where these assumptions lead to very poor (high cost) solutions to the real problem. To avoid making simplifying assumptions, this paper presents a genetic algorithm approach to learning low cost routes and schedules in large transportation networks.

Genetic algorithms provide an ideal platform for learning routes and schedules because they work with evaluation functions of arbitrary complexity. Cost function complexity is tolerated by genetic algorithms because they treat the cost function as a "black box." All that is required of the black box is a measure of performance for the candidate solution. Hence, the contents of the evaluation function black box can contain the nonlinearities of the actual problem.

Genetic algorithms can also tolerate the interactions between routes and schedules; however, the way in which genetic algorithms represent these interactions is the central problem for the algorithm designer. There is no single, obvious representation of the routing and scheduling problem that effectively captures all of the underlying interactions of the real problem. Hence, the design of a representation scheme for routing and scheduling problems is an important research question.

This paper presents two contributions. First, it shows that genetic algorithms can successfully learn low cost routes and schedules for realistic transportation problems. These are problems that have gone unsolved using other techniques. Second it provides a useful representation scheme for general routing and scheduling problems.

The next section describes the specific routing and scheduling problem that we considered – routing and scheduling trains for CSX Transportation (CSXT). Section 3 provides our genetic algorithm approach and representation scheme. Section 4 contains results from the application of this approach to the CSXT problem and Section 5 has conclusions and research directions.

2 ROUTING AND SCHEDULING TRAINS

Routing and scheduling comprise a class of transportation problems. This section defines routing and scheduling in the context of a rail freight transportation network and indicates the difficulties to traditional approaches.

The fundamental element of a rail freight network is a *car*, which transports goods from an origin to a destination. A *block* is a collection of cars which remain together from the originating yard to the destination yard. Each block has an associated weight. A *train* consists of one or more locomotives, a string of blocks, and an end-of-train device. Each train departs at a specified time from a yard and travels along the rails to a given termination yard. The problem to be addressed in this paper is to build trains from a given set of blocks and then route and schedule those trains through the rail freight network.

The *routing* within a network assigns a sequence of yards in the network to each block. The general strategy is to classify blocks with similar destinations together along the same route. Some important issues involved in developing a low cost routing strategy include the use of run-through trains (where a block is placed on a dedicated train between origin and destination yards, thereby avoiding costly intermediate switching and yard time), the capacity of the rail network, and the power available on the train.

The *schedule* for a network specifies the departure yard, the termination yard, and the departure time for each train. Any particular schedule imposes delay times, as well as costs, for blocks waiting in intermediate yards. It also determines other costs, including the cost for locomotives used by the train, crew cost, and total rental cost. The scheduling strategy attempts to minimize these costs, maximizing the profit contributed by each car on the train.

For example, Figure 1 depicts a typical problem involving the Atlanta, Montgomery, Chicago, and St. Louis rail yards. The figure shows the rails and intermediate yards that exist between these rail stations and the number of cars to be transported in each group (i.e. one block of 5 cars is to be transported from Atlanta to Chicago). In this simple network, all of the cars must pass through Nashville and Evansville, but they do not necessarily have to stop at those yards. The questions that must be answered include:

1) How many trains are required?
2) How often should the trains depart?
3) Should the trains stop at Nashville and/or Evansville?
4) Should blocks be joined at Nashville and/or Evansville?

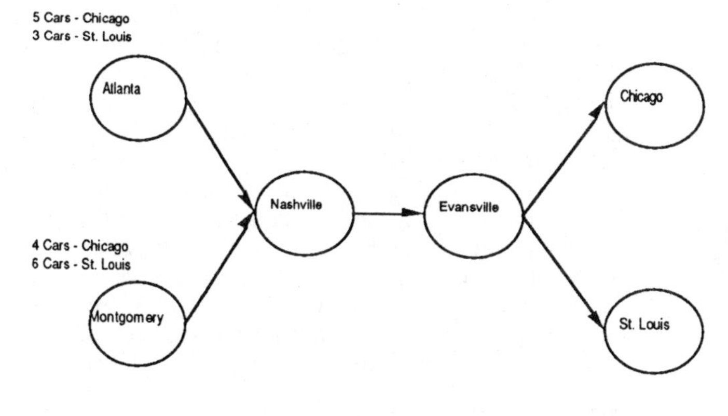

Figure 1: Example Problem

Routes and schedules are evaluated using operational costs. The cost of train operations has four components: freight car rental cost, fuel cost, crew cost, and locomotive cost. Freight car rental cost imposes a simple per hour and per mile rate on each car. However, the calculation of the transit time is a disjoint linear function based on the power used to pull the train and the weight of the train. Fuel cost is a function of the ratio between the number of locomotives required to pull the train and the weight of the train. The crew cost imposes a per hour rate for each crew on the train. A train requires one crew for every 12 hours of transit time. Finally, the locomotive cost imposes a per hour rate on each locomotive required to pull the train. A train requires one locomotive for every 3400 tons of freight.

The routing and scheduling problem we have described is quite complicated. Difficulties arise from the effects of freight car interactions on rental, fuel, crew and locomotive costs, which are each nonlinear. In fact, a simple version of our problem is NP-hard, which means that it is among the most difficult of all optimization problems. In the simple version, all but one of the blocks have been routed optimally, and the goal is to create the optimal route for the last block. Assume that rental, fuel, and crew costs are a function of mileage only. As an added simplification ignore locomotive costs, unless the weight of a train exceeds 3400 tons; then the cost of the locomotive is the weight minus 3400 tons. The problem is to minimize the mileage of this last block, subject to constraining the total weight-overage for the trains encountered along the block's route to zero. This is an example of the Shortest Weight-Constrained Path Problem (SWCPP) which is NP-hard [4].

Previous work in rail freight networks has not fully addressed routing and scheduling. The details of rail yard operations and the costs associated with these operations can be found in [2]. Several studies have been done to simulate rail operations using various strategies (e.g., [1]). However, little work has been done to integrate routing and scheduling into one optimization framework [3]. Also, since the variable costs associated with each process traditionally have been imbedded within the simulation work itself, they have not been studied as separate components. The emphasis of our work is on the cost impact of routing and scheduling.

The integration of both the routing and scheduling strategies during the analysis of rail transportation problems is new. This integration is important since the optimal route for a block of cars is heavily dependent upon the schedule of trains, and the optimal schedule of trains depends upon the routes for the shipments. These dependencies exist because the trains are operating throughout a network where the decisions at one yard affect the operations at neighboring yards in the network. The impact of these decisions is considerable. For example, CSXT runs over 1,200 scheduled trains, and at any point in time, CSXT has 3,000 locomotives and over 200,000 freight cars on its lines, representing an asset base of over $8 billion. The routing and scheduling plans for these cars are critical in insuring prompt service to CSXT customers at the lowest possible cost.

Genetic algorithms provide an environment for learning routes and schedules in an integrated way with realistic cost functions. However, this learning requires careful design of the representation scheme. The next section provides details of our approach.

3 AN APPROACH USING GENETIC ALGORITHMS

A Genetic Algorithm (GA) was incorporated in a system that allows the user to modify various parameters in the cost function in order to determine optimal routes and schedules under different scenarios. As do all GAs, this one operates on a set of strings, where each position (i.e., allele) in a string corresponds to the route taken by a block. The algorithm keeps a list of K different routes through the network, and each allele in a string is an index into this master list of routes. These routes define the set of trains taken to link a given source and destination. They are determined using a greedy local optimizer, which can be initialized with iterative improvements on a set of random routes. There may be several different routes in the list which link any given source to a given destination.

For example, assume that there are only three blocks and K = 10. Then, a "legal" string R might look as in Figure 2, where the large box is the master list of routes defining the trains used to travel from origin to destination. Note that each train in a route has a fixed origin, destination, and departure time. Hence, the path of the block and the schedule of the trains are implicitly represented within each solution string. If train #1 travels from Tampa to Waycross departing at 6 and train #40 travels from Waycross to Montgomery departing at 12, then block 2 will depart Tampa at 6 and depart Waycross at 12.

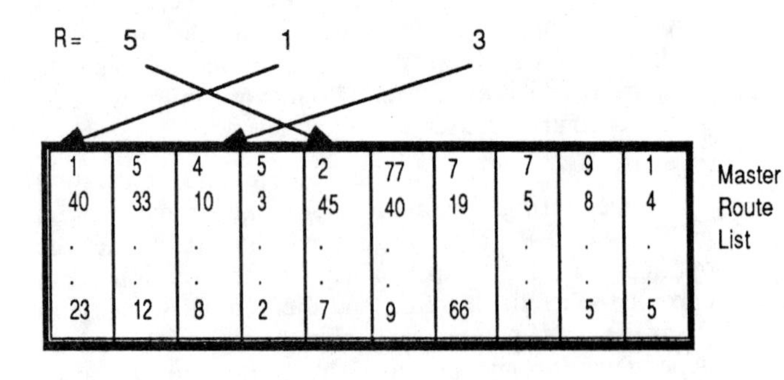

Figure 2: Example String

The GA was implemented with a population of 300 solution strings and a mutation rate of 0.05. Each generation produced 50 children using uniform crossover [Swerda, 1989]. Each run iterated through 1,000 generations. One feature that distinguishes this GA approach from the standard GA implementation is the use of local optimizers in searching for the global optimal routing and scheduling of given blocks. A fixed number of children are selected each generation for local optimization in a manner similar to that used for parental selection. Therefore, mutation proceeds in two ways: route generation and route reassignment. In the first way the process goes in four steps:

1. a block is selected from within a string,
2. a new route is constructed for it by a local optimizer,
3. the new route is added to the master list, and
4. the route index for that block (i.e., its allele value) is set to the new route.

This first kind of mutation happens with some probability m_1 for each group. The second type of mutation is the reassignment of an index to another legal route in the master list (e.g., reassigning group 2 to point to route 5). This happens with probability $m_2 > m_1$. Hence, local optimization only occurs occasionally, with the GA spending most of its time trying to find the right "mix" of routes.

4 RESULTS OF ROUTING AND SCHEDULING TRAINS ON CSX NETWORK

Results have been obtained on a 128 node iPSC/860 for a rail network consisting of 10 fully connected yards. The yards are fully connected to provide run through trains between any two yards in the network. These yards were chosen to reflect a sufficient segment of the CSX rail network and to provide interesting and useful routing and scheduling problems. The problem is further constrained by restricting the departure times of trains to every 4th hour during a 60 hour time window. Hence, for the 10 yard network there are 1,350 possible trains. The major question to be answered by this system is which trains out of the 1,350 possible trains should be used for any given configuration of blocks within the network. The example problem designates the origin, destination, ready time, and weight, for 770 cars in 15 blocks resulting in over 100,000 possible routes over the network. Because this problem was small enough, the system was able to search over all possible routes without requiring any local optimization.

One aspect of the GA approach that CSXT is interested in is the sensitivity of the routes and schedules to changes in the cost model. Hence, results were obtained for two cost

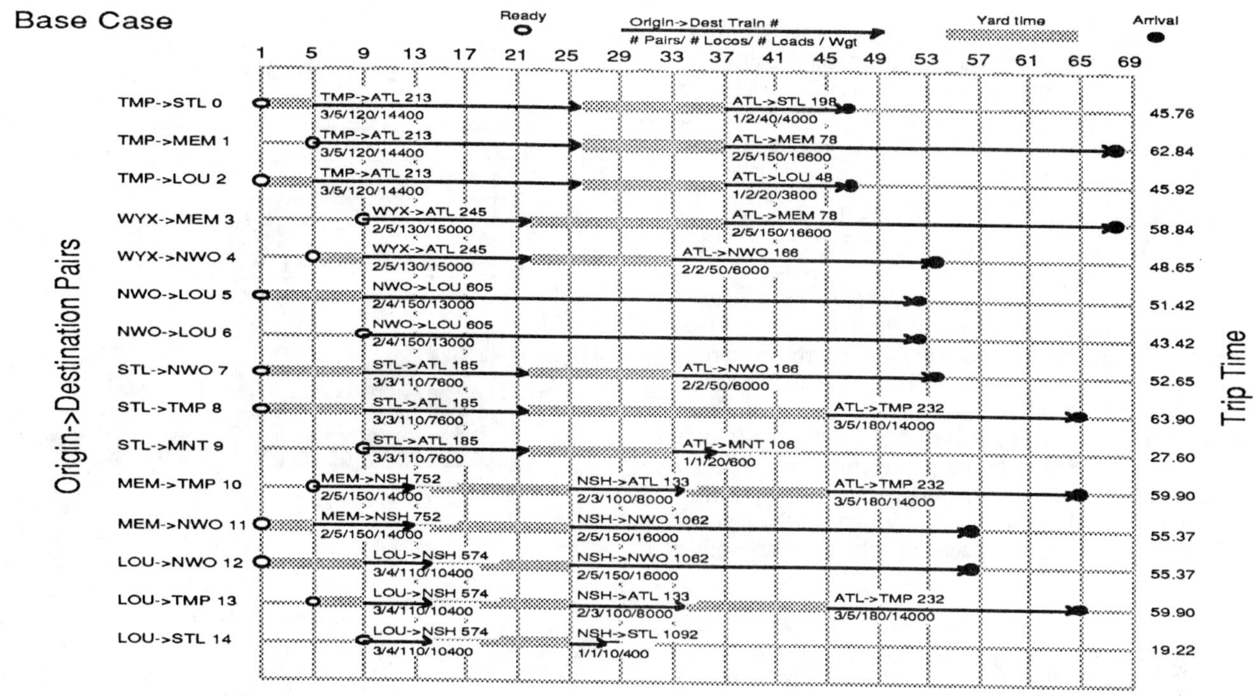

Figure 3a: Base Case Schedule

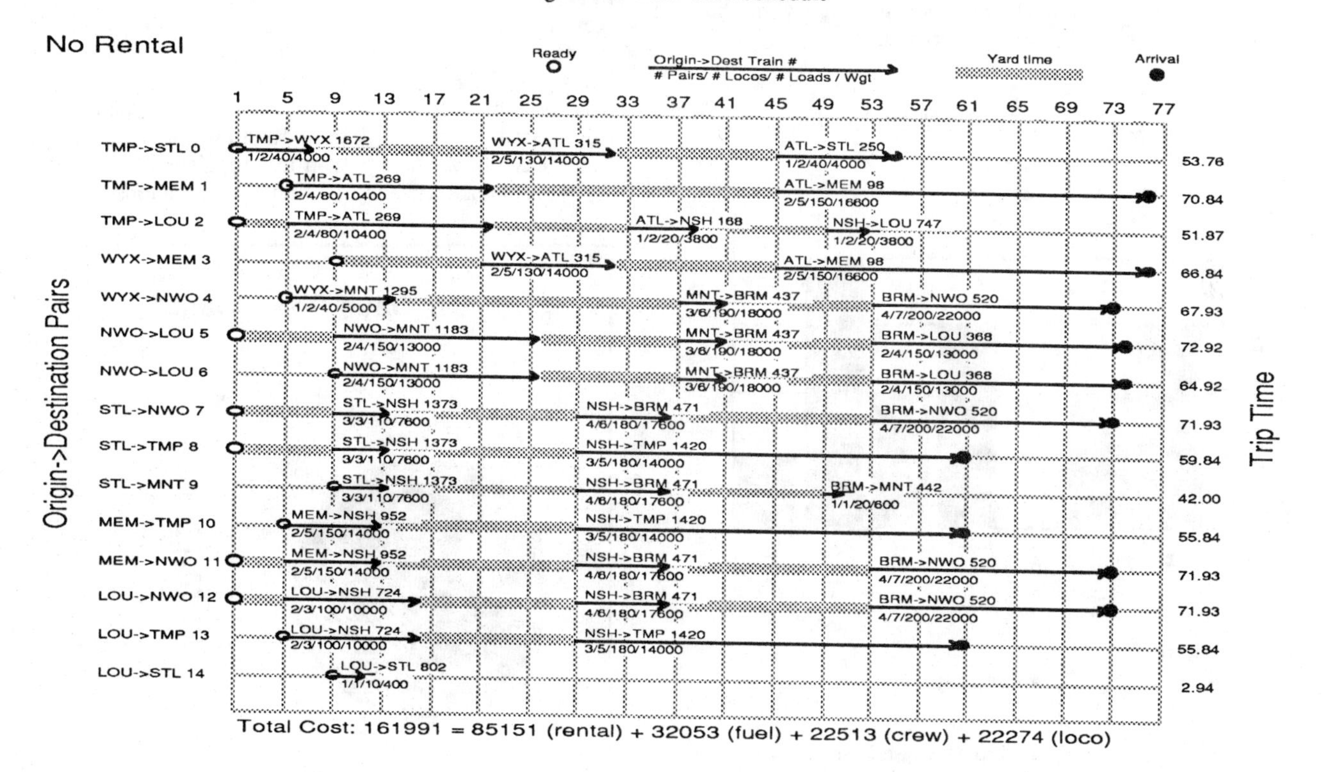

Figure 3b: No-Rental Case Schedule

scenarios. The first, called the Base case, includes all four costs in the train economics model. The second, called the No-Rental case, ignores car rental cost. Because genetic algorithms do not assume anything about the form of the evaluation function, it provides a way of looking at these types of scenarios.

The train schedules returned by the GA for the two cost scenarios are quite different (see Figure 3). For the Base case, there was a tendency toward a small number of very long trains. The No-Rental case, on the other hand, tended to drive the algorithm toward shorter trains. For example, block 12 has 50 cars to be routed from Louisville to New Orleans. In the Base case, the loads are placed on a run through train that goes non-stop from Nashville to New Orleans. However, in the No-Rental case the trip time is no longer as important, and and the cars took a less direct route, stopping in Birmingham to meet up with blocks 6, 7, and 11. Although the layover increased the trip time, it decreased the crew and other costs.

At a higher level, various statistics can be compared. Figure 4 shows the distribution of train lengths for the 15 trains scheduled for the Base case and the 22 trains scheduled for the No-rental case. Again, this figure shows how the Base case tended towards longer trains while the No-rental case tended towards shorter trains.

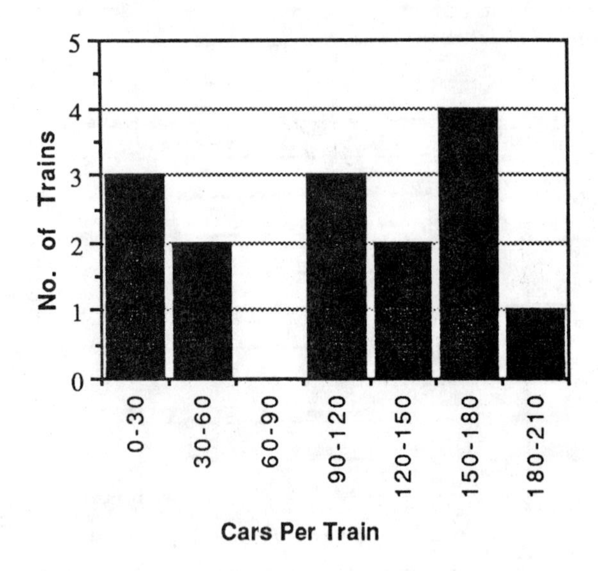

Figure 4a: Base Case

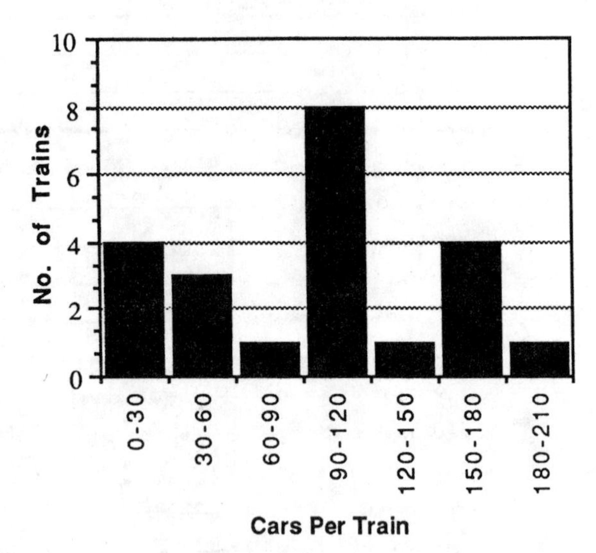

Figure 4b: No-Rental Case

In addition, the differences between the schedules can be seen when comparing car days, or the number of cars times the number of days in transit. Figure 5a shows that in the Base case, 72% of the trip time was spent in transit, 22% was spent in intermediate yards, and 6% was spent in the origin yard. However, Figure 5b shows that for the No-rental case, 35% was spent in intermediate yards waiting for more blocks traveling to similar destinations.

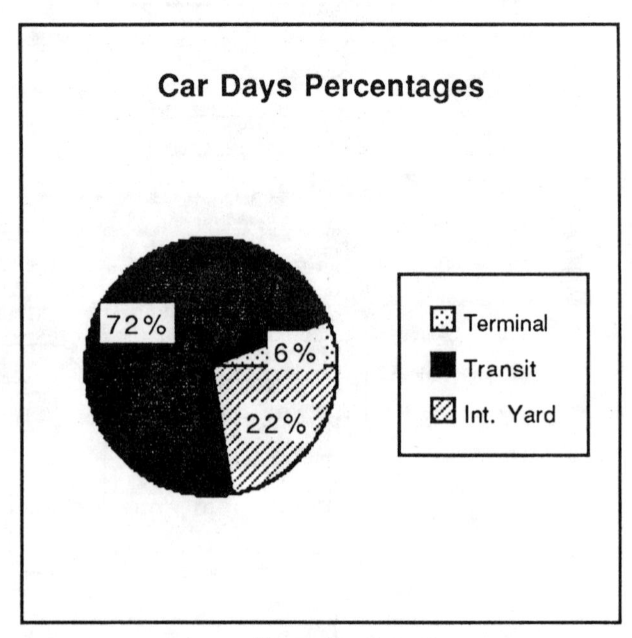

Figure 5a: Base Case

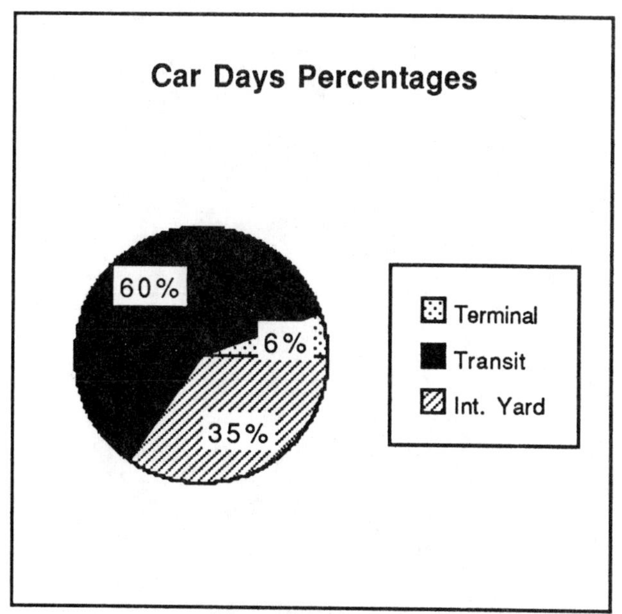

Figure 5b: No-Rental Case

We can also look at the distribution of the locomotives per train. Each 3400 tons on a train requires one locomotive. In the Base case trains tended to be longer, requiring more locomotives as shown in Figure 6a. The trains for the No-Rental case were shorter and required fewer locomotives as shown in Figure 6b. As cars are added to a train, the capacity of the locomotives is reached and the train speed will decrease. Hence, adding a block of cars may increase the transit time of the train. These complex relationships are captured within the GA.

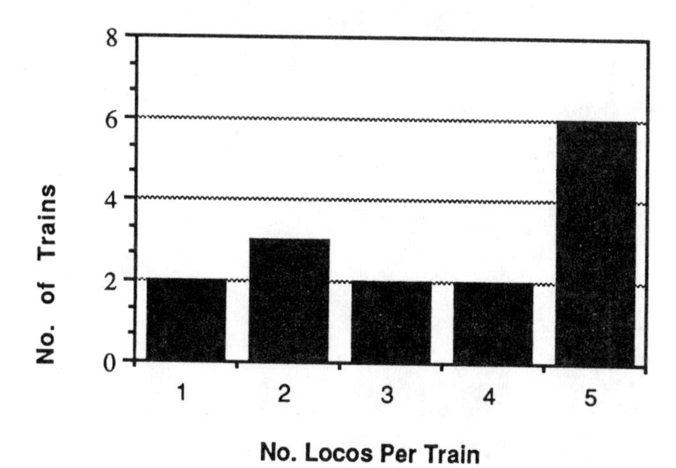

Figure 6a: Base Case

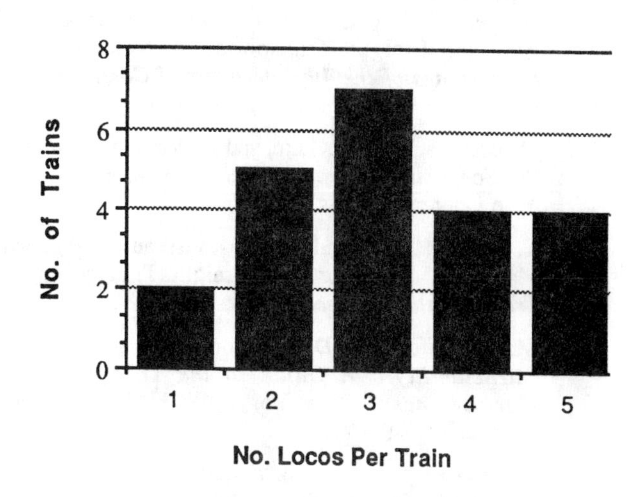

Figure 6b: No-Rental Case

5 CONCLUSIONS

The GA approach for rail transportation problems is successful for routing and scheduling trains along a network. Unlike current approaches, complex cost models are used and can be modified for sensitivity analysis. Important results have been obtained for CSXT which can easily be extended to many different types of transportation applications which share similar characteristics with the rail industry.

Extensions are necessary for handling larger networks and longer time windows. Current research is investigating various alternatives for local optimization and route generation for problems where all of the routes can not be stored. In addition, we are researching various alternatives for decomposing the routing and scheduling problem using new and innovative clustering techniques within the GA. It is important for the GA to learn not only the appropriate routes and schedules, but also the classes of traffic which are traveling on the network so that future blocks of cars can be accommodated quickly within the routes.

Acknowledgements

This research was supported by a grant from the 1990 CSX Corporation Technology Research & Development Program.

References

[1] W. Allman. A Computer Simulation Model of Railroad Freight Transportation Systems. In *Proceedings of the 4th International Conference on Operations Research*. Wiley, NY, 1966.

[2] M. Beckmann, C. McGuire, and C. Winsten. *Studies in the Economics of Transportation*. New Haven, CT.: Yale University Press, 1956.

[3] T. Crainic, J. Ferland, and J. Rousseau. A Tactical Planning Model for Rail Freight Transportation. *Transportation Science*, **18**, 165-184, 1984.

[4] M. R. Garey and D. Johnson. *Computers and Intractability: A Guide to the Theory of NP-Completeness*. New York: W. H. Freeman and Company, 1979.

[5] G. Syswerda. Uniform crossover in genetic algorithms. In J.D. Schaffer (Ed.). *Proceedings of the Third International Conference on Genetic Algorithms*. San Mateo: Morgan Kaufmann Publishers, Inc., pp. 2-9, 1989.

Genetic Algorithms and Computer-Assisted Music Composition

Andrew Horner and David E. Goldberg
University of Illinois at Urbana-Champaign
Urbana, IL 61801

Abstract

Genetic algorithms have been used with increasing frequency and effectiveness in a variety of problems. This paper investigates the application of genetic algorithms to music composition. A technique of thematic bridging is presented that allows for the specification of thematic material and delegates its development to the genetic algorithm. A look at the effects of building block linkage subsequently establishes a basis for implementing a GA-optimizable operation set for the problem. Some preliminary results are then discussed with an eye toward future work in GA-assisted composition.

1 Introduction

Genetic algorithms [6,10] have been applied to a wide array of problem domains from medical image registration [5] to stack filter design [1]. The accessibility of their interface has contributed to experimentation on problems from many diverse disciplines. Unlike most traditional optimization techniques, GAs don't rely on a particular problem structure or problem-specific knowledge. The general effectiveness of GAs in blindly optimizing strings in no small part accounts for their growing popularity.

It is this accessibility and effectiveness that suggests that GAs might provide a powerful tool in the computer-assisted composition of music. Various approaches to computer-assisted composition have been employed in generating music. These range from the relatively straightforward mapping of functions to compositional parameters [4] through elaborate stochastic feedback models [12,14]. Recent work using neural network composition methods to learn and generalize structures from existing musical examples is also of particular interest [13].

The approach adopted in this paper is one of thematic bridging. Thematic bridging is the transformation of an initial musical pattern to some final pattern over a specified duration. This method is often characteristic of *phase* or *minimalist* music and was chosen because of the first author's compositional approach to music and the fact that it lends itself so naturally to GA encoding.

The bridging is brought about through modifying or reordering elements of intervening patterns through a sequence of operations. The elements themselves may be discrete pitch, amplitude, or duration values. The resulting musical output consists of the concatenation of each partially developed pattern (or some variation of it) through the final pattern. The initial pattern itself is not part of the output, but could be a part of a preceding section as the final pattern.

As a simple example, with an initial note pattern (Gb,Bb,F,Ab,Db), a final pattern (F,Ab,Eb), and a note duration of 17, a bridge could be as follows:

operation description	resulting pattern
delete the last note:	Gb Bb F Ab
rotate the pattern:	Bb F Ab Gb
delete the last note:	Bb F Ab
mutate the first note:	Eb F Ab
rotate the pattern:	F Ab Eb

musical output: Gb Bb F Ab Bb F Ab Gb Bb F Ab Eb F Ab F Ab Eb

In addition to the thematic patterns and bridge duration, the basic operation set for transforming patterns must be specified. In the example above, the operation set includes atomic operations that change and delete notes, and an operation for rotating the pattern as a whole.

Some operation sets may not be capable of bridging the initial and final patterns over the duration specified, regardless of the sequencing. This suggests an iterative approach, where the problem is run to convergence and subsequently rerun with modifications if results are unsatisfactory (figure 1). For example, if the thematic patterns are dissimilar and the bridge duration is short, relatively high-level operations may be needed to bring off the transformation.

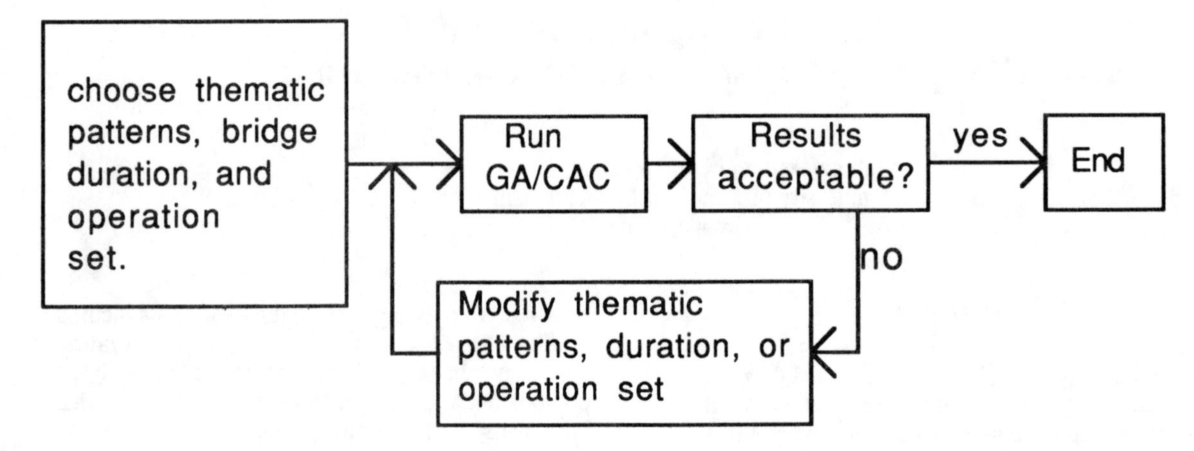

Figure 1: Iterative modification in GA-assisted composition.

Thematic bridging contrasts with more conventional algorithmic composition techniques in that the transition rules are not explicitly applied. Instead, the GA selects a sequence of operations or atomic transition rules to apply from a user-supplied operation set in constructing the bridge. This bridging bears some resemblance to the interpolation of melodies in neural network algorithmic composition [13], though learning is not a part of the bridging process.

2 GAs and Thematic Bridging

Thematic bridging bears a close resemblance to another GA application, job shop scheduling [2,3]. In job shop scheduling a group of tasks must be scheduled so that the total waiting time of the tasks is minimized. We can draw an analogy between the number of notes produced by a bridging operation and the wait of a corresponding job. Further, pattern bridging corresponds to constraints imposed upon job sequencing. This analogy may be useful in visualizing the problem.

The following section examines a particular operation set, and looks at its encoding. A given encoded operation sequence can be evaluated by simply executing the sequence and comparing its results to those desired. A simple 3-operator GA with binary tournament selection and 1-point crossover was used in all the examples presented in this paper. Only minor modifications of the GA were required to accommodate bridging operators and operands as a genic ordered pair.

2.1 An Operation Set and Encoding

The following is a general description of the operation set used in generating the results presented in this paper. Other operation sets can include any variety of customized operations. Note that the **Mutate** operation below should not be confused with the GA mutation operator.

Operation	Function
No-op	"do nothing" operator (takes up space)
Add	add an element to the pattern (either a needed element or one specified by the operand)
Delete	delete an element of the pattern (either an extraneous element or one specified by the operand)
Mutate	change an element of the pattern (tries to change an extraneous element to a needed one, otherwise as specified by the operands)
Rot	rotate the pattern (number of rotations and direction specified by the operands)
Exch	exchange elements of the pattern (moves first element to the position specified by the operand through a series of exchanges with neighboring elements)
IRot	rotate incorrectly positioned elements of the pattern
IExch	exchange incorrectly positioned elements of the pattern

The chromosome encodes the operation sequence from left to right with one operation per gene. Each gene is an operator that may or may not be accompanied by operands. In the actual implementation auxiliary structures were used to store the operands, which were crossed in parallel with the chromosome.

Each operand is represented by a real number between 0.0 and 1.0. Initially, a random number in this range is selected; when mutation occurs another random number is chosen as the replacement. These raw operands are subsequently scaled to appropriate values according to their operator's function. The use of real codings has a long if controversial history in the history of evolutionary methods. A recent paper [7] has put that usage on firmer theoretical footing.

Since simple GAs optimize fixed-length chromosomes, an appropriate length estimate for potential operation sequences must be found. This figure can be safely overestimated, since **No-ops** can effectively fill out extra positions. Without control flow instructions, execution proceeds in a simple left-to-right scan of the chromosome operations. In general, the length estimate should be based on the particular thematic patterns, operation set, and bridge duration under consideration. The examples tried to date have varied from a handful of operations up to lengths of 40. If control flow operations are included in the instruction set, their effect should also be considered.

Returning to the simple example presented earlier, the operation sequence would look something like:

(**Delete** 5) (**Rot** 1 forward) (**Delete** 4) (**Mutate** 1 3) (**Rot** 1 forward)

The **Mutate** and **Delete** operations use the first operand to specify the position affected. The second operand of the **Mutate** operation specifies the position in the final pattern to get the new value. The rotation operations indicate that a single forward rotation should be performed.

2.2 A Reasonable Fitness Function

The fitness function of a given operation sequence consists of a two-part hierarchy. The initial part is based on how close the developed pattern matches the final pattern. If there is an exact match, the fitness is set to a value derived by comparison of the resulting and desired output durations (i.e. the number of notes produced verses the number of notes desired).

The pattern-matching aspect of the fitness is computed from two components, the contents of the patterns and the ordering of their elements. These two checks insure that the right elements appear in the resulting pattern and in the correct order. The fitness function for this part is given by (maximizing):

$$\text{fitness} = 0.5 * (P_{common} + P_{correct})$$

where P_{commom} is the proportion of elements in the resulting pattern also in the final pattern, and $P_{correct}$ is the proportion of elements in the resulting pattern which match their corresponding element in the final pattern.

With an exact match of the resulting and final patterns (fitness=1.0), the fitness is augmented through a calculation which compares the resulting and desired musical output durations. The resulting fitness is maximal in the case of an exact duration match. The evaluation of the second part of the fitness is as follows:

$$\text{fitness} = 1.0 + N_d - \text{abs}(N_d - N_r)$$

where N_d is the desired duration, and N_r is the resulting duration.

In practice, a resulting duration slightly greater than the desired duration is often acceptable as a satisficing solution. In these cases, the extra output is simply truncated to the desired length.

3 Musical Results and Interpretation

This section considers some musical results and their relation to GA theory. Several levels of refinement were necessary in the implementation to arrive at these results. Problems with building block linkage arose in the initial implementation, but were subsequently overcome.

3.1 Musical Results

So what does genetic music sound like? This depends on the operation set and, of course, the fitness function.

The problem, operation set, and GA have been programmed in Smalltalk. A short piece has been created by the GA using thematic bridging for pitch and amplitude development. The piece is entitled "Epistasis" and consists of five voices in canon (the voices play the same material at different times) over six bridge passages. The GA was run for each of the passages independently, given the thematic patterns and duration of each. The musical output was then scheduled among the five voices so that the passages would overlap. The thematic material itself is quite simple, which is something of a challenge since its development must sustain interest. The results are musically pleasing to the authors with the usual qualifications regarding personal taste. Readers may contact the first author to obtain ordering information to receive a tape of GA-generated music.

3.2 First Results

The initial runs performed using the operation set presented above were not able to find useful solutions to even fairly trivial test problems. In these cases the run progressed to find a good collection of elements, but was unable to correctly order it. Inspection of the operations encoded in bad chromosomes revealed that no single operation change was sufficient to correct the ordering; generally, several changes were needed. Moreover, these changes were often widely spaced within the chromosome, a case of poor linkage.

The **Add** and **Delete** operations seemed to contribute most to these conditions. Generally, these operations used their operands to determine the position of the element to be added or deleted, which means picking the right element for the pattern while putting it in the right place. The nonlinearities introduced by simultaneously determining the correct collection and positioning introduced numerous local optima in the search space with poor linkage.

3.3 Improved Results

An alternative approach was adopted in an attempt to reduce the problem's nonlinearities to a reasonable level and improve linkage. This entailed a modification of operations such as **Add** and **Delete**. With the **Add** operation, an operand again determined the position of the added element, but the selection itself was made from elements in the final pattern not in the current pattern. This heuristically moves selection toward the correct collection of elements, and leaves room for their positioning. This significantly improves linkage since useful combinations of operator changes can occur closer together. All runs to date have converged to useful solutions using this strategy.

Runs using a hill-climbing GA (a GA with only selection and mutation operators) with the heuristic operation set have also been successful, but have been more susceptible to stalling. About one in five hill-climbing runs encounters problems, with long waits ensuing. These problems again occur when any single operation change to a suboptimal chromosome results in decreased fitness. Unlike hill-climbing, the normal GA with crossover recombines parts from distinct individuals to break out of this problem. With heuristic operations, these critical crosses are fairly likely since elements are generally added and deleted as needed, which makes operations easier to mix and match. Thus, a recombinative GA generally fares much better than a hill-climbing GA on the same problem.

4 Future Work

Several extensions to the work presented in this paper follow quite naturally. The thematic material used to date has been relatively simple; more sophisticated material and textures are needed to gage the general effectiveness of the technique. As mentioned earlier, experiments using control flow and other customized operations are likely to be fruitful. Beyond pitch, amplitude, and rhythmic development, manipulation of timbre (sound quality) via music synthesis techniques or MIDI control changes promises further compositional control using the same basic technique (interpolation may be used between discrete values for continuous changes).

In addition, there are other aspects of algorithmic composition that might lend themselves to GA optimization. These include the generation of coherent sets of material for thematic bridging and the subsequent scheduling of the GA-generated passages. Moreover, the application of messy GAs [8,9] to the problem would be a natural encoding, since the operation sequence is inherently variable-length. Genetic programming [11] might also be used in conjunction with hierarchical operation trees.

5 Conclusions

This paper has applied genetic algorithms to music composition through a technique of thematic bridging. The technique is only one method of using GAs in computer-assisted composition, but it raises many of the issues common to other approaches. For instance, any approach will have to consider the operation set, encoding, and fitness function. Similarly, consideration of building block linkage is useful in constructing a GA-friendly operation set. Finally, several directions for further study have been outlined and discussed which promise interesting musical results in future genetic composition.

Acknowledgments

This material is based upon work supported by the National Science Foundation under Grants CTS-8451610 and ECS-9022007. The first author also acknowledges support provided by the CERL Sound Group at the University of Illinois. Musical realization of this work was facilitated by the algorithmic composition tools in Symbolic Sound's Kyma Workstation.

References

[1] Chu, C. (1989). A genetic algorithm approach to the configuration of stack filters. *Proceedings of the Third International Conference on Genetic Algorithms and Their Applications*, 112-120.

[2] Cleveland, G., & Smith, S. (1989). Using genetic algorithms to schedule flow shop releases. *Proceedings of the Third International Conference on Genetic Algorithms and Their Applications*, 160-169.

[3] Davis, L. (1985). Job shop scheduling with genetic algorithms. *Proceedings of an International Conference on Genetic Algorithms and Their Applications*, 136-140.

[4] Dodge, C. (1988). Profile: A musical fractal. *Computer Music Journal 12*(3): 10-14.

[5] Fitzpatrick, J. M., Grefenstette, J. J., & Van Gucht, D., (1984). Image registration by genetic search. *Proceedings of IEEE Southeast Conference*, 460-464.

[6] Goldberg, D. E. (1989). *Genetic algorithms in search, optimization, and machine learning*. Reading, MA: Addison-Wesley.

[7] Goldberg, D. E. (1990). *Real-coded genetic algorithms, virtual alphabets, and blocking* (IlliGAL Report No. 90001). Urbana: University of Illinois, Illinois Genetic Algorithms Laboratory.

[8] Goldberg D. E., Deb, K., & Korb B. (1990). Messy genetic algorithms revisited: Studies in mixed size and scale. *Complex Systems*, *4*, 415-444.

[9] Goldberg D. E., Korb B., & Deb, K. (1989). Messy genetic algorithms: motivation, analysis, and first results. *Complex Systems*, *3*, 493-530.

[10] Holland, J. H. (1975). *Adaptation in natural and artificial systems*. Ann Arbor: The University of Michigan Press.

[11] Koza, J. (1990). *Genetic programming: A paradigm for genetically breeding populations of computer programs to solve problems* (Report No. STAN-CS-90-1314). Stanford: Stanford University, Department of Computer Science.

[12] Tipei, S. (1989). The computer: A composer's collaborator. *Leonardo*, *22*(2): 189-195.

[13] Todd, P. (1989). A connectionist approach to algorithmic composition.*Computer Music Journal 13*(4): 27-43.

[14] Xenakis, I. (1971). *Formalized music*. Bloomington: Indiana University Press.

Solving Partitioning Problems with Genetic Algorithms

Donald R. Jones and **Mark A. Beltramo**
Operating Sciences Department
General Motors Research Laboratories
Warren, MI 48090-9055
djones@gmr.com, beltramo@gmr.com

Abstract

We consider several approaches for solving partitioning problems with genetic algorithms. Two test problems are used to evaluate the approaches. The best approach encodes partitions as permutations and decodes them using a greedy adding heuristic. This approach performs well on both test problems; the other approaches give mixed results.

1 INTRODUCTION

A common problem in applied research is partitioning objects into a fixed number of groups to optimize an objective function. Solving this problem is difficult because the number of partitions of N objects into K groups increases with K^N [6]. In special cases, exact solutions are possible in polynomial time [5], but in general near-optimal solutions must be found using heuristics. This paper considers solving partitioning problems with genetic algorithms (GAs).

To solve partitioning problems with GAs, one must encode partitions in a way that allows manipulation by genetic operators. We consider three encoding methods. The first encodes a partition as an N-string whose i^{th} element is the group number assigned to object i. For example, the partition $\{AE\}\ \{CD\}\ \{BF\}$ could be represented by the string (1 3 2 2 1 3).

The second method encodes a partition as a permutation of N objects and $K-1$ group separators. For example, the partition $\{AE\}\ \{CD\}\ \{BF\}$ would be encoded by the permutation $(A\ E\ \mid\ C\ D\ \mid\ B\ F)$. In the computer, the permutation would be stored as (1 5 7 3 4 8 2 6) where the 1 corresponds to A, the 2 corresponds to B, etc., and the 7 and 8 correspond to the group separators. This encoding method has also been considered by J. Bhuyan, V. Raghavan, and V. Elayavalli [1].

The third method, suggested to us by L. Davis (personal communication), encodes a partition as a permutation of the N objects and decodes this permutation with a greedy adding heuristic. The greedy heuristic uses the first K objects in the permutation to initialize K groups. The remaining objects are then added to these groups in the order they appear in the permutation, always adding the object to the group that yields the best objective value. Davis's encoding method requires more knowledge about the objective function than the others, since the greedy heuristic requires the ability to evaluate partitions of less than N objects into K groups. For most problems, however, this is feasible.

We consider several crossover operators for each of these encoding methods, resulting in a total of nine distinct approaches. These approaches are compared on two test functions using a steady-state GA [11].

The rest of this paper is organized as follows. Section 2 presents the test problems and describes the Monte Carlo experiment used to evaluate the approaches. Sections 3 and 4 then consider the group-numbers and permutation encodings, respectively. Section 5 discusses some theoretical issues that may explain the performance of the different encodings.

2 TEST PROBLEMS

The first test problem is to divide N numbers x_i ($i = 1, \ldots, N$) into K groups to minimize differences among the group sums. Specifically, the objective is to minimize

$$f(G_1, \ldots, G_K) = \sum_{k=1}^{K} \mid s_k - \bar{s} \mid$$

where G_k is the kth group in the partition,

$$s_k = \sum_{i \in G_k} x_i, \quad \text{and} \quad \bar{s} = \frac{1}{K} \sum_{k=1}^{K} s_k .$$

We call this the "equal piles problem" since the idea is to create K "piles" of numbers of equal sum. Data for N=34, K=10, and $\bar{s}$=10000 is given in Table 1; an optimal solution with f=0 is also shown.

Table 1: The Equal Piles Problem

Data

i	x_i	i	x_i
1	3380	18	3433
2	1824	19	3519
3	1481	20	1363
4	2060	21	1824
5	1225	22	3305
6	836	23	2156
7	1363	24	3305
8	2705	25	3049
9	4635	26	3980
10	6481	27	2787
11	2588	28	4635
12	3380	29	4068
13	1952	30	2992
14	3832	31	5932
15	3176	32	528
16	2316	33	3304
17	2479	34	4107

Solution

k	$i \in$ Group k
1	1, 22, 27, 32
2	2, 3, 11, 34
3	14, 15, 30
4	4, 9, 24
5	16, 17, 23, 25
6	29, 31
7	5, 6, 28, 33
8	7, 8, 13, 26
9	10, 19
10	12, 18, 20, 21

The second test problem is to color the US map so that no two bordering states are the same color. That is, we want to partition the 48 states of the continental US into 4 groups (corresponding to colors) to minimize the number of "bad borders" — the number of bordering state pairs that are in the same group. Formally, let b_{ij}=1 if states i and j border and b_{ij}=0 otherwise. The objective is to minimize

$$f(G_1, \ldots, G_4) = \sum_{k=1}^{4} \sum_{\{(i,j) \mid i,j \in G_k,\, i<j\}} b_{ij}.$$

The optimal solution to this problem is f=0; the b_{ij} data can be read from any US map.

Partitioning and coloring problems are NP-complete [8]. However, the US map coloring problem appears to be easy since it can be solved quickly using simple heuristics. On the other hand, we designed the equal piles problem to be hard. We constructed the numbers in a way that we believe makes the solution unique (except for exchanges of identical values like x_7 and x_{20}). We also chose a large number of groups so that greedy heuristics would have some difficulty.

In sections 3 and 4 we will introduce several approaches for solving partitioning problems with GAs. The approaches are evaluated using a Monte Carlo experiment. We first draw 30 integers to serve as seed numbers for the GA's random number generator. For each approach and test problem, 30 steady-state GA runs are made using these seeds (see Appendix A for a description of the GA). For a given test problem, let f_i^a denote the solution at convergence for approach a using seed i. To test whether methods a and b have the same mean solution value, we compute the following statistic:

$$z = \frac{\sqrt{n}\,\bar{x}}{\sqrt{\frac{1}{n-1}\sum_{i=1}^{n}(x_i - \bar{x})^2}},$$

where $n = 30$,

$$x_i = f_i^a - f_i^b, \quad \text{and} \quad \bar{x} = \frac{1}{n}\sum_{i=1}^{n} x_i.$$

The central limit theorem implies that the asymptotic distribution of z is standard normal. We can therefore reject the null hypothesis that $\bar{x} = 0$ in favor of the alternative that $\bar{x} < 0$ with significance level p, where $p = \Phi(z)$ and $\Phi(\cdot)$ is the standard normal distribution function. In what follows, we will say method a is "significantly better" than method b if $p \leq 0.05$ and "not significantly better" otherwise.

Comparing mean solution values *at convergence* is somewhat arbitrary, since some approaches take longer to converge than others. For this reason, we will also show graphs of best solution value (averaged over 30 runs) versus iteration number; these graphs should give the reader an idea of how the approaches compare before convergence.

3 GROUP-NUMBERS ENCODING

3.1 METHODS

Encoding partitions as strings of group numbers allows the use of standard single-point and uniform crossover operators. These operators, however, create two problems. First, the child can have fewer groups than the parents. For example, if we cross strings (1 2 2 3 3 1) and (1 3 3 2 2 1) after the third position with single-point crossover, the child will be (1 2 2 2 2 1) which contains two groups instead of three. This example also illustrates the second problem. Both parents encode the same partition, $\{AF\}$ $\{BC\}$ $\{DE\}$; the only difference is in how the groups are numbered. Their child, on the other hand, encodes an entirely different partition. Normally we would like identical parents to have identical children (ignoring mutation).

Single-point crossover can be modified to eliminate these problems. To ensure that the child has the desired K groups, we use a rejection method. There are $2(N-1)$ children that can result from single-point crossover, since there are $N-1$ cross points and the cut strings can be juxtaposed in two ways (first part of parent 1 with the second part of parent 2, or first part of parent 2 with the second part of parent 1). For the rejection method, we select one of these $2(N-1)$ children at random. If this child has fewer than K groups, we select another child from those that remain. We continue in this way until we get a child with K groups or until no children remain. In the latter case, we set the child equal to one of the parents chosen at random. Note that, following Whitley [11], we produce only *one* child per crossover.

To ensure that parents encoding the same partition have a child that also encodes this partition, we can canonically renumber the parents. A convenient method renumbers each parent so the first group to appear is a 1, the second distinct group is a 2, and so on. Thus, (1 3 3 2 2 1) would become (1 2 2 3 3 1). With renumbering, parents representing the same partition will have the same *renumbered* string of group numbers and, hence, the child will be the same.

When the parents are renumbered, there will always be a cross point that gives a child with K groups; hence, the rejection method will never resort to setting the child equal to one of the parents. To see why, suppose we have cut the parent strings at some point and let j_i ($i = 1, 2$) be the highest group number in parent i to the left of the cut point. If $j_2 \leq j_1$, then a child with K groups can be formed by juxtaposing the first part of parent 1 with the second part of parent 2. The first part of parent 1 contains groups numbers $\{1, \ldots, j_1\}$. The second part of parent 2 contains group numbers $\{j_2 + 1, \ldots, K\}$ which, since we are assuming $j_2 \leq j_1$, must contain $\{j_1 + 1, \ldots, K\}$. Hence the child contains all group numbers. If $j_2 > j_1$, we juxtapose the first part of parent 2 with the second part of parent 1.

Uniform crossover [10] can also be used to cross strings of group numbers. Uniform crossover considers each object separately, giving the child the group number of either parent 1 or parent 2 depending on the flip of a simulated fair coin. Rejection and renumbering can also be used. However, because there are 2^N possible children, we do not maintain a list of possible children and sample them without replacement. Instead, we repeat uniform crossover until we get a child with K groups or until a limit on the number of attempted crosses is reached. (We use a limit of 200.) If the limit is reached without finding a child with K groups, the child is set to one of the parents chosen at random. The rejection method is unlikely to resort to setting the child equal to a parent; in fact, this never occurs in our runs.

A final crossover method is *edge-based crossover* [7]. The motivation for this operator comes from a similar edge-based operator used by Whitley et. al. [12] for the traveling salesman problem. Here, we describe edge-based crossover using an example; Appendix B gives details.

Consider the following parent partitions:

$$\begin{aligned} \text{Parent 1} &= \{A\}\{CDE\}\{BF\} \\ \text{Parent 2} &= \{C\}\{BDF\}\{AE\}. \end{aligned}$$

Two objects are connected by an edge if they are in the same group. Thus, parent 1 contains edges *cd*, *ce*, *de*, and *bf*, whereas parent 2 contains edges *bd*, *bf*, *df*, and *ae*. Edges are not directed, so that edge *bf* is the same as edge *fb*. The basic idea of edge-based crossover is to construct a child by combining edges from the parents. One child that can be constructed is

$$\text{Child} = \{AE\}\{CD\}\{BF\}$$

This child contains edges *ae*, *cd*, and *bf*, all of which it inherits from at least one parent. Edge *ae* is inherited from parent 2, edge *cd* from parent 1, and edge *bf* from both parents. Other children are possible. At an intuitive level, edge-based crossover can be thought of as selecting one of these children at random.

Although the intuition of edge-based crossover is clear, the child cannot be constructed by simply selecting some edges from parent 1 and some from parent 2. For example, we cannot take edge *ed* from parent 1 and edge *db* from parent 2. If we choose these edges, it would follow that E, D, and B were all in the same group, and the child would contain edge *eb*. But the child cannot inherit this edge, since neither parent has it. The algorithm for edge-based crossover is given in Appendix B.

When using the group-numbers encoding, some care must be taken in initializing the population and implementing mutation. Strings can be generated by setting each object's group number to a random integer between 1 and K. However, it is possible that the resulting string will be missing one or more of the K group numbers (i.e., it will encode a partition with less than K groups). In these cases, one must reject the string and generate another one. In mutation, objects in a group by themselves must not be mutated, since doing so will empty a group.

3.2 RESULTS

For the group-numbers encoding, we evaluated five crossover operators: single-point with rejection; single-point with rejection and renumbering; uniform with rejection; uniform with rejection and renumbering; and edge-based. Figure 1 shows the results for the equal piles problem. The figure graphs the best solution value (averaged over 30 runs) as a function of the number of iterations. At convergence, edge-based crossover is significantly better than the other crossover operators. Renumbering slightly improves the performance of single-point and uniform crossover, but the effect is insignificant. Differences between single-point and uniform crossover are also insignificant, though uniform takes more iterations to converge.

Figure 2 shows the results for the map coloring problem. Edge-based crossover is the only operator that finds the optimal solution in all 30 runs; it also converges in the least number of iterations. However, edge-based crossover is only *significantly* better

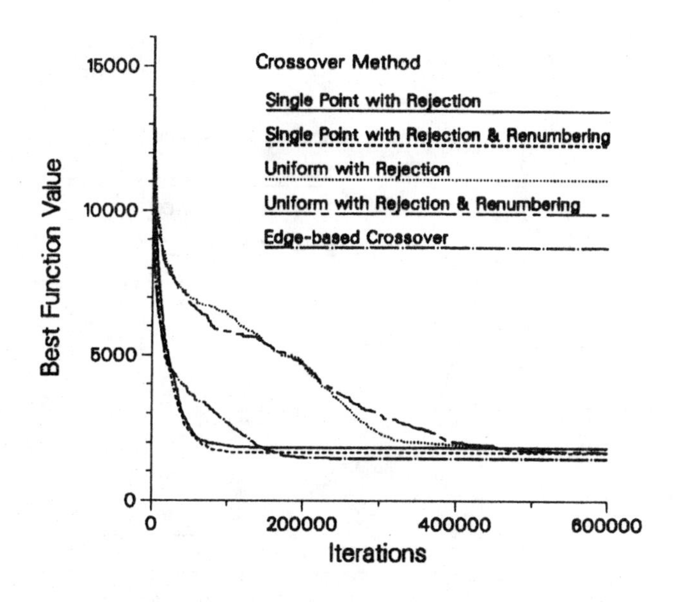

Figure 1: The Equal Piles Problem Encoded with Strings of Group Numbers

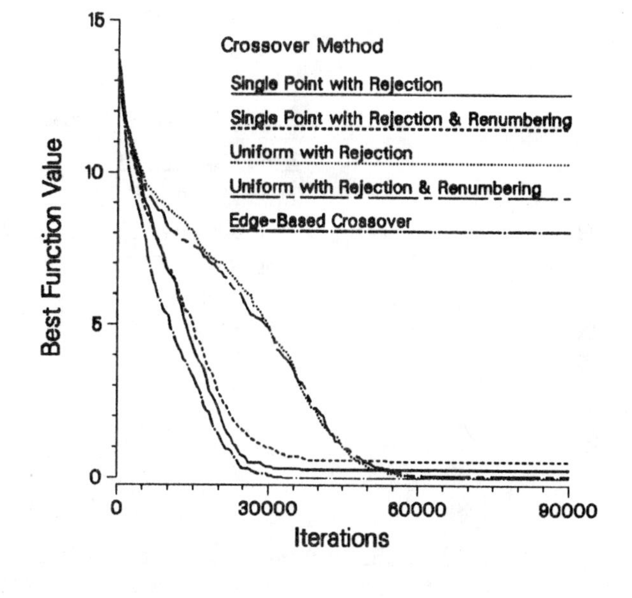

Figure 2: The Map Coloring Problem Encoded with Strings of Group Numbers

than single-point crossover (with and without renumbering); differences with uniform (with and without renumbering) are insignificant. In the map coloring problem, renumbering degrades the performance of single-point and uniform crossover, though the effects are insignificant. As in the equal piles problem, uniform crossover takes more iterations to converge than single-point crossover; here, however, uniform crossover does significantly better than single-point at convergence.

For the class of approaches using strings of group numbers, edge-based crossover appears best. However, edge-based crossover has $O(K^4)$ complexity [7] and is impractical for large K. For our test problems, it takes 2-5 times more computation time per iteration than the other crossover methods. Uniform crossover also takes a long time to converge. When speed is a concern, therefore, it may be best to use single-point crossover with rejection. Whether one should use renumbering is not clear; the effect varies for the two problems and is always insignificant.

4 PERMUTATION ENCODINGS

4.1 METHODS

Encoding partitions as permutations allows the use of any crossover operator designed for permutations. A good deal of work has been done on such crossover operators, focusing mainly on the traveling salesman problem. Here, we consider order crossover (OX) [3,9] and partially matched crossover (PMX) [4].

For the permutation encoding using group separators, care must be exercised in initializing the population and in implementing crossover and mutation. The reason is that not all permutations encode valid K-partitions (a permutation will encode less than K groups if a separator is in the first or last position or if two separators are together). The population is best initialized by generating random strings of group numbers and then converting these into permutations. For example, the string (3 2 1 3 2 1) would first be converted into the permutation ($C\ F \mid B\ E \mid A\ D$) by listing the objects in each group separated by separators. Objects within groups would then be randomly permuted, resulting in a permutation like ($F\ C \mid B\ E \mid D\ A$). In the computer, this permutation would be stored as (6 3 8 2 5 7 4 1) where 7 and 8 represent the separators (which also appear in a random order). In crossover, we repeatedly cross the parents until we get a child that decodes into a partition with K groups. Finally, we limit mutation to swapping objects, not group separators (see Appendix A); this ensures that mutation will not empty a group and will only cause a small change to the associated partition.

For the permutation encoding with greedy decoding (Davis's method), the greedy adding heuristic ensures that every permutation encodes a valid partition. Hence, one may as well initialize the population with

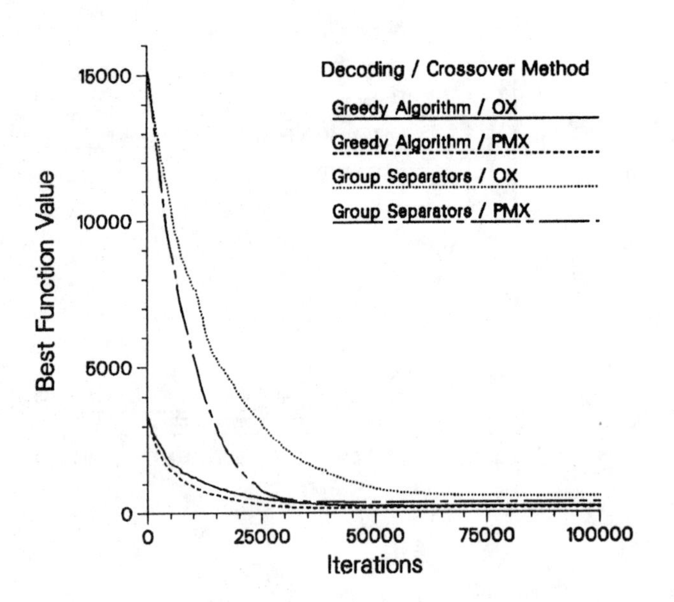

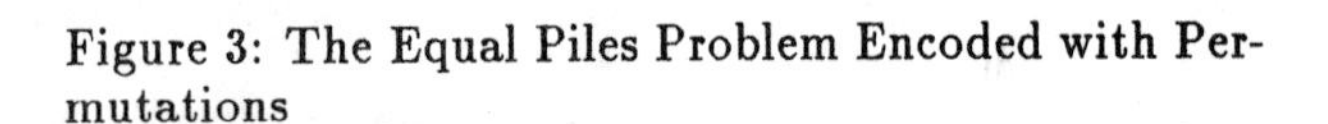
Figure 3: The Equal Piles Problem Encoded with Permutations

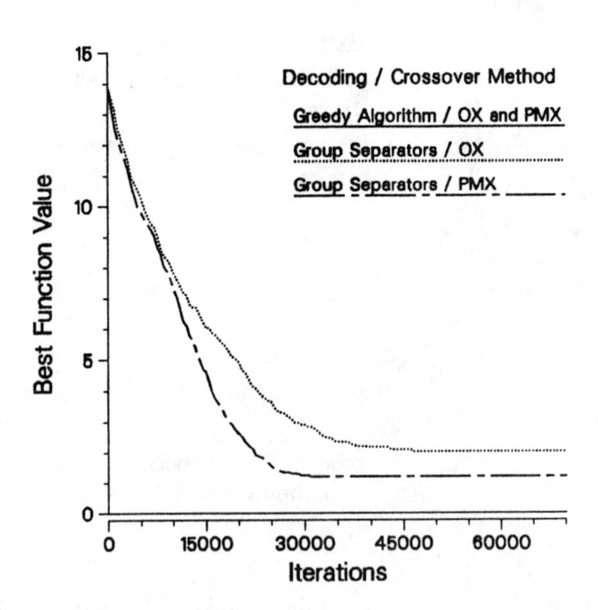

Figure 4: The Map Coloring Problem Encoded with Permutations

random permutations. Crossover and mutation of the permutations can be carried out normally.

4.2 RESULTS

Figure 3 shows the results for the equal piles problem. For both crossover operators, the "greedy" decoding significantly outperforms the "separators" decoding. In addition, for both decodings, partially matched crossover (PMX) significantly outperforms order crossover (OX).

Figure 4 shows the results for the map coloring problem. Here the greedy heuristic performs so well that the optimal solution is always found in the initial population! The decoding with group separators does poorly, though PMX is again significantly better than OX crossover.

For completeness, Table 2 gives a complete set of comparisons among the all of the approaches, based on solution value at convergence.

5 DISCUSSION

The clear winner in these experiments is the permutation encoding with greedy decoding and PMX crossover. This approach is a hybrid between a GA and the greedy heuristic. Like most hybrid methods, we should expect it to outperform a pure GA since it incorporates additional problem knowledge—in this case, knowledge about the value of partitions with less than N objects. The greedy heuristic is good choice for a hybrid method; for a specific problem, Cornuejols et. al. [2] show that the greedy heuristic has excellent worst-case performance.

We have not systematically compared GAs to other partitioning heuristics. But GAs can outperform some standard heuristics on some problems. For example, we have applied a switching heuristic [6] to the equal piles problem. Switching starts with a given partition and then iteratively improves it. In each iteration, it considers all $N(K-1)$ ways to move one of the objects from its current group to another group. It then makes the one move (or "switch") that most improves the objective function. The process stops when no further moves improve the objective function. We ran switching 30 times on the equal piles problem, each run starting with a randomly-generated partition. The best solution was 1692 and the mean solution was 3610 (standard deviation = 1220). In contrast, the GA/greedy hybrid with PMX had a best solution of 2 and a mean solution of 171 (standard deviation = 91). Another example where a GA outperforms standard clustering heuristics is given in [7].

Although the GA/greedy hybrid worked best on our test problems, the method can only be used if one can evaluate partitions with less than N objects. When this is not possible, only the pure GA methods can be used. The results for these methods were mixed: the "separators" encoding worked best for the equal piles

Table 2: Comparisons of Mean Solution Values at Convergence

Equal Piles Problem

Encoding/Cross	pgo	pgp	pso	psp	gs	gsr	gu	gur	ge
perm-greedy/OX (pgo)			**	**	**	**	**	**	**
perm-greedy/PMX (pgp)	*		**	**	**	**	**	**	**
perm-separators/OX (pso)					**	**	**	**	**
perm-separators/PMX (psp)			**		**	**	**	**	**
group/single (gs)									
group/single & renum (gsr)									
group/uniform (gu)									
group/uniform & renum (gur)									
group/edge-based (ge)					**	*	**	*	

US Map Coloring Problem

Encoding/Cross	pgo	pgp	pso	psp	gs	gsr	gu	gur	ge
perm-greedy/OX (pgo)			**	**	**	**			
perm-greedy/PMX (pgp)			**	**	**	**			
perm-separators/OX (pso)									
perm-separators/PMX (psp)			*						
group/single (gs)			**	**					
group/single & renum (gsr)			**	**					
group/uniform (gu)			**	**	**	**			
group/uniform & renum (gur)			**	**	*	**			
group/edge-based (ge)			**	**	**	**			

Note: The number of asterisks in a cell indicates the significance level at which we may accept the alternative hypothesis that the row approach is better than the column approach: one asterisk indicates $p \leq 0.05$ and two indicates $p \leq 0.01$.

problem, while the group-numbers encoding worked best for the map coloring problem. In the remainder of this section we discuss factors that may explain the relative performance of the pure GA methods.

Most of the encodings we have considered are redundant in that the number of possible encodings exceeds $P(N, K)$, the number of partitions of N objects into K groups. The value of $P(N, K)$ is given by the recursive formula

$$P(N, K) = K P(N-1, K) + P(N-1, K-1)$$

where

$$P(N, 1) = P(N, N) = 1$$

for all N (see [6]). Now let $E(N, K)$ denote the number of possible encodings. For the group-numbers encoding, every partition can be represented by $K!$ strings since there are $K!$ ways to number the groups; hence, $E(N, K) = K!\, P(N, K)$. Renumbering eliminates this redundancy, but has a negative side effect (discussed later). Using edge-based crossover also eliminates redundancy, since the operator works upon the decoded partitions, not the group numbers per se. Finally, for permutations of objects and group separators, the number of possible encodings is[1]

$$E(N, K) = N! \binom{N-1}{K-1} (K-1)!$$

Let us measure redundancy using the index $R \equiv \log_{10}[E(N, K)/P(N, K)]$. Using the base 10 log means that R can be interpreted as orders of magnitude. Values of R for the different methods and test problems are given in Table 3.

All else equal, we expect redundancy to degrade a GA's performance. Intuitively, redundancy causes the objective function to have many more optima in the encoded space than in the space of partitions. We can therefore expect several subpopulations to develop, each searching the niche around one of these optimal encodings. Crosses between subpopulations will be tend to be unproductive and wasteful. Eventually,

[1] The logic behind the expression for $E(N, K)$ is as follows. Start with a permutation of the objects without any separators; there are $N!$ such permutations. Now consider where to put the group separators; there are $N-1$ possible locations and we must choose $K-1$ (hence the second term). Finally, the group separators can be numbered in (K-1)! ways.

Table 3: Redundancy Index for Different Encodings and Test Problems

Encoding	Crossover	Redundancy Index: Equal Piles Problem	Redundancy Index: US Map Coloring Problem
Separators	OX, PMX	24.3	38.6
Group numbers	single/uniform no renumbering	6.6	1.4
Group numbers	single/uniform with renumbering	0.0	0.0
Group numbers	edge-based	0.0	0.0

one subpopulation will get closer to the optimum than the others and expand to include the entire population. But most of the final population will probably be descended from the small original subpopulation and, hence, may lack the diversity required to find an accurate solution.

As mentioned earlier, renumbering eliminates redundancy for the group numbers encoding. However, renumbering also has a potentially disruptive interaction with mutation. For example, consider the string (1 1 2 2 3 3). If the first group number is mutated to a 3, we will get (3 1 2 2 3 3) which, after renumbering, becomes (1 2 3 3 1 1). Thus, the mutated *and renumbered* string differs by five out of six group numbers! Such large amounts of mutation could seriously interfere with schema processing. Due to this interaction with mutation, the effects of redundancy cannot be measured by observing the effects of renumbering in Figures 1 and 2.

There is, however, one special case when the effect of redundancy *can* be measured. This occurs when we use single-point crossover and set the mutation rate equal to zero. Single-point crossover has the property that, if the parents are canonically numbered, then the child will also be canonically numbered (this is not true for uniform crossover). Thus, with zero mutation, we only need to renumber the initial population; all subsequent individuals will automatically be canonically numbered. Of course, with no mutation, the disruptive interaction between mutation and renumbering is eliminated. Hence, in this special case, the effect of redundancy is equivalent to that of renumbering.

Based on the above observation, we made some additional runs using single-point crossover and zero mutation. In particular, for each of our 30 seeds, two GAs were run: one in which the initial population was renumbered, and one in which it was not. For the equal piles problem, renumbering improved performance in 22 out of the 30 comparisons. The mean solution with renumbering was better than that without renumbering, and the difference was significant at the 0.01 level. For the map coloring problem, the renumbering solution was better in 11 comparisons, the same in 8, and worse in 11. The mean solution with renumbering was better than that without renumbering, but the difference was not significant. That renumbering has a smaller impact on the map coloring problem is consistent with the fact that the redundancy index is smaller for that problem (see Table 3).

For the values of N and K in our test problems, the separators encoding has much more redundancy than the group-numbers encoding (see Table 3). The separators approach may, however, have a feature that partially offsets the negative impact of redundancy. Consider the following permutation of objects and separators:

$$(A\ B \mid C\ D\ E\ F\ G\ H \mid I\ J\ K).$$

On the one hand, one might think that this permutation encodes nothing more than the partition $\{A\ B\}\ \{C\ D\ E\ F\ G\ H\}\ \{I\ J\ K\}$. On the other hand, one could argue that the partition also encodes information like "I am more confident that E and F are in the same group than I am about C and H." The reason is that crossover is more likely to leave E and F in the same group than it is to leave C and H in the same group. Thus, a kind of linkage among objects can develop. We hypothesize that this linkage is stronger for PMX than OX crossover, and that this explains why PMX outperforms OX on our test problems.

Putting it all together, the best method for our test problems is the permutation encoding with greedy decoding and PMX crossover. If a pure GA method is required, the results are mixed and suggest topics for future research. For example, further work could be done on the effects of redundancy. One could also study how much the "separators" encoding benefits from the hypothesized linkage effect. If linkage seems important, it may be possible to modify PMX crossover to enhance linkage and improve performance.

APPENDIX A

Our steady-state GA is similar to Whitley's GENITOR algorithm [11]. The population is initialized by randomly generating M candidate solutions. Each iteration begins by selecting two parents. Selection probability depends linearly on rank; if we let b denote the bias, then we select the best population member with probability b/M and the worst member with probability $(2-b)/M$. The parents are crossed to form a *single* child, which is then mutated in a way depending on the encoding. For strings of group numbers, we consider each position in the string and, with probability p_{mut}, change the group number; objects in a group by themselves are not mutated since this would empty the group. For permutations, with probability p_{mut} we randomly select two elements in the permutation and swap them. An iteration ends by inserting the child in the population and deleting the worst member. All runs use a population size of 1000 and a bias of 2. For strings of group numbers, we set $p_{mut} = 1/N$ where N is the number of objects. For permutations, we set $p_{mut} = 0.1$. The GAs are terminated when population minimum and maximum fitness are the same (complete convergence).

APPENDIX B

Let G_{pi} denote the ith group $(i = 1, \ldots, K)$ in the partition of parent p $(p = 1, 2)$. The algorithm for edge-based crossover is as follows:

1. *Initialize the Child*
   Find all intersections of the form $G_{1i} \bigcap G_{2j}$ for $i, j \in \{1, 2, \ldots, K\}$. Let L denote the number of these intersections that are nonempty, and set child partition equal to these L groups.
2. *Check Stopping Criterion*
   If $L=K$, stop. Otherwise, go to step 3.
3. *Join Two Groups in the Child*
   For each pair of groups in the child, compute the number of between-group edges not present in either parent; we call these "noninherited edges." Select the pair of groups with the minimum number of noninherited edges, breaking ties at random. Join this pair of groups, set $L = L - 1$, and go to step 2.

Step 1 ensures that edges present in both parents will be inherited by the child. Suppose both parents have an edge between A and B. This means that A and B are grouped together by parent 1 (say, in group G_{1i}) and by parent 2 (say, in G_{2j}). Therefore, A and B will be in the intersection $G_{1i} \bigcap G_{2j}$. Since this intersection will be contained in a group of the child, the child will also have an edge between A and B.

Step 3 attempts to ensure that every edge in the child comes from at least one parent. It does so by attempting to avoid noninherited edges when joining. For large K, however, there will usually come a point in the joining procedure when the minimum number of noninherited edges is nonzero. Still, based on our computational experiments, edge-based crossover always creates children with less non-inherited edges than single-point or uniform crossover. The random tie-breaking in step 3 makes it possible for several different children to be created from the same parents.

References

[1] J. Bhuyan, V. Raghavan, and V. Elayavalli. *Genetic-Based Clustering*. Technical Report 90-4-1, The Center for Advanced Computer Studies, University of Southwestern Louisiana, Lafayette, LA, March 1990.

[2] G. Cornuejols, M. Fisher, and G. Nemhauser. Location of bank accounts to optimize float: an analytic study of exact and approximate algorithms. *Management Science*, 23:789–810, 1977.

[3] L. Davis. Job shop scheduling with genetic algorithms. In *Proc. First Intl. Conf. on Genetic Algorithms*, pages 136–140, 1985.

[4] D. Goldberg and R. Lingle. Alleles, loci, and the travelling salesman problem. In *Proc. First Intl. Conf. on Genetic Algorithms*, pages 154–159, 1985.

[5] P. Hansen, B. Jaumard, and O. Frank. Maximum sum-of-splits clustering. *Journal of Classification*, 6:177–193, 1989.

[6] J. Hartigan. *Clustering Algorithms*. John Wiley & Sons, New York, 1975.

[7] D. R. Jones and M. A. Beltramo. *Clustering with Genetic Algorithms*. Research Publication GMR-7156, General Motors Research Laboratories, Warren, MI, September 24, 1990.

[8] F. Maffioli. The complexity of combinatorial optimization problems and the challenge of heuristics. In N. Christofides, A. Mingozzi, P. Toth, and C. Sandi, editors, *Combinatorial Optimization*, chapter 5, John Wiley & Sons, New York, 1979.

[9] I. Oliver, D. Smith, and J. Holland. A study of permutation crossover operators on the travelling salesman problem. In *Proc. Second Intl. Conf. on Genetic Algorithms*, pages 224–230, 1987.

[10] G. Syswerda. Uniform crossover in genetic algorithms. In *Proc. Third Intl. Conf. on Genetic Algorithms*, pages 2–9, 1989.

[11] D. Whitley. The GENITOR algorithm and selection pressure: why rank-based allocation of reproductive trials is best. In *Proc. Third Intl. Conf. on Genetic Algorithms*, pages 116–123, 1989.

[12] D. Whitley, T. Starkweather, and D. Fuquay. Scheduling problems and traveling salesmen: the genetic edge recombination operator. In *Proc. Third Intl. Conf. on Genetic Algorithms*, pages 133–140, 1989.

Design of an Adaptive Fuzzy Logic Controller Using a Genetic Algorithm

Charles L. Karr
U. S. Bureau of Mines
Tuscaloosa Research Center
Tuscaloosa, AL 35486-9777

Abstract

The U.S. Bureau of Mines has developed a powerful technique which uses genetic algorithms (GAs) to design fuzzy logic controllers (FLCs). FLCs are rule-based systems capable of mimicking the "rule-of-thumb" approach used by humans in process control. This human rule-of-thumb approach is modeled using rules that incorporate fuzzy linguistic terms, such as expressions like "high" and "low." These linguistic terms are described by membership functions. Although FLCs have been used to control a number of systems, the selection of acceptable fuzzy membership functions has been a subjective and time-consuming task. In this paper, GAs are used to select high-performance membership functions for FLCs that control a computer simulated cart-pole balancing system. In two examples presented, the development time required to design the FLCs is reduced substantially by employing a GA. More importantly, in both instances the GA-designed FLCs outperform human-designed FLCs. These results lead to the conclusion that GAs are both effective and robust when used in FLC design.

1 INTRODUCTION

Rule-based systems have become increasingly popular in practical applications of artificial intelligence. Although they have performed as well as humans in several problem domains, their performance in process control problems has been limited by a lack of flexibility in representing the subjective nature of human decision-making. The flexibility inherent in human decision-making can be incorporated into rules via fuzzy set theory [12]. In fuzzy set theory, abstract or subjective concepts can be represented with linguistic terms, such as expressions like "very cold" and "not quite hot." Linguistic terms have been incorporated into rule-based systems to form fuzzy logic controllers (FLCs).

FLCs use rules that are of the form common to many rule-based systems: IF [condition] THEN [action]. These rules prescribe a definite action to be taken for a particular condition existing in the physical environment. The rules used in FLCs incorporate fuzzy linguistic terms to describe both the condition and the action portion of the rule and, thus model a human's "rule-of-thumb" approach to process control. FLCs are increasing in popularity and have been successfully used in a number of control problems [11]. These fuzzy rule-based systems include rules to direct the decision process and membership functions to convert linguistic terms into precise numeric values needed by the computer. The rule set is gleaned from a human expert's knowledge and experience, and the membership functions are chosen by the FLC developer to represent the human expert's conception of the linguistic terms. The selection of high-performance membership functions is commonly the most time-consuming phase of FLC development. Changes in the membership functions alter the performance of the controller because the membership functions determine the contribution each rule makes to the choice of a final action. Thus, the performance of the FLC is directly related to the developer's choice of membership functions.

Procyk and Mamdani [10] introduced an iterative procedure for modifying membership functions, but in general, little has been done to develop techniques for choosing membership functions that improve FLC performance. In fact, the design of FLCs has often been a trial-and-error process in which most of the development time is devoted to the quest for efficient membership functions. An efficient method is needed for selecting membership functions that maximize FLC performance. Moreover, complex time-variant systems require adaptive FLCs: those that are able to account for changes in the characteristic parameters of the systems before the changes cause the physical systems to fail. A new algorithm is needed that is robust enough to alter membership functions in "real-time," i.e., as changes are actually occurring in the physical system.

In this paper, two examples are discussed in which GAs are used to design FLCs that oversee a cart-pole balancing system. The cart-pole system consists of a rolling cart with an attached inverted pendulum. The objective is to keep the pole balanced while applying horizontal forces that position the cart at the center of the track. The cart-pole system, which has received much attention in the literature, is a challenging control problem and has become a benchmark in the artificial intelligence community for testing control strategies [1,2,3,9].

Initially, an example is presented in which a simple GA (SGA) [4] is used to design a non-adaptive cart-pole FLC. In this paper, a non-adaptive FLC is considered to be one in which the membership functions do not vary once they have been established. In this first example, the characteristic parameters of the cart-pole system remain constant. Next, a second example is presented in which a GA is used to design an adaptive cart-pole FLC, one whose membership functions are altered in real-time. This adaptive FLC proves to be adept at controlling the cart-pole system even though the characteristic parameters, such as the cart mass, vary with time. Since this FLC must alter membership functions in real-time, it uses a rapidly converging, small population micro-GA (MGA) [6,7]. The MGA, based on an investigation of small population GAs by Goldberg [5], locates high-performance membership functions fast enough to provide the FLC with the necessary real-time adaptive capabilities.

2 FUZZY LOGIC CONTROLLERS

There are numerous approaches to FLC development. In this paper, a "streamlined" approach to FLC development is introduced and presented in a step-by-step procedure. This approach circumvents much of the fuzzy mathematics generally associated with FLCs. Although the particular application of interest is a cart-pole balancing system, the approach to FLC development is discussed in generalized terms to facilitate its application to physical systems other than the cart-pole.

The first step in developing a FLC is to determine which conditions occurring in the physical system will be important in choosing an effective control action. These "condition variables" are commonly the "state variables" associated with the physical system. Any number of condition variables may be considered, but the more condition variables considered, the larger the required rule set. The condition variables are, simply stated, the variables used to decide upon control actions. Also, as a part of this first step in FLC development, the membership functions defining the fuzzy linguistic terms used to describe the condition variables must be specified.

The linguistic terms used in FLCs are commonly vague; terms are used that mean different things to different people. Therefore, some mechanism must exist for describing the conditions in the system with linguistic terms. This mechanism is the fuzzy membership function. Fuzzy membership functions allow the precise numeric values of the condition variables in the physical system to be described by linguistic terms. Actually, fuzzy membership functions can be thought of as approximations to the confidence with which a precise numeric value is described by a linguistic term, and fuzzy membership function values (μ) are numeric representations of these confidences. When a fuzzy membership function has a value of $\mu = 1$, there is maximum confidence that the precise numeric value is being accurately described by the linguistic term. Figure 1 shows the commonly used Gaussian membership function form.

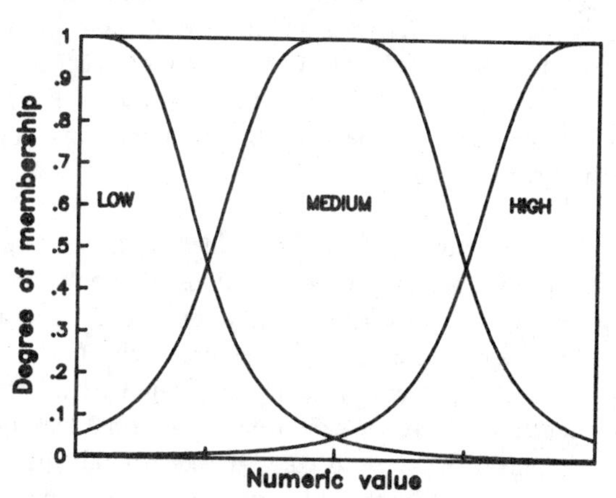

Figure 1: Guassian Membership Functions

Once the condition variables have been chosen, the second step in FLC development is to identify the "action variables," and describe them with fuzzy membership functions. The action variables represent control actions which can be taken to bring about a change in the physical system. Unlike the number of condition variables, the number of action variables chosen has no effect on the size of the rule set because the choice of actions is assigned by the developer.

The use of linguistic terms facilitates the third step in FLC development which is the production of the rule set. A FLC's rule set is formed by writing a rule for every possible condition that could exist in the physical system as described by the fuzzy membership functions set down for the condition variables (the rule set grows with the number of linguistic terms used to describe the condition variables). Each individual rule is of the form:

IF [c_1 IS C_1 AND c_2 IS C_2 AND ...]
THEN [a_1 IS A_1 AND a_2 IS A_2 AND ...],

where c_i is a condition variable, a_i is an action variable, C_i is a linguistic term describing the condition variable, and A_i is a linguistic term describing the action variable. To clarify, an example rule follows:

IF [(c_1 IS HIGH) AND (c_2 IS LOW)]
THEN [(a_1 IS VERY LARGE)].

Although determining an action to be associated with all of the possible conditions existing in the physical system as described by the fuzzy membership functions (writing the rule set) seems at first to be an imposing task, the incorporation of fuzzy linguistic terms into the rules makes their development much easier than the development of rules for conventional rule-based systems. This is because the rules are now more akin to the rules-of-thumb humans are comfortable using.

The above procedure has provided a means for converting a precise set of conditions existing in the physical system to a set of "fuzzy conditions," and a set of fuzzy rules has been developed each of which prescribes a "fuzzy action" associated with its particular set of fuzzy conditions. The final task in developing a FLC is to convert each of the fuzzy actions provided by the individual rules into a single, precise set of actions to be taken on the physical system. Larkin [8] found that a center of area (COA) procedure is an efficient method for determining these precise actions. In the COA method, the fuzzy membership functions describing the control actions are used in a weighted summing procedure to find one precise action. The scheme for finding the weighted sum can be easily thought of in graphical terms as seen in Figure 2. The membership function for the fuzzy action associated with each rule is plotted with a height equal to the minimum confidence (degree of membership) present in the condition portion of the rule. The single "crisp" control action to be applied to the physical system is the single value defined by the center of area of the scaled membership functions plotted for all of the rules. The rules in which one has the greatest confidence produce shapes with the largest areas, and thus have the greatest effect on the selected action.

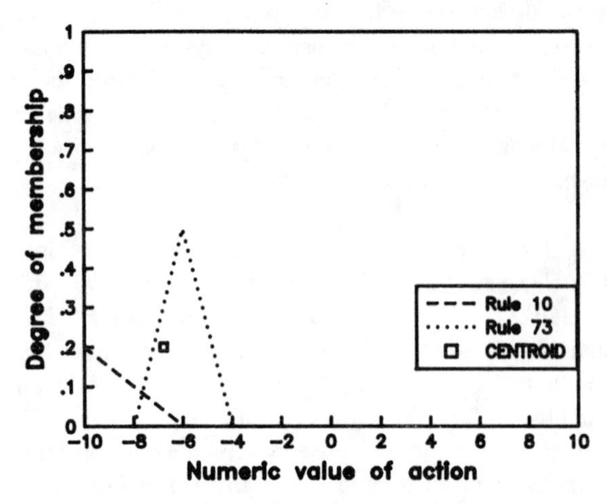

Figure 2: The Center of Area Method

This has been a brief description of one approach to FLC development. It is a relatively simple chore to produce a FLC that controls a particular system. However, to produce an efficient controller, the FLC must be properly designed. The rules must prescribe logical actions to be taken under all conditions, and the membership functions must be properly specified to ensure efficient FLC performance. The next section of this paper describes the use of a SGA for learning membership functions in a non-adaptive cart-pole FLC. A following section describes the use of a MGA for learning membership functions in an adaptive cart-pole FLC.

3 DESIGN OF A NON-ADAPTIVE CART-POLE FLC USING A SGA

In this section, the SGA detailed by Goldberg [4] learns membership functions that improve the efficiency of a FLC designed to control a cart-pole balancing system. The membership functions learned by the SGA produce a FLC that is more efficient than the FLC that uses membership functions conceived by the author. Before results are presented demonstrating the ability of a SGA to learn high-performance membership functions, the physical system is described and some details of the GA-FLC application are addressed.

The problem of interest in this application is the control of a cart-pole balancing system. A cart is free to move along a one-dimensional track while a pole is free to rotate only in the vertical plane of the cart and track. A multivalued force, F, can be applied at discrete time intervals in either direction to the center of mass of the cart. A schematic of the physical system is shown in Figure 3. The objective of the control problem is to apply forces to the cart until it is stationary at the center of the track and the pole is balanced in a vertical position. This task of centering a cart on a track while balancing a pole is often used as an example of the inherently unstable, multiple-output, dynamic systems present in many balancing situations, e.g., two-legged walking and the aiming of a rocket thruster.

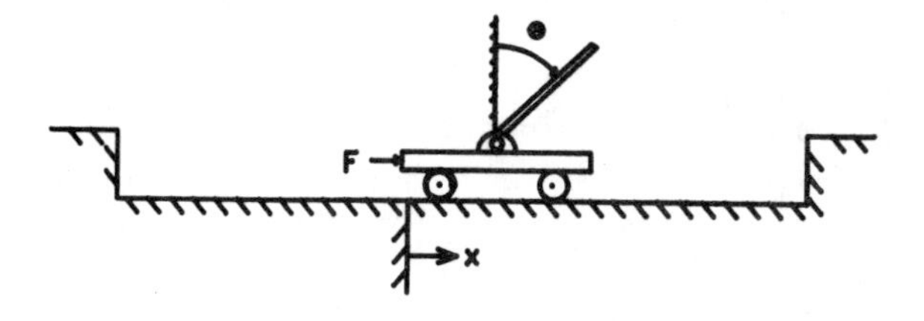

Figure 3: The Cart-pole Balancing System

The state of the cart-pole system at any time is described by four real-valued variables:

x = position of the cart,

$\dot{x}$ = linear velocity of the cart,

θ = angle of the pole with the vertical,

$\dot{\theta}$ = angular velocity of the pole.

The system is simulated on a computer by solving a set of ordinary differential equations [2]. The solution of the equations of motion is approximated using Euler's method with a time step of 0.02 seconds. This time step struck a balance between the accuracy of the solution and the computational time required to find the solution. As will be seen later, the time required for a simulation is an important consideration when using a GA to design a FLC because the GA's fitness function depends heavily on the time required to run a simulation.

A FLC was designed by the author to control the cart-pole system. This FLC used the four state variables listed above as its condition variables. Only one action variable was appropriate, the force applied to the cart, F. Three linguistic terms were used to characterize each of the four condition variables. Eighty-one rules were required because there were $3^4 = 81$ possible combinations of the linguistic terms used to describe the condition variables. Thus, one of seven linguistic terms used to characterize the action variable were assigned to the condition portion of each rule. Figure 4 shows the basic form of the fuzzy membership functions used to define the linguistic terms.

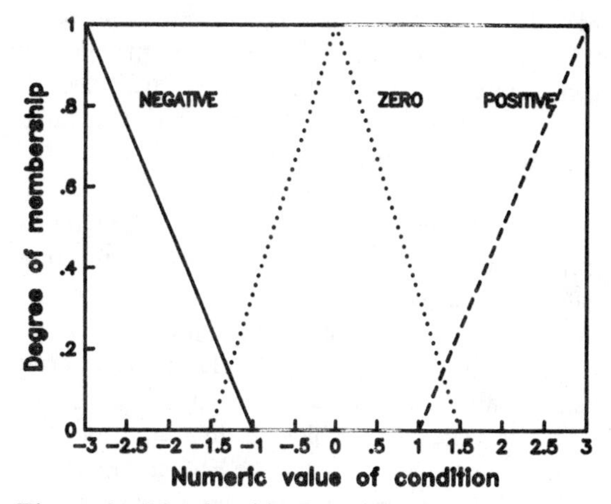

Figure 4: Membership Functions for Condition Variables

The AD-FLC was able to use the 81 rules and the associated membership functions to center the cart and balance the pole (see Figure 5). The membership functions were defined to represent the author's conception of the linguistic terms, and the use of these linguistic terms made the formation of the rule set a straightforward task. However, the fuzzy membership functions selected by the author through a trial-and-error procedure did not provide optimal FLC performance. In fact, a SGA was able to demonstrate that these author-developed membership functions were not optimal by learning membership functions that improved the performance of the cart-pole FLC.

Applying a SGA to the problem of learning membership functions is not difficult. When applying a three-operator GA to a search problem there are two basic decisions to be made: (1) how to code the possible solutions to the problem as finite bit strings and (2) how to evaluate the merit of each string.

Consider first the coding issue. Conventional concatenated, mapped, unsigned binary coding proves to be well suited for representing membership functions as finite bit strings. The desired parameters in this search problem were the anchor points that located the triangles defining the linguistic variables. The SGA was used to move and to expand or shrink the base of each triangle. The "extreme" linguistic variables, like "NEGATIVE" and "POSITIVE BIG," required the definition of only one anchor point since one side of the triangle was fixed at the limiting values of the variables. Two anchor points, the points defining the location of the triangles, were required to describe the "interior" triangles defining linguistic variables, like "ZERO" and "NEGATIVE SMALL." Only two anchor points were required for the interior triangles because their membership functions were constrained to isosceles triangles. In this manner, the design of the cart-pole FLC required the selection of 15 parameters.

Now, what about the second decision? How are the strings describing the potential membership functions evaluated? In judging the performance of the cart-pole FLC, it is important for the controller to center the cart, balance the pole, and hold this equilibrium position, all in the shortest time possible. These objectives can be achieved by adjusting the membership functions so that the FLC minimizes the squared difference between the cart position and the center of the track, summed over some finite time, while keeping the pole balanced (requiring the current value of θ to be accounted for in the fitness function). Mathematically this objective is described by the following equation for the fitness function:

$$f = \sum_{i=case\ 1}^{case\ 4} \left(\sum_{j=0\ sec}^{30\ sec} w_1 x_{ij}^2 + w_2 \theta_{ij}^2 \right)$$

where f is the fitness function and w_1 and w_2 are weighting constants. This fitness function offers a means for evaluating the performance of a cart-pole FLC using different fuzzy membership functions. For a given set of membership functions, a squared error was calculated for a particular set of initial conditions with the intent of using a GA to minimize this error term. So that a general purpose set of membership functions would be developed, four different sets of initial conditions were considered in the evaluation of each bit string. The four sets of initial conditions were chosen to ensure that the FLC could center the cart over a spectrum of system states.

One hundred, thirty-two bit long, strings were used by the SGA to learn efficient membership functions. After having viewed only a small portion of the search space (approximately 32000 of the $2^{132} = 5.44*10^{39}$ possible points), the SGA was able to learn membership functions that provided for much better control than those defined by the author. These results are summarized in Figure 5. For the set of initial conditions presented, the SGA-FLC controlled the cart-pole system more efficiently than did the AD-FLC. It is worthy to note that the set of initial conditions represented in Figure 5 was not included in the definition of the fitness function used by the SGA to learn the membership functions. This is a testimony to the robustness of the technique because a GA has improved the performance of the controller for a situation about which it had no prior information.

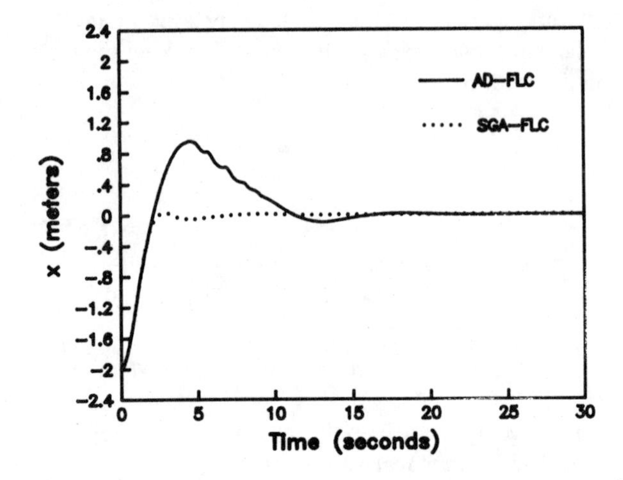

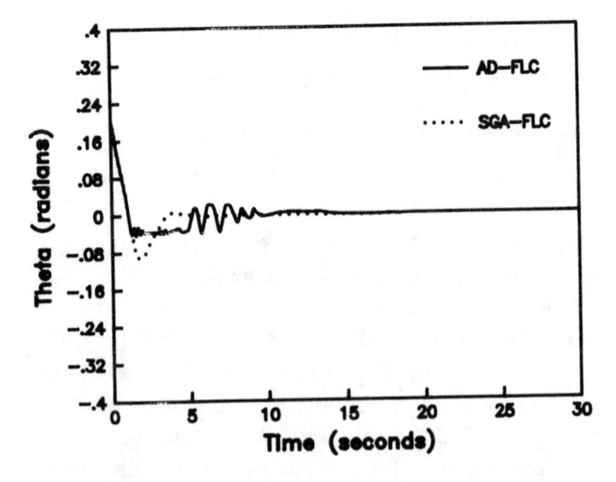

Figure 5: (a) Cart Position (b) Pole Angle

This section has provided the reader with the details of a technique for combining GAs and FLCs. A non-adaptive cart-pole FLC has been designed using a SGA. The SGA-FLC consistently outperforms the AD-FLC across a spectrum of initial conditions. In the next section, the basic control problem is made more complex. The mass of the cart is considered to be time-variant, thus necessitating an adaptive FLC to adequately manipulate the physical system. The alteration of membership functions in real-time requires a GA that converges even faster than the SGA. Therefore, a MGA [6,7] is employed.

4 DESIGN OF AN ADAPTIVE CART-POLE FLC USING A MGA

Consider a problem that is significantly more difficult than the problem addressed in the last section; the problem in which a characteristic parameter of the physical system, specifically the mass of the cart, changes with time. Figure 6 shows the specific mass-time profile to be considered. For this time-variant physical system, a satisfactory controller must be adaptive; it must be able to discern that changes have occurred, analyze those changes, and take an adequate action in real-time or the physical system will have actually changed characteristics and quite possibly have failed. To make the problem even more difficult, the controller does not include a mechanism in its rule set for considering changes in cart mass; it is only able to use information it ascertains from its past performance to even realize the mass has changed. Figure 7 shows a schematic of an adaptive controller. Fortunately, as will be demonstrated in this section, a MGA is well suited for the task of designing an adaptive FLC.

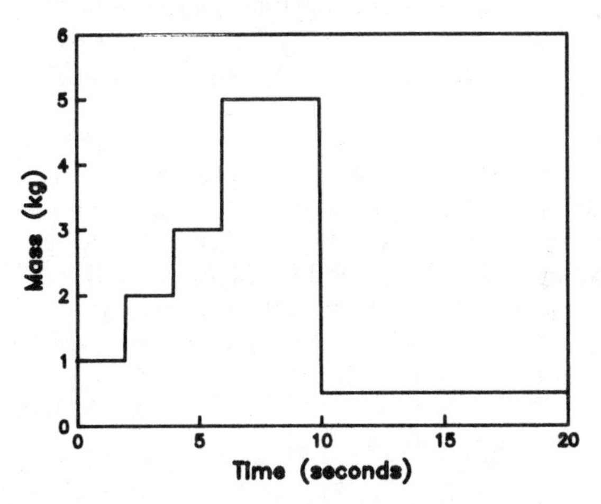

Figure 6: Changing Cart Mass

Historically, the rule-of-thumb for GA population sizes has been "the bigger the better." However, small population GAs have been theoretically shown to be more effective [5], and have proven to be efficient in non-stationary optimization problems [7]. A particular small population GA, a MGA, is well suited for the task of learning high-efficiency membership functions for an adaptive FLC in real-time.

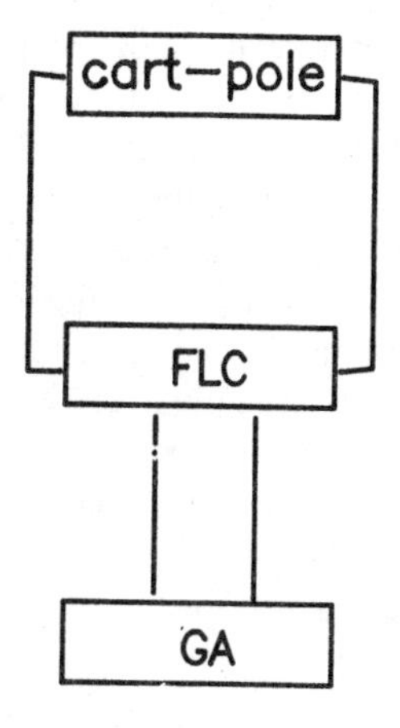

Figure 7: Adaptive Controller Using GA

When applying a MGA, a user must address the same two issues faced when applying a SGA: (1) coding the possible solutions to the problem as finite bit strings and (2) evaluating the merit of each string. The fact that the mass of the cart is no longer constant does not disrupt the mechanisms introduced in the last section for addressing either of these issues. Again, a concatenated, mapped, unsigned binary coding is appropriate for representing the membership functions as bit strings. As for the issue of evaluating the merit of the strings, the basic form of the fitness function employed to design the non-adaptive FLC presented in the last section can again be used. In this example, it is unnecessary to consider four sets of initial conditions; only the conditions at the current time must be considered. However, the duration of the simulations now becomes even more important. If the simulations utilized in the definition of the fitness function require too much time, the mass of the cart is liable to have changed, and the MGA will no longer be working on the correct problem. This fact must be weighed against the knowledge that the quality of the membership functions selected by the MGA improves with longer simulations. The simulations used in the fitness function evaluations for this study lasted 10 seconds, as compared to 30 seconds used by the fitness functions for the SGA.

The effectiveness of the MGA in this problem environment is demonstrated by the results presented in Figure 8. For point of comparison, an adaptive

FLC was developed using a SGA. Figure 8 presents a comparison of the performance of the original non-adaptive AD-FLC, the adaptive SGA-FLC, and the adaptive MGA-FLC, in the time-variant cart-pole environment. Notice that the non-adaptive AD-FLC was unable to maintain control of the system: the controller was unable to factor in the changing cart mass to maintain balance of the pole. The adaptive SGA-FLC was able to control the cart-pole balancing system but, its performance was deemed unacceptable; it could not center the cart quickly enough. On the other hand, when compared to the other two FLCs, the adaptive MGA-FLC did a remarkable job of controlling the cart-pole balancing system: it centered the cart and balanced the pole well before the other two systems. These results are especially impressive when one recalls the fact that the controllers had no way of knowing the exact value of the cart mass at any given time. The adaptive FLCs were able only to ascertain that their performances were declining, and that they must therefore make changes in the membership functions.

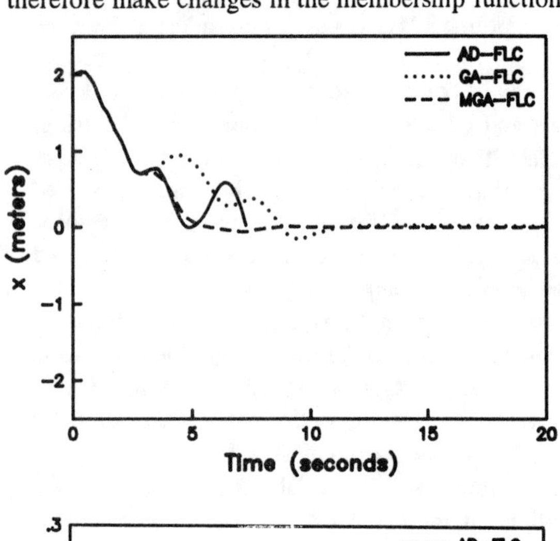

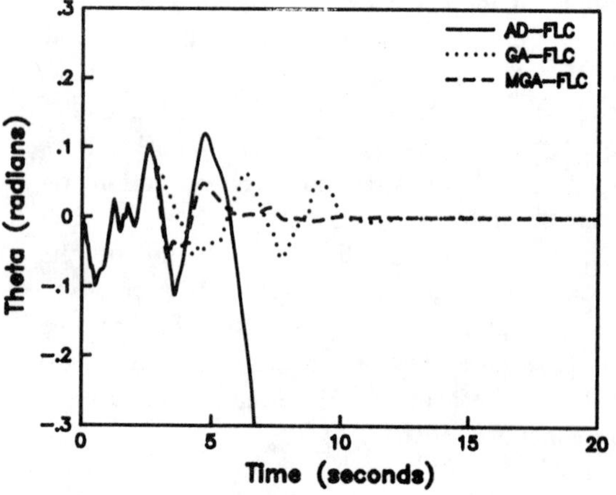

Figure 8: (a) Cart Position (b) Pole Angle

5 CONCLUSIONS

A technique was established and discussed in which GAs are used to design FLCs. Initially the mechanics of a FLC were described. Next, the technique combining GAs and FLCs was applied to two FLCs used to control a cart-pole balancing system. The performance of the FLCs, one non-adaptive and one adaptive, was enhanced via the use of GAs to select fuzzy membership functions.

The non-adaptive cart-pole FLC designed with a SGA outperformed an AD-FLC for the case in which all of the characteristic parameters of the cart-pole system remained constant. This example demonstrates the fact that GAs are potentially powerful tools for designing FLCs. This is significant because FLCs are becoming increasingly popular in the area of process control. And, the use of GAs for designing FLCs offers the potential for improving FLC performance and decreasing FLC development time.

Next, a MGA was used to design an adaptive cart-pole FLC. This adaptive MGA-FLC was capable of altering its membership functions in real-time to account for changes in the cart mass despite the fact that it received no feedback information on the magnitude of the changes in mass. The adaptive MGA-FLC outperformed both a non-adaptive AD-FLC and an adaptive SGA-FLC. The use of a GA to produce an adaptive FLC is quite significant. It opens the door to attaining the smooth control capabilities and efficient performance generally associated with FLC use in rapidly changing environments.

GAs improved the performance of two FLCs. The search problems associated with the design of these two FLCs presented the GA with different challenges. Because of the dramatic improvements in FLC performance in both of the applications, it is concluded that GAs are robust and are a powerful tool in FLC design.

Acknowledgements

The author would like to acknowledge Ed Gentry, a Computer Clerk at the U.S. Bureau of Mines' Tuscaloosa Research Center. Mr. Gentry did the computer programming associated with this research and was a valuable source of ideas.

References

[1] C. W. Anderson. Strategy learning with multilayer connectionist representations. *Proc. of the Fourth Intl. Workshop on Machine Learning*, 1987.

[2] A. G. Barto, R. S. Sutton, and C. W. Anderson. Neuronlike adaptive elements that can solve difficult learning control problems. *IEEE Trans. on Systems, Man, and Cybernetics*, 13, 1983.

[3] Y. Chen. Stability analysis of fuzzy control--A Lyapunov approach. *Proc. of the IEEE Intl. Conf. on Systems, Man, and Cybernetics*, 1987.

[4] D. E. Goldberg. *Genetic algorithms in search, optimization, and machine learning*. Addison-Wesley, Reading, MA, 1989.

[5] D. E. Goldberg. Sizing populations for serial and parallel genetic algorithms. In J. J. Grefenstette, editor, *Proc. Intl. Conf. on Genetic Algorithms and Their Applications*, 1978.

[6] C. L. Karr. Analysis and optimization of an air-injected hydrocyclone. Doctoral dissertation, The University of Alabama, Tuscaloosa, AL, 1989.

[7] K. S. Krishnakumar. Micro genetic algorithms for stationary and nonstationary function optimization. *Proc. of the SPIE Intelligent Control and Adaptive Systems Conf.*, 1989.

[8] L. I. Larkin. A fuzzy logic controller for aircraft flight control. In M. Sugeno, editor, *Industrial Applications of Fuzzy Control*, pages 87-104. North-Holland, Amsterdam, 1985.

[9] M. O. Odetayo, and D. R. McGregor. Genetic algorithm for inducing control rules for a dynamic system. In J. J. Grefenstette, editor, *Proc. Intl. Conf. on Genetic Algorithms and Their Applications*, 1987.

[10] T. J. Procyk, and E. H. Mamdani. A linguistic self-organising process controller, *Automatica*, 15, 1978.

[11] M. Sugeno, editor. *Industrial applications of fuzzy control*, Elsevier, Amsterdam, 1985.

[12] L. A. Zadeh. Outline of a new approach to the analysis of complex systems and decision processes. *IEEE Trans. on Systems, Man and Cybernetics*, 3, 1973.

A Parallel Genetic Algorithm for Network-Diagram Layout

Corey Kosak   Joe Marks   Stuart Shieber
Division of Applied Sciences
Harvard University
Cambridge, MA 02138

Abstract

Network-diagram layout (NDL) is one of the most difficult problems encountered in automating the design of informational graphics. We consider generalized network-diagram layout (GNDL), a more general version of the standard NDL problem in which aesthetic considerations are subordinated to perceptual-organization requirements. GNDL is computationally intractable, and existing heuristic methods for specialized instances of NDL cannot be adapted to this problem. We present here a genetic algorithm for GNDL that runs in parallel on a Connection Machine. The genetic algorithm is tailored to both the GNDL problem and the SIMD architecture of the Connection Machine. It has proven to be a robust heuristic method for GNDL. We utilize the algorithm in the ANDD+ system, a system for fully automating the design of network diagrams. We also describe several experiments designed to improve our understanding of the algorithm and to tune its parameters.

1 INTRODUCTION

The *network diagram* is one of the standard graphical forms used for information presentation.[1] Many network diagrams continue to be designed and drawn essentially by hand, though computers are now often used to automate some of the tasks involved in network-diagram design. In this paper we consider *network-diagram layout* (NDL), the most demanding subtask in the design of network diagrams.

Network diagram layout is the process of determining, subject to certain constraints, the two-dimensional positions of the symbols in a network diagram. Primary among these are constraints of *syntactic validity*; node symbols must not overlap, and link symbols must not cross through node symbols. In addition, criteria of *aesthetics* may be imposed, global syntactic criteria such as minimization of the number of link crossings, the sum of link lengths, or the total area; maximization of diagram symmetry; or evenness of the spatial distribution of nodes or the link lengths.

Many layout algorithms have been reported in the literature; Eades and Tammasia [3] provide an extensive bibliography. Much of the previous research on NDL has concentrated on special cases of the problem. For example, specialized algorithms have been developed for network diagrams that have tree-like, acyclic, or planar topologies. Furthermore, all previously reported algorithms focus on the problem of *aesthetic layout*—layout relative to syntactic validity constraints and various aesthetic criteria, as described above.

Our approach to NDL differs from most previous research in two respects. First, our algorithm computes layouts for all kinds of network diagrams, regardless of their topology. Second, we emphasize the role of *perceptual organization* [8, 6] in diagram layout as a new source of constraints on the layout task. Graphic designers and perceptual psychologists have long recognized the utility of perceptual organization in the design of effective informational graphics [1, 9, 13]. The most important kinds of layout-related perceptual groupings for network diagrams are those that group nodes by proximity, by "good continuation" (e.g., evenly-spaced alignment), and by closure (e.g., the use of enclosure boxes to surround and group nodes). We use the term *generalized network-diagram layout* (GNDL) to refer to a more general version of the NDL problem in which aesthetic considerations are subordinated to perceptual-organization requirements. The algorithm we describe is for the GNDL problem.

Our layout algorithm takes as input not only a specification of the nodes and links that comprise a network diagram, but also a set of *visual-organization features* (VOFs) that specify a desired perceptual organization [12]. Various kinds of perceptual grouping can be described in terms of VOFs. Some of the VOFs supported by our algorithm are:

- *Zoning*—nodes in the same zone are positioned within some rectangular region of the display from which other nodes are excluded;

[1] Network diagrams are also called node-link diagrams, circle-and-arrow diagrams, charts, or graphs. *Chart graphics* (e.g., bar charts and pie charts) and *maps* are the other major graphical forms for information presentation [1].

- *Clustering*—nodes in the same cluster are positioned in mutual spatial proximity;
- *Sequential Layout*—nodes are laid out horizontally or vertically, with an option for even spacing along the layout axis.
- *Alignment*—nodes are laid out in horizontal or vertical alignment, with an option for even spacing along the layout axis.
- *Symmetry*—nodes are positioned so that they have a horizontal or vertical axis of symmetry;
- *T-Shape*—nodes are laid out in a 'T' shape (as, e.g., a parent and its immediate children in a tree);
- *Hub-Shape*—nodes are laid out evenly along the perimeter of a circle, with a specified node at the center of the circle.

(An exhaustive list is given by Marks [12].) We restrict our attention to layouts that embed nodes in an integer grid and that have single-segment links. The GNDL problem is the problem of computing a layout that (i) is syntactically valid, (ii) exhibits the desired perceptual organization expressed as a set of VOFs, and (iii) is aesthetically optimal (according to some or all of the aesthetic criteria mentioned above). Further, this *priority ordering* of design considerations should be respected by any layout algorithm. For instance, degrading the perceptual organization in order to improve aesthetics should be avoided.

Like many less general NDL tasks, GNDL is computationally intractable: it is NP-complete [4, 12], and the solution space is enormous (a 15-node layout can be embedded in more than 10^{39} different ways in a 25×25 grid). Existing heuristic methods for aesthetic layout [3] cannot be adapted easily to the GNDL problem. In our first attempt at developing a satisfactory heuristic method for GNDL we took a rule-based approach [12]. Although fast, the rule-based method was not sufficiently robust. Our search for a more powerful heuristic method led us to consider stochastic optimization techniques. Initial experiments with an algorithm based on simulated annealing [7] were encouraging,[2] but we finally settled on a genetic algorithm because of this method's greater potential for parallelization.

We begin with a description of our algorithm. We utilize this algorithm in the ANDD+ system, which designs all aspects of a network diagram (not just the layout of nodes) automatically [13, 11, 12]. We next describe our experiences with the algorithm and the results of several experiments we have performed. We conclude by summarizing our plans for future work.

[2] Davidson and Harel [2] have developed a simulated-annealing algorithm for the standard NDL problem independently. We are indebted to David Johnson for bringing this paper to our attention.

2 A GA FOR GNDL

Development of a genetic algorithm for the GNDL problem requires only that layouts be encodable in a suitable data structure—a bit string encoding, for instance—and that they be evaluable by a function that ranks layouts as to their desirability. The familiar general form for such an algorithm is as follows: A set of layout encodings are randomly chosen. Iteratively, elements of this set are then subject to mutation (by altering bits in the encoding), crossover (by swapping bit substrings with another element of the set), and reproduction according to a schedule dependent on the relative quality of the layout encoded.

The key factor in enabling such a genetic algorithm to solve a layout problem is the ability to evaluate layouts according to their syntactic validity, perceptual organization, and aesthetics. In order for the genetic algorithm to effect an optimization process, the evaluation metric should vary continuously in such a way that the evaluation improves as the various factors become more closely satisfied. The evaluation function that we use is given by a weighted sum of penalties imposed for deviation from satisfaction of the factors; the relative weights force the priority ordering of design considerations described earlier. For syntactic validity criteria, a penalty accrues to each node overlap, each node cut by a link, and so forth. However, computing suitable penalties for some of the VOFs is considerably more complex. A measure of the degree of misalignment of a set of nodes subject to an alignment VOF is computed by calculating the difference in either the x or the y coordinates (depending on the specified alignment axis) of every pair of nodes in the set, and then summing those differences. To measure the degree of asymmetry of a set of nodes, the location of the requested axis of symmetry is computed from the rectangular extent of the nodes. Every node in the set is reflected around this axis, and the distances from these positions to the nearest (unreflected) nodes in the set are found. The asymmetry measure is then the sum of these distances. In a diagram exhibiting the requested symmetries there will already be nodes in all of the reflected positions; thus (as would be expected) the asymmetry measure is zero. Other VOFs are evaluated similarly. In this way, the evaluation tends to decrease as sets of nodes come to satisfy the appropriate VOFs. Finally, terms proportional to the area of the layout or the number of link crossings can be included in the evaluation formula so as to promote the tertiary considerations of global aesthetics.

Given an evaluation function such as this to optimize, we can envision solving the optimization by a stochastic process such as a genetic algorithm. However, we have modified the standard formulation of genetic algorithms in order to improve the efficiency of the search process by matching the genetic algorithm more closely both to a particular parallel computational architecture and to the problem being solved.

2.1 THE GA AND SIMD MACHINES

The paradigm of genetic algorithms seems appropriate for parallelization on multiprocessor computers (as described, for instance, by Hillis [5] or Mühlenbein [14]), which holds the promise of dramatic improvements in speed by dispersing the population among the processors. In order to take advantage of this possibility, we chose to implement the layout genetic algorithm on the Connection Machine, a parallel computer utilizing a SIMD architecture (single instruction, multiple data). In such an architecture, all processors are simultaneously executing the same instruction (though they may execute it on different data, or may idle for a portion of the execution). Therefore, the algorithm was modified to reflect this synchrony of operation. At each step, all of the processors are performing a mutation step, or a crossover step, or a reproduction. However, individual processors may perform the step in a different way, or, in the case of crossover and mutation, may even idle, skipping the opportunity to perform the step.

Another property of this parallel architecture (and other ones) is that communication among processors is, in general, very expensive. A general operation of crossover between layouts stored at arbitrary processors would consequently be extremely expensive. In particular cases, however, the cost of communication is reduced, namely when the communication is local with respect to the topological connectivity of the processors. For this reason, we restrict genetic operations to occur only among layouts in adjacent processors. The topology of the processor connections was chosen to be a two-dimensional, toroidally connected, square grid. As the machine has 4096 processors, the grid is 64 by 64 processors.[3]

In a single generation, a layout may be reproduced, then, only within its local neighborhood of nine processors. During the reproduction phase, a processor replaces its layout with that of one of the nine layouts in its neighborhood; a reproduction schedule specifies probabilities of replacement with the best of the nine, the second best, and so on. Similarly, crossover works by importing from a randomly chosen neighbor a portion of its layout. Note that crossover is not symmetric under this scheme. If a processor contributes information about the positions of a certain set of nodes to a neighboring processor, the neighbor does not necessarily contribute its specification of those same nodes to the first processor. Again, this asymmetric crossover allows for more uniform operation of the processors, thereby facilitating SIMD implementation.

There are two disadvantages to localization of genetic operations in this way. First, there is no advantage given to the globally best layouts. In particular, the single best layout derived in a particular generation may well be lost by the next generation before it has a chance to propagate. Mühlenbein [14] proposes adding the globally best solution as a participant in the reproduction process in order to prevent it from getting lost. We have used a different method, of reseeding the globally best solution in the processor that generated it if it would otherwise get lost. (We have just begun experimenting with Mühlenbein's method.)

The second disadvantage is that after several generations, reproduction will have created local "colonies" of layouts that share similar structure. Crossover within a colony is likely to be ineffective, as it is not combining different genetic material; the system comes to be incestuous. (A similar problem is noted by Hillis [5].) To remedy this problem, we allow for a phase in which a certain parametrically determined percentage of the processors send their layouts to other randomly chosen processors. This "scatter" process has the effect of distributing genetic material more uniformly throughout the processor topology, thereby mitigating the effect of incest. As scattering requires global communication, it is more expensive than the local operations, but it can involve fewer processors and still perform well at eliminating colony formation.

2.2 THE GA AND THE GNDL PROBLEM

Finally, in order to accelerate the convergence process, we do not restrict the genetic algorithm to the "pure" variant described initially, using bit string encodings and genetic operations. The mutation and crossover operations take advantage of the structure of the problem by being sensitive to the groupings of nodes as manifest in the VOFs. For instance, a mutation oper-

[3] Since the Connection Machine allows for the simulation of multiple processors by a single processor, larger topologies can be run on the same machine. We have, however, not used this method in our experiments.

ation may modify the location of a single node directly by adding a vector offset to its position, and not just through the intermediary of modifying its bit string encoding. A mutation may also uniformly move a *set* of nodes that participate in a single VOF constraint. Similarly, crossover replaces the locations of nodes or sets of nodes that participate in constraints. Another problem-specific modification is the use of a variant of the mutation operator that is restricted to move nodes only small distances.

To enable these kinds of problem-specific genetic operations, the encoding of layouts is as a structured object, rather than a simple bit string. A layout is encoded as a vector of x and y coordinates in the bounded plane. The genetic operations, then, operate directly on this vector, updating coordinates of nodes or sets of nodes. To allow this, the processors must have access to information about the problem structure, namely which sets of nodes participate in constraints. Similar ideas have been widely used to improve the rate of convergence of genetic algorithms, for instance, the use of λ-opt for mutation in solving the traveling salesman problem [10, 14]. Here also, a vector of positions is used as the encoding, and problem structure, in the form of connectivity information, is utilized by the genetic operators.

2.3 SUMMARY OF THE ALGORITHM

In summary, the algorithm can be described at a high level by the following pseudo-code.

```
parallel {initialization}
repeat
    parallel
    {   evaluation;
        reproduction;
        scatter?;
        crossover?;
        mutation?;
        find-best        }
until the algorithm has converged
return the best layout found
```

The *initialization* phase assigns a random layout at each processor. The *evaluation* process computes the evaluation function at each processor in parallel. The evaluation is used to govern the *reproduction* phase, in which the i-th best layout of a processor in the neighborhood is copied into the processor, where i is determined stochastically based on a reproduction schedule. The next three phases are each executed in a random subset of the processors (the probability being parametrically determined), as indicated by the '?' suffix. *Scatter* distributes a certain percentage of the layouts to other processors in order to prevent incest. *Crossover* copies from a neighboring processor locations of a node or set of nodes participating in a constraint. *Mutation* randomly translates a node or set of nodes participating in a constraint. Finally, a *find-best* phase determines and stores the best layout in the set so that it may be maintained across generations. If after a generation all processors have layouts inferior to the previous best, it can be seeded into a subset of the processors so that its genetic material is not lost. Alternatively, the best layout can be broadcast to all of the processors so that they may include it as an option during the next reproduction phase, as described earlier.

A final issue concerns deciding when the algorithm has converged. For this purpose, we keep running averages of improvement in the globally best layout (another reason to extract this information) over a multiple generation period (currently 50). We stop when this running average drops below a threshold.

3 EXPERIENCE AND RESULTS

In initial experimentation, we have used the genetic algorithm to compute layouts for more than 30 different network diagrams. We have gathered extensive performance statistics for two representative diagrams from the thirty. Example layouts for each of them are shown in Figures 1 and 2.[4] These diagrams evolved over 128 and 105 generations, respectively, with each generation comprising 4,096 individual layouts. On a 4,096-processor Connection Machine a new generation for either of these diagrams is generated in less than a second.

The diagram in Figure 1 exhibits zoning, clustering, and symmetry VOFs. The diagram in Figure 2 exhibits different VOFs: evenly spaced sequential layout and evenly spaced alignment. The following layout aesthetics were considered for both diagrams: number of link crossings, diagram area, and global symmetry.

Our purpose in gathering performance statistics for a representative selection of network diagrams is to identify optimal values for the various genetic-algorithm parameters. So far the only performance measure we have considered is quality of the final generated layout as measured by its evaluation. The number of generations required for a layout to evolve is another important performance measure, but we have assigned a sec-

[4] These network diagrams communicate the same information conveyed in a diagram designed by Sanden [15].

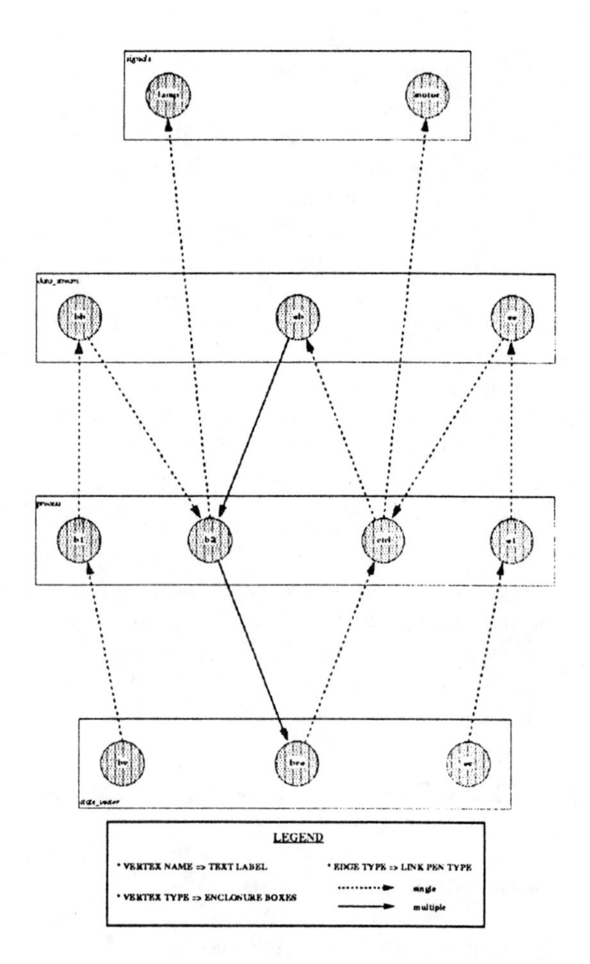

Figure 1: A layout computed by the GNDL GA.

ondary priority to this measure because convergence is relatively rapid. For instance, in the two example diagrams, convergence occurs almost universally within 200 generations.

In our first set of experiments we varied the parameters affecting reproduction, mutation, crossover, and scatter, and recorded the final layout evaluation for a single run at each distinct set of parameter values. These very preliminary experiments provided the following results: crossover and scatter had no noticeable effect on final layout evaluation; reproduction and mutation parameters were correlated weakly to final layout evaluation; and substantial stochastic variation existed in the computed layout evaluations.

On the basis of these initial results, we decided to run a more systematic set of experiments with scatter disabled. In order to reduce the degree of stochastic variation in final layout evaluation, each setting was run three times and the final evaluations averaged. The results of this second round of experiments for the network diagram shown in Figure 1 are illustrated in

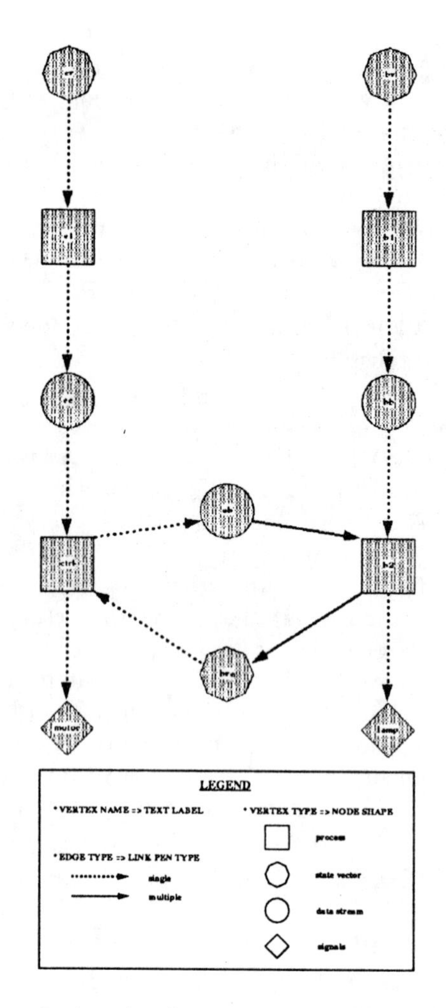

Figure 2: Another layout computed by the GA.

Figure 3. (Results for the other diagram were similar.) The x and y axes in this chart represent the mutation and reproduction parameters, respectively, expressed as percentages. The z axis (shown as the horizontal axis in the small bar charts) represents the crossover parameter, again expressed as a percentage. The length of the bars is a linear function of layout evaluation.[5] The layouts depicted in Figures 1 and 2 were generated with our standard settings (based

[5]The reproduction parameter indicates the probability of copying the best layout in a neighborhood. Given a reproduction parameter R, the probability $P(n)$ (for $1 \leq n \leq 9$) of choosing the n-th best layout is

$$P(n) = \begin{cases} min\left(\frac{1-\sum_{i=1}^{n-1} P(i)}{2}, P(n-1)\right) & \text{if } n > 1 \\ R & \text{if } n = 1 \end{cases}$$

The mutation parameter indicates the probability of a layout mutating from one generation to the next. The crossover parameter indicates the probability of a layout importing genetic material from a neighbor from one generation to the next.

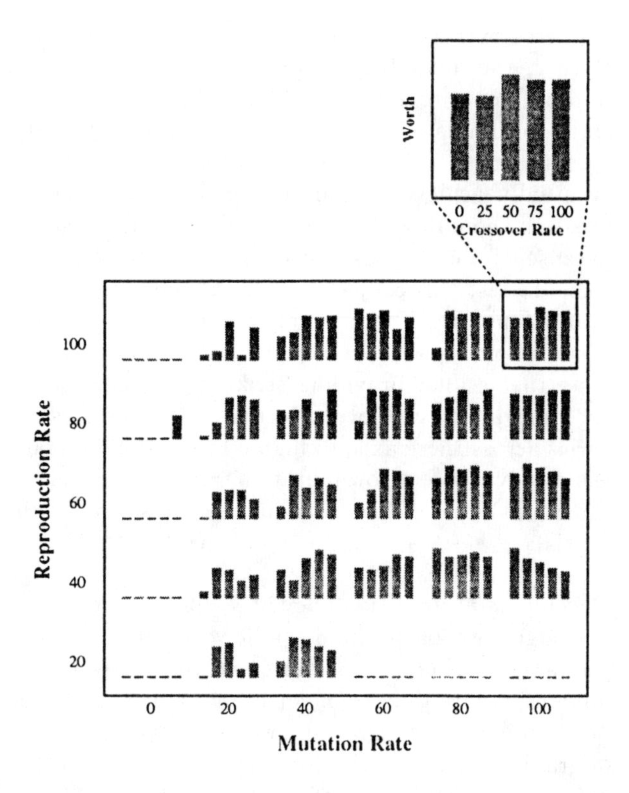

Figure 3: Layout evaluation at convergence for diagram of Figure 1 as a function of the reproduction, mutation, and crossover parameters.

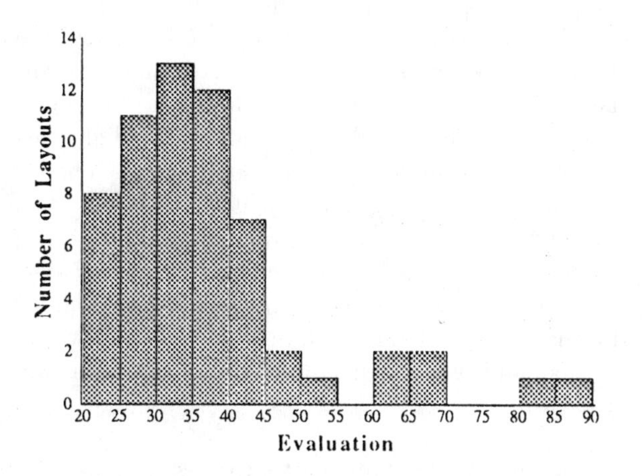

Figure 4: Histogram of 60 runs of the GA.

on these experiments) with reproduction .80, crossover .50, and mutation .80.

As an indication of the quality of the inferior parameter settings, Figure 5 shows a layout generated by one of the worst settings as determined by the data of Figure 3 (reproduction, .60; crossover, .25; mutation, 0; convergence at generation 145). Although a syntactically valid diagram has been generated, and some

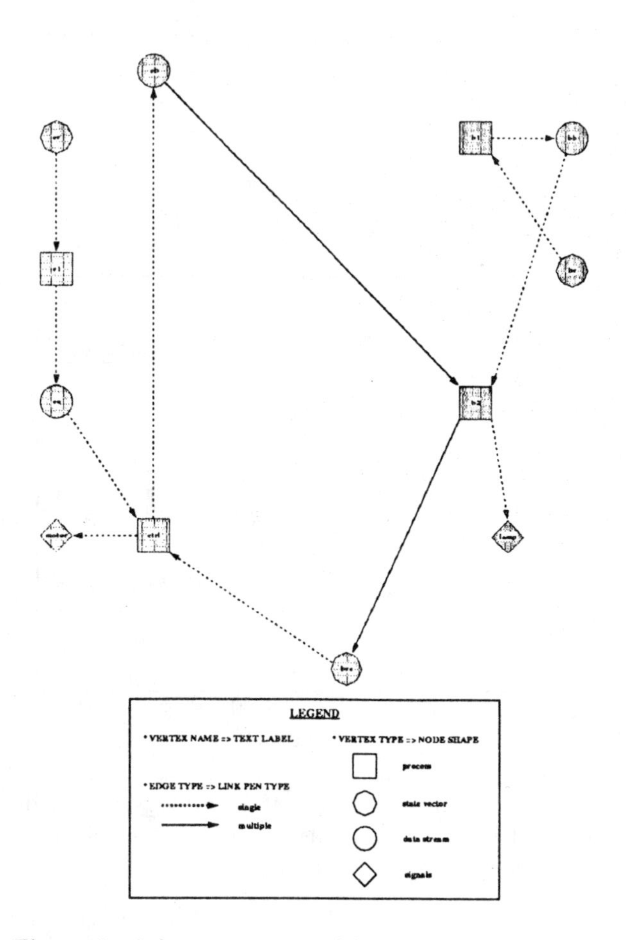

Figure 5: A layout generated by poor parameter settings.

progress has been made toward satisfying the evenly spaced alignment and sequential layout VOFs, not all have been completely satisfied. Furthermore, its aesthetic quality—as measured by the number of edge crossing, its area, and its degree of global symmetry—is poor.

Although these results are preliminary, some conclusions are obvious from Figure 3. It is clear that low values for the mutation rate (x axis) are extremely undesirable; low values for the reproduction (y axis) parameter are relatively undesirable; crossover (z axis) has little effect on final evaluation; and substantial stochastic variation in final layout evaluation can be expected, regardless of the actual parameter values used. Statistical analysis of the data verifies that the effects of reproduction and mutation were statistically significant, but those of crossover were not. The phenomena of stochastic variation and the apparently ineffectual nature of crossover are of particular interest. We turn now to some experiments performed to examine these phenomena.

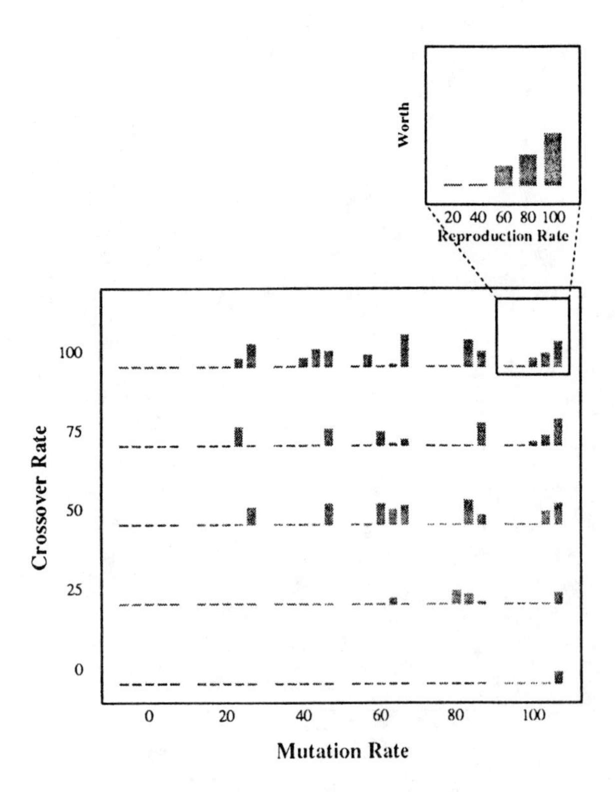

Figure 6: Layout evaluation at 50 generations for diagram of Figure 1 as a function of the reproduction, mutation, and crossover parameters.

Two approaches can be taken to eliminate stochastic variation: the population size per run can be increased, or the number of runs can be increased. To develop a sense of the degree of stochastic variation, we generated 60 runs each of the two diagrams, at the standard parameter settings. A histogram of the results for Figure 1 is presented in Figures 4. These experiments show that considerable variation is possible in the evaluation of the final layout. The layouts in all but the tail of the distribution are deemed acceptable. Taking the best layout from multiple runs may be required to reduce the probability of ending up in the tail to an acceptable level.

The relative uselessness of crossover was surprising. We speculated that crossover is more likely to affect the number of generations required for a layout to evolve than final layout evaluation for the following reason. The primary role of the crossover operator is to permit one individual to benefit from the advances of another. In a pure GA, this is especially important because the only way an improvement that is dependent on the problem structure (as determined implicitly by the evaluation function) can be shared is by such a crossover. But in an impure system in which mutation is already adapted to the structure of the problem, crossover is not needed to "learn" this structure. Rather, the advantage of crossover is to allow such improvements to be made more efficiently, as good traits accumulate in the pool (through highly evaluated schemata). Although the mutation operator can eventually achieve similar results without crossover, progress should be made more quickly with crossover because of the increased cooperation among processors. To test this conjecture, we examined the layout evaluations after 50 generations for the various settings of mutation, reproduction, and crossover. Figure 6 depicts the results, providing striking evidence for the speedup due to crossover. In this diagram, crossover varies across the y axis, to highlight the trend. Again, the disadvantage of low mutation rates can be seen (in the x axis), and of low reproduction rates (now in the z axis).

Developing an intuition about the operation of a parallel algorithm of this kind can be very difficult, due to the large number of processors and the overwhelming amount of data that are involved. To aid us in our understanding of the algorithm, we have developed a pair of graphic visualization tools. The first is designed to provide an overall view of what is occurring. On the Connection Machine's frame buffer, we display a grid of enlarged gray pixels in the same 64 by 64 geometric configuration as the processors in the machine, with one pixel per processor. The intensity of a pixel corresponds to the score of its related processor—the better the score, the brighter the pixel. Additionally, when the score crosses an arbitrary threshold, the hue of its pixel changes to red; this tends to highlight those processors whose scores are unusually good. The best score in the current generation is colored yellow, unless it also represents an improvement over all previous scores, in which case it is colored blue. This tool allows us to see the growth of the colonies that develop due to the the locality of the reproduction and crossover operators and it gives us a global sense of the algorithm's progress.

The second tool is intended to give us a closeup view on what is occurring in individual processors. It displays a simplified graphic form of the best layout in each generation. Optionally, the population can be partitioned into subsets, and the best diagram of the generation in each of the subsets can be displayed.

Although these visualization tools do not substitute for quantitative measures of the algorithm's performance, they have been invaluable in contributing to our understanding of the algorithm's behavior.

4 CONCLUSION

We have developed a parallel genetic algorithm for the GNDL problem, a computationally intractable problem that arises in the automated design of network diagrams. Although the current algorithm performs very well, we are investigating four strategies for improving its performance.

The first strategy is simply the continued refinement of algorithm parameters along the lines described in the previous section.

The second strategy involves the use of a gradient-descent method to achieve rapid convergence to locally optimal points in the search space.[6] Although the genetic algorithm has proven to be an excellent technique for identifying neighborhoods of locally optimal points in the search space, it finds the actual optimal points within neighborhoods inefficiently. We hope that a hybrid genetic algorithm that uses a gradient-descent method to move quickly to optimal points within a neighborhood will make for a more efficient technique.

Our third strategy for improving the genetic algorithm involves the use of dynamic algorithm parameters. A central component of any simulated-annealing technique [7] is the annealing schedule, which specifies how a key parameter (temperature) varies as the algorithm proceeds. In our current implementation we do not vary the algorithm parameters over time. It seems likely that different parameter values are preferable at different times, so we plan to investigate various approaches to dynamically varying the genetic-algorithm parameters as the algorithm progresses.

The final potential improvement strategy concerns the initial seeding of the processors. All processors are seeded initially with a random layout. By seeding some processors with non-random layouts, we might hope to achieve faster convergence to satisfactory solutions. A possible source of seed layouts are those generated by our rule-based heuristic method [12].

Acknowledgments

This research was supported in part by research contracts from U S WEST Advanced Technologies and Lockheed Space and Missile Company. We are happy to acknowledge the support and advice of Tony Cox, Dave Davis, Mark Friedell, and Barbara Grosz. Steve Sistare suggested the idea of a scatter operation.

[6] This is joint work with Mark Friedell, Peter McMurry, and Steve Sistare.

References

[1] J. Bertin. *Semiology of Graphics: Diagrams, Networks, Maps*. Univ. of Wisconsin Press, 1983. Translated by W. J. Berg.

[2] R. Davidson and D. Harel. Drawing graphs nicely using simulated annealing. In review, 1989.

[3] P. Eades and R. Tamassia. Algorithms for drawing graphs: An annotated bibliography. Technical report CS-89-09, Brown University, Dept. of Computer Science, 1989. Revised version.

[4] M. R. Garey and D. S. Johnson. Crossing number is NP-complete. *SIAM Journal of Albegraic and Discrete Methods*, 4(3):312–316, 1983.

[5] W. D. Hillis. Co-evolving parasites improve simulated evolution as an optimization procedure. Unpublished manuscript, 1989.

[6] L. Kaufman. *Sight and Mind: An Introduction to Visual Perception*. Oxford University Press, New York, 1974.

[7] S. Kirkpatrick, C. D. Gelatt Jr., and M. P. Vecchi. Optimization by simulated annealing. *Science*, 220:671–680, 1983.

[8] W. Köhler. *Gestalt Psychology*. The New American Library, New York and Toronto, 1947.

[9] S. M. Kosslyn. Understanding charts and graphs. *Applied Cognitive Psychology*, 3:185–226, 1989.

[10] S. Lin. Computer solutions of the traveling salesman problem. *Bell System Technical Journal*, 44:2245–2269, 1965.

[11] J. Marks. A syntax and semantics for network diagrams. In *Proceedings of the IEEE 1990 Workshop on Visual Languages*, pages 104–110, Skokie, Illinois, 1990.

[12] J. Marks. *Automating the Design of Network Diagrams*. PhD thesis, Harvard University, Cambridge, Massachusetts, 1991. In preparation.

[13] J. Marks and E. Reiter. Avoiding unwanted conversational implicatures in text and graphics. In *Proceedings of the Eighth National Conference on Artificial Intelligence (AAAI '90)*, pages 450–456, Boston, Massachusetts, 1990.

[14] H. Mühlenbein, M. Gorges-Schleuter, and O. Krämer. Evolution algorithms in combinatorial optimization. *Parallel Computing*, 7:65–85, 1988.

[15] B. Sanden. An entity-life modeling approach to the design of concurrent software. *CACM*, 32(3):337, 1989.

A Hybrid Genetic Algorithm for Task Allocation in Multicomputers

Nashat Mansour
School of Computer Science
Center for Computational Science
Syracuse University
Syracuse New York 13244

Geoffrey C. Fox
School of Computer Science
Department of Physics
Center for Computational Science
Northeast Parallel Architectures Ctr.
Syracuse University

Abstract

A hybrid genetic algorithm (HGATA) is proposed for the task allocation problem in parallel computing. It includes elitist ranking selection, variable rates for the genetic operators, the inversion operator and hill-climbing of individuals. Hill-climbing is done by a simple heuristic procedure tailored to the application. HGATA minimizes the likelihood of premature convergence and finds good solutions in a reasonable time. It also makes use of problem-specific knowledge to evade some computational costs and to reinforce some favorable aspects of the genetic search. The experimental results on realistic test cases support the HGATA approach for task allocation.

1 INTRODUCTION

Parallel computers offer a high computational power, which makes them useful for many problems in science, engineering and other areas. Generally, they are classified as Single Instruction Multiple Data and Multiple Instruction Multiple Data (MIMD). Distributed-memory MIMD computers will henceforth be called multicomputers. While offering high raw computational power, parallel computers can suffer from low utilization and show poor performance if the load is not distributed as equally as possible among the processors. This crucial issue leads to the task allocation problem. In multicomputers, task allocation aims at the minimization of the total execution time of a problem by balancing the calculations among the processors and minimizing the interprocessor communication. Task allocation may be based on partitioning the operations in the algorithm or the data set. In this work, data partitioning is considered.

The task allocation problem is a computationally intractable combinatorial optimization problem. Several heuristic methods have been proposed, such as mincut-based heuristics, orthogonal recursive bisection, scattered decomposition, neural networks and simulated annealing [7, 9, 10, 11, 12, 13, 19]. The deterministic methods are strong methods with predictable and low execution time. However, they, naturally, either make restrictive assumptions or tend to be biased towards particular instances of the problem. The stochastic methods make no assumptions about the problem considered. However, their execution time is currently unpredictable and is still an open question. A parallel version of the classic genetic algorithm has been suggested in [21] for the specific case of allocating matrix rows to a hypercube for the Gaussian elimination problem. The work in [21] is difficult to generalize to other allocation problems and assumes a computational model that is different from the model considered in this paper. The theory of complex systems has been suggested as a framework within which concurrency issues, such as task allocation can be studied [14]. It should be emphasized here that all the approaches mentioned above, as well as our approach, aim at producing good sub-optimal solutions, and not necessarily the optimal, in a reasonable time.

In this work, we propose a hybrid genetic algorithm for the task allocation problem (HGATA). HGATA enhances the classic genetic algorithm (GA) with a number of features in order to alleviate the problem of premature convergence and to improve the search efficiency. These features include a combination of design choices for the selection scheme, the genetic operators and the rates of the operators. The incorporation of a problem-specific hill-climbing procedure is also an essential feature and is responsible for the hybrid adjective.

This paper is organized as follows. Section 2 defines the task allocation problem and presents an objective function. Section 3 presents HGATA and explains the choices involved. The experimental results are reported and discussed in section 4. In section 5, conclusions are given.

2 THE TASK ALLOCATION PROBLEM

Task allocation consists of partitioning the problem into tasks, i.e., subproblems, and allocating these tasks to the processors of the multicomputer so that an objective function is minimized. An objective function associated with the total execution time required for solving a problem is given below. The computational model is explained first, then exact and approximate objective functions are presented and discussed. Some parameters, which will be utilized by HGATA, are also given.

The model of computation considered here is that of loosely synchronous parallel algorithms [14], where calculation and communication do not overlap. Processors run the same code (algorithm) and repeat a calculate-communicate cycle, where each processor performs calculations on its subproblem (task) and then communicates with other processors to exchange necessary boundary information.

To formulate an objective function representing the cost of task allocation, both the problem domain and the multicomputer are represented by graphs. The vertices of the problem graph are the data elements and the edges refer to the calculation dependency. The vertices of the multicomputer graph are the processors and the edges are given by the interconnections. Task allocation becomes a mapping of subsets of the vertices of the problem graph to vertices in the multicomputer graph. Let W(p) and C(p) denote the amount of calculation and communication for processor p, respectively. W(p) is proportional to the number of data elements allocated to p. C(p) is a function of the amount of information communicated by p and the distance it travels. The total execution time, T, for a parallel program is determined by the processor with the greatest load of calculation and communication, that is,

$$T = \max_p \{W(p) + C(p)\} \quad \text{.......... (1)}$$

Equation (1) represents the exact objective function to be minimized in task allocation and is the basis for evaluating the results of HGATA. However, the use of this minimax criterion is computationally expensive, because the calculation of a new T caused by any change in the mapping of elements to processors requires the recalculation of the load of all processors.To avoid this complexity, a quadratic objective function has been proposed [9, 12] to approximate the cost of task allocation. It can be expressed as

$$r^2 \sum_p N^2(p) + v\left(\frac{tcomm}{tcalc}\right) \sum_{p,q} d(p,q) \quad \text{....(2)}$$

where r is the amount of calculation per data element, N(p) is the number of elements allocated to processor p, (tcomm/tcalc) is a machine-dependent communication to calculation time ratio, *v* is a constant scaling factor expressing the relative importance of communication with respect to calculation, and d(p,q) is the Hamming distance between processors p and q. The main advantage of using this quadratic cost function is that it enjoys the locality property. Locality means that a change in the cost due to a change in the allocation of elements to processors is determined by the reallocated elements only. Since HGATA employs a hill-climbing scheme based on incremental reallocation of elements, the locality property becomes very important for keeping hill-climbing as fast as possible. Another important consideration in using the objective function in (2) is the choice of the weight *v*. In this work, values for *v* are chosen in harmony with the behavior of HGATA for the purpose of generating better quality solutions. This is elaborated in the next section within the HGATA context.

Two parameters which can be derived from the objective function in (2) are utilized by HGATA. The first is the degree of clustering (DOC) of the data elements in a task allocation instance. The maximum DOC, *DOC(max)*, corresponds to an optimal allocation. The second parameter is an estimate of the value of the optimal objective function. This estimate involves the problem size, the multicomputer size, and the scaling factor *v*. It is henceforth referred to as *OBJ(opt)*. The derivation for both parameters is omitted here. However, we note that DOC and *OBJ(opt)* are employed by HGATA for evading some computational costs and reinforcing some aspects of the search.

3 HYBRID GENETIC ALGORITHM

Genetic algorithms represent powerful weak methods for solving optimization problems, such as task allocation. However, the implementation of an efficient GA often encounters the problem of premature convergence to local optima, otherwise a long time may be required for the GA search to reach an optimal or a good suboptimal solution. Methods for overcoming the two problems of premature convergence and inefficiency would be conflicting, and a compromise is usually required. The incorporation of problem-specific knowledge has been proposed to direct the blind GA search to the fruitful regions of the search space for improving the efficiency [15, 16]. The resultant schemes are referred to as hybrid schemes. To address the problem of premature convergence, a number of techniques have been suggested. Some selection schemes have been proposed for reducing the stochastic sampling errors [2, 15]. Other techniques have been incorporated into the reproduction scheme to control the level of competition among individuals and to maintain diversity. Examples of these are prescaling, ranking and the use of sharing functions or crowding factors [1, 5, 6, 15]. Reduced-surrogate crossover and two-point crossover operators have been suggested for enhancing exploration and improving the search [3]. Adaptive rates for crossover and mutation have been found useful [3, 4]. The variation in these rates is usually inversely proportional to the level of diversity in the population.

The advantages of the techniques mentioned above have been demonstrated by comparing the resultant performance with that of the classical GA [17]. Often, the performance verification is carried out for DeJong's testbed of functions [6] or for other specific applications, such as the traveling salesperson problem. In this work, a number of techniques dealing with selection and genetic operators have been combined for improving the quality of the solutions for the task allocation problem. Also, a simple problem-specific hill-climbing procedure is added for improving the efficiency of the search. The techniques and the procedure comprise HGATA, which is outlined in Figure 1. Four objectives guide the design of HGATA. These are the minimization of the likelihood of premature convergence, increasing the search efficiency, keeping computational costs low, and utilizing domain knowledge, wherever possible, for satisfying the first three objectives. In the remainder of this section, HGATA is explained. The stages of the genetic search are described first as a prelude to the description of some design choices in the following subsections.

```
Read (problem graph and multicomputer graph);
Random Generation of initial population P(0) of size POP;
Evaluate fitness of individuals in P(0);
For (gen = 1 to maxgen) OR until convergence do
  Set (v, operator rates, flags);
  Rank individuals in P(gen-1), and
        allocate reproduction trials stored in MATES[];
  /* produce new generation P(gen) */
  For (i = 1 to POP step 2) do
    Randomly select 2 parents from MATES [];
    Apply genetic operators (2-pt xover,mutation,inversion);
    Hill-climbing by new individuals;
  endfor
  Evaluate fitness of individuals in P(gen);
  Retain the better of {fittest(gen) , fittest(gen-1)};
endfor
```

Figure 1: An Outline of HGATA.

3.1 THREE STAGES OF HGATA SEARCH

In the beginning of the search, the allocation of data elements to processors is almost random, and thus, the communication among processors is heavy and far from optimal, regardless of the distribution of the number of elements. In the successive generations, clusters of elements are expected to be grown gradually and allocated to processors so that interprocessor communication is constantly reduced, at least in the fitter individuals in the population. Then, at some point in the search, the balancing of the calculational load becomes more significant for increasing the fitness. Therefore, two stages of the search can be distinguished. The first stage is the clustering stage, which lays down the foundation of the basic pattern of the interprocessor communication. The second stage will be referred to as the calculation-balancing stage. Obviously, the two successive stages overlap.

A third stage in the search can also be identified when the population is near convergence. In this advanced stage, the average DOC of the population approaches *DOC(max)* and the clusters of elements crystallize. If these clusters are broken, the fitness of the respective individual would drop significantly, and its survival becomes less likely. At this point, crossover becomes less useful for introducing new building blocks, mutation of elements in the middle of the clusters is useless, and a fruitful search is that which concentrates on the adjustment of the boundaries of the clusters in the processors. This stage will henceforth be referred to as the tuning stage. Boundary adjustment can be accomplished mainly by the hill-climbing of individuals, which is explained below, aided by the probabilistic mutation of the boundary elements. The main responsibility of crossover becomes the propagation and the inheritance of high-performance building blocks and the maintenance of the drive towards convergence for the sake of search efficiency. For hill-climbing and boundary mutation to take on their role in this stage, it is necessary to increase the relative weight of the calculation term in the fitness function. This is elaborated below with the description of hill-climbing. It is worth noting here that the tuning stage constitutes a relatively small number of generations in comparison with the first two stages.

3.2 CHROMOSOMAL REPRESENTATION

An instance of task allocation is encoded by a chromosome whose length is equal to the number of data elements (vertices) in the problem graph. The value of an allele is an integer representing the processor to which a data element is allocated. The element is, therefore, the index (locus) of the processor (gene) to which it is assigned. For example, if we have a graph of four data elements and two processors, the genotype (1,1,2,1) indicates that elements 1,2 and 4 are allocated to processor 1 and element 3 to processor 2.

3.3 FITNESS EVALUATION

The fitness of an individual is evaluated as the reciprocal of the objective function in expression (2). As pointed out in section 2, the choice of v is of particular interest. Its value should be chosen in accordance with the properties of the HGATA search illustrated above. v should be so large that the communication term in the fitness function acquires sufficient importance in the clustering stage. But, v should not be too large, otherwise it will swamp the effect of the calculational term in the later stages. In other words, v is chosen to favor the fitness of the individuals whose structure involves nearest-neighbor interprocessor communication in the clustering stage. In the later phases of the search, the value of v should allow the emphasis to shift to the calculation term in the fitness. A value that satisfies these requirements can be determined from the ratio of the calculation and communication terms of *OBJ(opt)*, which is defined in section 2. In subsection 3.7, it will be argued that v has to be decreased in the tuning stage.

3.4 REPRODUCTION SCHEME

The reproduction scheme adopted in HGATA is elitist ranking followed by random selection of mates from the list of reproduction trials, or copies, allocated to the ranked individuals. In ranking [1], the individuals are sorted by their fitness values and are allocated a number of copies according to a predetermined scale of equidistant values for the population, and not according to their relative fitness. In HGATA, the ranks assigned to the fittest and the least fit individuals are 1.2 and 0.8, respectively. Individuals with ranks bigger than 1 are first assigned single copies. Then, the fractional part of their ranks and the ranks of the lower half of individuals are treated as probabilities for assignment of copies. This scheme has been found to produce a percent involvement value of 92% to 98% in different generations. It offers a suitable way for controlling the selective pressure, and hence, the inversely related population diversity [23]. This results in the control of premature convergence, which is the main reason for using ranking-based reproduction in HGA-

TA. The control of premature convergence by ranking outweighs the loss due to ignoring knowledge about the relative fitness, especially that the expression used for the fitness in our application is only an approximation to the exact one anyway. Furthermore, ranking dispenses with prescaling, which is usually necessary for fitness proportionate reproduction schemes. From an efficiency point of view, ranking provides a computationally cheap method for controlling the population diversity in comparison with expensive methods needed with fitness proportionate selection, such as sharing functions or DeJong's crowding schemes [5, 6, 15].

Elitism in the reproduction scheme refers to the preservation of the fittest individual. In HGATA, the preceding fittest individual is passed unscathed to the new generation, but it is forced to compete with the new fittest, and only the better of the two is retained. The purpose of elitism and its current implementation is the exploitation of good building blocks and ensuring that good candidate solutions are saved if the search is to be truncated at any point. To patch up a part of the loophole created by the use of the approximate objective function, the criterion for choosing between the current fittest and the preceding fittest individuals is changed in the tuning stage. The exact expression for fitness is used and has been found beneficial.

3.5 GENETIC OPERATORS

The Genetic operators employed in HGATA are crossover, mutation and inversion. The two-point ring-like crossover is used because it offers less positional bias than the one-point standard crossover without introducing any distributional bias [8]. Other more complex and presumably higher-performance crossover operators, such as shuffle crossover [8], are not used in this work in order to avoid excessive computations.

The standard mutation operator is employed throughout the search. In the tuning stage of the search, for the reason explained in subsection 3.1, mutation is restricted to elements at the boundaries of the clusters

Inversion is used in the standard biological way, where a contiguous section of the chromosome is inverted. In HGATA, the chromosome is considered as a ring. Inversion at a low rate helps in introducing new building blocks into the population for an application, such as task allocation.

3.6 OPERATOR RATES

It has become widely recognized that variable operator rates are useful for maintaining diversity in the population, and hence, for alleviating the premature convergence problem [3, 4]. Rates are varied in the direction that counteracts the drop in diversity. Several measures have been suggested for the detection of diversity, such as lost alleles, entropy, percent involvement, and others [1, 4, 15, 16]. The evaluation of these measures invariably requires considerable computations. In HGATA, the cost of computing measures of diversity is not incurred. Instead, the degree of clustering of elements is used to guide the variation of the rates. This design decision is based upon the observation that diversity is reduced in the population as the DOC increases. The current implementation uses a simple stepwise change in the rates. The smallest and largest rates are associated with the DOC of the first generation and the *DOC (max)* estimate, respectively.

3.7 HILL CLIMBING

Knowledge about the application can direct the blind genetic search to more profitable regions in the adaptive space. In HGATA, individuals carry out a simple problem-specific hill-climbing procedure that can increase their fitness. The procedure is greedy, and its inclusion improves the efficiency of the search significantly.

Hill-climbing for an individual is performed by considering only the boundary data elements allocated to the processors one at a time. A boundary element e is an element that is allocated to a processor p1 and has at least one neighboring element (in the problem graph) allocated to a different processor p2. Such an element is transferred from p1 to p2, if and only if, the transfer causes the objective function to drop or stay the same. It can be shown that the Change in Objective Function, COF, due to a transfer of element e is given by

$$2r^2 \left[1 + N(p2) - N(p1)\right] + 2vR\,(CCD)$$

where N(x) is the number of elements allocated to processor x before the transfer, R is the (tcomm/tcalc) ratio, and CCD is the change in communication cost (sum of distances) for element e. From this expression, it can easily be seen that a transfer of an element can only take place from overloaded processors to underloaded processors. It should be emphasized here that the formulation of COF, which leads to a simple implementation of hill-climbing, is a direct result of the locality property of the approximate objective function, as mentioned in section 2.

In the tuning stage of the evolution, a procedure for removing isolated elements is invoked as a part of hill-climbing. This amounts to eliminating noise components, which manifest themselves as artificial additions to both the calculation and communication loads of processors in a task allocation instance.

Hill-climbing plays a distinctive role in the tuning stage, where it fine-tunes the structures by adjusting the boundaries of the sizeable clusters assigned to the processors. In this advanced stage, the basic pattern of interprocessor communication can not be significantly changed, and the search ceases to offer significant gains. For these reasons, the emphasis upon balancing the calculational load should be artificially increased for the purpose of facilitating the boundary adjustment. This is achieved by decreasing the value of the weight v in the objective function gradually

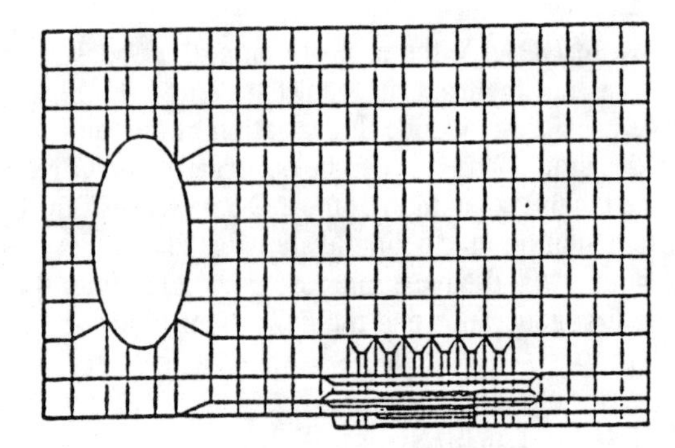

Figure 2: 551-Element Mesh1.

Figure 3: 301-Element Mesh2.

from the fixed value used throughout the search to a small suitable value determined by the COF expression. The smallest useful value for v is one that makes COF negative or zero when the following conditions are both true. The first condition is that an overloaded processor has two elements more than the underloaded processor. The second condition is that the transfer of an element e does not increase the sum of communication distances of e by more than one.

4 EXPERIMENTAL RESULTS

The experiments describe the solutions that can be obtained by HGATA for realistic problems. They also illustrate the design choices and parameters of HGATA. The experimental set-up is presented first, then the results are given and discussed.

4.1 EXPERIMENTAL DESIGN

A genetic algorithm is considered to be a 6-tuple of variables GA = (REP, XOV, INV, OPRATE, POP, MRANK), where REP and XOV refer to the reproduction scheme and the crossover operator, respectively, INV indicates whether inversion is included, OPRATE indicates either variable or fixed rates for the genetic operators, POP is the population size, and MRANK is the maximum rank for the ranking-based reproduction scheme. Other parameters are assumed to be the same as in the classical GA. POP has been empirically determined by extrapolation from small test cases. It has been found that a population size approximately equal to the size of the problem graph is adequate for HGATA, as long as the multicomputer graph is much smaller. Fixed rates for the genetic operators are 0.6 for crossover, 0.002 for mutation and 0.02 for inversion. Variable rates vary in a stepwise fashion as follows. The crossover rate increases from 0.5 to 1.0; the mutation rate increases from 0.002 to 0.004, and the inversion rate decreases from 0.03 to 0.0.

Several test cases have been used. For small and regular problems, HGATA has always found an optimal task allocation efficiently. These results are not be presented here. Instead, two irregular problems with realistic sizes are considered. These are shown in Figures 2 and 3, and are henceforth referred to as Mesh1 and Mesh2, respectively. Most of the results presented below are the averages of three runs. This small number of runs is satisfactory to illustrate HGATA's features, except for paragraph (iii), where 20 runs have been carried out.

In all experiments, a solution obtained at a certain point in the search refers to the fittest individual in the respective generation. The performance measures are the (exact) multicomputer's efficiency and the average fitness of the population. The efficiency is defined as the ratio of the sequential execution time to the product of T (equation 1) and the number of processors in the multicomputer. Both measures are plotted below with respect to the number of generations, which, in turn, is used to assess the efficiency of the search. For clarity, the results are given as ratios, where efficiency is normalized with respect to the (exact) optimum, and fitness is normalized with respect to the (approximate) fitness of the optimal solution. It should be understood that the use of exact efficiency and approximate fitness for expressing the quality of the solutions will obviously exhibit a discrepancy in the results for the two measures.

4.2 RESULTS

The first experiment only refers to Mesh1. All the following experiments refer to allocating Mesh2 to an 8-processor hypercube.

(i) For Mesh1 and a 16-processor hypercube, HGATA1 = (ranking, 2-point, yes, var, 500, 1.2) yields the allocation configuration depicted in Figure 4. The efficiency of this allocation is 0.93 of the optimum, and its fitness is 0.998 of the optimal fitness. This solution is obtained after 280 generations. Each generation takes about 30 seconds on a SPARC workstation.

(ii) HGATA2 = (ranking, 2-point, yes, var, 300, 1.2) applied to Mesh2 for a 3-cube finds a solution shown in Figure 5. The efficiency and fitness are shown in Figure 6, where the relative average loads of calculation and communication are also depicted. After generation 118, the search converges to a solution with efficiency and fitness ratios 0.97 and 0.998,

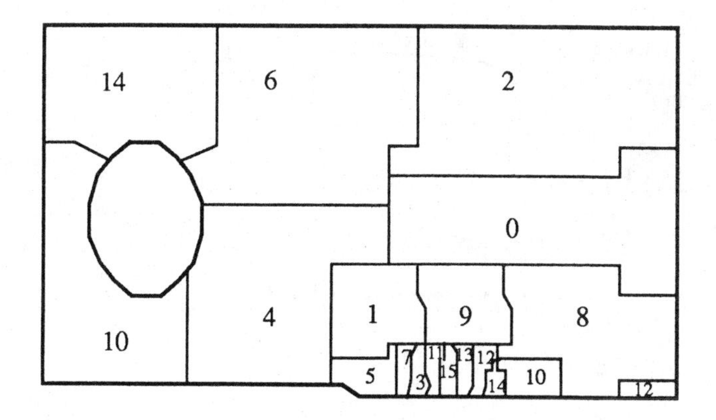

Figure 4: Allocation of Mesh1 to 4-Cube.

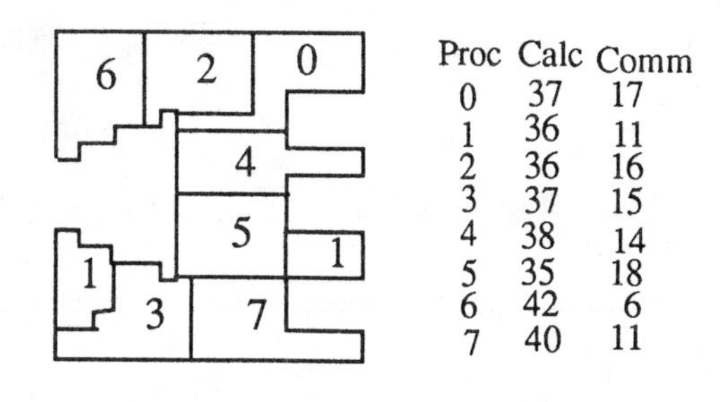

Proc	Calc	Comm
0	37	17
1	36	11
2	36	16
3	37	15
4	38	14
5	35	18
6	42	6
7	40	11

Figure 5: Allocation of Mesh2 to 3-Cube by HGATA2, and Processor Loads.

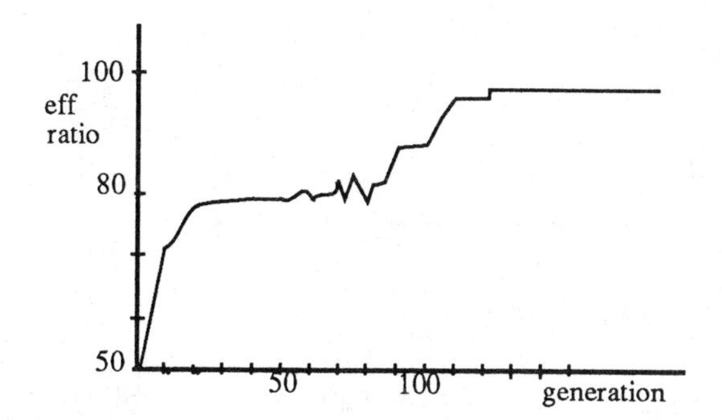

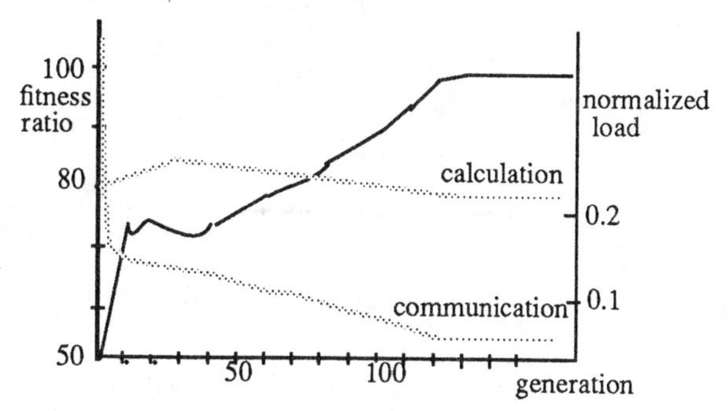

Figure 6: Efficiency and Fitness Ratios for HGATA2.

respectively. Each generation takes about 12 seconds.

The three stages of the search can be identified in the fitness and load curves in Figure 6. Roughly, their overlapping points are generations 50 and 100. It can be seen that in the first stage, the communication load drops steadily regardless of the calculation load, which happens to increase. In the second stage, both loads decrease, and the fitness rises. Decreasing v in the tuning stage enhances HGATA's tendency to reduce the calculation load. If v had not been decreased at this advanced stage, the efficiency would have been trapped at 89%.

(iii) The averages of 20 runs are shown in Figure 7 for comparing HGATA2 with a classical GA1 = (RSIS, 1-point, no, fixed, 300, -). GA1, however, still includes hill climbing, for speed, and the problem-specific features in the tuning stage, for improving the final solution. RSIS is Remainder Stochastic Independent Sampling [2] implemented here with prescaling. Figure 7 shows that GA1 converges before generation 80 to a fitness of 99% and an efficiency of 89%. The efficiency is later improved to 92.5% in generation 130 under the effect of mutation and tuning. HGATA2 takes 45 more generations to converge to 99% fitness and 96% efficiency in generation 125. The best solutions found in the 20 runs are 94.2% and 97.2% efficiency by GA1 and HGATA2, respectively. The worst is 90.4% and 93.5% for GA1 and HGATA2, respectively. The mean square deviation of the efficiency results are 1.18 for HGATA2 and 1.1 for GA1. Clearly, GA1 (without expensive sharing functions or crowding factors) results in a higher selection pressure and lacks the capability of controlling convergence. This explains the lower quality solutions produced by the classical GA1 and highlights the advantages of the combination of choices adopted in HGATA.

(iv) The effect of increasing the selection pressure is explored by increasing MRANK in HGATA3 = (ranking, 2-point, yes, var, 300, 2.0). This results in an early convergence, as shown in Figure 8. HGATA3 finds a good solution (96% efficiency ratio) in only 66 generations, which is 61% of the time required by HGATA2 to find a solution of the same quality. However, the large percentage of individuals (up to 20%) that die every generation, makes a maximum rank of 2.0 too high to be reliable, in general, for producing good solutions. This highlights the trade-off that exists between the solution quality and the search efficiency.

(v) Without hill-climbing, the efficiency of the search deteriorates tremendously. HGATA2, for example, becomes more than a hundred times slower.

(vi) The amount of improvement in the solution quality acquired in the tuning stage of the evolution has been found somewhat sensitive to the parameter that triggers this stage. If tuning is triggered too early, the time allowed for the first two stages of the evolution might be insufficient for produc-

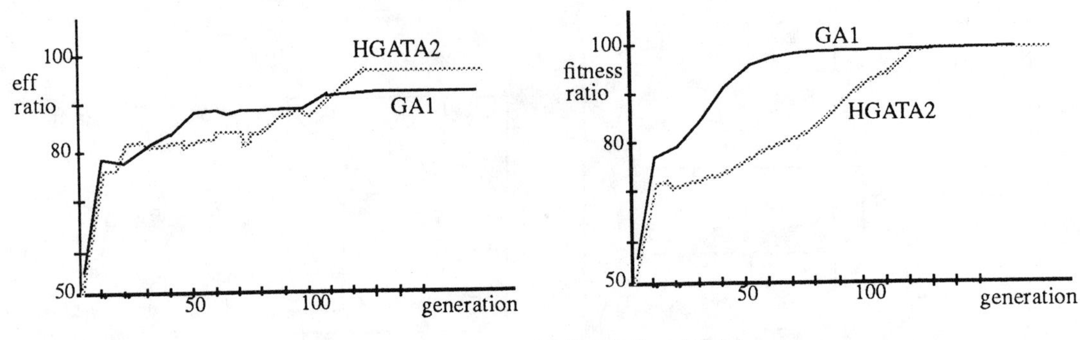

Figure 7: Comparison of HGATA2 and GA1.

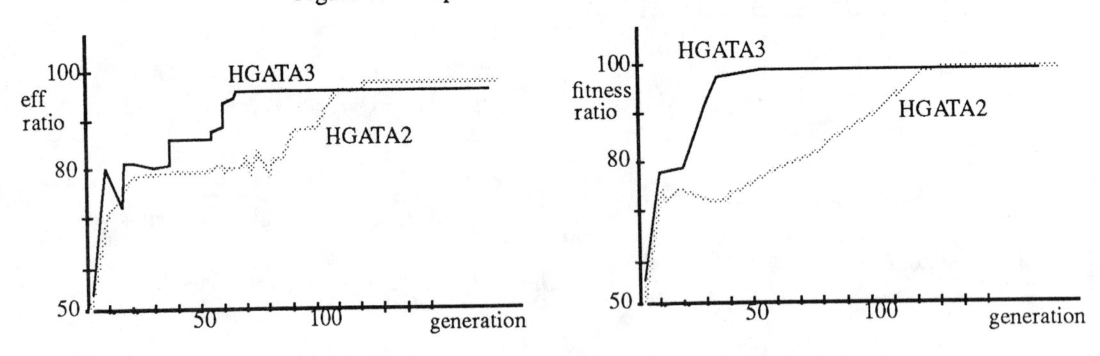

Figure 8: Comparison of HGATA2 and HGATA3.

ing near-optimal building blocks. If the tuning stage is invoked too late, convergence to a local optimum might have already prevailed in the population as a result of the first two stages.

4.3 DISCUSSION

The results obtained for Mesh1 and Mesh2 are good suboptimal task allocations. They are considerably better than the results obtained by other faster task allocation techniques. For example, recursive bisection [11] produces a solution for Mesh2 whose efficiency is 87% of the optimum. Scattered decomposition [19] yields an efficiency 61% of the optimum. The best result of 500 runs of the hill-climbing algorithm, each starting with different initial random configuration, has been found to be 83%.

HGATA is not restricted to the loosely synchronous model of computation described in section 2. It can be easily adapted to other models by modifying the objective function module. Furthermore, most of the constituents of HGATA can be employed for solving related combinatorial optimization problems such as graph partitioning and quadratic assignment.

The trade-off between the solution quality and the computational cost is worthwhile emphasizing. The search can be made less expensive by resorting to measures such as, for example, increasing the selection pressure by some proportion as in HGATA3. But, in such cases the solution quality is likely to be sacrificed, although at a smaller proportion. The determination of a suitable population size is another important and difficult issue affecting the solution quality and amount of computations. The use of theoretically derived estimates makes the search time impractical. Since we are interested in suboptimal results, heuristic estimates would be adequate. In this work, a population size of the order of the size of the problem graph has been found satisfactory; when the multicomputer size is many times smaller.

The hill-climbing procedure enables qualified individuals to rapidly climb the adaptive peaks, which speeds up the evolution. This improvement in the efficiency of the search may seem to cause the exploitation feature to gain an upper hand over exploration; contributing to premature convergence. However, although hill-climbing does fuel the exploitation aspect of the search, the experimental results do not reveal any negative effects. Hill-climbing enables exploration to be carried out in the space of genotypes representing local fitness optima. Further, It seems that it plays a role similar to that of a knowledgeable mutation operator and does not lead the search to be trapped in local optima.

5 CONCLUSIONS

The combined constituents of HGATA have been shown to provide a good balance between exploratory forces and exploitation forces for the task allocation problem. HGATA greatly reduces the causes of premature convergence and has found near-optimal solutions in a reasonable time, although the objective function used is only an approximation to the exact one. The use of the degree of clustering of data elements has obviated expensive diversity detection mechanisms. Also, it has been found that setting the weighting factor v in harmony with the properties of the search in different phases leads to better results.

The performance of HGATA can be improved in several ways. Firstly, the frequencies of the genetic operators can be adaptively varied according to a measure of the population diversity. Secondly, more fruitful crossover operators, such as the reduced surrogate operator [3] and the shuffle crossover [8], can be used to enable the search to concentrate on useful work. However, it should be clear that the computational cost per generation will increase for both suggestions. Nonetheless, high performance crossover can lead to a smaller number of generations. Thirdly, a better heuristic estimate for the population size needs to be worked out for our specific application. Fourthly, the search efficiency is likely to increase and better solutions might be produced by starting the hill climbing procedure at a randomly chosen gene instead of the first gene in the chromosome. Fifthly, faster execution can be obtained by parallel algorithms based on HGATA [18, 20, 22]. The parallel algorithms can also reduce the sensitivity to design parameters.

Acknowledgment

This work was supported by the Joint Tactical Fusion Program Office, and the National Science Foundation under Cooperative Agreement No. CCR-8809165.

References

[1] J. E. Baker. Adaptive Selection Methods for Genetic Algorithms. *ICGA*'85, 101-111.

[2] J. E. Baker. Reducing Bias and Efficiency in the Selection Algorithm. *ICGA*'87, 14-21.

[3] L. Booker. Improving Search in Genetic Algorithms. In L. Davis, ed, *Genetic Algorithms and Simulated Annealing*, Morgan Kaufmann, 1987, 61-73.

[4] L. Davis. Adapting Operator Probabilities on Genetic Algorithms. *ICGA*'89, 61-69.

[5] K. Deb, and D. E. Goldberg. An Investigation of Niche and Species Formation in Genetic Function Optimization. *ICGA*'89, 42-50.

[6] K. A. DeJong. *An Analysis of the Behavior of a Class of Genetic Adaptive Systems*. Doctoral Dissertation, University of Michigan, 1975.

[7] F. Ercal. *Heuristic Approaches to Task Allocation For Parallel Computing*. Doctoral Dissertation, Ohio State University, 1988.

[8] L. J. Eshelman, R. A. Caruana, and J. D. Schaffer. Biases in the Crossover Landscape. *ICGA*'89, 10-19.

[9] J. Flower, S. Otto, and M. Salama. Optimal Mapping of Irregular Finite Element Domains to Parallel Processors. *Caltech C3P* #292b, 1987.

[10] G. C. Fox. A Review of Automatic Load Balancing and Decomposition Methods for the Hypercube. In M. Shultz, ed., *Numerical Algorithms for Modern Parallel Computer Architectures*, Springer-Verlag, 1988, 63-76.

[11] G. C. Fox. A Graphical Approach to Load Balancing and Sparse Matrix Vector Multiplication on the Hypercube. *Caltech C3P* #327b, 1986.

[12] G. C. Fox, A. Kolawa, and R. Williams. The Implementation of a Dynamic Load Balancer. *Proc. 2nd Conf. Hypercube Multiprocessors*, 1987, 114-121.

[13] G. C. Fox and W. Furmanski. Load Balancing Loosely Synchronous Problems with a Neural Network. *Proc 3rd Conf. Hypercube Concurrent Computers, and Applications*, 1988, 241-278.

[14] G. C. Fox, M. Johnson, G. Lyzenga, S. Otto, J. Salmon, and D. Walker. *Solving Problems on Concurrent Processors*. Prentice Hall, 1988.

[15] D. E. Goldberg. *Genetic Algorithms in Search, Optimization and Machine Learning*. Addison-Wesley, 1989.

[16] J. J. Grefenstette. Incorporating Problem Specific Knowledge into Genetic Algorithms. In L. Davis, ed., *Genetic Algorithms and Simulated Annealing*, Morgan Kaufmann, 1987, 42-60.

[17] J. H. Holland. *Adaptation in Natural and Artificial Systems*. Univ. of Michigan Press, 1975.

[18] N. Mansour and G.C. Fox. Parallel Genetic Algorithms with Application to Load Balancing. *In Preparation*.

[19] R. Morison and S. Otto. The Scattered Decomposition for Finite Elements. *Caltech C3P* #286, 1985.

[20] H. Muhlenbein. Parallel Genetic Algorithms, Population Genetics, and Combinatorial Optimization. *ICGA*'89, 416-421.

[21] C. C. Pettey and M.R. Leuze. Parallel Placement of Parallel Processes. *Proc. 3rd Conf.Hypercube Concurrent Computers, and Applications*, 1988, 232-238.

[22] R. Tanese. Distributed Genetic Algorithms. *ICGA*'89, 434-440.

[23] D. Whitley. The GENITOR Algorithm and Selection Pressure: Why Rank-Based Allocation of Reproductive Trials is Best. *ICGA*'89, 116-123.

Conventional Genetic Algorithm for Job Shop Problems

Ryohei Nakano
Takeshi Yamada
Comm. and Info. Proc. Labs, NTT
1-2356 Take, Yokosuka, 238-03, Japan

Abstract

The job shop problem (JSP) is NP-hard, much harder than the traveling salesman problem. This paper shows how a conventional Genetic Algorithm (GA) can efficiently solve the JSP. We introduce unique ideas in representation, evaluation, and survival. A solution is succinctly represented as a binary genotype even though the JSP is an ordering problem. Mostly a genotype **g** produced by conventional crossover is illegal, i.e., represents no feasible schedule. Therefore an evaluation function first finds a legal genotype **g'** as similar to **g** as possible, and then evaluates **g'** to determine the fitness of **g**. The fitness of **g** is evaluated as the total elapsed time of the corresponding schedule. In survival of genotypes, we introduce a new treatment, called forcing, that replaces the genotype **g** with **g'** when **g** is selected as the survivor. Forcing both quickens convergence of GAs and drastically improves solution quality. A conventional GA using the three ideas is applied to three well-known JSP benchmarks. The solution quality approaches those obtained by branch and bound methods.

1 Introduction

The job shop problem (JSP) is among the hardest combinatorial problems. Not only is it NP-complete [Garey and Johnson, 1979], but it is one of the worst NP-complete class members. The flow shop problem (FSP), a restricted version of JSP, can be reduced to the traveling salesman problem (TSP) [Reddi and Ramamoorthy, 1972]; hence, the JSP is much harder than the TSP.

Research on the JSP has been the subject of much significant literature [Muth and Thompson, 1963; Balas, 1969; McMahon and Florian, 1975; Barker and McMahon, 1985; Carlier and Pinson, 1989]. The primary algorithms to solve JSP's are the branch and bound methods. The performance of existing algorithms has been evaluated through the widely known JSP benchmarks [Muth and Thompson, 1963]. The main historical progress in solution quality will be shown later together with our results.

Research on the application of Genetic Algorithms (GAs) to JSP's is relatively recent [Davis, 1985; Liepins and Hilliard, 1987; Cleveland and Smith, 1989; Whitley *et al.*, 1989]. Moreover, all investigated the FSP, although some discussed a more realistic problem than the FSP defined in the next section. Anyway the problems investigated so far are rather simple versions of the JSP.

This paper is organized as follows. Section 2 defines the JSP addressed and Section 3 describes a binary representation of a solution genotype. Section 4 presents how to evaluate a genotype **g**; **g** is mostly illegal, i.e., represents no feasible schedule. When **g** is illegal, an evaluation function finds a legal genotype **g'** as similar to **g** as possible, and then evaluates **g'** to assess the fitness of **g**. The fitness of **g** is equal to the total elapsed time of the corresponding schedule. Section 5 discusses the forcing which means the replacement of the genotype **g** with **g'**, when **g** is a survivor. Section 6 shows the results of experiments conducted on three well-known JSP benchmarks.

```
machine1:      111    44444333333333 6666666666222222222222555
machine2:2222222244444666111111555  3
machine3:333331 2222255555555544444                        6
machine4:     3333       666         4441111111           22225
machine5:             2222222222  5555533333334444444446666111111
machine6:         33333333  66666666662222222222225555111444444444
```

(a) Schedule (total elapsed time = 55)

```
machine1 :  1  4  3  6  2  5      job1 < job2 :  110100
machine2 :  2  4  6  1  5  3      job1 < job3 :  011000
machine3 :  3  1  2  5  4  6      job1 < job4 :  110010
machine4 :  3  6  4  1  2  5      job1 < job5 :  111100
machine5 :  2  5  3  4  6  1      job1 < job6 :  110000
machine6 :  3  6  2  5  1  4      job2 < job3 :  101000
                                  job2 < job4 :  111100
 (b) Symbolic representation      job2 < job5 :  111111
                                  job2 < job6 :  111000
job1    :  3  1  2  4  6  5       job3 < job4 :  111001
job2    :  2  3  5  6  1  4       job3 < job5 :  111100
job3    :  3  4  6  1  2  5       job3 < job6 :  111101
job4    :  2  1  3  4  5  6       job4 < job5 :  110100
job5    :  3  2  5  6  1  4       job4 < job6 :  111010
job6    :  2  4  6  1  5  3       job5 < job6 :  101000

(d) Machine sequences (given)     (c) Binary representation
```

Figure 1: Representations of Schedule for 6 × 6 Problem

2 Job Shop Problem

The job shop problem (JSP) to be solved has N jobs to be processed on M machines and assumes the following:

- A machine can process only one job at a time.
- The processing of a job on a machine is called an *operation*.
- An operation cannot be interrupted.
- A job consists of at most M operations.
- An operation sequence within a job, called *machine sequence*, and processing times for operations are given.
- An operation sequence on a machine, called *job sequence*, is unknown. The full set of job sequences is called a symbolic representation.
- A feasible symbolic representation is called a *schedule*.

The JSP is to find a schedule which minimizes the total elapsed time.

The flow shop problem (FSP) is a restricted JSP, where machine sequences are identical for all jobs.

3 Representation

There are at least two ways of representing a solution: symbolic and binary. As is true in the case of the TSP [Goldberg and Lingle, 1985; Grefenstette *et al.*, 1985; Oliver *et al.*, 1987], symbolic representation is also more straightforward in the JSP [Liepins and Hilliard, 1987; Cleveland and Smith, 1989; Whitley *et al.*, 1989]. Conventional GAs [Holland, 1975; Goldberg, 1986], however, are less suited to symbolic representation. For example, conventional crossover or mutation cannot be applied to symbolic representation. Hence binary representation is utilized below.

We focus our attention on a job pair [j,k]. Let the machine sequences within j and k be $[o_{j1}, o_{j2}, ..., o_{jM}]$ and $[o_{k1}, o_{k2}, ..., o_{kM}]$ respectively. Consider the following function assuming operations o1 and o2 are executed on the same machine: prior(o1,o2) = 1 if operation o1 is executed prior to o2, otherwise prior(o1,o2) = 0. Then we can get a bit vector for the job pair [j,k]:

$$[prior(o_{j1}, o_{k*}), ..., prior(o_{jM}, o_{k*})].$$

Note that operations o_{ji} and o_{k*} are executed

on the same machine. For N jobs, there exist N(N-1)/2 job pairs. Hence for N jobs and M machines, total MN(N-1)/2 bits are required to represent a solution; for example, 90 bits for N=M=6, 450 bits for N=M=10, and 950 bits for N=20, M=5.

Figure 1 shows symbolic and binary representations of a schedule for the 6 $\times$ 6 (N=6 jobs, M=6 machines) problem [Muth and Thompson, 1963]. In the binary matrix B, the rows represent the 15 job pairs and the columns represent the bit vectors. For example, B(1,1)=1 means job 1 be executed prior to job 2 on machine 3, and B(1,3)=0 means job 2 be prior to job 1 on machine 2. In the symbolic matrix, the i-th row represents the job sequence to be executed on machine i. For example, the first row indicates that on machine 1, operations are to be executed in the sequence: job1 $\rightarrow$ job4 $\rightarrow$ job3 $\rightarrow$...

Note that the above schedule in Figure 1 is an optimal schedule. The binary representation shows that bits have a tendency to continue within a machine. That is, the results confirm the heuristic that says a good schedule tends to keep the processing priority for each job pair. We can make use of the heuristic in both genotype initialization and conventional crossover. Note also that, in a binary representation, each bit has its own meaning, as seen in nature.

4 Evaluation

As stated above, a symbolic representation can be represented in a binary form. The inverse, however, does not hold. This means that the space of binary representation properly includes the space of symbolic representation. In fact, for the 6 $\times$ 6 problem, the number of elements of the former amounts to $2^{90} \simeq 10^{27}$, while that of the latter amounts to $(6!)^6 \simeq 10^{17}$. Hereafter, a binary representation is interchangeably called a genotype. Any evaluation of a genotype should take its wider space into consideration.

The idea behind our evaluation is as follows. In general, a genotype **g** produced either initially or by conventional crossover is illegal, i.e., represents no schedule. Therefore, our evaluation function first finds a legal genotype **g'** as similar to **g** as possible, and then evaluates **g'** to determine the fitness of **g**. The fitness of **g** is evaluated as the total elapsed time of the corresponding schedule.

The procedure that creates a legal genotype **g'** from an illegal genotype **g** is called the *harmonization algorithm*. The Hamming distance is used to assess the similarity of **g'** to **g**. The harmonization algorithm goes through two phases: local harmonization and global harmonization. The former creates a symbolic representation from **g**, removing local inconsistency within each machine. The symbolic representation may contain global inconsistencies between machines. By removing all global inconsistencies, the latter creates a schedule from the symbolic representation. The legal genotype **g'** represents the schedule.

							sum
job1:	*	0	0	1	1	0	2
job2:	1	*	0	0	1	1	3
job3:	1	1	*	1	1	0	4
job4:	0	1	0	*	0	0	1
job5:	0	0	0	1	*	1	2
job6:	1	0	1	1	0	*	3

(a) Original priority

							sum
job1:	*	0	0	1	1	0	2
job2:	1	*	0	0	1	1	3
job3:	1	1	*	1	1	1	5
job4:	0	1	0	*	0	0	1
job5:	0	0	0	1	*	1	2
job6:	1	0	0	1	0	*	2

(b) After selecting job 3

							sum
job1:	*	0	0	1	1	0	2
job2:	1	*	0	1	1	1	4
job3:	1	1	*	1	1	1	5
job4:	0	0	0	*	0	0	0
job5:	0	0	0	1	*	1	2
job6:	1	0	0	1	0	*	2

(c) After selecting job 2

							sum
job1:	*	0	0	1	1	1	3
job2:	1	*	0	1	1	1	4
job3:	1	1	*	1	1	1	5
job4:	0	0	0	*	0	0	0
job5:	0	0	0	1	*	1	2
job6:	0	0	0	1	0	*	1

(d) Final priority

Figure 2 : Local Harmonization Algorithm

The *local harmonization* algorithm works for each machine separately; hence, the following description addresses just one machine. The algorithm determines a job sequence, making use of an illegal genotype **g**. Since **g** states the priority by indicating which operation is to be executed prior to the other for all operation pairs on all machines, from **g** we can directly get a priority matrix for each machine. Figure 2 (a) shows a priority matrix for one machine directly gained from some illegal genotype. The element (2,1) = 1 indicates that job 2 is to be executed prior to job 1 on the machine. The algorithm searches for the operation having the highest priority, and finds the job 3 operation. When there is more than one such operation, one of them is selected. After selecting the operation having the highest priority, priority inconsistency is removed, as shown in (b). By repeating the above, the algorithm next selects the job 2 operation, resulting in (c). Thus repeating the process, the algorithm finally gets the consistent priority matrix, as shown in (d). It states the job sequence should be job3 $\rightarrow$ job2 $\rightarrow$ job1 $\rightarrow$ job5 $\rightarrow$ job6 $\rightarrow$ job4. On the whole, the algorithm changed three bits of the genotype.

It is rather easy to see that the local harmonization algorithm can find a valid job sequence while changing the minimum number of bits in a genotype. The algorithm goes halfway to get a legal genotype **g'** from an illegal genotype **g**.

The *global harmonization* algorithm is embedded in a simple scheduling algorithm. First, we describe the scheduling algorithm. The following notation is introduced:

$jnext(j)$: next operation to be executed within job j,
$jnext(j).machine$: machine to execute $jnext(j)$,
$mnext(m)$: next operation to be executed on machine m.

The simple scheduling algorithm inputs a symbolic representation and given conditions, i.e. machine sequences and processing time. It polls jobs checking if any job can be scheduled, and stops when no job can be scheduled. The job j can be scheduled if the following holds:

$$jnext(j) = mnext(jnext(j).machine).$$

The algorithm unconditionally schedules a job that can be scheduled. If a symbolic representation is a schedule, the algorithm always creates the schedule. Otherwise, it terminates when it meets a deadlock. Now the scheduling algorithm is modified to call the global harmonization algorithm whenever it meets a deadlock.

The global harmonization algorithm works as follows. Each job is checked to determine how far it is between jnext(j) and mnext(jnext(j).machine) in the job sequence list. The job with the minimum distance d is selected, and the job sequence is permuted to make mnext(jnext(j).machine) jnext(j). In the permutation, d operations are shifted right by one position each, resulting in d bits of change in the genotype. The global harmonization algorithm returns control to the scheduling algorithm, which in turn continues scheduling.

Thus the simple scheduling algorithm creates a schedule in cooperation with the global harmonization algorithm. It is not always guaranteed that the input (symbolic representation) yields the output (schedule) with the minimum number of shifts. We can expect, however, they are reasonably close.

On the whole, the harmonization algorithm creates a legal **g'** as similar to **g** as possible in the sense of Hamming distance. Developing optimal solutions for large problems suffers from computational explosion, and should be avoided since the evaluation will be repeated so many times. Note that usually the total number of bits of change is less than the sum of changes incurred by local harmonization and global harmonization.

5 Forcing

The framework of the conventional GA [Holland, 1975; Goldberg, 1986] is pursued. That is, conventional crossover and mutation are used. An unusual treatment called forcing, however, is introduced in this paper. Forcing means the replacement of an illegal genotype g with a quite similar legal genotype g'.

An original illegal genotype can be considered as an inherited character, while a refined legal genotype can be considered as an acquired one. Forcing can be considered as the inheritance of an acquired character, although it is not widely believed that an acquired character is inherited in nature. Since too much forcing may destroy the whole ecology and cause premature convergence, it is limited to the cases when **g** is selected as the survivor.

Forcing brings about at least two merits for GAs. One is to help them converge quickly. The other is to greatly improve the solution quality. The two advantages are demonstrated in the next section.

Table 1 : Main Results for Job Shop Problems

Papers	Algorithm	6 × 6 prob.	10 × 10 prob.	20 × 5 prob.
Balas1969	BAB	55	1177	1231
McMahon1975	BAB	55	972	1165
Barker1985	BAB	55	960	1303
Carlier1989	BAB	55	930	1165
Nakano1991	GA	55	965	1215

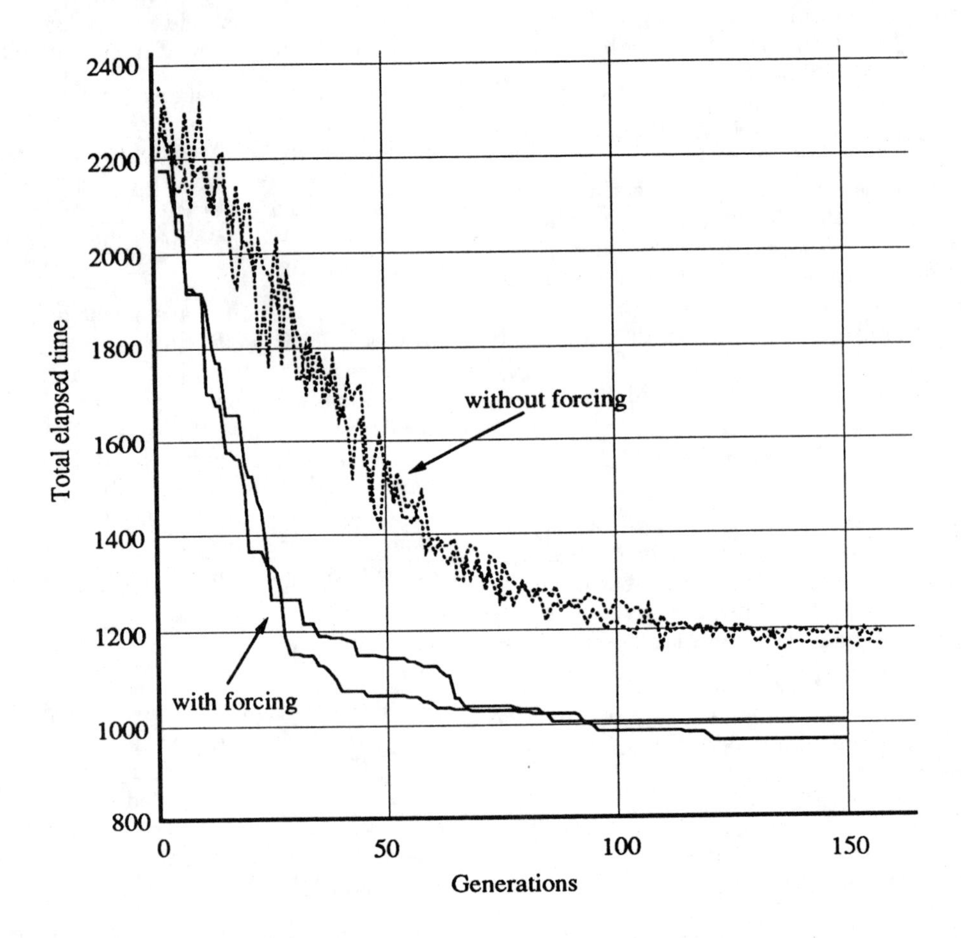

Figure 3 : Convergence of GAs for 10 × 10 problem

6 Experiments

This section shows the results of experiments conducted using three well-known JSP benchmarks [Muth and Thompson, 1963]:

- 6 × 6 (6 jobs, 6 machines) problem
- 10 × 10 (10 jobs, 10 machines) problem
- 20 × 5 (20 jobs, 5 machines) problem

Our best results (total elapsed time) are shown in Table 1 together with the historical progress in branch and bound methods.

The solution quality obtained in this paper is almost comparable with those obtained through more than 20 years of research into branch and bound methods. A population size of 1,000 is frequently used and this size is large enough to ensure a reasonably good solution for these problems.

Figure 3 shows convergence of GAs for the 10 × 10 problem. Convergence with and without forcing is displayed. The figure clearly shows that forcing both quickens convergence and drastically improves solution quality.

7 Conclusion

This paper shows how a conventional GA can effectively solve a tough combinatorial problem, the job shop problem. Three unique ideas are introduced in representation, evaluation, and survival. A binary representation pursued here makes it possible to apply conventional GAs. Even though a genotype produced by conventional crossover is usually illegal, the evaluation method presented here can evaluate it by finding a similar legal genotype. Forcing, the replacement of an illegal genotype with a legal one, improves convergence rates and solution quality. Experiments using well-known JSP benchmarks showed the solutions generated by the present approach were as good as those obtained by branch and bound methods.

Future work will include investigation of the complexity of this approach. The ideas presented in this paper can be used in symbolic representation, and such symbolic approach is also worth investigating.

Acknowledgments

The authors wish to thank Dr.Fumio Kanaya for discussion about theoretical issues, Yasuhiro Inooka for helpful information about the job shop problem, and Dr.Stephen I. Gallant for helpful comments.

References

[Balas, 1969] E. Balas. Machine sequencing via disjunctive graphs: an implicit enumeration algorithm. *Oper. Res.*, 17:941–957, 1969.

[Barker and McMahon, 1985] J.R. Barker and G.B. McMahon. Scheduling the general job-shop. *Manage. Sci.*, 31(5):594–598, 1985.

[Carlier and Pinson, 1989] J Carlier and E. Pinson. An algorithm for solving the job-shop problem. *Manage. Sci.*, 35(2):164–176, 1989.

[Cleveland and Smith, 1989] G.A. Cleveland and S.F. Smith. Using genetic algorithms to schedule flow shop releases. In *Proc. 3rd Int. Conf. on Genetic Algorithms and their Applications (Arlington, Va.)*, pages 160–169, 1989.

[Davis, 1985] L. Davis. Job shop scheduling with genetic algorithms. In *Proc. 1st Int. Conf. on Genetic Algorithms and their Applications (Pittsburgh,PA)*, pages 136–140, 1985.

[Garey and Johnson, 1979] M.R. Garey and D.S. Johnson. *Computers and Intractability - A Guide to the Theory of NP-Completeness.* Freeman and Company, New York, 1979.

[Goldberg and Lingle, 1985] D.E. Goldberg and R.Jr. Lingle. Alleles, loci, and the traveling salesman problem. In *Proc. 1st Int. Conf. on Genetic Algorithms and their Applications (Pittsburgh,PA)*, pages 154–159, 1985.

[Goldberg, 1986] D.E. Goldberg. *Genetic Algorithms in Search, Optimization, and Machine Learning.* Addison-Wesley, Reading,Mass., 1986.

[Grefenstette *et al.*, 1985] J. Grefenstette, R. Gopal, B. Rosmaita, and D.V. Gucht. Genetic algorithms for the traveling salesman problem. In *Proc. 1st Int. Conf. on Genetic Algorithms and their Applications (Pittsburgh,PA)*, pages 160–168, 1985.

[Holland, 1975] J.H. Holland. *Adaptation in Natural and Artificial Systems.* Unuv. of Michigan Press, 1975.

[Liepins and Hilliard, 1987] G.E. Liepins and M.R. Hilliard. Greedy genetics. In *Proc. 2nd Int. Conf. on Genetic Algorithms and their Applications (Cambridge,MA)*, pages 90–99, 1987.

[McMahon and Florian, 1975] G. McMahon and M. Florian. On scheduling with ready times and due dates to minimize maximum lateness. *Oper. Res.*, 23(3):475–482, 1975.

[Muth and Thompson, 1963] J.F. Muth and G.L. Thompson. *Industrial Scheduling.* Prentice-Hall, Englewood Cliffs, N.J., 1963.

[Oliver *et al.*, 1987] I.M. Oliver, D.J. Smith, and J.R.C. Holland. A study of permutation crossover operators on the traveling salesman problem. In *Proc. 2nd Int. Conf. on Genetic Algorithms and their Applications (Cambridge,MA)*, pages 224–230, 1987.

[Reddi and Ramamoorthy, 1972] S.S. Reddi and C.V. Ramamoorthy. On the flow-shop sequencing problem with no wait in process. *Operational Research Quarterly*, 23(3):323–331, 1972.

[Whitley *et al.*, 1989] D. Whitley, T. Starkweather, and D. Fuquay. Scheduling problems and traveling salesman: The genetic edge recombination operator. In *Proc. 3rd Int. Conf. on Genetic Algorithms and their Applications (Arlington, Va.)*, pages 133–140, 1989.

Genetic Algorithms And Their Potential For Use In Process Control: A Case Study

Jean-Pierre Nordvik and Jean-Michel Renders
COMMISSION OF THE EUROPEAN COMMUNITIES
Joint Research Centre, Institute for Systems Engineering and Informatics
I-21020 ISPRA (Va), Italy

Abstract

This paper presents three different approaches to applying Genetic Algorithms (GA) for process control, namely: Pure GA - the classical approach of GA to process control; Hybrid GA which combines a conventional controller and a GA-based system; and the Tesselation approach based on the tesselation of the state space. These approaches are discussed in the light of two initial requirements, namely of "on-line performance" and of "no *a priori* knowledge". To illustrate the different approaches, results of their application to a case study - the regulation of a simple hydraulic system - are presented in detail, showing in particular ability to adapt under environmental changes.

1. Introduction

Genetic Algorithms (GA), Neural Networks, and Immune Networks are three techniques or paradigms derived from adaptive mechanisms of biological systems which have shown potential for application in the field of Machine Learning [7]. A property common to these three paradigms is that of adapting to widely different environments - they reconfigure themselves to an unspecified environment in an incremental and robust fashion. In the domain of Process Control - the manipulation of certain parameters of complex industrial processes in order to achieve stable optimal operation - there appears to be a need for such adaptive performance.

In fact, simple adaptive control of time-evolving systems remains a problem not fully solved by traditional means. The amount of *a priori* information needed about the process concerned is considerable [1]: even the extension of traditional Control Theory (CT) known as adaptive CT requires precise knowledge of both the disturbances possible and the process to be regulated (e.g. in the form of a linear difference equation), as well as qualitative knowledge about the process (e.g. maximum order, stability and minimum-phase, sign of the instantaneous gain, time-delay ...). Moreover this knowledge has to be very precise and complete: any slight imprecision can degrade dramatically the quality of the control. Adaptive CT further relies on slow deviation of the process parameters with respect to the adaptive capacity of the controller.

The three "biological metaphors" mentioned above exhibit adaptive behaviour while having requirements less stringent than those of CT or adaptive CT - or at least different ones. They could therefore be used to provide new control strategies or to enhance conventional ones. Given the differences between the three approaches, it is appropriate to compare their performance when applied to process control. This laboratory is participating along with others in a research action called ERBAS - Exploratory Research on Biological Adaptive Systems - which started in 1990 and aims to carry out such a "benchmark" comparative study. The benchmark consists of a number of case studies, to be tackled one by one, in increasing order of complexity, by the laboratories involved in the research. After each case study has been finished, the results from the various laboratories are compared. This paper presents three different methods of applying the concept of GA to the first case study, the "Tank" case study, and the results of these three methods.

2. GA AND PROCESS CONTROL

To exploit GA for process control, two problems have to be faced: (a) to find a way to encode, in an efficient and generic manner, all relevant control parameters into one or more individuals; (b) to define a fitness function, based on the encoding schema, which measures the quality of control performed. Both encoding and fitness score procedures have to be as "domain-independent" as possible (so that no *a priori* information about the process structure is needed) in order to address a large range of control processes. Moreover, the procedures must be carried out "on-line", so that the adaptive control can be performed in real time.

The first and simplest way to implement the GA approach would be simply to use GA techniques [6] to optimise the parameters of traditional control algorithms, such as a PID regulator [2]. However the parameter space of such algorithms is typically Euclidian, continuous, smooth and of low dimensionality. GA could therefore not be expected to perform better than classical search methods, and might well indeed perform worse. Moreover the resulting controller will always be an instance of a traditional controller and would therefore suffer from the same limitations as CT. For these reasons, we have not pursued this approach.

Three other methods which do not suffer from these drawbacks have been investigated:

(a) Pure GA - the "classical application" of GA to process control;
(b) Hybrid GA - the combined use of a conventional controller and a GA-based system;
(c) the GA-based tesselation of the state space.

These three approaches are here described in turn, and the results of their application to the Tank case study are presented subsequently.

2.1. PURE GA

The "production systems" approach is one traditional way to apply in a simple fashion the concepts of GA to the problem of process control, avoiding the drawbacks mentioned above. Production systems are made up of a collection of production rules of the form "IF <premise> THEN <conclusion>". In our case, these rules represent the chunks of control knowledge which have to be discovered by the GA in order to regulate adequately the process under control. There exist two general encoding mechanisms [3]: either each individual encodes a single rule (the Michigan approach) [5,8,9], or each individual encodes one version of the whole set of rules about the process (the Pitt approach) [12]. In the first case, many individuals are needed for any control strategy, and GA converges to an optimum set of individuals. In the second case, in principle there could be one optimum individual, and GA represents a mechanism for constructing and choosing such an individual. While each mechanism has its own advantages, in this paper we consider only the Pitt approach.

A control rule takes the general form "IF <system state> THEN <control action>". A method for having each individual represent the complete set of control rules is to perform the following steps: (a) the state-space is partitioned into regions (as in the "boxes" of Michie and Chambers) with an index number assigned to each region; (b) all different control actions are listed and an encoding procedure for these actions is defined; (c) an individual is represented as a sequence of codes, one for each region in the partition, where each code corresponds to the control action prescribed for that region. This method has already been shown to control adequately various physical systems, although it depends on *a priori* knowledge in the form of the partition of the state space.

Moreover, it does not fulfil our on-line requirement, because each individual of the population has to "control", independently from the others, the physical process for a certain time in order to have its fitness score measured. The scores are then computed serially and use to derive a new generation. A certain number of such generations are needed initially, constituting a "training stage", which ends when an individual emerges which controls the process adequately. This individual is then "frozen" and used to control the process subsequently.

2.2. HYBRID GA

Another possibility is the use of a GA-based Controller (GAC) in combination with a conventional controller (CC). Figure 1 gives the block-diagram of such a hybrid controller. While CC performs in a classical manner, the GAC modifies the CC output so as to enhance the global performance of the system. The resulting combined system can no longer be thought of as a direct implementation of CT. By comparison with the previous method, this approach has two advantages: (a) it is compatible with CT and can therefore be applied to existing control systems; (b) it can be implemented in such a way that the performance of the resulting system is always at least as good as the CC alone (in the case where no improvement is possible, the GAC response is to leave the CC output unmodified).

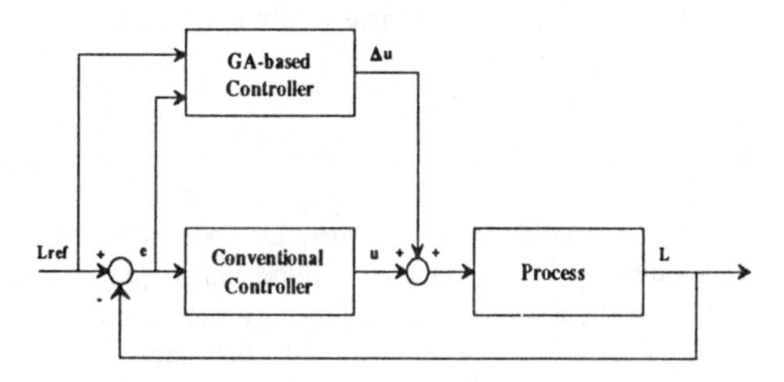

Figure 1 - Block Diagram of the Hybrid Controller

The same encoding procedure as described for the previous case can be applied to derive the individuals during the training stage. The only difference is that the output of the GAC no longer represents the global control variable, but only the modifications that have to be made to the CC output. For the same reasons as in the previous approach, both the requirements of "on-line performance" and "no *a priori* knowledge" are not completely fulfilled.

2.3. TESSELATION

A way to fulfil these two requirements is to change the way of representing the control knowledge, and to split the control problem into two stages: a modelisation and prediction stage, and a strategy-building stage based on the models thus constructed. The following rule pattern is used: "IF <state(t)> and <control action> THEN <state(t+1)> AND <system outputs>". Each individual now represents a physical model of the process, and therefore the fitness function no longer measures the quality of control, but rather measures the quality of a model by assessing its ability to predict future states. This enables computation of the fitness scores to be carried out in parallel (each individual matches its own predictions to the process and derives its fitness score) and therefore fulfils the on-line requirement. The individual with the best score is used as input for the second stage which determines the action to be undertaken through an analysis of the model represented (e.g. by computing some kind of inverse model).

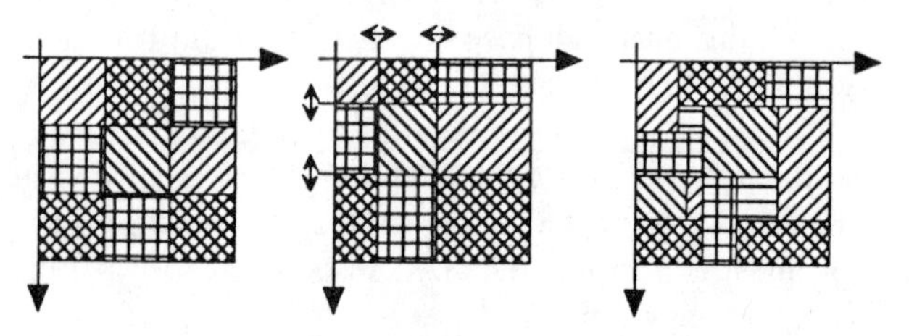

Figure 2 - Three types of tesselation in a 2-d space
(a) fixed partition (b) dynamic partition (c) general tesselation

As for the first two methods, a partition or "tesselation" of the state space (and in this case also of the control space) is required to describe the components of the rules which define the model. Whether the requirement of no *a priori* knowledge is fulfilled or not depends on how this tesselation is obtained: the tesselation can be derived dynamically from a GA-mechanism, or it can be fixed by the user prior to the experiment. The second case corresponds to the arbitrary partitioning of the domain of state and control variables (as in the previous two methods) which induces through orthogonal extension a tesselation of the product space; in this case, the cells are pre-defined and GA is used only to discover the rules of transition between them [4,11]. Figure 2(a) presents an example of such a tesselation.

A second possibility is to encode the partition of state and control variable domains explicitly in the genotype of the individuals, as well as the transition rules as defined above. By this means the requirement of no *a priori* knowledge is fulfilled, since both the static and the dynamic aspects of the model (both state definition and transition rules) are processed by GA. The advantage of this approach is that the partition of the domains changes dynamically (figure 2(b)), and eventually should converge on an optimum. Yet the resulting tesselation must still be one obtainable by orthogonal extension, and may therefore not be an overall optimum.

To expand the set of possible tesselations, it is necessary to define each sub-region explicitly. This permits the "general tesselation" of figure 2(c). To achieve such a general tesselation, each sub-region has to be independently encoded in the genotype. Unfortunately this causes the encoding of each individual to become much longer, which leads to computational difficulties.

Our present work involves several forms of tesselation. In this paper we present preliminary results from the "dynamic partition" (figure 2(b)) approach.

A further interesting development would be the use of Fuzzy Sets to define the tesselation. This can be seen either as extending the GA-based tesselation approach defined here by allowing fuzzy boundaries, or alternatively as the use of GA techniques to optimize a critical stage in the design of a fuzzy controller, the definition of a partition of the state space.

3. THE TANK CASE STUDY

Figure 3 shows a schematic representation of a tank and the related hydraulic system. The tank is filled up by an inflow, while a valve at the bottom of the tank is used to regulate the outflow. The regulation consists of adjusting this outflow in order to maintain a given level of liquid inside the tank while the inflow is changing. This inflow is thus considered as a perturbation upon the system. The control parameter is the valve opening position requested. The valve's response to a command is not instantaneous: after a request for a new position is sent, the valve moves asymptotically towards that new position. The valve itself is modelled by a non-linear relationship which relates the area of valve open to the valve position. Complete description of the terms of reference of the Tank case study can be found in [10].

It is assumed that the characteristics of the valve (the time constant of the control response and the relationship between area open and valve position), the perturbation upon the system (the inflow), and the physical equations describing the hydraulic systems are not known to the controller.

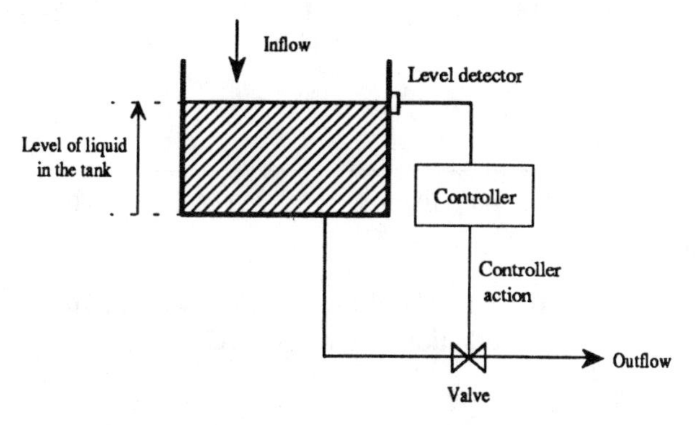

Figure 3 - Layout of the tank and hydraulic system

Various sets of simulation conditions have been used to assess the adaptiveness of the different control methods. The sets differ in the way the reference level and the perturbation change over time (e.g. step-wise or triangular function), and in the relationship which defines the valve (sigmoid, parabolic, etc.).

4. EXPERIMENTS AND RESULTS

4.1. PURE GA

It was decided to use a "bang-bang" control strategy, instead of a continuous control, in order to simplify the representation of the control variable (one allele is enough to code the variable). The two commands that the valve can receive were coded as follows: 0 for fully-closed and 1 for fully-open. The error space (i.e. the difference between the level of reference and the actual level in the tank) was divided arbitrarily into 28 intervals (table 1) which results in individuals of length 28 bits. Population size was 40. A 1-point crossover (p_c = 0.6) and a mutation operator (p_m = 0.02) were used in combination with a ranking-based selection scheme (pressure of selection = 1.5) and an elitist strategy (reproduction in each generation of the best individual from the previous generation).

The training stage consisted of the regulation of the tank for a certain period at a first reference level, followed by a period at a second (lower) reference level, with the inflow remaining constant. The fitness function was defined as the sum over time of the absolute value of the error between the reference level and the actual level in the tank.

Table 1 - Partition of the error space

Allele	Interval [cm]
1	[-Lmax, -100 [
2	[-100 , -50 [
3	[-50 , -25 [
4	[-25 , -10 [
5	[-10 , -9 [
6	[-9 , -8 [
...	...
27	[50 , 100 [
28	[100 , Lmax [

Figure 4 presents the mean and best performances, averaged over 10 experiments. After 50 generations, the pure GA controller has learned to control the tank adequately (in the sense that the strategy is nearly optimal, given the partition chosen and the restricted control signals permitted). Figure 5 presents a plot of the reference level over time during the training stage and a typical output of the process under control after that stage.

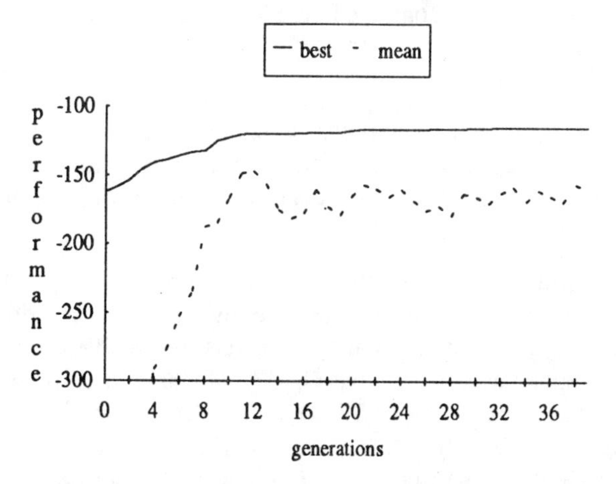

Figure 4 - Pure GA. Mean and best performances averaged over 10 experiments

The resulting controller was run under each set of simulation conditions, and demonstrated its ability to control the tank even in new situations. Figure 6 illustrates the robustness of this control, showing its functioning in the case of a constant reference level, but inflow (perturbation) varying over time. The performance of the control is degraded but it still works.

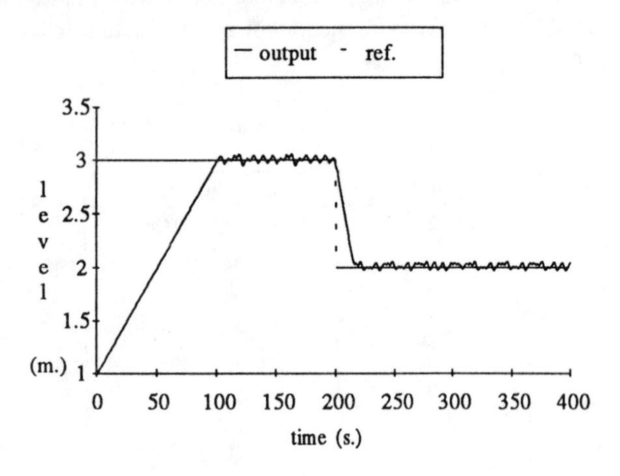

Figure 5 - Pure GA. Typical output of the process at the end of the training stage

An alternative implementation of this approach was to use all sets of simulation conditions in the training stage; the fitness function was defined as the absolute value of the error, summed over time and over all sets of conditions. Similar convergence performance was shown during this extended training stage. As expected this controller performed better across the range of simulation conditions (compare figure 7 with figure 6).

4.2. HYBRID GA

A near-optimal Proportional-Integral (PI) controller was obtained by hand tuning, using the same two-step reference level as above. This controller was shown to control the tank very well.

To achieve greater robustness of the overall system, both the error space and the reference level space were arbitrarily divided into 14 and 3 intervals, respectively. The command variable of the GAC could have 5 different values representing a -5%, -1%, 0%, 1% and 5% absolute modification of the PI output. This variable was described by a 5-value allele. The length of resulting individuals was therefore 42 alleles. Population size was 90. A 1-point crossover (p_c = 0.6) and a mutation operator (p_m = 0.02; symmetrical over all allele values) were used in combination with a ranking-based selection scheme (pressure of selection = 1.5) and an elitist strategy. Initially, an individual representing a "do not modify" control strategy (i.e. with a phenotype always equal to a 0% output) was introduced into the population to make sure that the GAC always had this possibility.

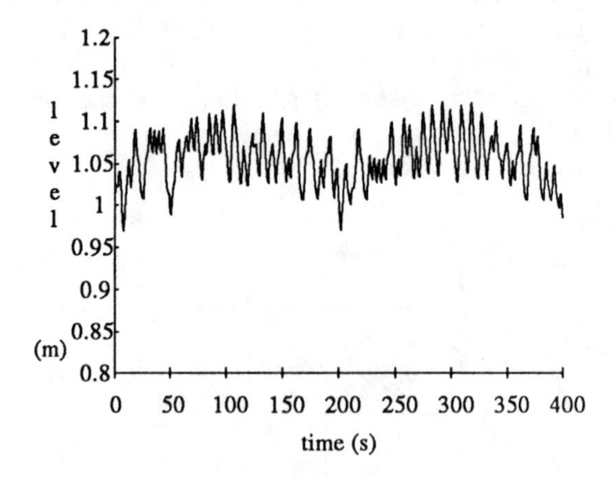

Figure 6 - Pure GA. Process output with varying perturbation (reference level = 1m)

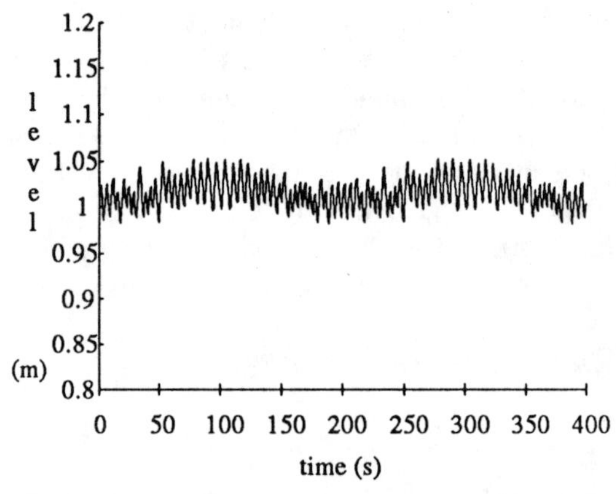

Figure 7 - Pure GA. Process output with varying perturbation (ext. training stage; ref. level =1m)

The training stage consisted of all sets of simulation conditions; the fitness function was defined as above. About 50 generations were needed to converge. After the training stage, the hybrid controller achieved better control than the PI alone, with all the sets of simulation conditions which involved variation over time of the reference level - one of the variables taken into account in the genotype. With the other sets, there was no significant change in the performance. Figure 8 presents, for both the PI alone and the hybrid controller once trained, a plot over time of the error between level and reference level covering a period shortly after a sudden change in the reference level. It illustrates the improved performance of control using the GAC: considerable reduction of the magnitude of the oscillation and faster damping.

The addition of a GA-based adaptive mechanism to a conventional controller thus performs better than the conventional controller alone. This hybrid controller is no longer a member of the PI-controller family. It is not an "Optimal PI-controller"; it is a PI-based controller enhanced and adapted to the simulation conditions for which it has been trained. Thus this hybrid approach combines the advantages of GA robustness and adaptiveness with the use of existing process knowledge and conventional CT.

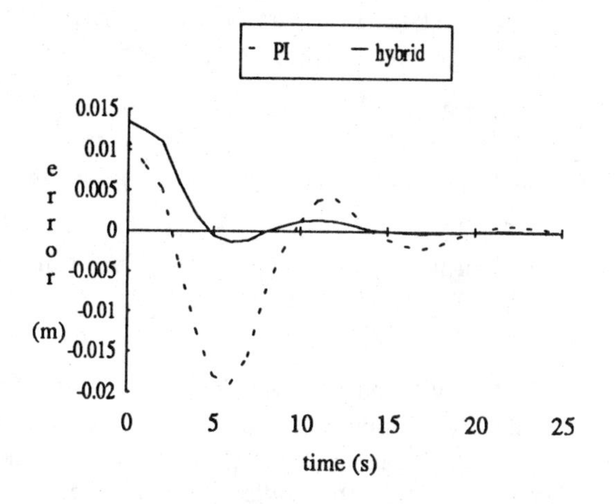

Figure 8 - Comparison of the error of the PI alone and the hybrid controller once trained (detail)

A limitation of this approach, however, is that the hybrid controller only adapts to modifications of the variables present in the genotype definition - which can only be measurable variables of the process. Such a controller cannot therefore achieve robustness to unknown perturbations such as change of inflow or valve characteristic.

4.3. TESSELATION

The GA-based tesselation approach is still under investigation and only preliminary results are available. The tesselation space used is 2-dimensional, one dimension representing the present level in the tank and one the control action. For the sake of simplicity a "bang-bang" control strategy was used. This substantially reduces the complexity of the problem: instead of a complete tesselation of the space, only a partition of two 1-dimensional spaces is needed, each space corresponding to one possible action. In this case, with all other parameters fixed, the model of the process can be described by equation (1). The objective of the GA mechanism, in the modelisation stage of this approach, is to produce a model which approximates f_0 and f_1.

$$(1)\quad \Delta L_{t+1} = \begin{cases} f_0(L_t) \text{ if control action is "fully closed"} \\ f_1(L_t) \text{ if control action is "fully open"} \end{cases}$$

An individual, which encodes a partition of the space, is mapped onto such a model by constructing two continuous piecewise linear functions, using a least squares approximation based on historical data from the process. The fitness score is defined as the error in this approximation.

For the subsequent strategy-building stage, a very simple technique was adopted: using the model derived from the best-scored individual, the consequences of each possible control action were calculated, and the action chosen was that which produced a level closest to the reference level.

Figures 9 and 10 presents the process output using a four-interval division of the level space, under the two sets of simulation conditions presented above.

These preliminary results shows that even with certain simplistic assumptions this approach achieves satisfactory control while fulfilling the two requirements of "on-line performance" and "no *a priori* knowledge", without the need for an explicit modelling stage beforehand.

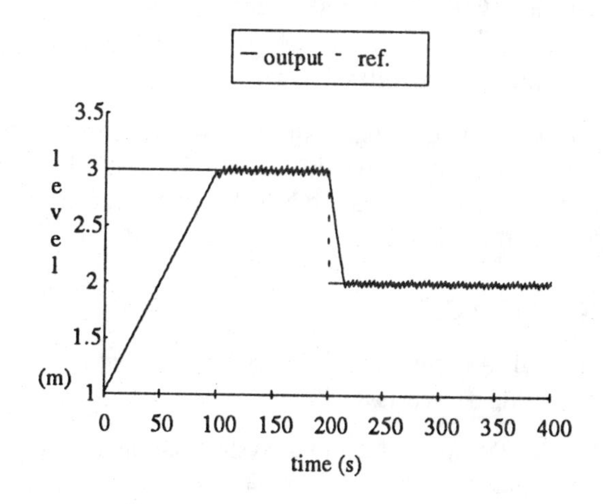

Figure 9 - Tesselation. Process output with step-wise reference level

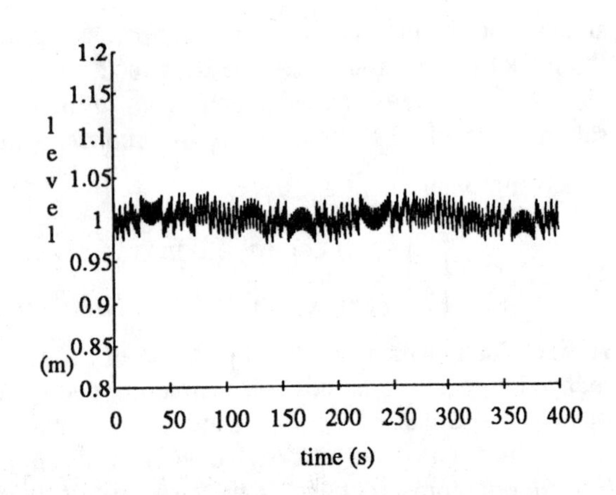

Figure 10 - Tesselation. Process output with varying perturbation (reference level = 1m)

5. CONCLUSION

The GA-based tesselation approach appears very promising. Even the somewhat restricted implementation presented here has been shown to achieve satisfactory control under on-line performance constraints, even with changing problem parameters. Such an on-line performance opens the way to the use of GA techniques to achieve robust adaptive control of processes in a changing environment, thus offering the prospect of GA-based control systems both more efficient and more secure.

Further work is needed to investigate the performance of such systems under difficult operating conditions, and to show whether the promise of "graceful degradation" under such conditions is really kept.

Acknowledgments

We would like to thank the ERBAS group for fruitful discussions on adaptive control and biological metaphors. Thanks also to Neil Mitchison for his help in writing this paper.

REFERENCES

[1] K.J. Astrom and B. Wittenmark. *Adaptive Control.* Addison-Wesley, 1989.

[2] K. De Jong. Adaptive System Design: A Genetic Approach. *IEEE Transactions on Systems, Man and Cybernetics,* 10, pages 566-574, 1980.

[3] K. De Jong. Learning with Genetic Algorithms: An Overview. *Machine Learning,* 3, pages 121-138, 1988.

[4] L.J. Fogel, A.J. Owens and M.J. Walsh. *Artificial Intelligence through Simulated Evolution.* J. Wiley & Sons, 1966.

[5] D.E. Goldberg. Genetic Algorithms and Rule Learning in Dynamic System Control. *Proc. of the First Conf. on Genetic Algorithms and Their Applications,* pages 8-15, 1988.

[6] D.E. Goldberg. *Genetic Algorithms in Search, Optimization, and Machine Learning.* Addison-Wesley, 1989.

[7] D.E. Goldberg and J.H. Holland. Genetic Algorithms and Machine Learning. *Machine Learning,* 3, pages 95-99, 1988.

[8] J.J. Grefenstette. Credit Assignment in Rule Discovery Systems Based on Genetic Algorithms. *Machine Learning,* 3, pages 225-245, 1988.

[9] J.J. Grefenstette. A System for Learning Control Strategies with Genetic Algorithms. *Proc. of the Third Conf. on Genetic Algorithms,* pages 183-190, 1989.

[10] J.P. Nordvik and J.M. Renders. The Tank Case Study: Terms of Reference - ERBAS JRC - Ispra. *TN I.90.97,* 1990.

[11] J.P. Nordvik and J.M. Renders J.M. Genetic Algorithms and Finite-State Automaton Discovery: Potential Application to Adaptive Process. Forthcoming in *IFAC Distributed Intelligence Systems,* Arlington, USA, 1991.

[12] M.O. Odetayo and D.R. McGregor. Genetic Algorithm for Inducing Control Rules for a Dynamic System. *Proc. of the Third Conf. on Genetic Algorithms,* pages 177-182, 1989.

A Genetic Algorithm for Primitive Extraction

Gerhard Roth
Institute for Information Technology
National Research Council of Canada
Ottawa, Canada

Martin D. Levine
McGill Research Center for
Intelligent Machines
Montreal, Canada

Abstract

Extracting geometric primitives from geometric sensor data is an important problem in computer vision and its solution is essential if intelligent machines are to be created. A novel representation of a primitive by a minimal subset of member points is proposed. By using this representation, crossover and mutation operations are defined to create a genetic algorithm for primitive extraction. We show that the genetic approach is an improvement over other algorithms that use this representation which rely on random search. Some experimental results are provided to justify this claim, and some conclusions regarding the place of genetic algorithms in computer vision are made.

1 Introduction

Extracting geometric primitives from geometric data is an important problem in computer vision. A geometric primitive is a curve or surface which can be described by an equation with a number of free parameters; geometric data are an unordered list of points in two-or three-dimensional Cartesian space. Such data are obtained by a variety of methods, such as edge detection [10], stereo vision [11], or by using active sensors such as laser rangefinders [12].

In this paper we describe a method for primitive extraction which uses a genetic algorithm (GA). It is based on a novel representation of geometric primitives not by their parameter vector, but by the minimal set of points necessary to define such a primitive. Such a representation enables the standard operations of crossover and mutation to be easily defined. While we have suggested this representation previously, it was always applied to primitive extraction with a purely random search algorithm [13]. As expected, the genetic approach is superior to random search in terms of execution time, with the difference becoming more significant as the difficulty of extracting the primitives increases.

The remainder of the paper is organized as follows: Section 2 defines the extraction problem more precisely and describes the concept of minimal subsets. Section 3 contains a description of the chromosome structure that uses minimal subsets, along with the definition of the crossover and mutation operations. Section 4 presents the experimental results. Section 5 contains the conclusions in which the merits of this approach over more traditional algorithms are discussed, and some comments are made on the place of GAs in computer vision.

2 Primitive Extraction

In the paradigm of model based computer vision the simplifying assumption is made that the environment consists of manufactured objects whose geometry is known a priori. This assumption is necessary if practical computer vision systems are to be built in the near future, and is reasonable given the fact that most manufactured parts are created with a computer aided design package. The surface of such parts can be described as a union of patches, where each patch is a particular geometric primitive. For example, the surface of a cube can be described as the union of six planar patches. More complex parts can be described by using more complex geometric primitives as surface patches. An important first step for an intelligent machine to understand its environment is the extraction of the underlying geometric primitives defining each surface patch.

More formally, we assume that the input consists of N points in Cartesian space, along with the equation defining the particular geometric primitive to be extracted. These points need not be given in any particular order, and can come from a variety of sources. Examples of geometric primitives are lines, circles and ellipses in 2D, and planes, spheres, cones, cylinders and tori in 3D. This does not by any means exhaust the list of possible primitives; all that is necessary is that the geometric primitive has an associated defining equation. A single application of the extraction process should produce some subset of the input points belonging to a single geometric primitive. If the input contains more than one such primitive, or contains a number of different geometric primitives, then the extraction process must be repeatedly applied with the appropriate primitive definition.

The minimal subset principle is based on the observation that a minimal subset of the points described by a geometric primitive is often a good representation of the primitive [2, 9, 13]. This is trivially true when the primitive is a perfect fit to the points, and becomes less true as the quality of the fit decreases. While this requirement for a good fit may seem restrictive, a closer study of the primitive extraction problem shows that this is not the case. The reason is that the resolution of modern sensors is usually high enough to make quantization noise insignificant. Therefore if the part model is correct, then the individual geometric primitives are good fits to the data. In Figure 1 we show an example of this representation; part (a) shows that three points define a unique circle, and part (b) shows that five points define a unique conic (an ellipse in this case). In both cases the primitive passes through these points exactly with no error in the fit.

In order for the minimal subset representation to be useful for primitive extraction it must be possible to produce the parameter vector of a given primitive from this minimal subset. The parameter vector is necessary to define the implicit equation, and we will show that the implicit equation is essential for the evaluation function of the GA. A geometric primitive is defined by an implicit equation in the form $f(\overline{p};\overline{a}) = 0$. In this notation $\overline{p}$ is a datum point, and $\overline{a}$ defines the parameter vector for this particular primitive. For example, a 2D line is described by the implicit equation $a_0 + a_1x + a_2y = 0$, where the parameter vector is (a_0, a_1, a_2) and the datum point is (x, y). It has been shown that it is possible to map a minimal subset of points to the parameter vector for a wide variety of geometric primitives, so this requirement is not a practical restriction on this approach [13]. The number of points in the minimal subset is one less than the size of the parameter vector of the implicit equation. The geometric primitive is an exact fit to this number of points which we label as P, and Table 1 shows the minimal number of points necessary to define a variety of geometric primitives.

It is easy to create an extraction algorithm based on the minimal subset representation. The algorithm operates by choosing random minimal subsets, creating a geometric primitive from each subset, evaluating the goodness of each primitive by an evaluation function, and choosing the best such primitive [13, 15]. The set of points belonging to the best primitive is then removed from the input data and this process is repeated till no more input points remain. The main difficulty with this method is that a large number of random samples is often necessary so the execution time is excessive, and the GA version of the extraction algorithm specifically addresses this issue.

Model Primitive	Number of Defining Points P
Line	2
Circle	3
Ellipse	5
Plane	3
Cylinder	5
Cone	6
Sphere	4
Torus	7

Table 1: The minimal number of points necessary to define different geometric primitives

The idea of using a GA for primitive extraction comes from the understanding that primitive extraction can be thought of as an optimization problem [14]. The objective is to find the parameter vector $\overline{a}$, which maximizes the scoring function for a given geometric primitive. In fact, previous methods of solving the primitive extraction that use the Hough transform [1, 7] and other methods from the field of robust statistics [4] can all be recast in this optimization framework, albeit with slightly different scoring functions. To our knowledge, this understanding of primitive extraction is fairly recent, and it immediately suggests the use of a GA. The obvious approach would be to use a GA that operates on a direct encoding of the parameter vector $\overline{a}$. The minimal subset principle says that this is inefficient because it is only necessary to look at the values of the parameter vector defined by all the possible minimal subsets. The reason is that for low noise levels it is very likely that the minimum found at the value of the parameter vector defined by one of the minimal subsets is very close to the global minimum. Besides being much more efficient in terms of the number of evaluations of the scoring function, this representation enables the crossover and mutation operators to be simply and naturally defined.

3 Genetic Algorithm Approach

A GA is a method of solving difficult optimization problems in a way that attempts to mimic nature. It requires a representation of an individual by a chromosome which is a list of genes, where each gene takes on a single value from a given set of tokens [5]. We represent a geometric primitive by a minimal set of points as described in the previous section. Therefore a chromosome consists of P genes, each of which is one of the P points in the minimal subset that defines the primitive. Thus the set of possible tokens for each

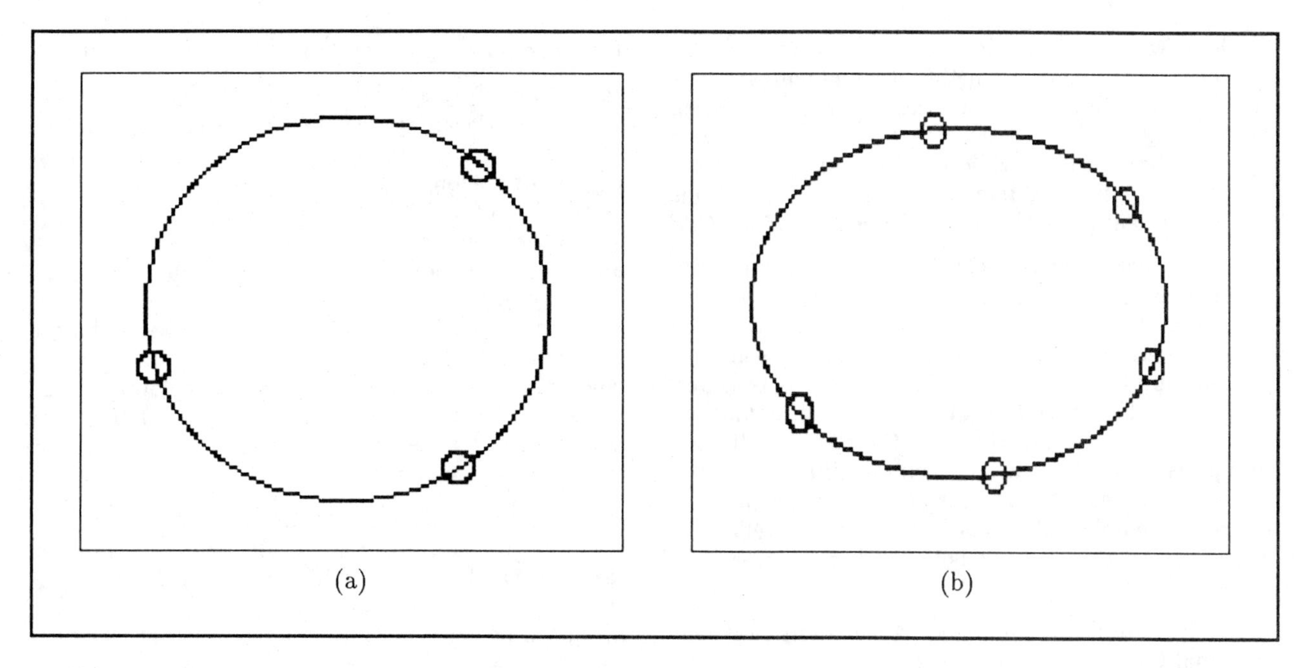

Figure 1: Representing geometric primitives by minimal point subsets (a) Three points define a unique circle (b) Five point define a unique conic (in this case an ellipse)

gene is a number from 1 to N which labels the point, where there are a total of N input points.

The GA takes two randomly chosen individuals (parents), and applies a crossover operation, followed by a mutation operation, to create two new population members (children). With our representation defining these operations is not a difficult task. The crossover operation creates two children from the two parents by making the children's chromosomes a combination of the genes of each parent's chromosome. The mutation operation is applied on each child's chromosome by replacing the token for a single gene (the index of one of the defining minimal points) by the index of a random input point. Such a natural definition of these operators is not obvious if a geometric primitive is represented by its parameter vector $\overline{a}$, instead of a minimal set of member points.

The next requirement for the GA is a way of scoring each individual in the population. While many scoring functions can be used to evaluate a geometric primitive [14], the simplest is a fixed band around each geometric primitive. This fixed band creates a template whose application to the input data computes the score of the geometric primitive. The score is simply the total number of input points contained in the fixed band. This is a sensible way to score a geometric primitive, since the more points belonging to it, the less likely that this alignment of points is random, and the better the chance that these points truly belong to the geometric primitive. The band size must be set by a careful study of the resolution of the particular sensor that created the input. Since modern sensors have very high resolution this band size is usually quite small. Given any implicit equation $f(\overline{p};\overline{a}) = 0$, a good approximation to the minimal distance of a point $\overline{p}$ to f is $f(\overline{p};\overline{a})/|\nabla f(\overline{p};\overline{a})|$. This formula enables the template matching to be done for any input function by simply cycling through all N input points, and counting those whose minimal distance puts them inside the fixed band which defines the template [13]. The only requirement is the ability to produce the parameter vector $\overline{a}$ from the minimal point that represents the primitive, which as discussed in the previous section, can be done for a wide variety of primitives.

In order for a genetic algorithm to work well the representation must enable partial solutions to be used as building blocks which are combined to produce even better new solutions [3]. Figure 2 shows how crossover accomplishes this task when using the minimal subset representation. In this figure all the points belong to a single circle which is not a perfect fit to these points since a small amount of noise has been added. In part (a) of the figure are two circles, described by two chromosomes, each consisting of three points. The points that make up the chromosome of the first circle are shaded in black, the second circle are shaded cross-hatched and the remaining points are not shaded. It is clear from the figure that neither circle is a good description of all the points. In part (b) of the figure, we see one of the possible children produced by a crossover of the two parents shown in part (a), with the new chromosome being a combination of the parents. The new chromosome points are the three filled points (two cross-hatched and one black) that exactly fit the new circle. Two chromosomes from one parent and one chromosome from the other parent are partial solutions which have combined to produce an even better solution. This demonstrates that the crossover operation can produce children that improve on their parents' score by combining partial solutions, and this ability is essential for the successful operation of a GA.

The final ingredient of a GA is a way of creating the initial population. This is done by selecting a small portion of the image from which a randomly chosen set of points is drawn to define the chromosomes of a small number of parents. This process is repeated over the entire image to create the initial population. In order to accomplish this task the image is divided into parts containing an equal number of points by the use of a k-d tree [16]. The results are not sensitive to the actual number of points in the lowest level of the k-d tree, and this number is set rather arbitrarily in our examples to 150.

A steady state GA (as opposed to the traditional generational GA) is used, along with a uniform crossover operation [17]. In the examples shown the initial population size is 50, the crossover probability is .8 and the mutation probability is .05. We have not found the results to be overly sensitive to these parameter settings, though more experimentation is necessary to validate this conclusion. The algorithm runs till 70 new children are created by the steady state GA, which is rather arbitrary. It is possible to set this stopping criterion in a more principled fashion, by modelling the input as consisting of good data corrupted by noise. Then we could stop when the GA obtained a statistically significant result, and this is an area of future research.

4 Experimental Results

The example we have chosen has as input a set of edge pixels extracted from an intensity image produced by a digitizer. The edge pixels were created using a public domain image processing package with the default threshold settings, and no attempt was made to optimize the results in any way by experimenting with these settings. The edge pixels are places where there is a significant image discontinuity and consist of an unordered list of 2D points. The particular image is of a number of coins, and the task is to extract the circles defined by these coins. In Figure 3, part (a) is the original intensity image of the coins. In part (b) of this figure are the edge pixels produced by the image processing package. The algorithms necessary to create such as an edge map can be found in any introductory book on computer vision [10]. Note that there are gaps in the circles, along with spurious points produced by the markings on the coins. However, this quality of input is generally typical of what any intelligent machine must deal with, and it is unreasonable for a computer vision system to expect any better.

Part (c) of the same figure shows the k-d tree decomposition of the image, with each of the smallest boxes contributing to the initial population. The resulting points that belong to all the chromosomes of the initial population of 50 circles are shown in part (d). Part (e) shows the circles that have been extracted by the GA, and they are superimposed on the original edge pixels in part (f). The circles were obtained by repeatedly running the GA, removing the points that belong to the highest scoring population member, and stopping when the number of input points is below a certain threshold. It can be seen by inspection that the extracted circles are essentially correct, though there are some minor deviations.

In our experiments a direct implementation of this extraction procedure that does not use a GA, but uses random search is noticeably slower than the GA. Further experiments must be done to quantify the speedup produced by the GA. Our experience is that the difference between the GA and random searching increases as the percentage of points on a single primitive decreases. This is not surprising, since as this percentage decreases the number of random subsets necessary for successful extraction also increases. A simple formula has been derived which shows the number of necessary random samples as a function of this percentage [15]. Thus the more complex the scene, the better the GA will perform when compared to random search. When a single geometric primitive contains a significant percentage (more than 30%) of all the data points, a GA does not decrease the execution time noticeably over random search.

We chose the above example because circle extraction is still an active area of research in computer vision [8, 19]. The Hough transform (HT) can be adapted to this task, but at the cost of using local edge directions at each pixel, which are usually not accurate. Because of the space requirements of the HT it must be customized for different primitive by splitting the parameter space into parts, and processing each part separately. It is not obvious how this is to be done in general, and it is also less accurate than using the entire parameter space. By contrast, the GA can be directly applied to many different geometric primitives. The only change that needs to be made is to use the appropriate function to map the gene points to the parameter vector of the primitive. An example of this flexibility is shown in Figure 4, where instead of extracting circles, we use the GA to extract an ellipse. In the same fashion as for circles, a minimal subset of points (in this case five) is used to represent an ellipse. Part (a) of this figure shows the initial edge pixels, while part (b) shows the extracted ellipse superimposed on the edge pixels. The ability to use the same algorithm without significant modification for a variety of complex geometric primitives is simply im-

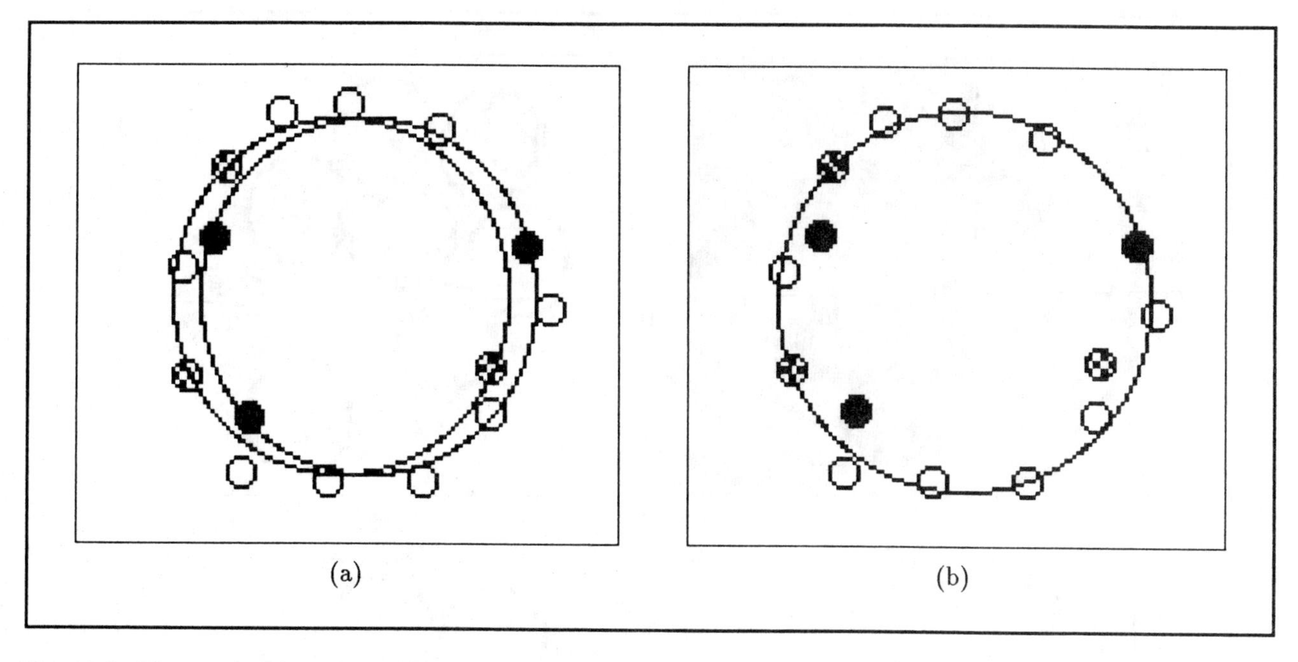

Figure 2: The scoring function and crossover operations (a) Two different genes and their associated circles (b) The new gene from crossover and the new circle

possible using the HT. Again it should be noted that extracting ellipses is also still an active area of research in computer vision [18, 6].

5 Conclusions

The GA succeeds because it uses local information defined by the initial population as building blocks to find a good global solution. A direct random search method using the minimal subset representation could be modified to do something similar by the use of a traditional computer vision merging algorithm, but we believe that the result would be inferior to the genetic approach. The reason is that the GA explores a number of possible alternatives in parallel. If the initial population is large enough, then in our experience the GA does not suffer from the problem of premature commitment. This occurs when a single possible answer is explored to the detriment of others and is the weakness of many computer vision algorithms. Further experimentation with the parameter settings of the GA are necessary since many are rather arbitrary. However, our experience so far is that the GA is not unduly sensitive to these settings. We can say with confidence that the GA approach is better than random search in terms of quality of results and execution time.

The GA requires a good chromosome representation to function properly. In our case this representation encodes a geometric primitive by the minimal set of points necessary to define a unique instance of that primitive. This representation has been previously described in [14], but has not to our knowledge been used in a GA. In order to evaluate a population member it is necessary that the chromosome representation be transformed into a parameter vector, which can be done for a wide variety of geometric primitives. We have shown examples of circle and ellipse extraction, but are also working on extracting planes, spheres, cylinders and tori from 3D data using the same basic GA approach.

The GA shares with the HT considerable robustness in terms of its ability to find the correct geometric primitive in the face of missing and spurious input data. This robustness is not without its computational cost, and we have shown elsewhere that adding robustness to an extraction process dramatically increases the number of local minima of the scoring function to be optimized [13]. Such optimization problems are the perfect domain for GAs, which have the ability to find a good solution very quickly. It should be noted that robustness is one of the distinguishing features of biological systems, and it is reassuring that a GA for primitive extraction has this characteristic.

On a more philosophical level GAs are one member of a family of so called "weak methods" which are currently enjoying a resurgence in the field of Artificial Intelligence. Another member of this family are the connectionist approaches typified by neural networks. Historically, these weak methods fell out of favour because of the inability to extend them to more complex situations. This produced the generally held hypothesis that the creation of intelligence requires a large amount of knowledge, and gave birth to the so called

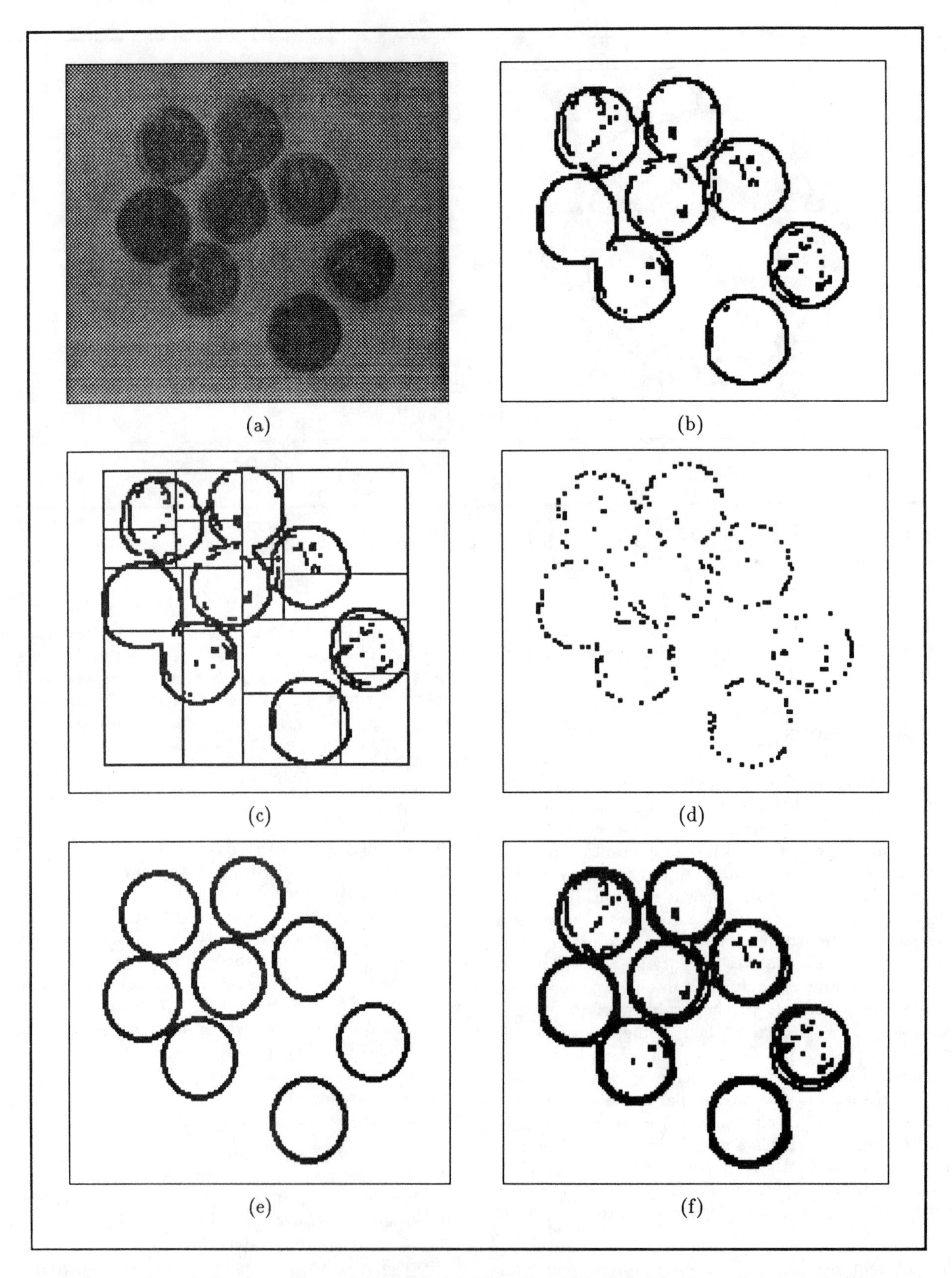

Figure 3: Extracting circles from a complex image (a) The original image (b) The edge pixels (c) The k-d tree used to create the starting population (d) The points in the initial population (d) The extracted circles (f) The extracted circles superimposed on the edge pixels

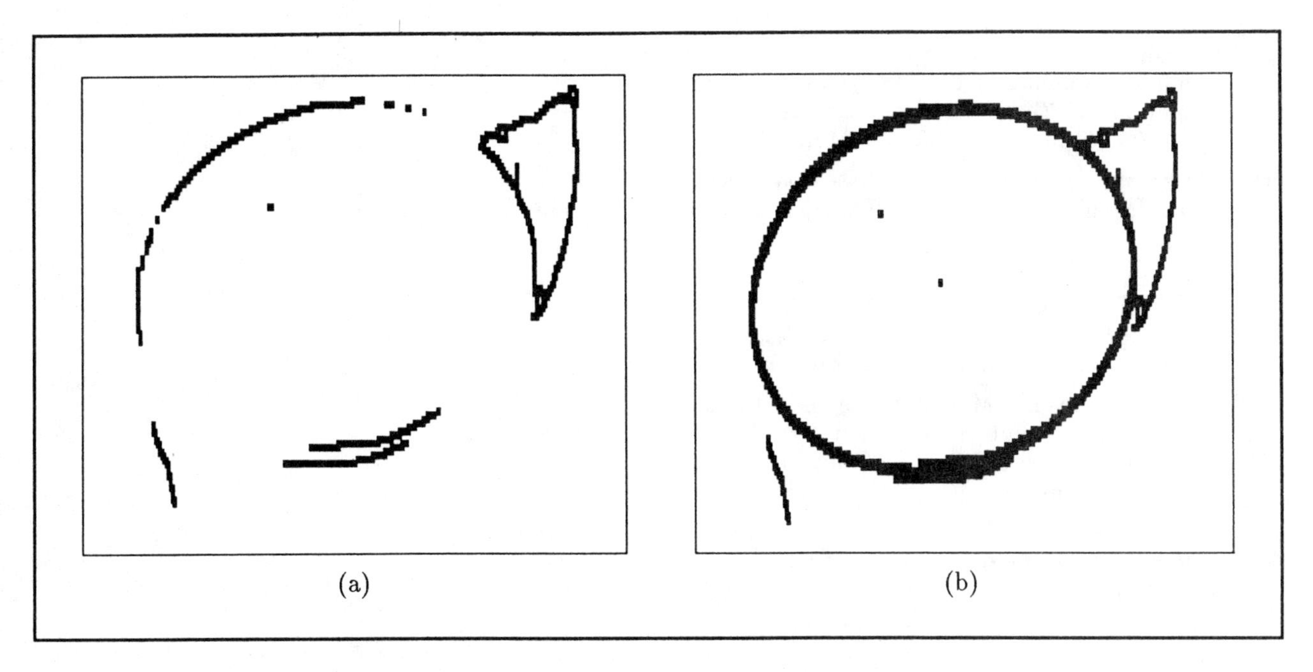

Figure 4: Extracting ellipses from a complex image (a) The edge pixels (b) The extracted ellipses

"strong methods" of Artificial Intelligence, sometimes referred to as the knowledge-based approaches. However, the difficulty of extracting and encoding knowledge, not to mention the complexity of the algorithms necessary to use such knowledge, has cast a pale on the entire knowledge based paradigm. It may be that early failure with weak methods was not due to any inherent fault in the paradigm, but simply due to the lack of appropriate hardware. The increasing computer power available from modern parallel architectures suggest that very simple approaches like GAs may work much better than was initially expected. In our case such hardware would enable the scoring function to be computed very rapidly. This could make it possible to use GAs for real time primitive extraction. This would be an important step towards producing robust and practical algorithms for such computer vision tasks as tracking and object identification. These basic tasks will be essential requirements for any intelligent machine.

Acknowledgements

The authors would like to thank one of the anonymous reviewers for the helpful comments on the manuscript. Gerhard Roth would like to thank the National Research Council of Canada (NRCC) for providing education leave for graduate studies. Martin D. Levine would like to thank the Canadian Institute for Advanced Research and PRECARN for its support. This work was partially supported by the National Science and Engineering Council of Canada.

References

[1] R.O. Duda and P.E. Hart. The use of the Hough transform to detect lines and curves in pictures. *Communications of the ACM*, 15:11–15, 1971.

[2] M.A. Fischler and R. C. Bolles. Random sample consensus. *Communications of the ACM*, 24(6):381–395, June 1981.

[3] David Goldberg. *Genetic algorithms in search, optimization and machine learning.* Addison-Wesley, Reading, Mass., 1988.

[4] F.R. Hampel, E.M. Ronchetti, P.J. Rousseeuw, and W.A. Stahel. *Robust statistics: the approach based on influence functions.* Wiley, New York, 1986.

[5] John H. Holland. *Adaptation in natural and artificial systems.* University of Michigan Press, 1975.

[6] C.L. Huang. Elliptical feature extraction via an improved Hough transform. *Pattern Recognition Letters*, 10:93–100, 1989.

[7] J. Illingworth and J. Kittler. The adaptive Hough transform. *IEEE Transactions On Pattern Analysis and Machine Intelligence*, 9:690–698, 1987.

[8] C. Kimme, D. Ballard, and J. Slansky. Find circles by an array of accumulators. *Communications of the ACM*, 18(2):120–122, 1975.

[9] P. Kultanen, E. Oja, and L. Xu. Randomized Hough transform (rht). In *10th International Conference on Pattern Recognition*, pages 631–635, Atlantic City, June 1990.

[10] Martin D. Levine. *Vision in man and machine.* McGraw-Hill, New York, 1985.

[11] Y. Ohta and T. Kanade. Stereo by intra- and inter-scanline search using dynamic programming. *IEEE Trans. Pattern Analysis and Machine Intelligence*, 7(2):139–154, Mar. 1985.

[12] M. Rioux. Laser rangefinders based on synchronized scanning. *Applied Optics*, 23:3837–3844, 1985.

[13] G. Roth and M.D. Levine. Segmentation of geometric signals using robust fitting. In *10th International Conference on Pattern Recognition*, pages 826–831, Atlantic City, June 1990.

[14] Gerhard Roth and Martin D. Levine. Random sampling for primitive extraction. In *International Workshop on Robust Computer Vision*, Seattle, Washington, Oct. 1990.

[15] P.J. Rousseeuw and A.M. Leroy. *Robust regression and outlier detection.* Wiley, 1987.

[16] H. Samet. The quadtree and related hierarchical data structure. *ACM Computing Surveys*, 16:187–260, 1984.

[17] Gilbert Syswerda. Uniform crossover in genetic algorithms. In *Proceedings of the third international conference on genetic algorithms*, pages 2 – 9, George Mason University, Fairfax, Virginia, July 1989.

[18] H.K. Yuen, J. Illingworth, and J. Kittler. Detecting partially occluded ellipses using the Hough transform. *Image and Vision Computing*, 7(1), Feb. 1989.

[19] H.K. Yuen, J. Princen, J. Illingworth, and J. Kittler. Compartive study of the Hough transform methods for circle finding. *Image and Vision Computing*, 8(1):71–77, Feb. 1990.

Genetic Algorithms for the 1-D Fractal Inverse Problem

R. Shonkwiler F. Mendivil A. Deliu
School of Mathematics
Georgia Institute of Technology
Atlanta, GA. 30332
e-mail: shenk@math.gatech.edu

Abstract

We describe the basics of one-dimensional IFS type fractals including their generation, the forward problem, and their encoding, the inverse problem. We also describe the details of a solution of the latter by a Genetic Algorithm. The resultant algorithm converges remarkably well, and, when parallelized as described, achieves a 600 fold speed-up using 12 processors.

1 Fractal Sets Generated by Iterated Function Systems

Let $\mathcal{W} = \{w_1, w_2, \ldots, w_n\}$ be a finite set of affine maps of the unit interval $I = [0,1]$ into itself, that is maps of the form

$$w(x) = sx + a, \qquad 0 \le x \le 1.$$

Here the parameter s is the scale factor and the parameter a is the translation. Alternatively, putting $l = a$ and $r = s + a$, then

$$w(x) = l(1-x) + rx. \tag{1.1}$$

In this form the unit interval is seen to map into the interval between l and r and so $0 \le l, r \le 1$. We impose the condition that the image set $w(I)$ be a strict subset of I, i.e. that w be a contraction map and $|s| < 1$. In this case such a map w has a fixed point in I.

Associated with every such collection $\mathcal{W}$ is the *attractor* A, that is the unique subset $A \subset I$ characterized by the selfcovering property that

$$A = \bigcup_{i=1}^{n} w_i(A). \tag{1.2}$$

For the proof of the existence and uniquness of an attractor which satisfies (1.2) see [1]. It is however easy to obtain points in A. Obviously the fixed point x_i^* of each map w_i in $\mathcal{W}$ belongs to A by (1.2). Further this equation provides that if $x \in A$ then also $w_i(x) \in A$ for each $i = 1, 2, \ldots, n$. It follows that for every composition $f = w_{i_1} w_{i_2} \cdots w_{i_k}$, where $i_j \in \{1, \ldots, n\}$, for all $j = 1, \ldots, k$, if $x \in A$ then also $f(x) \in A$. It is in this way that $\mathcal{W}$ is an iterated function system (IFS).

Moreover the observation above provides a method to visualize an attractor on a computer screen. Starting from the fixed point x^*, say of map w_1, choose $i_1 \in \{1, 2, \ldots, n\}$ at random and plot $x_1 = w_{i_1}(x^*)$. (To make the image easier to see, plot a short vertical line at x_1.) Now repeat this step with x_1 in place of x^* and $x_2 = w_{i_2}(x_1)$ in place of x_1, then repeat with x_2 in place of x_1, e.t.c. until say 10,000 points have been plotted. This construction is known as the *Random Iteration Algorithm* for constructing the attractor.

A difficulty is that the mapping $\mathcal{W} \to A_{\mathcal{W}}$, which assigns to an IFS W its attractor $A_{\mathcal{W}}$, is not one-to-one. As an example the Cantor set is the attractor of the IFS

$$w_1(x) = \frac{1}{3}x, \qquad w_2(x) = \frac{1}{3}x + \frac{2}{3},$$

as well as the attractor of the distinct IFS

$$w_1'(x) = -\frac{1}{3}x + \frac{1}{3}, \qquad w_2'(x) = \frac{1}{3}x + \frac{2}{3}.$$

See also fig.1.

Typically the attractor of an IFS is a fractal set.

2 Fractal Measure

Given an IFS $\mathcal{W}$ having n maps, let $\mathcal{P} = \{p_1, p_2, \ldots, p_n\}$ be a set of positive numbers summing to 1, i.e. a list of probabilities. Associated to such a pair $(\mathcal{W}, \mathcal{P})$ is a Borel probability measure μ on I, supported on the attractor A of the IFS $\mathcal{W}$. Moreover μ satisfies the characterizing property (*invariance*)

$$\mu(S) = \sum_{i=1}^{n} p_i \mu(w_i^{-1}(S)), \tag{2.1}$$

for all Borel subsets $S \subset A$.

Just as the Random Iteration Algorithm can be used to visualize the attractor, likewise it can be used to visualize the invariant measure. Proceed as described above except that when choosing the next map to apply, w_i is selected with probability p_i, $i = 1, 2, \ldots, n$. In this way the parts of the attractor having more mass will appear denser, especially during the initial iterations of the algorithm.

Alternatively the measure may be visualized by means of a histogram. Choose a suitable refinement of the unit interval, say 1024, and run the Random Iteration Algorithm with the modification as above, on the kth iteration the map w_i is chosen with probability p_i. Then obtain $x_k = w_{i_k}(x_{k-1})$ and increment the cell of the possible 1024 into which x_k falls. Some typical histograms are shown in fig. 4-11.

In addition to histograms, another basic way to describe a measure is to compute its moments m_k, $k = 0, 1, \ldots$. The kth moment is defined by

$$m_k = \int x^k \, d\mu.$$

From the equation (2.1) we deduce the following recursive formula for integrating a continuous function f over A with respect to the invariant measure ([1,p356]),

$$\int f \, d\mu = \sum_{i=1}^{n} p_i \int (f \circ w_i) \, d\mu.$$

If we let $f(x) = x^k$ in the previous formula, then by binomial expansion we get

$$(f \circ w_i)(x) = (s_i x + a_i)^k = \sum_{j=0}^{k} \binom{k}{j} s_i^{k-j} x^{k-j} a_i^k.$$

Hence

$$m_k = \int f(x) \, d\mu = \sum_{i=1}^{n} p_i \sum_{j=0}^{k} \binom{k}{j} s_i^{k-j} a_i^k m_{k-j},$$

or, solving for m_k,

$$\left(1 - \sum_{i=1}^{n} p_i s_i^k\right) m_k = \sum_{j=1}^{k} \binom{k}{j} m_{k-j} \left(\sum_{i=1}^{n} p_i s_i^{n-j} a_i^j\right). \qquad (2.2)$$

The coefficient on the left is not zero and hence this equation may be used to compute the moments m_k recursively. Of course $m_0 = 1$ since μ is a probability measure.

3 The Inverse Problem

The *inverse attractor problem* consists in finding an IFS $\mathcal{W}$ whose attractor A is given. Similarly, the *inverse measure problem* consists in finding a pair $(\mathcal{W}, \mathcal{P})$ whose invariant measure μ is given. These are also known as *encoding* an attractor or a measure respectively.

To solve the inverse problem for the attractor and for the measure we need a notion of closeness or distance between two subsets A and B of I, respectively between two positive measures μ and ν on I. In the former case we use the *Hausdorff* metric $h(A, B)$, see [3], which is defined as

$$h(A, B) = \sup_{x \in A} \inf_{y \in B} |x - y| + \sup_{y \in B} \inf_{x \in A} |y - x|.$$

In the latter case there are two candidates, the first is the L_2 distance between their moments, thus

$$M(\mu, \nu) = \left(\sum_{i=1}^{\infty} |m_i - n_i|^2\right)^{1/2}$$

where $(m_1, m_2, \ldots)$ are the moments of μ and $(n_1, n_2, \ldots)$ are those of ν. Another possible distance between measures is the *Hutchinson metric* $H(\mu, \nu)$, see [5], defined by

$$H(\mu, \nu) = \sup\{ \int_I f \, d\mu - \int_I f \, d\nu \; : \; |f(x) - f(y)| \leq |x - y| \}$$

The supremum is taken over all continuous real-valued functions satisfying the stated Lipschitz condition. The Hutchinson metric may be computed for an N-cell discretized measure on I (a histogram) by, (see [2])

$$H(\mu, \nu) = \sum_{i=1}^{N-1} |S_k|$$

where S_k is the kth partial sum

$$S_k = \sum_{i=1}^{k} (\mu(i) - \nu(i)).$$

The level surfaces which result from these distance functions are very pathological even for small numbers of maps. The surfaces rise and fall abruptly with small changes in parameter values and have numerous local minima, see [6]. The moment surface is especially bad; we'll have more to say on this below.

4 Solution by Genetic Algorithms

The aforementioned distance functions may be used to serve, in a reciprocal way, in the construction of fitness functions. We take the distance between a goal fractal and a potential solution fractal as the *error* in that attempted solution. For the purposes of genetic algorithms, the fitness can be inversely related to this error. Our exact mechanism for obtaining a fitness is given below.

After fitness, the next step toward the construction of a genetic algorithm is the invention of good genetic operators. Before discussing that we must describe the data structure upon which genetic operators would act.

4.1 Genes

First consider the inverse attractor problem. In this case a solution will be an IFS W comprising a set of n affine 1-variable maps each of which has two parameters. We choose the second form namely (1.1) i.e. $l(1-x)+rx$, for these maps because in this form the only constraints on the parameters l and r are that they lie in I,

$$0 \leq l \leq 1, \qquad \text{and} \qquad 0 \leq r \leq 1.$$

A lingering problem arises in connection with the parameter n, the number of maps in the collection. It is desirable to obtain a satisfactory solution with n as small as possible. This stems from one of the potential uses of fractal encoding, namely as a data compression technique. Indeed in a practical application one may want to hold n fixed (at 4 say) and pose the problem of finding the best n-map IFS representing a given attractor. Further, given a subset A of I, there is, in advance, no guarantee that it is the attractor of some IFS. In this case the problem is to find the nearest IFS attractor using a certain number of maps to the given set.

Consequently we choose to define a sub-problem to be that of finding the best fractal fit using a fixed number of maps. Then this sub-problem can be embedded in the larger problem of finding the best value of n if we so desire. This will be explained in more detail below.

Thus for a fixed value of n, our data structure, which we call a "bug", consists of a string

$$(l_1, r_1, l_2, r_2, \ldots, l_n, r_n)$$

of $2n$ real numbers each in the range 0 to 1. One may think of the string as the "genome" of the bug and each single number of the string as a "gene". The binary bits that make-up a floating point approximation to a real number can be thought of as "base-pair nucleotides" of the gene.

4.2 Matings and Mutations

Our genetic operators act on genes, i.e. the real numbers of the string. We use two operators, one a unary operator (it acts on a single string) which we term a "mutation", and the other a binary operator we refer to as a "mating". Both operators are stochastic.

The details of a mutation are as follows. A gene is selected at random equally likely. Suppose its present value is x, $0 \leq x \leq 1$. The new value of the gene, x', will be a random perturbation of the old value up to a built-in maximum, say Δ. Thus let

$$\begin{aligned} a &= \max\{0, x-\Delta\} \\ b &= \min\{1, x+\Delta\}. \end{aligned}$$

Put

$$x' = a + \text{rand}() * (b-a)$$

where the function *rand()* returns a uniformly generated random number in [0,1).

The details of the mating operation are as follows. Designate the "parents" for the mating as α and β and recall that each is a string of $2n$ genes. The offspring γ is formed by selecting for each gene position either the corresponding value of the α parent or the β parent with equal likelyhood.

For $i = 1$ to $2n$

if(rand() < 0.5) $\gamma_i = \alpha_i$

else $\gamma_i = \beta_i$

4.3 Seasonally Varying Population Size – A Generation Cycle

The individual bugs on which our genetic operators work are selected members of a *population* of such potential solution strings. We allow the population size to vary from generation to generation, stochastically, as follows. When a mutation occurs, a new bug is added to the population but the old one is not removed. Likewise with mating, the offspring is added to the population but neither parent is removed. This continues thoughout a generation.

At all times the population is kept in rank order of fitness. This is easy to do in that when a new member is added, it is simply merged into the population at its proper place according to its fitness.

A generation itself consists of 2 mutations and 10 matings. The operand for each mutation is selected at random from the present population equally likely. The parents for a mating are selected as follows. One parent, the α parent, will be from the best 10 members of the population in order, i.e. first the best, next the second best, and so on through the 10th best. The second parent, or β parent, for each mating is selected from the entire population at random.

The *mating/mutation* phase is followed by a *removal* phase. There will be 12 new members of the population integrated in according to fitness rank. In the removal phase some member may be removed (permanently) from the population before a new cycle begins. To determine if members will be removed and if so how many, we first select the desired population size for the next generation according to (an approximate) normal distribution with a mean of 16 individuals. For the sake of speed, the algorithm for this selection is only approximately normal. Moreover the selected size for the next generation is contrained to lie between 8 and 32. If the present population (with its 12 new members) is less than the target population for the next generation, no removals take place and the generation cycle is complete. If the present population is too high, then removals occur one by one until the target population size is reached.

4.4 Removals

Removals are done stochastically with the less fit bugs at greater risk. Specifically, a *living probability* $L = 0.85$ is applied to the population in reverse rank order something like Russian Roulette until one is selected. Thus the probability that the kth worst fit bug is removed will be (approximately) L^k. Approximately because if the entire present population survives the gauntlet, then an individual is selected at random for removal.

4.5 Stopping Criterion

We maintain an ongoing log of the best string to be obseved. We break out of the generation cycle loop and report this best value when no improvement has taken place in a fixed number of generation cycles (600) provided a specified minimum number of cycles have occurred (4000).

4.6 Treating Arbitrary Numbers of Maps

The algorithm so far optimizes on a fixed number of maps. To deal with a variable number of maps we take the process above as a subroutine and drive it in a loop with the number of maps as the incrementing variable. As n increases we expect the error as given by the appropriate distance function to decrease. This generally occurs, however as the method is stochastic, it can happen that the error for $n + 1$ maps exceeds that for n of them. Therefore we choose to terminate the search when no improvement is observed twice in succession, that is when the error for $n + 1$ and $n + 2$ maps exceeds that for n maps.

4.7 Fitness

From the foregoing we note that the only use made of the error function is to maintain a rank order of our artificial colony of strings. In turn however matings and removals depend on this order. One virtue of this approach is the avoidance of scaling problems – it makes no difference by how much one string outperforms another, only that it does.

$w_1(x) = .21x + .05$, $w_2(x) = .32x + .68$, $w_3(x) = .14x + .40$
3 MAP GOAL ATTRACTOR resolution 512
$w_1(x) = .21x + .051$, $w_2(x) = .321x + .679$, $w_3(x) = .140x + .400$
3 MAP SOLUTION Error = 4 (see text)
$w_1(x) = -.214x + .259$, $w_2(x) = -.326x + .994$, $w_3(x) = -.161x + .567$
3 MAP SOLUTION Error = 487 (see text)

Figure 1:

5 Results

5.1 The Inverse Attractor Problem

The algorithm was run both on parallel platforms and on a single processor computer. Our technique for parallelization is very simple and very effective. We simply run a copy of the algorithm on each of the multiple processors independently of each other except that one process gathers the final results of them all. (See Parallelizing Genetic Algorithms, this conference.)

The runs were made on randomly generated IFS attractors discretized to a resolution of 512. Only 2 and 3 map attractors were tested (4 and 6 parameter problems). Despite the small number of maps, the resulting attractors are quite complex and difficult to solve, see figures 1a-c. The figure 2 shows typical convergence curves for a single cpu and a parallel cpu run. The run times are on a Sun Sparc workstation and take about 1 hour for 10,000 iterations.

For the parallel algorithm the natural statistic to show is speed-up,

$$\text{speed-up} = \frac{\text{time for a single process}}{\text{time for the parallel process}}$$

However for a stochastic algorithm there is an inherent difficulty. For any time T however large (expressed in iterations until solution say), there is a non-zero probability the algorithm will take T time to finish. So

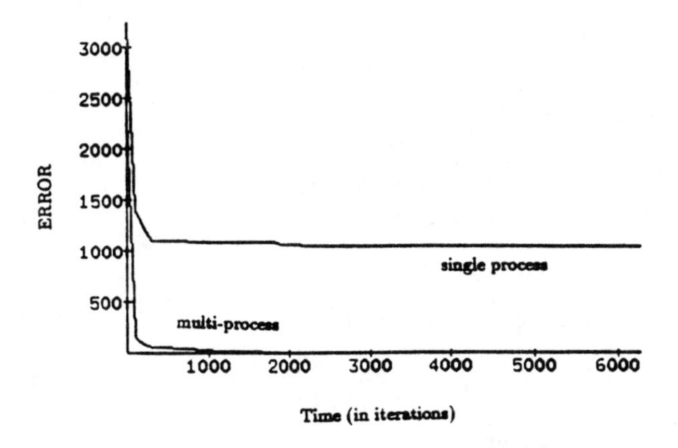

Figure 2: CONVERGENCE SCHEDULE

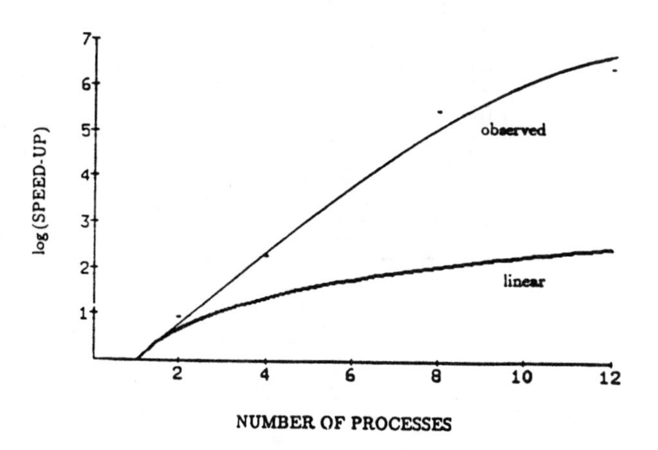

Figure 3: PARALLEL SPEEDUP

it is that in 24 runs on the problem above, 10 required an unknown number of iterations exceeding 100,000 in order to solve it. A solution was defined to be the achievement of a Hausdorff distance of less than 500.

Nevertheless by using the 100,000 value for these runs, we obtain an estimate of the expected run time which is too small. Hence our calculated speed-ups are under estimates of the actual ones.

The observed speed-ups on the order of hundreds for 10 to 20 processes is typical for this problem. As shown in [8], for a "deceptive" problem (in the terminology of [4]), the expected time to reach the goal can be infinite. In such cases the parallel speed-up will exhibit the noted behavior (cf Parallelizing Genetic Algorithms, this conference).

5.2 The Inverse Measure Problem

The first objective or error function tried here was *moment matching* as mentioned in §3. Its appeal consists in the speed with which the moments can be computed recursively by (2.2). This objective, however, proved to be very unsatisfactory. First it was discoverd that as many as 60 moments of two fractal measures could

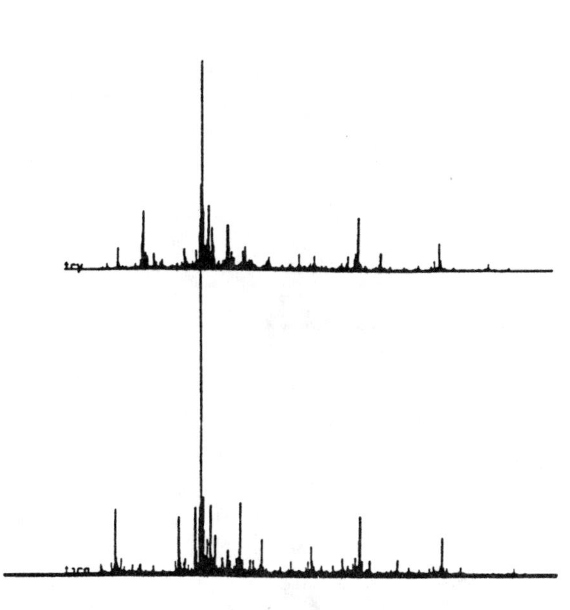

Figure 4:

agree through 10 decimal places and yet their two histograms were far apart. More disturbing is that perturbing the parameters of an IFS even slightly, gives rise to another IFS whose moments may differ considerably from those of the original IFS. Hence we sought another error function.

By contrast the Hutchinson metric worked splendidly. The only drawback to its use is that the histogram of the trial IFS must first be constructed. For this we use the Random Iteration Algorithm, a time consuming step. Nevertheless, in execution times on the order of 1 hour we can achieve results such as those depicted in figures 4-11. The lower histogram in each figure is the target, the upper is the best approximation. In these runs, an IFS with probability $\mathcal{W}, \mathcal{P}$, is randomly generated and its histogram computed, lower graph. Then the GA attemptes to find an IFS with probabilities whose histogram matches it, upper graph. Note that an n-map IFS with probability has a total of $3n-1$ free parameters. Runs were done on IFS's with up to 15 maps.

Of interest are figures 8-10 which show that a given fractal measure generated using a 15 map IFS can be reasonably well approximated by a 10 maps IFS. Figure 11 shows the result of attempting to fit an arbitrarily generated histogram. Such a histogram is highly unlikely to be the discretized measure arising from an IFS with probability having a small number of maps. The figures show that likewise the fit is much less satisfactory.

No parallelization was tried on this problem.

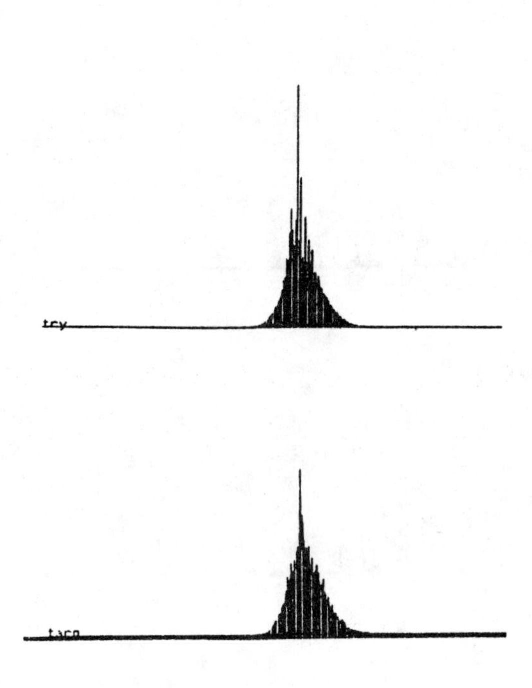

Figure 5:

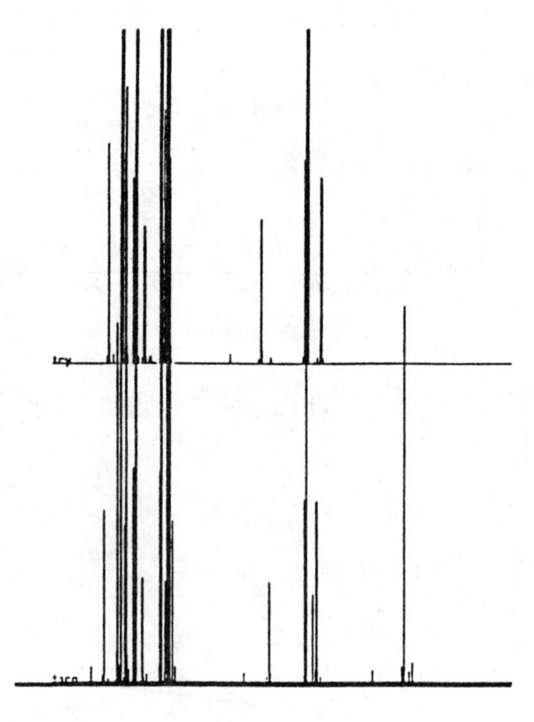

Figure 7:

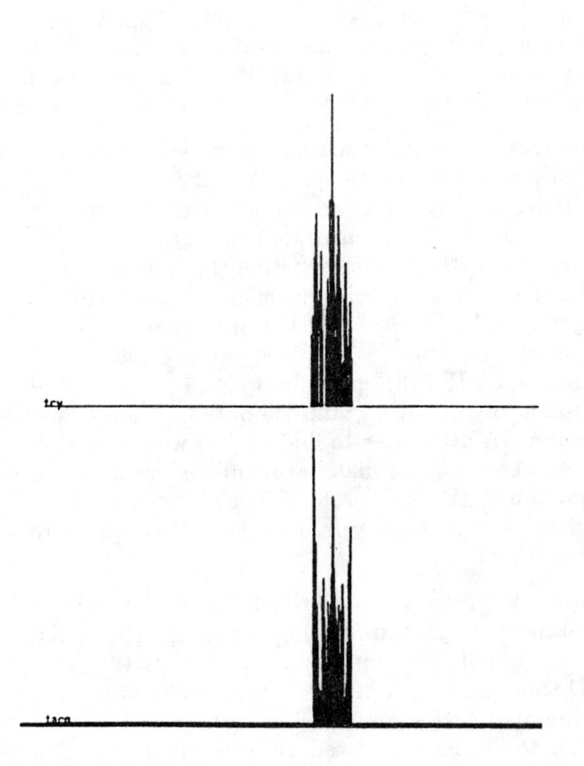

Figure 6:

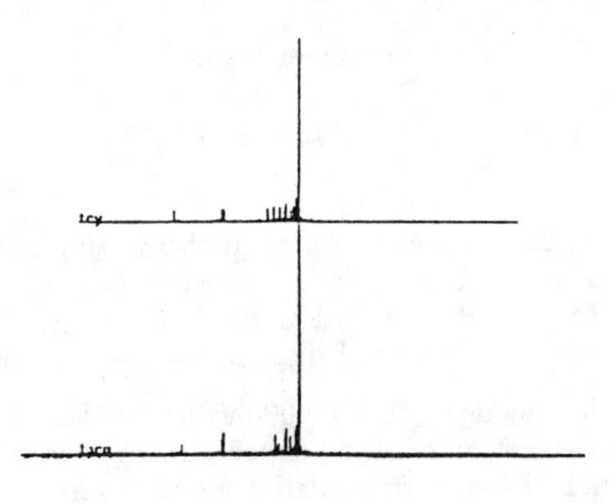

Figure 8:

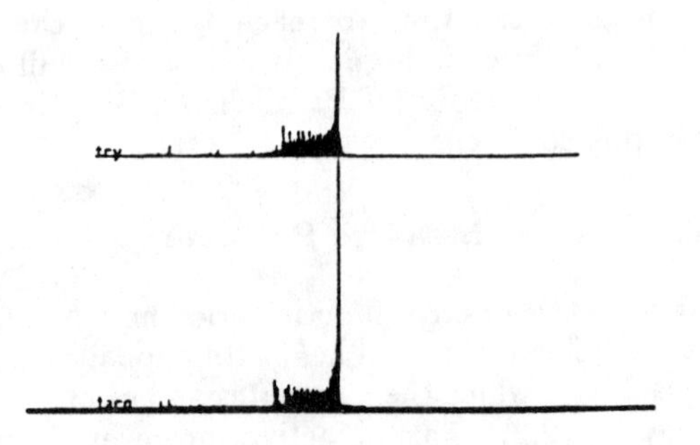

Figure 9:

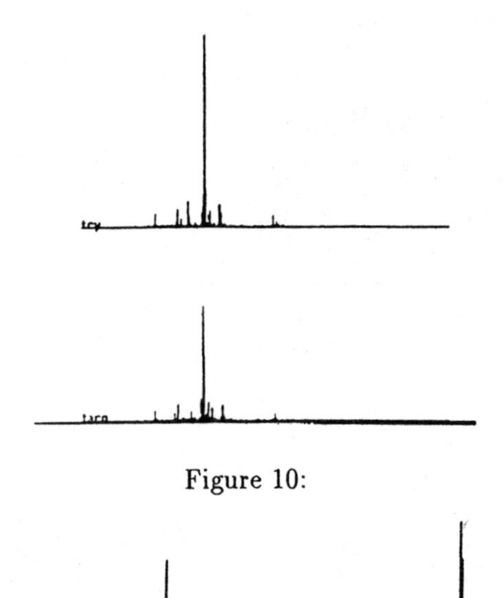

Figure 10:

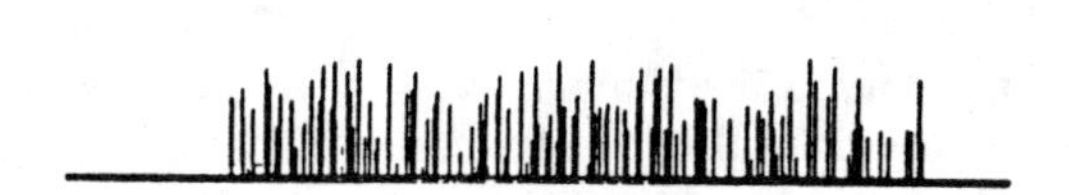

Figure 11:

6 Conclusion

In summary, the described Genetic Algorithms work well for these inverse problems. It would be nice if they worked much faster however. The fault lies with the expense of the function evaluation, namely an application of the Random Iteration Algorithm. Possibly there is an alternative mehod for generating and comparing two attractors.

Most encouraging is the vast speed-up of the parallelized code. While we have been able to theoretically prove that superlinear speed-up is possible and numerically confirm it on tractable examples, [8], a speed-up on the order of 100's using 12 processors is surprising.

References

1. Barnsley, M.F., *Fractals Everywhere*, Academic Press, NY, 1988.
2. Brandt,J., Cabrelli, C., and Molter, U., *An algorithm for the computation of the Hutchinson distance*, preprint (1990)
3. Falconer, K.J., *The geometry of fractal sets*, Cambridge University Press, 1985.
4. Goldberg, *Genetic Algorithms in Search, Optimization, and Machine Learning*, Addison-Wesley, Reading, Mass. (1989)
5. Hutchinson, J., *Fractals and selfsimilarity*, Indians Univ. J. Math., **30**, 713-747 (1981)
6. Mantica, G., and Sloan, A., *Chaotic optimization and the construction of fractals: solution of an inverse problem*, Complex Systems, **3**, 37-62 (1989)
7. Shonkwiler, R., *An image algorithm for computing the Hausdorff distance efficiently in linear time*, Inf. Proc. Lett. **30**, 87-89 (1988)
8. Shonkwiler, R., and Van Vleck, E., *Parallel speed-up of Monte Carlo methods for global optimization*, to appear in SIAM Journal of Scientific and Statistical Computing
9. Vrscay, R., *Moment and collage methods for the inverse problem of fractal construction with iterated function systems*, Fractal 90 Conference, Lisbon, June 6-8, 1990.

The Application of Genetic Algorithms to Resource Scheduling

Gilbert Syswerda
Jeff Palmucci

BBN Laboratories
BBN Systems and Technologies Corporation
10 Moulton Street
Cambridge, MA 02138

Abstract

We present a detailed report of the construction of an optimizer for a resource scheduling application. The optimizer is a combination of local expert search and global search provided by a genetic algorithm. We present the issues confronted in solving this real-world scheduling problem, show how we addressed these issues, and describe how we isolated the genetic algorithm portion of the system from these details. The resulting optimizer provides a good trade-off between quickly providing good results and improving schedules with further search.

1 INTRODUCTION

In this paper we present the messy details of a real-world application of genetic algorithms (GAs) to a scheduling problem, and show how to construct a GA-based optimizer that cleanly and effectively isolates the GA from these details.

The approach we have taken towards this scheduling problem has been described in [Syswerda 1991a]. We review the framework here, but the reader is referred to this book chapter for a more general treatment of the approach taken towards this particular schedule optimization problem.

2 THE SCHEDULING PROBLEM

The scheduling problem we address concerns the System Integration Test Station laboratory (SITS laboratory) of the U.S. Navy, which provides development support for the F-14 jet fighter. The laboratory consists of F-14 airframes, complete with cockpit controls, avionics, radar, and weapon control systems, embedded in a simulation environment that causes the systems to behave as if they are in a plane flying at a specified altitude and speed. All cockpit controls and indicators are active and respond as if the plane is flying. The radar environment is also simulated, causing the radar and weapon control systems to behave as if they are sensing other planes and missiles in the space around them. Numerous pieces of support equipment, such as computers, radios, and recorders, are available as well.

The laboratory is made available to developers of the F-14 jet fighters. Multiple users can use the laboratory at once, creating a scheduling problem. The resources to be scheduled are the F-14 frames, flight and radar simulation environment generators, and numerous pieces of support equipment. Rescheduling is also an important aspect of this scheduling problem. Once a schedule is in place, it is treated only as a working guideline, since high-priority emergency jobs are assigned preemptively, cancellations occur, and equipment breaks.

Users request time and resources via a task request. A task specifics the amount of time needed, time preferences, equipment needs, and the personnel and organizations involved. Since tasks play such a central role and since the personnel involved are represented by a task, it will be convenient to treat tasks somewhat anthropomorphically.

The major components of this scheduling problem are time, resources, and tasks.

2.1 TIME

Schedules are created one week in advance. Each week is divided into 168 discrete pieces of 1 hour each; more than one task can be scheduled during any hour. Some of the 168 hours are usually not available because the laboratory is typically open only Monday through Friday from 5 AM to 9 PM.

Laboratory users and administrators can specify time preferences. These time preferences can be local to a task (e.g. a task has a preference to be scheduled on Tuesday afternoon), local to a laboratory activity such as scheduled maintenance, and global to the schedule (a general preference to have any free time on a frame accumulated at the end of the day).

2.2 RESOURCES

The resources to be scheduled are pieces of equipment in the laboratory, including the frames, flight and radio environment simulators, and support equipment such as computers and recorders. The types and amounts of equipment change over time and is currently in the neighborhood 40 types of equipment pieces. There can be more than one instance of each type of equipment and this number can change as equipment becomes unavailable due to repair or assignment to other activities.

2.3 TASKS

A task is a request for both laboratory time and resources. The time requested can be one continuous block of time, or several discontinuous blocks.

Time preferences can be specified for each task. There are five preference levels: preemptory, which means the task must receive the exact time slots requested; preferred, meaning the system should attempt to assign a task within the preferred time; neutral, indicating no particular preference; preferred not, indicating a time block that should be used only if no other times are available; and illegal, indicating a time that cannot be assigned to the task. Preemptory time requests are handled outside of the optimization process, since they *must* be scheduled and cannot be moved.

Tasks may request resources as disjunctions. For example, a task may require equipment *A* and (*B* or *C*). A task typically specifies about 10 pieces of equipment. Tasks with conflicting equipment requests cannot be scheduled at the same time.

Tasks have personnel associated with them, and a person can be involved with more than one task. Tasks should not be scheduled on the same day if they have personnel in common, although this condition is treated as a weak constraint.

Tasks are assigned a priority. If there are more task requests than there is time to honor them, tasks with the lowest priority should be the ones left out of the schedule.

3 CONSTRAINTS

There are two kinds of constraints: weak and strong. A weak constraint is a preference that can be violated. It is generally better to violate a task's weak constraints than to not schedule the task. A strong constraint is a requirement; strong constraints may not be violated.

3.1 STRONG CONSTRAINTS

Time. Tasks may specify times for which they cannot be scheduled. Also, there are global times during which no tasks may be scheduled (times when the laboratory is not open).

Resources. Resources cannot be shared between tasks. If a resource has been assigned to a task for a certain time, other tasks that use the same resource cannot be scheduled during the same time.

Scheduled Maintenance. The amount of time allocated each day to scheduled maintenance of the equipment is flexible. The initial empty schedule is assigned an overabundance of scheduled maintenance time, and this time is gradually eroded as the schedule is filled. However, there is a minimum amount of time needed below which the deallocation of scheduled maintenance time becomes a strong constraint.

Setup Conditions. Laboratory equipment can be configured or *setup* for different uses. Setup requires between one and three hours to perform, during which time the equipment involved is not available (and a fee is not charged to a user). Once a particular setup has been performed, the setup condition persists until a different setup is done or the end of the day occurs. Tasks which require a particular setup cannot be scheduled at times for which the setup condition does not exist. In addition, tasks to perform a setup cannot be scheduled if a task has been already been scheduled later in the day that depends on a different setup condition.

Subtasks. The time requests of a task can be split into several portions (e.g. 3 blocks of 2 hours each). These subtasks cannot be scheduled at the same time.

3.2 WEAK CONSTRAINTS

Weak constraints are conditions that should be avoided, but are not illegal.

Time Preferences. Tasks can specify a time during which the task should not be scheduled, if possible. Tasks can also specify a time during which they prefer to be scheduled.

Personnel. Each task has associated personnel. Two tasks involving the same person should not be scheduled on the same day. This is a weak constraint since personnel substitutions can often be made.

Subtasks. The time requests for a task can be split into several portions (e.g. 3 blocks of 2 hours each). If possible, each subtask should be scheduled on a different day.

4 OVERALL SYSTEM DESIGN

The job of building a scheduling system involves more than just building an optimizer. There may exist special cases of constraints that are not common enough to warrant inclusion in the optimizer. The system must be able to combine manual scheduling of special cases with automatic scheduling based on more general criteria.

We have constructed a system that smoothly combines manual scheduling with schedule optimization. Manual scheduling is accomplished by the use of a graphical interface that allows users to simply pick up tasks with a mouse and place them into the schedule. The interface is intelligent in that it understands all the well-defined constraints of the scheduling problem, and advises the user about where to place tasks while disallowing the construction of illegal schedules. At any time in the scheduling process, the user may choose a set of tasks to schedule and a block of time to schedule them in (e.g. Thursday afternoon) and invoke the optimizer on just that portion of the schedule. This is most useful in rescheduling portions of a broken schedule.

From the perspective of schedule optimization, we have had to consider the basic time and resource constraints, interactions between tasks, global laboratory considerations such as scheduled maintenance and setup conditions, and global optimizations such as distribution of free time. In the next section, we present a GA-based optimizer that accommodates all these factors.

5 OPTIMIZER DESIGN

The GA-based optimizer we have constructed has three major components: a schedule builder, a schedule evaluator, and a genetic algorithm.

5.1 THE SCHEDULE BUILDER

The schedule builder receives as input an ordering of tasks from the GA. Using knowledge about the constraints of the problem, the schedule builder uses this ordering to construct a *legal* schedule without worrying about the much harder problem of producing a *good* schedule. It works by taking the first task from the list of tasks, and placing that task into the schedule by following a deterministic set of rules. The rules essentially specify that the task is to be placed into the first available legal position. The next task is then selected from the list, and is placed into the first legal place in the schedule, given the presence of the first task. This process is continued for the entire list. The resulting schedule is guaranteed to be legal, but because of task interaction, the quality of the schedule is largely dependent on the order of the tasks presented to the schedule builder.

5.2 THE SCHEDULE EVALUATOR

Once the schedule builder has constructed a schedule, it is evaluated. The evaluation procedure need only consider the weak constraints and global restrictions, since the schedule builder has constructed a schedule that satisfies all strong constraints. This greatly simplifies the evaluator and avoids the use of devices like penalty functions for illegal schedules.

5.3 THE GENETIC ALGORITHM

The schedule builder requires as input a permutation of the list of tasks to be scheduled. It uses this list to construct a schedule, which the evaluator then ranks. As a result, the schedule builder and evaluator have isolated the GA from all problem-specific details; the GA has only to work with permutations of a list of things.

We have previously shown [Syswerda 1991a] that two operators in particular are well suited for this optimization task. They are a swap mutation operator and position-based crossover.

The swap operator[1] is simple: choose two positions in the list of tasks, and swap the two tasks.

Position-based crossover works by randomly selecting a set of list positions, and imposing the positions of the tasks at those locations on the other parent, and vice versa. For the remaining positions, the tasks not selected are used to fill in the vacancies, in the order the tasks occurred in the parent. For example:

```
Parent 1:             a b c d e f g h i j
Parent 2:             e i b d f a j g c h
Selected positions:     *  *    *     *
Child 1:              a i b c f d e g h j
Child 2:              i b c d e f a h j g
```

6 DESIGN TRADEOFFS

The design of the optimizer gives us clean separation of knowledge. The schedule builder need know only about the strong constraints of the problem; its concern is constructing legal schedules. The schedule evaluator need consider only the weak constraints; it has a guarantee that the schedules it considers are legal. The GA need know nothing about the problem; it works in a domain-independent way with simple lists of things.

However, performance is an important issue. Integration of an appropriate amount of domain knowledge into a GA-based optimizer can often lead to a large increase in performance. We have three possible places in which to use domain knowledge: the schedule builder, schedule evaluator, and the GA.

Integration of domain knowledge into the schedule builder is a reasonable place to do so. Some knowledge of the weak constraints or global evaluation factors can make a large difference in the quality of the schedule generated. Instead of simply searching for the first legal position, the schedule builder can perform a local search for

[1] The swap mutation operator was called position-based mutation in [Syswerda 1991a].

the best position, given the current partially constructed schedule and its knowledge of weak constraints. It could go even further and move around tasks that have already been scheduled, or even change the chromosome to provide some local optimization. Taken to its logical extreme, the schedule builder could view the list of tasks presented to it as an unordered collection of things to be scheduled, and produce a schedule after performing some (probably prohibitive) amount of search.

Obviously, some care must be taken in applying domain knowledge. The GA works by modifying a population according to the individual fitnesses of each population member. As the schedule builder is made smarter, more and more chromosomes will result in the same schedule and thus the same evaluation. This hides information from the GA, since a chromosome and a somewhat worse chromosome can both result in the same evaluation. Where to draw the line will depend on how much faith one has in the abilities of a GA to perform effective global search, and how important it is to quickly generate good schedules. People working on this project who had relatively little experience with genetic algorithms tended to want to make the schedule builder very smart. The section on implementation details describes the current tradeoffs we have made between local and global search.

Use of domain knowledge in the schedule evaluator is necessary and tricky. The evaluation function must of course evaluate major characteristics such as high-priority tasks not being placed into the schedule. However, it must also capture what is is about schedules that make users of the system prefer one schedule over another. This can be a rather subtle affair. The evaluation function must return a single value, and since it computes over a number of criteria, the weighting of these criteria becomes an issue. Adjustment of these weights is an ongoing process; some control over the weights is given to the system's users so they can adjust the weights themselves.

Insertion of domain knowledge into the GA is usually done by adding operators that know about the domain. In our case this is difficult to do because of the representation we have chosen: a list of tasks. Suppose we wanted to create an operator that would look at the list of unscheduled tasks and attempt to schedule an unscheduled high-priority task by removing from the schedule a lower-priority task. To do this, it would need the actual schedule, which could be provided. However, to bring about the change, it would have to modify the permutation of tasks in the chromosome in such a way that the schedule builder would build a schedule with the desired change. This is difficult to do, requiring an inverse of the schedule builder.

If we felt strongly that such operators were necessary, we would change the representation of the chromosome from a simple list of tasks to something more closely resembling actual schedules. This is possible and has been explored in the context of job shop scheduling [Davis 1985]. However, performing recombination on actual schedules requires complicated operators, particularly if illegal schedules are to be avoided.

7 IMPLEMENTATION DETAILS

The SITS Scheduling system is written in CommonLisp, using the CLOS object system. The GA is a derivative of OOGA[2], a package developed by Lawrence Davis and Dan Cerys and described in [Davis 1991].

In the process of constructing the optimizer, there were many parameters decisions to make and performance hacks to devise. The next three sections describe the system as it existed to obtain the results presented in this paper.

7.1 THE SCHEDULE BUILDER

The scheduler builder receives a list of tasks, and a partially filled *base schedule*. If the schedule is being constructed completely from scratch, the base schedule contains only default scheduled maintenance and setup operations conducted during the early morning scheduled maintenance periods.

The base schedule can also contain tasks which have already been scheduled. These tasks have typically either been placed manually by the user prior to invoking the optimizer, or represent a schedule that has been disrupted and must have a portion rescheduled.

For each unscheduled task, the system computes a fitness value for each hour in the week, given the base schedule. These values are sorted first according to the computed fitness value and then according to a predetermined ordering of the hours of the week, resulting in a *preference vector* for each task. The hours of the week are ordered by the hour within the day across all days. Hours with hard constraint violations at this point are not included in the preference vector.

The schedule is constructed one task at a time. For each task in the unscheduled task list, starting with the first and working down, possible placement times are examined in the order specified by the preference vector. Each hour examined has a fitness value computed for it, similar to the fitness value previously computed for the preference vector. Hours are scanned until a fixed number of legal positions have been found. This number is the *depth of local search*. The hour with the highest fitness value is chosen, and the task is scheduled starting at that hour. If no valid time locations are found, the task is not scheduled.

7.2 SCHEDULE EVALUATION

Each task has a priority value, which typically ranges from 40 to 100. If a task has been placed into the schedule, its

[2] OOGA is available for a nominal fee from TSP, P.O. Box 991, Melrose MA 02176.

priority value is added to the overall evaluation figure. If some constraints are violated by this placement, a value (nominally 1.0) is subtracted from the fitness value for every constraint violation. In addition, any tasks whose constraints are violated by the placement also cause a value to be subtracted from the evaluation figure.

Each of the factors that can add to or subtract from the fitness value has an associated weight. These weights can be adjusted to change the emphasis of the evaluation function. The values used for the data presented in this paper are 1.0 for adding the priority value, and 10.0 for all constraint violations.

7.3 GA PARAMETERS

We used a steady-state GA[3] with a population size of 30. The two operators, crossover and mutation, were used independently in each generation. Over the length of a trial (3000 evaluations) the initial probabilities of using crossover/mutation were 0.8/0.2 and the ending probabilities were 0.2/0.8 with a linear interpolation between. No duplicates were allowed in the population. Parent selection was by scaled proportional fitness; deletion by exponentially decreasing ranking, with the probability that the population best would be deleted at any time set to one in one million.

8 PERFORMANCE

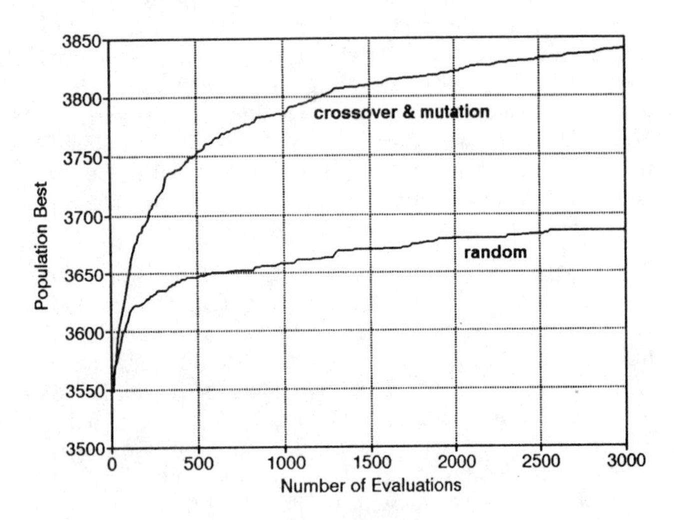

Figure 1: Random list generation versus position-based crossover and mutation. The random "operator" simply generates random task sequences.

Figure 1 presents a comparison of two optimization runs: random list generation compared with crossover and mutation. All data presented is the average of 25 runs.

Random: This plot shows the results of using only one operator: random list generation. This provides the baseline performance of the system, and is the best we can expect for the particular schedule builder we are using. We are using a fairly sophisticated schedule builder, and the schedules constructed using only random task lists are actually pretty good.

Crossover and mutation: A combination of the swap mutation operator and position based crossover. Crossover is used 80 percent of the time initially, and 20 percent at the end of the 3000 evaluations, while mutation starts at 20 percent and increases to 80.

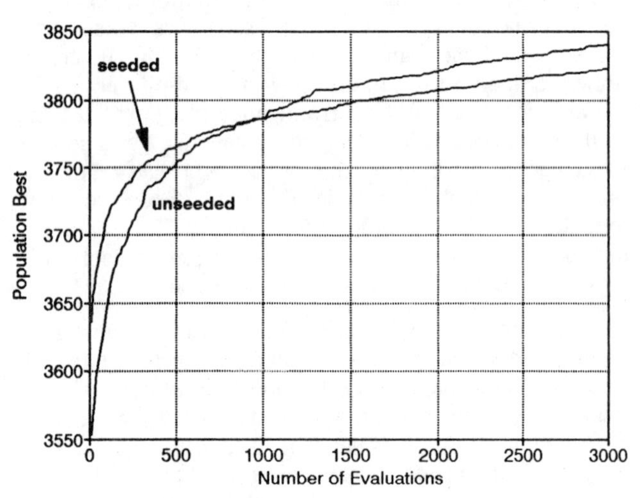

Figure 2: Unseeded versus seeded (bold) initial populations.

To see if significant improvements could be made by loading the initial population with non-random task orderings, we created an initial population by sorting the task list according to task priority, and within task priority sorting by the amount of time required. Using this initial member, we then filled half the population by randomly and locally perturbing the order of tasks. The other half of the population was filled with random permutations of tasks.

The results are given in figure 2, compared against the results for a random initial population. As the plots show, the initial population is better, but the difference rapidly disappears, and the runs with a random population do better in the end. Seeding a population by using a heuristic can sometimes drive the population to converge on a sub-optimal local minimum. It appears that is the case here.

The schedule builder has built into it a look-ahead parameter called the depth of local search. This parameter controls how many legal time slots will be considered for each task before it is placed. For values greater than one, this parameter increases a multiplier on an $O(N^2)$ algorithm, making it computationally expensive if made too large. Figure 3 presents a comparison of look ahead

[3] A more detailed discussion of steady state GAs and the various parameter settings can be found in [Syswerda 1991b].

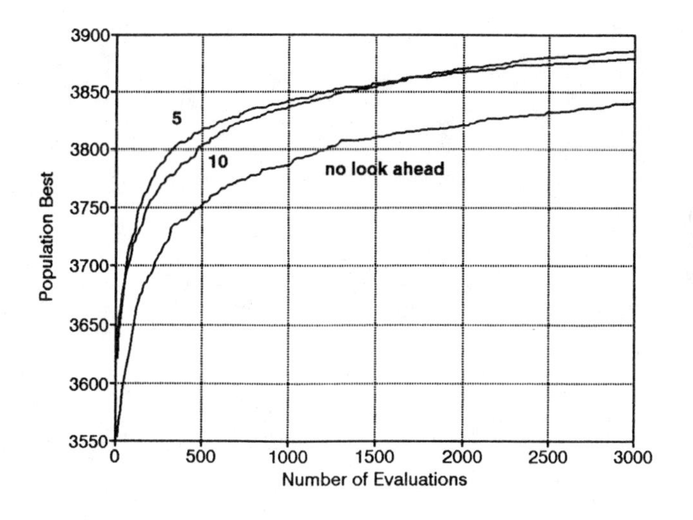

Figure 3: No local look ahead versus depths of 5 and 10.

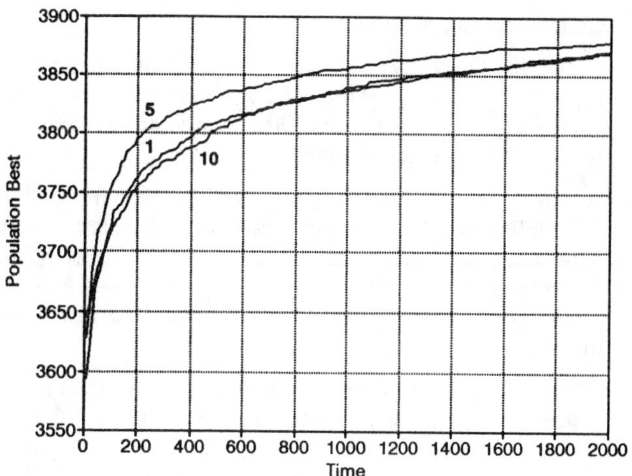

Figure 4: Look-ahead data normalized by time.

values of 1 (no look ahead), 5, and 10. At first glance, look ahead runs of 5 and 10 do significantly better than no look ahead. However, looking ahead is costly. Figure 4 depicts the same data as in figure 3, normalized with respect to computing time. As we can see, a look ahead value of 5 seems to be the most appropriate choice. This is not unexpected, since looking ahead by 5 results in the schedule builder scanning across the five working days in a week. The appropriate look ahead value to use would likely change if we changed the order of the preference vector, or perhaps even simply the code to perform the constraint computations.

9 SUMMARY

We have presented in some detail the construction of a GA-based resource scheduling application. The results to date indicate that we have made good compromises between greedy search and global optimization. The schedule builder is smart enough that, given a set of random task orderings (the initial GA population), it comes up with fairly good schedules. From that point, the GA continues to steadily improve the schedules. This arrangement strikes a balance between the needs of initially constructing the best schedules possible and rescheduling, where the major requirement is speed.

The basic architecture of the optimizer, which provides for a mix of local and global search and a clean separation of the GA from the details of the problem, promises to be a powerful approach for optimization problems of this kind.

Acknowledgements

This work was sponsored by the U.S. Navy, under the direction of Sam Wilson at the Pacific Missile Test Center, Pt. Mugu, CA.

References

Cleveland, G. and S. Smith (1989). Using genetic algorithms to schedule flow shop releases. *Proceedings of the Third International Conference on Genetic Algorithms and Their Applications.* San Mateo, Calif.: Morgan Kaufmann.

Coombs, S., and L. Davis (1987). Genetic algorithms and communication link speed design: constraints and operators. *Proceedings of the Second International Conference on Genetic Algorithms and Their Applications.* Hillsdale, N.J.: Lawrence Erlbaum Associates.

Davis, L. (ed.), (1991). *Handbook of Genetic Algorithms.* New York, New York: Van Nostrand Reinhold.

Davis, L. (1989). Adapting operator probabilities in genetic algorithms. *Proceedings of the Third International Conference on Genetic Algorithms and Their Applications.* San Mateo, Calif.: Morgan Kaufmann.

Davis, L. and S. Coombs (1987). Genetic algorithms and communication link speed design: theoretical considerations. *Proceedings of the Second International Conference on Genetic Algorithms and Their Applications.* Hillsdale, N.J.: Lawrence Erlbaum Associates.

Davis, L. (1985). Job shop scheduling with genetic algorithms. *Proceedings of an International Conference on Genetic Algorithms and Their Applications.* Hillsdale, N.J.: Lawrence Erlbaum Associates.

Syswerda, G. (1991a). Schedule optimization using genetic algorithms. Chapter 21 of the *Handbook of Genetic Algorithms*, New York, New York, Van Nostrand Reinhold.

Syswerda, G. (1991b). A study of reproduction in steady state and generational genetic algorithms. To be published in the Foundations of Genetic Algorithms, San Mateo,

Calif.: Morgan Kaufmann.

Syswerda, G. (1989). Uniform crossover in genetic algorithms. *Proceedings of the Third International Conference on Genetic Algorithms and Their Applications.* San Mateo, Calif.: Morgan Kaufmann.

Whitley, D., T. Starkweather, and D. Fuquay (1989). Scheduling problems and the traveling salesman: the genetic edge recombination operator. *Proceedings of the Third International Conference on Genetic Algorithms and Their Applications.* San Mateo, Calif.: Morgan Kaufmann.

Whitley, D., T. Starkweather, and D. Shaner (1991). The Traveling Salesman and Sequence Scheduling: Quality Solutions Using Genetic Edge Recombination. Chapter 22 of the *Handbook of Genetic Algorithms*, New York, New York, Van Nostrand Reinhold.

Fuzzy Logic Synthesis with Genetic Algorithms

Philip Thrift
Central Research Laboratories
Texas Instruments
P.O. Box 655936 M.S. 134
Dallas, Texas 75265
214 995-7906 thrift@resbld.ti.com

Abstract

This paper considers the application of a genetics-based learning algorithm to systems based on fuzzy logic. One of the more active areas in the application of fuzzy logic is fuzzy controllers. A fuzzy logic controller (FLC) is based on linguistic control strategies (or rules) that interface with real sensor and activator signals by means of fuzzification and defuzzification algorithms. The discrete nature of fuzzy strategies make them prime candidates for discovery by genetic algorithms. This approach is explored in this paper. Some general directions for genetics-based machine learning in fuzzy systems are outlined.

1 INTRODUCTION

There has been much recent activity in the use of fuzzy logic in the design of controllers from braking systems for trains to washing machines [10]. These systems are for the most part designed by a knowledge engineering process based on subjective experience and trial-and-error experimentation. There is much interest in the use of learning algorithms (including genetic algorithms [4,5]) to automatically synthesize such systems. This paper considers the use of genetics-based learning as a general approach to synthesizing fuzzy logic strategies. A genetic algorithm applied to a simple control example (cart centering) is presented.

2 FUZZY LOGIC CONTROLLERS

In this section we briefly review the basic concepts of fuzzy logic, and fuzzy logic controllers (FLCs) in particular. A general survey of the field as well as a fuzzy logic background can be found in [9].

Fuzzy logic is based on the concept of fuzzy sets [11]. A fuzzy set is a generalization of a classical set in that memberships are graded between 0 and 1 as opposed to being purely boolean. If x is some variable over some domain of discourse U, and X is a fuzzy set over U, then $\mu_X(x)$ is defined to be the degree of membership of x in X. As an example, U could be some measurable parameter of a system, such as **temperature**. X could be a fuzzy set, such as **cool**. Then $\mu_{\mathbf{cool}}(x)$ would be a number between 0 and 1 inclusive as indicated in Figure 1.

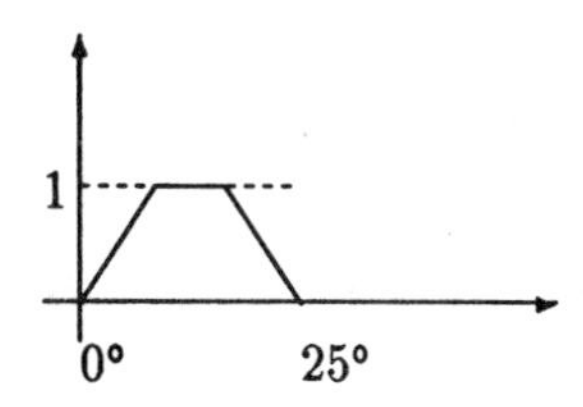

Figure 1: The fuzzy set **cool** over the domain **temperature**

Fuzzy set operations are a generalization of classical set operations. In particular,

$$\mu_{X_1 \cap X_2}(x) = \min\{\mu_{X_1}(x), \mu_{X_2}(x)\}$$
$$\mu_{X_1 \cup X_2}(x) = \max\{\mu_{X_1}(x), \mu_{X_2}(x)\}$$
$$\mu_{\overline{X}}(x) = 1 - \mu_X(x)$$

are definitions for fuzzy intersection, union, and complement. These are not the only ones (there is a product rule for intersection for example), but these are fairly standard.

Given a domain of discourse U, fuzzy sets over U can be identified with a set of names of linguistic variables. For example if U is identified with the parameter measuring **temperature**, then fuzzy sets over U can be

$$\{\textbf{cold}, \textbf{cool}, \textbf{moderate}, \textbf{warm}, \textbf{hot}\}.$$

Basic fuzzy sets may be modified by certain operators, such as *very*, or *slightly*, etc. These essentially change the shape of the membership function they modify.

A system of fuzzy sets over a domain can form a fuzzy partition of the domain as indicated in

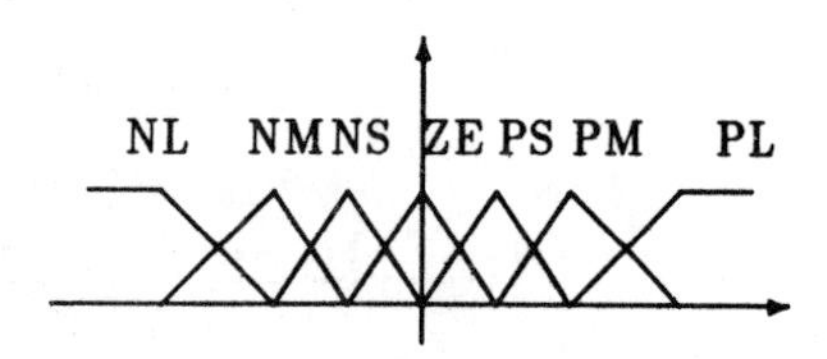

Figure 2: Fuzzy partitions of a domain

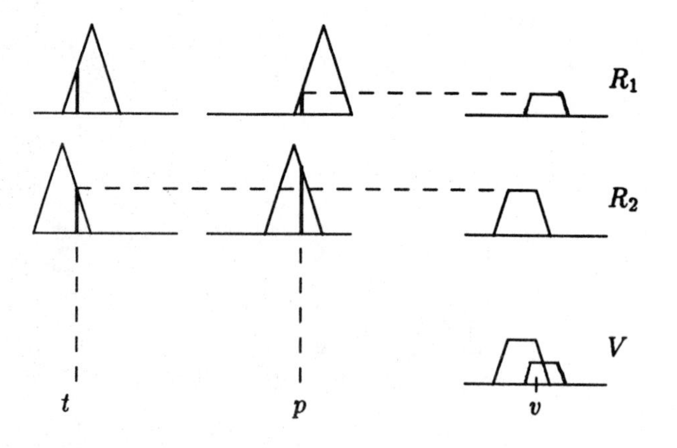

Figure 3: Fuzzy inference

	NL	NM		ZE	PS	PM	PL
NL	NB	NM	PL		ZE	PS	ZE
NM	NB		PL	NM	ZE	PL	ZE
NS	NB	NM	PL		NL	PM	PS
	NM	PS	NS	ZE	NL	ZE	PM
PS		PS		ZE	NS	NL	PL
PM	ZE	PS	ZE	PS	NS	NL	ZE
	ZE	NS	ZE	PL	PM	NM	ZE

Figure 4: Fuzzy decision table

Figure 2. Generically for linearly ordered parameter spaces, the fuzzy sets of a partition can be identified with lingustic variables such as **ZE**(zero), **PS**(positive small), **NS**(negative small), **NM**(negative medium), **PM**(positive medium), **NL**(negative large), **PL**(positive large). A subset of these can be chosen for a coarser resolution.

Given domains of discourse for a set of input and output variables for a system S, rules can be written in terms of fuzzy sets over these domains. Let us suppose that **temperature** (t) and **pressure** (p) are two input variables and the output variable is a **valve location** (v), then a fuzzy rule could be

if t is **hot** and p is **low** then turn v to **medium**

A set of rules of this type constitute a fuzzy rule-base. Once a fuzzy rule-base has been specified, either by an expert or an adaptive procedure (such as a genetic algorithm), the system can map actual (crisp) input values to output values by means of fuzzification, fuzzy inference, and defuzzification. The system uses a method of inference called *sup-star composition*. This procedure is indicated in Figure 3. The fuzzification is performed by evaluating every input parameter with respect to the fuzzy sets in the premise of rules. For example in the above example, $\mu_{\mathbf{hot}}(t)$ and $\mu_{\mathbf{low}}(p)$ would be evaluated. They are combined by

$$s = \mu_{\mathbf{hot}}(t) * \mu_{\mathbf{low}}(p)$$

where $*$ is either the multiplication or minimum operator. This gives the degree to which that particular rule is selected. The output of the rule is the fuzzy set R defined by the function

$$\mu_R(v) = \mu_{\mathbf{medium}}(v) * s.$$

This procedure produces a fuzzy set for each rule ($R_1(v)$,$R_2(v)$,...). A single fuzzy set V is produced by taking the fuzzy union (or max): $V = R_1 \cup R_2 \cup \ldots$. A single value v is then produced by a defuzzification operator

$$v = \text{defuzzify}(V).$$

In practice, several defuzzification stategies have been used. A typical strategy is to take the centroid of area under the curve specified by the membership function of V. This is the strategy used in the example below.

Fuzzy logic systems (such as fuzzy controllers) thus allow conflicting rules to apply allowing consensus answers to be formulated. This effect can be acheived by using partial matching and confidence factors in standard expert systems, but the fuzzy approach is a less ad-hoc approach.

An FLC consists of three components: a *fuzzification interface* with sensors, a *fuzzy rule-base* for inferencing, and a *defuzzication interface* for activators (or decison variables). The role of fuzzification is to map a sensor or input signal x into fuzzy membership values – one for each fuzzy set in the universe of discourse of x. For low dimensional systems, the fuzzy control logic can be specified by a table as shown in Figure 4. Entries in the table can be blank indicating no fuzzy set output for the corresponding rule.

3 LEARNING FUZZY RULES AND MEMBERSHIP FUNCTIONS

There are a number of previous approaches to learning fuzzy rule sets for implementing control strategies. These are referenced in [9], as well as a discussion of their general concept. A genetic algorithm approach is presented in [4]. In [5], a genetic algorithm is used to find both the memberships functions (parameterized by the endpoints of the triangular shapes) as well as the fuzzy rules. In this paper only the learning of fuzzy rules are considered.

4 FLC SYNTHESIS WITH A GA

One of the main applications of genetic algorithms and genetics-based machine learning (GBML) systems (such as classifier systems) is control. The use of GAs in controller synthesis has ranged from using low level binary pattern languages [2] to high level rule languages as the expression of control stratgies [3,7]. The use of fuzzy rule expressions is somewhat of a middle ground between these two.

We shall only consider fuzzy control synthesis for decision table forms. We will consider a table as a genotype with alleles that are fuzzy set indicators over the output domain. The phenotype is produced by the behavior produced by the fuzzification, max-* composition, and defuzzification operations. The eight gene values can be written {NL, NM, NS, ZE, PS, PM, PL, _}. The _ symbol indicates there is no fuzzy set entry at a position that it appears. A chromosome (genotype) is formed from the decision table by going rowwise and producing a string of numbers from the code set. Standard crossover and mutation operators can act on these strings. In the example below we will only take a subset of the fuzzy partition.

5 EXAMPLE

We consider the example of cart centering. This problem involves a cart with mass m that moves on a one dimensional track, as indicated in Figure 5. The state variable for this system are the **position** (x), and **velocity** (v). We assume in the dynamics of the model that the track is frictionless. The output of the controller is a **force** (F). The objective is, given an initial position and speed, to move the cart to zero position and velocity in minimum time.

The simulation for the cart is given by

$$x(t+\tau) = x(t) + \tau v(t)$$

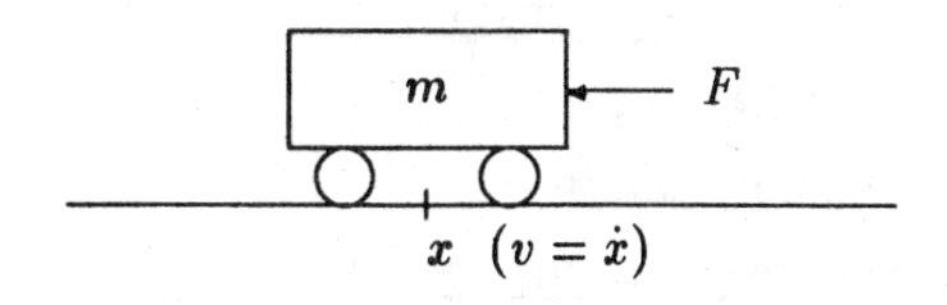

Figure 5: Cart centering example

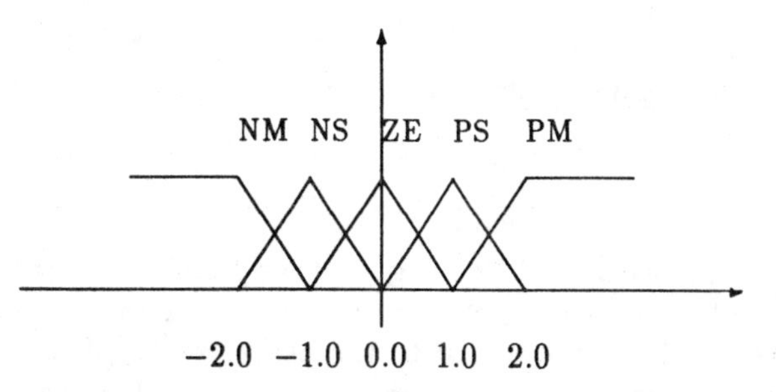

Figure 6: Fuzzy partitions for **position**, **velocity**, **force** (in meters, meters/sec, and newtons respectively)

$$v(t+\tau) = v(t) + \tau \frac{F(t)}{m}$$

Following [7], $\tau = .02$ sec, $m = 2.0$ kilograms. Position (velocity) is chosen randomly between -2.5 and 2.5 meters (meters/sec). For the first experiment, only five fuzzy partitions of each parameter were chosen. The fuzzy partitions for all three parameters (**position**, **velocity**, **force**) are chosen as indicated in Figure 6. The individual control strategies are are 5×5 tables, coded as chromosomes with alleles in $\{0, 1, 2, 3, 4, 5\}$. (corresponding to NM,NS,ZE,PS,PM, and _ respectively. Based on the sup-* algorithm above, a force F can be computed for a given (x, v). The defuzzification used is a simpilification of the centroid operator which computes a weighted average of the central points of the output fuzzy sets.

6 RESULTS

The fitness for an individual is determined by running a simulation of the cart for 500 steps (corresponding to 10 seconds for $\tau = 0.02$) with starting points (x_0, v_0) selected from 25 equally spaced positions. The fitness of an individual control strategy is measured by $500 - T$, where T is average time required to be sufficiently close to $(0,0)$ in (x, v) coordinates (chosen as $\max(|x|, |v|) < .5$). If, for a given starting point (x_0, v_0), more than 500 steps are required, the process "times out", recording 500 steps.

These are the particular features of the genetic al-

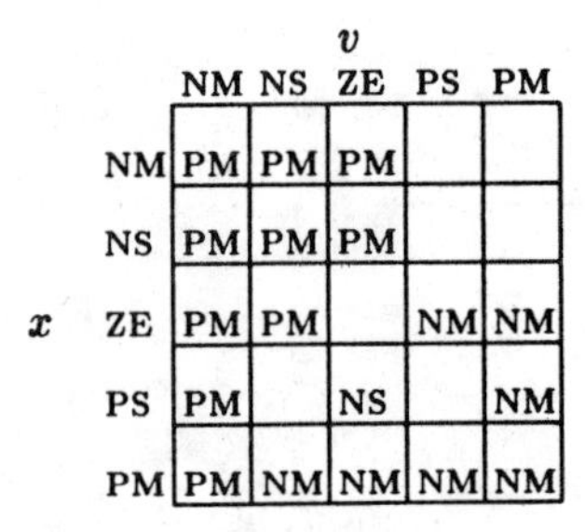

x \ v	NM	NS	ZE	PS	PM
NM	PM	PM	PM		
NS	PM	PM	PM		
ZE	PM	PM		NM	NM
PS	PM		NS		NM
PM	PM	NM	NM	NM	NM

Figure 7: Fuzzy control strategy found in generation 100

gorithm based on the coding strategy described above. A mutation operator changes a fuzzy code either up a level or down a level, or to the blank code. (if it is already blank, then it chooses a non-blank code at random). The crossover operator is the standard two-point crossover [1]. A mutation rate of 0.01 and a crossover (two-point) rate of 0.7 were chosen. A elite strategy was taken whereby the best solution at a given generation is promoted directly to the next.

A simulation of 100 generations with a population of 31 individuals produces a solution which translates into the fuzzy decision table shown in Figure 7. This rule set compares well with the optimal "bang-bang" control rule [7], which is defined as follows. If $F(t)$ is chosen to be either F or $-F$, where F is some positive constant, then choose $F(t)$ to be F if

$$\frac{v^2 \operatorname{sgn}(v)}{2|\frac{F}{m}|} < -x$$

and $-F$ otherwise. In this simulation $F = m = 2.0$.

The control strategy in Figure 7 was compared with the optimal control strategy over 100 runs with random starting points in $-2.5 < x < 2.5, -2.5 < v < 2.5$. The average number of time steps (< 500) were 164 (corresponding to 3.28 sec) for the fuzzy controller and 143 (2.86 sec) for the optimal controller. A strategy succeeds when the cart is within the tolerance as specified above (0.5 in x and v). Keeping the same fuzzy control strategy and reducing the tolerance to 0.2 resulted and allowing 2000 steps before time-out resulted in an average of 242 (4.8 sec) for the fuzzy controller and 161 (3.2 sec) for the optimal controller. The controller could be further optimized by using a GA to place the endpoints of the triangular membership functions shown in Figure 6.

7 CONCLUSION

In this paper we have examined one aspect of using GAs and GBML in fuzzy systems, namely to design a simple fuzzy controller. Fuzzy systems research and applications have been experiencing a recent acceleration [10], so the usefulness of genetic approaches as these systems become more complex could become more apparent. Future directions remain: more complex rule structures with modifiers and quantification, the relation of fuzzy logic and classifier systems, and adaptive generation of the shapes of fuzzy membership functions themselves. In regards to classifier systems, a classifer of the form 751 / 3 could code a rule of the form

$$\text{if } x_2 \text{ is } \mathbf{PM} \text{ and } x_3 \text{ is } \mathbf{NM} \text{ then } u \text{ is } \mathbf{ZE}$$

Rule activation would be by fuzzification and defuzzification instead of the standard pattern matching. Also fuzzy algorithms [11] expressed in a programing language form could be more robust than with non-fuzzy programs (the problem raised in [8] notes the radical change of phenotype due to minor change in genotypes based on standard programs with non-fuzzy decisions). The interface between genetic algorithms and fuzzy systems should prove to produce useful results.

In the languange of genetic algorithms, the alleles are fuzzy set indicators. This allows discrete specification to have continuous and robust interactions with the environment via the *fuzzy phenotype*.

References

[1] Davis, L. ed. *Handbook of Genetic Algorithms.* Van Nostrand Reinhold 1991.

[2] Goldberg, D.E. *Genetic Algorithms in Search, Optimization, and Machine Learning.* Addison-Wesley. 1989.

[3] Grefenstette, J.J. A System for Learning Control Strategies with Genetic Algorithms. *International Conference on Genetic Algorithms.* 1989.

[4] Karr, C. Genetic Algorithms for Fuzzy Controllers. *AI Expert.* February 1991.

[5] Karr, C. Applying Genetics to Fuzzy Logic. *AI Expert.* March 1991.

[6] Kong, S-G. Kosko, B. Comparison of Fuzzy and Neural Truck Backer-Upper Control Systems. *IJCNN* 1990. III:349-358.

[7] Koza, J.R. Keane, M.A. Cart Centering and Broom Balancing by Genetically Breeding Populations of Control Strategy Programs. *IJCNN-90.* 1990.

[8] Koza, J.R. Genetic Programming: A Paradigm for Genetically Breeding Populations of Computer Populations of Computer Programs to Solve Problems. Stanford University Department of Computer Science Report No. STAN-CS-90-1314. 1990.

[9] Lee, C.C. Fuzzy Logic in Control Systems: Fuzzy Logic Controller – Parts I & II. *IEEE Transaction on Systems, Man, and Cybernetics.* **20**:2. 1990. pages 404-435.

[10] Self, K. Designing with fuzzy logic. *IEEE Spectrum.* November 1990.

[11] Zadeh, L.A. Outline of a New Approach to the Analysis of Complex Systems and Decision Processes. *IEEE Transactions on Systems, Man, and Cybernetics* Vol. SMC-3, No. 1, January 1973.

Context Free Grammar Induction Using Genetic Algorithms

Peter Wyard
Speech and Language Technology,
BT Laboratories,
Martlesham Heath,
Ipswich, IP5 7RE,
England.

Abstract

A genetic algorithm was developed for the purpose of inferring context free grammars. Results are reported on the inference of two grammars in this class. Various forms of the grammar to generate the language of correctly balanced and nested brackets were successfully inferred, but more complex grammars were not learnt with the resources available. The paper also discusses various issues such as the representation of the grammars, the method of evaluation of the chromosomes and means of maintaining diversity in the population which will be important in future efforts to extend this work.

1 INTRODUCTION

The general nature of the grammatical inference problem is well described by [8]. A source is assumed given that generates strings, consisting entirely of symbols from a finite set of terminal symbols. These strings are assumed to possess some unique structural features that are characterised by a grammar G, which can be used to model the source. All of the strings that can be generated by the source are contained in the set L(G), the language generated by G, while all the strings that cannot be generated are contained in the complement set L′(G). An observer is given a finite set S+ of strings from L(G), called a positive sample, and possibly another finite set S− of strings from L′(G), a negative sample. Using this information the observer must infer the syntactic rules of the unknown grammar G.

The set S+ can be obtained by observing the output of the source. The set S− can only be defined if an external teacher is available with extra information about the properties of G, which it can use to define strings belonging to S−. In this paper, the algorithm has knowledge of the target grammars, and is therefore able to generate both positive and negative samples.

Many practical grammar inference methods have been devised. Some of these are surveyed in [1] and [2]. Many of these methods suffer from one or more of the following limitations:

1. The class of grammar inferred is limited, e.g. to regular grammars or a subset of regular grammars, and the inference method is often tailored to the characteristics of that class.

2. More information is required than a set of positive and negative strings, e.g. [3].

3. They do not run in polynomial time, so the inference of large-scale grammars is computationally infeasible.

Genetic algorithms (GAs) provide a general method of inferring any class of grammar, given only a set of positive and negative strings. Whether a GA can be designed to overcome the time problem in general is still unclear.

As far as the author is aware, GAs have not previously been applied to the problem of grammar induction, although they have been used for the acquisition of rules in rule-based systems, e.g. [16], [9].

Grammar induction has several practical applications outside the field of theoretical linguistics, such as structural pattern recognition (in both visual images and more general patterns) [5—7], automatic computer program synthesis [15] and programming by example [14]. When the samples from which a grammar is to be inferred come from a natural source, the class of grammar cannot easily be specified *a priori*, and an inference method such as a GA is highly appropriate.

For the purpose of this paper, the class of grammars to be inferred was restricted to context free. All grammars were written in Greibach Normal Form (GNF) to make parsing easier, but this was not a restriction, since any context free language (CFL) without the empty symbol epsilon can be generated by a GNF grammar [12]. Two languages were chosen for study. These were:

1. "Brackets", the language of correctly balanced and nested brackets, including strings such as (), (()), ()(), etc.

2. "AB", the language of strings containing equal numbers of a's and b's.

The target grammars, in GNF, which generate these languages are listed below:

Brackets:	S	→	(R
	S	→	(RS
	R	→	(RR
	R	→	)
AB:	S	→	aB
	S	→	bA
	A	→	a
	A	→	aS
	A	→	bAA
	B	→	b
	B	→	bS
	B	→	aBB

In the grammars above, lower-case letters represent terminal symbols and capital letters are non-terminals. These grammars were used to generate sentences for the inference algorithm. For each experiment, equal numbers of positive and negative strings were produced by generating strings at random, with an exponentially decaying length distribution, using the known set of terminal symbols, and then parsing them using the appropriate grammar to determine whether they were positive or negative.

The task of the GA was to induce a grammar which generated one of the two languages. This induced grammar may or may not contain the same set of rules as the target grammar — what matters is to infer any grammar in the same equivalence class.

2. THE GENETIC ALGORITHM USED

2.1 REPRESENTATION

Each chromosome in the population represented a candidate grammar. GAs usually use a bit string representation. This is partly a matter of simplicity, but Holland [10] advanced theoretical justification for this choice. If bit strings are used, each gene has only two alleles, but the length of the chromosome is large in order to carry a given amount of information. If a higher level representation is used, each gene may have many alleles, but the chromosome is shorter. The first representation gives a much greater number of schemata (subsets of alleles) per chromosome. Since GAs follow many schemata in parallel, the more schemata per chromosome, the greater the rate at which the search proceeds.

The present study uses a much higher-level representation for the chromosomes than bit strings: in fact, each chromosome is a list of production rules barely encoded at all. This has the advantage of ease of implementation, in that no difficult encoding has to be devised. It also allows one to define the genetic operators in such a way that they do not create illegal production rules. Thirdly, it enables one to follow the progress of the GA transparently, i.e. without having to do a lot of decoding.

In the light of Holland's theory presented above, further consideration ought to be given to devising a bit string representation for the grammars, but this would have to be more subtle than simply rewriting each symbol as a binary code. An example chromosome is given below, which represents the target grammar Brackets:

[(S,"(",[R]), (S,"(",[R,S]), (R,"(", [R,R]), (R,")",[])].

It can be seen that each chromosome is a list of rules, each rule being a triple with the format (Left Variable, Right Terminal, List of Right Variables), which is the format of rules dictated by GNF.

2.2 GENERATION OF THE INITIAL POPULATION

Chromosomes were generated with the same number of rules as in the target grammar (although this is not a real restriction since chromosome length can change under the influence of crossover). Each rule in each chromosome was generated randomly, with one left variable, one right terminal, and a random list of right variables of length randomly chosen between 0 and 4. Domain knowledge was used here, in that the set of variables and terminals which could be used was taken from the target grammar. This saved the GA from experimenting with an unnecessarily large number of variable symbols. Domain knowledge was also used in setting the maximum length of a right variable list equal to 4, since it was known that no target grammar had more than 3 right variables in any of its rules, and most had a maximum of 2 right variables. This choice was important to the size of the search space — see Section 4.1.

2.3 THE GENETIC OPERATORS

Mutation: the whole population is mutated once every generation, and each rule is mutated with a probability of 1 in 200. If a rule is selected for mutation, then just one of its symbols is randomly changed to another symbol, either variable or terminal as appropriate.

Reproduction: every generation, the fittest 10% of the population was selected to reproduce, and the chromosomes in this set were paired off at random to mate. (It was hypothesised that this method of selecting chromosomes for reproduction would not yield very different results in this experiment from the more common method of selecting chromosomes with a probability proportional to their fitness, but this may have been incorrect, since results are often quite sensitive to selection methods). Mating was achieved by means of the crossover operator. Crossover was allowed to occur at any point along the length of the chromosome, with one exception: crossover was not allowed to break an individual right variable list. This was done for convenience in aligning the two chromosomes. It could have been allowed, and would probably have been beneficial in generating greater variety.

Inversion: this was not used, because a lot of restrictions would have been necessary to maintain the correct format of the chromosome, although it could have been used in a limited way to invert a right variable list.

2.4 EVALUATION OF THE CHROMOSOMES

The correct choice of evaluation function is very important to the success of the GA. The method chosen was based on the ability of a candidate grammar (chromosome) to correctly accept positive strings and reject negative strings. Using the target grammar, a sample of equal numbers of positive and negative strings was generated. The more strings in the sample, the better the evaluation, but the longer the computation. This experiment used between 20 and 100 strings for evaluating grammars. A new sample of strings was used each generation, in order that the chromosomes should not evolve towards a false target because of insufficient information in the sample. This use of a moving target may have made learning more difficult; it is possible that using a fixed, but very large, sample would have been better. However, a very large sample would make the time for evaluation very long, and other workers have reported good results with a moving targets approach.

A chromosome scores +1 for accepting a positive string or rejecting a negative string, and scores −1 for rejecting a positive string or accepting a negative string. Thus with 50 strings in the sample, scores range from −50 to +50, but only even scores are possible. For example, a score of 40/50 would mean that just 5 of the 50 strings had been wrongly classified.

Chromosomes can change length in the course of evolution, and there is a tendency after a number of generations for high-scoring chromosomes to be longer than average, particularly when scores are not much above zero, which represents random guessing. This is because the more rules a grammar has, even if some of them are incorrect, the better the chance of parsing at least some of the positive string in the sample. This was not a problem with the two languages used in this study, because the scores quickly became quite high, but in general one would wish to introduce a bias into the evaluation function to favour chromosomes with fewer rules, and rules with fewer right variables.

2.5 THE OVERALL OPERATION OF THE GA

An initial population of chromosomes was produced as in Section 2.2. It was then mutated, and scored as described in Section 2.4. The fittest chromosomes were selected for reproduction as described in Section 2.3, and an equal number of chromosomes (the least fit) were eliminated from the population. Parent chromosomes were carried through to the next generation along with their offspring and the other survivors, and the procedure repeated. After each generation, the scores of the chromosomes were printed out.

It is difficult to know when to halt the genetic search, for a number of reasons. Firstly, a perfect score (e.g. 50/50) does not mean that the correct grammar has necessarily been inferred, because a sample of 50 strings contains limited information. Secondly, scores do not increase steadily, or even monotonically, so one never knows if a particular run will suddenly jump to the correct solution, or whether it has got stuck in a local maximum. In practice, a number of heuristic methods were used for deciding when to end the search, and it was necessary to monitor the progress of the search manually for best results.

3. EXPERIMENTS AND RESULTS

3.1 INFERRING A GRAMMAR FOR THE LANGUAGE "BRACKETS"

Five runs were performed with a population of 1000 chromosomes, three using 20 test strings, one using 50 test strings and one using 100 test strings. Two of these runs evolved to a correct grammar, one in 16 generations (one of the runs using 20 test strings) and one in only 3 generations. These correct grammars are not identical to the target grammar, but generate the same language, and are slightly more concise. An example is given below:

S → (R
R →)S
R →)
R → (RR or S → (SR

The other three runs all converged to a (different) wrong grammar, although each managed to achieve a 100% score on the evaluation function used. ('Convergence' means that all chromosomes in the population had become the same as each other). One example of such an incorrect grammar is:

R →)
S → (SS
S →)
S → (R

This is clearly wrong, since it generates the string ")". Another example is:

S → (RS
S → (SR
S → (R
R →)

This grammar is nearly correct: it generates a large subset of the language "Brackets", but fails to generate strings like "(())()", "(())(())", i.e. strings where there is a complete set of 4 or more brackets followed by a new set of brackets.

3.2 INFERRING A GRAMMAR FOR THE LANGUAGE "AB"

Three runs were performed, using population sizes ranging from 300 to 1000, and test string sample size ranging from 20 to 50. In none of these cases was the correct grammar inferred. The highest score obtained was 48/50, but the corresponding grammar generated quite a lot of erroneous strings, and the score appeared somewhat misleading. The runs were continued for 40 generations, and at the end of this time the population of chromosomes was fairly uniform, i.e. it had nearly converged to a local maximum.

4. DISCUSSION

4.1 SEARCH TIME AND SIZE OF SEARCH SPACE

Relatively limited computer resources were available to the author. Experiments were run on a SUN 3 workstation, using a fairly slow programming language (ML). This limited the opportunity for taking measures to prevent premature convergence (see Section 4.2 below).

In order to speed up the search, the size of the search space could have been reduced. de Jong [4] found that GAs performed very well at the task of function optimisation in search spaces of size $>> 10^{30}$. Although a direct comparison with grammar induction may be rash, the size of the search spaces in this study may be calculated. A candidate grammar for 'Brackets' consists of 4 triples. The first member of each triple is either S or R, the second is either (or), while the third is a list of 0 to 4 variables, selected from either R or S. This gives a total of $2 \times 2 \times 31 = 124$ variants per triple, and so a total of $124^4 \sim 10^8$ variants in total, if each grammar has 4 rules. Thus 10^8 is the size of the search space. The search space for 'AB' is of size $(3 \times 2 \times 121)^8 \sim 10^{23}$.

Although both these figures are lower than de Jong's 10^{30}, search space size could be reduced in two ways. If the maximum number of right variables per rule was reduced to 3, which is actually sufficient for these grammars, the search space sizes would become 10^7 for 'Brackets' and 10^{19} for "AB". Secondly, if it were possible to avoid treating two chromosomes with the same set of rules but merely in a different order as being different points in the search space, the size of the search space would be reduced by 4! for "Brackets' and 8! for "AB", giving search space sizes of 10^5 and 10^{14} respectively. This requires a more sophisticated representation of the grammar, but is well worth investigating.

4.2 AVOIDING PREMATURE CONVERGENCE

Premature convergence, or getting stuck in a local maximum, was a serious problem in this study. Altering various parameters of the GA such as the mutation rate was tried, without success. Increasing the population size to 1000 was more useful, but as mentioned in Section 3.2, after about 40 generations the population had converged on the wrong answer. A still larger population size was beyond the resources available.

Some preliminary experiments were done, following the general method of Mauldin [13], which forcibly maintained diversity in the population by killing off any duplicate chromosomes. In this experiment, any duplicates found were mutated. Unfortunately, although this method prevented premature convergence, it appeared to prevent any convergence at all. If a correct grammar appeared in the population at any stage, and was not spotted, it could not generate any copies of itself and was vulnerable to subsequent mutation. Although success was not achieved, the author believes that further investigation of ways to avoid premature convergence is necessary.

4.3 EVALUATION OF THE CHROMOSOMES

It is felt that the scoring function used was not optimal. High or perfect scores could be obtained with grammars that were some way from the target in terms of the strings which they could generate. One idea to improve the scoring function is to look at the generative capabilities of each candidate grammar, as well as its recognition capabilities. For example, 50 strings could be generated from each candidate, and these could then be parsed by the target grammar. Such an approach goes beyond the normal rules of the game of grammatical inference, where one is supposed to use only the strings in the presented sample, although it is being increasingly used in the machine learning field.

Another quite different approach is to use a bucket-brigade algorithm [11], where the population would consist of rules instead of complete grammars, and the scoring would take account of each rule's ability to correctly classify the test set of strings in conjunction with the other rules in the population. This approach has worked well with the induction of rule-based systems, and might well be suited to the problem of grammar induction.

4.4 REPRESENTATION OF THE GRAMMARS

As discussed in Section 2.1 and 4.1, this is a key issue. It would be worth investigating a range of representations from bit strings to fully symbolic representations, using different coding methods, designed to work well in conjunction with the genetic operators.

5. CONCLUSION

Genetic algorithms have been used to successfully infer the context free grammar (CFG) for the language of correctly nested and balanced brackets. They have shown promise at inferring a grammar for the language of strings with equal numbers of a's and b's. Further work is required to determine whether or not GA's are a good method for more general CFG induction, using larger computer resources than were available for this study, and investigating some of the modifications discussed in Section 4.

Acknowledgement

The author wishes to thank Alan MacDonald of Brunel University, where the work for this paper was done, for many helpful ideas and comments.

References

[1] D. Angluin and C.H. Smith. Inductive inference: theory and methods. Computing Surveys, Vol 15, No. 3, September 1983.

[2] P.R. Cohen and E.A. Feigenbaum. The handbook of artificial intelligence: volume 2. Los Altos, California. Publ. Morgan Kaufmann 1983.

[3] S. Crespi-Reghizzi. An efficient model for grammar inference. In "Information Processing 71", B. Gilchrist, Ed. Elsevier North-Holland, New York, pp. 524—529 1972.

[4] K.A. de Jong. Adaptive system design: a genetic approach. IEEE Trans. on Syst. Man and Cybern., Vol. SMC-10, No.9, pp. 566—574, September 1980.

[5] K.S. Fu. Syntactic methods in pattern recognition. Academic Press, New York 1975.

[6] K.S. Fu. Syntactic pattern recognition: applications. Springer-Verlag, New York 1977.

[7] K.S. Fu. Syntactic pattern recognition and applications. Prentice-Hall, New York 1982.

[8] K.S. Fu and T.L. Booth. Grammatical inference: introduction and survey, parts 1 and 2. IEEE Trans. Syst. Man Cybern. SMC-5, 95-111, 409-423 1975.

[9] D.E. Goldberg. Proc. 9th Intl. Jnt. Conf. on Artif. Intell. Los Angeles, Calif. pp 588-592 1983.

[10] J.H. Holland. Adaptation in natural and artificial systems, Univ. of Michigan Press, Ann Arbor 1975.

[11] J.H. Holland and J. Reitman. Cognitive systems based on adaptive algorithms, in "Pattern-directed inference systems", Waterman and Hayes-Roth, Eds., Academic Press 1978.

[12] J.E. Hopcroft and J.D. Ullman. Introduction to automata theory, languages and computation. Publ. Addison-Wesley 1979.

[13] M.L. Mauldin. Maintaining diversity in genetic search. Proc. Natl. Conf. on AI, AAAI-84, pp. 247—250 1984.

[14] T. Shinohara. Polynomial time inference of pattern languages and its applications. In Proc. 7th IBM Symposium on Mathematical Foundations of Computer Science 1982.

[15] D.R. Smith. A survey of the synthesis of LISP programs from examples. In "Automatic Program Construction Techniques", Biermann et al, Eds. Macmillan, New York 1982.

[16] S.F. Smith. A learning system based on genetic adaptive algorithms, PhD thesis, Dept. of Computer Science, Univ. of Pittsburgh 1980.

Connectionism and Artificial Life

The Immune Recruitment Mechanism: A Selective Evolutionary Strategy

Hugues Bersini
IRIDIA - Universite Libre de Bruxelles
CP 194/6
50, Av. Franklin Roosevelt
1050-Bruxelles

Francisco J. Varela
CREA - Ecole Polytechnique
1, rue Descartes
75005 Paris

Abstract

The recruitment mechanism of the Immune System has inspired a technique of optimization both in real (IRM) and in hamming spaces (GIRM). An eventual candidate is generated from very random structural manipulations within the current population. In order to be integrated in this population, the candidate must succeed a test of affinity with its neighbouring. This test requires the candidate to be more similar with its best neighbours. It is shown that in comparison with GA and ESs, this technique aims to explicit and to accelerate the parallel local hill-climbing inherent to them.

1 INTRODUCTION

The immune network model proposed and developed by Varela et al. [16] [14] [15] comprises two major aspects. The first aspect concerns what has been called the dynamics of the system i.e the differential equations governing the increase or decrease of the concentration of a fixed set of lymphocite clones and the corresponding immunoglobins. The network view relates to the immunoglobins interactions by mutual binding. The binding of two species is defined by an affinity value between these two species. Such value is function of the species physical and chemical properties and then does not change. The second aspect concerns what has been called the meta-dynamics of the system. Only this aspect will be emphasized and exploited here for problem solving ends. The meta-dynamics governs the recruitment of new species from an enormous pool of lymphocites freshly produced by the bone-marrow. This recruitment process selects for the generation of a new species on the basis of the current global state of the system i.e. according to the sensitivity of the network for this candidate species. This complementary process is fundamental because it modifies continuously the actors in presence like a neural net whose structure (the number and the nature of neurons) would change in time.

A possible candidate k will be recruited if $\Sigma_i m(k,i) f_i > T$, with i indexing the different species already present in the system, f_i being the concentration of species i and m(k,i) indicating the affinity between species k and i. T is the recruitment threshold. Two ways have been explored for characterizing a species i. The first one is due to Stewart and Varela [13]. Since each species is described by a specific value for several chemical and physical properties, it can be naturally represented by a point in a low dimension real space R^n called the "shape space". Then the affinity between two species becomes function of a certain distance between these two "point-species". The second way is due to Perelson et al. [10] [11]. Since each species can be characterized at the genotypic level, a binary coding has been preferred with the affinity related to the hamming distance.

The recruitment mechanism envisaged in the R^n shape space has inspired an algorithm of function optimization IRM (Immune Recruitment Mechanism) which will be described and compared with the Evolution Strategies (ESs, [7]) techniques (which share the same field of application). The recruitment mechanism envisaged with a binary coding has inspired an algorithm of combinatorial analysis GIRM (Genetic IRM) which will be described and compared with GA (which share the same field of application). What the paper aims to show is that in comparison with ESs and GA techniques, IRM and GIRM add a further test of selection which reinforces and then accelerates the parallel local hill-climbing intrinsic to these techniques. This test is based on the distance or similarity of a potential new candidate with the best candidates existing in the neighbourhood. In consequence, this new biologically-inspired mechanism should not be viewed as challenging any other existing mechanisms but rather as a complementary test at the generation level, that can be switched or not (eventually parametrized), intended to accelerate the search when possible and when needed.

2 IRM AND FUNCTION OPTIMIZATION

Let's assume a trivial optimization problem where you try to find the maximum of an one-dimensional unimodal function f(x). You possess already the values of the function at two points f(x1) and f(x2) with f(x2) > f(x1). You decide that the next point xk to recruit must satisfy a test very similar to the immune recruitment test when 1) equating the concentration to the fitness: $f_i = f(xi)$ 2) defining the the affinity by:

$$m(i,k) = m(xi, xk) = 1 - |xi - xk|/scaling$$

(scaling is the radius of affinity of each point, $0 \le m(xi,xk) \le 1$), and 3) with the threshold T given by $T=(f_1 + f_2)/2$. To succeed this new test, xk must satisfy:

$$\frac{\sum_{i=1}^{2} m(xi,xk) f_i}{\sum_{i=1}^{2} m(xi,xk)} \geq T$$

It is immediate to show that the only points that can be recruited are in the zone indicated in figure 1.

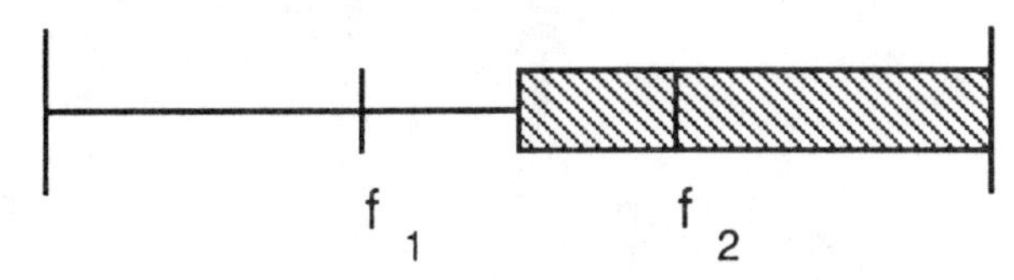

Figure 1

This resembles greatly to the deterministic two-points classic methods of search except the random aspect of the generation. Now suppose you know the values of the function at three points f_1, f_2, f_3 with $f_1 < f_3 < f_2$. If you apply the test on the three points but with the threshold still defined by the two best points: T = (f2 + f3)/2, it is less immediate but still easy to show that the only points to be recruited are in the zone indicated in figure 2.

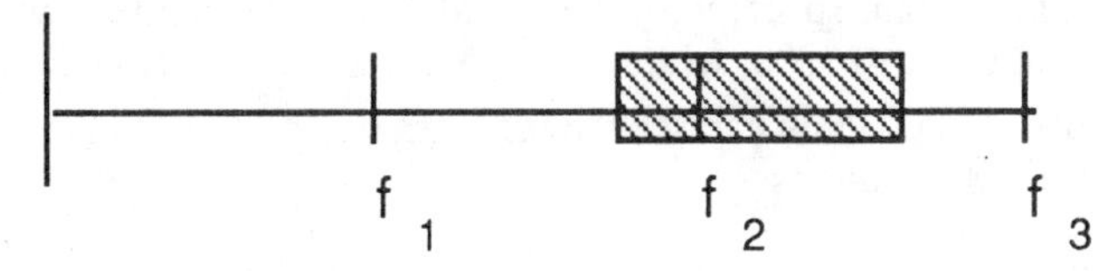

Figure 2

The length and the exact position of the zone are now function of f_1, f_2, f_3 and of the scaling. This slightly reminds the quadratic technic of search (which will place the next point in the 2-3 zone), except here again the random aspect of the generation. Applying this 3-2-points strategy in a 2-dimensional space shows some similitudes with the very popular simplex technique of search.

These two simple and illustrating problems will help to grasp the first part of a simple function optimization algorithm that we are still developing and testing. Suppose a function to optimize in a n-dimensional space, you need first to assign several parameters:

1) the sub-space containing the initial random generation of the Nt points population:

$$\{i = 1..n : x_i^{min} \le x_i \le x_i^{max}\}$$

2) the number Nb best points extracted from the population to calculate the threshold: $T = \frac{\sum_{i=1}^{Nb} f_i}{Nb}$

3) the number Np of points with which a potential candidate x_k will determine its affinity (j : 1..Nb...Np ; j indexes the ordered list of best points)

4) the number Nn of new points to recruit at each generation. To be recruited a proposed point x_k will have to satisfy the test:

$$\frac{\sum_{i=1}^{Np} m(xi,xk) f_i}{\sum_{i=1}^{Np} m(xi,xk)} \geq \frac{\sum_{i=1}^{Nb} f_i}{Nb}$$

5) the choice of an affinity function m. Different functions of the distance offering different advantages have been tested: a Gaussian, a quadratic, an inverse (see [2]). But here for sake of clarity, we limit ourselves to the simple similarity function:

$$m(xj,xk) = 1 - dist(xj,xk)/scaling$$

with dist being the euclidian distance and scaling being the radius of the domain of affinity ($0 \le m \le 1$).

The choice of an affinity function which decreases when the distance increases supposes implicitly that the fitness of the problem locally increases when getting close to the maxima. A lot of interesting functions does not show this property and we are studying other profiles of affinity no more inversely related to the euclidian distance.

6) a zone of proposition for each coordinate i of an eventual point, di. Following the initial random generation, all the subsequent propositions of point k will be done in a sub-space defined by: w = 1.. Nb (x^w = the list of the Nb best points in the current population):

$$\min(x_i^1, x_i^{Nb}) - di \le x_i^k \le \max(x_i^1, x_i^{Nb}) + di.$$

Like in classical optimization methods, we make this interval of search decrease with time.

We have tested IRM for the Schwefel's problem ($f(x)=\Sigma_{i=1,n} (\Sigma_{j=1,i}\, x_j)^2$, [7]) in very low dimensional space (n = 2). In fig.3 you can see the progression of the best point of the successive generations with : Np = 5,

Nb = 2, Nn = 1, the initial sub-space: $-2000 \leq x_i \leq 2000$, and d(i) = 100. After ± 40 generations, (0,0) is reached.

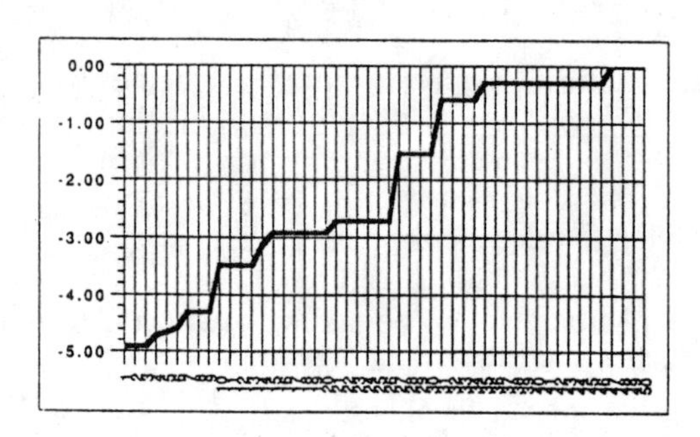

Figure 3

For higher dimensional space, the algorithm does not converge any more towards the function optimum because necessary information about the direction of progression completely lack. Two manipulations can contribute to improve the performance. The first one is to increase Np. Indeed, each one of the Np points which are not among the Nb best ones plays a kind of repulsion role, and the "intersection" of these various repulsions roughly indicates a good zone of search. However, this first manipulation remains insufficient for great dimension space and the addition of a second test becomes necessary leading, like in classic optimization theory and like in ESs, to continuously adapt the search of direction. To each recruited point k are now associated Nd directions of affinity: $\vec{\aleph}_k(j=1,Nd)$. This multi-affinity appears to be an important biological aspect of the immune system [13]. Here, each $\vec{\aleph}_k(j)$ simply is a normalized vector. To be recruited, an eventual point k needs now to succeed two tests: the first one, just described previously, a positional test, and the additional one, a directional test. This second test for a point k is still:

$$\frac{\sum_{i=1}^{Np} md(xi,xk) f_i}{\sum_{i=1}^{Np} md(xi,xk)} \geq \frac{\sum_{i=1}^{Nb} f_i}{Nb}$$

but now with md the directional affinity given by:

$md(x_i,x_k) = \max_j (\vec{x_i x_k} . \vec{\aleph}_i(j))$ obtained for j=max-j ($\vec{x_i x_k}$ has been normalized)

This test ensures that to be recruited a point k must move in the direction associated to the best points already recruited. When a point k satisfies this second test, it is recruited and one of its Nd directions of affinity is automatically fixed in function of the directions of affinity of the best points:

$$\vec{\aleph}_k(1) = \Sigma_{i=1,Nb}\, m(x_i,x_k)\, \vec{\aleph}_i(\text{max-j})$$

The Nd - 1 other directions are established randomly. This will allow to keep the good directions of search for subsequent points and to change them in case they drive to nowhere. If after a certain amount of directional tests, no more point is recruited, the Nb best points directions of search are inverted. The figure 4 illustrates the mechanism of this test in a 2-dimensional space. Suppose that 1 and 2 are the current best points, each one with two directions. Although satisfying the positional test, 3 will not be recruited. 4 will be recruited because the move is in the direction indicated by 1 and 2. The first direction of 4 is automatically fixed (in bold), the second one is random.

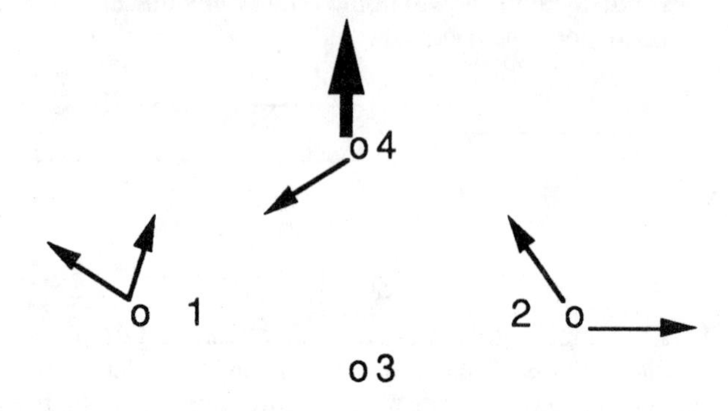

Figure 4: The directional test

The addition of this second directional test leads to considerable improvements for the optimization of functions in great dimensional space. In figure 5, IRM tries to optimize the Schwefel's function in a 10 dimensional space, with the initial sub-space defined by $-10000 \leq x_i \leq 10000$, with Np = 7, Nb = 2, Nd = 5, Nn = 5, d(i) = 3000. You can see the progression of the best point for 200 iterations.

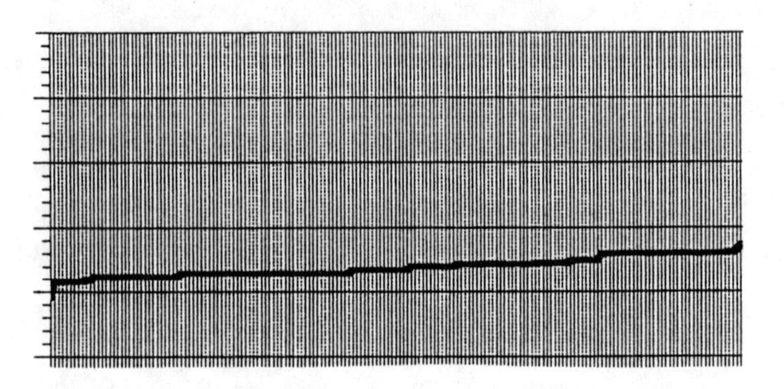

Figure 5

3 IRM AND ESs

Evolutionary Strategies are mainly dedicated to function optimization problems [7]. However when a new point is generated, no test is performed on the relative position of this new point with the best points of the current population located in the neighbourhood. A new point is generated by some moves (due to various specific mutation and recombination operators) from very selected points contained in the current population. All the efforts have been devoted to the generation step trying to improve these successive moves and to allow a gradual self-adaptation of their length and their direction. We believe that ESs show very appealing ideas for the generation step of potential candidates (this generation being equally the first step of IRM), but could be complemented in certain cases by the addition of the IRM positional and directional tests.

Since the objective of ESs is explicitly the realization of some kind of local hill-climbing in high-dimensional space (the greater exploitation of the local surface makes ESs to converge faster than GA for local search), reinforcing this hill-climbing by making the selection still more selective should nicely fit and complement the ESs goals.

Another sensible difference is that IRM relaxes the selection performed at the proposal level. Rather, the real selection follows a largely random proposition and is realized by means of the recruitment test.

4 GIRM AND GA

The GIRM algorithm results from the adoption of a binary coding in order to represent the different species. The similarities between the immune generation of new species and GA as well as between the whole immune network and classifiers system [8] have already been remarked and discussed [9] [10] [15]. Some works aiming to model the immune response i.e the recognition of antigens, the learning of these different encounters and the consequent continuous improvement of this response, have been achieved adopting a genotypic binary coding [11]. In a previous work [2], we discussed the three essential qualities that can be derived from the immune network comprehension and be likened to their specific counterparts in classifiers system: an interesting search algorithm, a large adaptive capacity and an endogenous selective memory. Here we deliberately restrict our attention on the first quality aware of the substantial impoverishment of the biological inspiration.

We have realized GIRM, a simple algorithm of combinatorial optimization within the hamming space. It presents two basic differences with classical GA [5]:

- At each iteration, the proposal of a new candidate presents a greater diversity in comparison with the basic two parents crossover. We rather allow a kind of multi-crossover for multi-parents (obviously there is no notion of parents in the genetic generation of new species). The composition of a new candidate string is achieved by selecting randomly some sub-strings of any length extracted from whatever string chosen randomly in the population. Additional mutations also take place during this generation. The huge diversity of antibodies is one of the most remarkable feature of the immune system [12]. In the biological reality new species are synthetized by somatic mutations and recombinations of genes coming from an enormous library. Our combinatorial mechanism, in increasing largely the pool of potential candidates ready for the test at each iteration, nicely fits with this well-known antibodies diversity. There again, the selectivity is postponed after a proposal step which is grounded indifferently in the entirety of the population.

- The heart of IRM i.e the recruitment test must be succeeded by an eventual candidate to be recruited. Now, instead of the euclidian distance, the affinity between two candidates is a function of their hamming distance dh. Different functions have been tested, but here again for sake of clarity, we limit ourselves to the simple linear function : $m(i,k) = 1 - dh(i,k)/L$ (here L is the length of the string, $0 \leq m \leq 1$).

To illustrate rapidly the effect of this intermediary test at the recruitment level, let's imagine a population of two strings S1 = 111000 and S2 = 000111 with respective fitness f1 > f2. The test (T = (f1 + f2)/2) will oblige a newly recruited string to have more in common (to be more similar) with the first string than with the second one. Then 111011 is possible but not 110111 which are both possible with GA. The test will make large schema already present in the best strings to have a greater chance to remain in subsequent generations. The effect of the recruitment test amounts to intensify the role of the best strings in the crossover and mutations operations (like with a probability distribution which would still more greatly favour the best strings to the detriment of the other ones during these operations).

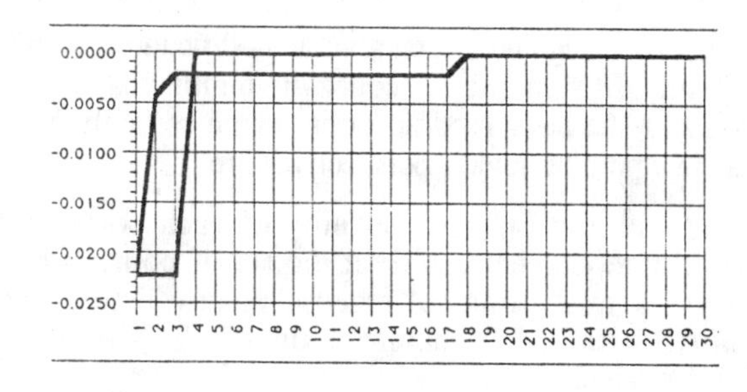

Figure 6 : The left (right)-side curve shows GA (GIRM) progression

In figure 6, you can see a comparison of GIRM and the basic GA for the optimization of $f(x)=x^{10}$, with binary strings of length 30 and population of 30 individuals. GIRM (Np = 30, Nb = 5, Nn = 30) finds the maximum two times faster than GA (15 vs 30 iterations).

In this simple case, increasing the selectivity, GIRM makes it faster in number of tests to perform (i.e it decreases the number of population generations) but eventually slower in total process time, because for each recruitment of a new point a certain number of possible candidates must be tested to overpass the recruitment threshold (± 20 in the case of fig.6). This appears to be an important weakness to take into account and limits the use of GIRM to applications where the tests on the function should not be too numerous and their quality more important than the processing time.

Nevertheless GA privileged field of applications is certainly not unimodal and smooth functions for which classical optimization techniques already propose a large set of performant algorithms, but rather (like biological realistic environment) multimodal functions with difficult and rugged profiles. So, since IRM and GIRM substantially reinforce the hill-climbing intrinsic to these techniques (GA and ESs), will they not fall within local-minima traps whose certainly GA (but less ESs) substantial goal is to avoid as far as possible?

5 IRM AND PARALLEL SEARCH

The aim of these evolutionary-inspired mechanisms is to face difficult situations characterized by a huge number of local solutions distributed in a rugged landscape. The mechanisms should simultaneously explore local solutions and not prematurely converge towards the first local minima that will be met. GA and ESs adepts could, with reason, criticize IRM that reinforcing the hill-climbing leads rapidly to local-minima attraction side-effect. However we know that the immune system has to face a highly complex natural environment with a lot of problems to simultaneously tackle. Moreover, biology makes us familiar with largely distributed systems in which the responsibility and the optimization of each actor are very localised.

We are achieving and testing several modifications both in IRM and GIRM in order to cope with multimodal rugged functions. The basic modifications, which certainly make sense in a pure biological perspective, are:

- to narrow the radius of affinity of each point with obvious consequences to limit the set of points which present a mutual affinity. A proposal point will only have affinity with other points situated in the strict neighbourhood.

- to calculate the threshold only in function of the best points having an affinity with the eventual candidate to test for recruitment. In consequence, the value of the threshold will not be constant in the whole search space and a local hill-climbing will not be hampered by another local, but already higher, climbing.

- to decrease gradually the radius of affinity of each point and, if we want to focus on the global optimum, to decrease gradually the size of the population.

- now when an eventual candidate is proposed, the first step is to locate the points with which it has some affinity. Then the recruitment test will take place only in relation with these neighbouring points.

- the recruitment will be done locally and the new point will replace the worst point among the affine ones. This goes not without recalling the De Jong's crowding scheme [3] where a new point replaces an highly similar one.

These modifications lead to a natural localisation and parallelisation of the hill-climbing inherent to the recruitment test. Each sub-set of mutually-affine points defines a natural niche in which evolution operates locally. The number of niches ideally should gradually equate the number of maxima of the multimodal function. In GA literature, the notion of implicit parallelism [6] often appears. Here, the localisation of the recruitment test by a reduction of the domain of affinity, by the variability of the threshold and by the local replacement, tends rather to a gradual "explicitation" of the parallelism concealed in the search space.

6 PRELIMINARY RESULTS FOR MULTIMODAL FUNCTIONS

We are currently testing both the performances of IRM and GIRM for multimodal functions. We applied IRM on the function f(x,y) represented in figure 7. In the graph, x is comprised in [-10,10] and y in [0,8].

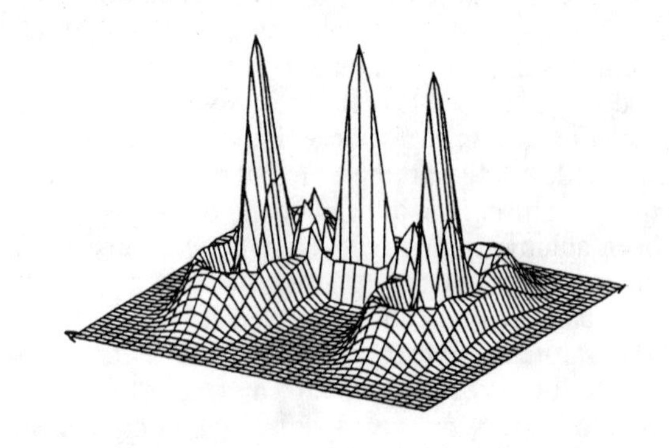

Figure 7

The version of IRM we executed initiates with a population of 50 points distributed randomly in $-10 \leq x,y \leq 10$. 5 new points are generated at each iteration. The initial radius of affinity is 1. To each point are associated 3 directions of affinity. The algorithm finds one of the two greatest peaks (the left or the right) in 90 iterations i.e 450 calculations of the function.

We applied GIRM on the f_5 - Shekel's foxholes function (figure 8) described in [7] in order to allow some comparisons with GA (which performs better than ESs for this problem).

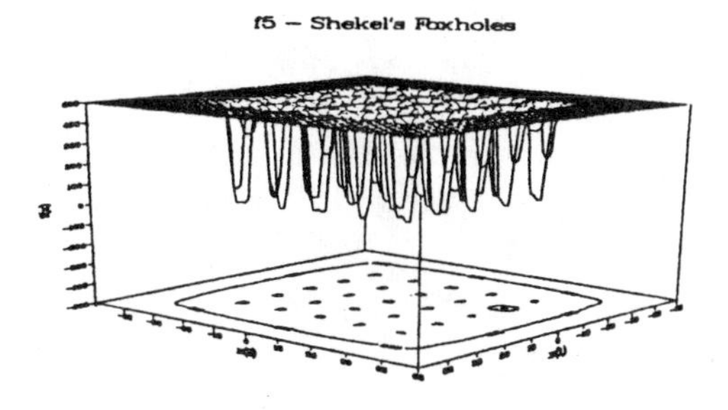

Figure 8

Like GA, GIRM manipulates and improves a population of 30 strings. Unlike GA, the structural manipulations responsible for the generation of a new string concern indifferently the whole population. A new "offspring" can be composed of blocks coming from all parents. Only 10 new strings are recruited at each generation. The radius of affinity can be defined either in the hamming space or in the real space. Here, it was defined in the real space. The function of affinity is quadratic :

$$m(i,k) = 1-(dh(i,k)/scale)^2$$

Preliminary results indicate that the global optimum is found in at least as few function calculations as the best GA variant described in [7] (GA elitist, ranking 2). However, a lot of improvements are still envisageable in this first version of GIRM i.e adjusting the affinity function, the decreasing of the population size and the radius of affinity, the structural manipulations for the generation of proposals, etc... We are still trying and testing these modifications on various multimodal functions.

7 CONCLUSIONS

In a last paper, Hoffmeister et al. [7] enlarge and abstract the notion of evolution strategies in order to encompass their ESs techniques and GA. Respecting the various aspects which characterize this large family, IRM might well apply for being another member of this biologically-grounded family. Indeed, the generation of a new population of points relies on structural manipulations of points present in the previous population. During this step and in order to reinforce the presence of benefic random, IRM relaxes the selective strategy which increases the influence of better points. However nothing in this family analysed explicitly (during the new population generation) the relative similarity of a new point with the previous ones situated in the neighbourhood. They suppose this similarity to result naturally from the selective and generative steps.

The recruitment test, which constitutes the basic novelty of IRM in comparison with GA or ESs, is responsible for this analysis. It intensifies the final selectivity by forcing the resemblance with the best points to the detriment of the less good ones. The consequence is an acceleration of the local hill-climbing(s). We are still carrying on numerous experiments in order to better understand what are or could be the real improvements (perhaps degradations) brought by IRM for problems whose profile range from simple to complicate fitness function.

Finally, the recruitment test fulfils a further function for distributed systems like the immune network. Suppose the performance of a system to be realized by a set of little operators whose responsibility is limited to a certain zone of action. All the zones of action form a partition of the problem state space (a good example of distributed system is the control of the cart-pole in [1]). We easily admit that two contiguous zones should contain similar operators in order to respect constraints of continuity. Then the optimization (based on the recruitment test) of one operator in a certain zone depends not only on the affinity with operators existing in the same zone but also with operators existing in the contiguous zones. The recruitment test integrates the network configuration of the system in its own optimization.

Acknowledgements

This work would have not been possible without the financial and intellectual support of the Shell Recherche SA France. We would like to thank the ERBAS group for fruitful discussions on adaptive control and biological metaphors. We acknowledge John Stewart, our colleagues from the Unité d'immubiologie at the Pasteur Institute of Paris and from the IRIDIA Laboratory of Brussels, for their essential collaboration.

References

[1] Barto, A., Sutton, R. and C. Anderson (1983): Neuronlike adaptive elements that can solve difficult learning control problems. *IEEE Transactions on Systems, Man and Cybernetics*, 13(5).

[2] Bersini, H. and F. Varela (1990) : Hints for adaptive problem solving gleaned from immune networks. In *proceedings of the first workshop on Parallel Problem Solving from Nature* - Springer Verlag.

[3] De Jong, K. (1975): *An analysis of the behaviour of a class of genetic adaptive systems*. PhD Thesis, University of Michigan, 1975. Diss. Abstr. Int. 36(10), 5140B, University Microfilms No. 76-9381.

[4] Doyne Farmer, J., A. Packard and A. Perelson (1987): The immune system, adaptation and machine learning. *Physica 22D.*

[5] Goldberg, D.E. (1989): *Genetic Algorithms in search, optimization and machine learning.* Addison-Wesley Publishing Company, Inc.

[6] Grefenstette, J.J. and J. E. Baker (1989): How genetic algorithms work: A critical look at implicit parallelism. In J. David Schaffer, editor, *Proceedings of the third international conference on genetic algorithms and their applications.* Morgan Kaufmann Publishers.

[7] Hoffmeister, F. and T. Bäck (1990): *Genetic Algorithms and Evolutions Strategies: Similarities and Differences.* Internal Report of Dortmund University - Bericht Nr. 365.

[8] Holland, J.H., Holyoak, K.J., Nisbett, R.E. & Thagard, P.R. (1986): *Induction: Processes of inference, learning and discovery.* Cambridge: MIT Press.

[9] Kauffman, S.A. (1989): Principles of Adaptation in Complex Systems, in D. Stein (Ed), *Lectures in the Sciences of Complexity*, SFI Series on the Science of Complexity, Addisson Wesley.

[10] Perelson, A. (1990): Theoretical Immunology. In *1989 Lectures in Complex Systems, SFI Studies in the Sciences of Complexity*, Lect. Vol. II, Ed. Erica Jen, Addison-Wesley.

[11] Perelson, A. and S. Forrest (1990): Genetic Algorithms and the Immune System. In *Proceedings of the 1990 Workshop on Parallel Problem Solving from Nature*, Springer-Verlag

[12] Roitt, I., Brostoff, J. and D. Male (1989): *Immunology.* Grover Medical Publishing Ltd 34-42 Cleveland Street, London, GB

[13] Stewart, J. and F. Varela (1990): Morphogenesis in Shape Space: Elementary meta-dynamics in a model of the immune network. Submitted to: *J. Theoret. Biol.*

[14] Varela, F., A. Coutinho, B. Dupire and N. Vaz (1988): Cognitive networks: Immune, neural and otherwise, in A. Perelson (Ed.), *Theoretical Immunology, Vol.2* SFI Series on the Science of Complexity, Addisson Wesley, New Jersey.

[15] Varela, F., V. Sanchez and A. Coutinho (1989): Adaptive strategies gleaned from immune networks, in B. Goodwin and P. Saunders (Eds.), *Evolutionary and epigenetic order from complex systems: A Waddington Memorial Volume.* Edinburgh U. Press

[16] Varela, F. and Stewart, J. (1990): Dynamics of a class of immune networks. I) Global behaviour. *J. theoret. Biol.* Vol. 144.

Is It Alive Or Is It GA?

Thomas S. Ray
School of Life & Health Sciences
University of Delaware
Newark, Delaware 19716
ray@brahms.udel.edu

Abstract

Synthetic organisms have been created based on a computer metaphor of organic life in which CPU time is the "energy" resource and memory is the "material" resource. Memory is organized into informational patterns that exploit CPU time for self-replication. Mutation generates new forms, and evolution proceeds by natural selection as different genotypes compete for CPU time and memory space. Diverse ecological communities have emerged. The genomes of these organisms consist of the machine code of a virtual computer. This machine code is robust to the genetic operations of mutation and recombination. This potentially makes it possible to evolve optimal solutions to problems expressed in machine code.

1 INTRODUCTION

Most evolutionary simulations are not open-ended. Their potential is limited by the structure of the model, which generally endows each individual with a genome consisting of a set of pre-defined genes, each of which may exist in a pre-defined set of allelic forms [5,2,7,1]. The object being evolved is generally a data structure representing the genome, which the simulator program mutates and/or recombines, selects, and replicates according to criteria designed into the simulator. The data structures do not contain the mechanism for replication, they are simply copied by the simulator if they survive the selection phase.

Self-replication is critical to evolution by natural selection, because without it, the mechanisms of selection must also be pre-determined by the simulator. Such artificial selection can never be as creative as natural selection. The organisms are not free to invent their own fitness functions. Freely evolving creatures will discover means of mutual exploitation and associated implicit fitness functions that we would never think of. Simulations constrained to evolve with pre-defined genes, alleles and fitness functions are dead ended, not alive.

The approach used here has generated rapidly diversifying communities of self-replicating organisms exhibiting open-ended evolution by natural selection. From a single rudimentary ancestral creature containing only the code for self-replication, interactions such as parasitism, immunity, hyper-parasitism, sociality and cheating have emerged spontaneously. This paper presents a methodology and some first results.

Apart from its value as a tool for the study or teaching of ecology and evolution, this model may have commercial applications. Evolution of machine code provides a new approach to the design and optimization of computer programs. In an analogy to genetic engineering, pieces of application code may be inserted into the genomes of digital organisms, and then evolved to new functionality or greater efficiency.

Genetic algorithms exploit the power of evolution to find optimal solutions to problems. Genetic operations such as mutation and recombination combined with replication and selection drive the process. One of the greatest difficulties in applying the method is in finding a representation of the space of possible solutions to a problem, in a form that can be subjected to the process. Problems are generally represented as a fixed length bit strings, although more flexible methods have been developed [6].

Machine instructions comprise a very powerful set of operations which are perhaps the most ideal general purpose representation of solutions. It is very easy to write solutions in the form of computer programs. Large numbers of persons are trained in the art, and there exists a great deal of software dedicated to assisting and automating the process.

However, when treated as genetic algorithms, computer programs have the shortcoming that solutions represented by sequences of machine instructions are not robust to the genetic operations of mutation and recombination. Computer programs of this class are notoriously brittle, in the sense that the ratio of viable programs to possible programs is virtually zero. The novel instruction set presented here overcomes the brittleness, and is demonstrated to support substantial evolution and optimization.

2 METHODS

2.1 THE METAPHOR

Organic life is viewed as utilizing energy, mostly derived from the sun, to organize matter. By analogy, digital life can be viewed as using CPU (central processing unit) time, to organize memory. Organic life evolves through natural selection as individuals compete for resources (light, food, space, etc.) such that genotypes which leave the most descendants increase in frequency. Digital life evolves through the same process, as replicating algorithms compete for CPU time and memory space, and organisms evolve strategies to exploit one another. CPU time is thought of as the analog of the energy resource, and memory as the analog of the spatial resource.

The memory, the CPU and the computer's operating system are viewed as elements of the "abiotic" (physical) environment. A "creature" is then designed to be specifically adapted to the features of the computational environment. The creature consists of a self-replicating machine language program (Appendix C).

All programs, regardless of the language they are written in, are converted into machine code before they are executed. Machine code is the natural language of the machine, and machine instructions are viewed by this author as the "atomic units" of computing. It is felt that machine instructions provide the most natural basis for an artificial chemistry of creatures designed to live in the computer.

In the biological analogy, the machine instructions are considered to be more like the amino acids than the nucleic acids, because they are "chemically active". They actively manipulate bits, bytes, CPU registers, and the movements of the instruction pointer. The digital creatures discussed here are entirely constructed of machine instructions. They are considered analogous to creatures of the RNA world, because the same structures bear the "genetic" information and carry out the "metabolic" activity.

A block of RAM memory in the computer is designated as a "soup" which can be inoculated with creatures. The "genome" of the creatures consists of the sequence of machine instructions that make up the creature's self-replicating algorithm. The prototype creature consists of 80 machine instructions, thus the size of the genome of this creature is 80 instructions, and its "genotype" is the specific sequence of those 80 instructions (Appendix C).

2.2 THE VIRTUAL COMPUTER—TIERRA

The work described here takes place on a virtual computer known as Tierra (Spanish for Earth). Tierra is a parallel computer of the MIMD (multiple instruction, multiple data) type, with a processor (CPU) for each creature. Parallelism is imperfectly emulated by allowing each CPU to execute a small time slice in turn. Each CPU of this virtual computer contains two address registers, two numeric registers, a flags register to indicate error conditions, a stack pointer, a ten word stack, and an instruction pointer (Appendix A). Computations performed by the Tierran CPUs are probabilistic due to flaws that occur at a low frequency (see Mutation below).

2.3 THE TIERRAN LANGUAGE

In developing this new virtual language, which is called "Tierran", close attention has been paid to the structural and functional properties of the informational system of biological molecules: DNA, RNA and proteins. Two features have been borrowed from the biological world which are considered to be critical to the evolvability of the Tierran language.

First, the instruction set of the Tierran language has been defined to be of a size that is the same order of magnitude as the genetic code. Information is encoded into DNA through 64 codons, which are translated into 20 amino acids. In its present manifestation, the Tierran language consists of 32 instructions, which can be represented by five bits, *operands included.*

Emphasis is placed on this last point because some instruction sets are deceptively small. Some versions of the redcode language of Core Wars [3,8] for example are defined to have ten operation codes. It might appear on the surface then that the instruction set is of size ten. However, most of the ten instructions have one or two operands. Each operand has four addressing modes, and then an integer. When we consider that these operands are embedded into the machine code, we realize that they are in fact a part of the instruction set, and this set works out to be about 10^{11} in size. Inclusion of numeric operands will make any instruction set extremely large in comparison to the genetic code.

In order to make a machine code with a truly small instruction set, we must eliminate numeric operands. This can be accomplished by allowing the CPU registers and the stack to be the only operands of the instructions. When we need to encode an integer for some purpose, we can create it in a numeric register through bit manipulations: flipping the low order bit and shifting left. The program can contain the proper sequence of bit flipping and shifting instructions to synthesize the desired number, and the instruction set need not include all possible integers.

A second feature that has been borrowed from molecular biology in the design of the Tierran language is the addressing mode, which is called "address by template". In most machine codes, when a piece of data is addressed, or the IP jumps to another piece of code, the exact numeric address of the data or target code is specified in the machine code. Consider that in the biological system by contrast, in order for protein molecule A in the cytoplasm of a cell to interact with protein molecule B, it does not specify the exact coordinates where B is located. Instead, molecule A presents a template on its surface which is complementary to some surface on B. Diffusion brings the two together, and the complementary conformations allow them to interact.

Addressing by template is illustrated by the Tierran JMP (jump) instruction. Each JMP instruction is followed by a sequence of NOP (no-operation) instructions, of which there are two kinds: NOP_0 and NOP_1. Suppose we have a piece of code with five instruction in the following order: JMP NOP_0 NOP_0 NOP_0 NOP_1. The system will search outward in both directions from the JMP instruction looking for the nearest occurrence of the complementary pattern: NOP_1 NOP_1 NOP_1 NOP_0. If the pattern is found, the instruction pointer will move to the end of the complementary pattern and resume execution. If the pattern is not found, an error condition (flag) will be set and the JMP instruction will be ignored (in practice, a limit is placed on how far the system may search for the pattern).

The Tierran language is characterized by two unique features: a truly small instruction set without numeric operands, and addressing by template. Otherwise, the language consists of familiar instructions typical of most machine languages, e.g., MOV, CALL, RET, POP, PUSH etc.

2.4 THE TIERRAN OPERATING SYSTEM

The Tierran virtual computer needs a virtual operating system that will be hospitable to digital organisms. The operating system will determine the mechanisms of interprocess communication, memory allocation, and the allocation of CPU time among competing processes. Algorithms will evolve so as to exploit these features to their advantage. More than being a mere aspect of the environment, the operating system together with the instruction set will determine the topology of possible interactions between individuals, such as the ability of pairs of individuals to exhibit predator-prey, parasite-host or mutualistic relationships.

2.4.1 Memory Allocation — Cellularity

The Tierran computer operates on a block of RAM of the real computer which is set aside for the purpose. This block of RAM is referred to as the "soup". In most of the work described here the soup consisted of about 60,000 bytes, which can hold the same number of Tierran machine instructions. Each "creature" occupies some block of memory in this soup.

Tierran creatures are considered to be cellular in the sense that they are protected by a "semi-permeable membrane" of memory allocation. The Tierran operating system provides memory allocation services. Each creature has exclusive write privileges within its allocated block of memory. The "size" of a creature is just the size of its allocated block (e.g., 80 instructions). This usually corresponds to the size of the genome. This "membrane" is described as "semi-permeable" because while write privileges are protected, read and execute privileges are not. A creature may examine the code of another creature, and even execute it, but it can not write over it. Each creature may have exclusive write privileges in at most two blocks of memory: the one that it is born with which is referred to as the "mother cell", and a second block which it may obtain through the execution of the MAL (memory allocation) instruction. The second block, referred to as the "daughter cell", may be used to grow or reproduce into.

When Tierran creatures "divide", the mother cell loses write privileges on the space of the daughter cell, but is then free to allocate another block of memory. At the moment of division, the daughter cell is given its own instruction pointer, and is free to allocate its own second block of memory.

2.4.2 Time Sharing — The Slicer

The Tierran operating system must be multi-tasking in order for a community of individual creatures to live in the soup simultaneously. The system doles out small slices of CPU time to each creature in the soup in turn. The system maintains a circular queue called the "slicer queue".

The number of instructions to be executed in each time slice may be set proportional to the size of the genome of the creature being executed, raised to a power. If the "slicer power" is equal to one, then the slicer is size neutral, the probability of an instruction being executed does not depend on the size of the creature in which it occurs. If the power is greater than one, large creatures get more CPU cycles per instruction than small creatures. If the power is less than one, small creatures get more CPU cycles per instruction. The power determines if selection favors large or small creatures, or is size neutral. A constant slice size selects for small creatures.

2.4.3 Mortality — The Reaper

Self-replicating creatures in a fixed size soup would rapidly fill the soup and lock up the system. To prevent this from occurring, it is necessary to include mortality. The Tierran operating system includes a "reaper" which begins "killing" creatures from a queue when the memory fills. Creatures are killed by deallocating their memory, and removing them from both the reaper and slicer queues. Their "dead" code is not removed from the soup.

In the present system, the reaper uses a linear queue. When a creature is born it enters the bottom of the queue. The reaper always kills the creature at the top of the queue. However, individuals may move up or down in the reaper queue according to their success or failure at executing certain instructions. When a creature executes an instruction that generates an error condition, it moves one position up the queue, as long as the individual ahead of it in the queue has not accumulated a greater number of errors. Two of the instructions are somewhat difficult to execute without generating an error (mal, div), therefore successful execution of these instructions moves the creature down the reaper queue one position, as long as it has not accumulated more errors than the creature below it.

The effect of the reaper queue is to cause algorithms which are fundamentally flawed to rise to the top of the queue and die. Vigorous algorithms have a greater longevity, but in general, the probability of death increases with age.

2.4.4 Mutation

Mutations occur in two circumstances. At some background rate, bits are randomly selected from the entire soup (e.g., 60,000 instructions totaling 300,000 bits) and flipped. This is analogous to mutations caused by cosmic rays, and has the effect of preventing any creature from being immortal, as it will eventually mutate to death. The background mutation rate has generally been set at about one bit flipped for every 10,000 Tierran instructions executed by the system.

In addition, while copying instructions during the replication of creatures, bits are randomly flipped at some rate in the copies. The copy mutation rate is the higher of the two, and results in replication errors. The copy mutation rate has generally been set at about one bit flipped for every 1,000 to 2,500 instructions moved. In both classes of mutation, the interval between mutations varies randomly within a certain range to avoid possible periodic effects.

In addition to mutations, the execution of Tierran instructions is flawed at a low rate. For most of the 32 instructions, the result is off by plus or minus one at some low frequency. For example, the increment instruction normally adds one to its register, but it sometimes adds two or zero. The bit flipping instruction normally flips the low order bit, but it sometimes flips the next higher bit or no bit. The shift left instruction normally shifts all bits one bit to the left, but it sometimes shifts left by two bits, or not at all. In this way, the behavior of the Tierran instructions is probabilistic, not fully deterministic.

It turns out that bit flipping mutations and flaws in instructions are not necessary to generate genetic change and evolution, once the community reaches a certain state of complexity. Genetic parasites evolve which are sloppy replicators, and have the effect of moving pieces of code around between creatures, causing rather massive rearrangements of the genomes. The mechanism of this ad hoc sexuality has not been worked out, but is likely due to the parasites' inability to discriminate between live, dead or embryonic code.

2.5 THE TIERRAN ANCESTOR

The Tierran language has been used to write a single self-replicating program which is 80 instructions long. This program is referred to as the "ancestor", or alternatively as genotype 0080aaa (Fig. 1). The ancestor is a minimal self-replicating algorithm which was originally written for use during the debugging of the simulator. No functionality was designed into the ancestor beyond the ability to self-replicate, nor was any specific evolutionary potential designed in.

3 RESULTS

The results of running creatures on the Tierran computer are presented in somewhat more detail in [9]. The dynamics of the process are illustrated in [10].

3.1 EVOLUTION

Evolutionary runs of the simulator are begun by inoculating the soup of about 60,000 instructions with a single individual of the 80 instruction ancestral genotype. The passage of time in a run is measured in terms of how many Tierran instructions have been executed by the simulator. The original ancestral cell executes 839 instructions in its first replication, and 813 for each additional replication. The initial cell and its replicating daughters rapidly fill the soup memory which starts the reaper. Typically, the system executes about 400,000 instructions in filling up the soup with about 375 individuals of size 80 (and their gestating daughter cells).

3.1.1 Micro-Evolution

Mutations in and of themselves, can not result in a change in the size of a creature, they can only alter the instructions in its genome. However, by altering the genotype, mutations may affect the process whereby the creature examines itself and calculates its size, potentially causing it to produce an offspring that differs in size from itself.

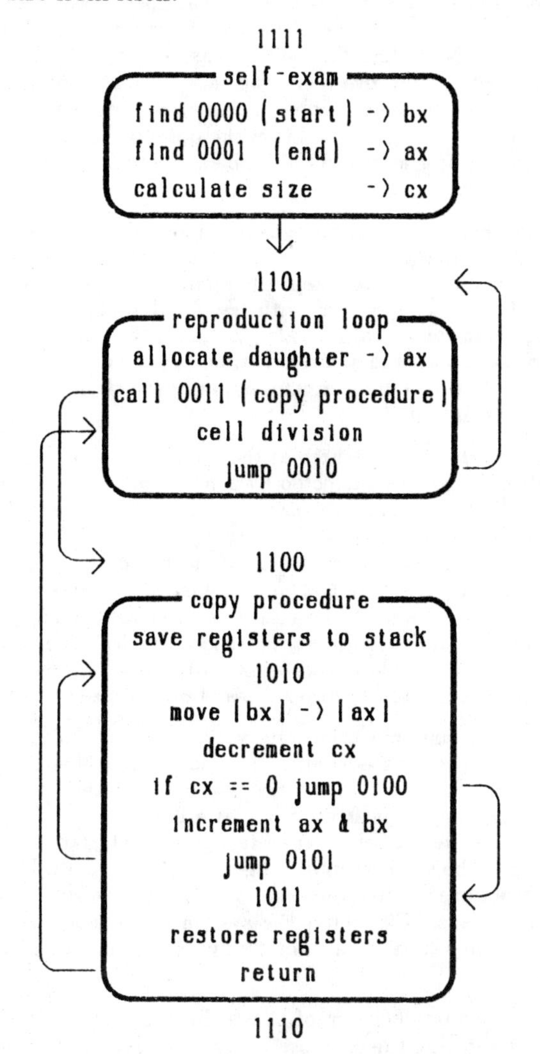

Figure 1: Metabolic flow chart for the ancestor: ax, bx and cx refer to CPU registers where location and size information are stored. [ax] and [bx] refer to locations in the soup indicated by the values in the ax and bx registers. Patterns such as 1101 are complementary templates used for addressing. Arrows outside of boxes indicate jumps in the flow of execution of the programs.

Parasites An example of the kind of error that can result from a mutation in a template is a mutation of the low order bit of instruction 42 of the ancestor. Instruction 42 is a NOP_0, the third component of the copy procedure template. A mutation in the low order bit would convert it into NOP_1, thus changing the template from 1 1 0 0 to: 1 1 1 0. This would then be recognized as the template used to mark the end of the creature, rather than the copy procedure.

A creature born with a mutation in the low order bit of instruction 42 would calculate its size as 45. It would allocate a daughter cell of size 45 and copy only instructions 0 through 44 into the daughter cell. The daughter cell then, would not include the copy procedure. This daughter genotype, consisting of 45 instructions, is named 0045aaa.

Genotype 0045aaa is not able to self-replicate in isolated culture. However, the semi-permeable membrane of memory allocation only protects write privileges. Creatures may match templates with code in the allocated memory of other creatures, and may even execute that code. Therefore, if creature 0045aaa is grown in mixed culture with 0080aaa, when it attempts to call the copy procedure, it will not find the template within its own genome, but if it is within the search limit (generally set at 200–400 instructions) of the copy procedure of a creature of genotype 0080aaa, it will match templates, and send its instruction pointer to the copy code of 0080aaa. Thus a parasitic relationship is established. Typically, parasites begin to emerge within the first few million instructions of elapsed time in a run.

Immunity to Parasites At least some of the size 79 genotypes demonstrate some measure of resistance to parasites. If genotype 45aaa is introduced into a soup, flanked on each side with one individual of genotype 0079aab, 0045aaa will initially reproduce somewhat, but will be quickly eliminated from the soup. When the same experiment is conducted with 0045aaa and the ancestor, they enter a stable cycle in which both genotypes coexist indefinitely. Freely evolving systems have been observed to become dominated by size 79 genotypes for long periods, during which parasitic genotypes repeatedly appear, but fail to invade.

Circumvention of Immunity to Parasites Occasionally these evolving systems dominated by size 79 were successfully invaded by parasites of size 51. When the immune genotype 0079aab was tested with 0051aao (a direct, one step, descendant of 0045aaa in which instruction 39 is replaced by an insertion of seven instructions of unknown origin), they were found to enter a stable cycle. Evidently 0051aao has evolved some way to circumvent the immunity to parasites possessed by 0079aab. The fourteen genotypes 0051aaa through 0051aan were also tested with 0079aab, and none were able to invade.

Hyper-Parasites Hyper-parasite have been discovered, (e.g., 0080gai, which differs by 19 instructions from the ancestor). Their ability to subvert the energy metabolism of parasites is based on two changes. The copy procedure does not return, but jumps back directly to the proper address of the reproduction loop. In this way it effectively seizes the instruction pointer from the parasite. However it is another change which delivers the coup de grâce: after each reproduction, the hyper-parasite re-examines itself, resetting the bx register with its location and the cx register with its size. After the instruction pointer of the parasite passes through this code, the CPU of the parasite contains the location and size of the hyper-parasite and the parasite thereafter replicates the hyper-parasite genome.

Social Hyper-Parasites Hyper-parasites drive the parasites to extinction. This results in a community with a relatively high level of genetic uniformity, and therefore high genetic relationship between individuals in the community. These are the conditions that support the evolution of sociality, and social hyper-parasites soon dominate the community. Social hyper-parasites appear in the 61 instruction size class. For example, 0061acg is social in the sense that it can only self-replicate when it occurs in aggregations. When it jumps back to the code for self-examination, it jumps to a template that occurs at the end rather than the beginning of its genome. If the creature is flanked by a similar genome, the jump will find the target template in the tail of the neighbor, and execution will then pass into the beginning of the active creature's genome. The algorithm will fail unless a similar genome occurs just before the active creature in memory. Neighboring creatures cooperate by catching and passing on jumps of the instruction pointer.

Cheaters: Hyper-Hyper-Parasites The cooperative social system of hyper-parasites is subject to cheating, and is eventually invaded by hyper-hyper-parasites. These cheaters (e.g., 0027aab) position themselves between aggregating hyper-parasites so that when the instruction pointer is passed between them, they capture it.

3.1.2 Macro-Evolution

The most thoroughly studied case for long runs is where selection, as determined by the slicer function, is size neutral. The longest runs to date (as much as 2.86 billion Tierran instructions) have been in a size neutral environment, with a search limit of 10,000, which would allow large creatures to evolve if there were some algorithmic advantage to be gained from larger size. These long runs illustrate a pattern which could be described as periods of stasis punctuated by periods of rapid evolutionary change, which appears to parallel the pattern of punctuated equilibrium [4].

Initially these communities are dominated by creatures with genome sizes in the eighties. This represents a period of relative stasis, which has lasted from 178 million to 1.44 billion instructions in the several long runs conducted to date. The systems then very abruptly (in a span of 1 or 2 million instructions) evolve into communities dominated by sizes ranging from about 400 to about 800. These communities have not yet been seen to evolve into communities dominated by either smaller or substantially larger size ranges.

The communities of creatures in the 400 to 800 size range also show a long-term pattern of punctuated equilibrium. These communities regularly come to be dominated by one or two size classes, and remain in that condition for long periods of time. However, they inevitably break out of that stasis and enter a period where no size class dominates. These periods of rapid evolutionary change may be very chaotic. Close observations indicate that at least at some of these times, no genotypes breed true. Many self-replicating genotypes will coexist in the soup at these times, but at the most chaotic times, none will produce offspring which are even their same size. Eventually the system will settle down to another period of stasis dominated by one or a few size classes which breed true.

3.2 EVOLUTIONARY OPTIMIZATION

In order to compare the process of evolution between runs of the simulator, a simple objective quantitative measure of evolution is needed. One such measure is the degree to which creatures improve their efficiency through evolution. This provides not only an objective measure of progress in evolution, but also sheds light on the potential application of synthetic life systems to the problem of the optimization of machine code.

The efficiency of the creature can be indexed in two ways: the size of the genome, and the number of CPU cycles needed to execute one replication. Clearly, smaller genomes can be replicated with less CPU time, however, during evolution, creatures also decrease the ratio of instructions executed in one replication, to genome size. The number of instructions executed per instruction copied, drops substantially.

Mutation rates are measured as the inverse fraction of individuals per generation affected by a move mutation. At the highest two sets of rates tested, one and two, the system is unstable. The genomes melt under the heat of the high mutation rates. The community often dies out, although some runs survived the 500 million instruction runs used in this study. The next lower rate, four, yields the highest rate of optimization without the risk of death of the community. At the five lower mutation rates, 8, 16, 32, 64 and 128, we see successively lower rates of optimization.

Runs at the same mutation rate vary in some details such as whether progress is continuous and gradual, or comes in bursts. Also, each run decreases to a size limit which it can not proceed past even if it is allowed to run much longer. However, different runs reach different plateaus of efficiency. The smallest limiting genome size seen has been 22 instructions, while other runs reached limits of 27 and 30 instructions. Evidently, the system can reach a local optima from which it can not easily evolve to the global optima.

The increase in efficiency of the replicating algorithms is even greater than the decrease in the size of the code. The ancestor is 80 instructions long and requires 839 CPU cycles to replicate. The creature of size 22 only requires 146 CPU cycles to replicate, a 5.75–fold difference in efficiency. These optimizations are illustrated in [11].

4 SUMMARY

4.1 GENERAL BEHAVIOR OF THE SYSTEM

In addition to an increase in the raw diversity of genotypes and genome sizes, there is an increase in the ecological diversity. Obligate commensal parasites evolve, which are not capable of self-replication in isolated culture, but which can replicate when cultured with normal (self-replicating) creatures. These parasites execute some parts of the code of their hosts, but cause them no direct harm, except as competitors. Some potential hosts have evolved immunity to the parasites, and some parasites have evolved to circumvent this immunity.

In addition, facultative hyper-parasites have evolved,

which can self-replicate in isolated culture, but when subjected to parasitism, subvert the parasites energy metabolism to augment their own reproduction. Hyper-parasites drive parasites to extinction, resulting in complete domination of the communities. The relatively high degrees of genetic relatedness within the hyper-parasite dominated communities leads to the evolution of sociality in the sense of creatures that can only replicate when they occur in aggregations. These social aggregations are then invaded by hyper-hyper-parasite cheaters.

Mutations and the ensuing replication errors lead to an increasing diversity of sizes and genotypes of self-replicating creatures in the soup. Within the first 100 million instructions of elapsed time, the soup evolves to a state in which about a dozen more-or-less persistent size classes coexist. The relative abundances and specific list of the size classes varies over time. Each size class consists of a number of distinct genotypes which also vary over time.

4.2 APPLICATIONS

A new version of the Tierran language is currently under development which will retain those features of the language that make it evolvable, while at the same time expanding its functionality to the point that it should be possible to develop cross-assemblers between it and real assembler languages. Application code written and compiled to run on real machines could be cross-assembled into the new Tierran language. Each procedure could then be inserted into the genome of a creature. Creatures could be rewarded with CPU time in proportion to the efficacy and efficiency of the evolving inserted code. In this way, artificial selection would lead to the optimization of the inserted code, which could then be cross-assembled back into the real machine code. If this proved to be practical, it would be worthwhile to render the Tierran virtual instruction set in silicon, thereby greatly accelerating the process. If machine code could evolve that quickly, then there is the possibility of using it as a generative process in addition to an optimization procedure.

Acknowledgements

I thank Robert Eisenberg, Doyne Farmer, Walter Fontana, Stephanie Forrest, Chris Langton, Dan Pirone, Stephen Pope, and Steen Rasmussen, for their discussions or readings of the manuscripts. Contribution No. 149 from the Ecology Program, School of Life and Health Sciences, University of Delaware.

References

[1] Ackley, D. H. & Littman, M. S. "Learning from natural selection in an artificial environment." In: *Proc. IJCNN*, Washington, DC. Hillsdale, NJ: Lawrence Erlbaum, Winter 1990.

[2] Dawkins, R. *The blind watchmaker.* New York: W. W. Norton & Co., 1987.

[3] Dewdney, A. K. "Computer recreations: In the game called Core War hostile programs engage in a battle of bits." *Sci. Amer.* **250** (1984): 14–22.

[4] Gould, S. J. & Eldredge, N. "Punctuated equilibria: the tempo and mode of evolution reconsidered." *Paleobiology* **3** (1977): 115–151.

[5] Holland, J. H. *Adaptation in natural and artificial systems.* Ann Arbor: Univ. of Michigan Press, 1975.

[6] Koza., J. R. *Genetic programming: a paradigm for genetically breeding populations of computer programs to solve problems.* Report No. STAN-CS-90-1314. Stanford, CA: Stanford U., 1990.

[7] Packard, N. H. "Intrinsic adaptation in a simple model for evolution." In: *Artificial life*, edited by C. Langton. Redwood City, CA: Addison-Wesley, 1989, 141–155.

[8] Rasmussen, S., Knudsen, C., Feldberg, R. & Hindsholm, M. "The coreworld: emergence and evolution of cooperative structures in a computational chemistry" *Physica D* **42** (1990): 111–134.

[9] Ray, T. "An approach to the synthesis of life." In: *Artificial Life II*, edited by Farmer, J. D., C. Langton, S. Rasmussen, & C. Taylor. Redwood City, CA: Addison-Wesley, (In press, 1991).

[10] Ray, T. "Population dynamics of digital organisms." In: *Artificial Life II Video Proceedings*, edited by C. G. Langton. Redwood City, CA: Addison-Wesley, (In press, 1991).

[11] Ray, T. "Synthetic Life: evolution and optimization of digital organisms." In: *1990 IBM Supercomputing Competition: Large Scale Computing Analysis and Modeling Conference Proceedings*, edited by H. U. Brown. Cambridge, MA: MIT Press (In press).

A Appendix

Structure definition of the Tierra virtual CPU. The executable or source code for the Tierra Simulator can be obtained by contacting the author by email or snail mail.

```
struct cpu {  /* registers of virtual cpu */
    int   ax; /* address register, % soup size */
    int   bx; /* address register, % soup size */
    int   cx; /* numerical register */
    int   dx; /* numerical register */
    char  fl; /* flag */
    char  sp; /* stack pointer */
    int   st[10]; /* stack */
    int   ip; /* instruction pointer */
    } ;
```

B Appendix

Instructions set for the Tierra Simulator:

nop_0, no operation; nop_1, no operation; or1, flip low order bit of cx, cx ^= 1; shl, shift left cx register, cx <<= 1; zero, set cx register to zero, cx = 0; if_cz, if cx==0 execute next instruction; sub_ab, subtract bx from ax, cx = ax - bx; sub_ac, subtract cx from ax, ax -= cx; inc_a, increment ax, ax++; inc_b, increment bx, bx++; dec_c, decrement cx, cx−−; inc_c, increment cx, cx++; push_ax, push ax on stack; push_bx, push bx on stack; push_cx, push cx on stack; push_dx, push dx on stack; pop_ax, pop top of stack into ax; pop_bx, pop top of stack into bx; pop_cx, pop top of stack into cx; pop_dx, pop top of stack into dx; jmp, move ip to nearest complementary template; jmpb, move ip backward to template; call, call a procedure; ret, return from a procedure; mov_cd, move cx to dx, dx = cx; mov_ab, move ax to bx, bx = ax; mov_iab, move instruction at address in bx to address in ax; adr, search both directions for nearest complementary template put address in ax; adrb, search backward for template; adrf, search forward for template; mal, allocate memory for daughter cell; divide, cell division (give daughter a cpu, cause mother to loose write privelages on daughter cell).

C Appendix

Assembler source code for the ancestral creature. This code requires 839 cpu cycles to replicate the first time, 813 cycles for each additional replication.

```
nop_1   ; 01  0 beginning template
nop_1   ; 01  1 beginning template
nop_1   ; 01  2 beginning template
nop_1   ; 01  3 beginning template
zero    ; 04  4 put zero in cx
or1     ; 02  5 put 1 in first bit of cx
shl     ; 03  6 shift left cx
shl     ; 03  7 shift left cx, (cx=4, templ size)
mov_cd  ; 18  8 move template size to dx
adrb    ; 1c  9 get (backward) addr. of beg. templ
nop_0   ; 00 10 compliment to beginning template
nop_0   ; 00 11 compliment to beginning template
nop_0   ; 00 12 compliment to beginning template
nop_0   ; 00 13 (ax = start of mother + 4)
sub_ac  ; 07 14 ax -= cx (ax = start of mother)
mov_ab  ; 19 15 bx = ax (bx = start of mother)
adrf    ; 1d 16 get (forward) addr. of end templ.
nop_0   ; 00 17 compliment to end template
nop_0   ; 00 18 compliment to end template
nop_0   ; 00 19 compliment to end template
nop_1   ; 01 20 (ax = end of mother)
inc_a   ; 08 21 incl dummy inst to sep creatures
sub_ab  ; 06 22 cx = ax - bx = size
nop_1   ; 01 23 reproduction loop template
nop_1   ; 01 24 reproduction loop template
nop_0   ; 00 25 reproduction loop template
nop_1   ; 01 26 reproduction loop template
mal     ; 1e 27 allocate memory for daughter
call    ; 16 28 call template (copy procedure)
nop_0   ; 00 29 (ax = start of daughter)
nop_0   ; 00 30 copy procedure compliment
nop_1   ; 01 31 copy procedure compliment
nop_1   ; 01 32 copy procedure compliment
divide  ; 1f 33 create independent daughter cell
jmp     ; 14 34 jump to template (reproduct loop)
nop_0   ; 00 35 reproduction loop compliment
nop_0   ; 00 36 reproduction loop compliment
nop_1   ; 01 37 reproduction loop compliment
nop_0   ; 00 38 reproduction loop compliment
if_cz   ; 05 39 dummy instr. to sep. templates
nop_1   ; 01 40 (begin copy procedure)
nop_1   ; 01 41 copy procedure template
nop_0   ; 00 42 copy procedure template
nop_0   ; 00 43 copy procedure template
push_ax ; 0c 44 push ax onto stack
push_bx ; 0d 45 push bx onto stack
push_cx ; 0e 46 push cx onto stack
nop_1   ; 01 47 copy loop template
nop_0   ; 00 48 copy loop template
nop_1   ; 01 49 copy loop template
nop_0   ; 00 50 copy loop template
mov_iab ; 1a 51 move [bx] to [ax]
dec_c   ; 0a 52 decrement cx
if_cz   ; 05 53 if cx == 0 perform next instr.
jmp     ; 14 54 jump to template
nop_0   ; 00 55 copy proc. exit compl.
nop_1   ; 01 56 copy proc. exit compl.
nop_0   ; 00 57 copy proc. exit compl.
nop_0   ; 00 58 copy proc. exit compl.
inc_a   ; 08 59 increment ax
inc_b   ; 09 60 increment bx
jmp     ; 14 61 jump to template
nop_0   ; 00 62 copy loop compliment
nop_1   ; 01 63 copy loop compliment
nop_0   ; 00 64 copy loop compliment
nop_1   ; 01 65 copy loop compliment
if_cz   ; 05 66 dummy instr., to sep. templ.
nop_1   ; 01 67 copy procedure exit template
nop_0   ; 00 68 copy procedure exit template
nop_1   ; 01 69 copy procedure exit template
nop_1   ; 01 70 copy procedure exit template
pop_cx  ; 12 71 pop cx off stack
pop_bx  ; 11 72 pop bx off stack
pop_ax  ; 10 73 pop ax off stack
ret     ; 17 74 return from copy procedure
nop_1   ; 01 75 end template
nop_1   ; 01 76 end template
nop_1   ; 01 77 end template
nop_0   ; 00 78 end template
if_cz   ; 05 79 dummy inst. to sep. creatures
```

D Appendix

Assembler source code for the smallest self-replicating creature. This code requires 146 cpu cycles to replicate the first time, 142 cycles for each additional replication.

```
nop_0   ; 00  0
adrb    ; 1c  1 find beginning
nop_1   ; 01  2
divide  ; 1f  3 fails the first time it is executed
sub_ac  ; 07  4
mov_ab  ; 19  5
adrf    ; 1d  6 find end
nop_0   ; 00  7
inc_a   ; 08  8 to include final dummy statement
sub_ab  ; 06  9 calculate size
mal     ; 1e 10
push_bx ; 0d 11 save beginning address on stack
nop_0   ; 00 12   in order to 'return' there
mov_iab ; 1a 13
dec_c   ; 0a 14
if_cz   ; 05 15
ret     ; 17 16 jump to beginning
inc_a   ; 08 17 (address saved on stack)
inc_b   ; 09 18
jmpb    ; 15 19
nop_1   ; 01 20
mov_iab ; 1a 21 dummy inst. to terminate template
```

Designing Biomorphs with an Interactive Genetic Algorithm

Joshua R. Smith*
jrs@cs.williams.edu
Department of Computer Science
Williams College
Williamstown, MA 01267

Abstract

We describe an interactive version of the Genetic Algorithm which can be used to evolve Fourier series based "biomorphs." The user plays the role of a binary-valued objective function. The usual Genetic operators of Selection, Mutation, and Crossover are applied normally. We discuss the Interactive Genetic Algorithm's use as a teaching tool and in domains in which measures of utility are difficult or impossible to specify mathematically. We argue that the evolution of biomorphs can be considered an ergonomic design problem, and with the evolution of biomorphs as an example, we argue that a large class of design problems can be cast as "imprecise optimization" and "solved" with Interactive Genetic Algorithms. Thus IGAs can be used for applications that fall somewhere between AI, optimization, and CAD.

1 Introduction

When does a curve look like a bug? "Looks-like-a-bug" is not the sort of relation that can easily be described in mathematical terms. Thus the Genetic Algorithm, which usually requires a precisely formulated fitness function, may not seem particularly "fit" for evolving bug-shaped curves. But if the objective function is replaced by a human expert, the GA can be applied in this and other imprecise domains. For example, if one were trying to evolve a model of a material with specific "squishy" or "fleshy" properties, it might be simpler to have a human expert tell the system which samples "felt" right than to try to specify mathematically the desired "feel" of the material.

A curve "looks-like-a-bug" when it interfaces with the human visual system in some particular but difficult to specify way, not when it satisfies some easily enumerated set of mathematical criteria. Thus the problem of evolving curves which look like bugs can be considered a "human-factors" or *ergonomic* design problem. Simon[7] pointed out that certain design problems, particularly in engineering, can be cast as optimization problems. Interactive Genetic Algorithms enlarge the class of problems that can be treated in this way. Ergonomic design is particularly well suited to IGA techniques because at some level, humans are the only possible measure of the quality of human-factors design. We maintain that the problem of evolving a curve which looks like a bug has much in common with other ergonomic design problems and that it may often be feasible to treat relations such as "feels-comfortable" in the same way we treat "looks-like-a-bug" here.

Like many versions of the Genetic Algorithm, the Interactive Genetic Algorithm has evolution-inspired "select-and-mutate" ancestors which do not achieve the GA's celebrated implicit parallelism[5] [4] because they do not maintain a large breeding population and do not use the crossover operator. The most famous of these ancestors is of course Richard Dawkins' *Blind Watchmaker,* which used a select-and-mutate approach to evolve tree-based forms which look surprisingly life-like. Dawkins christened these life-like shapes *biomorphs.* Obviously, we have inherited more than just the *term* biomorph from Dawkins: he originated the project of using interactive evolution techniques to grow them. Our variation on the process is the crossover operator and the maintenance of a large breeding population: in Dawkins' program, the next generation inherits its genetic material from a single parent in the current generation[2][1].

In a more practical vein, Oppenheimer [6] used interactive select-and-mutate methods to evolve impressive three dimensional images of trees (the biological variety, not computer science "trees"). Like Dawkins, Oppenheimer used a breeding population of size one and did not employ the crossover operator.

Yet Holland's Schema Theorem holds as well for inter-

*Emmanuel College, Cambridge, as of Oct. 1991

active GAs in which the objective function has been replaced by a human operator as it does for standard GAs. The difference is simply that it is impossible to specify *a prioi* which schemata will have above average fitness. Those short, low-order schemata with above-average fitness (no longer a precisely defined quantity) will still receive an exponentially increasing number of trials as long as the human operator behaves consistently. In interactive evolution problems, implicit parallelism is especially desirable, since time spent evolving a solution is expensive human time. Thus it makes sense to use the Genetic Algorithm for interactive evolution problems.

2 BUGS

Our example system for solving an interactive evolution problem enables the user to design biomorphs with the Genetic Algorithm. A biomorph's genotype consists of two sets of real numbers[1] which serve as Fourier coefficients in parametric equations specifying X and Y coordinates as a function of the parameter t. We call these two sets of numbers chromosomes. In the equation for $X(t)$, we let the coefficients of the sin terms equal zero; in the $Y(t)$ equation, we let the cos coefficients equal zero. That is why our curves are specified with two sets of numbers, not the four that would generally be required to specify two Fourier series. To put it succinctly, the first chromosome is the set of A_is and the second is the set of B_is in parametric equations of the form

$$X = \sum_{i=0}^{n} A_i \cos it$$

and

$$Y = \sum_{i=0}^{n} B_i \sin it.$$

We "grow" biomorphs from their genetic material, the Fourier coefficients, by graphing these parametric equations. Figure 1 shows some sample biomorphs. Their bilateral symmetry is due to the periodicity properties of the sin and cos functions.

We also graph each biomorph's genetic material below its 'portrait'. For each chromosome there is a baseline, on which genes with value zero fall. Genes above the line are positive; those below are negative. The graphs of the genes are visible below each biomorph in the Figure.

[1]Although we used real codings for the genes, there was no particular reason for this, and we could have just as easily used binary codings. Considering our earlier arguments about implicit parallelism, we probably should have.

3 An IGA in Action

As in most Genetic Algorithms, the population is initialized randomly. Then the user indicates which biomorphs may reproduce by clicking on-screen buttons. When the user gives the signal, the next generation is produced by fairly standard GA operators. The biomorphs which the user has marked as "fit to breed" all receive equal, positive fitness values; those not marked fit by the user receive fitness zero. At this point we chose pairs for breeding using the standard "roulette wheel" selection operator, stochastic sampling with replacement. Since the user is playing the role of a binary objective function, not of the selection operator, s/he can specify that certain biomorphs do *not* breed but cannot ensure that particular biomorphs *will* breed, and cannot specify which will breed with which. If the user does not indicate that any biomorphs are fit enough to breed, we assume they are all equally good; it is as if the user had indicated that all the biomorphs were fit.

We use one-point crossover, though we perform it twice, once per chromosome. Since we use real codings, we perform mutation by probabilistically adding Gaussian noise to each gene. The variance of the noise is specified as a parameter to the program.[2]

4 The IGA as a Teaching Tool

Using an interactive GA is a good way to gain an intuitive understanding of some features of GA operation. Since the genome is graphed on screen, the user can get some feeling for the operation of the crossover and mutation operators. The user may also acquire an understanding of stochastic selection and its perils: organisms marked fit by the user may nonetheless be passed up by chance. These stochastic sampling errors lead to the phenomena of genetic drift and premature convergence. If the user chooses a very small breeding population, the population converges almost immediately, which is not surprising to most people. But even if all biomorphs are given a chance to breed every time, that is, even in the absence of selective pressure, stochastic sampling errors will cause the population to converge after a time, a phenomenon known as genetic drift.[3] This may also teach a valuable lesson about biological evolution: not all inherited traits are adaptive. IGAs are useful both for improving one's own understanding of these phenomena and for explaining them to others.

[2]We would like to reiterate that these deviations from GA orthodoxy had nothing to do with the interactivity of our system; we could have just as easily used the standard crossover and mutate operators, or any others for that matter.

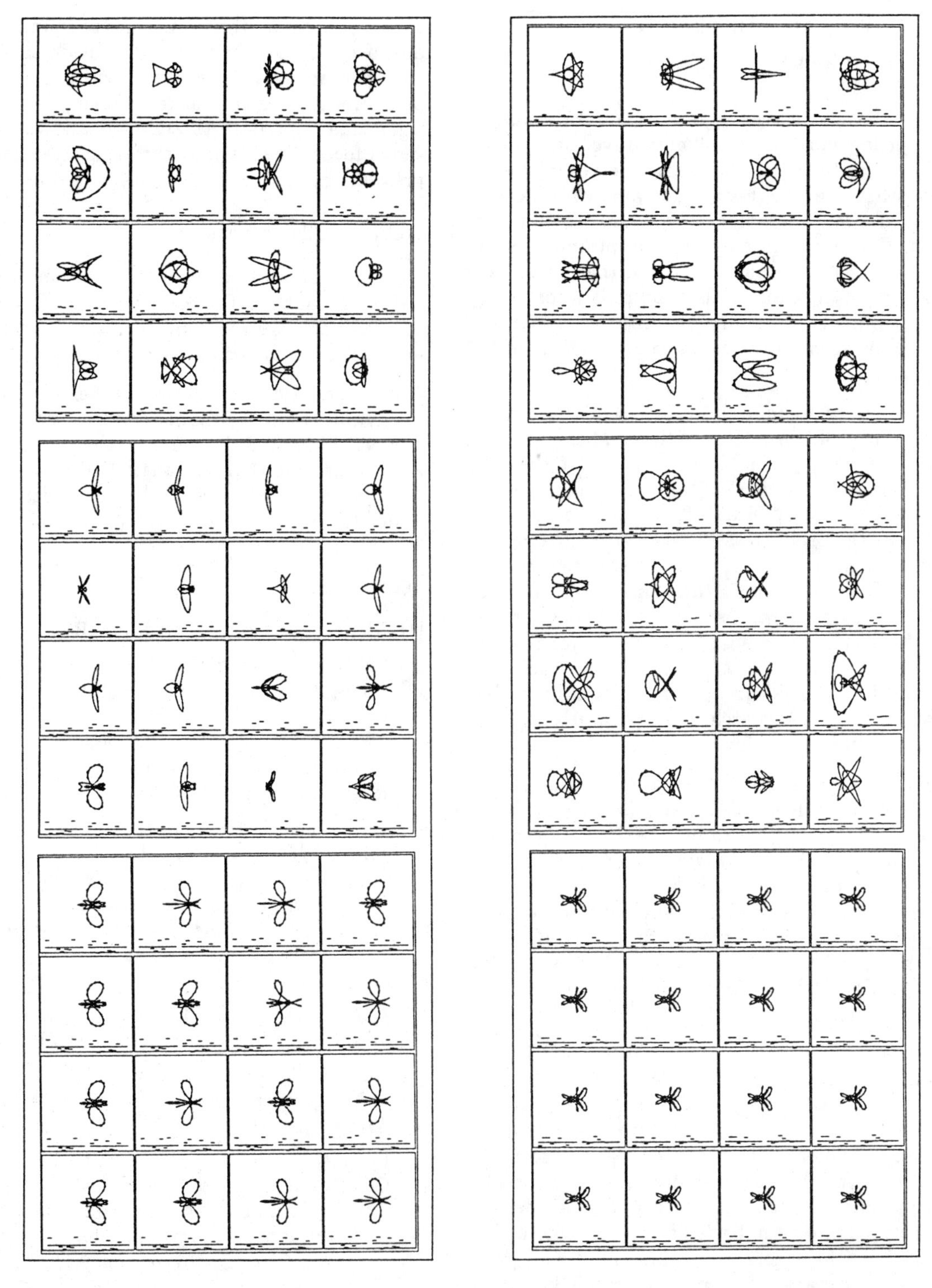

Figure 1: INTERACTIVELY GUIDED EVOLUTION of two biomorph populations. Below each biomorph is a graph of its genes. Each small horizontal segment represents a single gene. The large horizontal segments indicate the origin: a gene above its origin line has positive value; genes below are negative. In both trials shown here, the initial population (*top*) is generated randomly. After several generations of evolution with an interactive genetic algorithm, the curves become more bug-like. Diversity diminishes throughout this process, but the population does not converge immediately (*middle*). The time to convergence depends strongly on how selective the user is, that is, on the fraction of each population designated fit to breed. The final form to which the population converges (*bottom*) depends on the user's preference, on stochastic factors such as the make up of the initial population, and on the Fourier series-based "embryology" function which maps genes into curves.

5 The IGA and Imprecise Optimization

But, as we have already suggested, the IGA's value is not only pedagogic. Using an "interactive" or "human-expert" fitness function may allow the GA to be applied in otherwise inaccessible domains. We have already given a few examples of problems for which we think an IGA approach may be appropriate: the evolution (or "design") of biomorphs and the evolution of models of "squishy" materials, materials whose properties are difficult to specify quantitatively. We will call problems such as these "imprecise optimization" problems.

We will now try to map more precisely the class of IGA-solvable problems. Most of the conditions for GA applicability, with the notable exception of the need for a precisely formulated objective function, are also conditions for IGA applicability. It must be possible to formulate the problem as a search through a parameter space. That means the relevant parameters must be known. This is an important constraint: the need for this knowledge is what separates true design from mere optimization. A designer must discover relevant parameters; an optimizer, given this knowledge, must find good values for them. Again, IGAs, since they are based on the GA, search given parameter spaces. IGAs are optimizers in this sense.

Interactivity imposes an additional constraint on the class of IGA solvable problems: it must be possible to produce a new generation in near real time. While it might be acceptable to let a standard GA chug away for weeks at a time on a problem, a human is liable to lose patience if called upon periodically to babysit an IGA process over a long period of time. The IGA will be most useful when the expert or designer can see the next generation almost immediately (hence the name *interactive* GA).

So the IGA is likely to be applicable to optimization problems in which i) candidate solutions can be generated in near real time and ii) the utility of candidate solutions can be compared by humans but not (practically) by means of a precisely specified formula. This encompasses a class of design problems broader than engineering design (in which candidate designs can be compared on the basis of numbers like weight and cost) but which does not include all design problems, since the parameter space must be known in advance. Again, this second condition suggests that the IGA may be particularly well suited for ergonomic design.[3]

In these contexts, the IGA starts to sound more like an advanced CAD tool than an optimization technique. And perhaps that is where it will prove most useful: as a semi-intelligent assistant to suggest ideas the designer may not have thought of, and to help the designer focus the design by throwing away bad ideas and presenting good ones for frequent iterative refinement.

Acknowledgements

This work supported by the Bronfman Science Foundation. Thanks to Donald House and Duane Bailey of Williams College for guidance, and to Mike Donofrio and Bill Lenhart for their comments on this paper.

Note: The software discussed here is available via anonymous ftp at Williams College (`/pub/BUGS.tar.Z @cs.williams.edu`) and the `alife` server (`/pub/alife/software/unix/bugs/BUGS.tar.Z @iuvax.cs.indiana.edu`). It runs under Suntools and under the X Window System with XView.

References

[1] Richard Dawkins. *The Blind Watchmaker.* Longman, Harlow, 1986.

[2] Richard Dawkins. The evolution of evolvability. In Christopher G. Langton, editor, *Artificial Life: The Proceedings of an Interdisciplinary Workshop on the Synthesis and Simulation of Living Systems*, Redwood City, California, 1989. Addison-Wesley.

[3] David E. Goldberg. Finite markov chain analysis of genetic algorithms. In John J. Grefenstette, editor, *Genetic Algorithms and Their Applications: Proceedings of the Second International Conference on Genetic Algorithms*, Hillsdale, New Jersey, 1987. Lawrence Erlbaum Associates.

[4] David E. Goldberg. *Genetic Algorithms in Search, Optimization, and Machine Learning.* Addison-Wesley, Reading, Massachusetts, 1989.

[5] John H. Holland. *Adaptation in Natural and Artificial Systems.* The University of Michigan Press, Ann Arbor, 1975.

[6] Peter Oppenheimer. The artificial menagerie. In Christopher G. Langton, editor, *Artificial Life: The Proceedings of an Interdisciplinary Workshop on the Synthesis and Simulation of Living Systems*, Redwood City, California, 1989. Addison-Wesley.

[7] Herbert A. Simon. *The Sciences of the Artificial.* MIT Press, Cambridge, Massachusetts, second edition, 1985.

[3] For an example of ergonomic design that is close to the hearts of computer scientists, consider user interface design. IGAs, perhaps in conjunction with classifier systems, might find a home in adaptive user interfaces in which designing the interface became an ongoing process of interaction between the user and the machine. In the competition to do the user's bidding, some agents would make mistakes. The user would indicate that an error had occurred, and the responsible agents would receive low fitness ratings. In this way, one might evolve agents which can, to some extent, do "what-I-mean".

An approach to the analysis of the basins of the associative memory model using Genetic Algorithms

Keiji SUZUKI and Yukinori KAKAZU

Department of Precision Engineering, Faculty of Engineering, Hokkaido University, North-13, West-8, Sapporo 060, JAPAN

Abstract

In this paper, an approach to the analysis of the basin of a correlational associative memory model using the Genetic Algorithms and a new training algorithm for this model is described. The recalling process of a model described by direction cosine is insufficient for the better understanding of the dynamical behavior of the model. In order to know the characteristics of memorized states, the methodology of the Genetic Algorithms applied to analyze the recalling process concerned with each memorized state is proposed. Furthermore, before the analyzing, the LU-algorithm is proposed to give the model the ability of keeping a wide basin in both highly memorized rates and mutually non-orthogonal states. Finally, results of experiments related to the basin analysis are shown.

1. Introduction

In this paper, an approach to the analysis of the basin of a correlational associative memory model using Genetic Algorithms and new training algorithm for this model are described.

The study of the correlational associative memory models is one of the major fields in artificial neural models researches [2] [3]. The correlational associative memory models are characterized by its large number of units, dynamical behaviors and mechanism of storing information by learning. A large number of simple processing units, called neurons, are connected to each other to act asynchronously and to perform computational tasks in a collective manner as well as storing information in a distributive way. Through a training process using a learning algorithm, the models will have basins which are the sets of points in the state space which are attracted to one of the memorized points by the dynamical behavior of the models. This dynamical behavior of the models is characterized as a motion in time through the state space. Therefore, the memory-recalling process can be viewed as the dynamical behavior of the models searching for a stable state that is caused by being stimulated by an initial or external input vector.

In the recalling process, it is well known that the dynamical behavior shows the strange phenomena which are observed by computer simulations [2] [3]: Starting with an initial state close to a memorized state, the state monotonously approaches the memorized one. Whereas, starting with an initial state which is not so close to the memorized one, the initial state at first approaches the latter but then moves away from it. These phenomena are observed under the following conditions: the memorized states are mutually orthogonal and the distance between a memorized state and a state in the recalling process is measured by the direction cosine. From these phenomena, basins of attractors of a model seem to form strange shapes such as circles having splits towards the centers. That is, when a state closes to a center of a basin, the state in a split once approaches the center but then moves away at the bottom of the split; While only a state on the basin is able to directly approach the center. For the reason that the shapes of the basins can be viewed as forming strange shapes, it appears problematic to analyze the dynamical behaviors theoretically.

In addition, the dynamical behaviors show another phenomenon under the same conditions. That is, the dynamical behaviors always show the threshold phenomenon; There exits a threshold value of the direction cosine relating to an initial state in the recall process. Therefore, when a value of an initial state is greater than the threshold, the state always converges to a center of a basin, and when a value of an initial state is smaller than the threshold, the state goes away from the center of the basin.

However, if stored states in the model are no longer orthogonal under the new conditions, the threshold phenomenon disappears and the dynamical behaviors usually show more complex phenomena. Therefore, the theoretical analysis of the dynamical behaviors in the model is very difficult in practice and also numerical analysis will be difficult without a powerful analyzing method.

Although the above discussions are based on the distance of direction cosine between a state and a stored one, now, let us consider the reason why an initial state is attracted to a specific stable state and not attracted to the other stable states based on pattern recognition. Firstly, let's establish that each of the patterns of the states in a basin has a characteristic related to the stored pattern of the basin. Moreover, such characteristics of the patterns in a basin seem to vary with a set of memorized states since the shape of the basin concerned with the characteristics interacts with the shapes of other basins.

In the observation of the model, these characteristics will appear in the dynamical behaviors of the model starting from the attracted patterns to the stored pattern. Therefore, if the dynamical behaviors relating to a stored state can be analyzed, the result of the analysis could be applied to pattern recognition and clustering of a state space correlating to the memorized states.

In order to analyze the basins, because the performance of a training method determines the shapes of the basins concerned with memorized states, not only a powerful analyzing method but also a powerful learning method are required. That is, if a learning method only provides a poor ability for a model such as low memory capacity and acquiring many spurious states, the number of basin would be quite limited and the extents of the basins would be small and distorted. Therefore, the analyzing of the basins would be meaningless in practice.

In this paper, in order to achieve an appropriate condition of a model and analyze the basin, we propose the Learning-Unlearning (LU) algorithm for training and we use the Genetic Algorithms (GAs) for analyzing them. In the following sections, firstly, LU-algorithm is proposed and its learnability is described. Secondly, the method of applying the GAs to the analysis of basins is proposed. Finally, results of experiments related to the basin analysis are shown.

2. Associative Memory Model

Let V be a column vector having N components: $v_1, v_2, \ldots, v_N$ that represents a state of the model which has N units. Each unit receives N input signals and asynchronously determines the next state of itself, v_i, from the sum of weighted inputs, u_i (Eq.(1)).

$$v_i = \frac{1}{1+\exp(-u_i/\lambda)} = f(W_i V) \tag{1}$$

In Eq.(1), W_i represents a row vector which is an i-th row in the N×N weight matrix $W = (w_{ij})$, and λ is a positive coefficient. From Eq.(1), the type of the model in this paper is assumed as having a continuous state, $0 \le v_i \le 1$, and as having the discrete time type of the Hopfield neural model.

We define one unit time of the asynchronous dynamical behavior of the model when all units estimate themselves at once. This is formulated as the following Eq.(2), where k is the number of the count.

$$V' = \Phi^k(W, V) \tag{2}$$

In the characteristics of the model, the discrete time and the asynchronousness are not essential to the training algorithm but on the other hand the assumption of $0 \le v_i \le 1$ is essential to the algorithm. The reason will be shown in the following section. The dynamical behavior of the model is represented as the function which has the property of monotonously decreasing the energy (Eq.(3)).

$$E = -\frac{1}{2} W V V^T . \tag{3}$$

Strictly speaking, it is possible that the dynamical behavior of the model using continuous state doesn't appear to be monotonously decreasing the energy. The effect of this dynamical behavior is in practicality relatively small. Therefore, the effect is negligible.

Let $S = \{ S^j \mid j=1,\ldots,M \}$ be a set of memorized states. Each of the states in the set is N dimensional vector and the component of the vector is either 1 or 0. In order to memorize the states to the model, the outer product algorithm (OP-algorithm) for training the model is familiar and easy. The outer product formulation is described as follows (Eq.(4)):

$$W = \sum_i^M \varepsilon_{op} S^{i'} S^{i'T} , \tag{4}$$

$$\text{where } s_j^{i'} = 2s_j^i - 1,\ S^{i'} = (s_j^{i'}),\ j=1,\cdots,N$$

In Eq.(4), ε_{OP} is a positive coefficient. The formulation requires the assumption that *S* consists of the orthogonal vectors of each other. In relating the properties of the model given by the formulation, it is said that the critical memory capacity is about 0.15. Generally speaking, the memory rate M/N closely relates to the extent of the basins. That is, the extent of the basins becomes narrower when the memory rate increases. Furthermore, there is another important problem with the model trained with this formulation; the reverse states of the memorized states must be stored into the model and such states often have larger basins themselves than those of the memorized ones. Therefore, the extent of basins of the model is suppressed to a lower value.

If these spurious states can be decreased in the model or if it is possible to protect the model from storing these states, the extent of basins would increase dramatically. In the next section, it is shown that the scheme using the unlearning method and continuous state of 1 to 0 can not only protect the model by storing the reversing states of the memorized ones but can decrease the other types of spurious states and increase the extent of basins of the model.

3. Learning-Unlearning Algorithm

The unlearning method [1] can be formulated using a state V, learning time t and ε is a positive coefficient, as follows (Eq.(5)):

$$W(t) = W(t-1) - \Delta W = W(t-1) - \varepsilon V V^T \tag{5}$$

The property of the unlearning method can be understood from the behavior of the energy function with the state implemented in this scheme. That is, the energy of the state, V, decreases according to the application of the operation in Eq.(6).

$$E = -\frac{1}{2}W(t)VV^T = -\frac{1}{2}\left\{ W(t-1)VV^T - \varepsilon \sum_i^N \sum_j^N (v_i v_j)^2 \right\} \quad (6)$$

If the unlearning method is implemented to a stable state repeatedly, the equilibrium of the state becomes progressively weaker. Hence the state will at last lose stability in the end.

The algorithm using this method must avoid disturbing the stabilities of the memorized states. According to the this requirement, the proposed algorithm, which we name a Learning-Unlearning algorithm (LU-algorithm), has two phases similar to the learning algorithm for a Boltzmann Machine. One phase, called Learning-mode, is implemented so as to memorize the states. Another phase, called Unlearning-mode, is implemented so as to decrease the stabilities of the spurious states and separate the memorized states. Here, learning equation of the Learning-Unlearning algorithm is described as the following Eq.(7), where *t* is the learning time.

$$W(t) = W(t-1) + \Delta W^+ - \Delta W^- \quad (7)$$

The formulation of each phase is described as follows:

(I) Learning-mode... The sum of the matrixes, which is derived from the outer product with the memorized binary vector S^m (j=1,...,M), is written as

$$\Delta W^+ = \sum_{m=1}^{M} \varepsilon_L S^m S^{mT} = \sum_{m=1}^{M} \varepsilon_L \begin{bmatrix} \vdots \\ I_i(S^m)\, S^{mT} \\ \vdots \end{bmatrix} \quad (8)$$

In Eq.(8), $I_i(S^m)$ is the projection S^m on the i-th component of S^m, $I_i(S^m)S^{mT}$ is an i-th row of the matrix, and ε_L is a positive coefficient. The matrix includes the nonnegative value because the memorized vectors are all binary vectors. Therefore, if any other methods can not be used, this method only trains the model to memorize a state which overlaps the memorized states. Each state will be separated by the following method.

(II) Unlearning-mode... The method makes an outer product with the state X^m (m=1,...,M), which is a stable state or a transitional state starting from a randomly generated state $P = \{ P^r \mid r=1,..., tM \}$. The Unlearning-mode is formulated as follows Eq.(9). This formulation uses the transitional states in order to decrease the weight.

$$\Delta W^- = \sum_{r=q}^{q+M} \sum_{j=1}^{K} \varepsilon_U X_j^r X_j^{rT} \quad , \quad (9)$$

where $X_j^r = \Phi^1(W, X_{j-1}^r),\ X_0^r = P^r,\ \varepsilon_L > \varepsilon_U > 0.$

Learnability of the model trained with the algorithm can be shown using the changing direction of the sum of the inputs, u_i, in the training process. If a state S is an equilibrium, the i-th unit gets a sufficient positive value from the state S in the case of i-th component of S being 1 or gets a sufficient negative value from the state S in another case of i-th component of S being 0. The learnability will exist under the condition that the changing direction of u_i is always given toward getting the sufficient value along with the state of the i-th unit of the memorized one by implementing the training algorithm. The following illustrates the learnability of this algorithm.

Let $\boldsymbol{S} = \{ S^j \mid j=1,...,M \}$ be a set of memorized states, and let $\boldsymbol{X} = \{X^j \mid j=1,...,M\}$ be a set of unlearning states which is generated from the dynamical behaviors of the model starting with the random states. The following assumption is settled by the relationship of inner product between S^i and $\boldsymbol{S} \cup \boldsymbol{X}$ as described in Eq.(10).

$$S^i \cdot S^i = \gamma,\ S^i \cdot S^k = \alpha,\ S^i \cdot X^q = \beta$$

$$\rightarrow \alpha < \beta \le \gamma,\ \sum_{j \ne i,k}^{M} S^i \cdot S^j - \sum_{j \ne i,k}^{M} S^i \cdot X^j \cong 0 \quad (10)$$

In Eq.(10), the inner product between A and B is denoted as A•B. The weight, W, can be given with the last training using $\boldsymbol{S}$ and $\boldsymbol{X}$ as follows (Eq.(10)).

$$W = \begin{bmatrix} \vdots \\ W_i \\ \vdots \end{bmatrix} + \varepsilon_L \begin{bmatrix} \vdots \\ I_i(S^i)\, S^{iT} \\ \vdots \end{bmatrix} + \varepsilon_L \begin{bmatrix} \vdots \\ I_i(S^k)\, S^{kT} \\ \vdots \end{bmatrix} - \varepsilon_U \begin{bmatrix} \vdots \\ I_i(X^q)\, X^{qT} \\ \vdots \end{bmatrix} + \sum_{j \ne k,i}^{M} \varepsilon_L \begin{bmatrix} \vdots \\ I_i(S^j)\, S^{jT} \\ \vdots \end{bmatrix} - \sum_{j \ne k,i}^{M} \varepsilon_U \begin{bmatrix} \vdots \\ I_i(X^j)\, X^{jT} \\ \vdots \end{bmatrix} \quad (11)$$

The sum of the inputs with the model that takes the state S^i and uses W as its weight matrix is represented as the following Eq.(12). In Eq.(12), the inner product between S^i and the 5-th term in Eq.(11), and also between S^i and the 6-th term in Eq.(11), can be neglected from the assumptions held in Eq.(10).

$$U(S^i) = \begin{bmatrix} \vdots \\ W_i \cdot S^i \\ \vdots \end{bmatrix} + \varepsilon_L \begin{bmatrix} \vdots \\ I_i(S^i)\, \gamma \\ \vdots \end{bmatrix} + \varepsilon_L \begin{bmatrix} \vdots \\ I_i(S^k)\, \alpha \\ \vdots \end{bmatrix} - \varepsilon_U \begin{bmatrix} \vdots \\ I_i(X^q)\, \beta \\ \vdots \end{bmatrix} \quad (12)$$

The changing direction of the sum of the inputs, $\Delta u_i = I_i(U(S^i))$, is determined by the combination of $I_i(S^i)$, $I_i(S^k)$, and $I_i(X^q)$. The possible cases of the combination are illustrated in Table.1.

$(I_i(S^k)\ I_i(X^q))$	$I_i(S^i)=1$	$I_i(S^i)=0$
(1 v)	$\varepsilon_L\ (\gamma+\alpha)-\varepsilon_U\ v\beta>0$	$\varepsilon_L\ \alpha-\varepsilon_U\ v\beta<0$
(0 v)	$\varepsilon_L\ \gamma-\varepsilon_U\ v\beta>0$	$-\varepsilon_U\ v\beta<0$

Table-1. *The changing direction of the sum of the inputs, $\Delta u_i=I_i(U(S^i))$, in the case of using continuous state, $0\leq v\leq 1$.*

Under the assumption, $\alpha < \beta \leq \gamma$, the changing direction of Δu_i is always positive when $I_i(S^i)>0$ and always negative when $I_i(S^i) < 0$ except in the case when the combination is $(I_i(S^i), I_i(S^k), I_i(X^q)) = (0,1,v)$ from Table.1. In the case of (0, 1, v), Δu_i becomes negative when $v>\varepsilon_L\alpha / \varepsilon_U\beta$ but it becomes positive when $v<\varepsilon_L\alpha / \varepsilon_U\beta$. The value, $v>\varepsilon_L\alpha / \varepsilon_U\beta$, is the critical value for the learnability of this algorithm and this value effects the separability of memorized states. Namely, If the value is settled low, the model can learn and separate S^i with S^k.

With regard to the previous assumption, the relation, $\alpha < \gamma$, can be always held because the inner product of the vectors, which contains the values such that $0\leq v\leq 1$, has the relationship such that,

$$\max_V (V\cdot S) = S\cdot S \quad (13)$$

The assumption, however, will not exist when the components of these vectors ranges $-1\leq v\leq 1$. The continuous value function in Eq.(1) is also essential to the training of the model with this algorithm. When the model using a step value function is trained with this algorithm, the training will fail because the changing direction of Δu_i is insufficient to memorize the states in the case of $(I_i(S^i), I_i(S^k), I_i(X^q)) = (0, 1, 0)$ and $\Delta u_i = -\varepsilon_L\alpha>0$.

Experimental results of the LU-algorithm and the OP-algorithm are shown in both Fig.1 and Fig.2, where the model parameter is $N=100$, $\varepsilon_L=0.1$, $\varepsilon_U=0.08$, $K=10$, $T=200$, $\varepsilon_{op}=2.0$.

Fig.1 shows the result of the separability of the closely memorized states S. S corresponds to the relationship between the states as shown in the following Eq.(14), where the direction cosine, a_t, is defined in Eq.(15).

$$S = \{ S^m \mid \max_j (a(S^m, S^j)) = a_{max},\ m=1\cdots M \} \quad (14)$$

$$a_t = a(V^m, V^j) = \frac{1}{N}\sum_i^N v'_{mi}v'_{ji}, \quad \text{where } v'_{ji} = 2v_{ji} - 1,\ V^j = (v_{ji}) \quad (15)$$

Fig.2 shows the result of the maximum radii of the basins of memorized states, which are prepared at random, along with the memorizing rate. In Fig.2, each dot represents the maximum radius of each of the memorized states. The maximum radius is indicated by the direction cosine between a random starting state and a memorized state which is found by the starting state under the condition; $N=200$, $T=200$, $e_L=0.5$, $e_U=0.08$ and $K=5$.

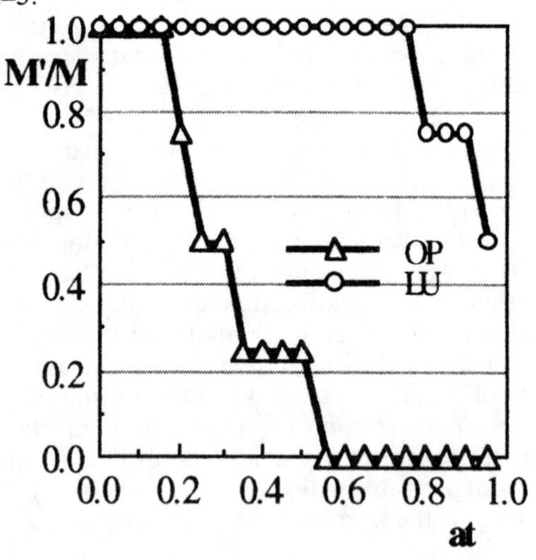

Fig.1 *The separability of the memorized states*

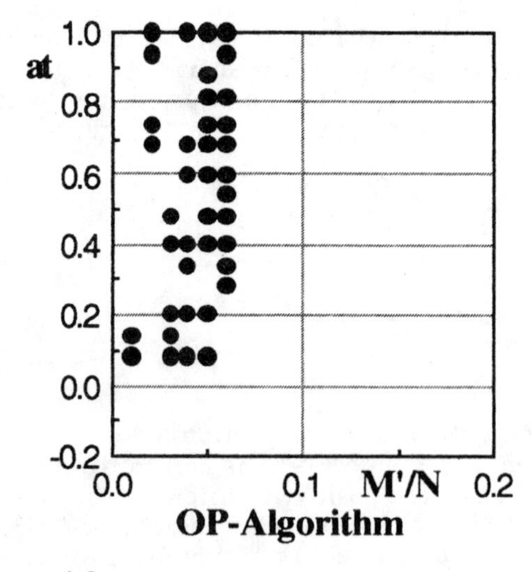

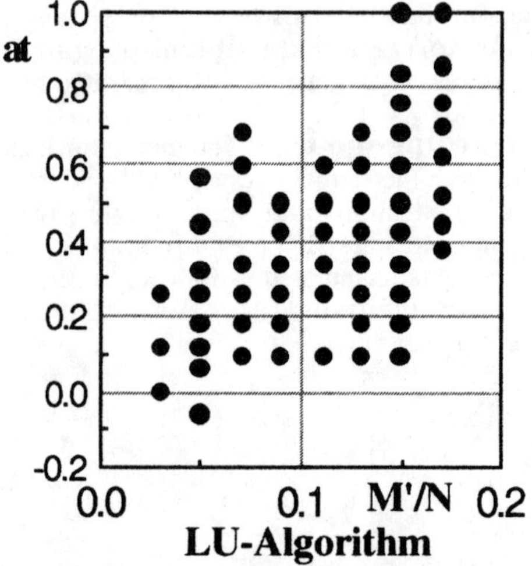

Fig.2 *The maximum radius of the basins of the memorized state along with the memory rate.*

All results of the experiments show this algorithm having higher performance than that of the OP-algorithm. Practically speaking, the results of the separability and the maximum radius of the basin show a difference of character between the LU-algorithm and the OP-algorithm. The LU-algorithm gives the model a high memory capacity, an ability to keep wide basins in high memorized rate and to memorize states which closely resemble each other.

4. Dynamical Behavior of the Model

Fig.3 shows a typical result of the dynamical behaviors in recall process, where N=64, M=4, learning time T=200.

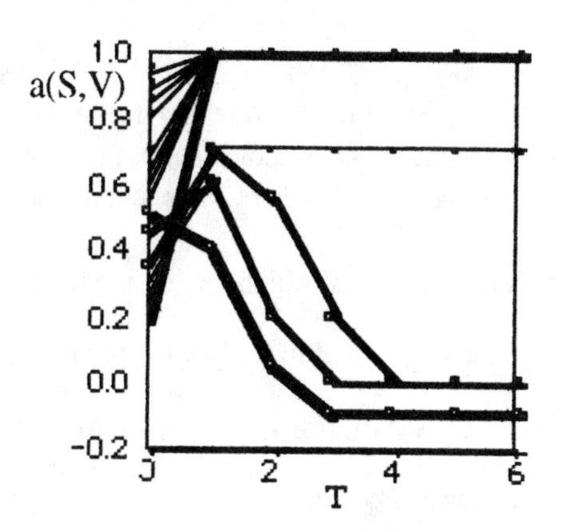

Fig.3 *Typical result of the dynamical behavior in recall process.*

In this figure, the ordinate is time *t*, the abscissa is the direction cosine a(S,V) where S is a memorized state, and the curves show the dynamical process of recalling, starting from various initial a(S,V). Under the condition that the memorized states don't hold the mutual orthogonal relationship, a basin of the memorized state is in close proximity to the other basins which include some spurious basins. Therefore, the dynamical behaviors behave as if only some initial states can approach the memorized state and the other initial states can not choose which states they should approach. Because of these phenomena, the basins of attractors of a model form strange shapes as illustrated in Fig.4. That is, when the states approach the basin center, the states in the splits at first approach the center but then move away to the bottoms of the splits. On the other hand, only states on the basin are able to approach the center of the basin. Since the basins can be seen as having strange shapes, it seems difficult to analyze related dynamical behaviors.

These dynamical behaviors are the particular results of measuring the distance between memorized state and transition states by the direction cosine. If other measurements are applied to the behaviors, the recalling process would quite possibly be easily understood. The phenomena relating to the dynamical behaviors that appear simple, such as seen in Fig.5, imply that these phenomena are observed to have threshold phenomenon and monotonous transition phenomena. That is, a transitional state approaches the center of the basin when its initial state of a measured value becomes greater than a threshold and, on the contrary, a transitional state moves away from the border of the basin when the initial state value becomes smaller than the threshold. If such ideal behaviors can be obtained by an analysis of the basins, the shape of the basins can be regarded as a complete sphere in a state space, which is illustrated in Fig.6.

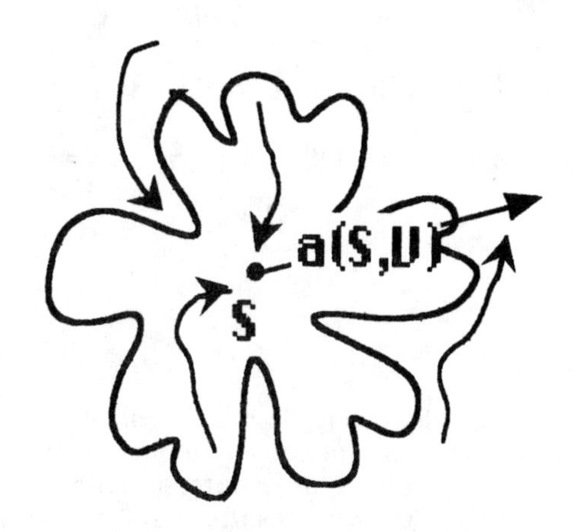

Fig.4 *Schematic view of the basin of the attraction.*

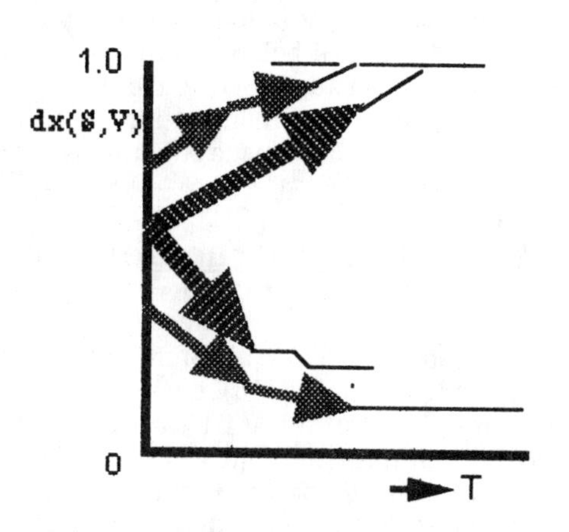

Fig.5 *Ideal dynamical behavior in recall process.*

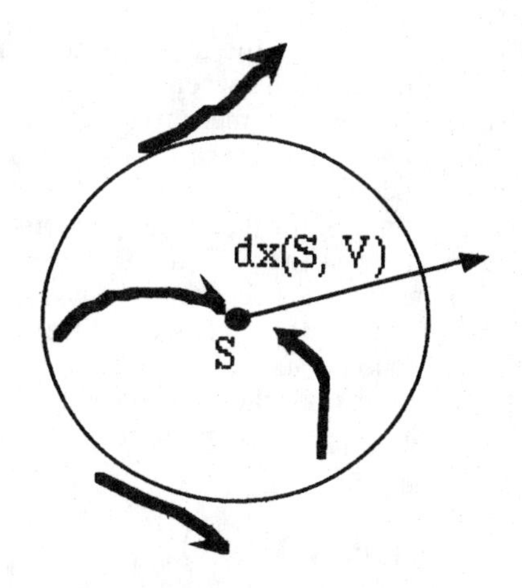

Fig.6 *Schematic view of the basin using the characteristic measure.*

The dynamical behaviors relating to the basins of the memorized states are different from each other thereby requiring the measurment of each basin. Here, the measurement of a basin is named characteristic measurement. However, not only the determination of the characteristic measurement itself is very difficult but the determination of the type of the characteristic measurements is also difficult. In this paper, it is assumed that the type of the characteristic measurements can be represented by a polynomial function of degree 2 as follows (Eq.(16)):

$$d_x(S, V) = \frac{1}{c}\sum_{k=1}^{c} a_k y_1^{\zeta_1} y_2^{\zeta_2} \cdots y_n^{\zeta_n}, \tag{16}$$

where $y_i = (2v_i-1)(2s_i-1)$, $S=(s_i)$, $V=(v_i)$,

$\zeta_i \in \{0, 1\}$, $\sum_i^n \zeta_i \leq 2$, $a_k \in \{0, 1\}$,

$c = \sum_{i=1} a_i$.

In Eq.(16), each characteristic measurement can be represented by the sequence of the coefficients $\{a1, a2, a3, \ldots, a\ c\}$. Therefore, determining these characteristic measurements results in an optimizing problem which searches the sequence of the coefficients of the characteristic measurements that can show the threshold phenomenon and monotonous phenomena concerned with the dynamical behaviors in basins. In the next section, the strategy of searching for combinations of the coefficients representing the ideal phenomenon is shown using the Genetic Algorithms.

5. Basin Analysis using the Genetic Algorithms

In order to apply the GAs, a search space must be encoded into strings. From Eq.(16), the search space can be given by the sequence of coefficients $\{a1, a2, a3, \ldots, a\,c\}$. Therefore, the sequence can directly be encoded into the string, d_i, for the GAs. For example, a sequence {0,1,1,1,0,0,...} is encoded into a string d_1=(011100...). The generation is represented by a set of strings, D, where $d_i \in D$ and $|D|$ represents its population.

In order to search the characteristic measurement which represents the ideal phenomenon, the amount of the difference between a phenomenon represented by a characteristic measurement and the ideal phenomenon must be estimated by an evaluation function which corresponds to the fitness value of the string to the ideal phenomenon. The ideal phenomenon has three properties, that is, (a) a threshold phenomenon, (b) a monotonous decreasing phenomenon when the value of an initial state is smaller than threshold and (c) a monotonous increasing phenomenon when the value of an initial state is greater than the threshold. Each of the properties can be formulated as follows:

(a) $d_x(S^i, V^x(\infty)) = 1,\ d_x(S^i, V^y(\infty)) \neq 1$

$\Rightarrow d_x(S^i, V^x(0)) > d_x(S^i, V^y(0))$

(b) $d_x(S^i, V^x(\infty)) = 1$

$\Rightarrow d_x(S^i, V^x(k)) \geq d_x(S^i, V^x(k-1))$,

$0 \leq k \leq \infty$

(c) $d_x(S^i, V^x(\infty)) \neq 1$

$\Rightarrow d_x(S^i, V^x(k)) \leq d_x(S^i, V^x(k-1))$,

$0 \leq k \leq \infty$ (17)

Furthermore, the total number of terms of the characteristic measurement should be small so that *c* of Eq.(16) should be minimized while searching. Therefore, the evaluation function can be composed with both the penalty value of the phenomena related to Eq.(17) observed by a characteristic measurement and the sum of the number of terms as follows (Eq.(18)):

$$F(d_i) = (\sigma_1\theta_a + \sigma_2\theta_b + \sigma_3\theta_c)^{-1}, \tag{18}$$

where $d_i \in D$, D is a set of the characteristic measures,

θ_a is the penaly value of the phenomenon of (a),

θ_b is sum of the penalty values of the phenomena of both (b) and (c),

$\theta_c = \sum_{k=1} a_k$ is the sum of the number of all the terms,

$\sigma_1, \sigma_2, \sigma_3$ are the weight coefficients for the penalties.

Hence, the objective of the string searching that minimizes the value of the evaluation function, can be formulated as follows (Eq.(19)):

$$\text{maximize } F(d_i) \text{ subject to } d_i \in D. \tag{19}$$

For the string manipulation, we use the following four types of the genetic operators, i.e., reproduction, crossover, mutation and reverse. The reproduction operation determines the survival rate of a string in the next generation according to the evaluation function. When a string d_i takes a value $E_i=F(d_i)$, survival rate R_i of the strings determined as:

$$R_i = \frac{e_i}{\sum_{j=1}^{|D|} e_j}, \tag{20}$$

where $e_i = (E^{min} - E_i)^2$, $E^{min} = \mathrm{MIN}_i^{|D|} [\, E_i \,]$

Using this R_i, the number of the string N_i in the next generation is determined as:

$$N_i = |D| \times R_i \tag{21}$$

Following the reproduction, the two crossover operations are applied which are called one-point crossover and two-point crossover in order to reconstruct contents of the strings. The one-point crossover consists of two steps. First, the members of the generation are mated at random. Second, each pair of strings undergoes crossing over as follows: an integer position *p* along the string is selected informally at random between $[1, \kappa]$ where κ is the length of the string.

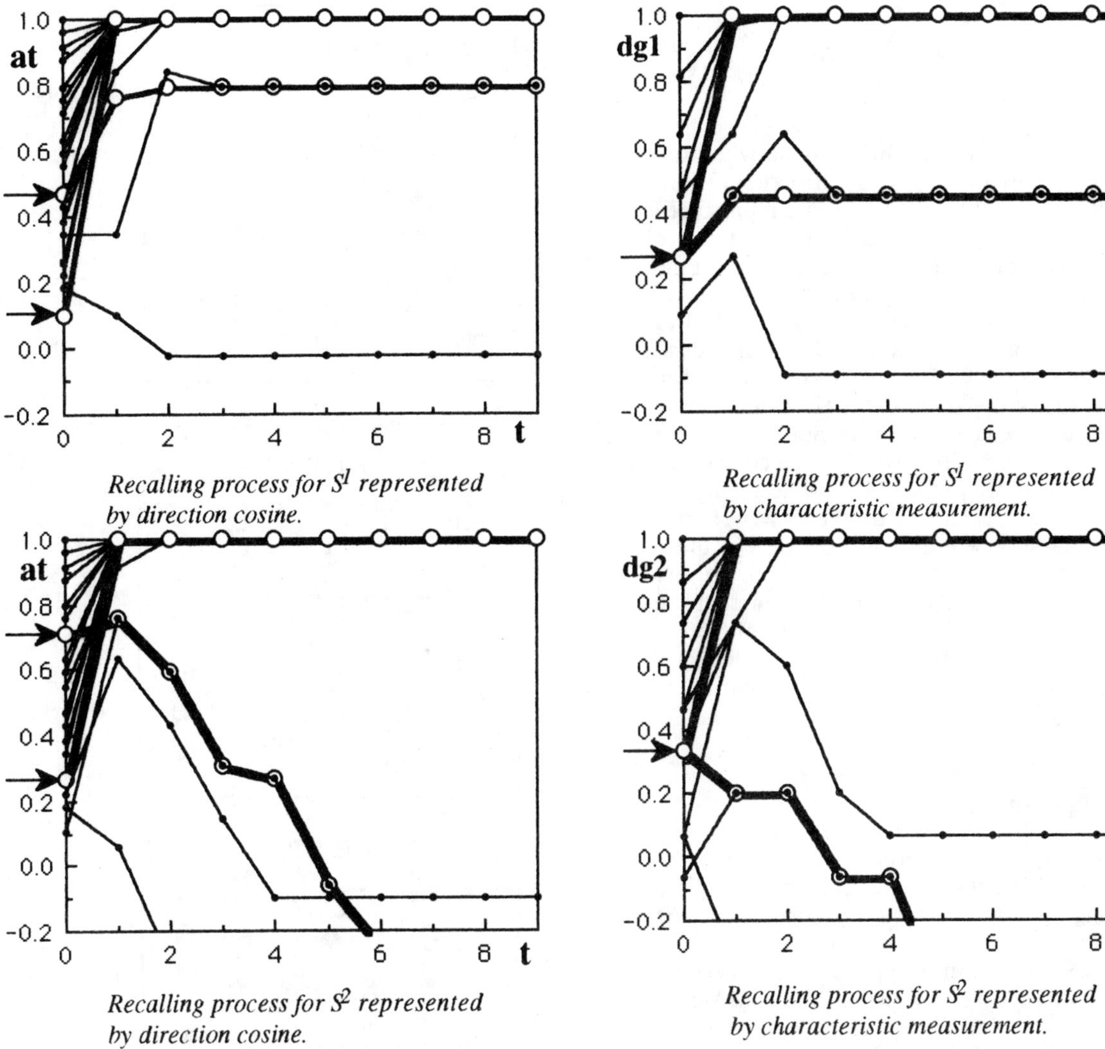

Recalling process for S^1 represented by direction cosine.

Recalling process for S^1 represented by characteristic measurement.

Recalling process for S^2 represented by direction cosine.

Recalling process for S^2 represented by characteristic measurement.

Fig.7 Recalling process of memorized state S^1 and S^2 that are illustrated by direction cosine.

Fig.8 Recalling process of memorized state S^1 and S^2 that are illustrated by characteristic measurements which are the results of searching by the GAs.

Two new strings are generated by swapping all the characters between position $p+1$ and κ inclusively. The two-point crossover consists of almost the same procedure. Two integer position p and q along the string are selected in the same manner. Two new strings are generated by swapping all the characters between position $p+1$ and q. In the following experiments, the one-point operator is carried to odd number generation and the two-point operator is carried out to even number generation.

The mutation is an occasional random alternation of the value of the contents of a string. In other words, a string and a location of the mutation are randomly selected according to the mutation rate, and the value of the corresponding bit position is reversed.

The reverse is also an occasional random alternation of strings. A string is selected randomly according to the reverse rate and all the values of the contents of the string are reversed. This operation can decrease the average of θc in the generation.

Hence, using these strings, evaluation function and the operation, the GAs can search the characteristic measurement for each basin of the memorized states.

6. Experiments

Under the condition that N=49, M=4 and that the model is trained with the LU under T=200, e_L=0.1, e_U=0.02, K=10, the GAs are applied to analyze each basin of the memorized state. The search parameters of the GAs are the following: |D|=200, σ_1=3000, σ_2=1500, σ_3=1, mutation rate 0.08=(number of mutation bit) / |D|, reverse rate 0.015 = (number of reversed string) / |D| . In Fig.7 and Fig.8, the result of the recalling processes for the memorized state S^1 and S^2 are represented by the direction cosine and by the characteristic measurements. The dynamical behaviors represented by the characteristic measurements show the threshold phenomenon but do not show the monotonous phenomena. Also the results of other characteristic measurements for the memorized state, S^3 and S^4, show the phenomena that are same as that of S^1 and S^2. The characteristic measurements cannot exactly represent the ideal behaviors, but the extents of the basins can be represented by these characteristic measurements as follows (Eq.(21)):

$$B(S^i) = \{ V^j(0) \mid d_x(S^i, V^j(0)) \geq \sigma^i \}, \quad (21)$$

where σ^i is the threshold of S^i.

In Fig.9 showing the relations between the memorized patterns and the characteristic measurements, the memorized patterns are represented by symbols `□' on the girds, and the characteristic measurements are represented by the dots `•' and the pairs of symbols `Δ' connected by the lines. The symbol `□' on the grid corresponds to a unit that takes state 1 of the memorized state and no marked grid corresponds to the units that takes state 0 of the memorized state. The dot `•' represents the term of degree one in the characteristic measurement and the position of this dot corresponds to the unit position which is the component of this term. The pair of symbols `Δ' connected by the line represent the term of degree two in the characteristic measurement and the positions of these dots correspond to the unit positions which are the components of this term. These results show that the characteristic measurements can be thought of as representing the characteristics of the memorized states and these characteristics seem to be able to be utilized for pattern recognition.

7. Conclusions

In this paper, one approach to analyzing the basin of a correlational associative memory model are described using the GAs and the LU-algorithm for this model, is proposed.

The LU-algorithm is expected to give the model an ability of keeping a wide basin in both highly memorized rate and mutually non-orthogonal states. For the recalling process of the model trained with this algorithm, the results are summarized as follows:

(a) The recalling process described by direction cosine is insufficient for the better understanding of the dynamical behavior of the model.

(b) The analyzing method for the recalling process using the Genetic Algorithms is proposed.

(c) Through the experiments, this analyzing method is proved to be useful for acquiring the characteristic measurement which indicates the extent of the basin.

Through the experiments, the contents of the characteristic measurements seem to show the characteristics of the memorized states and we can expect these measurements to be able to be successfully applied to pattern recognitions.

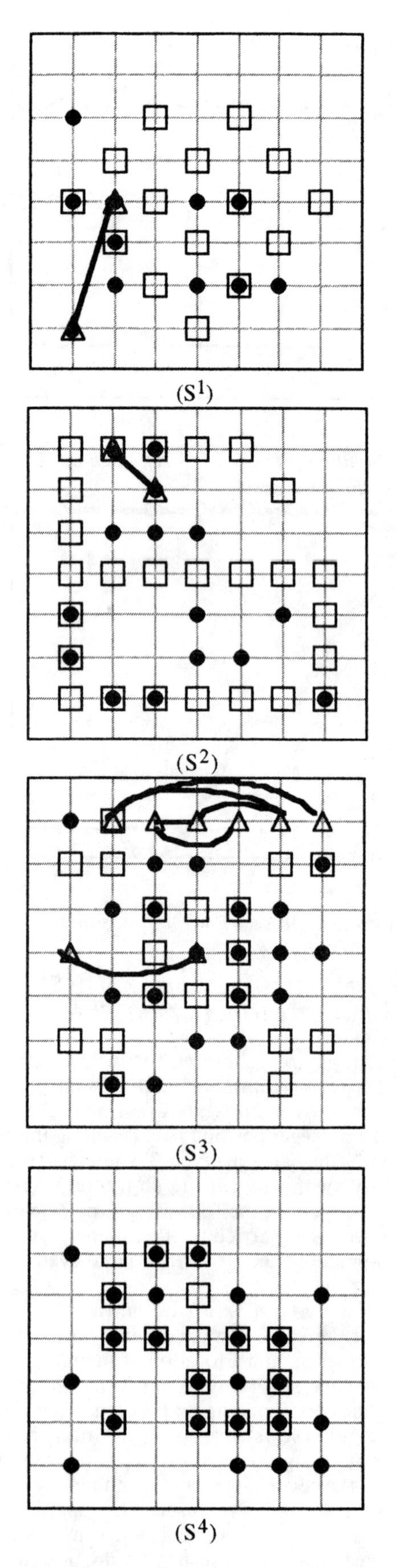

Fig.9 *Memorized states (`□') and characteristic measurements (`•', `Δ')*

References:

[1] J.J.Hopfield, D.I.Feinstein, and R.G.Palmer: 'Unlearning' has a stabilizing effect in collective memories, *Nature* Vol.304 14 JULY, 1983.

[2] S. Amari, and K.Maginu: Statistical Neurodynamics of associative Memory, *NeuralNetworks*, Vol.1, pp.63-73, 1988.

[3] M. Morita, S.Yoshizawa, and K.Nakano: Analysis and Improvement of the Dynamics of Autocorrelation Associative Memory, *Trans. IEICE*, Vol.73-D-II, No.2, pp.232-242. 1990. (in Japanese)

On the Sympatric Origin of Species: Mercurial Mating in the Quicksilver Model

Peter M. Todd
Department of Psychology
Jordan Hall, Building 420
Stanford University
Stanford, CA 94305
todd@psych.stanford.edu

Geoffrey F. Miller
Department of Psychology
Jordan Hall, Building 420
Stanford University
Stanford, CA 94305
geoffrey@psych.stanford.edu

Abstract

Traditional models of how interbreeding populations split apart into reproductively isolated populations (species) require the intervention of geographic barriers to mating or disruptive selection. We develop an alternate Quicksilver Model of speciation, and show through simulation that sympatric (barrierless) speciation can occur spontaneously, frequently, and robustly even in the absence of external divisive forces given certain broad conditions: (1) individuals have evolvable mate preferences based on degree of similarity to oneself along certain phenotypic dimensions, and (2) individuals compete to match the mate preferences of other individuals, and to have appropriate mate preferences themselves (i.e. sexual selection exists). Our models' success defends the notion of sympatric speciation against charges that it is impossible, implausible, or unlikely. It also offers an new vision of macroevolution based on appreciating the way modest psychological mechanisms of mate choice can have strong emergent effects on macroevolutionary dynamics.

1 SPECIATION IN BIOLOGY

The macroevolutionary record shows two central themes: *diversification* and *discontinuity* (Eldredge, 1989, p.99). Diversification represents the gradual accumulation of various phenotypic adaptations; discontinuity represents disruptions in the smooth continuum of diversity that would be expected from the basic accumulation of adaptations. Diversification is readily explained by normal Darwinian processes like natural selection, sexual selection, and genetic drift. But discontinuity remains a problem: why the gaps? Speciation theory is the study of such discontinuity. It seeks to explain why the macroevolutionary record resembles a branching bush with distinct trunks and twigs rather than a homogenous cone of biological diversity.

Mayr (1942, p.120) defined species as "groups of actually or potentially interbreeding natural populations, which are reproductively isolated from other such groups." Although in macroevolutionary theory species are understood as reproductively isolated populations, in paleontological practice they must often be identified by the empirical criterion of phenotypic similarity. Fortunately for taxonomists, morphological gaps between phenotypically similar groups correlate strongly with reproductive gaps between species. A central question in speciation emerges: does the morphological divergence cause the reproductive divergence, or vice-versa? Or are both caused by something else?

Most theories of speciation answer that both are caused by some external divisive force. These 'cleaver models' propose that some external dividing instrument or force cleaves a population in twain genetically and phenotypically, and then reproductive barriers arise afterwards through genetic drift or through selection against hybridization. In Mayr's (1942) model of *allopatric* speciation, the cleaver is a new geographic barrier arising to separate previously interbreeding populations. In Dobzhansky's (1937) model of *sympatric* speciation, the cleaver is more abstract: it is a low-fitness valley in an adaptive landscape. In both cases, the cleaver models aim to identify a single, simple, deterministic, causal force, external to the speciating population, driving the speciation process.

Since cleaver models form the backbone of contemporary speciation theory, they deserve closer chiropractic scrutiny. In Mayr's (1942) allopatric model, a physical, geographic barrier arises, like a river shifting course to isolate one population from another. Genetic drift, the founder effect, or disruptive selection then causes the two newly isolated groups to diverge phenotypically, and once enough phenotypic divergence accumulates, the groups can no longer interbreed even without the barrier, and so are recognized as separate species.

Sympatric models are generally analogous, but imagine the barrier to arise in an abstract adaptive landscape (Wright, 1931, 1932) rather than in geographic space. For example, an adaptive landscape might develop two high-fitness 'peaks' (econiches) separated by a low-fitness 'valley', thereby establishing disruptive selection (Mather, 1953) that drives an original population to diverge towards the separate peaks and thereby to become polymorphic. Dobzhansky (1937) further suggested that after divergence, reproductive isolation evolves through selection against hybridization: since hybrids will usually fall in the lower-fitness 'valley', mechanisms to prevent cross-breeding between the separate populations will tend to evolve. Thus the evolution of reproductive isolation (speciation itself) is viewed as a conservative process of consolidating adaptive change rather than a radical process of differentiation. Vrba (1985) and Futuyama (1987) concur that speciation serves a conservative function, acting like a 'ratchet' in macroevolution: only reproductive isolation allows a newly diverged population to effectively consolidate its adaptive differentiation; otherwise, the parent species will tend to genetically re-absorb it.

An important recent development in sympatric models is Paterson's (1985) concept of *specific mate recognition systems* (SMRSs: can be pronounced "s'mores" to highlight their sexual-appetitive function). SMRSs are phenotypic mechanisms each species uses to maintain itself as a self-defining reproductive community. A species is thus considered the largest collection of organisms with a shared SMRS. In Paterson's view, sympatric disruption and divergence of these SMRSs themselves (through some unspecified processes) leads to speciation. Eldredge (1989, p.120) emphasizes the potential macroevolutionary significance of SMRSs: "significant adaptive change in sexually reproducing lineages accumulates only in conjunction with occasional disruptions of the SMRSs."

The debate over allopatric versus sympatric speciation continues unabated and unresolved to the present. The debate's importance stems from its implications: if reproductive isolation is an adaptation in itself (as in sympatric theory) rather than a side-effect of drift (as in allopatric theory), then (1) species have more legitimacy as real biological entities because their boundaries are real evolved adaptations rather than by-products of phenotypic divergence, (2) sexual selection becomes almost as important as natural selection in macroevolution, and (3) species selection and macro-selection of lineages for their ability to speciate (e.g. Vrba, 1980, 1985) become central to macroevolutionary dynamics.

Allopatric speciation appears more immediately plausible, but its frequency and generality might be doubted. Geographic barriers are clearly sufficient to impose reproductive isolation between populations. But to explain the number of similar species that now live sympatrically (in the same geographic area), barriers would have to suddenly arise, persist long enough to allow divergence, and then subside, allowing the divergent groups to live together again (Pimentel, Smith, and Soans, 1967).

Reactions to the possibility of sympatric speciation have ranged from outright denial that it is possible (Mayr, 1942) to hesitant support based on a hunch that it might work (Dobzhansky, 1937). Yet sympatric speciation remains the only mechanism that can hope to explain the real number and diversity of species occupying the same geographic area. Historically, the acceptability of sympatric models has depended on the perceived ability of disruptive selection to generate stable polymorphisms and eventually reproductive isolation. A large number of experiments reviewed by Thoday (1972) show that disruptive selection is sufficient to generate phenotypic divergence even in the face of maximal gene flow between populations (which Mayr, 1963, p.472, saw as the Achilles' heel of sympatric speciation models), and that mechanisms of reproductive isolation would evolve to avoid hybrids and consolidate that divergence. Mathematical models by Maynard Smith (1966) and computer models by Crosby (1970) showed that sympatric speciation could occur when populations choose different micro-habitats, evolve stable polymorphisms through disruptive selection, and then evolve reproductive barriers to avoid hybridization. But, historically, the speciation debate has ground down to a question of whose cleaver is bigger: Mayr's (1942) geographic barriers or Dobzhansky's (1937) fitness valleys. Into the cogs of this debate, which has been churning away for fifty years, we hope to throw a monkey wrench, not just another cleaver.

2 GENETIC ALGORITHM MODELS OF SPECIATION

Most macroevolutionary change occurs just before, during, and after speciation events, not during the relatively gradual selection within stable species. This view is supported by the observed inhomogeneity in rates of morphological change in the fossil record (Eldredge and Gould, 1972). Since the arena of speciation is where much of evolution's adaptive action happens, understanding the dynamics of speciation should be of great importance in developing genetic algorithms capable of modelling complex biological adaptation and of solving difficult real-world problems. Conversely, since much of the ongoing debate in macroevolution centers around the feasibility of sympatric speciation, and simulation can be of great use in deciding feasibility issues and overcoming arguments from lack of imagination, genetic algorithm models of speciation may help illuminate evolutionary theory itself.

Most speciation work in the genetic algorithms (GA) community has bought into the cleaver model: speciation requires an imposed barrier of some sort, whether concrete (obstacles to migration) or abstract (valleys in the adaptive landscape). Mayr's (1942) allopatric model translates into Grosso's (1985) work on subpopulation formation and migration operators, and into the work of Muhlenbein (1989) and Gorges-Schleuter (1989) on parallel genetic algorithms. Implementing different GA populations on different computer processors is analogous to imposing geographic barriers between biological populations. The migration rules, the distribution of populations over processors, and the fitness landscape itself are all imposed rather than emergent, in accord with the cleaver model.

Dobzhansky's (1937) sympatric model informs most of the other GA work on speciation and niche formation. Goldberg (1989, p.189), for example, reasons that if there are multiple peaks in an adaptive landscape, then speciation will be useful in allowing separate sub-populations to explore those peaks independently, without the cost of repeatedly evaluating low-fitness hybrids between peaks. Sharing functions (Goldberg & Richardson, 1987; Deb & Goldberg, 1989) and crowding functions (De Jong, 1975) can be used to reinforce the disruptive selection emerging in a multi-peaked adaptive landscape, and to insure a reasonable allocation of individuals across different adaptive peaks.

A persistent question though has been how to allow reproductive isolation itself to evolve. Perry (1984) simply pre-defined a set of *external schemata*, that identify each individual as belonging to a particular reproductively isolated community: the species and niches are essentially fixed in advance. Deb and Goldberg (1989) developed a phenotypic mating restriction scheme such that individuals within a certain threshold distance in phenotype space were allowed to mate and individuals outside that distance were not. The threshold distance must be pre-set by the experimenter according to the structure of the adaptive landscape, again in accord with a cleaver model.

So far, very little GA research has allowed individuals to evolve their mating preferences themselves – to decide for themselves how much reproductive isolation they 'want'. The only example we know is Booker's (1985) use of evolvable mating templates. These genetic templates are matched against the functional (phenotype-encoding) portions of other individual's genotypes to determine who mates with whom. Booker's mating templates are essentially SMRSs that specify a particular genotypic ideal that each individual would like to mate with. Different sub-populations can in principle evolve different templates, hence different SMRSs, and thus evolve into distinct species. But for reasons discussed in later sections, we believe that mating templates defined in terms of absolute genotypic or phenotypic preferences are unlikely to promote speciation.

3 THE QUICKSILVER MODEL

In this paper we advance a view of speciation profoundly different from the cleaver models that underlie most current speciation theory in biology and GA research. We were inspired by three converging observations. First, the widespread existence of 'sibling species' (phenotypically very similar species occupying the same geographic area and perhaps even the same econiche) suggests that in many cases reproductive isolation can evolve *before* geographic barriers to mating or disruptive selection on phenotypic attributes. Sibling species are difficult to explain under any existing allopatric or sympatric model. Second, our previous experience with neural networks, artificial life, and genetic algorithms (e.g. Miller, Todd, & Hegde, 1989; Miller & Todd 1990; Todd & Miller 1991a, 1991b) led us to consider whether speciation might occur as a kind of spontaneous, emergent phenomenon. Third, we had an intuition that the evolution of psychological mechanisms to assess potential mates might in itself substantially increase the likelihood and alter the dynamics of speciation. We believe it is crucial for the fields of psychology, cognitive science, and even artificial life to recognize that evolution constructed all existing psychological adaptations and behavioral repertoires (Miller & Todd, 1990). But we are also intrigued by instances where psychological and behavioral phenomena can have profound reciprocal influences on evolutionary dynamics (as in sexual selection itself, and in the Baldwin effect: see Hinton & Nowlan, 1985).

Our speciation model derives metaphoric inspiration from the behavior of mercury ("quicksilver") shaken gently on a flat plate: sometimes large cohesive blobs will split into separate smaller blobs. The mercury's behavior depends on two opposing forces: the random shaking forces that tend to jostle blobs apart, and the cohesive, surface-tension forces that tend to keep blobs intact. In our metaphor, the flat plate represents a fitness landscape defined over a space of possible phenotypes (in this case, a simple 2-D space). The shaking force represents the random effects of mutation (and to some extent the disruptive effects of crossover), and the cohesive force represents mate-choice preferences evolved to favor phenotypically somewhat similar mates. The Quicksilver Model suggests that, given some reasonable balance between these forces of mutation and mate-choice, spontaneous sympatric speciation can and will occur. Note that no cleavers are required: no physical barriers intercede to limit mating, and the fitness landscape has no topography leading to disruptive selection. Sometimes speciation just happens.

Since most of the arguments against spontaneous sympatric speciation were essentially arguments from failure of imagination, we endeavored to develop a genetic algorithm simulation model to demonstrate the feasibility of sympatric speciation under the following conditions: (1) no geographic barriers, (2) no disruptive selection pressures (a 'flat' adaptive landscape), and (3) no pre-defined assortative mating preferences or SMRSs. Paterson's (1985) SMRS concept did seem to offer an alternate to cleaver models of speciation, but we wondered exactly how and why SMRSs might be disrupted spontaneously in a way capable of accounting for sibling species.

Our key observation was that although mate preferences can always be *represented* as probability-of-mating distributions over a phenotype space, they need not be *defined* directly in terms of phenotypic coordinates. Booker's (1985) mating templates represent a kind of mate preference defined directly in phenotypic coordinates: what we call a *space-relative mate preference*. But Deb and Goldberg's (1989) threshold mating distance represents a fundamentally different kinds of mate preference, defined relative to the individual's current position in phenotype space. This we call an *individual-relative mate preference*. Space-relative and individual-relative mate preferences have very different implications for speciation and macroevolutionary dynamics.

Paterson's (1985) and Eldredge's (1989) assumption that sympatric speciation would require a split in a population's SMRS would thus hold true only if each individual's mate preferences happen to be space-relative. For spontaneous sympatric speciation to happen with space-relative mate preferences through something analogous to the Quicksilver Model, the frequency distribution of actual phenotypes and the frequency distribution of mate preferences would have to both jostle (through mutation and crossover) into the same bimodal form at the same time. This seems an extremely unlikely event.

But *if a populations' mate preferences are individual-relative, sub-populations that are splitting apart can retain exactly the same mate preferences.* This makes the Quicksilver model more plausible because speciation requires only that the frequency distribution of actual phenotypes go temporarily bimodal; the frequency distribution of mate preferences need not change at all. So we postulated that individual-relative mate preferences could generate the cohesive forces needed in the Quicksilver model of speciation. The remainder of this paper demonstrates that indeed freely evolvable individual-relative mate preferences, in conjunction with sexual competition for mates, promotes fast, robust, consistent, and spontaneous sympatric speciation.

4 IMPLEMENTING THE QUICKSILVER MODEL IN A GA

To demonstrate the feasibility of our Quicksilver Model, we devised a very simple GA scenario which we hoped still had enough richness to allow sympatric speciation. Each genotype includes just three genes: two genes coding for phenotypic attributes, and one gene coding for a mating preference. Using just two phenotypic genes makes visualization and representation easy: the phenotype of each individual can be represented as a single x,y point on a 2-D plot, and the phenotype frequency distribution of an entire population can be represented as a set of points in the same space. In this case, we interpret phenotype space as a 1x1 toroid (where the top edge connects to the bottom, and left edge to right). It is imperative to remember that positions in this abstract 2-D phenotype space are not spatial locations in physical space, and that separation in this space cannot be interpreted allopatrically. Mate preferences permitting, it is as easy for two individuals far apart in phenotype space to mate as for two individuals close in this space.

All of the genes are Gray-coded. For the two phenotypic genes corresponding to x and y positions, Gray-coding has the nice property of making each edge of the phenotype-space just one mutation away from the corresponding opposite (and toroidally connected) edge; for example with four bits, 0000, which codes for 0.0, is one mutation away from 1000, which codes for the identical 1.0. The Gray-code representation is thus smooth in our toroidal phenotype space (which normal binary coding would not be), eliminating center-biasing and edge effects.

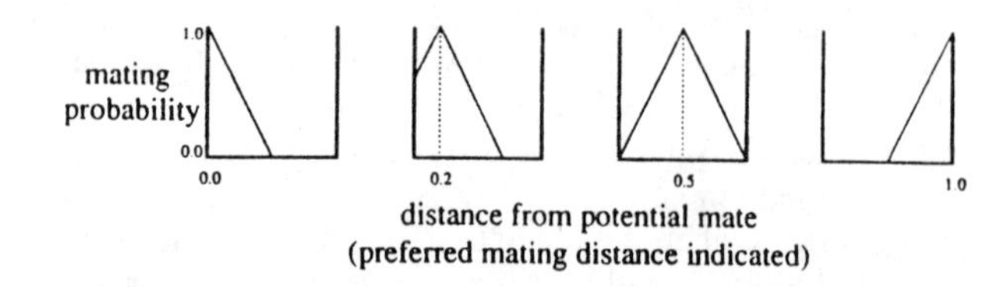

Figure 1: Mating Probability Function

In addition to the two phenotypic genes, every genotype includes one gene specifying a preferred mating distance, which in turn determines an individual-relative mate preference. We determined the general shape of a mate preference function: the single-peaked "triangular" function shown in Figure 1. This function maps relative phenotypic distance between the individual and some potential mate on to a probability that the individual would actually choose to indulge in this mating opportunity. The peak of this function represents the individual's preferred mating distance, and mating probabilities fall off linearly with equal slopes on both sides of this peak.

(The lower probability of mating with another individual closer than one's ideal can be thought of as an 'incest taboo'; the lower probability of mating with another farther than one's ideal can be thought of as 'xenophobia'.) Hence, different genetically encoded preferred mating distances essentially "slide" this triangular function back and forth, as can be seen in Figure 1, often yielding distances where the chance of mating is reduced to zero. By squaring this probability function, or raising it to a higher power, we can sharpen its peak and tighten up the preferred range of acceptable mating distances – we square the function for the simulations presented here (though the model is robust with respect to the peakiness of the function).

Implementing sexual selection with mating preferences in fairly straightforward. To create the next generation, a "mom" individual is first picked from the old population using normal roulette-wheel selection based on scaled fitness. Since our fitness landscape is flat, this amounts to uniform random sampling. Since it takes two to tango, a "dad" is next selected from the population using the same kind of selection. (Of course, individuals do not have a sex per se; how each individual is picked determines whether it plays the mom or dad role in each mating.) The phenotypic distance between the mom and the dad is computed (using the Euclidean metric in the 2-D toroidal phenotype space), and each individual computes its probability of mating with the other based on this mutual distance and its individual mate preference function. The two probabilities are likely to be different (mom might want dad more than the reverse), so they are multiplied together to yield a single probability that this pair will mate. Multiplying the two probabilities gives equal weight to each individual's choice in the matter, and mating happens only if both want to. Given this product of both their mating probabilities, a die is thrown, and if their number comes up, the pair gets to mate, and they are crossed over and their offspring put in the next population. If mating fails, then a new dad is chosen at random (with replacement), and the process is repeated, until the initially-selected mom finds a mate. Once the mom mates, a new mom is picked randomly (with replacement), and the search for a suitable dad begins again. This continues until the next generation is filled. (If a mom cannot find a suitable mate after five times through the population, she is deemed hopelessly picky, and a new mom is chosen.)

This moms-sample-dads scheme is unbiased in the sense that, although it enforces sexual selection, it does tend to preserve the frequency distribution of phenotypes from one generation to the next. It allows different phenotypically-separate and reproductively-isolated clusters to maintain their relative numeric proportions, by tending to pick mating pairs in just those proportions. This is important in promoting speciation. In contrast to our scheme, a random-pair-sampling scheme in which a mom and a dad are picked independently and randomly at each step, and thrown back if they didn't like each other, will bias the next generation's frequency distribution of phenotypes towards currently more common phenotypes. Smaller, reproductively-isolated phenotypic clusters would be at a great disadvantage, because there would be a much smaller probability of choosing both parents from the same small cluster simultaneously, so potential new species would be eliminated prematurely.

We use traditional bit-wise mutation and two-point crossover, which seemed sufficient to generate the mercurial jostling required in the Quicksilver model. Mutation rate should be high enough to shake things around (to produce the quicksilver effect), so we typically used a value of 0.01. The crossover rate is 1.0 – the mom and dad always cross over – because the very use of sexual selection implies that sexual reproduction will be employed. To allow fairly fine-grained structure to emerge in the phenotype space, we typically use 15 bits to encode each phenotype gene, and 30 bits for the preferred mating distance gene, yielding a total genotype length of 60 bits. Population size is usually 100.

Since speciation is the process of an existing reproductive community diverging out into the surrounding phenotype space, it would be inappropriate here to start out with a random initial population uniformly distributed throughout phenotype space (the default for most GA applications). Instead, we give the initial population some elbow room to branch out into by starting them out as a small random cluster in the middle of a much larger space of potential phenotypes. Initial x and y phenotype genes were constrained to be in a range from 0.45 to 0.55, and the initial preferred mating distances were constrained to be in a range from 0.0 to 0.1. This initial population (and all later generations) can be graphically displayed in a 1x1 grid with a dot at each x,y phenotypic position, surrounded by a circle of radius equal to the individual's preferred mating distance. Figure 2 shows such a representation for Generation 0, with the centered square of initial phenotypic positions, surrounded by the overlapping halos of their preferred mating distances.

5 RESULTS

With this instantiation of the Quicksilver Model in a GA, speciation is immediate and obvious, as can be seen in the progression of generations shown in Figure 2. After 10 generations, several "outpost" individuals have been shaken loose from the central blob, but no outlying phenotypic region has built up enough concentration of individuals to keep from being reabsorbed for long. The preferred mating distances of some individuals in the central blob have in fact grown to recapture a few of the errant individuals, as can be seen particularly in the circles extending to the left of the center. By 20 generations, though, a new species can be seen in the blobs above and below the original central one (these two new blobs represent just one new species, because of the

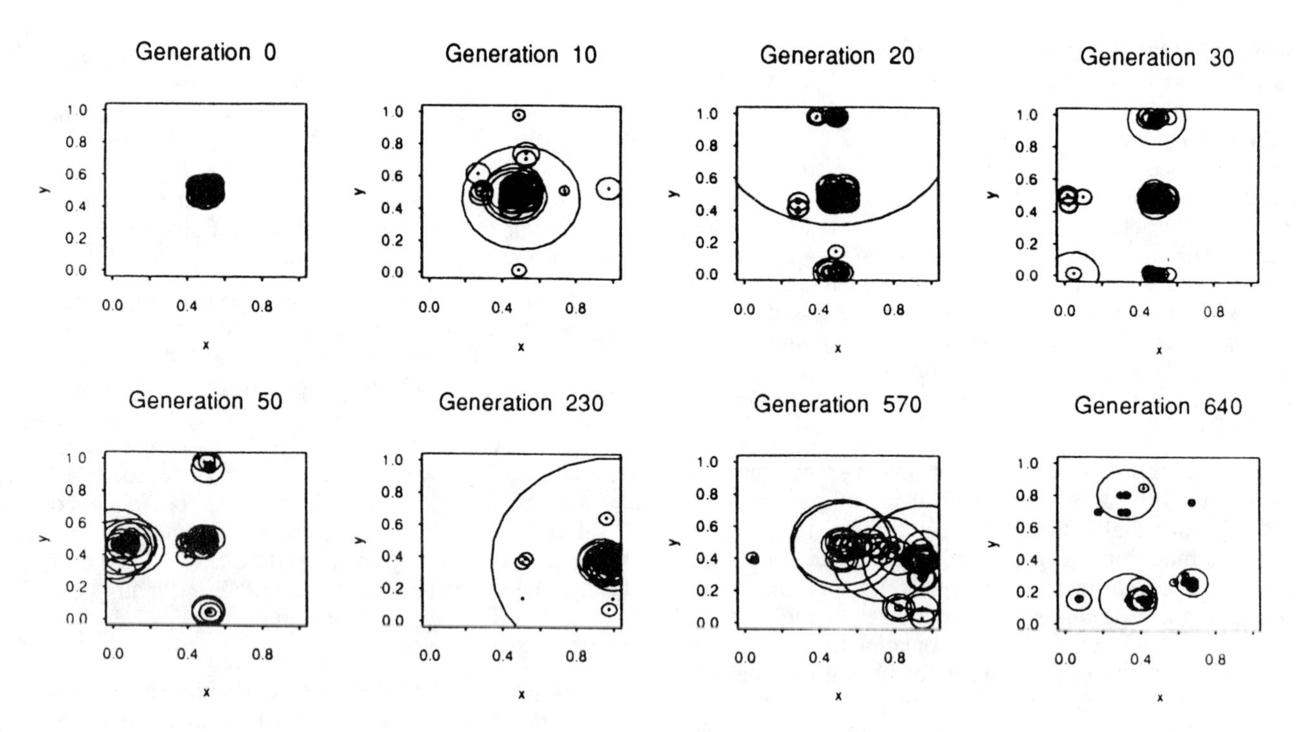

Figure 2: Sympatric Speciation Over Successive Generations

wraparound in this toroidal phenotype space). These individuals are far enough away from the ones in the center, and tightly enough clustered in phenotypes and mate preferences, to achieve reproductive isolation from the original species. Thus, this new species keeps from being directly reabsorbed by the original species. and it remains stable for many generations.

After 30 generations, the nascent uprising of a new species can be seen in the middle left edge of the space (x=0.0), "hopeful monsters" flung to this new phenotypic position and seeking others with similar inclinations; by 50 generations, this reproductively isolated subpopulation has grown significantly, at the expense of the second species. The species in the center hangs on for a long while, losing and recapturing individuals in a stochastic manner, with the advantage of having started out with the greatest number of members. But by 230 generations, even this founding population has dwindled to nothingness, being replaced by another single species in yet a new region of phenotype space.

About the messiest the population ever gets is shown at generation 570, where two relatively densely-populated regions of phenotype space are bridged by a succession of hybrids, and several individuals have larger preferred mating distances to take advantage of this spread. But by 640 generations, the preferred mating distances have again shrunk to highly selective values, and tight phenotypic clusters are evident. These shifting patterns of species formation, growth, dwindling, death, and rebirth are played out endlessly as the generations march on.

The fact that species come and go in this model, splitting and recombining, should not be seen as a drawback. In more complex biological systems, once speciation has been achieved through sexual isolation via mate preferences, divergent evolution of other phenotypic aspects (niche preference, identifying calls, mating apparatus, etc.) can serve to consolidate the species differences and prevent them from recombining. Removing the GA restriction used here of fixed population size will also serve to allow multiple species to coexist for longer, since growth of one will no longer automatically require the demise of another. Furthermore, the frequent emergence of new species relatively distant from already-existing ones, rather than a gradual budding and drifting apart, can be interpreted as an artifact of the particular scale we are witnessing the phenotypic changes on. If we "zoomed back" and took in a much large range of phenotype space, it is entirely possible that the formation of new species would have just this sort of budding-and-splitting look.

But at this phenotype-space scale, other interesting features emerge. In particular, we often see a sort of "sexual dimorphism" *within* species, where two (or sometimes more) subgroups within a species will each cluster tightly together, but will only mate with members from the *other* subgroup. This can be seen most clearly here in Figure 2 at generation 20 in the small group of individuals just to the left of the central blob. The individuals in the lower tiny cluster all have their preferred mating distances evolved so that they coincide exactly with the position of the members of the upper tiny cluster, and vice versa. The resulting "interlocking rings" structure can commonly be seen in these simulations, arising as a geometrically stable configuration. Interestingly, the separate "sexes" never diverge and become their own independent species; but we are looking for instances in which these sexually dimorphic species may divide *across* sex-lines to form new species (half the "males" and "females" go to one new species, the rest to another).

These speciation results are remarkably robust across parameter changes. A wide variety of mutation rates all yielded more or less messy but similar speciation, with larger mutation rates resulting in greater cluster spread and hence correspondingly larger preferred mating distances to exploit that spread. Different crossover rates seem merely to affect the amount of speciation (that is, the ease and frequency with which new species form); even in the absence of crossover, speciation still occurs, though at a slower rate, since mutation must then provide all of the scattering and jiggling forces. Different population sizes similarly just affect the amount of speciation, in ways that we are continuing to characterize. The length of the genes used also has effects we are investigating; we hope that by increasing the length of the preferred mating distance gene, relative to the length of the phenotype-component genes, we will be able to alter the rate of preference evolution with respect to phenotype evolution. In this way we hope to make it easier or harder for the preferred mating distances to change to track diverging phenotype clusters, and thus correspondingly harder or easier for new species to break away. Over the ranges we have investigated so far (from 1:2 ratio to 2:1), though, we have not found significant effects.

In work to be described more fully elsewhere, we have applied the Quicksilver Model to non-flat fitness landscapes, to investigate its ability to guard against low-fitness hybrids and increase speed of convergence. Our evolvable mating distances generally performed at least as well as traditional search without sexual selection, and populations often converged faster on an adaptive peak and stayed perched at the peak more steadily. Without crowding or sharing functions, however, separate species did not generally form on the different peaks. We intend to extend these comparisons to more complex GA-deceptive problems and see whether a more robust sexual selection scheme can be found that allows speciation to separate adaptive peaks without crowding or sharing functions.

6 CONCLUSIONS

Spontaneous sympatric speciation is easier than most biologists think. According to our Quicksilver Model, if a population's SMRS is composed of individual-relative mate preferences evolved for phenotypically similar individuals (cohesiveness), spontaneous sympatric speciation requires only that the phenotype frequency distribution itself change through mutation and crossover (jostling). These results support the feasibility of sympatric speciation in real biological systems, help clarify the long-standing debate between Mayr and Dobzhansky, and offer the possibility of developing a deeper understanding of sexual selection itself. Our results also point towards the potential for a more general class of genetic algorithms capable of navigating adaptively through very complex fitness landscapes.

We have already developed a general theory of mate preference schemes, including a method for representing how natural-selective and sexual-selective fitness landscapes interact with each other and with evolving mating preferences, which will appear elsewhere. In future work, we hope to show that if populations are allowed to evolve different kinds of mate preferences for different phenotypic dimensions, then populations will tend to evolve space-relative mate preferences along those phenotypic dimensions that have stable natural-selective fitness peaks, but will evolve individual-relative mate preferences along those dimensions with more unstable fitness peaks. Next, we intend to develop a very large macro-evolutionary simulation with natural-selective and sexual-selective pressures that tend to select lineages for their ability to speciate; perhaps speciation ability can be understood as a population-level 'adaptation' instantiated in the mate preferences of individuals in the population. Finally, as evolutionary psychologists, we will apply the theory to human mate selection and sexual attraction, by (1) analyzing the adaptive landscapes underlying different classes of phenotypic cues, (2) based on that analysis, predicting which phenotypic cues would be expected to be judged according to space-relative, population-relative, or individual-relative mate preferences, and (3) testing those predictions experimentally. If successful, this would be the first general theoretical framework for making sense of how human sexual attraction works.

Natural selection and sexual selection are the twin engines that drive the evolution of complex adaptations. Although GA simulations of natural selection are beginning to enter their maturity, GA simulations of sexual selection are only now beginning to gestate. Since sexual selection logically depends on evolvable mate preferences, we hope our Quicksilver Model of spontaneous sympatric speciation is taken as more than just another footnote to the debate on allopatric versus sympatric speciation. We hope it inspires other researchers to explore the rich, surprising, and illuminating dynamics that can emerge when we let our bit-strings evolve their own mate preferences.

References

Booker, L.B. (1985). Improving the performance of genetic algorithms in classifier systems. In *Proc. of an Int'l Conf. on Genetic Algorithms and their Applications*, pp. 80-92.

Crosby, J.L. (1970) The evolution of genetic discontinuity: Computer models of the selection of barriers to interbreeding between subspecies. *Heredity*, 25: 253-297.

De Jong, K.A. (1975) An analysis of the behavior of a class of genetic adaptive systems. (Doctoral dissertation, Univ. of Michigan). *Dissertation Abstracts International*, 36(10): 5140B.

Deb, K. & Goldberg, D.E. (1989) An investigation of niche and species formation in genetic function optimization. In J.D. Schaffer (Ed.), *Proc. of the Third Int'l Conf. on Genetic Algorithms*. San Mateo, CA: Morgan Kaufmann. pp. 42-50.

Dobzhansky, T. (1937). *Genetics and the origin of species.* (Reprinted 1982). New York: Columbia University Press.

Eldredge, N. (1989). *Macroevolutionary dynamics: Species, niches, and adaptive peaks.* New York: McGraw-Hill.

Eldredge, N., & Gould, S.J. (1972). Punctuated equilibria: An alternative to phyletic gradualism. In T.J.M. Schopf (Ed.), *Models in paleobiology.* San Francisco: Freeman, Cooper. pp. 82-115.

Futuyama, D.J. (1987). On the role of species in anagenesis. *American Naturalist* 130: 465-473.

Goldberg, D.E. (1989). *Genetic algorithms in search, optimization, and machine learning.* Reading, MA: Addison-Wesley.

Goldberg, D.E., & Richardson, J. (1987) Genetic algorithms with sharing for multimodal function optimization. *Genetic algorithms and their applications: Proc. of the Second Int'l Conf. on Genetic Algorithms*, pp. 41-49.

Gorges-Schleuter, M. (1989) ASPARAGOS: An asynchronous parallel genetic optimization strategy. In *Proc. of the Third Int'l Conf. on Genetic Algorithms*, pp. 422-427.

Grefenstette, J.J. (1987). *A user's guide to GENESIS* (Tech. Rep.). Washington, DC: Naval Research Laboratory, Navy Center for Applied Research in Artificial Intelligence.

Grosso, P.B. (1985) Computer simulation of genetic adaptation: Parallel subcomponent interaction in a multilocus model. (Doctoral dissertation, Univ. of Michigan.)

Hinton, G.E., & Nowlan, S.J. How learning guides evolution. *Complex systems*, 1: 495-502.

Mather, K. (1953) The genetical structure of populations. *Symp. Soc. exp. Biology*, 7: 66-95.

Maynard Smith, J. (1966) Sympatric speciation. *American Naturalist*, 100(916): 637-650.

Mayr, E. (1942). *Systematics and the origin of species.* (Reprinted 1982). New York: Columbia University Press.

Mayr, E. (1963) *Animal species and evolution.* Cambridge, MA: Harvard U. Press.

Miller, G.F., & Todd, P.M. (1990). Exploring adaptive agency I: Theory and methods for simulating the evolution of learning. In D.S. Touretzky, J.L. Elman, T.J. Sejnowski, & G.E. Hinton (Eds.), *Proc. of the 1990 Connectionist Models Summer School.* San Mateo, CA: Morgan Kaufmann. pp. 65-80.

Miller, G.F., Todd, P.M., & Hegde, S.U. (1989). Designing neural networks using genetic algorithms. In *Proc. of the Third Int'l Conf. on Genetic Algorithms*, pp. 379-384.

Muhlenbein, H. (1989) Parallel genetic algorithms, population genetics, and combinatorial optimization. In *Proc. of the Third Int'l Conf. on Genetic Algorithms*, pp. 416-421.

Paterson, H.E.H. (1985). The recognition concept of species. In E.S. Vrba (Ed.), *Species and Speciation. Transvaal Museum Monograph 4*, pp. 21-29.

Perry, Z.A. (1984) Experimental study of speciation in ecological niche theory using genetic algorithms. (Doctoral dissertation, Univ. of Michigan.) *Dissertation Abstracts International*, 45(12): 3870B.

Pimental, D., Smith, G.J.C., & Soans, J. (1967) A population model of sympatric speciation. *American Naturalist* 101(922): 493-504.

Thoday, J.M. (1972) Disruptive selection. *Proc. of the Royal Soc. of London B*, 182: 109-143.

Todd, P.M., & Miller, G.F. (1991a). Exploring adaptive agency II: Simulating the evolution of associative learning. In S.W. Wilson & J.-A. Meyer (Eds.), *From animals to animats: Proceedings of the First Int'l Conf. on Simulation of Adaptive Behavior.* Cambridge, MA: MIT Press/Bradford Books. pp. 306-315.

Todd, P.M., & Miller, G.F. (1991b). Exploring adaptive agency III: Simulating the evolution of habituation and sensitization. In *Proc. of the First Int'l Conf. on Parallel Problem Solving from Nature.* Berlin: Springer-Verlag.

Vrba, E.S. (1980). Evolution, species, and fossils: How does life evolve? *S. Afr. J. Sci.*, 76: 61-84.

Vrba, E.S. (1985). Environment and evolution: Alternative causes of the temporal distribution of evolutionary events. *South African Journal of Science*, 81: 229-236.

Wright, S. (1931). Evolution in Mendelian populations. *Genetics*, 16: 97-159.

Wright, S. (1932). The roles of mutation, inbreeding, crossbreeding, and selection in evolution. *Proce. of the Sixth Int'l Congress on Genetics*, 356-366.

Temporal Processing with Recurrent Networks: An Evolutionary Approach

Jan Torreele

Artificial Intelligence Laboratory
Free University of Brussels
Pleinlaan 2, 1050 Brussels
Belgium
e-mail: jant@arti.vub.ac.be

Abstract

In this paper we present an evolutionary approach to the problem of temporal processing with recurrent networks. A genetic algorithm is used to evolve both structure and weights, so as to alleviate the design and learning problem recurrent networks suffer from. The viability of this approach is demonstrated by successfully solving two nontrivial temporal processing problems. The important technique of teacher forcing is identified and its influence on the performance of the algorithm is empirically demonstrated.

1 INTRODUCTION

Many problems in areas such as signal processing, control and speech have a temporal extent. They often involve learning about temporal sequences with a priori unknown temporal properties. The use of recurrent neural networks has proven to be a promising approach to the problem of time-dependent processing [6], [9], [10], [13]. The recurrent connections in such networks allow the (intermediate) results of processing at time $t-\Delta t$ to influence the (intermediate) results of processing at time t, thereby shaping the network's subsequent behaviour by previous responses.

1.1 THE PROBLEM OF RECURRENT LEARNING

Despite the fact that elegant and powerful algorithms — such as the Backpropagation Algorithm [11] — have been developed for feedforward networks, no such algorithms are known today for their recurrent counterparts. Indeed, many of the learning algorithms devised for recurrent networks possess undesirable features such as the inability to perform on-line learning, growing memory or computational requirements, or the need for global information. See [10] for an overview.

One of the main reasons for this is that in contrast with feedforward networks, recurrent networks have a complex dynamics. Even the smallest recurrent networks can engage in very complex and unanticipated behaviours. This makes it very hard to characterize and analyze them, and therefore also extremely difficult to devise appropriate learning algorithms. Due to this lack of insight into the networks' dynamics it is also virtually impossible for the designer to handcraft good initial network architectures.

1.2 AN EVOLUTIONARY APPROACH

In this paper, we have studied an alternative approach to the problem of time-dependent processing with recurrent networks. We have used a Genetic Algorithm (GA) [5] to simulate an evolving population of recurrent networks which, under the influence of selective pressure, converges to a set of networks that successfully solve the task at hand. This evolutionary approach to the problem of time-dependent processing is interesting for several reasons:

1. It is well known that the error surface over the weight space of a neural network, given a nontrivial problem to be trained on, is typically highly degenerate. This property is probably even more pronounced for recurrent networks and poses serious challenges to any gradient-based technique for finding optimal weights. GAs are powerful stochastic optimization techniques, capable of searching large and complex problem spaces. They offer therefore a promising alternative to gradient-based techniques.
2. The use of a GA also frees us from the need to design good initial architectures for the networks. By allowing the GA to modify the structure of the networks, network architectures appropriate for solving the problem under consideration can emerge as the evolution proceeds.

In the remainder of this paper we will demonstrate how an evolutionary strategy can be employed to evolve recurrent networks that perform time-dependent processing. The structure of the paper is as follows: first we outline the architecture and the dynamics of the recurrent networks investigated. Then we describe how these networks can be cast in an evolutionary framework so as to evolve both structure and weights. In a next section we present results of our approach using two examples: The problem of processing a sequence of inputs with a temporal information content, and the problem of generating a particular temporal pattern. Finally, the validity of the approach and its relation with other research is discussed.

2 RECURRENT NETWORKS WITH LINEAR THRESHOLD ELEMENTS

The recurrent networks investigated in this paper are composed out of Linear Threshold Elements (LTE's), operating in a discrete-time, synchronous manner (figure 1). LTE's are simple processing elements that linearly sum up their inputs, and compute their output by thresholding this sum:

$$x_k(t) = \sum_j w_{kj} y_j(t-1) \quad (1)$$

$$y_k(t) = \begin{cases} 1 & \text{for } x_k(t) \geq \theta_k \\ 0 & \text{for } x_k(t) < \theta_k \end{cases} \quad (2)$$

where $x_k(t)$ is the total input of unit k at time t, $y_k(t)$ is the output of unit k at time t, w_{kj} is the weight of the connection from unit j to unit k and θ_k is the threshold of unit k. In our model, we have constrained the weights to be either +1 or −1. That is, the weights can be excitatory or inhibitory, but with the same amount. As a consequence, the threshold values of the units can be thought of as integers.

The units that receive external input are referred to as input units, and those that are observed from outside the network are output units. The remaining units are referred to as hidden, since they can only exchange signals with other parts of the network. External input is supplied by clamping the input units' output to zero or one. Notice that the set of input units can be empty. Indeed, even in the absence of any external input, the networks can show a variety of complex behaviours.

3 AN EVOLVING POPULATION OF RECURRENT NETWORKS

In order to simulate an evolving population of networks, an adequate representation scheme has to be established. This representation scheme has to allow genetic operators to act upon the networks so as to produce better adapted ones during the course of evolution. Different representation strategies for neural networks exist, with varying degree of developmental specification. Weak specification schemes use loose descriptions of the networks as genotypes [2], [3]. Strong representation schemes on the other hand, describe network architectures as well as other relevant parameters with a significant degree of detail [7].

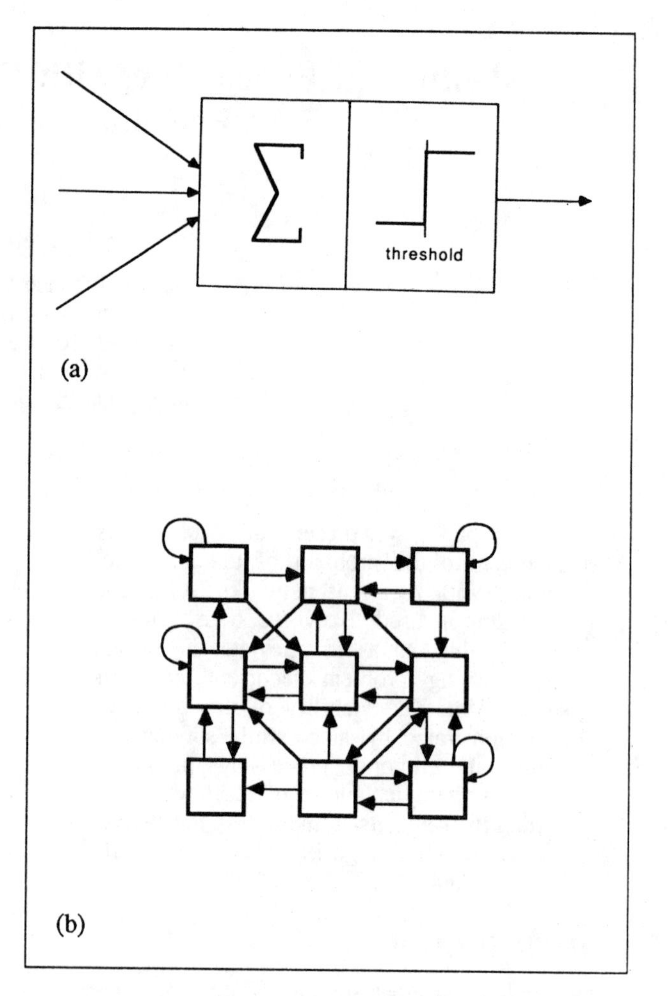

Fig. 1: (a) A linear threshold element (LTE). (b) A recurrent network with linear threshold elements.

3.1 A REPRESENTATION SCHEME

As the networks investigated in this paper are very sensitive to only minor changes in their parameters or architecture, it is clear that weak specification schemes — such as the stochastic specification scheme used by Bergman [2] — are not appropriate. We have chosen instead for a very strong specification scheme, in which networks are described in all their detail.

To accomplish this, the units of a network are arranged in a rectangular lattice (as in figure 1). A neighbourhood structure is then defined over the lattice. This neighbourhood structure can be viewed as a repertoire of connections, that is, as a collection of other units from which a particular unit can eventually receive input. It is specified as a list of relative positions on the lattice. The neighbourhood structure can be problem dependent, but it is the same for all units in the network and once chosen, remains fixed during the course of evolution. A typical (local) neighbourhood structure would be the list (0,1), (1,0), (0,–1), (–1,0), indicating that a unit at position (x,y) in the lattice can eventually get input from units located at positions $(x,y+1)$, $(x+1,y)$, $(x,y-1)$ and $(x-1,y)$.

The networks are converted to bit string genotypes by concatenating the encodings of the network units on the lattice in a row-major order. Each individual unit has a binary code of M bits:

$$M = 2N + \lceil \log_2(2N) \rceil$$

where N is the number of neighbours in the neighbourhood structure. Of those M bits, 2N bits are used to encode the connectivity of the unit: for each of the N possible incoming connections, one bit is used to indicate the presence of that connection, and a second bit is used to encode the weight of the connection (+1 or −1). The remaining $\log_2(2N)$ bits represent the local threshold of the unit, encoded as an integer in the range $\{-N, \ldots, N-1\}$. This representation of the networks is closed: any bit string of length $K \times L \times M$ is a valid description of a $K \times L$-unit network on a $K \times L$ lattice.

3.2 THE GENETIC ALGORITHM

We have used a rank based genetic algorithm with adaptive mutation rate [12]. The fitness of an individual in the population is simply its performance as a neural network, as computed with some standard error measure. Starting off with a population of randomly generated networks, the algorithm first assigns a ranking to each individual, based on their fitness. Next, the algorithm engages in a series of reproductive trials. Each cycle, two individuals are selected from the population in a probabilistic way, based on their ranking: the higher ranked individuals have a higher probability of being selected. A recombination of those individuals by means of two genetic operators, crossover and mutation, yields an offspring whose fitness is calculated. The newly generated individual is then inserted in the population, and ranked according to its fitness. The worst ranked individual thereby disappears. Offspring with fitnesses that are inferior to the fitness of the worst ranked individual are discarded.

4 EXPERIMENTS AND RESULTS

We have successfully tested our approach on a number of small though interesting problems. In all of these experiments the networks had to learn to represent useful information about the temporal properties of the problem at hand. We report here the results of the approach on two different temporal problems. The first problem involves the extraction of time-dependent information from an input stream in order to categorize it. The second problem demonstrates how the networks can be taught to generate a particular temporal pattern in the absence of any inputs.

4.1 THE TEMPORAL MULTIPLEXER

The multiplexer problem is a categorization task for binary strings of length l. Each string belongs either to category "0" or to category "1". The category a string belongs to can be computed by using the first n bits of the string as an unsigned address that points to one of the remaining 2^n bits as the category (where $l = n + 2^n$). The multiplexer is a difficult problem that has been widely used as a testbed for neural network algorithms (e.g. by Barto [1]).

The multiplexer problem is typically not of a temporal nature. It can be transformed though into a nontrivial temporal problem for a recurrent network in the following way. An instance of the multiplexer is given as input to the network one bit at a time, i.e. the l bits of the binary string are sequentially clamped onto the input unit, from time t up to time $t+l-1$. The network is then expected to output the correct categorization of the string at time $t+l$. This temporal version of the multiplexer incorporates several difficulties. In order to solve it, a network first has to decode the address bits. Next it has to figure out how many time steps it should wait before the appropriate data bit appears on the input unit, and finally it has to transmit this data bit to its output unit.

We have tested the algorithm on a simple version of this problem, the temporal 3-multiplexer. Networks with 4 units arranged on a 2x2 lattice are used. The neighbourhood structure imposed on the networks is one with full connectivity: each unit is allowed to get an incoming connection from any other unit. Two different units are chosen to be the input and the output unit of the networks respectively. An evaluation of a network consists of running it on the 8 different binary strings that have to be categorized. The fitness of a network is simply the number of bit strings for which it produces the correct classification on its output unit. In between the evaluations of each of the binary strings, the networks are re-initialized by setting the output of all its units to zero.

Figure 2 shows the performance of the algorithm on the 3-multiplexer, together with a typical solution found. In this experiment we have used a population size of 100, 2-point crossover and a mutation rate which varied linearly with the hamming distance between the parents, up to a maximum of 0.25.[1] The networks in this example are coded by bit strings of length 44 (3 threshold bits plus 8 connectivity bits per unit), leaving us with a search space of size 2^{44}.

As can be seen from the figure, the algorithm has used its ability to evolve the structure of the networks in a purposeful way. From the 4x4 possible connections between the units, only 9 are effectively used. It is

[1] In this experiment, and in the following, no attempt was made the optimize the parameters of the algorithm.

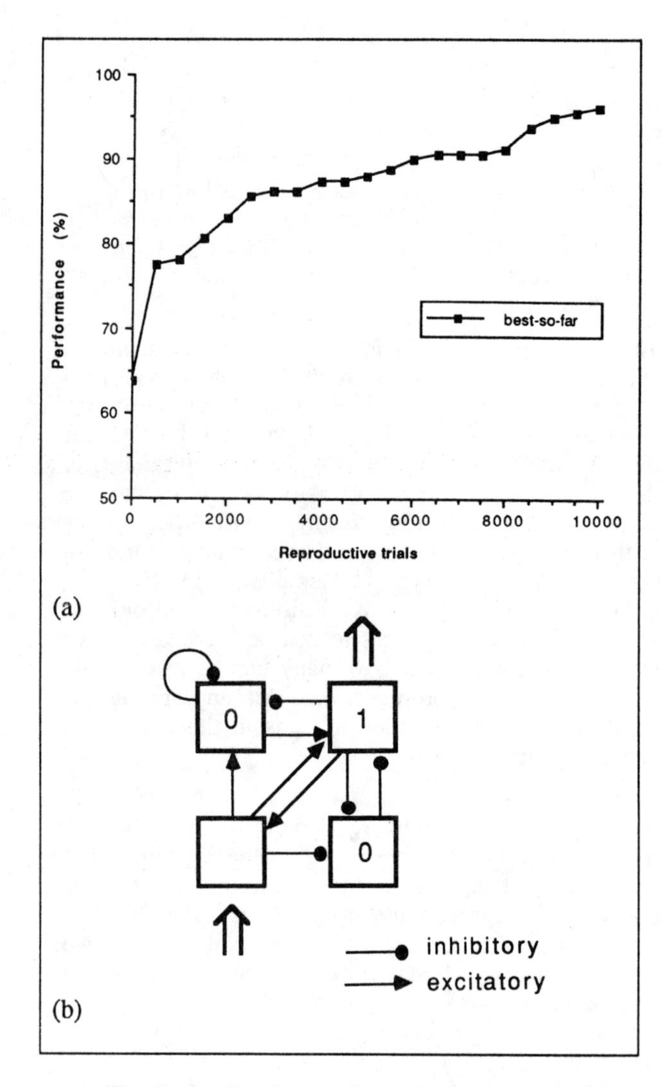

Fig. 2: (a) The best-so-far performance of the algorithm on the temporal 3-multiplexer, averaged over 20 runs. (b) A typical solution found. The threshold values are displayed inside the units.

instructive also to pay attention to the complexity of the design. Even with a relatively simple problem such as the 3-multiplexer, it becomes very hard to see how this network actually manages to solve it. An a priori design of the network structure would have been challenging.

The performance of the algorithm is satisfying. A benchmark over 20 different runs reveiled that in approximately 75% of the cases, the optimum was found within 10,000 reproductive trials. Furthermore, in all of our experiments a solution was always found within 20,000 reproductive trials.

4.2 THE BLINKER PROBLEM

Even in the absence of any external inputs, recurrent networks are able to engage into particular temporal behaviours. Training a network to do so is difficult, because no external cues (i.e. inputs) are available to guide its internal dynamics.

Consider the problem of an output unit of a network required to stay active (output 1) for n time steps and then to become inactive (output 0) for another n time steps, repeating this cycle forever. We have named this problem the n-blinker. The problem is clearly of a non-trivial temporal nature. A possible solution would be for a network to develop a counter from 0 to n in its hidden units, together with a flip-flop-like element in its output unit. We have performed a number of experiments to observe the algorithm's performance on the 4-blinker problem, using a population of 2×2 networks with fully connected neighbourhood structure.

In a first attempt to tackle the problem, a standard error measure was used to evaluate a network's behaviour: the network was run for a number of time steps (24 in this example) and its fitness was computed by counting the number of correct outputs.[2] This evaluation function will be referred to as *F1*. Results were not encouraging. Many of the networks in the population immediately acquire a fitness of about 50% by just keeping their output unit at a constant value. Networks of this kind tend to take over the complete population, which results in a loss of diversity, before the hidden units can even attempt to do something useful. Neither increasing the mutation rate nor softening the selection pressure alleviates this problem, since those "constant" networks already prevail in any randomly generated initial population. Enlarging the population size so as to diversify the behaviours of the networks eventually helps, but only because networks with fitnesses of up to 70% then start to appear as members of the starting population.

One of the difficulties with the blinker problem is that the output of a network has to be constant most of the time. In order to promote more activity in the networks' dynamics, the algorithm was tested using a more "severe" evaluation function *F2*. Function *F2* computes the fitness of a network by counting the number of successive time steps for which a correct output is generated, up to the first mistake. The idea is to force the networks to generate the desired temporal activity in a gradual way, thereby inhibiting the evolution of "constant" networks. Counter to our intuitions however, the algorithm performs very poorly using function *F2*. "Constant" networks still appear, although now with a smaller fitness. Out of 20 different test runs performed, only one found the optimum.

[2] The population size, mutation rate and crossover operator were set as in the previous example.

A third evaluation function, *F3*, provides satisfying results. *F3* can be viewed as an interesting variant of *F1*. The idea is to replace the actual output of a network by the desired output in the subsequent computations. That is, the future activity of a network is based on the desired state of the output unit rather than on its free-running state. This technique, which is called "teacher forcing", originates from neural network research where it was found useful during the training of recurrent networks (e.g. [6], [13]).

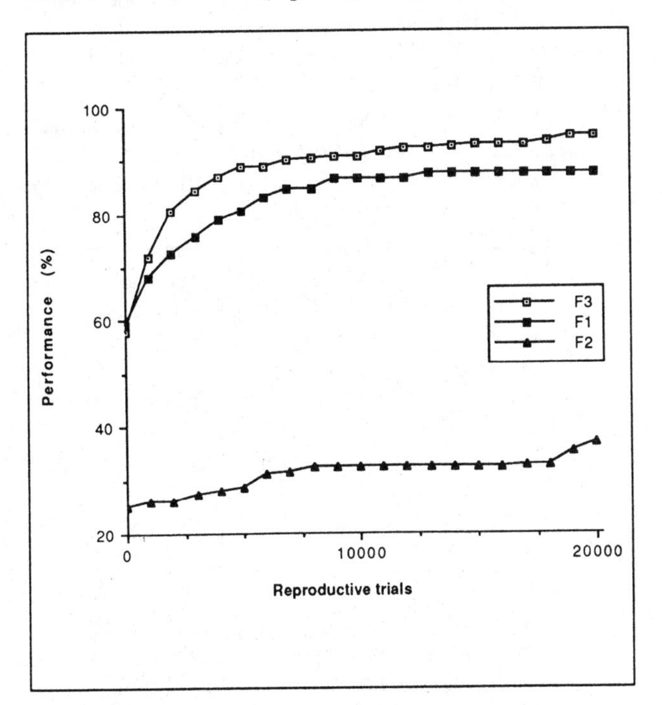

Fig. 3: The best-so-far performance of the algorithm on the 4-blinker problem, averaged over 20 runs. F1 is the result for the standard evaluation function, F2 uses a severe evaluation function and F3 uses teacher forcing (details in text).

Teacher forcing can be made valuable also in an evolutionary approach to recurrent learning, by integrating the technique with the evaluation function. This is achieved by modifying evaluation function *F1* such that it sets the output at its correct value after each time step. By doing so, the networks are encouraged (1) to work towards an appropriate pattern of activity in response to a number of temporal sub-sequences, and (2) to chain these short patterns of activity into a globally desired behaviour. Figure 3 shows the results of the algorithm for the 4-blinker problem, averaged over 20 different runs, using the 3 different evaluation functions. It is clear from the figure that teacher forcing (*F3*) produces significantly better results than a standard evaluation function (*F1*).

A typical solution found for the 4-blinker problem is shown in figure 4. Figure 4.a. shows the architecture and the weights of the network. Notice that the architecture found is remarkably simple: no extraneous connections are present. Figure 4.b. displays a trace of the activations of the network units during 8 successive time steps. One can clearly see that the network has developed a counter in its hidden units.

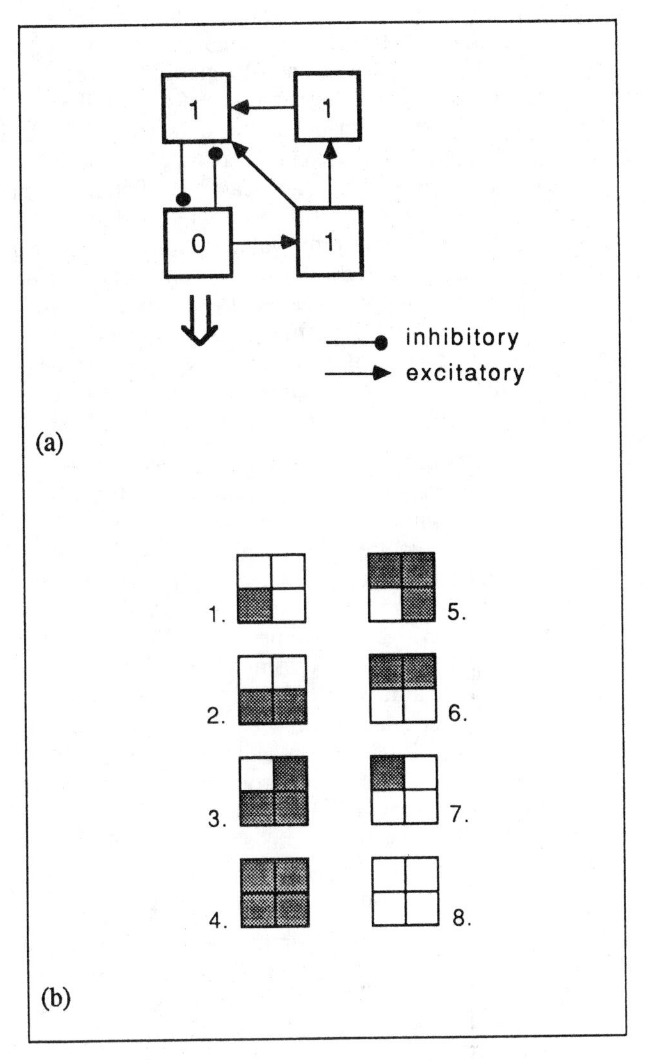

Fig. 4: (a) A typical solution found for the 4-blinker problem. Thresholds are displayed inside the units. (b) The activations of the network units during 8 successive time steps. The unit on the bottom left is the output unit.

5 DISCUSSION

The idea of applying genetic algorithms to neural networks is not new. Various approaches have been explored, some of them focusing on learning the weights of networks with predefined architectures, e.g. [12], others attempting to find appropriate initial architectures for the networks [7], [4]. The problem of evolving a population of recurrent networks to exhibit a particular temporal behaviour has received less attention in the past. Bergman [2] has used an evolutionary approach to learning both structure and weights of recurrent networks. The task for which the networks are trained however is of a non-temporal nature. The recurrent pathways in those networks, if any, will therefore serve to perform spatial rather than temporal processing. A completely different approach to learning with recurrent networks is advocated by Patarnello et al. [8]. They use simulated annealing instead of a genetic algorithm to perform global optimization over a space of boolean networks. Using this approach, a simple artificial creature was learned to set up a "clever" strategy for moving around in an environment, collecting food as fast as it can. The creature's strategy involves the use of several temporal cues present in the environment.

Recurrent networks suffer from particularly severe *design* and *training* problems. Motivated by this observation, the emphasis in our work has been laid on learning both structure and weights of the networks without using any prior information about the temporal nature of the problem to be solved. The use of a GA to simulate an evolving population of recurrent networks in order to optimize their performance on a given temporal problem is attractive for several reasons: being a powerful stochastic optimization technique, a GA is capable of searching large and complex problem spaces. It offers therefore a promising alternative to gradient-based techniques for the *recurrent learning problem*. By allowing the GA to also modify the structure of the networks, the *design problem* for recurrent networks can be alleviated.

Our primary goal in this paper has been to investigate the possibilities of such an evolutionary approach to the problem of temporal processing with recurrent networks. These initial investigations have led us to conclude the following:

1. The application of a GA to the problem of temporal processing with recurrent networks is a new approach. Its main *advantage* is that it alleviates both the design and the learning problems recurrent networks suffer from.

2. The *viability* of the approach has been demonstrated by successfully solving two nontrivial temporal processing problems. The algorithm thereby uses its ability to evolve the structure of the networks in a purposeful way.

3. Preliminary investigations have made us identify the important technique of *teacher forcing*. The way to integrate this technique with the genetic algorithm was described. The important influence of the technique of teacher forcing on the performance of the algorithm was demonstrated empirically.

Acknowledgements

Discussions with, and comments from Tony Bell, Bernard Manderick and Piet Spiessens have improved the quality of this paper. This work is sponsored by NATO grant CRG 90031 and by the Belgian Government under IMPULS contract no. RFO-AI-10.

References

1. A. Barto, "Learning by Statistical Cooperation of Self-Interested Neuron-like Computing Elements," COIN Technical Report 85-11, University of Massachusetts, 1985.

2. A. Bergman, "Self-Organization by Simulated Evolution," in *Lectures in Complex Systems, Proceedings of the 1989 Complex Systems Summerschool, Santa Fe*, ed. E. Jen, 1989.

3. A. Guha, S.A. Harp, and T. Samad, "Genetic Synthesis of neural networks," Technical Report CSDD-88-I4852-CC-1, Honeywell Corporate Systems Development Division, 1988.

4. S.A. Harp, T. Samad, and A. Guha , "Towards the Genetic Synthesis of Neural Networks," in *Proceedings of the Third International Conference on Genetic Algorithms*, ed. D. Schaffer, 1989.

5. J.H. Holland, *Adaptation in Natural and Artificial Systems*, University of Michigan Press, Ann Arbor, 1975.

6. M.I. Jordan, "Attractor Dynamics and Parallelism in a Connectionist Sequential Machine," in *Proceedings of the Eighth Annual Conference of the Cognitive Science Society*, 1986.

7. G.F. Miller, P.M. Todd, and S.U. Hegde, "Designing Neural Networks using Genetic Algorithms," in *Proceedings of the Third International Conference on Genetic Algorithms*, ed. D. Schaffer, 1989.

8. S. Patarnello and P. Carnevali, "Learning Capabilities of Boolean Networks," in *Neural Networks, from Models to Applications*, ed. L. Personnaz and G. Dreyfus, 1988 .

9. B.A. Pearlmutter, "Learning State Space Trajectories in Recurrent Neural Networks," Technical Report CMU-CS-88-191, Carnegie Mellon University, Department of Computer Science, 1988.

10. F.J. Pineda, "Recurrent Backpropagation and the Dynamical Approach to Adaptive Computation," *Neural Computation*, vol. 1, no. 2, 1989.

11. D.E. Rumelhart, G.E. Hinton, and R.J. Williams, "Learning Internal Representations by Error Propagation," in *Parallel Distributed Processing: Explorations in the Microstructure of Cognition*, Bradford Books, Cambridge, MA, 1986.

12. D. Whitley and T. Hanson, "Optimizing Neural Networks Using Faster, More Accurate Genetic Search," in *Proceedings of the Third International Conference on Genetic Algorithms*, ed. D. Schaffer, 1989.

13. R.J. Williams and D. Zipser, "Experimental Analysis of the Real-time Recurrent Learning Algorithm," *Connection Science*, vol. 1, no. 1, 1989.

Genetic Reinforcement Learning with Multilayer Neural Networks

Darrell Whitley, Stephen Dominic and Rajarshi Das
Computer Science Department
Colorado State University
Fort Collins, CO 80523, U.S.A.
whitley@cs.colostate.edu

Abstract

Empirical tests indicate that the genetic algorithms which have produced good performance for neural network weight optimization are really genetic hill-climbers, with a strong reliance on mutation rather than hyperplane sampling. Initial results are presented using genetic hill-climbers for reinforcement learning with multilayer neural networks for the control of a simulated cart-centering and pole-balancing dynamical system. "Genetic reinforcement learning" produces competitive results with AHC, a well-known reinforcement learning paradigm for neural networks that employs temporal difference methods.

1 INTRODUCTION

Attempts to apply genetic algorithms to neural network optimization problems have largely met with modest results. Most of the problems that have been solved are relatively small; though genetic algorithms have been successfully applied to weight optimization for some large problems, recent advances in neural network training algorithms (such as cascade correlation, Fahlman 1990) threaten to overshadow these successes. There are domains, however, where genetic algorithms can make a unique contribution to neural network learning.

We briefly review different strategies for applying genetic algorithms to neural networks and present evidence which suggests that genetic hill-climbers are currently the most effective genetic algorithms for neural net weight optimizations. Furthermore we argue that researchers should seek neural network applications where gradient information is unavailable or hard to obtain. As an example, a genetic hill-climber is used to train a neural network to control an inverted pendulum. We compare the results obtained with the genetic hill-climber to the "adaptive heuristic critc" (AHC), which uses a temporal difference method to learn to predict failure. The genetic hill-climbing algorithm displays comparable learning times and appears to be robust over a wide range of different learning conditions relative to the nonlinearity of the problem, and the chosen criteria for "failure."

These results are not intended as a general comparison to reinforcement learning using temporal difference methods, or even as a test of "genetic reinforcement learning" versus the "adaptive heuristic critic." Our comparative tests do show, however, that genetic reinforcment learning does offer a viable approach to "neurocontrol" problems.

2 GENETIC HILL-CLIMBERS

In previous work, we have found that genetic algorithms which predominately rely on the recombination of binary encoded strings to drive search can be used to optimize the weights in smaller neural networks, but we have not been able to duplicate this ability on problems with encodings larger than 300 bits (Whitley and Hanson 1989). Our analyses suggest that this "failure" is partially due to the fact that multiple symmetric representations exist for any single neural network. Recombining encodings for functionally dissimilar neural networks can result in inconsistent feedback to the genetic algorithm in the form of high variance associated with hyperplane samples (Whitley, Starkweather and Bogart 1990). Nevertheless, we have been able to "scale-up" a new version of the genetic algorithm to handle relatively large problems. Our results are similar to some of the results reported by Montana and Davis (1989) indicating that larger neural networks can be optimized using genetic algorithms.

Only three major implementation differences exist between the algorithms that have failed to optimize larger nets and those we have used to produced positive results. First, the problem encoding is real-valued instead of binary. This means that each parameter (weight) is represented by a single real value and that recombination can only occur between weights. Second, a much higher level of mutation is used; traditional genetic algorithms are largely driven by recombination, not mutation. Third, a small population is used (e.g. 50 individuals). We have found that a small population reduces the exploration of the multiple (representationally dissimilar) solutions for the same net. The stronger reliance on mutation also helps to avoid this problem since no recombination is involved when mutation occurs. The genetic algorithm we used is a variant of GENITOR (Whitley and Kauth 1988) which uses *one-at-a-time* recombination and *ranking*. These same attributes also characterized the Montana and Davis genetic algorithm. Thus, although the implementation details are not so different from conventional genetic algorithms, the result is a type of stochastic hill-climbing algorithm which we refer to as a "genetic hill-climber."

Using a population of 50, the genetic hill-climber converges to a solution in 90% of 50 runs on a net that adds two two-bit numbers; search times are roughly comparable with, but not superior to back propagation with momentum. The same approach (population of 50, real-value encodings, increased rate of mutation) was used to optimize a large signal detection network. The application is to identify a signal pulse in one of several channels that span a frequency range. This signal detection problem is complicated by 1) a valid signal causes "false signals" to appear in surrounding channels and 2) more than one valid signal may simultaneously exist across multiple channels. Approximately 300 signal samples are present in the training set while several thousand samples exist in the testing set. The genetic algorithm also produced results on this problem competitive with back propagation.

Our analyses indicate that this success does not stem exclusively from the hyperplane sampling processes that are normally viewed as the driving force of a genetic search, but rather involve stochastic hill-climbing. As already noted, the genetic hill-climber has a stronger reliance on mutation and uses a relatively small population. Mutation in the context of one-at-a-time recombination can be thought of as a hill-climbing operator: randomly change some portion of the string encoding the problem; if performance improves, retain the change, otherwise continue mutation. The strong reliance of the genetic hill-climber on mutation and stochastic hill-climbing is demonstrated by incrementally removing the genetic components of the algorithm until it is exclusively driven by stochastic hill-climbing. This is done by shrinking the population size and reducing the use of recombination until only one string is left in the population and mutation is the only operator. Surprisingly, this conversion consistently shows that as the role of the genetic algorithm is reduced the speed of the algorithm increases (the number of evaluations needed to reach convergence is reduced by an order of magnitude on the adder problem), but the rate of successful convergence decreases. These results are summarized in Table 1. These results (and other tests) suggest that the global efforts of the genetic algorithm do result in a higher convergence rate, but these same efforts require more evaluations to converge to a solution.

PopSize	Trials	Convergence
50	42,500	90 %
5	24,000	80 %
1	4,900	72 %

Table 1: *Results for the 2 2-bit adder problem. As population size (PopSize) is reduced faster hill-climbing occurs, but the convergence rate to acceptable solutions (error < 0.0025) decreases. Trials (rounded to the nearest 100) refers to the average trials to convergence only on those cases that did converge.*

Future researchers should be careful to distinguish between hill-climbing genetic algorithms and hyperplane sampling genetic algorithms since the theoretical foundations of the two are radically different, as are the problems to which they are suited. Goldberg's work (1991) also suggests that there are limitations to the search power of "real-coded coded genetic algorithms" on certain multimodal problems; this should not be a problem, however, when training neural networks.

3 REINFORCEMENT LEARNING

The need for reinforcement learning is based on the fact that for some problems knowledge about what actions are correct or incorrect is not immediate available. Feedback about performance may also be sparse. These kinds of situations preclude the use of relatively simple supervised training algorithms for neural networks such as back propagation. A important application domain of reinforcement learning are control tasks that require unsupervised learning without reliance on any known, a priori, control law. *Back propagation through time* has been applied to such problems, but this is much more complex than simple supervised learning. "Neurocontrol" can be used for the design of automatic controllers that improve their performance by learning from experience.

3.1 The Inverted Pendulum Problem

The inverted pendulum probem is a classic contr[illegible] problem that involves both pole-balancing and [illegible] centering. This is a well studied control probl[illegible] represents an inherently unstable mechani[illegible] a cart and a pole constrained to mov[illegible] cal plane. At any point in time, [illegible] information includes the angle [illegible] angular velocity of the po[illegible] of the cart, ρ, and the velocity of the cart, $\dot{\rho}$. The cart is placed on a track of finite length. Using θ, $\dot{\theta}$, ρ and $\dot{\rho}$ as inputs, the output of the neural net is an action to be applied to the cart: either full-push left or full-push right. One of these two actions occurs at each time step.

$$\ddot{\theta}_t = \frac{mg \sin \theta_t \; - \; \cos \theta_t \; [F_t \; + \; m_{pl} \, \dot{\theta}_t^2 \sin \theta_t]}{(4/3)ml \; - \; m_{pl} \cos^2 \theta_t} \quad (1)$$

$$\ddot{\rho}_t = \frac{F_t \; + \; m_{pl} \, [\dot{\theta}_t^2 \sin \theta_t \; - \; \ddot{\theta}_t \cos \theta_t]}{m} \quad (2)$$

where:

ρ is the cart position, with a range of $\pm 2.4m$
$\dot{\rho}$ is the cart velocity, with a range of ± 1.5
θ is the pole angle
$\dot{\theta}$ is the angular velocity of the pole
m_p is the mass of the pole $= 0.1kg$
m is the total mass of the system $= 1.1kg$
l is the length of the pole $= 0.5m$
F is the control force $= \pm\, 10N$
g is the acceleration due to gravity $= 9.8m/sec^2$

The system was simulated by numerically approximating the equations of motion using Euler's method with a time step of $\tau = 0.02$ seconds and discrete time equations of the form $\theta(t+1) = \theta(t) + \tau\dot{\theta}(t)$. The sampling rate of the system's state variables was the same as the rate of application of the control force (50 Hz).

There are several interesting aspects of this simulated control task that are relevant for analytical purposes. A failure signal is of course generated when the cart crashes into one end of the track, but a failure signal also is usually associated with a particular angle. For example, in some experiments a failure signal is generated when the pole falls beyond 12 degrees from vertical. The system dynamics can be linearly approximated when θ is less than 12 degrees. For larger angles product expansion terms for θ and $\dot{\theta}$ can not be linearly approximated and th[illegible] stem is nonlinear. The 12 degree limit is c[illegible] d; our experiments attempt to bal[illegible] much larger range of angles [illegible] nonlinear versions of [illegible]

[illegible] problem. Of par[illegible] which rely on the [illegible] such as Michie [illegible] on and Ander[illegible] to (1985) and [illegible] here, we were [illegible] layer neural [illegible] ion experi[illegible] performed [illegible] adaptation

3.2 Genetic Reinforcement Learning

Supervised training for neural nets implies that for each input in the training set there is a known desired output. But in the case of certain tasks, the correct output may not be known in advance. Consider a sequence of actions on the cart, followed by a failure signal which means the pole has fallen. Let a 1 indicate a push to the right and a 0 a push to the left; "F" indicates failure. Consider the following sequence: 100011110000F. A classic credit assignment problem exists: which of the actions contributed to success and which actions contributed to failure?

The attraction of using genetic algorithms to train neural networks for reinforcement learning is due to the fact that genetic algorithms do not use gradient information. They also compete with algorithms which use only a failure signal to learn. Genetic reinforcement learning can be done using a single net (other approaches use two nets) with performance information as feedback. Furthermore, the genetic based learning algorithm does not have to explicitly evaluate a state in relation to any other state in order to learn. The approach is therefore quite general.

On the pole-balancing problem each real-valued string in the population is decoded to form a network with five input units (four state variables and a constant bias input), five hidden units and one output unit. Five hidden units are used for compatibility with Anderson's (1989) network. The network is fully connected between the input layer and the hidden layer; the input layer is also fully and directly connected to the output unit. All five hidden units also feed into the output unit. Since there are 35 links in the network, each string used by the genetic search includes 35 real values concatenated together.

Before any input is applied to the network, the four state variables are normalized between 0 and 1. Each state variable feeds into one input unit. The action for a particular set of inputs is determined from the activation of the output unit.

A random start state is supplied to the net and an initial action is then applied to the system. The output of the neural net is a value from 0.0 (push left) to 1.0 (push right) using a bang-bang control. The output of the system is a new state which is then reintroduced as a new input to the net. This continues until a failure occurs. In this case, the only monitor or "critic" is an accumulator which determines how long a particular neural network is able to avoid failure; this length of time is a direct measure of fitness. We stopped the genetic search when a net was found that was able to maintain the system without failure for 120,000 time steps (40 minutes of simulated time).

One potential problem with such a simple evaluation criterion is that a favorable or unfavorable start state may bias the fitness ranking of an individual net. We would like to assign a fitness value to a string based on its ability to perform across all possible start states. In reality this is not practical. In our initial experiments we started from one start state and interpreted the output action deterministically; the nets learned, but performance was not roboust over a large sample of possible start states. The learning behavior of the net can be improved by interpreting the output "action" probabilistically: in other words, an output of +0.75 did not automatically mean that the action should be to push right, but rather that the probability of pushing right is 0.75. This interpretation of the output action allows the net to visit more of the state space and hence learn about more of the problem space. Anderson (1989) also uses a probabilistic interpretation of the output action.

Every algorithm faces the same generalization problem: it must build a decision model for all possible inputs given only a sample of all possible inputs. This problem is not restricted to genetic algorithms, but in genetic algorithms the "generalization problem" is also a noisy evaluation problem. Some set of initial states variables guarantee failure for any sequence of control actions. Networks which receive poor starting states may therefore be ranked lower than networks which receive good starting positions. This creates noise in the ranking function and may result in some nets being lost which are in fact very good competitors. Despite this noise GENITOR performs well in comparison to other learning algorithms. We did try averaging over three trials from different initial starting positions to obtain a better estimate of fitness for strings in the population. This appears to reduce worst case behavior, but does not dramatically change the average learning time. The results reported in this paper therefore are the result of a single learning trial. Methods to deal with this noise in the evaluation function remains a topic of interest; the work of Fitzpatrick and Grefenstette (1988) is particularly relevant.

3.3 Other Genetic Approaches to Control

Genetic Algorithms for learning to control a simulated pole-balancing system have been studied by several researchers. One approach uses predefined partitions of a discrete (binary) state space. When applied to this discretized problem, the genetic algorithm is used to find an appropriate action for each state space partition. The genetic encoding is merely a binary string, where each bit represents an action (push left or right) for each partition of the space. While we are largely interested in reinforcement learning for multilayered neural nets, some aspects of these discrete space approaches are interesting. Odetayo and McGregor (1989) have used this type of approach for pole balancing and cart centering as has Thierens and Vercauteren (1990). The first notable point is that the entire "vector" of binary values does not have to be correct for the entire space to keep the pole up; the pole need only be started in a favorable position and kept in a favorable position. In Thierens and Vercauteren (1990) experiments only the "critical" bits in the encodings took on correct values; bits for more marginal states were never learned. Another relevant observation is that for problems with a large number of variables it becomes impossible to discretize the space. Consider a problem with 30 variables; even if each variable is binary, the discretized space of 2^{30} partitions is unreasonable.

The multilayer network employed here does not require a discretized state space. This frees the experimenter from having to define the size, shape and placement of partitions in the state space. Furthermore, properly discretizing the state space is part of the task of designing a set of features that are sufficient to learn the control task (Anderson 1986). Our interest is in reinforcement learning algorithms which are capable of learning this feature representation and therefore do not require a discretized state space in order to learn difficult control problems.

4 ADAPTIVE HEURISTIC CRITIC

Anderson's (1989) approach to reinforcement learning uses an adaptive heuristic critic (AHC) composed of an evaluation network, an action network and a learning algorithm. The AHC algorithm attempts to associate "states" and failure predictions using temporal difference methods (Sutton 1988). The action net takes information about the system as input and outputs an action (push left or push right). The evaluation network uses temporal differences methods to learn to predict failure based on the current state of the system. The output of the evaluation net is used to "correct" the output of the action net using a form of back propagation. Temporal differences methods are similar to bucket brigage algorithms; information is passed from one state to the next each time a particular sequence of events occurs (Sutton 1988). Information only gradually reaches those states that are more removed from the failure signal.

The output of the evaluation net is a prediction of f
ure, with the strength of the output being rel
the expected time to failure. The action
tion networks learn simultaneously, re
initial assignments of credit and bl
tion net becomes more reliable
used by the "action" net to
in such a way as to av

The pole-balanc
raphy were
training
rate

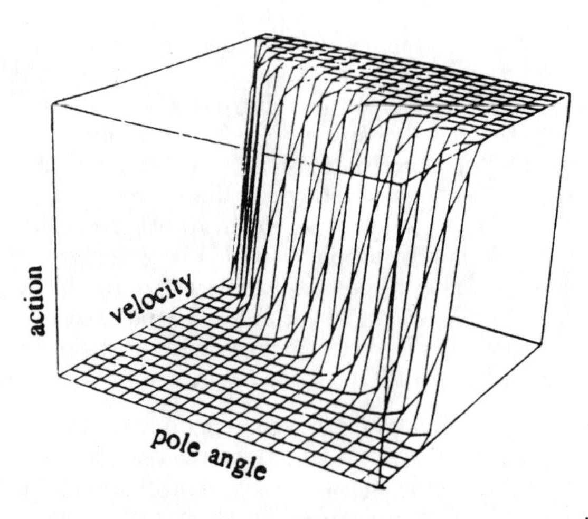

Figure 1: *Plot of pole angle θ, angular velocity $\dot{\theta}$ and the output action on the pole-balancing/cart centering problem using a failure at 12 degrees.*

Seeded Group					
Pop	Best	Worst	Median	Mean	S D
5	1846	17,366	5373	6340	3761
50	16,115	89,615	21,212	24,631	12,699
100	33,979	54,261	40,674	40,815	5031
Restart Group					
Pop	Best	Worst	Median	Mean	S D
5	87	13,104	869	2068	2735
50	225	14,103	3920	4089	2977
100	644	32,057	6223	9380	8457
Normal Group					
Pop	Best	Worst	Median	Mean	S D
5	251	121,988	1614	5971	21,604
50	268	8093	2809	3148	2036
100	332	16,413	3077	3690	2888

Table 2: *Effect of seeding, restarts on learning rate using population sizes (Pop) of 5, 50 and 100. "Normal" implies no restarts or seeding was used.*

5 EXPERIMENTAL RESULTS

Experiments were carried out with failure signals occurring at three different positions of the pole: 12 degrees, 35 degrees and 74 degrees. The 12 degree restriction means that problem has a solution which is is approximately linear. Figure 1 shows the decision boundary learned using the genetic algorithm for the 12 degree problem ploted as a function of the pole angle and the angular velocity. The decision boundaries appear to be very similar to those learned by AHC.

At 35 degrees the problem is nonlinear and contains many start states where it is impossib'e to balance the pole; empirically we have found it ... ible to balance the pole for states up to 39 ... vertical, but only if other all othe... e favorable. At 74 degrees th... tart states from which ... ole. The genetic ... ns of the pr... ncrease

time steps. This "seeded" strategy involved randomly generating strings until enough strings satisfying the "entry" requirement are obtained. This is analogous to using a initial population large enough to guarantee that X strings satisfy some basic threshold of performance; then the population is reduced to the X best strings. This helped to eliminate the extreme worst case behavior, but uniformly increases the costs of each individual run.

The second strategy (the "restart" strategy) was to pick a random initial population, but if sufficient progress is not displayed early on the population is completely regenerated and the run is restarted. A "restart" occurred if a net had not learned to balance for 600 time steps after 2500 evaluations. This strategy avoids uniform overhead costs, since if a particular run results in fast learning no action is required.

The third strategy shown in Table 2 was to do nothing: run the "normal" genetic algorithm on the initial random population to completion. These tests support the notion that genetic algorithms are fairly robust when left alone. Tests of seeded populations and restarts when using a failure signal at 35 and 74 degrees produced results consistent with those in Table 2; these results are also base on 30 experiments. In subsequent tests we did not use restarts or seeding.

In other tests, an evaluation function based upon the verage evaluation resulting from 3 (random) starting sitions was used in order to reduce the bias of favor-e starting positions. This did not significantly affect performance of the network. In subsequent exper-s we did find that using a harder convergence d more impact on the consistency of what was continuing training until the system is able to he pole 3 times in a row noticeably improves performance of the resulting nets. Both

Learning Rates at 12 degrees					
Method	Best	Worst	Median	Mean	SD
AHC	3866	15,000	4781	6606	3713
Genetic	268	8093	2809	3148	2036

Learning Rates at 35 degrees					
Method	Best	Worst	Median	Mean	SD
AHC	2229	45,000	3192	12,207	16,521
Genetic	602	42,297	4246	6618	7197

Learning Rates at 74 degrees					
Method	Best	Worst	Median	Mean	SD
Genetic	2089	80,268	12,092	17,409	17,378

Table 3: *Comparison of learning rates for AHC and the genetic algorithm (using a population of 50) using 12, 35 and 74 degree failure signals. Results are averaged over 30 experiments.*

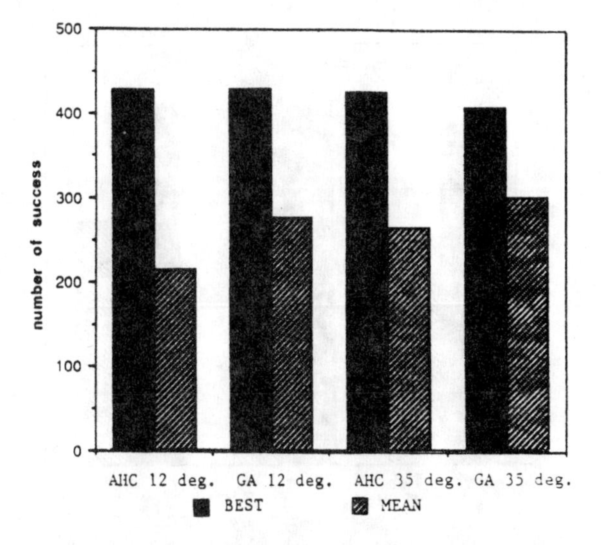

Trained and tested with 12 deg. failure signal					
Method	Best	Worst	Median	Mean	SD
AHC	428	97	179	214	92
Genetic	427	136	267	276	86

Trained and tested with 35 deg. failure signal					
Method	Best	Worst	Median	Mean	SD
AHC	424	65	257	265	96
Genetic	406	122	300	300	75

Figure 2: *Performance of learned function over 625 random start positions for both 12 and 35 degree problems; the genetic algorithm used a population of 50.*

worst case and average case performance is improved by extending the "stopping" criteria from a single success to multiple successes.

While the genetic algorithm did not show the same lack of convergence for the pole balancing problem as it did on the two two-bit adder problem, the results are nevertheless similar. On the pole balancing problem the median training times are lower when using a population of 5, but average and worst case training times are inferior to those achieved with larger populations. In both cases, small population often learn very quickly, but sometimes fail completely (as on the two two-bit adder) or result in very long training times (as on the pole balancing problem).

5.2 Comparative tests

One difference between the genetic algorithm and the AHC algorithm is in their convergence rate. Out of 30 attempts AHC learned a control strategy in only 26 cases (86% convergence) using a failure signal at 12 degrees. The average convergence time (for those cases that did converge) was 5315 trials which is roughly competitive with the average learning time for the various genetic experiments at 12 degrees. The AHC experiments at 12 degrees were run to a maximum training time of 15,000 trials. The average training time for all experiments was 6,606 which is very close to the 6,900 training time reported by Anderson (Anderson 1989) for a single experiment. The genetic algorithm learned 100% of the time and had an average training time of 3,148 trials using a population of 50 strings.

Anderson's original experiments only attempted to learn to balance the pole within the 12 degree range. We also ran experiments using both algorithms with the failure signal extended to 35 and 74 degrees to increase the difficulty of the problem. AHC successfully converged in 24 out of 30 attempts for the 35 degree problem (80%). The experiments were run out to a maximum of 45,000 trials. These convergence rates are better than those we achieved in initial tests of AHC; working with Charles Anderson we have made some minor changes in the implementation to improve its performance. When the problem was extended to use a failure signal at 74 degrees AHC converged to a successful solution in only 1 out of 30 attempts (learning time, 8261 trials) using upto 100,000 learning trials. There are several parameters (e.g., learning rate, etc.) that must be tuned for AHC and the algorithm is fairly sensitive to these parameters. According to Charles Anderson (personal communication) the algorithm may be overly tuned to the 12 degree problem. Population size is the only parameter we adjusted on the genetic algorithm; our only other parameter–the selective pressure–was set to a linear bias of 1.5 and never tuned.

Table 3 shows the best, average, and worst learning rates for the genetic reinforcement algorithm and AHC when the failure signal is generated at 35 degrees and, for the genetic algorithm, 74 degrees. The genetic algorithm reliably learned in all experiments, even when a failure signal is not generated until the pole reaches 74 degrees. Learning times are slower at 74 degrees, but the increases are reasonably modest.

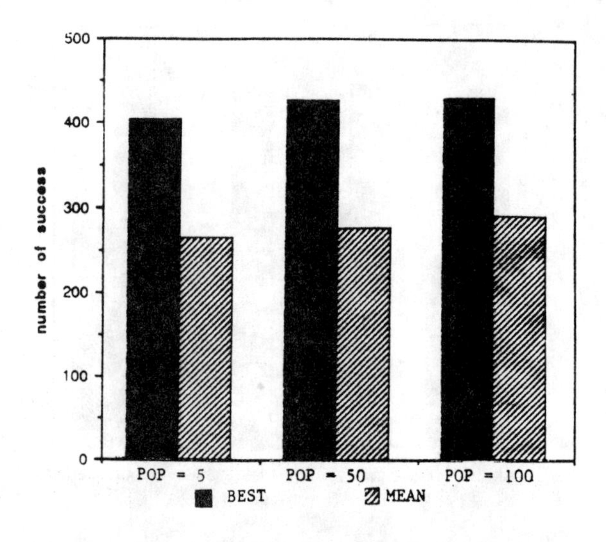

Population size vs. function generalization					
Pop Size	Best	Worst	Median	Mean	S D
5	404	87	260	263	93
50	427	136	267	276	86
100	432	111	297	292	88

Figure 3: *Effect of population size (Pop Size) on performance of the learned function at 12 degrees.*

To further compare the learned function found by the AHC method to the network discovered by genetic search, each system was tested for its ability to balance the pole from 625 random initial state variables. The nets tested were trained on either the 12 or 35 degree version of the pole-balancing problem. Best, average and worse case performance is given for the 30 nets trained by the genetic algorithm; only those nets that learned were tested for AHC. Each net was tested for up to 1000 time steps. Results are given in Figure 2. It is perhaps surprising that the best nets result in failure about 1/3 of the time. This is because many of these initial positions are irrecoverable no matter what control algorithm is used. Figure 3 shows the effect of population size on the ability of the network to balance the pole from 625 random starting positions. The larger population sizes of 50 and 100 tend to both reduce the worst case performance and improve the best case performance. Overall, the larger population sizes produced better results. This observation is consistent with the notion that larger populations should yield better performance when the evaluation (or "fitness") function is noisy (Whitley, Dominic, Das and Anderson 1991).

6 DISCUSSION

There are subtle differences in the information used by the two algorithms, at least in the learning component. In both cases, the "action" net requires input information about the current state in order to produce an output. But one distinction we wish to stress is in the information required for learning. For AHC the evaluation net also requires state information to learn, since without state information there is no prediction of failure. But the genetic algorithm (as opposed to the AHC) does not require state information to update the neural net; it only needs feedback about how long the pole stayed up in order to rank competing sets of weights. This may have important implications for learning to control real-world systems where only time-delayed noisy sensor data is available.

Another difference is that the genetic algorithm only continues to learn as long as failures occur. The reinforcement back propagation that occurs in the action net means that learning continues in the AHC net even when failures are not occurring. Further, the use of temporal difference methods means that the evaluation net is also being updated, even when failures are not occurring. In the genetic approach, updates to the action net occur only after one or more failures. In the genetic algorithm, two networks that avoid failure for an equal number of time steps are equally evaluated. However, the AHC algorithm evaluates networks by the trajectory of states that are experienced. Even though two networks avoid failure for the same number of steps, the evaluations associated with the two nets would differ when the AHC training algorithm is used, favoring the network that drives the cart-pole through more highly-valued states.

Another difference between the genetic algorithm and AHC algorithm is the lack of AHC success when the failure signal occurs at wider pole bounds. This may result from the combination of the incremental learning algorithm used to adjust the weights and the way AHC generalizes. A good prediction of failure is hard to learn when the majority of start states lead to failure. Without a good failure prediction function, a successful control strategy cannot be learned using the reinforcement driven back propagation. The genetic algorithm, because it ranks each network based on performance, is able to basically ignore those cases where the pole cannot be balanced; only the successful cases will obtain the chance to engage in genetic reproduction. For AHC however, these cases provide misleading information which may be learned by the system.

There have been several attempts to apply genetic algorithms to neural network problems. Many of these applications work well for small problems but do not scale up to larger problems. The genetic hill-climbing algorithms described here and used by Montana and Davis work well on large supervised weight optimization problems, but are not as effective as methods such as cascade correlation. We argue that researchers should seek novel applications of genetic algorithms for neural networks such that they are applied in domains where gradient methods are difficult to use. Reinforcement learning for neurocontrol is an application area where it is difficult to directly apply gradient descent methods and where genetic hill-climbing has been shown to produce interesting results. Future work should look at more difficult reinforcement learning problems; genetic reinforcement learning should be also compared to methods such as back propagation through time.

Our purpose in the current study is to demonstrate that genetic hill-climbing algorithms can be used for training neural networks for control problems. The differences between "genetic" reinforcement learning and and other methods for training neural networks using only performance information must be explored more carefully, especially in other application domains. Anderson (1990) discusses several challenging control problems that could provide an initial test bed. We are just finishing the construction of an actual inverted pendulum and will use this to further test the algorithms. We are also interested in ways in which these algorithms might be combined or hybridized. Recently, Ackley and Littman (1991) have combined genetic methods and neural nets for control problems in a different way. They use a genetic algorithm to train the evaluation net, then use the output of the evaluation net to do reinforcement error propagation on the action net. There are also many other ways in which a hybrid system might be developed that uses ideas from both genetic reinforcement learning, AHC and other reinforcement learning paradigms.

Acknowledgements

Our thanks to Charles Anderson for supplying his AHC code and for helpful discussions of the work. This research was supported in part by NSF grant IRI-9010546 and by a grant from the Colorado Institute of Artificial Intelligence (CIAI). CIAI is sponsored in part by the Colorado Advanced Technology Institute (CATI), an agency of the State of Colorado.

References

Ackley, D. and Littman, M. (1990) Interactions Between Learning and Evolution. Submitted to *Proc. 2nd Conf. on Artificial Life.* C.G. Langton, ed. Addison-Wesley.

Anderson, C. W. (1986) Learning and Problem Solving with Multi-layer Connectionist Systems, PhD Dissertation, University of Massachusetts.

Anderson, C. W. (1989) Learning to Control an Inverted Pendulum Using Neural Networks, *IEEE Control Systems Magazine*, vol. 9, no. 3, pp. 31-37.

Anderson, C. W. (1990) A Challenging Set of Control Problems. In: *Neural Networks for Control*, W.T. Miller, R. Sutton and P. Werbos, eds. MIT Press.

Barto, A. G., Sutton, R. S., and Anderson, C. W. (1983) Neuronlike Adaptive Elements That Can Solve Difficult Learning Control Problems. *IEEE Trans. Syst., Man, Cybern.*, SMC-13:834-846.

Fahlman, S. and C. Lebiere (1990) The Cascade Correlation Learning Architecture. *Advances in Neural Information Processing Systems 2.* Morgan Kaufmann.

Fitzpatrick, J.M. and J. Grefenstette (1988) Genetic Algorithm in Noisy Environments. *Machine Learning.* 3:101-120.

Goldberg, D. (1991) Real Coded Genetic Algorithms, Virtual Alphabets and Blocking. *Parallel Problem Solving from Nature* Springer/Verlag, 1991.

Michie, D. and Chambers, R. (1968) BOXES: An Experiment in Adaptive Control. *Machine Learning 2*, E. Dale and D. Michie, Eds., Edinburgh: Oliver and Boyd, pp. 137-152.

Montana, D. and Davis, L. (1989) Training Feedforward Neural Networks Using Genetic Algorithms. *Proc. IJCAI-89*, 1:762-767.

Odetayo, M. and McGregor, D. (1989) Genetic Algorithm for Inducing Control Rules for a Dynamic System. *Proc. Third International Conf. on Genetic Algorithms.* Morgan Kaufmann.

Selfridge, O.G., Sutton, R.S., and Barto, A.G. (1985) Training and Tracking in Robotics, *Proc. IJCAI-85*, pp. 670-672.

Sutton, R. S. (1984) Temporal aspects of credit assignment in reinforcement learning, PhD Dissertation, University of Massachusetts, 1984.

Sutton, R. S. (1988) Learning to Predict by the Methods of Temporal Differences. *Machine Learning* 3:9-44.

Thierens, D and Vercauteren, L. (1990) Incremental Reinforcement Learning with Topology Perserving Maps to Control Dynamic Systems. *Parallel Problem Solving from Nature* Springer/Verlag, 1991.

Whitley, D. and Kauth, J. (1988) GENITOR: A Different Genetic Algorithm. *Proceeding of the 1988 Rocky Mountain Conference on Artificial Intelligence.*

Whitley, D. and Hanson, T. (1989) Optimizing Neural Nets Using Faster, More Accurate Genetic Search. *Proc. Third International Conf. on Genetic Algorithms.* Morgan Kaufmann.

Whitley, D., Starkweather T., and Bogart, C. (1990) Genetic Algorithm and Neural Networks: Optimizing Connections and Connectivity. *Parallel Computing.* 14:347-361.

Whitley, D., Dominic, S., Das, R. and Anderson, C. (1991) Genetic Reinforcement Learning for Neurocontrol Problems. Tech. Report. Dept. Computer Science. Colorado State University.

INTERNATIONAL SOCIETY FOR GENETIC ALGORITHMS CLASSIFICATION TAXONOMY (PROPOSED)

8. Database/Information retrieval 400, 416

J. Relation to AI
   1. Knowledge representation 10, 509
   2. Heuristic search 474
   3. Planning 264
   4. Machine learning 303, 362, 514
      a. Connectionist networks, (*see also* I.H.5) 370, 539, 555, 562
      b. Simulated annealling 174
      c. Version space

II. CLASSIFIER SYSTEMS

A. Condition matching

B. Apportionment of credit
   1. Classifier utility measures 354
   2. Bucket brigade 324

C. Rule discovery 288, 303, 311, 339, 346
   1. Genetic algorithm, (*see also* I.A.4)
   2. Triggered operations

D. Classifier chaining

E. Parallel implementation 296, 311

F. Mathematical models
   1. Dynamical systems
   2. PAC learning; VC-dimension 318
   3. Complexity theory
   4. Parallel computation
   5. Optimization

G. Modeling
   1. Biological 354
   2. Economic
   3. Psychology 324

H. Applications
   1. Diagnosis 288, 334
   2. Robotics/adaptive control 296, 303, 346

I. Relation to AI
   1. Knowledge representation
   2. Machine learning 288
      a. Connectionist networks 324
      b. Simulated annealling
      c. Version space
   3. Planning

KEY WORD INDEX

AUTHOR INDEX